Lecture Notes in Computer Science 5823

Commenced Publication in 1973
Founding and Former Series Editors:
Gerhard Goos, Juris Hartmanis, and Jan van Leeuwen

Abraham Bernstein David R. Karger
Tom Heath Lee Feigenbaum Diana Maynard
Enrico Motta Krishnaprasad Thirunarayan (Eds.)

The Semantic Web - ISWC 2009

8th International Semantic Web Conference, ISWC 2009
Chantilly, VA, USA, October 25-29, 2009
Proceedings

Springer

Volume Editors

Abraham Bernstein
Universität Zürich, 8050 Zurich, Switzerland
E-mail: bernstein@ifi.uzh.ch

David R. Karger
Massachusetts Institute of Technology, Cambridge, MA 02139, USA
E-mail: karger@mit.edu

Tom Heath
Talis Information Ltd., Birmingham Business Park, B37 7YB, UK
E-mail: Tom.Heath@talis.com

Lee Feigenbaum
Cambridge Semantics Inc., Cambridge, MA 02142, USA
E-mail: lee@cambridgesemantics.com

Diana Maynard
University of Sheffield, Sheffield, S1 4DP, UK
E-mail: diana@dcs.shef.ac.uk

Enrico Motta
The Open University, Milton Keynes, MK7 6AA, UK
E-mail: e.motta@open.ac.uk

Krishnaprasad Thirunarayan
Wright State University, Dayton, OH 45435, USA
E-mail: t.k.prasad@wright.edu

Library of Congress Control Number: 2009935764

CR Subject Classification (1998): H.3.5, I.2.4, H.2.4, H.5.1, C.2

LNCS Sublibrary: SL 3 – Information Systems and Application, incl. Internet/Web and HCI

ISSN 0302-9743
ISBN-10 3-642-04929-X Springer Berlin Heidelberg New York
ISBN-13 978-3-642-04929-3 Springer Berlin Heidelberg New York

springer.com

Printed in Germany

Typesetting: Camera-ready by author, data conversion by Scientific Publishing Services, Chennai, India
Printed on acid-free paper SPIN: 12774173 06/3180 5 4 3 2 1 0

Preface

As the Web continues to grow, increasing amounts of data are being made available for human and machine consumption. This emerging Semantic Web is rapidly entering the mainstream and, as a result, a variety of new solutions for searching, aggregating and the intelligent delivery of information are being produced, both in research and commercial settings. Several new challenges arise from this context, both from a technical and human–computer interaction perspective – e.g., as issues to do with the scalability and usability of Semantic Web solutions become particularly important.

The International Semantic Web Conference (ISWC) is the major international forum where the latest research results and technical innovations on all aspects of the Semantic Web are presented. ISWC brings together researchers, practitioners, and users from the areas of artificial intelligence, databases, social networks, distributed computing, Web engineering, information systems, natural language processing, soft computing, and human–computer interaction to discuss the major challenges and proposed solutions, success stories and failures, as well the visions that can advance the field.

This volume contains the main proceedings of ISWC 2009, which we are excited to offer to the growing community of researchers and practitioners of the Semantic Web. We received a tremendous response to our call for research papers from a truly international community of researchers and practitioners from all over the world, who submitted 360 abstracts and 250 research papers. Each paper received more than 3 (some up to 5) reviews as well as a recommendation by one of the Vice Chairs, who read the papers under investigation as well as the comments made by Program Committee members. After the first round of reviews, authors had the opportunity to submit a rebuttal, leading to further discussions among the reviewers and — where needed — to additional reviews. All papers, where at least one of the reviewers or Vice Chairs argued for acceptance, were discussed in a Program Committee meeting held simultaneously in Zurich, Switzerland and Palo Alto, USA, which could be also attended through video conferencing.

As the Semantic Web develops, we found that a variety of subjects of interest have emerged. This year the keywords of the accepted papers were distributed as follows (frequency in parentheses): Interacting with Semantic Web data (3); Semantic Web content creation and annotation (2); mashing up Semantic Web data and processes (2); Semantic Web applications to Web-2.0 sites (1); Information visualization (1); beyond description logic (3); lightweight semantics (4); ontology modeling, reuse, extraction, and evolution (9); ontology mapping, merging, and alignment (3); searching and ranking ontologies (4); ontology evaluation (2); applications with clear lessons learned or evaluations (1); Semantic Web for large-scale applications (4); Semantic Web for multimedia, sensors, and situational awareness (1); Semantic Web technologies for P2P, services, agents, grids and middleware (4); Semantic Web technologies for software and systems engineering (3); mobile Semantic Web (1); Semantic Web technologies for life sciences and healthcare (3); languages, tools and methodologies for representing and managing Semantic Web data (6); database, IR and AI technologies for the Semantic Web (7); search, query, integration, and analysis on the Semantic Web (13); robust and scalable knowledge management and reasoning on the Web (8); machine learning and information extraction for the Semantic Web (3); cleaning, assurance, trust, and provenance of Semantic Web data, services and processes (3); principles & applications of very large semantic databases (4); social networks and processes on the Semantic Web (2); Semantic Web technologies for collaboration and cooperation (2); representing and reasoning about trust, privacy, and security (2); reasoning (10); Semantic Web for desktops or personal information management (1); and Semantic Web technology applications in other areas (1).

Overall, as the field continues to mature, the ISWC Program Committee members have adopted high expectations as to what constitutes high-quality Semantic Web research and what a paper must deliver in terms of theory, practice, and evaluation in order to be accepted in the research track. Correspondingly, the Program Committee accepted only 43 papers, 17.2% of the submissions. Four submissions were accepted for the in-use track after further discussion with its track chairs.

The Semantic Web In-Use Track received 59 submissions, more than double the submissions in previous years. Each paper was reviewed by three members of the In-Use Track Program Committee. We accepted 12 papers, along with 4 papers referred from the Research Track. The large number of submissions this year demonstrated the increasingly diverse breadth of applications which take advantage of Semantic Web technologies. In particular, many papers highlighted the increasing importance of Linked Data, while others focused on ontology engineering, vocabulary mapping, logic and rules, semantic search, and semantic query. The disciplines in which Semantic Web technologies are used are wide-ranging as well, covering education, social networking, supply chains, customer data management, e-discovery, clinical systems integration, geospatial features, and much more.

For the fifth consecutive year, ISWC also contained a Doctoral Consortium track for PhD students within the Semantic Web community, giving them the opportunity not only to present their work but also to discuss in detail their research topics and plans, and to receive extensive feedback from leading scientists in the field, from both academia and industry. Out of 19 submissions, 7 were accepted as full papers, and a further 6 were accepted for the poster session. The standard of submissions this year was particularly high, and covered a wide range of topics. Each student was assigned a mentor who led the discussions following the presentation of the work, and provided detailed feedback and comments, focusing on the PhD proposal itself and presentation style, as well as on the actual work presented.

Another unique aspect of the International Semantic Web Conferences is the Semantic Web Challenge. In this competition practitioners and scientists are encouraged to showcase useful and leading-edge applications of Semantic Web technology. This year the Semantic Web Challenge was organized by Chris Bizer and Peter Mika. For the second time it was extended to include the Billion Triple Challenge. Here the focus was not so much on having a sleek and handsome application, but rather on managing a large heterogeneous data set made available by the challenge organizers.

The ISWC program was further enriched by Keynote Talks given by leading figures from both the academic and business world. Specifically, Pat Hayes, a Senior Research Scientist at the Institute for Human and Machine Cognition discussed the fundamental implications for logical reasoning, which derive from the emergence of the linked data cloud; Nova Spivack, CEO and Founder of Radar Networks, talked about the paradigm shift caused by the emergence of semantic approaches to search; and finally Tom Mitchell, Fredkin Professor of Artificial Intelligence and Machine Learning at Carnegie Mellon University, addressed the issue of large-scale data generation from unstructured sources.

As in previous ISWC editions, the conference program also included an extensive Tutorial Program, which was co-ordinated by Marta Sabou and Jennifer Golbeck and comprised 11 tutorials, and a Workshop Program, which was organized by Jeff Heflin and Philippe Cudré-Mauroux and consisted of 19 workshops.

We would also like to thank Harith Alani, Tania Tudorache and Oscar Corcho for chairing an excellent Poster and Demo Session, and Matt Fisher and John Callahan for co-ordinating the Industry Track, a forum for latest discussions and demonstrations of semantic applications in the commercial world. The Industry Track serves as a complement to the In Use Track and shows just how far semantics are expanding through the enterprise field.

The conference also included a Lightning Talk session, where ISWC attendees could at very short notice get five minutes of attention from the audience, to report on anything they have done, plan to do, like or dislike about the Semantic Web.

We are also much indebted to Krishnaprasad Thirunarayan, our Proceedings Chair, who provided invaluable support in compiling the printed proceedings and exhibited super-human patience in allowing the other Chairs to stretch deadlines to the absolute limit. Many thanks also to Joel Sachs, our Fellowships Chair and to Todd Schneider, Publicity Chair, both for their excellent handling of their portfolios, as well as for their wider contribution to the collaborative decision-making process within the Organizing Committee.

As has been the case for the past few years, ISWC 2009 also contributed to the linked data cloud, by providing semantically characterized data about aspects of the conference. This would not have been possible without the efforts of our Metadata Chairs, Knud Möller and Richard Cyganiak.

We would like to give a special thank you to the Local Organization Chairs, Mike Dean and Leo Obrst and their team from Washington DC, who did a brilliant job in taking care of the local arrangements and ensuring that anything the Organizing Committee needed was promptly made available. In this context we would also like to acknowledge the excellent contributions of Valerie Sumner and Jimmy Ervin. We would also like to thank the generous contribution from our sponsors and the fine work of the Sponsorship Chairs, Pete Pflugrath and Fabio Ciravegna. Finally, we are also indebted to James Stewart from Precision Conference Solutions for his prompt support with the conference system.

October 2009

Abraham Bernstein and David R. Karger
Program Committee Co-chairs, Research Track

Diana Maynard
Doctoral Consortium Chair

Lee Feigenbaum and Tom Heath
Program Committee Co-chairs, Semantic Web In Use Track

Enrico Motta
Conference Chair

Conference Organization

General Chair

Enrico Motta	Open University

Program Chairs

Abraham Bernstein	Universität Zürich
David R. Karger	M.I.T.

Semantic Web In Use Chairs

Tom Heath	Talis
Lee Feigenbaum	Cambridge Semantics Inc.

Semantic Web Challenge Chairs

Peter Mika	Yahoo! Research
Chris Bizer	Free University of Berlin

Doctoral Consortium Chair

Diana Maynard	University of Sheffield

Proceedings Chair

Krishnaprasad Thirunarayan	Wright State University

Local Chairs

Mike Dean	BBN Technologies
Leo Obrst	MITRE

Workshop Chairs

Jeff Heflin	Lehigh University
Philippe Cudré-Mauroux	M.I.T.

Tutorial Chairs

Marta Sabou	Open University
Jennifer Golbeck	University of Maryland

Industry Track Chairs

Matt Fisher	Progeny Systems
John Callahan	Johns Hopkins University

Poster and Demos Chairs

Harith Alani	Open University
Tania Tudorache	Stanford University
Oscar Corcho	Universidade Politecnica de Madrid

Sponsor Chairs

Pete Pflugrath	BBN Technologies
Fabio Ciravegna	University of Sheffield

Metadata Chairs

Knud Möller	DERI, National University of Ireland, Galway
Richard Cyganiak	DERI, National University of Ireland, Galway

Publicity Chair

Todd Schneider	Raytheon

Fellowship Chair

Joel Sachs	University of Maryland Baltimore County

Vice Chairs - Research Track

Ed Chi
Philipp Cimiano
Claudia d'Amato
Stefan Decker
Steven Drucker
Jérôme Euzenat
Jennifer Golbeck
Claudio Gutierrez
Siegfried Handschuh
David Huynh
Georg Lausen
Thomas Lukasiewicz
David Martin
Sheila McIlraith
Peter Mika
Natasha Noy

Bijan Parsia
M. C. Schraefel
Umberto Straccia
Heiner Stuckenschmidt
Valentina Tamma

Program Committee – Research Track

Mark Ackerman
Harith Alani
Paul André
Anupriya Ankolekar
Chutiporn Anutariya
Kemafor Anyanwu
Hassan Aït-Kaci
Lora Aroyo
Medha Atre
Sören Auer
Jie Bao
Steve Battle
Michael Benedikt
Michael Bernstein
Eric Bier
Chris Bizer
Sebastian Blohm
Fernando Bobillo
Olivier Bodenreider
Kalina Bontcheva
Amancio Bouza
John Breslin
Christopher Brewster
Dan Brickley
Tom Briggs
Adriana Budura
Paul Buitelaar
Mark Burstein
Michael Cafarella
Andrea Cal
Diego Calvanese
Vinay Chaudhri
Mariano Consens
Olivier Corby
Oscar Corcho
Paulo Costa
Melanie Courtot
Isabel Cruz
Philippe Cudré-Mauroux
Bernardo Cuenca Grau
Richard Cyganiak
Mathieu d'Aquin
Carlos Damasio
Jérôme David
Brian Davis
David De Roure
Mike Dean
Thierry Declerck
Duane Degler
Klaas Dellschaft
Tommaso Di Noia
Li Ding
John Domingue
Mira Dontcheva
Michel Dumontier
Martin Dzbor
Thomas Eiter
Daniel Elenius
Robert H.P. Engels
Nicola Fanizzi
Cristina Feier
Pablo Fillottrani
Tim Finin
Sergio Flesca
Tim Furche
Fabien Gandon
Aldo Gangemi
Lise Getoor
Chiara Ghidini
Yolanda Gil
Birte Glimm
Carole Goble
Asunción Gómez-Pérez
Mark Greaves
Marko Grobelnik
Volker Haarslev
Peter Haase
Harry Halpin

Andreas Harth
Michael Hausenblas
Pat Hayes
Tom Heath
Jeff Heflin
Nathalie Henry Riche
Martin Hepp
Kaoru Hiramatsu
Pascal Hitzler
Matthew Horridge
Andreas Hotho
Jane Hunter
Carlos Hurtado
Eero Hyvönen
Giovambattista Ianni
Antoine Isaac
Leif Isaksen
Manfred Jaeger
Anthony Jameson
Vana Kalogeraki
Aditya Kalyanpur
Marcel Karnstedt
Jörg-Uwe Kietz
Sheila Kinsella
Matthias Klusch
Mitch Kokar
Yiannis Kompatsiaris
Jacek Kopecky
Spyros Kotoulas
Manolis Koubarakis
Sebastian Kruk
Markus Krötzsch
Mounia Lalmas
Patrick Lambrix
Kathryn Laskey
Ken Laskey
Ora Lassila
Faith Lawrence
Domenico Lembo
Nicola Leone
Holger Lewen
Jiwen Li
Francesca Alessandra Lisi
Gergely Lukacsy
Tiziana Margaria
Paolo Massa
Tara Matthews
Wolfgang May
Diana Maynard
Francis McCabe
Pankaj Mehra
Christian Meilicke
Eduardo Mena
Thomas Meyer
Sebastian Michel
Matthew Michelson
Maja Milicic
Malgorzata Mochol
Ralf Moeller
Leora Morgenstern
Boris Motik
Mark Musen
Knud Möller
Wolfgang Nejdl
Matthias Nickles
Mathias Niepert
John O'Donovan
Daniel Oberle
Georgios Paliouras
Jeff Pan
Alexandre Passant
Terry Payne
Adam Perer
Maria S. Perez
Charles Petrie
Antonio Picariello
Axel Polleres
Alun Preece
Wolfgang Prinz
Andrea Pugliese
Jorge Pérez
Guilin Qi
Azzurra Ragone
Gerald Reif
Vinny Reynolds
Andrea Rodriguez
Riccardo Rosati
Sebastian Rudolph
Lloyd Rutledge
Marta Sabou

Matthias Samwald
Kai-Uwe Sattler
Simon Scerri
Sebastian Schaffert
Thomas Scharrenbach
Anne Schlicht
Stefan Schlobach
Michael Schmidt
Lars Schmidt-Thieme
Thomas Schneider
Guus Schreiber
Daniel Schwabe
Luciano Serafini
Amit Sheth
Pavel Shvaiko
Wolf Siberski
Elena Simperl
Michael Sintek
Evren Sirin
Sergej Sizov
Michael Smith
Daniel Alexander Smith
Shirin Sohrabi
Steffen Staab
Robert Stevens
V.S. Subrahmanian
Bongwon Suh
York Sure
Vojtech Svatek
Kerry Taylor
Herman ter Horst
VinhTuan Thai
Yannis Theoharis
Thanh Tran
Volker Tresp
Cassia Trojahn
Raphael Troncy
Tania Tudorache
Giovanni Tummarello
Anni-Yasmin Turhan
Victoria Uren
Alejandro Vaisman
Petko Valtchev
Ludger van Elst
Willem Robert van Hage
Frank van Harmelen
Jacco van Ossenbruggen
Kunal Verma
Tomas Vitvar
Denny Vrandecic
Johanna Völker
Holger Wache
Haofen Wang
Rob Warren
Fang Wei
Cathrin Weiss
Max L. Wilson
Katy Wolstencroft
Andreas Wombacher
Peter Wood
Yeliz Yesilada
Filip Zelezny
Cai-Nicolas Ziegler
Antoine Zimmermann

Program Committee – Semantic Web In Use

Harith Alani
Dean Allemang
Melli Annamalai
Phil Archer
Sören Auer
Michael Bergman
Mike Bevil
Chris Bizer
David Booth
Christopher Brewster
Sam Chapman
Huajun Chen
Kendall Clark
Chris Clarke
Christine Connors
Mathieu d'Aquin
John Davies
Leigh Dodds
Catherine Dolbear
Bob DuCharme

Robert H.P. Engels
Mark Feblowitz
Paul Gearon
Hugh Glaser
John Goodwin
Mark Greaves
Siegfried Handschuh
Michael Hausenblas
William Hayes
Ivan Herman
David Huynh
Eero Hyvönen
Renato Iannella
Simon Johnston
Mitch Kokar
Mike Lang
Ora Lassila
Joanne Luciano
Dickson Lukose
Christopher Matheus
Brian McBride
Peter Mika
Knud Möller
Marco Neumann
Eric Neumann
Lyndon Nixon
Chimezie Ogbuji
Daniel Olmedilla
Jeff Pan
Valentina Presutti
Eric Prud'hommeaux
Yves Raimond
Gerald Reif
Dave Reynolds
Yuzhong Qu
Marta Sabou
Andy Seaborne
Kavitha Srinivas
Susie Stephens
Sam Tunnicliffe
Jan Wielemaker
Gregory Todd Williams
David Wood

Program Committee – Doctoral Consortium

Saartje Brockmans
Paul Buitelaar
Irene Celino
Philipp Cimiano
Emanuele Della Valle
Martin Dzbor
Adam Funk
Marko Grobelnik
Peter Haase
Siegfried Handschuh
Andreas Harth
Natasha Noy
Lyndon Nixon
Joel Sachs
Vojtech Svatech
Pavel Shvaiko
Elena Simperl
Holger Wache
Willem Robert van Hage

External Reviewers

Liaquat Ali
Carlo Allocca
Renzo Angles
Christian Becker
Ravish Bhagdev
Carlos Bobed
Amancio Bouza
Arina Britz
Markus Bundschus
Krisztian Buza
Gianluca Correndo
Minh Dao-Tran
Frithjof Dau
Tommaso di Noia

Laura Dragan
Renaud Delbru
Cristina Feier
Pablo Fillottrani
Haizhou Fu
Sidan Gao
Raul Garcia Castro
Zeno Gantner
Aurona Gerber
Jose Manuel Gomez-Pérez
Giovanni Grasso
Brynjar Gretarsson
Stephan Grimm
Gunnar Grimnes
Tudor Groza
Parisa Haghani
Benjamin Heitmann
Matthias Hert
Aidan Hogan
Yi Huang
Sergio Ilarri
Patrick Kapahnke
Szymon Klarman
Georgi Kobilarov
Stasinos Konstantopoulos
Thomas Krennwallner
Anastasia Krithara
Artus Krohn-Grimberghe
Vita Lanfranchi
Jens Lehmann
Gergely Lukacsy
Marco Manna
Giuseppe M. Mazzeo
Michael Martin
Michael Meier
Manuel Möller
Jakub Moskal
Stefan Nesbigall
Barry Norton
Kow Weng Onn
Rafael Peñaloza
Giuseppe Pirrò
Antonella Poggi
Jeff Pound
Jörg Pührer
Achim Rettinger
Francesco Ricca
Marco Ruzzi
Ratnesh Sahay
Manuel Salvadores
Florian Schmedding
Tan Yew Seng
Rob Shearer
Anastasios Skarlatidis
Giorgos Stoilos
Carmen Suarez de Figueroa Baonza
Christopher Tuot
Juergen Umbrich
Shenghui Wang
Gang Wu
Fouad Zablith
Mohammad Reza Beik Zadeh

Sponsors

Platinum Sponsors

- BBN Technologies
- Lockheed Martin
- Raytheon

Gold Sponsors

- Clark & Parsia
- LarKC
- ManTech Intl. Corp.
- NeOn
- Orbis Technologies Inc.
- Progeny Systems
- SRI International
- X-Media

Silver Sponsors

- Franz Inc.
- Talis
- Vistology
- Yahoo Labs

Table of Contents

Research Track

Semantic Web In Use

Doctoral Consortium

Invited Talks

Queries to Hybrid MKNF Knowledge Bases through Oracular Tabling

José Júlio Alferes, Matthias Knorr, and Terrance Swift

CENTRIA, Dep. Informática, Faculdade de Ciências e Tecnologia
Univ. Nova de Lisboa, 2825-516 Caparica, Portugal

Abstract. An important issue for the Semantic Web is how to combine open-world ontology languages with closed-world (non-monotonic) rule paradigms. Several proposals for hybrid languages allow concepts to be simultaneously defined by an ontology and rules, where rules may refer to concepts in the ontology and the ontology may also refer to predicates defined by the rules. Hybrid MKNF knowledge bases are one such proposal, for which both a stable and a well-founded semantics have been defined. The definition of Hybrid MKNF knowledge bases is parametric on the ontology language, in the sense that non-monotonic rules can extend any decidable ontology language. In this paper we define a query-driven procedure for Hybrid MKNF knowledge bases that is sound with respect to the original stable model-based semantics, and is correct with respect to the well-founded semantics. This procedure is able to answer conjunctive queries, and is parametric on an inference engine for reasoning in the ontology language. Our procedure is based on an extension of a tabled rule evaluation to capture reasoning within an ontology by modeling it as an interaction with an external oracle and, with some assumptions on the complexity of the oracle compared to the complexity of the ontology language, maintains the data complexity of the well-founded semantics for hybrid MKNF knowledge bases.

1 Introduction

Ontologies and Rules offer distinctive strengths for the representation and transmission of knowledge in the Semantic Web. Ontologies offer the deductive advantages of first-order logics with an open domain while guaranteeing decidability. Rules offer non-monotonic (closed-world) reasoning that can be useful for formalizing scenarios under (local) incomplete knowledge; they also offer the ability to reason about fixed points (e.g. reachability) which cannot be expressed within first-order logic. Interest in both and their combination is demonstrated by the pervasive interest in Ontology languages for the Semantic Web and the growing interest on Rule languages for the Semantic Web, cf. the RIF and the RuleML initiatives.

The two most common semantics for rules are the well-founded semantics (WFS) [19] and the answer-sets semantics [6]. Both semantics are widely used; both offer closed-world reasoning and allow the representation of fixed points;

A. Bernstein et al. (Eds.): ISWC 2009, LNCS 5823, pp. 1–16, 2009.

furthermore the relationship between the two semantics has been fully explored. Of the two, the well-founded semantics is weaker (in the sense that it is more skeptical), but has the advantage that its lower complexity allows it to be integrated into the general-purpose programming language Prolog. Thus in addition to its features for knowledge representation, WFS rules can provide a reactive or procedural component missing from ontologies. Several formalisms have concerned themselves with combining decidable ontologies with WFS rules [3,5,9]. Among these, the Well-Founded Semantics for Hybrid MKNF knowledge bases ($MKNF_{WFS}$), introduced in [9] and overviewed in Section 2 below, is the only one which allows knowledge about instances to be fully inter-definable between rules and an ontology that is taken as a parameter of the formalism. Using this parameterized ontology, $MKNF_{WFS}$ is defined using a monotonic fixpoint operator that computes in each iteration step, besides the usual immediate consequences from rules, the set of all atoms derivable from the ontology whose ABox is augmented with the already proven atomic knowledge. The least fixpoint of the $MKNF_{WFS}$ operator coincides with the original WFS [19] if the DL-component is empty, and when dealing with tractable description logics $MKNF_{WFS}$ retains a tractable data complexity. Furthermore, $MKNF_{WFS}$ is sound wrt. to that of [12] for MKNF knowledge bases, which is based on answer-set semantics and coincides with the answer-sets semantics if the DL-part is empty.

In one sense, the fixpoint operator of $MKNF_{WFS}$ provides a way to compute, in a naive bottom-up fashion, all consequences of a knowledge base. However, such an approach is far from practical for large knowledge bases, as in the Semantic Web context. As a concrete example, consider a medical knowledge base about patients in a large research study. Such a knowledge base might use a standard OWL-ontology representing pathologies, treatment procedures, pharmaceuticals, and so on (e.g. http://www.mindswap.org/2003/CancerOntology). At the same time rules may be used to represent complex temporal constraints that a research study imposes on a given patient, to interface with a patient's electronic health record, and even to extend the ontology with local procedures or policies. To be practical this requires efficient techniques to answer queries about patients, health-care workers, and other objects of interest.

This paper presents a querying mechanism, called **SLG($\mathcal{O}$)**, that is sound and complete for $MKNF_{WFS}$, and sound for MKNF knowledge bases of [12]. **SLG($\mathcal{O}$)** accepts DL-safe conjunctive queries, (i.e. conjunctions of predicates with variables where queries have to be ground when processed in the ontology), returning all correct answer substitutions for variables in the query. To the best of our knowledge, this is the first query-driven, top-down like, procedure for knowledge bases that tightly combine an ontology with non-monotonic rules.

The Gist of the Approach

The main element of our approach addresses the interdependency of the ontology and rules. In particular, our program evaluation method SLG($\mathcal{O}$), presented in Section 4, extends SLG resolution [2], which evaluates queries to normal logic programs (i.e. sets of non-disjunctive non-monotonic rules) under WFS. SLG is a form of tabled resolution that handles loops within the program, and does not change the data complexity of WFS. It does that by resorting to already computed results, in a forest of derivation trees, a technique also known as *tabling*. To adjoin an ontology to rules, the first thing that needs to be done is to allow an SLG evaluation to make calls to an inference engine for the ontology. Since MKNF is parametric on any given decidable ontology formalism[1], the inference engine is viewed in SLG as an oracle. In fact, every time SLG selects an atom that is (perhaps jointly) defined in the ontology, the oracle's inference engine must be called, in case the atom is not provable by the rules alone. Such a queried atom, say $P(a)$, might thus be provable but only if a certain set of atoms in turn is provable via rules. Our approach captures this by allowing the oracle to return a new program clause, say $P(a)$:- *Goals*, which has the property that (possibly empty) *Goals*, in addition to the axioms in the ontology and the atoms already proven by the program would be sufficient to prove $P(a)$. $\mathbf{SLG}(\mathcal{O})$ then treats these new clauses just as if they were program clauses. Note that, getting these conditional answers does not endanger decidability (or tractability, if it is the case) of reasoning in the ontology alone. In fact, it is easy to conceive a modification of a tableaux based inference engine for an ontology, that is capable of returning these conditional answers and is decidable if the tableaux algorithm is: add all those atoms that are defined in the program to the ABox; then proceed with the tableaux as usual, but collect all those added facts that have been used in the proof. Under some assumptions on the complexity of the oracle, it is shown (in Section 5 along with some other properties) that $\mathbf{SLG}(\mathcal{O})$ also retains tractability.

The other element of our approach arises from the need to combine the classical negation of an ontology with the non-monotonic negation of rules. This problem is similar to the issue of *coherence* that arises when adding strong negation to logic programs [6,13,14]: the strong (or classical) negation must imply negation by default. In our case, if the ontology entails that some atom A is false, then perforce the default negation *not* A must hold in the program. The derivation must accordingly be modified since the proof of *not* A cannot simply rely on the failure of the proof of A as it is usual in logic programming. For simplicity, instead of modifying $\mathbf{SLG}(\mathcal{O})$, our proposal (in Section 3) transforms the original knowledge base $\mathcal{K}$ to ensure coherence. $\mathbf{SLG}(\mathcal{O})$ is then applied to the transformed $\mathcal{K}$. This transformation itself provides an alternative formulation of $MKNF_{WFS}$ and is another original result of the paper.

2 Preliminaries

2.1 Syntax of Hybrid MKNF Knowledge Bases

We presume a basic understanding of the well-founded semantics [19] and first-order logics, in particular notions related to logic programming and resolution

[1] The limitation to decidability is theoretically not strictly necessary but a choice to achieve termination and complexity results in accordance with the decidable ontology languages like OWL (`http://www.w3.org/2004/OWL/`).

(see e.g. [11]). Hybrid MKNF knowledge bases as introduced in [12] are essentially formulas in the logics of minimal knowledge and negation as failure (MKNF) [10], i.e. first-order logics with equality and two modal operators **K** and **not** allowing inspection of the knowledge base: intuitively, given a first-order formula φ, $\mathbf{K}\varphi$ asks whether φ is known while $\mathbf{not}\varphi$ is used to check whether φ is not known. Hybrid MKNF knowledge bases consist of two components, a decidable description logics (DL) knowledge base[2], translatable into first-order logics, and a finite set of rules.

Definition 1. *Let $\mathcal{O}$ be a DL knowledge base built over a language $\mathcal{L}$ with distinguished sets of countably infinitely many variables N_V, and finitely many individuals N_I, and predicates N_C. An atom $P(t_1, \ldots, t_n)$ where $P \in N_C$ and $t_i \in N_V \cup N_I$ is called a* DL-atom *if P occurs in $\mathcal{O}$, otherwise it is called* non-DL-atom. *An MKNF rule r has the following form where H_i, A_i, and B_i are atoms: $\mathbf{K}H \leftarrow \mathbf{K}A_1, \ldots, \mathbf{K}A_n, \mathbf{not}B_1, \ldots, \mathbf{not}B_m$. H is called the* (rule) head *and the sets $\{\mathbf{K}A_i\}$, and $\{\mathbf{not}B_j\}$ form the* (rule) body. Literals[3] *are* positive literals $\mathbf{K}A$ *or* negative literals $\mathbf{not}A$. *A rule r is* positive *if $m = 0$ and a* fact *if $n = m = 0$. A* program $\mathcal{P}$ *is a finite set of MKNF rules and a* hybrid MKNF knowledge base $\mathcal{K}$ *is a pair $(\mathcal{O}, \mathcal{P})$.*

We will usually omit the modal operators **K** in the rule head and the positive body, though they remain implicit however. Furthermore, we sometimes also omit the terms t_i of an atom as well (in the context of description logics).

For decidability DL-safety is applied which basically constrains the use of rules to individuals actually appearing in the knowledge base under consideration. Formally, an MKNF rule r is *DL-safe* if every variable in r occurs in at least one non-DL-atom $\mathbf{K}B$ occurring in the body of r. A hybrid MKNF knowledge base $\mathcal{K}$ is *DL-safe* if all its rules are DL-safe[4]. Likewise, to ensure decidability, grounding the knowledge base, i.e. its rules, is restricted to individuals appearing in the knowledge base and not to the whole infinite domain[5]. Therefore, given a hybrid MKNF knowledge base $\mathcal{K} = (\mathcal{O}, \mathcal{P})$, the *ground instantiation of* $\mathcal{K}$ is the KB $\mathcal{K}_G = (\mathcal{O}, \mathcal{P}_G)$ where $\mathcal{P}_G$ is obtained from $\mathcal{P}$ by replacing each rule r of $\mathcal{P}$ with a set of rules substituting each variable in r with constants from $\mathcal{K}$ in all possible ways (for more details we refer to [12] and [9]). DL-safety is also imposed on (conjunctive) queries:

Definition 2. *A* conjunctive query *q is a non-empty set, i.e. conjunction, of literals where each variable in q occurs in at least one non-DL atom in q. We also write q as a rule $q(X_i) \leftarrow A_1, \ldots, A_n, \mathbf{not}B_1, \ldots, \mathbf{not}B_m$ where X_i is the (possibly empty) set of variables, appearing in the body, which are requested.*

---

[2] For a thorough introduction on description logics we refer to [1].

[3] In [9], the term modal atom is used and modal atoms in MKNF are in general not restricted to first-order atoms but in this paper it is essentially their only appearance.

[4] In the following all hybrid MKNF knowledge bases are assumed to be DL-safe.

[5] As well-known, description logics semantics usually require an infinite domain to admit the intended semantics for statements involving unknown individuals.

The restriction of conjunctive queries to DL-safety is not always necessary: for DLs like SHIQ conjunctive query answering is possible ([7]) and we may also make use of such existing algorithms, however, when there is no algorithm for conjunctive query answering yet or it is even not decidable (like for $\mathcal{EL}^{++}$ [15]) then the limitation is required to achieve decidability in the combined approach.

2.2 Well-Founded Semantics of Hybrid MKNF Knowledge Bases

The well-founded MKNF semantics as presented in [9] is based on a complete three-valued extension of the original MKNF semantics. However, here we are not interested in obtaining the entire semantics where a model consists of two sets of sets of first-order interpretations. Instead we limit ourselves here to the computation of what is called the well-founded partition in [9]: basically the literals which are true and false. For that reason, and in correspondence to logic programming, we will name this partition the well-founded model. At first, we recall some notions from [9] which will be useful in the definition of the operators for obtaining that well-founded model.

Definition 3. *Consider a hybrid MKNF knowledge base $\mathcal{K} = (\mathcal{O}, \mathcal{P})$. The set of* **K**-atoms *of $\mathcal{K}$, written $\mathsf{KA}(\mathcal{K})$, is the smallest set that contains (i) all positive literals occurring in $\mathcal{P}$, and (ii) a literal $\mathbf{K}\xi$ for each literal $\mathbf{not}\xi$ occurring in $\mathcal{K}$. Furthermore, for a set of literals S, S_{DL} is the subset of DL-atoms of S, and $\widehat{S} = \{\xi \mid \mathbf{K}\xi \in S\}$.*

Basically all literals appearing in the rules are collected in $\mathsf{KA}(\mathcal{K})$ as a set of positive literals and the other two notions provide restrictions on such a set.

To guarantee that all atoms that are false in the ontology are also false by default in the rules, we introduce new positive DL atoms which represent first-order false DL atoms, and another program transformation making these new literals available for reasoning in the respective rules.

Definition 4. *Let $\mathcal{K}$ be a hybrid MKNF knowledge base. We obtain $\mathcal{K}^+ = (\mathcal{O}^+, P)$ from $\mathcal{K}$ by adding an axiom $\neg P \sqsubseteq NP$ for every DL atom P which occurs as head in at least one rule in $\mathcal{K}$ where NP is a new predicate not already occurring in $\mathcal{K}$. Moreover, we obtain $\mathcal{K}^*$ from $\mathcal{K}^+$ by adding $\mathbf{not}NP(t_1, \ldots, t_n)$ to the body of each rule with a DL atom $P(t_1, \ldots, t_n)$ in the head.*

By $\mathcal{K}^+$, NP represents $\neg P$ (with its corresponding arguments) and $\mathcal{K}^*$ introduces a restriction on each rule with such a DL atom in the head saying intuitively that the rule can only be used to conclude the head if the negation of its head does not hold already[6].

We continue now by defining an operator $T_{\mathcal{K}}$ which allows to draw conclusions from positive hybrid MKNF knowledge bases.

[6] Note that $\mathcal{K}^+$ and $\mathcal{K}^*$ are still hybrid MKNF knowledge bases, so we only refer to $\mathcal{K}^+$ and $\mathcal{K}^*$ explicitly when it is necessary.

Definition 5. *For $\mathcal{K}$ a positive hybrid MKNF knowledge base, $R_\mathcal{K}$, $D_\mathcal{K}$, and $T_\mathcal{K}$ are defined on the subsets of $\mathsf{KA}(\mathcal{K}^*)$ as follows:*

$$\begin{aligned}
R_\mathcal{K}(S) &= S \cup \{\mathbf{K}H \mid \mathcal{K} \text{ contains a rule of the form (1) such that } \mathbf{K}A_i \in S \\
&\qquad \text{for each } 1 \leq i \leq n\} \\
D_\mathcal{K}(S) &= \{\mathbf{K}\xi \mid \mathbf{K}\xi \in \mathsf{KA}(\mathcal{K}^*) \text{ and } \mathcal{O} \cup \widehat{S}_{DL} \models \xi\} \cup \{\mathbf{K}Q(b_1,\ldots,b_n) \mid \\
&\qquad \mathbf{K}Q(a_1,\ldots,a_n) \in S \setminus S_{DL},\ \mathbf{K}Q(b_1,\ldots,b_n) \in \mathsf{KA}(\mathcal{K}^*), \text{ and} \\
&\qquad \mathcal{O} \cup \widehat{S}_{DL} \models a_i \approx b_i \quad \text{for } 1 \leq i \leq n\} \\
T_\mathcal{K}(S) &= R_\mathcal{K}(S) \cup D_\mathcal{K}(S)
\end{aligned}$$

$R_\mathcal{K}$ derives consequences from the rules while $D_\mathcal{K}$ obtains knowledge from the ontology $\mathcal{O}$, respectively from non-DL-atoms and the equalities occuring in $\mathcal{O}$.

The operator $T_\mathcal{K}$ is shown to be monotonic in [9] so, by the Knaster-Tarski theorem, it has a unique least fixpoint, denoted $\mathsf{lfp}(T_\mathcal{K})$, which is reached after a finite number of iteration steps.

The computation follows the alternating fixpoint construction [18] of the well-founded semantics for logic programs which neccesitates turning a hybrid MKNF knowledge base into a positive one to make $T_\mathcal{K}$ applicable.

Definition 6. *Let $\mathcal{K}_G = (\mathcal{O}, \mathcal{P}_G)$ be a ground hybrid MKNF knowledge base and let $S \subseteq \mathsf{KA}(\mathcal{K}_G)$. The* MKNF transform *$\mathcal{K}_G/S = (\mathcal{O}, \mathcal{P}_G/S)$ is obtained by $\mathcal{P}_G/S$ containing all rules $H \leftarrow A_1, \ldots, A_n$ for which there exists a rule $\mathbf{K}H \leftarrow \mathbf{K}A_1, \ldots, \mathbf{K}A_n, \mathbf{not}B_1, \ldots, \mathbf{not}B_m$ in $\mathcal{P}_G$ with $\mathbf{K}B_j \notin S$ for all $1 \leq j \leq m$.*

This resembles the transformation known from answer-sets [6] of logic programs and the following two operators are defined.

Definition 7. *Let $\mathcal{K}$ be a hybrid MKNF knowledge base and $S \subseteq \mathsf{KA}(\mathcal{K}^*)$.*

$$\Gamma_\mathcal{K}(S) = \mathsf{lfp}(T_{\mathcal{K}_G^+/S}) \qquad \Gamma'_\mathcal{K}(S) = \mathsf{lfp}(T_{\mathcal{K}_G^*/S})$$

Both operators are shown to be antitonic [9] and form the basis for defining the well-founded MKNF model. Here we present its alternating computation.

$$\begin{aligned}
\mathbf{P}_0 &= \emptyset & \mathbf{N}_0 &= \mathsf{KA}(\mathcal{K}^*) \\
\mathbf{P}_{n+1} &= \Gamma_\mathcal{K}(\mathbf{N}_n) & \mathbf{N}_{n+1} &= \Gamma'_\mathcal{K}(\mathbf{P}_n) \\
\mathbf{P}_\omega &= \bigcup \mathbf{P}_n & \mathbf{N}_\omega &= \bigcap \mathbf{N}_n
\end{aligned}$$

Note that by finiteness of the ground knowledge base the iteration stops before reaching ω. It was shown in [9] that the sequences are monotonically increasing, decreasing respectively, and that $\mathbf{P}_\omega$ and $\mathbf{N}_\omega$ form the well-founded MKNF model.

Definition 8. *Let $\mathcal{K} = (\mathcal{O}, \mathcal{P})$ be a hybrid MKNF knowledge base and let $\mathbf{P}_\mathcal{K}, \mathbf{N}_\mathcal{K} \subseteq \mathsf{KA}(\mathcal{K})$ with $\mathbf{P}_\mathcal{K}$ being $\mathbf{P}_\omega$ and $\mathbf{N}_\mathcal{K}$ being $\mathbf{N}_\omega$, both restricted to the literals only occurring in $\mathsf{KA}(\mathcal{K})$. Then $M_{WF} = \{\mathbf{K}A \mid A \in \mathbf{P}_\mathcal{K}\} \cup \{\mathbf{K}\pi(\mathcal{O})\} \cup \{\mathbf{not}A \mid A \in \mathsf{KA}(\mathcal{K}) \setminus \mathbf{N}_\mathcal{K}\}$ is the* well-founded MKNF model *of $\mathcal{K}$.*

All literals in M_{WF} are true, its counterparts are false (e.g. if $\mathbf{K}H$ is true then $\mathbf{not}H$ is false) and all other literals from $\mathsf{KA}(\mathcal{K})$ are undefined. Note that $\mathbf{K}\pi(\mathcal{O})$ appears in the model for conciseness with [9].

3 Alternative Computation of $MKNF_{WFS}$

As we have seen, the bottom-up computation of the well-founded MKNF model requires essentially two operators each with its own transformation of the knowledge base. Using directly this as a basis for the top-down procedure, would complicate it, in that we would have to consider two different copies of the program, and use them alternately in different parts of the procedure. This is why, in this section, we define that computation in a different way. Namely, we double the rules in $\mathcal{K}$ using new predicates, transform them appropriately, and double the ontology, so that we can apply just one operator and still obtain the well-founded MKNF model.

Definition 9. *Let $\mathcal{K} = (\mathcal{O}, \mathcal{P})$ be a hybrid MKNF knowledge base. We introduce new predicates, i.e. a predicate A^d for each predicate A appearing in $\mathcal{K}$, and define $\mathcal{K}^d$ as the knowledge base obtained by adding $\mathcal{O}^+$ to $\mathcal{O}$ where each predicate A in $\mathcal{O}$ is substituted by A^d, and transforming each $H(t_i) \leftarrow A_1, \ldots, A_n, \mathbf{not} B_1, \ldots, \mathbf{not} B_m$ occuring in $\mathcal{P}$, t_i representing the arguments of H, into the following two rules:*

(1) $H(t_i) \leftarrow A_1, \ldots, A_n, \mathbf{not} B_1^d, \ldots, \mathbf{not} B_m^d$ *and either*
(2a) $H^d(t_i) \leftarrow A_1^d, \ldots, A_n^d, \mathbf{not} B_1, \ldots, \mathbf{not} B_m, \mathbf{not} NH(t_i)$ *if H is a DL-atom; or*
(2b) $H^d(t_i) \leftarrow A_1^d, \ldots, A_n^d, \mathbf{not} B_1, \ldots, \mathbf{not} B_m$ *otherwise*

Note that the predicate $\mathbf{not} NH$ is in fact the one introduced by $\mathcal{K}^+$.

We can now define a new operator Γ^d on $\mathcal{K}^d$ only[7].

Definition 10. *Let $\mathcal{K} = (\mathcal{O}, \mathcal{P})$ be a hybrid MKNF knowledge base and $S \subseteq \mathsf{KA}(\mathcal{K}^d)$. We define $\Gamma^d_{\mathcal{K}}(S) = \mathsf{lfp}(T_{\mathcal{K}^d_G/S})$ and $\Upsilon_{\mathcal{K}}(S) = \Gamma^d_{\mathcal{K}}(\Gamma^d_{\mathcal{K}}(S))$.*

The operator $\Gamma^d_{\mathcal{K}}$ is antitonic just like $\Gamma_{\mathcal{K}}$, and so $\Upsilon_{\mathcal{K}}$ is a monotonic operator. Therefore $\Upsilon_{\mathcal{K}}$ also has a least and a greatest fixpoint and we can formulate their iteration in the same manner as for $\mathbf{P}_\omega$ and $\mathbf{N}_\omega$.

$$\mathbf{P}^d_0 = \emptyset \qquad \mathbf{N}^d_0 = \mathsf{KA}(\mathcal{K}^d)$$
$$\mathbf{P}^d_{n+1} = \Gamma^d_{\mathcal{K}}(\mathbf{N}^d_n) \qquad \mathbf{N}^d_{n+1} = \Gamma^d_{\mathcal{K}}(\mathbf{P}^d_n)$$
$$\mathbf{P}^d_\omega = \bigcup \mathbf{P}^d_n \qquad \mathbf{N}^d_\omega = \bigcap \mathbf{N}^d_n$$

We can now state the relation of the least and the greatest fixpoint of $\Upsilon_{\mathcal{K}}$ to $\mathbf{P}_\omega$ and $\mathbf{N}_\omega$, from which the well-founded MKNF model is obtained.

Theorem 1. *Let $\mathcal{K} = (\mathcal{O}, \mathcal{P})$ be a hybrid MKNF knowledge base and let $\mathbf{P}^d_\omega$ be the least fixpoint of $\Upsilon_{\mathcal{K}}$ and $\mathbf{N}^d_\omega$ be the greatest fixpoint of $\Upsilon_{\mathcal{K}}$. We have:*

- $A \in \mathbf{P}_\omega$ *if and only if* $A \in \mathbf{P}^d_\omega$
- $B \notin \mathbf{N}_\omega$ *if and only if* $B^d \notin \mathbf{N}^d_\omega$

It follows immediately from this theorem that we can use $\Upsilon_{\mathcal{K}}$ to compute the well-founded MKNF model. We also derive from the theorem that we have to use the new predicates A^d if we query for negative literals.

[7] Note that the operators in Definition 5 are now defined for subsets of $\mathsf{KA}(\mathcal{K}^d)$.

4 Tabled SLG($\mathcal{O}$)-Resolution for Hybrid MKNF

Now we present the new SLG-resolution, **SLG($\mathcal{O}$)**, for Hybrid MKNF Bases which extends [2]. We should mention that the definition of **SLG($\mathcal{O}$)** is quite involved and requires that certain definitions are interlinked with each other (links are provided in each of these cases). It is based on sets of trees of derivations (forests). Tree nodes contain sets of literals which we also call *goals* (cf. Def.12). To deal with termination, some literals must be delayed during the tabled resolution, and so we have to define delay literals (cf. Def.11). Also, a notion of completely evaluated goals in a forest is needed for termination (cf. Def.14). The trees, and the delaying of literals, are constructed according to the set of operations in Def.18. Some of these operations require resolution of selected literals with program rules and, since delayed literals may exist, a slightly changed resolution is required (cf. Def.13). For dealing with the ontology, derivation steps must also take into account an oracle; thus, a suitable definition of what is expected from the oracle is required (cf. Def.17). We begin the presentation of **SLG($\mathcal{O}$)** by defining the delay literals, and then forests:

Definition 11. *A* negative delay literal *has the form* **not** *A, where A is a ground atom.* Positive delay literals *have the form A^{Call}_{Answer}, where A is an atom whose truth value depends on the truth value of some literal $Answer$ for the literal $Call$. If θ is a substitution, then $(A^{Call}_{Answer})\theta = (A\theta)^{Call}_{Answer}$.*

Positive delay literals can only appear as a result of resolution. Its special form as indicated by the names is meant to keep track of the answer and call used for that resolution and possible substitutions are thus only applied to A itself.

Definition 12. *A* node *has the form*

$$Answer\_Template \text{ :- } Delays|Goals \qquad or \qquad fail.$$

In the first form, $Answer\_Template$ is an atom, $Delays$ is a sequence of (positive and negative) delay literals and $Goals$ is a sequence of literals. The second form is called a failure node. *A* program tree *T is a tree of nodes whose root is of the form S :- $|S$ for some atom S: S* is the root node for *T and T* is the tree for *S. An* SLG forest *$\mathcal{F}$ is a set of program trees. A node N is an* answer *when it is a leaf node for which $Goals$ is empty. If $Delays$ of an answer is empty it is termed an* unconditional answer*, otherwise, it is a* conditional answer*. Program trees T may be marked as complete.*

Whenever $Goals$ contains various elements we effectively have to select one of them, by using a *selection function.* The only requirement for such a selection function, is that DL-atoms are not selected until they are ground (which is always possible given DL-safety).

The definition of answer resolution is slightly different from the usual one to take delay literals in conditional answers into account.

Definition 13. *Let N be a node A :- $D|L_1, ..., L_n$, where $n > 0$. Let $Ans = A'$:- $D'|$ be an answer whose variables have been standardized apart from N.*

N is SLG resolvable *with Ans if* $\exists i$, $1 \leq i \leq n$, *such that* L_i *and* A' *are unifiable with an mgu* θ. *The SLG resolvent of N and Ans on* L_i *has the form:*

$$(A \text{ :- } D|L_1, ..., L_{i-1}, L_{i+1}, ..., L_n)\theta$$

if D' *is empty; otherwise the resolvent has the form:*

$$(A \text{ :- } D, {L_i}^{L_i}_{A'}|L_1, ..., L_{i-1}, L_{i+1}, ..., L_n)\theta$$

We delay L_i rather than propagating the answer's delay list. This is necessary, as shown in [2], to ensure polynomial data complexity[8].

At a certain point in SLG($\mathcal{O}$) resolution, a set of goals may be completely evaluated, i.e. it can produce no more answers.

Definition 14. *A set* $\mathcal{S}$ *of literals is* completely evaluated *if at least one of the conditions holds for each* $S \in \mathcal{S}$

1. *The tree for S contains an answer S :- |; or*
2. *For each node N in the tree for S:*
   (a) *The underlying subgoal*[9] *of the selected literal of N is completed; or*
   (b) *The underlying subgoal of the selected literal of N is in* $\mathcal{S}$ *and there are no applicable* New Subgoal, Program Clause Resolution, Oracle Resolution, Equality Resolution, Positive Return, Negative Return *or* Delaying *operations (Definition 18) for N.*

Once a set of literals is determined to be completely evaluated, the Completion operation marks the trees for each literal (Definition 12). Such completely evaluated trees can then be used to simplify other trees in the evaluation.

According to Definition 13, if a conditional answer is resolved against the selected literal in the set *Goals* of a node, the information about the delayed literals in the answer is not propagated. However, in certain cases, the propagation of conditional answers can lead to a set of *unsupported answers* — conditional answers that are false in the well founded model (see e.g. Example 1 of [17])[10].

Definition 15. *Let* $\mathcal{F}$ *be an SLG forest, S a root of a tree in* $\mathcal{F}$*, and Answer be an atom that occurs in the head of some answer of S. Then Answer is supported by S in* $\mathcal{F}$ *if and only if:*

1. *S is not completely evaluated; or*
2. *there exists an answer node Answer :- Delays| of S such that for every positive delay literal* D^{Call}_{Ans}, *Ans is supported by Call.*

We can obtain an interpretation from an SLG forest representing the truth values of the roots of its trees. This interpretation will later also correspond to M_{WF} (cf. Theorem 4).

---

[8] If we propagated the delay lists, we would propagate all derivations which could be exponential in bad cases.

[9] The *underlying subgoal* of literal L is L if L is positive and S if $L = not\ S$.

[10] As an aside, we note that unsupported answers appear to be uncommon in practical evaluations which minimize the use of delay such as [16].

Definition 16. *Let $\mathcal{F}$ be a forest. Then the* interpretation induced by $\mathcal{F}$, $I_{\mathcal{F}}$, *is the smallest set such that:*

- *A (ground) atom $A \in I_{\mathcal{F}}$ iff A is in the ground instantiation of some unconditional answer Ans :- | in $\mathcal{F}$.*
- *A (ground) literal not $A \in I_{\mathcal{F}}$ iff A is in the ground instantiation of a completely evaluated literal in $\mathcal{F}$, and A is not in the ground instantiation of any answer in a tree in $\mathcal{F}$.*

An atom S is successful *(*failed*) in $I_{\mathcal{F}}$ if S' (not S') is in $I_{\mathcal{F}}$ for every S' in the ground instantiation of S. A negative delay literal not D is successful (failed) in a forest $\mathcal{F}$ if D is (failed) successful in $\mathcal{F}$. Similarly, a positive delay literal D^{Call}_{Ans} is successful (failed) in a $\mathcal{F}$ if $Call$ has an unconditional answer Ans :- | in $\mathcal{F}$.*

In order to describe a tabled evaluation that is parameterized by an oracle, we need to characterize the behavior of an abstract oracle, $\mathcal{O}$[11] that computes entailment according to a theory, i.e. the ontology. For that purpose, we define an oracle transition function that in just one step computes all possible atoms required to prove the goal. In other words, such an oracle, when posed a query S non-deterministically returns in one step a set of atoms defined in the program (i.e. atoms for which there is at least one rule with it in the head) such that, if added to the oracle theory, immediately derives S.

Definition 17. *Let $\mathcal{K} = (\mathcal{O}, \mathcal{P})$ be a hybrid MKNF knowledge base, S a goal, and L a set of ground atoms which appear in at least one rule head in $\mathcal{P}_G$. The* complete transition function *for $\mathcal{O}$, denoted $compT_{\mathcal{O}}$, is defined by*

$$compT_{\mathcal{O}}(I_{\mathcal{F}_n}, S, L) \text{ iff } \mathcal{O} \cup I_{\mathcal{F}_n} \cup L \models S$$

We are now able to characterize SLG($\mathcal{O}$) operations.

Definition 18 (SLG($\mathcal{O}$) Operations). *Let $\mathcal{K} = (\mathcal{O}, \mathcal{P})$ be a hybrid MKNF knowledge base. Given a forest $\mathcal{F}_n$ of an SLG($\mathcal{O}$) evaluation of $\mathcal{K}$, $\mathcal{F}_{n+1}$ may be produced by one of the following operations.*

1. New Subgoal: *Let $\mathcal{F}_n$ contain a tree with non-root node*

$$N = Ans \text{ :- } Delays|G, Goals$$

where G is the selected literal S or not S. Assume $\mathcal{F}_n$ contains no tree with root S. Then add the tree S :- $|S$ to $\mathcal{F}_n$.

2. Program Clause Resolution: *Let $\mathcal{F}_n$ contain a tree with root node $N = S$:- $|S$ and C be a rule $Head$:- $Body$ such that $Head$ unifies with S with mgu θ. Assume that in $\mathcal{F}_n$, N does not have a child $N_{child} = (S \text{ :- } |Body)\theta$. Then add N_{child} as a child of N.*

[11] We overload $\mathcal{O}$ syntactically to represent the oracle and the ontology, since semantically they are the same anyway.

3. ORACLE RESOLUTION: *Let $\mathcal{F}_n$ contain a tree with root node $N = S$:- $|S$ and S and all $G \in Goals$ be DL-atoms. Assume that $compT_{\mathcal{O}}(I_{\mathcal{F}_n}, S, Goals)$. If N does not have a child $N_{child} = S$:- $|Goals$ in $\mathcal{F}_n$ then add N_{child} as a child*
4. EQUALITY RESOLUTION: *Let $\mathcal{F}_n$ contain a tree with root node $N = S$:- $|S$ where S and $G \in Goal$ are ground non-DL-atoms with the identical predicate. Assume that $compT_{\mathcal{O}}(I_{\mathcal{F}_n}, S, Goal)$. If N does not have a child $N_{child} = S$:- $|Goal$ in $\mathcal{F}_n$ then add N_{child} as a child*
5. POSITIVE RETURN: *Let $\mathcal{F}_n$ contain a tree with non-root node N whose selected literal S is positive. Let Ans be an answer for S in $\mathcal{F}_n$ and N_{child} be the SLG resolvent of N and Ans on S. Assume that in $\mathcal{F}_n$, N does not have a child N_{child}. Then add N_{child} as a child of N.*
6. NEGATIVE RETURN: *Let $\mathcal{F}_n$ contain a tree with a leaf node, whose selected literal not S is ground*

$$N = Ans \text{ :- } Delays|not\ S, Goals.$$

   (a) NEGATION SUCCESS: *If S is failed in $\mathcal{F}$ then create a child for N of the form: Ans :- $Delays|Goals$.*
   (b) NEGATION FAILURE: *If S succeeds in $\mathcal{F}$, then create a child for N of the form* fail.
7. DELAYING: *Let $\mathcal{F}_n$ contain a tree with leaf $N = Ans$:- $Delays|not\ S, Goals$, such that S is ground, in $\mathcal{F}_n$, but S is neither successful nor failed in $\mathcal{F}_n$. Then create a child for N of the form Ans :- $Delays, not\ S|Goals$.*
8. SIMPLIFICATION: *Let $\mathcal{F}_n$ contain a tree with leaf node $N = Ans$:- $Delays|$, and let $L \in Delays$*
   (a) *If L is failed in $\mathcal{F}$ then create a child* fail *for N.*
   (b) *If L is successful in $\mathcal{F}$, then create a child Ans :- $Delays'|$ for N, where $Delays' = Delays - L$.*
9. COMPLETION: *Given a completely evaluated set $\mathcal{S}$ of literals (Definition 14), mark the trees for all literals in $\mathcal{S}$ as completed.*
10. ANSWER COMPLETION: *Given a set of unsupported answers $\mathcal{UA}$, create a failure node as a child for each answer $Ans \in \mathcal{UA}$.*

The only thing now missing is the formalization of the initialization of an SLG evaluation process.

Definition 19. *Let $\mathcal{K}$ be a hybrid MKNF knowledge base and let q be a query of the form $q(X_i) \leftarrow A_1, \ldots, A_n, \textbf{not}B_1, \ldots, \textbf{not}B_m$ where X_i is the (possibly empty) set of requested variables. We set $\mathcal{F}_0 = \{q(X_i) : - \mid q(X_i)\}$ to be the initial forest of an SLG($\mathcal{O}$) evaluation of $\mathcal{K}^d$ for q.*

Of course, if the query is atomic we can usually simply start with that atomic query. Note that since we use $\mathcal{K}^d$, the technically correct way to query negative literals is to use $\textbf{not}B^d$ instead of $\textbf{not}B$ for any atom B.

Example 1. In order to illustrate the actions of SLG($\mathcal{O}$) we consider a derivation of an answer to the query `?- discount(bill)` to the KB due to [12][12]:

[12] We adopt that only DL-atoms start with a capital letter. Also, to ease the reading, and since it has no influence in this example, instead of $\mathcal{K}^d$ we operate on $\mathcal{K}$ directly.

$NonMarried \equiv \neg Married$ $\qquad \neg Married \sqsubseteq HighRisk$
$\exists Spouse.T \sqsubseteq Married$ $\qquad bill \in (\exists Spouse.michelle)$
$NonMarried(X) \leftarrow$ **not** $Married(X)$.
$discount(X) \leftarrow$ **not** $HighRisk(X)$

Note that both TBox and ABox information are each distributed over both the description logic and the program. Figure 1 shows the final forest for this evaluation, where elements are marked in the order they are created. The initial forest for the evaluation consists of node 0 only. Since the selected literal of node 0, `discount(bill)` is a non-DL-atom and there are no equalities in the KB, we can only apply PROGRAM CLAUSE RESOLUTION which produces node 1, followed by a NEW SUBGOAL to produce node 2. Node 2 is a DL-atom, there are no rules applicable for `HighRisk(bill)`, but an ORACLE RESOLUTION operation can be applied to derive $bill \in NonMarried$ (node 3). Then via a NEW SUBGOAL operation node 4 is obtained. The selected literal for node 4, `NonMarried(bill)` is a DL-atom that also is the head of a rule, so the oracle and the program evaluation may both try to derive the atom. On the program side, PROGRAM CLAUSE RESOLUTION produces nodes 5 and 6. The selected literal of node 6, `Married(bill)`, is a DL-atom that is not the head of a program rule, so once again the only possibility is to use ORACLE RESOLUTION, and derive `Married(bill)`; using this a NEGATIVE RETURN operation produces node 8, and the tree for `Married(bill)` can be early completed. The tree for `NonMarried(bill)` which does not have an answer must be completed (step 10), and the same for `HighRisk(bill)` (step 11). Once this occurs, a NEGATIVE RETURN operation is enabled to produce node 12.

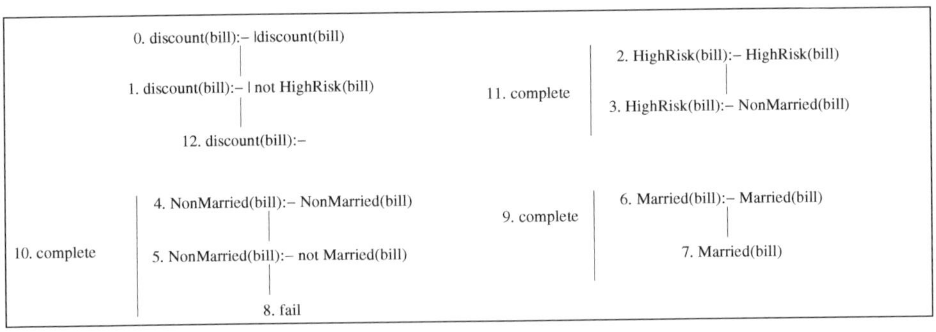

Fig. 1. Final Forest for query `?- discount(bill)` to $\mathcal{K}_1$

The evaluation illustrates several points. First, the evaluation makes use of classical negation in the ontology along with closed world negation in the rules. From an operational perspective, the actions of the description logic prover and the program are interleaved, with the program "calling" the oracle by creating new trees for DL-atoms, and the oracle "calling" the rule system through ORACLE RESOLUTION operations. As a result, trees for DL-atoms must either be early-completed, or explicitly completed by the tabulation system.

5 Properties

Theorem 2. *Let $q = L$ be a query to a hybrid MKNF knowledge base $\mathcal{K}$. Then any* **SLG($\mathcal{O}$)** *evaluation of q will terminate after finitely many steps, producing a finite final forest.*

The way **SLG($\mathcal{O}$)** is defined there is no real order in which to apply any of the operations possible in a forest $\mathcal{F}_i$. Some orders of application are in general more efficient than others but it was shown in [2] that any order yields the same outcome for any query. We adopt this statement here to **SLG($\mathcal{O}$)**.

Theorem 3. *Let $\mathcal{E}_1$ and $\mathcal{E}_2$ be two* **SLG($\mathcal{O}$)** *evaluations of a query $q = L$ to a hybrid knowledge based $\mathcal{K}_G^d$. Let $\mathcal{F}_1$ be the final forest of $\mathcal{E}_1$ and $\mathcal{F}_2$ be the final forest of $\mathcal{E}_2$. Then, $I_{\mathcal{F}_1} = I_{\mathcal{F}_2}$.*

This theorem will also be helpful when it comes to proving that **SLG($\mathcal{O}$)** is in fact a query procedure for $MKNF_{WFS}$ and may terminate within the same complexity bounds as the semantics defined in [9]. At first, we will show that the procedure presented in the previous section coincides with $MKNF_{WFS}$. Intuitively, what we have to show is that the well-founded MKNF model, as presented in section 2 and based on the computation presented in Section 3, and the interpretation $I_\mathcal{F}$ induced by $\mathcal{F}_n$ for some query q to $\mathcal{K}^d$ coincide for each ground literal appearing in $\mathcal{K}_G^d$. We can simplify that by showing for each literal L appearing in $\mathcal{K}_G^d$ that $L \in M_{WF}$ if and only if $L \in I_\mathcal{F}$ with query $q = L$ and $\mathcal{F}_n$ for some n. Additionally, we prove that the same correspondence holds for (positive)[13] atoms only appearing in the ontology.

Theorem 4. *Let $\mathcal{K}$ be a hybrid MKNF knowledge base and L be a modal atom which appears in $\mathcal{K}_G^d$.* **SLG($\mathcal{O}$)** *resolution is correct and complete wrt.* MKNF$_{WFS}$, *i.e. $L \in M_{WF}$ if and only if $L \in I_\mathcal{F}$ where $I_\mathcal{F}$ is induced by the forest $\mathcal{F}$ of an* **SLG($\mathcal{O}$)** *evaluation of $\mathcal{K}_G^d$ for query $q = L$ and, for atoms P not appearing in any rule, $M_{WF} \models P$ if and only if $P \in I_\mathcal{F}$.*

Given the soundness of $MKNF_{WFS}$ wrt. the semantics of MKNF knowledge bases of [12], it follows easily that:

Corollary 1. *Let $\mathcal{K}$ be a consistent hybrid MKNF knowledge base and L be a modal atom which appears in $\mathcal{K}_G^d$. If $L \in I_\mathcal{F}$, where $I_\mathcal{F}$ is induced by the forest $\mathcal{F}$ of an* **SLG($\mathcal{O}$)** *evaluation of $\mathcal{K}_G^d$ for query $q = L$, then L belongs to all MKNF two-valued models (as in [12]) of $\mathcal{K}$.*

In addition to the interpretation of the final forest $I_\mathcal{F}$ being sound with respect to the 2-valued MKNF model, the conditional answers in $\mathcal{F}$ can be seen as a well-founded reduct of the rules in $\mathcal{K}$, augmented with conditional answers derived through Oracle Resolution and Equality Resolution operations. As a

---
[13] We cannot query directly for explicit negated atoms, however, a simple transformation similar to the one yielding $\mathcal{K}^+$ provides a solution to that problem.

result, the final forest can be seen as a *residual program*: a sound transformation not only of the rules, but of information from the oracle, and can be used to construct a partial 2-valued stable model.

Regarding complexity, it is clear that the complexity of the whole $\mathbf{SLG}(\mathcal{O})$ depends on the complexity of the oracle, and also on the number of results returned by each call to the oracle. Clearly, the complexity associated to the computation of one result of the oracle function is a lower-bound of the complexity of $\mathbf{SLG}(\mathcal{O})$. Moreover, even if e.g. the computation of one result of the oracle is tractable, if exponentially many solutions are generated by the oracle (e.g. returning all supersets of a solution), then the complexity of $\mathbf{SLG}(\mathcal{O})$ becomes exponential. This is so, because our definition of the oracle is quite general, and in order to prove interesting complexity results some assumptions most be made about the oracle. We start by defining a correct partial oracle:

Definition 20. *Let $\mathcal{K} = (\mathcal{O}, \mathcal{P})$ be a hybrid MKNF knowledge base, S a goal, and L a set of ground atoms which appear in at least one rule head in $\mathcal{P}_G$ (called program atoms). A* partial oracle *for $\mathcal{O}$, denoted $pT_{\mathcal{O}}$, is a relation $pT_{\mathcal{O}}(I_{\mathcal{F}_n}, S, L)$ such that if $pT_{\mathcal{O}}(I_{\mathcal{F}_n}, S, L)$ then $\mathcal{O} \cup I_{\mathcal{F}_n} \cup L \models S$.*

A partial oracle $pT_{\mathcal{O}}$ is correct *iff when replacing $compT_{\mathcal{O}}$ in* $\mathbf{SLG}(\mathcal{O})$ *succeeds for exactly the same set of queries.*

Note that the complete oracle is indeed generating unnecessarily many answers, and it can be replaced by a partial one which assures correctness. E.g. consider a partial oracle that does not return supersets of other results. Such a partial oracle is obviously correct. Making assumptions on the complexity and number of results of an oracle, complexity results of $\mathbf{SLG}(\mathcal{O})$ are obtained.

Theorem 5. *Let $pT_{\mathcal{O}}$ be a correct partial oracle for the hybrid MKNF knowledge base $\mathcal{K} = (\mathcal{O}, \mathcal{P})$, such that for every goal S, the cardinality of $pT_{\mathcal{O}}(I_{\mathcal{F}_n}, S, L)$ is bound by a polynomial on the number of program atoms. Moreover, assume that computing each element of $pT_{\mathcal{O}}$ is decidable with data complexity $\mathcal{C}$. Then, the* $\mathbf{SLG}(\mathcal{O})$ *evaluation of a query in $\mathcal{K}_G^d$ is decidable with data complexity* $\mathrm{P}^{\mathcal{C}}$.

In particular, this means that if the partial oracle is tractable, and only with polynomial many results, then $\mathbf{SLG}(\mathcal{O})$ is also tractable. Clearly, for an ontology part of the knowledge base which is a tractable fragment, it is possible to come up with a correct partial oracle that is also tractable. Basically, all it needs to be done is to proceed with the usual entailment method, assuming that all program atoms hold, and collecting them for the oracle result. To guarantee that the number of solutions of the oracle is bound by a polynomial, and still keeping with correctness, might be a bit more difficult. It amounts to find a procedure that returns less results, and at the same time does not damage the completeness proof (similar to that of Theorem 4). At least for the tractable case this is possible, albeit the oracle being the (polynomial complexity) bottom-up procedure that defines $MKNF_{WFS}$.

6 Discussion and Conclusions

Together with the alternate computation method of Section 3, **SLG($\mathcal{O}$)** provides a sound and complete querying method for hybrid MKNF knowledge bases, that unlike others (cf. below) freely allows bidirectional calls between the ontology and the rules, and that does not impose a burden of complexity beyond that of the ontology. As such it presents a significant step towards making hybrid MKNF knowledge bases practically usable for the Semantic Web. In fact, work has begun on a prototype implementation of the **SLG($\mathcal{O}$)** method presented here using XSB Prolog and its ontology management library CDF [8]. Because the CDF theorem prover is implemented directly using XSB, the ORACLE RESOLUTION and EQUALITY RESOLUTION operations of Section 4 are more easily implemented than they would be using a separate prover, as is the detection of when a mutually dependent set of subgoals is completely evaluated (Definition 14), and the guarantee of the polynomial size of the oracle. The resulting implementation will enable further study into how hybrid MKNF knowledge bases can be practically used and will indicate needed optimizations. For instance, since XSB supports constraint processing, temporal or spatial constraints can be added to the ABox. From a systems perspective, the multi-threading of XSB can allow for the construction of hybrid MKNF knowledge servers that make use of either Prolog rules or F-logic rules (via FLORA-2, which is implemented using XSB). As mentioned in Section 5 the final forest of a **SLG($\mathcal{O}$)** evaluation produces a well-founded reduct of the rules and oracle information. This reduct, which is materialized in a table in XSB, can be sent to a stable model generator through XSB's XASP library to obtain a partial stable MKNF model of [12].

There are two other semantics which define a well-founded model for a combination of rules and ontologies, namely [5] and [3]. The approach of [5] combines ontologies and rules in a modular way, i.e. keeps both parts and their semantics separate, thus having similarities with our approach. The interface is done by the dlv hex system [4]. Though with identical data complexity to the well-founded MKNF semantics for a tractable DL, it has a less strong integration, having limitations in the way the ontology can call back program atoms (see [5] for details). Hybrid programs of [3] are even more restrictive in the combination: in fact it only allows to transfer information from the ontology to the rules and not the other way around. Moreover, the semantics of this approach differs from MKNF [12,9] and also[5] in that if an ontology expresses $B_1 \vee B_2$ then the semantics in [3] derives p from rules $p \leftarrow B_1$ and $p \leftarrow B_2$, p while MKNF and [5] do not.

While queries posed to KBs without an ontology are handled in the same way as in SLG, strictly speaking the queries posed to the (oracle) DL fragment, are not conjunctive queries in the sense of [7] where boolean queries may contain anonymous variables which never get instantiated. Here we ask whether a ground atom holds when querying the oracle. We nevertheless obtain conjunctive queries up to a certain extent in the sense of [7] only wrt. the entire KB, and our queries are not limited to fit the tree-shaped models there. One line of future work will thus be an extension to such queries which is supported by possible anonymous variables in XSB, the system in which the semantics is currently implemented.

References

1. Baader, F., Calvanese, D., McGuinness, D.L., Nardi, D., Patel-Schneider, P.F. (eds.): The Description Logic Handbook: Theory, Implementation, and Applications, 2nd edn. Cambridge University Press, Cambridge (2007)
2. Chen, W., Warren, D.S.: Tabled Evaluation with Delaying for General Logic Programs. J. ACM 43(1), 20–74 (1996)
3. Drabent, W., Małuszynski, J.: Well-founded semantics for hybrid rules. In: Marchiori, M., Pan, J.Z., de Marie, C.S. (eds.) RR 2007. LNCS, vol. 4524, pp. 1–15. Springer, Heidelberg (2007)
4. Eiter, T., Ianni, G., Schindlauer, R., Tompits, H.: Effective integration of declarative rules with external evaluations for semantic web reasoning. In: Sure, Y., Domingue, J. (eds.) ESWC 2006. LNCS, vol. 4011, pp. 273–287. Springer, Heidelberg (2006)
5. Eiter, T., Lukasiewicz, T., Schindlauer, R., Tompits, H.: Well-founded semantics for description logic programs in the semantic web. In: Antoniou, G., Boley, H. (eds.) RuleML 2004. LNCS, vol. 3323, pp. 81–97. Springer, Heidelberg (2004)
6. Gelfond, M., Lifschitz, V.: Logic programs with classical negation. In: Warren, Szeredi (eds.) ICLP 1990. MIT Press, Cambridge (1990)
7. Glimm, B., Horrocks, I., Lutz, C., Sattler, U.: Conjunctive query answering for the description logic SHIQ. J. of Artificial Intelligence Research 31, 151–198 (2008)
8. Gomes, A.S.: Derivation methods for hybrid knowledge bases with rules and ontologies. Master's thesis, Univ. Nova de Lisboa (2009)
9. Knorr, M., Alferes, J.J., Hitzler, P.: A coherent well-founded model for hybrid MKNF knowledge bases. In: Ghallab, M., Spyropoulos, C.D., Fakotakis, N., Avouris, N. (eds.) Proceedings of the 18th European Conference on Artificial Intelligence, ECAI2008, pp. 99–103. IOS Press, Amsterdam (2008)
10. Lifschitz, V.: Nonmonotonic databases and epistemic queries. In: International Joint Conferences on Artifical Intelligence, IJCAI 1991, pp. 381–386 (1991)
11. Lloyd, J.W.: Foundations of Logic Programming. Springer, Berlin (1984)
12. Motik, B., Rosati, R.: A faithful integration of description logics with logic programming. In: 20th Int. Joint Conf on Artificial Intelligence (IJCAI), Hyderabad, India, January 6–12 2007, pp. 477–482. AAAI Press, Menlo Park (2007)
13. Pearce, D., Wagner, G.: Reasoning with negative information I: Strong negation in logic programs. In: Haaparanta, L., Kusch, M., Niiniluoto, I. (eds.) Language, Knowledge and Intentionality, Acta Philosophica Fennica, vol. 49, pp. 430–453 (1990)
14. Pereira, L.M., Alferes, J.J.: Well founded semantics for logic programs with explicit negation. In: European Conference on Artifical Intelligence, ECAI, pp. 102–106 (1992)
15. Rosati, R.: On conjunctive query answering in EL. In: DL 2007.CEUR Electronic Workshop Proceedings (2007)
16. Sagonas, K., Swift, T., Warren, D.S.: The limits of fixed-order computation. Theoretical Computer Science 254(1-2), 465–499 (2000)
17. Swift, T., Pinto, A., Pereira, L.: Incremental answer completion. In: Hill, P.A., Warren, D.S. (eds.) ICLP 2009. LNCS, vol. 5649, pp. 519–524. Springer, Heidelberg (2009)
18. van Gelder, A.: The alternating fixpoint of logic programs with negation. In: Principles of Database Systems, pp. 1–10. ACM Press, New York (1989)
19. van Gelder, A., Ross, K.A., Schlipf, J.S.: Unfounded sets and well-founded semantics for general logic programs. Journal of the ACM 38(3), 620–650 (1991)

Automatically Constructing Semantic Web Services from Online Sources

José Luis Ambite, Sirish Darbha, Aman Goel, Craig A. Knoblock, Kristina Lerman, Rahul Parundekar, and Thomas Russ

University of Southern California
Information Sciences Institute
4676 Admiralty Way
Marina del Rey, CA 90292
{ambite,darbha,amangoel,knoblock,lerman,parundek,tar}@isi.edu

Abstract. The work on integrating sources and services in the Semantic Web assumes that the data is either already represented in RDF or OWL or is available through a Semantic Web Service. In practice, there is a tremendous amount of data on the Web that is not available through the Semantic Web. In this paper we present an approach to automatically discover and create new Semantic Web Services. The idea behind this approach is to start with a set of known sources and the corresponding semantic descriptions and then discover similar sources, extract the source data, build semantic descriptions of the sources, and then turn them into Semantic Web Services. We implemented an end-to-end solution to this problem in a system called DEIMOS and evaluated the system across five different domains. The results demonstrate that the system can automatically discover, learn semantic descriptions, and build Semantic Web Services with only example sources and their descriptions as input.

1 Introduction

Only a very small portion of data on the Web is available within the Semantic Web. The challenge is how to make Web sources available within the Semantic Web without the laborious process of manually labeling each fact or converting each source into a Semantic Web Service. Converting an existing Web service into a Semantic Web Service requires significant effort and must be repeated for each new data source. We have developed an alternative approach that starts with an existing set of known sources and their descriptions, and then goes on to automatically discover new sources and turn them into Semantic Web Services for use in the Semantic Web.

The system starts with a set of example sources and their semantic descriptions. These sources could be Web services with well-defined inputs and outputs or even Web forms that take a specific input and generate a result page as the output. The system is then tasked with finding additional sources that are similar, but not necessarily identical, to the known source. For example, the system

A. Bernstein et al. (Eds.): ISWC 2009, LNCS 5823, pp. 17–32, 2009.

may already know about several weather sources and then be given the task of finding new ones that provide additional coverage for the world. To do this it must build a semantic description of these new weather sources to turn them into Semantic Web Services. In general, the type of source that we focus on in this paper are information-producing sources where there is a web form that takes one or more input values and produces a result page that has the same format across all output pages. We have found that this type of source is much more common than Web services.

The overall problem can be broken down into the following subtasks. First, given an example source, find other similar sources. Second, once we have found such a source, extract data from it. For a web service, this is not an issue, but for a Web site with a form-based interface, the source might simply return an HTML page from which the data has to be extracted. Third, given the syntactic structure of a source (i.e., the inputs and outputs), identify the semantics of the inputs and outputs of that source. Fourth, given the inputs and outputs, find the function that maps the inputs to the outputs. Finally, given the semantic description, construct a wrapper that turns the source into a Semantic Web Service that can be directly integrated into the Semantic Web.

In previous work we have developed independent solutions to each of these subtasks. Here, we describe the integration of these separate components into a single unified approach to discover, extract from, and semantically model new online sources. In the previous work each of these components made assumptions that were not consistent with the other components. We had to address these issues to build an end-to-end system. This work provides the first general approach to automatically discovering and modeling new sources of data. Previous work, such as the ShopBot system [16], did this in a domain-specific way where a significant amount of knowledge was encoded into the problem (e.g., shopping knowledge in the case of ShopBot).

In this paper we present DEIMOS, a system that provides an end-to-end approach to discovering and building Semantic Web Services. First, we review the previous work on which the system is built (Section 2). Second, we describe the architecture of DEIMOS (Section 3), and describe how we built on the previous work to discover new sources (Section 3.1), invoke and extract data from the discovered sources (Section 3.2), semantically type the inputs and outputs of these sources (Section 3.3), semantically model the function performed by these sources (Section 3.4), and then use this semantic model to turn the web source into a Semantic Web Service (Section 4). Third, we present results of an end-to-end evaluation in five different information domains, where the only input to the system is an example of a source in that domain and a semantic description of that source (Section 5). Finally, we compare with related work (Section 6) and conclude with a discussion and directions for future research (Section 7).

2 Prior Work

In previous work, we have developed the core technologies used by the integrated system. Plangprasopchok & Lerman [15] developed an automatic *source discovery* method that mines a corpus of tagged Web sources from the social bookmarking site del.icio.us to identify sources similar to a given source. For example, given a weather service that returns current weather conditions at a specified location, the method can identify other weather services by exploiting the tags used to describe such sources on del.icio.us. Tags are keywords from an uncontrolled personal vocabulary that users employ to organize bookmarked Web sources on del.icio.us. We use topic modeling techniques [4,10] to identify sources whose tag distribution is similar to that of the given source.

Gazen & Minton [7] developed an approach to automatically structure Web sources and *extract data* from them without any previous knowledge of the source. The approach is based on the observation that Web sources that generate pages dynamically in response to a query specify the organization of the page through a page template, which is then filled with results of a database query. The page template is therefore shared by all pages returned by the source. Given two or more sample pages, we can derive the page template and use it to automatically extract data from the pages.

Lerman *et al.* [13] developed a domain-independent approach to *semantically label* online data. The method learns the structure of data associated with each semantic type from examples of that type produced by sources with known models. The learned structure is then used to recognize examples of semantic types from previously unknown sources.

Carman and Knoblock [5] developed a method to learn a *semantic description* of a source that precisely describes the relationship between the inputs and outputs of a source in terms of known sources. This is done as a logical rule in a relational query language. A data integration system can then use these source descriptions to access and integrate the data provided by the sources [14].

3 End-to-End Discovery, Extraction, and Modeling

The overall architecture of the DEIMOS system is shown in Figure 1. DEIMOS starts with a known source and background knowledge about this source. It then invokes each module to discover and model new related sources. Our techniques are domain-independent, but we will illustrate them with examples from the weather domain.

The background knowledge required for each domain consists of the semantic types, sample values for each type, a domain input model, the known sources (seeds), and the semantic description of each seed source. For the weather domain, the background knowledge consists of: (1) Semantic types: e.g., TempF, Humidity, Zip; (2) Sample values for each type: e.g., "88 F" for TempF, and "90292" for Zip; (3) Domain input model: a weather source may accept Zip or a combination of City and State as input; (4) Known sources (seeds): e.g., http://wunderground.com; (5) Source descriptions: specifications of the functionality of the source in a formal language of the kind used by data integration

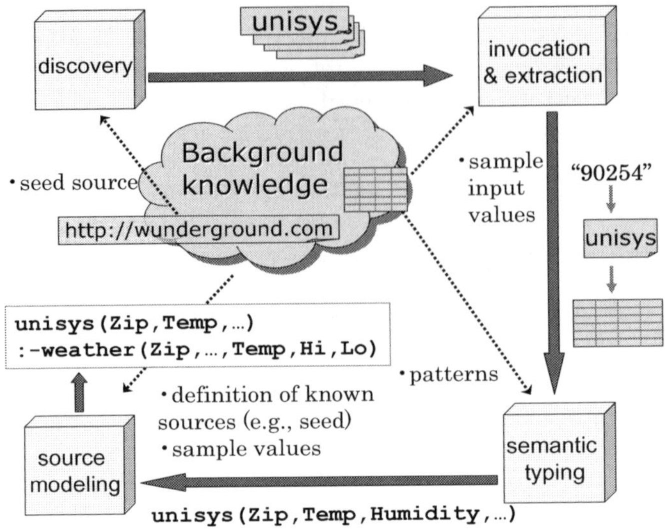

Fig. 1. DEIMOS system architecture

systems. For example, the following Local-as-View [14] Datalog rule[1] specifies that wunderground returns current weather conditions and five day forecast for a given zip code:

```
wunderground($Z,CS,T,F0,S0,Hu0,WS0,WD0,P0,V0,
            FL1,FH1,S1,FL2,FH2,S2,FL3,FH3,S3,FL4,FH4,S4,FL5,FH5,S5) :-
  weather(0,Z,CS,D,T,F0,_,_,S0,Hu0,P0,WS0,WD0,V0)
  weather(1,Z,CS,D,T,_,FH1,FL1,S1,_,_,_,_,_),
  weather(2,Z,CS,D,T,_,FH2,FL2,S2,_,_,_,_,_),
  weather(3,Z,CS,D,T,_,FH3,FL3,S3,_,_,_,_,_),
  weather(4,Z,CS,D,T,_,FH4,FL4,S4,_,_,_,_,_),
  weather(5,Z,CS,D,T,_,FH5,FL5,S5,_,_,_,_,_).
```

which has an input attribute (denoted by "$") Z (of type Zip) and outputs CS (CityState), T (Time), FHi and FLi high and low temperatures in Farenheit degrees (TempInF) on the ith forecast day (0= today, 1= tomorrow, ...), D (Date), S (Sky conditions), Hu (Humidity), WS (Wind speed in MPH), WD (WindDirection), P (Pressure in inches), and V (Visibility in miles). The semantics of the source are specified by the conjunctive formula in the body of the rule that uses predicates from a domain ontology (weather() in this example).

[1] We use the usual rule syntax for LAV rules common in the data integration literature. However, logically this rule should be interpreted as:

$$\text{wunderground}(\ldots) \rightarrow \text{weather}(\ldots) \wedge \text{weather}(\ldots) \wedge \ldots$$

This means that every tuple from wunderground satisfies the formula over the domain predicates (weather), but not viceversa. That is, the source is not complete.

DEIMOS first uses the discovery module to identify sources that are likely to provide functionality similar to the seed. Once a promising set of target sources has been identified, DEIMOS uses the invocation and extraction module to determine what inputs are needed on Web forms and how to extract the returned values. DEIMOS then invokes the semantic typing module to automatically infer the semantic types of the output data. Once DEIMOS constructs a type signature for a new source, it then invokes the source modeling module to learn its source description. We will describe each of these modules in turn, along with the challenges in building an end-to-end solution.

3.1 Source Discovery

This module identifies sources likely to provide functionality similar to the seed. DEIMOS first collects popular tags annotating the seed from the social bookmarking site del.icio.us. As of October 2008, http://wunderground.com has been tagged by over 3200 people. Among popular tags are useful descriptors of the service: "weather," "forecast," and "meteo." Next, DEIMOS retrieves all other sources that were annotated with those tags on del.icio.us. By analogy to document topic modeling, we view each source as a document, and treat the tags created by users who bookmarked it as words.

The system uses Latent Dirichlet Allocation (LDA) [4] to learn a compressed description, or 'latent topics', of tagged sources [17]. The learned topics form the basis for comparing similarity between sources. If a source's topic distribution is similar to the seed's, it is likely to have similar functionality. We rank retrieved sources according to their similarity to the seed and pass the 100 top-ranked sources to the next module. In the weather domain, among sources similar to http://wunderground.com are weather sources such as http://weather.yahoo.com and http://weather.unisys.com

3.2 Source Invocation and Extraction

To retrieve data from the discovered Web sources, DEIMOS has to figure out how to invoke the source. These sources typically use standard HTML forms for input and return a result page. During the invocation step, DEIMOS analyzes the sources's document model and extracts forms and form elements. For each of the forms, DEIMOS identifies the input fields, which can be text (input) or menu (select) fields. DEIMOS relies on background knowledge to constrain the search for valid inputs. The background knowledge contains information about the typical input types expected by sources in the domain and sample values for each input type: e.g., *weather* sources expect zipcodes or city and state combinations, while *mutual funds* sources typically expect a fund symbol as input.

DEIMOS uses a brute force approach, trying all permutations of input types in the input form's fields. We do allow for optional input fields, leaving some input fields blank. Since our domains generally have only a small number of possible input types, the combinatorics of the brute force approach are manageable. However, some types of sources, such as hotel booking sites, had so many possible form inputs that searching all combinations of possible inputs was intractable. We believe the solution to this problem is to exploit more of the context information on the form, such as the form label and variable name to narrow the possible inputs to each field. This will be a direction for future work so that we will be able to model information domains that have more numerous input fields in the forms.

DEIMOS repeatedly invokes the source with the different permutations of domain input values, looking for a set of mappings that yields results pages from which it can successfully extract data.

Next, DEIMOS extracts data from pages returned by the source in response to a query. For this, DEIMOS uses the Autowrap algorithm, described in [7], which exploits the regularity of dynamically generated pages. It assumes that the organization of dynamically generated page is specified through a *page template* that is shared by all pages returned by the source. Given two or more sample pages, we can derive the page template and use it to extract data from the pages.

A *template* is a sequence of alternating stripes and slots. Stripes are the common substrings and slots are placeholders for data. Autowrap uses the Longest Common Subsequence algorithm to induce a template from sample pages. The common substrings are the template stripes and the gaps between stripes are the slots. Given snippets from two pages, "`HI:65<br>LO:50`" and "`HI:73<br>LO:61`", it induces the template "`HI:*<br>LO:*`" where "`*`" marks a slot. The induced template can be used to extract data from new pages that share the same template. This involves locating the stripes of the template on the new page. Substrings that lie between the stripes are extracted as field values. Applying the template above to the snippet "`HI:50<br>LO:33`" results in two values: "`50`" and "`33`".

We modified the basic approach described above to deal with the challenges encountered while integrating the technology within DEIMOS. One extraction problem we encountered was that some of the strings that were useful for disambiguating data values, such as units on numbers, ended up being considered part of the page template by Autowrap. Consider, for example, a temperature value of '10 C' and a wind speed value of '10 mph', which look very similar once you remove the units. The extraction module finds strings that *change* across pages, and in structured sources such as these, the units will not be extracted because they rarely change. Since units are typically a single token that comes immediately following the value, we built a post-processor that generated additional candidates for semantic typing that included tokens that were most likely to capture unit information or other context. This is done by checking the document object model (DOM) of the page and appending tokens immediately following a value if it occurs at the same level in the DOM tree, which means that it likely occurs immediately after the value on the page. For '10 mph', the system would generate both '10' and '10 mph' and the next step would attempt to determine the semantic type of each of them.

Another challenge was that for seemingly minor variations across pages, there were significant difference in the page structure, which prevented the system from finding the page template. An example of this was in a weather source where some of the cities had a weather advisory on the page. This resulted in a different underlying DOM structures and Autowrap failed to find the shared portion of the structure. To address this problem requires searching a much larger space to find the page template, so for the current set of results DEIMOS fails on some sources that should be learnable.

3.3 Semantic Typing of Sources

This module semantically types data extracted from Web sources using the approach described in [13]. This approach represents the structure of a data field as a sequence of tokens and syntactic types, called a pattern [12]. The syntactic types, e.g., alphabetic, all-capitalized, numeric, one-digit, have regular expression-like recognizers. The patterns associated with a semantic type can be efficiently learned from example values of the type, and then used to recognize instances of a semantic type by evaluating how well the patterns describe the new data. We developed a set of heuristics to evaluate the quality of the match. These heuristics include how many of the learned patterns match data, how specific they are, and how many tokens in the examples are matched [13].

The output of this module is a semantically typed signature of a source with its input and output parameters assigned to semantic types in the domain. For example, a subset of the type signature learned for source weather.unisys.com is:

```
unisys($Zip,TempF,TempC,Sky,Humidity, ...)
```

The most significant challenge encountered in this module is that the typing component did not always have enough information to distinguish between two alternative types and chose the incorrect one. We plan to improve the semantic typing by using additional features of the values, such as numeric ranges, which will allow the system to make finer-grained semantic-typing distinctions.

3.4 Source Modeling

The typed input/output signature of a new source offers only a partial description of the source's behavior. What we need is a semantic characterization of its functionality—the relationship between its input and output parameters. We use the approach described in Carman & Knoblock [5] to learn a Local-as-View (LAV) description of the source (a Datalog rule) [14]. We illustrate the main ideas of the inference algorithm using our running example.

Consider the following *conjunctive* LAV source description for weather.unisys.com:

```
unisys($Z,CS,T,F0,C0,S0,Hu0,WS0,WD0,P0,V0,
      FL1,FH1,S1,FL2,FH2,S2,FL3,FH3,S3,FL4,FH4,S4,FL5,FH5,S5):-
  weather(0,Z,CS,D,T,F0,_,_,S0,Hu0,P0,WS0,WD0,V0)
  weather(1,Z,CS,D,T,_,FH1,FL1,S1,_,_,_,_,_),
  weather(2,Z,CS,D,T,_,FH2,FL2,S2,_,_,_,_,_),
  weather(3,Z,CS,D,T,_,FH3,FL3,S3,_,_,_,_,_),
  weather(4,Z,CS,D,T,_,FH4,FL4,S4,_,_,_,_,_),
  weather(5,Z,CS,D,T,_,FH5,FL5,S5,_,_,_,_,_),
  centigrade2farenheit(C0,F0).
```

A *domain model/ontology* (consisting of predicates weather and centigrade2farenheit in the example) assigns precise semantics to sources (such as unisys) in an application domain.

The Source Modeling module of DEIMOS learns these definitions by combining *known* sources to emulate the input/output values of a new *unknown* source. For the weather domain, the system already knows the description of wunderground (cf. Section 3) and the following temperature conversion service:

```
convertC2F($C,F) :- centigrade2farenheit(C,F)
```

Using these known sources, the system learns the following join, which describes some of the input/output values of the previously unknown unisys source:

```
unisys($Z,_,_,_,_,_,_,F9,_,C,_,F13,F14,Hu,_,F17,_,_,_,_,
        S22,_,S24,_,_,_,_,_,_,_,_,_,S35,S36,_,_,_,_,_,_,_,_) :-
  wunderground(Z,_,_,F9,_,Hu,_,_,_,_,F14,F17,S24,_,_,S22,_,_,
               S35,_,_,S36,F13,_,_),
  convertC2F(C,F9)
```

Replacing the known sources, wunderground and convertC2F, by their definitions yields a version of the above LAV source description for unisys (cf. Section 4 for a Semantic Web version of this source description).

Learning this definition involves searching the space of possible hypotheses (Datalog conjunctive rules) that could explain the observed inputs and outputs. DEIMOS uses an approach based on Inductive Logic Programming to enumerate the search space in an efficient, best-first manner and finds the most specific rule that best explains the observed data. During this search the system uses the learned semantic types (for the unknown source) and the already known types of the background sources to prune candidate hypotheses. The system considers only conjunctive queries that join on variables of compatible types.

DEIMOS evaluates each candidate hypothesis (conjunctive query) over a set of sample input tuples, generating a set of *predicted* output tuples. It then compares the generated output tuples with those actually produced by the source being modeled to see if the predicted and actual outputs are similar. As part of its background knowledge, DEIMOS associates a similarity function with each semantic type. For numbers, the similarity is an absolute or a relative (percent) difference. For text fields, it uses string similarity metrics (e.g., Levenshtein distance). DEIMOS uses the Jaccard similarity to rank different hypotheses according to the amount of overlap between the predicted output tuples and the observed ones. For some types we found a large variation in the values returned for the same inputs by different sources. In the weather domain, for example, the temperature and humidity values reported for the same location had a high variance. We had to allow for larger differences in the similarity function used by the source modeling module in order to discover any matches on these fields.

We encountered several challenges when integrating the source modeling component within DEIMOS. First, there are often synonyms for values that are critical to invoking sources and comparing resulting values. For example, in the flight domain some sources take the airline name as input and others the corresponding

3-letter airline code. We addressed the problem of synonyms and functionally-equivalent values by providing synonym mapping tables as additional sources that can be used in the source modeling step.

The second challenge is that sometimes closely-related attributes would be typed incorrectly due to precision errors on the values. For example, in the weather domain the forecast for the high temperature on the 3rd day would get confused with the high temperature for the 5th day. The problem arose because the 3rd-day and 5th-day high temperature values were very close for the set of sample input cities. This problem can be addressed by using additional input examples that can disambiguate between the attributes. However, a larger number of examples sometimes entails a greater variability of the resulting pages, which makes the extraction task harder (e.g., recall the page structure change due to weather advisory events discussed in Section 3.2).

4 Automatically Building Semantic Web Services

After the source modeling phase, DEIMOS constructs a semantic web service (SWS) encapsulating the discovered web source. The SWS accepts RDF input and produces RDF output according to the domain ontology. Internally, the SWS calls the discovered web form using the input values from the input RDF to the semantic web service. It then extracts the data from the resulting HTML using the learned *page template* (cf. Section 3.2). The output data obtained by applying the page template is filtered according to the learned source description (cf. Section 3.4). In this way the system is certain of the semantics of the extracted values. Finally, the extracted values are converted to RDF according to the specification of the source description.

We describe the construction of the semantic web service using our running unisys weather source example. For brevity and convenience earlier in the paper, we have used a domain model with n-ary predicates (such as weather()). However, since we are interested in producing RDF-processing semantic web services, in our source descriptions we actually use a domain ontology composed of unary and binary predicates, which can be straightforwardly translated to RDF. For example, the definition for wunderground is:

```
wunderground($Z,CS,T,F0,C0,S0,Hu0,WS0,WD0,P0,V0,
              FL1,FH1,S1,FL2,FH2,S2,FL3,FH3,S3,FL4,FH4,S4,FL5,FH5,S5) :-
 Weather(@w0),hasForecastDay(@w0,0),hasZIP(@w0,Z),hasCityState(@w0,CS),
  hasTimeWZone(@w0,T),hasCurrentTemperatureFarenheit(@w0,F0),
  hasCurrentTemperatureCentigrade(@w0,C0),hasSkyConditions(@w0,S0),
  hasHumidity(@w0,Hu0),hasPressure(@w0,P0),hasWindSpeed(@w0,@ws1),
  WindSpeed(@ws1),hasWindSpeedInMPH(@ws1,WS0),hasWindDir(@ws1,WD0),
  hasVisibilityInMi(@w0,V0),
 Weather(@w1),hasForecastDay(@w1,1),hasZIP(@w1,Z),hasCityState(@w1,CS),
  hasLowTemperatureFarenheit(@w1,FL1),hasHighTemperatureFarenheit(@w1,FH1),
  hasSkyConditions(@w1,S1), ...
```

Thus, the RDF-like source description learned for unisys (cf. Section 3.4) is:

```
unisys($Z,_,_,_,_,_,_,_,F9,_,C,_,F13,F14,Hu,_,F17,_,_,_,_,
        S22,_,S24,_,_,_,_,_,_,_,_,_,_,S35,S36,_,_,_,_,_,_,_,_,_) :-
  Weather(@w0),hasForecastDay(@w0,0),hasZIP(@w0,Z),
   hasCurrentTemperatureFarenheit(@w0,F9),centigrade2farenheit(C,F9),
   hasCurrentTemperatureCentigrade(@w0,C),hasHumidity(@w0,Hu0),
  Weather(@w1),hasForecastDay(@w1,1),hasZIP(@w1,Z),hasCityState(@w1,CS),
   hasTimeWZone(@w1,T),hasLowTemperatureFarenheit(@w1,F14),
   hasHighTemperatureFarenheit(@w1,F17),hasSkyConditions(@w1,S24),
  Weather(@w2),hasForecastDay(@w2,2),hasZIP(@w2,Z),hasSkyConditions(@w2,S22),
  Weather(@w3),hasForecastDay(@w3,3),hasZIP(@w3,Z),hasSkyConditions(@w3,S35),
  Weather(@w4),hasForecastDay(@w4,4),hasZIP(@w4,Z),hasSkyConditions(@w4,S36),
  Weather(@w5),hasForecastDay(@w5,5),hasZIP(@w5,Z),
   hasLowTemperatureFarenheit(@w5,F13).
```

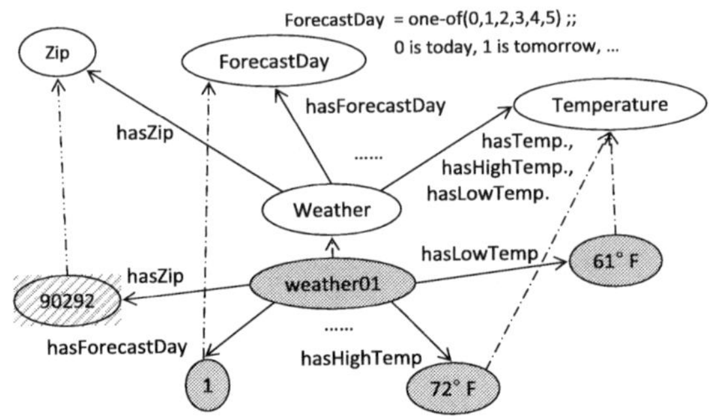

Fig. 2. Results from invoking the Semantic Web Service generated for the Unisys source

This rule means that given an RDF object Z (of type zip) as input, the SWS generated by DEIMOS produces as output an RDF graph consiting of 6 new objects (@w0 ... @w5) with literals for some of their properties as extracted from the web form. For example, the learned SWS for unisys produces the current temperature in centigrade and farenheit degrees, as well as the low temperature of the fifth forecast day among other data. Note that all of the weather forecast objects refer to the same zip code. This learned source description can be seen as a *lifting* rule à la SA-WSDL [11]. Figure 2 illustrates the input/output behaviour of the unisys SWS. The input RDF instance is the 90292 zip (shown with diagonal shading) and the output RDF graph of weather objects (solid shading).

5 Results on Discovering and Modeling New Services

We performed an end-to-end evaluation of DEIMOS on the *geospatial*, *weather*, *flight*, *currency converter*, and *mutual fund* domains. The seeds for these domains are, respectively: geocoder.us, which returns geographic coordinates of a specified address; wunderground.com, which returns weather conditions for a specified location; flytecomm.com, which returns the status of a specified flight; xe.com, which converts amounts from one currency to another based on conversion rates; and finance.yahoo.com, which provides information about mutual funds.

DEIMOS starts by crawling del.icio.us to gather sources possibly related to each seed according to the following strategy. For each seed we (i) retrieve the 20 most popular tags that users applied to this resource; (ii) for each of the tags, retrieve

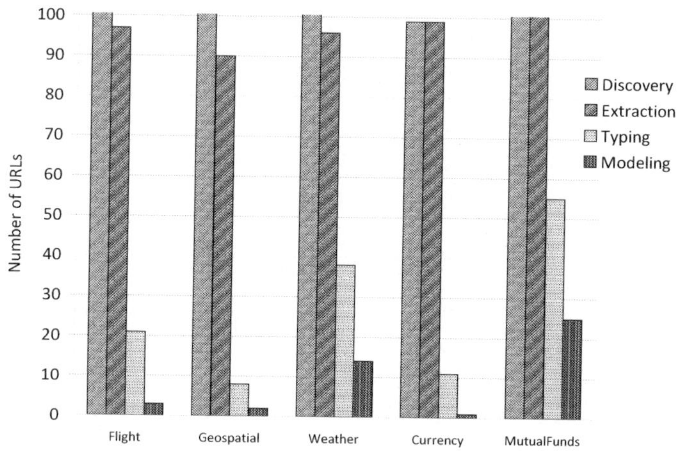

Fig. 3. URL filtering by module for all domains

other sources that have been annotated with that tag; and (iii) collect all tags for each source. We removed low (< 10) and high (> 10,000) frequency tags, and applied LDA, with the number of topics fixed at 80 to learn the hidden topics in each domain. We then ranked sources according to how similar their topic distributions are to the seed.

The 100 top-ranked URLs from the discovery module are passed to the invocation & extraction module, which tries to (1) recognize the form input parameters and calling method on each URL, and (2) extract the resulting output data. For the successful extractions, the semantic typing module, produces a typed input/ouput signature that allows DEIMOS to treat the web sources as web services. Finally, for each typed service, the source modeling module learns the full semantic description. In these experiments, DEIMOS invoked each target source with 10–30 sample inputs.

Figure 3 shows the number of target sources returned by each DEIMOS module.[2] The Invocation & Extraction module provides very little filtering because it is often able to build a template, even in cases where there is no useful data to extract. This happens in some cases when it turns out that the site really is not a good domain source. It can also occur if sample pages from the site have some differences in the DOM structure that cannot be handled with our current heuristics (for example, a weather source which dynamically inserts a severe weather alert into results for some queries). In these cases, the extracted data often contains chunks of HTML that the Source Typing module cannot recognize.

The Semantic Typing and Semantic Modeling modules provide most of the filtering. The Semantic Typing filters a source if it cannot recognize any semantic types other than the input types (which often appear in the output). The Source

[2] Note that we started with only 99 sources for the currency domain because one source was dropped from the experiments at the request of a site administrator.

Table 1. Confusion matrices (A=Actual, P=Predicted, T=True, F=False) for each domain associated with (a) the top-ranked 100 URLs produced by the discovery module, and (b) for the descriptions learned by the semantic modeling module

Geospatial

	PT	PF
AT	8	8
AF	8	76

Weather

	PT	PF
AT	46	15
AF	15	24

Flight

	PT	PF
AT	4	10
AF	10	76

Currency

	PT	PF
AT	56	15
AF	15	14

Mutual funds

	PT	PF
AT	21	16
AF	16	47

(a) Source Discovery

Geospatial

	PT	PF
AT	2	0
AF	0	6

Weather

	PT	PF
AT	15	4
AF	8	14

Flight

	PT	PF
AT	2	0
AF	5	6

Currency

	PT	PF
AT	1	10
AF	0	0

Mutual funds

	PT	PF
AT	17	4
AF	8	26

(b) Source Modeling

Modeling filters a source it fails to build a model that describes any of the source outputs. The primary reasons for failing to find a source model are one of following: (a) the source was not actually a domain source, (b) the semantic typing module learned an incorrect type signature, (c) the source extraction module extracted extraneous text following the extracted data value, or (d) there was a mismatch in the attribute values.

We use two check-points, at the first and last module's output, to evaluate the system by manually checking the retained URLs. We judge the top-ranked 100 URLs produced by the discovery module to be relevant if they provide an input form that takes semantically-similar inputs as the seed and returns domain-relevant outputs. The *geospatial* had $n = 16$ relevant sources, *weather* $n = 61$, *flight* $n = 14$, *currency* $n = 71$ and *mutual funds* $n = 37$.

Table 1(a) shows the confusion matrices associated with the top-ranked 100 sources in each domain. The numbers in the column *PT* show how many of the top-ranked n sources were relevant (*AT*) and not relevant (*AF*) to the domain in question. The R-precision[3] for each domain is 50%, 75%, 29%, 79%, and 57%, respectively (with the same recall values). Although there is a similar number of *geospatial* and *flight* sources, there were twice as many relevant *geospatial* sources (measured by R-precision) among the top-ranked results compared to the *flight* sources. We suspect that the reason for this is less consistency in the vocabulary of users tagging the flight sources.

At the second check-point, we count the services for which DEIMOS learned a semantic description. Table 1(b) presents confusion matrices for this test. In the *geospatial* domain DEIMOS learned source descriptions for 2 out of the 8 semantically-typed sources, namely geocoder.ca and the seed. We manually checked the remaining 6 sources and found out that although some were related to geospatial topics, they were not geocoders. Similarly, in the *weather* domain

[3] R-precision is the precision of the n top-ranked sources, where n is the number of relevant sources in our set of 100 sources.

Table 2. Precision, Recall and F1-measure for actual sources in each domain for which DEIMOS learned descriptions

domain	Precision	Recall	F_1-measure
weather	0.64	0.29	0.39
geospatial	1.00	0.86	0.92
flights	0.69	0.35	0.46
currency	1.00	1.00	1.00
mutualfund	0.72	0.30	0.42

DEIMOS correctly identified 15 relevant (true positives) and 14 not relevant (true negatives) sources; it failed to recognize 4 weather sources and proposed descriptions for 8 sources that were not actual weather sources. The false positives (where the system found a description for a non-weather source) consisted of very short descriptions with only a few attributes modeled. These were the result of invoking a search form, which returned the input, and one of the numeric values on the page randomly matched a seed attribute with a weak pattern for its semantic type. In the *currency* domain, DEIMOS learned the description of one source accurately. It failed to learn description for most of the other sources because the resultant currency value after conversion could not be extracted from them because of their use of Javascript to perform the conversions without generating a new result page. In the *mutualfund* domain, DEIMOS correctly learned source descriptions for 17 sources. There were 8 sources that were incorrectly identified to be from this domain (false positives) because their forms returned a result page where the reported time taken to process the query (e.g., 0.15 s) was incorrectly typed as the change in net value of the fund over a day.

We are ultimately interested in learning logical source descriptions, not just identifying sources input/ouputs types. Therefore, we evaluated the quality of the learned semantic source descriptions. We do this by comparing the learned description to the model a human expert would write for the source. We report precision (how many of the learned attributes were correct), and recall (how many of the actual attributes were learned). The average precision, recall, and F_1-measure for the attributes in the source descriptions learned by DEIMOS for actual services in each domain are shown in Table 2. As an example of our evaluation methodology consider the description learned for geocoder.ca:

```
geocoder.ca(A,_,SA,_,Z,S,_,La,Lo) :- geocoder.us(A,S,C,SA,Z,La,Lo).
```

with attributes A (of semantic type Address), S (Street), C (City), SA (State), Z (ZIP), La (Latitude), and Lo (Longitude). Manually verifying the *attributes* of geocoder.ca yields a precision of 100% (6 correct attributes out of 6 learned) and recall of 86% (6 correct out of 7 present in the *actual* source). Similarly, the conjunctive source description learned for unisys.com, which is shown in Section 3.4, has a precision of 64% (7/11) and a recall of 29% (7/24).

We used strict criteria to judge whether a learned attribute was correct. In one case, for example, the semantic typing component mistakenly identified the field containing flight identifiers such as "United 1174" as Airline, which led to a description containing the Airline attribute. We labeled this attribute as not correct, even though the first component was the airline name. In the *weather* domain, DEIMOS incorrectly labeled the 3rd-day forecast as a 5th-day forecast, because the values of these attributes were sufficiently close. Learning using more sample inputs would reduce the chance of a fortuitous value match.

Overall, we consider these results quite promising. DEIMOS was able to discover Web sources, convert them into programmatically accessible services and learn semantic descriptions of these services in a completely automated fashion. We would like to improve the precision and recall of the learned source models and we believe this can be done largely by improving the semantic typing module and learning over more data.

6 Related Work

Early work on learning semantic descriptions of Internet sources was the *category translation problem* of Perkowitz *et al.* [16]. That problem can be seen as a simplification of the source induction problem, where the known sources have no binding constraints or definitions and provide data that does not change over time. Furthermore, it is assumed that the new source takes a single value as input and returns a single tuple as output. There has also been a significant amount of work on extracting and labeling data found on structured web pages (e.g., the work on Lixto [2]), but this work assumes that a user provides examples of the data to extract and a label for the extracted data, while the approach in this paper requires no labeling.

More recently, there has been work on classifying web services into different domains [8] and on clustering similar services [6]. These techniques can indicate that a new service is likely a *weather* service based on similarity to other weather services. This knowledge is useful for service discovery, but too abstract for automating service integration. We learn more expressive descriptions of web services—view definitions that describe how the attributes of a service relate to one another. Hess & Kushmerick [9] developed an approach that helps users to semantically annotate Web services for data integration. It uses an ensemble of classifiers to predict how various elements of the WSDL should be annotated. The goal of this work is similar, but we handle the more general problem of supporting web sources and our approach works in a completely unsupervised fashion.

Within the bioinformatics space, where web services are widely used, there is a pressing need to build semantic descriptions of existing bioinformatics services. Belhajjame et al. [3] exploit the fact that many of these Web services have been composed into workflows and the connections in the parameters of the workflows can be used to infer constraints on the semantic types of the inputs and outputs of each of these Web services. This is a clever way to infer semantics for Web service parameters, but this method does not provide a complete semantic description of a Web service. Afzal et al. [1] developed an NLP-based approach to learning descriptions of bioinformatics Web services that attempts to extract both the type and the function performed by a service. This approach can provide broad coverage since it can be applied to a wide variety of services, however, it can only provide a high level classification of services (e.g., algorithm, application, data, etc.) and a limited description of the function. In contrast, the goal of DEIMOS is to build a semantic description that is sufficiently detailed to support automatic retrieval and composition.

7 Conclusion

We presented a completely automatic approach to discover new online sources, invoke and extract the data from those sources, learn the semantic types of their inputs and outputs, and learn a semantic description of the function performed by the source. These results allow us to turn an online source into a Semantic Web Service. We also presented empirical results showing that the system can learn semantic models for previously unknown sources. Our approach is general and only requires a small amount of background knowledge for each domain. This work makes it possible to automatically take existing online sources and make them available for use within the Semantic Web.

A limitation of the current work is that it can only learn a new source if it already has models of sources that contain the same information. In future work, we plan to learn models of sources that cover information for which the system has no previous knowledge. In particular, we will focus on learning models of sources for which the current system can already learn partial models. For example, the system might only learn a small subset of the attributes of a particular source. We plan to develop an approach that can learn new semantic types (e.g., barometric pressure), new attributes (e.g., 10th-day forecasted high temperature), new relations that convert between new semantic types and known types (e.g., converting Fahrenheit to Celsius; converting state names to two-letter abbreviations), and learning more accurate descriptions of the domain and ranges of sources (e.g., distinguishing between a weather source that provides information for the US versus one that provides information for the world). The ability to learn models of sources that go beyond the current knowledge within a system will greatly expand the range of sources that the system can discover and model automatically.

Acknowledgments

We would like to thank Mark Carman, Cenk Gazen, Jeonhyung Kang, Steve Minton, Maria Muslea, and Anon Plangprasopchok for their contributions to the system and ideas presented in this paper. This research is based upon work supported in part by the National Science Foundation under award numbers IIS-0324955 and CMMI-0753124, in part by the Air Force Office of Scientific Research under grant number FA9550-07-1-0416, and in part by the Defense Advanced Research Projects Agency (DARPA) under Contract No. FA8750-07-D-0185/0004.

References

1. Afzal, H., Stevens, R., Nenadic, G.: Mining semantic descriptions of bioinformatics web services from the literature. In: Aroyo, L., Traverso, P., Ciravegna, F., Cimiano, P., Heath, T., Hyvönen, E., Mizoguchi, R., Oren, E., Sabou, M., Simperl, E. (eds.) ESWC 2009. LNCS, vol. 5554, pp. 535–549. Springer, Heidelberg (2009)
2. Baumgartner, R., Flesca, S., Gottlob, G.: Declarative information extraction, web crawling, and recursive wrapping with lixto. In: Eiter, T., Faber, W., Truszczyński, M. (eds.) LPNMR 2001. LNCS (LNAI), vol. 2173, pp. 21–41. Springer, Heidelberg (2001)
3. Belhajjame, K., Embury, S.M., Paton, N.W., Stevens, R., Goble, C.A.: Automatic annotation of web services based on workflow definitions. ACM Trans. Web 2(2), 1–34 (2008)
4. Blei, D.M., Ng, A.Y., Jordan, M.I.: Latent dirichlet allocation. Journal of Machine Learning Research 3, 993–1022 (2003)
5. Carman, M.J., Knoblock, C.A.: Learning semantic definitions of online information sources. Journal of Artificial Intelligence Research (JAIR) 30, 1–50 (2007)
6. Dong, X., Halevy, A.Y., Madhavan, J., Nemes, E., Zhang, J.: Simlarity search for web services. In: Proceedings of VLDB (2004)
7. Gazen, B., Minton, S.: Autofeed: an unsupervised learning system for generating webfeeds. In: KCAP 2005: Proceedings of the 3rd international conference on Knowledge capture, pp. 3–10. ACM, New York (2005)
8. Heß, A., Kushmerick, N.: Learning to attach semantic metadata to Web services. In: Fensel, D., Sycara, K., Mylopoulos, J. (eds.) ISWC 2003. LNCS, vol. 2870, pp. 258–273. Springer, Heidelberg (2003)
9. Heß, A., Kushmerick, N.: Iterative ensemble classification for relational data: A case study of semantic web services. In: Boulicaut, J.-F., Esposito, F., Giannotti, F., Pedreschi, D. (eds.) ECML 2004. LNCS (LNAI), vol. 3201, pp. 156–167. Springer, Heidelberg (2004)
10. Hofmann, T.: Probabilistic latent semantic analysis. In: Proc. of UAI, pp. 289–296 (1999)
11. Kopecky, J., Vitvar, T., Bournez, C., Farrell, J.: Sawsdl: Semantic annotations for WSDL and XML schema. IEEE Internet Computing 11(6), 60–67 (2007)
12. Lerman, K., Minton, S., Knoblock, C.: Wrapper maintenance: A machine learning approach. Journal of Artificial Intelligence Research 18, 149–181 (2003)
13. Lerman, K., Plangprasopchok, A., Knoblock, C.A.: Semantic labeling of online information sources. International Journal on Semantic Web and Information Systems, Special Issue on Ontology Matching 3(3), 36–56 (2007)
14. Levy, A.Y.: Logic-based techniques in data integration. In: Minker, J. (ed.) Logic-Based Artificial Intelligence. Kluwer Publishers, Dordrecht (2000)
15. Plangprasopchok, A., Lerman, K.: Exploiting social annotation for resource discovery. In: AAAI workshop on Information Integration on the Web, IIWeb 2007 (2007)
16. Perkowitz, M., Doorenbos, R.B., Etzioni, O., Weld, D.S.: Learning to understand information on the Internet: An example-based approach. Journal of Intelligent Information Systems 8, 133–153 (1999)
17. Plangprasopchok, A., Lerman, K.: Modeling social annotation: a bayesian approach. Technical report, Computer Science Department, University of Southern California (2009)

Exploiting User Feedback to Improve Semantic Web Service Discovery

Anna Averbakh, Daniel Krause, and Dimitrios Skoutas

L3S Research Center
Hannover, Germany
{averbakh,krause,skoutas}@l3s.de

Abstract. State-of-the-art discovery of Semantic Web services is based on hybrid algorithms that combine semantic and syntactic matchmaking. These approaches are purely based on similarity measures between parameters of a service request and available service descriptions, which, however, fail to completely capture the actual functionality of the service or the quality of the results returned by it. On the other hand, with the advent of Web 2.0, active user participation and collaboration has become an increasingly popular trend. Users often rate or group relevant items, thus providing valuable information that can be taken into account to further improve the accuracy of search results. In this paper, we tackle this issue, by proposing a method that combines multiple matching criteria with user feedback to further improve the results of the matchmaker. We extend a previously proposed dominance-based approach for service discovery, and describe how user feedback is incorporated in the matchmaking process. We evaluate the performance of our approach using a publicly available collection of OWL-S services.

1 Introduction

Web services have emerged as a key technology for implementing Service Oriented Architectures, aiming at providing interoperability among heterogeneous systems and integrating inter-organization applications. At the same time, users increasingly use the Web to search not only for pages or other multimedia resources, but also to find services for fulfilling a given task. For example, a service typically either returns some information to the user, such as a weather forecast for a given location and time period, or it performs some task, such as flight booking for a given date, departure and destination. Hence, as both the user needs and the number of available services and service providers increases, improving the effectiveness and accuracy of Web service discovery mechanisms becomes a crucial issue.

A Web service description is a document written in a formal, standardized language, providing information about what the service does and how it can be used. Given a service request, expressed as the description of a desired service, the task of matchmaking refers to identifying and selecting from a service repository those services whose description closely matches the request, under

A. Bernstein et al. (Eds.): ISWC 2009, LNCS 5823, pp. 33–48, 2009.

one or more matching criteria. Typically, the matchmaking process is based on computing a degree of match between the parameters of the requested and advertised service, which may include both functional parameters (i.e., inputs, outputs, pre-conditions, and effects), as well as non-functional parameters (i.e., QoS attributes, such as price, execution time, availability).

Several approaches, reviewed in Section 2, exist for this task, ranging from methods relying on techniques commonly used in Information Retrieval [1], to methods employing logic-based matching [2]. Also, hybrid methods have been proposed [3], and, more recently, methods that combine multiple matching criteria simultaneously [4]. Nevertheless, all these works rely solely on the parameters involved in the service descriptions, thus facing an important drawback. A service description may not always capture completely and accurately all the aspects related to the functionality and usage of the service. Additional information which may be useful for determining how appropriate a service is for a given request, often remains implicit, not being encoded in the formal description. Moreover, different services may be more relevant to a specific request for different users and in different contexts.

On the other hand, with the advent of Web 2.0, active user participation and collaboration has become an increasingly popular trend. Users provide valuable information by tagging, rating, or grouping similar resources, such as bookmarks (e.g., Del.icio.us), music (e.g., Last.fm) or videos (e.g., YouTube). Essentially, these activities allow collecting user feedback, which can then be exploited to enhance the accuracy and effectiveness of the search [5].

In this paper, we propose a method for leveraging user feedback to improve the results of the service discovery process. Given a service request, the matchmaker searches the repository for available services and returns a ranked list of candidate matches. Then, the system allows the user posing the query to rate any of these matches, indicating how relevant or appropriate they are for this request. The provided ratings are stored in the system for future use, when the same or a similar request is issued. Designing intuitive, easy-to-use user interfaces, can help the process of collecting user feedback. In this work, we do not deal with this issue; instead, our focus is on how the collected feedback is processed and integrated in the matchmaking process to improve the results of subsequent searches. Notice, that it is also possible to collect user feedback automatically, assuming that the system can track which service(s) the user actually used; however, this information would typically be incomplete, since not all relevant services are used.

Our main contributions are summarized below.

- We consider the problem of employing user feedback to improve the quality of search for Semantic Web services.
- We propose a method for processing user feedback and incorporating it in the matchmaking process.
- We experimentally evaluate our approach on a publicly available collection of Semantic Web services, applying standard Information Retrieval evaluation metrics.

The rest of the paper is organized as follows. Section 2 reviews related work. Section 3 presents our architecture for incorporating user feedback in the service discovery process. Section 4 describes the basic matchmaking approach, employing multiple matching criteria. Applying user feedback to improve the search results is presented in Section 5. Finally, Section 6 presents our experimental results, while Section 7 concludes the paper.

2 Related Work

WSDL and UDDI are existing industry standards for describing and discovering Web services. However, their focus lies on specifying the structure of the service interfaces and of the exchanged messages. Thus, they address the discovery problem relying on structural, keyword-based matching, which limits their search capabilities. Other earlier works have also focused on applying Information Retrieval techniques to the service discovery problem. For example, the work presented in [1] deals with similarity search for Web services, using a clustering algorithm to group names of parameters into semantically meaningful concepts, which are then used to determine the similarity between input/output parameters. An online search engine for Web services is seekda[1], which crawls and indexes WSDL files from the Web. It allows users to search for services by entering keywords, by using tag clouds, or by browsing using different facets, such as the country of the service provider, the most often used services or the most recently found ones.

To deal with the shortcomings of keyword search, several approaches have been proposed for exploiting ontologies to semantically enhance the service descriptions (WSDL-S [6], OWL-S [7], WSMO [8]). These so-called Semantic Web services can better capture and disambiguate the service functionality, allowing for formal, logic-based matchmaking. Essentially, a logic reasoner is employed to infer subsumption relationships between requested and provided service parameters [2,9]. Along this line, several matching algorithms assess the similarity between requested and offered inputs and outputs by comparing the positions of the corresponding classes in the associated domain ontology [10,11,12]. Similarly, the work in [13] semantically matches requested and offered parameters, modeling the matchmaking problem as one of matching bipartite graphs. In [14], OWL-S services are matched using a similarity measure for OWL objects, which is based on the ratio of common RDF triples in their descriptions. An approach for incorporating OWL-S service descriptions into UDDI is presented in [15], focusing also on the efficiency of the discovery process. Efficient matchmaking and ranked retrieval of services is also studied in [16].

Given that logic-based matching can often be too rigid, hybrid approaches have also been proposed. In an earlier work [17], the need for employing many types of matching has been discussed, proposing the integration of multiple external matching services to a UDDI registry. The selection of the external matching service to be used is based on specified policies, e.g., selecting the

[1] http://seekda.com/

first available, or the most successful. If more than one matching services are invoked, again the system policies specify whether the union or the intersection of the results should be returned. OWLS-MX [3] and WSMO-MX [18] are hybrid matchmakers for OWL-S and WSMO services, respectively. More recently, an approach for simultaneously combining multiple matching criteria has been proposed [4].

On the other hand, some approaches already exist about involving the user in the process of service discovery. Ontologies and user profiles are employed in [19], which then uses techniques like query expansion or relaxation to better satisfy user requests. The work presented in [20] focuses on QoS-based Web service discovery, proposing a reputation-enhanced model. A reputation manager assigns reputation scores to the services based on user feedback regarding their performance. Then, a discovery agent uses the reputation scores for service matching, ranking and selection. The application of user preferences, expressed in the form of soft constraints, to Web service selection is considered in [21], focusing on the optimization of preference queries. The approach in [22] uses utility functions to model service configurations and associated user preferences for optimal service selection. In [1], different types of similarity for service parameters are combined using a linear function, with manually assigned weights. Learning the weights from user feedback is proposed, but it is left as an open issue for future work.

Collecting and exploiting user feedback as a form of interaction between a user and an application is a key concept of Web 2.0 [23]. Users are more than ever before willing to actively contribute by tagging, rating or commenting any kind of content or offered service – even better, their feedback is very valuable to improve search and retrieval algorithms [24,25]. In this work we employ a simple user feedback model, focusing on how this feedback can be exploited in Semantic Web service discovery, to improve the quality and accuracy of the search results. More fine-grained forms of user feedback, as well as exploiting additional information from social networks of users [26] are considered as possible extensions of our approach.

In a different line of research, relevance feedback has been extensively considered in Information Retrieval [27]. The main idea is that the user issues a query, the system returns an initial set of results, and the user marks some of the returned documents as relevant or non-relevant; then, based on this user feedback, the system revises the set of retrieved results to better capture the user's information need. However, approaches in this area typically rely on the vector space model and term re-weighting, and, hence, they are not suitable for service matchmaking.

3 Architecture

Typical service matchmaking systems are based on a unidirectional information flow. First, an application that needs a specific Web Service to perform a task creates a service request, containing the requirements that a service should fulfil. This service request is then delivered to a matchmaking component that utilizes

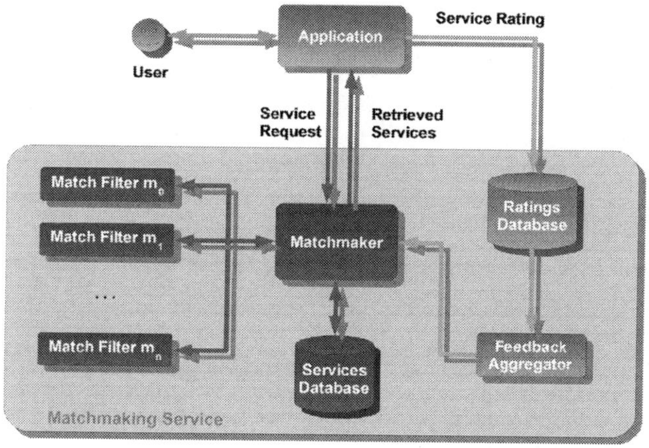

Fig. 1. Matchmaking service with feedback component

one or more match filters to retrieve the best-matching services from a repository of Semantic Web service descriptions. These services are finally returned to the application which invoked the matchmaker. The drawback in this scenario is that if a service is not appropriate or sufficient for any reason to perform the original task, the application has no option to inform the matchmaker about the inappropriateness of this match result.

Hence, our matchmaking architecture is extended by a feedback loop, as illustrated in Figure 1, enabling the matchmaking mechanism to use previously provided user feedback in order to improve the quality of the retrieved results.

Enabling this feedback loop, relies on the assumption that the application users can assess the quality of retrieved Web services. This is a common principle in Web 2.0 applications, where users can rate available resources. One possibility is that users can rate services explicitly. If it is not possible or easy for the users to rate services directly, the application can still infer implicit ratings for a service through user behavior. For example, if an applications uses services to generate music recommendations, then users can be asked whether they consider the given recommendations appropriate. Based on the assumption that services delivering high quality recommendations are better matches for this task, the application can infer the relevance of a service, and pass this information as a user rating to the matchmaking service.

The user ratings are stored in an RDF triple store (e.g., SESAME [28]). As user ratings refer to a given service request, each `Rating` instance contains the user who performed the rating, the service request, the rated service, and finally a rating score that ranges from 0 to 1 (with higher scores denoting higher rating). For example, a rating from Bob about a request X and a service Y would be stored as:

```
<r:Rating>
  <foaf:Person rdf:about="#bob"/>
  <r:Request rdf:about="#requestX"/>
  <r:Service rdf:about="#serviceY"/>
  <r:Score rdf:datatype="&xsd;double">0.90</r:score>
</r:Rating>
```

This RDF database, which contains the user feedback in form of ratings, is exploited by the user feedback component. This component aggregates previous ratings provided by different users, to determine the relevance between a service request and an actual service.

Then, given a service request, the matchmaker component combines the relevance score from the feedback component with the similarity scores calculated by the match filter(s) to assess the degree of match for each available service, and returns a ranked list of match results to the application.

4 Service Matchmaking

In this section we describe the basic service matchmaking and ranking process, without taking into account user feedback. For this task, we adopt the approach from [4], because, as shown in Section 5, it allows us to integrate user feedback in a more flexible and seamless way. In the following, we give a brief overview of how the matchmaking and ranking of services is performed.

For simplicity, we focus on input and output parameters, annotated by ontological concepts, but other types of parameters can be handled similarly. Let R be a service request with a set of input and output parameters, denoted by R_{IN} and R_{OUT}, respectively. We use $R.p_j$ to refer to the j-th input parameter, where $p_j \in R_{IN}$ (similarly for outputs). Also, assume an advertised service S with inputs and outputs S_{IN} and S_{OUT}, respectively. Note that S can be a match to R, even when the cardinalities of their parameter sets differ, i.e., when a service advertisement requires less inputs or produces more outputs than requested.

The matchmaking process applies one or more matching functions to assess the degree of match between requested and offered parameters. Each matching function produces a score in the range $[0, 1]$, where 1 indicates a perfect match, while 0 indicates no match. Typical examples are the matching functions provided by the OWLS-MX service matchmaker [3]. These comprise a purely logic-based match (M0), as well as hybrid matches based on string similarity measures, namely loss-of-information (M1), extended Jaccard similarity coefficient (M2), cosine similarity (M3), and Jensen-Shannon information divergence based similarity (M4). Given a request R, a service S, and a matching function m_i, the *match instance* of S with respect to R is defined as a vector s_i such that

$$s_i[j] = \begin{cases} \max\limits_{p_k \in S_{IN}} \{m_i(S.p_k, R.p_j)\}, & \forall j : p_j \in R_{IN} \\ \max\limits_{p_k \in S_{OUT}} \{m_i(S.p_k, R.p_j)\}, & \forall j : p_j \in R_{OUT} \end{cases} \tag{1}$$

Table 1. Example of the match object for the request `book_price_service.owls` and the service `novel_price_service.owls`

Match Filter	Book	Price
M0	0.88	1.00
M1	0.93	1.00
M2	0.69	1.00
M3	0.72	1.00
M4	0.93	1.00

The match instance s_i has a total of $d = |R_{IN}| + |R_{OUT}|$ entries that correspond to the input and output parameters of the request. Intuitively, each s_i entry quantifies how well the corresponding parameter of the request R is matched by the advertisement S, under the matching criterion m_i. Clearly, an input (output) parameter of R can only match with an input (output) parameter of S.

Let $\mathcal{M}$ be a set of matching functions. Given a request R and an advertisement S, each $m_i \in \mathcal{M}$ results in a distinct match instance. We refer to the set of instances as the *match object* of the service S. In the following, and in the context of a specific request R, we use the terms service and match object interchangeably, denoted by the same uppercase letter (e.g., S). On the other hand we reserve lowercase letters for match instances of the corresponding service (e.g., s_1, s_2, etc.). The notation $s_i \in S$ implies that the match instance s_i corresponds to the service S. Hence, a match object represents the result of the match between a service S and a request R, with each contained match instance corresponding to the result of a different match function.

As a concrete example, consider a request `book_price_service.owls`, with input `Book` and output `Price`, and a service `novel_price_service.owls`, with input `Novel` and output `Price`, matched applying the five aforementioned functions M0–M4 of the OWLS-MX matchmaker. The resulting match object is shown in Table 1, comprising an exact match for the outputs, while the degree of match between `Book` and `Novel` varies based on the similarity measure used.

Next, we describe how services are ranked based on their match objects. Let $\mathcal{I}$ be the set of all match instances of all services. Given two instances $u, v \in \mathcal{I}$, we say that u *dominates* v, denoted by $u \succ v$, iff u has a higher or equal degree of match in all parameters and a strictly higher degree of match in at least one parameter compared to v. Formally

$$u \succ v \quad \Leftrightarrow \quad \forall i \; u[i] \geq v[i] \wedge \exists j \; u[j] > v[j] \tag{2}$$

If u is neither dominated by nor dominates v, then u and v are incomparable.

Given this dominance relationship between match instances, we proceed with defining *dominance scores* that are used to rank the available service descriptions with respect to a given service request. Intuitively, a service should be ranked highly in the list if

- its instances are dominated by as few other instances as possible, and
- its instances dominate as many other instances as possible.

To satisfy these requirements, we formally define the following dominance scores, used to rank the search results for a service request.

Given a match instance u, we define the *dominated score* of u as

$$u.dds = \frac{1}{|\mathcal{M}|} \sum_{V \neq U} \sum_{v \in V} |v \succ u| \tag{3}$$

where $|u \succ v|$ is 1 if $u \succ v$ and 0 otherwise. Hence, $u.dds$ accounts for the instances that dominate u. Then, the dominated score of a service U is defined as the (possibly weighted) average of the dominated scores of its instances:

$$U.dds = \frac{1}{|\mathcal{M}|} \sum_{u \in U} u.dds \tag{4}$$

The dominated score of a service indicates the average number of services that dominate it, i.e., a *lower* dominated score indicates a better match result.

Next, we look at the instances that a given instance dominates. Formally, given a match instance u, we define the *dominating score* of u as

$$u.dgs = \frac{1}{|\mathcal{M}|} \sum_{V \neq U} \sum_{v \in V} |u \succ v| \tag{5}$$

Similarly to the case above, the dominating score of a service U is then defined as the (possibly weighted) average of the dominating scores of its instances:

$$U.dgs = \frac{1}{|\mathcal{M}|} \sum_{u \in U} u.dgs \tag{6}$$

The dominating score of a service indicates the average number of services that it dominates, i.e., a *higher* dominating score indicates a better match result.

Finally, we define the *dominance score* of match instances and services, to combine both of the aforementioned criteria. In particular, the dominance score of a match instance u is defined as

$$u.ds = u.dgs - \lambda \cdot u.dds \tag{7}$$

where the parameter λ is a scaling factor. This promotes u for each instance it dominates, while penalizing it for each instance that dominates it. Then, the dominance score of a service U is defined as the (possibly weighted) average of the dominance scores of its instances:

$$U.ds = \frac{1}{M} \sum_{u \in U} u.ds \tag{8}$$

The ranking process comprises computing the aforementioned scores for each service, and then sorting the services in descending order of their dominance score. Efficient algorithms for this computation can be found in [4].

5 Incorporating User Feedback

In this section, we present our approach for processing user feedback and incorporating it in the dominance-based matchmaking and ranking method described in Section 4.

Our approach is based on the assumption that the system collects feedback from the users by allowing them to rate how appropriate the retrieved services are with respect to their request (see feedback component in Section 3). Assume that the collected user ratings are stored as a set $\mathcal{T} \subseteq \mathcal{U} \times \mathcal{R} \times \mathcal{S} \times F$ in the Ratings Database, where $\mathcal{U}$ is the set of all users that have provided a rating, $\mathcal{R}$ is the set of all previous service requests stored in the system, $\mathcal{S}$ is the set of all the available Semantic Web service descriptions in the repository, and $F \in [0, 1]$ denotes the user rating, i.e., how relevant a particular service was considered with respect to a given request (with higher values representing higher relevance). Thus, a tuple $T = (U, R, S, f) \in \mathcal{T}$ denotes that a user U considers the service $S \in \mathcal{S}$ to be relevant for the request $R \in \mathcal{R}$ with a score f.

To aggregate the ratings from different users into a single feedback score, different approaches can be used. For example, [29] employs techniques to identify and filter out ratings from spam users, while [30] proposes the aging of feedback ratings, considering the more recent ratings as more relevant. It is also possible to weight differently the ratings of different users, assigning, for example, higher weights to ratings provided previously by the same user as the one currently issuing the request, or by users that are assumed to be closely related to him/her, e.g., by explicitly being included in his/her social network or being automatically selected by the system through techniques such as collaborative filtering or clustering. However, as the discussion about an optimal aggregation strategy for user ratings is orthogonal to our main focus in this paper, without loss of generality we consider in the following all the available user ratings as equally important, and we calculate the feedback value as the average of all user ratings of the corresponding service. Hence, the feedback score fb between a service request $R \in \mathcal{R}$ and a service advertisement $S \in \mathcal{S}$ is calculated as:

$$fb(R, S) = \frac{\sum\limits_{(U,R,S,f) \in \mathcal{T}} f}{|\{(U, R, S, f) \in T\}|} \tag{9}$$

However, it may occur that for a given pair of a request R and a service S, no ratings (U, R, S, f) exist in the database. This may be because the request R is new, or because the service S has been recently added to the database and therefore has been only rated for a few requests. Moreover, even if some ratings exist, they may be sparse and hence not provide sufficiently reliable information for feedback. In these cases, Equation (9) is not appropriate for determining the feedback information for the pair (R, S). To address this issue, we generalize this method to consider not only those ratings that are directly assigned to the current service requests R, but also user ratings that are assigned to requests that are similar to R. Let $SIM(R)$ denote the set of requests which are considered to be similar to R. Then, the feedback can be calculated as:

$$fb(R,S) = \frac{\sum_{(U,Q,S,f)\in\mathcal{T}:Q\in SIM(R)} f * sim(R,Q)}{|\{(U,Q,S,f)\in\mathcal{T} : Q \in SIM(R)\}|} \tag{10}$$

In Equation (10), $sim(R,Q)$ is the match instance of Q with respect to R, calculated by a matching function m_i, as discussed in Section 4. Notice that $sim(R,Q)$ is a vector of size equal to the number of parameters of R, hence in this case $fb(R,S)$ is also such a vector, i.e., similar to a match instance. Also, Equation (9) can be derived as a special case of Equation (10), by considering $SIM(R) = \{R\}$. By weighting the given feedback by the similarity between the requests, we ensure that feedback from requests which are more similar to the considered one, is taken more into account.

A question that arises is how to select the similar requests for a given request R, i.e., how to determine the set $SIM(R)$. This choice involves a trade-off. Selecting a larger number of similar queries, allows the use of more sources of information for feedback; however, if the similarity between the original request and the selected ones is not high enough, then the information from this feedback is also not highly appropriate, and may eventually introduce noise in the results. On the other hand, setting a very strict criterion for selecting similar queries, reduces the chance of finding enough feedback information. As a solution to this trade-off, we use a top-k query with constraints: given a request R, we select the top-k most similar requests from the database, given that the values of their match instances are above a specified threshold.

The process described above results in a *feedback instance* $fb(R,S)$ for the given request R and a service S. The next step is to integrate this instance to the match object of the service S, comprising the other instances obtained by the different matching functions m_i. We investigate two different strategies for this purpose:

1. *Feedback instance as an additional match instance.* In this case we add the feedback information to the match object of the service as an additional instance (combined with the average of the previous values). That is, this method treats the feedback mechanism as an extra matchmaking function.
2. *Feedback instance integrated with match instances.* In this case we update the values of the match instances by adding the values of the feedback instance. That is, this method adjusts the results of the matchmaking functions applying the feedback information.

As a concrete example, consider the match object presented in Table 1. Assume that the feedback instance for the pair (`book_price_service.owls`, `novel_price_service.owls`) is $fb = [0.77 \quad 1.00]$. Then this match object will be modified as shown in Table 2.

Table 2. Example of the match object for the request `book_price_service.owls` and the service `novel_price_service.owls` updated using feedback information

(a) Method 1

Match Filter	Book	Price
M0	0.88	1.00
M1	0.93	1.00
M2	0.69	1.00
M3	0.72	1.00
M4	0.93	1.00
AVG(M_i)+FB	1.60	2.00

(b) Method 2

Match Filter	Book	Price
M0+FB	1.65	2.00
M1+FB	1.70	2.00
M2+FB	1.46	2.00
M3+FB	1.49	2.00
M4+FB	1.70	2.00

6 Experimental Evaluation

In this section, we evaluate the quality of our feedback-based matchmaking approach in comparison to state-of-the-art matchmaking algorithms. In Section 6.1, we discuss implementation issues and the data collections we have used. The evaluation metrics and the experimental results are then reported in Section 6.2.

6.1 Experimental Setup

We have implemented the feedback-based matchmaking and ranking process described in Sections 4 and 5. The implementation utilizes the OWLS-MX service matchmaker [3], to process service requests and advertisements described in OWL-S, and to compute the pairwise similarities between parameters. In particular, OWLS-MX provides five different matching filters. The first performs a purely logic-based match (M0), characterising the result as exact, plug-in, subsumes, or subsumed-by. The other four perform hybrid match, combining the semantic-based matchmaking with the following measures: loss-of-information (M1), extended Jaccard similarity coefficient (M2), cosine similarity (M3), and Jensen-Shannon information divergence based similarity (M4). Notice, that for each pair (R, S) of a service request and service advertisement, OWLS-MX applies one of the filters M0–M4, and calculates a single score denoting the degree of match between R and S. We have modified this functionality to get all the individual degrees of match between the compared parameters of R and S (i.e., a vector); also, we have applied for each pair (R, S) all the similarity measures M0–M4, to get the individual match instances, as described in Section 4. Finally, our implementation includes also the process described in Section 5 for processing and using the available feedback information.

For our experiments, we have used the publicly available service retrieval test collection OWLS-TC v2[2]. This collection comes in two versions, an original one containing 576 services, and an extended one, containing 1007 services. To better assess the performance of our method, we have conducted our experiments on

[2] This collection is available at `http://projects.semwebcentral.org/projects/owls-tc/`. Before running the experiments we have fixed some typos that prevented some services from being processed and/or retrieved.

Table 3. Characteristics of the test collections

Collection	# of requests	# of services	# of rel. services per req. (average)
OWL-S TC I	28	576	15.2
OWL-S TC II	28	1007	25.4

both versions, denoted in the following as OWLS-TC I and OWLS-TC II, respectively. The contained service descriptions are based on real-world Web services, retrieved mainly from public IBM UDDI registries, covering 7 different domains, such as economy, education, and travel. Also, the collection comprises a set of 28 sample requests. Notice that the extended version of the collection comprises one extra request, namely `EBookOrder1.owls`; however, in our experiments, we have excluded this request, so that in both cases the set of queries used for the evaluation is the same. For each request, a relevance set is provided, i.e., the list of services that are considered relevant to this request, based on human judgement. The characteristics of the two data sets are summarized in Table 3.

To evaluate our feedback-based mechanism, there needs to be, for each request, at least one similar request for which some services have been rated as relevant. As this was not the case with the original data set, due to the small number of provided requests, we have extended both of the aforementioned collections by creating a similar query for each of the 28 original ones. This was done by selecting a request, then selecting one or more of its input and/or output parameters, and replacing its associated class in the ontology with one that is a superclass, subclass or sibling. Then, for each of these newly created queries, some of the services in the collection were rated as relevant. To simplify this task, we have restricted our experimental study in binary ratings, i.e., the value of the user rating was either 1 or 0, based on whether the user considered the service to be relevant to the request or not. The new queries and the ratings, provided in the form of corresponding relevance sets, are made available for further use at: `http://www.l3s.de/~krause/collection.tar.gz`

6.2 Experimental Results

In the following, we evaluate the performance of our approach, including both strategies described in Section 5. For this purpose, we compare the retrieved results to the ones produced without taking user feedback into consideration. In particular, we have implemented and compared the following 5 methods:

- *NF1*: No feedback is used; one match instance per service is considered. The values of the match instance are the degrees of match between the request and service parameters, computed applying the Jensen-Shannon similarity measure, i.e., the filter M4 from OWLS-MX, which is shown in [3] to slightly outperform the other measures.
- *NF5*: No feedback is used; five match instances per service are considered. The values of the match instances are the degrees of match between the request and service parameters computed by the filters M0–M4 of OWLS-MX.

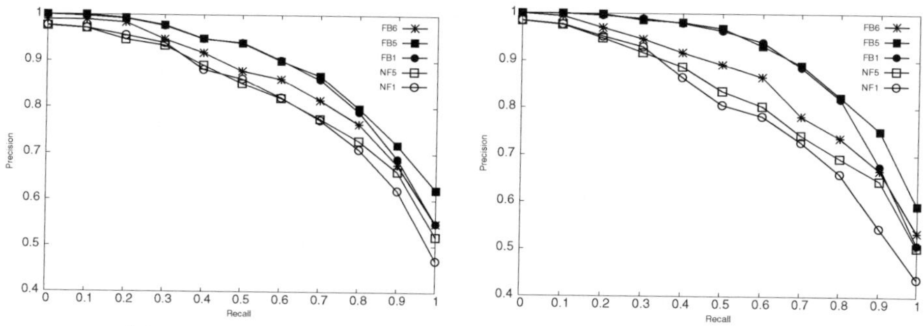

(a) OWLS-TC I (b) OWLS-TC II

Fig. 2. Precision-Recall curve for the OWLS test collections

- *FB1*: Feedback is used; one match instance per service is considered. The values of the match instance are the sum of the degrees of match between the request and service parameters computed by M4 and the feedback values calculated by Equation (10).
- *FB5*: Feedback is used; five match instances per service are considered. The value of each match instance is the sum of the degrees of match between the request and service parameters computed by one of the measures M0–M4 and the feedback values calculated by Equation (10).
- *FB6*: Feedback is used; six match instances per service are considered. The values of the first five match instances are the degrees of match between the request and service parameters computed by the filters M0–M4. The values of the sixth match instance are computed as the averages of the previous ones plus the feedback values calculated by Equation (10). Notice, that the reason for using also the average values of the initial instances, instead of only the feedback values, is mainly to avoid penalizing services that constitute good matches but have not been rated by users.

To measure the effectiveness of the compared approaches, we apply the following standard IR evaluation measures [27]:

- *Interpolated Recall-Precision Averages*: measures precision, i.e., percent of retrieved items that are relevant, at various recall levels, i.e., after a certain percentage of all the relevant items have been retrieved.
- *Mean Average Precision (MAP)*: average of precision values calculated after each relevant item is retrieved.
- *R-Precision (R-prec)*: measures precision after all relevant items have been retrieved.
- *bpref*: measures the number of times judged non-relevant items are retrieved before relevant ones.

Table 4. IR metrics for the OWLS test collections

(a) OWLS-TC I

Method	MAP	R-prec	bpref	R-rank	P@5	P@10	P@15	P@20
FB6	0.8427	0.7772	0.8206	0.9762	0.9214	0.8357	0.7690	0.6589
FB5	0.8836	0.7884	0.8600	1.0000	0.9714	0.8857	0.7952	0.6696
FB1	0.8764	0.7962	0.8486	1.0000	0.9786	0.8786	0.7929	0.6625
NF5	0.8084	0.7543	0.7874	0.9405	0.9071	0.7964	0.7500	0.6393
NF1	0.8027	0.7503	0.7796	0.9405	0.9214	0.8143	0.7357	0.6357

(b) OWLS-TC II

Method	MAP	R-prec	bpref	R-rank	P@5	P@10	P@15	P@20
FB6	0.8426	0.7652	0.8176	1.0000	0.9714	0.8964	0.8476	0.7875
FB5	0.9090	0.8242	0.8896	1.0000	0.9857	0.9679	0.9214	0.8536
FB1	0.8960	0.8024	0.8689	1.0000	0.9857	0.9607	0.9167	0.8411
NF5	0.8007	0.7388	0.7792	0.9643	0.9429	0.8607	0.8119	0.7536
NF1	0.7786	0.7045	0.7499	0.9643	0.9357	0.8607	0.7976	0.7268

- *Reciprocal Rank (R-rank)*: measures (the inverse of) the rank of the top relevant item.
- *Precision at N(P@N)*: measures the precision after N items have been retrieved.

Figure 2 plots the precision-recall curves for the 5 compared methods, for both considered test collections. Overall, the main observation is that the feedback-aware methods clearly outperform the other two ones in both test collections. The best overall method in both collections is $FB5$, because it provides two advantages: a) it utilizes user feedback, and b) it combines all the available similarity measures for matchmaking service parameters. The method $FB1$, which combines feedback information with the Jensen-Shannon hybrid filter, also demonstrates a very high accuracy. The method $FB6$, which treats the feedback information as an additional match instance, achieves lower precision, but it still outperforms the non-feedback methods. This behavior is due to the fact that although feedback is utilized, its impact is lower since it is not considered for the 5 original match instances, but only as an extra instance. Regarding $NF5$ and $NF1$, the former exhibits better performance, which is expected as it combines multiple similarity measures. Another interesting observation is that $FB5$ and $FB1$ follow the same trend as $NF5$ and $NF1$, respectively, which are their non-feedback counterparts, however having considerably higher precision values at all recall levels. Finally, for the collection OWLS-TC II, which comprises an almost double number of services, the trends are the same as before, but with the differences between the feedback-aware and the non-feedback methods being even more noticeable. Another interesting observation in this case is that after the recall level 0.8 the precision of $FB1$ drops much faster than that of $FB6$; thus, although $FB1$ has an overall higher performance than $FB6$, the latter appears to be more stable, which is due to having more instances per match object, i.e., taking into account more similarity measures.

Table 4 presents the results for the other IR evaluation metrics discussed above. These results again confirm the aforementioned observations. For all the considered metrics, $FB5$ and $FB1$ perform better, followed by $FB6$.

7 Conclusions

In this paper we have dealt with the problem of Semantic Web service discovery, focusing on how to exploit user feedback to improve the quality of the search results. This relies on the idea that modern Web 2.0 applications allow users to explicitly express their opinion by giving feedback about available resources, in the form of rating, tagging, etc. We have presented an architecture that allows to collect user feedback on retrieved services and incorporate it in the Semantic Web service matchmaking process. We have proposed different methods to combine this user feedback with the output of matchmaking algorithms in order to improve the quality of the match results. Further, we have discussed how to overcome the problems of a limited amount of feedback or of previously unknown requests (i.e., where no previous feedback is available for the request), by utilizing information from similar requests. We have conducted an experimental evaluation using a publicly available collection of OWL-S services. We have compared our feedback-aware matchmaking strategies to state-of-the-art matchmaking algorithms that do not take feedback into account. Our experimental results show that user feedback is a valuable source of information for improving the matchmaking quality.

Our current and future work focuses mainly on two directions: a) investigating in more detail the social aspects involved in the process of collecting and aggregating user ratings for services, and b) extending our experimental setup with additional scenarios.

Acknowledgements. This work was partially supported by the European Commission in the context of the FP7 projects SYNC3 and GRAPPLE.

References

1. Dong, X., Halevy, A.Y., Madhavan, J., Nemes, E., Zhang, J.: Similarity Search for Web Services. In: VLDB, pp. 372–383 (2004)
2. Paolucci, M., Kawamura, T., Payne, T.R., Sycara, K.P.: Semantic Matching of Web Services Capabilities. In: Horrocks, I., Hendler, J. (eds.) ISWC 2002. LNCS, vol. 2342, pp. 333–347. Springer, Heidelberg (2002)
3. Klusch, M., Fries, B., Sycara, K.P.: Automated Semantic Web service discovery with OWLS-MX. In: AAMAS, pp. 915–922 (2006)
4. Skoutas, D., Sacharidis, D., Simitsis, A., Kantere, V., Sellis, T.: Top-k Dominant Web Services under Multi-criteria Matching. In: EDBT, pp. 898–909 (2009)
5. Bao, S., Xue, G.-R., Wu, X., Yu, Y., Fei, B., Su, Z.: Optimizing Web Search Using Social Annotations. In: WWW, pp. 501–510 (2007)
6. Akkiraju, R., et al.: Web Service Semantics - WSDL-S. In: W3C Member Submission (November 2005)
7. Burstein, M., et al.: OWL-S: Semantic Markup for Web Services. In: W3C Member Submission (November 2004)
8. Lausen, H., Polleres, A., Roman, D. (eds.): Web Service Modeling Ontology (WSMO). W3C Member Submission (June 2005)
9. Li, L., Horrocks, I.: A Software Framework for Matchmaking based on Semantic Web Technology. In: WWW, pp. 331–339 (2003)

10. Cardoso, J.: Discovering Semantic Web Services with and without a Common Ontology Commitment. In: IEEE SCW, pp. 183–190 (2006)
11. Skoutas, D., Simitsis, A., Sellis, T.K.: A Ranking Mechanism for Semantic Web Service Discovery. In: IEEE SCW, pp. 41–48 (2007)
12. Skoutas, D., Sacharidis, D., Kantere, V., Sellis, T.: Efficient Semantic Web Service Discovery in Centralized and P2P Environments. In: Sheth, A.P., Staab, S., Dean, M., Paolucci, M., Maynard, D., Finin, T., Thirunarayan, K. (eds.) ISWC 2008. LNCS, vol. 5318, pp. 583–598. Springer, Heidelberg (2008)
13. Bellur, U., Kulkarni, R.: Improved Matchmaking Algorithm for Semantic Web Services Based on Bipartite Graph Matching. In: ICWS, pp. 86–93 (2007)
14. Hau, J., Lee, W., Darlington, J.: A Semantic Similarity Measure for Semantic Web Services. In: Web Service Semantics Workshop at WWW (2005)
15. Srinivasan, N., Paolucci, M., Sycara, K.P.: An Efficient Algorithm for OWL-S Based Semantic Search in UDDI. In: Cardoso, J., Sheth, A.P. (eds.) SWSWPC 2004. LNCS, vol. 3387, pp. 96–110. Springer, Heidelberg (2005)
16. Constantinescu, I., Binder, W., Faltings, B.: Flexible and Efficient Matchmaking and Ranking in Service Directories. In: ICWS, pp. 5–12 (2005)
17. Colgrave, J., Akkiraju, R., Goodwin, R.: External Matching in UDDI. In: ICWS, p. 226 (2004)
18. Kaufer, F., Klusch, M.: WSMO-MX: A Logic Programming Based Hybrid Service Matchmaker. In: ECOWS, pp. 161–170 (2006)
19. Balke, W.-T., Wagner, M.: Cooperative Discovery for User-Centered Web Service Provisioning. In: ICWS, pp. 191–197 (2003)
20. Xu, Z., Martin, P., Powley, W., Zulkernine, F.: Reputation-Enhanced QoS-based Web Services Discovery. In: ICWS, pp. 249–256 (2007)
21. Kießling, W., Hafenrichter, B.: Optimizing Preference Queries for Personalized Web Services. In: Communications, Internet, and Information Technology, pp. 461–466 (2002)
22. Lamparter, S., Ankolekar, A., Studer, R., Grimm, S.: Preference-based selection of highly configurable web services. In: WWW, pp. 1013–1022 (2007)
23. O'Reilly, T.: O'Reilly Network: What is Web 2.0 (September 2005)
24. Abel, F., Henze, N., Krause, D.: Ranking in Folksonomy Systems: Can Context Help?. In: CIKM, pp. 1429–1430 (2008)
25. Hotho, A., Jäschke, R., Schmitz, C., Stumme, G.: Information Retrieval in Folksonomies: Search and Ranking. In: Sure, Y., Domingue, J. (eds.) ESWC 2006. LNCS, vol. 4011, pp. 411–426. Springer, Heidelberg (2006)
26. Mislove, A., Gummadi, K.P., Druschel, P.: Exploiting Social Networks for Internet Search. In: Workshop on Hot Topics in Networks. (2006)
27. Manning, C.D., Raghavan, P., Schütze, H.: An Introduction to Information Retrieval. Cambridge University Press, Cambridge (2008)
28. Broekstra, J., Kampman, A., van Harmelen, F.: Sesame: A Generic Architecture for Storing and Querying RDF and RDF Schema. In: Horrocks, I., Hendler, J. (eds.) ISWC 2002. LNCS, vol. 2342, pp. 54–68. Springer, Heidelberg (2002)
29. Whitby, A., Josang, A., Indulska, J.: Filtering Out Unfair Ratings in Bayesian Reputation Systems. In: AAMAS (2004)
30. Yu, B., Singh, M.P., Sycara, K.: Developing trust in large-scale peer-to-peer systems. In: IEEE Symposium on Multi-Agent Security and Survivability, pp. 1–10 (2004)

A Generic Approach for Large-Scale Ontological Reasoning in the Presence of Access Restrictions to the Ontology's Axioms

Franz Baader[1], Martin Knechtel[2], and Rafael Peñaloza[1]

[1] Theoretical Computer Science TU Dresden, Germany
{baader,penaloza}@tcs.inf.tu-dresden.de
[2] SAP AG, SAP Research CEC Dresden, Germany
martin.knechtel@sap.com

Abstract. The framework developed in this paper can deal with scenarios where selected sub-ontologies of a large ontology are offered as views to users, based on criteria like the user's access right, the trust level required by the application, or the level of detail requested by the user. Instead of materializing a large number of different sub-ontologies, we propose to keep just one ontology, but equip each axiom with a label from an appropriate labeling lattice. The access right, required trust level, etc. is then also represented by a label (called user label) from this lattice, and the corresponding sub-ontology is determined by comparing this label with the axiom labels. For large-scale ontologies, certain consequence (like the concept hierarchy) are often precomputed. Instead of precomputing these consequences for every possible sub-ontology, our approach computes just one label for each consequence such that a comparison of the user label with the consequence label determines whether the consequence follows from the corresponding sub-ontology or not.

In this paper we determine under which restrictions on the user and axiom labels such consequence labels (called boundaries) always exist, describe different black-box approaches for computing boundaries, and present first experimental results that compare the efficiency of these approaches on large real-world ontologies. Black-box means that, rather than requiring modifications of existing reasoning procedures, these approaches can use such procedures directly as sub-procedures, which allows us to employ existing highly-optimized reasoners.

1 Introduction

Assume that you have a large ontology $\mathcal{T}$, but you want to offer different users different views on this ontology, i.e., each user can see only a subset of the actual ontology, which is selected by an appropriate criterion. This criterion could be the access right that this user has, the level of trust (in the axioms of the ontology) that the user requires, the level of details that is deemed to be appropriate for this user, etc. In principle, you could explicitly create a sub-ontology for each (type of) user, but then you might end up with exponentially many different

A. Bernstein et al. (Eds.): ISWC 2009, LNCS 5823, pp. 49–64, 2009.

ontologies, where each is a subset of $\mathcal{T}$. Instead, we propose to keep just the big ontology $\mathcal{T}$, but label the axioms in $\mathcal{T}$ such that a comparison of the axiom label with the user criterion determines whether the axiom belongs to the sub-ontology for this user or not. To be more precise, we use a labeling lattice $(L, \leq)$, i.e., a set of labels L together with a partial order $\leq$ on these labels such that a finite set of labels always has a join (supremum, least upper bound) and a meet (infimum, greatest lower bound) w.r.t. $\leq$.[1]. All axioms $t \in \mathcal{T}$ are now assumed to have a label $\mathsf{lab}(t) \in L$, and the user also receives a label $\ell \in L$ (which can be read as access right, required level of trust, etc.). The sub-ontology that a user with label ℓ can see is then defined to be[2]

$$\mathcal{T}_\ell := \{t \in \mathcal{T} \mid \mathsf{lab}(t) \geq \ell\}.$$

Of course, the user of an ontology should not only be able to see its axioms, but also the consequences of these axioms. Thus, a user with label ℓ should be able to see all the consequences of $\mathcal{T}_\ell$. For large ontologies, certain relevant consequences are often pre-computed. The goal of the pre-computation is that certain user queries can be answered by a simple look-up in the pre-computed consequences, and thus do not require expensive reasoning during the deployment phase of the ontology. For example, in the version of the large medical ontology SNOMED CT[3] that is distributed to hospitals, all the subsumption relationships between the concept names occurring in the ontology are pre-computed. For a labeled ontology as introduced above, it is not enough to pre-compute the relevant consequences of $\mathcal{T}$. In fact, if the relevant consequence α follows from $\mathcal{T}$, then we also need to know for which user labels ℓ it still follows from $\mathcal{T}_\ell$. Otherwise, if a user with label ℓ asks whether α holds, the system could not simply look this up in the pre-computed consequences, but would need to compute the answer on-the-fly by reasoning over the sub-ontology $\mathcal{T}_\ell$. Our solution to this problem is to compute a so-called *boundary* for the consequence α, i.e., an element μ_α of L such that α follows from $\mathcal{T}_\ell$ iff $\ell \leq \mu_\alpha$.

There are basically two approaches for computing a boundary. The *glass-box approach* takes a specific reasoner (or reasoning technique) for an ontology language (e.g., a tableau-based reasoner for OWL DL [20]) and modifies it such that it can compute a boundary. Examples for the application of the glass-box approach to specific instances of the problem of computing a boundary are tableau-based approaches for reasoning in possibilistic Description Logics [15,13] (where the lattice is the interval $[0, 1]$ with the usual order) and glass-box approaches to axiom pinpointing in Description Logics [19,14,12,3,4] (where the lattice consists of (equivalence classes of) monotone Boolean formulae with implication as order [4]). The problem with glass-box approaches is that they

---

[1] Figure 1 in Section 3 shows a small lattice. A detailed introduction to lattices and orders can, e.g., be found in [9].

[2] To define this sub-ontology, an arbitrary partial order would be sufficient. However, the existence of suprema and infima will be important for the computation of a boundary of a consequence (see below).

[3] http://www.ihtsdo.org/snomed-ct/

have to be developed for every ontology language and reasoning technique anew and that optimizations of the original reasoning technique do not always apply to the modified reasoners. In contrast, the *black-box approach* can re-use existing optimized reasoners without modifications, and it can be applied to arbitrary ontology languages: one just needs to plug in a reasoner for this language.

In this paper, we introduce three different black-box approaches for computing a boundary, and compare their performance on real-world ontologies. The first approach uses an axiom pinpointing algorithm as black-box reasoner, whereas the second one modifies the Hitting-Set-Tree-based black-box approach to axiom pinpointing [11,21]. The third uses binary search and can only be applied if the labeling lattice is a linear order. It can be seen as a generalization of the black-box approach to reasoning in possibilistic Description Logics described in [16]. All the proofs omitted in this paper can be found in [2].

2 Basic Definitions and Results

To stay as general as possible, we do not fix a specific ontology language. We just assume that *ontologies* are finite sets of *axioms* such that every subset of an ontology is again an ontology. If $\mathcal{T}'$ is a subset of the ontology $\mathcal{T}$, then $\mathcal{T}'$ is called a *sub-ontology* of $\mathcal{T}$. The ontology language determines which sets of axioms are admissible as ontologies. For a fixed ontology language, a *monotone consequence relation* $\models$ is a binary relation between ontologies $\mathcal{T}$ of this language and *consequences* α such that, for every ontology $\mathcal{T}$, we have that $\mathcal{T}' \subseteq \mathcal{T}$ and $\mathcal{T}' \models \alpha$ imply $\mathcal{T} \models \alpha$. If $\mathcal{T} \models \alpha$, then we say that α *follows from* $\mathcal{T}$ and that $\mathcal{T}$ *entails* α. For instance, given a Description Logic $\mathcal{L}$ (e.g., the DL $\mathcal{SHIN}(D)$ underlying OWL DL), an ontology is an $\mathcal{L}$-TBox, i.e., a finite set of general concept inclusion axioms (GCIs) of the form $C \sqsubseteq D$ for $\mathcal{L}$-concept descriptions C, D. As consequences we can, e.g., consider subsumption relationships $A \sqsubseteq B$ for concept names A, B.

We consider a lattice $(L, \leq)$ and respectively denote by $\bigoplus_{\ell \in S} \ell$ and $\bigotimes_{\ell \in S} \ell$ the *join* (least upper bound) and *meet* (greatest lower bound) of the finite set $S \subseteq L$. A *labeled ontology with labeling lattice* $(L, \leq)$ is an ontology $\mathcal{T}$ together with a labeling function lab that assigns a *label* $\mathsf{lab}(t) \in L$ to every element t of $\mathcal{T}$[4] We denote with L_{lab} the set of all labels occurring in the labeled ontology $\mathcal{T}$, i.e., $L_{\mathsf{lab}} := \{\mathsf{lab}(t) \mid t \in \mathcal{T}\}$. Every element of the labeling lattice $\ell \in L$ defines a sub-ontology $\mathcal{T}_\ell$ that contains the axioms of $\mathcal{T}$ that are labeled with elements greater than or equal to ℓ:

$$\mathcal{T}_\ell := \{t \in \mathcal{T} \mid \mathsf{lab}(t) \geq \ell\}.$$

Conversely, every sub-ontology $\mathcal{S} \subseteq \mathcal{T}$ defines an element $\lambda_\mathcal{S} \in L$, called the *label* of $\mathcal{S}$: $\lambda_\mathcal{S} := \bigotimes_{t \in \mathcal{S}} \mathsf{lab}(t)$. The following lemma states some simple relationships between these two notions.

Lemma 1. *For all $\ell \in L$, $\mathcal{S} \subseteq \mathcal{T}$, it holds that $\ell \leq \lambda_{\mathcal{T}_\ell}$, $\mathcal{S} \subseteq \mathcal{T}_{\lambda_\mathcal{S}}$ and $\mathcal{T}_\ell = \mathcal{T}_{\lambda_{\mathcal{T}_\ell}}$.*

[4] An example of a labeled ontology is given in Example 2 in Section 3.

Notice that, if a consequence α follows from $\mathcal{T}_\ell$ for some $\ell \in L$, it must also follow from $\mathcal{T}_{\ell'}$ for every $\ell' \leq \ell$, since then $\mathcal{T}_\ell \subseteq \mathcal{T}_{\ell'}$. A maximal element of L that still entails the consequence will be called a margin for this consequence.

Definition 1 (Margin). *Let α be a consequence that follows from the ontology $\mathcal{T}$. The label $\mu \in L$ is called a $(\mathcal{T},\alpha)$-*margin *if $\mathcal{T}_\mu \models \alpha$, and for every ℓ with $\mu < \ell$ we have $\mathcal{T}_\ell \not\models \alpha$.*

If $\mathcal{T}$ and α are clear from the context, we usually ignore the prefix $(\mathcal{T},\alpha)$ and call μ simply a *margin*. The following lemma shows three basic properties of the set of margins that will be useful throughout this paper.

Lemma 2. *Let α be a consequence that follows from the ontology $\mathcal{T}$. We have:*

1. *If μ is a margin, then $\mu = \lambda_{\mathcal{T}_\mu}$;*
2. *if $\mathcal{T}_\ell \models \alpha$, then there is a margin μ such that $\ell \leq \mu$;*
3. *there are at most $2^{|\mathcal{T}|}$ margins for α.*

If we know that μ is a margin for the consequence α, then we know whether α follows from $\mathcal{T}_\ell$ for all $\ell \in L$ that are comparable with μ: if $\ell \leq \mu$, then α follows from $\mathcal{T}_\ell$, and if $\ell > \mu$, then α does not follow from $\mathcal{T}_\ell$. However, the fact that μ is a margin gives us no information regarding elements that are incomparable with μ. In order to obtain a full picture of when the consequence α follows from $\mathcal{T}_\ell$ for an arbitrary element of l, we can try to strengthen the notion of margin to that of an element ν of L that accurately divides the lattice into those elements whose associated sub-ontology entails α and those for which this is not the case, i.e., ν should satisfy the following: for every $\ell \in L$, $\mathcal{T}_\ell \models \alpha$ iff $\ell \leq \nu$. Unfortunately, such an element need not always exist, as demonstrated by the following example.

Example 1. Consider the distributive lattice $(S_4, \leq_4)$ having the four elements $S_4 = \{0, a_1, a_2, 1\}$, where 0 and 1 are the least and greatest elements, respectively, and a_1, a_2 are incomparable w.r.t. $\leq_4$. Let $\mathcal{T}$ be the set formed by the axioms ax_1 and ax_2, which are labeled by a_1 and a_2, respectively, and let α be a consequence such that, for every $\mathcal{S} \subseteq \mathcal{T}$, we have $\mathcal{S} \models \alpha$ iff $|\mathcal{S}| \geq 1$. It is easy to see that there is no element $\nu \in S_4$ that satisfies the condition described above. Indeed, if we choose $\nu = 0$ or $\nu = a_1$, then a_2 violates the condition, as $a_2 \not\leq \nu$, but $\mathcal{T}_{a_2} = \{\mathrm{ax}_2\} \models \alpha$. Accordingly, if we choose $\nu = a_2$, then a_1 violates the condition. Finally, if $\nu = 1$ is chosen, then 1 itself violates the condition: $1 \leq \nu$, but $\mathcal{T}_1 = \emptyset \not\models \alpha$.

It is nonetheless possible to find an element that satisfies a restricted version of the condition, where we do not impose that the property must hold for every element of the labeling lattice, but only for those elements that are *join prime* relative to the labels of the axioms in the ontology.

Definition 2 (Join prime). *Let $(L, \leq)$ be a lattice. Given a finite set $K \subseteq L$, let $K_\otimes := \{\bigotimes_{\ell \in M} \ell \mid M \subseteq K\}$ denote the closure of K under the meet operator. An element $\ell \in L$ is called* join prime relative to K *if, for every $K' \subseteq K_\otimes$, $\ell \leq \bigoplus_{k \in K'} k$ implies that there is an $k_0 \in K'$ such that $\ell \leq k_0$.*

In Example 1, all lattice elements with the exception of 1 are join prime relative to $\{a_1, a_2\}$.

Definition 3 (Boundary). *Let $\mathcal{T}$ be an ontology and α a consequence. An element $\nu \in L$ is called a $(\mathcal{T}, \alpha)$-*boundary *if for every element $\ell \in L$ that is join prime relative to L_{lab} it holds that $\ell \leq \nu$ iff $\mathcal{T}_\ell \models \alpha$.*

As with margins, if $\mathcal{T}$ and α are clear from the context, we will simply call such a ν a *boundary*. In Example 1, the element 1 is a boundary. Indeed, every join prime element ℓ relative to $\{a_1, a_2\}$ (i.e., every element of L except for 1) is such that $\ell < 1$ and $\mathcal{T}_\ell \models \alpha$. From a practical point of view, our definition of a boundary has the following implication: we must enforce that user labels are always join prime relative to the set L_{lab} of all labels occurring in the ontology.

3 Computing a Boundary

In this section, we describe three black-box approaches for computing a boundary. The first two approaches are based on Lemma 3 below, and the third one, a modification of binary search, can be used if the labeling lattice is a linear order.

Lemma 3. *Let $\mu_1, \ldots, \mu_n$ be all $(\mathcal{T}, \alpha)$-margins. Then $\bigoplus_{i=1}^n \mu_i$ is a boundary.*

By Lemma 2, a consequence always has finitely many margins, and thus Lemma 3 shows that a boundary always exists. Note, however, that a consequence may have boundaries different from the one of Lemma 3. To identify the particular boundary of Lemma 3, we will call it the *margin-based boundary*.

3.1 Using Full Axiom Pinpointing

From Lemma 3 we know that the set of all margins yields sufficient information for computing a boundary. The question is now how to compute this set. In this subsection, we show that all margins (and thus the margin-based boundary) can be computed through *axiom pinpointing*. Axiom-pinpointing refers to the task of computing *MinAs* [6]: minimal (w.r.t. set inclusion) sub-ontologies from which a consequence α still follows. More formally, $\mathcal{S} \subseteq \mathcal{T}$ is called a *MinA* for $\mathcal{T}$ and α if $\mathcal{S} \models \alpha$, and $\mathcal{S}' \not\models \alpha$ for every $\mathcal{S}' \subset \mathcal{S}$. The following lemma shows that every margin can be obtained from some MinA.

Lemma 4. *For every margin μ for α there is a MinA $\mathcal{S}$ such that $\mu = \lambda_{\mathcal{S}}$.*

Notice that this lemma does not imply that the label of any MinA $\mathcal{S}$ corresponds to a margin. However, as the consequence follows from every MinA, point 2 of Lemma 2 shows that $\lambda_{\mathcal{S}} \leq \mu$ for some margin μ. The following theorem is an immediate consequence of this fact together with Lemma 3 and Lemma 4.

Theorem 1. *If $\mathcal{S}_1, \ldots, \mathcal{S}_n$ are all MinAs for $\mathcal{T}$ and α, then $\bigoplus_{i=1}^n \lambda_{\mathcal{S}_i}$ is the margin-based boundary for α.*

Thus, to compute a boundary, it is sufficient to compute all MinAs. Several methods exist for computing the set of all MinAs, either directly [19,11,7] or through a so-called pinpointing formula [6,4,5], which is a monotone Boolean formula encoding all the MinAs. The main advantage of using the pinpointing-based approach for computing a boundary is that one can simply use existing implementations for computing all MinAs, such as the ones offered by the ontology editor Protégé 4[5] and the CEL system.[6]

3.2 Label-Optimized Axiom Pinpointing

From Lemma 4 we know that every margin is of the form $\lambda_{\mathcal{S}}$ for some MinA $\mathcal{S}$. In the previous subsection we have used this fact to compute a boundary by first obtaining the MinAs and then computing their labels. This process can be optimized if we directly compute the labels of the MinAs, without necessarily computing the actual MinAs. Additionally, not all the labels of MinAs are necessary, but only the maximal ones. We present here a black-box algorithm that uses the labels of the axioms to find the boundary in an optimized way. Our algorithm is a variant of the Hitting-Set-Tree-based [17] method (HST approach) for axiom pinpointing [11,21]. First, we briefly describe the HST approach for computing all MinAs, which will serve as a starting point for our modified version.

The HST algorithm computes one MinA at a time while building a tree that expresses the distinct possibilities to be explored in the search of further MinAs. It first computes an arbitrary MinA $\mathcal{S}_0$ for $\mathcal{T}$, which is used to label the root of the tree. Then, for every axiom t in $\mathcal{S}_0$, a successor node is created. If $\mathcal{T} \setminus \{t\}$ does not entail the consequence, then this node is a dead end. Otherwise, $\mathcal{T} \setminus \{t\}$ still entails the consequence. In this case, a MinA $\mathcal{S}_1$ for $\mathcal{T} \setminus \{t\}$ is computed and used to label the node. The MinA $\mathcal{S}_1$ for $\mathcal{T} \setminus \{t\}$ obtained this way is also a MinA of $\mathcal{T}$, and it is guaranteed to be distinct from $\mathcal{S}_0$ since $t \notin \mathcal{S}_1$. Then, for each axiom s in $\mathcal{S}_1$, a new successor is created, and treated in the same way as the successors of the root node, i.e., it is checked whether $\mathcal{T} \setminus \{t, s\}$ still has the consequence, etc. This process obviously terminates, and the end result is a tree, where each node that is not a dead end is labeled with a MinA, and every MinA appears as the label of at least one node of the tree (see [11,21]).

An important ingredient of the HST algorithm is a procedure that computes a single MinA from an ontology. Such a procedure can, for example, be obtained by going through the axioms of the ontology in an arbitrary order, and removing redundant axioms, i.e., ones such that the ontology obtained by removing this axiom from the current sub-ontology still entails the consequence (see [6] for a description of this and of a more sophisticated logarithmic procedure). As said before, in our modified HST algorithm, we are now not interested in actually computing a MinA, but only its label. This allows us to remove all axioms having a "redundant" label rather than a single axiom. Algorithm 1 describes a black-box method for computing $\lambda_{\mathcal{S}}$ for some MinA $\mathcal{S}$ that is based on this idea. In

[5] http://protege.stanford.edu/

[6] http://code.google.com/p/cel/

Algorithm 1. Compute a minimal label set of one MinA.

Procedure min-lab($\mathcal{T}, \alpha$)
Input: $\mathcal{T}$: ontology; α: consequence
Output: $M_L \subseteq L$: minimal label set for a MinA

1: **if** $\mathcal{T} \not\models \alpha$ **then**
2:     **return** no MinA
3: $\mathcal{S} := \mathcal{T}$
4: $M_L := \emptyset$
5: **for** every $k \in L_{\mathsf{lab}}$ **do**
6:     **if** $\bigotimes_{l \in M_L} l \not\leq k$ **then**
7:         **if** $\mathcal{S} - k \models \alpha$ **then**
8:             $\mathcal{S} := \mathcal{S} - k$
9:         **else**
10:             $M_L := (M_L \setminus \{l \mid k < l\}) \cup \{k\}$
11: **return** M_L

fact, the algorithm computes a *minimal label set* of a MinA $\mathcal{S}$, a notion that will also be useful when describing our variant of the HST algorithm.

Definition 4 (Minimal label set). *Let $\mathcal{S}$ be a MinA for α. A set $K \subseteq \{\mathsf{lab}(t) \mid t \in \mathcal{S}\}$ is called a* minimal label set *of $\mathcal{S}$ if distinct elements of K are incomparable and $\lambda_{\mathcal{S}} = \bigotimes_{\ell \in K} \ell$.*

Algorithm 1 removes all the labels that do not contribute to a minimal label set. If $\mathcal{T}$ is an ontology and $\ell \in L$, then the expression $\mathcal{T} - \ell$ appearing at Line 7 denotes the sub-ontology $\mathcal{T} - \ell := \{t \in \mathcal{T} \mid \mathsf{lab}(t) \neq \ell\}$. If, after removing all the axioms labeled with k, the consequence still follows, then there is a MinA none of whose axioms is labeled with k. In particular, this MinA has a minimal label set not containing k; thus all the axioms labeled with k can be removed in our search for a minimal label set. If the axioms labeled with k cannot be removed, then all MinAs of the current sub-ontology need an axiom labeled with k, and hence k is stored in the set M_L. This set is used to avoid useless consequence tests: if a label is greater than or equal to $\bigotimes_{\ell \in M_L} \ell$, then the presence or absence of axioms with this label will not influence the final result, which will be given by the infimum of M_L; hence, there is no need to apply the (possibly complex) decision procedure for the consequence relation.

Theorem 2. *Let $\mathcal{T}$ and α be such that $\mathcal{T} \models \alpha$. There is a MinA $\mathcal{S}_0$ for α such that Algorithm 1 outputs a minimal label set of $\mathcal{S}_0$.*

Once the label of a MinA has been found, we can compute new MinA labels by a successive deletion of axioms from the ontology using the HST approach. Suppose that we have computed a minimal label set $\mathcal{M}_0$, and that $\ell \in \mathcal{M}_0$. If we remove all the axioms in the ontology labeled with ℓ, and compute a new minimal label set $\mathcal{M}_1$ of a MinA of this sub-ontology, then $\mathcal{M}_1$ does not contain ℓ, and thus $\mathcal{M}_0 \neq \mathcal{M}_1$. By iterating this procedure, we could compute all minimal label sets, and hence the labels of all MinAs. However, since our goal is to compute

Algorithm 2. Hitting set tree (HST) algorithm for computing the boundary

Procedure hst-boundary($\mathcal{T}, \alpha$)
Input: $\mathcal{T}$: ontology; α: consequence
Output: boundary ν for α

Global : $\mathbf{C}, \mathbf{H} := \emptyset; \nu$
$\mathcal{M} :=$ min-lab($\mathcal{T}, \alpha$)
$\mathbf{C} := \{\mathcal{M}\}$
$\nu := \bigotimes_{\ell \in \mathcal{M}} \ell$
for each label $\ell \in \mathcal{M}$ **do**
  expand-hst($\mathcal{T}_{\not\leq \ell}, \alpha, \{\ell\}$)
return ν

Procedure expand-hst($\mathcal{T}, \alpha, H$)
Input: $\mathcal{T}$: ontology; α: consequence; H: list of lattice elements
Side effects: modifications to $\mathbf{C}$, $\mathbf{H}$ and ν

if there exists some $H' \in \mathbf{H}$ such that $\{h \in H' \mid h \not\leq \nu\} \subseteq H$ **or**
  H' contains a prefix-path P with $\{h \in P \mid h \not\leq \nu\} = H$ **then**
  return (early path termination $\diamond$)
if there exists some $\mathcal{M} \in \mathbf{C}$ such that for all $\ell \in \mathcal{M}, h \in H$, $\ell \not\leq h$ and $\ell \not\leq \nu$ **then**
  $\mathcal{M}' := \mathcal{M}$ (MinLab reuse)
else
  $\mathcal{M}' :=$ min-lab($\mathcal{T}_{\not\leq \nu}, \alpha$)
if $\mathcal{T}_{\not\leq \nu} \models \alpha$ **then**
  $\mathbf{C} := \mathbf{C} \cup \{\mathcal{M}'\}$
  $\nu := \bigoplus\{\nu, \bigotimes_{\ell \in \mathcal{M}'} \ell\}$
  for each label $\ell \in \mathcal{M}'$ **do**
    expand-hst($\mathcal{T}_{\not\leq \ell}, \alpha, H \cup \{\ell\}$)
else
  $\mathbf{H} := \mathbf{H} \cup \{H\}$ (normal termination $\odot$)

the supremum of these labels, the algorithm can be optimized by avoiding the computation of MinAs whose labels will have no impact on the final result. Based on this we can actually do better than just removing the axioms with label ℓ: instead, all axioms with labels $\leq \ell$ can be removed. For an element $\ell \in L$ and an ontology $\mathcal{T}$, $\mathcal{T}_{\not\leq \ell}$ denotes the sub-ontology obtained from $\mathcal{T}$ by removing all axioms whose labels are $\leq \ell$. Now, assume that we have computed the minimal label set $\mathcal{M}_0$, and that $\mathcal{M}_1 \neq \mathcal{M}_0$ is the minimal label set of the MinA $\mathcal{S}_1$. For all $\ell \in \mathcal{M}_0$, if $\mathcal{S}_1$ is not contained in $\mathcal{T}_{\not\leq \ell}$, then $\mathcal{S}_1$ contains an axiom with label $\leq \ell$. Consequently, $\bigotimes_{m \in \mathcal{M}_1} m = \lambda_{\mathcal{S}_1} \leq \bigotimes_{m \in \mathcal{M}_0} m$, and thus $\mathcal{M}_1$ need not be computed. Algorithm 2 describes our method for computing the boundary using a variant of the HST algorithm that is based on this idea.

In the procedure hst-boundary, three global variables are declared: $\mathbf{C}$ and $\mathbf{H}$, initialized with $\emptyset$, and ν. The variable $\mathbf{C}$ stores all the minimal label sets computed so far, while each element of $\mathbf{H}$ is a set of labels such that, when all the axioms with a label less than or equal to any label from the set are removed from the ontology, the consequence does not follow anymore; the variable ν stores the supremum of the labels of all the elements in $\mathbf{C}$ and ultimately corresponds to the boundary that the method computes. The algorithm starts by computing a first minimal label set $\mathcal{M}$, which is used to label the root of a tree. For each element of $\mathcal{M}$, a branch is created by calling the procedure expand-hst.

The procedure expand-hst implements the ideas of HST construction for pinpointing [11,21] with additional optimizations that help reduce the search space as well as the number of calls to min-lab. First notice that each $\mathcal{M} \in \mathbf{C}$ is a minimal label set, and hence the infimum of its elements corresponds to the label of some MinA for α. Thus, ν is the supremum of the labels of a set of MinAs for α. If this is not yet the boundary, then there must exist another MinA $\mathcal{S}$ whose label is not less than or equal to ν. This in particular means that no element of $\mathcal{S}$ may have a label less than or equal to ν, as the label of $\mathcal{S}$ is the infimum of the labels of the axioms in it. When searching for this new MinA we can then exclude all axioms having a label $\leq \nu$, as done in Line 6 of expand-hst. Every time we expand a node, we extend the set H, which stores the labels that have been removed on the path in the tree to reach the current node. If we reach normal termination, it means that the consequence does not follow anymore from the reduced ontology. Thus, any H stored in $\mathbf{H}$ is such that, if all the axioms having a label less than or equal to an element in H are removed from $\mathcal{T}$, then α does not follow anymore. Lines 1 to 4 of expand-hst are used to reduce the number of calls to the subroutine min-lab and the total search space. We describe them now in more detail. The first optimization, *early path termination*, prunes the tree once we know that no new information can be obtained from further expansion. There are two conditions that trigger this optimization. The first one tries to decide whether $\mathcal{T}_{\not\leq \nu} \models \alpha$ without executing the decision procedure. As said before, we know that for each $H' \in \mathbf{H}$, if all labels less than or equal to any in H' are removed, then the consequence does not follow. Hence, if the current list of removal labels H contains a set $H' \in \mathbf{H}$ we know that enough labels have been removed to make sure that the consequence does not follow. It is actually enough to test whether $\{h \in H' \mid h \not\leq \nu\} \subseteq H$ since the consequence test we need to perform is whether $\mathcal{T}_{\not\leq \nu} \models \alpha$. The second condition for early path termination asks for a prefix-path P of H' such that $P = H$. If we consider H' as a list of elements, then a prefix-path is obtained by removing a final portion of this list. The idea is that, if at some point we have noticed that we have removed the same axioms as in a previous portion of the search, we know that all possibilities that arise from that search have already been tested before, and hence it is unnecessary to repeat the work. Hence we can prune the tree here.

The second optimization avoids a call to min-lab by *reusing* a previously computed minimal label set. Notice that our only requirement on min-lab that it produces a minimal label set. Hence, any minimal label set for the ontology obtained after removing all labels less than or equal to any $h \in H$ or to ν would work. The MinLab reuse optimization checks whether there is such a previously computed minimal label set. If this is the case, it uses this set instead of computing a new one by calling min-lab.

Theorem 3. *Let $\mathcal{T}$ and α be such that $\mathcal{T} \models \alpha$. Then Algorithm 2 computes the margin-based boundary of α.*

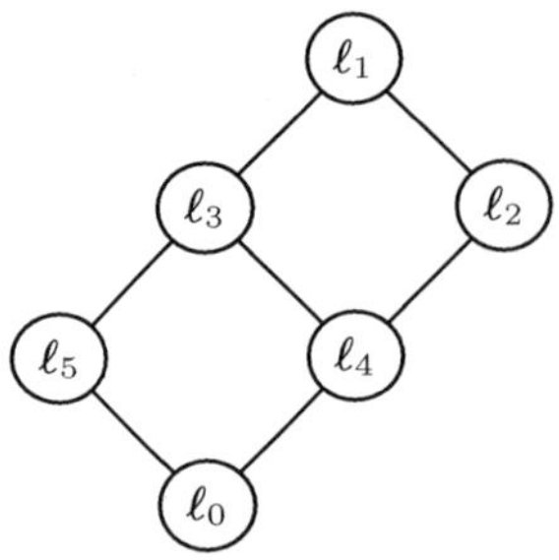

Fig. 1. A lattice

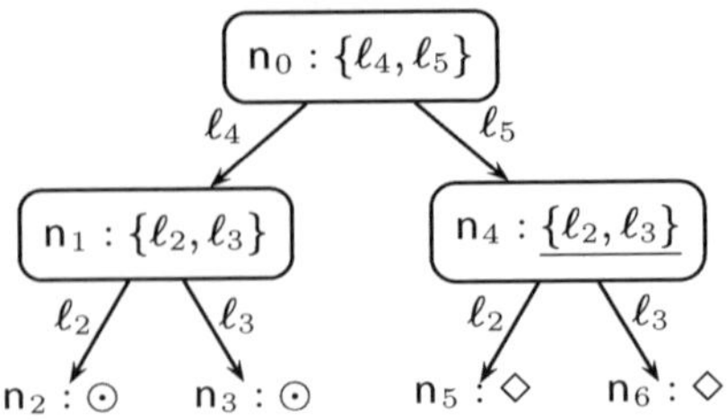

Fig. 2. An expansion of the HST method

A proof of this theorem can be found in [2]. Here, we just illustrate how it works by a small example.

Example 2. Consider the lattice in Figure 1, and let $\mathcal{T}$ be the (Description Logic) ontology consisting of the following five axioms:

$$t_1 : A \sqsubseteq P_1 \sqcap Q_1, \quad t_2 : P_1 \sqsubseteq P_2 \sqcap Q_2, \quad t_3 : P_2 \sqsubseteq B,$$
$$t_4 : Q_1 \sqsubseteq P_2 \sqcap Q_2, \quad t_5 : Q_2 \sqsubseteq B,$$

where each axiom t_i is labeled with $\mathsf{lab}(t_i) = \ell_i$. There are four MinAs for the subsumption relation $A \sqsubseteq B$ w.r.t. $\mathcal{T}$, namely $\{t_1, t_2, t_3\}, \{t_1, t_2, t_5\}, \{t_1, t_3, t_4\}$, and $\{t_1, t_4, t_5\}$. All the elements of the labeling lattice except ℓ_1 and ℓ_3 are join prime relative to L_{lab}. Figure 2 shows a possible run of the hst-boundary algorithm. The algorithm first calls the routine min-lab($\mathcal{T}, A \sqsubseteq B$). Consider that the **for** loop of min-lab is executed using the labels $\ell_1, \ldots, \ell_5$ in that order. Thus, we try first to remove t_1 labeled with ℓ_1. We see that $\mathcal{T} - \ell_1 \not\models A \sqsubseteq B$; hence t_1 is not removed from $\mathcal{T}$, and M_L is updated to $M_L = \{\ell_1\}$. We then see that $\mathcal{T} - \ell_2 \models A \sqsubseteq B$, and thus t_2 is removed from $\mathcal{T}$. Again, $\mathcal{T} - \ell_3 \models A \sqsubseteq B$, so t_3 is removed from $\mathcal{T}$. At this point, $\mathcal{T} = \{t_1, t_4, t_5\}$. We test then whether $\mathcal{T} - \ell_4 \models A \sqsubseteq B$ and receive a negative answer; thus, ℓ_4 is added to M_L; additionally, since $\ell_4 < \ell_1$, the latter is removed from M_L. Finally, $\mathcal{T} - \ell_5 \not\models A \sqsubseteq B$, and so we obtain $M_L = \{\ell_4, \ell_5\}$ as an output of min-lab.

The minimal label set $\{\ell_4, \ell_5\}$, is used as the root node n_0, setting the value of $\nu = \ell_4 \otimes \ell_5 = \ell_0$. We then create the first branch on the left by removing all the axioms with a label $\leq \ell_4$, which is only t_4, and computing a new minimal label set. Assume, for the sake of the example, that min-lab returns the minimal label set $\{\ell_2, \ell_3\}$, and ν is accordingly changed to ℓ_4. When we expand the tree from this node, by removing all the axioms below ℓ_2 (left branch) or ℓ_3 (right branch), the subsumption relation $A \sqsubseteq B$ does not follow any more, and hence we have a normal termination, adding the sets $\{\ell_4, \ell_2\}$ and $\{\ell_4, \ell_3\}$ to **H**. We then create the second branch from the root, by removing the elements below ℓ_5. We see that the previously computed minimal axiom set of node n_1 works also as a minimal axiom set in this case, and hence it can be reused (MinLab reuse), represented as an underlined set. The algorithm continues now by calling expand-hst($\mathcal{T}_{\not\leq \ell_2}, A \sqsubseteq B, \{\ell_5, \ell_2\}$). At this point, we detect that there is $H' = \{\ell_4, \ell_2\}$

Algorithm 3. Compute a boundary by binary search.

Input: $\mathcal{T}$: ontology; α: consequence
Output: ν: $(\mathcal{T}, \alpha)$-boundary

if $\mathcal{T} \not\models \alpha$ **then**
  return no boundary
$\ell := \mathbf{0}_{\mathsf{lab}}; h := \mathbf{1}_{\mathsf{lab}}$
while $l < h$ **do**
  set $m, \ell < m \leq h$ such that $\delta(\ell, m) - \delta(m, h) \leq 1$.
  if $\mathcal{T}_m \models \alpha$ **then**
    $\ell := m$
  else
    $h := \mathsf{pred}(m)$
return $\nu := \ell$

satisfying the first condition of early path termination (recall that $\nu = \ell_4$), and hence the expansion of that branch at that point. Analogously, we obtain an early path termination on the second expansion branch of the node n_4. The algorithm then outputs $\nu = \ell_4$, which can be easily verified to be a boundary.

3.3 Binary Search for Linear Ordering

In this subsection, we assume that the labeling lattice $(L, \leq)$ is a linear order, i.e., for any two elements ℓ_1, ℓ_2 of L we have $\ell_1 \leq \ell_2$ or $\ell_2 \leq \ell_1$.

Lemma 5. *Let $\mathcal{T}$ and α be such that $\mathcal{T} \models \alpha$. Then the unique boundary of α is the maximal element μ of L_{lab} with $\mathcal{T}_\mu \models \alpha$.*

A direct way for computing the boundary in this restricted setting thus consists of testing, for every element in $\ell \in L_{\mathsf{lab}}$, in order (either increasing or decreasing) whether $\mathcal{T}_\ell \models \alpha$ until the desired maximal element is found. This process requires in the worst case $n := |L_{\mathsf{lab}}|$ iterations. This can be improved using binary search, which requires a logarithmic number of steps measured in n. Algorithm 3 describes the binary search algorithm. In the description of the algorithm, the following abbreviations have been used: $\mathbf{0}_{\mathsf{lab}}$ and $\mathbf{1}_{\mathsf{lab}}$ represent the minimal and the maximal elements of L_{lab}, respectively; for $\ell_1 \leq \ell_2 \in L_{\mathsf{lab}}$, $\delta(\ell_1, \ell_2) := |\{\ell' \in L_{\mathsf{lab}} \mid \ell_1 < \ell' \leq \ell_2\}|$ is the *distance* function in L_{lab} and for a given $\ell \in L_{\mathsf{lab}}$, $\mathsf{pred}(\ell)$ is the maximal element $\ell' \in L_{\mathsf{lab}}$ such that $\ell' < \ell$.

The variables ℓ and h are used to keep track of the relevant search space. At every iteration of the **while** loop, the boundary is between ℓ and h. At the beginning these values are set to the minimum and maximum of L_{lab} and are later modified as follows: we first find the *middle* element m of the search space; i.e., an element whose distance to ℓ differs by at most one from the distance to h. We then test whether $\mathcal{T}_m \models \alpha$. If that is the case, we know that the boundary must be larger or equal to m, and hence the lower bound ℓ is updated to the value of m. Otherwise, we know that the boundary is strictly smaller than m as m itself cannot be one; hence, the higher bound h is updated to the maximal element of L_{lab} that is smaller than $m : \mathsf{pred}(m)$. This process terminates when the search space has been reduced to a single point, which must be the boundary.

4 Empirical Evaluation

4.1 Test Data and Test Environment

We test on a PC with 2GB RAM and Intel Core Duo CPU 3.16GHz. We implemented all approaches in Java and used Java 1.6, CEL 1.0, Pellet 2.0.0-rc5 and OWL API trunk revision 1150. The boundary computation with full axiom pinpointing (FP in the following) uses log-extract-mina() (Alg. 2 from [7], which is identical to Alg. 8 from [21]) and the HST based hst-extract-all-minas() (Alg. 9 from [21]). The set of extracted MinAs is then used to calculate the label of the consequence. We break after 10 found MinAs in order to limit the runtime, so there might be non-final label results. The boundary computation with label-optimized axiom pinpointing (LP in the following) with min-lab() and hst-boundary() are implementations of Alg. 1 and Alg. 2 of the present paper. The boundary computation with binary search for linear ordering (BS in the following) implements Alg. 3 of the present paper.

Although we focus on comparing the efficiency of the presented algorithms, and not on practical applications of these algorithms, we have tried to use inputs that are closely related to ones encountered in applications. The two labeling lattices $(L_d, \leq_d)$ and $(L_l, \leq_l)$ are similar to ones encountered in real-world applications. The labeling lattice $(L_d, \leq_d)$ was already introduced in Fig. 1. Lattices of this structure (where the elements correspond to hierarchically organized user roles) can be obtained from a real-world access matrix with the methodology presented in [8]. The set of elements of L_d that are allowed to represent user roles if all elements of the lattice can be used as axiom labels are the elements that are join prime relative to the whole lattice, i.e., $\ell_0, \ell_2, \ell_4, \ell_5$. The labeling lattice $(L_l, \leq_l)$ is a linear order with 6 elements $L_l = L_d = \{\ell_0, \ldots, \ell_5\}$ with $\leq_l := \{(\ell_n, \ell_{n+1}) \mid \ell_n, \ell_{n+1} \in L_l \wedge 0 \leq n \leq 5\}$, which could represent an order of trust values as in [18] or dates from a revision history.

We used the two ontologies O^{Snomed} and O^{Funct} with different expressivity and types of consequences for our experiments. The Systematized Nomenclature of Medicine, Clinical Terms (Snomed ct) is a comprehensive medical and clinical ontology which is built using the Description Logic (DL) $\mathcal{EL}+$. Our version O^{Snomed} is the January/2005 release of the DL version, which contains 379,691 concept names, 62 object property names, and 379,704 axioms. Since more than five million subsumptions are consequences of O^{Snomed}, testing all of them was not feasible and we used the same sample subset as described in [7], i.e., we sampled 0.5% of all concepts in each top-level category of O^{Snomed}. For each sampled concept A, all positive subsumptions $A \sqsubseteq_{O^{\text{Snomed}}} B$ with A as subsumee were considered. Overall, this yielded 27,477 positive subsumptions. Following the ideas of [7], we precomputed the reachability-based module for each sampled concept A with CEL and stored these modules. This module for A was then used as the start ontology when considering subsumptions with subsumee A.

Table 1. Emprical results of FP and LP with lattice $(L_d, \leq_d)$ on a sampled set of 21,001 subsumptions from O^{Snomed} and on a set of 307 consequences from O^{Funct} with less than 10 MinAs (time in ms)

			♯early termination	♯reuse	♯calls to extract MinA (MinLab)	♯MinA (♯MinLab)	♯axioms (♯labels) per MinA (MinLab)	lattice operations time	total labeling time
O^{Snomed}	FP	avg	81.05	9.06	26.43	2.07	5.40	0.25	143.55
		max	57,188.00	4,850.00	4,567.00	9.00	28.67	45.00	101,616.00
		stddev	874.34	82.00	90.48	1.86	3.80	0.86	1,754.03
	LP	avg	0.01	0.00	2.76	1.03	1.73	0.35	4.29
		max	2.00	1.00	6.00	3.00	3.00	57.00	70.00
		stddev	0.13	0.02	0.59	0.16	0.56	0.98	3.62
O^{Funct}	FP	avg	43.59	29.52	26.56	4.26	3.05	0.49	3,403.56
		max	567.00	433.00	126.00	9.00	6.50	41.00	13,431.00
		stddev	92.16	64.04	30.90	2.84	1.01	2.38	3,254.25
	LP	avg	0.09	0.02	2.80	1.33	1.40	0.76	207.32
		max	2.00	1.00	7.00	4.00	3.00	22.00	1,295.00
		stddev	0.34	0.13	0.90	0.54	0.48	1.56	87.29

O^{Funct} is an OWL ontology for functional description of mechanical engineering solutions presented in [10]. It has 115 concept names, 47 object property names, 16 data property names, 545 individual names, 3,176 axioms, and the DL expressivity used in the ontology is $\mathcal{SHOIN}(\mathbf{D})$. Its 716 consequences are 12 subsumption and 704 instance relationships (class assertions).

To obtain labeled ontologies, axioms in both labeled ontologies received a random label assignment of elements from $L_l = L_d$. As black-box subsumption and instance reasoner we used the reasoner Pellet since it can deal with the expressivity of both ontologies. For the expressive DL $\mathcal{SHOIN}(\mathbf{D})$ it uses a tableau-based algorithm and for $\mathcal{EL}+$ it uses an optimized classifier for the OWL 2 EL profile, which is based on the algorithm described in [1].

4.2 Results

The results for O^{Snomed} and $(L_d, \leq_d)$ are given in the upper part of Table 1. LP computed all labels, but since we limit FP to $<$10 MinAs, only 21,001 subsumptions have a final label, which is guaranteed to be equal to the boundary. The 6,476 remaining subsumptions (31%) have a non-final label which might be too low in the lattice since there might be further MinAs providing a higher label. The overall labeling time for all 21,001 subsumptions with FP was 50.25 minutes, for LP 1.50 minutes which means that LP is about 34 times faster than FP, but again this is only for the subset of subsumptions which were finished by FP. An estimation for the time needed to label all of the more than 5 million subsumptions in O^{Snomed} with LP would be approximately 6 hours.

The final labels of FP and LP (i.e., the computed boundaries) were identical, the non-final labels of FP were identical to the final labels of LP (i.e., the boundaries) in 6,376 of the 6,476 cases (98%), i.e., in most cases the missing MinAs would not have changed the already computed label. Table 2 provides results for the subsumptions with more than 10 MinAs: FP took 2.5 hours on this set without final results (since it stopped after 10 MinAs), whereas LP took 0.6% of that time and returned final results after 58 seconds. We started a test series

Table 2. Emprical results of FP and LP with lattice $(L_d, \leq_d)$ on a sampled set of 6,476 subsumptions from O^{Snomed} and on a set of 409 class assertions from O^{Funct} with at least 10 MinAs (time in ms)

			♯early termination	♯reuse	♯calls to extract MinA (MinLab)	♯MinA (♯MinLab)	♯axioms (♯labels) per MinA (MinLab)	lattice operations time	total (non-final) labeling time
O^{Snomed}	FP	avg	432.11	42.25	126.54	10.20	16.38	0.30	1,378.66
		max	42,963.00	5,003.00	4,623.00	16.00	37.80	14.00	148,119.00
		stddev	1,125.06	121.15	186.33	0.49	5.00	0.54	3,493.02
	LP	avg	0.04	0.00	3.12	1.06	2.05	0.32	8.88
		max	3.00	2.00	6.00	3.00	3.00	46.00	86.00
		stddev	0.21	0.04	0.50	0.25	0.44	1.04	4.26
O^{Funct}	FP	avg	30.01	16.00	26.44	10.04	4.41	0.56	8,214.91
		max	760.00	511.00	411.00	11.00	6.50	3.00	25,148.00
		stddev	85.33	47.79	40.61	0.20	1.08	0.55	3,428.97
	LP	avg	0.09	0.01	2.76	1.38	1.32	0.77	200.55
		max	3.00	2.00	7.00	4.00	2.00	16.00	596.00
		stddev	0.33	0.12	0.91	0.64	0.43	1.40	61.11

Table 3. Emprical results of LP and BS on a sampled set of 27,477 subsumptions in O^{Snomed}/ all 716 consequences of O^{Funct} with lattice $(L_l, \leq_l)$ (time in ms)

		LP							BS	
		♯early termination	♯reuse	♯calls to extract MinLab	♯MinLab	♯labels per MinLab	lattice operations time	total labeling time	iterations	total labeling time
O^{Snomed}	avg	0.03	0.00	2.24	1.03	1.23	0.37	4.75	2.41	2.81
	max	1.00	0.00	5.00	3.00	2.00	329.00	330.00	3.00	75.00
	stddev	0.18	0.00	0.45	0.19	0.42	4.85	6.37	0.49	2.94
O^{Funct}	avg	0.09	0.00	2.50	1.27	1.24	0.82	186.98	2.55	95.80
	max	1.00	0.00	5.00	3.00	2.00	62.00	1147.00	3.00	877.00
	stddev	0.28	0.00	0.72	0.49	0.40	2.74	69.55	0.50	45.44

limiting runs of FP to <30MinAs, which did not terminate after 90 hours, with 1,572 labels successfully computed and 30 subsumptions skipped since they had ≥30MinAs. Interestingly, in both consequence sets, LP can rarely take advantage of the optimizations early termination and MinA reuse, which might be due to the simple structure of the lattice.

For O^{Funct} the comparison between FP and LP is given in the lower part of Tables 1 and 2. Again, the computation of FP was restricted to <10 MinAs. This time, only 363 out of 409 (88%) non-final labels of FP were equal to the final labels of LP (i.e., the boundary). Although the ontology is quite small, LP again behaves much better than FP. The reason could be that in this ontology consequences frequently have a large set of MinAs. From Tables 1 and 2, one can see that LP requires at most three MinLabs for O^{Snomed}, at most four for O^{Funct}, and usually just one MinLab whereas FP usually requires more MinAs.

Table 3 provides results for LP vs. BS with the total order $(L_l, \leq_l)$ as labeling lattice. For O^{Snomed}, LP takes 130.4 and BS takes 77.1 seconds to label all 27,477 subsumptions. For O^{Funct}, LP takes 133.9 and BS takes 68.6 seconds to label all 716 consequences. So BS is about twice as fast as LP. Interestingly, labeling all consequences of O^{Funct} and O^{Snomed} takes roughly the same time, perhaps due to a tradeoff between ontology size and expressivity.

5 Conclusion

We have considered a scenario where ontology axioms are labeled and user labels determine views on the ontology, i.e., sub-ontologies that are obtained by comparing the user label with the axiom labels. Our approach can be used for large-scale ontologies since, on the one hand, it allows to precompute consequences without having to do do this separately for all possible views: once we have computed a boundary for the consequence, checking whether this consequence entailed by a sub-ontology is reduced to a simple label comparison. On the other hand, the fact that we employ a black-box approach for computing the boundary allows us to use existing highly-optimzed reasoners, rather than having to implement a new reasoner from scratch.

Our general framework allows to use any restriction criterion that can be represented using a lattice, such as user roles, levels of trust, granularity, or degrees of uncertainty. In the presence of access restrictions, each user label defines a sub-ontology containing the axioms visible to this user. In the presence of trust restrictions, the user label specifies the trust level required for the ontology axiom. This supports scenarios with axioms from different sources, like company-internal with high trust level and public Web with low trust level. In the presence of uncertainty, e.g. in possibilistic reasoning, each axiom has an associated certainty degree in the interval $[0, 1]$. The user label then specifies the certainty degree required for the axioms and the consequences. Similarly, granularity restrictions (i.e., on how much details the ontology should provide for the user) can be expressed by a total order.

Our experiments have shown that this framework can be applied to large ontologies. From the two black-box algorithms that can deal with arbitrary lattices, the Full Axiom Pinpointing approach is clearly outperformed by the Label-Optimized Axiom Pinpointing approach. For the special case where the labeling lattice is a total order, the latter is again outperformed by the Binary Search approach.

References

1. Baader, F., Brandt, S., Lutz, C.: Pushing the $\mathcal{EL}$ envelope. In: Proc. of 19th Int. Joint Conf. on Art. Int. IJCAI 2005, Edinburgh, UK. Morgan-Kaufmann, San Francisco (2005)
2. Baader, F., Knechtel, M., Peñaloza, R.: Computing boundaries for reasoning in sub-ontologies. Technical Report 09-02, LTCS (2009), `http://lat.inf.tu-dresden.de/research/reports.html`
3. Baader, F., Peñaloza, R.: Axiom pinpointing in general tableaux. In: Olivetti, N. (ed.) TABLEAUX 2007. LNCS (LNAI), vol. 4548, pp. 11–27. Springer, Heidelberg (2007)
4. Baader, F., Peñaloza, R.: Automata-based axiom pinpointing. In: Armando, A., Baumgartner, P., Dowek, G. (eds.) IJCAR 2008. LNCS (LNAI), vol. 5195, pp. 226–241. Springer, Heidelberg (2008)
5. Baader, F., Peñaloza, R.: Axiom pinpointing in general tableaux. Journal of Logic and Computation (To appear, 2009)

6. Baader, F., Peñaloza, R., Suntisrivaraporn, B.: Pinpointing in the description logic $\mathcal{EL}^+$. In: Hertzberg, J., Beetz, M., Englert, R. (eds.) KI 2007. LNCS (LNAI), vol. 4667, pp. 52–67. Springer, Heidelberg (2007)
7. Baader, F., Suntisrivaraporn, B.: Debugging SNOMED CT using axiom pinpointing in the description logic $\mathcal{EL}^+$. In: Proc. of the International Conference on Representing and Sharing Knowledge Using SNOMED (KR-MED 2008), Phoenix, Arizona (2008)
8. Dau, F., Knechtel, M.: Access policy design supported by FCA methods. In: Dau, F., Rudolph, S. (eds.) Proc. of the 17th Int. Conf. on Conceptual Structures, ICCS (2009)
9. Davey, B.A., Priestley, H.A.: Introduction to Lattices and Order, 2nd edn. Cambridge University Press, Cambridge (2002)
10. Gaag, A., Kohn, A., Lindemann, U.: Function-based solution retrieval and semantic search in mechanical engineering. In: Proc. of the 17th Int. Conf. on Engineering Design, ICED 2009 (to appear, 2009)
11. Kalyanpur, A., Parsia, B., Horridge, M., Sirin, E.: Finding all justifications of OWL DL entailments. In: Aberer, K., Choi, K.-S., Noy, N., Allemang, D., Lee, K.-I., Nixon, L.J.B., Golbeck, J., Mika, P., Maynard, D., Mizoguchi, R., Schreiber, G., Cudré-Mauroux, P. (eds.) ASWC 2007 and ISWC 2007. LNCS, vol. 4825, pp. 267–280. Springer, Heidelberg (2007)
12. Kalyanpur, A., Parsia, B., Sirin, E., Hendler, J.A.: Debugging unsatisfiable classes in OWL ontologies. J. Web Sem. 3(4), 268–293 (2005)
13. Lesot, M.-J., Couchariere, O., Bouchon-Meunier, B., Rogier, J.-L.: Inconsistency degree computation for possibilistic description logic: An extension of the tableau algorithm. In: Proc. of NAFIPS 2008, pp. 1–6. IEEE Comp. Soc. Press, Los Alamitos (2008)
14. Meyer, T., Lee, K., Booth, R., Pan, J.Z.: Finding maximally satisfiable terminologies for the description logic ALC. In: Proc. of the 21st Nat. Conf. on Artificial Intelligence (AAAI 2006). AAAI Press/The MIT Press, Menlo Park (2006)
15. Qi, G., Pan, J.Z.: A tableau algorithm for possibilistic description logic. In: Domingue, J., Anutariya, C. (eds.) ASWC 2008. LNCS, vol. 5367, pp. 61–75. Springer, Heidelberg (2008)
16. Qi, G., Pan, J.Z., Ji, Q.: Extending description logics with uncertainty reasoning in possibilistic logic. In: Mellouli, K. (ed.) ECSQARU 2007. LNCS (LNAI), vol. 4724, pp. 828–839. Springer, Heidelberg (2007)
17. Reiter, R.: A theory of diagnosis from first principles. Artificial Intelligence 32(1), 57–95 (1987)
18. Schenk, S.: On the semantics of trust and caching in the semantic web. In: Sheth, A., Staab, S., Dean, M., Paolucci, M., Maynard, D., Finin, T., Thirunarayan, K. (eds.) ISWC 2008. LNCS, vol. 5318, pp. 533–549. Springer, Heidelberg (2008)
19. Schlobach, S., Cornet, R.: Non-standard reasoning services for the debugging of description logic terminologies. In: Gottlob, G., Walsh, T. (eds.) Proc. of the 18th Int. Joint Conf. on Artificial Intelligence (IJCAI 2003), Acapulco, Mexico, pp. 355–362. Morgan Kaufmann, Los Altos (2003)
20. Sirin, E., Parsia, B.: Pellet: An OWL DL reasoner. In: Proc. of the 2004 Description Logic Workshop (DL 2004), pp. 212–213 (2004)
21. Suntisrivaraporn, B.: Polynomial-Time Reasoning Support for Design and Maintenance of Large-Scale Biomedical Ontologies. PhD thesis, Fakultät Informatik, TU Dresden (2009), `http://lat.inf.tu-dresden.de/research/phd/#Sun-PhD-2008`

OntoCase-Automatic Ontology Enrichment Based on Ontology Design Patterns*

Eva Blomqvist

StLab, ISTC-CNR, via Nomentana 56, 00161 Roma, Italy
eva.blomqvist@istc.cnr.it

Abstract. OntoCase is a framework for semi-automatic pattern-based ontology construction. In this paper we focus on the retain and reuse phases, where an initial ontology is enriched based on content ontology design patterns (Content ODPs), and especially the implementation and evaluation of these phases. Applying Content ODPs within semi-automatic ontology construction, i.e. ontology learning (OL), is a novel approach. The main contributions of this paper are the methods for pattern ranking, selection, and integration, and the subsequent evaluation showing the characteristics of ontologies constructed automatically based on ODPs. We show that it is possible to improve the results of existing OL methods by selecting and reusing Content ODPs. OntoCase is able to introduce a general top structure into the ontologies, and by exploiting background knowledge the ontology is given a richer overall structure.

1 Introduction

Ontology engineering, for the Semantic Web and other application fields, is a tedious and error-prone process requiring expertise in knowledge modelling and logical languages. With the emergence of the Semantic Web we have seen an increase in popularity of light-weight ontologies that provide just a bit of formal semantics to a data set. Additionally, general web developers have become more and more interested in ontology engineering. To exploit the full benefits of the Semantic Web, ontologies need to be easy to construct, perhaps automatically. In line with this, we have focused on automatic methods (so called ontology learning - OL) for light-weight ontology construction, and especially how the ontology quality can be improved by applying ontology design patterns (ODPs).

Typically, learnt ontologies are quite shallow (in a taxonomical sense), sparse (with respect to the number of relations) and contain a large set of unconnected concepts. An inherent problem when OL is attempted based on text corpora is the fact that most information is actually implicit in the text

* This research was primarily conducted at the Information Engineering research group, Jönköping University (SE). Financially supported through projects SEMCO (KK-foundation grant no. 2003/0241), MediaIlLOG (grant from Carl-Olof och Jenz Hamrins Stiftelse), and DEON (STINT). Evaluations were also conducted within the NeOn project, funded by the European Commission, grant no. IST-2005-027595.

A. Bernstein et al. (Eds.): ISWC 2009, LNCS 5823, pp. 65–80, 2009.

(e.g. as observed in [1]). The consequence is that learnt ontologies lack an overall general structure representing this background knowledge. We believe that some of this knowledge can be added by means of ODPs, in particular Content ODPs. Such patterns can also assist in structuring the existing learnt knowledge.

In this paper we show a particular method for enriching learnt ontologies, by means of Content ODPs. The method is not language dependent as such, but is implemented based on current de facto standards, e.g. OWL. Evaluations attempt to show the two main features of this method: 1) The ability to add a (taxonomical) top structure to the ontology, representing background knowledge implicit in the texts that were the basis of the input ontology. 2) The ability to add more structure to the learnt ontology, i.e. to increase the taxonomical depth and the relation-to-concept ratio. In the following sections we present some background and related work with respect to OL and ODPs. In section 3 we discuss the OntoCase methods, and section 4 presents experiments validating the OntoCase approach. We conclude and discuss future work in section 5.

1.1 Ontology Learning

So far OL approaches have largely dealt with element extraction (see overviews [2,3]), i.e. extraction of single concepts or relations from text corpora, such as in [4]. Approaches have mostly focused on adapting techniques from NLP, computational linguistics, machine learning, and text mining. An example of a state-of-the-art OL tool is Text2Onto[1] [5]. A few recent approaches, e.g. [6,7], have attempted to extract complex axioms, but primarily based on structured input, such as dictionary entries. An approach sharing our goal of providing a more structured output is [8]. The approach uses semantic frames, from linguistics, to extract knowledge from text and transform the frames into small ontologies.

Even though many of the techniques used in OL have been around for many years, they remain subjects of research. The quality of the ontologies is far from perfect; without manual revision the ontologies are not directly usable. An important issue is improving the output ontology quality of OL systems, i.e. the main focus of OntoCase, but still we are not attempting to replace the ontology engineers, merely provide a better starting point for further development.

1.2 Content Ontology Design Patterns

There exist different types of ODPs having different characteristics, for details on ODP types see [9], but in this paper we focus on Content ODPs. Content ODPs are small ontologies with explicit documentation of design rationales, which can be used as building blocks in ontology design, as shown in [10,11].

As an example we describe a Content ODP that is called *Action*. It represents the relations between different types of actions, the state of the actions, and

[1] http://ontoware.org/projects/text2onto/

defines the relation between a plan and a set of proposed actions, see Fig. 1. Content ODPs are collected and presented in different catalogues, such as the *ODP portal*[2]. In addition to their diagrammatic representation Content ODPs are described using a number of catalogue entry fields (c.f. software pattern templates), such as *name*, *intent*, *consequences*, and *building block* (linking to an OWL realization of the pattern). Reusing Content ODPs is a special case of ontology reuse, when the elements of the Content ODP are specilized.

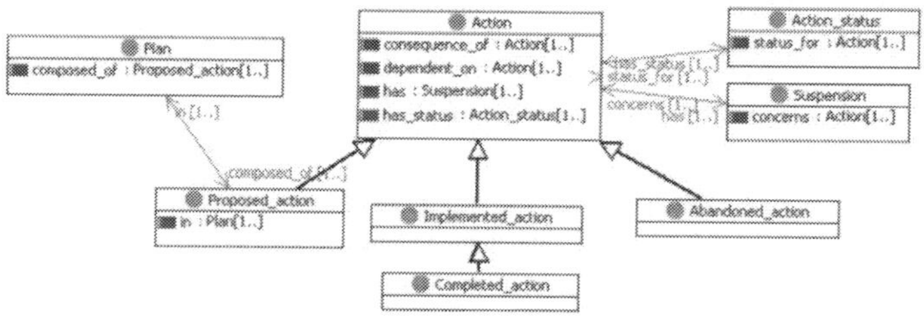

Fig. 1. The Action Content ODP's graphical representation in UML

2 Related Work

Our research is inspired by early AI approaches in analogical reasoning, e.g. scripts and frames, and case-based reasoning (CBR), see [12] for a discussion related to CBR, as well as earlier work in knowledge engineering (KE) and reuse of problem-solving methods. However, recent developments in the KE and Semantic Web fields have led to a situation where a lot of general background knowledge and other knowledge resources, such as ODPs, are now readily available. Consequently we are now able to realize the true potential of such techniques.

Within the ontology engineering field, this research is strongly related to ontology reuse, since Content ODPs are in essence small ontologies. However, no previous approaches specifically target the automatic selection and integration of Content ODPs. Related work exists in ontology search and ranking, e.g. search engines such as Watson[3], SWOOGLE[4], and Sindice[5]. Tools such as the NeOn toolkit in combination with the Watson plug-in let a developer integrate parts of the retrieved ontologies into the one being built, but the matching is simple keyword matching. The problems of selecting the right ontology, e.g. Content ODP, specializing it, and composing several ontologies are largely unaddressed.

An elaborate ranking approach that has inspired our research is AktiveRank [13]. However, this approach incorporates measures such as the centrality of the keywords in the ontology, which are not suitable for small patterns where the notion of centrality is not applicable. The methods in this paper are similar

[2] http://www.ontologydesignpatterns.org
[3] http://watson.kmi.open.ac.uk
[4] http://swoogle.umbc.edu/
[5] http://sindice.com/

to ontology matching, see [14], and methods for analyzing names in ontologies [15]. However, they have been tailored to the specific case of matching a large diverse input ontology on one hand to a small Content ODP on the other hand. To the best of our knowledge no such specific methods have been proposed previously.

3 OntoCase - Retrieve and Reuse

OntoCase is a general framework for pattern-based semi-automatic ontology construction, but in this paper we focus on the retrieve and reuse phases that have been implemented for OWL ontologies. The overall framework can be seen in Fig. 2. The third and fourth phases, revise and retain, are still future work.

3.1 Assumptions and Input

The process is initiated by the input of a data set from which to bootstrap the ontology; "Input" in Fig. 2. The "Element extraction" step uses state-of-the-art OL, i.e. deriving ontology elements from some non-ontological input, such as a text corpus. Since such extraction has been treated in previous research, we simply assume that a preliminary OWL ontology is present. Many OL tools provide a notion of extraction confidence; if present, such values can be used by OntoCase. The methods in this paper cover the "Pattern matching", "Pattern selection", "Pattern adaptation", and "Pattern composition" steps in Fig. 2.

A catalogue of Content ODPs is assumed to be available, "Pattern base" in Fig. 2. Currently the OWL building block itself is used for retrieval and reuse. The current implementation of OntoCase does not treat the pattern construction problem, although the problem is recognized as future work, i.e. the fourth phase (retain). This is not an unreasonable assumption, since recently catalogues of Content ODPs have emerged, such as the ODP portal[6] and catalogues re-engineered from other sources (such as data model patterns [16]), as in [12].

For the matching procedure terms representing both concepts and properties are used, i.e. names or labels (the best match is chosen disregarding if it is a label or a local name). We are assuming that both Content ODPs and the input ontology have human-readable local names (the local part of the resource URI), and/or labels (defined through `rdfs:label`), defined for all classes and properties. Content ODPs represent best practices, hence it should hold for all patterns since this is in itself a good practice, and additionally most OL methods apply some form of term extraction when forming concepts. When matching properties, in the current implementation only the domains and ranges of the properties are used. To explicitly define domain and range is also a best practice, hence most pattern properties have explicit domain and range definitions. For learnt ontologies this may not hold, but again due to the methods commonly applied in OL, such as relation extraction through term co-occurrence, it is very common that domains and ranges are actually present for most properties.

[6] http://ontologydesignpatterns.org

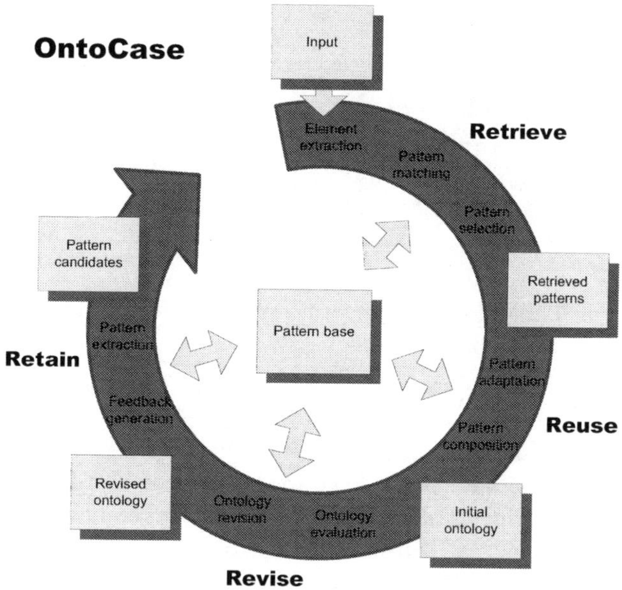

Fig. 2. The overall OntoCase framework

3.2 Example Scenario

As an example, assume that we are constructing a software engineering ontology, and we have extracted the concepts (with local names as follows) "project plan" (with a confidence of 0.5), "schedule" (with a confidence of 0.5), "execution" (with confidence 1.0), and "software engineering task" (with confidence 0.5) from a text corpus. Additionally there is one extracted property named "compose", with domain "software engineering task" and range "project plan". Let us also assume that we have the pattern called *Action* in our pattern catalogue, see section 1.2. This example will be used throughout the description of the method.

3.3 Pattern Ranking

Our pattern ranking and selection method (realizing the "Pattern matching" and "Pattern selection" steps in Fig. 2) takes as input a catalogue of Content ODPs and a preliminary OWL ontology, possibly extracted using some OL tool. The ranking is based on matching between the patterns and the input ontology. Content ODPs are inherently *small* ontologies, hence computationally expensive graph operations can be applied without risking the performance. Content ODP reuse is done through specialization, hence generalizations of pattern concepts are disregarded. The ranking scheme contains three main parts; concept coverage, relation coverage, and utility measures, which are applied as in Fig. 3. Current formulas (details in [12]) are based on related work in ontology ranking, and on analyzing what features intuitively impact pattern suitability.

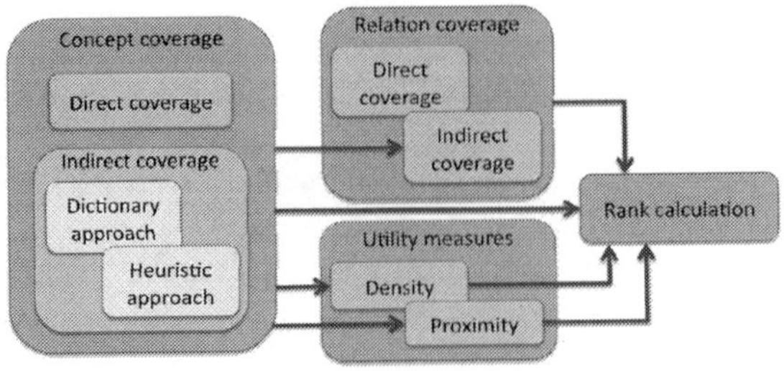

Fig. 3. Parts of the ranking scheme and their dependencies

Concept Coverage. Concept coverage is computed based on direct and indirect term matching. To determine the direct coverage, i.e. discover possible candidates for equivalent concepts, string matching of concept names and labels is used. In order to find only possible equivalences, the matching threshold has to be set quite high. A future extension of this method would be to also study partial inclusion of strings, which would indicate different kinds of relationships between the concepts, or to use "naming patterns" as proposed by [15]. The direct coverage is computed based on the fraction of the pattern concepts that match terms representing a concept in the input ontology O_{input}. The string matching between terms produces a similarity value representing the degree of similarity between two strings; any common normalised string matching measure could be used. These values are then composed into a weighted matching value for each discovered match, using the string matching score and the confidence (if present). For each concept the direct coverage score is computed as the maximum weighted matching value, i.e. the matching score of the best match.

For our example (see section 3.2) we assume the string matching metric is simple string inclusion. Only one term, i.e. "project plan", will match any term, i.e. "Plan", in the *Action* pattern. The term "Plan" constitutes one third of the character string "project plan", hence the resulting score is 0.33. To arrive at the weighted matching score we multiply this with the confidence value, i.e. 0.5, and arrive at a final direct coverage of the "plan"-concept of 0.17.

For the indirect matching, i.e. hints of subclass relations, clues can be found among hypernym relations between terms. Hyponymy/hypernymy denotes a hierarchical relation between terms indicating the specificity of the terms; such a hierarchy is for instance present in the WordNet dictionary. Two approaches are used; hypernym chains in WordNet and the "head heuristic". The "head heuristic" states that a compound term is more specific than the head of the term, i.e. "graduate student" is a specialization of "student".

The dictionary approach starts with a concept in O_{input}. The terms representing this concept are matched against the WordNet dictionary, through exact string matching, and corresponding WordNet terms t_i, if any, are retrieved. The same is done for all pattern concepts. The hypernyms of each t_i are searched for pattern terms. Since t_i can have several senses, there can be several paths, not all leading to a pattern term. The dictionary coverage of a pattern concept is computed as the sum of the contributions by each O_{input} concept, which in turn depend on the number of paths, the length of the shortest path, and the number of senses. The intuition is that the score increases if several paths are found, while it decreases with an increased length of the shortest path, and with an increased number of senses (i.e. an increased uncertainty).

In our example the terms "schedule" and "execution" can be found in WordNet. The noun "execution" has 7 senses, and for two of the senses we can find a path of hypernym relations connecting them to the term "Action". One path is of length one and the other of length 6. When computing the dictionary coverage of the "Action" concept the number of paths (i.e. two) is divided by the length of the shortest path (i.e. one) and the number of senses (i.e. 7), and weighted with the confidence, arriving at the value 0.29. "Schedule" has two senses in WordNet, and one path of length one leads to the pattern term "Plan". The dictionary coverage of "Plan" is thereby 0.25.

For a concept in O_{input} represented by a multi-word term also the "head heuristic" is applied. The number of modifiers, in our case defined as the number of additional words (disregarding the possibility of multi-word terms being individual modifiers) preceding the head word, are treated analogously to a step in the hypernym chains above. Since we have no information about senses, this is disregarded. In our example the only multi-word term matching a pattern term is "project plan", which matches "Plan" with one modifier (i.e. the word "project"). The head coverage (weighted by the confidence) of "Plan" is 0.5 .

Although more elaborate weighting mechanisms could be applied, currently the total concept coverage of each pattern concept is computed as the sum of the three scores, with a maximum score set to 1. In our example this means that the direct coverage, the dictionary coverage and the head coverage of "Plan" are added, resulting in 0.92. The total concept coverage score of a pattern is the average over all concepts. In the example we have 8 concepts in the pattern, where two of them had matches, resulting in a total concept coverage of 0.15.

Relation Coverage. Matching of properties can be done both based on property names and labels, and on the direct concept matches. The intuition behind this is that neither the property name nor the domain and range is sufficient to determine equivalence (or similarity) of the property, but in combination they give a strong indication. For each pattern property from concept c_d to concept c_r, the best match (if any) is selected from the extracted properties. The score is calculated based on the individual matching scores of the direct concept matches (see above). When matching property names and labels a string similarity measure is used. The score of one pattern property with respect to all properties in O_{input}, is the maximum matching value for any property, which in turn is computed as the average of the string matching score and the combined direct concept coverages (multiplied and weighted by the property confidence).

In our example there is only one direct concept match, hence no pair of such matches connected by a property can be found. The name of the extracted property "compose" can be found similar to the pattern property "composed_of" (with the score 0.64, based on string inclusion). The total relation coverage of the "composed_of" property is then 0.32. The total relation coverage is the average of the individual matching scores. In our example there are three properties, hence the total coverage is 0.11. A more elaborate strategy for matching relation names could include also common "naming patterns" to identify the possibility of "composed_of" being the inverse of "compose", rather than the same relation.

Utility Measures. The intuition is to assess the "utility" of enriching the input ontology with particular pattern, based on the concept and relation matches. We remind the reader of the aim to give the input ontology a richer structure, thereby utility is interpreted as the ability to add structure. Two utility measures are used; density and proximity, which are inspired by [13].

Density refers to the amount of "structure" that surrounds a certain concept. OntoCase currently considers the number of sub- and superclasses, taxonomical siblings, and concepts directly related through object properties (explicitly defined domains and ranges). In our example, the "Action" concept has three direct subclasses and is related to two other concepts, resulting in the density 5 divided by 8 (the total number of concepts), i.e. 0.63. The density of "Plan" is 0.13. The density values are then weighted using the concept coverage; the weighted density of "Plan" is 0.12 (0.92∗0.13). The complete density of a pattern is the normalized sum of the densities of matched concepts.

The proximity measure considers the distance $dist(c_i, c_j)$ between two matched concepts c_i and c_j in a pattern, which is computed as the length of the shortest path between the concepts, taking into account all relations (subclass and properties explicitly defined with domain and range, disregarding the direction), except paths passing `owl:Thing`. The maximum distance between any two concepts in a pattern is denoted the *pattern diameter*, which is used for normalization together with the fraction of matched concepts. The total proximity value of a pattern is the normalized sum of all the individual proximities.

Ranking and Selection. The three parts are then aggregated into one ranking value. Although more advanced combinations could be imaginable, for simplicity reasons a linear combination is used, currently with equal weights on all measures. The simplest approach for selection is to let the user set a threshold on the ranking value. A more elaborate approach would be to study the total coverage of the patterns over the input.

3.4 Pattern-Based Enrichment

The reuse phase is concerned with adapting, i.e. specializing, and composing the selected Content ODPs, and integrating them into an enriched version of the input ontology (steps "Pattern adaptation" and "Pattern composition" in Fig. 2). More specifically, pattern specialization means adding all the matching results (or those with a score above a certain threshold) as equivalence axioms and subclass relations respectively. In our previous example this means that "execution" would be added as a subclass of "Action" with a confidence of 0.29 (see above), "schedule" added as a subclass of "Plan", and so on. "Project plan" is ambiguous, i.e. we have evidence of a subclass relation to "Plan" and also an equivalence, hence both relations with their associated confidences are added.

We would additionally like to add the "composed_of"-property, however the "Proposed_action" concept was not matched, only its superconcept "Action". The default heuristic is to only include those parts of a pattern that had some match, not to introduce any unnecessary concepts, but the enrichment can be enhanced by a set of additional heuristics, to create a more well-structured ontology. Firstly, the composition process can be performed in two different modes, pruning or pure enrichment. The pruning mode implies that the input ontology is pruned and only those parts that can be connected to any selected pattern will be included. The pure enrichment leaves the input ontology as it is and only adds the parts of the selected patterns that were matched. Secondly, the (slightly overlapping) heuristics currently available include (some being applicable in only one of the two modes; pruning or pure enrichment):

1. Add all properties (even if not matched) between included concepts.
2. Use the transitive property of subclass relations; if an intermediate concept is missing then add the child directly at the level of the missing concept.
3. Add all extracted subconcepts of concepts in the input ontology, where the parent-concept matched a pattern concept.
4. An object property that originally relates two concepts is added even if one of the concepts is missing if there is a subconcept of the missing concept, which can replace the missing one, present in the ontology.
5. Add all superclasses of matched pattern concepts.

The first heuristic is proposed since relations are harder to extract than terms, hence it is very likely that there will be a number of properties missing in the input ontology. Additionally, we are reluctant to decrease the structural density of the pattern concepts. The second and fourth heuristics are intended to preserve the structure even if there are "gaps" in the matched taxonomy. An alternative strategy would be to include the complete taxonomy, even if only some parts were matched, as suggested in heuristic three and five. Heuristic three is based on the intuition that if a structure exists in the input ontology, this is based on "real-world" evidence (e.g. texts), hence we do not attempt to change this. The intuition for heuristic five, is that we want to use the patterns to add the abstract knowledge that is often missing, even though it might be hard to match it to the input ontology. Adding relations based on the first heuristic even if a concept in the taxonomy is missing, is the focus of the fourth heuristic.

Pattern composition is another task. In the current implementation only simple composition is performed, i.e. relations are not explicitly discovered between patterns. During pattern composition the focus is instead on overlaps between patterns, which are handled using heuristics, for example assuming that two concepts are equivalent if they have the same name and no conflicting axioms. The confidence values resulting from the matching are stored with the ontology.

3.5 Implementation

A first version of the OntoCase method, including the two first phases, has been implemented as a command-line research prototype. The Jena API[7] was used for handling ontologies. Additional external software is the WordNet[8] lexical database. The pattern base is currently deployed as a simple database. The actual pattern ontologies are locally stored as OWL-files, or directly linked to online ontologies on the web. The initial text processing can be done through the Text2Onto[9] tool (an interface to an older version of the tool is provided). The alternative is to provide an input ontology represented as an OWL-file. At the moment a graphical user interface is not provided; plug-in implementations for both Protégé[10] and the NeOn toolkit[11] will be considered in the future.

4 Evaluation

The OntoCase retrieval and reuse phases have been evaluated in three independent settings; the SEMCO project's requirements engineering ontology, the JIBSNet university intranet ontology, and an agricultural ontology of the FAO. All ontologies were constructed using the current OntoCase implementation, set in its pruning mode (described in section 3.4).

4.1 Evaluation Setup

Ontology evaluation methods were used for studying the quality and characteristics of the output of OntoCase. In [17] an overall framework for ontology evaluation is described, consisting of three levels; structural, functional, and usability evaluations. Structural evaluations analyze the quality of the syntax and semantics of the ontology as it is represented. Functional evaluations analyze how well the ontology conforms to the intended conceptualization, i.e. the requirements. Usability evaluations concern the understandability and reusability of the ontology, as well as user satisfaction. We have aimed to cover all levels, although due to practical reasons it has not been feasible for all experiments.

The structural level was analyzed within all experiments, based on measures such as number of concepts, number of concepts at the top level (i.e. root concepts, with no other superclass but `owl:Thing`), number of subclass relations and properties, and average depth of the taxonomy, as suggested in [17,18]. We chose not to apply any formal measure of tangledness, but to evaluate this by inspecting the ontology graphically. The two most well-known approaches for taxonomic evaluation, presented in [19] and OntoClean in [20], were used when feasible (these evaluations were conducted manually by two ontology engineers).

7 http://jena.sourceforge.net/
8 http://wordnet.princeton.edu/
9 http://ontoware.org/projects/text2onto/
10 http://protege.stanford.edu/
11 http://www.neon-toolkit.org/

A weakness is that, for two of experiments, only a sample of the elements were evaluated, resulting in the unfeasibility to evaluate, for example, improper semantic leveling, level of detail, and other issues concerning the overall structure.

To evaluate functional characteristics, i.e. the content of the ontologies, a subset of the OntoMetric framework suggested in [21] was used in the SEMCO experiments. Only the dimension *Content* was deemed interesting. This evaluation was performed by two domain experts at the enterprise in question. In the JIBSNet and FAO cases the evaluation was performed using a random sample[12] of classes and properties, whereby the same factors were not applicable. Instead we applied individual assessment of the concepts and properties by domain experts (or ontology engineers in the FAO case). Through a graphical illustration of the concepts, their placement in the taxonomy, and their properties, the experts were asked to classify them into one of five categories; "essential" (i.e. highly relevant for inclusion in the ontology), "accept" (i.e. correct but not essential), "not sure" (i.e. confusing or hard to assess), "not correctly modeled" (i.e. elements that should be included but not in their current form, e.g. too general concepts, or elements wrongly placed in the ontology structure), and "incorrect" (i.e. not to be included). Six domain experts from JIBS participated, representing different roles in the organization and different educational backgrounds. The individual opinions were weighted together, categorizing the elements as "correct" (representing the *essential* and *accept* judgements), "uncertain", and "incorrect" (representing the *not correctly modeled* and *incorrect* judgements). For JIBSNet this can be seen as a usability evaluation, since it was performed by end-users of the application. While, for the FAO case the evaluation was performed by two ontology engineers, using FAO knowledge sources for the evaluation, hence it was a functional evaluation.

4.2 The SEMCO Requirements Engineering Ontology

Within the research project SEMCO an ontology was constructed with the aim to support structuring and retrieval of information and artefacts during the software development process of an enterprise, focussing on the requirements engineering phase. Initially the ontology was aimed at structuring of artefacts within a tool, i.e. ArtifactManager [22]. The aims of this set of experiments were to compare the implementation of OntoCase to both manual ontology engineering, and to an alternative implementation using naive methods.

The naive implementation applies only existing tools, such as string matching and basic heuristics for the enrichment, and resulted in the ontology OA_{naive} as seen in Table 1 (all numbers represent absolute counts). Additionally the resulting ontologies were compared to two versions of a manually constructed ontology, an initial version, $OM_{initial}$, constructed manually based on the same sources used by OntoCase, i.e. a set of documents, and an enriched version also

[12] The sample sizes were between 3 and 72% of the total number of elements (the aim was to stay above 10% but due to practical limitations on subject availability this had to be reduced for the largest ontologies, hence reducing reliability).

refined based on interviews with domain experts, OM_{final}. $OA_{improved}$ is an ontology constructed using the same pattern catalogue but using the current OntoCase implementation, and OA_{final} is the final version of the automatically constructed ontology, constructed using an extended pattern catalogue (including both a domain specific catalogue and the complete set of patterns at that time available from *ontologydesignpatterns.org*).

Table 1. General characteristics of the SEMCO ontologies

Characteristic	$OM_{initial}$	OA_{naive}	OM_{final}	$OA_{improved}$	OA_{final}
Number of concepts	224	85	379	90	150
Number of root concepts	8	35	5	21	13
Number of properties	15	34	246	37	48
Number of subclass relations	224	48	380	95	243
Average depth	2,52	1,95	3,5	2,10	1,62

4.3 The JIBSNet Information Structure Ontology

The second set of experiments were performed in the university domain. JIBSNet is an intranet present at Jönköping International Business School (JIBS). The intranet contains internal documents of all kinds, from personnel instructions to meeting minutes and information for students. An ontology could help to improve classification, presentation and retrieval of information from JIBSNet. The aim of this experiment was to thoroughly evaluate an ontology constructed by means of OntoCase with actual domain experts, i.e. end-users of a tentative application, but also to compare the resulting ontology to the input ontology.

The ontologies were first evaluated using structural measures, see Table 2 (all numbers are absolute counts within each ontology), and then the usability evaluation was performed together with domain experts at JIBS. In Table 2 O_{in} denotes the input ontology, constructed with the OL tool Text2Onto, and O_{out} denotes the output from OntoCase when applying patterns on top of O_{in}.

4.4 The FAO Agricultural Ontology

The third evaluation was set in the agriculture domain, with focus on concepts related to growing rice. The intention of this experiment was mainly to further establish the previous results showing that OntoCase is in fact able to improve the results of existing OL methods. The setting was the Food and Agriculture

Table 2. General characteristics of the JIBSNet ontology

Characteristic	O_{in}	O_{out}
Number of concepts	6535	2576
Number of root concepts	6368	15
Number of properties	218	147
Number of subclass relations	189	4714
Average depth	1.03	2.72

Organization (FAO) of the United Nations and their work on improving the use of agricultural resources around the globe. The organization is trying to improve their processes by moving from simple structures, such as thesauri, to more complex definitions of concepts, such as ontologies.

An existing manually engineered light-weight ontology was used as a starting point, denoted O_1 in Table 3 (all values represent absolute counts within each ontology). Ontologies were also constructed directly from text corpora, using the OL tool Text2Onto; O_2, O_3, and O_4. Two of the text corpora were generated by extracting DBPedia[13] abstracts and comments respectively, collected by using the concepts in O_1 as search terms. A third text corpus was produced by FAO based on article abstracts connected to the AGROVOC thesaurus[14], related to the term "rice". The two final ontologies, O_5 and O_6, were constructed as combinations of O_1 and O_2, and O_1 and O_3, respectively. "+OC" in the table denotes the output ontologies, after applying OntoCase.

Table 3. General characteristics of the agricultural ontologies

Characteristic	O_1	O_{1+OC}	O_2	O_{2+OC}	O_3	O_{3+OC}	O_4	O_{4+OC}	O_5	O_{5+OC}	O_6	O_{6+OC}
No of concepts	266	280	1086	812	365	321	3575	2823	1256	1293	536	570
No of root conc.	155	112	1018	22	290	22	1822	27	1080	758	352	248
No of properties	37	46	17	28	3	16	49	63	53	73	39	62
No of subclass rel.	110	245	89	1471	189	628	1954	4162	199	1679	299	921
Avg depth	1.68	2.19	1.06	2.37	1.34	2.39	1.68	2.84	1.20	3.36	1.57	3.06

4.5 Result Summary

From the SEMCO experiment we see a clear difference between OntoCase and naive methods. Primarily related to the ability to include abstract knowledge, due to the more elaborate matching methods used, hence the ontology is given a more abstract top structure. Although it is not inherently a positive feature to include more abstract concepts, in the specific case of enriching very shallow ontologies, such as the ones learnt by OL tools (even containing unconnected concepts), this is actually an improvement. When compared to manual methods the automatic approach of course does not perform as well, but it has some merits, especially compared to manually constructing an ontology only based on similar (textual) sources. For example, more properties were included in the ontologies constructed by OntoCase. The manually constructed ontologies had even fewer root concepts, i.e. an even more abstract top structure, but it remains to be determined if they are in fact too abstract. Evaluation of the taxonomy (see [19,20]) showed a comparable level of correctness between all ontologies.

The results of the functional and usability evaluations in the JIBSNet and FAO experiments show that with respect to concepts OntoCase performs on the same level of accuracy as Text2Onto, a state-of-the-art OL method. While, with respect to the structure OntoCase considerably improves on the results

[13] http://dbpedia.org/

[14] http://www.fao.org/agrovoc/

of Text2Onto. The top concepts added give a more intuitive structure to the ontology and the relations are deemed correct to a larger extent than the relations of the input ontologies. Table 4 presents the results for the JIBSNet ontology (O_{in} being the input ontology and O_{out} the output of OntoCase). Randomly selected sets were used in the evaluation, for concepts and taxonomic relations one set representing a selection from the top level of the taxonomy and one representing the overall ontology. The reason for this distinction was to show that even the top structure, where most of the pattern concepts and relations were added, is reasonable. The results of the correctness assessment of the FAO agricultural ontologies can be seen in Table 5. The most significant increase in correctness can be seen in the relation assessment (here subclass relations and properties were assessed together).

An increased tangledness of the taxonomy, together with increased redundancy, e.g. a direct subclass relation being present while simultaneously being derivable from a chain of other subclass relations, could be noted in all of the cases. It is important to point out that this might actually be an advantage, if exploited in the right way. As we saw in the example previously, this is due to the ambiguity of the matching results, but it also provides options for an ontology engineer when continuing to refine the ontology, and guidance is given in the form of confidence values. Intuitively the reader may agree that providing some guidance and choices to the user is most likely better than to select one alternative automatically, whereas wrong choices would sometimes be made.

Table 4. Results for JIBSNet concepts and relations

Evaluation set	Assessment	% of concepts	% of subclass relations	% of properties
O_{in} (top structure)	Correct	85.1%	33.4%	N/A
	Uncertain	11.7%	17.5%	N/A
	Incorrect	3.2%	49.2%	N/A
O_{in} (total)	Correct	75.3%	26.8%	50.7%
	Uncertain	15.9%	15.8%	16.2%
	Incorrect	9.2%	57.4%	33.1%
O_{out} (top structure)	Correct	80.6%	58.1%	N/A
	Uncertain	16.1%	38.7%	N/A
	Incorrect	3.2%	3.2%	N/A
O_{out} (total)	Correct	73.7%	53.4%	65.0%
	Uncertain	16.1%	24.9%	11.9 %
	Incorrect	10.2%	21.7%	23.1%

Table 5. Correctness of the agricultural ontologies

Measure	O_1	O_{1+OC}	O_2	O_{2+OC}	O_3	O_{3+OC}	O_4	O_{4+OC}	O_5	O_{5+OC}	O_6	O_{6+OC}
Concepts												
Correct	92.8	97.8	85.5	87.9	94.2	93.0	84.9	88.5	87.8	88.1	92.0	93.6
Unsure	5.4	2.2	3.4	4.6	2.9	4.4	5.7	5.8	5.6	5.6	4.3	3.6
Incorrect	1.8	0.0	11.1	7.6	2.9	2.6	9.4	5.8	6.7	6.4	3.7	2.7
Relations												
Correct	90.0	92.8	61.6	77.5	64.1	86.1	65.9	79.3	76.2	79.1	78.4	88.2
Unsure	5.7	2.9	12.8	2.5	11.5	6.3	13.2	6.9	10.7	8.1	4.9	5.9
Incorrect	4.3	4.4	25.6	20.0	24.4	7.6	20.9	13.8	13.1	12.8	16.7	5.9

To summarize these results we can conclude that OntoCase in fact provides an added structure to the ontologies, and does connect unconnected parts of the ontologies produced by other OL methods. A general top structure is introduced, adding some of the missing general background knowledge not found explicitly in a text corpus. These improvements, conforming to our hypotheses presented in section 1, are achieved without increasing the error-rate of the ontologies.

5 Conclusions and Future Work

Open issues within the OntoCase framework include to cover the complete cycle, i.e. also address the revision and pattern construction problems. A graphical user interface for the revision of OntoCase ontologies would be a great benefit to the user, in order to study the alternative modelling choices present, and to exploit the confidence values as decision support. This would also give the opportunity to experimentally study the amount of manual work required by ontology engineers in different settings. More specific improvements can be found in the use of background knowledge, where WordNet could for example be replaced by domain specific knowledge or knowledge sources available on the Semantic Web, e.g. online ontologies. Some of the rank calculations currently apply quite naive numerical combinations of values, whereas a certain tuning most likely would be beneficial. Also the composition step can benefit from future refinement, for example by explicitly finding relations between patterns, and testing consistency of hypotheses before adding them to the ontology. We do no envision that OntoCase will replace manual editing of the ontologies, aiming for total correctness would require reasoning mechanisms that will most likely not scale. However, as an aid for providing assistance to the user when selecting and integrating Content ODPs we find OntoCase quite useful, despite the simplicity of some of the methods and heuristics currently implemented.

There are many challenges when semi-automatically constructing ontologies from sources such as text corpora. A prime challenge is how to incorporate the "missing information" that is not explicitly stated in domain specific texts. This is both common-sense knowledge and domain knowledge that is assumed and not stated explicitly. We have addressed this challenge by introducing Content ODPs into semi-automatic ontology construction, and we have proposed a general framework for pattern-based semi-automatic ontology construction called OntoCase. Experiments have shown that the ontologies produced are reasonable with respect to their intended domain, and improve the quality of output compared to typical OL methods, primarily with respect to the ontology structure.

References

1. Brewster, C., Ciravegna, F., Wilks, Y.: Background and foreground knowledge in dynamic ontology construction. In: Proc. Semantic Web Workshop, SIGIR (2003)
2. Cimiano, P.: Ontology Learning and Population from Text - Algorithms, Evaluation and Applications. Springer, Heidelberg (2006)

3. Cimiano, P., Buitelaar, P., Magnini, B. (eds.): Ontology Learning and from Text: Methods, Evaluation and Applications. IOS Press, Amsterdam (2005)
4. Ciaramita, M., Gangemi, A., Ratsch, E., Rojas, I., Saric, J.: Unsupervised learning of semantic relations between concepts of a molecular biology ontology. In: Proceedings of IJCAI 2005 (2005)
5. Cimiano, P., Völker, J.: Text2onto - a framework for ontology learning and data-driven change discovery. In: Montoyo, A., Muńoz, R., Métais, E. (eds.) NLDB 2005. LNCS, vol. 3513, pp. 227–238. Springer, Heidelberg (2005)
6. Voelker, J., Vrandecic, D., Sure, Y., Hotho, A.: Learning disjointness. In: Proceedings of the 4th European Semantic Web Conference, Innsbruck (2007)
7. Völker, J., Haase, P., Hitzler, P.: Learning expressive ontologies. In: Ontology Learning and Population: Bridging the Gap between Text and Knowledge. Frontiers in Artificial Intelligence and Applications. IOS Press, Amsterdam (2008)
8. Coppola, B., Gangemi, A., Gliozzo, A., Picca, D., Presutti, V.: Frame detection over the semantic web. In: Proc. of ESWC 2009. Springer, Heidelberg (2009)
9. Gangemi, A., Presutti, V.: Ontology design patterns. In: Handbook on Ontologies, 2nd edn. International Handbooks on Information Systems. Springer, Heidelberg (2009)
10. Gangemi, A.: Ontology Design Patterns for Semantic Web Content. In: Proc. of the Fourth International Semantic Web Conference, Galway. Springer, Heidelberg (2005)
11. Presutti, V., Gangemi, A.: Content ontology design patterns as practical building blocks for web ontologies. Proc. of ER 2008, 128–141 (2008)
12. Blomqvist, E.: Semi-automatic Ontology Construction based on Patterns. PhD thesis, Linköping University, Department of Computer and Information Science at the Institute of Technology (2009)
13. Alani, H., Brewster, C.: Ontology Ranking based on the Analysis of Concept Structures. In: Proceedings of KCAP 2005, Banff, Alberta, Canada (October 2005)
14. Euzenat, J., Shvaiko, P.: Ontology Matching. Springer, Heidelberg (2007)
15. Svab-Zamazal, O., Svatek, V.: Analysing ontological structures through name pattern tracking. In: Gangemi, A., Euzenat, J. (eds.) EKAW 2008. LNCS (LNAI), vol. 5268, pp. 213–228. Springer, Heidelberg (2008)
16. Hay, D.C.: Data Model Patterns - Conventions of Thought. Dorset House (1996)
17. Gangemi, A., Catenacci, C., Ciaramita, M., Lehmann, J.: Qood grid: A metaontology-based framework for ontology evaluation and selection. In: Proc. of the 4th International EON Workshop, Located at WWW (2006)
18. Yao, H., Orme, A.M., Etzkorn, L.: Cohesion Metrics for Ontology Design and Application. Journal of Computer Science 1(1), 107–113 (2005)
19. Gómez-Pérez, A.: Evaluation of Taxonomic Knowledge in Ontologies and Knowledge Bases. In: Proc. of KAW 1999, Banff, vol.2 (1999)
20. Guarino, N., Welty, C.: Evaluating Ontological Decisions with OntoClean. Communications of the ACM 45(2), 61–65 (2002)
21. Lozano-Tello, A., Gómez-Pérez, A.: ONTOMETRIC: A Method to Choose the Appropriate Ontology. Journal of Database Management 15(2) (April-June 2004)
22. Billig, A., Sandkuhl, K.: Enterprise ontology based artefact management. GI Jahrestagung P134, 681–687 (2008)

Graph-Based Ontology Construction from Heterogenous Evidences

Christoph Böhm[1], Philip Groth[2], and Ulf Leser[2]

[1] Hasso-Plattner-Institut, Prof.-Dr.-Helmert-Str. 2-3, 14482 Potsdam, Germany
christoph.boehm@hpi.uni-potsdam.de
[2] Humboldt-Universität zu Berlin, Unter den Linden 6, 10099 Berlin, Germany
{groth,leser}@informatik.hu-berlin.de

Abstract. Ontologies are tools for describing and structuring knowledge, with many applications in searching and analyzing complex knowledge bases. Since building them manually is a costly process, there are various approaches for bootstrapping ontologies automatically through the analysis of appropriate documents. Such an analysis needs to find the concepts and the relationships that should form the ontology. However, since relationship extraction methods are imprecise and cannot homogeneously cover all concepts, the initial set of relationships is usually inconsistent and rather imbalanced - a problem which, to the best of our knowledge, was mostly ignored so far. In this paper, we define the problem of extracting a consistent as well as properly structured ontology from a set of inconsistent and heterogeneous relationships. Moreover, we propose and compare three graph-based methods for solving the ontology extraction problem. We extract relationships from a large-scale data set of more than 325K documents and evaluate our methods against a gold standard ontology comprising more than 12K relationships. Our study shows that an algorithm based on a modified formulation of the dominating set problem outperforms greedy methods.

1 Introduction

A primary use of ontologies in information systems is the description and structuring of shared knowledge [11]. For instance, ontologies are used extensively in Life Science databases to describe properties of biomedical objects, such as the function of a gene [6] or properties of a biological sequence [4]. The most prominent such ontology, the Gene Ontology [6], consists of more than 28,000 terms describing the molecular function, biological processes, and cellular locations of genes. It is used by dozens of databases throughout the world, is constantly extended and revised, and has established itself as a major research tool in functional genomics.

Building and maintaining a high-quality ontology is costly. Creating the first version of the GO has taken more than two years and involved several people from various research groups. Its maintenance and further development since then requires a constant funding of 10-20 researchers. Due to these high costs, there have been various proposals to automatically learn ontologies from a set

A. Bernstein et al. (Eds.): ISWC 2009, LNCS 5823, pp. 81–96, 2009.

of domain-specific documents (ontology induction, e.g., [15]). Results of such an endeavor may either be used directly, e.g., to support semantic search in documents [12], or may serve as a basis for a subsequent manual verification and extension process (ontology bootstrapping). However, ontology induction is complex and involves a series of subproblems. First, the set of terms (concepts) that are to be included in the ontology must be determined. Second, all occurrences of those concepts in the available corpus must be found. This is not trivial, because concepts may consist of several tokens, which may or may not appear in different orders. Furthermore, morphological variations, abbreviations, and usage of synonymous words are common. Next, evidences for the semantic relationships between concepts must be found (relationship extraction), for instance by searching for specific grammatical patterns involving two concepts. Finally, the ontology itself must be constructed from the set of all extracted relationships, a step we call *ontology extraction*.

Our work focusses on ontology extraction. This is an important step in ontology induction, because in large-scale projects extracted relationships are often redundant, contradicting, or circular. Such cases destroy the semantic consistency of an ontology and need to be resolved. More precisely, we study the following problem: Given a set S of ISA-relationships between concepts, find a subset $S' \subset S$ such that S' is cycle-free and also fulfills certain other criteria. A natural first choice for additional criteria might be that S' has maximal size, i.e., that it is computed from S by removing the least number of edges. But this is not necessarily the best choice in ontology extraction. It is common-sense that an ontology should exhibit certain topological properties to make it better accessible to humans (see, for instance, [1, 23, 24]). These properties include:

1. The topology should roughly resemble a tree, i.e., the number of nodes with more than one parent should be low (trees are easier to grasp),
2. Nodes should not also be parents of their siblings (this is often perceived as semantic nonsense),
3. Nodes with only one child should be avoided (merging them would be appropriate),
4. Leaves should have a comparable depth (to avoid imbalanced subparts and varying semantic granularity of leaves),
5. Grossly imbalanced numbers of children should be avoided (to prevent varying semantic granularity in inner nodes),
6. Nodes should not have (additional) parents that are more than one level away (which goes beyond property 2 to avoid links completely inconsistent with the tree-like ontology backbone), and
7. The ontology should have a single root node.

Consider Fig. 1 as an example. The left part shows a set of ISA-relationships between concepts a, b, c, d, and e as it could emerge from a relationship extraction phase (edges are not weighted to keep the example simple). The set contains cycles and is thus inconsistent. There are various ways to break the cycles by removing one or more edges. Some resulting (consistent) ontologies are shown in the second to fifth part of the figure. Option 1 requires only one edge to

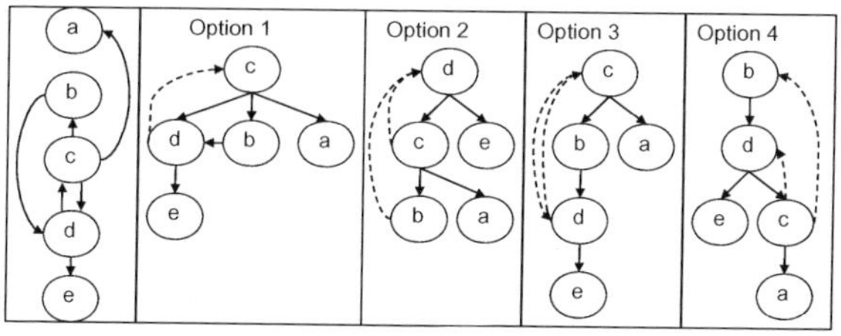

Fig. 1. Extracting a consistent ontology from a set of inconsistent relationships. The left part shows concepts and extracted (cyclic) relationships. The second to fifth part depict different ontologies, which can be derived. Option 1: removal of edge $d \rightarrow c$; 2: $c \rightarrow d, b \rightarrow d$; 3: $c \rightarrow d, d \rightarrow c$; 4: $c \rightarrow d, c \rightarrow b$.

be removed, while Options 2, 3, and 4 require two edge removals. Most of the options have some property that might be considered unfortunate for a 'clean' ontology. For instance, in Option 1, node b is sibling and parent of node d, in Option 3, the depth of the leaves varies greatly, and Options 3 and 4 contain many nodes with only one child.

Certainly one should take into account that extracted relationships often are simply false. The probably most important, yet also most difficult to evaluate, criterium for S' is whether it contains a maximal number of true relationships and a minimal number of false relationships. Finally, it is important that relationship evidences usually carry method-specific confidence values. These values must be considered when choosing the relationships to form the ontology.

In this work, we study the problem of constructing a semantically consistent, correct and well-formed ontology from a given set of heterogeneous, weighted, and possibly inconsistent evidences. For this, we use two different methods for extracting ISA-relationships among concepts from a domain-specific corpus resulting in a set of $> 29K$ relationships. We propose three graph-based algorithms for selecting a semantically consistent subset of relationships from this set. We evaluate our methods by trying to (re-)construct a phenotype ontology (a phenotype is, broadly speaking, an observable property of an organism attributed to the (mal-)function of a gene [8]), using as corpus a set of 327,200 phenotype descriptions downloaded from PhenomicDB [7] as of 02/2008. We converted the texts to lower case, removed URIs, punctuation, and tokens consisting only of numbers. For evaluation we leverage as our gold standard the Mammalian Phenotype Ontology (MPO) [22], consisting of 11,700 concepts (plus synonyms) and 6,830 direct ISA-relationships (03/2008). For the evaluation we use all transitive relationships from the MPO; we consider two concepts a and b to be in transitive relationship if there is a path from a to b or a and b are synonyms. MPO contains 172.134 transitive relationships.

2 Related Work

Ontology induction in general is well studied (e.g. [25]). Systems covering the entire process are, e.g., OntoEdit [15] or OntoLearn [16]. However, to the best of our knowledge, the problem of ontology extraction with inconsistent evidences has achieved little attention in the scientific literature so far, probably due to the fact that most studies in ontology construction are rather small-scale and therefore do not face this problem on a notable scale (e.g., [3, 9]).

We are aware of only few papers that explicitly deal with this problem. Schmitz reported on a study for creating term hierarchies from *flickr* tags [21]. He used subsumption for relationship extraction (see below) and suggested filtering techniques for the resulting relationships, such as a required minimum occurrence of tags. The paper also proposed a pruning strategy that eliminates all relationships that would form relations within the same hierarchy level. This approach is not comparable to ours, because no global consistency of the resulting ontology is targeted or guaranteed and structural properties are not considered. Krishna and Krishnapuram presented a clustering-based approach to hierarchy construction in [13]. However, this algorithm is used to cluster documents for improved browsing and works on entire documents, not extracted relations. This paper also mentions desirable properties for concept hierarchies, which are more usage-oriented than ours. We refrained from reusing these properties, because their formulations is vague and an evaluation is only possible through usage observation. [14] described a method for constructing a concept hierarchy based on a probabilistic language model derived from term co-occurrences. The authors acknowledge the existence of inconsistent relationships and describe a pruning strategy based on the Dominating Set Problem (DSP). This paper largely influenced our work; however, we go beyond their proposal by refining the DSP approach and by comparing it to two other strategies; furthermore, we provide a thorough, large-scale evaluation of both methods, which is lacking in [14] where only 500 texts were used as input (compared to 327,200 in this work).

In sharp contrast to ontology extraction, relationship extraction has been researched extensively over the last decades. In this work, we use Hearst-style POS pattern [10] and subsumption [20], because of their simplicity and expected coverage. Furthermore, we wanted to evaluate how our methods for ontology extraction deal with heterogeneous sources of evidences. Using multiple sources of evidence for relationship extraction is not a new idea; for instance, [3] combined Hearst-style patterns, WordNet relations, head-modifier properties, and term co-occurrence. Note that this work has a much smaller scale than our study and does not cover ontology extraction.

3 Concept Occurrences and Relationship Extraction

In this work, we consider the set of concepts predefined by the gold standard ontology, i.e., MPO, to allow a comparative evaluation. However, we still need to spot occurrences of such concepts in the corpus, which we discuss first. We then show how we extract weighted relationships between those concepts.

3.1 Concept Occurrences

We need to locate all occurrences of all concepts in our corpus. Note that this step is non-trivial for biomedical ontologies, especially because concepts often consist of multiple tokens, which only rarely appear as such in a text. For instance, in the MPO the average number of tokens per concept is 3.5 and only 5.5% of the concepts consist of only one token. Thus, exact matching of concepts would yield a low recall.

We apply approximative concept matching using various meanings of 'approximative'. We investigated the following options to determine whether a set of tokens in a text should be considered as a match with a concept: (1) tokens of a concept appear in a window of w consecutive tokens (integer parameter $w > 0$) in the text; (2) tokens of the concept must or need not appear in the same order in the text (boolean parameter *order*); (3) concepts and corpus are stemmed before matching (boolean parameter *stemming*); (4) stop words are removed from the concepts before matching (boolean parameter *stopwords*). Stemming is accomplished using Porter Stemmer [17]. The list of stop words was taken from [18] (320 terms). In the following, we describe the actual setting of parameters using a vector $\boldsymbol{p} = (order, stemming, stopwords)$. We searched all occurrences of 11,700 MPO concepts (incl. synonyms) using different parameter settings over a range of values for w. We do evaluate our methods only by counting the number of distinct concepts matched, because checking whether matches are actually true would be extremely time-consuming and is not in the focus of our work. We are not aware of any MPO-annotated corpus, which we could use to compute classical IR metrics.

For $w = 2...6$ we find 4,500-6,500 concepts, which is ≈50% of all MPO concepts. The highest number of different concepts is matched when token order is ignored, *stemming* is used, and no stop words are removed ($\boldsymbol{p} = (0, 1, 0)$). Although, we expect $\boldsymbol{p} = (1, 0, 0)$ to produce the least number of false positives, manual reviews have shown that $\boldsymbol{p} = (0, 1, 0)$ performs comparably. Considering the number of transitive MPO relations, which only use matched concepts, we find 35,000-61,500 MPO relations, which amounts to at most 35.5% of all transitive MPO relationships. Note that these numbers form an upper bound to all subsequent steps, since a relationship including a concept that is never found in the corpus cannot be inferred by any method.

3.2 Relationship Extraction

We use two methods for extracting relationships between concepts spotted in the previous step: *subsumption* [20] and Hearst-style *POS-patterns* [10]. Both are explained and evaluated on our corpus in the following subsections. Evaluation is carried out in terms of precision and coverage. We omit figures for recall for the following reason: Imagine a gold standard like $a_1 \rightarrow a_2 \rightarrow \ldots \rightarrow a_{10}$. These nine direct relationships induce 45 transitive relationships. If a method would recover all direct relationships but $a_5 \rightarrow a_6$, it would miss 25 transitive relationships and its recall would already drop down to 45%. Therefore, we consider recall values under our evaluation scheme as somewhat misleading and instead give absolute numbers of correctly extracted relationships (coverage).

Subsumption. The underlying hypothesis for subsumption is that if a concept a always occurs close to a concept b, then a is related to b. If furthermore a occurs more often than b, then a is considered to be more general than b. This intuition is formalized as follows: Let a and b be two concepts and let $p(a|b)$ denote the relative frequency with which a occurs in a window of tokens of length v that also contains b. We say that b ISA a if $p(a|b) \geq t$ and $p(b|a) < 1$, where t is a threshold. Such a relationship is assigned the score $s_{Sub}(a,b) = p(a|b)$.

For evaluating subsumption, we took six parameters into account: v, t and the four parameters from the concept matching (w, *order*, *stemming*, *stopwords*). The impact of varying t is generally as expected: the higher t, the higher precision and the lower coverage. We found $t = 0.9$ to be a good compromise; results for varying t are omitted for brevity. Also, changing w between $2 \ldots 6$ has no significant influence on precision and coverage; we show data for $w = 5$.

Results for various other parameter settings are shown in Fig. 2. The main tendency is that increasing v causes a decrease in precision but an increase in coverage. $\boldsymbol{p} = (1,0,0)$ reaches the best precision $(0.75 - 0.6)$ at average coverage, while $\boldsymbol{p} = (0,1,0)$ yields a precision which is only 0.05 less but has a much better coverage. To keep precision high, we fixed this setting ($\boldsymbol{p} = (0,1,0)$, $t = 0.9$, $w = 5$) at $v = 10$ for all subsequent experiments.

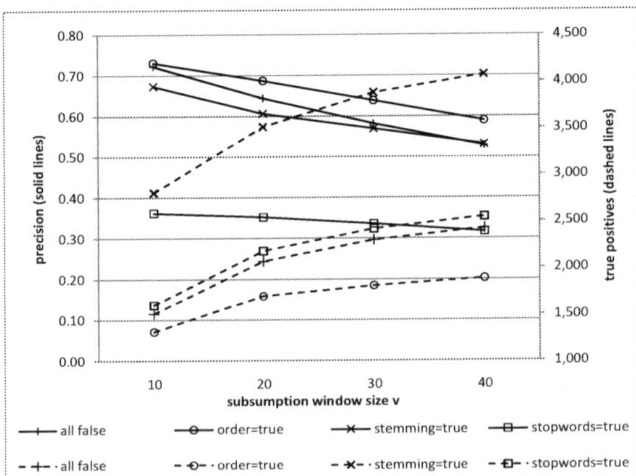

Fig. 2. Performance of subsumption for different parameter settings and $w = 5, t = 0.9$: precision (left axis, solid lines) and coverage (right axis, dashed lines)

POS-patterns. Hearst-style patterns are fixed patterns of word forms or word types that hint to a certain relationship. Consider the following sentence: "*The occurrence of cleft lip and palate in association with skeletal changes such as absent radius suggests Roberts syndrome.*" From the key phrase "*such as*" one can conclude that *absent radius* is a *skeletal change*. In this work, we used the following patterns: *[a such as b]*, *[a includes b]*, *[a especially b]*, *[a like b]* and *[a for example b]*. However, we have to pay special attention to determining meaningful borders of those a and b. We therefore perform a series of preprocessing steps on the corpus before searching the patterns. First, we run a sentence splitter [2] to exclude matches across sentences boundaries. We discard sentences that do not contain a key phrase. Finally, we used a chunker [2] to divide each sentence into syntactically correlated phrases, especially noun phrases. Chunking is important, because the concepts connected in a pattern are not always the ones that are immediately before or after the key phrase.

We considered x noun phrases before and x noun phrases after a key phrase as being connected. The precision obviously decreases for increasing x since a relation discovered by x_j is also discovered by any $x_k > x_j$ but not vice versa (of course, recall increases). We use these precision values to score extracted relationships in the following way: Let $p_1 ... p_i$ be the precision values (in descending order) the process achieves for $x_1 ... x_i$ ($x_1 \leq ... \leq x_i$). We assign scores $s_P(a,b) = \frac{p_j}{p_1}$ if $a \rightarrow b$ was discovered using x_j ($1 \leq j \leq i$) to all generated relationships.

We ran experiments with different values for $\boldsymbol{p}$ and x (at $w = 5$); see Fig. 3. Generally, precision decreases and coverage increases with increasing x. $\boldsymbol{p} = (1,0,0)$ achieves the highest precision followed by $\boldsymbol{p} = (0,1,0)$. Coverage ranges greatly, from 30 ($\boldsymbol{p} = (1,0,0), x = 1$) to 1,900 ($\boldsymbol{p} = (0,1,0), x = 5$). $\boldsymbol{p} = (0,1,0)$ reaches the highest number of true positives at a precision that does not differ significantly from the other settings for values $x > 2$. Accordingly, we used $\boldsymbol{p} = (0,1,0)$ and $x = 1 \ldots 5$ for subsequent experiments.

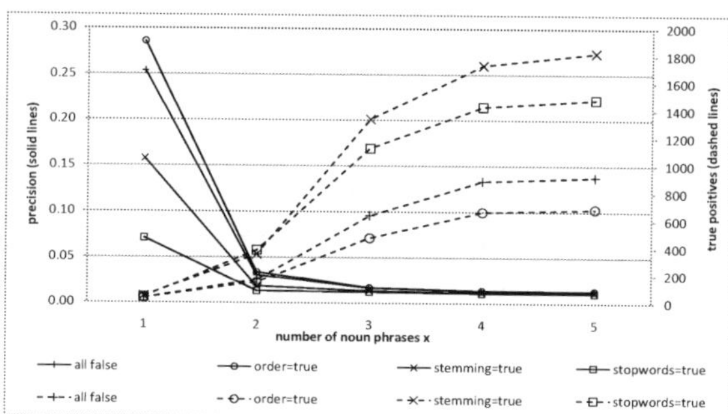

Fig. 3. Performance of POS pattern discovery for different parameter settings and $w = 5$: precision (left axis, solid lines) and coverage (right axis, dashed lines)

4 Graph-Based Ontology Extraction

The result of the previous phase are lists of relationships with a score, either resulting from subsumption or from Hearst-style patterns. This set of weighted relationships does not yet form a consistent and well-formed ontology. First, relationships often are found by both methods, which leads to redundant links; those should be merged in a way that the weights of the original links are properly considered. Second, relationships may be directly contradicting, even when we look at only one source of evidence. Third, relationships may form cycles. Both, cycles and contradicting relationships (essentially cycles of length two), destroy the semantic consistency of an ontology.

Removing these inconsistencies requires choices onto which relationships to exclude and which to include. These choices should be guided by the confidence in the respective relationships, their influence on topological properties of the resulting ontology (as discussed in the introduction), and their semantic correctness. In this section, we devise graph-based algorithms for solving this problem. All operate on a so-called concept graph, which unifies matched concepts and all extracted relationships into a single graph. Note that a nice property of this approach is that adding new sources of relationship evidence is trivial.

Definition 1. *A* concept graph *is a directed graph $G = (V, E)$, where V (nodes) is the set of concepts and E (edges) is constructed from the extracted, weighted evidences (see s_{Sub}, s_P from Sec. 3.2). Let $a, b \in V$ and $0 \leq w_S, w_P \leq 1$ be weighting factors. We define edges E,* edge weights $s(a,b)$, *and* node weights $s(a)$ *as follows:*

$$E = \{(a,b) | s_{Sub}(a,b) > 0 \vee s_P(a,b) > 0\}$$

$$s(a,b) = w_S * s_{Sub}(a,b) + w_P * s_P(a,b)$$

$$s(a) = \sum_{(a,b) \in E} s(a,b)$$

4.1 The Problem

As previously described, the concept graph itself cannot be used as an ontology (in most cases), because it is semantically inconsistent. However, any cycle-free subgraph of the concept graph could in principle be used as an ontology. Therefore, we define the problem of *ontology extraction* as follows.

Definition 2. *Let $G = (V, E)$ be a concept graph. We call a subgraph $G' = (V', E')$ of G with $V' = V$ and $E' \subseteq E$* consistent *if G' is cycle-free. The problem of* ontology extraction *is to choose a consistent subgraph of a given concept graph.*

Consider again Fig. 1 and assume this concept graph would have edge weights as given in Fig. 4. One way to further specify the ontology extraction problem is to require the extracted subgraph to be consistent and *maximal*, i.e., that the sum of the weights of extracted edges is maximal under all consistent subgraphs (by definition $V = V'$). This would result in option 4 with a total weight of 2.6. Using such a formulation would disregard many other criteria to judge good from bad ontologies. Furthermore, the corresponding optimization problem is too complex to be solved for concept graphs of non-trivial size. Actually it is already NP-hard to solve the f-DAG problem, which is to find, given a weighted digraph G and a parameter f, the heaviest DAG in G with out-degree $\leq f$ (by reduction to Minimum Feedback Arc Set [5]).

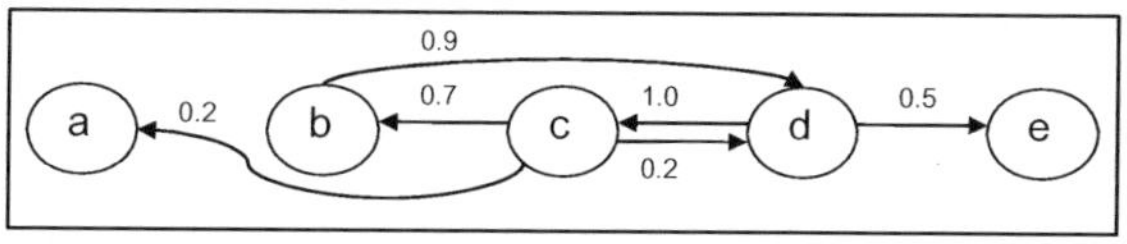

Fig. 4. Weighted Concept Graph with inconsistent relationships from Fig. 1

In the following, we propose and compare three algorithms to solve the ontology extraction problem. The first, which we call *Greedy Edge Inclusion* (*GEI*), adds edges to an empty graph in the order of their weights. It is a heuristic for finding cycle-free graphs that disregards the specific nature of the ontology extraction problem. In contrast, the second and third solution, which we call *Hierarchical Greedy Expansion* (*HGE*) and *Weighted Dominating Set Problem* (*wDSP*), build balanced tree-like structures by iteratively adding *strong* nodes to an empty graph, i.e, nodes that have strong evidences to be a super-concept of many other nodes. They differ in the exact definition of *strong*. In the following, we explain each of these algorithms in more depth.

4.2 Greedy Edge Inclusion (GEI)

The GEI algorithm works as follows: It first copies all nodes from the concept graph into the emerging ontology graph. It then adds edges from the concept graph to the ontology graph sorted by their weights. Every edge that introduces a cycle is omitted.

The main advantage of this method is its simplicity. However, it has many disadvantages. First, it completely disregards topological properties of the resulting structure, which leads to DAGs with many nodes with many parents, grossly imbalanced numbers of children, and many edges spanning large distances (see Sec. 5). Furthermore, due to its many cycle-checks it is rather costly to be computed. Finally, it does not produce a rooted DAG (and therefore has no distinct leaves). Rooting such a structure could be approached by a topological sort, which would open further choices since a topological sort is not unique. Therefore, we evaluate the GEI algorithm only in its unrooted form.

4.3 Strong Node Sets

The following two algorithms share a common idea captured by the notion of *strong* node sets. We first explain this idea in more depth before we discuss the *HGE* and *wDSP* algorithms.

Both algorithms iteratively build an ontology. They start from an artificial root node and add, level by level, *strong* nodes from the concept graph. Therein, they frequently need to solve the following problem: Given a node from the concept graph, choose a subset of its children such that (a) the resulting ontology is consistent, (b) that the overall sum of evidences added to the graph is high, and (c) that it obeys the requirements formulated in the introduction as much as possible. We approximate these requirements by the notion of a *strong node set*. Intuitively, a *strong node set* is a set of nodes that have strong evidences to be a super-concept for many other concepts. This idea leads to the algorithmic framework shown in Listing 1. We use the following notations (let $G = (V, E)$):

- $strong(G, f)$ computes a *strong node set* from G with maximally f members, i.e., a set $V' \subseteq V$, $|V'| \leq f$ and all $v \in V'$ have high evidence to be super-concepts of many other nodes. The concrete definition of *strong* is different in HGE (Sec. 4.4) and wDSP (Sec. 4.5).
- $subgraph(G, S)$ is the subgraph of G induced by $S \subseteq V$.
- (n, l, D) is a triple of a node n, a level l (an integer), and a (strong) set D of nodes.

Listing 1. Ontology extraction from a concept graph

```
1   last_level = 0
2   to_be_removed = {}
3   V' = {root} // resulting nodes
4   E' = {} // resulting edges
5   strong_set = strong(G, f)
6   V' = V' ∪ strong_set
7   for each child ∈ strong_set
8       E' = E' ∪ {(root, child)}
9       add (child, 1, strong_set) to queue
10  while ( queue not empty )
11      (n, current_level, S) = next from queue
12      if ( current_level > d )
13          break
14      if ( last_level < current_level )
15          last_level = current_level
16          G = subgraph(G, V − to_be_removed)
17          to_be_removed = {}
18      to_be_removed = to_be_removed ∪ {n}
19      subgraph_nodes = {c|(n, c) ∈ E} − S // n's children without S
20      strong_set = strong(subgraph(G, subgraph_nodes), f)
21      V' = V' ∪ strong_set
22      for each child ∈ strong_set
23          E' = E' ∪ {(n, child)}
24          add (child, l + 1, strong_set) to queue
25  return G' = (V', E')
```

The first run of *strong* determines nodes for the first level of abstraction (line 6). To obtain a hierarchy-like structure with a root the algorithm creates edges from an artificial root node to the first-level-nodes (line 9). In a next step, it inserts the first-level-nodes, the current level (here 1), and the current *strong* set into a queue (line 10). The queue allows for a breadth-first processing of the nodes. The current *strong* set is stored to avoid that nodes in the set are part of following *strong* sets and thereby cause a cycle. The while loop (line 11) successively processes the nodes in the queue. The algorithm exits the loop if the maximum number d of levels is extracted from the concept graph (line 13-14). The following steps in the while loop determine the descendants (in the hierarchy) of the nodes in the queue. This is achieved by applying *strong* to the children of the nodes in the queue (line 20-21). There is a set of candidate nodes that may be direct descendants of a node n in the resulting hierarchy, i.e., $subgraph\_nodes = \{c|(n, c) \in E\} - S$ (line 20). The application of *strong* to

$subgraph(G, subgraph\_nodes)$ chooses a subset to form a part of the next level of abstraction (line 21). A level l is complete when all nodes from level $l-1$ have been processed and are thus removed from the queue. When a level has been completely processed (line 15) the algorithm removes the nodes from the input graph (line 16-18). To connect levels the process creates only edges from a node n to nodes in the *strong* (sub)set of its children (line 24).

Regarding the requirements stated above, this algorithm has the following properties. (a) It is consistent, because edges that cause cycles are removed actively. (b) It tries to maximize the global sum of edges by maximizing the local sum of edges emerging from each node. (c) The properties stated in Sec. 1 are considered as follows:

1. By repeating the strong node set computation in a breadth-first manner we inherently extract a hierarchy-like structure.
2. We actively remove edges between members of a chosen strong node set and neighboring strong node sets on each level.
3. Avoidance of nodes with only one child is not embedded in the algorithm; however, we shall see that both algorithms create such cases rather rarely (compared to the gold standard; see Sec. 5, Tab. 1).
4. We restrict the number of iterations and thus avoid leaves of grossly different depths.
5. We fix a global upper bound for the size of a strong node set, which leads to a relatively evenly distributed fanout.
6. We only add edges from nodes to strong (sub)set of its children and therefore remain consistent with the tree-like ontology backbone across multiple levels of abstraction.
7. The ontology has a single root by construction.

4.4 Hierarchical Greedy Expansion (HGE)

As stated above, *strong* node sets $V' \subseteq V$ ($|V'| \leq f$) contain nodes that have a strong evidence to be super-concepts for many other nodes. Therefore, a *strong* node set wants to maximize $\sum_{v \in V'} s(v)$ to gather as much evidence as possible; however, it also needs to consider the sheer number of children added to the graph, i.e., $|\bigcup_{v \in V'} \{w | v \rightarrow w \in E\}|$, because many children likely lead to many grand-children and so forth.

We implemented two ways of identifying such *strong* node sets from a concept (sub)graph. Our first approach is called Hierarchical Greedy Expansion and works as follows: For a given concept graph $G = (V, E)$ and an upper bound f it adds nodes n to a strong node set R in the order of their scores $s(n)$ (from Def. 1) until $|R| = f$. After adding a node, it removes the added node and all its edges from G and adjusts the scores of the remaining nodes in V. Recall that the score $s(n)$ of a node n is the sum of the weights of its outgoing edges.

4.5 Weighted Dominating Set Problem (wDSP)

Our second approach models $strong(G, f)$ as a slightly modified instance of the Dominating Set Problem (DSP). Formally, the DSP is the following: Given a graph $G = (V, E)$, find a subset $D \subset V$ (the superconcepts), such that for each node $v \in V - D$ a node $d \in D$ exists, where $(d, v) \in E$, and $|D|$ is minimal. Thus, the DSP is quite similar to the problem of selecting a set of strong nodes in every iteration of our framework. This was first observed by Lawrie et al. in [14]. We reuse and refine their method in this section. Note that we need to modify the original proposal to consider weights of nodes.

Definition 3. *Let $G = (V, E)$ be a concept graph and f be an upper bound. Let $d(D) := \{v|v \in V - D \wedge \exists d \in D|(d, v) \in E\}$ denote the subset of $V - D$ dominated by $D \subset V$. The* ***weighted Dominating Set Problem (wDSP)*** *is to find a set of nodes $D \subset V$, which satisfies the following requirements:*

1. $|D| \leq f$,
2. $\sum_{d \in D} s(d)$ *is maximal,*
3. $\nexists \hat{D}$ *such that $\hat{D}$ satisfies (1) and (2) as well as* $|d(\hat{D})| > |d(D)|$.

However, solving DSP is NP-hard [5], and so is solving wDSP. Therefore, we use an approximation called *GreedyVote* (adapted to wDSP), which was shown to be the best performing heuristic for solving DSP in [19]. *GreedyVote* takes the neighborhood of a node n into account when it decides whether n should be added to a dominating set or not. It uses a score $gv(n)$ that captures for every node how 'difficult' or 'easy' it can be dominated; for details, we refer the reader to [19]. We modified the algorithm to take node weights into account. For each node, we compute a new score $score(n) = gv(n) + s(a)$ and thus include both maximization criteria, $|\bigcup_{v \in V'} \{w|v \rightarrow w \in E\}|$ as well as $\sum_{v \in V'} s(v)$.

5 Evaluation

We run all three algorithms on the set of relationships extracted from the phenotype corpus using the methods explained in Sec. 3.2. For the HGE and wDSP algorithm, we show results for varying values of f (max. size of strong node sets) and d (max. depth of ontology); all other parameters were fixed as described in Sec. 3.2. We compare all resulting ontologies to the MPO. In the following, we first concentrate on precision and coverage and then discuss differences in topological properties. Remember that we consider a relationship $a \rightarrow b$ to be true if concepts a and b are in transitive relationship in MPO, i.e., there is a path from a to b or a and b are synonyms.

We also ran experiments with different weights w_S, w_P for subsumption and pattern discovery (see Definition 1). For brevity, we do not show results here. In general, coverage and precision is better when both weights are set equally compared to using only one evidence, but the increase is small ($< 5\%$ on coverage). All further results were produced with $w_S = w_P = 0.5$. Another interesting information is the precision and coverage of the concept graph itself. Its coverage serves as an upper bound to all algorithms. We considered all relationships with $s_{Sub}(a, b) > 0.5$ or $s_P(a, b) > 0$. Then, precision is 0.47 and coverage is 5,070.

We first compare precision and coverage of HGE and wDSP for varying values of f and d. Fig. 5 gives coverage values for HGE and wDSP for increasing values of f. Additionally, it shows the fraction of true positives that are direct relations in MPO. The figure shows that $wDSP$ produces 10-25% more true positives than HGE, both, for transitive and direct relations. At the same time, precision (data not show) slightly decreases from 0.54 to 0.51 for wDSP and remains almost constant ≈ 0.5 for HGE. Though the increase of true positives slows down for increasing f, it is not bounded, i.e., the number of true positives grows monotonously. This monotony implies that better coverage values than reported here are possible, but only at the expense of less readable ontologies (many nodes with > 50 children).

Fig. 6 show coverage of HGE and wDSP for different values of d. wDSP has a steep increase until $d = 6$ and then roughly levels out, since the fluctuation for $d = 7...20$ probably can be contributed to the random factor in $GreedyVote$. In contrast, HGE's number of true positives is considerable lower but increases constantly. As in the previous experiment, precision of wDSP is better than for HGE (e.g., ≈ 0.54 versus ≈ 0.50 at $d = 2 \ldots 20$). The difference in coverage is mostly due to the fact that wDSP manages to include more relationships into the ontology than HGE without loosing precision.

Compared to HGE and wDSP, the (parameter-free) GEI algorithm has a much higher coverage (4,376 true positive relationships). However, this is at the cost of a considerably lower precision (0.45) and a less well-formed topology (see below). Fig. 5 and 6 indicate that the coverage of HGE and wDSP likely could be increased by giving extremely high limits for f and d, but this would probably lead to equally twisted structures as in the case of the GEI algorithm.

Table 1 compares the performance of the GEI, HGE, and wDSP algorithm with regard to the topological properties stated in Sec. 1. We also computed those number for the gold standard itself, i.e., the MPO. GEI clearly performs the worst, since the standard deviation of both in- and out degree is much higher than for the other two algorithms (recall the GEI produces an unrooted DAG and therefore properties 4 and 6 are meaningless). The comparison between HGE and

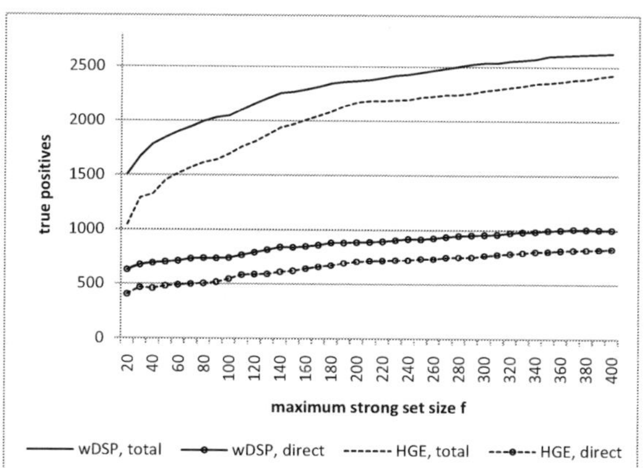

Fig. 5. Coverage of HGE and wDSP for different maximum *strong* set sizes f ($d = 10$)

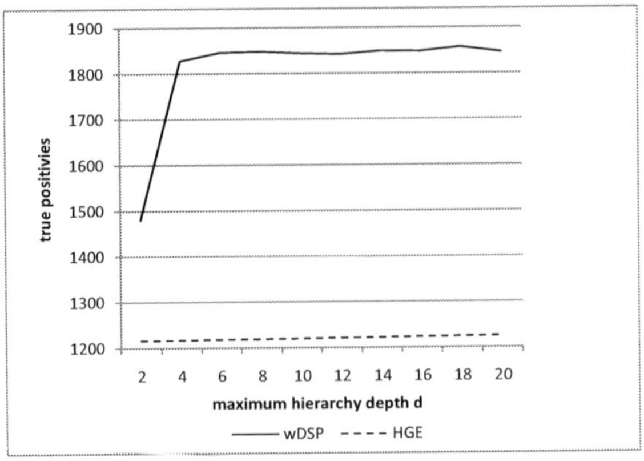

Fig. 6. Coverage of HGE and wDSP for different maximum hierarchy depths d ($f = 50$)

wDSP is less clear. wDSP has a much more homogeneous out-degree distribution, but a slightly worse in-degree distribution. Its ontology is deeper and has a much lower fan-out on average; furthermore, its values overall are considerable closer to the gold standard than those of the HGE algorithm. Interestingly, MPO does not quite adhere to Prop.3, i.e., it has 661 concepts that have only one child (which could therefore be merged). We attribute this to the still immature nature of the MPO where many concepts might still be placeholders for further expansion; however, this problem is also "reconstructed" by wDSP.

Overall, the evaluation shows that wDSP outperforms both, GEI and HGE. It has the highest precision of all methods and a better coverage than HGE. The topology of its computed ontology is closest to what one would consider as desirable for ontologies. However, it is also important to state that the simple GEI algorithm produces by far the best coverage at a small expense of precision. Possibly, this expense could be lowered when an evidence cut-off value for edges in the concept graph is applied. Therefore, we believe that further research on greedy methods is valuable, which would also need to include a method to root the ontology.

6 Discussion

In this paper, we defined and studied the problem of constructing a consistent and well-formed ontology from a set of inconsistent and heterogenous concept relationships. We presented and analyzed three different strategies ranging from a simple greedy algorithm to a sophisticated reformulation of the DSP. We evaluated our approaches on a large-scale real-world scenario from the Life Sciences. Algorithms were judged based on the semantic correctness and certain topological properties of the created ontologies. Our results indicate that the wDSP approach outperforms greedy solutions in terms of precision and topology of the ontology, but it also has a a considerably lower coverage.

Table 1. Topological properties for resulting ontologies compared to MPO

		abs	min	max	avg	stdev
GEI	1) in degree	-	1	53	2.83	2.42
	2) sibling/parent nodes	11,271	-	-	-	-
	3) single-child-nodes	1101	-	-	-	-
	4) depth	-	-	-	-	-
	5) out-degree	-	1	1129	4.92	28.35
	6) multiple-level-parents	-	-	-	-	-
	7) rooted hierarchy	no	-	-	-	-
HGE	1) in degree	-	1	7	1.32	0.61
	2) sibling/parent nodes	0	-	-	-	-
	3) single-child-nodes	163	-	-	-	-
	4) depth	-	2	7	2.63	0.81
	5) out-degree	-	1	50	7.17	11.70
	6) multiple-level-parents	0	-	-	-	-
	7) rooted hierarchy	yes	-	-	-	-
wDSP	1) in degree	-	1	7	1.45	0.83
	2) sibling/parent nodes	0	-	-	-	-
	3) single-child-nodes	529	-	-	-	-
	4) depth	-	2	8	3.09	0.92
	5) out-degree	-	1	50	3.76	7.14
	6) multiple-level-parents	0	-	-	-	-
	7) rooted hierarchy	yes	-	-	-	-
MPO	1) in degree	-	1	4	1.18	0.41
	2) sibling/parent nodes	0	-	-	-	-
	3) single-child-nodes	661	-	-	-	-
	4) depth	-	1	13	5.53	1.41
	5) out-degree	-	1	38	3.15	3.18
	6) multiple-level-parents	0	-	-	-	-
	7) rooted hierarchy	yes	-	-	-	-

In the future, we will look into ways to directly incorporate topological requirements into the extraction algorithm. Additionally, we will work on increasing coverage without scarifying precision. This can be achieved by either improving the extraction algorithms or by refining the concept matching strategy or by simply adding further sources of relationship evidences. Certainly, the first of these options is the most interesting, but also most challenging one.

References

[1] Brank, J., Grobelnik, M., Mladenić, D.: A Survey of Ontology Evaluation Techniques. In: Proc. Conf. Data Mining and Data Warehouses (2005)

[2] Buyko, E., Wermter, J., Poprat, M., Hahn, U.: Automatically Adapting an NLP Core Engine to the Biology Domain. In: Proc. of the ISMB 2006: BioLink & Bio-Ontoligies SIG Meeting (2006)

[3] Cimiano, P., Pivk, A., Schmidt-Thieme, L., Staab, S.: Learning Taxonomic Relations from Heterogeneous Sources of Evidence, ch. II.4, pp. 59–76. IOS Press Publication, Amsterdam (2003)

[4] Eilbeck, K., Lewis, S.E., Mungall, C.J., Yandell, M., Stein, L., Durbin, R., Ashburner, M.: The Sequence Ontology: a tool for the unification of genome annotations. Genome Biol. 6 (2005)

[5] Garey, M.R., Johnson, D.S.: Computers and Intractability; A Guide to the Theory of NP-Completeness. W. H. Freeman & Co., New York (1990)

[6] Gene Ontology Consortium. Creating the gene ontology resource: design and implementation. Genome Res 11, 1425–1433 (2001)
[7] Groth, P., Pavlova, N., Kalev, I., Tonov, S., Georgiev, G., Pohlenz, H.-D., Weiss, B.: PhenomicDB: a new cross-species genotype/phenotype resource. Nucl. Acids Res. 35, 696–699 (2007)
[8] Groth, P., Weiss, B.: Phenotype Data: A Neglected Resource in Biomedical Research? Current Bioinformatics, vol. 1, pp. 347–358 (2006)
[9] Gulla, J.A., Brasethvik, T.: A Hybrid Approach to Ontology Relationship Learning. In: Proc. of the 13th Int. Conf. on Natural Language and Information Systems, pp. 79–90. Springer, Heidelberg (2008)
[10] Hearst, M.A.: Automatic acquisition of hyponyms from large text corpora. In: Proc. of the 14th Conf. on Computational Linguistics. Association for Computational Linguistics (1992)
[11] Jurisica, I., Mylopoulos, J., Yu, E.: Ontologies for Knowledge Management: An Information Systems Perspective. Knowl. Inf. Syst. 6, 380–401 (2004)
[12] Kasneci, G., Suchanek, F.M., Ifrim, G., Ramanath, M., Weikum, G.: NAGA: Searching and Ranking Knowledge. In: Proc. of the 24th Int. Conf. on Data Engineering, pp. 953–962 (2008)
[13] Krishna, K., Krishnapuram, R.: A clustering algorithm for asymmetrically related data with applications to text mining. In: Proc. of the 10th Int. Conf. on Information and Knowledge Management, pp. 571–573. ACM, New York (2001)
[14] Lawrie, D., Croft, W.B., Rosenberg, A.: Finding topic words for hierarchical summarization. In: Proc. of the 24th Annu. Int. ACM SIGIR Conf. on Research and Development in Information Retrieval, pp. 349–357. ACM Press, New York (2001)
[15] Maedche, A., Staab, S.: Ontology Learning for the Semantic Web. IEEE Intelligent Systems 16, 72–79 (2001)
[16] Navigli, R., Velardi, P.: Learning Domain Ontologies from Document Warehouses and Dedicated Web Sites. Comput. Linguist. 30, 151–179 (2004)
[17] Porter, M.F.: An algorithm for suffix stripping. Readings in Information Retrieval, 313–316 (1997)
[18] Rijsbergen, C.J.v.: Information retrieval. Butterworths (1979)
[19] Sanchis, L.A.: Exoerimental Analysis of Heuristic Algorithms for the Dominating Set Problem. Algorithmica 33, 3–18 (2002)
[20] Sanderson, M., Croft, B.: Deriving concept hierarchies from text. In: Proc. of the 22nd Annu. Int. ACM SIGIR Conf. on Research and Development in Information Retrieval, pp. 206–213. ACM Press, New York (1999)
[21] Schmitz, P.: Inducing Ontology from Flickr Tags. In: Proc. of the Collaborative Web Tagging Workshop (WWW 2006). IW3C2 (2006)
[22] Smith, C.L., Goldsmith, C.A., Eppig, J.T.: The Mammalian Phenotype Ontology as a tool for annotating, analyzing and comparing phenotypic information. Genome Biol. 6 (2005)
[23] Supekar, K., Patel, C., Lee, Y.: Characterizing Quality of Knowledge on Semantic Web. In: Proc. of the Seventeenth Int. Florida Artificial Intelligence Research Society Conf., (2004)
[24] Tartir, S., Arpinar, I.B., Moore, M., Sheth, A.P., Aleman-Meza, B.: OntoQA: Metric-Based Ontology Quality Analysis. In: Proc. of IEEE Workshop on Knowledge Acquisition from Distributed, Autonomous, Semantically Heterogeneous Data and Knowledge Sources (2005)
[25] Zhou, L.: Ontology learning: state of the art and open issues. Information Technology and Management 8, 241–252 (2007)

DOGMA: A Disk-Oriented Graph Matching Algorithm for RDF Databases

Matthias Bröcheler[1], Andrea Pugliese[2], and V.S. Subrahmanian[1]

[1] University of Maryland, USA
[2] Università della Calabria, Italy

Abstract. RDF is an increasingly important paradigm for the representation of information on the Web. As RDF databases increase in size to approach tens of millions of triples, and as sophisticated graph matching queries expressible in languages like SPARQL become increasingly important, scalability becomes an issue. To date, there is no graph-based indexing method for RDF data where the index was designed in a way that makes it disk-resident. There is therefore a growing need for indexes that can operate efficiently when the index itself resides on disk. In this paper, we first propose the DOGMA index for fast subgraph matching on disk and then develop a basic algorithm to answer queries over this index. This algorithm is then significantly sped up via an optimized algorithm that uses efficient (but correct) pruning strategies when combined with two different extensions of the index. We have implemented a preliminary system and tested it against four existing RDF database systems developed by others. Our experiments show that our algorithm performs very well compared to these systems, with orders of magnitude improvements for complex graph queries.

1 Introduction

RDF is becoming an increasingly important paradigm for Web knowledge representation. As more and more RDF database systems come "online" and as RDF gets increasing emphasis from both established companies like HP and Oracle, as well as from a slew of startups, the need to store and efficiently query massive RDF datasets is becoming increasingly important. Moreover, large parts of query languages like SPARQL increasingly require that queries (which may be viewed as graphs) be matched against databases (which may also be viewed as graphs) – the set of all possible "matches" is returned as the answer.

For example, the GovTrack database [1] tracks events in the US Congress and stores the data in RDF. RDF triple stores primarily store triples (s, p, v) where s is a subject, p is a property, and v is a value. Fig. 1(a) shows a small portion of the GovTrack dataset (we have changed the names of individuals identified in that dataset). The reader can readily see that the RDF data forms a graph where the nodes correspond to subjects and values, and the edges linking them are labeled with a property. For instance, in Fig. 1(a), we see that Jeff Ryster sponsored Bill B0045 whose subject is Health Care. This corresponds to two triples (Jeff Ryster, sponsor, Bill B0045) and (Bill B0045, subject, Health Care). A user who is using such a database might wish to ask queries

A. Bernstein et al. (Eds.): ISWC 2009, LNCS 5823, pp. 97–113, 2009.

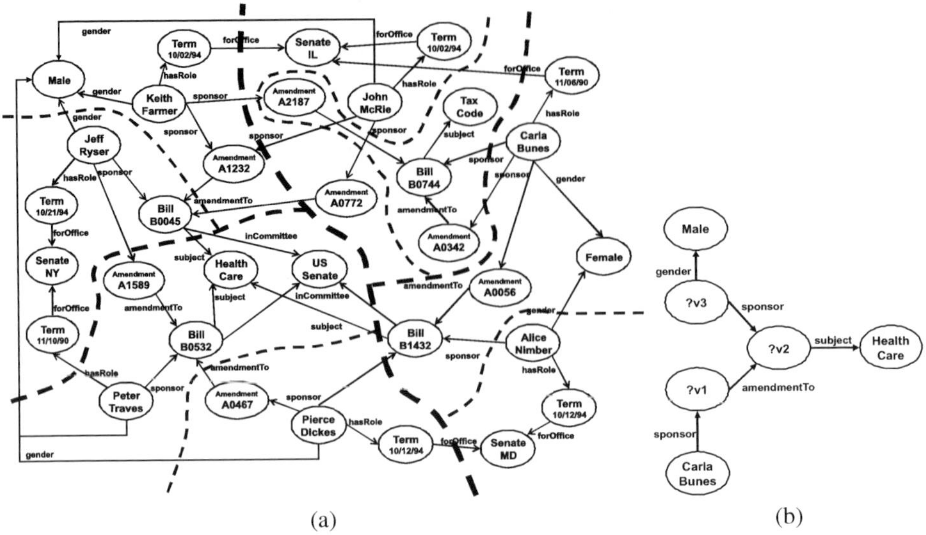

Fig. 1. Example RDF graph (a) and query (b)

such as that shown in Fig. 1(b). This query asks for all amendments ($?v1$) sponsored by Carla Bunes to bill ($?v2$) on the subject of health care that were originally sponsored by a male person ($?v3$). The reader can readily see that when answering this query, we want to find *all* matches for this query graph in the original database. The reader who tries to answer this very simple query against this very tiny database will see that it takes time to do so, even for a human being!

In this paper, we propose a *graph-based* index for RDF databases called DOGMA, that employs concepts from graph theory to efficiently answer queries such as that shown above. DOGMA is tuned for scalability in several ways. First, the index itself can be stored on disk. This is very important. From experiences in relational database indexing, it is clear that when the data is large enough to require disk space, the index will be quite large and needs to be disk resident as well. DOGMA, defined in Section 3, is the first graph-based index for RDF that we are aware of that is specifically designed to reside on disk. We define the DOGMA data structure and develop an algorithm to take an existing RDF database and create the DOGMA index for it. In Section 4, we develop algorithms to answer graph matching queries expressible in SPARQL [2] (we emphasize that we do not claim DOGMA supports all SPARQL queries yet). Our first algorithm, called DOGMA_basic, uses the index in a simple manner. Subsequently, we provide the improved algorithm DOGMA_adv and two extensions of the index called DOGMA_ipd and DOGMA_epd, that use sophisticated pruning methods to make the search more efficient without compromising correctness. Third, in Section 5, we show the results of an experimental assessment of our techniques against four competing RDF database systems (JenaTDB, Jena2, Sesame2, and OWLIM). We show that DOGMA performs very well compared to these systems.

2 Preliminaries

In this section, we briefly explain our notation. We assume the existence of a set $\mathcal{S}$ whose elements are called *subjects*, a set $\mathcal{P}$ whose elements are called *properties* and a set $\mathcal{V}$ whose elements are called *values*. Throughout this paper, we assume that $\mathcal{S}, \mathcal{P}, \mathcal{V}$ are all arbitrary, but fixed sets. If $s \in \mathcal{S}, p \in \mathcal{P}$ and $v \in \mathcal{V}$, then (s, p, v) is called an *RDF triple*. Note that $\mathcal{V}$ and $\mathcal{S}$ are not required to be disjoint. An *RDF database* is a finite set of RDF triples. For example, as mentioned earlier, (Jeff Ryster, sponsor, Bill B0045) and (Bill B0045, subject, Health Care) are RDF triples. Every RDF database $\mathcal{R}$ has an associated *RDF graph* $G_\mathcal{R} = (V_\mathcal{R}, E_\mathcal{R}, \lambda_\mathcal{R})$ where $V_\mathcal{R} = \mathcal{S} \cup \mathcal{V}, E_\mathcal{R} \subseteq \mathcal{S} \times \mathcal{V}$, and $\lambda_\mathcal{R} : E_\mathcal{R} \to \mathcal{P}$ is a mapping such that for all $(s, p, v) \in \mathcal{R}$, $\lambda_\mathcal{R}(s, v) = p$. [1] As there is a one-one correspondence between RDF graphs and RDF databases, we will often use the terms synonymously.

In this paper, we only focus on graph matching queries. In order to define such queries, we assume the existence of some set VAR of *variable symbols*. In this paper, all variable symbols will start with a "?". A *graph query* is any graph $Q = (V_Q, E_Q, \lambda_Q)$ where $V_Q \subseteq \mathsf{VAR} \cup \mathcal{S} \cup \mathcal{V}$, $E_Q \subseteq V_Q \times V_Q$, and $\lambda_Q : E_Q \to \mathcal{P}$ is a mapping. Suppose Q is a query. A *substitution* for query Q is a mapping $V_Q \cap \mathsf{VAR} \to \mathcal{S} \cup \mathcal{V}$. In other words, a substitution maps all variable vertices in query Q to either a subject or a value. For instance, in Fig. 1, the mapping θ which assigns B0744 to $?v1$, B0744 to $?v2$ and Jeff Ryster to $?v3$ is a substitution. If θ is a substitution for query Q, then $Q\theta$ denotes the replacement of all variables $?v$ in V_Q by $\theta(?v)$. In other words, the graph structure of $Q\theta$ is exactly like that of Q except that nodes labeled with $?v$ are replaced by $\theta(?v)$. A substitution θ is an *answer* for query Q w.r.t. database $\mathcal{R}$ iff $Q\theta$ is a subgraph of $G_\mathcal{R}$. The answer set for query Q w.r.t. an RDF database $\mathcal{R}$ is the set $\{\theta \mid Q\theta \text{ is a subgraph of } G_\mathcal{R}\}$.

Example 1. Consider the example query and RDF database in Fig. 1. In this case, the substitution θ such that $\theta(?v1)$ = Amendment A0056, $\theta(?v2)$ = Bill B1432, and $\theta(?v3)$ = Pierce Dickes is the only answer substitution for this query. □

3 The DOGMA Index

In this section, we develop the DOGMA index to efficiently answer graph queries in situations where the index itself must be very big (which occurs when $\mathcal{R}$ is very big). Before we define DOGMA indexes, we first define what it means to merge two graphs.

Suppose G is an RDF graph, and G_1 and G_2 are two RDF graphs such that $V_1, V_2 \subseteq V_\mathcal{R}$ and k is an integer such that $k \leq \max(|V_1|, |V_2|)$. Graph G_m is said to be a *k-merge of graphs G_1, G_2 w.r.t. G* iff: *(i)*$|V_m| = k$; *(ii)* there is a *surjective* (i.e. onto) mapping $\mu : V_1 \cup V_2 \to V_m$ called the *merge mapping* such that $\forall v \in V_m, rep(v) = \{v' \in V_1 \cup V_2 \mid \mu(v') = v\}$, and $(v_1, v_2) \in E$ iff there exist $v'_1 \in rep(v_1), v'_2 \in rep(v_2)$ such that $(v'_1, v'_2) \in E$. The basic idea tying k-merges to the DOGMA index is that we want

[1] For the sake of simplicity, we ignore many features in RDF such as reification, containers, blank nodes, etc. Moreover, we define $E_\mathcal{R} \subseteq \mathcal{S} \times \mathcal{V}$ for notational convenience; our implementation allows for multiple edges between vertices.

DOGMA to be a binary tree each of whose nodes occupies a disk page. Each node is labeled by a graph that "captures" its two children in some way. As each page has a fixed size, the number k limits the size of the graph so that it fits on one page. The idea is that if a node N has two children, N_1 and N_2, then the graph labeling node N should be a k-merge of the graphs labeling its children.

A DOGMA index for an RDF database $\mathcal{R}$ is a generalization of the well known binary-tree specialized to represent RDF graph data in a novel manner.

Definition 1. A DOGMA index of order k ($k \geq 2$) *is a binary tree* $\boldsymbol{D}_{\mathcal{R}}$ *with the following properties:*

1. *Each node in* $\boldsymbol{D}_{\mathcal{R}}$ *equals the size of a disk page and is labeled by a graph.*
2. $\boldsymbol{D}_{\mathcal{R}}$ *is balanced.*
3. *The labels of the set of leaf nodes of* $\boldsymbol{D}_{\mathcal{R}}$ *constitute a partition of* $G_{\mathcal{R}}$*.*
4. *If node* N *is the parent of nodes* N_1, N_2*, then the graph* G_N *labeling node* N *is a* k*-merge of the graphs* G_{N_1}, G_{N_2} *labeling its children.*

Note that a single RDF database can have many DOGMA indexes.

Example 2. Suppose $k = 4$. A DOGMA index for the RDF graph of Fig. 1(a) might split the graph into the 8 components indicated by dashed lines in Fig. 1(a) that become the leaf nodes of the index (Fig. 2). Consider the two left-most leaf nodes. They can be 4-merged together to form a parent node. Other leaf nodes can also be merged together (due to space constraints, the results of k-merging are not shown in the inner nodes). □

Even though many different DOGMA indexes can be constructed for the same RDF database, we want to find a DOGMA index with as few "cross" edges between subgraphs stored on different pages as possible. In other words, if node N is the parent of nodes N_1, N_2, then we would like relatively fewer edges in $\mathcal{R}$ between some node

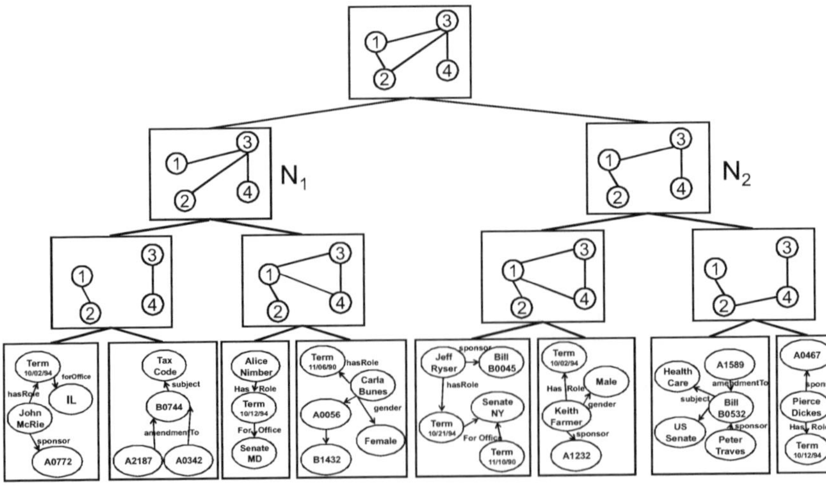

Fig. 2. A DOGMA index for the RDF database of Fig. 1(a)

in G_{N_1} and some node in G_{N_2}. The smaller this number of edges, the more "self-contained" nodes N_1, N_2 are, and the less likely that a query will require looking at both nodes N_1 and N_2. In the description of our proposed algorithms, we employ an external graph partitioning algorithm (many of which have been proposed in the literature) that, given a weighted graph, partitions its vertex set in such a way that (*i*) the total weight of all edges crossing the partition is minimized and (*ii*) the accumulated vertex weights are (approximately) equal for both partitions. In our implementation, we employ the *GGGP* graph partitioning algorithm proposed in [3].

Fig. 3 provides an algorithm to build a DOGMA index for an RDF graph $G_{\mathcal{R}}$. The BuildDOGMAIndex algorithm starts with the input RDF graph, which is set to G_0. It assigns an arbitrary weight of 1 to each vertex and each edge in G_0. It iteratively coarsens G_0 into a graph G_1 that has about half the vertices in G_0, then coarsens G_1 into a graph G_2 that has about half the vertices as G_1, and so forth until it reaches a G_j that has k vertices or less.

Algorithm BuildDOGMAIndex
Input: RDF graph $G_{\mathcal{R}}$, page size k
(level L, colors C)
Output: DOGMA index $\mathbf{D}_{\mathcal{R}}$

$G_0 \leftarrow G_{\mathcal{R}}$
for all $v \in V_{\mathcal{R}}$
  $weight(v) \leftarrow 1$
for all $e \in E_{\mathcal{R}}$
  $weight(e) \leftarrow 1$
$i \leftarrow 0$
while $|G_i| > k$
  $i \leftarrow i + 1$
  $G_i, \mu_i \leftarrow$ CoarsenGraph(G_{i-1})
$root(\mathbf{D}_{\mathcal{R}}) \leftarrow$ a new "empty" node R
BuildTree(R, i, G_i)
ColorRegions(L, $\mathbf{D}_{\mathcal{R}}$, C) /* Only required for the DOGMA_epd index discussed later */
return $\mathbf{D}_{\mathcal{R}}$

Algorithm CoarsenGraph
Input: RDF graph $G_{\mathcal{R}}$
Output: Coarsened graph $G'_{\mathcal{R}}$, merge mapping μ

$G'_{\mathcal{R}} \leftarrow G_{\mathcal{R}}$
$\mu \leftarrow (V_{\mathcal{R}} \rightarrow V'_{\mathcal{R}})$ /* identity map */
while $2 \times |V'_{\mathcal{R}}| > |V_{\mathcal{R}}|$
  $v \leftarrow$ uniformly random chosen vertex from $V'_{\mathcal{R}}$
  $N_v \leftarrow \{u \mid (u, v) \in E'_{\mathcal{R}}\}$
  $m \leftarrow x \in N_v$ s.t. $x \succeq y \ \forall y \in N_v$
  $weight(m) \leftarrow weight(m) + weight(v)$
  for all $(v, u) \in E'_{\mathcal{R}}$
    if $(m, u) \in E_{\mathcal{R}}$
      $weight((m, u)) \leftarrow weight((m, u))$
        $+ weight((v, u))$
    else
      $E'_{\mathcal{R}} \leftarrow E'_{\mathcal{R}} \cup \{(m, u)\}$
      $weight((m, u)) \leftarrow weight((v, u))$
  $V'_{\mathcal{R}} \leftarrow V'_{\mathcal{R}} \setminus \{v\}$
  $\mu(\mu^{-1}(v)) \leftarrow m$
  $E'_{\mathcal{R}} \leftarrow E'_{\mathcal{R}} \setminus \{(v, u) \in E'_{\mathcal{R}}\}$
return $G'_{\mathcal{R}}, \mu$

Algorithm BuildTree
Input: Binary tree node N, level i,
subgraph S at level i
Output: Graph merge hierarchy $\{G_j\}_{j \geq 0}$
and merge mappings $\{\mu_j\}_{j > 0}$

$label(N) \leftarrow S$
if $|S| > k$
  $S_1, S_2 \leftarrow$ GraphPartition(S)
  $L \leftarrow leftChild(N)$
  $R \leftarrow rightChild(N)$
  $S_L \leftarrow$ induced subgraph in G_{i-1}
    by vertex set $\{v \mid \mu_i(v) \in V_{S_1}\}$
  $S_R \leftarrow$ induced subgraph in G_{i-1}
    by vertex set $\{v \mid \mu_i(v) \in V_{S_2}\}$
  BuildTree(L, $i-1$, S_L)
  BuildTree(r, $i-1$, S_R)
$P_N \leftarrow \{v \mid \mu_i(\mu_{i-1}(\dots\mu_1(v))) \in V_S\}$
for all $v \in P_N$ /* Only for DOGMA_ipd */
  $ipd(v, N) \leftarrow \min_{u \in V_0 \setminus P_N} d_{G_0}(u, v)$

Fig. 3. BuildDOGMAIndex, CoarsenGraph, and BuildTree algorithms

The coarsening is done by invoking a CoarsenGraph algorithm that randomly chooses a vertex v in the input graph, then it finds the immediate neighbors N_v of v, and then finds those nodes in N_v that are best according to a total ordering $\succeq$. There are many ways to define $\succeq$; we experimented with different orderings and chose to order by increasing edge weight, then decreasing vertex weight. The CoarsenGraph algorithm appropriately updates node and edge weights and then selects a maximally weighted node, denoted m, to focus the coarsening on. The coarsening associated with the node v merges neighbors of the node m and m itself into one node, updates weights, and removes v. Edges from m to its neighbors are removed. This process is repeated till we obtain a graph which has half as many vertices (or less) than the graph being coarsened. The result of CoarsenGraph is a k-merge where we have merged adjacent vertices. The BuildDOGMAIndex algorithm then uses the sequence $G_0, G_1, \ldots, G_j$ denoting these coarsened graphs to build the DOGMA index using the BuildTree subroutine. Note that Line 12 in the BuildDOGMAIndex algorithm (where L denotes the level at which to color the subgraphs and C is a list of unique colors) is only needed for the DOGMA_epd index introduced later, as well as lines 12–14 in BuildTree are for the DOGMA_ipd index. They are included here to save space.

Proposition 1. *Algorithm BuildDOGMAIndex correctly builds a DOGMA index for an RDF graph $G_{\mathcal{R}}$. Moreover, the worst-case time complexity of Algorithm BuildDOGMAIndex is $O(|E_{\mathcal{R}}| + \Lambda(k)\frac{|V_{\mathcal{R}}|}{k})$ where $\Lambda(k)$ is the worst-case time complexity of Algorithm GraphPartition over a graph with k vertices and $O(k)$ edges.*

4 Algorithms for Processing Graph Queries

In this section, we first present the DOGMA_basic algorithm for answering queries against a DOGMA index stored on external memory. We then present various extensions that improve query answering performance on complex graph queries.

4.1 The DOGMA_basic Query Processing Algorithm

Fig. 4 shows our basic algorithm for answering graph matching queries using the DOGMA index. In the description of the algorithm, we assume the existence of two index retrieval functions: retrieveNeighbors($\mathbf{D}_{\mathcal{R}}, v, l$) that retrieves from DOGMA index $\mathbf{D}_{\mathcal{R}}$ the unique identifiers for all vertices v' that are connected to vertex v by an edge labeled l, i.e., the neighbors of v restricted to label l, and retrieveVertex($\mathbf{D}_{\mathcal{R}}, v$) that retrieves from $\mathbf{D}_{\mathcal{R}}$ a complete description of vertex v, i.e., its unique identifier and its associated metadata. Note that retrieveVertex implicitly exploits locality, since after looking up neighboring vertices, the probability is high that the page containing the current vertex's description is already in memory.

DOGMA_basic is a recursive, depth-first algorithm which searches the space of all substitutions for the answer set to a given query Q w.r.t an RDF database $\mathcal{R}$. For each variable vertex v in Q, the algorithm maintains a set of constant vertices $R_v \subseteq V_{\mathcal{R}}$ (called result candidates) to prune the search space; for each answer substitution θ for Q, we have $\theta(v) \in R_v$. In other words, the result candidates must be a superset of the set of all matches for v. Hence, we can prune the search space by only considering those

```
Algorithm DOGMA_basic
Input: Graph query Q, DOGMA index D_R, partial substitution θ, candidate sets {R_z}
Output: Answer set A, i.e. set of substitutions θ s.t. Qθ is a subgraph of G_R
1  if ∀z ∈ V_Q ∩ VAR : ∃c : (z → c) ∈ θ
2    A ← A ∪ {θ}
3    return /* done - a correct answer substitution has been found */
4  if θ = ∅
5    for all z ∈ V_Q ∩ VAR
6      R_z ← null /* no candidate substitutions for any vars in the query initially */
7    for all c ∈ V_Q ∩ (S ∪ V)
8      for all edges e = (c, v) incident on c and some v ∈ V_Q ∩ VAR
9        if R_v = null
10         R_v ← retrieveNeighbors(D_R, c, λ_Q(e)) /* use index to retrieve all nbrs of c with same label as e */
11       else
12         R_v ← R_v ∩ retrieveNeighbors(D_R, c, λ_Q(e)) /* restrict space of possible subst. for z */
13 R_w ← argmin_{R_z ≠ null, s.t. z ∈ V_Q ∩ V \ dom(θ)} |R_z|
14 if R_w = ∅
15   return "NO"
16 else
17   for all m ∈ R_w
18     retrieveVertex(D_R, m)
19     θ' ← θ ∪ {w → m}
20     for all z ∈ V_Q ∩ VAR
21       R'_z ← R_z
22     for all edges e = (w, v) incident on w and some v ∈ V_Q ∩ VAR \ dom(θ)
23       if R_v = null
24         R'_v ← retrieveNeighbors(D_R, m, λ_Q(e))
25       else
26         R'_v ← R_v ∩ retrieveNeighbors(D_R, m, λ_Q(e))
27     DOGMA_basic(θ'(Q), D_R, θ', {R'_z})
```

Fig. 4. DOGMA_basic algorithm

substitutions θ for which $\theta(v) \in R_v$ for all variable vertices v in Q. DOGMA_basic is called initially with an empty substitution and uninitialized result candidates (lines 4-6). We use uninitialized result candidates $R_v = \mathbf{null}$ to efficiently denote $R_v = V_{\mathcal{R}}$, i.e., the fact that there are no constraints on the result candidates yet. The algorithm then initializes the result candidates for all variable vertices v in Q which are connected to a constant vertex c in Q through an edge labeled by l (lines 7-12). Here we employ the fact that any answer substitution θ must be such that $\theta(v)$ is a neighbor of c, and thus the set of all neighbors of c in $G_{\mathcal{R}}$ reachable by an edge labeled l are result candidates for v. We use the DOGMA index $\mathbf{D}_{\mathcal{R}}$ to efficiently retrieve the neighborhood of c. If v is connected to multiple constant vertices, we take the intersection of the respective constraints on the result candidates.

At each recursive invocation, the algorithm extends the given substitution and narrows down the result candidates for all remaining variable vertices correspondingly. To extend the given substitution θ, we greedily choose the variable vertex w with the smallest set of result candidates (line 13). This yields a locally optimal branching factor of the search tree since it provides the smallest number of extensions to the current substitution. In fact, if the set of result candidates is empty, then we know that θ cannot be extended to an answer substitution, and we thus directly prune the search (lines 14-15). Otherwise, we consider all the possible result candidates $m \in R_w$ for w by deriving extended substitutions θ' from θ which assign m to w (lines 17-19) and then calling DOGMA_basic recursively on θ' (line 27). Prior to this, we update the result candidates for all remaining variable vertices (lines 20-26). By assigning the constant vertex m to w we can constrain the result candidates for all neighboring variable vertices as discussed above.

Note that our description of the algorithm assumes that edges are undirected, to simplify the presentation. Obviously, our implementation takes directionality into account and thus distinguishes between outgoing and incoming edges when determining vertex neighborhoods.

Example 3. Consider the example query and RDF database in Fig. 1. Fig. 5(a) shows the initial result candidates for each of the variable vertices $?v_1, ?v_2, ?v_3$ in boxes. After initialization, DOGMA_basic chooses the smallest set of result candidates to extend the currently empty substitution $\theta = \emptyset$. We have that $|R_{v_1}| = |R_{v_2}| = 3$; suppose R_{v_2} is chosen. We can now extend θ by assigning each of the result candidates (Bill B0045, Bill B0532, Bill B1432) to $?v_2$. Hence, we first set $\theta'(?v_2) =$ Bill B0045. This introduces a new constant vertex into the query and we thus constrain the result candidates of the two neighbor variable vertices v_1, v_3 by the "amendmentTo" and "sponsor" neighborhood of Bill B0045 respectively. The result is shown in Fig. 5(b); here we call DOGMA_basic recursively to encounter the empty result candidates for v_1. Hence we reached a dead end in our search for an answer substitution and the algorithm backtracks to try the remaining extensions for θ. Eventually, DOGMA_basic considers the extension $v_2 \rightarrow$ Bill B1432 which leads to the query answer. □

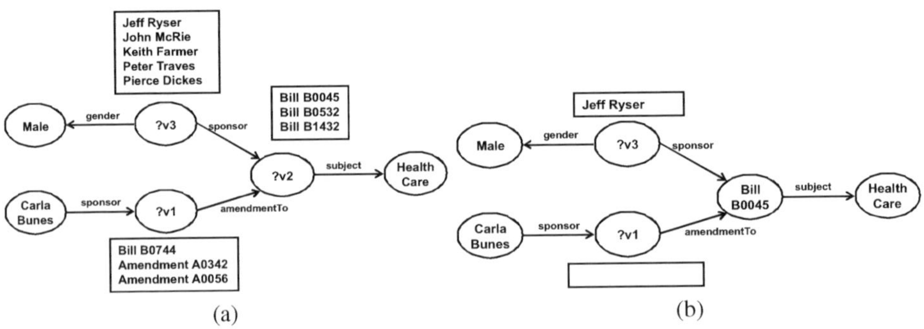

Fig. 5. Execution of DOGMA_basic on the example of Fig. 1

Proposition 2. *Suppose $\boldsymbol{D}_{\mathcal{R}}$ is a DOGMA index for an RDF database $\mathcal{R}$ and Q is a graph query. Then: DOGMA_basic($Q, \boldsymbol{D}_{\mathcal{R}}, \{\}, null$) returns the set of all correct answer substitutions for query Q w.r.t. $\mathcal{R}$. Moreover, the worst-case complexity of the DOGMA_basic algorithm is $O(|V_{\mathcal{R}}|^{|V_Q \cap VAR|})$.*

The algorithm is therefore exponential in the number of variables in the query in the worst case. However, the algorithm is efficient in practice as we will show in Section 5. Furthermore, we propose two extensions of the DOGMA index that improve its performance.

4.2 The DOGMA_adv Algorithm

The basic query answering algorithm presented in the previous section only uses "short range" dependencies, i.e., the immediate vertex neighborhood of variable vertices, to constrain their result candidates. While this suffices for most simple queries, considering "long range" dependencies can yield additional constraints on the result candidates and thus improve query performance. For instance, the result candidates for v_1 in our example query not only must be immediate neighbors of "Carla Bunes": in addition, they must be at most at a distance of 2 from "Health Care". More formally, let $d_{\mathcal{R}}(u, v)$ denote the length of the shortest path between two vertices $u, v \in V_{\mathcal{R}}$ in the undirected counterpart of a RDF graph $G_{\mathcal{R}}$, and let $d_Q(u, v)$ denote the distance between two vertices in the undirected counterpart of a query Q; a long range dependency on a variable vertex $v \in V_Q$ is introduced by any constant vertex $c \in V_Q$ with $d_Q(v, c) > 1$.

We can exploit long range dependencies to further constrain result candidates. Let v be a variable vertex in Q and c a constant vertex with a long range dependency on v. Then any answer substitution θ must satisfy $d_Q(v, c) \geq d_{\mathcal{R}}(\theta(v), c)$ which, in turn, means that $\{m \mid d_{\mathcal{R}}(m, c) \leq d_Q(v, c)\}$ are result candidates for v. This is the core idea of the DOGMA_adv algorithm shown in Fig. 6, which improves over and extends DOGMA_basic. In addition to the result candidates sets R_v, the algorithm maintains sets of distance constraints C_v on them. As long as a result candidates set R_v remains uninitialized, we collect all distance constraints that arise from long range dependencies on the variable vertex v in the constraints set C_v (lines 15-16 and 34-35). After the result candidates are initialized, we ensure that all elements in R_v satisfy the distance constraints in C_v (lines 17-18 and 37-38). Maintaining additional constraints therefore reduces the size of R_v and hence the number of extensions to θ we have to consider (line 23 onward).

DOGMA_adv assumes the existence of a *distance index* to efficiently look up $d_{\mathcal{R}}(u, v)$ for any pair of vertices $u, v \in V_{\mathcal{R}}$ (through function retrieveDistance), since computing graph distances at query time is clearly inefficient. But how can we build such an index? Computing all-pairs-shortest-path has a worst-case time complexity $O(|V_{\mathcal{R}}|^3)$ and space complexity $O(|V_{\mathcal{R}}|^2)$, both of which are clearly infeasible for large RDF databases. However, we do not need to know the *exact* distance between two vertices for DOGMA_adv to be correct. Since all the distance constraints in DOGMA_adv are *upper bounds* (lines 18, 31, and 38), all we need is to ensure that $\forall u, v \in V_{\mathcal{R}}$, retrieveDistance($\mathbf{D}_{\mathcal{R}}, u, v) \leq d_{\mathcal{R}}(u, v)$.

Thus, we can extend the DOGMA index to include distance information and build two "lower bound" distance indexes, DOGMA_ipd and DOGMA_epd, that use approximation techniques to achieve acceptable time and space complexity.

4.3 DOGMA_ipd

For building the DOGMA index, we employed a graph partitioner which minimizes cross edges, to ensure that strongly connected vertices are stored in close proximity on disk; this implies that distant vertices are likely to be assigned to distinct sets in the partition. We exploit this to extend DOGMA to a distance index.

As seen before, the leaf nodes of the DOGMA index $\mathbf{D}_{\mathcal{R}}$ are labeled by subgraphs which constitute a partition of $G_{\mathcal{R}}$. For any node $N \in \mathbf{D}_{\mathcal{R}}$, let P_N denote the union

```
Algorithm DOGMA_adv
Input: Graph query Q, DOGMA Index D_R, partial substitution θ, candidate sets {R_z}, constraint sets {C_z}
Output: Answer set A, i.e. set of substitutions θ s.t. θ(Q)⊆~G
if ∀z ∈ V_Q ∩ VAR : ∃c : (z → c) ∈ θ
   A ← A ∪ {θ}
   return
if θ = ∅
   for all z ∈ V_Q ∩ VAR
      R_z ← null
   for all c ∈ V_Q ∩ (S ∪ V)
      for all edges e = (c, v) incident on c and some v ∈ V_Q ∩ VAR
         if R_v = null
            R_v ← retrieveNeighbors(D_R, c, λ_Q(e))
         else
            R_v ← R_v ∩ retrieveNeighbors(D_R, c, λ_Q(e))
   for all c ∈ V_Q ∩ (S ∪ V)
      for all variable vertices v ∈ V_Q ∩ VAR s.t. d_Q(c, v) > 1
         if R_v = null
            C_v ← C_v ∪ {(c, d_Q(c, v))}
         else
            R_v ← {u ∈ R_v | retrieveDistance(D_R, c, u) ≤ d_Q(c, v)}
R_w ← argmin_{R_z ≠ null, s.t. z ∈ V_Q ∩ VAR \ dom(θ)} |R_z|
if R_w = ∅
   return
else
   for all m ∈ R_w
      retrieveVertex(D_R, m)
      θ' ← θ ∪ {w → m}
      for all z ∈ V_Q ∩ VAR
         R'_z ← R_z
         C'_z ← C_z
      for all edges e = (w, v) incident on w and some v ∈ V_Q ∩ VAR \ dom(θ)
         if R_v = null
            R'_v ← {u ∈ retrieveNeighbors(D_R, m, λ_Q(e)) | ∀(c, d) ∈ C_v : retrieveDistance(D_R, c, u) ≤ d}
         else
            R'_v ← R_v ∩ retrieveNeighbors(D_R, m, λ_Q(e))
      for all variable vertices v ∈ V_Q ∩ VAR \ dom(θ) s.t. d_Q(w, v) > 1
         if R_v = null
            C_v ← C_v ∪ {(m, d_Q(w, z))}
         else
            R_v ← {w ∈ R_v | retrieveDistance(D_R, m, v) ≤ d_Q(w, v)}
      DOGMA_basic(θ'(Q), D_R, θ', {R'_z}, {C'_z})
```

Fig. 6. DOGMA_adv algorithm

of the graphs labeling all leaf nodes reachable from N. Hence, P_N is the union of all subgraphs in $G_{\mathcal{R}}$ that were eventually merged into the graph labeling N during index construction and therefore corresponds to a larger subset of $G_{\mathcal{R}}$. For example, the dashed lines in Fig 1(a) mark the subgraphs P_N for all index tree nodes N of the DOGMA index shown in Fig. 2 where bolder lines indicate boundaries corresponding to nodes of lower depth in the tree.

The DOGMA *internal partition distance* (DOGMA_ipd) index stores, for each index node N and vertex $v \in P_N$, the distance to the outside of the subgraph corresponding to P_N. We call this the *internal partition distance* of v, N, denoted $ipd(v, N)$, which is thus defined as $ipd(v, N) = \min_{u \in V_{\mathcal{R}} \setminus P_N} d_{\mathcal{R}}(v, u)$. We compute these distances during index construction as shown in Fig. 3 (BuildTree algorithm at lines 12-14). At query time, for any two vertices $v, u \in V_{\mathcal{R}}$ we first use the DOGMA tree index to identify those distinct nodes $N \neq M$ in $\mathbf{D}_{\mathcal{R}}$ such that $v \in P_N$ and $u \in P_M$, which are at the same level of the tree and closest to the root. If such nodes do not exist (because v, u are associated with the same leaf node in $\mathbf{D}_{\mathcal{R}}$), then we set $d_{ipd}(u, v) = 0$. Otherwise we set $d_{ipd}(u, v) = \max(ipd(v, N), ipd(u, M))$. It is easy to see that d_{ipd}

is an admissible lower bound distance, since $P_N \cap P_M = \emptyset$. By choosing those distinct nodes which are closest to the root, we ensure that the considered subgraphs are as large as possible and hence $d_{ipd}(u, v)$ is the closest approximation to the actual distance.

Proposition 3. *Building the* DOGMA_ipd *index has a worst-case time complexity* $O(\log \frac{|V_\mathcal{R}|}{k}(|E_\mathcal{R}| + |V_\mathcal{R}| \log |V_\mathcal{R}|))$ *and space complexity* $O(|V_\mathcal{R}| \log \frac{|V_\mathcal{R}|}{k})$.

Example 4. Consider the example of Fig. 1. As shown in Fig. 7(a), there is a long range dependency between "Carla Bunes" and variable vertex v_2 at distance 2. The boldest dashed line in Fig. 1(a) marks the top level partition and separates the sets P_{N_1}, P_{N_2}, where N_1, N_2 are the two nodes directly below the root in the DOGMA index in Fig. 2. We can determine that ipd(Carla Bunes, N_2) $= 3$ and since Bill B0045 and B0532 lie in the other subgraph, it follows that d_{ipd}(Carla Bunes, B0045/B0532) $= 3$ and therefore we can prune both result candidates. □

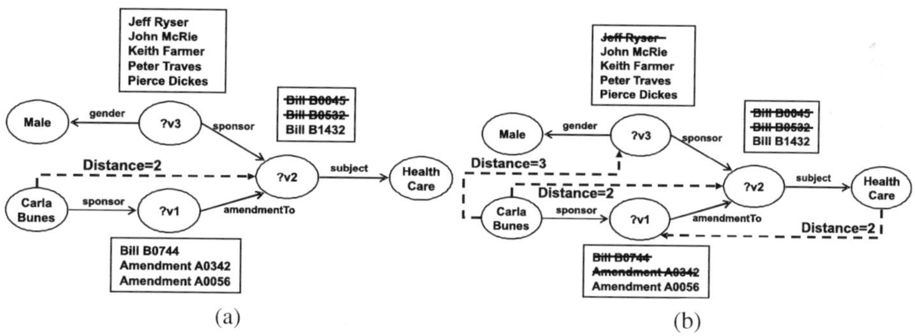

Fig. 7. Using DOGMA_ipd and DOGMA_epd for query answering

4.4 DOGMA_epd

The DOGMA *external partition distance* (DOGMA_epd) index also uses the partitions in the index tree to compute a lower bound distance. However, it considers the distance to *other* subgraphs rather than the distance within the *same* one. For some fixed level L, let $\mathcal{N}_L$ denote the set of all nodes in $\mathbf{D}_\mathcal{R}$ at distance L from the root. As discussed above, $P = \{P_N\}_{N \in \mathcal{N}_L}$ is a partition of $G_\mathcal{R}$. The idea behind DOGMA_epd is to assign a color from a fixed list of colors C to each subgraph $P_N \in P$ and to store, for each vertex $v \in V_\mathcal{R}$ and color $c \in C$, the shortest distance from v to a subgraph colored by c. We call this the *external partition distance*, denoted $epd(v, c)$, which is thus defined as $epd(v, c) = \min_{u \in P_N, \phi(P_N)=c} d_\mathcal{R}(v, u)$ where $\phi : P \rightarrow C$ is the color assignment function. We store the color of P_N with its index node N so that for a given pair of vertices u, v we can quickly retrieve the colors c_u, c_v of the subgraphs to which u and v belong. We then compute $d_{epd}(v, u) = \max(epd(v, c_u), epd(u, c_v))$. It is easy to see that d_{epd} is an admissible lower bound distance.

Ideally, we want to assign each partition a distinct color but this exceeds our storage capabilities for large database sizes. Our problem is thus to assign a limited number of colors to the subgraphs in such a way as to maximize the distance between subgraphs of the same color. Formally, we want to minimize the objective function $\sum_{P_N \in P} \sum_{P_M \in P, \phi(P_N)=\phi(P_M)} \frac{1}{d(P_N, P_M)}$ where $d(P_N, P_M) = \min_{u \in P_N, v \in P_M} d_{\mathcal{R}}(u, v)$. Inspired by the work of Ko and Rubenstein on peer-to-peer networks [4], we designed a probabilistic, locally greedy optimization algorithm for the maximum distance coloring problem named ColorRegions, that we do not report here for reasons of space. The algorithm starts with a random color assignment and then iteratively updates the colors of individual partitions to be locally optimal. A *propagation radius* determines the neighborhood that is analyzed in determining the locally optimal color. The algorithm terminates if the cost improvement falls below a certain threshold or if a maximum number of iterations is exceeded.

Proposition 4. *Computing the external partition distance has a worst-case time complexity $O(|C|\,(|E_{\mathcal{R}}| + |V_{\mathcal{R}}| \log |V_{\mathcal{R}}|))$ and space complexity $O(|V_{\mathcal{R}}|\,|C|)$.*

Example 5. Consider the example of Fig. 1(a) and assume each set in the lowest level of the DOGMA index in Fig. 2 is colored with a different color. Figure 7(b) indicates some long range dependencies and shows how the external partition distance can lead to additional prunings in the three result candidates sets which can be verified against Fig. 1(a). □

5 Experimental Results

In this section we present the results of the experimental assessment we performed of the DOGMA_adv algorithm combined with DOGMA_ipd and DOGMA_epd indexes.

We compared the performance of our algorithm and indexes with 4 leading RDF database systems developed in the Semantic Web community that are most widely used and have demonstrated superior performance in previous evaluations [5]. *Sesame2* [6] is an open source RDF framework for storage, inferencing and querying of RDF data, that includes its own RDF indexing and I/O model and also supports a relational database as its storage backend. We compare against Sesame2 using its native storage model since initial experiments have shown that Sesame2's performance drops substantially when backed by a relational database system. *Jena2* [7] is a popular Java RDF framework that supports persistent RDF storage backed by a relational database system (we used PostgreSQL [8]). SPARQL queries are processed by the ARQ query engine which also supports query optimization [9]. *JenaTDB* [10] is a component of the Jena framework providing persistent storage and query optimization for large scale RDF datasets based on a native indexing and I/O model. Finally, *OWLIM* [11] is a high performance semantic repository based on the Sesame RDF database. In the experiments, we compared against the internal memory version of OWLIM which is called *SwiftOWLIM* and is freely available. SwiftOWLIM loads the entire dataset into main memory prior to query answering and therefore must be considered to have an advantage over the other systems.

Moreover, we used 3 different RDF datasets. *GovTrack* [1] consists of more than 14.5 million triples describing data about the U.S. Congress. The *Lehigh University Benchmark* (LUBM) [12] is frequently used within the Semantic Web community as the basis

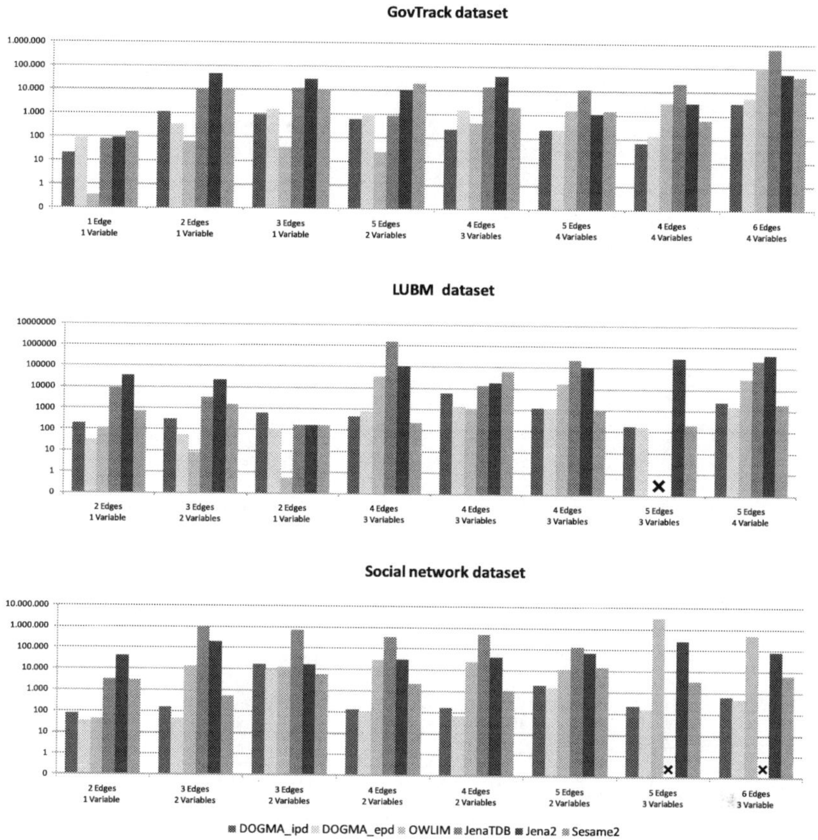

Fig. 8. Query times (ms) for graph queries of low complexity

for evaluation of RDF and ontology storage systems. The benchmark's RDF generator employs a schema which describes the university domain. We generated a dataset of more than 13.5 million triples. Finally, a fragment of the Flickr social network [13] dataset was collected by researchers of the MPI Saarbrücken to analyze online social networks [14] and was generously made available to us. The dataset contains information on the relationships between individuals and their memberships in groups. The fragment we used for the experiments was anonymized and contains approximately 16 million triples. The GovTrack and social network datasets are well connected (with the latter being denser than the former), whereas the dataset generated by the LUBM benchmark is a sparse and almost degenerate RDF graph containing a set of small and loosely connected subgraphs.

In order to allow for a meaningful comparison of query times across the different systems, we designed a set of graph queries with varying complexity, where constant vertices were chosen randomly and queries with an empty result set were filtered out. Queries were grouped into classes based on the number of edges and variable vertices.

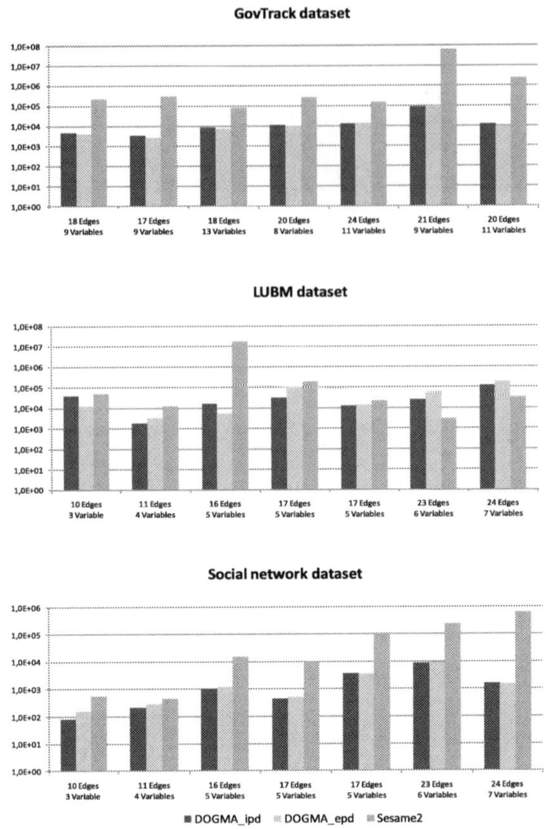

Fig. 9. Query times (ms) for graph queries of high complexity

We repeated the query time measurements multiple times for each query, eliminated outliers, and averaged the results. Finally, we averaged the query times of all queries in each class. All experiments were executed on a machine with a 2.4Ghz Intel Core 2 processor and 3GB of RAM.

In a first round of experiments, we designed several relatively simple graph queries for each dataset, containing no more than 6 edges, and grouped them into 8 classes.The results of these experiments are shown in Fig. 8 which reports the query times for each query class on each of the three datasets. Missing values in the figure indicate that the system did not terminate on the query within a reasonable amount of time (around 20 mins). Note that the query times are plotted in logarithmic scale to accommodate the large discrepancies between systems. The results show that OWLIM has low query times on low complexity queries across all datasets. This result is not surprising, as OWLIM loads all data into main memory prior to query execution. The performance advantage of DOGMA_ipd and DOGMA_epd over the other systems increases with query complexity on the GovTrack and social network dataset, where our proposed

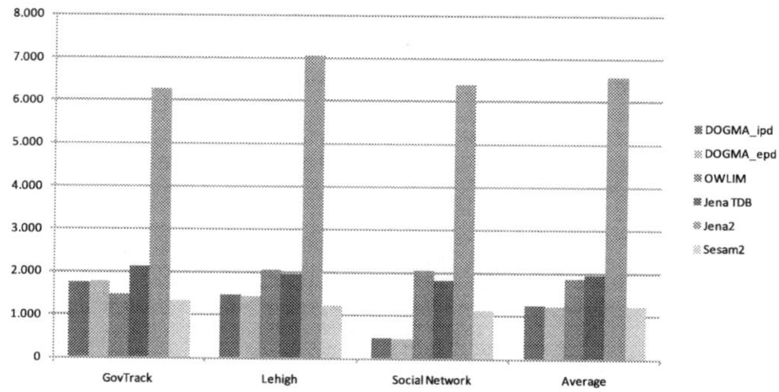

Fig. 10. Index size (MB) for different datasets

techniques are orders of magnitude faster on the most complex queries. On the LUBM dataset, however, Sesame2 performs almost equally for the more complex queries. Finally, DOGMA_epd is slightly faster on the LUBM and social network dataset, whereas DOGMA_ipd has better performance on the Govtrack dataset.

In a second round of experiments, we significantly increased the complexity of the queries, which now contained up to 24 edges. Unfortunately, the OWLIM, JenaTDB, and Jena2 systems did not manage to complete the evaluation of these queries in reasonable time, so we exclusively compared with Sesame2. The results are shown in Fig. 9. On the GovTrack and social network dataset, DOGMA_ipd and DOGMA_epd continue to have a substantial performance advantage over Sesame2 on all complex graph queries of up to 40000%. For the LUBM benchmark, the picture is less clear due to the particular structure of the generated dataset explained before.

Finally, Fig. 10 compares the storage requirements of the systems under comparison for all three datasets. The results show that DOGMA_ipd,DOGMA_epd and Sesame2 are the most memory efficient.

To wrap up the results of our experimental evaluation, we can observe that both DOGMA_ipd and DOGMA_epd are significantly faster than all other RDF database systems under comparison on complex graph queries over non-degenerate graph datasets. Moreover, they can efficiently answer complex queries on which most of the other systems do not terminate or take up to 400 times longer, while maintaining a satisfactory storage footprint. DOGMA_ipd and DOGMA_epd have similar performance, yet differences exist which suggest that each index has unique advantages for particular queries and RDF datasets. Investigating these is subject of future research.

6 Related Work

Many approaches to RDF storage have been proposed in the literature and through commercial systems. In Section 5 we briefly reviewed four such systems that we used in the performance comparison. Discussing all prior work on RDF storage and retrieval in detail is beyond the scope of this paper. Approaches differ with respect to their storage regime, index structures, and query answering strategies. Some systems use relational databases as their back-end [15]; for instance by inferring the relational schema of the given RDF data [16,17], or using a triple based denormalized relational schema [7], whereas others propose native storage formats for RDF [11]. To efficiently retrieve triples, RDF databases typically rely on index structures, such as the popular B-tree and its generalizations, over subjects, predicates, objects or any combination thereof [18]. Query answering is either handled by the relational database back-end after a SPARQL query is translated into its SQL equivalent or employs existing index structures to retrieve stored triples that match the query. [19] does some additional tuning through B^+-tree page compression and optimized join processing. Recent work on query optimization for RDF uses triple selectivity estimation techniques similar to those used in relational database systems [9].

Despite these differences, the great majority of RDF databases are *triple oriented* in the sense that they focus on the storage and retrieval of individual triples. In contrast, our work is *graph oriented* because we analyze the graph spanned by RDF data and exploit graph properties, such as connectedness and shortest path lengths, for efficient storage and, more importantly, retrieval. This explains DOGMA's performance advantage on complex queries. GRIN [20] was the first RDF indexing system to use graph partitioning and distances in the graphs as a basis for indexing for SPARQL-like queries. However, GRIN did not operate on disk and the authors subsequently found errors in the experimental results reported in that paper. There is also some related work in other communities. LORE [21], a database system for semi-structured data, proposed path indexes based on the assumption that the input data can be accurately represented as a tree. This assumption clearly does not hold for RDF data. Furthermore, there is a lot work on approximate query answering over graph datasets in the bioinformatics community [22]. However, the biological datasets are small enough to fit into main memory and hence storage and retrieval are not being addressed. Finally, [19,23] focus on the physical data structures to optimally store RDF triples. Their work is thus orthogonal to ours, since a DOGMA index could be built on the physical data structures proposed in these papers in order to additionally exploit graph distance locality.

7 Conclusions and Future Work

In this paper, we proposed the DOGMA index for fast subgraph matching on disk and developed algorithms to answer queries over this index. The algorithms use efficient (but correct) pruning strategies and can be combined with two different extensions of the index. We tested a preliminary implementation of the proposed techniques against four existing RDF database systems, showing very good query answering performance. Future work will be devoted to an in-depth study of the advantages and disadvantages of each of the proposed indexes when dealing with particular queries and RDF datasets. Moreover, we plan to extend our indexes to support efficient updates, also trying to improve over usual index maintenance schemes such as those based on a partial use of the space in index nodes.

References

1. GovTrack dataset: `http://www.govtrack.us`
2. Seaborne, A., Prud'hommeaux, E.: SPARQL query language for RDF. W3C recommendation (January 2008)
3. Karypis, G., Kumar, V.: A fast and high quality multilevel scheme for partitioning irregular graphs. SIAM Journal on Scientific Computing 20, 359–392 (1999)
4. Ko, B., Rubenstein, D.: Distributed self-stabilizing placement of replicated resources in emerging networks. Networking, IEEE/ACM Transactions on 13(3), 476–487 (2005)
5. Lee, C., Park, S., Lee, D., Lee, J., Jeong, O., Lee, S.: A comparison of ontology reasoning systems using query sequences. In: Proceedings of the 2nd international conference on Ubiquitous information management and communication, Suwon, Korea, pp. 543–546. ACM, New York (2008)
6. Sesame2: `http://www.openrdf.org`
7. Wilkinson, K., Sayers, C., Kuno, H., Reynolds, D.: Efficient RDF storage and retrieval in Jena2. In: Proceedings of SWDB, vol. 3, pp. 7–8 (2003)
8. PostgreSQL: `http://www.postgresql.org`
9. Stocker, M., Seaborne, A., Bernstein, A., Kiefer, C., Reynolds, D.: SPARQL basic graph pattern optimization using selectivity estimation. In: Proceeding of the 17th international conference on World Wide Web, Beijing, China, pp. 595–604. ACM, New York (2008)
10. JenaTDB: `http://jena.hpl.hp.com/wiki/TDB`
11. Kiryakov, A., Ognyanov, D., Manov, D.: OWLIM - a pragmatic semantic repository for OWL. In: WISE Workshops, pp. 182–192 (2005)
12. The Lehigh University Benchmark: `http://swat.cse.lehigh.edu/projects/lubm`
13. Flickr: `http://www.flickr.com`
14. Mislove, A., Marcon, M., Gummadi, K.P., Druschel, P., Bhattacharjee, B.: Measurement and analysis of online social networks. In: Proceedings of the 7th ACM SIGCOMM conference on Internet measurement, pp. 29–42. ACM, New York (2007)
15. Theoharis, Y., Christophides, V., Karvounarakis, G.: Benchmarking database representations of RDF/S Stores, pp. 685–701 (2005)
16. Broekstra, J., Kampman, A., van Harmelen, F.: Sesame: An architecture for storing and querying RDF data and schema information. In: Spinning the Semantic Web, pp. 197–222 (2003)
17. Sintek, M., Kiesel, M.: RDFBroker: A signature-based high-performance RDF store. In: Sure, Y., Domingue, J. (eds.) ESWC 2006. LNCS, vol. 4011, pp. 363–377. Springer, Heidelberg (2006)
18. Harth, A., Decker, S.: Optimized index structures for querying RDF from the Web. In: Proceedings of the 3rd Latin American Web Congress, pp. 71–80 (2005)
19. Neumann, T., Weikum, G.: RDF-3X: a RISC-style engine for RDF. PVLDB 1(1), 647–659 (2008)
20. Udrea, O., Pugliese, A., Subrahmanian, V.S.: GRIN: A graph based RDF index. In: AAAI, pp. 1465–1470 (2007)
21. Goldman, R., McHugh, J., Widom, J.: From semistructured data to XML: migrating the Lore data model and query language. In: Proceedings of the 2nd International Workshop on the Web and Databases (WebDB 1999), pp. 25–30 (1999)
22. Tian, Y., McEachin, R.C., Santos, C.: SAGA: a subgraph matching tool for biological graphs. Bioinformatics 23(2), 232 (2007)
23. Neumann, T., Weikum, G.: Scalable join processing on very large RDF graphs. In: SIGMOD Conference, pp. 627–640 (2009)

Semantically-Aided Business Process Modeling

Chiara Di Francescomarino, Chiara Ghidini, Marco Rospocher, Luciano Serafini, and Paolo Tonella

FBK-irst, Via Sommarive 18 Povo, I-38123, Trento, Italy
{dfmchiara,ghidini,rospocher,serafini,tonella}@fbk.eu

Abstract. Enriching business process models with semantic annotations taken from an ontology has become a crucial necessity both in service provisioning, integration and composition, and in business processes management. In our work we represent semantically annotated business processes as part of an OWL knowledge base that formalises the business process structure, the business domain, and a set of criteria describing correct semantic annotations. In this paper we show how Semantic Web representation and reasoning techniques can be effectively applied to formalise, and automatically verify, sets of constraints on Business Process Diagrams that involve both knowledge about the domain and the process structure. We also present a tool for the automated transformation of an annotated Business Process Diagram into an OWL ontology. The use of the semantic web techniques and tool presented in the paper results in a novel support for the management of business processes in the phase of process modeling, whose feasibility and usefulness will be illustrated by means of a concrete example.

1 Introduction

Semantic Business Process Management [11,6] has the main objective of improving the level of automation in the specification, implementation, execution, and monitoring of business processes by extending business process management tools with the most significant results from the area of semantic web. When the focus is on process modeling, i.e. the activity of specification of business processes at an abstract level (descriptive and non executable), annotating process descriptions with labels taken from a set of domain ontologies provides additional support to the business analysis (see e.g. [16]).

A crucial step in process modeling is the creation of valid diagrams, which not only comply with the basic requirements of the process semantics, but also satisfy properties that take into account the domain specific semantics of the labels of the different process elements. For instance, an important requirement for a valid on-line shopping process should be the fact that *the activity of providing personal data is always preceded by an activity of reading the policy of the organization*. As the notion of semantically annotated processes becomes more and more popular, and business experts start to annotate elements of their processes with semantic objects taken from a domain ontology, there is an increasing potential to use Semantic Web technology to support business experts in their modeling activities, including the modeling of valid diagrams which satisfy semantically enriched and domain specific constraints. A clear demonstration of this, is the stream of recent work on the introduction and usage of formal semantics to support Business Process Management [19,20,13,2,5,18,15].

A. Bernstein et al. (Eds.): ISWC 2009, LNCS 5823, pp. 114–129, 2009.

Analyzing this stream of work we can roughly divide the different approaches into two groups: (i) those adding semantics to specify the *dynamic* behavior exhibited by a business process, and (ii) those adding semantics to specify the meaning of the entities of a business process in order to improve the automation of business process management. In this paper we place ourselves in the second group and we focus on the usage of Semantic Web technology to specify and verify *structural* constrains, that is, constraints that descend from structural requirements which refer to *descriptive* properties of the annotated process diagram and not to its execution. We focus on structural requirements for two fundamental reasons: first, structural requirements complement behavioral properties, as they can be used to express properties of the process which cannot be detected by observing the execution of a process. Second, structural requirements provide an important class of expressions whose satisfiability can be directly verified with existing Description Logic (DL) reasoners.

Thus, the purpose of this paper is to propose a concrete formalization of typical classes of structural requirements over annotated BPMN processes [4], and to show how DL reasoners can be used to provide the verification services to support modeling activities. More in detail: we first recall how to represent semantically annotated BPMN processes within an OWL-DL Business Process Knowledge Base (BPKB), firstly introduced in [7]; with respect to [7], we extend the structure of BPKB to incorporate constraints used to formalize structural requirements (Section 3); then we provide concrete examples of how to encode typical classes of structural requirements in BPKD (section 4); finally, we show how to automatically translate an annotated BPMN process into a set of assertions of the Business Process Knowledge Base and we evaluate the usage of Description Logic reasoners to validate structural requirements (Section 5). An example, introduced in Section 2, is used throughout the paper to illustrate the proposed formalism, while a comparison with related approaches is contained in Section 6.

2 A Motivating Example

In this section, we describe a portion of an on-line shopping process which we use throughout the paper as a motivating and explanatory example. The annotated process we refer to is illustrated in Figure 1. The Business Process Modeling Notation (BPMN) [17] is the (graphical) language used to draw Business Process Diagrams (BPD). In our semantic variant of BPMN we allow for the annotation of objects of a BPD with concept descriptions taken from domain ontologies, i.e. shared formalizations of a specific domain. Annotations are preceded in a BPD by the "@" symbol. Due to space limitations we consider here only the initial steps of the on-line shopping process (e.g. the product presentation and selection and the customer authentication), leaving out the last phases, e.g. the checkout.

The realization of the on-line shopping process depicted in Figure 1 can involve a team of business experts, who may wish to impose requirements (constraints) on the process itself. These requirements could cover different aspects of the process, ranging from the correct annotation of business processes to security issues, as in the following examples:

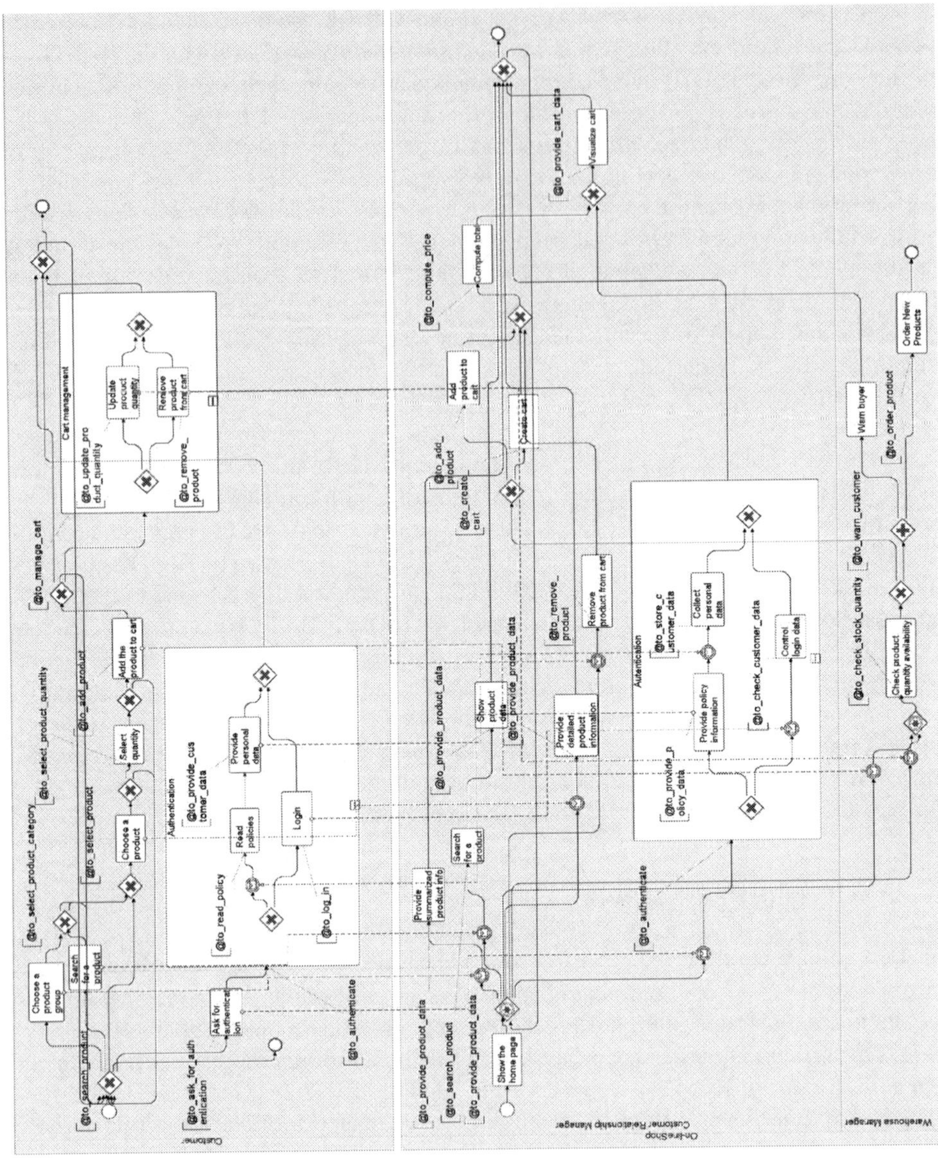

Fig. 1. A portion of the On-line shopping business process diagram

- issues related to the semantic annotation:
  (a) *"to_manage" is a complex action and can be used only to annotate BPMN sub-processes (and not atomic activities).*
- privacy issues:
  (b) *the activity of providing personal data is always preceded by an activity of reading the policy of the organization;*

(c) *the activity of reading the policy of the organization is activated by an event generated from an activity of providing these policies to the customer itself*;

– security issues:

(d) *the customer pool must contain an authentication sub-process which, in turn, contains a log-in activity and an insertion of personal data activity*;

– general issues :

(e) *in the on-line shopping process there must be a "Customer" pool and an "On-line shop" pool*;

(f) *inclusive gateway cannot be used in the on-line shopping process (to force all the alternative actions to be mutually exclusive)*;

(g) *each gateway must have at most 2 outgoing gates (to keep the process simple)*;

(h) *each pool must contain a single authentication activity / sub-process (to ease maintenance)*;

(i) *the activity of managing a shopping cart is a sub-process which contains an activity of removing products from the cart.*

All these constraints are examples of structural requirements as they relate to the descriptive properties of the annotated process diagram and complement properties which may refer to the process execution. While some of the requirements listed above can bear some similarity with behavioral properties, it is important to note here that expressing them as structural requirements allows to give them a different meaning, as it is well known that the same process behavior can be obtained by completely different process diagrams. To make a simple example we could "rephrase" constraint (h) in the apparently equivalent *the execution paths of all pools must contain a single authentication activity*. Nevertheless while this requirement is satisfied by both diagram in Figure 2, requirement (h) is only satisfied by diagram 2(b), which is definitely easier to maintain if changes to the authentication sub-process are foreseen. Thus, structural requirements are the appropriate way to express static properties of the diagram, which may even not be discovered by analyzing the behavior of the process.

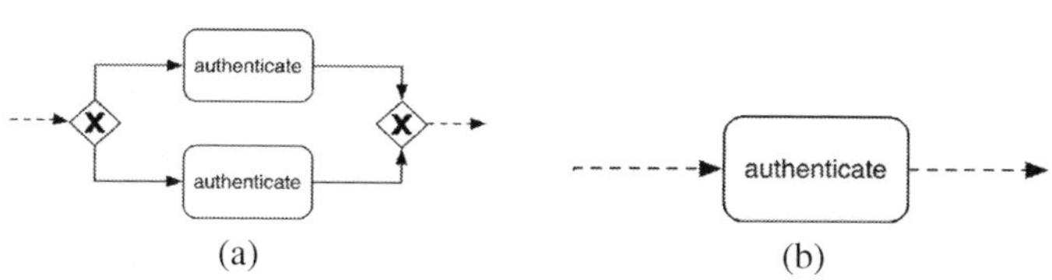

Fig. 2. Diagrams with equivalent behavior

3 Representing Semantically Annotated Processes

In order to represent semantically annotated BPDs, and to support automated reasoning on the requirements that can be expressed on them, we extend the notion of *Business*

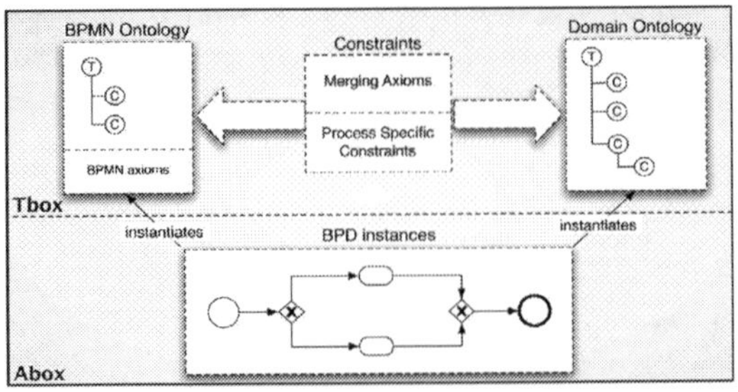

Fig. 3. The Business Processes Knowledge Base

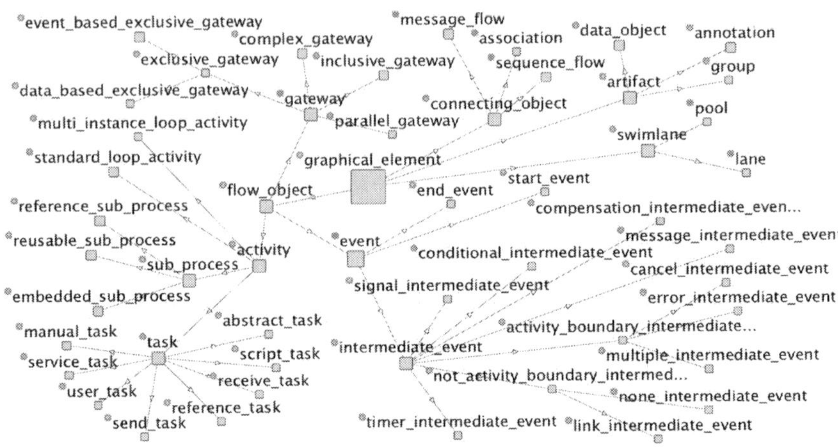

Fig. 4. The graphical elements of BPMNO

Processes Knowledge Base (BPKB), firstly introduced in [7] and schematized in Figure 3. The extension, described in details in Section 4, allows to define process specific constraints which are used to state specific structural requirements on the process to be modelled.

A BPKB is composed of four modules: a BPMN ontology, a domain ontology, a set of constraints and the BPD instances.

The BPMN Ontology. The BPMN ontology, hereafter called BPMNO, formalizes the structure of a BPD. It is a formalization of the BPMN standard as described in Annex B of [17], and consists of a set of axioms that describe the BPMN elements and the way in which they can be combined for the construction of BPDs. The taxonomy of the graphical elements of BPMNO is illustrated in Figure 4. The ontology has currently the expressiveness of $\mathcal{ALCHOIN}(\mathcal{D})$ and a detailed description is contained in [10]. We remark again that BPMNO provides a formalization of the structural part of BPDs, i.e. which are the basic elements of a BPD and how they are (can be) connected. BPMNO is not intended to model the dynamic behavior of BPDs (that is, how the flow proceeds within a process). Ontology languages are not particularly suited to specify behavioral semantics. This part can be better modeled using formal languages for Workflow or Business Process Specification based on Petri Nets, as proposed in [13].

The Domain Ontology. The domain ontology component, hereafter called BDO, consists of a (set of) OWL ontology(es) that describes a specific business domain. It allows to give a precise semantics to the terms used to annotate business processes. The BDO can be an already existing business domain ontology (e.g. RosettaNet or similar standard business ontologies), a customization of an existing ontology, or an artefact developed on purpose. Top level ontologies such as DOLCE [8] can be included as "standard" components of the domain ontology and used to provide typical annotation patterns to the BPD objects.

The Constraints. Constraints are used to ensure that important semantic structural requirements of process elements are satisfied. We distinguish among two different kinds of constraints: **merging axioms** and **process specific constraints**. Merging axioms state the correspondence between the domain ontology and the BPMN ontology. They formalize the criteria for correct/incorrect semantic annotations. Process specific constraints are expressions used to state specific structural requirements that apply to the process under construction. Differently from merging axioms, these expressions can have many different forms to match a variety of different properties of the process.

The BPD Instances. The BPD instances (or BPD objects) contain the description of a set of annotated BPDs in terms of instances of the BPMN ontology and the domain ontology. Roughly speaking, the BPD instances obtained from an annotated BPD β are all the graphical objects g of β. The assertions on these elements can be divided into three groups: *BPM-type assertions*, *BPM-structural assertions* and *BPM-semantic assertions*. The first two groups of assertions involve concepts from BPMNO only, while the third group involves concepts from BDO only. BPM-type assertions are used to store informations on the type of graphical object g. Thus we represent the fact that g is an exclusive gateway with the BPM-type assertion data_based_exclusive_gateway(g). Analogously the assertion sequence_flow(s) states that the BPM object s is of type sequence_flow. BPM-structural assertions are used to store information on how the graphical objects are connected. Thus for every connecting object c of β that goes from a to b, we generate two structural assertions of the form SourceRef(c,a) and TargetRef(c,b). For instance, the assertion has_sequence_flow_source_ref(c,a) states that the sequence flow c originates from gateway a. Finally BPM-semantic assertions are used to represent annotation of graphical elements. For instance the assertion to_search_product(t) states that task t is an instance of concept to_search_product and is obtained from the semantic annotation to_search_product of the BPD in Figure 1.

We have implemented BPKB using the standard semantic web language OWL-DL based on Description Logics (DL) [1]. The terminological part (Tbox), which is the stable description of a given domain, is provided by the upper level modules of Figure 3. Instead, the changeable part, which corresponds to a specific process description, is provided in the form of assertional knowledge (Abox).

4 Specifying Structural Requirements

To ensure that important structural requirements are satisfied, we make use of constraints. We distinguish between two different kinds of constraints: **merging axioms** and **process specific constraints**.

4.1 Merging Axioms

Though belonging to different ontologies, concepts from the BMPN ontology and the domain ontology are not totally unrelated. Indeed, precise correspondences often exist which define criteria for correct / incorrect semantic annotations. Examples of these criteria, which may hold in many application domains, are:

A BPMN activity ***can be annotated only*** *with actions of the domain ontology (and not e.g., with objects).* (1)

A BPMN data-object ***cannot be annotated*** *with actions or events of the domain ontology (but e.g., with objects).* (2)

A BPMN Event ***can be annotated only*** *with events of the domain ontology (and not e.g., with objects).* (3)

An additional domain specific criterion, which refer to the particular business process or domain ontology at hand, is requirement (a) in Section 2.

To allow the business designer to specify the kind of positive and negative constraints described above, in [7] we have proposed the usage of four relations: "*annotatable only by*" ($\xrightarrow{\mathsf{AB}}$) and "*not annotatable by*" ($\xrightarrow{\mathsf{nAB}}$) from BPMNO concepts to BDO concepts, and the symmetrical "*annotates only*" ($\xrightarrow{\mathsf{A}}$) and "*cannot annotate*" ($\xrightarrow{\mathsf{nA}}$) from BDO concepts to BPMNO concepts. In the following table we report the intuitive meaning, and the formalization as DL axioms, of these four constructs. We use x to denote a concept of BPMNO and y to denote a concept of BDO.

Merging Axiom	Intuitive meaning	DL axiom[1]
$x \xrightarrow{\mathsf{AB}} y$	a BPMN element of type x can be annotated only with a concept equivalent or more specific than y	$x \sqsubseteq y$
$x \xrightarrow{\mathsf{nAB}} y$	a BPMN element of type x cannot be annotated with a concept equivalent or more specific than y	$x \sqsubseteq \neg y$
$y \xrightarrow{\mathsf{A}} x$	any concept equivalent or more specific than y can be used to denote only BPMN elements of type x	$y \sqsubseteq x$
$y \xrightarrow{\mathsf{nA}} x$	any concept equivalent or more specific than y can not be used to denote BPMN elements of type x	$y \sqsubseteq \neg x$

The formalization of the four constructs as DL axioms is the basis for the translation of the informal expressions such as (1)–(3) and (a) into a formal set of expressions, denoted with $\mathsf{Merging\_Axioms(BPMNO,BDO)}$. Note that though the meaning of $x \xrightarrow{\mathsf{nAB}} y$ and $y \xrightarrow{\mathsf{nA}} x$ coincide, we provide both primitives as, depending on the case to be modeled, one may result more intuitive than the other.

Merging axioms can describe "domain independent" criteria, such as (1)–(3), and "domain specific" criteria, such as requirement (a). Domain independent criteria, may hold in many application domains, as they relate elements of BPMN, such as data-objects, activities or events to very general concepts, like the elements of a top-level ontology, e.g. DOLCE [8]. These kinds of constraints can be thought of as "default" criteria for correct / incorrect semantic annotations, and in this case DOLCE can be provided as a "default" component of the domain ontology in the workspace. The advantage of having these criteria already included in the BPKB is that in many situations it might be the case that the analysts, which are usually very focused on their application domain, forget to add them explicitly while they may tend to add more domain-specific constraints; note however that these "default" criteria could still be modified by the analysts to reflect the actual annotation criteria for the specific domain at hand.

To support the creation of merging axioms, we have implemented a first library of domain independent merging axioms between BPMN and DOLCE (see [9] for a detailed description). Based on this work, expression (1) can be represented with the merging axiom $\mathsf{activity} \xrightarrow{\mathsf{AB}} \mathsf{process}$ (identifying action with class process in DOLCE) which in turn is formally represented with the DL statement $\mathsf{BPMNO{:}activity} \sqsubseteq \mathsf{BDO{:}process}$, expression (2) can be represented with the merging axiom $\mathsf{data\_object} \xrightarrow{\mathsf{nAB}} \mathsf{perdurant}$ (where DOLCE class perdurant is a general class covering both processes and events) which in turn is represented with $\mathsf{BPMNO{:}data\_object} \sqsubseteq \neg\mathsf{BDO{:}perdurant}$, and similarly with the other expressions.

4.2 Process Specific Constraints

These constraints are expressions used to state specific properties that apply to the process under construction. Differently from merging axioms, these expressions can have many different forms to match a variety of different properties of the process. In this paper we focus on three types of process specific constraints that can be expressed over the Business Process Diagrams: (i) containment constraints (including existence constraints), (ii) enumeration constraints, and (iii) precedence constraints.

Containment Constraints. Containment constraints are of the form *X contains Y* or *X is contained in Y* and are used to represent the fact that the BPD or certain graphical elements contain other graphical elements. As such they can be used to express also informal statements of the form *exists X* and *non exists X*, which are rephrased in the containment constraint *diagram X contains Y* and *diagram X does not contain Y*. A simple containment constraint of the form *X contains Y* which can be expressed over the on-line shopping process is provided by requirement (i). A constraint of the form *exists X* is instead provided by requirements (e), while a constraint of the form *non exists X* is given by requirement (f).

Containment constraints can be encoded in Description Logics using specific BPMNO roles which formalise the containment relations existing between different BPD objects as described by specific attributes in [17]. Examples of these roles, used in DL to represent object properties and data properties, are:

- has_embedded_sub_process_sub_graphical_elements. This role corresponds to the GraphicalElement attribute of an Embedded Sub-Process, as described in [17], and represents all of the objects (e.g., Events, Activities, Gateways, and Artifacts) that are contained within the Embedded Sub-Process;
- has_pool_process_ref which corresponds to the ProcessRef attribute and is used to represent the process that is contained within a pool;
- has_process_graphical_element which corresponds to the GraphicalElements attribute of BPMN and identifies all of the objects that are contained within a process;
- has_business_process_diagram_pools which allows to relate a BPD with the pools it contains.

Using r to indicate one of the roles above, containment constraints are typically expressed as statements of the form $X \sqsubseteq \exists r.Y$ or $X \sqsubseteq \forall r.Y$ which use the basic existential and universal quantification constructs $\exists r.Y$ and $\forall r.Y$. Requirement (i) can therefore be formalized as follows, where we use has_embedded as a shorthand for has_embedded_sub_process_sub_graphical_elements for the sake of readability:

$$\text{BDO:to\_manage\_cart} \sqsubseteq \text{BPMN:embedded\_sub\_process} \tag{4}$$

$$\text{BDO:to\_manage\_cart} \sqsubseteq \exists \text{BPMN:has\_embedded.}(\text{BPMN:activity} \sqcap \text{BDO:to\_remove\_product}) \tag{5}$$

Similarly, requirement (e) can be encoded as follows:

$$\begin{aligned} &\text{BPMNO:business\_process\_diagram} \sqsubseteq \\ &\quad \exists \text{BPMNO:has\_business\_process\_diagram\_pools.BDO:customer} \sqcap \\ &\quad \exists \text{BPMNO:has\_business\_process\_diagram\_pools.BDO:on-line\_shop} \end{aligned} \tag{6}$$

while requirement (f) can be formalized by asserting:

$$\begin{aligned} &\text{BPMNO:process} \sqsubseteq \\ &\quad \forall \text{BPMNO:has\_process\_graphical\_element.} \neg \text{BPMNO:inclusive\_gateway} \end{aligned} \tag{7}$$

A more complex example of containment constraint is provided by requirement (d). The formalization of this constraint is omitted for lack of space.

Enumeration Constraints. Enumeration constraints further refine containment constraints by stating that *X contains (at least / at most / exactly) n objects of type X*. A simple example of enumeration constraint which concern a plain BPMN element is provided by requirements (g). An enumeration constraint which also involves semantically annotated objects is provided by requirement (h). Enumeration constraints can be encoded in Description Logics using the constructors of *number restriction* and *qualified number restriction* [1]. Number restrictions are written as $\geq nR$ (at-least restriction) and $\leq nR$ (at-most restriction), with n positive integer, while qualified number restrictions are written as $\geq nR.C$ and $\leq nR.C$. The difference between the two is that number restriction allows to write expressions such as, e.g., $\leq 3 hasChild$, which characterise the set of individuals who have at most 3 children, while qualified number restriction allows to write expressions such as, e.g., $\leq 3hasChild.Female$, which characterise the set of individuals who have at most 3 female children. At-least and at-most operators can be combined to obtain statement of the form $= nR$.

Thus, a formalization of requirements (g) can be provided by the DL statement:

$$\text{BPMNO:gateway} \sqsubseteq (\leq 2)\text{BPMNO:has\_gateway\_gate} \tag{8}$$

while a formalization of requirement (h) is given by:

$$\begin{aligned} &\text{BPMN:pool} \sqsubseteq \forall \text{BPMN:has\_pool\_process\_ref}. \\ &\quad (= 1)\text{BPMN:has\_process\_graphical\_element.BDO:to\_authenticate} \end{aligned} \tag{9}$$

Precedence Constraints. Precedence constraints are used to represent the fact that certain graphical objects appear before others in the BPD. They can be of several forms. Significant examples are: *X is always preceded by Y* in all possible paths made of sequence flows and *X is once preceded by Y* in at least a path composition of sequence flows. Particular cases of these constraints are *X is always immediately preceded by Y* and *X is once immediately preceded by Y*. These constraints also require that *X* is a graphical object immediately preceded by *Y* by means of a sequence flow. Finally the precedence constraint *X is activated by Y* requires that *X* is activated by *Y* by means of a message flow. Two simple examples of precedence constraint are provided by requirements (b) and (c).

Precedence constraints can be encoded in Description Logics using specific BPMNO roles which formalize the connection between graphical objects. In particular the key roles we can use are:

- has_sequence_flow_source_ref and has_sequence_flow_target_ref.
- has_message_flow_source_ref and has_message_flow_target_ref.

These roles represent the SourceRef and TargetRef attributes of BPMN and identify which graphical elements the connecting object is connected from and to respectively. The first two roles refer to connecting object which are sequence flow, while the other two roles refer to message flow.
Constraint (b) can be formalized in DL by means of two statements

$$\begin{aligned} &\text{BDO:to\_provide\_sensible\_data} \sqsubseteq \forall \text{BPMN:has\_sequence\_flow\_target\_ref}^{-}. \\ &\quad \forall \text{BPMN:has\_sequence\_flow\_source\_ref.BDO:to\_read\_policy}^{*} \end{aligned} \tag{10}$$

$$\begin{aligned} &\text{BDO:to\_read\_policy}^{*} \equiv \neg \text{BPMN:start\_event} \sqcap ((\text{BDO:to\_read\_policy} \sqcap \\ &\quad \text{BPMN:activity}) \sqcup \forall \text{BPMN:has\_sequence\_flow\_target\_ref}^{-}. \\ &\quad \forall \text{BPMN:has\_sequence\_flow\_source\_ref.BDO:to\_read\_policy}^{*}) \end{aligned} \tag{11}$$

The statements above use has_sequence_flow_source_ref and has_sequence_flow_target_ref, together with an auxiliary concept BDO:to_read_policy*. In a nutshell the idea is that the concept BDO:to_provide_sensible_data is immediately preceded, in all paths defined by a sequence flow, by a graphical object of type BDO:to_read_policy*. This new concept is, in turn, defined as a graphical object which is not the start event and either it is an activity of type BDO:to_read_policy or it is preceded in all paths by

BDO:to_read_policy*. By replacing BDO:to_provide_sensible_data, BDO:to_read_policy, and BDO:to_read_policy* with X, Y and Y^* in (10) and (11) we can obtain a general encoding of constraints of the form *X is always preceded by Y*. In addition by replacing $\forall$ with $\exists$ we can obtain an encoding of *X is once preceded by Y*.

Note that, if we replace the constraint (b) with a simpler constraint of the form *"the activity of providing personal data is always* ***immediately*** *preceded by an activity of reading the policy of the organization"* then, we do not need to introduce the auxiliary concept Y^* and the encoding is directly provided by the statement

$$\begin{aligned}&\mathsf{BDO{:}to\_provide\_sensible\_data} \sqsubseteq \forall \mathsf{BPMN{:}has\_sequence\_flow\_target\_ref}^{-}.\\&\quad\forall \mathsf{BPMN{:}has\_sequence\_flow\_source\_ref.BDO{:}to\_read\_policy}\end{aligned} \tag{12}$$

To formalise (c) we need to check that the activities annotated with BDO:to_read_policy are activated by an intermediate event (message) which refers to a message flow which originates by an activity of type BDO:to_provide_policy_data:

$$\begin{aligned}&\mathsf{BDO{:}to\_read\_policy} \sqsubseteq \exists \mathsf{BPMN{:}has\_sequence\_flow\_target\_ref}^{-}.\\&\quad\exists \mathsf{BPMN{:}has\_sequence\_flow\_source\_ref}.(\\&\qquad\mathsf{BPMN{:}intermediate\_event} \sqcap \exists \mathsf{BPMN{:}has\_message\_flow\_target\_ref}^{-}.\\&\qquad\exists \mathsf{BPMN{:}has\_message\_flow\_source\_ref}.(\\&\qquad\quad\mathsf{BDO{:}to\_provide\_policy\_data} \sqcap \mathsf{BPMN{:}activity}))\end{aligned} \tag{13}$$

Again by replacing the specific BDO concepts with X and Y we can obtain a schematic encoding of constraints of the form *X is activated by Y*.

Combining different constraints. By combining containment, enumeration and precedence constraints, we can encode more complex requirements. An example is provided by the following requirement

All the paths that originate from the exclusive gateways contained in the
On-line shop pool must start with an event which comes from the Customer pool.

Due to expressiveness limitation imposed by Description Logics and by the fact that we want to remain in a decidable version of OWL, there are also constraints on the static part of the BPMN diagram which are are not represented in our approach. In particular all the properties that, once translated into first order logic, require more than two variables. A typical example of this kind of constraint is the fact that *X mutually_excludes Y*, that is that X and Y are always preceded by the same exclusive gateway. In fact we can express this property as follows, where $<$ is used to indicate the precedence relation between graphical elements:

$$\begin{aligned}&\forall X.\forall Y.\exists Z(xor(Z) \wedge precede(Z,X) \wedge precede(Z,Y) \wedge \forall W.(precede(Z,W) \wedge\\&\qquad precede(W,X) \wedge precede(W,Y)) \rightarrow \neg gateway(W))\end{aligned} \tag{14}$$

We can instead represent a weaker version of the constraint above, which states that X and Y are immediately preceded by an exclusive XOR gateway as in our BPD in Figure 1, by using precedence constraints. Similar limitations apply to constraints involving parallel gateways.

5 Checking Constraints as a Satisfiability Problem

Given an Abox A_β which contains the OWL representation of a semantically annotated BPD β, we can reduce the problem of checking constraints over a BPD to a satisfiability problem in DL. In particular we can reformulate the fact that A_β represents a business process correctly annotated according to Merging_Axioms(BPMNO, BDO) and which satisfies a set of process specific constraints Constraints(BPMNO,BDO) as the fact that

$$\text{BPMNO} \cup \text{BDO} \cup \text{Merging\_Axioms}(\text{BPMNO}, \text{BDO}) \cup \text{Constraints}(\text{BPMNO},\text{BDO}) \cup A_\beta$$

is a consistent knowledge base. In this section we illustrate how we can support the automatic transformation of a semantically annotated BPD β into the corresponding Abox A_β. We also provide an experimental evaluation of the performance of the tool, and of reasoning systems used to check structural constraints over the BPD.

5.1 Automatically Encoding a BPD into an Abox

We developed a tool for the automated transformation of a BPD into an OWL Abox. Given BPMNO, BDO, Merging_Axioms(BPMNO,BDO) and an annotated BPD β, the tool creates the Abox A_β and populates the ontology with instances of BPMN elements belonging to the specific process.

The input BPMN process is currently described in a `.bpmn` file, one of the files generated by both the Eclipse SOA Tools Platform and the Intalio Process Modeler tools. The `.bpmn` file is an XML file that contains just the structural description of the process, leaving out all the graphical details. The ontology is populated by parsing the file and instantiating the corresponding classes and properties in the BPKB Tbox.

The mapping between the XML elements/attributes used in the `.bpmn` file of the Eclipse tool and concepts and properties in the BPKB Tbox is realized by means of a mapping file. It associates each XML element of the `.bpmn` file to the corresponding concept in BPMNO and each of its attributes and XML elements (with a correspondence in the BPMNO) to the corresponding concept or property. The fragment of mapping file in Figure 5 shows the correspondences between the pool process element (i.e. the XML element having type `bpmn:Pool` in the `.bpmn` file) and the BPMNO. Each XML element of this type in the `.bpmn` file will be translated into an instance of the concept BPMNO:Pool. The values of its attributes `name` and `documentation` will be the target of the two data properties of the concept BPMNO:Pool, respectively has_swimlane_name and has_BPMN_element_documentation. Moreover, the two target objects (instances of the classes BPMNO:Object and BPMNO:Process) of the BPMNO:Pool object properties has_BPMN_element_id and has_pool_process_ref, will be instatiated by exploiting the unique id of the process element `pool`. Finally, the XML elements contained in the element `pool` and annotated with the XML tags `lanes` and `vertices` will respectively be the target objects of the BPMNO:Pool object property has_pool_lanes and of the BPMNO:Process object property has_process_graphical_elements.

The BPMN process descriptions currently generated by the Eclipse tool or the Intalio tool do not exhaustively cover the features provided by the BPMN specification and, therefore, the full ontology potential. Thus the mapping file is limited to the subset of the specification actually implemented by the considered tools and is also based on

```
<bpmn:Pool class="pool">
    <attributes>
        <this prop="has_BPMN_element_id" range="object"/>
        <this.object prop="this.object->has_object_id"/>
        <this prop="has_pool_process_ref" range="process"/>
        <this.process prop="this.process->has_process_name"/>
        <documentation prop="has_BPMN_element_documentation"/>
        <name prop="has_swimlane_name"/>
    </attributes>
    <relations>
        <lanes prop="has_pool_lanes" range="lane"/>
        <vertices prop="this.process->has_process_graphical_elements"
                  range="graphical_element"/>
    </relations>
</bpmn:Pool>
```

Fig. 5. A fragment of the mapping file

Table 1. Evaluation Results

Process	Process Elements	Semantic Annotations	Abox Creation	Classification of BPKB	Classification of BPKB with Constraints
PROCESS_A	79	13	185s	5s	205s
PROCESS_B	196	30	601s	20s	143s
PROCESS_C	324	56	1377s	28s	338s
PROCESS_D	629	112	4075s	750s	1141s

assumptions implicitly made by the tools. Similarly, the mapping file depends on the particular process representation adopted by these tools and must be adjusted whenever a different tool is used for process editing.

Finally the semantic annotations added to process elements and contained in the `.bpmn` file as XML elements of type `bpmn:TextAnnotation` are also used for populating the BDO concepts. By parsing the file, the process element associated to each XML element having type `bpmn:TextAnnotation` will be added as an instance of the BDO concept corresponding to the value of the semantic annotation.

Our tool uses the `org.w3c.dom` XML parsing library to manage the `.bpmn` input file, Protégé libraries to populate the resulting OWL Abox, and Pellet for reasoning.

5.2 An Experimental Evaluation

We performed some experiments in order to evaluate the performances of our tool in populating the BPKB Abox and the performance of DL reasoners for checking the satisfiablity of structural constraints. We considered four processes of increasing size and complexity. PROCESS_A is composed by the "Customer" pool of the process in Figure 1, PROCESS_B is the process reported in Figure 1, PROCESS_C extends PROCESS_B by describing the entire on-line shopping process, including the checkout part, and finally PROCESS_D is composed by doubling PROCESS_C. The number of BPMN graphical elements and of semantic annotations contained in the four BPDs is contained in Table 1. The processor used for the experiments is a 1.86GHz Intel Core 2, with 2 Gb of RAM and running Windows XP. The reasoner was Pellet 1.5.

Table 1 reports, for all the processes, the time spent by our tool to generate the corresponding Abox. It also reports the time spent to classify the $\mathcal{ALCHOIN}(\mathcal{D})$ BPKBs

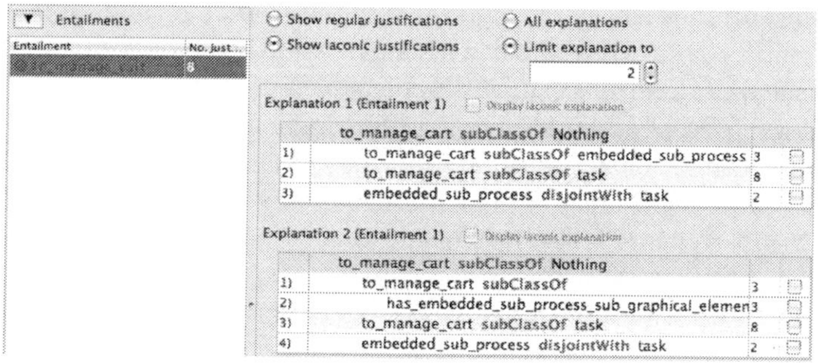

Fig. 6. Explanation generation

which encodes the annotated processes without constraints and the $\mathcal{ALCHOIQ}(\mathcal{D})$ BPKBs enriched with a set of constraints similar to the ones described in the paper by the Pellet reasoner. By looking at the results the time seems compatible with an off-line usage of the tools on medium size processes. Note that, in the case of unsatisfiable classes, which originate from conflicting requirements, an off-line usage of DL reasoners and explanation techniques similar to the ones described in [12] can be also useful to provide justifications to the business experts. Figure 6 shows an explanation obtained by using the Explanation Workbench plugin for Protege-4 of the unsatisfiable concept to_manage_cart in the case the assertion BDO:to_manage_cart $\sqsubseteq$BPMN:task is added to the knowledge base, together with constraint (4).

6 Related Work

We can roughly divide the existing proposals for adding semantics to business processes into two groups: (1) those adding semantics to specify the dynamic behavior exhibited by a business process [19,20,13], and (2) those adding semantics to specify the meaning of the entities of a BPD in order to improve the automation of business process management [2,5,18,15]. We clearly belong to the second group.

Thomas and Fellmann [18] consider the problem of augmenting EPC process models with semantic annotations. They propose a framework which joins process model and ontology by means of properties (such as the "semantic type" of a process element). Markovic [14] considers the problem of querying and reasoning on business process models. He presents a framework for describing business processes which integrates functional, behavioral, organizational and informational perspectives: the elements of the process are represented as instances of an ontology describing the process behavior (based on π-calculus), and the annotations of these elements with respect to the ontologies formalizing the aforementioned perspectives are described as relation instances. Born *et al.* [3] propose to link the elements of a business process to the elements of an ontology describing objects, states, transitions, and actions. These proposals differ substantially from ours, which establishes a set of subsumption (aka subclass or is-a) relations between the classes of the two ontologies being integrated (BPMN metamodel and domain ontology), instead of associating annotation properties to the process instances. This difference has a direct impact on the kind of constraints that can be automatically enforced (e.g. BPMN elements annotatable by domain concepts).

De Nicola *et al.* [15] propose an abstract language (BPAL) that bridges the gap between high-level process description (e.g. in BPMN) and executable specification (e.g. in BPEL). The formal semantics offered by BPAL refers to notions such as activity, decision, etc., while the problem of integrating process model and domain ontology is not their focus. In Weber *et al.* [19], semantic annotations are introduced for validation purposes, i.e. to verify constraints about the process execution semantic. In their work, semantic annotations with respect to a background ontology are used to ensure that an executable process model behaves as expected in terms of preconditions to be fulfilled for execution and its effects. In the SUPER project [5], the SUPER ontology is used for the creation of semantic annotations of both BPMN and EPC process models in order to support automated composition, mediation and execution. Our work represents an extension of the existing literature in that we provide an ontology based approach that supports automated verification of semantic constraints defining the correctness of semantic process annotations and rich structural requirements.

7 Conclusions

In this paper we have presented an ontology-based framework to verify sets of structural constraints that involve both knowledge about the domain and the process structure. We have also described a tool for the automated transformation of an annotated business process into an OWL ontology and evaluated how standard DL reasoners can be used to automatically verify these constraints as ontology consistency violations.

Although some effort is required to create all the components of the BPKB, we envisage situations in which the business designer is mainly concerned with the specification of constraints and the validation of this specification. In particular, we foresee a situation in which the domain ontology (or an adaptation of an existing one) is available to the business designer, pre-defined high level merging axioms can be directly plugged in the system, and (candidate) annotations can be provided automatically by means of matching algorithms.

In our future work, we will work towards this objective by extending the library of "domain independent" merging axioms by examining different upper-level ontologies, including business domain ontologies. We will also investigate how to simplify the tasks of constraint specification and checking for business experts by means of more user friendly notations and tools. Finally, we will validate the approach further, on larger case studies.

References

1. Baader, F., Calvanese, D., McGuinness, D.L., Nardi, D., Patel-Schneider, P.F. (eds.): The Description Logic Handbook: Theory, Implementation, and Applications. Cambridge University Press, Cambridge (2003)
2. Beeri, C., Eyal, A., Kamenkovich, S., Milo, T.: Querying business processes. In: VLDB 2006, pp. 343–354 (2006)

3. Born, M., Dörr, F., Weber, I.: User-friendly semantic annotation in business process modeling. In: Weske, M., Hacid, M.-S., Godart, C. (eds.) WISE Workshops 2007. LNCS, vol. 4832, pp. 260–271. Springer, Heidelberg (2007)
4. Business Process Management Initiative (BPMI). Business process modeling notation: Specification (2006), http://www.bpmn.org
5. Dimitrov, M., Simov, A., Stein, S., Konstantinov, M.: A bpmo based semantic business process modelling environment. In: Proceedings of the Workshop on Semantic Business Process and Product Lifecycle Management at the ESWC. CEUR-WS, vol. 251 (2007)
6. Wetzstein, B., et al.: Semantic business process management: A lifecycle based requirements analysis. In: Proc. of the Workshop on Semantic Business Process and Product Lifecycle Management. CEUR Workshop Proceedings, vol. 251 (2007)
7. Di Francescomarino, C., Ghidini, C., Rospocher, M., Serafini, L., Tonella, P.: Reasoning on semantically annotated processes. In: Bouguettaya, A., Krueger, I., Margaria, T. (eds.) ICSOC 2008. LNCS, vol. 5364, pp. 132–146. Springer, Heidelberg (2008)
8. Gangemi, A., Guarino, N., Masolo, C., Oltramari, A., Schneider, L.: Sweetening Ontologies with DOLCE. In: Gómez-Pérez, A., Benjamins, V.R. (eds.) EKAW 2002. LNCS (LNAI), vol. 2473, pp. 166–181. Springer, Heidelberg (2002)
9. Ghidini, C., Hasan, M.K., Rospocher, M., Serafini, L.: A proposal of merging axioms between bpmn and dolce ontologies. Technical report, FBK-irst (2009), https://dkm.fbk.eu/index.php/BPMN_Related_Resources
10. Ghidini, C., Rospocher, M., Serafini, L.: A formalisation of BPMN in description logics. Technical Report TR 2008-06-004, FBK-irst (2008), https://dkm.fbk.eu/index.php/BPMN_Related_Resources
11. Hepp, M., Leymann, F., Domingue, J., Wahler, A., Fensel., D.: Semantic business process management: A vision towards using semantic web services for business process managemen. In: ICEBE 2005: Proceedings of the IEEE International Conference on e-Business Engineering, pp. 535–540. IEEE Computer Society, Los Alamitos (2005)
12. Horridge, M., Parsia, B., Sattler, U.: Laconic and precise justifications in owl. In: Sheth, A.P., Staab, S., Dean, M., Paolucci, M., Maynard, D., Finin, T., Thirunarayan, K. (eds.) ISWC 2008. LNCS, vol. 5318, pp. 323–338. Springer, Heidelberg (2008)
13. Koschmider, A., Oberweis, A.: Ontology based business process description. In: Proceedings of the CAiSE 2005 Workshops. LNCS, pp. 321–333. Springer, Heidelberg (2005)
14. Markovic, I.: Advanced querying and reasoning on business process models. In: Abramowicz, W., Fensel, D. (eds.) BIS. LNBIP, vol. 7, pp. 189–200. Springer, Heidelberg (2008)
15. De Nicola, A., Lezoche, M., Missikoff, M.: An ontological approach to business process modeling. In: Proceedings of the 3rd Indian International Conference on Artificial Intelligence (IICAI), December 2007, pp. 1794–1813 (2007)
16. Thomas, M.F.O.: Semantic epc: Enhancing process modeling using ontology languages. In: Proc. of the Workshop on Semantic Business Process and Product Lifecycle Management at the ESWC. CEUR-WS, vol. 251 (2007)
17. OMG. Business process modeling notation, v1.1., www.omg.org/spec/BPMN/1.1/PDF
18. Thomas, O., Fellmann, M.: Semantic epc: Enhancing process modeling using ontology languages. In: Proceedings of the Workshop on Semantic Business Process and Product Lifecycle Management (SBPM), June 2007, pp. 64–75 (2007)
19. Weber, I., Hoffmann, J., Mendling, J.: Semantic business process validation. In: Proceedings of the Workshop on Semantic Business Process and Product Lifecycle Management (SBPM) (June 2008)
20. Wong, P.Y.H., Gibbons, J.: A relative timed semantics for bpmn. In: Proceedings of 7th International Workshop on the Foundations of Coordination Languages and Software Architectures, FOCLASA 2008 (2008)

Task Oriented Evaluation of Module Extraction Techniques

Ignazio Palmisano, Valentina Tamma, Terry Payne, and Paul Doran

Department of Computer Science, University of Liverpool,
Liverpool L69 3BX, United Kingdom
{ignazio,V.Tamma,T.R.Payne,P.Doran}@liverpool.ac.uk

Abstract. Ontology Modularization techniques identify coherent and often reusable regions within an ontology. The ability to identify such modules, thus potentially reducing the size or complexity of an ontology for a given task or set of concepts is increasingly important in the Semantic Web as domain ontologies increase in terms of size, complexity and expressivity. To date, many techniques have been developed, but evaluation of the results of these techniques is sketchy and somewhat ad hoc. Theoretical properties of modularization algorithms have only been studied in a small number of cases. This paper presents an empirical analysis of a number of modularization techniques, and the modules they identify over a number of diverse ontologies, by utilizing objective, task-oriented measures to evaluate the fitness of the modules for a number of statistical classification problems.

1 Introduction

One of the emerging areas of ontology engineering that has received considerable attention by the community is that of ontology modularization (OM) [1,2,3,4,5,6]. OM is the collective name of approaches for fragmenting ontologies into smaller, coherent components (*modules*), each of which are themselves ontologies. The approaches can be broadly divided into two categories: *ontology partitioning*, whereby the ontology is partitioned into a number of modules such that the union of all the modules is semantically equivalent to the original ontology; or *module extraction techniques*, whereby concepts that form a coherent fragment of an ontology are extracted to form a module, such that it covers a given vocabulary (based on an initial *module signature*). OM techniques, have been proposed for a wide variety of tasks, such as ontology design, maintenance, and reuse; knowledge selection; and integration [2,7], and hence the precise definition of modularization can vary depending on the technique used, with respect to the types of concepts, axioms and properties (found in the ontology) that are desirable within the module, and on the characteristics that the modules should exhibit [7].

The diversity of approaches can therefore make a comparative evaluation of the different modularization processes extremely challenging. Most evaluation approaches in the literature propose the use of the *size* of the module, which

A. Bernstein et al. (Eds.): ISWC 2009, LNCS 5823, pp. 130–145, 2009.

includes the number of named concepts and properties in the module [2], as an objective metric. However, size is a crude measure that does not fully reflect the *content* of a module (for example it ignores the effect of restrictions over concepts or properties). More recently, other approaches [8,9] propose various ways to capture different aspects of the *content* of a module, in terms of its structural properties, or information content. Whilst the criteria proposed by Schlicht and Stuckenschmidt [8] focus on the structure of the ontology modules produced and attempt to assess the trade-off between maintainability and efficiency of reasoning in distributed systems, the entropy-inspired approach [9] attempts to assess the difference between modules by measuring the information content carried by axioms modelling domain and language entities in the ontology. This latter approach models the *combined effect* of both domain and language entities, by aggregating the quantitative estimates of the dimensions represented by the criteria. Whilst these different techniques capture to some extent the structural differences between modules, they are based on objective measures that fail to consider fully the suitability of a module for a specific task, and therefore are limited in the level of assistance they can provide in supporting ontology engineers and users alike in choosing the most appropriate modularization technique for a specific task.

The focus of this paper is to provide a systematic and extensive empirical evaluation of various module extraction approaches, from the perspective of their suitability for a specific task. The aim is to identify the broad suitability of an approach for a given task, and to detect possible boundary cases where other approaches might be more suitable. In this way, we provide a guide that ontology engineers and users alike can use to tailor the modularization process to their problem. Three related problems have been identified that support a number of common tasks such as query answering or service retrieval: *Instance retrieval*, *Subclass retrieval*, and *Superclass retrieval*. In this paper, we provide a systematic evaluation of a number of different modularization techniques for these problems. In each case, a query is constructed and used as the *signature* (described below) of an identified module. The resulting modules are then evaluated to determine their utility for query answering, when compared to using the original ontologies.

The paper is therefore structured as follows: Section 2 reviews different ontology modularization approaches and the different evaluation techniques. Section 3 presents the evaluation of the reviewed modularization approaches. The evaluation problems are formally defined in terms of statistical classification problems, we then present and discuss the results. Finally, concluding remarks are illustrated in Section 4.

2 Ontology Modularization

Ontology modularization [1,2,3,10,5,6] refers to the process of fragmenting existing ontologies into a set of smaller, and possibly interconnected parts, or *modules*. Broadly speaking, modularization approaches aim to identify the minimal set of necessary concepts and definitions for different parts of the original ontology.

Whilst size could be considered a factor, the modules themselves should be *fit for purpose* for some task; and hence the algorithms proposed can differ significantly depending on this intended task. Likewise, the reasons for modularizing can be different and range from ontology reuse in order to support the work of ontology engineers [1,3,10] to information integration [2], or to support efficient agent communication [11]. Thus, whilst size is often quoted for some modularization techniques, it unsuitable as an objective indicator of the quality of a module or the modularisation approach. This section reviews the different approaches for modularizing ontologies, focussing in particular on module extraction techniques, and presents the different techniques proposed in the literature for evaluating the result of modularization approaches.

An ontology O is defined as a pair $O = (Ax(O), Sig(O))$, where $Ax(O)$ is a set of axioms (intensional, extensional and assertional) and $Sig(O)$ is the signature of O[1]. This signature consists of the set of entity names used by O, i.e., its vocabulary. Ontology modularization is the process of defining a module $M = (Ax(M), Sig(M))$, where M is a subset of O, $M \subseteq O$, such that $Ax(M) \subseteq Ax(O)$ and $Sig(M) \subseteq Sig(O)$. No assumptions beyond this are made here about the nature of a module.

Approaches for modularizing ontologies belong to two main categories: ontology partitioning and ontology module extraction. *Ontology partitioning* is the process of fragmenting an ontology O into a set of (not necessarily disjoint[2]) modules $\mathcal{M}= \{M_1, M_2, ..., M_n\}$, such that the union of all the modules should be equivalent to the original ontology O; i.e. $\{M_1 \cup M_2 \cup ... \cup M_n\} = O$. Thus, a function $partition(O)$ can be formally defined as follows:

$$partition(O) \rightarrow \mathcal{M} = \{\{M_1, M_2, ..., M_n\} | \{M_1 \cup M_2 \cup ... \cup M_n\} = O\}$$

Ontology module extraction refers to the process of extracting a module M from an ontology O that covers a specified signature $Sig(M)$, such that $Sig(M) \subseteq Sig(O)$. M is the relevant part of O that is said to cover the elements defined by $Sig(M)$, as such $M \subseteq O$. M is now an ontology itself and could elicit further modules, depending on the signatures subsequently used. Thus, a function $extract(O, Sig(M))$ can be defined as follows:

$$extract(O, Sig(M)) \rightarrow \{M | M \subseteq O\}$$

This paper focusses on ontology module extraction approaches, since the concept queried can form the basis of the signature used to extract modules. Ontology partitioning approaches are independent from any specific signature in input, and thus would not reflect a query answering task. The techniques for ontology module extraction in the literature can be further subdivided into

[1] This definition is agnostic with respect to the ontology language used to represent the ontology, but it should be noted that the modularization techniques detailed in this section assume a description logic representation.

[2] This is in contrast to the mathematical definition of partitioning that requires partitions to be disjoint.

two distinct groups: *traversal approaches* and *logical approaches*. Traversal approaches [3,1,4,5] represent the extraction as a graph traversal, with the module being defined by the conditional traversal, which implicitly considers the ontological semantics, of the graph. Logical approaches [2,12] focus on maintaining the logical properties of coverage and minimality; as such, they explicitly consider the ontological semantics when extracting an ontology module.

2.1 Traversal Based Extraction

All of the following methods for ontology module extraction can be considered as traversal based extraction techniques. Each represents the ontology as a graph and the ontology module is defined as a conditional traversal over this graph.

d'Aquin *et al.* [3] address the specific task of extracting modules related to components found in a given web page. Their ontology module extraction technique is integrated within a larger *knowledge selection* process. The specific aim is to dynamically retrieve the relevant components from online ontologies to annotate the webpage currently being viewed in the browser. The knowledge selection process comprises three phases: (i) selection of relevant ontologies, (ii) modularization of selected ontologies and (iii) merging of the relevant ontology modules. The principle used for the extraction of an ontology module (i.e. phase (ii)) is to include all the elements that participate in the definition, either directly or indirectly, of the entities (similar to the approach proposed by Seidenberg and Rector [5]). There are two distinct characteristics of this approach:

- **Inferences** are used during the extraction, rather than assuming an inferred model (as is the case with techniques such as Doran *et al.* [1]), i.e. that all inferences are made a priori to the module extraction process. For example, the transitivity of the *subClassOf* edge allows new subclass relations to be inferred in the input ontology.
- '**Shortcuts**' are taken in the class hierarchy by including only the named classes that are the most specific common super-classes of the included classes. This is done by restricting the possible values of the *Least Common Subsumer(LCS)* algorithm [13] to the classes in the ontology.

Doran *et al.* [1] tackle the problem of ontology module extraction from the perspective of an Ontology Engineer wishing to reuse part of an existing ontology. The approach extracts an ontology module corresponding to a single user-supplied concept that is self-contained, concept centred and consistent. This approach is agnostic with respect to the language the ontology is represented in, provided that the ontology language itself can be transformed into the *Abstract Graph Model*. A conditional traversal descends down the *is-a* hierarchy from the signature concept via two edge sets: one set of edges to traverse and a second set containing edges that are not traversed. Exceptions to these traversal sets are permitted during the first iteration of the algorithm. For example, when extracting an ontology module from an OWL ontology, `owl:disjointWith` edges are not traversed during the first iteration, but are considered in subsequent iterations (to prevent relevant definitions from being skipped).

Noy and Musen [4] define the notion of *traversal view extraction*, which defines an *ontology view* of a specified concept, which is analogous to an ontology module. Starting from one class of the ontology being considered, relations from this class are recursively traversed to include the related entities. These relations are selected by the user, and for each relation selected, a depth of traversal (or *traversal directive*) is assigned. This traversal directive is used to halt the traversal of the corresponding relation when the specified depth is reached. A *traversal view* consists of a set of traversal directives. This flexible approach (which was incorporated into PROMPT [14]) allows an Ontology Engineer to iteratively construct the ontology module that they require by extending the current 'view'. However, this can require the Ontology Engineer to have a deep understanding of the ontology that is being used.

We do not consider this approach in our evaluation since it has a high degree of interactivity with the ontology engineer, that can affect the detemination of a module.

Seidenberg and Rector [5] developed a technique specifically for extracting an ontology module for a given signature, $Sig(M)$, from the Galen ontology. Their technique identifies all elements that participate (even indirectly) to the definition of the signature, or other elements in the extracted module. The algorithm can be decomposed as follows: assuming we have a $Sig(M) = \{A\}$. Firstly the hierarchy is upwardly traversed (analogous to Upper Cotopy defined in [15]), so all of the A's superclasses are included. Next the hierarchy is downwardly traversed so that all the A's subclasses are included. It should be noted that the sibling classes of A are not included, they could be included by explicitly adding them to the $Sig(M)$. The restrictions, intersection, union and equivalent classes of the already included classes can now be added to the module. Lastly, links across the hierarchy from the previously included classes are traversed; the target of these links are also upwardly traversed.

Whilst the degree of generality for this approach is high (with respect to other ontologies), the focus on GALEN introduces certain features that may be less suitable for other ontologies. For example, result of property filtering can lead to class definitions becoming equivalent, whilst this is not incorrect it does introduce unnecessary definitions that can be transformed into primitive classes.

Table 1 compares the features of the traversal based extraction techniques.

Table 1. Comparison of features for traversal based ontology module extraction

	Interactive	Traversal Direction	Property Filtering	Least Common Subsumer	Assume Inferred Model
Whole Ontology	✗	Up & Down	✗	✗	✗
d'Aquin	✗	Up & Down	✗	✓	✓
Doran	✗	Down	✗	✗	✗
Noy and Musen	✓	Up & Down	✗	✗	✗
Seidenberg and Rector	✗	Up	✓	✗	✗

2.2 Logical Based Extraction

The logical based extraction techniques are based on the notion of conservative extension [16]. An ontology module extracted from a given ontology is a conservative extension if the entailments regarding the ontology module are captured totally within its signature. More formally Lutz *et al.* [16]present the following definition:

Definition 1. *Conservative Extension Let $\mathcal{T}_1$ and $\mathcal{T}_2$ be TBoxes formulated in a DL $\mathcal{L}$, and let $\Gamma \subseteq sig(\mathcal{T}_1)$ be a signature. Then $\mathcal{T}_1 \cup \mathcal{T}_2$ is a Γ-conservative extension of $\mathcal{T}_1$ if for all $C_1, C_2 \in \mathcal{L}(\Gamma)$, we have $\mathcal{T}_1 \models C_1 \sqsubseteq C_2$ iff $\mathcal{T}_1 \cup \mathcal{T}_2 \models C_1 \sqsubseteq C_2$.*

Thus, all the entailments regarding the signature of the ontology module are equivalent to using the ontology module with the ontology it was taken from. Unfortunately, Lutz *et al.* [16] also show that deciding if an O is a conservative extension is undecidable for OWL DL. However, Konev *et al.* [12] have developed an algorithm, MEX, for extracting conservative extensions from acyclic terminologies formulated in $\mathcal{ALCI}$ or $\mathcal{ELI}$. Whilst these restrictions limit the use of this approach, it can be successfully applied to large, real world ontologies such as SNOMED CT. The experimental evaluation presented later in Section 3 does not include the method by Konev and colleagues since it has been shown to be undecidable for ontologies of complexity higher than $\mathcal{EL}$.

Grau *et al.* [2] overcome the limitations of conservative extensions for more expressive description logics by utilizing approximations; they term these modules as locality-based modules. Coverage and safety are the properties that locality-based modules can guarantee, but this is done at the expense of minimality which is also guaranteed by conservative extensions. Coverage and safety [17] are defined in terms of a module being imported by a local ontology ($\mathcal{L}$) as follows:

Coverage. Extract everything the ontology defines for the specified terms. The module O' covers the ontology O for terms from, some signature, X if for all classes A and B built from terms in X, such that if $\mathcal{L} \cup O \models A \sqsubseteq B$ then $\mathcal{L} \cup O' \models A \sqsubseteq B$.

Safety. The meaning of the extracted terms is not changed. $\mathcal{L}$ uses the terms from X safely if for all classes A and B built from terms in X, such that if $\mathcal{L} \cup O' \models A \sqsubseteq B$ then $O' models A \sqsubseteq B$.

Two different variants of locality are described by Grau *et al.* [18]. Syntactic locality can be computed in polynomial time, but semantic locality is PSPACE-complete. Syntactic locality is computed based on the syntactic structure of the axiom whereas semantic locality is computed based on the interpretation ($\mathcal{I}$) of the axiom. The issue concerning the syntactic locality is that syntactically different (but semantically equivalent) axioms can be treated differently. For example, Borgida and Giunchiglia [19] raise this issue of the syntactic approximation via the following example; consider the two sets of axioms $\{A \sqsubseteq (B \sqcap C)\}$

and $\{A \sqsubseteq B, A \sqsubseteq C\}$. These axioms are semantically equivalent, but the syntactic difference will effect the extraction process. The syntactic locality also can not handle tautologies, but this is unlikely to affect real world applications as ontologies with tautologies would be considered badly engineered.

Table 2 compares the features of the traversal based extraction techniques.

Table 2. Comparison of features for logical based ontology module extraction

	Coverage	Minimality	DL Expressivity	Tractable
Whole Ontology	✓	✗	Any	✓
Locality Based	✓	✗	OWL 1 (SHOIN)	✓
MEX	✓	✓	EL++	✓
Conservative Extension	✓	✓	Any	✗

2.3 Evaluation of Ontology Modularization Approaches

Whilst a number of different approaches for modularizing ontologies have been proposed in the literature, few efforts have addressed the problem of providing objective measures for the evaluation of the outcome of modularization techniques [7]. The prevalent measure to discriminate between ontology modules is the module *size*, that is the number of named concepts and properties that compose the module [18]. Other criteria have been proposed [8,20] that look at the structure of the ontology modules produced and attempt to assess the trade-off between maintainability and efficiency of reasoning in distributed systems. These criteria include:

- *Redundancy*: The extent to which ontology modules overlap. The inclusion of redundancy in modules improves efficiency and robustness but in contrast it requires a higher effort to maintain the modules;
- *Connectedness*: The number of edges shared by the modules generated by a modularization approach. This criterion assumes that the modules are represented as a graph, where the vertices are the axioms in the ontology, and the edges are the properties (roles) connecting two axioms with a shared symbol[3]. Connectedness estimates the degree of independence of the set of modules generated by the modularization approach.
- *Distance*: The process of modularization can simplify the structure of the module wrt the ontology in input. Two different distance measures, *inter-module distance* and *intra-module distance* have been defined that count the number of modules that relate to entities, and the number of relations in the shortest path from two entities in a module, respectively.

All the above criteria (including size) assess a specific aspect of the module obtained that depends on the task for which modularization is carried out [7]. However, there is no measure that attempts to capure the *combined effect* and

[3] The use of graph-based representations of ontologies has frequently been used for ontology evaluation [21], and the transformation of an OWL ontology into a graph has been defined (`http://www.w3.org/TR/owl-semantics/mapping.html`).

aggregates the quantitative estimates of the dimensions represented by the criteria. In contrast, [9] proposes an entropy measure, based on the analogous notion used in information theory [22], in order to provide a quantitative measure of the information contained in a message. A variation of the entropy measure, presented in [9], captures the information content of a module, *i.e.* the amount of *definition* in a module, that is how precisely the concepts in the modules are defined, thus providing a profile for the module.

A module's information content accounts for both *language* and *domain* related content, whereby the former depends on the constructors allowed by the ontology language used to represent a module; the domain-related content depends on the domain entities represented in the module. This differentiation is needed in order to capture the difference in meaning carried by the different types of knowledge represented in the ontology, and their effect on the modelling. For instance, a relationship between two concepts (represented by the OWL statement `<owl:ObjectProperty>`) should not be considered in the same way as an equivalence between two concepts (represented in OWL by the statement `<owl:equivalentClass>`), because these two notions carry different meanings, and have different consequences with respect to a modularization technique: equivalent concepts should always be grouped together in a module, whilst this is not necessarily the case with object properties.

The proposed reformulation of Shannon's entropy measure accounts for the different types of relationships that can exist between concepts by separating the notion of *language level entropy*, from *domain level entropy*. *Language level entropy* thus estimates the information content carried by the edges that represent language level constructs. These constructs are part of the ontological representation that is being used, for instance the OWL statements `<owl:equivalentClass>` or `<rdfs:subClassOf>` are language level constructs. The notion of *domain level entropy* is concerned with the domain specific relationships; these are the constructs that allow an Ontology Engineer to tailor the ontology to their domain. Such a construct in OWL would be the definition of an object property through the `<owl:ObjectProperty>` statement. In contrast, domain level entropy captures the information content that a relationship contributes to an ontology or to a module. These two types of entropy are then aggregated in the *integrated entropy*, that is a function of the *domain level entropy* and *language level entropy*, and that the authors postulate can be used to evaluate the outcome of the modularization process. Given two modules $\mathcal{M}_1$ and $\mathcal{M}_2$, that have the same size and the same signature, if the integrated entropy of $\mathcal{M}_1$ is greater than the integrated entropy of $\mathcal{M}_2$ this indicates that the number of definitions of $\mathcal{M}_1$ is greater than the number of definitions of $\mathcal{M}_2$. The integrated entropy measure is a more accurate indicator than size only, because it takes into account not only the number of named concepts in the module, but also the degree of the nodes (thus accounting also for connectedness).

Although these evaluation techniques try to capture the structural features and the content of a module, they do not help users in deciding which modularization technique is the best suited for a task at hand. The reason for this is that these techniques do not bear any relation with the task for which the modularization process is performed. Thus, in this paper we present a systematic evaluation of the modularization techniques with respect to the task of query answering and we analyse the features of the module produced in an attempt to correlate these features with the suitability for a task.

3 Evaluation

The main contribution of this paper is to provide a systematic evaluation of the different modularization techniques presented in the previous section. We identify three problems that support a number of common tasks such as query answering or service retrieval: *Instance retrieval*, *Subclass retrieval*, and *Superclass retrieval*.

These three tasks are considered as statistical classification problems over the original ontology O, and the module M_i computed using a modularization technique whose input signature is $S_{m_i} = \{C_i\}$. Let $O = \langle \mathcal{T}, \mathcal{A} \rangle$ be a knowledge base; let $Ind(\mathcal{A})$ be the set of all individuals occurring in $\mathcal{A}$, and let $C = \{C_1, C_2, \cdots, C_s\}$ the set of both primitive and defined concepts in O.

The $InstanceRetrieval_{(O,C_i)}$ problem can be defined as follows: **given** an individual $a \in (A)$, and a class $C_i \in C$ **determine** the set $I_O = \{Instance(C_i) \subseteq Ind(\mathcal{A})$ such that $O \models C_i(a)\}$.

For the purpose of this evaluation M_i is to be considered an ontology itself, so the evaluation distinguishes between the task of instance retrieval in the original ontology O, $InstanceRetrieval_{(O,C_i)}$, from the task of instance retrieval in the module M_i where the signature of the module $S_{m_i} = \{C_i\}$, therefore the problem becomes determining $I_{M_i} = \{Instance(C_i) \subseteq Ind(\mathcal{A})$ such that $M_i \models C_i(a)\}$.

We can also define the tasks of subclass and superclass retrieval as follows:

- $SubclassRetrieval_{(O,C_i)}$
  given a class $C_i \in C$ **determine** the set $Sub_O = \{X_1, X_2, \cdots, X_m\} \subseteq C$ such that $\forall X_j, j = 1, \cdots m : O \models \mathcal{I}^{X_j} \subseteq \mathcal{I}^{C_i}$
- $SuperclassRetrieval_{(O,C_i)}$
  given a class $C_i \in C$ **determine** the set $Sup_O = \{X_1, X_2, \cdots, X_m\} \subseteq C$ such that $\forall X_j, j = 1, \cdots m : O \models \mathcal{I}^{C_i} \subseteq \mathcal{I}^{X_j}$

Analogously, we define the problems $SubclassRetrieval_{(M_i,C_i)}$, with the set Sub_{M_i}; and $SuperclassRetrieval_{(M_i,C_i)}$ for the module M_i by substituting to the definitions above M_i to any occurrence of O.

To evaluate the performance of a modularization technique, a named concept C_i is selected in O; the corresponding module M_i is then built. Then we consider the problems $InstanceRetrieval_{(O,C_i)}$ and $InstanceRetrieval_{(M_i,C_i)}$ obtaining respectively I_O and I_{M_i} (set of instances of C_i in O and in M_i), Sub_O and Sub_{M_i} (set of subclasses of C_i in O and in M_i), and Sup_O and Sup_{M_i} (set of superclasses of C in O and in I_M).

We utilise these three pairs of results, in order to compute the number of *true positive*, *false positive* and *false negative* for the sets generated by the problems of

Instance retrieval, *Subclass retrieval*, and *Superclass retrieval*. The number of *true positive* is the number of entities (classes or instances) that are classified correctly (*i.e.* as Subclass, Superclass or Instance of a class C_i); the number of *false positive* is the number of entities that were incorrectly classified as positive, whilst the number of *false negative* is the number of correct entities that were missed (*i.e.* entities that were not classified as belonging to the true positive class but should have been). *False negative* can occur when the modularization approach does not preserve all the constraints on class definitions that were present in the original ontology, thus generating new class definitions that are included in the module, for instance a disjunction axiom is lost in the modularization process (this can occur those modularization approaches that do not guarantee *safety*, that is to leave the concept definitions unchanged, as discussed in 2). We discuss this case more in detail in Section 3.2.

1. $truepositive = |I_O \cap I_{M_i}|$: the number of instances of C_i in both O and M_i;
2. $falsepositive = |I_O \setminus I_M|$: the number of instances of C_i in O which are not instances of C_i in M_{C_i};
3. $falsenegative = |I_M \setminus I_O|$: the number of instances of C_i in M_i which are not instances of C_i in O.

The same values are computed for (Sub_O, Sub_M) and (Sup_O, Sup_M), substituting instances with subclasses and superclasses.

These values enable us to compute classical precision and recall measures for the three problems in consideration, where precision and recall are defined as follows:

$$precision = \frac{truepositive}{truepositive + falsepositive}$$

$$recall = \frac{truepositive}{truepositive + falsenegative}$$

To synthetize the values of precision and recall in a single value, we use the F-measure:

$$F - measure = \frac{2 \times precision \times recall}{precision + recall}$$

3.1 Evaluation Setup

The dataset for this evaluation consists of eleven ontologies from the OAEI 2007 Conference Track[4]; the full list, as well as some metrics for these ontologies, such as expressivity, number of named concepts and roles, and number of anonymous concepts, is available in Table 3.

The modularization techniques available for the experiment are:

1. Cuenca Grau *et al.* [2], *lower variant*, shortened in the following as CG-L;
2. Cuenca Grau *et al.* [2], *upper variant*, shortened as CG-U;
3. d'Aquin *et al.* [3], shortened as DAQ;

[4] `http://oaei.ontologymatching.org/2007/conference/`

Table 3. Classes, properties, and expressivity for each of the OAEI ontologies

Ontology	Named Classes	Object Properties	Datatype Properties	Anonymous Classes	Expressivity
cmt	29	49	10	11	$\mathcal{ALCHIF}(\mathcal{D})$
Conference	59	46	18	33	$\mathcal{ALCHIF}(\mathcal{D})$
confOf	38	13	23	42	$\mathcal{SHIF}(\mathcal{D})$
crs_dr	14	15	2	0	$\mathcal{ALCHIF}(\mathcal{D})$
edas	103	30	20	30	$\mathcal{ALCHIF}(\mathcal{D})$
ekaw	73	33	0	27	$\mathcal{SHIN}$
MICRO	31	17	9	33	$\mathcal{ALCHOIF}(\mathcal{D})$
OpenConf	62	24	21	63	$\mathcal{ALCHOI}(\mathcal{D})$
paperdyne	45	58	20	109	$\mathcal{ALCHOIF}(\mathcal{D})$
PCS	23	24	14	26	$\mathcal{ALCHIF}(\mathcal{D})$
sigkdd	49	17	11	15	$\mathcal{ALCHI}(\mathcal{D})$

Table 4. Comparison of the Module Size (in terms of named entities) for each of the different modularization approaches. Both the number of modules generated containing more than two named concepts, and this value as a percentage of all modules for each ontology are given.

Ontology Name	#	DAQ Modules (> 1)		DOR Modules (> 1)		SEID Modules (> 1)		CG-L Modules (> 1)		CG-U Modules (> 1)	
	Cl.	Num	% Cl.	Num	% Cl.	Num	% Cl.	Num	% Cl.	Num	% Cl.
Conference	59	26	44.1%	24	40.7%	49	83.1%	35	59.3%	35	59.3%
cmt	29	14	48.3%	11	37.9%	22	75.9%	9	31.0%	9	31.0%
confOf	38	7	18.4%	8	21.1%	33	86.8%	25	65.8%	25	65.8%
crs-dr	14	9	64.3%	3	21.4%	10	71.4%	0	0.0%	0	0.0%
edas	103	18	17.5%	25	24.3%	89	86.4%	102	99.0%	102	99.0%
ekaw	73	14	19.2%	23	31.5%	67	91.8%	15	20.5%	15	20.5%
MICRO	31	14	45.2%	6	19.4%	28	90.3%	24	77.4%	24	77.4%
OpenConf	62	12	19.4%	25	40.3%	60	96.8%	62	100.0%	62	100.0%
paperdyne	45	22	48.9%	8	17.8%	41	91.1%	36	80.0%	36	80.0%
PCS	23	13	56.5%	10	43.5%	17	73.9%	7	30.4%	7	30.4%
sigkdd	49	11	22.4%	14	28.6%	43	87.8%	8	16.3%	8	16.3%
Mean values:			**36.7%**		**29.7%**		**85.0%**		**52.7%**		**52.7%**

4. Doran *et al.* [1], shortened as DOR;
5. Seidenberg and Rector [5], shortened as SEID.

For each one of these techniques, the implementation made available by the authors has been used to guarantee the behaviour of each approach as intended by the original authors. Wrappers have been used to fit the techniques into the testing framework, such as changes to the expected input method, from a URL for the ontology to load to a local file. However, these do not modify the data, and have no affect on the results of the evaluation.

For each ontology, the set of named concepts has been considered. For each named concept, each technique has been used to produce the related module; the modularization signature was in each case the single named concept.

Table 5. Results broken down by ontology and technique; all the modules for a single ontology and a single technique are averaged together; this table only reports the *instances* results

	CG-L			CG-U			DAQ			DOR			SEID		
	P	R	FM	P	R	FM	P	R	FM	P	R	FM	P	R	FM
Conference	1.00	0.84	0.912	1.00	0.84	0.912	1.00	0.80	0.888	1.00	1.00	1.000	0.83	0.72	0.773
cmt	1.00	0.76	0.862	1.00	0.76	0.862	1.00	0.78	0.875	1.00	1.00	0.998	0.76	0.60	0.670
confOf	1.00	0.83	0.909	1.00	0.83	0.909	1.00	0.83	0.909	1.00	1.00	1.000	0.87	0.74	0.801
crs_dr	1.00	0.84	0.911	1.00	0.84	0.911	1.00	0.84	0.911	1.00	1.00	1.000	0.71	0.71	0.714
edas	1.00	0.86	0.926	1.00	0.86	0.926	1.00	0.83	0.907	1.00	0.99	0.995	0.87	0.81	0.838
ekaw	1.00	0.76	0.865	1.00	0.76	0.865	1.00	0.76	0.865	1.00	1.00	1.000	0.92	0.73	0.813
MICRO	1.00	0.87	0.928	1.00	0.87	0.928	1.00	0.82	0.903	1.00	0.97	0.987	0.94	0.81	0.869
OpenConf	0.90	0.78	0.835	0.90	0.78	0.835	1.00	0.73	0.841	0.89	1.00	0.942	0.97	0.81	0.885
paperdyne	0.96	0.82	0.884	0.96	0.82	0.884	0.99	0.87	0.924	0.84	1.00	0.910	0.91	0.82	0.862
PCS	1.00	0.75	0.860	1.00	0.75	0.860	1.00	0.76	0.864	1.00	1.00	1.000	0.74	0.61	0.667
sigkdd	1.00	0.81	0.893	1.00	0.81	0.893	1.00	0.81	0.893	0.99	1.00	0.996	0.88	0.76	0.816
Average	0.987	0.810	0.889	0.987	0.810	0.889	0.999	0.802	0.889	0.974	0.997	0.984	0.854	0.739	0.792

Table 6. Results broken down by ontology and technique; all the modules for a single ontology and a single technique are averaged together; this table only reports the *subclasses* results

	CG-L			CG-U			DAQ			DOR			SEID		
	P	R	FM	P	R	FM	P	R	FM	P	R	FM	P	R	FM
Conference	1.00	0.87	0.928	1.00	0.87	0.928	1.00	0.84	0.910	1.00	1.00	1.000	0.83	0.74	0.784
cmt	1.00	0.80	0.892	1.00	0.80	0.892	1.00	0.82	0.901	1.00	1.00	0.999	0.76	0.64	0.692
confOf	1.00	0.86	0.924	1.00	0.86	0.924	1.00	0.86	0.924	1.00	1.00	1.000	0.87	0.76	0.813
crs_dr	1.00	0.87	0.929	1.00	0.87	0.929	1.00	0.87	0.929	1.00	1.00	1.000	0.71	0.71	0.714
edas	1.00	0.88	0.939	1.00	0.88	0.939	1.00	0.87	0.929	1.00	1.00	1.000	0.87	0.82	0.845
ekaw	1.00	0.80	0.889	1.00	0.80	0.889	1.00	0.80	0.889	1.00	1.00	1.000	0.92	0.77	0.835
MICRO	1.00	0.89	0.941	1.00	0.89	0.941	1.00	0.88	0.934	1.00	1.00	1.000	0.94	0.83	0.882
OpenConf	0.90	0.81	0.852	0.90	0.81	0.852	1.00	0.83	0.907	0.98	1.00	0.992	0.97	0.90	0.931
paperdyne	0.96	0.84	0.896	0.96	0.84	0.896	0.99	0.89	0.937	0.98	1.00	0.990	0.91	0.83	0.871
PCS	1.00	0.81	0.893	1.00	0.81	0.893	1.00	0.81	0.896	1.00	1.00	1.000	0.74	0.63	0.682
sigkdd	1.00	0.84	0.912	1.00	0.84	0.912	1.00	0.84	0.912	0.99	1.00	0.997	0.88	0.78	0.826
Average	0.987	0.842	0.909	0.987	0.842	0.909	0.999	0.845	0.915	0.996	1.000	0.998	0.854	0.766	0.807

The total number of modules obtained in this way is the sum of the *Concept* column in Table 3, multiplied by the number of techniques; this gives a total of 2630 modules, of which 526 were generated for each technique.

For each module, precision, recall and F-measure have been computed, as outlined in Section 3. The results[5] have then been presented by ontology and technique (Tables 5, 6 and 7).

Table 4 lists the results for the modules generated by each of the modularization techniques across each of the ontologies, as well as the number of named classes defined by each ontology. For each technique, the number of modules containing two or more named concepts is given. Cases where no modules can be generated, or where modules of size one are generated are not given. This is in part due to the fact that some techniques (such as *DOR* and *SEID*) are guaranteed to generate a module containing at least the concept specified in the

[5] The complete set of results are available at:
`http://www.csc.liv.ac.uk/semanticweb/`

Table 7. Results broken down by ontology and technique; all the modules for a single ontology and a single technique are averaged together; this table only reports the *superclasses* results

	CG-L			CG-U			DAQ			DOR			SEID		
	P	R	FM	P	R	FM	P	R	FM	P	R	FM	P	R	FM
Conference	0.87	0.82	0.845	1.00	0.53	0.689	0.98	0.70	0.821	0.71	0.54	0.612	1.00	0.64	0.783
cmt	1.00	0.61	0.758	1.00	0.55	0.711	1.00	0.53	0.693	1.00	0.67	0.806	1.00	0.75	0.860
confOf	0.74	0.65	0.689	0.99	0.56	0.718	0.88	0.80	0.837	0.83	0.71	0.767	1.00	0.77	0.871
crs_dr	1.00	0.55	0.707	1.00	0.72	0.839	1.00	0.77	0.871	1.00	0.56	0.718	1.00	0.61	0.758
edas	1.00	0.52	0.684	1.00	0.83	0.906	0.97	0.83	0.892	1.00	0.53	0.689	1.00	0.51	0.678
ekaw	1.00	0.66	0.792	0.94	0.91	0.922	0.99	0.57	0.722	0.92	0.77	0.837	1.00	0.84	0.912
MICRO	0.76	0.63	0.687	0.95	0.79	0.863	1.00	0.47	0.643	1.00	0.60	0.752	1.00	0.51	0.676
OpenConf	1.00	0.84	0.912	1.00	0.55	0.708	1.00	0.57	0.724	0.90	0.85	0.877	1.00	0.57	0.723
paperdyne	0.90	0.73	0.810	1.00	0.89	0.940	1.00	0.55	0.711	1.00	0.55	0.708	1.00	0.55	0.708
PCS	1.00	0.89	0.940	0.99	0.61	0.759	0.87	0.69	0.769	0.90	0.73	0.810	0.99	0.65	0.786
sigkdd	1.00	0.56	0.718	1.00	0.55	0.706	1.00	0.60	0.753	1.00	0.64	0.783	0.95	0.79	0.863
Average	0.934	0.677	0.777	0.988	0.680	0.796	0.972	0.644	0.767	0.933	0.651	0.760	0.994	0.654	0.783

module signature, whereas Cuenca Grau *et al.*'s two techniques (CG^L and CG^U) can generate empty modules (i.e. of size zero). Finally, the number of modules (of size > 1) generated for each ontology is given as a percentage of the total number of named concepts. The mean values across all the ontologies for the percentage of modules considered.

3.2 Monotonicity in DL and False Negatives

Some of the extracted modules cause some instances to be misclassified with respect to the whole ontology, i.e., there is some concept C_i for which there are instances in M_i that are not instances of C in O; this seems counterintuitive given the monotonic nature of Description Logics. According to monotonicity, all the inferences that can be drawn from a subset of axioms can be drawn against the whole set of axioms in a knowledge base, so no inferences that can be drawn from a module M_i should be lost when considering O. This intuition, however, is not well founded; there are theoretical constraints on the assumptions that must be made, i.e., O must be a conservative extension of M_i (or at least guarantee safety, as defined in [2]); this is not guaranteed by all the techniques considered.

The same problem can arise with subclasses and superclasses; in the current evaluation, these problems have been detected in 31 cases, involving two ontologies (paperdyne and openconf).

3.3 Discussion

The results presented demonstrate that all the modularization techniques evaluated perform reasonably well in terms of precision and recall across all but two of the considered ontologies. However, all the approaches but *SEID* experienced some degradation in performance when applied to OpenConf and Paperdyne. This could be due to the fact that these two ontologies are both very complex and interconnected ontologies that cause all the approaches to degrade.

However, we note that Seidenberg's technique seems to have the greatest degree of variation with respect to the considered ontology, with many cases in which the precision is either 0 or 100%. This result seems to indicate that some of the heuristics used by Seidenberg's modularization approach might have been overly specific to the characteristics of the GALEN ontology, and thus are not so well suited for ontologies that have different characteristics with respect to GALEN.

One interesting result is that there is no discernible difference between logic based approaches and traversal based approaches in terms of precision of the results and recall. However the modules differ in sizes, and percentages of modules with 0 or one concept only. This seems to indicate that users need to look at the characteristics of the task they have in mind in order to choose the most appropriate modularization approach. Hence, for instance, we might want to distinguish the task of single instance retrieval from the more generic task of Instance retrieval. The former is typical of queries where a single instance of a concept is required. For example, in service provision, where the request of a service that is of a certain class as *Give me a service that is an instance of Weather service.* The latter provides *all* the instances of a class. In the first case, any of the modularization approaches with high precision results (Cuenca Grau upper and lower variants, DAQ and DOR) would perform equally well; whilst DOR has the lowest precision, it is still within a 0.05% error. Recall, in this scenario it would not be as important as finding just one correct instance which would suffice to satisfy the user request.

Conversely, if we are looking at the problem of generalized instance retireval, then recall becomes important, and in this case DOR has a better recall (whilst maintanining a good performance) followed by DAQ, and Cuenca Grau's variants, whose recall values are very similar.

If the problem consists of retrieving all the subclasses, then DOR once again performs better than the others. This is an artefact of the type of traversal used by the approach, that traverse mainly from the signature concept down. Interestingly enough, the results for subclass retrieval and superclass retrieval on this dataset seem to indicate that the content of a module is defined by the definition of the subclasses of the signature concept, whilst the superclasses seem to provide a lesser contribution to the module definition. For this reason Doran's approach, that includes only the constraints from the superclasses of the signature that are inherited down the hierarchy, performs as well as other approaches like d'Aquin or Cuenca Grau.

Other considerations that a user might want to take into account when choosing a module are related to the specific characteristics of the task. If the module produced is to be used for extensive reasoning, then Cuenca Grau's approach is to be preferred as it is the only one amongst those considered that guarantees safety. If safety is not a concern, then Doran and d'Aquin are good candidates.

4 Conclusions

Whilst a number of modularization techniques have been proposed to date, there has been little systematic analysis and comparison of these approaches with respect to common tasks. Objective, measures such as size or Integrated Entropy

[9] give some information about a module, but fail to capture task-related information, such as whether the module is fit for purpose, or can lose information (with respect to using the original ontology). To this end, we have presented a systematic and extensive empirical evaluation of various module extraction approaches, from the perspective of their suitability for a specific task. Three related problems have been identified that support a number of common tasks such as query answering or service retrieval: *Instance retrieval*, *Subclass retrieval*, and *Superclass retrieval*.

The results suggest that pragmatic, heuristic approaches such as those that assume graph traversal may be as good as logical-based approaches for most scenarios. Whilst better for tasks that may require safety guarantees or extensive reasoning, logical based approaches may not offer many benefits when used for generalized instance retrieval. However, in nearly all cases, little utility is gained by considering the definition of concepts that are more general than those appearing in the signature. Future work will extend this analysis to better identify boundary cases whereby certain techniques may be more suitable than others.

References

1. Doran, P., Tamma, V., Iannone, L.: Ontology module extraction for ontology reuse: an ontology engineering perspective. In: CIKM, pp. 61–70. ACM, New York (2007)
2. Cuenca Grau, B., Horrocks, I., Kazakov, Y., Sattler, U.: Modular reuse of ontologies: Theory and practice. J. of Artificial Intelligence Research (JAIR) 31, 273–318 (2008)
3. d'Aquin, M., Sabou, M., Motta, E.: Modularization: a key for the dynamic selection of relevant knowledge components. In: First Int.Workshop on Modular Ontologies, ISWC, Athens, Georgia, USA (2006)
4. Noy, N.F., Musen, M.A.: Specifying ontology views by traversal. In: Int. Semantic Web Conference (2004)
5. Seidenberg, J., Rector, A.: Web ontology segmentation: analysis, classification and use. In: WWW 2006: Proceedings of the 15th Int. Conference on World Wide Web (2006)
6. Stuckenschmidt, H., Klein, M.: Structure-based partitioning of large concept hierarchies. In: Proc. of the 3rd Int. Semantic Web Conference, Hiroshima, Japan (2004)
7. d'Aquin, M., Schlicht, A., Stuckenschmidt, H., Sabou, M.: Ontology modularization for knowledge selection: Experiments and evaluations. In: Wagner, R., Revell, N., Pernul, G. (eds.) DEXA 2007. LNCS, vol. 4653, pp. 874–883. Springer, Heidelberg (2007)
8. Schlicht, A., Stuckenschmidt, H.: Towards structural criteria for ontology modularization. In: Proceedings of the 1st International Workshop on Modular Ontologies, WoMO 2006, co-located with the International Semantic Web Conference, ISWC 2006, Athens, Georgia, USA, November 5 (2006)
9. Doran, P., Tamma, V., Payne, T.R., Palmisano, I.: An entropy inspired measure for evaluating ontology modularization. In: 5th International Conference on Knowledge Capture, KCAP 2009 (2009)

10. Noy, N.F., Musen, M.A.: Prompt: Algorithm and tool for automated ontology merging and alignment. In: Proc. of the 17th National Conference on Artificial Intelligence and 12th Conference on Innovative Applications of Artificial Intelligence (2000)
11. Doran, P., Tamma, V., Palmisano, I., Payne, T.R.: Dynamic selection of ontological alignments: a space reduction mechanism. In: Twenty-First International Joint Conference on Artificial Intelligence, IJCAI 2009 (2009)
12. Konev, B., Lutz, C., Walther, D., Wolter, F.: Semantic modularity and module extraction in description logics. In: Proceedings of ECAI 2008: 18th European conference on Artificial Intelligence (2008)
13. Cohen, W.W., Borgida, A., Hirsh, H.: Computing least common subsumers in description logics. In: AAAI, pp. 754–760 (1992)
14. Noy, N.F., Musen, M.A.: The PROMPT suite: interactive tools for ontology merging and mapping. International Journal of Human-Computer Studies 59, 983–1024 (2003)
15. Maedche, A., Staab, S.: Measuring similarity between ontologies. In: Gómez-Pérez, A., Benjamins, V.R. (eds.) EKAW 2002. LNCS (LNAI), vol. 2473, pp. 251–263. Springer, Heidelberg (2002)
16. Lutz, C., Walther, D., Wolter, F.: Conservative extensions in expressive description logics. In: IJCAI 2007, Proceedings of the 20th International Joint Conference on Artificial Intelligence, pp. 453–458 (2007)
17. Cuenca Grau, B., Horrocks, I., Kazakov, Y., Sattler, U.: Just the right amount: Extracting modules from ontologies. In: WWW 2007, Proceedings of the 16th International World Wide Web Conference, Banff, Canada, May 8-12, 2007, pp. 717–727 (2007)
18. Grau, B.C., Horrocks, I., Kazakov, Y., Sattler, U.: A logical framework for modularity of ontologies. In: IJCAI 2007, Proceedings of the 20th International Joint Conference on Artificial Intelligence, pp. 298–303 (2007)
19. Borgida, A., Giunchiglia, F.: Importing from functional knowledge bases - a preview. In: Cuenca-Grau, B., Honavar, V., Schlicht, A., Wolter, F. (eds.) WOMO (2007)
20. d'Aquin, M., Schlicht, A., Stuckenschmidt, H., Sabou, M.: Criteria and evaluation for ontology modularization techniques. In: Stuckenschmidt, H., Parent, C., Spaccapietra, S. (eds.) Modular Ontologies. LNCS, vol. 5445, pp. 67–89. Springer, Heidelberg (2009)
21. Gangemi, A., Catenacci, C., Ciaramita, M., Lehman, J.: Ontology evaluation and validation. an integrated formal model for the quality diagnostic task. Technical report, Laboratory for Applied Ontology, ISTC-CNR (2005)
22. Shannon, C.E.: A mathematical theory of communication. Technical Report 27:379-423, 623-656, Bell System Technical Report (1948)

A Decomposition-Based Approach to Optimizing Conjunctive Query Answering in OWL DL

Jianfeng Du[1,2], Guilin Qi[3,4], Jeff Z. Pan[5], and Yi-Dong Shen[2]

[1] Institute of Business Intelligence & Knowledge Discovery, Guangdong University of Foreign Studies, Guangzhou 510006, China
[2] State Key Laboratory of Computer Science, Institute of Software, Chinese Academy of Sciences, Beijing 100190, China
{jfdu,ydshen}@ios.ac.cn
[3] AIFB, Universität Karlsruhe, D-76128 Karlsruhe, Germany
[4] School of Computer Science and Engineering, Southeast University, Nanjing, China
[5] Department of Computing Science, The University of Aberdeen, Aberdeen AB24 3UE, UK

Abstract. Scalable query answering over Description Logic (DL) based ontologies plays an important role for the success of the Semantic Web. Towards tackling the scalability problem, we propose a decomposition-based approach to optimizing existing OWL DL reasoners in evaluating conjunctive queries in OWL DL ontologies. The main idea is to decompose a given OWL DL ontology into a set of target ontologies without duplicated ABox axioms so that the evaluation of a given conjunctive query can be separately performed in every target ontology by applying existing OWL DL reasoners. This approach guarantees sound and complete results for the category of conjunctive queries that the applied OWL DL reasoner correctly evaluates. Experimental results on large benchmark ontologies and benchmark queries show that the proposed approach can significantly improve scalability and efficiency in evaluating general conjunctive queries.

1 Introduction

Scalable query answering over Description Logic (DL) based ontologies plays an important role for the success of the Semantic Web (SW). On the one hand, the W3C organization proposed the standard Web Ontology Language (OWL)[1] to represent ontologies in the SW, which is based on DLs and provides shared vocabularies for different domains. On the other hand, ontology query engines are expected to be scalable enough to handle the increasing semantic data that the Web provides.

OWL DL is the most expressive species in the OWL family that is decidable in terms of consistency checking. Though the decidability of conjunctive query answering in OWL DL is still an open problem, many OWL DL reasoners implement decision procedures for some categories of conjunctive queries (CQs) for which decidability is known, e.g., for CQs that have a kind of tree-shape or CQs that do not contain non-distinguished variables (i.e. existentially quantified variables). To name a few, Pellet [14] is a well-known OWL DL reasoner that supports general CQs that have a kind of tree-shape

[1] http://www.w3.org/TR/owl-semantics/

A. Bernstein et al. (Eds.): ISWC 2009, LNCS 5823, pp. 146–162, 2009.

(i.e. do not contain cycles through non-distinguished variables); KAON2 [11] is another well-known OWL DL reasoner that supports CQs without non-distinguished variables. These reasoners still suffer from the scalability problem and call for optimizations to make them scale to larger ABoxes or more complex TBoxes.

To make existing OWL DL reasoners more scalable, we propose a decomposition-based approach to optimizing conjunctive query answering in OWL DL (see Section 4). Basically, the approach computes explicit answers (i.e. facts that satisfy the given CQ) first and then identifies candidate answers and target ontologies that are sufficient for checking whether candidate answers are indeed answers to the query. Different target ontologies have no common ABox axioms but may have common TBox axioms. The verification of whether a candidate answer is an answer is delegated to an existing OWL DL reasoner. This approach guarantees sound and complete results for the categories of CQs that the OWL DL reasoner correctly evaluates. For the categories of CQs that the OWL DL reasoner cannot handle, this approach still returns all candidates and results in an unsound but complete evaluation.

We implement the proposed approach and conduct experiments on LUBM [8] and UOBM [10] ontologies (see Section 5). Experimental results on all benchmark CQs given in [8,10] show that the proposed approach can significantly improve scalability and efficiency in evaluating general CQs.

Related Work. There are approaches to conjunctive query answering that have certain contributions to the scalability problem. Motik *et al.* [12] propose a resolution-based approach, implemented in KAON2 [11], to evaluating CQs without non-distinguished variables. This approach reduces the problem of conjunctive query answering to the problem of reasoning in disjunctive datalog programs; the latter problem has more scalable solutions for handling large ABoxes. Currently KAON2 does not support nominals, which are allowed in OWL DL. Dolby *et al.* [2] propose a summarization and refinement approach to instance retrieval, which is later adapted to evaluating CQs without non-distinguished variables by adding some optimizations for retrieving role instances [3]. This approach improves scalability because it works on a summarization of the ABox, but it does not support nominals either. Pan and Thomas [13] propose a semantic approximation approach to OWL DL. The approach converts an OWL DL ontology to a DL-Lite [1] ontology, which allows CQs to be evaluated in polynomial time. The above approaches, however, do not support or may not correctly evaluate CQs with non-distinguished variables.

The idea of decomposition has been exploited in managing large ontologies. The result of decomposing an ontology is usually a set of subsets of axioms in the ontology. Stuckenschmidt and Klein [15] propose a method for decomposing all concepts in an ontology to facilitate visualization of the ontology. This method does not concern ontology reasoning. Cuenca Grau *et al.* [6] propose a method for decomposing an ontology into a set of modules such that all inferences about the signature contained in a module can be made locally. The method focuses on TBoxes and does not concern conjunctive query answering. Guo and Heflin [7] propose a method for decomposing an ABox into possibly overlapped subsets. Only instance retrieval of atomic concepts/roles can be correctly performed in separate resulting subsets together with the whole TBox. Compared with the above methods, the primary distinction of our proposed approach is that it yields target ontologies without duplicated ABox axioms and ensures conjunctive query answering to be correctly performed in separate target ontologies.

2 Preliminaries

OWL DL and Conjunctive Query Answering. OWL DL corresponds to DL $\mathcal{SHOIN}$. We assume that the reader is familiar with OWL DL and thus we do not describe it in detail, but recall that an OWL DL ontology $\mathcal{O} = (\mathcal{O}_\mathcal{T}, \mathcal{O}_\mathcal{A})$ consists of a *terminological box* (*TBox*) $\mathcal{O}_\mathcal{T}$ and an *assertional box* (*ABox*) $\mathcal{O}_\mathcal{A}$. The TBox $\mathcal{O}_\mathcal{T}$ consists of a finite set of *concept inclusion axioms* $C \sqsubseteq D$, *transitivity axioms* $\mathsf{Trans}(R)$ and *role inclusion axioms* $R \sqsubseteq S$, where C and D are OWL DL concepts, and R and S roles. The ABox $\mathcal{O}_\mathcal{A}$ consists of a finite set of *concept assertions* $C(a)$, *role assertions* $R(a, b)$, *equality assertions* $a \approx b$ and *inequality assertions* $a \not\approx b$, where C is an OWL DL concept, R a role, and a and b individuals.

We briefly introduce the direct model-theoretic semantics for an OWL DL ontology $\mathcal{O}$. An *interpretation* $\mathcal{I} = (\Delta^\mathcal{I}, \cdot^\mathcal{I})$ consists of a *domain* $\Delta^\mathcal{I}$ and a function $\cdot^\mathcal{I}$ that maps every atomic concept A to a set $A^\mathcal{I} \subseteq \Delta^\mathcal{I}$, every atomic role R to a binary relation $R^\mathcal{I} \subseteq \Delta^\mathcal{I} \times \Delta^\mathcal{I}$, and every individual a to $a^\mathcal{I} \in \Delta^\mathcal{I}$. $\mathcal{I}$ is called a *model* of $\mathcal{O}$ if every axiom in $\mathcal{O}$ is satisfied by $\mathcal{I}$. $\mathcal{O}$ is *consistent* or *satisfiable* iff it has a model.

A conjunctive query (CQ) is of the form $q(\overrightarrow{x}) \leftarrow \exists \overrightarrow{y}.\mathsf{conj}(\overrightarrow{x}, \overrightarrow{y}, \overrightarrow{c})$ or simply $q(\overrightarrow{x}) \leftarrow \mathsf{conj}(\overrightarrow{x}, \overrightarrow{y}, \overrightarrow{c})$, where $q(\overrightarrow{x})$ is the *head* of q, $\mathsf{conj}(\overrightarrow{x}, \overrightarrow{y}, \overrightarrow{c})$ is the *body* of q, $\overrightarrow{x}$ are *distinguished variables*, $\overrightarrow{y}$ are *non-distinguished variables*, $\overrightarrow{c}$ are individuals, and $\mathsf{conj}(\overrightarrow{x}, \overrightarrow{y}, \overrightarrow{c})$ is a conjunction of atoms of the form $A(v)$ or $R(v_1, v_2)$ for A an atomic concept, R an atomic role, and v, v_1 and v_2 variables in $\overrightarrow{x}$ and $\overrightarrow{y}$ or individuals in $\overrightarrow{c}$. Here allowing only atomic concepts/roles is not a big issue in practice as querying against named relations is usual when people query over relational databases [13]. A CQ is called a *Boolean conjunctive query* (BCQ) if it has no distinguished variables.

A tuple $\overrightarrow{t}$ of individuals in an ontology $\mathcal{O}$ is called an *answer* of $q(\overrightarrow{x})$ in $\mathcal{O}$, denoted by $\mathcal{O} \models q[\overrightarrow{x} \mapsto \overrightarrow{t}]$, if every model of $\mathcal{O}$ satisfies $q[\overrightarrow{x} \mapsto \overrightarrow{t}]$, i.e. the body of q with every variable in $\overrightarrow{x}$ substituted by its corresponding individual in $\overrightarrow{t}$. A BCQ $q() \leftarrow \mathsf{conj}(\overrightarrow{y}, \overrightarrow{c})$ is said to have an answer $\langle\rangle$ in $\mathcal{O}$ if $\mathcal{O} \models q[\langle\rangle \mapsto \langle\rangle]$ (simply denoted by $\mathcal{O} \models q$). The problem of evaluating a CQ in $\mathcal{O}$, i.e. computing all answers of the CQ in $\mathcal{O}$, is called a problem of *conjunctive query answering*.

First-order Logic. We use the standard clausal form to represent a first-order logic program. *Terms* are variables, constants or functional terms of the form $f(t_1, \ldots, t_n)$, where f is a function symbol of arity n and t_1, ..., t_n are terms. Throughout this paper, we use (possibly with subscripts) x, y, z for variables, a, b, c for constants, and s, t for terms. We only consider unary function symbols because only unary function symbols occur in first-order logic programs that are translated from OWL DL ontologies. *Atoms* are of the form $T(t_1, \ldots, t_n)$ where T is a predicate symbol of arity n and $t_1, \ldots, t_n$ are terms. A *literal* is a positive or negative atom and a *clause* is a disjunction of literals. Terms, atoms and clauses that do not contain variables are called *ground*.

A *first-order logic program* is a set of clauses in which all variables are universally quantified. For a clause $cl = \neg A_1 \vee \ldots \vee \neg A_n \vee B_1 \vee \ldots \vee B_m$, the set of atoms

$\{A_1, \ldots, A_n\}$ is denoted by cl^-, whereas the set of atoms $\{B_1, \ldots, B_m\}$ is denoted by cl^+. By $|S|$ we denote the cardinality of a set S. A clause cl is called a *fact* if $|cl^-| = 0$, and said to be *definite* if $|cl^+| = 1$.

A *propositional program* Π is a first-order logic program consisting of only ground clauses. The set of ground atoms occurring in Π is denoted by $\mathsf{atoms}(\Pi)$.

For a first-order logic program P, the set of ground terms (resp. ground atoms) defined from the first-order signature of P is called the *Herbrand universe* (resp. *Herbrand base*) of P, denoted by $\mathrm{HU}(P)$ (resp. $\mathrm{HB}(P)$). The set of ground clauses obtained by replacing all variables occurring in each clause in P with ground terms from $\mathrm{HU}(P)$ is called the *primary grounding* of P, denoted by $\mathcal{G}(P)$. An *interpretation* M of P is a set of ground atoms in $\mathrm{HB}(P)$; it is a *model* of P if for any ground clause $cl \in \mathcal{G}(P)$ such that $cl^- \subseteq M$, $cl^+ \cap M \neq \emptyset$; it is a *minimal model* of P if there is no model M' of P such that $M' \subset M$. P is *satisfiable* iff it has a model. Given a CQ $q(\overrightarrow{x}) \leftarrow \mathsf{conj}(\overrightarrow{x}, \overrightarrow{y}, \overrightarrow{c})$, a tuple $\overrightarrow{t}$ of constants is called an *answer* of $q(\overrightarrow{x})$ in P, denoted by $P \models q[\overrightarrow{x} \mapsto \overrightarrow{t}]$, if every model of P satisfies $q[\overrightarrow{x} \mapsto \overrightarrow{t}]$.

The first-order logic program P translated from a $\mathcal{SHOIN}$ ontology may contain the equality predicate $\approx$, which is interpreted as a *congruence relation* and different from ordinary predicates. This difference is not captured by the above first-order semantics. However, the equality predicate $\approx$ can be explicitly axiomatized via a well-known transformation from [5]. Let $\mathcal{E}(P)$ denote the first-order logic program consisting of the following clauses: (1) $t \approx t$, for each ground term $t \in \mathrm{HU}(P)$; (2) $\neg(x \approx y) \vee y \approx x$; (3) $\neg(x \approx y) \vee \neg(y \approx z) \vee x \approx z$; (4) $\neg(x \approx y) \vee f(x) \approx f(y)$, for each function symbol f occurring in P; (5) $\neg T(x_1, \ldots, x_i, \ldots, x_n) \vee \neg(x_i \approx y_i) \vee T(x_1, \ldots, y_i, \ldots, x_n)$, for each predicate symbol T other than $\approx$ occurring in P and each position i. Appending $\mathcal{E}(P)$ to P allows to treat $\approx$ as an ordinary predicate, i.e., M is a model of P that interprets $\approx$ as a congruence relation, iff for any ground clause $cl \in \mathcal{G}(P \cup \mathcal{E}(P))$ such that $cl^- \subseteq M$, $cl^+ \cap M \neq \emptyset$.

3 The Proposed Decomposition-Based Approach

Throughout this section, by $\mathcal{O} = (\mathcal{O}_\mathcal{T}, \mathcal{O}_\mathcal{A})$ we denote a given OWL DL ontology. We assume that the given ontology $\mathcal{O}$ is consistent and treat $\mathcal{O}$ as a set of axioms.

3.1 The Basic Idea of the Proposed Approach

We use a BCQ $Q : q() \leftarrow p_1(\overrightarrow{y_1}, \overrightarrow{c_1}) \wedge ... \wedge p_n(\overrightarrow{y_n}, \overrightarrow{c_n})$ to show the basic idea of our approach. The approach to checking if $\mathcal{O} \models q$ consists of two phases.

In the first phase, we first translate $\mathcal{O}$ to a first-order logic program P such that $\mathcal{O} \models q$ iff $P \cup \{\neg p_1(\overrightarrow{y_1}, \overrightarrow{c_1}) \vee ... \vee \neg p_n(\overrightarrow{y_n}, \overrightarrow{c_n})\}$ is unsatisfiable (see Subsection 3.2), then consider transforming P to a proposition program Π such that $P \cup \{\neg p_1(\overrightarrow{y_1}, \overrightarrow{c_1}) \vee ... \vee \neg p_n(\overrightarrow{y_n}, \overrightarrow{c_n})\}$ is unsatisfiable iff $\Pi \cup \mathrm{Inst}_{\mathrm{BCQ}}(\Pi, Q) \cup \{\neg w_Q()\}$ is unsatisfiable, where w_Q a new predicate symbol corresponding to Q and not occurring in Π, and $\mathrm{Inst}_{\mathrm{BCQ}}(\Pi, Q)$ is a set of ground clauses instantiated from the clause $\neg p_1(\overrightarrow{y_1}, \overrightarrow{c_1}) \vee ... \vee \neg p_n(\overrightarrow{y_n}, \overrightarrow{c_n}) \vee w_Q()$ based on Π. We develop a basic method for extracting a target ontology $\mathcal{O}_{rel} \subseteq \mathcal{O}$ such that $\Pi \cup \mathrm{Inst}_{\mathrm{BCQ}}(\Pi, Q) \cup \{\neg w_Q()\}$ is unsatisfiable only if

$\mathcal{O}_{rel} \models q$ (note that this implies $\mathcal{O} \models q$ only if $\mathcal{O}_{rel} \models q$). Since Π may be infinite due to presence of function symbols in P, we instead transform P to a finite variant Π' of Π (see Subsection 3.3), such that the target ontology $\mathcal{O}'_{rel} \subseteq \mathcal{O}$ extracted from $\Pi' \cup \text{Inst}^{\dagger}_{\text{BCQ}}(\Pi', Q) \cup \{\neg w_Q()\}$ by using a similar method satisfies $\mathcal{O}_{rel} \subseteq \mathcal{O}'_{rel}$, where $\text{Inst}^{\dagger}_{\text{BCQ}}$ is a variant of Inst_{BCQ}. It should be noted that P and Π' are independent of any given query, so this phase (i.e., computing P and Π') can be performed offline.

In the second phase (see Subsection 3.4), we check if there exists a ground substitution σ such that $\{p_1(\overrightarrow{y_1}, \overrightarrow{c_1}), ..., p_n(\overrightarrow{y_n}, \overrightarrow{c_n})\}\sigma$ is a set of ground atoms occurring in definite ground facts in P. If such ground substitution exists, we conclude that $\mathcal{O} \models q$; otherwise, we extract the aforementioned target ontology $\mathcal{O}'_{rel}$ from $\Pi' \cup \text{Inst}^{\dagger}_{\text{BCQ}}(\Pi', Q) \cup \{\neg w_Q()\}$ and conclude that $\mathcal{O} \models q$ iff $\mathcal{O}'_{rel} \models q$.

From the above descriptions, we can see that our approach is correct, i.e., $\mathcal{O} \models q$ iff there is a ground substitution σ such that $\{p_1(\overrightarrow{y_1}, \overrightarrow{c_1}), ..., p_n(\overrightarrow{y_n}, \overrightarrow{c_n})\}\sigma$ is a set of ground atoms occurring in definite ground facts in P, or $\mathcal{O}'_{rel} \models q$.

Due to the space limitation, we do not provide proofs of lemmas and theorems in this paper, but refer the interested reader to our technical report[2].

3.2 Translating to First-Order Logic

Since a direct translation from $\mathcal{SHOIN}$ to first-order clauses may incur exponential blowup [9], we apply the well-known *structural transformation* [11,9] to $\mathcal{O}$ before translating $\mathcal{O}$ to first-order clauses. By $\Theta(ax)$ we denote the result of applying structural transformation to an axiom ax, and by $\Theta(\mathcal{O})$ we denote $\bigcup_{ax\in\mathcal{O}} \Theta(ax)$. As structural transformation is well-known, we do not give its definition here but refer the reader to [11,9] or our technical report[2].

Throughout this section, we use $\mathcal{O}^{\dagger}$ to denote $\Theta(\mathcal{O})$ if not otherwise specified and treat $\mathcal{O}^{\dagger}$ as a set of axioms as well. By $\Xi(ax)$ we denote the result of translating an axiom ax in $\mathcal{O}^{\dagger}$ to a set of first-order clauses using the standard methods (see [9] or our technical report[2] for details). By $\Xi(\mathcal{O}^{\dagger})$ we denote $\bigcup_{ax\in\mathcal{O}^{\dagger}} \Xi(ax)$, and by $\Xi'(\mathcal{O}^{\dagger})$ we denote $\Xi(\mathcal{O}^{\dagger}) \cup \mathcal{E}(\Xi(\mathcal{O}^{\dagger}))$ if some equational atom $s \approx t$ occurs positively in $\Xi(\mathcal{O}^{\dagger})$, or $\Xi(\mathcal{O}^{\dagger})$ otherwise. Recall that $\mathcal{E}(\Xi(\mathcal{O}^{\dagger}))$ is used to axiomatize the equality predicate in $\Xi(\mathcal{O}^{\dagger})$ (see Section 2).

The following lemma shows that the problem of evaluating a BCQ in $\mathcal{O}$ can be reduced to a satisfiability problem about $\Xi'(\mathcal{O}^{\dagger})$. This lemma is similar to existing results given in [11,9].

Lemma 1. *For a BCQ $q() \leftarrow p_1(\overrightarrow{y_1}, \overrightarrow{c_1}) \wedge ... \wedge p_n(\overrightarrow{y_n}, \overrightarrow{c_n})$ in $\mathcal{O}$, $\mathcal{O} \models q$ iff $\Xi'(\mathcal{O}^{\dagger}) \cup \{\neg p_1(\overrightarrow{y_1}, \overrightarrow{c_1}) \vee ... \vee \neg p_n(\overrightarrow{y_n}, \overrightarrow{c_n})\}$ is unsatisfiable.*

Example 1. In our running example, we consider an ontology $\mathcal{O} = \{\mathsf{Man} \sqsubseteq \leq_1 \mathsf{hasFather} \sqcap \exists \mathsf{hasFather.Man} \sqcap \mathsf{Human}, \mathsf{Man}(a_1), \mathsf{hasFather}(a_1, a_2)\}$. By applying the structural transformation, we obtain $\mathcal{O}^{\dagger} = \{\mathsf{Man} \sqsubseteq \leq_1 \mathsf{hasFather}, \mathsf{Man} \sqsubseteq \exists \mathsf{hasFather.Man}, \mathsf{Man} \sqsubseteq \mathsf{Human}, \mathsf{Man}(a_1), \mathsf{hasFather}(a_1, a_2)\}$. By translating $\mathcal{O}^{\dagger}$ to

[2] `http://www.aifb.uni-karlsruhe.de/WBS/gqi/DecomBaR/decompose-long.pdf`

first-order clauses, we obtain $\Xi(\mathcal{O}^\dagger) = \{cl_1, ..., cl_6\}$. Since $y_1 \approx y_2$ occurs positively in $\Xi(\mathcal{O}^\dagger)$, we have $\Xi'(\mathcal{O}^\dagger) = \{cl_1, ..., cl_{13}\} \cup \{t \approx t \mid t \in \mathrm{HU}(\Xi(\mathcal{O}^\dagger))\}$.

$cl_1: \neg\mathsf{Man}(x) \vee \neg\mathsf{hasFather}(x, y_1) \vee \neg\mathsf{hasFather}(x, y_2) \vee y_1 \approx y_2$

$cl_2: \neg\mathsf{Man}(x) \vee \mathsf{hasFather}(x, f(x))$ $\quad cl_3: \neg\mathsf{Man}(x) \vee \mathsf{Man}(f(x))$

$cl_4: \neg\mathsf{Man}(x) \vee \mathsf{Human}(x)$ $\quad cl_5: \mathsf{Man}(a_1)$ $\quad cl_6: \mathsf{hasFather}(a_1, a_2)$

$cl_7: \neg(x \approx y) \vee y \approx x$ $\quad cl_8: \neg(x \approx y) \vee \neg(y \approx z) \vee x \approx z$ $\quad cl_9: \neg(x \approx y) \vee f(x) \approx f(y)$

$cl_{10}: \neg\mathsf{Man}(x) \vee \neg(x \approx y) \vee \mathsf{Man}(y)$ $\quad cl_{11}: \neg\mathsf{hasFather}(x, y) \vee \neg(x \approx z) \vee \mathsf{hasFather}(z, y)$

$cl_{12}: \neg\mathsf{hasFather}(x, y) \vee \neg(y \approx z) \vee \mathsf{hasFather}(x, z)$ $\quad cl_{13}: \neg\mathsf{Human}(x) \vee \neg(x \approx y) \vee \mathsf{Human}(y)$

3.3 Approximate Grounding of the First-Order Logic Program

According to Lemma 1, we need to address a satisfiability problem about $\Xi'(\mathcal{O}^\dagger)$. This can be done by considering a propositional program that is transformed from $\Xi'(\mathcal{O}^\dagger)$ and has the same set of minimal models as $\Xi'(\mathcal{O}^\dagger)$ has. We extend the well-known intelligent grounding (IG) technique [4] which computes, in a fixpoint-evaluation manner, a semantically equivalent propositional program containing only derivable ground atoms from a function-free first-order logic program. By generalizing the idea of the IG technique, we define a method for grounding a general first-order logic program, called *reduced grounding* and defined below.

Definition 1 (Reduced Grounding). For a first-order logic program P, the *reduced grounding* of P, denoted by $\mathcal{G}_r(P)$, is the union of two sets of ground clauses $\Pi_1 \cup \Pi_2$, where $\Pi_1 = \{cl \in P \mid cl \text{ is a definite ground fact}\}$, and Π_2 is the least fixpoint of $\Pi^{(n)}$ such that $\Pi^{(0)} = \emptyset$ and for $n > 0$, $\Pi^{(n)} = \{cl\,\sigma \mid cl \in P, \sigma \text{ is a ground substitution such that } cl^-\sigma \subseteq \mathsf{atoms}(\Pi^{(n-1)} \cup \Pi_1), cl^+\sigma \subseteq \mathrm{HB}(P) \text{ and } cl^+\sigma \cap \mathsf{atoms}(\Pi_1) = \emptyset\}$.

Lemma 2. *Let P be a first-order logic program in which the equality predicate $\approx$ has been axiomatized. Then $\mathcal{G}_r(P)$ is a subset of $\mathcal{G}(P)$ and has the same set of minimal models as P has.*

In the following theorem, we show a method that uses $\mathcal{G}_r(\Xi'(\mathcal{O}^\dagger))$ to check if $\mathcal{O} \models q$ for a BCQ $Q: q() \leftarrow p_1(\vec{y_1}, \vec{c_1}) \wedge ... \wedge p_n(\vec{y_n}, \vec{c_n})$. By $\mathrm{Inst}_{\mathrm{BCQ}}(\Pi, Q)$ we denote the result of instantiating the clause $cl: \neg p_1(\vec{y_1}, \vec{c_1}) \vee ... \vee \neg p_n(\vec{y_n}, \vec{c_n}) \vee w_Q()$ based on a propositional program Π, i.e., $\mathrm{Inst}_{\mathrm{BCQ}}(\Pi, Q) = \{cl\,\sigma \mid \sigma \text{ is a ground substitution such that } cl^-\sigma \subseteq \mathsf{atoms}(\Pi)\}$, where w_Q is a predicate symbol corresponding to Q and not occurring in Π. The introduction of $\mathrm{Inst}_{\mathrm{BCQ}}(\mathcal{G}_r(\Xi'(\mathcal{O}^\dagger)), Q)$ enables the checking of $\mathcal{O} \models q$ to be performed in a propositional program, while the introduction of w_Q facilitates extracting a target ontology w.r.t. Q.

Theorem 1. *For a BCQ $Q: q() \leftarrow p_1(\vec{y_1}, \vec{c_1}) \wedge ... \wedge p_n(\vec{y_n}, \vec{c_n})$ in $\mathcal{O}$, $\mathcal{O} \models q$ iff $\mathcal{G}_r(\Xi'(\mathcal{O}^\dagger)) \cup \mathrm{Inst}_{\mathrm{BCQ}}(\mathcal{G}_r(\Xi'(\mathcal{O}^\dagger)), Q) \cup \{\neg w_Q()\}$ is unsatisfiable.*

Based on the above theorem, we develop a basic method that could improve the performance in evaluating a BCQ $Q: q() \leftarrow p_1(\vec{y_1}, \vec{c_1}) \wedge ... \wedge p_n(\vec{y_n}, \vec{c_n})$ in $\mathcal{O}$. This method first extracts a relevant subset Π_{rel} of $\mathcal{G}_r(\Xi'(\mathcal{O}^\dagger)) \cup \mathrm{Inst}_{\mathrm{BCQ}}(\mathcal{G}_r(\Xi'(\mathcal{O}^\dagger)), Q) \cup \{\neg w_Q()\}$ such that Π_{rel} is unsatisfiable iff $\mathcal{G}_r(\Xi'(\mathcal{O}^\dagger)) \cup \mathrm{Inst}_{\mathrm{BCQ}}(\mathcal{G}_r(\Xi'(\mathcal{O}^\dagger)), Q) \cup \{\neg w_Q()\}$ is unsatisfiable, then identifies a subset $\mathcal{O}_{rel}$ of axioms in $\mathcal{O}^\dagger$ from Π_{rel} such that Π_{rel} is unsatisfiable only if $\mathcal{O}_{rel} \models q$, and finally checks if $\mathcal{O}_{rel} \models q$. By Theorem 1, we have $\mathcal{O} \models q$ only if $\mathcal{O}_{rel} \models q$. Since $\mathcal{O}_{rel} \subseteq \mathcal{O}^\dagger$ and $\mathcal{O} \models q$ iff $\mathcal{O}^\dagger \models q$, we also have $\mathcal{O} \models q$ if $\mathcal{O}_{rel} \models q$.

However, the basic method cannot be realized in general as $\mathcal{G}_r(\Xi'(\mathcal{O}^\dagger))$ can be infinite. We therefore consider a mapping function on ground terms occurring in a propositional program Π such that the range of this function is finite. We call a mapping function $\lambda : \text{terms}(\Pi) \mapsto \text{terms}(\Pi)$, where $\text{terms}(\Pi)$ is the set of ground terms occurring in Π, an ***equality-and-functional-term-collapsed mapping function*** (simply *eft-mapping function*) for Π, if for every functional term $f_1(...f_n(a))$ (where $n > 1$) occurring in Π, $\lambda(f_1(...f_n(a))) = \lambda(f_n(a))$, and for every equational atom $s \approx t$ occurring positively in Π, $\lambda(s) = \lambda(t)$.

We naturally extend a mapping function λ on ground terms to other first-order objects, i.e., by $\lambda(\alpha)$, $\lambda(cl)$, $\lambda(\mathcal{A})$ and $\lambda(P)$ we respectively denote the results obtained from an atom α, a clause cl, a set $\mathcal{A}$ of atoms and a first-order logic program P by replacing every ground term t occurring in it with $\lambda(t)$.

It is clear that, when a propositional program Π is infinite but the number of constants, predicate symbols and function symbols occurring in Π is finite, $\lambda(\Pi)$ is finite for any eft-mapping function λ for Π. Even when Π is finite, $\lambda(\Pi)$ can be much smaller than Π because the subset of ground clauses in Π that form a congruence relation is collapsed in $\lambda(\Pi)$.

By $\text{Inst}'_{\text{BCQ}}(\Pi, Q, \lambda)$ we denote the result of instantiating the clause $cl : \neg p_1(\vec{y_1}, \vec{c_1}) \vee ... \vee \neg p_n(\vec{y_n}, \vec{c_n}) \vee w_Q()$ based on a propositional program Π and a mapping function λ, i.e., $\text{Inst}'_{\text{BCQ}}(\Pi, Q, \lambda) = \{cl\,\sigma \mid \sigma \text{ is a ground substitution such that } \lambda(cl^-)\sigma \subseteq \mathsf{atoms}(\Pi)\}$, where w_Q is a predicate symbol corresponding to Q and not occurring in Π. We revise the basic method by replacing $\mathcal{G}_r(\Xi'(\mathcal{O}^\dagger))$ with a finite superset Π_{sup} of $\lambda(\mathcal{G}_r(\Xi'(\mathcal{O}^\dagger)))$, where λ is an eft-mapping function for $\mathcal{G}_r(\Xi'(\mathcal{O}^\dagger))$. The revised method first extracts a relevant subset Π_{rel} of $\Pi_{sup} \cup \text{Inst}'_{\text{BCQ}}(\Pi_{sup}, Q, \lambda) \cup \{\neg w_Q()\}$, then computes the set $\mathcal{O}_{rel}$ of axioms ax in $\mathcal{O}^\dagger$ such that $\lambda(cl\,\sigma) \in \Pi_{rel}$ for some clause $cl \in \Xi(ax)$ and some ground substitution σ, and finally checks if $\mathcal{O}_{rel} \models q$.

Consider the extraction of a relevant subset Π_{rel} mentioned above. Our extraction method is based on the notion of *connected component* (see Definition 2 below). Simply speaking, a connected component of a propositional program Π is a subset of Π such that any two clauses in it have common ground atoms. This notion has been used to confine the search space in solving SAT problems because an unsatisfiable propositional program must have a maximal connected component that is unsatisfiable.

Definition 2 (Connected Component). Let Π be a propositional program. Two ground clauses cl and cl' are called *connected* in Π if there exists a sequence of clauses $cl_0 = cl, cl_1, \ldots, cl_n = cl'$ in Π such that cl_{i-1} and cl_i have common ground atoms for any $1 \leq i \leq n$. A *connected component* Π_c of Π is a subset of Π such that any two clauses cl and cl' in Π_c are connected in Π_c. Π_c is called *maximal* if there is no connected component Π'_c of Π such that $\Pi_c \subset \Pi'_c$.

The basic idea for extracting Π_{rel} is that when $\Pi_{sup} \cup \text{Inst}'_{\text{BCQ}}(\Pi_{sup}, Q, \lambda) \cup \{\neg w_Q()\}$ is unsatisfiable, the maximal connected component of $\Pi_{sup} \cup \text{Inst}'_{\text{BCQ}}(\Pi_{sup}, Q, \lambda) \cup \{\neg w_Q()\}$ where $w_Q()$ occurs is also unsatisfiable. To obtain a smaller unsatisfiable subset, we extend the basic idea by removing a subset Π_{ur} from Π_{sup} first and then extracting the maximal connected component Π_{rel} of $(\Pi_{sup} \cup \mathrm{Inst}'_{\mathrm{BCQ}}(\Pi_{sup}, Q, \lambda) \cup \{\neg w_Q()\}) \setminus \Pi_{ur}$ where $w_Q()$ occurs. The detailed description and the correctness of the method are shown in the following theorem.

Theorem 2. *Let $Q : q() \leftarrow p_1(\overrightarrow{y_1}, \overrightarrow{c_1}) \wedge ... \wedge p_n(\overrightarrow{y_n}, \overrightarrow{c_n})$ be a BCQ, λ be an eft-mapping function for $\mathcal{G}_r(\Xi'(\mathcal{O}^\dagger))$, Π_{sup} be a superset of $\lambda(\mathcal{G}_r(\Xi'(\mathcal{O}^\dagger)))$ and $\Pi' = \Pi_{sup} \cup \mathrm{Inst}'_{\mathrm{BCQ}}(\Pi_{sup}, Q, \lambda) \cup \{\neg w_Q()\}$. Let Π_{ur} be a subset of Π' such that for all clauses $cl \in \Pi_{ur}$, cl^+ contains at least one ground atom not occurring in $\Pi' \setminus \Pi_{ur}$ and Π_{rel} be the maximal connected component of $\Pi' \setminus \Pi_{ur}$ where $w_Q()$ occurs. Let $\mathcal{O}_{rel} = \{ax \in \mathcal{O}^\dagger \mid$ there exists a clause $cl \in \Xi(ax)$ and a ground substitution σ such that $\lambda(cl\,\sigma) \in \Pi_{rel}\}$. Then $\mathcal{O} \models q$ iff $\mathcal{O}_{rel} \models q$.*

Based on the above theorem, we develop an algorithm to compute a superset of $\lambda(\mathcal{G}_r(\Xi'(\mathcal{O}^\dagger)))$ for some eft-mapping function λ. This algorithm, denoted by Approx-Ground($\mathcal{O}^\dagger$), accepts $\mathcal{O}^\dagger$ and returns a triple $(\Pi, \mathcal{S}_{df}, \mathcal{S})$, where $\mathcal{S}_{df}$ and $\mathcal{S}$ are two sets of sets of ground atoms and Π is a superset of $\lambda(\mathcal{G}_r(\Xi'(\mathcal{O}^\dagger)))$ for some eft-mapping function λ that can be constructed from $\mathcal{S}_{df}$ and $\mathcal{S}$. Since the algorithm is rather technical, we only explains the basic idea here and refer the interested reader to our technical report[2] for technical details.

The output $\mathcal{S}_{df}$ is actually the set of sets of constants such that for any constant a in any $\mathcal{C} \in \mathcal{S}_{df}$, there exists an equality assertion $a \approx b$ in $\mathcal{O}^\dagger$ for some constant $b \in \mathcal{C}$. The output $\mathcal{S}$ is actually the set of sets of ground terms whose functional depth is at most one, such that for any ground term s in any $\mathcal{C} \in \mathcal{S}$, there exists a ground term $t \in \mathcal{C}$ such that the equational atom $s \approx t$ appears in the execution of the algorithm. Both $\mathcal{S}_{df}$ and $\mathcal{S}$ are used to merge ground terms that may occur in the same equational atom occurring positively in $\mathcal{G}_r(\Xi'(\mathcal{O}^\dagger))$, making the output Π smaller. We call an element of $\mathcal{S}_{df}$ or $\mathcal{S}$ a *congruence class*.

The algorithm does not directly compute an eft-mapping function λ for $\mathcal{G}_r(\Xi'(\mathcal{O}^\dagger))$, because such mapping function can be constructed from $\mathcal{S}_{df}$ and $\mathcal{S}$. By $\mathsf{map}(t, \mathcal{S}_{df}, \mathcal{S})$ we denote a function $\mathrm{HU}(\Xi'(\mathcal{O}^\dagger)) \mapsto \mathrm{HU}(\Xi'(\mathcal{O}^\dagger))$ based on $\mathcal{S}_{df}$ and $\mathcal{S}$, recursively defined as follows, where a and b are constants, and s and t ground terms.

- $\mathsf{map}(f_1(...f_n(a)), \mathcal{S}_{df}, \mathcal{S}) = \mathsf{map}(f_n(a), \mathcal{S}_{df}, \mathcal{S})$, where $n > 1$;
- $\mathsf{map}(f(a), \mathcal{S}_{df}, \mathcal{S}) = \mathsf{map}(f(b), \emptyset, \mathcal{S})$, where $b = \min(\mathcal{C})$ if $a \in \mathcal{C}$ for some $\mathcal{C} \in \mathcal{S}_{df}$, or $b = a$ otherwise;
- $\mathsf{map}(a, \mathcal{S}_{df}, \mathcal{S}) = \mathsf{map}(b, \emptyset, \mathcal{S})$, where $b = \min(\mathcal{C})$ if $a \in \mathcal{C}$ for some $\mathcal{C} \in \mathcal{S}_{df}$, or $b = a$ otherwise;
- $\mathsf{map}(s, \emptyset, \mathcal{S}) = t$, where $t = \min(\mathcal{C})$ if $s \in \mathcal{C}$ for some $\mathcal{C} \in \mathcal{S}$, or $t = s$ otherwise.

We naturally extend the function map to other first-order objects, i.e., by $\mathsf{map}(\alpha, \mathcal{S}_{df}, \mathcal{S})$, $\mathsf{map}(cl, \mathcal{S}_{df}, \mathcal{S})$, $\mathsf{map}(\mathcal{A}, \mathcal{S}_{df}, \mathcal{S})$ and $\mathsf{map}(P, \mathcal{S}_{df}, \mathcal{S})$ we respectively denote the results obtained from an atom α, a clause cl, a set $\mathcal{A}$ of atoms and a first-order logic program P by replacing every ground term t occurring in it with $\mathsf{map}(t, \mathcal{S}_{df}, \mathcal{S})$.

We call a mapping function $\lambda : \mathrm{HU}(\Xi'(\mathcal{O}^\dagger)) \mapsto \mathrm{HU}'(\Xi(\mathcal{O}^\dagger))$ *induced from* the function map w.r.t. $\mathcal{S}_{df}$ and $\mathcal{S}$ if $\lambda(t) = \mathsf{map}(t, \mathcal{S}_{df}, \mathcal{S})$ for all ground terms $t \in \mathrm{HU}(\Xi'(\mathcal{O}^\dagger))$. The first goal of the algorithm is to ensure the mapping function λ induced from map w.r.t. $\mathcal{S}_{df}$ and $\mathcal{S}$ to be an eft-mapping function for $\mathcal{G}_r(\Xi'(\mathcal{O}^\dagger))$,

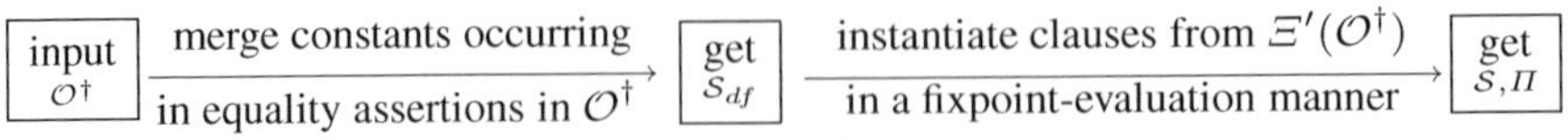

Fig. 1. The main steps for approximately grounding $\Xi'(\mathcal{O}^\dagger)$

i.e., ensure $\mathsf{map}(s, \mathcal{S}_{df}, \mathcal{S}) = \mathsf{map}(t, \mathcal{S}_{df}, \mathcal{S})$ for all equational atoms $s \approx t$ occurring positively in $\mathcal{G}_r(\Xi'(\mathcal{O}^\dagger))$. The second goal of the algorithm is to return a superset of $\lambda(\mathcal{G}_r(\Xi'(\mathcal{O}^\dagger)))$. To achieve the above two goals, the algorithm works in two main steps, as shown in Figure 1.

In the first step, the algorithm places any two constants a and b that occur in the same equality assertion in $\mathcal{O}^\dagger$ into the same congruence class $\mathcal{C}$ and adds $\mathcal{C}$ to $\mathcal{S}_{df}$. After this step, $\mathcal{S}_{df}$ will not be changed anymore.

In the second step, the algorithm instantiates clauses from $\Xi'(\mathcal{O}^\dagger)$ in a fixpoint-evaluation manner to generate a superset of $\lambda(\mathcal{G}_r(\Xi'(\mathcal{O}^\dagger)))$, where λ denotes a mapping function induced from map w.r.t. $\mathcal{S}_{df}$ and $\mathcal{S}$.

Before giving a fixpoint-like characterization of a superset of $\lambda(\mathcal{G}_r(\Xi'(\mathcal{O}^\dagger)))$, we need to introduce a restriction on $\mathcal{O}^\dagger$. We call $\mathcal{O}^\dagger$ *congruence-complete* if (1) $a \approx b \in \mathcal{O}^\dagger$ implies $b \approx a \in \mathcal{O}^\dagger$; (2) $a \approx b \in \mathcal{O}^\dagger$ and $b \approx c \in \mathcal{O}^\dagger$ imply $a \approx c \in \mathcal{O}^\dagger$; (3) $a \approx b \in \mathcal{O}^\dagger$ and $A(a) \in \mathcal{O}^\dagger$ imply $A(b) \in \mathcal{O}^\dagger$; (4) $a \approx b \in \mathcal{O}^\dagger$ and $R(a, c) \in \mathcal{O}^\dagger$ imply $R(b, c) \in \mathcal{O}^\dagger$; and (5) $a \approx b \in \mathcal{O}^\dagger$ and $R(c, a) \in \mathcal{O}^\dagger$ imply $R(c, b) \in \mathcal{O}^\dagger$.

Let Π_1 denote the set of definite ground facts in $\Xi'(\mathcal{O}^\dagger)$. By induction on the level n of $\Pi^{(n)}$ given in Definition 1, we can show that, if Π is a subset of $\lambda(\mathcal{G}(\Xi'(\mathcal{O}^\dagger)))$ such that (*) $\lambda(cl\,\sigma) \in \Pi$ for any clause $cl \in \Xi'(\mathcal{O}^\dagger)$ and any ground substitution σ such that $\lambda(cl^-\sigma) \subseteq \mathsf{atoms}(\Pi \cup \lambda(\Pi_1))$ and every ground atom in $\lambda(cl^+\sigma) \cap \mathsf{atoms}(\lambda(\Pi_1))$ contains ground terms that are functional or occur in $\mathcal{S}$, then $\Pi \cup \lambda(\Pi_1)$ is a superset of $\lambda(\mathcal{G}_r(\Xi'(\mathcal{O}^\dagger)))$ when $\mathcal{O}^\dagger$ is congruence-complete. We refer the interested reader to our technical report[2] to see the proof of the above conclusion, a counterexample on why the restriction on congruence-completeness is needed, as well as a simple method for making $\mathcal{O}^\dagger$ congruence-complete when $\Theta(\mathcal{O})$ is not congruence-complete.

We in what follows assume that $\mathcal{O}^\dagger$ is congruence-complete. Under this assumption, we refine the second goal of the algorithm to finding a subset Π of $\lambda(\mathcal{G}(\Xi'(\mathcal{O}^\dagger)))$ that satisfies the above condition (*). To achieve this goal, the second step of the algorithm adds $\lambda(cl\,\sigma)$ to Π for any clause $cl \in \Xi'(\mathcal{O}^\dagger)$ and ground substitution σ such that (i) $\lambda(cl^-\sigma) \subseteq \mathsf{atoms}(\Pi_P \cup \lambda(\Pi_1))$ and (ii) every ground atom in $\lambda(cl^+\sigma) \cap \mathsf{atoms}(\lambda(\Pi_1))$ contains ground terms that are functional or occur in $\mathcal{S}$. Meanwhile, any equational atom $s \approx t$ occurring positively in $cl\,\sigma$ is handled by placing s and t into the same congruence class $\mathcal{C}$ and by adding $\mathcal{C}$ to $\mathcal{S}$. In order to achieve the first goal of the algorithm, $f(s)$ and $f(t)$ for f a function symbol occurring in $\Xi(\mathcal{O}^\dagger)$ are merged similarly as s and t, because the clause $\neg(s \approx t) \vee f(s) \approx f(t)$, instantiated from the clause of the form (4) in Section 2, may belong to $\mathcal{G}_r(\Xi'(\mathcal{O}^\dagger))$.

The following lemma shows the correctness and the complexity of the algorithm.

Lemma 3. *Let $(\Pi, \mathcal{S}_{df}, \mathcal{S})$ be returned by* ApproxGround($\mathcal{O}^\dagger$) *and λ be a mapping function induced from the function* map *w.r.t. $\mathcal{S}_{df}$ and $\mathcal{S}$. Then: (1) $\lambda(\mathcal{G}_r(\Xi'(\mathcal{O}^\dagger))) \subseteq \Pi$*

and λ is an eft-mapping function for $\mathcal{G}_r(\Xi'(\mathcal{O}^\dagger))$; (2) ApproxGround($\mathcal{O}^\dagger$) *works in time polynomial in s^m and $|\Pi|$ is polynomial in s, where m is the maximum number in number restrictions in $\mathcal{O}$ and s is the size of $\mathcal{O}$.*

Based on the above lemma, we can use the result $(\Pi, \mathcal{S}_{df}, \mathcal{S})$ of ApproxGround($\mathcal{O}^\dagger$) to compute a subset $\mathcal{O}_{rel}$ of $\mathcal{O}^\dagger$ such that $\mathcal{O} \models q$ iff $\mathcal{O}_{rel} \models q$ for a BCQ $q() \leftarrow \mathsf{conj}(\overrightarrow{y}, \overrightarrow{c})$, by applying the method given in Theorem 2.

Before ending this subsection, we show the result of ApproxGround($\mathcal{O}^\dagger$) for the ontology $\mathcal{O}^\dagger$ given in Example 1 (note that $\mathcal{O}^\dagger$ is congruence-complete).

Example 2 (Example 1 continued). ApproxGround($\mathcal{O}^\dagger$) returns a triple $(\Pi, \mathcal{S}_{df}, \mathcal{S})$, where $\Pi = \{cl'_1, ..., cl'_{30}\}$, $\mathcal{S}_{df} = \emptyset$ and $\mathcal{S} = \{\{a_2, f(a_1)\}\}$.

$cl'_1:\ \mathsf{Man}(a_1)\quad cl'_2:\ \mathsf{hasFather}(a_1, a_2)$
$cl'_3:\ \neg\mathsf{Man}(a_1) \vee \neg\mathsf{hasFather}(a_1, a_2) \vee \neg\mathsf{hasFather}(a_1, a_2) \vee a_2 \approx a_2$
$cl'_4:\ \neg\mathsf{Man}(a_2) \vee \neg\mathsf{hasFather}(a_2, f(a_2)) \vee \neg\mathsf{hasFather}(a_2, f(a_2)) \vee f(a_2) \approx f(a_2)$
$cl'_5:\ \neg\mathsf{Man}(f(a_2)) \vee \neg\mathsf{hasFather}(f(a_2), f(a_2)) \vee \neg\mathsf{hasFather}(f(a_2), f(a_2)) \vee f(a_2) \approx f(a_2)$
$cl'_6:\ \neg\mathsf{Man}(a_1) \vee \mathsf{hasFather}(a_1, a_2)\quad cl'_7:\ \neg\mathsf{Man}(a_2) \vee \mathsf{hasFather}(a_2, f(a_2))$
$cl'_8:\ \neg\mathsf{Man}(f(a_2)) \vee \mathsf{hasFather}(f(a_2), f(a_2))\quad cl'_9:\ \neg\mathsf{Man}(a_1) \vee \mathsf{Man}(a_2)$
$cl'_{10}:\ \neg\mathsf{Man}(a_2) \vee \mathsf{Man}(f(a_2))\quad cl'_{11}:\ \neg\mathsf{Man}(f(a_2)) \vee \mathsf{Man}(f(a_2))$
$cl'_{12}:\ \neg\mathsf{Man}(a_1) \vee \mathsf{Human}(a_1)\quad cl'_{13}:\ \neg\mathsf{Man}(a_2) \vee \mathsf{Human}(a_2)\quad cl'_{14}:\ \neg\mathsf{Man}(f(a_2)) \vee \mathsf{Human}(f(a_2))$
$cl'_{15}:\ a_1 \approx a_1\quad cl'_{16}:\ a_2 \approx a_2\quad cl'_{17}:\ f(a_2) \approx f(a_2)$
$cl'_{18}:\ \neg(a_2 \approx a_2) \vee a_2 \approx a_2\quad cl'_{19}:\ \neg(a_2 \approx a_2) \vee f(a_2) \approx f(a_2)$
$cl'_{20}:\ \neg(f(a_2) \approx f(a_2)) \vee f(a_2) \approx f(a_2)\quad cl'_{21}:\ \neg(a_2 \approx a_2) \vee \neg(a_2 \approx a_2) \vee a_2 \approx a_2$
$cl'_{22}:\ \neg(f(a_2) \approx f(a_2)) \vee \neg(f(a_2) \approx f(a_2)) \vee f(a_2) \approx f(a_2)$
$cl'_{23}:\ \neg\mathsf{Man}(a_2) \vee \neg(a_2 \approx a_2) \vee \mathsf{Man}(a_2)\quad cl'_{24}:\ \neg\mathsf{Man}(f(a_2)) \vee \neg(f(a_2) \approx f(a_2)) \vee \mathsf{Man}(f(a_2))$
$cl'_{25}:\ \neg\mathsf{hasFather}(a_1, a_2) \vee \neg(a_2 \approx a_2) \vee \mathsf{hasFather}(a_1, a_2)$
$cl'_{26}:\ \neg\mathsf{hasFather}(a_2, f(a_2)) \vee \neg(a_2 \approx a_2) \vee \mathsf{hasFather}(a_2, f(a_2))$
$cl'_{27}:\ \neg\mathsf{hasFather}(a_2, f(a_2)) \vee \neg(f(a_2) \approx f(a_2)) \vee \mathsf{hasFather}(a_2, f(a_2))$
$cl'_{28}:\ \neg\mathsf{hasFather}(f(a_2), f(a_2)) \vee \neg(f(a_2) \approx f(a_2)) \vee \mathsf{hasFather}(f(a_2), f(a_2))$
$cl'_{29}:\ \neg\mathsf{Human}(a_2) \vee \neg(a_2 \approx a_2) \vee \mathsf{Human}(a_2)$
$cl'_{30}:\ \neg\mathsf{Human}(f(a_2)) \vee \neg(f(a_2) \approx f(a_2)) \vee \mathsf{Human}(f(a_2))$

3.4 Computing All Answers with the Help of the Grounding

In this subsection, we present a method for computing all answers of a general CQ by using $\Xi(\mathcal{O}^\dagger)$ and the result of ApproxGround($\mathcal{O}^\dagger$).

For a propositional program Π, two sets $\mathcal{S}_{df}$ and $\mathcal{S}$ of sets of ground terms occurring in Π, and a CQ $Q : q(\overrightarrow{x}) \leftarrow p_1(\overrightarrow{x_1}, \overrightarrow{y_1}, \overrightarrow{c_1}) \wedge ... \wedge p_n(\overrightarrow{x_n}, \overrightarrow{y_n}, \overrightarrow{c_n})$, where $\overrightarrow{x}$ is the union of $\overrightarrow{x_1}, ..., \overrightarrow{x_n}$, by $\text{Inst}_{\text{CQ}}(\Pi, Q, \mathcal{S}_{df}, \mathcal{S})$ we denote the result of instantiating the clause $cl : \neg p_1(\overrightarrow{x_1}, \overrightarrow{y_1}, \overrightarrow{c_1}) \vee ... \vee \neg p_n(\overrightarrow{x_n}, \overrightarrow{y_n}, \overrightarrow{c_n}) \vee w_Q(\overrightarrow{x})$ based on Π, $\mathcal{S}_{df}$ and $\mathcal{S}$, where w_Q is a predicate symbol corresponding to Q and not occurring in Π, i.e., $\text{Inst}_{\text{CQ}}(\Pi, Q, \mathcal{S}_{df}, \mathcal{S}) = \{cl\,\sigma \mid \sigma$ is a ground substitution such that all ground atoms in $\mathsf{map}(cl^-, \mathcal{S}_{df}, \mathcal{S})\sigma$ occur in Π and $\overrightarrow{x}\sigma$ is a tuple of constants$\}$. For example, for $\Pi, \mathcal{S}_{df}$ and $\mathcal{S}$ given in Example 2 and a CQ $Q : q(x) \leftarrow \mathsf{Man}(x) \wedge \mathsf{hasFather}(x, y)$, $\text{Inst}_{\text{CQ}}(\Pi, Q, \mathcal{S}_{df}, \mathcal{S}) = \{\neg\mathsf{Man}(a_1) \vee \neg\mathsf{hasFather}(a_1, a_2) \vee w_Q(a_1)\} \cup \{\neg\mathsf{Man}(a_2) \vee \neg\mathsf{hasFather}(a_2, f(a_2)) \vee w_Q(a_2)\}$. With the above denotation, the following lemma gives a necessary condition that an answer of a CQ in $\mathcal{O}$ should satisfy.

Lemma 4. *Let $(\Pi, \mathcal{S}_{df}, \mathcal{S})$ be returned by* ApproxGround($\mathcal{O}^\dagger$). *For a CQ Q : $q(\overrightarrow{x}) \leftarrow \mathsf{conj}(\overrightarrow{x}, \overrightarrow{y}, \overrightarrow{c})$ in $\mathcal{O}$, a tuple of constants $\overrightarrow{t}$ is an answer of Q in $\mathcal{O}$ only if $\mathsf{map}(w_Q(\overrightarrow{t}), \mathcal{S}_{df}, \mathcal{S})$ occurs in* $\text{Inst}_{\text{CQ}}(\Pi, Q, \mathcal{S}_{df}, \mathcal{S})$.

The following lemma gives a sufficient condition that ensures a tuple $\overrightarrow{t}$ of constants to be an answer of a CQ $Q : q(\overrightarrow{x}) \leftarrow \mathsf{conj}(\overrightarrow{x}, \overrightarrow{y}, \overrightarrow{c})$ in $\mathcal{O}$, where $q[\overrightarrow{x} \mapsto \overrightarrow{t}, \overrightarrow{y} \mapsto \overrightarrow{s}]$ denotes the body of q with every variable in $\overrightarrow{x}$ and $\overrightarrow{y}$ respectively substituted by its corresponding ground terms in $\overrightarrow{t}$ and $\overrightarrow{s}$.

Lemma 5. *For a CQ $Q : q(\overrightarrow{x}) \leftarrow \mathsf{conj}(\overrightarrow{x}, \overrightarrow{y}, \overrightarrow{c})$ in $\mathcal{O}$, a tuple $\overrightarrow{t}$ of constants is an answer of Q in $\mathcal{O}$ if there exists a tuple $\overrightarrow{s}$ of ground terms such that every ground atom occurring in $q[\overrightarrow{x} \mapsto \overrightarrow{t}, \overrightarrow{y} \mapsto \overrightarrow{s}]$ is satisfied by all models of $\Xi'(\mathcal{O}^\dagger)$.*

Based on the above two lemmas, we can identify a set of candidate answers of a CQ $Q : q(\overrightarrow{x}) \leftarrow \mathsf{conj}(\overrightarrow{x}, \overrightarrow{y}, \overrightarrow{c})$ in $\mathcal{O}$. Afterwards, we may, for every candidate answer $\overrightarrow{t}$, compute a subset of axioms in $\mathcal{O}^\dagger$ to check if $\mathcal{O} \models q[\overrightarrow{x} \mapsto \overrightarrow{t}]$ by applying the method given in Theorem 2. However, handling candidate answers one by one is inefficient because it needs to extract, for each candidate answer $\overrightarrow{t}$, a relevant subset of $\Pi \cup \mathrm{Inst}_{\mathrm{CQ}}(\Pi, Q, \mathcal{S}_{df}, \mathcal{S}) \cup \{\mathsf{map}(\neg w_Q(\overrightarrow{t}), \mathcal{S}_{df}, \mathcal{S})\}$; such computation is quite costly. To improve the efficiency, we extract all relevant subsets from $\Pi \cup \mathrm{Inst}_{\mathrm{CQ}}(\Pi, Q, \mathcal{S}_{df}, \mathcal{S}) \cup \{\mathsf{map}(\neg w_Q(\overrightarrow{t}), \mathcal{S}_{df}, \mathcal{S}) \mid \overrightarrow{t}$ is a candidate answer$\}$ in one pass, then from each extracted subset, identify a subset $\mathcal{O}_{rel}$ of axioms in $\mathcal{O}^\dagger$ and evaluate Q over $\mathcal{O}_{rel}$ by applying an OWL DL reasoner.

The algorithm for query answering is given in Figure 2, where the input $\mathcal{A}$ can be the set of ground atoms occurring in definite ground facts in $\Xi(\mathcal{O}^\dagger)$. We explain how the algorithm works. Lines 1–2 respectively compute a set Ans of explicit answers and a set $Cands$ of candidate answers of Q in $\mathcal{O}$ based on Lemma 5 and Lemma 4. Line 3 decomposes $\Pi^\dagger = \Pi \cup \mathrm{Inst}_{\mathrm{CQ}}(\Pi, Q, \mathcal{S}_{df}, \mathcal{S}) \cup \{\mathsf{map}(\neg w_Q(\overrightarrow{t}), \mathcal{S}_{df}, \mathcal{S}) \mid \overrightarrow{t} \in Cands\}$ to a set of disjoint subsets from which target ontologies can be extracted. The subprocedure Decompose($\Pi^\dagger$) first filters the largest subset Π_0 of $\Pi^\dagger$ such that for all clauses $cl \in \Pi_0$, cl^+ has at least one ground atom not occurring in $\Pi^\dagger \setminus \Pi_0$, then returns the set of maximal connected components of $\Pi^\dagger \setminus \Pi_0$. Basically, Π_0 is the greatest fixpoint of $\Pi_0^{(n)}$ such that $\Pi_0^{(0)} = \Pi^\dagger$ and for $n > 0$, $\Pi_0^{(n)} = \{cl \in \Pi_0^{(n-1)} \mid cl^+ \setminus \mathsf{atoms}(\Pi^\dagger \setminus \Pi_0^{(n-1)}) \neq \emptyset\}$ (see our technical report[2] for how to realize the decomposition process). Lines 4–6 handle every maximal connected component Π_{rel} of $\Pi^\dagger \setminus \Pi_0$: if any ground atom over w_Q does not occur in Π_{rel}, Π_{rel} is irrelevant to Q and thus neglected; otherwise, a subset $\mathcal{O}_{rel}$ of $\mathcal{O}^\dagger$ is extracted from Π_{rel}, and all answers of Q in $\mathcal{O}_{rel}$ are added to Ans (the evaluation of $\mathcal{O}_{rel}$ of $\mathcal{O}^\dagger$ is realized in TraditionalEvaluate, which applies an OWL DL reasoner). Note that different extracted ontologies have no common ABox axioms because ABox axioms correspond to ground atoms in $\Pi^\dagger \setminus \Pi_0$ and different maximal connected components of $\Pi^\dagger \setminus \Pi_0$ have no common ground atoms.

Example 3 (Example 2 continued). Given a CQ $Q : q(x) \leftarrow \mathsf{Man}(x) \wedge \mathsf{hasFather}(x, y)$ in $\mathcal{O}$ given in Example 1, we show how DecompBasedEvaluate($Q, \mathcal{A}, \Pi, \mathcal{S}_{df}, \mathcal{S}$) works, where $\mathcal{A}$ is the set of ground atoms occurring in definite ground facts in $\Xi(\mathcal{O}^\dagger)$, i.e., $\mathcal{A} = \{\mathsf{Man}(a_1), \mathsf{hasFather}(a_1, a_2)\}$, and $\Pi, \mathcal{S}_{df}, \mathcal{S}$ are given in Example 2. Line 1 in Figure 2 sets Ans as $\{a_1\}$. Line 2 sets $Cands$ as $\{a_2\}$. Line 3 computes $\Pi^\dagger = \Pi \cup \mathrm{Inst}_{\mathrm{CQ}}(\Pi, Q, \mathcal{S}_{df}, \mathcal{S}) \cup \{\mathsf{map}(\neg w_Q(\overrightarrow{t}), \mathcal{S}_{df}, \mathcal{S}) \mid \overrightarrow{t} \in Cands\} = \{cl'_1, \ldots, cl'_{33}\}$ and calls Decompose($\Pi^\dagger$), where $cl'_1, \ldots, cl'_{30}$ are given in Example 2, cl'_{31} is

Algorithm 1. DecompBasedEvaluate$(Q, \mathcal{A}, \Pi, \mathcal{S}_{df}, \mathcal{S})$

In: A CQ $Q : q(\overrightarrow{x}) \leftarrow p_1(\overrightarrow{x_1}, \overrightarrow{y_1}, \overrightarrow{c_1}) \wedge \ldots \wedge p_n(\overrightarrow{x_n}, \overrightarrow{y_n}, \overrightarrow{c_n})$, a set $\mathcal{A}$ of ground atoms satisfied by all models of $\Xi'(\mathcal{O}^\dagger)$ and the results $\Pi, \mathcal{S}_{df}, \mathcal{S}$ returned by ApproxGround$(\mathcal{O}^\dagger)$.
Out: The set of answers of Q in $\mathcal{O}$.
1. $Ans := \{\overrightarrow{x}\sigma \mid \sigma$ is a ground substitution such that $\{p_1(\overrightarrow{x_1}, \overrightarrow{y_1}, \overrightarrow{c_1}), \ldots, p_n(\overrightarrow{x_n}, \overrightarrow{y_n}, \overrightarrow{c_n})\}\sigma \subseteq \mathcal{A}\}$ and $\overrightarrow{x}\sigma$ is a tuple of constants$\}$;
2. $Cands := \{\overrightarrow{x}\sigma \mid \sigma$ is a ground substitution such that $\mathsf{map}(w_Q(\overrightarrow{x}\sigma), \mathcal{S}_{df}, \mathcal{S})$ occurs in $\mathrm{Inst}_{\mathrm{CQ}}(\Pi, Q, \mathcal{S}_{df}, \mathcal{S})$ and $\overrightarrow{x}\sigma$ is a tuple of constants$\} \setminus Ans$;
3. $RSets :=$ Decompose$(\Pi \cup \mathrm{Inst}_{\mathrm{CQ}}(\Pi, Q, \mathcal{S}_{df}, \mathcal{S}) \cup \{\mathsf{map}(\neg w_Q(\overrightarrow{t}), \mathcal{S}_{df}, \mathcal{S}) \mid \overrightarrow{t} \in Cands\})$;
4. **for** each $\Pi_{rel} \in RSets$ such that some ground atoms over w_Q occur in Π_{rel} **do**
5. $\quad \mathcal{O}_{rel} := \{ax \in \mathcal{O}^\dagger \mid$ there exists a clause $cl \in \Xi(ax)$ and a ground substitution σ such that $\mathsf{map}(cl\,\sigma, \mathcal{S}_{df}, \mathcal{S}) \in \Pi_{rel}\}$;
6. $\quad Ans := Ans \cup$ TraditionalEvaluate$(Q, \mathcal{O}_{rel})$;
7. **return** Ans;

Fig. 2. A decomposition-based algorithm for evaluating a CQ

$\neg\mathsf{Man}(a_1) \vee \neg\mathsf{hasFather}(a_1, a_2) \vee w_Q(a_1)$, cl'_{32} is $\neg\mathsf{Man}(a_2) \vee \neg\mathsf{hasFather}(a_2, f(a_2)) \vee w_Q(a_2)$ and cl'_{33} is $\neg w_Q(a_2)$. It can be checked that the filtered set Π_0 of $\Pi^\dagger$ is $\{cl'_{12}, cl'_{13}, cl'_{14}, cl'_{29}, cl'_{30}, cl'_{31}\}$ and the remaining set has a single maximal connected component. Hence Decompose$(\Pi^\dagger)$ returns $\{\{\Pi_{rel}\}\}$, where $\Pi_{rel} = \Pi^\dagger \setminus \Pi_0$. Since $w_Q(a_2)$ occurs in Π_{rel}, line 5 in Figure 2 is executed, yielding $\mathcal{O}_{rel} = \{\mathsf{Man} \sqsubseteq \leq_1 \mathsf{hasFather}, \mathsf{Man} \sqsubseteq \exists\mathsf{hasFather}.\mathsf{Man}, \mathsf{Man}(a_1), \mathsf{hasFather}(a_1, a_2)\}$. Note that $\mathcal{O}_{rel}$ is a subset of $\mathcal{O}^+$ from which the axiom $\mathsf{Man} \sqsubseteq \mathsf{Human}$ is removed. By applying an OWL DL reasoner, we can check that a_2 is the unique answer of Q in $\mathcal{O}_{rel}$, so Ans is updated to $\{a_1, a_2\}$ in line 6 and finally returned by DecompBasedEvaluate$(Q, \mathcal{A}, \Pi, \mathcal{S}_{df}, \mathcal{S})$.

In the remainder of this section, we assume that the OWL DL reasoner applied in our approach is sound and complete for the category of given CQs, i.e., the subprocedure TraditionalEvaluate is correct. The following theorem shows the correctness of DecompBasedEvaluate.

Theorem 3. *Let $Q : q(\overrightarrow{x}) \leftarrow \mathsf{conj}(\overrightarrow{x}, \overrightarrow{y}, \overrightarrow{c})$ be a CQ, $\mathcal{A}$ be a set of ground atoms satisfied by all models of $\Xi'(\mathcal{O}^\dagger)$, $(\Pi, \mathcal{S}_{df}, \mathcal{S})$ be returned by* ApproxGround$(\mathcal{O}^\dagger)$. *Then* DecompBasedEvaluate$(Q, \mathcal{A}, \Pi, \mathcal{S}_{df}, \mathcal{S})$ *returns the set of answers of Q in $\mathcal{O}$.*

3.5 Optimization by Computing More ABox Entailments

Based on Figure 2 (line 1), we can see that if we obtain more definite ground facts of $\Xi(\mathcal{O}^\dagger)$, we can compute more explicit answers of a given CQ; thus, we can further improve the performance of our approach. We therefore present an important optimization that computes more entailments of $\mathcal{O}^\dagger$ before calling ApproxGrounding$(\mathcal{O}^\dagger)$.

Basically, the optimization computes a set $\mathcal{A}$ of ground atoms from the set of definite clauses in $\Xi(\Theta(\mathcal{O}))$ such that the functional depth of every ground term occurring in $\mathcal{A}$ is at most one. Recall that $\Theta(\mathcal{O})$ is the result of applying structural transformation to $\mathcal{O}$. We call such subset $\mathcal{A}$ the *bounded entailment set* of $\Theta(\mathcal{O})$, which is defined as the least fixpoint of $\mathcal{A}^{(n)}$ such that $\mathcal{A}^{(0)} = \emptyset$ and for $n > 0$, $\mathcal{A}^{(n)} = \bigcup\{cl^+\sigma \mid cl \in \mathrm{DS}(\Xi'(\Theta(\mathcal{O}))), \sigma$ is a ground substitution such that

$cl^{-}\sigma \subseteq \mathcal{A}^{(n-1)}$, $cl^{+}\sigma \subseteq \mathrm{HB}(\Xi(\Theta(\mathcal{O})))$ and $\mathsf{depth}(cl\,\sigma) \leq 1\}$, where $\mathsf{depth}(cl)$ denotes the maximum functional depth of all ground terms occurring in a ground clause cl, and $\mathrm{DS}(\Xi'(\Theta(\mathcal{O})))$ denotes the set of all definite clauses in $\Xi'(\Theta(\mathcal{O}))$.

Let $\mathcal{A}$ be the bounded entailment set of $\Theta(\mathcal{O})$. Since $\mathcal{A}$ is a set of ground atoms in the least model of $\mathrm{DS}(\Xi'(\Theta(\mathcal{O}))$ and the least model of $\mathrm{DS}(\Xi'(\Theta(\mathcal{O})))$ is a subset of every model of $\Xi'(\Theta(\mathcal{O}))$, every ground atom in $\mathcal{A}$ is satisfied by all models of $\Xi'(\Theta(\mathcal{O}))$. Let $\mathcal{A}_C$ be the subset of $\mathcal{A}$ that contains only constants. When $\Theta(\mathcal{O}) \cup \mathcal{A}_C$ has equality assertions, some equational atoms must occur positively in $\Xi(\Theta(\mathcal{O}))$, so $\mathcal{E}(\Xi(\Theta(\mathcal{O}))) \subseteq \mathrm{DS}(\Xi'(\Theta(\mathcal{O}))$. It follows that $\Theta(\mathcal{O}) \cup \mathcal{A}_C$ is congruence-complete. In the remainder of this section, we assume that $\mathcal{O}^{\dagger} = \Theta(\mathcal{O}) \cup \mathcal{A}_C$. Since every ground atom in $\mathcal{A}_C$ is satisfied by all models of $\Theta(\mathcal{O})$, Lemma 1 and whatever follows from Lemma 1 still hold when $\mathcal{O}^{\dagger} = \Theta(\mathcal{O}) \cup \mathcal{A}_C$.

To evaluate a CQ Q in $\mathcal{O}$, we now call DecompBasedEvaluate(Q, $\mathcal{A}$, Π, $\mathcal{S}_{df}$, $\mathcal{S}$), where $\mathcal{A}$ is the bounded entailment set of $\Theta(\mathcal{O})$ and $(\Pi, \mathcal{S}_{df}, \mathcal{S})$ are returned by ApproxGround($\mathcal{O}^{\dagger}$). Theorem 4 shows that the optimization also guarantees sound and complete results. Example 4 illustrates that the optimization does take effect in our running example.

Theorem 4. *Let $Q : q(\overrightarrow{x}) \leftarrow \mathsf{conj}(\overrightarrow{x}, \overrightarrow{y}, \overrightarrow{c})$ be a CQ, $\mathcal{A}$ be the bounded entailment set of $\Theta(\mathcal{O})$, $(\Pi, \mathcal{S}_{df}, \mathcal{S})$ be returned by* ApproxGround($\mathcal{O}^{\dagger}$) *and Ans be returned by* DecompBasedEvaluate($Q, \mathcal{A}, \Pi, \mathcal{S}_{df}, \mathcal{S}$). *Then Ans is the set of answers of Q in $\mathcal{O}$.*

Example 4 (Example 3 continued). For the ontology $\mathcal{O}$ given in Example 1, since $\Xi'(\Theta(\mathcal{O})) = \{cl_1, ..., cl_{13}\} \cup \{t \approx t \mid t \in \mathrm{HU}(\Xi(\mathcal{O}^{\dagger}))\}$, where $cl_1, ..., cl_{13}$ are given in Example 1, we have $\mathrm{DS}(\Xi'(\Theta(\mathcal{O}))) = \Xi'(\Theta(\mathcal{O}))$. We can compute the bounded entailment set $\mathcal{A}$ of $\Theta(\mathcal{O})$ as $\{\mathsf{Man}(a_1), \mathsf{Man}(f(a_1)), \mathsf{Man}(a_2), \mathsf{Man}(f(a_2)), \mathsf{Human}(a_1), \mathsf{Human}(f(a_1)), \mathsf{Human}(a_2), \mathsf{Human}(f(a_2)), \mathsf{hasFather}(a_1, a_2), \mathsf{hasFather}(a_1, f(a_1)), \mathsf{hasFather}(a_2, f(a_2)), \mathsf{hasFather}(f(a_1), f(a_2)), a_2 \approx f(a_1), f(a_1) \approx a_2, a_1 \approx a_1, f(a_1) \approx f(a_1), a_2 \approx a_2, f(a_2) \approx f(a_2)\}$. By appending to $\Theta(\mathcal{O})$ the set $\mathcal{A}_C$ of ground atoms in $\mathcal{A}$ that contains only constants, we obtain $\mathcal{O}^{\dagger} = \{\mathsf{Man} \sqsubseteq_{\leq 1} \mathsf{hasFather}, \mathsf{Man} \sqsubseteq \exists \mathsf{hasFather}.\mathsf{Man}, \mathsf{Man} \sqsubseteq \mathsf{Human}, \mathsf{Man}(a_1), \mathsf{Man}(a_2), \mathsf{Human}(a_1), \mathsf{Human}(a_2), \mathsf{hasFather}(a_1, a_2), a_1 \approx a_1, a_2 \approx a_2\}$. ApproxGround($\mathcal{O}^{\dagger}$) returns $(\Pi, \mathcal{S}_{df}, \mathcal{S})$, where $\Pi = \{cl'_1, \ldots, cl'_{11}, cl'_{13}, \ldots, cl'_{30}\}$ (see $cl'_1, ..., cl'_{30}$ in Example 2), $\mathcal{S}_{df} = \emptyset$ and $\mathcal{S} = \{\{a_2, f(a_1)\}\}$. Consider again the CQ $Q : q(x) \leftarrow \mathsf{Man}(x) \wedge \mathsf{hasFather}(x, y)$ given in Example 3. By calling DecompBasedEvaluate(Q, $\mathcal{A}$, Π, $\mathcal{S}_{df}$, $\mathcal{S}$), we get $Ans = \{a_1, a_2\}$ in line 1 and $Cands = \emptyset$ in line 2, thus we obtain the set of answers of Q in $\mathcal{O}$ without calling OWL DL reasoners.

To summarize, our decomposition-based approach to conjunctive query answering works as follows. In the offline phase, we compute the bounded entailment set $\mathcal{A}$ of $\Theta(\mathcal{O})$ and set $\mathcal{O}^{\dagger}$ as $\Theta(\mathcal{O}) \cup \{ax \in \mathcal{A} \mid ax$ contains only constants$\}$, then call ApproxGround($\mathcal{O}^{\dagger}$), obtaining $(\Pi, \mathcal{S}_{df}, \mathcal{S})$. In the online phase, for every coming CQ Q, we call DecompBasedEvaluate(Q, $\mathcal{A}$, Π, $\mathcal{S}_{df}$, $\mathcal{S}$), obtaining all answers of Q.

4 Experimental Evaluation

We implemented the proposed approach in GNU C++, using MySQL as the back-end SQL engine. The implemented system[3] is called DecomBaR (abbr. for **Decom**position-**Ba**sed **R**easoner). It maintains ABox axioms and ground atoms in the approximate grounding in databases, maintains ground clauses in the approximate grounding in disk files, and calls Pellet [14] (version 2.0.0-rc6)[4] to evaluate CQs over extracted ontologies. All our experiments were conducted on a machine with 2GHz Intel Pentium Dual CPU and 2GB RAM, running Windows XP, where the maximum Java heap size was set to (max) 1312MB.

4.1 Experimental Setup

We conducted experiments on Lehigh University Benchmark (LUBM) [8] and University Ontology Benchmark (UOBM) [10] (including UOBM-Lite and UOBM-DL) ontologies. We used the above benchmark ontologies because they all come with benchmark CQs, which can provide a reasonable assessment on the effectiveness of our proposed approach. By LUBMn, UOBM-Lite$n$ and UOBM-DLn we respectively denote the instances of LUBM, UOBM-Lite and UOBM-DL that contain axioms about n universities. We specifically tested on LUBM1, LUBM10, UOBM-Lite1, UOBM-Lite10, UOBM-DL1 and UOBM-DL10, where the former two were generated by the LUBM data generator[5], and the latter four were all downloaded from the UOBM Website[6].

Before testing our approach we stored ABoxes to MySQL databases. Table 1 lists the characteristics of the six test ontologies.

Table 1. The characteristics of test ontologies and the execution time in the offline phase

$\mathcal{O}$	$\|N_C\|$	$\|N_R\|$	$\|N_I\|$	$\|\mathcal{T}\|$	$\|\mathcal{A}\|$	$\|\mathcal{T}^\dagger\|$	$\|\mathcal{A}^\dagger\|$	Offline(sec)
LUBM1	59	16	50,253	94	100,543	105	100,543	17
LUBM10	59	16	629,568	94	1,272,575	105	1,272,575	219
UOBM-Lite1	51	43	95,010	130	245,864	151	245,864	101
UOBM-Lite10	51	43	820,208	130	2,096,973	151	2,096,973	1193
UOBM-DL1	112	44	96,081	151	260,540	210	260,587	242
UOBM-DL10	112	44	825,455	151	2,217,302	210	2,217,349	7103

Note: $\mathcal{O} = (\mathcal{T}, \mathcal{A})$ is a test ontology and $\Theta(\mathcal{O}) = (\mathcal{T}^\dagger, \mathcal{A}^\dagger)$. N_C, N_R and N_I are respectively the sets of concept names, role names and individuals in $\mathcal{O}$.

4.2 Experimental Results

We compared DecomBaR with the original Pellet reasoner (simply Pellet) and the KAON2 reasoner (simply KAON2) on evaluating benchmark CQs given in [8,10]. We

[3] `http://www.aifb.uni-karlsruhe.de/WBS/gqi/DecomBaR/`
[4] `http://clarkparsia.com/pellet/`
[5] `http://swat.cse.lehigh.edu/projects/lubm/index.htm`
[6] `http://www.alphaworks.ibm.com/tech/semanticstk/`

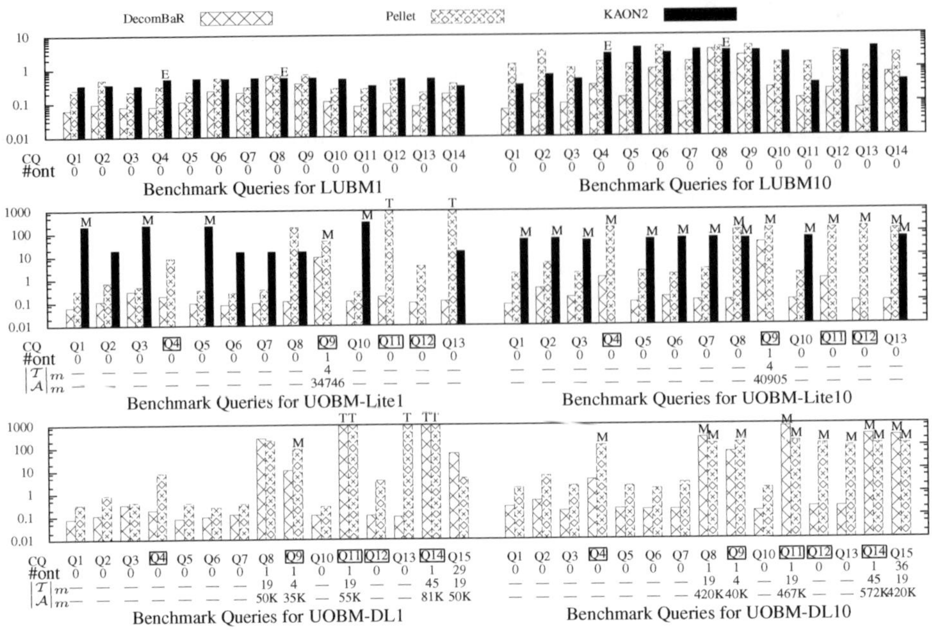

Fig. 3. The execution time (in seconds) for evaluating all benchmark CQs

implemented an interface to allow Pellet or KAON2 to read ABoxes from databases. We did not test KAON2 on UOBM-DLn ontologies nor CQs with non-distinguished variables as they are not supported by KAON2.

The execution time (in seconds) in the offline phase of DecomBaR is shown in the last column of Table 1. The results for evaluating every benchmark CQ are shown in Figure 3. The execution time about DecomBaR is the total evaluation time in the online phase, including the time for decomposing the propositional program compiled in the offline phase and the time for loading extracted ontologies to the called reasoner, waiting the called reasoner to return and handling the returned results. The execution time about Pellet or KAON2 is the time for query answering only, excluding the time for ontology loading and consistency checking (as we assume that the ontology is loaded and checked consistency offline) and the time for writing results.

Below the horizontal axis in Figure 3, "#ont" denotes the number of extracted ontologies over which Pellet is called to evaluate a test query, and "$|\mathcal{T}|_{max}$" (resp. "$|\mathcal{A}|_{max}$") denotes the maximum number of axioms in the TBox (resp. the ABox) of every extracted ontology. The name of a CQ is framed iff the CQ has non-distinguished variables. Above a bar, "M" means running out of memory after the displayed time, "T" means exceeding the time limit of 1000s, and "E" means that the set of computed answers is incorrect; we call any of these cases an unsuccessful evaluation. For every benchmark CQ that both DecomBaR and Pellet (resp. KAON2) successfully evaluate, the answers computed by DecomBaR and Pellet (resp. KAON2) coincide.

Comparing DecomBaR with Pellet, DecomBaR is more efficient than Pellet except for Q8 and Q15 on UOBM-DL1.[7] Such exception is due to that sometimes decomposition and interaction with the called reasoner introduce a significant overhead in the execution of DecomBaR. However, when DecomBaR does not generate any candidate answer (i.e. when #ont = 0), DecomBaR works very fast because it only needs to extract explicit and candidate answers by accessing the database through a SQL query. For example, DecomBaR spends about 0.1s for evaluating Q8 on both UOBM-Lite1 and UOBM-Lite10, while for evaluating the same CQ Pellet spends about 180s on UOBM-Lite1 and runs out of memory on UOBM-Lite10. Moreover, DecomBaR is more scalable than Pellet against increasing size of ABoxes. This is because accessing databases through SQL queries is relatively scalable (in case #ont = 0) and extracted ontologies could have similar sizes for different size of ABoxes (in case #ont > 0). For example, the UOBM-Lite benchmark query Q9 has an individual in the query body, which forces $\mathrm{Inst}_{\mathrm{CQ}}$ (defined in Subsection 3.4) to return similar ground clauses and then forces the extracted ontologies to have similar sizes for UOBM-Lite1 and UOBM-Lite10.

Comparing DecomBaR with KAON2, DecomBaR is generally more efficient (esp. for UOBM-Lite$n$ ontologies, by orders of magnitude more efficient) than KAON2. Moreover, the scalability of DecomBaR is comparable with that of KAON2 on LUBMn ontologies, and is much better than that of KAON2 on UOBM-Lite$n$ ontologies. This shows that DecomBaR is much more scalable than KAON2 against increasing complexity of TBoxes. It should also be mentioned that DecomBaR supports more expressive CQs than KAON2 does. In particular, KAON2 may not correctly evaluate CQs involving datatypes (e.g. the LUBM benchmark queries Q4 and Q8); this is a limitation of the resolution-based query mechanism [11] exploited in KAON2.

5 Conclusion and Future Work

In this paper, we have proposed a decomposition-based approach to optimize conjunctive query answering in OWL DL ontologies. The basic idea of the approach is to evaluate a CQ with the help of a precompiled propositional program: it computes explicit answers first and then computes other answers over separate ontologies that are extracted from the precompiled propositional program. Experimental results demonstrate the advantages of the proposed approach.

The proposed approach still has some limitations. First, it only works well on ontologies that rarely change as the offline phase is somewhat costly. We plan to upgrade the compilation method to an incremental one to copy with ontology changes. Second, the approach fails when some extracted ontologies are still too large to be handled by the called reasoner. This is the reason why our implemented system DecomBaR does not successfully evaluate six benchmark CQs in our experiments (see Figure 3). We will tackle this limitation by exploiting the idea of summarization [2,3].

[7] Here we do not compare DecomBaR and Pellet for those CQs that both DecomBaR and Pellet do not successfully evaluate.

Acknowledgments. Thanks all anonymous reviewers for their useful comments. Jianfeng Du is supported in part by NSFC grants 60673103 and 70801020. Yi-Dong Shen is supported in part by NSFC grants 60673103, 60721061, and by the 863 Program.

References

1. Calvanese, D., De Giacomo, G., Lembo, D., Lenzerini, M., Rosati, R.: DL – Lite: Tractable description logics for ontologies. In: Proc. of AAAI 2005, pp. 602–607 (2005)
2. Dolby, J., Fokoue, A., Kalyanpur, A., Kershenbaum, A., Schonberg, E., Srinivas, K., Ma, L.: Scalable semantic retrieval through summarization and refinement. In: Proc. of AAAI 2007, pp. 299–304 (2007)
3. Dolby, J., Fokoue, A., Kalyanpur, A., Ma, L., Schonberg, E., Srinivas, K., Sun, X.: Scalable grounded conjunctive query evaluation over large and expressive knowledge bases. In: Sheth, A.P., Staab, S., Dean, M., Paolucci, M., Maynard, D., Finin, T., Thirunarayan, K. (eds.) ISWC 2008. LNCS, vol. 5318, pp. 403–418. Springer, Heidelberg (2008)
4. Eiter, T., Leone, N., Mateis, C., Pfeifer, G., Scarcello, F.: A deductive system for non-monotonic reasoning. In: Proc. of LPNMR 1997, pp. 364–375 (1997)
5. Fitting, M.: First-order Logic and Automated Theorem Proving, 2nd edn. Springer, New York (1996)
6. Cuenca Grau, B., Parsia, B., Sirin, E., Kalyanpur, A.: Modularity and web ontologies. In: Proc. of KR 2006, pp. 198–209 (2006)
7. Guo, Y., Heflin, J.: A scalable approach for partitioning OWL knowledge bases. In: Proc. of SSWS 2006, pp. 47–60 (2006)
8. Guo, Y., Pan, Z., Heflin, J.: LUBM: A benchmark for OWL knowledge base systems. Journal of Web Semantics 3(2–3), 158–182 (2005)
9. Kazakov, Y., Motik, B.: A resolution-based decision procedure for $\mathcal{SHOIQ}$. Journal of Automated Reasoning 40(2-3), 89–116 (2008)
10. Ma, L., Yang, Y., Qiu, Z., Xie, G., Pan, Y., Liu, S.: Towards a complete OWL ontology benchmark. In: Sure, Y., Domingue, J. (eds.) ESWC 2006. LNCS, vol. 4011, pp. 125–139. Springer, Heidelberg (2006)
11. Motik, B.: Reasoning in Description Logics using Resolution and Deductive Databases. PhD thesis, Univesität karlsruhe, Germany (January 2006)
12. Motik, B., Sattler, U., Studer, R.: Query answering for OWL-DL with rules. Journal of Web Semantics 3(1), 41–60 (2005)
13. Pan, J.Z., Thomas, E.: Approximating OWL-DL ontologies. In: Proc. of AAAI 2007, pp. 1434–1439 (2007)
14. Sirin, E., Parsia, B., Cuenca Grau, B., Kalyanpur, A., Katz, Y.: Pellet: A practical OWL-DL reasoner. Journal of Web Semantics 5(2), 51–53 (2007)
15. Stuckenschmidt, H., Klein, M.: Structure-based partitioning of large concept hierarchies. In: McIlraith, S.A., Plexousakis, D., van Harmelen, F. (eds.) ISWC 2004. LNCS, vol. 3298, pp. 289–303. Springer, Heidelberg (2004)

Goal-Directed Module Extraction for Explaining OWL DL Entailments

Jianfeng Du[1,2], Guilin Qi[3,4], and Qiu Ji[3]

[1] Institute of Business Intelligence and Knowledge Discovery,
Guangdong University of Foreign Studies, Guangzhou 510006, China
[2] State Key Laboratory of Computer Science, Institute of Software,
Chinese Academy of Sciences, Beijing 100190, China
jfdu@ios.ac.cn
[3] AIFB, Universität Karlsruhe, D-76128 Karlsruhe, Germany
{gqi,qiji}@aifb.uni-karlsruhe.de
[4] School of Computer Science and Engineering, Southeast University, Nanjing, China

Abstract. Module extraction methods have proved to be effective in improving the performance of some ontology reasoning tasks, including finding justifications to explain why an entailment holds in an OWL DL ontology. However, the existing module extraction methods that compute a *syntactic locality-based module* for the sub-concept in a subsumption entailment, though ensuring the resulting module to preserve all justifications of the entailment, may be insufficient in improving the performance of finding all justifications. This is because a syntactic locality-based module is independent of the super-concept in a subsumption entailment and always contains all concept/role assertions. In order to extract smaller modules to further optimize finding all justifications in an OWL DL ontology, we propose a goal-directed method for extracting a module that preserves all justifications of a given entailment. Experimental results on large ontologies show that a module extracted by our method is smaller than the corresponding syntactic locality-based module, making the subsequent computation of all justifications more scalable and more efficient.

1 Introduction

As the Semantic Web (SW) is evolving, ontologies become more and more important because they provide formal representation of knowledge shared within the SW. To enable users to understand the knowledge in a consistent way, ontologies need to be logically constructed and have well-defined semantics. To this end, the W3C organization proposed the Web Ontology Language (OWL)[1] for representing ontologies, which is based on Description Logics (DLs) [1]. With well-defined semantics from DLs, some important ontology reasoning tasks emerged, including classical reasoning tasks like subsumption checking and instance retrieval [1], and non-standard reasoning tasks like finding *justifications* to explain why an entailment holds in a given ontology [12,17], where a justification of an entailment is a minimal subset of axioms in the given ontology that are responsible for the entailment.

[1] http://www.w3.org/TR/owl-semantics/

A. Bernstein et al. (Eds.): ISWC 2009, LNCS 5823, pp. 163–179, 2009.

Reasoning tasks in OWL ontologies are in general computationally hard, thus rely on optimization techniques. To improve the performance of ontology reasoning, module extraction methods are often employed and their effectiveness has been adequately verified. A module extraction method reduces a given ontology to a specific module depending on specific tasks. These tasks include ontology reuse [16,4], subsumption checking [21], incremental classification [7] and finding of justifications [2,22].

In this paper, we focus on the problem of extracting small modules to improve the performance of an important reasoning task, namely finding all justifications of a given entailment. The idea of using modules to optimize finding justifications is not new. It has been shown in [2] and [22] that the syntactic locality-based module introduced in [8] can be used to improve the performance of finding all justifications. As proved in [22], a syntactic locality-based module is a module that preserves all subsumption entailments $A \sqsubseteq B$ w.r.t. a concept name A and is independent of the super-concept B. It follows that the extraction of a syntactic locality-based module is not goal-directed; it may yield an unexpectedly large module as illustrated below.

Example 1. Let $\mathcal{O}$ be an ontology consisting of the following axioms.
ax_1: ChiefActress $\sqsubseteq$ Person ax_2: ChiefActress $\sqsubseteq$ Actress ax_3: Actress $\sqsubseteq$ Woman
ax_4: Person $\sqsubseteq$ Man $\sqcup$ Woman ax_5: ChiefActress $\sqsubseteq \neg$Man

Consider the above example. There are two justifications $\{ax_1, ax_4, ax_5\}$ and $\{ax_2, ax_3\}$ of the subsumption entailment ChiefActress $\sqsubseteq$ Woman in $\mathcal{O}$. The syntactic locality-based module w.r.t. ChiefActress in $\mathcal{O}$ is exactly $\mathcal{O}$. This module cannot be used to optimize finding all justifications of a subsumption entailment of the form ChiefActress $\sqsubseteq B$ no matter what the concept name B is. On the other hand, a syntactic locality-based module must contain all concept/role assertions because these assertions can never be *local* [8]. This means that extracting a syntactic locality-based module can hardly optimize finding justifications of a membership entailment $A(a)$ or $R(a, b)$, as all concept/role assertions in the given ontology must be involved.

In order to extract modules smaller than the syntactic locality-based ones to further optimize finding justifications, we propose a goal-directed method for module extraction in $\mathcal{SHOIQ}$ ontologies (see Section 4). The DL $\mathcal{SHOIQ}$ corresponds to OWL DL allowing qualifying number restrictions. Since the problem of finding justifications of a subsumption entailment can be reduced to the problem of finding justifications of a membership entailment by introducing new concept assertions, our method focuses on finding a module that preserves all justifications of a given membership entailment. The basic idea is to start from a given membership entailment and backward traverse a set of axioms that might be responsible for the given entailment. To ensure the traversed set to be an intended module, our method compiles a propositional program from the given ontology in which each clause may have a flag corresponding to an axiom in the given ontology, such that the traversal of clauses corresponds to the traversal of axioms, i.e., the set of flags appearing in the traversed clauses corresponds to an intended module.

We implement the proposed method and test on large ontologies (see Section 5). Experimental results show that our method can extract modules that are much smaller than syntactic locality-based modules. They also show that, with our proposed module extraction method, it is possible to efficiently find all justifications of a membership entailment in a large ontology with millions of ABox axioms.

2 Related Work

Module extraction methods form an important category in ontology modularization. Some methods rely on syntactically traversing axioms in the ontology and exploiting different heuristics to determine which axioms belong to the extracted module, such as those given in [16], [18] and [4]. Other methods formally specify the intended outputs by considering the semantics of the ontology language, such as the method proposed by Cuenca Grau et al. [8], which computes a so-called locality-based module that preserves all subsumption entailments $A \sqsubseteq B$ for a given concept name A. There are two kinds of locality-based modules, namely semantic ones and syntactic ones, both contain all concept/role assertions [8]. The method proposed by Suntisrivaraporn [21] computes so-called reachability-based modules in $\mathcal{EL}^+$ ontologies, which coincide with syntactic locality-based modules. The above methods do not concern extracting a suitable module to optimize finding all justifications.

Another important category in ontology modularization is formed by ontology partitioning methods, each of which divides a given ontology into a set of modules. For example, Stuckenschmidt and Klein [20] propose a method to partition all concepts in an ontology to facilitate visualization of the ontology; Cuenca Grau *et al.* [9] propose a method for dividing an ontology into a set of modules such that all inferences about the signature contained in a module can be made locally. These methods are less relevant to ours because they yield a set of modules other than a single one.

3 Preliminaries

$\mathcal{SHOIQ}$ Syntax and Semantics. A $\mathcal{SHOIQ}$ ontology $\mathcal{O}$ consists of a *terminological box (TBox)* $\mathcal{O}_\mathcal{T}$ and an *assertional box (ABox)* $\mathcal{O}_\mathcal{A}$. The TBox $\mathcal{O}_\mathcal{T}$ consists of a finite set of *concept inclusion axioms* $C \sqsubseteq D$, *transitivity axioms* $\mathrm{Trans}(R)$ and *role inclusion axioms* $R \sqsubseteq S$, where C and D are $\mathcal{SHOIQ}$ concepts, and R and S roles. A role is either a role name or an *inverse role* R^- for some role name R; it is called *simple* if it has not any transitive subrole. The set of $\mathcal{SHOIQ}$ concepts is the smallest set such that each concept name A (also called *atomic concept*) is a $\mathcal{SHOIQ}$ concept and, for C and D $\mathcal{SHOIQ}$ concepts, R a role, S a simple role, a an individual, and n a nonnegative integer, $\top$, $\bot$, $\{a\}$, $\neg C$, $C \sqcap D$, $C \sqcup D$, $\exists R.C$, $\forall R.C$, $\leq_n S.C$ and $\geq_n S.C$ are also $\mathcal{SHOIQ}$ concepts. The ABox $\mathcal{O}_\mathcal{A}$ consists of a finite set of *concept assertions* $C(a)$, *role assertions* $R(a,b)$ or $\neg R(a,b)$, *equality assertions* $a \approx b$ and *inequality assertions* $a \not\approx b$, where C is a $\mathcal{SHOIQ}$ concept, R a role, and a and b individuals. In this paper, both $\mathcal{O}_\mathcal{T}$ and $\mathcal{O}_\mathcal{A}$ are treated as sets of axioms, so is $\mathcal{O} = \mathcal{O}_\mathcal{T} \cup \mathcal{O}_\mathcal{A}$.

As a fragment of first-order logic, $\mathcal{SHOIQ}$ inherits its semantics from first-order logic by the standard translations known e.g. from [14]. Let π denote the operator for mapping a $\mathcal{SHOIQ}$ ontology (resp. axiom) to a first-order logic program (resp. clause) as given in [14]. Then $\mathcal{O}$ is called *consistent* or *satisfiable* if $\pi(\mathcal{O})$ has a model. An atomic concept A is called *satisfiable* in $\mathcal{O}$ if there exists a model M of $\pi(\mathcal{O})$ such that $A(a) \in M$ for some constant a. $\mathcal{O}$ is called *coherent* if all atomic concepts in $\mathcal{O}$ are satisfiable. When $\mathcal{O}$ has atomic concepts, $\mathcal{O}$ being coherent implies $\mathcal{O}$ being consistent. An axiom ax is called an *entailment* of $\mathcal{O}$, denoted by $\mathcal{O} \models ax$, if $\pi(ax)$ is satisfied by all models of $\pi(\mathcal{O})$. An entailment of $\mathcal{O}$ is called a *subsumption entailment* (resp. a *membership entailment*) if it is of the form $C \sqsubseteq D$ (resp. $C(a)$ or $R(a, b)$).

First-Order Logic. We use the standard clausal form to represent a first-order logic program. *Terms* are variables, constants or functional terms of the form $f(t_1, \ldots, t_n)$, where f is a function symbol of arity n and t_1, ..., t_n are terms. Throughout this paper, we use (possibly with subscripts) x, y, z for variables, a, b, c for constants, and s, t for terms. We only consider unary function symbols because only unary function symbols occur in first-order logic programs that are translated from $\mathcal{SHOIQ}$ ontologies.

Atoms are of the form $T(t_1, \ldots, t_n)$ where T is a predicate symbol of arity n and $t_1, \ldots, t_n$ are terms. A *literal* is a positive or negative atom and a *clause* is a disjunction of literals. Terms, atoms and clauses that do not contain variables are called *ground*.

A *first-order logic program* is a set of clauses in which all variables are universally quantified. For a clause $cl = \neg A_1 \vee \ldots \vee \neg A_n \vee B_1 \vee \ldots \vee B_m$, the set of atoms $\{A_1, \ldots, A_n\}$ is denoted by cl^-, whereas the set of atoms $\{B_1, \ldots, B_m\}$ is denoted by cl^+. By $|S|$ we denote the cardinality of a set S. A clause cl is called a *fact* if $|cl^-| = 0$.

A *propositional program* Π is a first-order logic program consisting of only ground clauses. The set of ground atoms occurring in Π is denoted by $\mathsf{atoms}(\Pi)$.

For a first-order logic program P, the set of ground terms (resp. ground atoms) defined from the first-order signature of P is called the *Herbrand universe* (resp. *Herbrand base*) of P, denoted by $\mathrm{HU}(P)$ (resp. $\mathrm{HB}(P)$). The set of ground clauses obtained by replacing all variables occurring in each clause in P with ground terms from $\mathrm{HU}(P)$ is called the *primary grounding* of P, denoted by $\mathcal{G}(P)$. An *interpretation* M of P is a set of ground atoms in $\mathrm{HB}(P)$; it is a *model* of P if for any ground clause $cl \in \mathcal{G}(P)$ such that $cl^- \subseteq M$, $cl^+ \cap M \neq \emptyset$; it is a *minimal model* of P if there is no model M' of P such that $M' \subset M$. P is said to be *satisfiable* if P has a model.

The first-order logic program P translated from a $\mathcal{SHOIQ}$ ontology may contain the equality predicate $\approx$, which is interpreted as a *congruence relation* and different from ordinary predicates. This difference is not captured by the above first-order semantics. However, the equality predicate $\approx$ can be explicitly axiomatized via a well-known transformation from [6]. Let $\mathcal{E}(P)$ denote the first-order logic program consisting of the following clauses: (1) $t \approx t$, for each ground term $t \in \mathrm{HU}(P)$; (2) $\neg(x \approx y) \vee y \approx x$; (3) $\neg(x \approx y) \vee \neg(y \approx z) \vee x \approx z$; (4) $\neg(x \approx y) \vee f(x) \approx f(y)$, for each function symbol f occurring in P; (5) $\neg T(x_1, \ldots, x_i, \ldots, x_n) \vee \neg(x_i \approx y_i) \vee T(x_1, \ldots, y_i, \ldots, x_n)$, for each predicate symbol T other than $\approx$ occurring in P and each position i. Appending $\mathcal{E}(P)$ to P allows to treat $\approx$ as an ordinary predicate, i.e., M is a model of P that interprets $\approx$ as a congruence relation, iff for any ground clause $cl \in \mathcal{G}(P \cup \mathcal{E}(P))$ such that $cl^- \subseteq M$, $cl^+ \cap M \neq \emptyset$.

4 Goal-Directed Module Extraction

We in this paper address the following problem: Given a coherent $\mathcal{SHOIQ}$ ontology $\mathcal{O}$ which contains atomic concepts (thus $\mathcal{O}$ is consistent too) and an entailment ax of the form $A \sqsubseteq B$, $A(a)$ or $R(a, b)$ in $\mathcal{O}$, where A and B are concept names, R a role name, and a and b individuals, extract a *just-preserving module* of $\mathcal{O}$ for ax, which preserves all *justifications* of ax.

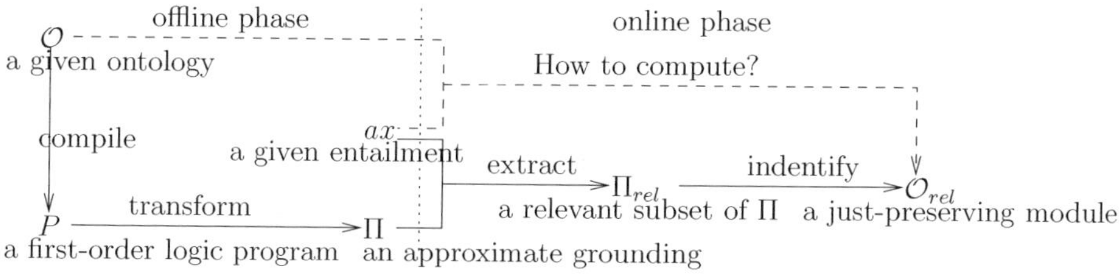

Fig. 1. Illustration of our proposed method

Definition 1 (Justification). For $\mathcal{O}$ an ontology and ax an axiom such that $\mathcal{O} \models ax$, a subset $\mathcal{J}$ of axioms in $\mathcal{O}$ is called a *justification* of ax in $\mathcal{O}$ if $\mathcal{J} \models ax$ and for all proper subsets $\mathcal{J}'$ of $\mathcal{J}$, $\mathcal{J}' \not\models ax$.

Definition 2 (Just-Preserving Module). For $\mathcal{O}$ an ontology and ax an axiom such that $\mathcal{O} \models ax$, a subset $\mathcal{M}$ of axioms in $\mathcal{O}$ is called a *just-preserving module* of $\mathcal{O}$ for ax if $\mathcal{J} \subseteq \mathcal{M}$ for all justifications $\mathcal{J}$ of ax in $\mathcal{O}$.

We present a goal-directed method for extracting just-preserving modules, depicted in Figure 1. The method consists of two phases. In the first phase, the method first compiles a given $\mathcal{SHOIQ}$ ontology $\mathcal{O}$ to a first-order logic program P (see Subsection 4.1), then transforms P to a propositional program Π. Both P and Π are independent of any given entailment, thus this phase can be performed offline. To show the correctness of the method in a clearer way, we present two algorithms for transforming P to Π, though only one of them is used in the method. One algorithm transforms P to a propositional program which has the same set of minimal models as P has (see Subsection 4.2). The other algorithm transforms P to a finite variant of the above propositional program (see Subsection 4.4); it is actually used in the proposed method. In the second phase, for every given entailment ax, the method extracts a relevant subset Π_{rel} from Π in a goal-directed manner and then identifies a just-preserving module for ax from Π_{rel} (see Subsection 4.3). Due to the space limitation, we do not provide proofs of lemmas and theorems in this paper, but refer the interested reader to our technical report[2].

4.1 Compiling a Diagnosing Program

Let $\mathcal{O}^\dagger$ denote $\mathcal{O} \cup \{A(a_A) \mid A$ is an atomic concept in $\mathcal{O}$ and a_A is a new globally unique individual corresponding to $A\}$. Recall that $\mathcal{O}$ is assumed coherent and having concept names in our addressing problem, so $\mathcal{O}^\dagger$ is consistent. Since the new individuals introduced in $\mathcal{O}^\dagger$ is unique, we have $\mathcal{O} \models A \sqsubseteq B$ iff $\mathcal{O}^\dagger \models B(a_A)$. Hence, $\mathcal{J}$ is a justification of $A \sqsubseteq B$ in $\mathcal{O}$ iff $\mathcal{J} \cup \{A(a_A)\}$ is a justification of $B(a_A)$ in $\mathcal{O}^\dagger$. It follows that $\mathcal{M}$ is a just-preserving module of $\mathcal{O}$ for $A \sqsubseteq B$ iff $\mathcal{M} \cup \{A(a_A)\}$ is a just-preserving module of $\mathcal{O}^\dagger$ for $B(a_A)$. That is, our addressing problem can be reduced to a problem of extracting a just-preserving module for a membership entailment. We therefore, in what follows, focus on membership entailments.

2 http://www.aifb.uni-karlsruhe.de/WBS/gqi/jp-module/module-long.pdf

Note that, for an concept/role assertion ax, $\mathcal{O}^\dagger \models ax$ iff $\mathcal{O}^\dagger \cup \{\neg ax\}$ is inconsistent, so $\mathcal{J}$ is a justification of ax in $\mathcal{O}^\dagger$ iff $\mathcal{J} \cup \{\neg ax\}$ is a minimally inconsistent subset of axioms in $\mathcal{O}^\dagger \cup \{\neg ax\}$ (simply called a *MIS* of $\mathcal{O}^\dagger \cup \{\neg ax\}$). It follows that $\mathcal{M}$ is a just-preserving module of $\mathcal{O}^\dagger$ for ax iff $S \subseteq \mathcal{M} \cup \{\neg ax\}$ for all MISs S of $\mathcal{O}^\dagger \cup \{\neg ax\}$. To extract a just-preserving module of $\mathcal{O}^\dagger$ for ax, we characterize all MISs of $\mathcal{O}^\dagger \cup \{\neg ax\}$ by a first-order logic program, called the *diagnosing program* of $\mathcal{O}^\dagger$, denoted by $\mathcal{D}(\mathcal{O}^\dagger)$, in which every clause may have a positive literal $\hbar_{ax}$ that corresponds to an axiom $ax \in \mathcal{O}^\dagger$. The atom $\hbar_{ax}$ is called the *decision atom* of ax. It is nullary and works as a flag to associate the clauses containing it with the axiom ax. We detail below the method for computing the diagnosing program of $\mathcal{O}^\dagger$.

We first translate $\mathcal{O}^\dagger$ to a first-order logic program. Since a direct translation from $\mathcal{SHOIQ}$ to first-order clauses may incur exponential blowup [13], we apply the well-known *structural transformation* [11,13] to an axiom before translating it to first-order clauses. By $\Theta(ax)$ we denote the result of applying structural transformation and clausification (i.e. translating to clauses) to an axiom ax. As both structural transformation and clausification are well-known, we do not give their definition here but refer the reader to [13] or our technical report[2].

Let $\Xi(ax)$ denote the set of clauses obtained from $\Theta(ax)$ by adding the decision atom $\hbar_{ax}$ to every clause in $\Theta(ax)$, i.e., $\Xi(ax) = \{cl \vee \hbar_{ax} \mid cl \in \Theta(ax)\}$. Let $\Xi(\mathcal{O}^\dagger) = \bigcup_{ax \in \mathcal{O}^\dagger} \Xi(ax)$. Then the *diagnosing program* of $\mathcal{O}^\dagger$, i.e. $\mathcal{D}(\mathcal{O}^\dagger)$, is defined as follows: if some equational atom $s \approx t$ occurs positively in $\Xi(\mathcal{O}^\dagger)$, then $\mathcal{D}(\mathcal{O}^\dagger) = \Xi(\mathcal{O}^\dagger) \cup \mathcal{E}(\Xi(\mathcal{O}^\dagger))$, otherwise $\mathcal{D}(\mathcal{O}^\dagger) = \Xi(\mathcal{O}^\dagger)$. Recall that $\mathcal{E}(\Xi(\mathcal{O}^\dagger))$ is used to axiomatize the equality predicate in $\Xi(\mathcal{O}^\dagger)$ (see Section 3).

Example 2 (Example 1 continued). Consider the ontology $\mathcal{O}$ in Example 1. We compute $\mathcal{O}^\dagger$ as $\mathcal{O} \cup \{ax_6 : \mathsf{ChiefActress}(a_1), ax_7 : \mathsf{Person}(a_2), ax_8 : \mathsf{Actress}(a_3), ax_9 : \mathsf{Woman}(a_4), ax_{10} : \mathsf{Man}(a_5)\}$, where $a_1, \ldots, a_5$ are all new individuals. We then compute $\Xi(\mathcal{O}^\dagger)$ as $\{cl_1, \ldots, cl_{10}\}$ given below.

$cl_1 : \neg\mathsf{ChiefActress}(x) \vee \mathsf{Person}(x) \vee \hbar_{ax_1}$
$cl_2 : \neg\mathsf{ChiefActress}(x) \vee \mathsf{Actress}(x) \vee \hbar_{ax_2}$
$cl_3 : \neg\mathsf{Actress}(x) \vee \mathsf{Woman}(x) \vee \hbar_{ax_3}$
$cl_4 : \neg\mathsf{Person}(x) \vee \mathsf{Man}(x) \vee \mathsf{Woman}(x) \vee \hbar_{ax_4}$
$cl_5 : \neg\mathsf{ChiefActress}(x) \vee \neg\mathsf{Man}(x) \vee \hbar_{ax_5}$
$cl_6 : \mathsf{ChiefActress}(a_1) \vee \hbar_{ax_6}$
$cl_7 : \mathsf{Person}(a_2) \vee \hbar_{ax_7}$
$cl_8 : \mathsf{Actress}(a_3) \vee \hbar_{ax_8}$
$cl_9 : \mathsf{Woman}(a_4) \vee \hbar_{ax_9}$
$cl_{10} : \mathsf{Man}(a_5) \vee \hbar_{ax_{10}}$

Since there is no equational atom occurring positively in $\Xi(\mathcal{O}^\dagger)$, the diagnosing program of $\mathcal{O}^\dagger$, i.e. $\mathcal{D}(\mathcal{O}^\dagger)$, is exactly $\Xi(\mathcal{O}^\dagger)$.

Before showing how to use the diagnosing program, we introduce some notions. Given a first-order logic program P, a set X of ground atoms never occurring negatively in P and a truth assignment Φ_X on X, we define the *reduction* of P w.r.t. Φ_X, denoted by $P \downarrow \Phi_X$, as a first-order logic program obtained from P by deleting every clause $cl \in P$ that contains a ground atom $\alpha \in X$ such that $\Phi_X(\alpha) = \text{true}$, and by removing all ground atoms in X from the remaining clauses. Since any ground atom in X does not occur negatively in P, it is clear that $P \downarrow \Phi_X$ is satisfiable iff there exists a model M of P such that $M \cap X = \{\alpha \in X \mid \Phi_X(\alpha) = \text{true}\}$.

In the remainder of this section, let X denote the set of decision atoms in $\mathcal{D}(\mathcal{O}^\dagger)$, i.e., $X = \{\hbar_{ax} \mid ax \in \mathcal{O}^\dagger\}$, and ax a membership entailment of $\mathcal{O}^\dagger$ on concept/role names if not otherwise specified. Consider an ontology $\mathcal{O}'$ such that $\mathcal{O}^\dagger \subseteq \mathcal{O}'$. For Φ_X a truth assignment on X, we define the *reduction* of $\mathcal{O}'$ w.r.t. Φ_X, denoted by $\mathcal{O}' \downarrow \Phi_X$, as $\mathcal{O}' \setminus \{ax' \in \mathcal{O}^\dagger \mid \Phi_X(\hbar_{ax'}) = \mathrm{true}\}$, i.e. the subset of $\mathcal{O}'$ where all axioms $ax' \in \mathcal{O}^\dagger$ such that $\Phi_X(\hbar_{ax'}) = \mathrm{true}$ are absent. We have the following relationship.

Lemma 1. *Let $P = \mathcal{D}(\mathcal{O}^\dagger) \cup \{\neg ax\}$ and $\mathcal{O}' = \mathcal{O}^\dagger \cup \{\neg ax\}$. For any truth assignment Φ_X on X, $P \downarrow \Phi_X$ is satisfiable iff $\mathcal{O}' \downarrow \Phi_X$ is satisfiable.*

The above relationship implies a correspondence between MISs of $\mathcal{O}^\dagger \cup \{\neg ax\}$ and *maximally unsatisfiable X-assignments* for $\mathcal{D}(\mathcal{O}^\dagger) \cup \{\neg ax\}$ (see Lemma 2), where a maximally unsatisfiable X-assignment for $\mathcal{D}(\mathcal{O}^\dagger) \cup \{\neg ax\}$ is a truth assignment Φ_X on X such that $(\mathcal{D}(\mathcal{O}^\dagger) \cup \{\neg ax\}) \downarrow \Phi_X$ is unsatisfiable and $\{\alpha \in X \mid \Phi_X(\alpha) = \mathrm{true}\}$ is maximal, formally defined below.

Definition 3 (Maximally Unsatisfiable X-Assignment). For P a first-order logic program and X a set of ground atoms never occurring negatively in P, a truth assignment Φ_X on X for P is called an *unsatisfiable X-assignment* for P if $P \downarrow \Phi_X$ is unsatisfiable; an unsatisfiable X-assignment Φ_X for P is called a *maximally unsatisfiable X-assignment* for P if there is no other unsatisfiable X-assignment Φ'_X for P such that $\{\alpha \in X \mid \Phi_X(\alpha) = \mathrm{true}\} \subset \{\alpha \in X \mid \Phi'_X(\alpha) = \mathrm{true}\}$.

Lemma 2. *Let $P = \mathcal{D}(\mathcal{O}^\dagger) \cup \{\neg ax\}$ and $\mathcal{O}' = \mathcal{O}^\dagger \cup \{\neg ax\}$. For a MIS S of $\mathcal{O}'$, the truth assignment Φ_X on X, such that $\Phi_X(\hbar_{ax'}) = \mathrm{true}$ iff $ax' \in \mathcal{O}' \setminus S$ for all $ax' \in \mathcal{O}^\dagger$, is a maximally unsatisfiable X-assignment for P. For a maximally unsatisfiable X-assignment Φ_X for P, $\mathcal{O}' \setminus \{ax' \in \mathcal{O}^\dagger \mid \Phi_X(\hbar_{ax'}) = \mathrm{true}\}$ is a MIS of $\mathcal{O}'$.*

According to Lemma 2, we can characterize a just-preserving module of $\mathcal{O}^\dagger$ for ax with the help of maximally unsatisfiable X-assignments for $\mathcal{D}(\mathcal{O}^\dagger) \cup \{\neg ax\}$.

Theorem 1. *$\mathcal{M}$ is a just-preserving module of $\mathcal{O}^\dagger$ for ax iff $\{ax' \in \mathcal{O}^\dagger \mid \Phi_X(\hbar_{ax'}) = \mathrm{false}\} \subseteq \mathcal{M}$ for all maximally unsatisfiable X-assignments Φ_X for $\mathcal{D}(\mathcal{O}^\dagger) \cup \{\neg ax\}$.*

4.2 Exact Grounding of the Diagnosing Program

According to Theorem 1, we need to consider satisfiability problems on $\mathcal{D}(\mathcal{O}^\dagger) \cup \{\neg ax\}$. This can be done by considering a propositional program that is transformed from $\mathcal{D}(\mathcal{O}^\dagger) \cup \{\neg ax\}$ and has the same set of minimal models as $\mathcal{D}(\mathcal{O}^\dagger) \cup \{\neg ax\}$ has. We extend the well-known intelligent grounding (IG) technique [5] which computes, in a fixpoint-evaluation manner, a semantically equivalent propositional program containing only derivable ground atoms from a function-free first-order logic program. By generalizing the idea of the IG technique, we define the so-called *bottom-up grounding* of a general first-order logic program P, denoted by $\mathcal{G}_{bu}(P)$, as the least fixpoint of $\Pi^{(n)}$ such that $\Pi^{(0)} = \emptyset$ and for $n \geq 1$, $\Pi^{(n)} = \{cl\,\sigma \mid cl \in P,\ \sigma$ is a ground substitution such that $cl^-\sigma \subseteq \mathsf{atoms}(\Pi^{(n-1)})$ and $cl^+\sigma \subseteq \mathrm{HB}(P)\}$.

Lemma 3. *Let P be a first-order logic program in which the equality predicate $\approx$ has been axiomatized. Then (1) $\mathcal{G}_{bu}(P)$ is the least subset S of $\mathcal{G}(P)$ such that $cl\,\sigma \in S$ for any clause $cl \in P$ and any ground substitution σ such that $cl^-\sigma \subseteq \mathsf{atoms}(S)$ and $cl^+\sigma \subseteq \mathrm{HB}(P)$; (2) $\mathcal{G}_{bu}(P)$ has the same set of minimal models as P has.*

Example 3 (Example 2 continued). This example shows the steps of computing the bottom-up grounding of $\mathcal{D}(\mathcal{O}^\dagger)$ given in Example 2. By applying the fixpoint-evaluation process for defining $\mathcal{G}_{bu}(\mathcal{D}(\mathcal{O}^\dagger))$, we have $\Pi^{(0)} = \emptyset$, $\Pi^{(1)} = \{cl'_1, ..., cl'_5\}$, $\Pi^{(2)} = \{cl'_1, ..., cl'_9\}$, $\Pi^{(3)} = \{cl'_1, ..., cl'_{11}\}$, $\Pi^{(4)} = \{cl'_1, ..., cl'_{12}\}$ and $\Pi^{(5)} = \Pi^{(4)}$, so $\mathcal{G}_{bu}(\mathcal{D}(\mathcal{O}^\dagger)) = \{cl'_1, ..., cl'_{12}\}$, where $cl'_1, \ldots, cl'_{12}$ are given below.

$cl'_1:$ $\mathsf{ChiefActress}(a_1) \vee \hbar_{ax6}$
$cl'_2:$ $\mathsf{Person}(a_2) \vee \hbar_{ax7}$
$cl'_3:$ $\mathsf{Actress}(a_3) \vee \hbar_{ax8}$
$cl'_4:$ $\mathsf{Woman}(a_4) \vee \hbar_{ax9}$
$cl'_5:$ $\mathsf{Man}(a_5) \vee \hbar_{ax10}$
$cl'_6:$ $\neg\mathsf{ChiefActress}(a_1) \vee \mathsf{Person}(a_1) \vee \hbar_{ax1}$
$cl'_7:$ $\neg\mathsf{ChiefActress}(a_1) \vee \mathsf{Actress}(a_1) \vee \hbar_{ax2}$
$cl'_8:$ $\neg\mathsf{Actress}(a_3) \vee \mathsf{Woman}(a_3) \vee \hbar_{ax3}$
$cl'_9:$ $\neg\mathsf{Person}(a_2) \vee \mathsf{Man}(a_2) \vee \mathsf{Woman}(a_2) \vee \hbar_{ax4}$
$cl'_{10}:$ $\neg\mathsf{Actress}(a_1) \vee \mathsf{Woman}(a_1) \vee \hbar_{ax3}$
$cl'_{11}:$ $\neg\mathsf{Person}(a_1) \vee \mathsf{Man}(a_1) \vee \mathsf{Woman}(a_1) \vee \hbar_{ax4}$
$cl'_{12}:$ $\neg\mathsf{ChiefActress}(a_1) \vee \neg\mathsf{Man}(a_1) \vee \hbar_{ax5}$

Since ax is a membership entailment of $\mathcal{O}^\dagger$, by lemma 3, $\mathcal{G}_{bu}(\mathcal{D}(\mathcal{O}^\dagger)) \models ax$. Hence ax occurs positively in some clauses in $\mathcal{G}_{bu}(\mathcal{D}(\mathcal{O}^\dagger))$. Note that $\mathcal{G}_{bu}(\mathcal{D}(\mathcal{O}^\dagger) \cup \{\neg ax\})$ can be computed by first grounding $\mathcal{D}(\mathcal{O}^\dagger)$ then adding the clause $\{\neg ax\}$, and that the adding of clause $\neg ax$ to $\mathcal{G}_{bu}(\mathcal{D}(\mathcal{O}^\dagger))$ does not introduce new ground atoms to $\mathcal{G}_{bu}(\mathcal{D}(\mathcal{O}^\dagger))$, so $\mathcal{G}_{bu}(\mathcal{D}(\mathcal{O}^\dagger) \cup \{\neg ax\}) = \mathcal{G}_{bu}(\mathcal{D}(\mathcal{O}^\dagger)) \cup \{\neg ax\}$. It follows from Theorem 1 and Lemma 3 that $\mathcal{M}$ is a just-preserving module of $\mathcal{O}^\dagger$ for ax iff $\{ax' \in \mathcal{O}^\dagger \mid \Phi_X(\hbar_{ax'}) = \text{false}\} \subseteq \mathcal{M}$ for all maximally unsatisfiable X-assignments Φ_X for $\mathcal{G}_{bu}(\mathcal{D}(\mathcal{O}^\dagger)) \cup \{\neg ax\}$. Subsequently, a just-preserving module of $\mathcal{O}^\dagger$ for ax can be computed as $\{ax' \in \mathcal{O}^\dagger \mid \hbar_{ax'} \in \mathsf{atoms}(\Pi_{rel})\}$, where Π_{rel} is a subset of $\mathcal{G}_{bu}(\mathcal{D}(\mathcal{O}^\dagger)) \cup \{\neg ax\}$ such that $\{\hbar_{ax'} \in X \mid \Phi_X(\hbar_{ax'}) = \text{false}\} \subseteq \mathsf{atoms}(\Pi_{rel})$ for all maximally unsatisfiable X-assignments Φ_X for $\mathcal{G}_{bu}(\mathcal{D}(\mathcal{O}^\dagger)) \cup \{\neg ax\}$. We call such Π_{rel} a *just-preserving relevant subset* of $\mathcal{G}_{bu}(\mathcal{D}(\mathcal{O}^\dagger)) \cup \{\neg ax\}$ for ax.

4.3 Extracting a Justification-Preserving Module

In this subsection, by Π we denote $\mathcal{G}_{bu}(\mathcal{D}(\mathcal{O}^\dagger)) \cup \{\neg ax\}$ if not otherwise specified. Our method for extracting a just-preserving relevant subset of Π is based on the notion of a *connected component* (see Definition 4 below), which is a subset of a propositional program such that any two clauses in it have common ground atoms. This notion has been used to confine the search space in solving SAT problems [3] because an unsatisfiable propositional program must have a maximal connected component that is unsatisfiable.

Definition 4 (Connected Component). Let Π be a propositional program. Two ground clauses cl and cl' are called *connected* in Π if there exists a sequence of clauses $cl_0 = cl, cl_1, \ldots, cl_n = cl'$ in Π such that cl_{i-1} and cl_i have common ground atoms for any $1 \leq i \leq n$. A *connected component* Π_c of Π is a subset of Π such that any two clauses cl and cl' in Π_c are connected in Π_c. Π_c is called *maximal* if there is no connected component Π'_c of Π such that $\Pi_c \subset \Pi'_c$.

The basic idea employed in our method is that the maximal connected component of Π where ax occurs is a just-preserving relevant subset of Π for ax. To obtain a smaller just-preserving relevant subset, our method extends the basic idea by first removing two subsets Π_{ur1} and Π_{ur2} from Π in turn, then extracting a maximal connected component from the remaining set. The description and the correctness of the method are shown in Lemma 4, while the idea is illustrated in Example 4.

Example 4 (Example 3 cont.). Let $\Pi = \mathcal{G}_{bu}(\mathcal{D}(\mathcal{O}^\dagger)) \cup \{\neg\mathsf{Person}(a_1)\}$ for $\mathcal{G}_{bu}(\mathcal{D}(\mathcal{O}^\dagger))$ given in Example 3. This example shows the sets Π_{ur1}, Π_{ur2} and Π_{rel} that may be computed by our method.

$cl'_2: \boxed{\mathsf{Person}(a_2)} \vee \hbar_{ax_7}$
$cl'_3: \boxed{\mathsf{Actress}(a_3)} \vee \hbar_{ax_8}$
$cl'_4: \boxed{\mathsf{Woman}(a_4)} \vee \hbar_{ax_9}$
$cl'_5: \boxed{\mathsf{Man}(a_5)} \vee \hbar_{ax_{10}}$
$cl'_7: \neg\mathsf{ChiefActress}(a_1) \vee \boxed{\mathsf{Actress}(a_1)} \vee \hbar_{ax_2}$
$cl'_8: \neg\mathsf{Actress}(a_3) \vee \boxed{\mathsf{Woman}(a_3)} \vee \hbar_{ax_3}$
$cl'_9: \neg\mathsf{Person}(a_2) \vee \boxed{\mathsf{Man}(a_2)} \vee \boxed{\mathsf{Woman}(a_2)} \vee \hbar_{ax_4}$
$cl'_{10}: \neg\mathsf{Actress}(a_1) \vee \boxed{\mathsf{Woman}(a_1)} \vee \hbar_{ax_3}$
$cl'_{11}: \neg\mathsf{Person}(a_1) \vee \mathsf{Man}(a_1) \vee \boxed{\mathsf{Woman}(a_1)} \vee \hbar_{ax_4}$

Π_{ur1} : for all clauses $cl \in \Pi_{ur1}$, cl^+ contains at least one ground atom not in $X \cup \mathsf{atoms}(\Pi \setminus \Pi_{ur1})$.

The set $M_0 = \bigcup_{cl \in \Pi_{ur1}} cl^+ \setminus \mathsf{atoms}(\Pi \setminus \Pi_{ur1})$, i.e. the set of framed ground atoms, is a model of Π_{ur1}.

$cl'_{12}: \neg\mathsf{ChiefActress}(a_1) \vee \neg\mathsf{Man}(a_1) \vee \hbar_{ax_5}$

Π_{ur2} : for all clauses $cl \in \Pi_{ur2}$,
$cl^- \not\subseteq \bigcup_{cl \in \Pi \setminus (\Pi_{ur1} \cup \Pi_{ur2})} cl^+$.

Note that $M_0 \cup \mathcal{A}$ is a model of Π_{ur2} for any subset $\mathcal{A}$ of ground atoms occurring in $\Pi \setminus (\Pi_{ur1} \cup \Pi_{ur2})$.

$\neg\mathsf{Person}(a_1)$
$cl'_6: \neg\mathsf{ChiefActress}(a_1) \vee \mathsf{Person}(a_1) \vee \hbar_{ax_1}$
$cl'_1: \mathsf{ChiefActress}(a_1) \vee \hbar_{ax_6}$

Π_{rel} = the maximal connected component of $\Pi \setminus (\Pi_{ur1} \cup \Pi_{ur2})$ where $\mathsf{Person}(a_1)$ occurs.
Π_{rel} is a just-preserving subset of Π for $\mathsf{Person}(a_1)$, and $\{ax' \in \mathcal{O}^\dagger \mid \hbar_{ax'} \in \mathsf{atoms}(\Pi_{rel})\} = \{ax_1, ax_6\}$ is a just-preserving module of $\mathcal{O}^\dagger$ for $\mathsf{Person}(a_1)$.

In this example, $\Pi = \Pi_{ur1} \cup \Pi_{ur2} \cup \Pi_{rel}$. The key point to prove that Π_{rel} is a just-preserving subset of Π is that for any maximally unsatisfiable X-assignment Φ_X of Π, $\Phi_X(\alpha) = \mathsf{true}$ for all $\alpha \in X \setminus \mathsf{atoms}(\Pi_{rel})$, otherwise $\Pi \downarrow \Phi'_X$ is satisfiable for Φ'_X the truth assignment on X such that $\Phi'_X(\alpha) = \Phi_X(\alpha)$ for all $\alpha \in X \cap \mathsf{atoms}(\Pi_{rel})$ and $\Phi'_X(\alpha) = \mathsf{true}$ for all $\alpha \in X \setminus \mathsf{atoms}(\Pi_{rel})$; this implies that $\Pi_{rel} \downarrow \Phi_X = \Pi_{rel} \downarrow \Phi'_X$ is satisfiable and $\Pi_{rel} \downarrow \Phi_X$ has a minimal model M, but then $M \cup M_0$ is model of $\Pi \downarrow \Phi_X$ (contradiction).

Algorithm 1. Extract(Π_{in}, X, α)
Input: A propositional program Π_{in} with decision atoms, the set X of decision atoms occurring in Π_{in}, and a ground atom α occurring in Π_{in}.
Output: A subset of $\Pi_{in} \cup \{\neg\alpha\}$.

```
 Π_ur1 := Π_in; Π'_ur1 := ∅;
 while Π'_ur1 ≠ Π_ur1 do
   Π'_ur1 := Π_ur1;
   for each clause cl ∈ Π_ur1 such that cl⁺ ⊆ X ∪ atoms(Π_in \ Π_ur1) do
     Π_ur1 := Π_ur1 \ {cl};
 Π_0 := Π_ur1; Π'_0 := ∅;
 while Π'_0 ≠ Π_0 do
   Π'_0 := Π_0;
   for each clause cl ∈ Π_in \ Π_0 such that cl⁻ ⊈ ⋃_{cl ∈ Π_in \ Π_0} cl⁺ do
     Π_0 := Π_0 ∪ {cl};
 Π_rel := {¬α}; Π'_rel := ∅;
 while Π'_rel ≠ Π_rel do
   Π'_rel := Π_rel;
   for each clause cl ∈ Π_in \ Π_0 such that (cl⁺ ∪ cl⁻) ∩ atoms(Π_rel) ≠ ∅ do
     Π_rel := Π_rel ∪ {cl};
 return Π_rel;
```

Fig. 2. The algorithm for extracting a just-preserving relevant subset of $\Pi_{in} \cup \{\neg\alpha\}$ for α

Lemma 4. *Let Π_{ur1} be a subset of Π such that for all clauses $cl \in \Pi_{ur1}$, cl^+ contains at least one ground atom not in $X \cup \mathsf{atoms}(\Pi \setminus \Pi_{ur1})$. Let Π_{ur2} be a subset of $\Pi \setminus \Pi_{ur1}$ such that $cl^- \not\subseteq \bigcup_{cl \in \Pi \setminus (\Pi_{ur1} \cup \Pi_{ur2})} cl^+$ for all clauses $cl \in \Pi_{ur2}$. Let Π_{rel} be the* *maximal connected component of* $\Pi \setminus (\Pi_{ur1} \cup \Pi_{ur2})$ *where* ax *occurs. Then* Π_{rel} *is a just-preserving relevant subset of* Π *for* ax*, i.e.,* $\{\hbar_{ax'} \in X \mid \Phi_X(\hbar_{ax'}) = \text{false}\} \subseteq \mathsf{atoms}(\Pi_{rel})$ *for all maximally unsatisfiable* X*-assignments* Φ_X *for* Π*.*

According to Lemma 4, we develop an algorithm for extracting a just-preserving relevant subset of Π for ax, shown in Figure 2. In the algorithm, lines 1–5 compute Π_{ur1} given in Lemma 4, lines 6–10 compute Π_{ur2} (where $\Pi_0 \setminus \Pi_{ur1}$ in the algorithm corresponds to Π_{ur2} given in Lemma 4), and lines 11-15 compute Π_{rel}. Note that Π_{rel} is computed in a goal-directed (backward traversal) manner.

Consider $\text{Extract}(\mathcal{G}_{bu}(\mathcal{D}(\mathcal{O}^\dagger)), X, \mathsf{Person}(a_1))$ for $\mathcal{G}_{bu}(\mathcal{D}(\mathcal{O}^\dagger))$ given in Example 3. It can be checked that the values of Π_{ur1}, Π_{ur2} (i.e. $\Pi_0 \setminus \Pi_{ur1}$ in the algorithm) and Π_{rel} are exactly those given in Example 4. By Lemma 4, Π_{rel} is a just-preserving subset of Π for $\mathsf{Person}(a_1)$, thus $\{ax' \in \mathcal{O}^\dagger \mid \hbar_{ax'} \in \mathsf{atoms}(\Pi_{rel})\} = \{ax_1, ax_6\}$ is a just-preserving module of $\mathcal{O}^\dagger$ for $\mathsf{Person}(a_1)$. Note that $\mathsf{ChiefActress}(a_1) \in \mathcal{O}^\dagger \setminus \mathcal{O}$, so $\{ax_1\}$ is a just-preserving module of $\mathcal{O}$ for $\mathsf{ChiefActress} \sqsubseteq \mathsf{Person}$. Recall that the syntactic locality-based module of ChiefActress in $\mathcal{O}$ is exactly $\mathcal{O}$. This shows that the just-preserving module extracted by our goal-directed method can be much smaller than the syntactic locality-based module.

The following theorem shows the correctness and complexity of the algorithm.

Theorem 2. $\text{Extract}(\mathcal{G}_{bu}(\mathcal{D}(\mathcal{O}^\dagger)), X, ax)$ *returns a just-preserving relevant subset of* $\mathcal{G}_{bu}(\mathcal{D}(\mathcal{O}^\dagger)) \cup \{\neg ax\}$ *for* ax *in time polynomial in* $|\mathcal{G}_{bu}(\mathcal{D}(\mathcal{O}^\dagger))|$.

It follows from the above theorem that $\{ax' \in \mathcal{O}^\dagger \mid \hbar_{ax'} \in \mathsf{atoms}(\Pi_{rel})\}$ is a just-preserving module of $\mathcal{O}^\dagger$ for ax, where Π_{rel} is the result of $\text{Extract}(\mathcal{G}_{bu}(\mathcal{D}(\mathcal{O}^\dagger)), X, ax)$. The remaining issue for extracting just-preserving modules is that $\mathcal{G}_{bu}(\mathcal{D}(\mathcal{O}^\dagger))$ can be infinite when there are function symbols occurring in $\mathcal{D}(\mathcal{O}^\dagger)$. We in the next subsection show that a just-preserving module of $\mathcal{O}^\dagger$ for ax can also be extracted from a finite variant of $\mathcal{G}_{bu}(\mathcal{D}(\mathcal{O}^\dagger))$ whose size is polynomial in the size of $\mathcal{O}$.

4.4 Approximate Grounding of the Diagnosing Program

To tackle the problem that the bottom-up grounding Π of a first-order logic program can be infinite, we consider a mapping function on ground terms occurring in Π such that the range of this function is finite and small. We call a mapping function $\lambda : \text{terms}(\Pi) \mapsto \text{terms}(\Pi)$, where $\text{terms}(\Pi)$ is the set of ground terms occurring in Π, a *convergent mapping function* for Π, if for every functional term $f_1(...f_n(a))$ (where a is a constant) occurring in Π, $\lambda(f_1(...f_n(a))) = \lambda(a)$, and for every equational atom $s \approx t$ occurring in Π, $\lambda(s) = \lambda(t)$. For example, given a propositional program $\Pi^\dagger = \{\neg\mathsf{hasFather}(a, b) \vee \neg\mathsf{hasFather}(a, f(a)) \vee b \approx f(a), \neg\mathsf{Person}(a) \vee \mathsf{hasFather}(a, f(a)), \neg\mathsf{Person}(a) \vee \mathsf{Person}(f(a))\}$, the mapping function $\lambda = \{a \mapsto a, b \mapsto a, f(a) \mapsto a\}$ is a convergent mapping function for $\Pi^\dagger$, but the mapping function $\lambda' = \{a \mapsto b, b \mapsto a, f(a) \mapsto a\}$ is not since $\lambda'(f(a)) \neq \lambda'(a)$.

Given a mapping function λ on ground terms, by $\lambda(\alpha)$, $\lambda(cl)$, $\lambda(\mathcal{A})$ and $\lambda(\Pi)$ we respectively denote the results obtained from a ground atom α, a clause cl, a set $\mathcal{A}$ of ground atoms and a propositional program Π by replacing every ground term t occurring in it with $\lambda(t)$. Take $\Pi^\dagger$ given above for example, for the mapping function $\lambda =$

Algorithm 2. $\text{ApproxGround}(\mathcal{O}^\dagger)$
Input: An $\mathcal{SHOIQ}$ ontology $\mathcal{O}^\dagger$.
Output: A propositional program and a set of sets of constants.

```
Let P be obtained from D(O†) by stripping all function symbols from functional terms;
Π := ∅; S := ∅;
repeat
   Π' := Π;
   for each clause cl ∈ P and each ground substitution σ such that cl⁻σ ⊆ atoms(Π) ∩
   HB(P) and cl⁺σ ⊆ HB(P) do
      cl_inst := cl σ;
      for each equational atom a ≈ b ∈ cl⁺_inst such that a and b are different constants do
         MergeConstants(a, b, S, cl_inst, P);
      Π := Π ∪ {cl_inst};
until Π' = Π;
Π := {cl ∈ Π | cl⁺ ∪ cl⁻ ⊆ HB(P)};
return (Π, S);
```

Subprocedure $\text{MergeConstants}(a, b, \mathcal{S}, cl, P)$

```
Let C_a be {a} if a does not occur in S, or the congruence class in S where a belongs to;
Let C_b be {b} if b does not occur in S, or the congruence class in S where b belongs to;
C_new := C_a ∪ C_b; rep(C_new) := b; S := (S \ {C_a, C_b}) ∪ C_new;
Update cl and P by replacing every occurrence of a with b;
```

Fig. 3. The algorithm for approximately grounding the diagnosing program $\mathcal{D}(\mathcal{O}^\dagger)$

$\{a \mapsto a, b \mapsto a, f(a) \mapsto a\}$, $\lambda(\Pi^\dagger) = \{\neg\mathsf{hasFather}(a, a) \vee \neg\mathsf{hasFather}(a, a) \vee a \approx a$, $\neg\mathsf{Person}(a) \vee \mathsf{hasFather}(a, a)$, $\neg\mathsf{Person}(a) \vee \mathsf{Person}(a)\}$.

It is obvious that for any convergent mapping function λ for Π, $\lambda(\Pi)$ is finite when Π is infinite but the number of constants and predicate symbols occurring in Π is finite. Even when Π is finite and does not contain function symbols, $\lambda(\Pi)$ can be much smaller than Π because the subset of ground clauses in Π that form a congruence relation is collapsed in $\lambda(\Pi)$.

Lemma 5. *Let λ be an convergent mapping function for $\mathcal{G}_{bu}(\mathcal{D}(\mathcal{O}^\dagger))$ where decision atoms, treated as nullary atoms, are not in the domain of λ, Π be a superset of $\lambda(\mathcal{G}_{bu}(\mathcal{D}(\mathcal{O}^\dagger)))$, and Π_{rel} be returned by* $\text{Extract}(\Pi, X, \lambda(ax))$. *Then $\{ax' \in \mathcal{O}^\dagger \mid \hbar_{ax'} \in \mathsf{atoms}(\Pi_{rel})\}$ is a just-preserving module of $\mathcal{O}^\dagger$ for ax.*

To employ the above lemma to extract a just-preserving module of $\mathcal{O}^\dagger$ for ax, we develop an algorithm, shown in Figure 3, to compute a superset of $\lambda(\mathcal{G}_{bu}(\mathcal{D}(\mathcal{O}^\dagger)))$ for some convergent mapping function λ for $\mathcal{G}_{bu}(\mathcal{D}(\mathcal{O}^\dagger))$.

In the algorithm, P is a first-order logic program obtained from $\mathcal{D}(\mathcal{O}^\dagger)$ by stripping all function symbols from functional terms, i.e., by replacing every functional term $f_1(...f_n(t))$ with t where t is a variable or a constant; $\text{HB}(P)$ is the Herbrand base of P, which does not contain any function symbol and is changed whenever P is changed; $\mathcal{S}$ is the set of sets of constants such that for any constant a in any element $\mathcal{C} \in \mathcal{S}$, there exists a constant $b \in \mathcal{C}$ such that the equational atom $a \approx b$ appears in the execution of the algorithm. We call an element in $\mathcal{S}$ a *congruence class*. Each congruence class $\mathcal{C}$ is associated with a *representative constant*, denoted by $\mathsf{rep}(\mathcal{C})$, which is an element of $\mathcal{C}$.

The algorithm does not directly compute a convergent mapping function λ for $\mathcal{G}_{bu}(\mathcal{D}(\mathcal{O}^{\dagger}))$, because such mapping function can be constructed from $\mathcal{S}$. By $\mathsf{map}(t, \mathcal{S})$ we denote the function for mapping a term $t \in \mathrm{HU}(\mathcal{D}(\mathcal{O}^{\dagger}))$ to a constant occurring in $\mathcal{D}(\mathcal{O}^{\dagger})$ based on $\mathcal{S}$, recursively defined as follows, where a is a constant.

- $\mathsf{map}(f_1(...f_n(a)), \mathcal{S}) = \mathsf{map}(a, \mathcal{S})$, where $n > 0$;
- $\mathsf{map}(a, \mathcal{S}) = b$, where $b = \mathsf{rep}(\mathcal{C})$ if $a \in \mathcal{C}$ for some $\mathcal{C} \in \mathcal{S}$, or $b = a$ otherwise.

We naturally extend the function map to (sets of) ground atoms and ground clauses, i.e., by $\mathsf{map}(\alpha, \mathcal{S})$, $\mathsf{map}(\mathcal{A}, \mathcal{S})$ and $\mathsf{map}(cl, \mathcal{S})$ we respectively denote the results obtained from a ground atom α, a set $\mathcal{A}$ of ground atoms, and a clause cl by replacing every ground term t occurring in it with $\mathsf{map}(t, \mathcal{S})$.

We call a mapping function $\lambda : \mathrm{HU}(\mathcal{D}(\mathcal{O}^{\dagger})) \mapsto \mathrm{HU}(\mathcal{D}(\mathcal{O}^{\dagger}))$ *induced from* the function map w.r.t. $\mathcal{S}$ if $\lambda(t) = \mathsf{map}(t, \mathcal{S})$ for all ground terms $t \in \mathrm{HU}(\mathcal{D}(\mathcal{O}^{\dagger}))$. One goal of the algorithm is to ensure the mapping function λ induced from map w.r.t. $\mathcal{S}$ to be an convergent mapping function for $\mathcal{G}_{bu}(\mathcal{D}(\mathcal{O}^{\dagger}))$, i.e., ensure $\mathsf{map}(s, \mathcal{S}) = \mathsf{map}(t, \mathcal{S})$ for all equational atoms $s \approx t$ occurring positively in $\mathcal{G}_{bu}(\mathcal{D}(\mathcal{O}^{\dagger}))$. Another goal of the algorithm is to return a superset Π of $\lambda(\mathcal{G}_{bu}(\mathcal{D}(\mathcal{O}^{\dagger})))$. Consider how to achieve this goal. By Lemma 3, $\mathcal{G}_{bu}(\mathcal{D}(\mathcal{O}^{\dagger}))$ is the least subset S of $\mathcal{G}(\mathcal{D}(\mathcal{O}^{\dagger}))$ such that $cl\,\sigma \in S$ for any clause $cl \in \mathcal{D}(\mathcal{O}^{\dagger})$ and any ground substitution σ such that $cl^{-}\sigma \subseteq \mathsf{atoms}(S)$ and $cl^{+}\sigma \subseteq \mathrm{HB}(\mathcal{D}(\mathcal{O}^{\dagger}))$. It follows that $\lambda(\mathcal{G}_{bu}(\mathcal{D}(\mathcal{O}^{\dagger})))$ is the least subset S' of $\lambda(\mathcal{G}(\mathcal{D}(\mathcal{O}^{\dagger})))$ such that $\lambda(cl\,\sigma) \in S'$ for any clause $cl \in \mathcal{D}(\mathcal{O}^{\dagger})$ and any ground substitution σ such that $\lambda(cl^{-}\sigma) \subseteq \mathsf{atoms}(S')$ and $\lambda(cl^{+}\sigma) \subseteq \lambda(\mathrm{HB}(\mathcal{D}(\mathcal{O}^{\dagger})))$. If Π is a subset of $\lambda(\mathcal{G}(\mathcal{D}(\mathcal{O}^{\dagger})))$ such that $\lambda(cl\,\sigma) \in \Pi$ for any clause $cl \in \mathcal{D}(\mathcal{O}^{\dagger})$ and any ground substitution σ such that $\lambda(cl^{-}\sigma) \subseteq \mathsf{atoms}(\Pi)$ and $\lambda(cl^{+}\sigma) \subseteq \lambda(\mathrm{HB}(\mathcal{D}(\mathcal{O}^{\dagger})))$, then we have $\lambda(\mathcal{G}_{bu}(\mathcal{D}(\mathcal{O}^{\dagger}))) \subseteq \Pi$. Hence we refine the second goal to the goal of finding the above subset Π.

In the following descriptions, we use λ to denote a mapping function induced from map w.r.t. $\mathcal{S}$. At the beginning (before line 3), it is clear that $\mathrm{HU}(P) = \{\mathsf{map}(t, \mathcal{S}) \mid t \in \mathrm{HU}(\mathcal{D}(\mathcal{O}^{\dagger}))\}$. Hence we can use $\mathrm{HU}(P)$ to represent $\{\mathsf{map}(t, \mathcal{S}) \mid t \in \mathrm{HU}(\mathcal{D}(\mathcal{O}^{\dagger}))\}$ in the algorithm. To this end, we should ensure $\{\mathsf{map}(t, \mathcal{S}) \mid t \in \mathrm{HU}(\mathcal{D}(\mathcal{O}^{\dagger}))\} = \mathrm{HU}(P)$ throughout the algorithm.

We now describe the main part of the algorithm. All clauses in P are instantiated iteratively until a fixpoint is reached (lines 3–10), making the instantiated set Π satisfy (†) $\lambda(cl\,\sigma) \in \Pi$ for any clause $cl \in \mathcal{D}(\mathcal{O}^{\dagger})$ and any ground substitution σ s.t. $\lambda(cl^{-}\sigma) \subseteq \mathsf{atoms}(\Pi)$ and $\lambda(cl^{+}\sigma) \subseteq \lambda(\mathrm{HB}(\mathcal{D}(\mathcal{O}^{\dagger})))$. Consider any clause $cl \in \mathcal{D}(\mathcal{O}^{\dagger})$ and any ground substitution σ. Let cl' be obtained from cl by stripping all function symbols from functional terms and by replacing every occurrence of constant a with $\lambda(a)$, and σ' be obtained from σ by replacing each mapping $x \mapsto t$ with $x \mapsto \mathsf{map}(t, \mathcal{S})$. Then $cl' \in P$ and $\lambda(cl\,\sigma) = \lambda(cl'\sigma')$. Hence we only need to consider adding $\lambda(cl\,\sigma)$ to Π for every clause $cl \in P$ and every ground substitution σ such that every ground term occurring in $cl\,\sigma$ also occurs in $\{\mathsf{map}(t, \mathcal{S}) \mid t \in \mathrm{HU}(\mathcal{D}(\mathcal{O}^{\dagger}))\} = \mathrm{HU}(P)$. Note that $\lambda(cl\,\sigma) = cl\,\sigma$ when every ground term occurring in $cl\,\sigma$ also occurs in $\mathrm{HU}(P)$ and that $\mathrm{HB}(P) = \lambda(\mathrm{HB}(\mathcal{D}(\mathcal{O}^{\dagger})))$, so in order to satisfy the condition (†), every clause $cl \in P$ and every ground substitution σ such that $cl^{-}\sigma \subseteq \mathsf{atoms}(\Pi) \cap \mathrm{HB}(P)$ and $cl^{+}\sigma \subseteq \mathrm{HB}(P)$ are handled as follows (lines 5–9). Let $cl_{inst} = cl\,\sigma$. Every ground term $a \approx b \in cl^{+}_{inst}$ such that a and b are different constants is handled by merging a and b to the same congruence class in $\mathcal{S}$ and by updating cl_{inst} and P accordingly (see the subprocedure MergeConstants). The merge of a and b is to ensure $\mathsf{map}(a, \mathcal{S}) = \mathsf{map}(b, \mathcal{S})$. The update of cl_{inst} is to ensure $\mathsf{map}(cl_{inst}, \mathcal{S}) = cl_{inst}$. The update of P is to ensure again $\{\mathsf{map}(t, \mathcal{S}) \mid t \in \mathrm{HU}(\mathcal{D}(\mathcal{O}^{\dagger}))\} = \mathrm{HU}(P)$. Since $\mathcal{S}$ is only updated in the subprocedure MergeConstants, $\{\mathsf{map}(t, \mathcal{S}) \mid t \in \mathrm{HU}(\mathcal{D}(\mathcal{O}^{\dagger}))\} = \mathrm{HU}(P)$ holds throughout the algorithm. After every ground term $a \approx b \in cl^{+}_{inst}$ is handled, the possibly updated cl_{inst} is added to Π. After Π reaches a fixpoint, since Π should be a subset of $\lambda(\mathcal{G}(\mathcal{D}(\mathcal{O}^{\dagger})))$, every ground term occurring in Π should be in $\{\lambda(t) \mid t \in \mathrm{HU}(\mathcal{D}(\mathcal{O}^{\dagger}))\} = \mathrm{HU}(P)$, so those clauses in Π that contain ground atoms not in $\mathrm{HB}(P)$ are removed from Π (line 11). Since the propositional program Π returned by the algorithm satisfies the above condition (†), we have the following lemma.

Lemma 6. *Let* $(\Pi, \mathcal{S})$ *be returned by* ApproxGround($\mathcal{O}^{\dagger}$) *and* λ *be a mapping function induced from the function* map *w.r.t.* $\mathcal{S}$. *Then (1)* $\lambda(\mathcal{G}_{bu}(\mathcal{D}(\mathcal{O}^{\dagger}))) \subseteq \Pi$ *and* λ *is a convergent mapping function for* $\mathcal{G}_{bu}(\mathcal{D}(\mathcal{O}^{\dagger}))$*; (2)* ApproxGround($\mathcal{O}^{\dagger}$) *works in time polynomial in* s^m *and* $|\Pi|$ *is polynomial in* s, *where* m *is the maximum number in all qualifying number restrictions in* $\mathcal{O}$ *and* s *is the size of* $\mathcal{O}$.

The following theorem shows the correctness of our proposed method for extracting a just-preserving module, which is immediately follows from Lemma 5 and Lemma 6.

Theorem 3. *Let* $(\Pi, \mathcal{S})$ *be returned by* ApproxGround($\mathcal{O}^{\dagger}$), *and* Π_{rel} *be returned by* Extract(Π, X, $\mathsf{map}(ax, \mathcal{S})$). *Then* $\{ax' \in \mathcal{O}^{\dagger} \mid \hbar_{ax'} \in \mathsf{atoms}(\Pi_{rel})\}$ *is a just-preserving module of* $\mathcal{O}^{\dagger}$ *for* ax.

To summarize, our proposed method works as follows. In the first and offline phase, we first compute $\mathcal{O}^{\dagger} = \mathcal{O} \cup \{A(a_A) \mid A$ is an atomic concept in $\mathcal{O}$ and a_A is a new globally unique individual corresponding to $A\}$, then compute $(\Pi, \mathcal{S})$ as the result of ApproxGround($\mathcal{O}^{\dagger}$). In the second and online phase, for every coming entailment ax, we first set $\alpha = B(a_A)$ if the given entailment ax is $A \sqsubseteq B$, or $\alpha = ax$ otherwise, then compute $\Pi_{rel} = \mathrm{Extract}(\Pi, X, \mathsf{map}(\alpha, \mathcal{S}))$, and finally compute a just-preserving module of $\mathcal{O}$ for ax as $\{ax' \in \mathcal{O}^{\dagger} \mid \hbar_{ax'} \in \mathsf{atoms}(\Pi_{rel})\} \cap \mathcal{O}$.

5 Experimental Evaluation

We implemented the proposed method in GNU C++, using MySQL as the back-end SQL engine. In the implementation, ABox axioms are maintained in databases, instantiated clauses in the course of grounding are maintained in disk files, and Pellet [19] (version 2.0.0-rc6)[3] is called to find all justifications of a given entailment. All our experiments were conducted on a machine with 2.33GHz Intel Xeon E5410 CPU and 8GB RAM, running Windows Server 2003, where the maximum Java heap size was set to (max) 1252MB. The implementation, test sets and complete test results are available at `http://www.aifb.uni-karlsruhe.de/WBS/gqi/jp-module/`.

[3] Pellet (`http://clarkparsia.com/pellet/`) employs a hitting set tree (HST) based algorithm [12] to find all justifications.

Table 1. The characteristics of test ontologies and the execution time in the offline phase

$\mathcal{O}$	$\|N_C\|$	$\|N_R\|$	$\|N_I\|$	$\|\mathcal{T}\|$	$\|\mathcal{A}\|$	Offline time(sec)
GALEN	2,748	412	0	4,529	0	1,431
GO	20,465	1	0	28,897	0	7,519
NCI	27,652	70	0	46,940	0	10,901
LUBM1	59	16	50,253	94	100,543	9
LUBM10	59	16	629,568	94	1,272,575	116
UOBM-Lite1	51	43	95,010	130	245,864	62
UOBM-Lite10	51	43	820,208	130	2,096,973	679

Note: $\mathcal{O} = (\mathcal{T}, \mathcal{A})$ is a test ontology. N_C, N_R and N_I are respectively the sets of concept names, role names and individuals in $\mathcal{O}$.

5.1 Experimental Setup

We conducted experiments on the GALEN Medical Knowledge Base[4], the Gene Ontology (GO)[5], the US National Cancer Institute Ontology (NCI)[6], as well as Lehigh University Benchmark (LUBM) [10] and University Ontology Benchmark (OWL Lite version) [15] (UOBM-Lite) ontologies. By LUBMn and UOBM-Lite$n$ we respectively denote the instances of LUBM and UOBM-Lite that contain axioms about n universities. We specifically tested on LUBM1, LUBM10, UOBM-Lite1 and UOBM-Lite10, where the former two were generated by the LUBM data generator[7], and the latter two were all downloaded from the UOBM Website[8].

Before testing our approach we stored ABoxes to MySQL databases. Table 1 lists the characteristics of the seven test ontologies.

In our experiments, we randomly selected 40 subsumption entailments for each of GALEN, GO and NCI, and randomly selected 40 membership entailments over concept names for each of LUBM1, LUBM10, UOBM-Lite1 and UOBM-Lite10. For each selected subsumption, we extracted the just-preserving module introduced in this paper; for comparison, we also extracted the corresponding syntactic locality-based module. In particular, for each selected subsumption entailment $A \sqsubseteq B$, we extracted the syntactic locality-based module for A; for each selected membership entailment $A(a)$, we extracted the syntactic locality-based module for $\{a\}$. We set a time limit of 1000 seconds for Pellet to find all justifications of a selected entailment.

5.2 Experimental Results

Regarding the offline phase of our method, the execution time (in seconds) in this phase is shown in the last column of Table 1. Though the offline phase is rather costly, it is reasonable due to the following reasons. First, the offline phase is independent of any given subsumption/membership entailment; i.e., once it is executed, the extraction

[4] `http://www.openclinical.org/prj_galen.html`
[5] `http://www.geneontology.org`
[6] `http://www.mindswap.org/2003/CancerOntology/nciOntology.owl`
[7] `http://swat.cse.lehigh.edu/projects/lubm/index.htm`
[8] `http://www.alphaworks.ibm.com/tech/semanticstk/`

Table 2. The test results on finding all justifications of a selected entailment

	Module Extr. by Our Method			Syn. Locality-based Module			# SH_{LB}	# Sz_{JP}
	#SH	SHT_{avg}(sec)	$Size_{avg}$	#SH	SHT_{avg}(sec)	$Size_{avg}$	$\setminus SH_{JP}$	$< Sz_{LB}$
GALEN	40	3.555	69.75	40	3.814	134.78	0	40
GO	40	7.314	9.55	40	11.985	32.25	0	40
NCI	40	4.065	7.23	40	7.518	70.95	0	40
LUBM1	40	69.061	22.15	20	201.481	100,596.00	0	40
LUBM10	40	95.721	20.48	0	MO_1	1,272,615.00	0	40
UOBM-Lite1	16	24.813	897.80	11	155.220	245,966.00	0	40
UOBM-Lite10	15	32.278	799.83	0	MO_2	2,097,047.00	0	40

Note: "#SH" is the number of selected entailments that are *successfully* handled, i.e., all justifications of the entailment can be computed over the extracted module without running out of time/memory. "SHT_{avg}" is the average execution time for finding all justifications of each selected entailment that is successfully handled. "$Size_{avg}$" is the average number of axioms in each extracted module (counting all 40 selected entailments). "$\#SH_{LB} \setminus SH_{JP}$" is the number of selected entailments that are successfully handled in syntactic locality-based modules but not successfully handled in modules extracted by our method. "$\#Sz_{JP} < Sz_{LB}$" is the number of selected entailments whose module extracted by our method is smaller than the corresponding syntactic locality-based module. "MO_1" (resp. "MO_2") means that all runs are out of memory when finding justifications (resp. when loading modules).

of just-preserving modules for any given subsumption/membership entailment can be done without executing it again. This mechanism is suitable for the situation where the given ontology is stable, which can happen because users may want to know why an unwanted/surprising entailment holds when they get some implicit results from ontology reasoning. Second, the offline phase is theoretically tractable. More precisely, it works in time polynomial in the size of the given ontology under the assumption that numbers in qualifying number restrictions are bounded by some constant (see Lemma 6).

Regarding the online phase of our method, the test results are reported in Table 2. For comparison, Table 2 also shows the results for applying Pellet to find all justifications over syntactic locality-based modules. We need to point out that the execution time for finding all justifications over a module extracted by our method includes the module extraction time, but the execution time for finding all justifications over a syntactic locality-based module excludes the module extraction time as we assume that all syntactic locality-based modules are extracted offline.

The test results have some implications. First, a module extracted by our method is smaller than the corresponding syntactic locality-based module; esp. for membership entailments, by orders of magnitude smaller. Second, finding all justifications over modules extracted by our method is generally more efficient than finding all justifications over syntactic locality-based modules. The efficiency improvement is rather small for test ontologies with large TBoxes because most of the selected entailments have only a small justification (which results in a small syntactic locality-based module) and the module extraction time is only included in the former case. Third, finding all justifications over modules extracted by our method is much more scalable than over syntactic locality-based modules against increasing number of ABox axioms. In particular, a good number of membership entailments are handled efficiently in modules that are extracted from a test ontology with millions of ABox axioms using our method. It seems that the size of a module extracted by our method does not depend on the number of ABox axioms, but mainly depends on the TBox complexity.

6 Conclusion and Future Work

In this paper, we have proposed a goal-directed method for extracting a just-preserving module for a given entailment. The basic idea of the method is to first extract a relevant subset of a finite propositional program compiled from the given ontology, then identify a just-preserving module for the given entailment from the extracted subset. Experimental results on large ontologies show that a module extracted by our method is smaller than the corresponding syntactic locality-based module, improving the performance of the subsequent computation of all justifications. For future work, we plan to adapt the method to optimize finding all justifications of the inconsistency/incoherence of an OWL DL ontology, and upgrade the compilation method to an incremental one to cope with ontology changes.

Acknowledgments. Thanks all anonymous reviewers for their useful comments. Jianfeng Du is supported in part by NSFC grants 60673103 and 70801020. Guilin Qi is supported in part by the EU under the IST project NeOn (IST-2006-027595).

References

1. Baader, F., Calvanese, D., McGuinness, D.L., Nardi, D., Patel-Schneider, P.F. (eds.): The Description Logic Handbook: Theory, Implementation, and Applications. Cambridge University Press, Cambridge (2003)
2. Baader, F., Suntisrivaraporn, B.: Debugging SNOMED CT using axiom pinpointing in the description logic $\mathcal{EL}^+$. In: Proc. of KR-MED 2008 (2008)
3. Biere, A., Sinz, C.: Decomposing SAT problems into connected components. Journal on Satisfiability, Boolean Modeling and Computation 2, 201–208 (2006)
4. Doran, P., Tamma, V.A.M., Iannone, L.: Ontology module extraction for ontology reuse: an ontology engineering perspective. In: Proc. of CIKM 2007, pp. 61–70 (2007)
5. Eiter, T., Leone, N., Mateis, C., Pfeifer, G., Scarcello, F.: A deductive system for non-monotonic reasoning. In: Dix, J., Fuhrbach, U., Nerode, A. (eds.) LPNMR 1997. LNCS (LNAI), vol. 1265, pp. 364–375. Springer, Heidelberg (1997)
6. Fitting, M.: First-order Logic and Automated Theorem Proving, 2nd edn. Springer, New York, Inc. (1996)
7. Cuenca Grau, B., Halaschek-Wiener, C., Kazakov, Y.: History matters: Incremental ontology reasoning using modules. In: Aberer, K., Choi, K.-S., Noy, N., Allemang, D., Lee, K.-I., Nixon, L.J.B., Golbeck, J., Mika, P., Maynard, D., Mizoguchi, R., Schreiber, G., Cudré-Mauroux, P. (eds.) ASWC 2007 and ISWC 2007. LNCS, vol. 4825, pp. 183–196. Springer, Heidelberg (2007)
8. Cuenca Grau, B., Horrocks, I., Kazakov, Y., Sattler, U.: Just the right amount: Extracting modules from ontologies. In: Proc. of WWW 2007, pp. 717–726 (2007)
9. Cuenca Grau, B., Parsia, B., Sirin, E., Kalyanpur, A.: Modularity and web ontologies. In: Proc. of KR 2006, pp. 198–209 (2006)

10. Guo, Y., Pan, Z., Heflin, J.: LUBM: A benchmark for OWL knowledge base systems. Journal of Web Semantics 3(2–3), 158–182 (2005)
11. Hustadt, U., Motik, B., Sattler, U.: Reasoning in description logics by a reduction to disjunctive datalog. Journal of Automated Reasoning 39(3), 351–384 (2007)
12. Kalyanpur, A., Parsia, B., Horridge, M., Sirin, E.: Finding all justifications of OWL DL entailments. In: Aberer, K., Choi, K.-S., Noy, N., Allemang, D., Lee, K.-I., Nixon, L.J.B., Golbeck, J., Mika, P., Maynard, D., Mizoguchi, R., Schreiber, G., Cudré-Mauroux, P. (eds.) ASWC 2007 and ISWC 2007. LNCS, vol. 4825, pp. 267–280. Springer, Heidelberg (2007)
13. Kazakov, Y., Motik, B.: A resolution-based decision procedure for $\mathcal{SHOIQ}$. Journal of Automated Reasoning 40(2-3), 89–116 (2008)
14. Motik, B., Sattler, U., Studer, R.: Query answering for OWL-DL with rules. Journal of Web Semantics 3(1), 41–60 (2005)
15. Ma, L., Yang, Y., Qiu, Z., Xie, G., Pan, Y., Liu, S.: Towards a complete OWL ontology benchmark. In: Sure, Y., Domingue, J. (eds.) ESWC 2006. LNCS, vol. 4011, pp. 125–139. Springer, Heidelberg (2006)
16. Noy, N.F., Musen, M.A.: Specifying ontology views by traversal. In: McIlraith, S.A., Plexousakis, D., van Harmelen, F. (eds.) ISWC 2004. LNCS, vol. 3298, pp. 713–725. Springer, Heidelberg (2004)
17. Schlobach, S., Cornet, R.: Non-standard reasoning services for the debugging of description logic terminologies. In: Proc. of IJCAI 2003, pp. 355–362 (2003)
18. Seidenberg, J., Rector, A.L.: Web ontology segmentation: Analysis, classification and use. In: Proc. of WWW 2006, pp. 13–22 (2006)
19. Sirin, E., Parsia, B., Cuenca Grau, B., Kalyanpur, A., Katz, Y.: Pellet: A practical OWL-DL reasoner. Journal of Web Semantics 5(2), 51–53 (2007)
20. Stuckenschmidt, H., Klein, M.: Structure-based partitioning of large concept hierarchies. In: McIlraith, S.A., Plexousakis, D., van Harmelen, F. (eds.) ISWC 2004. LNCS, vol. 3298, pp. 289–303. Springer, Heidelberg (2004)
21. Suntisrivaraporn, B.: Module extraction and incremental classification: A pragmatic approach for $\mathcal{EL}^+$ ontologies. In: Bechhofer, S., Hauswirth, M., Hoffmann, J., Koubarakis, M. (eds.) ESWC 2008. LNCS, vol. 5021, pp. 230–244. Springer, Heidelberg (2008)
22. Suntisrivaraporn, B., Qi, G., Ji, Q., Haase, P.: A modularization-based approach to finding all justifications for owl dl entailments. In: Domingue, J., Anutariya, C. (eds.) ASWC 2008. LNCS, vol. 5367, pp. 1–15. Springer, Heidelberg (2008)

Analysis of a Real Online Social Network Using Semantic Web Frameworks

Guillaume Erétéo[1], Michel Buffa[2], Fabien Gandon[3], and Olivier Corby[3]

[1] Orange Labs, Sophia Antipolis, 06921 France
guillaume.ereteo@orange-ftgroup.com
[2] KEWI, I3S, University of Nice, France
buffa@unice.fr
[3] INRIA - Edelweiss, 2004 rt des Lucioles, BP 93, 06902 Sophia Antipolis
{fabien.gandon,olivier.corby}@sophia.inria.fr

Abstract. Social Network Analysis (SNA) provides graph algorithms to characterize the structure of social networks, strategic positions in these networks, specific sub-networks and decompositions of people and activities. Online social platforms like Facebook form huge social networks, enabling people to connect, interact and share their online activities across several social applications. We extended SNA operators using semantic web frameworks to include the semantics of these graph-based representations when analyzing such social networks and to deal with the diversity of their relations and interactions. We present here the results of this approach when it was used to analyze a real social network with 60,000 users connecting, interacting and sharing content.

Keywords: semantic web, social network analysis, graph representations.

1 Introduction

We are witnessing the deployment of a social media landscape where "expressing tools allow users to express themselves, discuss and aggregate their social life", "sharing tools allow users to publish and share content", and "networking tools allow users to search, connect and interact with each other" [4]. Social platforms, like Facebook, Orkut, Hi5, etc., are at the center of this landscape as they enable us to host and aggregate these different social applications. You can publish and share your del.icio.us bookmarks, your RSS streams or your microblog posts via the Facebook news feed, thanks to dedicated Facebook applications. This integration of various means for publishing and socializing enables us to quickly share, recommend and propagate information to our social network, trigger reactions, and finally enrich it.

More and more social solutions (e.g. Socialtext[1]) are deployed in corporate intranets to reproduce information sharing success stories from the web into organizations' intranets. However, the benefit of these platforms is often hindered when the social network becomes so large that relevant information is frequently lost in an

[1] http://www.socialtext.com/

A. Bernstein et al. (Eds.): ISWC 2009, LNCS 5823, pp. 180–195, 2009.

overwhelming flow of activity notifications. Organizing this huge amount of information is one of the major challenges of web 2.0[2] for its acceptance in corporate contexts and to achieve the full potential of Enterprise 2.0, i.e., the efficient use of Web 2.0 technologies like blogs and wikis within the Intranet [17].

Social Network Analysis (SNA) proposes graph algorithms to characterize the structure of a social network, strategic positions, specific sub-networks and networking activities [22]. Collaborative applications now capture more and more aspects of physical social networks and human interactions in a decentralized way. Such rich and diffuse data cannot be represented using only raw graphs as in classical SNA algorithms without some loss of knowledge. Semantic Web frameworks answer this problem of representing and exchanging knowledge on such social networks with a rich typed graph model (RDF[3]), a query language (SPARQL[3]) and schema definition frameworks (RDFS[3] and OWL[3]). Reporting on the experiments of SNA on online social networks, we have shown in [10] the lack of techniques for applying SNA on these rich typed representations of social networks. We proposed a framework to exploit directly the RDF representations of social networks using semantic web search engines, in order to take advantage of the rich information they hold and in particular the typed relations that form these labeled graphs [1] [11].

Today, most common analyses of social networks rely directly on graph theory or on algebraic approaches. Mika [18] showed that folksonomies can be exploited using graph theory in order to identify user groups and interest emergence. An approach by [19] uses FOAF profiles in order to identify communities of interest from the network of LiveJournal.com. [14] studied trust propagation in social networks using semantic web frameworks. [12] verified the power law of the degrees and community structures in FOAF profiles. [15] worked on merging FOAF profiles and identities used on different sites. Other researchers like [8] (see [10] and [11]) have extended tools, e.g., the SPARQL query language, in order to find paths between semantically linked resources in RDF-based graphs. These works will be a basis for us to work on graph-based and ontology-based social network representation and analysis.

Many algorithms are available for detecting social structures, roles and positions. But one aspect of our work that differentiates us from other approaches is in going beyond the application of classical algorithms to social networks described with semantic web technologies. We propose to extend these algorithms and to create new ones, in order to manage communities' life cycles, taking into account not only the graph structure but also the semantics of the ontological primitives used to label its nodes and arcs.

We first introduce our framework for exploiting the graph models underlying RDF representations of social networks. We provide formal definitions in SPARQL of SNA operators parameterized by the ontologies underlying these representations. In addition we detail SemSNA, an ontology of SNA characteristics used to annotate social networks. Finally we present how we conducted a semantic social network analysis on an anonymized dataset from Ipernity.com and the results we obtained.

[2] http://oreilly.com/web2/archive/what-is-web-20.html
[3] Semantic Web, W3C, http://www.w3.org/2001/sw/

2 Semantic Social Network Analysis

We use the RDF graphs to represent social networks, and we type those using existing ontologies together with specific domain ontologies if needed. Some social data are already readily available in a semantic format (RDF, RDFa, μformats, etc.). However, today, most of the data are still only accessible through APIs (flickr, Facebook, etc.) or by crawling web pages and need to be converted. To annotate these social network representations with SNA indices, we designed SemSNA (Fig. 1), an ontology that describes SNA notions, e.g., centrality. With this ontology, we can (1) abstract social network constructs from domain ontologies to apply our tools on existing schemas by having them extend our primitives; and we can (2) enrich the social data with new annotations (Fig. 2) such as the SNA indices that will be computed. These annotations enable us to manage more efficiently the life cycle of an analysis, by pre-calculating relevant SNA indices and updating them incrementally when the network changes over time. On top of SemSNA we propose SPARQL formal definitions of SNA operators handling the semantics of the representations. The current tests use the semantic search engine Corese [7] that supports powerful SPARQL extensions particularly well suited for SNA features such as path computations [8].

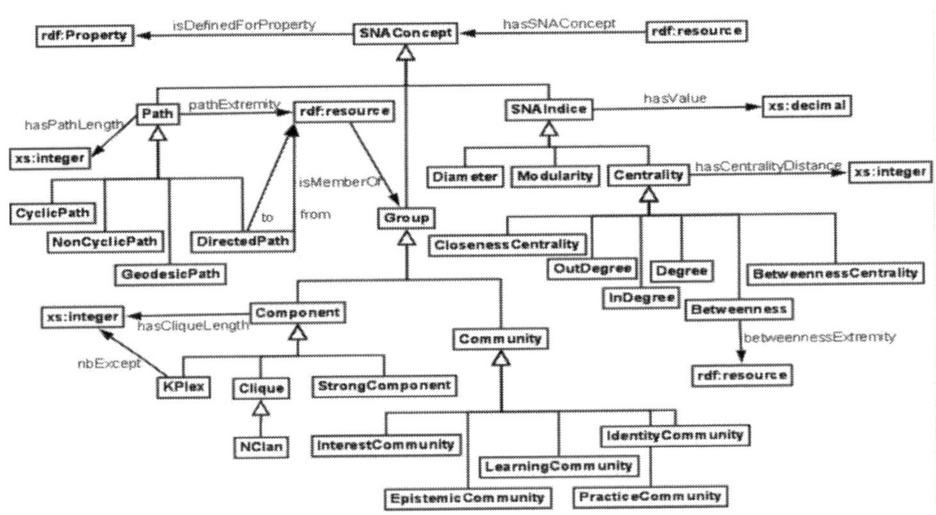

Fig. 1. Schema of SemSNA: the ontology of Social Netwtok Analysis

2.1 A New Version of SemSNA: The Ontology of Social Network Analysis

We have designed a new version of SemSNA[4], an ontology of Social Network Analysis. The first version [11] was focused on strategic position based on Freeman's definition of centrality [13]. Since then, we introduced many new primitives to annotate social data with different definitions of groups and useful indices to characterize their properties.

[4] http://ns.inria.fr/semsna/2009/06/21/voc

The main class SNAConcept is used as the super class for all SNA concepts. The property isDefinedForProperty indicates for which relationship, i.e., sub-network, an instance of the SNA concept is defined. An SNA concept is attached to a social resource with the property hasSNAConcept. The class SNAIndice describes valued concepts such as centrality, and the associated value is set with the property hasValue. We have slightly modified the modelling of centralities of the previous version [11]. We created a super class Centrality for all centralities defined by the classes Degree, InDegree, OutDegree, Betweenness, BetweennessCentrality and ClosenessCentrality. The property hasCentralityDistance defines the distance of the neighbourhood taken into account to measure the centrality.

We propose a set of primitives to define and annotate groups of resources linked by particular properties. The class Group is a super class for all classes representing an alternative concept of group of resources. The class Component represents a set of connected resources. The class StrongComponent defines a component of a directed graph where the paths connecting its resources do not contain any change of direction.

The Diameter, subclass of Indice, defines the length of the longest geodesics (shortest paths between resources) of a component. The property maximumDistance enables us to restrict component membership to a maximum path length between members. A clique is a complete sub graph, for a given property according to our model. An n-clique extends this definition with a maximum path length (*n*) between members of the clique; the class Clique represents this definition, and the maximum path length is set by the property maximumDistance. Resources in a clique can be linked by shortest paths going through non clique members. An NClan is a restriction of a clique that excludes this particular case. A KPlex relaxes the clique definition to allow connecting to *k* members with a path longer than the clique distance, *k* is determined by the property nbExcept. Finally the concept Community supports different community definitions: InterestCommunity, LearningCommunity, GoalOrientedCommunity, PraticeCommunity and EpistemicCommunity [16] [5]. These community classes are linked to more detailed ontologies like [23] used to represent communities of practice.

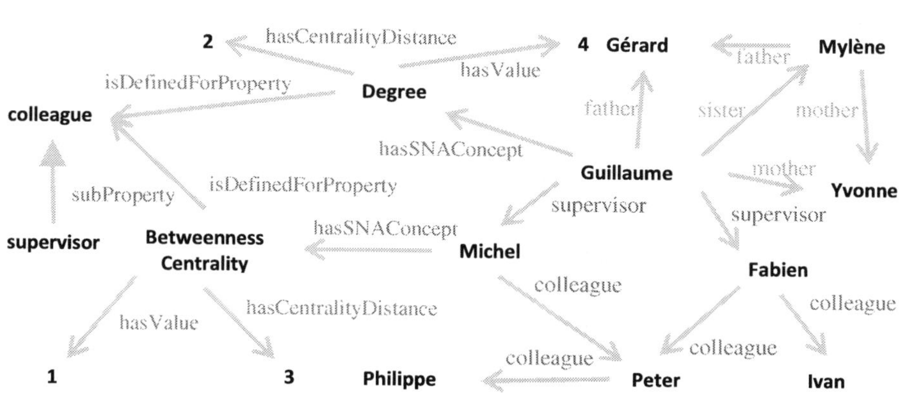

Fig. 2. A social network annotated with SemSNA indices (Degree, Betweenness)

2.2 Extract SNA Concepts with SPARQL

In [21], researchers have shown that SPARQL "is expressive enough to make all sensible transformations of networks". However, this work also shows that SPARQL is not expressive enough to meet SNA requirements for global metric querying (density, betweenness centrality, etc.) of social networks. Such global queries are mostly based on result aggregation and path computation which are missing from the standard SPARQL definition. The Corese search engine [7] provides such features with result grouping, aggregating function like `sum()` or `avg()` and path retrieving [8] [11]. We present here how to perform social network analysis combining structural and semantic characteristics of the network with queries performed with Corese, e.g., Figure 3 illustrates the calculation of a parameterized degree where only family relations are considered.

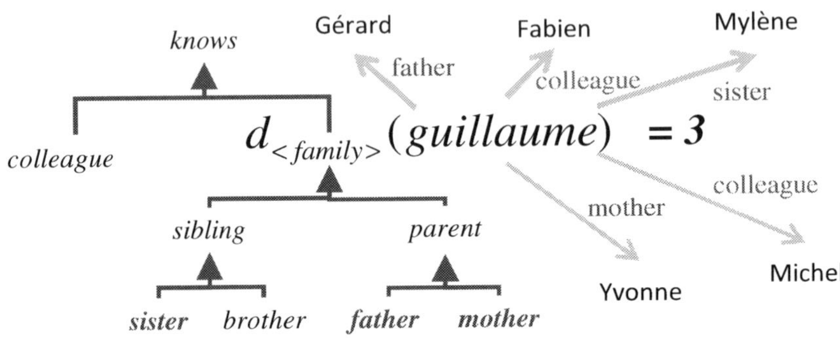

Fig. 3. A Parameterized degree that considers a hierarchy of relations

We introduced a new syntactic convention in Corese for path extraction. A regular expression is now used instead of the property variable to specify that a path is searched and to describe its characteristics. The regular expression operators are: `/` (sequence), `|` (or), `*` (0 or more), `?` (optional), `!` (not). We can bind the path with a variable specified after the regular expression. The following example retrieves a path between two resources `?x` and `?y` starting with zero or more `foaf:knows` properties and ending with the `rel:worksWith` property; the path length must be equal to or less than 3:

```
?x foaf:knows*/rel:worksWith::$path ?y
filter(pathLength($path) <= 3)
```

Path characteristics are defined by adding options before the regular expression: `'i'` to allow inverse properties, `'s'` to retrieve one shortest path, `'sa'` to retrieve all shortest paths, e.g., `?x i sa foaf:knows*/worksWith ?y`.

Path retrieval enables us to exploit the hierarchy of relation, by taking into account sub-properties at each step. Consequently we propose parameterized queries that accept as argument a regular expression of properties. Figure 3 shows a parameterized degree that only considers the hierarchy of family relationships.

Table 1. Definition of SNA notions in labeled oriented graphs and notations used

SNA indices and definition	Notation
Graph: defined as in [1] and [11]	$G=(E_G, R_G, n_G, l_G)$ where : • E_G and R_G are two disjoint finite sets respectively, of nodes and relations. • $n_G : R_G \rightarrow E_G^*$ associates to each relation a couple of entities called the arguments of the relation. If $n_G(r)=(e_1,e_2)$ we note $n_G^i(r)=e_i$ the ith argument of r. • $l_G : E_G \cup R_G \rightarrow L$ is a labelling function of entities and relations.
Number of actors: the number of actors of a given type.	$nb^{actor}_{<type>}(G)$
Number of actors: the number of actors involved in a given relation as subject or object.	$nb^{actor}_{<rel>}(G)$
Number of subject actors: the number of actors involved in a given relation as subject.	$nb^{subject}_{<rel>}(G)$
Number of object actors: the number of actors involved in a given relation as object.	$nb^{object}_{<rel>}(G)$
Number of relations: number of pairs of resources linked by a relation *rel*.	$nb^{relation}_{<rel>}(G)$
Path: a list of nodes of a graph *G* each linked to the next by a relation belonging to *G*.	$path\ p_{<rel>} =< x_0, x_1, x_2 ... x_n > \wedge \forall i < n \exists r \in R_G ; n_G(r) = (x_i, x_{i+1}) \wedge l_G(r) \leq rel$
Length: the number of relations/links/arcs involved in a path	$length(p =< x_0, x_1, x_2 ... x_n >) = n$
Density: proportion of the maximum possible number of relationships. It represents the cohesion of the social network.	$Den_{<rel>}(G) = \dfrac{nb^{relation}_{<rel>}(G)}{nb^{actor}_{<domain(rel)>}(G) * nb^{actor}_{<range(rel)>}(G)}$
Component: a connected sub graph for a given property (and sub-properties) with no link to resources outside the component	$Comp_{<rel>}(G) = (G_1, G_2, ..., G_k)$ where G_k is a subgraph of G such that for every pair of nodes n_i, n_j of G_k there exist a path from n_i to n_j in G_k.
Degree: number of paths of properties of type *rel* (or subtype) having *y* at one end and with a length smaller or equal to *dist*. It highlights local popularities.	$D_{<rel,dist>}(y) = \left\| \begin{Bmatrix} x_n ; \exists path\ p_{<rel>} =< x_0, x_1, x_2 ... x_n > \\ \wedge ((x_0 = y) \vee (x_n = y)) \wedge n \leq dist \end{Bmatrix} \right\|$

Table 1. *(continued)*

In-Degree: number of paths of properties of type *rel* (or subtype) ending by *y* and with a length smaller or equal to *dist*. It highlights supported resources.	$D^{in}_{<rel,dist>}(y) = \left\| \begin{Bmatrix} x_n ; \exists path\ p_{<rel>} = < x_0, x_1, x_2 \ldots x_n > \\ \wedge (x_n = y) \wedge n \leq dist \end{Bmatrix} \right\|$
Out-Degree: number of paths of properties of type *rel* (or subtype) starting by *y* and with a length smaller or equal to *dist*. It highlights influencing resources.	$D^{out}_{<rel,dist>}(y) = \left\| \begin{Bmatrix} x_n ; \exists path\ p_{<rel>} = < x_0, x_1, x_2 \ldots x_n > \\ \wedge (x_0 = y) \wedge n \leq dist \end{Bmatrix} \right\|$
Geodesic between *from* and *to*: a geodesic is a shortest path between two resources, an optimal communication path.	$g_{<rel>}(from, to) = < from, x_1, x_2 \ldots, x_n, to >$ such that: $\forall p_{<rel>} = < from, y_1, y_2 \ldots, y_m, to >$ n•m
Diameter: the length of le longest geodesic in the network.	$Diam_{<rel>}(G)$
Number of geodesics between *from* and *to*: the number of geodesics between two resources shows the dependency of their connectivity with their intermediaries.	$nb^{g}_{<rel>}(from, to)$
Number of geodesics between *from* and *to* going through node *b*: the number of geodesics between two resources going through a given intermediary node; it shows the connectivity dependency of these resources w.r.t. *b*.	$nb^{g}_{<rel>}(b, from, to)$
Closeness centrality: the inverse sum of shortest distances to each other resource. It represents the capacity of a resource to access or to be reached.	$C^{c}_{<rel>}(k) = \left[\sum_{x \in E_G} length(g_{<rel>}(k, x)) \right]^{-1}$
Partial betweenness: the probability for *k* to be on a shortest path of properties of type *rel* (or subtype) between *x* and *y*. It represent the capacity of *k* to be an intermediary or a broker between *x* and *y* for this type (or subtype) of property.	$B_{<rel}(b, x, y) = \frac{nb^{g}_{<rel>}(b, x, y)}{nb^{g}_{<rel>}(x, y)}$
Betweenness centrality: the sum of the partial betweenness of a node between each other pair of nodes. It represents the capacity of a node to be an intermediary in the whole network.	$C^{b}_{<rel>}(b) = \sum_{x, y \in E_G} B_{<rel>}(b, x, y)$

Table 2. Formal definition in SPARQL of semantically parameterized SNA indices

SNA indices	SPARQL formal definition

$nb^{actor}_{<type>}(G)$	select merge[5] count(?x) as ?nbactor from <G>
	where{
	?x rdf:type param[type][6]
	}

$nb^{actor}_{<rel>}(G)$	select merge count(?x) as ?nbactors from <G>
	where{
	{?x param[rel] ?y}
	UNION{?y param[rel] ?x}
	}

$nb^{subject}_{<rel>}(G)$	select merge count(?x) as ?nbsubj from <G>
	where{
	?x param[rel] ?y
	}

$nb^{object}_{<rel>}(G)$	select merge count(?y) as ?nbobj from <G>
	where{
	?x param[rel] ?y
	}

$nb^{relation}_{<rel>}(G)$	select cardinality(?p) as ?card from <G> where
	{
	{ ?p rdf:type rdf:Property
	filter(?p ^ param[rel]) }
	UNION
	{ ?p rdfs:subPropertyOf ?parent
	filter(?parent ^ param[rel]) }
	}

$Comp_{<rel>}(G)$	select ?x ?y from <G> where { ?x param[rel] ?y }group by any[7]
$D_{<rel,dist>}(y)$	select ?y count(?x) as ?degree where { {?x (param[rel])*::$path ?y filter(pathLength($path) <= param[dist])} UNION {?y param[rel]::$path ?x filter(pathLength($path) <= param[dist])} }group by ?y
$D^{in}_{<rel,dist>}(y)$	select ?y count(?x) as ?indegree where{ ?x (param[rel])*::$path ?y filter(pathLength($path)[8] <= param[dist]) }group by ?y

[5] The *merge* keyword merges all results in one with distinct values for each variable.
[6] Corese accepts a list of parameters at the execution of a query.
[7] The keyword *any* enables to group results having the same value for any result variables
[8] The *pathLength()* function returns the length a the path given in parameter.

Table 2. *(continued)*

$D^{out}_{<rel,dist>}(y)$	select ?x count(?y) as ?outdegree where { ?x (param[rel])*::$path ?y filter(pathLength($path) <= param[dist]) }group by ?x
$g_{<rel>}(from,to)$	select ?from ?to $path pathLength($path) as ?length where{ ?from sa (param[rel])*::$path ?to }group by ?from ?to
$Diam_{rel>}(G)$	select pathLength($path) as ?length from <G> where { ?y s (param[rel])*::$path ?to }order by desc(?length) limit 1
$nb^{g}_{<rel>}(from,to)$	select ?from ?to count($path) as ?count where{ ?from sa (param[rel])*::$path ?to }group by ?from ?to
$nb^{g}_{<rel>}(b,from,to)$	select ?from ?to ?b count($path) as ?count where{ ?from sa (param[rel])*::$path ?to graph $path{?b param[rel] ?j} filter(?from != ?b) optional { ?from param[rel]::$p ?to } filter(!bound($p)) }group by ?from ?to ?b
$C^{c}_{<rel>}(y)$	select distinct ?y ?to pathLength($path) as ?length (1/sum(?length)) as ?centrality where{ ?y s (param[rel])*::$path ?to }group by ?y
$B_{<rel>}(b,from,to)$	select ?from ?to ?b (count($path)/count($path2)) as ?betweenness where{ ?from sa (param[rel])*::$path ?to graph $path{?b param[rel] ?j} filter(?from != ?b) optional { ?from param[rel]::$p ?to } filter(!bound($p)) ?from sa (param[rel])*::$path2 ?to }group by ?from ?to ?b
$C^{b}_{<rel>}(b)$	Non SPARQL post-processing on shortest paths.

We propose a set of queries (Table 2) to compute SNA metrics adapted to the oriented labeled graphs described in RDF (Table 1). These queries exploit the path retrieval features as well as grouping and aggregating functions. We implemented and tested all the operators that we present.

3 Linking Online Interaction Data to the Semantic Web

Ipernity.com, the social network we analyzed, offers users several options for building their social network and sharing multimedia content. Every user can share pictures, videos, music files, create a blog, a personal profile page, and comment on other's shared resources. Every resource can be tagged and shared. To build the social network, users can specify the type of relationship they have with others: friend, family, or simple contact (like a favorite you follow). Relationships are not symmetric, *Fabien* can declare a relationship with *Michel* but *Michel* can declare a different type of relationship with *Fabien* or not have him in his contact list at all; thus we have a directed labeled graph. Users have a homepage containing their profile information and pointers to the resources they share. Users can post on their profile and their contacts' profiles depending on access rights. All these resources can be tagged including the homepage. A publisher can configure the access to a resource to make it public, private or accessible only for a subset of its contacts, depending on the type of relationship (family, friend or simple contact), and can monitor who visited it. Groups can also be created with topics of discussion with three kinds of visibility, public (all users can see it and join), protected (visible to all users, invitation required to join) or private (invitation required to join and consult).

3.1 SemSNI: Extending SIOC to Model Social Networking Interactions

Several Ontologies already exist to represent online social networks [10] [11], and we use them as a basis for our model. We use FOAF[9] to represent persons' profiles and their relationships in combination with the RELATIONSHIP[10] ontology that extends the `foaf:knows` property with sub-properties. RELATIONSHIP proposes many family properties (`parentOf`, `siblingOf`, etc.) but it does not have the super property we need for all these relations: family. Thus we extend the RELATIONSHIP ontology with the property `SemSNI:family` as a super property for all these family relationships. We modelled the favorite and friend relationship respectively with the property `knowsInPassing` and `friendOf` of this ontology. The SIOC[11] [3] ontology provides the basis for defining a user (class `User`), the content he produces (class `Item`, property `has_creator`) and the actions of others users on this content (property `has_reply`). SIOC types[12] extend `sioc:Item` to specify different types of resources produced online. We use `sioc:Post`, `sioct:ImageGallery`, `sioct:Weblog` and `sioct:Comment` to model respectively posts on homepages, photo albums, blogs and comments on resources. We use SCOT[13] to model tagging.

In order to model homepages, private messages, discussion topics, and documents that do not exist in SIOC types with the required semantics, we designed SemSNI (Semantic Social Network Interactions, Fig. 4). SemSNI defines the class `UserHome`, `PrivateMessage`, `Topic` and `Document` as subclasses of `sioc:Item`

9 http://www.foaf-project.org/
10 http://vocab.org/relationship/
11 http://sioc-project.org/
12 http://rdfs.org/sioc/types
13 http://scot-project.org/

(`SemSNI:Document` also extends `foaf:Document`). The class `Visit` and the properties `visitedResource` and `hasVisitor` enable us to describe the visits of a user to a resource. In order to infer new relationships between users from their interactions and the content they share, SemSNI defines the class `Interaction` and the property `hasInteraction` (domain: `sioc:User`, range: `sioc:User`). The classes representing exchanges on a content (`sioct:Comment`, `SemSNI:Visit` and `SemSNI:PrivateMessage`) are defined as subclasses of `SemSNI:Interaction` and we can infer a relation `hasInteraction` between the creator of the resource and its consumer. We did not type more precisely such relations, but we can extend this property in order to increase the granularity in the description of interactions between users. We use the types of access defined in AMO[14] (`Public`, `Private`, `Protected`) in combination with the property `SemSNI:sharedThroughProperty` to model the kind of sharing for resources.

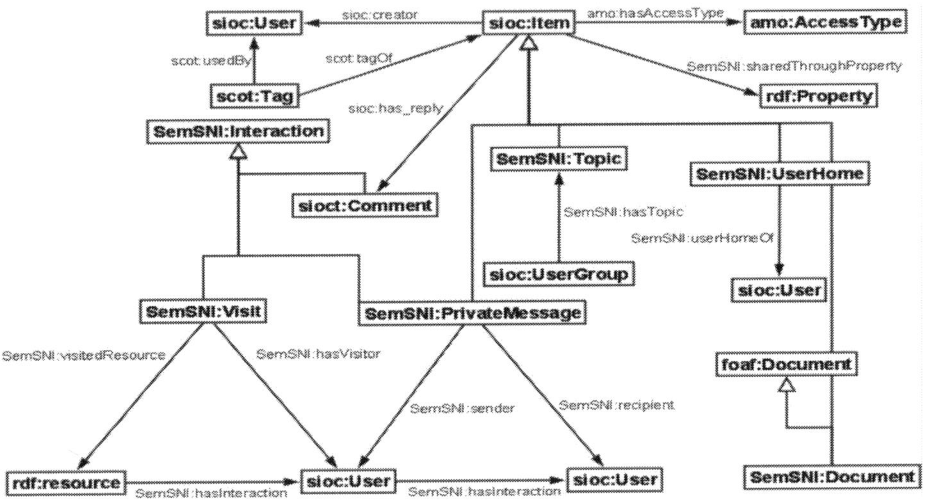

Fig. 4. Schema of SemSNI; an ontology of Social Network Interactions

3.2 Generating RDF from a Relational Database with Corese

We used another feature of Corese to transform the relational database of the Ipernity.com social network into RDF/XML. Indeed, Corese has an extension that enables us to nest SQL queries within SPARQL queries. This is done by means of the `sql()` function that returns a sequence of results for each variable in the SQL `select` clause. Corese has an extension to the standard SPARQL `select` clause that enables us to bind these results to a list of variables [9]. In the following example, we show how we retrieve the friend relationships from the relational database, using this `sql()` function and another one (`genIdUrl()`) that generates URIs from relational database primary keys (ids):

[14] Access Management Ontology, currently developed by the KEWI team at I3S.

```
construct { ?url_user1 rel:friendOf ?url_user2 }
select sql('jdbc:mysql://localhost:3306/mysql',
'com.mysql.jdbc.Driver', 'user', 'pwd',
'SELECT user1_id, user2_id from relations where rel = 2 limit
100000' ) as (?id1, ?id2)
fun:genIdUrl(?id1, 'http://semsni.fr/people/') as ?url_user1
fun:genIdUrl(?id2, 'http://semsni.fr/ people/') as ?url_user2
where {   }
```

Like in [21], once exported in RDF, this network can be processed to leverage the social network representation with a SPARQL query using a `construct` block. Corese can automate some transformations with inference rules [6]. As an example we can infer a property `SemSNI:hasInteraction` between two actors when one commented on the other's resource using the following rule (future improvements include using the Rule Interchange Format syntax):

```
<cos:if>
  { ?doc sioc:has_creator ?person1
    ?doc sioc:has_reply ?comment
    ?comment sioc:has_creator ?person2 }
</cos:if>
<cos:then>
  {?person1 semsni:hasInteraction ?person2   }
</cos:then>
```

4 Results

We tested our algorithms and queries on an bi-processor quadri-core Intel(R) Xeon(R) CPU X5482 3.19GHZ, 32.0Gb of RAM. We applied the defined queries on relations and interactions from Ipernity.com. We analyzed the three types of relations separately (*favorite*, *friend* and *family*) and also used polymorphic queries to analyze them as a whole using their super property: `foaf:knows`. We also analyzed the interactions produced by exchanges of private messages between users, as well as the ones produced by someone commenting someone else's documents.

We first applied quantitative metrics to get relevant information on the structure of the network and activities: the number of links and actors, the *components* and the *diameters*. 61,937 actors are involved in a total of 494,510 relationships. These relationships are decomposed in 18,771 *family* relationships between 8,047 actors, 136,311 *friend* relationships involving 17,441 actors and 339,428 *favorite* relationships

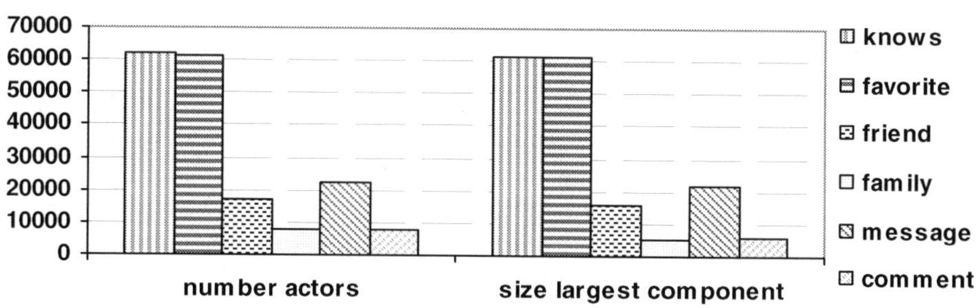

Fig. 5. Number of actors and size of the largest component of the studied networks

Table 3. Performance of queries

Indice	Relation	Query time	Nb of graph projections
$Diam_{rel>}(G)$	Knows	1 m 41.93 s	10,000,000
	Favorite	1 m 51.37 s	10,000,000
	Friend	1 m 42.62 s	10,000,000
	Family	1 m 54.28 s	10,000,000
	Comment	35.06 s	1,000,000
	Message	1 m 50.84 s	10,000,000
$nb_{actors<rel>}(G)$	Knows	1 m 9.62 s	989,020
	Favorite	2 m 35.29 s	678,856
	Friend	11.67 s	272,622
	Family	0.68 s	37,542
	Message	17.62 s	1,448,225
	Comment	8 m 27.25 s	7,922,136
$Comp_{<rel>}(G)$	Knows	0.71 s	494,510
	Favorite	0.64 s	339,428
	Friend	0.31 s	136,311
	Family	0.03 s	18,771
	Message	1.98 s	795,949
	Comment	9.67 s	2,874,170
$D_{<rel,1>}(y)$	Knows	20.59 s	989,020
	Favorite	18.73 s	678,856
	Friend	1.31 s	272,622
	Family	0.42 s	37,542
	Message	16.03 s	1,591,898
	Comment	28.98 s	5,748,340
Shortest paths used to calculate $C_{b<rel>}(b)$	Knows	Path length <= 2: 2h 56m 34.13s	1,000,000
	Favorite	Path length <= 2: 5h 33m 18.43s	2,000,000
	Friend	Path length <= 2: 1m 12.18 s	1,000,000
	Family	Path length <= 2 : 27.23 s Path length <= 3 : 1m 10.71 s	1,000,000 1,000,000

for 61,425 actors. These first metrics show that the semantics of relations are globally respected, as *family* relations are less used than *friend* and *favorite*. 7,627 actors have interacted through 2,874,170 comments and 22,500 have communicated through 795,949 messages. All these networks are composed of a largest *component* containing most of the actors (Fig. 5) and few very small components (less than 100 actors) that show "the effectiveness of the social network at doing its job" [20], in connecting people. The interaction sub networks have a very small diameter (3 for comments and 2 for messages) due to their high density. The *family* network has a high diameter

(19), consistent with its low density. However the *friend* and *favorite* networks have a low density and a low diameter revealing the presence of highly intermediary actors.

The betweenness and degree centralities confirm this last hypothesis. The *favorite* network is highly centralized, with five actors having a betweenness centrality higher than 0, with a dramatically higher value for one actor: one who has a betweenness centrality of 1,999,858 and the other 4 have a value comprised between 2.5 and 35. This highest value is attributed to the official animator of the social network who has a *favorite* relationship[15] with most actors of the network, giving him the highest degree: 59,301. In the *friend* network 1,126 actors have a betweenness centrality going from 0 to 96,104 forming a long tail, with only 12 with a value higher than 10,000. These actors do not include the animator, showing that the *friend* network has been well adopted by users. The *family* network has 862 actors with a betweenness centrality from 0 to 162,881 with 5 values higher than 10,000. Only one actor is highly intermediary in both *friend* and *family* networks. The centralization of these three networks presents significant differences showing that the semantics of relations have an impact on the structure of the social network. The betweenness centralities of all the relations, computed using the polymorphism in SPARQL queries with the `knows` property, highlight both the importance of the animator who has again the significantly highest centrality and the adoption by users with 186 actors playing a role of intermediary. The employees of Ipernity.com have validated these interpretations of the metrics that we computed, showing the effectiveness of a social network analysis that exploits the semantic structure of relationships.

The Corese engine works in main memory and such an amount of data is memory consuming. The 494,510 relations declared between 61,937 actors use a space of 4.9 Gb. Annotations of all messages use 14.7 Gb and the representation of documents with their comments use 27.2 Gb. On the other hand working in main memory allows us to process the network very rapidly. The path computation is also time and space consuming and some queries had to be limited to a maximum number of graph projections when too many paths could be retrieved. However, in that case approximations are sufficient to obtain relevant metrics (Table 3) on a social network, i.e., for centralities [2]. Moreover, we can limit the distance of the paths we are looking for by using others metrics. For example, we limit the depth of paths to be smaller or equal to the diameter of the components when computing shortest paths.

5 Conclusion

The huge amounts of social activities and user-generated contents have to be properly organized and filtered to preserve the benefits of online collaboration. While SNA provides relevant metrics to understand the structure of online activities, Semantic Web technologies enable us to represent, to mash and to query social data from applications spanning both internet and intranet networks. The directed labeled graph structure of RDF is well suited to represent such social knowledge and such socially produced metadata. Our framework allows analyzing these rich typed representations of social networks and handling the diversity of interactions and relationships with parameterized SNA metrics. Classical SNA ignores the semantics of richly typed

---

15 This animator is an employee of the company that animates the social network. He declares as *favorite* every user who just created an account and sends him welcome messages.

graphs like RDF and classical Relational Database approaches do not offer simple mechanisms for handling the semantics of type lattices. Subsumption relations are natively taken into account when querying the RDF graph in SPARQL with an engine like CORESE. Parameterized operators formally defined in SPARQL rely on this to allow us to adjust the granularity of the analysis of relations. Moreover, a new range of pre-processing can be used such as rules crawling the network to add types or relations whenever they detect a pattern, e.g., an actor frequently commenting on posts by another actor is linked to him by a relation "monitors".

New queries that compute new operators can be defined at anytime and SemSNA can be extended. Network assortativity [20] is an example of future operators that could both leverage the semantics of the schemas (e.g., similarity between two nodes) and extension mechanisms of SPARQL (e.g., counting the number of shared connections). In addition, using a schema to add the results of our queries (or rules) to the network also allows us to decompose complex processing into two or more stages and to factorize some computation among different operators, e.g., we can augment the network with in-degree calculation and betweenness calculation and then run a query on both criteria to identify nodes with an in-degree > y and a betweenness > x.

Furthermore we validated this framework on a real social network and revealed the importance of considering the diversity of relationships and their semantic links. The sub-networks we analyzed present different characteristics that highlight in particular the strategic actors and the partioning of the different activities. The approach is applied as batch processing on large RDF triple store (CORESE is a freeware handling millions of nodes but other engines with the same extensions could be used just as well). Consequently we annotate the social data with the results of these parameterized SNA metrics using SemSNA ontology to provide services based on this analysis (e.g. filter social activity notifications), to use them in the calculation of more complex indices or (in the future) to support iterative or parallel approaches in the computations. Computation is time consuming and even if CORESE runs in main memory, experiments reported in the paper show that handling a network with millions of actors is out of our reach today. We started to study different approaches for addressing that problem: (1) identifying computation techniques that are iterative, parallelizable, etc. (2) identifying approximations that can be used and under which conditions they provide good quality results (3) identifying graph characteristics (small worlds, diameters, etc.) that can help us cut the calculation space and time for the different operators. Our perspectives include also the development of a semantic based community detection algorithm and methods to manage the evolution of such ever-changing networks. More precisely, we plan to exploit these semantic based SNA metrics to structure overwhelming flows of corporate social data and to foster social interactions.

Acknowledgments. We thank the ANR for funding the ISICIL project ANR-08-CORD-011 that led to these results. We thank Peter Sander for proof-reading this article.

References

1. Baget, J.-F., Corby, O., Dieng-Kuntz, R., Faron-Zucker, C., Gandon, F., Giboin, A., Gutierrez, A., Leclère, M., Mugnier, M.-L., Thomopoulos, R.: Griwes: Generic Model and Preliminary Specifications for a Graph-Based Knowledge Representation Toolkit. In: Eklund, P., Haemmerlé, O. (eds.) ICCS 2008. LNCS (LNAI), vol. 5113, pp. 297–310. Springer, Heidelberg (2008)

2. Brandes, U., Pich, C.: Centrality estimation in large networks. Bifurcation and Chaos in Applied Sciences and Engineering 17(7), 2303–2318 (2007)
3. Breslin, J.G., Harth, A., Bojars, U., Decker, S.: Towards Semantically-Interlinked Online Communities. In: Gómez-Pérez, A., Euzenat, J. (eds.) ESWC 2005. LNCS, vol. 3532, pp. 500–514. Springer, Heidelberg (2005)
4. Cavazza, F.: `http://www.fredcavazza.net/2009/04/10/social-media-landscape-redux/` (2009)
5. Conein, B.: Communautés épistémiques et réseaux cognitifs: coopération et cognition distribuée. Revue D'Economie Politique 113, 141–159 (2004)
6. Corby, O., Faron-Zucker, F.: Corese: A Corporate Semantic Web Engine. In: Workshop on Real World RDF and Semantic Web Applications (2002)
7. Corby, O., Dieng-Kuntz, R., Faron-Zucker, C.: Querying the Semantic Web with the Corese Search Engine. In: ECA/PAIS 2004 (2004)
8. Corby, O.: Web, Graphs and Semantics. In: Eklund, P., Haemmerlé, O. (eds.) ICCS 2008. LNCS (LNAI), vol. 5113, pp. 43–61. Springer, Heidelberg (2008)
9. Corby, O., Kefi-Khelif, L., Cherfi, H., Gandon, F., Khelif, K.: Querying the Semantic Web of Data using SPARQL, RDF and XML. INRIA Research Report n°6847 (2009)
10. Erétéo, G., Buffa, M., Gandon, F., Grohan, P., Leitzelman, M., Sander, P.: A State of the Art on Social Network Analysis and its Applications on a Semantic Web. In: SDoW 2008, Workshop at ISWC 2008 (2008)
11. Erétéo, G., Gandon, F., Corby, O., Buffa, M.: Semantic Social Network Analysis. In: Web Science 2009 (2009)
12. Finin, T., Ding, L., Zou, L.: Social networking on the semantic web. Learning organization journal 5(12), 418–435 (2005)
13. Freeman, L.C.: Centrality in Social Networks: Conceptual Clarification. Social Networks 1, 215–239 (1979)
14. Golbeck, J., Parsia, B., Hendler, J.: Trust network on the semantic web. In: Proceedings of cooperative information agents (2003)
15. Goldbeck, J., Rothstein, M.: Linking social Networks on the web with FOAF. In: Proceedings of the twenty-third conference on artificial intelligence, AAA 2008 (2008)
16. Henri, F., Pudelko, B.: Understanding and analyzing activity and learning in virtual communities. Journal of Computer Assisted Learning 19, 474–487 (2003)
17. McAfee, A.-P.: Enterprise 2.0: The Dawn of Emergent Collaboration, MIT Sloan Management Review, Management of Technology and Innovation (April 1, 2006)
18. Mika, P.: Ontologies are Us: a unified Model of Social Networks and Semantics. In: Gil, Y., Motta, E., Benjamins, V.R., Musen, M.A. (eds.) ISWC 2005. LNCS, vol. 3729, pp. 522–536. Springer, Heidelberg (2005)
19. Paolillo, J.C., Wright, E.: Social Network Analysis on the Semantic Web: Techniques and Challenges for Visualizing FOAF. In: Book Visualizing the semantic Web Xml-based Internet and Information (2006)
20. Newman, M.E.J.: The Structure and Function of Complex Networks. SIAM rev. 45, 167–256 (2003)
21. San Martin, M., Gutierrez, C.: Representing, Querying and Transforming Social Networks with RDF / SPARQL. In: ESWC 2009 (2009)
22. Scott, J.: Social Network Analysis, a handbook, 2nd edn. Sage, Thousand Oaks (2000)
23. Vidou, G., Dieng-Kuntz, R., El Ghali, A., Evangelou, C., Giboin, A., Tifous, A., Jacquemart, S.: Towards an Ontology for Knowledge Management in Communities of Practice (2006)

Coloring RDF Triples to Capture Provenance

Giorgos Flouris[1], Irini Fundulaki[1], Panagiotis Pediaditis[1,2], Yannis Theoharis[1,2], and Vassilis Christophides[1,2]

[1] Institute of Computer Science, FORTH, Greece
[2] Computer Science Department, University of Crete, Greece
{fgeo,fundul,pped,theohari,christop}@ics.forth.gr

Abstract. Recently, the W3C Linking Open Data effort has boosted the publication and inter-linkage of large amounts of RDF datasets on the Semantic Web. Various ontologies and knowledge bases with millions of RDF triples from Wikipedia and other sources, mostly in e-science, have been created and are publicly available. Recording provenance information of RDF triples aggregated from different heterogeneous sources is crucial in order to effectively support trust mechanisms, digital rights and privacy policies. Managing provenance becomes even more important when we consider not only explicitly stated but also implicit triples (through RDFS inference rules) in conjunction with declarative languages for querying and updating RDF graphs. In this paper we rely on *colored RDF triples* represented as quadruples to capture and manipulate explicit provenance information.

1 Introduction

Recently, the W3C Linking Open Data [29] effort has boosted the publication and interlinkage of large amounts of RDF datasets on the Semantic Web [1]. Various ontologies and knowledge bases with millions of RDF triples from Wikipedia [26] and other sources have been created and are available online [25]. In addition, numerous data sources in e-science are published nowadays as RDF graphs, most notably in the area of life sciences [27], to facilitate community annotation and interlinkage of both scientific and scholarly data of interest. Finally, Web 2.0 platforms are considering RDF and RDFS as non-proprietary exchange formats for the construction of information mashups [16,28].

In this context, it is of paramount importance to be able to store the *provenance of a piece of data* in order to effectively support trust mechanisms, digital rights and privacy policies. *Provenance* means *origin* or *source* and refers to *from where* and *how* the piece of data was obtained [33]. In the context of scientific communities, provenance information can be used in the proof of the correctness of results and in general determines their quality. In some cases, provenance of data is considered more important than the result itself.

The popularity of the RDF data model [8] and RDF Schema language (RDFS) [2] is due to the flexible and extensible representation of information, independently of the existence or absence of a schema, under the form of *triples*. An RDF triple, $(subject,$-$property,object)$, asserts the fact that *subject* is associated with *object* through *property*. RDFS is used to add semantics to RDF triples, by imposing *inference rules* [15]

A. Bernstein et al. (Eds.): ISWC 2009, LNCS 5823, pp. 196–212, 2009.

(mainly related to the transitivity of subsumption relationships) which can be used to entail new *implicit* triples (i.e., facts) that are not explicitly asserted.

Currently, there is no adequate support for managing (i.e., querying and updating) provenance information of implicit RDF triples. There are two different ways in which such implicit knowledge can be viewed and this affects the semantics of update operations only. Under the *coherence* semantics [10], implicit knowledge does not depend on the explicit one but has a value on its own; therefore, there is no need for explicit "support" of some triple. Under this viewpoint, implicit triples are "first-class citizens", i.e., considered of equal value as explicit ones. On the other hand, under the *foundational* semantics [10] each implicit triple depends on the existence of the explicit triple(s) that imply it: implicit knowledge is only valid as long as the supporting explicit knowledge exists. It should be emphasized that the selected viewpoint is irrelevant as far as standard implication and querying is considered, but it affects the way updates should be performed; in particular, the coherence semantics corresponds to "belief set changes", whereas foundational semantics corresponds to "belief base changes", in the belief revision terminology [10].

In this paper we propose the use of *colors* (as in [4]) in order to capture the *provenance* of RDF *data* and *schema* triples. Intuitively, the color of an implicit and explicit triple represents the source from which the triple was obtained. We record the color of an RDF triple as a *fourth column*, hence obtaining an *RDF quadruple* as in [7,17].

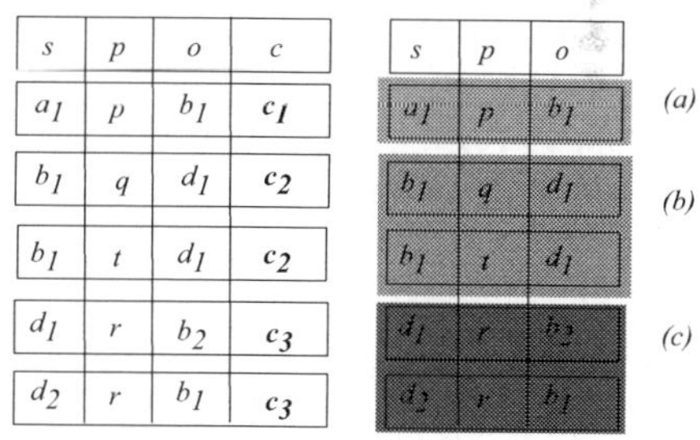

Fig. 1. Granularity Levels of Provenance

To provide the intuition of the granularity levels of provenance for RDF datasets that we capture with this work, we compare it with the representation of provenance information in the relational context. For instance, authors in [4] use colors to capture the provenance of relational tables, tuples and attributes. If we consider that a relational tuple of the form $[a_1{:}v_1, \ldots, a_k{:}v_k]$ with tuple identifier *tid* corresponds to a set of triples (tid, a_j, v_j), $j = 1, \ldots k$, then (see Fig. 1): a color assigned to *(a)* a single triple captures provenance at the level of *an attribute of the relational tuple*; *(b)* a collection of triples sharing the same subject captures provenance at the level of *the relational tuple* and finally *(c)* a set of triples whose subjects are instances of the same schema class, captures *provenance of the relational table*. The quadruples used to represent colored RDF/S triples leverage the syntax of RDF Named Graphs [5]: an RDF named graph can be modeled by arbitrary sets of triples sharing the same color.

The main contributions of this work are:

- We rely on the notion of *colors* to capture provenance information of explicit and implicit RDF triples. In particular, we employ a *semigroup* structure, defined by a set of colors to record and a binary operation "+" to reason over the provenance of RDF triples.

- We extend the RDFS inference rules [15] for computing the composite provenance of implicit RDF triples. In this respect, we devise an algorithm for determining *on the fly* the provenance of *non-materialized implicit* RDF triples.
- We study the semantics of *provenance propagation* and *querying* for the *subclass*, *subproperty* and *type* RDF hierarchies when colored RDF triples are represented as *quadruples*. In addition, we discuss atomic update operations (i.e., inserts and deletes) of RDF quadruples; updates are considered under the *coherence semantics* which allow us to preserve more knowledge during update operations [10].

The paper is structured as follows: in Section 2 we present the motivating example that we will use throughout this paper. Section 3 presents the basic RDF and RDFS notions. In Section 4 we introduce the notions of *color* and *quadruple* and discuss *inference* rules for sets of RDF quadruples. Section 5 discusses simple queries and atomic updates. We present related work in Section 6 and conclude in Section 7.

2 Motivating Example

We will use, for illustration purposes, an example taken from the News application domain. The RDFS schema of our News example captures information related to *newspapers* and *politicians* as well as the *relationships* between them and is contributed by different information sources.

	s	p	o	c
q_1	$\&NYT$	$endorses$	$\&B.\ Obama$	c_2
q_2	$\&NYT$	rdf:type	$Newspaper$	c_4
q_3	$Newspaper$	rdf:type	rdfs:Class	c_3
q_4	$Newspaper$	rdfs:subClassOf	$Mass\ Media$	c_3
q_5	$Mass\ Media$	rdfs:subClassOf	$Media$	c_5
q_6	$Candidate$	rdf:type	rdfs:Class	c_5
q_7	$\&B.\ Obama$	rdf:type	$Candidate$	c_5
q_8	$endorses$	rdf:type	rdf:Property	c_1
q_9	$endorses$	rdfs:domain	$Newspaper$	c_1
q_{10}	$endorses$	rdfs:range	$Candidate$	c_1
q_{11}	$Candidate$	rdfs:subClassOf	$Person$	c_1
q_{12}	$supports$	rdf:type	rdf:Property	c_1
q_{13}	$supports$	rdfs:domain	$MassMedia$	c_1
q_{14}	$supports$	rdfs:range	$Person$	c_1
q_{15}	$endorses$	rdfs:subPropertyOf	$supports$	c_2
q_{16}	$\&NYT$	$endorses$	$\&B.\ Obama$	c_1
q_{17}	$Media$	rdf:type	rdfs:Class	c_5

Fig. 2. Relation $\mathcal{Q}(s,p,o,c)$

Relation $\mathcal{Q}(s,p,o,c)$ that stores both schema and data triples is shown in Fig. 2. For $\mathcal{Q}(s,p,o,c)$, the s, p and o columns stand for the *subject*, *predicate* and *object* of an RDF triple. Column c is used to store the *provenance (color)* of a triple (s, p, o), which corresponds to the source from which this triple originates from and is represented by a URI (the URI of the source). We say that a triple (s, p, o) is *colored* c iff (s,p,o,c) is in $\mathcal{Q}(s,p,o,c)$. The set of triples (s, p, o) can be obtained by projecting on columns s, p and o of $\mathcal{Q}(s,p,o,c)$. In $\mathcal{Q}(s,p,o,c)$, a triple (s, p, o) can be assigned different colors (e.g., quadruples q_1 and q_{16}); this way, we can capture data integration scenarios in which the same piece of information originates from different sources.

Since RDF/S graphs can be seen as a kind of *node and edge labeled directed graphs*, we use the following graphical notation: *classes* and *properties* (binary relations between classes) of an RDFS vocabulary, are represented with boxes, and ovals respectively. *Instances* of classes contain their URI reference and to distinguish between individual resources and classes, we prefix the URI of the former with the "&" symbol. RDFS built-in properties rdfs:subClassOf, rdf:type and rdfs:subPropertyOf are represented by dashed, dotted and dotted-dashed arrows respectively. Fig. 3 shows the graph obtained from the quadruples in $\mathcal{Q}(s, p, o, c)$ of Fig. 2. For (s, p, o, c), we color the edge with c and we draw $s \text{—}p\ c\!\rightarrow o$ (except if p is one of the built-in RDF/S properties where we draw $s \text{—}c\!\rightarrow o$).

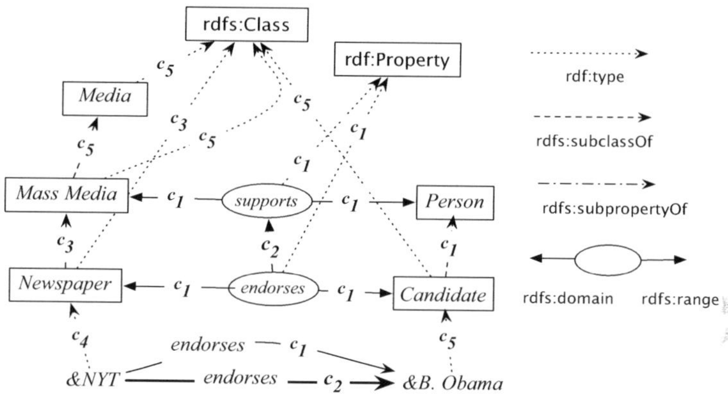

Fig. 3. Graph representation of $\mathcal{Q}(s, p, o, c)$

A great part of the information captured by a set of RDF triples can be inferred [15] by the *transitivity* of class and property subsumption relationships (rdfs:subClassOf and rdfs:subPropertyOf respectively) stated in the associated schemas. For instance, although not explicitly asserted in $\mathcal{Q}(s, p, o, c)$, we can infer that *Newspaper* is a subclass of *Media* because *(i) Newspaper* is a subclass of *Mass Media* (quadruple q_4) and *(ii) Mass Media* is a subclass of *Media* (quadruple q_5).

The question raised here is the following: *"what is the color of the implicit RDF triple?"*: the triple cannot be colored either c_3 or c_5 (colors of quadruples q_4 and q_5 respectively). In terms of provenance, we view the origin of this triple as *composite*: this triple should be colored c_3 *and* c_5. A possible way to represent this fact is to add quadruples (*Newspaper*, rdfs:subClassOf, *Media*, c_3) and (*Newspaper*, rdfs:subClassOf, -*Media*, c_5) in $\mathcal{Q}(s, p, o, c)$. But, in this case, the query *"return the triples colored c_3"* would falsely return (*Newspaper*, rdfs:subClassOf, *Media*). Hence, *composite origin cannot be captured by associating with the implicit triple separately the colors of its implying triples.*

Instead, we capture the provenance of implicit triples using colors defined by applying the special operation "+" on the colors of its implying triples. For instance, we color triple (*Newspaper*, rdfs:subClassOf, *Media*) with the color $c_{(3,5)} = c_3 + c_5$. Color $c_{(3,5)}$ can be seen as a *new, composite source of information*: it is a new URI and is assigned to the triples which are implied by triples colored c_3 and c_5 (as in the above example). Note that in this paper, we only consider *where* provenance [33], i.e., we record the sources that contributed in generating a particular triple, not the processes (e.g., the particular inference rules) that led to its generation (*how* provenance) [33]. In Section 4 we discuss the properties of operation "+" in more detail.

As with annotated databases [11,12], we aim at supporting both *provenance propagation* and *provenance querying* over the subclass, subproperty and type RDF hierarchies. In the case of *provenance propagation*, queries are targeted to the original data and provenance is propagated to the query results. The result of these queries are *explicit* and *implicit* colored RDF triples. For instance, one can ask the query *"return all instances of class Newspaper"*. The result of the query will be ($\&NYT$, rdf:type, $Newspaper$, c_4). Query *"return all subclasses of Media"* will return ($Newspaper$, rdfs:subClassOf, $Media$, $c_{(3,5)}$) and ($Mass\ Media$, rdfs:subClassOf, $Media$, c_5). Note that ($Newspaper$, rdfs:subClassOf, $Media$, $c_{(3,5)}$) is an *implicit* quadruple obtained by applying the transitive rdfs:subClassOf subsumption relationship on quadruples ($Mass\ Media$, rdfs:subClassOf, $Media$, c_5) and ($Newspaper$, rdfs:subClassOf, $Mass\ Media$, c_3) as previously discussed. On the other hand, in the case of *provenance querying*, queries are explicitly targeted at provenance information. Examples falling in this category are queries asking for the color of a given triple, or queries that filter triples given a color. For instance, the query *"return the color of (Newspaper,* rdfs:subClassOf, *Media)"*, will return $c_{(3,5)}$.

In this work, we support atomic updates; more specifically, one can either *(a)* insert a quadruple, or *(b)* delete an explicit or implicit quadruple. As previously stated, for our update operations we adhere to the coherence semantics [10], according to which we must explicitly retain all the implicit triples that will no longer be implied after the deletion of an explicit triple; we must also ensure that the resulting set of quadruples is redundant-free. As an example, consider the deletion of ($\&B.\ Obama$, rdf:type, $Candidate$, c_5). According to the coherence semantics we must retain the implicit information ($\&B.\ Obama$, rdf:type, $Person$, $c_{(1,5)}$) implied by quadruples ($\&B.\ Obama$, rdf:type, $Candidate$, c_5) and ($Candidate$, rdf:subClassOf, $Person$, c_1). Had we followed the foundational semantics, the implicit triple (and all the information it carries) would be lost, which would correspond to an unnecessary loss of information.

3 Preliminaries

In this work, we ignore non-universally identified resources, called *unnamed* or *blank nodes* [14]. In this respect, we consider two disjoint and infinite sets $\mathbf{U}$, $\mathbf{L}$, denoting the URIs and literals respectively.

Definition 1. *An RDF triple* $(subject, predicate, object)$ *is any element of the set* $\mathcal{T} = \mathbf{U} \times \mathbf{U} \times (\mathbf{U} \cup \mathbf{L})$.

The RDF Schema (RDFS) language [2] provides a built-in vocabulary for asserting user-defined schemas in the RDF data model. For instance, RDFS names rdfs:Resource (res), rdfs:Class (class) and rdf:Property (prop)[1] could be used as objects of triples describing *class* and *property* types. Furthermore, one can assert *instance of* relationships of resources with the RDF predicate rdf:type (type), whereas *subsumption* relationships among classes and properties are expressed with the RDFS rdfs:subClassOf (sc)

[1] In parenthesis are the terms we use to refer to the RDFS built-in classes and properties.

and rdfs:subPropertyOf (sp) predicates respectively. In addition, RDFS rdfs:domain (domain) and rdfs:range (range) predicates allow one to specify the domain and range of the properties in an RDFS vocabulary. In the rest of this paper, we consider two disjoint and infinite sets of URIs of classes ($\mathbf{C} \subset \mathbf{U}$) and property types ($\mathbf{P} \subset \mathbf{U}$).

4 Provenance for RDF/S Data

In the same spirit as in [4] for relational databases, we assign a *color* to an RDF triple in order to capture its provenance.

Definition 2. *Structure* ($\mathbf{I}$, *"+") is a* commutative semigroup *where:*

- $\mathbf{I} \subset \mathbf{U}$ *is the set of colors, disjoint from the sets of class* $\mathbf{C}$ *and property types* $\mathbf{P}$.
- *"+" is a binary operation with the following properties:* $\forall c_1, c_2, c_3 \in \mathbf{I}$

$$
\begin{array}{lll}
c_1 + c_1 & = c_1 & (\mathit{Idempotence}) \\
c_1 + c_2 & = c_2 + c_1 & (\mathit{Commutativity}) \\
c_1 + (c_2 + c_3) & = (c_1 + c_2) + c_3 & (\mathit{Associativity})
\end{array}
$$

Binary operation "+" is defined to capture the provenance of *implicit* RDF triples (see Table 1) and returns a *composite color* which is also in $\mathbf{I}$, i.e., if a triple t is implied by triple t_1 (whose color is c_1) and t_2 (whose color is c_2), then the color of t should be $c_1 + c_2$ and is denoted by $c_{(1,2)}$. In fact, $c_{(1,2)}$ is a new URI in $\mathbf{I}$ whose exact form is an implementation detail that we don't address in this paper. Note that, for n colors, there are $O(2^n)$ possible composite colors, so the storage of a composite color would require $O(n)$ bits in an efficient implementation. For an implicit RDF triple, its implying triples are those used to obtain it through the application of the inference rules that will be discussed in Section 4.1.

We say that color c_k is a *defining color* of $c_{(1,2,...,n)}$ and write $c_k \lessdot c_{1,2,...,n}$ iff $k \in \{1, 2, ..., n\}$, i.e., if $c_{(1,2,...,n)}$ can be obtained by an operation of the form $c_{(1,2,...,n)} = c_k + c$ for some color c. In other words, if a triple t is colored $c_{(1,2,...,n)}$ and c_k is a defining color of $c_{(1,2,...,n)}$, then t is an implicit triple which has been inferred using some triple colored c_k.

The intuition behind the properties of the "+" operation is the following: *(a)* an implicit RDF triple obtained from triples of the same color inherits the color of its implying triples (*idempotence*) *(b)* the color of an implicit triple is uniquely determined by the colors of the triples that imply it and not by the order of application of the inference rules (*commutativity* and *associativity*).

We should also note that $c_1 + c_2$ is a new color unless $c_1 = c_2$. In this manner, we keep a clear separation of what is given (explicit) and what can be implied (implicit). Moreover, the choice of idempotence is due to the semantics we give to colors as sources of information and the type of provenance we support: we need to know which sources participated in the creation of a new triple irrespective of which triples contributed to its implication. On the other hand, commutativity and associativity stem from the fact that the set of triples that are implied by a given triple set is the same irrespective of the order in which the inference rules are applied.

Definition 3. *An RDF quadruple* $(subject, predicate, object, color)$ *is any element of the set* $\mathcal{D} = \mathbf{U} \times \mathbf{U} \times (\mathbf{U} \cup \mathbf{L}) \times \mathbf{I}$.

Using this definition, we can define the notion of an RDF dataset featuring triples associated with their provenance information as follows:

Definition 4. *An RDF Dataset* d *is a finite set of quadruples in* $\mathcal{D}$ $(d \subseteq \mathcal{D})$.

Note that none of the existing approaches in the literature combine intentional and extensional assignment of triples to provenance information (colors). In [30] RDF Named Graphs, capturing the provenance of RDF triples, are defined intentionally through SPARQL [31] views and do not support the explicit assignment of triples to such graphs, whereas in [5] a purely extensional definition is followed. The notion of colors as introduced in this paper allows us to capture both the intentional and extensional aspects of RDF graphs (i.e., sets of RDF triples) that are useful to record and reason about provenance information in the presence of updates.

4.1 Inference

Similarly to RDF graphs (i.e., sets of triples) we define a consequence operator that abstracts a set of inference rules. These rules compute the *closure* of an RDF dataset d, denoted by $Cn(d)$, which is obtained using the inference rules of Table 1. We say that a dataset d entails a quadruple q, and write $d \vdash q$, iff $q \in Cn(d)$. We say that a triple $t = (s,\ p,\ o)$ is colored c iff $(s, p, o, c)\ \in Cn(d)$.

The inference rules shown in Table 1 extend those specified in [15] in a straightforward manner to take into account colors. They compute the color of an implicit triple, using the colors of its implying triples. For instance, for our motivating example of Section 2, quadruple $(Newspaper,$ rdfs:subClassOf, $Media,\ c_{(3,5)})$ is obtained by applying the *Transitivity of* sc $I_d^{(2)}$ rule on quadruples $(Newspaper,$ rdfs:subClassOf, $Mass\ Media,\ c_3)$ and $(Mass\ Media,$ rdfs:subClassOf, $Media,\ c_5)$.

4.2 Redundancy Elimination

In our work we consider that the RDF datasets are *redundant free*. An RDF dataset is redundant free if there does not exist a quadruple that can be implied by others when applying inference rules $I_d^{(1)} - I_d^{(6)}$ of Table 1. The detection and removal of redundancies is straightforward using these rules.

In the sequel, we assume that queries and updates are performed upon redundant free RDF datasets. In effect, this means that redundancies are detected (and removed) at update time rather than at query time. This choice was made because we believe that in real scale Semantic Web systems, query performance should prevail over update performance. Redundant free RDF datasets were chosen because they offer a number of advantages in the case of transaction management for concurrent updates and queries.

Table 1. A subset of the RDFS Inference Rules for RDF Datasets

Reflexivity of sc	Transitivity of sc
$I_d^{(1)}: \dfrac{(C, \mathsf{type}, \mathsf{class}, c_1)}{(C, \mathsf{sc}, C, c_1)}$	$I_d^{(2)}: \dfrac{(C_1, \mathsf{sc}, C_2, c_1), (C_2, \mathsf{sc}, C_3, c_2)}{(C_1, \mathsf{sc}, C_3, c_{(1,2)})}$
Reflexivity of sp	**Transitivity of sp**
$I_d^{(3)}: \dfrac{(P, \mathsf{type}, \mathsf{prop}, c_1)}{(P, \mathsf{sp}, P, c_1)}$	$I_d^{(4)}: \dfrac{(P_1, \mathsf{sp}, P_2, c_1), (P_2, \mathsf{sp}, P_3, c_2)}{(P_1, \mathsf{sp}, P_3, c_{(1,2)})}$
Transitivity of class instantiation	**Transitivity of property instantiation**
$I_d^{(5)}: \dfrac{(x, \mathsf{type}, C_1, c_1), (C_1, \mathsf{sc}, C_2, c_2)}{(x, \mathsf{type}, C_2, c_{(1,2)})}$	$I_d^{(6)}: \dfrac{(P_1, \mathsf{sp}, P_2, c_1), (x_1, P_1, x_2, c_2)}{(x_1, P_2, x_2, c_{(1,2)})}$

5 Querying and Updating RDF Datasets

5.1 Querying RDF Datasets

In this section we discuss a simple class of queries that allow one to express queries on the *subclass*, *subproperty* and *type* hierarchies of an RDF dataset. We consider $\mathcal{V}$, $\mathcal{C}$ to be two sets of variables for resources and colors respectively; $\mathcal{V}$, $\mathcal{C}$, $\mathbf{U}$ (URIs) and $\mathbf{L}$ (literals) are mutually disjoint sets. We define a simple form of a quadruple pattern, called *q-pattern* which is an element from $(\mathbf{U} \cup \mathcal{V}) \times \{\mathsf{type} \cup \mathsf{sc} \cup \mathsf{sp}\} \times (\mathbf{U} \cup \mathcal{V}) \times (\mathbf{I} \cup \mathcal{C})$. In our context, a query is of the form (H, B, C) where H (head) is a q-pattern, B (body) is an expression defined after the antecedent of the inference rules presented in Section 4.1, and C (constraints) is a conjunction of *atomic predicates*. Each atomic predicate has the form: *(1)* $v = const$ for $v \in \mathcal{V}$; *(2)* $v < v'$ for $v, v' \in \mathcal{C}$; *(3)* $c = c_1 + c_2 \ldots + c_k$ where $c \in \mathcal{C}$ and $c_i \in \mathbf{I}$.

According to the above definition, one can express constraints on resources *(1)*, on colors *(2)*, as well as to specify that a color considered in the query is defined by a set of other colors *(3)*. In addition, we require that all variables that appear in the head of the query (H) appear in the query's body (B). This restriction is imposed in order to have computationally desirable properties.

We denote variables with $?x$, $?y$, ... for resources and $?c_1$, $?c_2$, ... for colors. To define the query semantics, we use the notion of mapping (as in [21]) as follows: a *mapping* μ is a partial function $\mu : (\mathcal{V} \cup \mathcal{C}) \rightarrow \mathbf{U}$. The domain of μ ($dom(\mu)$) is the subset of $\mathcal{V} \cup \mathcal{C}$ where μ is defined. For a variable $?v$, $\mu(?v)$ denotes the resource or color to which $?v$ is mapped through μ.

To define the semantics of a *q-pattern* we must define first the *semantics of a property* r *over an RDF dataset* d, denoted by $[[r]]_d$. Given an RDF dataset d, $[[r]]_d$ for properties type, sc, sp, domain, range and user defined property p is given in Table 2. We write $d \vdash_S q$ to denote that dataset d entails quadruple q when the inference rules in S are applied on d.

Table 2. $[[r]]_d$ for type, sc, sp, domain, range and user defined property p

$$
\begin{aligned}
[[\mathsf{type}]]_d &= \{(x,\ y,\ c) \mid d \vdash_{\{I_d^{(2)}, I_d^{(5)}\}} (x, \mathsf{type}, y, c)\} \\
[[\mathsf{sc}]]_d &= \{(x,\ y,\ c) \mid d \vdash_{\{I_d^{(1)}, I_d^{(2)}\}} (x, \mathsf{sc}, y, c)\} \\
[[\mathsf{sp}]]_d &= \{(x,\ y,\ c) \mid d \vdash_{\{I_d^{(3)}, I_d^{(4)}\}} (x, \mathsf{sp}, y, c)\} \\
[[\mathsf{domain}]]_d &= \{(x,\ y,\ c) \mid d \vdash (x, \mathsf{domain}, y, c)\} \\
[[\mathsf{range}]]_d &= \{(x,\ y,\ c) \mid d \vdash (x, \mathsf{range}, y, c)\} \\
[[p]]_d &= \{(x,\ y,\ c) \mid d \vdash_{\{I_d^{(4)}, I_d^{(6)}\}} (x, p, y, c)\}
\end{aligned}
$$

Table 3. Semantics of *q-patterns*

$$
\begin{aligned}
[[(a, exp, ?y, ?c)]]_d &= \{\mu \mid dom(\mu) = \{?y, ?c\} \textit{ and } (a,\ \mu(?y),\ \mu(?c)) \in [[exp]]_d\} \\
[[(?x, exp, a, ?c)]]_d &= \{\mu \mid dom(\mu) = \{?x, ?c\} \textit{ and } (\mu(?x),\ a,\ \mu(?c)) \in [[exp]]_d\} \\
[[(?x, exp, ?y, c)]]_d &= \{\mu \mid dom(\mu) = \{?x, ?y\} \textit{ and } (\mu(?x),\ \mu(?y),\ c) \in [[exp]]_d\} \\
[[(a, exp, b, ?c)]]_d &= \{\mu \mid dom(\mu) = \{?c\} \textit{ and } (a,\ b,\ \mu(?c)) \in [[exp]]_d\} \\
[[(a, exp, b, c)]]_d &= \{\mu \mid dom(\mu) = \emptyset \textit{ and } (a,\ b,\ c) \in [[exp]]_d\}
\end{aligned}
$$

We write $\langle r \rangle_d$ to denote the semantics of property r when no inference rule is used. We can now define the semantics of a *q-pattern*. Consider an RDF dataset d and $q = (?X, exp, ?Y, ?c)$ a *q-pattern*, where exp is one of sc, sp, type, domain and range. Then the evaluation of q over d is defined as follows:

$$[[q]]_d = \{\mu \mid dom(\mu) = \{?X, ?Y, ?c\} \textit{ and } (\mu(?X),\ \mu(?Y),\ \mu(?c)) \in [[exp]]_d\}$$

In Table 3 we give the semantics of some *q-patterns* when URIs and colors are considered (where $a, b \in \mathbf{U}$ and $c \in \mathbf{I}$). Finally, given a mapping μ we say that μ satisfies an atomic predicate C, denoted by $\mu \vdash C$, per the following conditions:

$$
\begin{aligned}
&\mu \vdash (?x = const) && \textit{iff } \mu(?x) = const, const \in \mathbf{U}, ?x \in dom(\mu) \\
&\mu \vdash (?x = ?y) && \textit{iff } \mu(?x) = \mu(?y), ?x, ?y \in dom(\mu) \\
&\mu \vdash (?c = ?c') && \textit{iff } \mu(?c) = \mu(?c'), ?c, ?c' \in dom(\mu) \\
&\mu \vdash (?c \lessdot ?c') && \textit{iff } \mu(?c) \lessdot \mu(?c')\ ?c, ?c' \in dom(\mu) \\
&\mu \vdash (?c = c_1 + ... + c_k) && \textit{iff } \mu(?c) = c_1 + \ldots + c_k, ?c \in dom(\mu)
\end{aligned}
$$

For a query $Q = (H, B, C)$ an RDF dataset d and mapping μ, such that $\mu \in [[B]]_d$, and $\mu \vdash C$, $\mu(H)$ is the quadruple obtained by replacing every variable $?x$ in $dom(\mu)$ with $\mu(?x)$. The color variable (if any) is replaced by the color obtained by applying the "+" operator as specified by the inference rules $I_d^{(1)}$ to $I_d^{(6)}$ of Table 1. The answer to Q is the union of the quadruples $\mu(H)$ for each such mapping μ.

5.2 Updating RDF Datasets

In this section we discuss *atomic* update operations (*inserts* and *deletes*). Recall that in our work we follow the *coherence semantics* [10] according to which we need to retain implicit information that would be lost during a triple deletion. Moreover, we enforce that the resulting RDF datasets will remain valid with respect to the employed RDFS schema. The notion of validity has been described in various fragments of the RDFS language ([18,32]), and is used to overrule certain triple combinations. In the context of RDF datasets, the validity constraints are applied (and defined) at the level of the dataset, but the color-related part of the quadruple is not considered. The validity constraints that we consider in this work concern the *disjointness* between class, property and color names and the acyclicity of rdfs:subClassOf and rdfs:subPropertyOf subsumption relationships. An additional validity constraint that we consider in our work is that the subject and object of the instance of some property should be correctly classified under the domain and range of the property respectively. For a full list of the related validity constraints, see [19].

The semantics of each atomic update is specified by its corresponding *effects* and *side-effects*. The effect of an insert or delete operation consists of the straightforward insertion/deletion of the requested quadruples. The side-effects ensure that the resulting RDF dataset continues to be valid and non-redundant as discussed in [35]. Update semantics adhere to the principle of *minimal change* [9], per which a minimal number of insertions and deletions should be performed in order to restore a valid and non-redundant state of an RDF dataset. The effects and side-effects of insertions and deletions are determined by the kind of triple involved, i.e., whether it is a *class instance* or *property instance* insertion or deletion. Due to space restrictions we only describe *class instance* insertions and deletions.

INSERT **Operation.** A primitive insert operation is of the form: $\mathsf{insert}(s, p, o, i)$ where $s, p \in \mathbf{U}$, $o \in \mathbf{U} \cup \mathbf{L}$, $i \in \mathbf{I}$.

Algorithm 1. Class Instance Insertion Algorithm

Data: $\mathsf{insert}(x, \mathsf{type}, y, i)$, RDF dataset d
Result: Updated RDF dataset d
1 **if** $(\exists\ (x,\ y,\ i) \in [[\mathsf{type}]]_d)$ **then return** d;
2 **if** $(y \notin \mathbf{C})$ **then**
3 | **return** d;
4 **forall** $((x,\ z,\ i') \in \langle\mathsf{type}\rangle_d$ *s.t.* $\exists(y,\ z,\ i'') \in [[\mathsf{sc}]]_d$ *and* $i' = i + i'')$ **do**
5 | $d = d \setminus \{(x, \mathsf{type}, z, i')\}$;
6 **end**
7 $d = d \cup \{(x, \mathsf{type}, y, i)\}$;
8 **return** d

A formal description of the insertion of a quadruple (x, type, y, i) in an RDF dataset d along with its side-effects can be found in Algorithm 1. At line 1 we examine if the quadruple already belongs to the semantics of property type. If not, then we ensure that y is a class (lines 2–3). If it is, then we remove all class instantiation quadruples from the RDF dataset which can be implied through the quadruple to be inserted and the class subsumption relationships (lines 4–6), to guarantee that the result is redundant free. Finally, the quadruple is inserted (line 7). An example of a class instance insertion is shown in Figures 4(a) and 4(b).

Algorithm 2. Class Instance Deletion Algorithm

Data: $\mathsf{delete}(x, \mathsf{type}, y, i)$, RDF dataset d
Result: Updated RDF dataset d

1 **if** ($\nexists\ (x,\ y,\ i) \in [[\mathsf{type}]]_d$) **then return** d;
2 **forall** ($(x,\ y',\ i') \in [[\mathsf{type}]]_d, (y',\ y,\ i'') \in [[\mathsf{sc}]]_d$ *s.t.* $i = i' + i''$) **do**
3   **forall** $(y',\ z,\ k) \in \langle \mathsf{sc} \rangle_d$ *s.t.* $y'! = z$ **do**
4   **if** $\nexists (z,\ y,\ h) \in [[\mathsf{sc}]]_d$ **then** $d = d \cup \{(x, \mathsf{type}, z, i' + k)\}$
5   **end**
6     $d = d \setminus \{(x, \mathsf{type}, y', i')\}$
7 **end**
8 **forall** $(x,\ o,\ h) \in \langle q \rangle_d$, *s.t.* $(q,\ c,\ i) \in \langle \mathsf{domain} \rangle_d$ **do**
9   **if** $\nexists\ (x,\ c,\ j) \in [[\mathsf{type}]]_d$ **then**
10         $d = d \setminus \{(x, q, o, h)\}$;
11   **forall** q' *s.t.* $\exists\ (q,\ q',\ h'') \in [[\mathsf{sp}]]_d$ **do**
12   **if** $\exists\ (x,\ e,\ k) \in [[\mathsf{type}]]_d$ *s.t.* $\exists\ (q,\ e,\ k') \in \langle \mathsf{domain} \rangle_d$ **then** $d = d \cup \{(x, q', o, h + h'')\}$
13   **end**
14
15 **end**
16 **forall** $(o,\ x,\ h) \in \langle q \rangle_d$, *s.t.* $(q,\ c,\ i) \in \langle \mathsf{range} \rangle_d$ **do**
17   **if** $\nexists\ (x,\ c,\ j) \in [[\mathsf{type}]]_d$ **then**
18         $d = d \setminus \{(o, q, x, h)\}$;
19   **forall** q' *s.t.* $\exists\ (q,\ q',\ h'') \in [[\mathsf{sp}]]_d$ **do**
20   **if** $\exists\ (x,\ e,\ k) \in [[\mathsf{type}]]_d$ *s.t.* $\exists\ (q,\ e,\ k') \in \langle \mathsf{range} \rangle_d$ **then** $d = d \cup \{(o, q', x, h + h'')\}$
21   **end**
22
23 **end**
24 **return** d

DELETE **Operation.** A primitive delete operation is of the form: $\mathsf{delete}(s, p, o, i)$ where $s, p \in \mathbf{U}$, $o \in \mathbf{U} \cup \mathbf{L}$, $i \in \mathbf{I}$.

A formal description of the deletion of a quadruple (x, type, y, i) is given in Algorithm 2. At line 1 we examine if (x, type, y, i) belongs to the semantics of property type. If this is the case, then we must (1) insert all the quadruples that are implied by the quadruple that we wish to delete (per the coherence semantics) and (2) delete the quadruples that if retained would imply the quadruple we wish to delete (lines 2–7). To ensure that the RDF dataset is still valid after the updates, we must remove all properties originating from (or reaching resp.) x whose domain (or range resp.) is a class that x is no longer an instance of (lines 8–23). Examples of class instance deletions can be found in Figures 4(c) and 4(d).

5.3 Complexity Analysis

When working with colored RDF triples, one of the basic kinds of queries that we need to answer is *"what is the color of a triple"*. This is a *provenance query* and essentially boils down to finding, for a given triple $(s,\ p,\ o)$ and RDF dataset d, all quadruples of

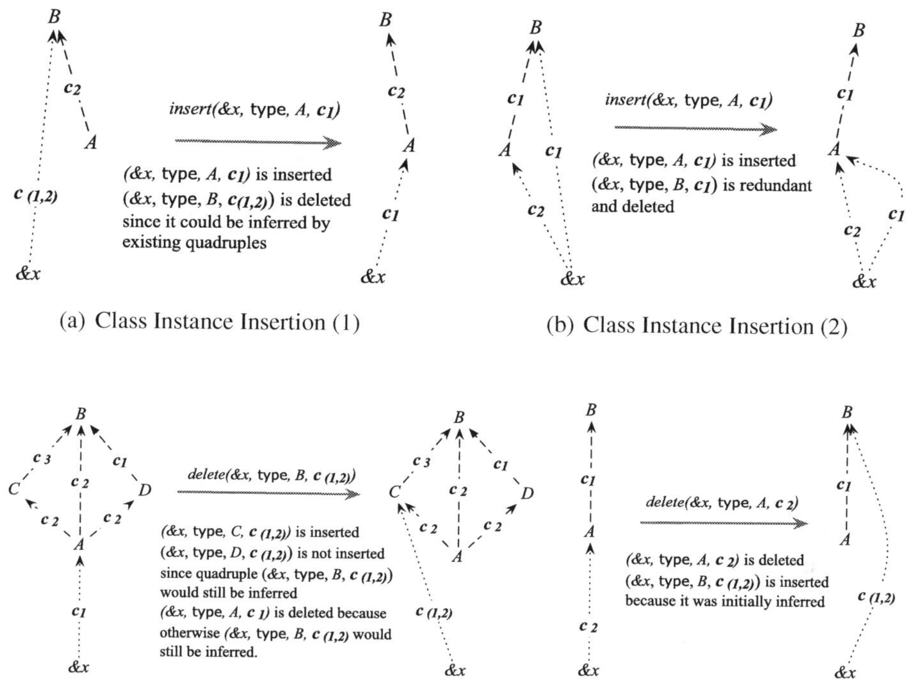

(a) Class Instance Insertion (1)

(b) Class Instance Insertion (2)

(c) Class Instance Deletion (1)

(d) Class Instance Deletion (2)

Fig. 4. Examples of Class Instance Insertions ((a)–(b)) and Deletions ((c)–(d))

the form (s, p, o, c) in $Cn(d)$. Given that we work with redundant free RDF datasets (so implicit triples are not materialized), we present here an algorithm that computes the color of RDF triples *without* materializing $Cn(d)$, and discuss its complexity.

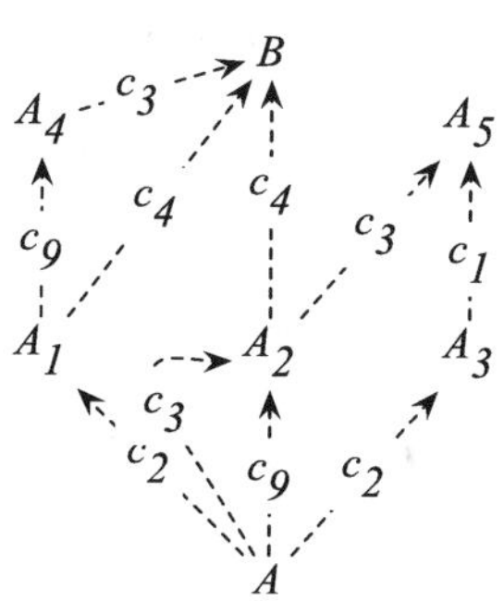

Fig. 5. sc Hierarchy

Consider an RDF dataset d whose size is N. In order to determine the color of $(s,\ p,\ o)$, without computing $Cn(d)$, we need to find all the possible ways in which a quadruple of the form (s, p, o, c) (for any c) can be inferred using quadruples from d. Using the algorithm below, this can be made in $O(N \log(N))$ time.

Certain quadruples are not involved in the inference rules in Table 1, so they cannot be implied by others: these are all the quadruples that *do not involve* the RDFS type, sc and sp relationships. Determining the color of such a quadruple is trivial, as we only need to check d (rather than $Cn(d)$), so it can be made in $O(\log(N))$ time by a simple search in d (using an appropriate index).

To determine the color of quadruples that involve the aforementioned built-in RDFS properties, we view the sc and sp hierarchies as *directed acyclic graphs*. Nodes in the graph are classes (properties resp.), and there exists an edge between nodes x and y iff x is a subclass (subproperty resp.) of y. The problem of obtaining the sequence of quadruples from which e.g., quadruple (A, sc, B, c) (for any c), can be implied is equivalent to discovering the path(s) (i.e., sequences of edges) in the DAG from node A to node B. For each of these sequences of edges, we keep the color of each involved quadruple. Recall that an edge may involve several quadruples (i.e., a triple can be colored with different colors). The algorithm uses a depth-first search and terminates when B is reached. At each step of the algorithm, we *(i)* find the superclasses of the context node (i.e., the node that we are currently looking at) that are also subclasses of the target node (e.g., B) and *(ii)* store the set of colors of the edges that we have already examined.

Algorithm 3. Traverse_Subsumption_Graph

```
    Data: Classes source and target, RDF Dataset d, Set of colors S;
    Result: Set of colors S;
 1  if source=target then
 2  |   return S;
 3  res ← ∅ ;
 4  foreach class x, where x is a superclass of source and subclass of target do
 5  |   Let (source, sc, x, c) ∈ d;
 6  |   Let S_x ← ∅ ;
 7  |   forall colors c_i in S do
 8  |   |   S_x ← S_x ∪ {c_i + c};
 9  |   end
10  |   res ← res ∪ Traverse_Subsumption_Graph(x,target,d,S_x);
11  end
12  return res
```

In Algorithm 3 in order to obtain all the classes that are superclasses of *source* and subclasses of *target* (line 4), we use the labeling scheme introduced in [6] that captures the subsumption relationships between classes and properties and allows us to determine whether a class (or property) is a subclass (or subproperty) of another in *constant time*. This is achieved by simply comparing the labels of the classes/properties. The labeling scheme is *solely used to prune irrelevant subclass and subproperty paths*, thereby limiting our search space significantly, without taking into consideration colors of triples.

As an example of application, consider the subclass hierarchy shown in Fig. 5. Suppose that we need to discover the color of triple (A, sc, B). We start with class A and in the first step we will keep classes A_1,A_2 and sets of colors $S_{A_1} = \{c_2\}$, $S_{A_2} = \{c_3, c_9\}$ since the edges that determine that A is a subclass of A_1, A_2 are formed by triples colored $\{c_2\}$, $\{c_3, c_9\}$ respectively. For each of the superclasses of A (i.e., A_1, A_2), we perform exactly the same process and extend the sets of colors that we have obtained using operation "+". Upon return of the recursion, these colors are placed into the set res to be returned. The output of the algorithm are the colors: $c_{(3,4)}$, $c_{(9,4)}$, $c_{(2,4)}$, $c_{(2,9,3)}$, where $c_{(1,\ldots,n)} = c_1 + \ldots + c_n$.

Given the fact that we prune subsumption paths that are not relevant in line 4 (using the labeling scheme), and that no cycles are allowed, the described process will, in the worst case, consider each quadruple in the RDF dataset d once. Given that each access requires a search in d, the cost of each access is $O(\log(N))$ (using an adequate indexing system). Therefore, the total complexity is $O(N \log(N))$.

For the other triples that may appear as implicit ones, the algorithm is similar. In particular, if t is of the form $(P, \ \mathsf{sp}, \ Q)$, then the process is identical to the above, except that properties and subproperty relationships are considered instead of class and subclass relationships. If t is of the form $(A, \ \mathsf{type}, \ B)$ then the process is almost identical, except that, in the first step of the recursion, we search for all classes whose explicit instance is A and are also (implicit or explicit) subclasses of B; the rest is the same. Finally, if t is of the form $(x, \ P, \ y)$, then, again, the process is identical, except that, in the first step of the recursion, we search for all explicit quadruples of the form (x, Q, y, c) such that Q is an explicit or implicit subproperty of P. The rest of the recursion steps are as in the above cases.

6 Related Work

So far, research on recording provenance for RDF data has focused on either associating triples with an *RDF named graph* [5,34] or by *extending* an RDF triple to a *quadruple* where the fourth element is a URI, a blank node or an identifier [7,17]. These works vary in the semantics of the fourth element which is used to represent *provenance*, *context* and *access control* information. *RDF Named graphs* have been proposed in [5,34] to capture *explicit provenance* information by allowing users to refer to specific parts of RDF/S graphs in order to decide *"how credible is"*, or *"how evolves"* a piece of information. An RDF named graph is a set of triples to which a URI has been assigned and can be referenced by other graphs as a normal resource; in this manner, one can assign explicit provenance information to this collection of triples.

Currently, there is no adequate support on how to manage provenance of *implicit* and explicit triples in the presence of *queries* and *updates*. Authors in [5] do not discuss RDFS inference, queries and updates in the presence of RDF named graphs. Unfortunately, existing declarative languages for querying and updating RDF triples have been extended either with RDF named graphs (such as SPARQL [23] and SPARQL Update [31]) or with RDFS inference support [22,24], but not with both. In this paper, we attempt to fill this gap by proposing a framework based on the use of colors to capture provenance of RDF triples and reason about the provenance of implicit triples for simple queries and atomic updates. In a previous work [20], we have introduced the notion of *RDF/S graphsets* which builds upon and extends the notion of RDF named graphs. In that paper we showed that the mechanism of RDF named graphs cannot capture the provenance of implicit RDF triples, and proposed RDF graphsets as a solution to this problem. In this paper, we use colors as an elegant and uniform way to capture the provenance of both explicit and implicit RDF triples. Colors are a generalization of

RDF named graphs [5]: a set of triples colored with a single color can be considered as belonging to the RDF named graph whose URI is the color (recall that colors are URIs). Colors obtained by applying the "+" operation on other colors simulate graph-sets [20]. The notion of colors as introduced in this paper allows us to capture both the intentional and extensional aspects of RDF graphs that are useful to record and reason about provenance information in the presence of updates whereas none of the existing approaches [5,30] combine intentional and extensional assignment of triples to provenance information.

On the other side of the spectrum, a significant amount of work on the issue has been done for relational and tree-structured databases [3,4,13,12]. Unlike these works, we consider both recursion and updates, whereas [13] does not consider updates, [3,4] supports updates but not recursion and [12] considers neither recursion or updates. Moreover, we do not focus on provenance propagation through relational operators but through inference rules that can be translated to relational unions and joins with bounded recursion. However, inference rules compute on the fly implicit triples without the need to materialize the provenance of intermediate results (per recursion step).

7 Conclusion

In this paper we proposed the use of colors to capture the provenance of RDF data and schema triples. We used the logical representation of quadruples to store the color of an RDF triple. The use of colors allows us to capture provenance at several granularity levels and can be considered as a generalization of RDF Named Graphs. One of the main contributions of the paper is the extension of RDFS inference rules to determine the provenance of *implicit* RDF triples, an extension which is not possible under the RDF named graphs approach and has been overlooked in the majority of approaches that deal with managing provenance information for RDF graphs. Note that the extended inference rules do not entail any additional complexity or scalability concerns to those already involving RDFS reasoning (see complexity analysis). As a future work we will study a more general algebraic structure (as in [13]) to capture the provenance of triples and mappings obtained by the SPARQL operators [23]. In addition, we plan to study the insertion and deletion of colors where the latter can be expressed as a batch deletion of quadruples sharing the same color.

Acknowledgements. This work was partially supported by the EU project CASPAR (FP6-2005-IST-033572).

References

1. Berners-Lee, T., Hendler, J., Lassila, O.: The Semantic Web. Scientific American-American Edition (2001)
2. Brickley, D., Guha, R.V.: RDF Vocabulary Description Language 1.0: RDF Schema (2004), `http://www.w3.org/TR/2004/REC-rdf-schema-20040210`
3. Buneman, P., Chapman, A.P., Cheney, J.: Provenance Management in Curated Databases. In: SIGMOD (2006)

4. Buneman, P., Cheney, J., Vansummeren, S.: On the Expressiveness of Implicit Provenance in Query and Update Languages. In: Schwentick, T., Suciu, D. (eds.) ICDT 2007. LNCS, vol. 4353, pp. 209–223. Springer, Heidelberg (2006)
5. Carroll, J., Bizer, C., Hayes, P., Stickler, P.: Named graphs, Provenance and Trust. In: WWW (2005)
6. Christophides, V., Plexousakis, D., Scholl, M., Tourtounis, S.: On labeling schemes for the semantic web. In: WWW (2003)
7. Dumbill, E.: Tracking Provenance of RDF Data. Technical report, ISO/IEC (2003)
8. McBride, B., Manola, F., Miller, E.: B.M.: RDF Primer (February 2004), `http://www.w3.org/TR/rdf-primer`
9. Gardenfors, P.: Belief Revision: An Introduction. Belief Revision (29), 1–28 (1992)
10. Gardenfors, P.: The dynamics of belief systems: Foundations versus coherence theories. Revue Internationale de Philosophie 44, 24–46 (1992)
11. Geerts, F., den Bussche, J.V.: Relational Completeness of Query Languages for Annotated Databases. In: Arenas, M., Schwartzbach, M.I. (eds.) DBPL 2007. LNCS, vol. 4797, pp. 127–137. Springer, Heidelberg (2007)
12. Geerts, F., Kementsietsidis, A., Milano, D.: MONDRIAN: Annotating and Querying Databases through Colors and Blocks. In: ICDE. (2006)
13. Green, T.J., Karvounarakis, G., Tannen, V.: Provenance semirings. In: PODS (2007)
14. Gutierrez, C., Hurtado, C.A., Mendelzon, A.O.: Foundations of Semantic Web Databases. In: PODS (2004)
15. Hayes, P.: RDF Semantics (February 2004), `http://www.w3.org/TR/rdf-mt`
16. Le-Phuoc, D., Polleres, A., Hauswirth, M., Tummarello, G., Morbidoni, C.: Rapid Prototyping of Semantic Mash-ups through Semantic Web Pipes. In: WWW (2009)
17. MacGregor, R., Ko, I.Y.: Representing Contextualized Data using Semantic Web Tools. In: Practical and Scalable Semantic Systems,conjunction with ISWC (2003)
18. Munoz, S., Perez, J., Gutierrez, C.: Minimal deductive systems for RDF. In: Franconi, E., Kifer, M., May, W. (eds.) ESWC 2007. LNCS, vol. 4519, pp. 53–67. Springer, Heidelberg (2007)
19. Pediaditis, P.: Querying and Updating RDF/S Named Graphs. Master's thesis, Computer Science Department, University of Crete (2008)
20. Pediaditis, P., Flouris, G., Fundulaki, I., Christophides, V.: On Explicit Provenance Management in RDF/S Graphs. In: TAPP (2009)
21. Perez, J., Arenas, M., Gutierrez, C.: Semantics and Complexity of SPARQL. In: Cruz, I., Decker, S., Allemang, D., Preist, C., Schwabe, D., Mika, P., Uschold, M., Aroyo, L.M. (eds.) ISWC 2006. LNCS, vol. 4273, pp. 30–43. Springer, Heidelberg (2006)
22. Perez, J., Arenas, M., Gutierrez, C.: nSPARQL: A Navigational Language for RDF. In: Sheth, A.P., Staab, S., Dean, M., Paolucci, M., Maynard, D., Finin, T., Thirunarayan, K. (eds.) ISWC 2008. LNCS, vol. 5318, pp. 66–81. Springer, Heidelberg (2008)
23. Prud'hommeaux, E., Seaborne, A.: SPARQL Query Language for RDF (January 2008), `http://www.w3.org/TR/rdf-sparql-query`
24. PSPARQL, `http://psparql.inrialpes.fr`
25. DBLP Comp. Science Bibliography, `http://www.informatik.uni-trier.de/~ley/db`
26. DBPedia, `http://www.dbpedia.org`
27. Gene Ontology., `http://www.geneontology.org`
28. RDFizers, `http://simile.mit.edu/wiki/RDFizers`
29. W3C Linking Open Data, `http://esw.w3.org/topic/SweoIG/TaskForces/CommunityProjects/LinkingOpenData`

30. Schenk, S., Staab, S.: Networked graphs: a declarative mechanism for SPARQL rules, SPARQL views and RDF data integration on the Web. In: WWW (2008)
31. Seaborne, A., Manjunath, G.: SPARQL/Update: A language for updating RDF graphs (April 2008), http://jena.hpl.hp.com/~afs/SPARQL-Update.html
32. Serfiotis, G., Koffina, I., Christophides, V., Tannen, V.: Containment and Minimization of RDF/S Query Patterns. In: Gil, Y., Motta, E., Benjamins, V.R., Musen, M.A. (eds.) ISWC 2005. LNCS, vol. 3729, pp. 607–623. Springer, Heidelberg (2005)
33. Tan, W.C.: Provenance in databases: Past, current, and future. Bulletin of the IEEE Computer Society Technical Committee on Data Engineering (2007)
34. Watkins, E., Nicole, D.: Named Graphs as a Mechanism for Reasoning About Provenance. In: Zhou, X., Li, J., Shen, H.T., Kitsuregawa, M., Zhang, Y. (eds.) APWeb 2006. LNCS, vol. 3841, pp. 943–948. Springer, Heidelberg (2006)
35. Zeginis, D., Tzitzikas, Y., Christophides, V.: On the Foundations of Computing Deltas Between RDF Models. In: Aberer, K., Choi, K.-S., Noy, N., Allemang, D., Lee, K.-I., Nixon, L.J.B., Golbeck, J., Mika, P., Maynard, D., Mizoguchi, R., Schreiber, G., Cudré-Mauroux, P. (eds.) ASWC 2007 and ISWC 2007. LNCS, vol. 4825, pp. 637–651. Springer, Heidelberg (2007)

TripleRank: Ranking Semantic Web Data by Tensor Decomposition

Thomas Franz, Antje Schultz, Sergej Sizov, and Steffen Staab

ISWeb - University of Koblenz-Landau, Germany
{franz,antjeschultz,sizov,staab}@uni-koblenz.de
http://isweb.uni-koblenz.de

Abstract. The Semantic Web fosters novel applications targeting a more efficient and satisfying exploitation of the data available on the web, e.g. faceted browsing of linked open data. Large amounts and high diversity of knowledge in the Semantic Web pose the challenging question of appropriate relevance ranking for producing fine-grained and rich descriptions of the available data, e.g. to guide the user along most promising knowledge aspects. Existing methods for graph-based authority ranking lack support for fine-grained latent coherence between resources and predicates (i.e. support for link semantics in the linked data model). In this paper, we present TripleRank, a novel approach for *faceted* authority ranking in the context of RDF knowledge bases. TripleRank captures the additional latent semantics of Semantic Web data by means of statistical methods in order to produce richer descriptions of the available data. We model the Semantic Web by a 3-dimensional tensor that enables the seamless representation of arbitrary semantic links. For the analysis of that model, we apply the PARAFAC decomposition, which can be seen as a multi-modal counterpart to Web authority ranking with HITS. The result are groupings of resources and predicates that characterize their authority and navigational (hub) properties with respect to identified topics. We have applied TripleRank to multiple data sets from the linked open data community and gathered encouraging feedback in a user evaluation where TripleRank results have been exploited in a faceted browsing scenario.

1 Introduction

Relevance ranking is a crucial component for a wide range of Semantic Web applications, such as semantic search and semantic browsing. Online services such as the dbpedia.org [4] provide rich descriptions of Semantic Web resources that render manual browsing extremely difficult. For instance, the resource The_Beatles has no less than 1228 links going in and out of it. In the context of RDF knowledge bases, data can be seen as a graph where nodes represent RDF resources and edges correspond to RDF predicates that link resources. Consequently, graph-based authority ranking algorithms known from Web retrieval, such as PageRank [8], HITS [23] or SALSA [27], can be adopted for the Semantic Web setting, too. Instead of ratings for Web pages they will then output ratings for RDF resources, with respect to one or more criteria, e.g. hub and authority scores in HITS. These scores reflect the centrality/importance of particular RDF resources in the knowledge representation and thus can be exploited for relevance estimation.

A. Bernstein et al. (Eds.): ISWC 2009, LNCS 5823, pp. 213–228, 2009.

Two important observations can be made about the authority ranking for RDF data. On one hand, the computational models of standard algorithms for Web analysis only consider structural information, i.e. the connectivity of graph nodes. Additional link semantics, e.g. knowledge about different link types, is not used. On the other hand, knowledge representation in the Semantic Web is heterogeneous. There is no single ontology for the Semantic Web that describes all of the available data in a concise way. Instead, there are many cases of overlapping, redundant, and conflicting ontologies describing similar information. Therefore, we may expect redundancies like the co-existence of different predicates with highly similar (or identical) meaning. For instance, even the well maintained data from dbpedia.org contains the redundant links http://dbpedia.org/ontology/writer and http://dbpedia.org/property/writer which have the same semantics. Common authority ranking algorithms provide no support for finding such groups of semantically coherent relationships.

For incorporating missing link semantics, various adaptations to the common authority ranking models can be constructed. However, changes to the computational model may alter the original behavior of an algorithm and impose new restrictions. For instance, the ObjectRank [6] approach adds link semantics to PageRank. However, it requires to assign static probabilities to each type of link before its application. Thus, ObjectRank lacks the flexibility of the original approach which can be easily applied to arbitrary Semantic Web graphs without such upfront assignments. Moreover, the modification of the original models and algorithms can impose side effects and hampers the comprehensibility of the approach. Interpretations on the impact of each modification are required to enable the evaluation of such modifications.

In this paper, we introduce a new approach TripleRank for authority ranking of linked data on the Semantic Web that naturally takes into account its additional semantics. We reconsider the paradigm of two-dimensional graph representation (e.g. adjacency matrix for HITS, probabilistic transition matrix for PageRank, a mixture of these models for SALSA, etc.) and represent semantic graphs by 3-dimensional tensors. Our model has a clear semantics and supports the seamless representation of RDF graphs, including link/predicate semantics. For finding authoritative sources, we apply the PARAFAC decomposition [18] which can be seen as a multi-dimensional correspondent to a singular value decomposition of a matrix. Tensor decomposition yields rich information on the resources and predicates of the analyzed Semantic Web graph beyond simple rankings. Rated groupings of RDF resources and RDF predicates with respect to their (latent) topic, authority, and navigational (hub) characteristics are the outcome of the analysis. These results can be exploited to support a variety of applications, such as semantic faceted navigation. We present encouraging results of TripleRank gathered by its application to multiple RDF data sets from the Wikipedia free encyclopedia and the DBLP research publications database.

The contribution is organized as follows. In Section 2 we distinguish our work from related approaches for rating Web pages and (semi-)structured data. Section 3 introduces the tensor-based semantic graph representation for RDF knowledge bases and its PARAFAC decomposition for authority ranking. Section 4 addresses design and architecture issues of TripleRank. Consequently, Section 5 shows results of systematic user studies that demonstrate the viability of TripleRank and its advantages in comparison with other methods. Section 6 summarizes and concludes the contribution.

2 Related Work

From the conceptual perspective, two topics can be seen as closely related to our TripleRank approach: authority ranking for Web contents and graph-based relevance ranking for semi-structured data. This section gives a short overview of these areas and distinguishes TripleRank from other existing solutions.

2.1 Rating Web Pages

PageRank [8], HITS [23] and SALSA [27] are prominent algorithms for ranking Web pages based on link analysis. PageRank builds upon a model of a random walk among Web pages, where the stationary probability of passing through a certain page is interpreted as measure of its importance. HITS is based on the notion of a mutual reenforcement between importance (authority) and connectivity (hub) scores of Web pages. SALSA can be seen as a more complex hybrid solution that integrates ideas of PageRank and HITS by combination of both link traversing directions (i.e. forward and backward) for constructing graph models. The conceptual generalization for this kind of methods is given in [14]. Unlike TripleRank, this family of methods provides no natural mechanisms for expressing and exploiting link semantics.

The contextualization of graph models can be achieved through different customizations of the mentioned models. Possible adaptations include various custom weightings of graph edges (e.g. based on appearance of particular terms in Web documents [31,29], content classification [13,19], structural properties like in-domain vs. out-domain linking [7], etc.) or joint probabilistic modeling for content and connectivity of Web pages [12]. In contrast to TripleRank, these solutions are designed for the Web setting and do not introduce distinguished link semantics. The solution presented in [26] uses for Web authority ranking the higher-order representation of the hyperlink graph by labeling the graph edges with the anchor text of the hyperlinks. This method is closely related to TripleRank, but addresses a fully different problem setting (links and anchors in the Web graph vs. linked data in Semantic Web RDF graphs). Additionally, our paper augments the introduction of TripleRank with a user evaluation that gives insight into the applicability of the overall authority ranking approach.

Another kind of contextualization for authority ranking models can be observed in the area of search personalization. For instance, Eirinaki and Vazirgiannis present a modification of the PageRank algorithm to compute personalized recommendations of Web pages given a path of visited pages [16]. Their approach requires access to web server logs that provide statistics about the paths browsed by other users. BrowseRank [28] is a further example of a page ranking approach that requires to collect statistics on user behavior such as the time spent on a web page. The generalized algorithm for personalized authority ranking is described in [22].

Our TripleRank approach is designed for a different scenario of resource recommendation when browsing linked data. As when browsing the Web, detailed statistical information about prior user visits is not available in our problem setting. Our TripleRank approach is conceptually more general and does not rely on user profiles and query logs. As when browsing the Web, detailed statistical information about prior user visits is often not available in our Semantic Web scenario. However, this information can be easily integrated with TripleRank, if necessary.

2.2 Rating (Semi-)Structured Data

ObjectRank [6] adds authority transfer weights for different types of links to the PageRank algorithm. Such weights influence the *random walk* of prospective users and are to be assigned by domain experts. Beagle++ [10] is an extension for the Beagle desktop search engine that applies ObjectRank to RDF meta data about desktop objects to improve their ranking in desktop search scenarios. TripleRank also considers the semantics of link types, however, it is an approach for computing ranks for RDF resources at runtime, does not rely on manually assigned link weights, and is based on the generalized HITS algorithm instead of PageRank.

Several research works have dealt with information retrieval in the Semantic Web, presenting approaches for ranking resources based on keyword input, e.g. Swoogle [15] and ReConRank [21]. Browsing-based search strategies, as targeted by TripleRank, differ from keyword search with respect to the search process, its perception by users, and the input and output data processed to implement browsing support [9]. Rocha et al. [32] presented an approach to semantic search that associates tf-idf, represented by semantic specificity and semantic similarity, to semantic relations to apply a spreading activation approach on this augmented graph.

Anyanwu and Sheth present a framework for query answering with respect to so called semantic associations [3]. A semantic association represents semantic similarity between paths connecting different resources in an RDF model. Aleman-Meza et al. [1] presented and evaluated methods for ranking semantic associations. As a continued work of [3], the presented methods target the identification of similar resources to apply it in scenarios like terror-prevention. Their approach involves ranking criteria considering graph structure, and user context. User context is defined statically by selecting ontology concepts that are considered as representative for a user's context. Ramakrishnan et al. present heuristics for weighting graph patterns connecting two nodes in a graph considering the differences of edges given by RDF graphs that include schema information encoded as RDFS ontologies [30]. Prior approaches on graph pattern analysis presented methods assuming that only one type of edge exists. Next to a presentation of the heuristics, they present an evaluation of them targeting the question which heuristic results in higher quality patterns.

3 TripleRank: The Semantic Web as Tensor

The authority ranking approach by TripleRank builds upon the analysis of a tensor. In this section, we first introduce our modeling approach for representing the Semantic Web as tensor. We then continue with details on its analysis.

3.1 The TripleRank Model

We define a Semantic Web graph as a graph $G = (V, L, E, linkType)$ where V is the set of RDF resources in the graph, L is the set of literals contained, and E is the set of links between RDF resources in V. Additionally, the function $linkType : E \rightarrow V$ returns the URI of the property that links two resources. Fig. 1(a) shows a small Semantic Web graph that contains seven resources (A,B,C,D,E,loves,hates) and ten links of two different types: *loves* and *hates*.

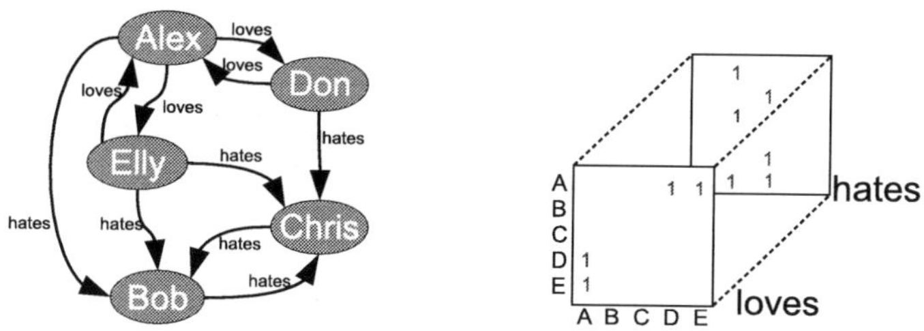

(a) Sample Semantic Web Graph (b) Tensor Representation

Fig. 1. Modeling Example

We represent Semantic Web graphs by a 3-dimensional tensor $\mathbf{T}$ where each of its slices represents an adjacency matrix for one RDF property. Figure 1(b) illustrates the tensor resulting from the transformation of the sample graph shown in Fig. 1(a). The first adjacency matrix $\mathbf{T}(:,:,1)$[1] models linkage by the property *loves*. An entry > 0 corresponds to the existence of a link by this property, empty entries are considered as zeroes. The second matrix $\mathbf{T}(:,:,2)$ models links by the property *hates*. For instance, the graph expresses that Alex *hates* Bob, which corresponds to $\mathbf{T}(1,2,2) = 1$.

3.2 PARAFAC for Authority Ranking

The Semantic Web graph can be described by an adjacency matrix. For a network graph matrix M the well known authority ranking methods like HITS [23] can be applied. HITS defines the authority ranking problem through mutual reinforcement between so-called hub and authority scores of graph nodes. The authority (relevance) score of each node is defined as the sum of hub scores of its predecessors. Analogously, the hub (connectivity) score of each node is defined as a sum of the authority scores of its successors. By applying the Singular Value Decomposition (SVD) to the adjacency matrix, we obtain hub and authority scores of graph nodes for each singular value of M, which can be interpreted as rankings regarding different themes or latent topics of interest. Formally, by this method, some arbitrary matrix $M \in \mathbf{R}^{k \times l}$ is splitted into three matrices $U \in \mathbf{R}^{k \times m}$, $S \in \mathbf{R}^{m \times m}$, $V \in \mathbf{R}^{l \times m}$. U and V represent the outlinks and inlinks with respect to the principal factor contained in S. Corresponding to our

[1] Throughout this paper we use the common Matlab-notation for addressing entries in tensors and vectors.

notation, M can be written as sum of rank-one-matrices by $M = \sum_{k=1}^{m} S^k \cdot U^k \circ V^k$. This 2-way decomposition yields authority and hub scores (cf. Fig. 2(a)) [25].

Modeling several link types by separate matrices results in very sparse and not connected matrices. Instead, the tensor model applied by TripleRank enables the representation of all adjacency matrices including information about the connections between link types. Tensor decomposition methods like PARAFAC can then detect further hidden dependencies.

These methods are regarded as higher-order equivalents to matrix decompositions. The PARAFAC tensor decomposition has the advantage of robustness and computational efficiency. These advantages are due to its uniqueness up to scaling and permutation of the produced component matrices [18]. By PARAFAC input tensors are transformed to so called Kruskal tensors, a sum of rank-one-tensors. Consequently, in TripleRank we derive authority and hub scores for particular latent aspects (topics) of the analyzed data from particular rank-one-tensors of the decomposition. In the context of this paper we focus on three-mode-tensors that represent connectivity between graph nodes together with semantics of links (predicates).

Formally, a tensor $\mathbf{T} \in \mathbf{R}^{k \times l \times m}$ is decomposed by n-Rank-PARAFAC into components matrices $U_1 \in \mathbf{R}^{k \times n}$, $U_2 \in \mathbf{R}^{l \times n}$, $U_3 \in \mathbf{R}^{m \times n}$ and n principal factors (pf) λ_i in descending order. Via these $\mathbf{T}$ can be written as a Kruskal tensor by $\mathbf{T} \approx \sum_{k=1}^{n} \lambda_k \cdot U_1^k \circ U_2^k \circ U_3^k$ where λ_k denotes the kth principal factor, U_i^k the kth column of U_i and $\circ$ the outer product [25]. U_i yields the ratio of the ith dimension to the principal factors. So, similar to SVD, PARAFAC derives hidden dependencies related to the pfs and expresses the dimensions of the tensor by relations to the pfs. Depending on the number of pfs PARAFAC decomposition can be loss-free. For a third-mode-tensor $\mathbf{T} \in \mathbf{R}^{k \times l \times m}$ a weak upper bound for this rank is known: $rank(\mathbf{T}) \leq \min\{kl, lm, km\}$ [25]. There is no proper way for estimating the optimal number of pfs for an appropriate decomposition but several indicators like residue analysis or core consistency exist [2].

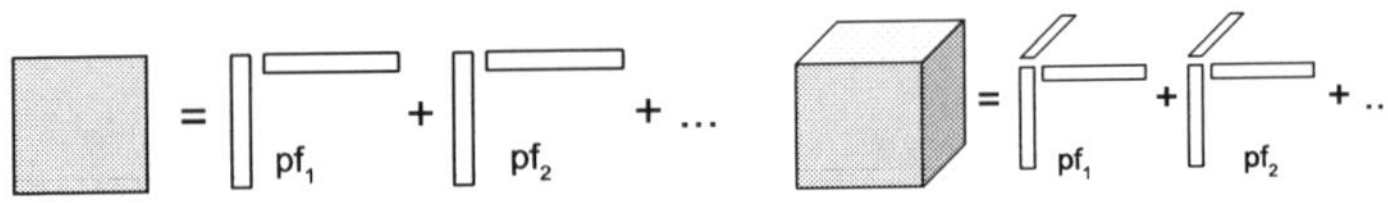

(a) Matrix Decomposition [25] (b) Tensor Decomposition [25]

The PARAFAC decomposition of a tensor derives authority and hub scores plus additional scores for the relevance of link types (cf. Fig. 2(b)). The tensor $\mathbf{T}$ in section 3.1 combines information about who loves and hates who. So the PARAFAC decomposition would yield U_1 with subject-pf relation, U_2 with object-pf relation and U_3 with property-pf relation. In other words U_1 keeps the hub scores as relevance of the subjects to the pfs, U_2 the authorities scores as relevance of the objects to the pfs and U_3 scores of the relevance of property types to the pfs. In line with HITS the largest entry of U_1^1 corresponds to the best hub for the first pf and the largest entry of U_2^1 to the best authority.

By the three matrices U_1, U_2 and U_3, subjects, objects, and properties can be compared in pairs regarding the pfs derived from all three dimensions. E.g. the relation of a property to specific subjects or objects is derived by multiplying the corresponding column of U_3 with those of U_1 or U_2 respectively.

3.3 Ranking Example

Applying the above transformation and analysis to the graph illustrated by Fig. 1 yields the results shown by the first four columns of Table 1. Two groups are identified, one where the predicate hates has a high score, and one where loves is scored highly. The authoritative resources for each group differ from each other. Bob and Chris have high scores with respect to hates. Don and Alex are the top authorities with respect to loves. The application of HITS results in the ranking shown by column 5 and 6. The HITS ranking corresponds to a ranking based on the indegrees of the resources. Notably, the rankings produced by the PARAFAC analysis are different from the HITS results as they provide rankings with respect to different knowledge aspects in the data.

Table 1. PARAFAC vs. HITS Results

PARAFAC				HITS	
Score	Predicate	Score	Resource	Score	Resource
Group 1				0.62	Chris
1.00	hates	0.71	Bob	0.56	Bob
-	-	0.70	Chris	0.50	Alex
Group 2				0.16	Don
1.00	loves	0.70	Don	0.16	Elly
0.001	hates	0.70	Alex		
-	-	0.10	Elly		

4 Implementation

Having introduced the theoretical background behind TripleRank, we present the implementation into an applicable system below. We describe the three core components of the TripleRank architecture, which encapsulate a 3-step process, namely i) the collection of data and its transformation to a tensor model, ii) its pre-processing, and iii) analysis.

4.1 Data Collection and Transformation

The first process step for the ranking of Semantic Web data is its collection. The TripleRank-collector requests RDF data from (linked open) data providers on the Semantic Web as follows. For a given starting point, i.e. some uniform resource identifier (URI), it executes a breadth first exploration of the surrounding resources. The exploration is parameterized by the maximal exploration depth, the maximal number of statements, and the maximal number of links to follow for each resource and link type. The collected data is then transformed into the tensor representation introduced in Section 3.1.

4.2 Pre-processing

The second component of the TripleRank architecture implements the pre-processing step on the collected data. Pre-processing is applied for two reasons: First, to reduce the amount of data to be analyzed. Second, to increase the quality of the collected data.

Information-theoretic notions ground the pre-processing as implemented by TripleRank. Predicates linking the majority of resources are pruned as they convey little information and *dominate* the data set. More precisely, for all of the results presented in this paper, a threshold of 40% has been used, i.e. predicates that occur in more than 40% of all statements are pruned. The predicate wikilink as used within the dbpedia.org [4] data set is an example of such a dominating predicate. It corresponds to links between pages of the Wikipedia encyclopedia, e.g. linking from a page describing a music band to associated band members, producers, tracks and further *different* types of information. Accordingly, the semantics of the relation established by this property are rather unclear.

A further pre-processing step is the weighting of the collected data to further remedy the negative effects of domination. We amplify statements based on their predicate frequency so that statements with less frequent predicates are amplified stronger than more common statements. As an effect, the adjacency indicators in the tensor have the following property:

$$\mathbf{T}(x, y, z) = \begin{cases} 1 + log\frac{\alpha}{links(z)}, & \text{if x points to y using property z} \\ 0, & \text{else} \end{cases}$$

The value α denotes the number of statements in which the most dominant predicate participates. The function $links(v)$ ($links : V \rightarrow \mathbb{N}_0$) returns the number of statements linked by property v.

We remark that the implemented pre-processing steps are valuable for generating ranking analyses in general. Notably, simple methods for authority ranking, e.g. the counting of inlink scores per resource and predicate, benefit more from such pre-processing than more complex methods like PARAFAC.

4.3 Analysis

The analysis step implements the PARAFAC decomposition of the tensor, as modeled and created by the previous process steps. We have integrated existing software packages [5] for this purpose. As indicated in Section 3.2, the number of factors for the PARAFAC decomposition is crucial for the quality of the results of the analysis. The determination of the optimal number of factors is a case of open research. However, heuristics for determining a suitable number of factors have been published, e.g. the core consistency diagnostic (CORCONDIA) [2]. The factor determination applied in TripleRank builds upon such research.

The result of the analysis is a Kruskal (cf. Sect. 3.2) tensor [25] that approximates the original tensor. As illustrated in Figure 2, the resulting vectors for the first (row), second (column), and third dimension are represented by three matrices. The columns of each of the matrices correspond to the scores calculated for the different factors $f_1 \ldots f_k$. Analogue to the SVD, entries in the column vectors correspond to authority scores, i.e. indicating the relevance of a resource with respect to its in-degree. Entries in the row vectors correspond to hub scores, i.e. indicating the relevance of a resource with respect to its out-degree. We refer to [23] for a thorough analysis of the correspondence between SVD and its interpretation for link analysis. Entries of the vectors in the third dimension indicate the relevance of a resource with respect to the hub and authority resources. Based on this notion, we interpret hub scores as indicative for the relevance of

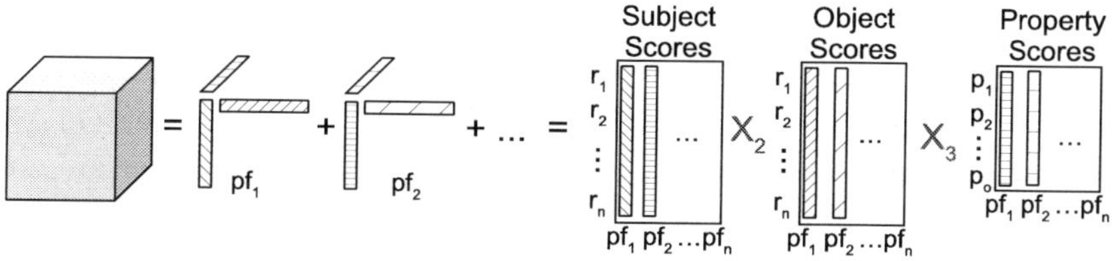

Fig. 2. Result of the analysis

resource when occurring as subject. Vice versa, authority scores indicate the relevance of a resource as object of a statement. As we modeled RDF properties by the third dimension, their relevance can be looked up in the vectors of the third dimension.

5 Evaluation

We have applied the implementation of TripleRank to existing Semantic Web data to investigate on the quality of the results and the runtime performance of the approach. Moreover, we have evaluated TripleRank in a user study where we have exploited it to support faceted navigation.

5.1 Data Sets

We have created an evaluation testbed by applying the extraction and transformation procedure introduced before. The testbed consists of multiple extracts around resources from dbpedia.org and the SWETO DBLP corpus, e.g. *The Beatles* and *Semantic Web*.

Table 2 describes the different data sets. Data sets created with a different parameterization are separated by a horizontal rule. The table shows, for each data set, the number of statements contained initially, the number of distinct properties, the top 3 properties with respect to the number of statements they occur in (shown in parentheses), and the number of triples after the pre-processing. The first group of data sets has been crawled by restricting the number of links per resource to 500 and the maximum exploration depth to 10. For the second group, containing the data sets on the SPARQL query language and the James Bond movie, a limit per predicate and resource has been set. For

Table 2. Description of Data Sets

Start Resource	Triples Before	Properties	Top 3 Link Types	Triples After
dbpr:The_Beatles	5084	85	dbpp:wikilink(1719), dbpp:origin(509), dbpp:artist(315)	5084
dbpr:HITS_algorithm	5145	70	dbpp:wikilink(2780), skos:subject(440), dbpp:wikiPageUsesTemplate(143)	5145
dbpr:Berlin	5044	82	dbpp:wikilink(1671), rdf:type(345), dbpp:birthplace(170)	5044
dbpr:The_Lord_of_the_Rings	5035	51	dbpp:wikiPageUsesTemplate(1234), skos:subject(1126), dbpp:wikilink(1063)	5035
dbpr:SPARQL	-	75	skos:subject(2640), rdf:type(426), owl:sameAs(288)	7853
dbpr:James_Bond	18421	93	dbpp:wikilink(13257), rdf:type(182), skos:subject(121)	5164
dbpr:The_Beatles	158608	421	dbpp:wikilink(105254), dbpo:birthPlace(11097), dbpp:countryofbirth(7067)	55354
dblp:semweb/2007	20421	7	rdf:type(9902), dc:publisher(5559), opus:in_series(4849)	10519

Namespaces: dbpr:<http://dbpedia.org/resource/>, dbpp:<http://dbpedia.org/property/>, skos:<http://www.w3.org/2004/02/skos/core#>, dblp:<http://dblp.uni-trier.de/rec/bibtex/>, dc:<http://purl.org/dc/elements/1.1/>, opus:<http://lsdis.cs.uga.edu/projects/semdis/opus#>

each resource, a maximum of 100 links per link type have been explored while the maximum exploration depth has been set to 8. The final two data sets have been created by restricting only the exploration depth. Here we see significant differences between the data sources dbpedia and dblp. The latter data set contains much less link types compared to the first. Among all[2] dbpedia data sets, we recognize the domination of the predicate wiklink. The crawling parameterization used for the first group of data sets already alleviated the domination, however, for less restrictive crawling parameters there is a strong domination. For instance, statements containing wikilink make up about 72% of all statements in the data set on James Bond. We provide download links for these data sets online at `http://isweb.uni-koblenz.de/Research/DataSets`

5.2 Performance

Applying TripleRank pre-processing and analysis steps to the data sets described above has led to reasonable and interesting results that provide a rich description of the analyzed data sets. As an example, Table 3 shows some of the results for the smaller Beatles data set that contains 5048 triples. As explained above, the PARAFAC analysis yields groupings of predicates and resources. A grouping corresponds to one box in Table 3. The first two columns within a box show for each grouping, the highest ranked predicates including their scores. The third and fourth column show the highest ranked statement objects (authorities) and associated scores. Predicates with a score below 0.1 and resources with a score below 0.0001 have been omitted from the results. For instance, the group on the top left shows the most authoritative resources for the topic described by the predicates skos:subject, dbpo:label, dbpp:label, and dbpp:producer. The top resources in this group are accordingly authoritative for the combination of these predicates and do not necessarily have to be linked by the top predicate, skos:subject for this group. For instance, the resource dbpr:Apple_Records is contained as authority. It describes a music label that produced songs of the Beatles.

As the quality of the results as illustrated by Table 3 seem sensible, we have also measured the runtime performance of the analysis step. The data collection and transformation steps have been excluded as these steps perform linear to the number of statements. Specifically, the performance of the data collection depends heavily on network bandwidth and the performance of the service providing the data. For all of the data sets, the analysis execution times have been within 2 to 18 seconds on a standard laptop computer with a dual core 2.0MHz processor and 2GB RAM.

5.3 User Evaluation: Faceted Browsing

We have evaluated TripleRank with 16 test persons to gather objective feedback on the sensibility of its results. The fine-grained results produced by TripleRank can be exploited for a number of applications ranging from similarity search for triples to facet identification in RDF data. We have chosen the scenario of faceted browsing for the user evaluation as we believe this scenario is well comprehensible for the participants of the study.

[2] The originally crawled data set on SPARQL has been accidently overwritten with the preprocessed data set so that statements containing wikilink are missing.

Table 3. TripleRank Results for the Beatles Data Set

Score	Predicate	Score	Resource	Score	Predicate	Score	Resource
0.66	skos:subject	0.35	dbpr:Category:Songs_produced-_by_George_Martin	0.70	dbpo:genre	0.73	dbpr:Rock_and_Roll
0.49	dbpo:label	0.32	dbpr:Category:1968_songs	0.70	dbpp:genre	0.45	dbpr:Beat_music
0.48	dbpp:label	0.27	dbpr:Category:Jazz_songs	0.16	dbpp:wikilink	0.23	dbpr:Psychedelic_rock
0.11	dbpp:producer	0.27	dbpr:Category:Psychedelic_songs	-	-	0.23	dbpr:Jazz_waltz
-	-	0.27	dbpr:Category:Folk_songs	-	-	0.23	dbpr:Folk_music
-	-	0.27	dbpr:Category:The_Beatles_songs-_sung_by_George_Harrison	-	-	0.13	dbpr:Folk_rock
-	-	0.27	dbpr:Category:George_Harrison-_songs	-	-	0.12	dbpr:Rock_music
-	-	0.16	dbpr:Apple_Records	-	-	0.09	dbpr:And_I_Love_Her
-	-	0.11	dbpr:Category:Apple_Records-_albums	-	-	0.06	dbpr:Capitol_Records
-	-	0.11	dbpr:Category:Double_albums	-	-	0.03	dbpr:Country_rock
0.95	dbpp:tracks	0.36	dbpr:Cry_Baby_Cry	0.69	dbpo:label	0.5	dbpr:Apple_Records
0.32	dbpp:wikilink	0.36	dbpr:Long	0.67	dbpp:label	0.26	dbpr:Capitol_Records
-	-	0.16	dbpr:Piggies	0.15	dbpp:wikilink	0.23	dbpr:EMI
-	-	0.16	dbpr:Ob-La-Di	-	-	0.21	dbpo:Resource
-	-	0.16	dbpr:I	-	-	0.18	dbpr:Template:infobox_album
-	-	0.16	dbpr:Rocky_Raccoon	-	-	0.15	dbpo:MusicalWork
-	-	0.16	dbpr:Good_Night_	-	-	0.15	dbpo:Work
-	-	0.16	dbpr:While_My_Guitar_Gently-_Weeps	-	-	0.14	dbpr:Parlophone
-	-	0.16	dbpr:Mother_Nature	-	-	0.14	dbpr:Template:succession_box
-	-	0.16	dbpr:Julia_	-	-	0.1	dbpo:Song
0.96	rdf:type	0.25	yago:TheBeatlesAlbums	0.70	dbpo:recordplace	0.93	dbpr:Abbey_Road_Studios
0.29	skos:subject	0.25	umbel:InstrumentalArtifact	0.70	dbpp:recorded	0.14	dbpr:1966
-	-	0.25	umbel:Artifact	0.14	dbpp:wikilink	0.11	dbpr:9_May
-	-	0.25	yago:AlbumsProducedBy-GeorgeMartin	-	-	0.11	dbpr:19_May
-	-	0.25	dbpo:Album	-	-	0.11	dbpr:16_May
-	-	0.25	yago:Album106591815	-	-	0.09	dbpr:27_April
-	-	0.12	yago:ParlophoneAlbums	-	-	0.09	dbpr:2_June
-	-	0.12	yago:AppleRecordsAlbums	-	-	0.09	dbpr:27_February
-	-	0.12	dbpo:Place	-	-	0.09	dbpr:1_July
-	-	0.12	umbel:Place	-	-	0.09	dbpr:16_July
1.00	dbpp:origin	0.99	dbpr:England				
-	-	0.11	dbpr:Liverpool				

Faceted browsing is a common means to ease the navigation in data by structuring associated data into so called facets [20]. For some resource, e.g. as viewed in a RDF browser, facets correspond to collections of associated resources. Resources within these collections have a commonality with respect to the currently viewed resource, e.g. the facet *band-members* for the resource *The Beatles* should contain only resources describing members of this band, e.g. John Lennon and Paul McCartney.

We have applied TripleRank to automatically select and order resources for given facets. Accordingly, the evaluation has been centered around the following question: *Given a currently viewed resource (subject) and a predicate (facet) associated to it, what are the most interesting, most related, most useful resources (objects) for the facet?*

We have created 10 of such queries using the data sets explained above and presented each of the 10 queries to each of the 16 test persons. As subject and predicate instances we have used the start URIs from the collected data sets (cf. Table 2) and associated properties. Test persons have been presented with candidate resources as produced by the randomly ordered union of the top 10 results from the TripleRank method and a baseline method. For each resource, they could indicate whether it is a good or bad match for the query, or whether they are undecided. Figure 3 illustrates how queries have been presented to test persons by the evaluation tool. Please note that the shown

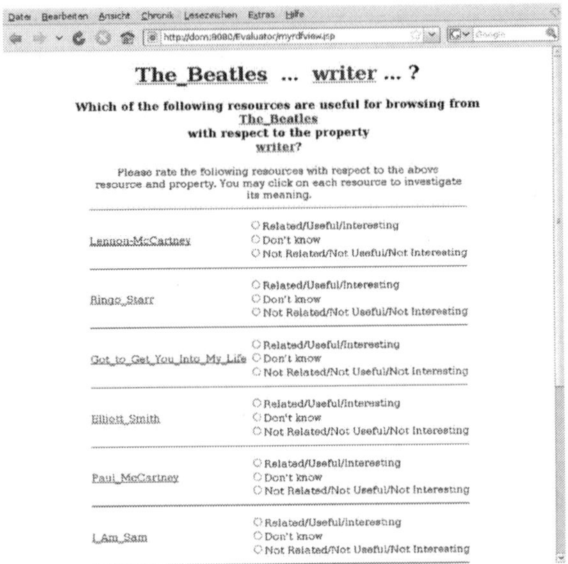

Fig. 3. Screenshot of the User Evaluation

resources (subject, predicate, objects) have been presented as hyperlinks to the service providing the data. Test persons could use these links to investigate on the resource by its complete description. For instance, the resource *The Beatles* links to the web page `http://dbpedia.org/resource/The_Beatles`

5.4 Baseline Method

The evaluation scenario requires to rank resources with respect to their authority for a given subject and facet. The baseline methods implement a straightforward strategy for computing scores for ranking using the tensor **T** as follows.

Baseline I. The straightforward and obvious baseline method for computing such a ranking given **T** follows a simple procedure:

1. Select statements matching the subject and predicate (facet) and project the objects.
2. Rank selected objects by their number of inlinks as known from the data set as inlinks indicate authority [23].

For data sets that are available completely and that contain little or no noise, this procedure performs very well with respect to runtime and the quality of results. It will select the relevant objects with high precision. The dbpedia and DBLP data sets obviously contain very little noise. Dbpedia data relies on the careful authoring by numerous users, whereas DBLP data maps to settled and rigid structures maintained by a database system. However, due to the fact that our data sets have been crawled from online services, they are incomplete and represent only a small portion of these large data sets. Accordingly, when applying this baseline method to our data sets, we have experienced no or very little recall due to the sparsity of the data. For instance, applying this baseline for answering the query *"What are relevant subjects for James Bond?"* it returns only one resource while the enhanced Baseline II method we present in the following returns 6 relevant resources. Nevertheless, we imagine that every method for ranking facet objects will include this baseline method as it produces high precision when data sets are complete and without noise.

Baseline II. We modified the Baseline I method in order to improve its competitiveness, i.e. improving its recall for incomplete data sets. Instead of considering only exact matches of subject-predicate patterns, we extended this criteria to consider also predicate matches only. As the data sets have been created around a URI that is central within the data sets, this extended baseline method turned out to produce more competitive results, specifically an increased recall. This method is also closer to the PARAFAC analysis as used by TripleRank as it corresponds to the authority ranking based on the in-degree (number of incoming links) of resources. In fact, this method can be seen as an approximation of authority scores in HITS, which are highly correlated with in-degrees of nodes [14] (see also the example given in Sect. 3.3). Accordingly, we have evaluated TripleRank against this improved Baseline II method.

5.5 TripleRank Method

The TripleRank method is similar to the Baseline II method with the exception that it builds upon the results of the PARAFAC decomposition of **T** instead of plain link-counting using **T**. First, the top factors for the predicate (facet) in question are selected. They are looked up from the matrix containing the predicate scores (cf. Figure 2). Those factors are selected where the predicate is ranked first or has a score of at least 0.6. Afterwards, the resource vectors of these factors, as given by the object-score matrix, are considered for the selection of the ten resources with the highest scores.

5.6 Results

We have received 1387 answers by 16 test persons. On average, the union of the results produced by the BaselineII and TripleRank method led to 8.669 candidate objects per query. While we asked for the top 10 results by each method, the average number of results is lower, as for several questions both methods produced less than 10 results.

We calculated the agreement among test persons using the index raw agreement [17]. It represents the ratio between the measured inter-rater agreement and the best possible agreement. Accordingly, a value of 1 corresponds to the maximal agreement, while 0 corresponds to no agreement. As expected, the evaluation shows a high inter-rater agreement. Considering the specific agreement on the relevance of results, test persons showed a higher agreement rate on positive candidates (agreement-ratio: 0.68) than on negative candidates (agreement-ratio: 0.57). For the overall agreement, we have measured an even higher ratio of 0.70 among the test persons.

The evaluation results presented in Table 4 built on these user ratings and give insight into the constitution of the union of the results. Table 4 lists the average number of results found exclusively by each method (BaselineII–TripleRank, TripleRank–BaselineII), and the average number of results suggested by both of the

Table 4. Results of the User Evaluation

	Avg Number of Results				Avg Precision	
	Total	Positive	Negative	Undecided	Macro	Micro
BaselineII−TripleRank	1.075	0.273	0.444	0.344	0.478	0.393
TripleRank−BaselineII	4.731	1.913	1.650	1.169	0.521	0.537
BaselineII∩TripleRank	2.863	1.338	0.763	0.763	0.659	0.637
BaselineII	3.948	1.626	1.207	1.107	0.607	0.574
TripleRank	7.594	3.251	2.413	1.932	0.557	0.574

methods, namely the intersection of their results (BaselineII∩TripleRank). In addition, Table 4 also presents the average numbers of ratings received for the results. Here, we notice that the TripleRank method succeeds in suggesting more positively rated results than negative ones. Comparing the total number of results produced by each method, the TripleRank approach nearly doubles the number of results (3.948 vs 7.594). For the number of results considered relevant by users, TripleRank achieves an increase from 1.626 to 3.251, i.e. the recall is increased by the factor of 1.999. For these results, we have also calculated additional confidence intervals with confidence degree 0.9. In most of the cases, TripleRank outperforms the baseline with statistical significance.

Table 4 also lists the precision of each approach. For a fine-grained analysis, we have calculated the macro precision, reflecting the users' perspective on the performance of the methods, and the micro precision, reflecting the systems view [11]. For the results produced exclusively by each method, TripleRank has higher precision than the baseline approach. Considering the overall results, the baseline method has equal precision on the micro-level and higher precision on the macro-level. Significance tests with confidence degree 0.9, however, have shown that all measured differences in precision are not statistically significant. Accordingly, we conclude that the TripleRank approach results in a substantially increased recall without loss of precision.

5.7 Lessons Learned: TripleRank Advantages

By further analysis of the evaluation results, we have investigated why the TripleRank approach outperforms the baseline method. Among the findings, we identified that the results of the analysis step implicate similarities among properties. For instance, it turned out that the dbpedia data sets contain predicates with similar semantics, e.g. the predicate `http://dbpedia.org/property/genre` and the predicate `http://dbpedia.org/ontology/genre` (cf. Table 3). Both properties associate bands or songs to a genre. The analysis step identified this similarity producing a factor where both properties are among the top two properties. The most authoritative resources for that factor are accordingly those resources having both or either one of these properties as inlink. As an effect, both precision and recall are increased when compared to baseline approaches that cannot identify the similarities of links. The advantage of TripleRank, however, is not only given when properties are directly grouped into one factor. The results of the tensor analysis also enable to increase recall by considering multiple factors where a predicate is ranked high. Within our experiments such increase of recall has been achieved without a loss of precision.

6 Conclusion and Future Work

In this paper we presented TripleRank, a novel approach for authority ranking in SemanticWeb applications. Conceptually, TripleRank is a SemanticWeb correspondent to authority ranking methods known from Web retrieval, such as PageRank or HITS. Our approach exploits the novel representational model for semantic RDF graphs, based on 3-dimensional tensors. This allows us to exploit in the natural way the available semantics of RDF predicates. By applying the PARAFAC tensor decomposition we identify authoritative sources in the knowledge base as well as groups of semantically coherent predicates and resources. The systematic evaluation shows the viability of TripleRank for a wide range of search/browsing applications in the Semantic Web, such as semantic faceted navigation and ranked retrieval. Therefore, TripleRank can be seen as a next step towards efficient and effective search/retrieval technology for Semantic Web. As next steps for this research, we pursue the integration of TripleRank into existing Linked Open Data browsers, e.g. our own Semantic Web browser LENA [24], to investigate on further enhancements of the method.We also envision the addition of semantic relations that go beyond link semantics. For instance, we consider the modeling of resource similarity based on the similarity of the types of resources.

Acknowledgements. This work was funded by the X-Media project (www.x-media-project.org) sponsored by the European Commission as part of the Information Society Technologies (IST) programme under EC grant number IST-FP6-026978.

References

1. Aleman-Meza, B., Halaschek-Wiener, C., Arpinar, I.B., Ramakrishnan, C., Sheth, A.P.: Ranking complex relationships on the semantic web. IEEE Internet Computing 9(3), 37–44 (2005)
2. Andersson, C.A., Bro, R.: The n-way toolbox for matlab. Chemometrics and Intelligent Laboratory Systems 52(1), 1–4 (2000)
3. Anyanwu, K., Sheth, A.P.: The p operator: Discovering and ranking associations on the semantic web. SIGMOD Record 31(4), 42–47 (2002)
4. Auer, S., Bizer, C., Kobilarov, G., Lehmann, J., Cyganiak, R., Ives, Z.G.: Dbpedia: A nucleus for a web of open data. In: Aberer, K., Choi, K.-S., Noy, N., Allemang, D., Lee, K.-I., Nixon, L.J.B., Golbeck, J., Mika, P., Maynard, D., Mizoguchi, R., Schreiber, G., Cudré-Mauroux, P. (eds.) ASWC 2007 and ISWC 2007. LNCS, vol. 4825, pp. 722–735. Springer, Heidelberg (2007)
5. Bader, B.W., Kolda, T.G.: Algorithm 862: MATLAB tensor classes for fast algorithm prototyping. ACM Transactions on Mathematical Software 32(4), 635–653 (2006)
6. Balmin, A., Hristidis, V., Papakonstantinou, Y.: Objectrank: Authority-based keyword search in databases. In: VLDB, pp. 564–575 (2004)
7. Bharat, K., Henzinger, M.R.: Improved Algorithms for Topic Distillation in a Hyperlinked Environment. In: 21st Annual International ACM SIGIR Conference, Melbourne, Australia, pp. 104–111 (1998)
8. Brin, S., Page, L.: The anatomy of a large-scale hypertextual web search engine. In: Seventh International World-Wide Web Conference, WWW 1998 (1998)
9. Broder, A.: A taxonomy of web search. SIGIR Forum 36(2), 3–10 (2002)
10. Chirita, P.A., Ghita, S., Nejdl, W., Paiu, R.: Beagle++: Semantically enhanced searching and ranking on the desktop. In: Sure, Y., Domingue, J. (eds.) ESWC 2006. LNCS, vol. 4011, pp. 348–362. Springer, Heidelberg (2006)

11. Manning, H.S.C., Raghavan, P.: Introduction to Information Retrieval. Cambridge University Press, Cambridge (2008)
12. Cohn, D.A., Hofmann, T.: The missing link - a probabilistic model of document content and hypertext connectivity. In: 13th Conference on Advances in Neural Information Processing Systems (NIPS), Denver, USA, pp. 430–436 (2000)
13. Diligenti, M., Gori, M., Maggini, M.: Web Page Scoring Systems for Horizontal and Vertical Search. In: 11th International World Wide Web Conference (WWW), Honolulu, USA, pp. 508–516 (2002)
14. Ding, C.H.Q., He, X., Husbands, P., Zha, H., Simon, H.D.: PageRank, HITS and a Unified Framework for Link Analysis. In: 25th Annual International ACM SIGIR Conference on Research and Development in Information Retrieval, Tampere, Finland, pp. 353–354 (2002)
15. Ding, L., Pan, R., Finin, T.W., Joshi, A., Peng, Y., Kolari, P.: In: International Semantic Web Conference, pp. 156–170 (2005)
16. Eirinaki, M., Vazirgiannis, M.: Usage-based pagerank for web personalization. In: IEEE International Conference on Data Mining, pp. 130–137 (2005)
17. Von Eye, A., Mun, E.Y.: Analyzing Rater Agreement: Manifest Variable Methods. Lawrence Erlbaum Associates, Mahwah (2004)
18. Harshman, R.A., Lundy, M.E.: Parafac: Parallel factor analysis. Computational Statistics & Data Analysis 18(1), 39–72 (1994)
19. Haveliwala, T.H.: Topic-sensitive PageRank. In: 11th International World Wide Web Conference (WWW), Honolulu, USA, pp. 517–526 (2002)
20. Hildebrand, M., van Ossenbruggen, J., Hardman, L.: /facet: A browser for heterogeneous semantic web repositories. In: international Semantic Web Conference, pp. 272–285 (2006)
21. Hogan, A., Harth, A., Decker, S.: Reconrank: A scalable ranking method for semantic web data with context. In: 2nd Workshop on Scalable Semantic Web Knowledge Base Systems (2006)
22. Jeh, G., Widom, J.: Scaling Personalized Web Search. In: 12th International World Wide Web Conference (WWW), Budapest, Hungary, pp. 271–279 (2003)
23. Kleinberg, J.M.: Authoritative sources in a hyperlinked environment. J. ACM 46(5), 604–632 (1999)
24. Koch, J., Franz, T., Staab, S.: Lena - browsing rdf data more complex than foaf. In: International Semantic Web Conference (ISWC) Demo Session (2008)
25. Kolda, T.G., Bader, B.W.: Tensor decompositions and applications. SIAM Review 51(3) (to appear) (September 2009)
26. Kolda, T.G., Bader, B.W., Kenny, J.P.: Higher-Order Web Link Analysis Using Multilinear Algebra. In: 5th IEEE International Conference on Data Mining (ICDM), Houston, USA, pp. 242–249 (2005)
27. Lempel, R., Moran, S.: SALSA: the Stochastic Approach for Link-Structure Analysis. ACM Transactions on Information Systems (TOIS) 19(2), 131–160 (2001)
28. Liu, Y.-T., Gao, B., Liu, T.-Y., Zhang, Y., Ma, Z., He, S., Li, H.: Browserank: letting web users vote for page importance. In: SIGIR, pp. 451–458 (2008)
29. Rafiei, D., Mendelzon, A.O.: What is this Page known for? Computing Web Page Reputations. Computer Networks 33(1-6), 823–835 (2000)
30. Ramakrishnan, C., Milnor, W.H., Perry, M., Sheth, A.P.: Discovering informative connection subgraphs in multi-relational graphs. SIGKDD Explor. Newsl. 7(2), 56–63 (2005)
31. Richardson, M., Domingos, P.: The Intelligent surfer: Probabilistic Combination of Link and Content Information in PageRank. In: 14th Conference on Advances in Neural Information Processing Systems (NIPS), Vancouver, Canada, pp. 1441–1448 (2001)
32. Rocha, C., Schwabe, D., de Aragão, M.P.: A hybrid approach for searching in the semantic web. In: WWW, pp. 374–383 (2004)

What Four Million Mappings Can Tell You about Two Hundred Ontologies

Amir Ghazvinian, Natalya F. Noy, Clement Jonquet, Nigam Shah, and Mark A. Musen

Stanford University, Stanford, CA 94305, US
{amirg,noy,jonquet,nigam,musen}@stanford.edu

Abstract. The field of biomedicine has embraced the Semantic Web probably more than any other field. As a result, there is a large number of biomedical ontologies covering overlapping areas of the field. We have developed BioPortal—an open community-based repository of biomedical ontologies. We analyzed ontologies and terminologies in BioPortal and the Unified Medical Language System (UMLS), creating more than 4 million mappings between concepts in these ontologies and terminologies based on the lexical similarity of concept names and synonyms. We then analyzed the mappings and what they tell us about the ontologies themselves, the structure of the ontology repository, and the ways in which the mappings can help in the process of ontology design and evaluation. For example, we can use the mappings to guide users who are new to a field to the most pertinent ontologies in that field, to identify areas of the domain that are not covered sufficiently by the ontologies in the repository, and to identify which ontologies will serve well as background knowledge in domain-specific tools. While we used a specific (but large) ontology repository for the study, we believe that the lessons we learned about the value of a large-scale set of mappings to ontology users and developers are general and apply in many other domains.

1 Why Create the Mappings?

The field of biomedicine has embraced the Semantic Web probably more than any other field. Ontologies in biomedicine facilitate information integration, data exchange, search and query of heterogeneous biomedical data, and other critical knowledge-intensive tasks [12]. As a result, there is a large number of biomedical ontologies covering overlapping areas of the field [3]. Creating mappings among ontologies by identifying similar concepts is a critical step in integrating data and applications that use different ontologies. With these mappings, for example, we can link resources annotated with terms in one ontology to resources annotated with related terms in another ontology, discovering new relations among the resources themselves (e.g., linking drugs and diseases).

As part of our work for the National Center for Biomedical Ontology (NCBO), we have developed BioPortal—an open community-based repository of biomedical ontologies [10].[1] This repository contains 140 ontologies with more than one million concepts among them. We view mappings between concepts in different ontologies as an

[1] http://bioportal.bioontology.org

A. Bernstein et al. (Eds.): ISWC 2009, LNCS 5823, pp. 229–242, 2009.

essential part of the ontology repository. Users can browse the mappings, create new mappings, upload mappings created with other tools, download the mappings stored in BioPortal, or comment on the mappings and discuss them [11]. Other NCBO tools, such as a service for automatic creation of ontology-based text annotations [7], rely on the mappings to annotate biomedical resources with terms from different ontologies and for linking these resources to one another.

Over the past year, our team and our collaborators have uploaded more than 30,000 mappings to BioPortal. However, these mappings constitute only a tiny subset of the mappings between concepts in BioPortal ontologies. Thus, one of our goals was to use simple methods to generate quickly a large number of high-precision mappings to include in the repository. Our earlier studies have shown that in the case of biomedical ontologies simple lexical techniques, such as comparing preferred names of concepts and their synonyms, are extremely effective in creating mappings [6]. We have reported elsewhere [6] that this simple method for generating mappings achieves extremely high precision for biomedical ontologies—where preferred names of concepts and synonyms are used extensively and represent a rich source of information. In addition, we compared our results with the gold standard produced by the Ontology Alignment Evaluation Initiative (OAEI) [4]. For the use case of biomedical ontologies, our method achieved levels of recall and precision comparable to the best tools in the competition.

We have applied this simple lexical matching of preferred names and synonyms to generate mappings between concepts in BioPortal ontologies. In addition, the Unified Medical Language System (UMLS) [9] is a large collection of biomedical ontologies and terminologies that the researchers at the US National Library of Medicine have integrated. We plan to include UMLS terminologies in BioPortal in the near future and thus, when creating the mappings, we included the UMLS terminologies as well.

We generated mappings across all concepts in 140 BioPortal ontologies and 67 UMLS terminologies with more than 4 million concepts among them. We had two goals in generating the mappings: (1) we wanted to create a large set of mappings for the BioPortal resource that other applications can access and use; and (2) we wanted to learn more about the characteristics of the ontologies and the relationships between them. Using a set of more than 4 million mappings generated over our ontology set, we would like to answer several practical questions with implications for ontology reuse and development. These questions include, but are not limited to, the following:

To what degree are the domains covered by different ontologies connected? Mapping out the connections between ontologies will help us understand which domains are closely related, which ontologies may serve as a bridge between domains, and so on.

If you are new to a domain, what are the important or representative ontologies with good coverage? Ontology developers seeking to develop or reuse knowledge from a particular domain will be interested to know which ontologies have good coverage within that domain.

If you want to build domain-specific tools for creating ontology mappings, what are good ontologies to use for background knowledge? Many tools that seek to provide mappings for the purpose of ontology alignment use background knowledge to improve their ability to produce valid and accurate mappings (e.g., [15,1]). Such background knowledge allows these tools to use information about the representation of the domain to identify equivalent concepts. Thus knowing which ontologies are optimal for use as a source of background knowledge may greatly improve these tools.

What can we learn about the characteristics of the ontologies themselves and the ontology repository from the mappings between them? A set of mappings can provide insight about the ontologies themselves, their importance to their respective domains, or their coverage.

Researchers have previously successfully applied network analysis to gain insights into the structure and connectedness of large data sets [8]. In this paper, we apply network-analysis methods to analyze the ontologies and their mappings, to answer the questions posed above, and to reason about the distribution of mappings among the ontologies.

Note that our analysis does not depend on the specific methods that was used to generate the mappings and relies only on the fact that we have large set of high-precision mappings.

This paper makes the following contributions:

- We demonstrate that large-scale mapping sets can be useful in understanding the structure of an ontology repository, by identifying the most pertinent ontologies, the domains of overlap among ontologies, and the missing parts in an ontology repository.
- We propose network-based analysis metrics of ontologies based on mappings between them.
- We produce a set of more than 4 million mappings for the repository of ontologies and terminologies in BioPortal and UMLS.

2 Materials and Methods: What's in a Link?

We will now describe how we created the mappings used in this study and what data we analyzed.

We define a **mapping** as a relationship between two classes from different ontologies. The mappings that we discuss in this paper are **similarity mappings**: we declare that two classes from different ontologies are **similar** if the meaning that one class represents is similar or identical to the meaning of the other. We use the term "mapping" throughout this paper to refer to similarity mapping.

2.1 The NCBO Ontology Set

To develop our mappings, we used a set of 207 ontologies in the domain of biomedicine. These ontologies include 140 ontologies in BioPortal and 67 terminologies in the UMLS. The ontologies in BioPortal come from two sources: (1) 70 ontologies are downloaded nightly from the OBO Foundry repository;[2] (2) 70 ontologies are submitted by their developers directly to BioPortal. Among them, the 207 ontologies and terminologies that we used in this study contained 4,021,662 concepts.

[2] http://obofoundry.org

2.2 Creating Lexical Mappings between Concepts

We created the mappings using the following steps, which we describe in detail in the rest of this section:

1. generate a database of *terms* used for preferred names and synonyms of ontology concepts;
2. *normalize* the strings in the database;
3. find pairs of *matching* terms;
4. create *mappings* between concepts based on the matching terms identified in the previous step.

In the first step—creating a database of preferred names and synonyms of all concepts in all ontologies—we needed to identify for each ontology which properties contained the strings representing these preferred names and synonyms. All UMLS terminologies have preferred names and synonyms clearly identified for all the concepts. Many of the BioPortal ontologies are represented in the OBO format, which also has designated properties to define preferred name and synonyms of a class. The OWL language does not itself provide any special annotation properties to store preferred names and synonyms (although many ontologies use `rdfs:label` for the former).[3] Thus, when users submit an OWL ontology to BioPortal, we ask them to indicate which OWL properties their ontology uses for preferred names and synonyms. We store the names of these properties as part of ontology metadata.

Extracting all terms for preferred names and synonyms resulted in a database of 7,637,125 terms. We then normalized all the strings, by converting them to lower case and removing all delimiters (e.g., spaces, underscores, parentheses, etc.). We used a mySQL database to store each term along with the ID for the ontology and the concept that it came from.

We used an SQL query to find pairs of matched terms among the normalized terms. From the database table of normalized terms, we utilized an SQL query to identify pairs of terms that matched exactly. To improve precision, we compared only strings with at least three characters and ignored the strings with three characters or less.

Since each term refers to a preferred name or synonym of a specific concept from a particular ontology, we used matching terms to connect concepts from different ontologies.

Consider the following example. The class "Myocardium" in Foundational Model of Anatomy (FMA) has the preferred name "Myocardium." The class "Heart myocardium" in the ontology of Mouse adult gross anatomy has a synonym "myocardium."[4] We will match the two normalized terms "myocardium" and therefore create a mapping between the two classes, "Myocardium" in FMA and "Heart myocardium" in Mouse adult gross anatomy ontology.

Such mapping between the mouse myocardium represented in Mouse adult gross anatomy ontology and the human myocardium represented in FMA can facilitate

---

[3] The Simple Knowledge Organization System (SKOS) provides the RDF-based vocabulary for defining preferred names and synonyms for concepts, but so far none of the BioPortal ontologies use SKOS.

[4] `http://bioportal.bioontology/org/ontologies`

cross-species data exploration and integration. Having created this mapping, we could then integrate data annotated with the concept "Heart myocardium" in a database describing mouse experiments and data annotated with "Myocardium" describing human-related data.

This process resulted in a set of 4,001,775 mappings, where each each mapping represents a class from one ontology that is similar to a class from another ontology. The mapping is bi-directional as the similarity relationship generated this way is symmetric.

Note that UMLS itself contains a large set of manually created mappings between terms in different ontologies and terminologies. In the work described in this paper, we did not include those mappings. Rather, we used only the lexical mappings that we have generated. In future work, we plan to include the mappings that UMLS provides for additional information.

2.3 Identifying Links between Ontologies

Because our goals include analysis of relations between *ontologies* and not individual concepts, we define a link between two ontologies based on a set of mappings between concepts from those ontologies. We use the notation $mapping(c_1, c_2)$ to describe a mapping between two concepts from different ontologies, such as the mappings that we described in Section 2.2. We denote a set of all concept-to-concept mappings between two ontologies S and T as $M(S,T)$, where $M(S,T) = \{mapping(c_s, c_t), c_s \in S, c_t \in T\}$.

Definition 1 (Mapping-Based Link between Ontologies). *Given two ontologies, the source ontology S and the target ontology T, and a set of mappings between them $M(S,T)$, we say that there exists a mapping-based link L between ontologies S and T iff M is not an empty set: $M \neq \emptyset$.*

If two ontologies have at least one pair of concepts with similar names or synonyms between them, there will be a mapping-based link between the two ontologies. However, a more meaningful measure is the *number* of links between two ontologies or, more precisely, the *fraction* of one ontology that is mapped to another. For instance, if two ontologies each have 1,000 concepts, then, intuitively, the two ontologies are much closer to each other if 700 of these concepts match than if 5 concepts do. Thus, we define the notion of a percent-normalized link between ontologies which reflects not only how many mappings one ontology has to another, but also normalizes this measure with respect to the ontology size.

Definition 2 (Percent-Normalized Link between Ontologies). *Given two ontologies, the source ontology S and the target ontology T, and a set of mappings $M(S,T)$ between them, we say that there is a percent-normalized link between S and T, $L_p(S,T)$ where $p \geq 0$ and $p \leq 100$, iff at least p% of the concepts in the ontology S are sources for the mappings in $M(S,T)$. We say that $L_0(S,T)$ holds if there is at least one mapping between concepts in S and T*

For instance, if an ontology S has 1,000 concepts, and 500 of these concepts are mapped to concepts in an ontology T, then $L_p(S,T)$ is true for all values of p from 0% to 50%.

Note that $L_p(S,T)$ is directional and it is entirely possible (and, in fact, common) for $L_p(S,T)$ to be true and for $L_p(T,S)$ to be false at the same time. If one ontology is much larger than another, a large fraction of the smaller ontology may be mapped to the larger ontology, but the set of mappings still constitutes a small portion of the larger ontology.

Intuitively, the percent-normalized link reflects how significant a set of mappings between ontologies is in the context of those ontologies. We evaluated the distribution of these links for several different values of p. We determined what percent of all ontology links were present at different values of p. Additionally, we implemented a graphical visualization of the links at each of these thresholds to analyze the clustering patterns and the link distribution for different values of p. Finally, we counted the number of links for each ontology at different values. We used this data to analyze the frequency with which an ontology has exactly k links.

3 Results

We used the data that we collected from the mappings to plot and analyze several metrics: First, we analyzed the number of links between ontologies at different values of p (i.e., at varying sizes of the mapped portion of the ontology, normalized by the ontology size) and the distribution of ontologies based on the number of links (Section 3.1). Second, we treated ontologies and the links between them as nodes and edges in a graph, again for several values of p. We analyzed the properties of these graphs as networks, using metrics such as the number of hubs and clusters (Section 3.2). Finally, we examined the overall similarity of the ontologies (Section 3.3). We discuss and analyze our results in Section 4.

3.1 How Many Links Do Ontologies Have?

Figure 1 shows a distribution of the number of links that ontologies have for two values of p: 20% (Figure 1a) and 1% (Figure 1b). Recall that when $p = 20\%$, we create a link between source ontology S and target ontology T iff at least 20% of concepts from S are mapped to concepts from T. In other words, ontologies that we count in the graph on the right have a looser connection to each other than the ontologies in the graph on the left. Due to lack of space, we do not present the graphs for values of $p > 20\%$.[5] The distribution for larger values of p is very similar to the distribution for $p = 20\%$.

The graphs in Figure 1 show that the links between ontologies follow a power-law distribution for $p = 20\%$ (and larger values of p): There is a small number of ontologies that have large number of links and a large number of ontologies with just a few links. For smaller values of p, however, such as $p = 1\%$, where we include ontologies with very little overlap, our network becomes essentially random.

We analyzed the average distance between two nodes in the graph for some values of p. We found that for small values of p, the network is quite well connected and

[5] This data is available at `http://www.bioontology.org/wiki/index.php/Mapping_Set`

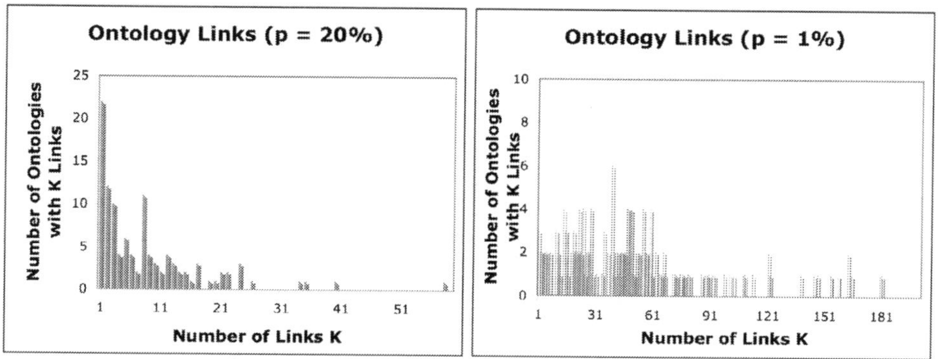

A B

Fig. 1. Number of Links between ontologies for(a) $p = 20\%$ **and (b)** $p = 1\%$**:** The x-axis represents a number of links to other ontologies that each ontology has. The y-axis represents the number of ontologies with that number of links. The graph demonstrates the power-law distribution for $p = 20\%$: there is a small number of ontologies that have a large number of links (the hubs) and a large number of ontologies with just a few links. If we use $p = 1\%$ (there is a link from one ontology to another if at least 1% of its concepts are mapped), the distribution becomes essentially random.

represents a small world: A network exhibits the small world property if any one node in the network is never more than a few hops away from any other [2]. For $p = 10\%$, the average distance between any two nodes is only 1.1 hops.

3.2 Hubs and Clusters

We constructed directed graphs of ontologies at several different thresholds of p (Figure 2). Nodes in the graphs are ontologies. There is a directed edge from a node corresponding to the ontology S to a node corresponding to an ontology T iff there is a link $L_p(S,T)$ between these ontologies for this values of p. We say that a particular node is a **hub** in this graph if it has more than twice the number of links (incoming and outgoing) than the average number of links for nodes in the graph. A set of connected ontologies forms a **cluster**.

We used these graphs to identify connections between different ontologies, based on varying levels of overlap. The graphs identified clear hubs, ontologies to which many other ontologies link, with the Systematized Nomenclature of Medicine - Clinical Terms (SNOMED-CT) ontology being the most prominent hub: We found that SNOMED-CT had the most links of any ontology, with 58 links (57 as "target", 1 as "source") at p = 20%. In other words, 58 out of 207 had at least 20% of their concepts mapped to concepts to SNOMED-CT. On further analysis, we found that more than 85% of SNOMED-CT concepts are mapped to at least one other concept in our repository.

Figure 3a shows variation of some key graph features (number of hubs, number of ontologies, number of clusters, and size of the largest cluster) as we change the value of p. Additionally, the plot in Figure 3b displays the percent of ontologies in the largest cluster.

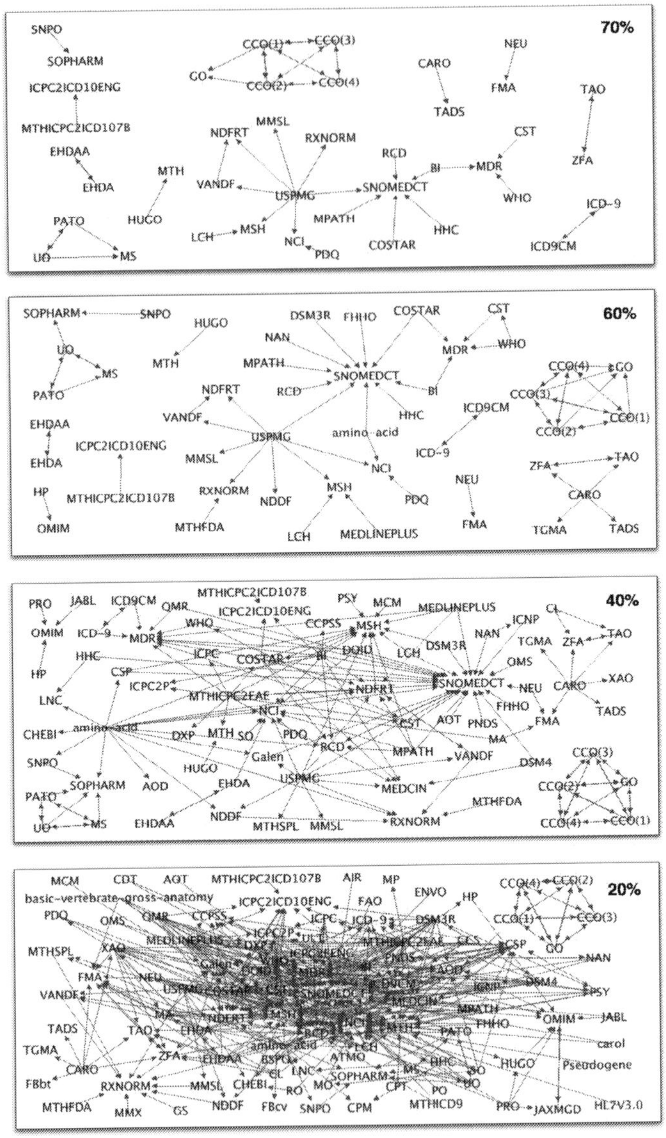

Fig. 2. The graphs show percent-normalized links between ontologies that are true for p = 20%, 40%, 60%, and 70%. Nodes represent ontologies in the repositories. Please refer to http://bioportal.bioontology.org/ontologies for the list of full ontology names corresponding to the acronyms that we use as labels in the graph. An edge from a node representing on ontology O_1 to a node representing an ontology O_2 means that at least p% of concepts from O_1 map to some concept in O_2.

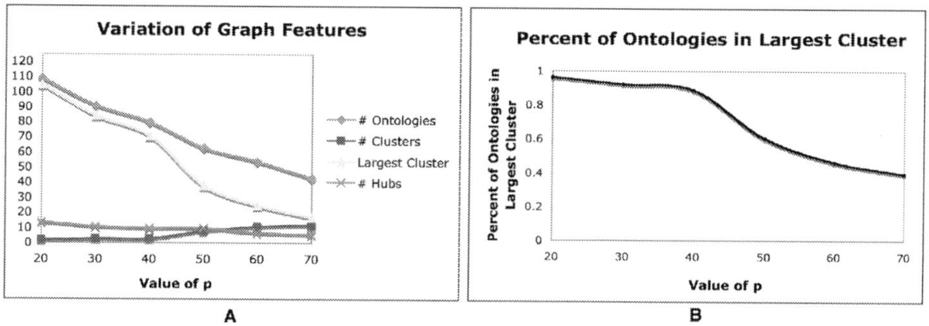

Fig. 3. Variation of graph features as p changes: (a) The graph shows how the number of ontologies, clusters, hubs, and the size of the largest cluster, all on the y-axis, change as p changes. As p decreases (from right to left), the number of clusters decreases slowly as clusters merge and the number of hubs increases slowly. Additionally, the largest cluster increases to include a larger fraction of the ontologies, which also increase in number because more ontologies are connected as the threshold for p decreases. (b) The graph shows the percent of ontologies that are in the largest cluster for different thresholds of p. As p decreases (right to left), almost all ontologies become connected very quickly.

Note that the graphs have two distinct types of hubs: (1) hubs in which almost all of the links (directed edges) were incoming and (2) hubs in which almost all of the directed edges were outgoing. That is, some, usually small, ontologies have a large fraction of their concepts covered by other ontologies. Other, usually larger, ontologies include concepts that are similar to a large fraction of concepts in many smaller ontologies in the repository.

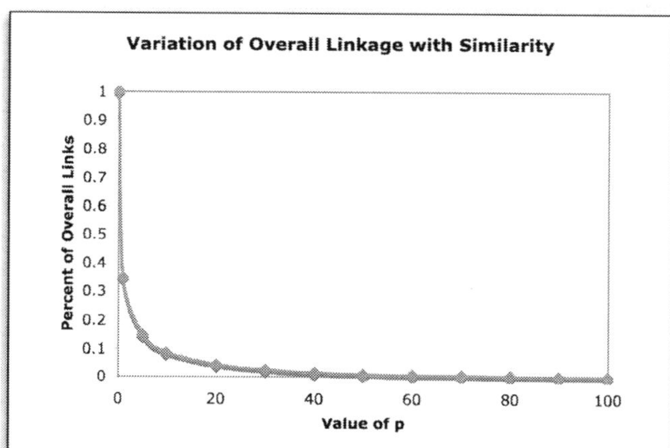

Fig. 4. Percentage of overall links: The x-axis is the value of p. The y-axis shows the fraction of all links between ontologies (L_O) that are present for a given value of p (i.e., L_p / L_O) as p changes.

3.3 How Similar Are the Ontologies?

Finally, Figure 4 shows a graph of the percentage of overall links that are included at each threshold value of p. This graph demonstrates that:

- 96% of ontology links are between ontologies that are less than 20% similar
- 91% between ontologies less than 10% similar
- 65% between ontologies less than 1% similar

And our final observation: Out of 207 ontologies, 68 ontologies (or 33%) have at least 50% of their terms mapped to terms in some other ontology.

4 Discussion and Analysis

The figures and data that we presented in Section 3 allow us to make several observations and answer several of the questions that we posed at the beginning of the paper.

To what degree are the domains covered by different ontologies connected?

- If we use lexical mappings as the basis for determining how connected the ontologies are, then the biomedical ontologies in our repository are very closely connected, with 33% of them having at least half of their concepts mapped to concepts in other ontologies.
- With such a large overlap among the ontologies, attempts to find "canonical" representations for concepts may be doomed: a large number of concepts in biomedicine are already represented in many different ways. One could argue that our data shows that given the state of the biomedical ontologies today, the most feasible way of integrating ontologies is by creating mappings, possibly complex ones, between them rather than trying to eliminate overlap (as the OBO Foundry initiative is trying to do [14]).
- Our study found that the small world property holds on our set of ontologies for low values of p, with the average distance between the nodes being as low as 1.1 hops. Other research has found the small world property to be true for other large data sets as well[8]. Further analysis on the properties of small world networks, such as strength of ties and k-core among others, may provide additional useful insight about the connectedness of ontologies.

If you are new to a domain, what are the important or representative ontologies with good coverage?

- The lexical mappings identified SNOMED-CT as the most prominent hub, and, indeed, SNOMED-CT is the largest and one of the most prominent and popular biomedical ontologies. Thus if we use mappings as a way of identifying prominent ontologies in a domain (i.e., an ontology with lots of mappings to other ontologies is an "important" one), then at least in this case, this approach would have identified correctly the ontology that a newcomer to the domain of biomedical ontologies must become familiar with.

- Hubs with many outgoing links show shared domains, particularly at high threshold values for p. For these hub ontologies, a large portion of their concepts is mapped to several different ontologies. Thus, ontologies that are linked through such a hub likely share the content that is represented in the hub ontology. For example, at p= 50%, the Common Anatomy Reference Ontology (CARO) is a hub with outgoing links to Foundational Model of Anatomy, Zebrafish anatomy and development, Tick gross anatomy, Teleost anatomy and development, and Mosquito gross anatomy—all ontologies in the anatomy domain. At p= 70%, the United States Pharmacopeia Model Guidelines ontology (USPMG) has outgoing links to Multum MediSource Lexicon, RxNorm Vocabulary, Veterans Health Administration National Drug File, National Drug File Reference Terminology, Medical Subject Headings, National Cancer Institute Thesaurus, and SNOMED-CT—all ontologies that describe drugs, among other things.

If you want to build domain-specific tools for creating ontology mappings, what are good ontologies to use for background knowledge?

- The two previous points lead to several practical uses of hubs identified through mappings: First, for ontology-mapping algorithms that require domain-specific background knowledge, hubs with many incoming links (such as SNOMED-CT) can serve as useful sources of such background knowledge. Second, these hubs are also good candidates for being representative ontologies for a domain.

What can we learn about the characteristics of the ontologies themselves and the ontology repository from the mappings between them?

- Links at a low value of p (1%) (i.e., when less than 1% of the concepts from the source ontology have a mapping to the target) do not say much about connectedness of ontologies. The domain of biomedicine is such that there is a little bit of overlap in everything, resulting in the extremely connected model we see at 1% mark. At 20%, however, we see a meaningful power-law distribution. At even higher thresholds, we can see ontologies that are very closely related. For example, we see that the Gene Ontology (GO) is very closely related to the cell cycle ontologies (CCO). 65% of links fall in a range lower than 1% similarity, which indicates that links below the p = 1% threshold are not as informative of connections between ontologies.
- If we were to use mappings between terms (identified in any way) as an indication of distance or similarity between ontologies in a repository, then the previous observation leads to the following practical implication: These links at low values of p are not very meaningful and should probably not be used as an indication of any relation between the ontologies.

5 Conclusions, Limitations, and Future Work

Our analysis does not depend on the specific method used to identify mappings between concepts. We used a simple method because it worked well and was very scalable (cf Section 2.2). Our earlier research has found that most of the openly available advanced mapping algorithms are simply not scalable to the size of biomedical ontologies in our repository [6]. One of the interesting directions for future work, when more scalable advanced algorithms become available, would be to perform similar analysis of relationships between ontologies taking a more advanced set of mappings as input.

Our main contribution in this paper is not the method for generating mappings between ontologies, but rather the analysis of these mappings. We believe that network analysis serves as a powerful tool for analyzing the structure of an ontology repository by providing insights into the characteristics of the ontologies and the structure of the repository. As the Semantic Web grows in popularity and use of ontologies expands, these methods may play an important role in understanding the connections among ontologies.

Our approach has certain critical limitations. Because we use a simple lexical matching method, our results are limited to the domain of biomedicine and other domains where such mapping method works well. In other domains, where concept definitions do not contain rich lexical information in the form of preferred names and synonyms, one will need to find scalable tools that would produce a large number of mappings that enable statistically significant analysis. Also, because of the way we generate the mappings, we do not account for the ontologies that have alternate lexical structures to represent the same concepts. Thus, we may miss a connection between two ontologies that actually have a significant amount of overlap in terms of the actual concepts they represent simply because these concepts have different lexical structures in the two ontologies. Our methods would work best for sets of ontologies that use similar naming conventions [13].

Another limitation of our work is that it gives no information as to the nature of the mappings or the actual specific relationship between linked ontologies. We cannot know whether one ontology simply has the same concept names as another ontology or if it imports terms from that ontology directly. Many BioPortal ontologies use OBO format and often simply copy the ontology that they import rather than use the import mechanism that recently has become available in OBO. As a result, a number of the mappings that we created are not actually mappings in the traditional sense. We plan to develop heuristics to identify this overlap and exclude it from the mappings set.

As part of our future work, we plan to compare our lexical mappings on the set of UMLS terminologies to the set of mappings provided by UMLS. The UMLS has a large number of mappings between the terminologies that were created manually. While the purpose of those mappings was slightly different from ours, comparing the results of the lexical mapping to the manual one will likely produce useful insights.

We also plan to implement automatic maintenance and updates on the set of mappings generated with this method. Not only does the number of BioPortal ontologies increase regularly, but also new versions of some of the ontologies are uploaded every night. These frequent updates makes manual creation of mappings among ontologies in the repository a daunting, if not impossible, task. By contrast, UMLS terminologies update twice a year.

In addition to the work outlined above, we plan to perform similar analysis on the mappings from other sources if the number of mappings is significant enough to make such analysis valid. For instance, the Alignment Server [5] in the NeON project could prove to be one such source of mappings in the future.

Finally, we uploaded the set of mappings between BioPortal ontologies to BioPortal. Users can browse the mappings through the BioPortal user interface and access them programmatically through REST services. We believe that the mappings should prove useful to developers of tools dependent on these mappings or for ontology developers looking at specific domain ontologies. All the data that we used for the analysis in this paper is available in raw spreadsheet form at `http://www.bioontology.org/wiki/index.php/Mapping_Set`

The study that we reported in this paper offered the first glimpse at the possibilities and challenges that large numbers of related ontologies bring to the fore. Our results show that using network analysis on a network defined by mappings between ontology terms helps us understand and navigate a world with a large number of ontologies. As more ontologies become available on the Semantic Web, such analysis will become more interesting, more useful, and more challenging.

Acknowledgements

This work was supported by the National Center for Biomedical Ontology, under roadmap-initiative grant U54 HG004028 from the National Institutes of Health. Nick Griffith has implemented the support for mappings in BioPortal.

References

1. Aleksovski, Z., Klein, M., ten Kate, W., van Harmelen, F.: Matching unstructured vocabularies using a background ontology. In: Staab, S., Svátek, V. (eds.) EKAW 2006. LNCS (LNAI), vol. 4248, pp. 182–197. Springer, Heidelberg (2006)
2. Barabsi, A.-L.: Linked: How Everything Is Connected to Everything Else and What It Means for Business, Science, and Everyday Life. Basic Books (2003)
3. Bodenreider, O., Stevens, R.: Bio-ontologies: current trends and future directions. Briefings in Bioinformatics 7, 256–274 (2006)
4. Caracciolo, C., Euzenat, J., Hollink, L., Ichise, R., Isaac, A., Malaise, V., Meilicke, C., Pane, J., Shvaiko, P., Stuckenschmidt, H., Svab Zamazal, O., Svatek, V.: Results of the ontology alignment evaluation initiative 2008. In: 3rd International Workshop on Ontology Matching (OM-2008) at ISWC 2008, Karlsruhe, Germany (2008)
5. Euzenat, J.: Alignment infrastructure for ontology mediation and other applications. In: Workshop on Mediation in Semantic Web Services (2005)
6. Ghazvinian, A., Noy, N.F., Musen, M.A.: Creating mappings for ontologies in biomedicine: Simple methods work. In: AMIA Annual Symposium (AMIA 2009), San Francisco, CA (2009)
7. Jonquet, C., Shah, N.H., Musen, M.A.: The open biomedical annotator. In: AMIA Summit on Translational Bioinformatics, San Francisco, CA, USA, pp. 56–60 (2009)
8. Leskovec, J., Horvitz, E.: Planetary-scale views on a large instant-messaging network. In: 17th International World Wide Web Conference (WWW 2008), Beijing, China (2008)
9. Lindberg, D., Humphreys, B., McCray, A.: The unified medical language system. Methods of Information in Medicine 32(4), 281 (1993)

10. Noy, N., Shah, N., Dai, B., Dorf, M., Griffith, N., Jonquet, C., Montegut, M., Rubin, D., Youn, C., Musen, M.: Bioportal: A web repository for biomedical ontologies and data resources. In: Demo session at 7th International Semantic Web Conference (ISWC 2008), Karlsruhe, Germany. Springer, Heidelberg (2008)
11. Noy, N.F., Griffith, N., Musen, M.A.: Collecting community-based mappings in an ontology repository. In: Sheth, A.P., Staab, S., Dean, M., Paolucci, M., Maynard, D., Finin, T., Thirunarayan, K. (eds.) ISWC 2008. LNCS, vol. 5318, pp. 371–386. Springer, Heidelberg (2008)
12. Rubin, D.L., Shah, N.H., Noy, N.F.: Biomedical ontologies: a functional perspective. Briefings in Bioinformatics 9(1), 75–90 (2008)
13. Schober, D., Smith, B., Lewis, S.E., Kusnierczyk, W., Lomax, J., Mungall, C., Taylor, C.F., Rocca-Serra, P., Sansone, S.-A.: Survey-based naming conventions for use in obo foundry ontology development. BMC Bioinformatics (2009)
14. Smith, B., Ashburner, M., Rosse, C., Bard, J., Bug, W., Ceusters, W., Goldberg, L.J., Eilbeck, K., Ireland, A., Mungall, C.J., Leontis, N., Rocca-Serra, P., Ruttenberg, A., Sansone, S.A., Scheuermann, R.H., Shah, N., Whetzel, P.L., Lewis, S.: The OBO Foundry: coordinated evolution of ontologies to support biomedical data integration. Nature Biotechnology 25(11), 1251–1255 (2007)
15. Zhang, S., Bodenreider, O.: Alignment of multiple ontologies of anatomy: Deriving indirect mappings from direct mappings to a reference. In: AMIA Annual Symposium, pp. 864–868 (2005)

Modeling and Query Patterns for Process Retrieval in OWL

Gerd Groener and Steffen Staab

ISWeb — Information Systems and Semantic Web,
Institute for Computer Science, University of Koblenz-Landau, Germany
{groener,staab}@uni-koblenz.de

Abstract. Process modeling is a core task in software engineering in general and in web service modeling in particular. The explicit management of process models for purposes such as process selection and/or process reuse requires flexible and intelligent retrieval of process structures based on process entities and relationships, i.e. process activities, hierarchical relationship between activities and their parts, temporal relationships between activities, conditions on process flows as well as the modeling of domain knowledge. In this paper, we analyze requirements for modeling and querying of process models and present a pattern-oriented approach exploiting OWL-DL representation and reasoning capabilities for expressive process modeling and retrieval.

1 Introduction

Process models are used in various applications like business process modeling, modeling of system processes and also the combination and interaction of web services are described as processes. In order to use existing processes it is necessary to support basic process model management tasks like the retrieval of processes [16].

There are various approaches for modeling processes and corresponding retrieval methods like a keyword search or retrieval methods based on data in- and output or process properties which are mainly process annotations. Retrieval with respect to the activities of a process and especially to the execution order and dependencies of them requires the consideration of the control flow of a process (cf. Section 2.2,[12]).

The problem of process retrieval based on control flow information depends on the internal process structure. The structure includes sequences, choices, parallel activities as well as activity specialization. Therefore, search capabilities for processes must include reasoning facilities that are able to match the heterogeneous requirements that exist at different levels of granularities of process descriptions between a query and a process. Moreover, this matching capability needs to distinguish modalities of matches, i.e. whether a specific condition on a process (part) must be always fulfilled or only for some of its instances. In order to tackle the described challenges, we consider two main tasks:

A. Bernstein et al. (Eds.): ISWC 2009, LNCS 5823, pp. 243–259, 2009.

- A formal description of the control flow of a process which explicitly models the various hierarchical and ordering relationships between activities.
- A characterization of query and reasoning tasks in order to retrieve processes with specified control flow characteristics and modalities.

OWL-DL allows for the description of terminology of processes and activities. DL reasoning enables matching of queries and processes (cf. Section 4). The available reasoning procedures account for the aggregation of activities and for different modalities.

OWL-DL and DL have been used for process modeling and retrieval before (cf. [4,11,16,17]). However, in these process models, the control flow is either not represented at all or represented by syntactic means that do not allow for reasoning as needed or the process models are too weak to express realistic control flows. A comprehensive survey is given in Section 6.

To remedy this situation, this paper provides a threefold contribution. First, we analyze modeling and querying requirements for process retrieval in general (Section 2). Second, we provide patterns for modeling process control flow in OWL-DL (Section 3). Third, we demonstrate how to exploit these semantic-based models for expressive retrieval of (parts of) process models by reasoning in OWL-DL (Section 4). The retrieval capabilities of our model are evaluated in Section 5.

2 Process Retrieval Problems

This section outlines requirements for process modeling and retrieval motivated by a retrieval scenario.

2.1 An Example Process Model as UML Activity Diagram

We use the widespread and standardized UML-Activity Diagrams for graphical process modeling. Figure 1 depicts three different process models describing *SalesOrder* processes with different characteristics. The elements of UML Activity Diagrams are described in Table 2. Activities are the tasks which are executed by the process. An edge is the connection from an activity to the follower activity.

The activities in the diagrams from Figure 1 are decomposed into more fine-grained descriptions which are depicted in Figure 2. The *Reorder* process contains a decision in combination with a condition. If the article is a standard article an internal reorder activity is executed otherwise an external reorder activity.

2.2 Requesting Process Descriptions

We consider a situation where a customer wants to use an online shop to buy a product. The selling procedure of the online shop is specified by a salesorder process description. Depending on the individual preferences of the customer, he may want to select a shopping service based on properties of the process control flow which is realized by the shop. For instance, he may want to search for a shop (Q) and expects a concrete response (R) fulfilling his request:

Q1: Which shop allows me to pay after delivery?
  R1: Process that executes *Delivery* before *Payment*.
Q2: Which shop allows me to pay by cash?
  R2: SalesOrder that offers cash payment as payment method.
Q3: Which shop accepts only cash payment?
  R3: SalesOrder process that only offers cash payment.

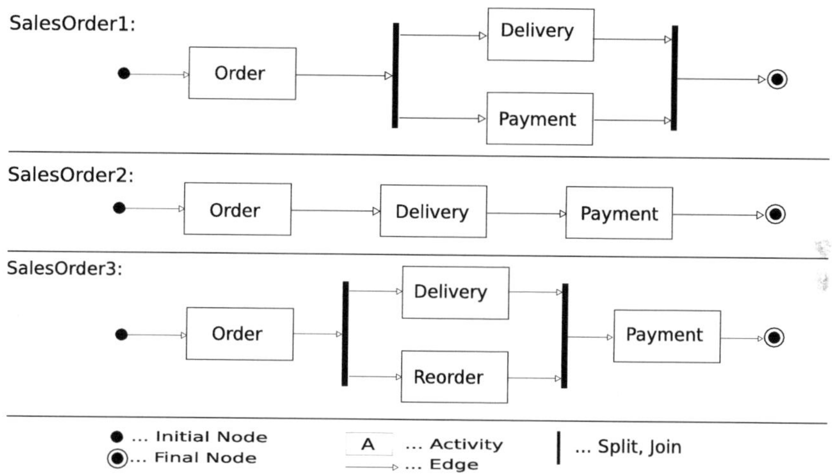

Fig. 1. UML-Activity Diagrams for Sales Order Processes

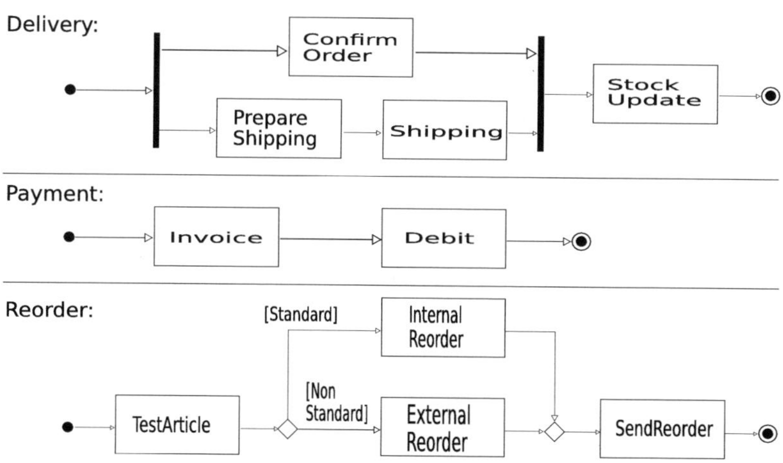

Fig. 2. UML-Activity Diagrams for the Subprocesses

In fact, quite often an online shop is requested that fulfills multiple constraints simultaneously. Besides the customer, the service provider may have further queries in order to find an appropriate process.

Q4: Which process executes at least all activities of $SalesOrder2$?
  R4: $SalesOrder3$ process contains all activities of $SalesOrder2$ and an additional activity.

Q5: Which process runs are instances of $SalesOrder2$ and $SalesOrder3$?
  R5: Process runs which are instances of both processes simultaneously.

2.3 Requirements for Process Modeling

In order to facilitate process retrieval we derived the requirements for process modeling from the demonstrated queries. Table 1 describes the derived requirements with respect to the questions. (1) A query must consider the execution order, e.g. that one activity happens (possibly indirectly) before another activity. (2) Process specializations and (3) activity decomposition involve the terminology, e.g. that one must consider the different specializations of payment types. (4) Queries have to cover modality, e.g. that a property like the type of payment is unavoidable in all possible process enactments. (5) Instance queries are relevant for all modeling dimensions. These requirements may be combined, e.g. queries concerning the execution order have also to cover terminological information. Obviously, these queries are general and imprecise process descriptions, e.g. only a description of relevant parts of the process. Hence the process retrieval has to account for incomplete process descriptions.

Table 1. Requirements derived from the queries

Requirements	Questions
Order of Activities and Subactivities	Q1
Specialization, Inheritance	Q4
Activity Decomposition	Q2, Q3
Modality	Q2, Q3
Instance Queries	Q5

3 Process Modeling in OWL

We have investigated two important design decisions for modeling processes in OWL-DL. The first method models occurrences of activities and their relative positioning in the control flow. The advantage is that each step can be modeled separately and - as we see later - some advantages with regard to refinements are implied. The second method models a process as one complex DL expression capturing all the steps. In this second model it is not fully possible to capture all temporal and refinement relationships simultaneously. However, the advantage of the second approach is that there are more possibilities for retrieval queries than in the first approach. In this paper we only focus on the second approach.

3.1 Design Principles and Transformation

A process model (sometimes called a process template) describes the set of all process runs it allows, i.e. a concept in OWL-DL. Process runs are instances in OWL-DL. A process run is a composition of activity instances. We translate the language primitives of the UML-Activity Diagram into an OWL-DL[1] representation. Table 2 lists our translation rules. A, B, C and D are activities, P is a process name.

Table 2. Transformation to Description Logic

Construct	UML Notation	DL Notation
1. Start		$Start_i$
2. End		End_i
3. Activity		A
4. Edge		TO_i
5. Process P (Axiom)	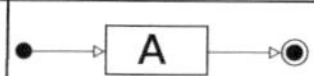	$P \equiv Start_i \sqcap \exists_{=1} TO_i.(A \sqcap \exists_{=1} TO_i.End_i)$
6. Flow		$A \sqcap \exists_{=1} TO_i.B$
7. Decision	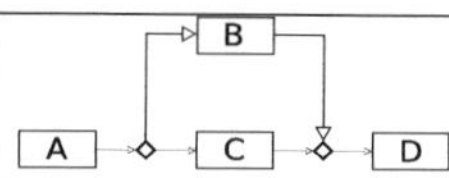	$A \sqcap \exists_{=1} TO_i.((B \sqcup C) \sqcap \exists_{=1} TO_i.D$
8. Condition		$A \sqcap \exists_{=1} TO_i.((B \sqcap \kappa_{Cond}) \sqcup (Stalled \sqcap \neg\kappa_{Cond}))$
9. Fork and Join	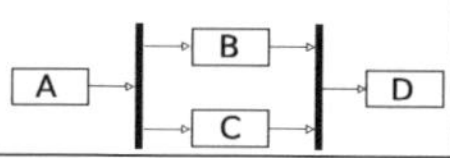	$A \sqcap \exists\, TO_i.(B \sqcap \exists_{=1} TO_i.D) \sqcap \exists\, TO_i.(C \sqcap \exists_{=1} TO_i.D) \sqcap = 2\, TO_i$
10. Loop	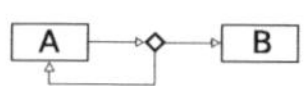	$Loop_j \sqcap \exists_{=1} TO_i.B$, $Loop_j \equiv A \sqcap \exists_{=1} TO_j.(Loop_j \sqcup End_j)$

The relation between an activity and the follower activity is represented by roles in DL, i.e. TO [2]. In order to allow for process composition and refinement in combination with cardinality restrictions the roles for each process (TO_i) are distinguished from each other. All roles TO_i are defined as subroles of TO in order to simplify query formulations. Therefore we may use TO for the retrieval of all processes. TOT is a transitive superrole of TO (cf. Table 3) which is similar to the relation pattern in DOLCE plan extension [9]. The combination of both roles enables process retrieval with respect to direct and indirect activity order.

A process is either atomic or composed. An atomic process is a non-decomposable activity. A composed process is built using axioms (No.5). An axiom defines a process starting with $Start_i$ followed by a sequence of composed activities. The last activity is the concept End_i. The follower relation TO_i refers

[1] We use the DL language $\mathcal{SHOIN}(D)$ which corresponds to OWL-DL.

[2] For sake of a short name, the role name TO is used instead of a more meaningful name like $FollowedBy$.

to an edge in the process diagram. A flow (No.6) is translated to DL with the expression $A \sqcap \exists\, TO_i.B$. This means that the process flow defines an execution of activity A which is directly followed by the activity B. The disjointness of A and B in a flow is here not required which accounts for a larger result set in the retrieval (cf. 5).

Decisions (No.7) are modeled in DL as a concept union $\exists\, TO_i.(B \sqcup C)$ representing a non-deterministic choice between two possible following activities. Both activities reach an activity D. The cardinality restrictions guarantee that there is only one follower activity. A condition can be assigned to a flow (No.8). Adding disjoint conditions to each flow of a decision leads to deterministic decisions. These deterministic decisions are used to model exclusive decisions (exclusive or), i.e. exactly one path after the decision is executed.

Disjointness of the activities B and C is also not required. We describe conditions as set of states that may be reached when the condition is fulfilled. The notation of No.8 describes that directly after the activity A there is an activity B and the condition $Cond$ holds, i.e. the activity B is only executed if the condition is satisfied. We add an additional activity $Stalled$ if the condition is not satisfied. The activity $Stalled$ means that the control flow of the process is stopped due to a violation of the condition. All conditions are subclasses of a general concept $Condition$ (Table 3).

Table 3. Process Model Axiomatization

Statement	DL Axiom
Stalled has no follower	$Stalled \sqcap \exists\, TO.Activity \sqsubseteq \bot$
All activities are subclasses of *Activity*	$A, B, C, D \sqsubseteq Activity$
Conditions are subclasses of the concept *Condition*	$\kappa_j \sqsubseteq Condition$
Domain and Range of TO is *Activity*	$\exists\, TO.\top \sqsubseteq Activity$ and $\top \sqsubseteq \forall\, TO.Activity$
TO is superrole of all TO_i	$TO_i \sqsubseteq TO$
TOT is the transitive superrole of TO	$TO \sqsubseteq TOT$, and $TO^+ \sqsubseteq TOT$
$Start_i$ is the first activity	$Start_i \sqcap \exists\, TO_i^-.Activity \sqsubseteq \bot$
End_i is the last activity	$End_i \sqcap \exists\, TO_i.Activity \sqsubseteq \bot$

A parallel execution (No.9) starts with a fork and ends with a join. The fork is described by the explicit statement that there exist multiple follower sequences, which is described by an intersection of the sequences. The join of sequences like for decisions and parallel executions is represented by activity (D) in the complex DL expression. Loops (No.10) are described by a subprocess $Loop_j$ which contains the activities of the loop (A) and the follower activity is either $Loop_j$ or End_j. This construct is a special decision with a self reference. In case of a nested loop structure, i.e. there is a loop within another loop, the transformation is straightforward. The inner loop is treated like a single activity of the outer loop, e.g. activity A in No.10 could be an inner loop.

For each process definition we require that $Start_i$ has no predecessor, End_i has no follower which is described in the second part of Table 3. In order to model process refinements these conditions are only valid for this (sub-) process. Therefore the start and end activities are distinguished from each other, using levels (i) for each process. Further axioms like domain and range restrictions, subclass relations between activities and conditions and the role hierarchy for the follower relations (TO_i) are described in the first part of Table 3.

In our OWL-DL model there are limitations in the accuracy. Some model constructs are non-deterministic, e.g. we use a disjunction in DL to model a non-deterministic control flow description and we do not require a general disjointness of the activities of a process. However, this supports query expressions using general and incomplete process descriptions with a maximal possible result set.

3.2 Process Transformation to OWL

Based on the described transformation pattern from UML process models to OWL, we describe the *SalesOrder* processes from Section 2 in OWL-DL. Figure 3 depicts the axioms that describe the *SalesOrder* processes. For being more readable, we omit here the cardinality restrictions in all process definitions and the *Stalled* activity for violated conditions.

$$
\begin{aligned}
SalesOrder1 \equiv\ & Start_1 \sqcap \exists\, TO_1.(Order \sqcap \exists\, TO_1.(Delivery \sqcap \exists\, TO_1.End_1) \\
& \sqcap \exists\, TO_1.(Payment \sqcap \exists\, TO_1.End_1)) \\
SalesOrder2 \equiv\ & Start_2 \sqcap \exists\, TO_2.(Order \sqcap \exists\, TO_2. \\
& (Delivery \sqcap \exists\, TO_2.(Payment \sqcap \exists\, TO_2.End_2))) \\
SalesOrder3 \equiv\ & Start_3 \sqcap \exists\, TO_3. \\
& (Order \sqcap \exists\, TO_3.\ (Delivery \sqcap \exists\, TO_3.(Payment \sqcap \exists\, TO_3.End_3)) \\
& \sqcap \exists\, TO_3.(Reorder \sqcap \exists\, TO_3.(Payment \sqcap \exists\, TO_3.End_3))) \\
Delivery \equiv\ & Start_4 \sqcap \exists\, TO_4.(ConfirmOrder \sqcap \exists\, TO_4.\ (StockUpdate \sqcap \exists\, TO_4.End_4)) \\
& \sqcap \exists\, TO_4(PrepareShipping \sqcap \exists\, TO_4.(Shipping \sqcap \\
& \exists\, TO_4.(StockUpdate \sqcap \exists\, TO_4.End_4))) \\
Payment \equiv\ & Start_5 \sqcap \exists\, TO_5.(\ Invoice \sqcap \exists\, TO_5.(Debit \sqcap \exists\, TO_5.End_5)) \\
Reorder \equiv\ & Start_6 \sqcap \exists\, TO_6.(TestArticle \sqcap \exists\, TO_6.\ ((InternalReorder \sqcap Standard \sqcup \\
& ExternalReorder \sqcap \neg Standard) \sqcap \exists\, TO_6. \\
& (SendReorder \sqcap \exists\, TO_6.End_6)))
\end{aligned}
$$

Fig. 3. The Sales Order Processes in DL

Domain knowledge (Figure 4) contains terminological information about activities and subactivities of the processes, e.g. a credit card payment is a subclass of payment. Figure 5 contains further process definitions also without cardinality restrictions. *CreditCardPayment* and *CashPayment* are specializations of the *Payment* activity. *SalesOrder2_Credit* and *SalesOrder2_Cash* refine the *SalesOrder2* process using the subconcepts *CreditCardPayment* and *CashPayment* instead of *Payment*. In the remainder of this paper, we refer to the set of axioms from Table 3, Figures 3, 4 and 5 as the knowledge base KB.

$$\begin{aligned}
CreditCardPayment &\sqsubseteq Payment \\
CashPayment &\sqsubseteq Payment \\
CashPayment \sqcap CreditCardPayment &\sqsubseteq \bot
\end{aligned}$$

Fig. 4. Domain Knowledge Axioms

$$\begin{aligned}
SalesOrder2_Credit \equiv\; & Start_2 \sqcap \exists\, TO_2.(Order \sqcap \exists\, TO_2. \\
& (Delivery \sqcap \exists\, TO_2.(CreditCardPayment \sqcap \exists\, TO_2.End_2))) \\
SalesOrder2_Cash \equiv\; & Start_2 \sqcap \exists\, TO_2.(Order \sqcap \exists\, TO_2. \\
& (Delivery \sqcap \exists\, TO_2.(CashPayment \sqcap \exists\, TO_2.End_2)))
\end{aligned}$$

Fig. 5. Additional SalesOrder-Processes in DL

3.3 Relations between Processes

Specialization and refinements are orthogonal relationships between processes. A specialization (or sometimes called an extension) is a relationships between processes in which the specialization of a process consists either of additional activities or some of the activities are specialized. An activity specialization or a subactivity is a subclass of the more general activity. Specializations are described as inheritance relationships in [20]. There is a distinction between four different inheritance relationships. One of them is the *projection inheritance* which refers to the notion of specialization in our process model. A formal definition is given in Definition 1.

As in [20] we use the term hidden activity to refer to an activity in a process which is not executed, i.e. this activity would be removed from the flow. The edges to and from the hidden activity remain in the process as a valid path but without these activity. For instance $SalesOrder3$ is a specialization of $SalesOrder2$. The additional activity is *Reorder*. If *Reorder* is hidden in $SalesOrder3$ the processes are equivalent. With equivalence we refer to bisimulation equivalence. The path via *Reorder* is still an existing path, but without any activity. Therefore a hidden activity is also called an activity without effect.

Definition 1. *A process P' is a specialization or extension of P if the following conditions hold: (1) Each activity from process P is either an activity in P' or there is a subactivity in P' with respect to the terminology. The subactivity is subsumed by the activity from P. (2) If all additional activities in P' are hidden then P' and P are bisimilar. Additional activities of P' are activities that are neither in P nor subactivities of them are in P.*

These definition refers to definitions like in [5,20]. In our model, we use DL concept subsumption to analyze process specializations. If a process specialization contains a parallel execution instead of a sequential or it contains additional activities or additional conditions, this specialization is subsumed, since it is modeled as concept intersection. These DL concept subsumptions conform to Definition 1, e.g. $SalesOrder3$ is subsumed by $SalesOrder2$ and $SalesOrder3$ is a specialization with respect to the definition.

A process containing a decision like *Reorder* is not subsumed by a process without the decision. For instance the *Reorder* process without *InteralReorder* activity does not subsume the *Reorder* process, since the decision is modeled as a concept union and therefore define a more general concept. However, decisions are not specializations with respect to our definition, since hidden additional activities do not lead to the same process. The path of the hidden activity (e.g. *InternalReorder*) still exists, and therefore the activity *ExternalReorder* could be omitted using the other path (with no effect) which is still available. For loops the DL subsumption also confirms to the definition. Adding a $Loop_j$ to a process is like adding a single activity to the more general process.

In [21] process specialization is defined using execution set semantics. The execution set of a process contains all process runs. Process P' is a specialization with respect to the minimal execution set semantics, if it contains at least all process runs from the general process. This definition is valid for processes P' containing decisions compared to a process P without decision. Under the maximal execution set the specialization consists of a subset of the process P. This definition refers to all other primitives except decisions that satisfy the concept subsumption $P' \sqsubseteq P$. Therefore this model can not be referred to only one of these specialization notions.

A refinement is an equivalent representation with another granularity (cf. [21]). The same process is described in a more fine-grained representation. The inverse of a refinement is an abstraction. A refinement replaces activities of a process with corresponding subprocesses, e.g. *Payment* is replaced in a *SalesOrder* process as described in Figure 2. A subprocess can also consist of a single activity.

4 Semantic Query Patterns for Process Retrieval

In Section 3 we have used the expressiveness of OWL to describe the control flow of processes, activity decomposition and specialization of activities and processes (Figure 4,5). In order to satisfy the requirements from Section 2.3 we express queries in DL and use DL reasoning to query the process models and derive process information and relationships.

Queries are general and incomplete process descriptions which specify the requested core functionality of the process. The query result contains all processes of the *KB* satisfying the query specification. The retrieval of a process with respect to the query depends on the axioms in the *KB*, i.e. the process information. We apply two different inference strategies in order to demonstrate how information or missing information in the *KB* may lead to different query results, depending on the applied inference method.

A strong inference method is the entailment of concept subsumption, i.e. the query subsumes the retrieved processes. The subsumption is used only in one direction. The result processes are specializations of the query. A weaker inference method is the satisfiability of concept conjunction, i.e. to test whether a process and the query process may have a common process run. This weaker condition leads to a higher number of results. The result processes are candidates for the query process. Missing information, e.g. missing disjointness axioms leads to a higher number of positive results. In general, adding further axioms to the KB reduces the result set for this inference method.

We consider three non-disjoint query patterns with respect to the control flow of a process. The first pattern mainly considers the retrieval of processes with respect to the execution order. The second pattern considers terminological queries. Queries for the modality are demonstrated with the third pattern. The query pattern consist of three components: query, inference and result. The *query* input P is a process description. *Inference* is the description of the applied reasoning strategy, i.e. entailment or satisfiability. For sake of a shorter presentation, we only depict the applied inference for one process (here $SalesOrder1$), but this is performed for all processes in the KB.

Pattern for Execution Order. A query for question Q1 from section 2.2 is described below. The query result contains all processes from the KB which conform to the described execution order.

Query. Which processes execute *Payment* after *Delivery*?
$\{P \equiv \exists\ TOT.(Delivery \sqcap \exists\ TOT.Payment)\}$

Inference. Test entailment of concept subsumption:
$KB \models SalesOrder1 \sqsubseteq P, \ldots$

Result. $\{SalesOrder2, SalesOrder2\_Cash, SalesOrder2\_Credit, SalesOrder3\}$
The result contains all processes that execute *Delivery* before *Payment* with an arbitrary number of activities between them. The result includes also processes which specialize the *Payment* activities. If the query should only retrieve processes with directly connected activities, the transitive role TOT is replaced by the role TO in the query. For instance the following query searches for all processes with *Payment* as direct follower of *Order*.

Query. Which processes execute *Payment* directly after *Order*?
$\{P \equiv\ \exists\ TOT.(Order\ \sqcap\ \exists\ TO.Payment)\}$

Inference. Test entailment of concept subsumption:
$KB \models SalesOrder1 \sqsubseteq P, \ldots$

Result. $\{SalesOrder1\}$

Pattern for Process Terminology. This pattern uses the terminological knowledge. This covers the extension of processes and the specialization of activities. The following query searches for all processes in which the *Delivery* activity is executed before the *Debit* activity. In all described processes there is at least the *Invoice* step between these activities.

Query. Which processes execute *Delivery* before *Debit*?
$\{P \equiv\ \exists\ TOT.(Delivery\ \sqcap\ \exists\ TOT.Debit)\}$

Inference. Test entailment of concept subsumption:
$KB \models SalesOrder1 \sqsubseteq P, \ldots$

Result: $\{SalesOrder2, SalesOrder2\_Cash, SalesOrder2\_Credit, SalesOrder3\}$
The query refers to the refinement of *Payment*, since only the decomposed process contains the *Debit* activity. If the query only uses the *TO* relationship the result-set is empty.

The next query searches for all extensions of the process *SalesOrder*2, as described in question Q4.

Query. Which processes extend *SalesOrder*2?
$\{P \equiv SalesOrder2\}$

Inference. Test entailment of concept subsumption:
$KB \models SalesOrder1 \sqsubseteq P, \ldots$

Result. $\{SalesOrder2\_Credit, SalesOrder2\_Cash, SalesOrder3\}$
The extensions *SalesOrder2_Credit* and *SalesOrder2_Cash* are generalized by *SalesOrder*2 since they contain a payment activity which is defined as a specialization of *Payment*. The extension relation between *SalesOrder*2 and *SalesOrder*3 is inferred since *SalesOrder*3 contains all activities from *SalesOrder*2 and additional activities (*Recorder*). For the entailment the same TO_i roles are required.

Pattern for Process Modality. This pattern considers queries which express the modality of processes that refer to questions like Q2 and Q3 and also terminological query Q4 from Section 2.2. The first query searches processes that offer a credit card payment.

Query. Which process offers *CreditCardPayment*?
$\{P \equiv \exists\, TOT.CreditCardPayment\}$

Inference. Test entailment of concept subsumption:
$KB \models SalesOrder1 \sqsubseteq P, \ldots$

Result. $\{SalesOrder2\_Credit\}$ There is only one process definition with an explicit statement that the payment method is *CreditCardPayment*. The next query searches for all processes that only offer credit card payment. Using concept subsumption there is no process that fulfills this condition, i.e. is subsumed by the query process. Since in no process definition the condition is explicitly stated. However, if the inference is changed to the weaker satisfiability of concept intersection, there are processes which are candidates for satisfying the query. The p in the query refers to a process run which is not in the KB.

Query. Which process only offers *CreditCardPayment*?
$\{P \equiv \forall\, TOT.(\neg Payment \sqcup CreditCardPayment)\}$

Inference. Test satisfiability of concept intersection: $KB \cup \{p : (SalesOrder1 \sqcap P)\}, \ldots$

Result. $\{SalesOrder1, SalesOrder2, SalesOrder2\_Credit, SalesOrder3\}$ The query description P describes that if there is a *Payment* activity in the process it must be a *CreditCardPayment*. The formula is valid for all SalesOrder processes except for the *SalesOrder2_Cash* processes since there is an axiom in the KB that the payment methods *CreditCardPayment* and *CashPayment* are disjoint.

5 Evaluation

Dataset: We evaluated the process retrieval with a KB of 182 different process models. These process models are based on ten real world processes (business processes) from the SAP Business Workflow library[3], from the Enterprise Services (ES) Workplace at the SAP Developer Network (SDN)[4] and from the process library [5]. We used multipliers in order to create from each basic process model between 16 and 25 different models. A multiplier modifies the processes using the following operators: (1) Add and remove activities (2) Change of activity order, (3) Replace activity with sub- or superactivity and (4) Add and remove flow conditions. These process models were transformed into OWL. The process models contain unconditional decisions (70%), conditional decisions (65%), split/fork (40%) and loops (25%). The DL expressivity is $\mathcal{SHIQ}$.

Methodology. For the evaluation we used the Pellet 2.0.0 reasoner in Java 1.6 running on a computer with 1.7 GHz CPU and 1 GB RAM. The retrieval is evaluated on four knowledge bases with different size. The first contains 52, the second contains 107, and the third and fourth contain 182 process models. In the last two KB the size of the processes (number of activities) is different. Due to this generation method the KB contains many similar process models that only differ in some activities or their ordering.

Depending on the size of the KB the evaluation consists of 25 until 35 queries like the demonstrated queries from Section 4. In the evaluation each query invokes the reasoner without caching. The queries for activity order also considered activities within non-sequential process parts, like in parallel executions or decisions. For terminology, we queried also for flow conditions and their complement.

Result. The evaluation result is depicted in Table 4. The first column refers to the number of processes in the KB. The second column contains the average number (`Av.`) of activities in a process and the third column the maximal (`Max.`) activities of a process. The number of axioms of the ontology is described in column four. The columns five and six describe the retrieval time (average and maximum) for simple queries in milliseconds (`ms`). Simple queries contain only one activity or a negated activity. All other queries with multiple activities (at least three) are referred to as complex queries, the retrieval time is in the columns

[3] http://help.sap.com
[4] http://esworkplace.sap.com/sdn
[5] http://bpt.hpi.uni-potsdam.de/Prozessbibliothek/WebHome

Table 4. Retrieval Evaluation Result

No.	KB Size	Process Size		Axioms	Simple Query Time [msec.]		Complex Query Time [msec.]		Concept Sat. Time [msec.]
		Av.	Max.		Av.	Max.	Av.	Max.	Av.
1	52	12.7	19	238	1420	1504	1508	1820	1390
2	107	12.9	19	416	1684	1720	1608	1754	1603
3	182	12.9	21	568	4548	4606	4391	4590	4141
4	182	21.2	28	587	4793	4890	4460	4557	4178

seven and eight. The average size of the result set (number of process models) of the queries using concept subsumption is 9.3 for the smallest KB, 9.8 for the KB with 107 models, 13.7 and 14.6 for the knowledge bases with 182 models. The retrieval time for all processes which satisfy the concept intersection with the query process (weak reasoning condition) are outlined in column nine (`Sat.`).

The performance evaluation from Table 4 indicates the following results. The influence of the size of the process models is negligible, as described in No.3, 4. In both settings there are 182 process models in the KB but the average size of the process models in No.4 is about 60% higher. For simple queries the retrieval time is even lower. The retrieval time increases by the number of processes in the KB. As indicated in the settings No.1 - 3 the retrieval time increases lightly from a KB with 52 processes compared with the KB with 107. If the KB is extended to 182 which is an increase of 70% the retrieval time is more than doubled. There is no performance difference between queries which contain activities either at the beginning of a process or at the end. The complexity of the query process (simple or complex) has also no influence. This indicates that only the number of process models in the KB affects the retrieval time.

The evaluation outlines the following qualitative results. (i) Processes form the KB are subsumed by the more general query process using the roles TO and TOT, except queries with decisions, since decisions are not considered as specializations and therefore not subsumed. (ii) However, if the query contains activities which are after a decision and before the merging of the flows the query has to cover either every sequence after the decision or use the disjoint flow conditions to retrieve the processes containing the decision. Since loops are special decisions, the retrieval of processes containing loops is realized in the same way.

(iii) The retrieval of refined processes depends on the refinement. If the refinement is realized by the decomposition of activities into subprocesses at the end of a process, the processes are retrieved using concept subsumption. Otherwise it requires two queries. For instance the query for a refined process $\exists\, TOT.(StockUpdate \sqcap \exists\, TOT.Payment)$ is transformed into a first query that selects the subprocess, the second query uses the start-activity of the refined process, e.g. $\exists\, TOT.(Start_i \sqcap \exists\, TOT.Payment)$, at which $Start_i$ is the start-activity (first activity) of the subprocess containing $StockUpdate$. The problem is that the order relation TO is not inferred for the subactivities (e.g. $StockUpdate$) with respect to the following activities of the original process (e.g. the relations

Table 5. Comparison of Modeling Dimensions

Modeling Characteristic	[3]	[8]	[10]	[11]	[6, 16]	[19]	Our Model
Control Flow Modeling	y	y	n	y	n	n	y
Realization of Process Retrieval	n	n	y	y	y	y	y
Activity Terminology in Modeling and Retrieval	n	n	n	y	y	n	y
Process Refinement	y	y	y	n	y	y	y
Modality	n	n	n	n	n	n	y
State Change	n	n	n	n	n	y	n

from *StockUpdate* to *Payment*). (iv) More complex refinements, e.g. change of the original execution order in combination with activity specialization is not retrieved by concept subsumption.

Distinction to other Technological Approaches. (i) Another way of using OWL for process modeling is the annotation of process models, i.e. other process models like UML are enriched with OWL annotations. However, the aim of our model is a generic process model in OWL with an explicit description of activity relations within a process. (ii) Non-ontological models like graph-based representations and queries [7] or tree-shaped diagram unraveling can manage the same ordering queries, but there is no reasoning benefit for process retrieval and a lack of representing terminological information, which is another aim of our approach.

Lessons Learned

Process Modeling. Modeling with cardinality restrictions leads to a more precise model. The difference between a decision and a parallel execution is emphasized with this restriction that there exists either exactly one follower or exactly n follower. There is a modeling tradeoff between accuracy and generic modeling. Generic descriptions account for a larger result set with respect to the demonstrated query mechanism. We do not model the join of decisions and parallel flows, which could be easily added with axioms like $D \sqsubseteq \exists TO_i^-.B \sqcap \exists TO_i^-.C$ for No.7 and 9 in Table 2. The join is described by the occurrence of the same activity (name) in all paths after the decision or split. This accounts for a maximal possible result set.

For the same reason we model decisions as union $\exists TO_i.(B \sqcup C)$ with only one TO_i relationship instead of $\exists TO_i.B \sqcup \exists TO_i.C$ in order to retrieve a process with this decision also for queries which contain both activities simultaneously, e.g. either B before C or vice versa. We do not require disjointness of activities in a flow and within decisions and parallel flows which could be simply added to the knowledge base by disjointness axioms.

Process Retrieval. Retrieval of refined processes is difficult due to the process definition using complex expressions. The relation from an activity to its successor is only defined for the process as a whole and not for the activities. However, in the demonstrated modeling methodology the reasoning complexity is lower than in the other modeling approach with explicit descriptions of the activity relations since the number of defined concepts in the ontology would be significantly higher. The query expressions are also more intuitive. Another benefit from the chosen modeling technique is the possibility to identify relationships between processes, e.g. process specializations. Since processes would be modeled by explicit relationships between activities, an activity can not occur in two different processes. Therefore it is required to use subclasses to denote the activities of the different processes.

6 Related Work

A comparison of our model with other process models is depicted in Table 5. The comparison contains modeling and retrieval.

For a core of our model, i.e. sequences only, this had already been accomplished by [14]. We significantly extend their model to account also for decisions, conditions, forks, joins, and loops. The matching of processes with different reasoning strategies is adopted from [13]. Process retrieval based on process information and annotation is realized in in [19]. A layered process representation is used. A service retrieval with additional functional information is described in [8,15] Other applications use search algorithms for process retrieval, like [6,16]. Similarity measure with respect to the process annotations and descriptions is used instead of logical reasoning. They use a process ontology and a corresponding query language. The retrieval in [10] also uses search algorithms instead of reasoning. DL based process models are described in [11] in order to enable process reuse. The model does not consider complex control flow.

The following approaches are focused on modeling and analyzing without retrieval. The OWL-S process model [1] is an OWL ontology for process description. Flow sequences are modeled as ordered lists. Like in OWL-S, processes are described as service interactions in WS-BPEL [2]. A XML messaging protocol is defined in order to specify the interoperability of services. The process specification language (PSL) [18] provides a mathematical model for activities, relations between activities and for data exchange. In [3] the control flow is analyzed according to the OWL-S process model. A combination of OWL and petri nets is demonstrated in [17]. OWL is only used to annotate the process models.

7 Conclusion

In this paper, we described process control flow modeling and retrieval in OWL-DL. For an application scenario we outlined the requirements of process retrieval with respect to the internal process structure. In order to fulfill these requirements it is necessary to describe explicitly the execution order of a process and to express the modality and terminology of process activities and structures. This also requires reasoning support to identify process extensions and terminological relationships between activities. Based on a process description with UML Activity Diagram the process models were transformed into OWL-DL. This contained a general description of the transformation and the design principles. The query patterns for process retrieval with respect to the control flow also cover the relationship between processes and demonstrate the usage of terminology and modality in the retrieval. In the evaluation the modeling and retrieval results, strengths and weaknesses are discussed.

Acknowledgements

This work has been supported by the European Projects MOST (ICT-FP7-2008 216691) and X-Media (IST-FP6-026978).

References

1. OWL-S: Semantic Markup for Web Services (2004), `http://www.w3.org/Submission/OWL-S`
2. Web Services Business Process Execution Language V. 2.0 (2007), `http://docs.oasis-open.org/wsbpel/2.0/OS/wsbpel-v2.0-OS.html`
3. Ankolekar, A., Paolucci, M., Sycara, K.: Spinning the OWL-S Process Model - Toward the Verification of the OWL-S Process Models. In: Proc. of ISWC Workshop on Semantic Web Services (2004)
4. Aslam, M.A., Auer, S., Shen, J., Herrmann, M.: Expressing Business Process Models as OWL-S Ontologies. In: Proc. of Int. Workshop on Grid and Peer-to-Peer based Workflows (GPWW). Springer, Heidelberg (2006)
5. Basten, T., van der Aalst, W.M.P.: Inheritance of Behavior. J. Log. Algebr. Program. 47(2), 47–145 (2001)
6. Bernstein, A., Klein, M.: Towards High-Precision Service Retrieval. In: Horrocks, I., Hendler, J. (eds.) ISWC 2002. LNCS, vol. 2342, p. 84. Springer, Heidelberg (2002)
7. Bildhauer, D., Ebert, J.: Querying Software Abstraction Graphs. In: Working on Query Technologies and Applications for Program Comprehension (2008)
8. Ferndriger, S., Bernstein, A., Dong, J.S., Feng, Y., Li, Y.-F., Hunter, L.: Enhancing Semantic Web Services with Inheritance. In: Sheth, A.P., Staab, S., Dean, M., Paolucci, M., Maynard, D., Finin, T., Thirunarayan, K. (eds.) ISWC 2008. LNCS, vol. 5318, pp. 162–177. Springer, Heidelberg (2008)
9. Gangemi, A., Borgo, S., Catenacci, C., Lehmenn, J.: Task Taxonomies for Knowledge Content D07. In: Metokis Deliverable, pp. 20–42 (2004)
10. Gil, Y., Gonzalez-Calero, P.A., Kim, J., Moody, J., Ratnakar, V.: Automatic Generation of Computational Workflows from Workflow Templates Using Distributed Data and Component Catalogs (2008)
11. Goderis, A., Sattler, U., Goble, C.: Applying DLs to workflow reuse and repurposing. In: Description Logic Workshop (2005)
12. Goderis, A., Sattler, U., Lord, P., Goble, C.: Seven Bottlenecks to Workflow Reuse and Repurposing. In: Gil, Y., Motta, E., Benjamins, V.R., Musen, M.A. (eds.) ISWC 2005. LNCS, vol. 3729, pp. 323–337. Springer, Heidelberg (2005)
13. Grimm, S., Motik, B., Preist, C.: Variance in e-Business Service Discovery. In: Proc. of the ISWC Workshop on Semantic Web Services (2004)
14. Hirsh, H., Kudenko, D.: Representing Sequences in Description Logics. In: Proc. of AAAI (1997)

15. Hull, D., Zolin, E., Bovykin, A., Horrocks, I., Sattler, U., Stevens, R.: Deciding Semantic Matching of Stateless Services. In: Proc. of AAAI (2006)
16. Kiefer, C., Bernstein, A., Lee, H.J., Klein, M., Stocker, M.: Semantic Process Retrieval with iSPARQL. In: Franconi, E., Kifer, M., May, W. (eds.) ESWC 2007. LNCS, vol. 4519, pp. 609–623. Springer, Heidelberg (2007)
17. Koschmider, A., Oberweis, A.: Ontology Based Business Process Description. In: EMOI-INTEROP (2005)
18. Menzel, C., Grüninger, M.: A formal Foundation for Process Modeling. In: Proc. of Int. Conf. on Formal Ontology in Information Systems, pp. 256–269 (2001)
19. Wolverton, M., Martin, D., Harrison, I., Thomere, J.: A Process Catalog for Workflow Generation. In: Sheth, A.P., Staab, S., Dean, M., Paolucci, M., Maynard, D., Finin, T., Thirunarayan, K. (eds.) ISWC 2008. LNCS, vol. 5318, pp. 833–846. Springer, Heidelberg (2008)
20. van der Aalst, W.M.P.: Inheritance of Business Processes: A Journey Visiting Four Notorious Problems. In: Petri Net Technology for Communication-Based Systems, pp. 383–408 (2003)
21. Wyner, G., Lee, J.: Defining Specialization for Process Models. In: Organizing Business Knowledge: The Mit Process Handbook. MIT Press, Cambridge (2003)

Context and Domain Knowledge Enhanced Entity Spotting in Informal Text

Daniel Gruhl[1], Meena Nagarajan[2], Jan Pieper[1], Christine Robson[1], and Amit Sheth[2]

[1] IBM Almaden Research Center
650 Harry Road, San Jose, CA
{dgruhl,jhpieper,crobson}@us.ibm.com
[2] Knoesis, 377 Joshi Research Center
3640 Colonel Glenn Highway, Dayton, OH
{meena,amit}@knoesis.org

Abstract. This paper explores the application of restricted relationship graphs (RDF) and statistical NLP techniques to improve named entity annotation in challenging Informal English domains. We validate our approach using on-line forums discussing popular music. Named entity annotation is particularly difficult in this domain because it is characterized by a large number of ambiguous entities, such as the Madonna album "Music" or Lilly Allen's pop hit "Smile".

We evaluate improvements in annotation accuracy that can be obtained by restricting the set of possible entities using real-world constraints. We find that constrained domain entity extraction raises the annotation accuracy significantly, making an infeasible task practical. We then show that we can further improve annotation accuracy by over 50% by applying SVM based NLP systems trained on word-usages in this domain.

1 Introduction

The semantic web and the plethora of relationships expressed as RDF files provide a wealth of information as to how entities in a document might relate. However, in the absence of a training corpus with in-line references to the entities (a "pre-annotated corpus"), it becomes difficult to identify and disambiguate named entities in text[13] to leverage these relationships in more complex tasks. The mapping of regions of text to entries in an ontology becomes harder when the regions are words used commonly in everyday language, such as "Yesterday," which could refer to the previous day, a Beatles song (one of 897 songs with that title), or a movie (there are three productions so named).

Sense disambiguation (the process of identifying which meaning of a word is used in any given context) becomes even more challenging when there is insufficient context surrounding the discourse; the language used is in the Informal English domain common to social networking sites – a blend of abbreviations, slang and context dependent terms delivered with an indifferent approach to grammar and spelling. Understanding the semantic relationships between entities in these challenging domains is necessary for a variety of information-centric

A. Bernstein et al. (Eds.): ISWC 2009, LNCS 5823, pp. 260–276, 2009.

applications, including the BBC SoundIndex [1]. This application, developed by the authors and others, provides a realtime "top 40" chart of music popularity based on sources such as MySpace and YouTube.

If we wish to utilize this type of content we need to transform it into a structured form by identifying and sense disambiguating particular entities such as mentions of artists, albums and tracks within the posts. In this paper we explore how the application of domain models (represented as a relationship graph, e.g., RDF) can complement traditional statistical NLP techniques to increase entity spotting[1] accuracy in informal content from the music domain. Semantic annotation of track and album name mentions are performed with respect to MusicBrainz RDF[2] - a knowledge base of instances, metadata and relationships in the music domain. An example snapshot of the MusicBrainz RDF is shown in Figure 1.

1.1 Challenging Features of the Music Domain

Availability of domain models is increasingly common with today's many Semantic Web initiatives. However, employing them for annotating Informal English content is non-trivial, more so in the music domain (see Table 1). Song titles are often short and ambiguous. Songs such as "The" (four songs), "A" (74 songs), "If" (413 songs), and "Why" (794 songs) give some idea of the challenges in spotting these entities. In annotating occurrence of these elements in text, for example, 'Yesterday' in "loved your song Yesterday!", we need to identify which entity 'Yesterday', among the many in the ontology, this one refers to.

Table 1. Challenging features of the music domain

Bands with a song "Merry Christmas"	60
Songs with "Yesterday" in the title	3,600
Releases of "American Pie"	195
Artists covering "American Pie"	31

Here, we present an approach that systematically expands and constrains the scope of domain knowledge from MusicBrainz used by the entity spotter to accurately annotate such challenging entity mentions in text from user comments. The MusicBrainz data set contains 281,890 artists who have published at least one track and 4,503,559 distinct artist/track pairs.

1.2 Our Approach and Contributions

We begin with a light weight, edit distance based entity spotter that works off a constrained set of potential entities from MusicBrainz. The entities we are interested in spotting in this work are track, album and song mentions. We constrain the size of the set of potential entities by manually examining some of the restrictions that can be applied on the MusicBrainz ontology. Restrictions are obtained using additional information from the context of the entity spot.

[1] We define spotting as finding a known list of named entities in text in real-time.

[2] http://wiki.musicbrainz.org/RDF

For example, when considering a spot in a comment from a discussion group on country music, we may only consider artists and songs from that genre.

Further improvement is needed to disambiguate the usage of song titles. For example, while Lilly Allen has a song titled 'Smile,' not all mentions of this word on her MySpace page refer to the song, for example, "your face lights up when you smile". We disambiguate the output of our naive spotter with more advanced NLP techniques using an SVM classifier that takes into account the characteristics of word usages.

We find that constraining the domain of possible entity matches before spotting can improve precision by several orders of magnitude over an admittedly poor baseline of the light weight spotter. We note that these improvements follow a Zipf distribution, where a reduction of possible entity matches by 50% equals a doubling of precision. We also find that use of our NLP system can improve accuracy by more than another 50%. These two steps, presented in the rest of this paper, can form the beginning of a processing pipeline to allow higher precision spot candidates to flow to upstream applications.

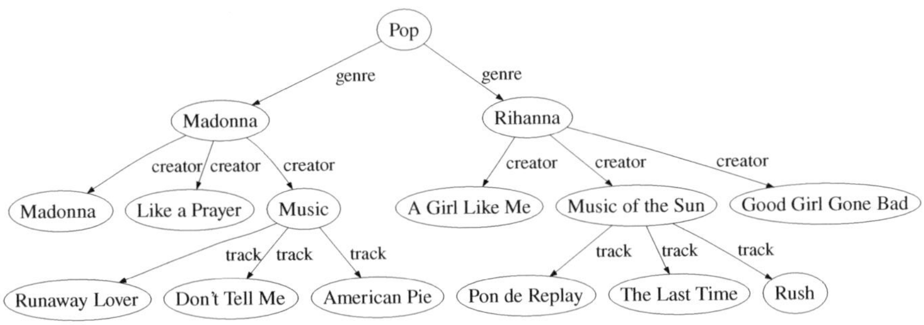


```
I went to <artist id=89>Madge's</artist> concert last night.
<artist id=262731>Rihanna</artist> is the greatest!
I love <artist id=357688>Lily's</artist> song <track id=8513722>smile</track>.
```

Fig. 1. RDF Snapshot of MusicBrainz and example of in-line annotations. These annotations illustrate how messages in our corpus can be tagged with universally unique identifiers (in this case the MusicBrainz id number) to facilitate searches both for individual mentions as well as Business Intelligence style roll-ups of aggregate statistics on mentions in the corpus.

2 Related Work

2.1 Named Entity Recognition and Use of Domain Knowledge

Named Entity Recognition (NER) is an important task in information extraction. Nadeau and Sekine present a comprehensive survey of NER since 1991 [15]. The KnowItAll Information Extraction system [8] makes use of entity recognition techniques, in a domain-independent fashion. Related work by Chieu and Ng has shown high performance in entity extraction with a single classifier and information from the whole document to classify each word [6].

Closely related to our work, domain dictionaries have been widely used in NER, including Wikipedia[4] and Wiktionary [14], DBLP [10], KAON [3], and MusicBrainz [1]. They have also been used for the task of disambiguating entity senses, an important step in accurately extracting entities. Work in [4] exploited the link and textual features of Wikipedia to perform named entity disambiguation. Entity disambiguation by gathering context from the document and comparing it with context in the knowledge base was also explored in [10].

These provide inspiration for our work, demonstrating that it is possible to do efficient and accurate NER on a document-by-document basis using domain knowledge supplemented with natural language processing techniques. Our work differs in how we constrain a domain knowledge base in order to annotate a set of known named entities in Informal English content.

2.2 Named Entity Recognition in Informal English

The challenge of NER in noisy and informal text corpora has been explored from several angles. Minkov et al. were the first to address NER in "informal text" such as bulletin board and newsgroup postings, and email [13]. Their work on recognizing personal names in such corpora is particularly relevant, as it uses dictionaries and constraining dictionary entries. They use a TF/IDF based approach for constraining the domain space, an approach we considered in early versions of our music miner. However, we found this approach to be problematic in the music domain, as song titles often have very low novelty in the TF/IDF sense (e.g. the Beatles song, "Yesterday"). Work by Ananthanarayanan et al. has also shown how existing domain knowledge can be encoded as rules to identify synonyms and improve NER in noisy text [2].

Our approach to NER in informal text differs in that it is a two step process. Given a set of known named entities from the MusicBrainz RDF, we first eliminate extraneous possibilities by constraining the domain model using available metadata and further use the natural language context of entity word-usages to disambiguate entities that appear as entities of interest and those that do not. Some word-usage features we employ are similar to those used in the past [13], while others are derived from our domain of discourse.

3 Restricted Entity Extraction

We begin our exploration of restricted RDF graphs or Ontologies to improve entity spotting by investigating the relationship between the number of entities (artists, songs and albums) considered for spotting and the precision of the entity spotter. The result is a calibration curve that shows the increase in precision as the entity set is constrained. This can be used to gauge the benefit of implementing particular real world constraints in annotator systems. For example, if detecting that a post is about an artist's recent album requires three weeks of work, but only provides a minor increase in precision, it might be deferred in favor of an "artist gender detector" that is expected to provide greater restriction in most cases.

3.1 Ground Truth Data Set

Our experimental evaluation focuses on user comments from the MySpace pages of three artists: Madonna, Rihanna and Lily Allen (see Table 2). The artists were selected to be popular enough to draw comment but different enough to provide variety. The entity definitions were taken from the MusicBrainz RDF (see Figure 1), which also includes some but not all common aliases and misspellings.

Table 2. Artists in the Ground Truth Data Set

Madonna	an artist with a extensive discography as well as a current album and concert tour
Rihanna	a pop singer with recent accolades including a Grammy Award and a very active MySpace presence
Lilly Allen	an independent artist with song titles that include "Smile," "Allright, Still", "Naive", and "Friday Night" who also generates a fair amount of buzz around her personal life not related to music

We establish a ground truth data set of 1858 entity spots for these artists (breakdown in Table 3). The data was obtained by crawling the artist's MySpace page comments and identifying all exact string matches of the artist's song titles. Only comments with at least one spot were retained. These spots were then hand scored by four of the authors as "good spot," "bad spot," or "inconclusive." This dataset is available for download from the Knoesis Center website[3].

Table 3. Manual scoring agreements on naive entity spotter results

Artist (Spots scored)	Good spots Agreement		Bad spots Agreement	
	100%	75 %	100%	75%
Rihanna (615)	165	18	351	8
Lily (523)	268	42	10	100
Madonna (720)	138	24	503	20

The human taggers were instructed to tag a spot as "good" if it clearly was a reference to a song and not a spurious use of the phrase. An agreement between at least three of the hand-spotters with no disagreement was considered agreement. As can be seen in Table 3, the taggers agreed 4-way (100% agreement) on Rihanna (84%) and Madonna (90%) spots. However ambiguities in Lily Allen songs (most notably the song "Smile"), resulted in only 53% 4-way agreement.

We note that this approach results in a recall of 1.0, because we use the naive spotter, restricted to the individual artist, to generate the ground truth candidate set. The precision of the naive spotter after hand-scoring these 1858 spots was 73%, 33% and 23% for Lilly Allen, Rihanna and Madonna respectively (see Table 3). This represents the best case for the naive spotter and accuracy drops quickly as the entity candidate set becomes less restricted. In the next Section we take a closer look at the relationship between entity candidate set size and spotting accuracy.

[3] http://knoesis.wright.edu/research/semweb/music

3.2 Impact of Domain Restrictions

One of the main contributions of this paper is the insight that it is often possible to restrict the set of entity candidates, and that such a restriction increases spotting precision. In this Section we explore the effect of domain restrictions on spotting precision by considering random entity subsets.

We begin with the whole MusicBrainz RDF of 281,890 publishing artists and 6,220,519 tracks, which would be appropriate if we had no information about which artists may be contained in the corpus. We then select random subsets of artists that are factors of 10 smaller (10%, 1%, etc). These subsets always contain our three actual artists (Madonna, Rihanna and Lily Allen), because we are interested in simulating restrictions that remove invalid artists. The most restricted entity set contains just the songs of one artist (≈0.0001% of the MusicBrainz taxonomy). In order to rule out selection bias, we perform 200 random draws of sets of artists for each set size - a total of 1200 experiments. Figure 2 shows that the precision increases as the set of possible entities shrinks. For each set size, all 200 results are plotted and a best fit line has been added to indicate the average precision. Note that the figure is in log-log scale.

We observe that the curves in Figure 2 conform to a power law formula, specifically a Zipf distribution ($\frac{1}{n^{R^2}}$). Zipf's law was originally applied to demonstrate the Zipf distribution in frequency of words in natural language corpora [18], and has since been demonstrated in other corpora including web searches [7]. Figure 2 shows that song titles in Informal English exhibit the same frequency characteristics as plain English. Furthermore, we can see that in the average case, a domain restrictions of 10% of the MusicBrainz RDF will result approximately in a 9.8 times improvement in precision of a naive spotter.

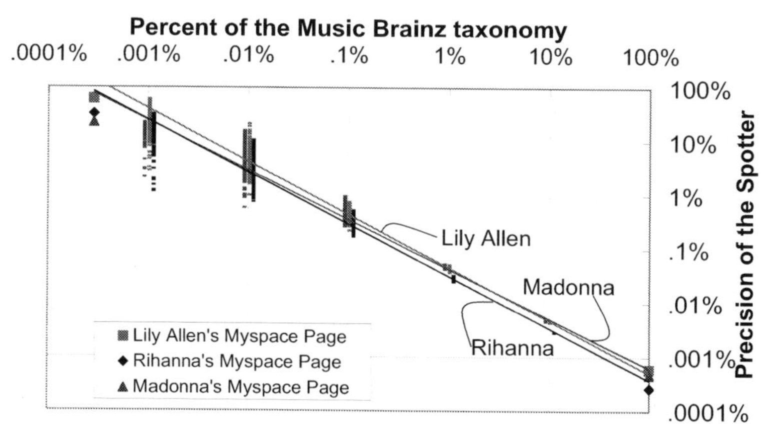

Fig. 2. Precision of a naive spotter using differently sized portions of the MusicBrainz Taxonomy to spot song titles on artist's MySpace pages

This result is remarkably consistent across all three artists. The R^2 values for the power lines on the three artists are 0.9776, 0.979, 0.9836, which gives a deviation of 0.61% in R^2 value between spots on the three MySpace pages.

4 Real World Constraints

The calibration results from the previous Section show the importance of "ruling out" as many artists as possible. We observe that simple restrictions such as gender that might rule out half the corpus could potentially increase precision by a factor of two. One way to impose these restrictions is to look for real world constraints that can be identified using the metadata about entities as they appear in a particular post. Examples of such real world constraints could be that an artist has released only one album, or has a career spanning more than two decades.

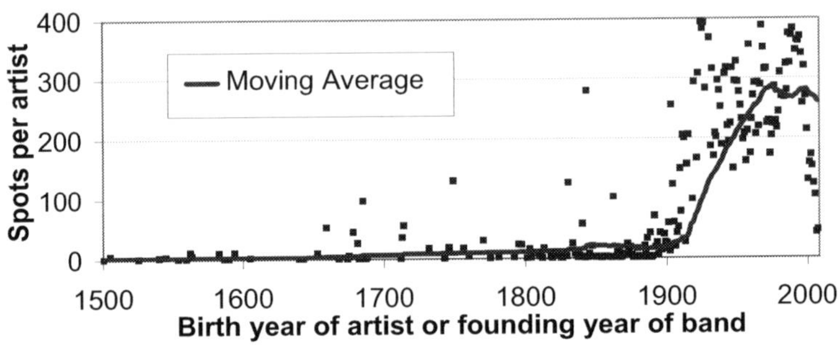

Fig. 3. Songs from all artists in our MySpace corpus, normalized to artists per year

We are interested in two questions. First, do real world constraints reduce the size of the entity spot set in a meaningful way? Second, by how much does the trivial spotter improve with these real world constraints and does this match with our predicted improvements from Figure 2? The effect of restricting the RDF by artist's age can be seen in Figure 3, which shows spots per artist by birth date. Interestingly, we can see a spike in the graph beginning around 1920 with the emergence of Jazz and then Rock and Roll, reflecting the use of common words as song titles, (e.g. "Blues" and "South" by Louis Armstrong). For all artists since this date (94% of the MusicBrainz Ontology, and 95.5% of the naive spots on our corpus), the increased use of natural language utterances as song titles is evidence that we should expect the Zipf distribution to apply to any domain restriction over the corpus.

Having established that domain restrictions do reduce spot size, we look for further constraints that can be inferred from the user-generated text. As an example, we observe that comments such as "Saw you last night in Denver!!!" indicate the artist is still alive. A more informational post such as "Happy 25th B-DAY!" would allow us to further narrow the RDF graph to 0.081% of artists in the Ontology, and 0.221% of the naive spots on Lily Allen's MySpace Page.

Table 4. The efficacy of various sample restrictions

Key	Count	Restriction
Artist Career Length Restrictions- Applied to Madonna		
B	22	80's artists with recent (within 1 year) album
C	154	First album 1983
D	1,193	20-30 year career
Recent Album Restrictions- Applied to Madonna		
E	6,491	Artists who released an album in the past year
F	10,501	Artists who released an album in the past 5 years
Artist Age Restrictions- Applied to Lily Allen		
H	112	Artist born 1985, album in past 2 years
J	284	Artists born in 1985 (or bands founded in 1985)
L	4,780	Artists or bands under 25 with album in past 2 years
M	10,187	Artists or bands under 25 years old
Number of Album Restrictions- Applied to Lily Allen		
K	1,530	Only one album, released in the past 2 years
N	19,809	Artists with only one album
Recent Album Restrictions- Applied to Rihanna		
Q	83	3 albums exactly, first album last year
R	196	3+ albums, first album last year
S	1,398	First album last year
T	2,653	Artists with 3+ albums, one in the past year
U	6,491	Artists who released an album in the past year
Specific Artist Restrictions- Applied to each Artist		
A	1	Madonna only
G	1	Lily Allen only
P	1	Rihanna only
Z	281,890	All artists in MusicBrainz

Our constraints are tabulated in Table 4, and are derived manually from comments such as, "I've been a fan for 25 years now," "send me updates about your new album," and "release your new album already! i'm getting tired of playing your first one on repeat!" Since we have chosen our corpus to represent three specific artists, the name of the artist is a further narrowing constraint.

We consider three classes of restrictions - career, age and album based restrictions, apply these to the MusicBrainz RDF to reduce the size of the entity spot set in a meaningful way and finally run the trivial spotter. For the sake of clarity, we apply different classes of constraints to different artists.

We begin with restrictions based on length of career, using Madonna's MySpace page as our corpus. We can restrict the RDF graph based on total length of career, date of earliest album (for Madonna this is 1983, which falls in the early 80's), and recent albums (within the past year or 5 years). All of these restrictions are plotted in Figure 4, along with the Zipf distribution for Madonna from Figure 2. We can see clearly that restricting the RDF graph based on career characteristics conforms to the predicted Zipf distribution.

For our next experiment we consider restrictions based on age of artist, using Lily Allen's MySpace page as our corpus. Our restrictions include Lily Allen's age of 25 years, but overlap with bands founded 25 years ago because of how dates are recorded in the MusicBrainz Ontology. We can further restrict using album information, noting that Lily Allen has only a single album, released in the past two years. These restrictions are plotted in Figure 4, showing that these restrictions on the RDF graph conform to the same Zipf distribution.

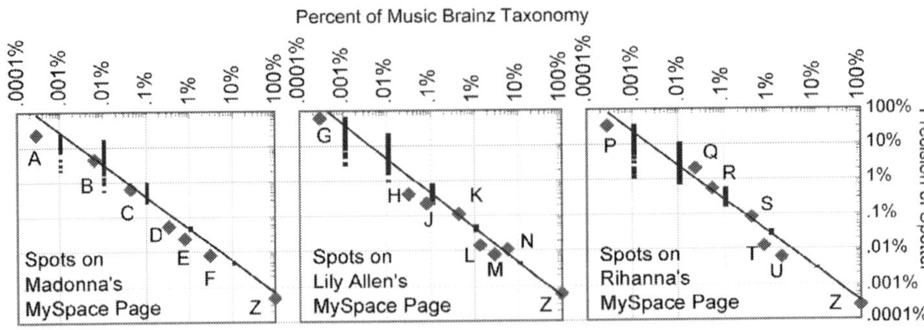

Fig. 4. Naive spotter using selected portions of the MusicBrainz RDF based on descriptive characteristics of Madonna, Lily Allen and Rihanna, respectively. The Key to the data points is provided in Table 4.

Finally, we consider restrictions based on absolute number of albums, using Rihanna's MySpace page as our corpus. We restrict to artists with three albums, or at least three albums, and can further refine by the release dates of these albums. These restrictions fit with Rihanna's short career and disproportionately large number of album releases (3 releases in one year). As can be seen in Figure 4, these restrictions also conform to the predicted Zipf distribution.

The agreement of the three types of restrictions from above with the random restrictions from the previous Section are clear from comparing the plots in Figure 4. This confirms the general effectiveness of limiting domain size to improve precision of the spotter, regardless of the type of restriction, as long as the restriction only removes off-target artists. A reduction in the size of the RDF graph results in an approximately proportionate increase in precision. This is a particularly useful finding, because it means that any restriction we can apply will improve precision, and furthermore we can estimate the improvement in precision.

5 NLP Assist

While reducing extraneous possibilities improved precision of the naive spotter significantly, false positives resulting from spots that appear in different senses still need attention (see Table 5). The widely accepted *'one sense per discourse'* notion[17] that the sense or meaning of a word is consistent within a discourse does not hold for this data given the use of common words as names for songs and albums.

Table 5. Spots in multiple senses

Valid:	Got your new album *Smile*. Loved it!
Invalid:	Keep your *SMILE* on. You'll do great!

The task is to assess whether a spot found is indeed a valid track or album. This is similar to the word sense disambiguation problem where the task is to resolve which one of many pre-defined senses is applicable to a word[9]. Here, we use a learning algorithm over local and surrounding word contexts, an approach similar in principle to several past efforts but adapted to our domain of discourse [16].

Formally, our task can be regarded as a binary classification problem. Consider the set of all spots found by the naive spotter. Each spot in this set can be labeled 1 if it is a track; and -1 if it is not, where the label is associated with a set of input features that characterize a spot s. This is implemented as a Support Vector Machine (SVM), a machine learning approach known to be effective for solving binary pattern recognition, named entity recognition and document classification problems[11].

5.1 Features

We trained and tested the SVM learner on two sets of features collectively observed in the tagged data (see Section 3.1); *basic features*, that characterize a spot and *advanced features* that are based on the context surrounding the spot.

Basic Features. We encode a set of spot-level boolean features (see Table 6) that include whether the spot is all capitalized, starts with capital letters or is enclosed in quotes. If the entire comment including the spot is capitalized, we do not record a 1 for *s.allCaps* or *s.firstCaps*. We also encode features derived from the part-of-speech (POS) tags and NP-chunking of comments (see syntactic features in Table 6)[4]. To encode syntactic features, we created a list of the Penn Treebank tag set[5] also used by the Stanford parser. If the parser returns a tag for the spot, we obtain the tag's index position in the list to encode this feature. If the sentence is not parsed this feature is not encoded.

Advanced features. We encode the following advanced features intended to exploit the local context surrounding every spot. We encode the POS tags of word tokens appearing before and after a spot in a sentence.

Sentiment expressions and domain-specific terms. We found that spots that co-occurred with sentiment expressions and domain-specific words such as 'music', 'album', 'song', 'concert', etc. were more likely to be valid spots. We encode these boolean features in the following manner.

First, we curated a sentiment dictionary of 300 positive and negative expressions from UrbanDictionary[6] (UD), given the use of slang by this poster demographic. Starting with expressions such as 'good', and 'bad', we obtained the top 10 related sentiment expressions for these words. We continued this process for the newly obtained words until we found no new words. Note that we are not concerned with the polarity, but mere co-occurrence of sentiment expressions with spots. A dictionary of 25 domain-specific terms, such as 'music', 'album', 'track', 'song' etc. was created manually by consulting MusicBrainz. These dictionaries are available for download from the Knoesis Center website[7].

[4] Obtained using the Stanford NL Parser http://nlp.stanford.edu/software/lex-parser.shtml

[5] http://www.cis.upenn.edu/~treebank/

[6] www.urbandictionary.com

[7] http://knoesis.wright.edu/research/semweb/music

Table 6. Features used by the SVM learner

Syntactic features	**Notation-S**
+POS tag of *s*	s.POS
POS tag of one token before *s*	$s.POS_b$
POS tag of one token after *s*	$s.POS_a$
Typed dependency between *s* and sentiment word	$s.POS\text{-}TD_{sent}$*
Typed dependency between *s* and domain-specific term	$s.POS\text{-}TD_{dom}$*
Boolean Typed dependency between *s* and sentiment	$s.B\text{-}TD_{sent}$*
Boolean Typed dependency between s and domain-specific term	$s.B\text{-}TD_{dom}$*
Word-level features	**Notation-W**
+Capitalization of spot *s*	s.allCaps
+Capitalization of first letter of *s*	s.firstCaps
+*s* in Quotes	s.inQuotes
Domain-specific features	**Notation-D**
Sentiment expression in the same sentence as *s*	$s.S_{sent}$
Sentiment expression elsewhere in the comment	$s.C_{sent}$
Domain-related term in the same sentence as *s*	$s.S_{dom}$
Domain-related term elsewhere in the comment	$s.C_{dom}$

+Refers to basic features, others are advanced features
*These features apply only to one-word-long spots.

If one or more sentiment expressions, domain-specific terms or their word forms were spotted in the same sentence as the spot, values for $s.S_{sent}$ and $s.S_{dom}$ are recorded as 1. Corresponding $s.C_{sent}$ and $s.C_{dom}$ features were also used to record similar values when these terms were found elsewhere in the comment. Encoding the actual number of co-occurring sentiment or domain expressions did not significantly change the classification result.

Typed Dependencies
We also captured the typed dependency paths (grammatical relations) via the $s.POS\text{-}TD_{sent}$ and $s.POS\text{-}TD_{dom}$ boolean features. These were obtained between a spot and co-occurring sentiment and domain-specific words by the Stanford parser[12] (see example in 7). We also encode a boolean value indicating whether a relation was found at all using the $s.B\text{-}TD_{sent}$ and $s.B\text{-}TD_{dom}$ features. This allows us to accommodate parse errors given the informal and often non-grammatical English in this corpus.

Table 7. Typed Dependencies Example

Valid spot: Got your new album **Smile**. Simply *loved* it! **Encoding:** nsubj(loved-8, Smile-5) implying that **Smile** is the nominal subject of the expression *loved*. **Invalid spot:** Keep your **smile** on. You'll do *great*! **Encoding:** No typed dependency between **smile** and *great*

Table 8. Classifier accuracy in percentages for different feature combinations. Best performers in bold.

	Features	Valid	Invalid Spots			Acc.
		Set1	Set2	Set3	Avg.	Split
(1)	W	45	88	84	**86**	45 - 86
(2)	W+S	74	43	37	40	74 - 40
(3)	W+D	62	85	83	**84**	62 - 84
(4)	D	70	50	62	56	70 - 56
(5)	D+S	72	34	36	35	72 - 35
(6)	W+D+s.POS	61	66	74	70	61 - 70
(7)	W+D+s.POS$_{b,a}$+s.POS-TDs	**78**	47	53	50	78 - 50
(8)	W+D+s.POS$_{b,a}$+s.B-TDs	**90**	33	37	35	90 - 35
(9)	W+D+only s.POS$_{b,a}$	62	81	87	**84**	62 - 84
(10)	W+D+only s.POS-TDs	60	79	91	**85**	60 - 85
(11)	W+D+only s.B-TDs	71	68	72	70	71 - 70
(12)	All features	42	89	93	**91**	42 - 91

5.2 Data and Experiments

Our training and test data sets were obtained from the hand-tagged data (see Table 3). Positive and negative *training examples* were all spots that all four annotators had confirmed as valid or invalid respectively, for a total of 571 positive and 864 negative examples. Of these, we used 550 positive and 550 negative examples for training. The remaining spots were used for test purposes.

Our positive and negative *test sets* comprised of all spots that three annotators had confirmed as valid or invalid spots, i.e. had a 75% agreement. We also included spots where 50% of the annotators had agreement on the validity of the spot and the other two were *not sure*. We further divided our negative test set into two disjoint equal sets that allowed us to confirm generality of the effect of our features. Finally, our test set of *valid spots*, Set 1, contained 120 spots and the two test sets for *invalid spots*, Set 2 and Set 3, comprised of 229 spots each.

We evaluated the efficacy of features shown in Table 6 in *predicting the labels assigned by the annotators.* All our experiments were carried out using the SVM classifier from [5] using 5-fold cross-validation. As one way of measuring the relative contribution of advanced contextual and basic spot-level features, we removed them one after another, trying several combinations. Table 8 reports those combinations for which the accuracy in labeling either the valid or invalid datasets was at least 50% (random labeling baseline). Accuracy in labeling valid and invalid spots refer to the percentage of true and false positives that were labeled correctly by the classifier. In the following discussion, we refer to the average performance of the classifier on the false positives, Sets 2 and 3 and its performance on the true positives, Set 1.

5.3 Usefulness of Feature Combinations

Our experiments revealed some expected and some surprising findings about the usefulness of feature combinations for this data. For valid spots, we found that the best feature combination was the word-level, domain-specific and contextual syntactic tags (POS tags of tokens before and after the spot) when used with the boolean typed dependency features. This feature combination labeled 90% of good spots accurately. The next best and similar combination of word-level, domain-specific and contextual tags when used with the POS tags for the typed dependency features yielded an accuracy of 78%. This suggests that *local word descriptors along with contextual features* are good predictors of valid spots in this domain.

For the invalid spots (see column listing average accuracy), the use of all features labeled 91% of the spots correctly. Other promising combinations included the word-level; word-level and domain-specific; word-level, domain-specific and POS tags of words before and after the spot; word-level, domain-specific and the typed dependency POS tags, all yielding accuracies around 85%.

It is interesting to note that the POS tags of the spot itself were not good predictors for either the valid or invalid spots. However, the POS typed dependencies were more useful than the boolean typed dependencies for the invalid spots. *This suggests that not all syntactic features are useless, contrary to the general belief that syntactic features tend to be too noisy to be beneficial in informal text.* Our current investigations to improve performance include the use of other contextual features like commonly used bi-grams and tri-grams and syntactic features of more tokens surrounding a spot.

Table 9. Average performance of best feature combinations on 6 sets of 500 invalid spots each

Feature Combination	Mean Acc.	Std.Dev Acc.
All Features	99.4%	0.87%
W	91.3%	2.58%
W+D	83%	2.66%
W+D+only s.POS-TDs	80.8%	2.4%
W+D+only s.POS$_{b,a}$	77.33%	3.38%

Accuracy in Labeling Invalid Spots. As further confirmation of the generality of effect of the features for identifying incorrect spots made by the naive spotter, we picked the best performing feature combinations from Table 8 and tested them on a dataset of 3000 known invalid spots for artist Rihanna's comments from her MySpace page. This dataset of invalid spots was obtained using the entire MusicBrainz taxonomy excluding Rihanna's song/track entries - effectively allowing the naive spotter to mark all invalid spots. We further split the 3000 spots into 6 sets of 500 spots each. The best feature combinations were tested on the model learned from the same training set as our last experiment. Table 9 shows the average and standard deviation performance of the feature combinations across the 6 sets. As we see, the feature combinations performed remarkably consistently for this larger test set. The combination of all features was the most useful, labeling 99.4% of the invalid spots correctly.

6 Improving Spotter Accuracy Using NLP Analysis

The last set of experiments confirmed the usefulness of the features in classifying whether a spot was indeed a track or not. In this next experiment, we sought to measure the improvement in the overall spotting accuracy - first annotating comments using the naive spotter, followed by the NLP analytics. This approach of boosting allows the more time-intensive NLP analytics to run on less than the full set of input data, as well as giving us a certain amount of control over the precision and recall of the final result.

Figure 5 shows the improvement in precision for spots in the three artists after boosting the naive spotter with the NLP component. Ordered by decreasing recall, we see an increase in precision for the different feature combinations. For example, the precision of the naive spotter for artist Madonna's spots was 23% and almost 60% after boosting with the NLP component and using the feature combinations that resulted in a $42-91$ split in accurately labeling the valid and invalid spots.

Although our classifier was built over the results of the naive spotter, i.e. it already knew that the spot was a potential entity, our experiments suggest that the features employed might also be useful for the traditional named entity recognition problem of labeling word sequences as entities.

Our experiments also suggest that although informal text has different characteristics than formal text such as news or scientific articles, simple and inexpensive learners built over a dictionary-based naive spotter can yield reasonable performance in accurately extracting entity mentions.

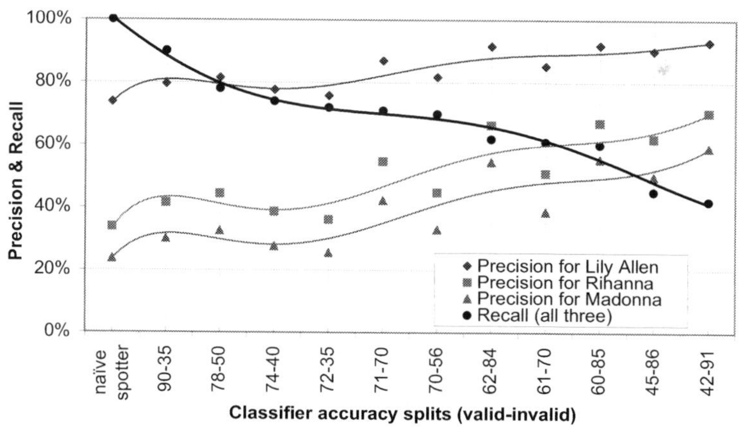

Fig. 5. NLP Precision-Recall curves for three artists and feature combinations

7 Conclusion and Future Work

Spotting music tracks in Informal English is a critical enabling technology for applications such as the BBC SoundIndex that allows real-time tracking of opinions in on-line forums. The presented approach is applicable to other domains as well. We are currently adopting our system to spot automobiles and find that car brands, makes and models provide a well formed ontology as well. We believe that such on-demand information will play an increasingly important role in business as companies seek to better understand their customers. Rapid detection of events (e.g. the artist "OK Go" ramping up the chart within hours of being featured on the popular TV show Big Brother) illustrate the possibilities of these systems.

There are several challenges in constructing these systems. Discussions of music produce millions of posts a day, which need to be processed in real-time, prohibiting more computational intensive NLP techniques. Moreover, since 1920, song titles based on common words or phrases have become very popular (see Figure 3), making it difficult to spot and disambiguate song titles.

In this paper, we presented a two stage approach - entity spotting based on scoping a domain model followed by SVM based NLP system to facilitate higher quality entity extraction. We studied the impact of restricting the size of the entity set being matched and noted that the spot frequency follows a Zipf distribution. We found that R^2 for this distribution is fairly consistent among a sample of artists. This allows a reasonable a priori evaluation of the efficacy of various restriction techniques using the calibration curve shown in Figure 2. We found that in many cases such restrictions can come from the language of the spotted text itself.

Given these potential spots, we show that simple classifiers trained on generic lexical, word and domain specific characteristics of a spot can effectively eliminate false positives in a manner that can improve accuracy up to a further 50%. Our experiments suggest that although informal text has different characteristics than formal text, learners that improve a dictionary-based naive spotter can yield reasonable performance in accurately extracting entity mentions.

7.1 Future Work

Future areas of interest include applying standard measures such as TF-IDF to predict the ambiguity of entities for use with the NLP component. One drawback of the current approach for scoping the linked data or RDF graph to a single artist occurs when references to multiple artists are made in text (e.g. your song "X" reminds me of Y's song "Z"). Even though these mentions are sparse, we are planning to include non-ambiguous artist names as "activators" in the base spotting set. If a post mentions another artist, the spotter would temporarily activate entities from the RDF belonging to that specific artist.

Another area of interest is to examine automatic constraint selection based on the posts themselves. For example a "birthdate note" detector, a gender of artist identifier, a recent album release detector, etc. Using the Zipf distribution in Figure 2 we can estimate how helpful each detector might be before we implement it. Once a robust set of post based constraint detectors are developed we can begin to experiment on "free domain" spotting - that is spotting in domains where less focused discussions are expected, e.g. Twitter messages.

We also plan to extend our work to other free text domains. The ability to achieve reasonable performance in this problem suggests that this approach will work well in other, less challenging domains where the entities are less overlapping (e.g. company name extraction) or the English is less informal (e.g. news releases).

Acknowledgements

We would like to thank the BBC SoundIndex team for their help and support. We would also like to thank Marti Hearst, Alexandre Evfimievski, Guozhu Dong, Amy Vandiver and our shepherd Peter Mika for their insightful reviews and suggestions.

References

1. Alba, A., Bhagwan, V., Grace, J., Gruhl, D., Haas, K., Nagarajan, M., Pieper, J., Robson, C., Sahoo, N.: Applications of voting theory to information mashups. In: ICSC, pp. 10–17. IEEE Computer Society, Los Alamitos (2008)
2. Ananthanarayanan, R., Chenthamarakshan, V., Deshpande, P.M., Krishnapuram, R.: Rule based synonyms for entity extraction from noisy text. In: ACM Workshop on Analytics for noisy unstructured text data, pp. 31–38 (2008)
3. Bozsak, E., Ehrig, M., Handschuh, S., Hotho, A., Maedche, A., Motik, B., Oberle, D., Schmitz, C., Staab, S., Stojanovic, L., Stojanovic, N., Studer, R., Stumme, G., Sure, Y., Tane, J., Volz, R., Zacharias, V.: KAON - towards a large scale semantic web. In: Bauknecht, K., Tjoa, A.M., Quirchmayr, G. (eds.) EC-Web 2002. LNCS, vol. 2455, pp. 304–313. Springer, Heidelberg (2002)
4. Bunescu, R.C., Pasca, M.: Using encyclopedic knowledge for named entity disambiguation. In: EACL. The Association for Computer Linguistics (2006)
5. Chang, C.-C., Lin, C.-J.: LIBSVM: a library for support vector machines (2001), `http://www.csie.ntu.edu.tw/~cjlin/libsvm`
6. Chieu, H.L., Ng, H.T.: Named entity recognition: A maximum entropy approach using global information. In: COLING (2002)
7. Cunha, C., Bestavros, A., Crovella, M.: Characteristics of www client-based traces. Technical report, Boston University, Boston, MA, USA (1995)
8. Etzioni, O., Cafarella, M., et al.: Web-scale information extraction in knowitall (preliminary results). In: WWW 2004, pp. 100–110. ACM Press, New York (2004)
9. Ide, N., Véronis, J.: Word sense disambiguation: The state of the art. Computational Linguistics 24, 1–40 (1998)
10. Aleman-Meza, B., Hassell, J., Arpinar, I.B.: Ontology-driven automatic entity disambiguation in unstructured text. In: Cruz, I., Decker, S., Allemang, D., Preist, C., Schwabe, D., Mika, P., Uschold, M., Aroyo, L.M. (eds.) ISWC 2006. LNCS, vol. 4273, pp. 44–57. Springer, Heidelberg (2006)
11. Joachims, T.: Text categorization with support vector machines: Learning with many relevant features. In: Machine Learning. LNCS, pp. 137–142. Springer, Heidelberg (1998)
12. Marneffe, M., Maccartney, B., Manning, C.: Generating typed dependency parses from phrase structure parses. In: Proceedings of LREC-2006, pp. 449–454 (2006)

13. Minkov, E., Wang, R.C., Cohen, W.W.: Extracting personal names from email: Applying named entity recognition to informal text. In: HLT/EMNLP. The Association for Computational Linguistics (2005)
14. Muller, C., Gurevych, I.: Using wikipedia and wiktionary in domain-specific information retrieval. In: Working Notes for the CLEF 2008 Workshop, Aarhus, Denmark (2008)
15. Nadeau, D., Sekine, S.: A survey of named entity recognition and classification. Linguisticae Investigationes (2007)
16. Doina, T.: Word sense disambiguation by machine learning approach: A short survey. Fundam. Inf. 64(1-4), 433–442 (2004)
17. Yarowsky, D.: Hierarchical Decision Lists for WSD. Kluwer Acadmic Publishers, Dordrecht (1999)
18. Zipf, G.K.: Human Behavior and the Principle of Least Effort: An Introduction to Human Ecology. Addison-Wesley, Cambridge (1949)

Using Naming Authority to Rank Data and Ontologies for Web Search

Andreas Harth[1,2], Sheila Kinsella[1], and Stefan Decker[1]

[1] National University of Ireland, Galway
Digital Enterprise Research Institute
[2] Institut AIFB, Universität Karlsruhe (TH), Germany

Abstract. The focus of web search is moving away from returning relevant documents towards returning structured data as results to user queries. A vital part in the architecture of search engines are link-based ranking algorithms, which however are targeted towards hypertext documents. Existing ranking algorithms for structured data, on the other hand, require manual input of a domain expert and are thus not applicable in cases where data integrated from a large number of sources exhibits enormous variance in vocabularies used. In such environments, the authority of data sources is an important signal that the ranking algorithm has to take into account. This paper presents algorithms for prioritising data returned by queries over web datasets expressed in RDF. We introduce the notion of naming authority which provides a correspondence between identifiers and the sources which can speak authoritatively for these identifiers. Our algorithm uses the original PageRank method to assign authority values to data sources based on a naming authority graph, and then propagates the authority values to identifiers referenced in the sources. We conduct performance and quality evaluations of the method on a large web dataset. Our method is schema-independent, requires no manual input, and has applications in search, query processing, reasoning, and user interfaces over integrated datasets.

1 Introduction

More and more structured interlinked data is appearing on the web, in the form of microformats, XML, and RDF (Resource Description Format). A common feature of these formats is that they take a loosely object-oriented view, describing objects such as people, events, or locations. Given that the information published in these formats exhibits more structure and a fine-grained description of objects, typical keyword-based search engines do not exploit the full potential that the more structured data offers. The established methods for information retrieval are not directly applicable to structured data, since i) the basic result units of search moves from documents to objects which may be associated with several sources and ii) the users are able, in addition to keyword searches, to more accurately state their information needs via precise queries.

The problem of search in hypertext collections has been extensively studied, with the web as the premier example. Given the large number of data providers

A. Bernstein et al. (Eds.): ISWC 2009, LNCS 5823, pp. 277–292, 2009.

compared to traditional database scenarios, an information retrieval system for hypertext on the web must be able to handle data with the following properties:

- Domain variety: the web contains data about many topics (e.g., from social networks to protein pathways to entertainment)
- Structural variety: aggregating data from many autonomous web sources, with no data curation or quality control of any sort, results in datasets with overlapping, redundant, and possibly contradictory information
- Noise: with the number of data contributors the amount of errors increase (e.g., syntax errors, spelling mistakes, wrong identifiers)
- Spam: when everybody can say anything about anything with little effort or cost, malicious activity emerges
- Scale: identifiers and documents on the web number in the trillions

We expect the properties identified above to also hold for structured information sources on the web, though they are typically not taken into account for classical relational query processing or data warehousing, where the number of autonomous sources is orders of magnitude lower. In traditional data integration systems, the schema of the integrated data is known in advance, and is hard-coded into applications to, for example, determine the order in which data elements are displayed. In this paper we focus on the problem of ranking in structured datasets which have been integrated from a multitude of disparate sources without a-priori knowledge on the vocabulary used. We assume a basic interlinked data model, enabling the definition of objects and relationships between those objects. Further, we assume the possibility of global identifiers which enable the reuse of identifiers across sources and thus the interlinkage of data. RDF, and to a limited extent XML and microformats fulfil these assumptions.

Based on the above scenario the contributions of this paper are as follows:

- We introduce the notion of naming authority which establishes a connection between an identifier (in our case a URI) and the source which has authority to assign that identifier (in our case a web source, also identified via a URI). The notion of naming authority can be generalised to other identifier schemes (e.g., trademarks) for which it is possible to establish a connection to the provenance of the identifier, such as a person or an organisation (Section 4).
- We present a method to derive a naming authority matrix from a dataset, and use the PageRank algorithm to determine rankings for sources. In a second step, an algorithm ranks individual identifiers based on the values assigned to their sources (Section 5).
- We provide an experimental evaluation on real-world web datasets containing up to 1.1bn data items from 6.5m web sources, and provide evidence for the quality of the rankings with a user study of 36 participants (Section 6).

We present an example scenario in Section 2, and provide an overview in Section 3. Section 7 outlines related approaches and Section 8 concludes.

2 Motivating Example

For the motivating example, we use social network information describing people, who they know and what they do, as there is large amounts of this data available. Data sources typically express person-related information in the Friend-of-a-Friend (FOAF) vocabulary, but also use their own schemas (i.e. sub-classing the general Person class with classes such as Professor or Ph.D. student). The personal data available on the web exhibits the properties we expect from any large-scale loosely coordinated distributed knowledge base.

In our running example URIs are used to identify both the objects (e.g. `http://danbri.org/foaf.rdf#danbri`) and the data sources (e.g. `http://danbri.org/foaf.rdf`). Identifiers might be reused across sources, for example, the source `http://danbri.org/foaf.rdf` uses the URI `http://www.-w3.org/People/BernersLee/card#i` to denote the Person Tim Berners-Lee. The reuse of identifiers provides a direct means to consolidate information about objects from different sources. In our experimental dataset, Dan's URI is referenced in at least 70 sources. Representing only a small subset of the available data, the graphs in Figure 1 depict object descriptions from two sources.

Several challenges arise when displaying data aggregated from the web:

1. How to prioritise results from keyword searches (i.e. in which order to show the resulting objects)?

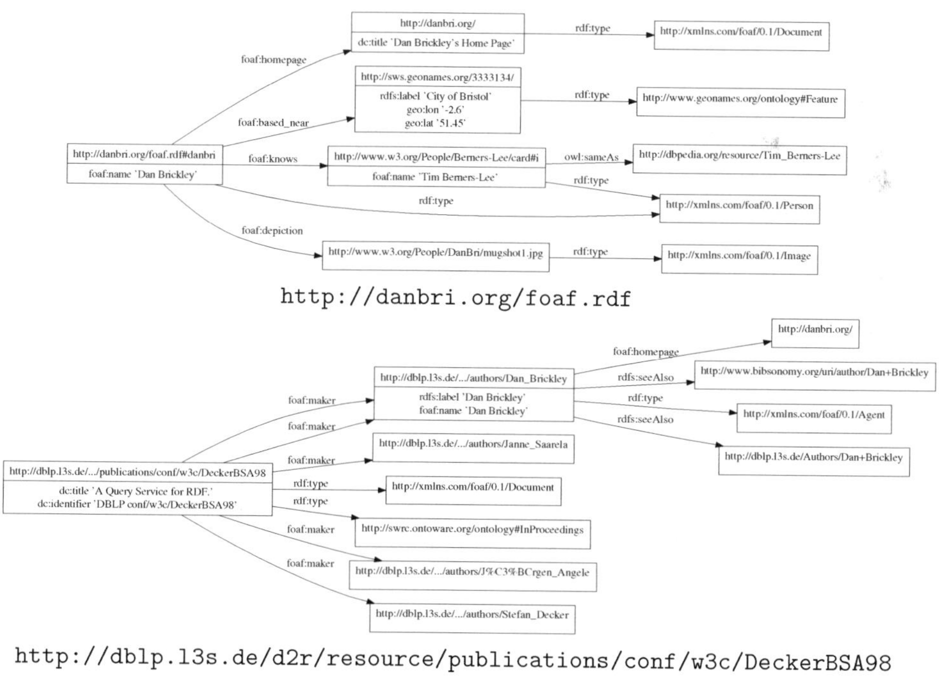

Fig. 1. A subset of RDF descriptions about Dan, taken from two sources

Fig. 2. Rendering of information pertaining to the object `http://www.danbri.org/-foaf.rdf#danbri` (datatype properties and values on the left, object properties and objects on the right, and data sources on the bottom). The decision on ordering of all data elements on the page has been taken based on rankings alone, without a priori schema knowledge.

2. How to prioritise predicates (i.e. which predicate/value pairs to show first)?
3. How to prioritise objects from multi-valued attributes (i.e. which image to show first in case there are multiple depictions of the person)?
4. How to prioritise sources that contribute data to an object (i.e. in which order to present the sources, from the most important to the least)?

Question 1) surfaces mainly in a web search scenario, where a search engine returns a list of identifiers matching the user-specified query (keywords or structured queries such as "return all instances of type `owl:Class`" or "return all instances of type `foaf:Person` that have `foaf:workplaceHomepage` `http://www.deri.org/`"). Questions 2) to 4) are also relevant for Linked Data browsers such as Disco[1], Data Explorer[2], or Tabulator[3] in case vocabularies

[1] `http://www4.wiwiss.fu-berlin.de/bizer/ng4j/disco/`
[2] `http://demo.openlinksw.com/rdfbrowser2/`
[3] `http://www.w3.org/2005/ajar/tab`

which are unknown to the browsers are used in the data. Figure 2 shows the rendering of information about an object using rankings derived with the method presented here in VisiNav[4], which also has Linked Data browsing functionality.

3 Method Overview

We introduce the data model, present the requirements for a ranking method on web data and outline our procedure for ranking data sources and data items.

3.1 Data Model

In general, our ranking method can be applied to datasets with i) global, reused identifiers, ii) tracking of provenance, and iii) correspondence between object identifiers and source identifiers. Specifically, we assume the following:

- a set of identifiers I, encompassing a set of global identifiers U, a set of local identifiers B, and a set of strings L (we omit datatypes such as integer or date for brevity)
- a set of data sources $S \subseteq U$
- a function *ids* which maps sets of global and local identifiers and literals $i \in I$ to the sources $s \in S$ in which they occur

This generic model applies to a wide variety of data models, such as hypertext, graph-structured data, and relational data.

3.2 Requirements

A ranking procedure operating on collaboratively-edited datasets should exhibit the following properties:

- The use of an identifier owned by source s_a by a source s_b indicates an acknowledgement of the authority of s_a and should benefit the ranking of s_a
- Data providers who reuse identifiers from other sources should not be penalised, i.e. their data sources should not lose any rank value.
- A data provider who simply links to other important identifiers (requiring no external effort) should not gain any rank from doing so. Using only the node-link graph without taking into account the source (e.g. [4]) makes the ranking method receptive for spam: by adding a triple pointing from a popular URI to a spam URI, the spam URI gains rank from the popular URI.
- We cannot make any assumptions of directionality of links between objects, since link direction is arbitrary (is u_a related to u_b or u_b related to u_a?). Thus we cannot use links occurring in the data graph as a vote.

3.3 Algorithm Outline

Our method for deriving a rank value for all data elements in a dataset consists of the following steps:

[4] `http://visinav.deri.org/`

1. Based on the occurrence of identifiers $u \in S$, construct the naming authority graph $S \times S$ that serves as input to a fixpoint calculation.
2. The naming authority graph is used to derive PageRank scores for the data sources S.
3. The source ranks are used to derive a rank value for both global identifiers $u \in U$ and data elements with local scope $b \in B$, $l \in L$.

4 Naming Authority

The main point of ranking on the web is to rate popular pages higher than unpopular ones. Indeed, PageRank[15] interprets hyperlinks to other pages as votes. A possible adaptation for structured data sources would rank popular data sources higher than unpopular ones. However data models such as RDF do not specify explicit links to other web sites or data sources. Therefore a straightforward adaptation of PageRank for structured data is not possible, since the notion of a hyperlink (interpreted as a vote for a particular page) is missing.

However, a closer examination of the data model leads to the following observation: a crucial feature of structured data sources is the use of global identifiers. Typically, these global identifiers – URIs in case of the web – have a specified syntax, and exploit naming mechanisms such as the domain name system.

Consider Dan's identifier `http://www.danbri.org/foaf.rdf#danbri`. Clearly the owner of the `danbri.org` domain can claim authority for creating this URI. Thus if the URI is used on `danbri.org` to denote Dan, the usage of the URI on other sites can be seen as a vote for the authority of the data source `danbri.org`.

To generalise this idea, one needs to define the notion of "naming authority" for identifiers. A naming authority is a data source with the power to define identifiers of a certain structure. Naming authority is an abstract term which could be applied to the provenance of a piece of information, be that a document, host, person, organisation or other entity. Data items which are denoted by unique identifiers may be reused by sources other than the naming authority.

We now define the general notion of naming authority:

Definition 1 (Naming Authority). *The naming authority of a global identifier $u \in U$ is the data source $s \in S$ which has the authority to mint the globally unique identifier u.*

4.1 Naming Authority for URIs

For naming authority in the RDF case, we assume a relation between $u \in U$ and $s \in S$ in the following way:

- if e contains a `#`, we assume the string before the `#` as the naming authority of the element, e.g. the naming authority for `http://danbri.org/foaf.rdf#-danbri` is `http://www.danbri.org/foaf.rdf`

– if e does not contain a #, we take the full element URI as the naming authority, e.g. the naming authority for `http://xmlns.com/foaf/0.1/maker` is `http://xmlns.com/foaf/0.1/maker`

The Hypertext Transfer Protocol (HTTP)[5], which is used to retrieve content on the web, allows the redirection of URIs, possibly multiple times, resulting in redirect chains. Redirect chains are unidirectional, i.e. the redirect relation does not follow the logical properties of the equivalence relation. To derive the correct naming authority we have to take HTTP redirects into account. Thus, for each naming authority, we check if the naming authority URI is redirected to another URI. If there is a redirect, we assume the redirected URI as the naming authority. E.g. `http://xmlns.com/foaf/0.1/-maker` redirects to `http://xmlns.com/foaf/spec/`, hence the naming authority becomes `http://xmlns.com/foaf/spec/`. This is in-line with the Linked Data principles[6].

4.2 Deriving the Naming Authority Matrix

As a first step, we derive the naming authority graph from the input dataset. That is, we construct a graph encoding links between data sources based on the implicit connections via identifiers.

Definition 2 (Naming Authority Matrix). *Given a data source $s_i \in S$ we define the naming authority matrix A as a square matrix defined as:*

$$a_{i,j} = \begin{cases} 1 & \textit{if } s_i \textit{ uses identifiers for which } s_j \textit{ has naming authority} \\ 0 & \textit{otherwise} \end{cases}$$

A naming authority graph for the example in Figure 1 is shown in Figure 3. In this example we assume the naming authority on a pay-level domain (PLD) level as described in the following.

4.3 Pay-Level Domains

In a variation of our algorithm we use the notion of pay-level domains (PLDs) as defined in [13]. A top-level domain (TLD) is a domain one level below the root in the DNS tree and appears as the label after the final dot in a domain name (e.g., `.com`, `.ie`). A pay-level domain is a domain that must be paid for at a TLD registrar. PLDs may be one level below the corresponding TLD (e.g., `livejournal.com`), but there are many exceptions where they are lower in the hierarchy (e.g., `cam.ac.uk`).

Bharat and Henzinger [5] suggest using a host rather than a web page as the unit of voting power in their HITS-based [11] ranking approach. PLD-level granularity is preferable to domain or host-level granularity because some sites

[5] `http://www.ietf.org/rfc/rfc2616.txt`

[6] `http://www.w3.org/DesignIssues/LinkedData.html`

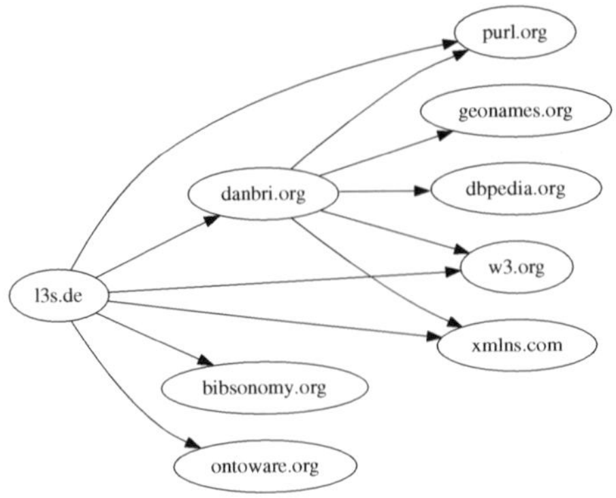

Fig. 3. Naming authority graph for data in Figure 1 on a pay-level domain granularity

like Livejournal assign subdomains to each user, which would result in large tightly-knit communities if domains were used as naming authorities. The use of PLDs reduces the size of the input graph to the PageRank calculation.

Previous work has performed PageRank on levels other than the page level, for example at the more coarse granularity of directories, hosts and domains [10], and at a finer granularity such as logical blocks of text [6] within a page.

4.4 Internal vs. External Links

Regardless of the level of granularity for naming authorities, there exist internal references occurring within a single naming authority, and external references occurring between different naming authorities. An internal reference is when a data source refers to an identifier which it has the authority to mint. An external reference is when a data source refers to an identifier which is under the authority of a different data source. Since our authority algorithm considers a reference as analogous to a vote for a source, it may be desirable to treat external references (from another source) differently to internal links (from the same source).

Similarly, in the HTML world, the HITS algorithm [11] considers only links which exist between different domains (which they call transverse links) and not links which occur within a single domain (which they call intrinsic links). The reason why only cross-domain links are taken into account in the HITS algorithm is that many intra-domain links are navigational and do not give much information about the authority of the target page. In variations of our algorithm we use either all links or take only external links into account.

5 Calculating Ranks

Having constructed the naming authority matrix, we now can compute scores for data sources, and in another step propagate data source scores to both global identifiers and identifiers and literals with a local scope which cannot be re-used in other sources. The algorithm can be implemented using a single scan over the dataset which derives both the naming authority matrix and a data structure that records the use of terms in data sources. As such, the method can be applied to streaming data. Subsequent calculations can then be carried out on the intermediate data structures without the use of the original data.

5.1 Calculating Source Ranks

For computing ranking scores we calculate PageRank over the naming authority graph. Essentially, we calculate the dominant eigenvector of the naming authority graph using the Power iteration while taking into account a damping factor.

In the input graph there may be sources which have no outlinks, referred to by the inventors of PageRank as dangling nodes [15]. The rank of these dangling nodes is split and distributed evenly across all remaining nodes. Conversely, there might be sources which have no inlinks, in the case where nobody uses the source's identifier, or identifiers the source speaks authoritatively for; these sources only receive the damping factor plus the rank of the dangling nodes.

5.2 Calculating Identifier Ranks

Based on the rank values for the data sources, we now calculate the ranks for individual identifiers. The rank value of the individual identifier $u \in U$ depends on the rank values of the data sources $s \in S$ where the identifier occurs. The identifier rank of a global identifier $u \in U$ is defined as the sum of the ranks of the sources $s \in S$ in which u occurs.

$$identifierrank(u) = \sum_{s \in \{s|u \in s; s \in S\}} sourcerank(s) \tag{1}$$

The identifier rank of local identifiers $b \in B$ and $l \in L$ are defined as the source rank of the source in which b or l occurs.

6 Experiments and Evaluation

In the following we evaluate our method on two real-world web datasets. We first introduce the datasets (one small and one large), then present runtime performance results, and finally present and analyse results of a quality evaluation of several variants of our method.

6.1 Datasets

We constructed two datasets:

- a small-scale RDF dataset crawled from a seed URI[7]. All unique URIs were extracted and their content downloaded in seven crawling rounds. The uncompressed dataset occupies 622MB of disk space.
- a large-scale RDF dataset derived from the 2008 Billion Triple Challenge datasets[8]. From these seed sets (around 450m RDF statements) we extracted

[7] `http://www.w3.org/People/Berners-Lee/card`

[8] `http://challenge.semanticweb.org/`; we used the Falcon, Swoogle, Watson, SWSE-1, SWSE-2 and DBpedia datasets.

all unique URIs and downloaded their contents during June and July 2008. The uncompressed dataset occupies 160GB of disk space.

Table 1 lists the properties of the datasets. In our experiments we followed one redirect, however, it is possible to take longer redirect chains into account.

Table 1. Dataset properties

Symbol	Small Dataset	Large Dataset
S	14k	6.5m
U	500k	74.3m
tuple	2.5m	1.1bn
Redirects	77k	4.5m

6.2 Evaluation Variants

The experiments were carried out on five different algorithms which are enumerated in in Table 2. The methods evaluated include four variants of our algorithm which differ according to the level of the naming authority (URI or PLD), and the references which we took into consideration for the authority calculation (all references, or external references only). We compared our method to a naive version of PageRank operating directly on the node-link graph without taking sources into account. We did not compare to ObjectRank [4] because ObjectRank requires manual assignment of weights to each of the thousands of properties in the dataset, which is infeasible.

On the small dataset, we compared performance of all the methods listed in Table 2, and carried out a quality evaluation on the results. On the large dataset, we conducted a scale-up experiment on the EU variant to demonstrate the scalability of our algorithm. We performed all experiments on a quad-core Xeon 2.33GHz machine with 8GB of main memory and a 500GB SATA drive.

Table 2. Ranking methods evaluated

Method	Description
AU	All links contribute authority URI-level naming authorities
EU	External links exclusively contribute authority URI-level naming authorities
AP	All links contribute authority PLD-level naming authorities
EP	External links exclusively contribute authority PLD-level naming authorities
PR	PageRank over the object graph

The implementation assures the uniqueness of links in the naming authority graph via sorting, and aggregates rank scores by writing the rank fractions to a file, sorting the file to group common identifiers together, and summing up the values via file scan. For the PageRank calculation, we fixed the number of iterations to ten rather than having the method converge in specified error bounds, which means we have to maintain only the current version of the rank vector rather than also maintaining the version from the previous iteration for comparison.

6.3 Performance Evaluation

The runtime of the five tested variants on the small dataset is plotted in Figure 4. Each processing step is plotted separately. The relatively large amounts of time spent on the AP and EP variants is due to the reduction of full URIs to pay-level domains. Given the coarser granularity of pay-level domains, the derived naming authority graph is quite small and thus accounts for the short running time of the PageRank calculation. In all approaches, a significant time is spent on the pre-processing of the data, processing steps which could be either carried out in-memory for small datasets or could be distributed across machines if required.

6.4 Scale-Up Experiment

We also conducted a scale-up experiment in which the algorithm performed ranking on a dataset with 1.1bn statements. In the scale-up experiment we used the Linux command `sort` with 6GB of main memory for sorting and uniquing instead of our standard Java implementation. The runtime of each processing step is listed in Table 3.

6.5 Quality Evaluation

In the information retrieval community, there are clearly defined processes and well-established metrics for evaluating how well an system performs in meeting

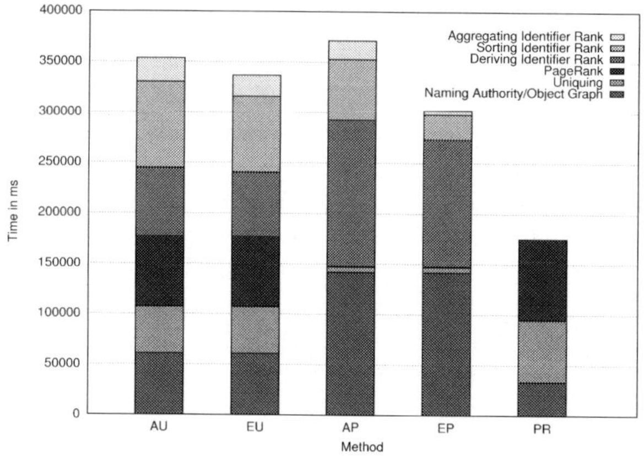

Fig. 4. Runtime of algorithm variations

Table 3. Runtime of processing steps for the large dataset

Processing Step	Duration
Deriving Naming Authority Graph	55h36m22s
Uniquing Naming Authority Graph	40h17m52s
PageRank Calculation	12h30m24s
Deriving Identifier Ranks	39h7m13s
Sorting Identifier Ranks	14h36m27s
Aggregating Identifier Ranks	1h56m43s

the information requirements of its users. Standard test collections consisting of documents, queries, and relevance judgements are available and are widely used. Search over Semantic Web data is a relatively new area however, and equivalent labelled collections do not yet exist. Therefore, given the lack of a labelled dataset, we use an alternative approach to quality assessment.

To compare the quality of the methods, we conducted a study in which we asked participants to manually rate results of queries for each algorithm. For every query, we presented the evaluators with five different ranked lists, each corresponding to one of the ranking methods. The result lists consisted of the top ten results returned by the respective method.

Table 4. Scenarios used for evaluation. N is the number of evaluators.

Scenario	Request	N
S1	Query for all persons in the dataset	11
S2	Keyword search: evaluator's own name	11
S3	Keyword search: "Tim Berners-Lee"	15
S4	Keyword search: "Dan Brickley"	15
S5	Keyword search: "John Breslin"	15

The evaluators were asked to order these lists from 1 to 5, according to which lists they deemed to represent the best results for each query. Our analysis covered the five scenarios listed in Table 4. For each scenario there were between 11 and 15 evaluators. Scenarios 1 and 2 were evaluated by the participants of a Semantic Web related workshop held in November 2008. During this evaluation, we presented each participant with results to a general query (S1) and with results of a query for their own name (S2), which was possible since all eleven participants have FOAF files and thus satisfactory data coverage. Scenarios 3 - 5 were evaluated by Ph.D. students and post-docs in the university. These final three scenarios were queries for people with whom the evaluators are familiar.

For each scenario we plotted the mean rank assigned by evaluators for each method. Error bars represent one standard deviation. We determine significance of the results using the Wilcoxon test with $p < .05$.

Figure 5 shows the average ranks which resulted for S1, a query for all people in the dataset. For this query, the top ten results of the methods AU and EU were identical. The differences between methods in this table are all statistically significant. The evaluators were almost unanimous in ranking this particular query which is why for three of the points there is no standard deviation marked.

Figure 6 shows the average ranks which were assigned by evaluators to a query for their own name (S2). In this table, the differences between each variant of our method and the PageRank method are all statistically significant. However the differences between the variants of our method are not statistically significant.

Figures 7, 8 and 9 show the average ranks which resulted for queries for three people involved in the Semantic Web community - scenarios S3, S4 and S5. The differences between each variant of our method and the PageRank method are statistically significant for all queries, with one exception (scenario S4, methods AP and OR). For S5, every evaluator rated PR as the worst method so for the corresponding point there is no standard deviation marked. The differences between the variants of our method are generally not statistically significant.

For scenarios 2 - 5 the average ranks follow similar patterns, showing that the evaluator's assessments of the methods were consistent over different queries.

We can conclude from the quality evaluation that our algorithm gives significantly better results than simply implementing PageRank on the object graph.

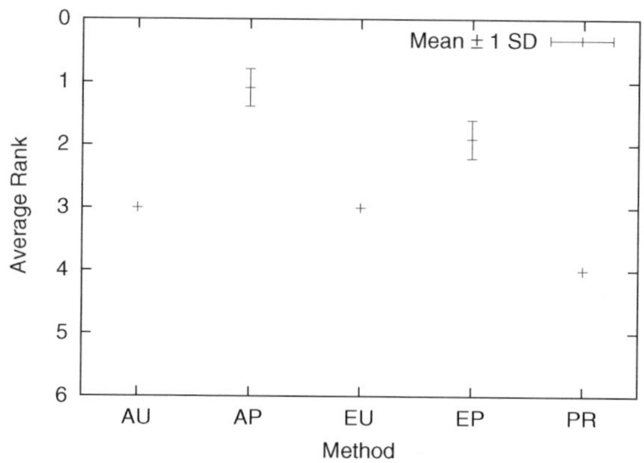

Fig. 5. Average ranks for scenario S1: Query for all persons in the dataset

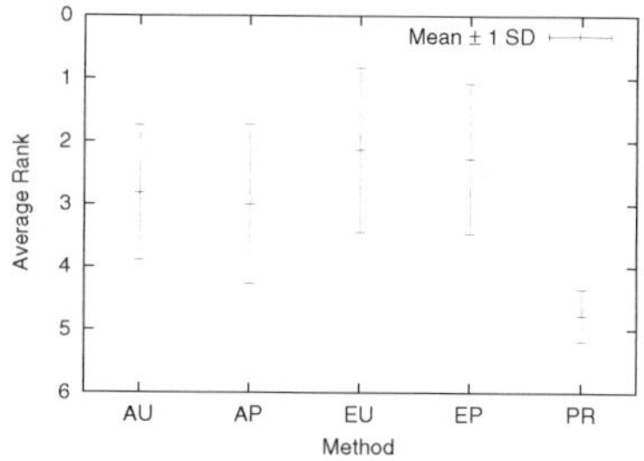

Fig. 6. Average ranks for scenario S2: Keyword search for the evaluator's own name

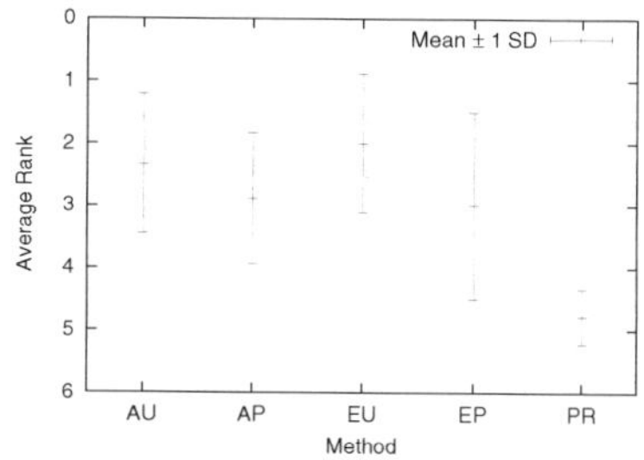

Fig. 7. Average ranks for scenario S3: Keyword search for "Tim Berners-Lee"

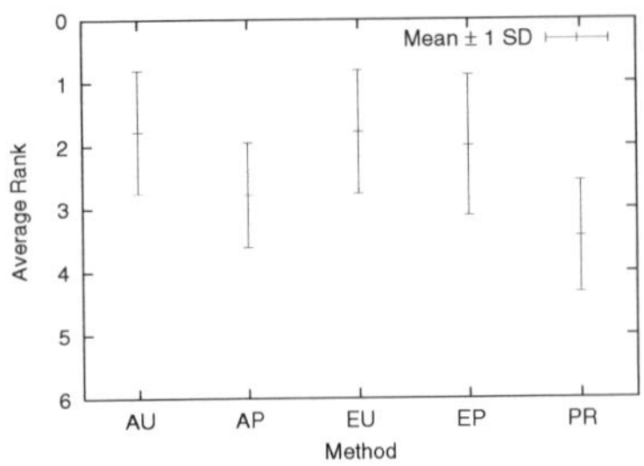

Fig. 8. Average ranks for scenario S4: Keyword search for "Dan Brickley"

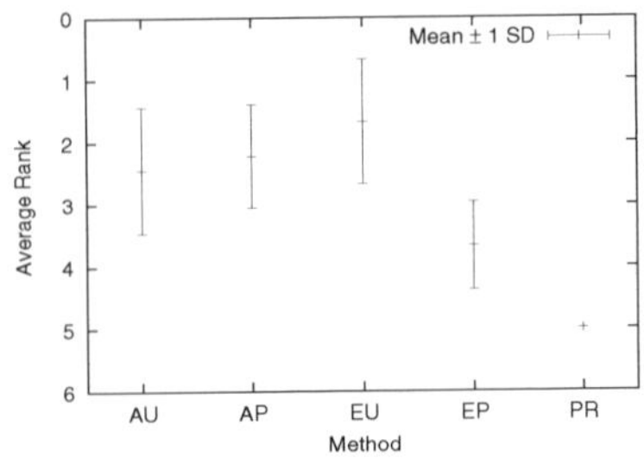

Fig. 9. Average ranks for scenario S5: Keyword search for "John Breslin"

We cannot determine the best variant of our algorithm with statistical significance. However with the exception of the query for all people in the dataset, the best method is always EU (where naming authorities are URIs, and only external links contribute to authority).

7 Related Work

Citation-based algorithms have been investigated in sociology in the area of social network analysis [16], which states as unresolved issue the mismatch between two-dimensional graph theory and multi-dimensional social networks. We have extended links-based connectivity algorithms such as PageRank [15] and HITS [11] to operate on data graphs. PageRank scales very well but only operates on two-dimensional matrices, the graph derived from the hyperlink structure.

An extension to PageRank and HITS to the multi-dimensional case is TOPHITS [12], a ranking procedure rooted in multilinear algebra which encodes the hypertext link structure (including link labels) as a three-way tensor. Rather, our approach operates on a general model for semistructured data, and is centred around the notion of trustworthiness of data sources.

ObjectRank [4] is an approach to rank a directed labelled graph using PageRank. The work includes a concept called authority transfer schema graphs, which defines weightings for the transfer of propagation through different types of links. ObjectRank relies on user input to weight the connections between nodes to describe their semantic weight, so that the three-way representation can be collapsed into a two-way matrix, on which a PageRank-style algorithm is applied. Our approach does not require any manual input, which is not feasible given the scale and heterogeneity of the input. In addition, omitting the provenance of data as in ObjectRank opens up the method to abuse - anyone could maliciously link to their own identifiers from well-known, highly ranked identifiers and therefore gain reputation by association. Using our notion of naming authority, reusing popular identifiers only results in a propagation of reputation from the containing sources to the popular source. As an added benefit, taking into account the naming authority results in a much smaller graph for PageRank calculation.

ReConRank [8] applies a PageRank-type algorithm to a graph which unifies the documents and resources in a dataset. The method generates scores for the documents and entities in a collection, but not for the properties. ReConRank does take data provenance into account, however because it simultaneously operates on the object graph, it is still susceptible to spamming.

SemRank [3] ranks relations and paths on Semantic Web data using information-theoretic measures. In contrast, we assign a rank value to all identifiers occurring in the data sources, based on a fixpoint calculation on the naming authority graph.

Swoogle [7] ranks documents using the OntoRank method, a variation on PageRank which iteratively calculates ranks for documents based on references to terms (classes and properties) defined in other documents. We extend the method described in [7] in several important ways: we generalise the notion of term use to naming authority which establishes a connection between identifier and source; we include the PLD abstraction layer which has been found to be advantageous for ranking in the web environment; and we extend our ranking scheme to not only cover vocabulary terms but instance identifiers as well, which is important in our Linked Data browser use-case scenario.

The notion of naming authority is related to that of authoritative sources as considered by the SAOR reasoning system [9]. SAOR uses authoritative sources to determine whether a source has authority to extend a class or property, while we use naming authority to rank sources and identifiers.

AKTiveRank [1] is a system for ranking ontologies based on how well they cover specified search terms. AKTiveRank combines the results of multiple analytic methods to rank each ontology. Individual instances and vocabulary terms are not ranked. Ontocopi [2] provides a way of locating instances in a knowledge base which are most closely related to a target instance. The Ontocopi tool uses a spreading activation algorithm and allows both manual and automatic tuning. However the source of data is not taken into consideration. Similarly, the SemSearch system [14] ranks entities according to how well they match the user query but does not consider the source of data.

8 Conclusion

Ranking provides an important mechanism to prioritise data elements and assuage the noise inherent in datasets which have been aggregated from disparate sources or have been created in a decentralised way. We have demonstrated a set of scalable algorithms for ranking over a general model of structured data collected from an open, distributed environment, based on the notion of naming authority. We adapted the general model to the case of RDF, taking the intricacies of RDF data from the web into account.

In comparison to using plain PageRank on a node-link graph representation of RDF, our methods exhibit similar runtime properties while improving on the quality of the calculated rankings. Contrary to methods which require manual input of a domain expert to specify schema weights, our method derives rankings for all identifiers in the dataset automatically.

We foresee our method having applications in search, query processing, reasoning, and user interfaces over integrated datasets from a large number of sources, an environment where assessing trustworthiness of sources and prioritising data items without a priori schema knowledge is vital.

References

1. Alani, H., Brewster, C., Shadbolt, N.: Ranking ontologies with AKTiveRank. In: Cruz, I., Decker, S., Allemang, D., Preist, C., Schwabe, D., Mika, P., Uschold, M., Aroyo, L.M. (eds.) ISWC 2006. LNCS, vol. 4273, pp. 1–15. Springer, Heidelberg (2006)
2. Alani, H., Dasmahapatra, S., O'Hara, K., Shadbolt, N.: Identifying communities of practice through ontology network analysis. IEEE Intelligent Systems 18(2), 18–25 (2003)
3. Anyanwu, K., Maduko, A., Sheth, A.: Semrank: ranking complex relationship search results on the semantic web. In: 14th International Conference on World Wide Web, pp. 117–127 (2005)
4. Balmin, A., Hristidis, V., Papakonstantinou, Y.: Objectrank: authority-based keyword search in databases. In: Proceedings of the 13th International Conference on Very Large Data Bases, pp. 564–575 (2004)
5. Bharat, K., Henzinger, M.R.: Improved algorithms for topic distillation in a hyperlinked environment. In: 21st International ACM SIGIR Conference on Research and Development in Information Retrieval, pp. 104–111 (1998)
6. Cai, D., He, X., Wen, J.R., Ma, W.Y.: Block-level link analysis. In: 27th International ACM SIGIR Conference on Research and Development in Information Retrieval, pp. 440–447 (2004)
7. Ding, L., Pan, R., Finin, T., Joshi, A., Peng, Y., Kolari, P.: Finding and ranking knowledge on the semantic web. In: Gil, Y., Motta, E., Benjamins, V.R., Musen, M.A. (eds.) ISWC 2005. LNCS, vol. 3729, pp. 156–170. Springer, Heidelberg (2005)
8. Hogan, A., Harth, A., Decker, S.: ReConRank: A scalable ranking method for semantic web data with context. In: 2nd Workshop on Scalable Semantic Web Knowledge Base Systems (2006)
9. Hogan, A., Harth, A., Polleres, A.: SAOR: Authoritative Reasoning for the Web. In: 3rd Asian Semantic Web Conference, pp. 76–90 (2008)
10. Jiang, X.-M., Xue, G.-R., Song, W.-G., Zeng, H.-J., Chen, Z., Ma, W.-Y.: Exploiting PageRank at Different Block Level . In: 5th International Conference on Web Information Systems, pp. 241–252 (2004)
11. Kleinberg, J.M.: Authoritative sources in a hyperlinked environment. Journal of the ACM 46(5), 604–632 (1999)
12. Kolda, T.G., Bader, B.W., Kenny, J.P.: Higher-order web link analysis using multilinear algebra. In: 5th IEEE International Conference on Data Mining, pp. 242–249 (2005)
13. Lee, H.-T., Leonard, D., Wang, X., Loguinov, D.: Irlbot: scaling to 6 billion pages and beyond. In: 17th International Conference on World Wide Web, pp. 427–436 (2008)
14. Lei, Y., Uren, V., Motta, E.: Semsearch: A search engine for the semantic web. In: 14th International Conference on Knowledge Engineering and Knowledge Management, pp. 238–245 (2006)
15. Page, L., Brin, S., Motwani, R., Winograd, T.: The PageRank citation ranking: Bringing order to the web. Technical report, Stanford Digital Library Technologies Project (1998)
16. Scott, J.: Trend report: Social network analysis. Sociology 22(1), 109–127 (1988)

Executing SPARQL Queries over the Web of Linked Data

Olaf Hartig[1], Christian Bizer[2], and Johann-Christoph Freytag[1]

[1] Humboldt-Universität zu Berlin
lastname@informatik.hu-berlin.de
[2] Freie Universität Berlin
firstname.lastname@fu-berlin.de

Abstract. The Web of Linked Data forms a single, globally distributed dataspace. Due to the openness of this dataspace, it is not possible to know in advance all data sources that might be relevant for query answering. This openness poses a new challenge that is not addressed by traditional research on federated query processing. In this paper we present an approach to execute SPARQL queries over the Web of Linked Data. The main idea of our approach is to discover data that might be relevant for answering a query during the query execution itself. This discovery is driven by following RDF links between data sources based on URIs in the query and in partial results. The URIs are resolved over the HTTP protocol into RDF data which is continuously added to the queried dataset. This paper describes concepts and algorithms to implement our approach using an iterator-based pipeline. We introduce a formalization of the pipelining approach and show that classical iterators may cause blocking due to the latency of HTTP requests. To avoid blocking, we propose an extension of the iterator paradigm. The evaluation of our approach shows its strengths as well as the still existing challenges.

1 Introduction

An increasing amount of data is published on the Web according to the Linked Data principles [1,2]. Basically, these principles require the identification of entities with URI references that can be resolved over the HTTP protocol into RDF data that describes the identified entity. These descriptions can include RDF links pointing at other data sources. RDF links take the form of RDF triples, where the subject of the triple is a URI reference in the namespace of one data source, while the object is a URI reference in the namespace of the other. The Web of Linked Data that is emerging by connecting data from different sources via RDF links can be understood as a single, globally distributed dataspace [3].

Querying this dataspace opens possibilities not conceivable before: Data from different data sources can be aggregated; fragmentary information from multiple sources can be integrated to achieve a more complete view. However, evaluating queries over the Web of Linked Data also poses new challenges that do not arise

A. Bernstein et al. (Eds.): ISWC 2009, LNCS 5823, pp. 293–309, 2009.

```
1  SELECT DISTINCT ?author ?phone WHERE {
2    <http://data.semanticweb.org/conference/eswc/2009/proceedings>
3                                                      swc:hasPart ?pub .
4    ?pub swc:hasTopic ?topic .
5    ?topic rdfs:label ?topicLabel .
6    FILTER regex( str(?topicLabel), "ontology engineering", "i" ) .
7
8    ?pub swrc:author ?author .
9    {?author owl:sameAs ?authAlt} UNION {?authAlt owl:sameAs ?author}
10
11   ?authAlt foaf:phone ?phone
12 }
```

Fig. 1. SPARQL query which asks for the phone numbers of people who authored an ontology engineering related paper at ESWC'09 (prefix declarations omitted)

in traditional federated query processing: Due to the openness of the dataspace, it is not possible to know all data sources that might be relevant for answering a query in advance.

Consider, for instance, the SPARQL query [4] in Figure 1 which asks for the phone number of people who authored an ontology engineering related paper at the European Semantic Web Conference 2009 (ESWC'09). This query cannot be answered from a single dataset but requires data from a large number of sources on the Web. For instance, the list of papers and their topics (cf. lines 2 to 4) is part of the Semantic Web Conference Corpus[1]; the names of the paper topics (cf. line 5) are provided by the sources authoritative for the URIs used to represent the topics; the phone numbers (cf. line 11) are provided by the authors. Hence, this kind of queries can only be answered using a method to execute queries without knowing the sources that contribute to the query result in advance. In this paper we present such a method.

The main idea of our approach is the asynchronous traversal of RDF links to discover data that might be relevant for a query during the query execution itself. Hence, the query engine executes each query over a growing set of potentially relevant data retrieved from the Web. Notice, in contrast to federated query processing [5] – which presumes each data source provides a query service – we do not distribute parts of the query evaluation. Instead, we only require the data sources to publish data following the Linked Data principles. This approach enables the execution of queries without knowing the sources that contribute to the query result in advance. Our main contributions are:

- an approach to execute SPARQL queries over the Web of Linked Data,
- a formal description of the realization of our approach with an iterator-based pipeline that enables an efficient query execution, and
- an extension of the iterator paradigm that avoids blocking of the query execution caused by waiting for data from the Web.

This paper is structured as follows. First, Section 2 gives an overview of our approach. In Section 3 we introduce an iterator-based evaluation of queries and

[1] http://data.semanticweb.org

present a formalism to describe this kind of evaluation. Section 4 discusses the application of an iterator-based query evaluation to our query execution approach and presents a strategy to execute these queries more efficiently. Even with these strategies, waiting for data from the Web may cause delays in query execution. Thus, in Section 5 we introduce an extension to the iterator paradigm that avoids blocking caused by these delays. An evaluation of our approach and the concepts to implement it is given in Section 6. Finally, we review related work in Section 7 and conclude in Section 8.

2 Overview of the Query Execution Approach

This section gives an informal overview of the proposed approach to execute SPARQL queries over the Web of Linked Data. SPARQL, the query language for RDF data [4], is based on graph patterns and subgraph matching. The basic building block from which more complex SPARQL query patterns are constructed is a basic graph pattern (BGP). A BGP is a set of triple patterns which are RDF triples that may contain query variables at the subject, predicate, and object position. During query evaluation solutions that bind values to the variables are determined.

To query the Web of Linked Data, we propose to intertwine query execution with the traversal of RDF links to discover data that might be relevant to answer the query. Using the data retrieved from looking up the URIs in a query as a starting point we evaluate parts of the query. The intermediate solutions resulting from this partial evaluation usually contain further URIs. These URIs link to additional data which may provide further solutions for the same or for other parts of the query. To determine results for the whole query we alternately evaluate query parts and dereference URIs. Hence, during query evaluation we continuously augment the queried dataset with potentially relevant data from the Web. The discovery of this data is driven by the URIs in intermediate results.

Example 1. The evaluation of the query in Figure 1 may start with RDF data retrieved from the Semantic Web Conference Corpus by dereferencing the URI identifying the ESWC'09 proceedings. This data contains a set of RDF triples that match the triple pattern in lines 2 and 3 of the query. The query engine generates a set of intermediate solutions from this set. Each of these solutions binds query variable `?pub` to the URI representing one of the papers in the ESWC'09 proceedings. Dereferencing these URIs yields RDF data about the papers including the topics of the publications. Hence, in this newly retrieved data the query engine finds matching triples for the pattern at line 4 with the given `?pub` binding. Based on these matches existing intermediate solutions can be augmented with bindings for variable `?topic`. Since the topics are also resources represented by URIs additional data will be added to the queried dataset. The query engine proceeds with the outlined strategy in order to determine solutions that are results of the whole query.

To consider all data that is available by traversing RDF links from matching triples our approach uses the following heuristic. Before we evaluate a triple pattern in order to find matching triples in the local dataset we ensure that the local dataset contains at least all data retrievable from dereferencing all URIs that are part of the triple pattern. For instance, the evaluation of the triple pattern in line 4 of our sample query using the intermediate solutions from the triple pattern in lines 2 and 3 comprises the evaluation of multiple triple pattern, actually – one for each `?pub` binding. As discussed before, we dereference each URI bound to variable `?pub` before we evaluate the corresponding triple pattern. Notice, in addition to the data retrieved by dereferencing the URIs in a query as an initial seed it is also possible to load further RDF data in the local dataset before executing the query. This possibility allows to explicitly ensure considering data that must not be missed during query execution. Furthermore, reusing the same local dataset for different queries may yield more complete results since an already filled local dataset may contain data relevant to a query that would not be discoverable by executing the query itself. Such a shared dataset, however, requires the application of caching strategies which identify formerly retrieved data that might be stale and that has to be requested again.

Due to the openness and the widely distributed nature of the Web we cannot assume to find all data that is relevant to answer a query with our approach. Hence, we should never expect complete results. The degree of completeness depends on the structure of the network of RDF links as well as on the number of links. In a Web sparsely populated with links chances are slight to discover relevant data. Nonetheless, we are convinced the Web of Linked Data will rapidly grow in the coming years and so will the number and density of links. Further limitations of our approach are i) the need for initial URIs in the queries to start the link traversal, ii) infinite link discovery, iii) the retrieval of unforeseeable large RDF graphs from the Web, iv) URI dereferencing that takes unexpectedly long, v) Linked Data servers that put restrictions to clients such as serving only a limited number of requests per second, and vi) the possibility to overload a server. Some of these issues might be addressed with user-configurable options such as seed URIs, timeouts, and limits on traversal depth and filesizes. However, further investigation is required to find suitable defaults for these options and to address the listed issues in general.

3 Pipelining-Based Basic Graph Pattern Matching

We propose to implement our query execution approach using an iterator-based pipeline that enables an efficient, parallelized execution of queries. Introducing the pipelined evaluation of BGP queries over a dataset that is continuously augmented with potentially relevant data from the Web requires an understanding of the static case. Therefore, this section presents a formalism to describe a pipelining-based evaluation of BGPs over a fixed set of RDF graphs. While the SPARQL specification introduces different types of graph patterns we focus on BGPs, the fundamental building block, in this paper; the application of the presented concepts to more complex query patterns is subject to further research.

3.1 Solutions in SPARQL Query Execution

A formal description of our approach requires an explicit definition of the notion of a solution. Solutions in SPARQL are defined in the context of BGP matching where each solution basically represents a matching subgraph in the queried RDF graph. To describe these solutions the SPARQL specification [4] introduces a *solution mapping* that binds query variables to RDF terms such as URIs and literals. Solution mappings can be understood as a set of variable-term-pairs where no two pairs contain the same variable. The application of such a solution mapping μ to a BGP b, denoted with $\mu[b]$, implies replacing each variable in the BGP by the RDF term it is bound to in the mapping; unbound variables must not be replaced. Based on these mappings the specification defines solutions of BGPs for RDF graphs. Since our approach is based on a local dataset that contains multiple RDF graphs – one from each dereferenced URI – we slightly adjust this definition and introduce solutions of BGPs for sets of RDF graphs:

Definition 1. *Let b be a BGP; let $\mathcal{G}$ be a set of RDF graphs. The solution mapping μ is a* **solution** *for b in $\mathcal{G}$ if i) $\mu[b]$ is a subgraph of $\bigcup_{G \in \mathcal{G}} G$ and ii) μ does not map variables that are not in b.*

3.2 An Algorithm to Evaluate Basic Graph Patterns

During query processing a query is represented by a tree of logical operators. From this operator tree the query engine generates a query execution plan that implements the logical operators by physical operations. A well established approach to realize query plans is *pipelining* in which each solution produced by one operation is passed directly to the operation that uses it [6]. The main advantage of pipelining is the rather small amount of memory that is needed compared to approaches that completely materialize intermediate results. Pipelining in query engines is typically realized by a tree of *iterators* that implement the physical operations [7]. An iterator is a group of three functions: `Open`, `GetNext`, and `Close`. `Open` initializes the data structures needed to perform the operation; `GetNext` returns the next result of the operation; and `Close` ends the iteration and releases allocated resources. An iterator allows a consumer to get the results of an operation separately, one at a time. In a tree of iterators the `GetNext` functions of an iterator typically call `GetNext` on the child(ren) of the iterator. Hence, a tree of iterators calculates solutions in a pull fashion.

Many in-memory RDF stores realize the evaluation of BGP queries by a tree of iterators as follows: Each iterator is responsible for a single triple pattern of the BGP $\{tp_1, \ldots, tp_n\}$. The iterators are chained together such that the iterator I_i which is responsible for triple pattern tp_i is the argument of the iterator I_{i+1} responsible for tp_{i+1}. Each iterator returns solution mappings that are solutions for the set of triple patterns assigned to it and to its predecessors. For instance, the iterator I_i returns solutions for $\{tp_1, \ldots, tp_i\}$ $(i \leq n)$. To determine these solutions the iterators basically execute the following three steps repeatedly: first, they consume the solution mappings from their direct predecessor; second, they apply these input mappings to their triple pattern; and, third, they try to

```
1  FUNCTION Open
2      LET G := the queried set of RDF graphs;
3      LET I_{i-1} := the iterator responsible for tp_{i-1};
4      CALL I_{i-1}.Open;
5
6      LET μ_cur := NotFound;
7      LET I_find := an iterator over an empty set of RDF triples;
8
9  FUNCTION GetNext
10     LET t := I_find.GetNext;
11     WHILE t = NotFound DO
12     {
13       LET μ_cur := I_{i-1}.GetNext;
14       IF μ_cur = NotFound THEN
15         RETURN NotFound;
16
17       LET I_find := an iterator over a set of all triples that match μ_cur[{tp_i}] in G;
18       LET t := I_find.GetNext;
19     }
20     LET μ' := the solution for μ_cur[{tp_i}] in G that corresponds to t;
21     RETURN μ_cur ∪ μ';
22
23 FUNCTION Close
24     CALL I_{i-1}.Close;
```

Fig. 2. Pseudo code notation of the `Open`, `GetNext`, and `Close` functions for an iterator I_i that evaluates triple pattern tp_i

find triples in the queried RDF data that match the triple patterns resulting from the applications of the input mappings. Figure 2 illustrates the corresponding algorithm executed by the `GetNext` function of the iterators. Notice, the third step is realized with another type of iterator that returns the matching triples. We do not discuss this *helper iterator* in detail because its realization is not important for the concepts presented in this paper. Hence, in the remainder we simply assume the helper iterator iterates over a set of all RDF triples that match a triple pattern in a set of RDF graphs.

3.3 A Formalization of Iterator-Based Pipelining

The set of solutions provided by an iterator I_i can be divided in subsets where each subset corresponds to one of the solutions consumed from the direct predecessor; i.e. each of these subsets contains all solutions that have been determined based on the same input solution. We denote these subsets with $Succ^{\mathcal{G}}(\mu, i)$.

$$Succ^{\mathcal{G}}(\mu, i) = \{\mu \cup \mu' \mid \mu' \text{ is a solution for } \mu[\{tp_i\}] \text{ in } \mathcal{G}\} \tag{1}$$

Notice, the μ's in Equation (1) correspond to the μ' in the algorithm (cf. line 20 in Figure 2). There is no μ' in Equation (1) that binds a variable which is already bound in μ because the application of μ to $\{tp_i\}$ yields $\{tp'_i\}$ where tp'_i does not contain a variable bound in μ. Hence, each μ' merely adds new bindings for variables not considered in μ. For this reason it holds, if μ is a solution for $\{tp_1, \dots, tp_{i-1}\}$ then each $\mu^* \in Succ^{\mathcal{G}}(\mu, i)$ is a solution for $\{tp_1, \dots, tp_i\}$.

With $\Omega_i^{\mathcal{G}}$ we denote all solutions determined by the ith iterator. It holds

$$\Omega_i^{\mathcal{G}} = \begin{cases} Succ^{\mathcal{G}}(\mu_0, 1) & ; \text{if } i = 1 \\ \bigcup_{\mu \in \Omega_{i-1}^{\mathcal{G}}} Succ^{\mathcal{G}}(\mu, i) & ; \text{else} \end{cases} \tag{2}$$

where $\mu_0 = \varnothing$ is the empty solution mapping consumed by the first iterator.

Proposition 1. *$\Omega_n^{\mathcal{G}}$ is the result for BGP $\{tp_1, \ldots, tp_n\}$ from RDF graph set $\mathcal{G}$.*

Proof. Each $\mu \in \Omega_1^{\mathcal{G}}$ is a solution from $\mathcal{G}$ for $\{tp_1\}$ because

$$\begin{aligned} \Omega_1^{\mathcal{G}} = Succ^{\mathcal{G}}(\mu_0, 1) &= \{\mu_0 \cup \mu' \mid \mu' \text{ is a solution for } \mu_0[\{tp_1\}] \text{ from } \mathcal{G}\} \\ &= \{\mu' \mid \mu' \text{ is a solution for } \{tp_1\} \text{ from } \mathcal{G}\} \end{aligned}$$

Let $i > 1$ and let $\Omega_{i-1}^{\mathcal{G}}$ all solutions from $\mathcal{G}$ for $\{tp_1, \ldots, tp_{i-1}\}$. Due to Equation (1) it holds for each $\mu \in \Omega_{i-1}^{\mathcal{G}}$ that each $\mu_i \in Succ^{\mathcal{G}}(\mu, i)$ is a solution from $\mathcal{G}$ for $\{tp_1, \ldots, tp_i\}$. Furthermore, $\Omega_i^{\mathcal{G}}$ is complete because it is the union of $Succ^{G}(\mu, i)$ for all possible $\mu \in \Omega_{i-1}^{\mathcal{G}}$. Hence, $\Omega_n^{\mathcal{G}}$ is the complete result from $\mathcal{G}$ for $\{tp_1, \ldots, tp_n\}$. □

4 Evaluating Basic Graph Patterns over the Web

In this section we formalize our approach to evaluate BGP queries over the Web of Linked Data and we introduce strategies to execute these queries more efficiently.

To query the Web of Linked Data we cannot evaluate BGP queries over a fixed set of RDF graphs as introduced in the previous section. Instead, the queried dataset grows during the evaluation since we continuously add further, potentially relevant data by following the heuristic outlined in Section 2. The heuristic is based on the assumption that RDF graphs retrieved by looking up the URIs in a triple pattern might contain triples that match the triple pattern. Hence, we require that the local dataset contains these RDF graphs before we evaluate the triple pattern. More formally, we have to guarantee the following requirement.

Requirement 1. *The calculation of $Succ^{\mathcal{G}}(\mu, i)$ requires that*

1. *$deref\big(subj(\mu[tp_i])\big) \in \mathcal{G}$ if $subj(\mu[tp_i])$ is a URI,*
2. *$deref\big(pred(\mu[tp_i])\big) \in \mathcal{G}$ if $pred(\mu[tp_i])$ is a URI, and*
3. *$deref\big(obj(\mu[tp_i])\big) \in \mathcal{G}$ if $obj(\mu[tp_i])$ is a URI*

where $\mu[tp]$ denotes the application of solution mapping μ to a triple pattern tp; $subj(tp)$, $pred(tp)$, and $obj(tp)$ denote the subject, predicate and object of a triple pattern tp, respectively; and $deref(u)$ represents the RDF graph that we retrieve by dereferencing the URI u.

```
1  FUNCTION GetNext
2      LET t := I_find.GetNext ;
3      WHILE t = NotFound DO
4      {
5          LET μ_cur := I_{i-1}.GetNext ;
6          IF μ_cur = NotFound THEN
7              RETURN NotFound;
8
9          CALL EnsureRequirement for μ_cur[{tp_i}] ;
10         LET I_find := an iterator over a set of all triples that match μ_cur[{tp_i}] in G ;
11         LET t := I_find.GetNext ;
12     }
13
14     LET μ' := the solution for μ_cur[{tp_i}] in G that corresponds to t ;
15     RETURN μ_cur ∪ μ' ;
```

Fig. 3. Pseudo code notation of the `GetNext` function for an iterator I_i that evaluates triple pattern tp_i over the Web of Linked Data

To guarantee Requirement 1 we adjust the `GetNext` function of our iterators as follows. Before the algorithm initializes the embedded helper iterator (cf. line 17 in Figure 2) it invokes a function called `EnsureRequirement`. This function checks the requirement and, if necessary, it dereferences URIs and waits until dereferencing has been finished. Figure 3 illustrates the adjusted `GetNext` function. Based on this adjustment the queried dataset $\mathcal{G}$ grows during the iterative calculation of $\Omega_n^{\mathcal{G}}$. The calculation of any solution subset $Succ^{\mathcal{G}}(\mu, i)$ uses the dataset $\mathcal{G}$ that is augmented with all RDF graphs retrieved for the calculation of previous subsets. However, each time the algorithm initializes I_{find} (cf. line 10 in Figure 3) it uses an isolated snapshot of $\mathcal{G}$ that cannot be augmented by other iterators. This isolation avoids conflicts and an endless query execution because once the calculation of any subset $Succ^{\mathcal{G}}(\mu, i)$ has been initiated later additions to $\mathcal{G}$ are ignored for that calculation. The downside of this isolation is that the completeness of a query result depends on the order by which the iterators for the patterns in a query are chained. The development of concepts to find an order that is optimal with respect to maximizing result completeness is subject to further research.

The dereferencing requests in function `EnsureRequirement` of the adjusted iterators should be implemented by asynchronous function calls such that multiple dereferencing tasks can be processed in parallel. However, waiting for the completion of the dereferencing tasks in function `EnsureRequirement` delays the execution of the `GetNext` function and, thus, slows down query execution times. It is possible to address this problem with the following prefetching strategy.

Instead of dereferencing each URI at the time when the corresponding RDF graph is required we suggest to initiate the dereferencing task as soon as the URI becomes part of a solution. Considering that dereferencing requests are implemented by asynchronous function calls the query engine can immediately proceed the evaluation while the dereferencing tasks are executed separately. Whenever a subsequent iterator requires the corresponding dereferencing result chances are high that the dereferencing task has already been completed.

```
1  FUNCTION GetNext
2      LET t := I_find.GetNext;
3      WHILE t = NotFound DO
4      {
5          LET μ_cur := I_{i-1}.GetNext;
6          IF μ_cur = NotFound THEN
7              RETURN NotFound;
8
9          CALL EnsureRequirement for μ_cur[{tp_i}];
10         LET I_find := an iterator over a set of all triples that match μ_cur[{tp_i}] in G;
11         LET t := I_find.GetNext;
12     }
13
14     LET μ' := the solution for μ_cur[{tp_i}] in G that corresponds to t;
15
16     FOR EACH (var, val) ∈ μ' DO
17         IF val is a URI AND deref(val) ∉ G THEN Request the retrieval of deref(val);
18
19     RETURN μ_cur ∪ μ';
```

Fig. 4. Pseudo code notation of the `GetNext` function for an iterator I_i that prefetches URIs during the evaluation of triple pattern tp_i

Figure 4 illustrates an adjusted `GetNext` function that realizes our URI prefetching strategy in lines 16 and 17. In Section 6.2 we analyze the impact of URI prefetching on query execution times.

5 Non-blocking Iterators

In this section we introduce an extension to the iterator paradigm. This extension prevents unnecessary long execution times caused by delays that cannot be avoided with URI prefetching.

URI prefetching as introduced in the previous section is an attempt to avoid delays during the execution of the `GetNext` function. However, this approach is not always sufficient as the following example demonstrates.

Example 2. Let tp_i = (?x,?p,?o) the ith triple pattern in a BGP $\{tp_1, \dots, tp_n\}$ where $1 < i \leq n$. Consider the iterator I_i which is responsible for tp_i is asked for the next solution by calling its `GetNext` function. Let the current helper iterator I_{find} of I_i be exhausted so that the algorithm enters the while loop (cf. line 3 in Figure 4) and requests the next input solution from the predecessor iterator I_{i-1} (cf. line 5). Let I_{i-1} return a solution mapping μ_{i-1} that binds variable ?x to URI *uri* where this binding has been determined by I_{i-1}; i.e. it holds (?x, *uri*) $\in \mu_{i-1}$ where $\mu_{i-1} \in Succ^{\mathcal{G}}(\mu, i-1) \subseteq \Omega^{\mathcal{G}}_{i-1}$ but (?x, *uri*) $\notin \mu$. Let $deref(uri) \notin \mathcal{G}$. In this case, I_{i-1} has requested the retrieval of $deref(uri)$ just before it has returned μ_{i-1} to I_i. Since I_i immediately calls `EnsureRequirement` for $\mu_{i-1}[\{tp_i\}]$ = (*uri*,?p,?o) it is very likely that the dereferencing of *uri* is still in progress. Hence, I_i has to wait before it can initialize the next helper iterator and return the next solution.

As can be seen from Example 2 the prefetching of URIs does not prevent delays in the `GetNext` function in general. Unfortunately, an iterator that waits during the execution of `GetNext` causes a blocking of the whole query execution because the subsequent iterators wait for the result of `GetNext` and the predecessor iterators are not requested to do anything either. This problem might be addressed with program parallelism and the use of asynchronous pipelines [8]. According to this approach all iterators work in parallel; each pair of connected iterators shares a buffering queue to which the providing iterator asynchronously adds its intermediate solutions; the consuming iterator dequeues these solutions to process them. However, the realization of this approach would require a major rewrite of existing query engines that are based on synchronous pipelines. For this reason we propose an extension to the iterator paradigm which is compatible with the commonly used synchronous pipelining approach presented in this paper.

The core idea of our extension is to enable iterators to temporarily reject a solution consumed from its predecessor. Given such a possibility an iterator that finds Requirement 1 is not fulfilled for an input solution could reject that solution and ask the predecessor for another solution. To enable this possibility we add a new function, called `Reject`, to the iterator paradigm and we slightly extend the semantics of the `GetNext` function. The new function `Reject` treats the result that has most recently been provided by the `GetNext` function as if this result has never been provided by `GetNext`. This means `Reject` takes the rejected result back and keeps it for later requests. The extended `GetNext` function either returns the next result of the operation performed by the iterator or it returns one of the rejected results. Once a formerly rejected result is not being rejected again it must not be kept any further. Notice, we allow `GetNext` to decide nondeterministically to return a newly calculated result or to return a formerly rejected result. This approach provides more flexibility for realizations of our extended iterator paradigm. Analogously, we do not prescribe which of the rejected results have to be returned. However, in most cases it would probably be unwise to immediately reoffer a recently rejected result.

Figure 5 illustrates an application of the extended iterator paradigm to the iterators that evaluate BGP queries over the Web of Linked Data. Notice, these iterators call a function `CheckRequirement` which is similar to the function `EnsureRequirement` introduced in Section 4. Both functions check Requirement 1 and, if necessary, request the dereferencing of URIs. However, in contrast to `EnsureRequirement` the function `CheckRequirement` does not wait until the requested data has been retrieved, but, it returns an indication that either Requirement 1 is fulfilled or that the retrieval of data has been requested. In the latter case the algorithm in function `GetNext` rejects the current input solution from the predecessor iterator (cf. line 39 in Figure 5). Hence, the extended iterator paradigm allows to temporarily reject input solutions. This possibility to postpone the processing of certain solutions has a significant impact on query execution times as our evaluation in Section 6.2 illustrates.

```
FUNCTION Open
      LET 𝒢 := the queried set of RDF graphs ;
      LET I_{i-1} := the iterator responsible for tp_{i-1} ;
      CALL I_{i-1}.Open ;

      LET μ_cur := NotFound ;
      LET I_find := an iterator over an empty set of RDF triples ;

      LET Ψ := an empty list ; // used to keep rejected solutions
      LET μ_last := NotFound ; // used to hold the most recently provided solutions


FUNCTION GetNext
      LET new := a randomly choosen element from {TRUE, FALSE} ;
      IF Ψ is not empty AND new = FALSE THEN
        LET μ_last := the first element in Ψ ;
        Remove μ_last from Ψ ;
        RETURN μ_last ;

      LET t := I_find.GetNext ;
      WHILE t = NotFound DO
      {
        LET μ_cur := I_{i-1}.GetNext ;
        IF μ_cur = NotFound THEN
        {
          IF Ψ is empty THEN
            LET μ_last := NotFound ;
            RETURN μ_last ;

          LET μ_last := the first element in Ψ ;
          Remove μ_last from Ψ ;
          RETURN μ_last ;
        }

        IF CheckRequirement for μ_cur[{tp_i}] returned TRUE THEN
          LET I_find := an iterator over a set of all triples that match μ_cur[{tp_i}] in 𝒢 ;
          LET t := I_find.GetNext ;
        ELSE
          CALL I_{i-1}.Reject ;
      }

      LET μ' := the solution for μ_cur[{tp_i}] in 𝒢 that corresponds to t ;

      FOR EACH (var, val) ∈ μ' DO
        IF val is a URI AND deref(val) ∉ 𝒢 THEN Request the retrieval of deref(val) ;

      LET μ_last := μ_cur ∪ μ' ;
      RETURN μ_last ;


FUNCTION Reject
      IF μ_last ≠ NotFound THEN Append μ_last to Ψ ;


FUNCTION Close
      CALL I_{i-1}.Close ;
```

Fig. 5. Pseudo code notation of the `Open`, `GetNext`, `Reject`, and `Close` functions for a non-blocking iterator I_i that evaluates triple pattern tp_i

6 Evaluation

As a proof-of-concept, we implemented our approach to query the Web of Linked Data in the Semantic Web Client Library[2] (SWClLib). This library is available as Free Software; it is portable to any major platform due to its implementation in Java. Based on SWClLib we evaluate our approach in this section; we present real-world use cases and we evaluate the iterator-based implementation in a controlled environment.

6.1 Real-World Examples

To demonstrate the feasibility of our approach we tested it with the following four queries that require data from multiple data sources to be answered:

$Q1$: Return phone numbers of authors of ontology engineering papers at ESWC09
$Q2$: What are the interests of the people Tim Berners-Lee knows?
$Q3$: What natural alternatives can be used instead of the drug "Varenicline"?
$Q4$: Return the cover images of soundtracks for movies directed by Peter Jackson.

Our demo page[3] provides the SPARQL representations of the test queries and it allows to execute these queries using SWClLib. In Table 1 we present average measures for these executions; the table illustrates the number of query results for each query, the number of RDF graphs retrieved during query execution, the number of servers from which the graphs have been retrieved, and the execution time. The first query, $Q1$, which is the sample query introduced in Section 1 (cf. Figure 1) has been answered using the Semantic Web Conference Corpus, DBpedia, and personal FOAF profiles. Answering query $Q2$ required data from Tim Berners-Lee as well as data from all the people he knows. Query $Q3$ was answered with data retrieved from the DrugBank database, the Diseasome dataset, and the Traditional Chinese Medicine dataset. The results for $Q4$ have been constructed with data from LinkedMDB and the MusicBrainz dataset.

In addition to the test queries we developed Researchers Map [9] to demonstrate that our approach is suitable for applications that consume Linked Data. Researchers Map is a simple mash-up that is solely based on the results of queries evaluated over the Web of Linked Data. The main feature of Researchers Map

Table 1. Statistics about the test queries

query	$Q1$	$Q2$	$Q3$	$Q4$
# of results	2	27	7	5
# of retrieved graphs	297	85	28	442
# of accessed servers	16	46	6	6
execution time	3min 47sec	1min 11sec	0min 46sec	1min 24sec

[2] http://www4.wiwiss.fu-berlin.de/bizer/ng4j/semwebclient/
[3] http://squin.informatik.hu-berlin.de/SQUIN/

is a map of professors from the German database community; the list of professors in the map can be filtered by research interests; selecting a professor opens a list of her/his publications. This application relies on data published by the professors as well as on data from DBpedia and the DBLP dataset.

6.2 The Impact of URI Prefetching and Non-blocking Iterators

Based on the SWClLib we conducted experiments to evaluate the presented concepts for an iterator-based realization of our query execution approach. For our tests we adopt the Berlin SPARQL Benchmark (BSBM) [10]. The BSBM executes a mix of 12 SPARQL queries over generated sets of RDF data; the datasets are scalable to different sizes based on a scaling factor. Using these datasets we set up a Linked Data server[4] which publishes the generated data following the Linked Data principles. With this server we simulate the Web of Linked Data in our experiments.

To measure the impact of URI prefetching and of our iterator paradigm extension we use the SWClLib to execute the BSBM query mix over the simulated Web. For this evaluation we adjust the SPARQL queries provided by BSBM in order to access our simulation server. We conduct our experiments on an Intel Core 2 Duo T7200 processor with 2 GHz, 4 MB L2 cache, and 2 GB main memory. Our test system runs a recent 32 bit version of Gentoo Linux with Sun Java 1.6.0. We execute the query mix for datasets generated with scaling factors of 10 to 60; these datasets have sizes of 4,971 to 26,108 triples, respectively. For each dataset we run the query mix 6 times where the first run is for warm up and is not considered for the measures.

Figure 6(a) depicts the average times to execute the query mix with three different implementations of SWClLib: i) without URI prefetching, ii) with prefetching, and iii) with the extended iterators that postpone the processing of input solutions. As can be seen from the measures URI prefetching reduces the query execution times to about 80%; our non-blocking iterators even halve the time.

The chart in Figure 6(b) puts the measures in relation to the time it takes to execute the query mixes without the need to retrieve data from the Web. We obtained this optimum by executing each query twice over a shared dataset; we measured the second executions which did not require to look up URIs because all data has already been retrieved in the first pass. These measures represent only the time to actually evaluate the queries as presented in Section 3. Hence, these times are a lower bound for possible optimizations to the iterator-based execution of our approach to query the Web of Linked Data. Using this lower bound we calculate the times required for data retrieval in the three implementations. These times are the differences between execution times measured for the three implementations and the lower bound, respectively. Figure 6(b) depicts these numbers which illustrate the significant impact of the possibility to

[4] Our server is based on RAP Pubby which is available from http://www4.wiwiss.fu-berlin.de/bizer/rdfapi/tutorial/RAP_Pubby.htm

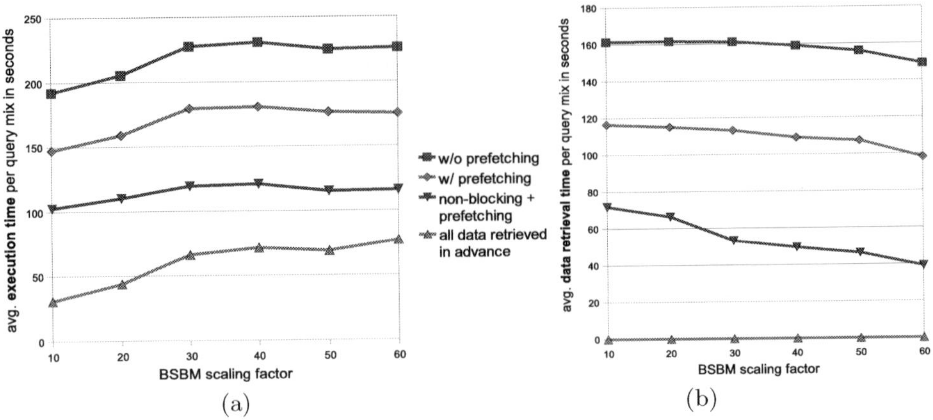

Fig. 6. Average times to execute the BSBM query mix with the SWClLib over a simulated Web of different sizes measured without URI prefetching, with prefetching, and with the extended iterators that temporarily reject input solutions, respectively

postpone the processing of certain input solutions in our non-blocking iterators. The chart additionally illustrates that the data retrieval times compared to the whole query execution time decreases for larger datasets, in particular in the case of the non-blocking iterators.

7 Related work

In this paper we present an approach to execute queries over the Web of Linked Data. Different solutions with a similar goal have been proposed before. These approaches can be classified in two categories: query federation and data centralization.

Research on federated query processing has a long history in database research. Sheth and Larson [5] provide an overview of the concepts developed in this context. Current approaches adapt these concepts to provide integrated access to distributed RDF data sources on the Web. The DARQ engine [11], for instance, decomposes a SPARQL query in subqueries, forwards these subqueries to multiple, distributed query services, and, finally, integrates the results of the subqueries. Very similar to the DARQ approach the SemWIQ [12] system contains a mediator service that transparently distributes the execution of SPARQL queries. Both systems, however, do not actively discover relevant data sources that provide query services.

In contrast to federation based systems the idea of centralized approaches is to provide a query service over a collection of Linked Data copied from different sources on the Web. For instance, OpenLink Software republishes the majority of the datasets from the Linking Open Data community project[5] on a

[5] http://esw.w3.org/topic/SweoIG/TaskForces/CommunityProjects/LinkingOpenData

single site[6]; users can issue SPARQL queries over the whole set of mirrored datasets. As with the federation based systems this approach executes queries only on selected datasets that are part of the collection. Another form of the centralized approach are search engines for the Web of Linked Data such as Sindice [13], Swoogle [14], and Watson [15]. These search engines crawl the Web by following RDF links, index discovered data, and provide query interfaces to their indexes. In contrast to our approach where the queried data is discovered and retrieved at query execution time the crawlers collect the queried data in advance. Due to the possibility to consider data that would not be discovered by our method a crawling-based approach might return more complete query results. However, setting up an index that covers large amounts of data on the Web requires much more resources (e.g. storage, compute power, administrative personnel) than is needed to apply our approach. Furthermore, our method is superior with respect to the timeliness of query results because we only use that data that is available at the time of query execution.

The idea of looking up URIs during application runtime as in our approach has first been proposed by Berners-Lee et al. [16]. The authors outline an algorithm that traverses RDF links in order to obtain more data about the resources presented in the Tabulator Linked Data browser. Our approach is based on the idea of Berners-Lee et al. We integrate this idea in the process of query execution and, thus, resolve the need to implement link traversal algorithms in each application that consumes Linked Data.

8 Conclusion

In this paper we introduce an approach to query the Web of Linked Data and we present concepts and algorithms to implement this approach. Our approach is based on traversing RDF links to discover data that might be relevant for a query during the query execution itself. We propose to implement this idea with an iterator-based pipeline and a URI prefetching approach to execute queries efficiently. To improve the performance of query execution even more the paper introduces an extension to the iterator paradigm that allows to temporarily reject certain input results. We provide the Semantic Web Client Library as a first prototype that implements our approach; an application that uses this library demonstrates the feasibility of our idea.

Our approach benefits from a high number of links in the Web of Linked Data; the more links exist the more complete results can be expected because more relevant data might be discovered. In addition to relying on a dense network of links, sharing the queried dataset and in advance crawling show great promise to improve the completeness of the results. In both approaches query execution does not have to start on an empty dataset. Instead, potentially relevant data might already be available. Hence, we are investigating possibilities to combine these techniques with our approach. Such a combination introduces the need

[6] http://lod.openlinksw.com

for a proper caching solution with suitable replacement strategies and refreshing policies. We are currently working on these issues.

The openness of the Web of Linked Data holds an enormous potential which could enable users to benefit from a virtually unbound set of data sources. However, this potential will become available not until applications take advantage of the characteristics of the Web which requires approaches to discover new data sources by traversing data links. This paper presents a proof of concept to enable this evolutionary step.

References

1. Berners-Lee, T.: Design Issues: Linked Data. (retrieved May 25, 2009), `http://www.w3.org/DesignIssues/LinkedData.html`
2. Bizer, C., Heath, T., Berners-Lee, T.: Linked data - the story so far. Journal on Semantic Web and Information Systems (in press, 2009)
3. Franklin, M.J., Halevy, A.Y., Maier, D.: From databases to dataspaces: A new abstraction for information management. SIGMOD Record 34(4), 27–33 (2005)
4. Prud'hommeaux, E., Seaborne, A.: SPARQL query language for RDF. W3C Recommendation (January 2008), `http://www.w3.org/TR/rdf-sparql-query/` (retrieved June 11, 2009)
5. Sheth, A.P., Larson, J.A.: Federated database systems for managing distributed, heterogeneous, and autonomous databases. ACM Computing Surveys 22(3), 183–236 (1990)
6. Garcia-Molina, H., Widom, J., Ullman, J.D.: Database Systems: The Complete Book. Prentice-Hall, Inc., Upper Saddle River (2002)
7. Graefe, G.: Query evaluation techniques for large databases. ACM Computing Surveys 25(2), 73–169 (1993)
8. Pirahesh, H., Mohan, C., Cheng, J., Liu, T.S., Selinger, P.: Parallelism in relational data base systems: Architectural issues and design approaches. In: Proceedings of the 2nd International Symposium on Databases in Parallel and Distributed Systems (DPDS), pp. 4–29. ACM, New York (1990)
9. Hartig, O., Mühleisen, H., Freytag, J.-C.: Linked data for building a map of researchers. In: Proceedings of 5th Workshop on Scripting and Development for the Semantic Web (SFSW) at ESWC (June 2009)
10. Bizer, C., Schultz, A.: Benchmarking the performance of storage systems that expose SPARQL endpoints. In: Proceedings of the Workshop on Scalable Semantic Web Knowledge Base Systems at ISWC (October 2008)
11. Quilitz, B., Leser, U.: Querying distributed RDF data sources with SPARQL. In: Bechhofer, S., Hauswirth, M., Hoffmann, J., Koubarakis, M. (eds.) ESWC 2008. LNCS, vol. 5021, pp. 524–538. Springer, Heidelberg (2008)
12. Langegger, A., Wöß, W., Blöchl, M.: A semantic web middleware for virtual data integration on the web. In: Bechhofer, S., Hauswirth, M., Hoffmann, J., Koubarakis, M. (eds.) ESWC 2008. LNCS, vol. 5021, pp. 493–507. Springer, Heidelberg (2008)
13. Oren, E., Delbru, R., Catasta, M., Cyganiak, R., Stenzhorn, H., Tummarello, G.: Sindice.com: A document-oriented lookup index for open linked data. International Journal of Metadata, Semantics and Ontologies 3(1) (2008)

14. Ding, L., Finin, T.W., Joshi, A., Pan, R., Cost, R.S., Peng, Y., Reddivari, P., Doshi, V., Sachs, J.: Swoogle: A search and metadata engine for the semantic web. In: Proceedings of the 13th ACM Conference on Information and Knowledge Management (CIKM), November 2004, pp. 652–659. ACM, New York (2004)
15. d'Aquin, M., Motta, E., Sabou, M., Angeletou, S., Gridinoc, L., Lopez, V., Guidi, D.: Toward a new generation of semantic web applications. IEEE Intelligent Systems 23(3), 20–28 (2008)
16. Berners-Lee, T., Chen, Y., Chilton, L., Connolly, D., Dhanaraj, R., Hollenbach, J., Lerer, A., Sheets, D.: Tabulator: Exploring and analyzing linked data on the semantic web. In: Proceedings of the 3rd Semantic Web User Interaction Workshop (SWUI) at ISWC (November 2006)

Dynamic Querying of Mass-Storage RDF Data with Rule-Based Entailment Regimes*

Giovambattista Ianni[1], Thomas Krennwallner[2], Alessandra Martello[1], and Axel Polleres[3]

[1] Dipartimento di Matematica, Università della Calabria, I-87036 Rende (CS), Italy
{ianni,a.martello}@mat.unical.it
[2] Institut für Informationssysteme 184/3, Technische Universität Wien, Austria
tkren@kr.tuwien.ac.at
[3] Digital Enterprise Research Institute, National University of Ireland, Galway
axel.polleres@deri.org

Abstract. RDF Schema (RDFS) as a lightweight ontology language is gaining popularity and, consequently, tools for scalable RDFS inference and querying are needed. SPARQL has become recently a W3C standard for querying RDF data, but it mostly provides means for querying simple RDF graphs only, whereas querying with respect to RDFS or other entailment regimes is left outside the current specification. In this paper, we show that SPARQL faces certain unwanted ramifications when querying ontologies in conjunction with RDF datasets that comprise multiple named graphs, and we provide an extension for SPARQL that remedies these effects. Moreover, since RDFS inference has a close relationship with logic rules, we generalize our approach to select a custom ruleset for specifying inferences to be taken into account in a SPARQL query. We show that our extensions are technically feasible by providing benchmark results for RDFS querying in our prototype system GiaBATA, which uses Datalog coupled with a persistent Relational Database as a back-end for implementing SPARQL with dynamic rule-based inference. By employing different optimization techniques like magic set rewriting our system remains competitive with state-of-the-art RDFS querying systems.

1 Introduction

Thanks to initiatives such as DBPedia or the Linked Open Data project,[1] a huge amount of machine-readable RDF [1] data is available, accompanying pervasive ontologies describing this data such as FOAF [2], SIOC [3], or YAGO [4].

A vast amount of Semantic Web data uses rather small and lightweight ontologies that can be dealt with rule-based RDFS and OWL reasoning [5,6,7], in contrast to the full power of expressive description logic reasoning. However, even if many practical use cases do not require complete reasoning on the terminological level provided

* This work has been partially supported by the Italian Research Ministry (MIUR) project Interlink II04CG8AGG, the Austrian Science Fund (FWF) project P20841, by Science Foundation Ireland under Grant No. SFI/08/CE/I1380 (Lion-2).

[1] http://dbpedia.org/ and http://linkeddata.org/

A. Bernstein et al. (Eds.): ISWC 2009, LNCS 5823, pp. 310–327, 2009.

by DL-reasoners, the following tasks become of utter importance. First, a Semantic Web system should be able to handle and evaluate (possibly complex) queries on large amounts of RDF instance data. Second, it should be able to take into account implicit knowledge found by ontological inferences as well as by additional custom rules involving built-ins or even nonmonotonicity. The latter features are necessary, e.g., for modeling complex mappings [8] between different RDF vocabularies. As a third point, joining the first and the second task, if we want the Semantic Web to be a solution to – as Ora Lassila formulated it – *those problems and situations that we are yet to define*,[2] we need triple stores that allow dynamic querying of different data graphs, ontologies, and (mapping) rules harvested from the Web. The notion of *dynamic* querying is in opposition to *static* querying, meaning that the same dataset, depending on context, reference ontology and entailment regime, might give different answers to the same query. Indeed, there are many situations in which the dataset at hand and its supporting class hierarchy cannot be assumed to be known upfront: think of distributed querying of remotely exported RDF data.

Concerning the first point, traditional RDF processors like Jena (using the default configuration) are designed for handling large RDF graphs in memory, thus reaching their limits very early when dealing with large graphs retrieved from the Web. Current RDF Stores, such as YARS [9], Sesame, Jena TDB, ThreeStore, AllegroGraph, or OpenLink Virtuoso[3] provide roughly the same functionality as traditional relational database systems do for relational data. They offer query facilities and allow to import large amounts of RDF data into their persistent storage, and typically support SPARQL [10], the W3C standard RDF query language. SPARQL has the same expressive power as non-recursive Datalog [11,12] and includes a set of built-in predicates in so called **`filter`** expressions.

However, as for the second and third point, current RDF stores only offer limited support. OWL or RDF(S) inference, let alone custom rules, are typically fixed in combination with SPARQL querying (cf. Section 2). Usually, dynamically assigning different ontologies or rulesets to data for querying is neither supported by the SPARQL specification nor by existing systems. Use cases for such dynamic querying involve, e.g., querying data with different versions of ontologies or queries over data expressed in related ontologies adding custom mappings (using rules or "bridging" ontologies).

To this end, we propose an extension to SPARQL which caters for knowledge-intensive applications on top of Semantic Web data, combining SPARQL querying with dynamic, rule-based inference. In this framework, we overcome some of the above mentioned limitations of SPARQL and existing RDF stores. Moreover, our approach is easily extensible by allowing features such as aggregates and arbitrary built-in predicates to SPARQL (see [8,14]) as well as the addition of custom inference and mapping rules. The contributions of our paper are summarized as follows:

- We introduce two additional language constructs to the normative SPARQL language. First, the directive **`using ontology`** for dynamically coupling a dataset with

[2] `http://www.lassila.org/publications/2006/SCAI-2006-keynote.pdf`

[3] See `http://openrdf.org/`, `http://jena.hpl.hp.com/wiki/TDB/`, `http://threestore.sf.net/`, `http://agraph.franz.com/allegrograph/`, `http://openlinksw.com/virtuoso/`, respectively.

an arbitrary RDFS ontology, and second *extended dataset clauses*, which allow to specify datasets with named graphs in a flexible way. The **`using ruleset`** directive can be exploited for adding to the query at hand proper rulesets which might used for a variety of applications such as encoding mappings between entities, or encoding custom entailment rules, such as RDFS or different rule-based OWL fragments.

- We present the GiaBATA system [15], which demonstrates how the above extensions can be implemented on a middle-ware layer translating SPARQL to Datalog and SQL. Namely, the system is based on known translations of SPARQL to Datalog rules. Arbitrary, possibly recursive rules can be added flexibly to model arbitrary ontological inference regimes, vocabulary mappings, or alike. The resulting program is compiled to SQL where possible, such that only the recursive parts are evaluated by a native Datalog implementation. This hybrid approach allows to benefit from efficient algorithms of deductive database systems for custom rule evaluation, and native features such as query plan optimization techniques or rich built-in functions (which are for instance needed to implement complex **`filter`** expressions in SPARQL) of common database systems.
- We compare our GiaBATA prototype to well-known RDF(S) systems and provide experimental results for the LUBM [16] benchmark. Our approach proves to be competitive on both RDF and dynamic RDFS querying without the need to pre-materialize inferences.

In the remainder of this paper we first introduce SPARQL along with RDF(S) and partial OWL inference by means of some motivating example queries which existing systems partially cannot deal in a reasonably manner in Section 2. Section 3 sketches how the SPARQL language can be enhanced with custom ruleset specifications and arbitrary graph merging specifications. We then briefly introduce our approach to translate SPARQL rules to Datalog in Section 4, and how this is applied to a persistent storage system. We evaluate our approach with respect to existing RDF stores in Section 5, and then conclusions are drawn in Section 6.

2 SPARQL and Some Motivating Examples

Similar in spirit to structured query languages like SQL, which allow to extract, combine and filter data from relational database tables, SPARQL allows to extract, combine and filter data from RDF graphs. The semantics and implementation of SPARQL involves, compared to SQL, several peculiarities, which we do not focus on in this paper, cf. [10,18,11,19] for details. Instead, let us just start right away with some illustrating example motivating our proposed extensions of SPARQL; we assume two data graphs describing data about our well-known friends Bob and Alice shown in Fig. 1(b)+(c). Both graphs refer to terms in a combined ontology defining the FOAF and Relationship[4] vocabularies, see Fig. 1(a) for an excerpt.

On this data the SPARQL query (1) intends to extract names of persons mentioned in those graphs that belong to friends of Bob. We assume that, by means of `rdfs:seeAlso` statements, Bob provides links to the graphs associated to the persons he is friend with.

[4] `http://vocab.org/relationship/`

```
@prefix foaf: <http://xmlns.com/foaf/0.1/>.
@prefix rel: <http://purl.org/vocab/relationship/>.
...
rel:friendOf rdfs:subPropertyOf foaf:knows.
foaf:knows rdfs:domain foaf:Person.
foaf:knows rdfs:range foaf:Person.
foaf:homepage rdf:type owl:inverseFunctionalProperty.
...
```

(a) Graph G_M (<http://example.org/myOnt.rdfs>), a combination of the FOAF&Relationship ontologies.

```
<http://bob.org#me> foaf:name "Bob"; a foaf:Person;
  foaf:homepage <http://bob.org/home.html>;
  rel:friendOf [ foaf:name "Alice";
                 rdfs:seeAlso <http://alice.org> ].
```

(b) Graph G_B (<http://bob.org>)

```
<http://alice.org#me> foaf:name "Alice"; a foaf:Person;
rel:friendOf [ foaf:name "Charles" ],
             [ foaf:name "Bob";
               foaf:homepage <http://bob.org/home.html>
].
```

(c) Graph G_A (<http://alice.org>)

Fig. 1. An ontology and two data graphs

```
select ?N from <http://example.org/myOnt.rdfs>
          from <http://bob.org>
          from named <http://alice.org>
where { <http://bob.org#me> foaf:knows ?X . ?X rdfs:seeAlso ?G .
        graph ?G { ?P rdf:type foaf:Person; foaf:name ?N } }        (1)
```

Here, the **from** and **from named** clauses specify an RDF dataset. In general, the *dataset* $DS = (G, N)$ of a SPARQL query is defined by (i) a *default graph G* obtained by the RDF merge [20] of all graphs mentioned in **from** clauses, and (ii) a set $N = \{(u_1, G_1), \ldots, (u_k, G_k)\}$ of *named graphs*, where each pair (u_i, G_i) consists of an IRI u_i, given in a **from named** clause, paired with its corresponding graph G_i. For instance, the dataset of query (1) would be $DS_1 = (\ G_M \uplus G_B, \{($<http://alice.org>$, G_A)\})$, where $\uplus$ denotes merging of graphs according to the normative specifications.

Now, let us have a look at the answers to query (1). Answers to SPARQL **select** queries are defined in terms of multisets of partial variable substitutions. In fact the answer to query (1) is empty when – as typical for current SPARQL engines – only simple RDF entailment is taken into account, and query answering then boils down to simple graph matching. Since neither of the graphs in the default graph contain any triple matching the pattern `<http://bob.org#me> foaf:knows ?X` in the **where** clause, the result of (1) is empty. When taking subproperty inference by the statements of the ontology in G_M into account, however, one would expect to obtain three substitutions for the variable ?N: {?N/"Alice", ?N/"Bob", ?N/"Charles"}. We will explain in the following why this is not the case in standard SPARQL.

In order to obtain the expected answer, firstly SPARQL's basic graph pattern matching needs to be extended, see [10, Section 12.6]. In theory, this means that the graph patterns in the **where** clause needs to be matched against an enlarged version of the original graphs in the dataset (which we will call the *deductive closure* $Cl(\cdot)$) of a given entailment regime. Generic extensions for SPARQL to entailment regimes other than simple RDF entailment are still an open research problem,[5] due to various problems: (i) for (non-simple) RDF entailment regimes, such as full RDFS entailment, $Cl(G)$ is infinite, and thus SPARQL queries over an empty graph G might already have infinite answers, and (ii) it is not yet clear which should be the intuitive answers to queries over

[5] For details, cf. http://www.polleres.net/sparqltutorial/, Unit 5b.

inconsistent graphs, e.g. in OWL entailment, etc. In fact, SPARQL restricts extensions of basic graph pattern matching to retain finite answers. Not surprisingly, many existing implementations implement finite approximations of higher entailment regimes such as RDFS and OWL [6,5,21]. E.g., the RDF Semantics document [20] contains an informative set of entailment rules, a subset of which (such as the one presented in Section 3.2 below) is implemented by most available RDF stores. These rule-based approximations, which we focus on in this paper, are typically expressible by means of Datalog-style rules. These latter model how to infer a finite closure of a given RDF graph that covers sound but not necessarily complete RDF(S) and OWL inferences. It is worth noting that Rule-based entailment can be implemented in different ways: rules could be either dynamically evaluated upon query time, or the closure wrt. ruleset $\mathcal{R}$, $Cl_{\mathcal{R}}(G)$, could be materialized when graph G is loaded into a store. Materialization of inferred triples at loading time allows faster query responses, yet it has drawbacks: it is time and space expensive and it has to be performed once and statically. In this setting, it must be decided upfront.

(a) which ontology should be taken into account for which data graph, and
(b) to which graph(s) the inferred triples "belong", which particularly complicates the querying of named graphs.

As for exemplifying (a), assume that a user agent wants to issue another query on graph G_B with only the FOAF ontology in mind, since she does not trust the Relationship ontology. In the realm of FOAF alone, `rel:friendOf` has nothing to deal with `foaf:knows`. However, when materializing all inferences upon loading G_M and G_B into the store, `bob:me foaf:knows _:a` would be inferred from $G_M \uplus G_B$ and would contribute to such a different query. Current RDF stores prevent to dynamically parameterize inference with an ontology of choice at query time, since indeed typically all inferences are computed upon loading time *once and for all.*

As for (b), queries upon datasets including named graphs are even more problematic. Query (1) uses G_B in order to find the IRI identifiers for persons that Bob knows by following `rdfs:seeAlso` links and looks for persons and their names in the *named* RDF graphs found at these links. Even if rule-based inference was supported, the answer to query (1) over dataset DS_1 is just $\{$?N/`"Alice"`$\}$, as "Alice" is the only (explicitly) asserted `foaf:Person` in G_A. Subproperty, domain and range inferences over the G_M ontology do not propagate to G_A, since G_M is normatively prescribed to be merged into the default graph, but not into the named graph. Thus, there is no way to infer that `"Bob"` and `"Charles"` are actually names of `foaf:Persons` within the named graph G_A. Indeed, SPARQL does not allow to merge, on demand, graphs into the named graphs, thus there is no way of combining G_M with the named graph G_A.

To remedy these deficiencies, we suggest an extension of the SPARQL syntax, in order to allow the specification of datasets more flexibly: it is possible to group graphs to be merged in parentheses in **`from`** and **`from named`** clauses. The modified query, obtaining a dataset $DS_2 = (\, G_M \uplus G_B, \{(\texttt{http://alice.org}, G_M \uplus G_A)\})$ looks as follows:

```
select ?N
  from (<http://example.org/myOnt.rdfs> <http://bob.org/>)
  from named <http://alice.org>
    (<http://example.org/myOnt.rdfs> <http://alice.org/>)              (2)
where { bob:me foaf:knows ?X . ?X rdfs:seeAlso ?G .
        graph ?G { ?X foaf:name ?N . ?X a foaf:Person . } }
```

For ontologies which should apply to the whole query, i.e., graphs to be merged into the default graph as well as any specified named graph, we suggest a more convenient shortcut notation by adding the keyword **using ontology** in the SPARQL syntax:

```
select ?N
  using ontology <http://example.org/myOnt.rdfs>
  from <http://bob.org/>
  from named <http://alice.org/>                                       (3)
where { bob:me foaf:knows ?X . ?X foaf:seeAlso ?G .
        graph ?G { ?X foaf:name ?N . ?X a foaf:Person. } }
```

Hence, the **using ontology** construct allows for coupling the entire given dataset with the terminological knowledge in the `myOnt` data schema. As our investigation of currently available RDF stores (see Section 5) shows, none of these systems easily allow to merge ontologies into named graphs or to dynamically specify the dataset of choice.

In addition to parameterizing queries with ontologies in the dataset clauses, we also allow to parameterize the ruleset which models the entailment regime at hand. Per default, our framework supports a standard ruleset that "emulates" (a finite subset of) the RDFS semantics. This standard ruleset is outlined in Section 3 below. Alternatively, different rule-based entailment regimes, e.g., rulesets covering parts of the OWL semantics á la ter Horst [5], de Bruijn [22, Section 9.3], OWL2 RL [17] or other custom rulesets can be referenced with the **using ruleset** keyword. For instance, the following query returns the solution {?X/<http://alice.org#me>, ?Y/<http://bob.org#me>}, by doing equality reasoning over inverse functional properties such as `foaf:homepage` when the FOAF ontology is being considered:

```
select ?X ?Y
 using ontology <http://example.org/myOnt.rdfs>
 using ruleset rdfs
 using ruleset <http://www.example.com/owl-horst>                      (4)
 from <http://bob.org/>
 from <http://alice.org/>
where { ?X foaf:knows ?Y }
```

Query (4) uses the built-in RDFS rules for the usual subproperty inference, plus a ruleset implementing ter Horst's inference rules, which might be available at URL `http://www.example.com/owl-horst`. This ruleset contains the following additional rules, that will "equate" the blank node used in G_A for "Bob" with <http://bob.org#me>:[6]

```
?P rdf:type owl:iFP . ?S1 ?P ?O . ?S2 ?P ?O .   → ?S1 owl:sameAs ?S2.
?X owl:sameAs ?Y                                 → ?Y owl:sameAs ?X.
?X ?P ?O . ?X owl:sameAs ?Y                      → ?Y ?P ?O.           (5)
?S ?X ?O . ?X owl:sameAs ?Y                      → ?S ?Y ?O.
?S ?P ?X . ?X owl:sameAs ?Y                      → ?S ?P ?Y.
```

[6] We use `owl:iFP` as shortcut for `owl:inverseFunctionalProperty`.

3 A Framework for Using Ontologies and Rules in SPARQL

In the following, we will provide a formal framework for the SPARQL extensions outlined above. In a sense, the notion of *dynamic querying* is formalized in terms of the dependence of BGP pattern answers $[\![P]\!]^{\mathcal{O},\mathcal{R}}$ from a variable ontology $\mathcal{O}$ and ruleset $\mathcal{R}$. For our exposition, we rely on well-known definitions of RDF datasets and SPARQL. Due to space limitations, we restrict ourselves to the bare minimum and just highlight some standard notation used in this paper.

Preliminaries. Let I, B, and L denote pairwise disjoint infinite sets of IRIs, blank nodes, and RDF literals, respectively. A *term* is an element from $I \cup B \cup L$. An *RDF graph* G (or simply *graph*) is defined as a set of *triples* from $I \cup B \times I \cup B \times I \cup B \cup L$ (cf. [18,12]); by $blank(G)$ we denote the set of blank nodes of G.[7]

A *blank node renaming* θ is a mapping $I \cup B \cup L \rightarrow I \cup B \cup L$. We denote by $t\theta$ the application of θ to a term t. If $t \in I \cup L$ then $t\theta = t$, and if $t \in B$ then $t\theta \in B$. If (s, p, o) is a triple then $(s, p, o)\theta$ is the triple $(s\theta, p\theta, o\theta)$. Given a graph G, we denote by $G\theta$ the set of all triples $\{t\theta \mid t \in G\}$. Let G and H be graphs. Let θ^G_H be an arbitrary blank node renaming such that $blank(G) \cap blank(H\theta^G_H) = \emptyset$. The *merge of* G *by* H, denoted $G \uplus H$, is defined as $G \cup H\theta^G_H$.

An *RDF dataset* $D = (G_0, N)$ is a pair consisting of exactly one unnamed graph, the so-called default graph G_0, and a set $N = \{\langle u_1, G_1\rangle, \ldots, \langle u_n, G_n\rangle\}$ of named graphs, coupled with their identifying URIs. The following conditions hold: (i) each G_i $(0 \leq i \leq n)$ is a graph, (ii) each u_j $(1 \leq j \leq n)$ is from I, and (iii) for all $i \neq j$, $\langle u_i, G_i\rangle, \langle u_j, G_j\rangle \in N$ implies $u_i \neq u_j$ and $blank(G_i) \cap blank(G_j) = \emptyset$.

The syntax and semantics of SPARQL can now be defined as usual, cf. [10,18,12] for details. For the sake of this paper, we restrict ourselves to `select` queries as shown in the example queries (1)–(4) and just provide an overview of the necessary concepts. A query in SPARQL can be viewed as a tuple $Q = (V, D, P)$, where V is the set of variables mentioned in the `select` clause, D is an RDF dataset, defined by means of `from` and `from named` clauses, and P is a *graph pattern*, defined in the `where` clause.

Graph patterns are in the simplest case sets of RDF triples (s, p, o), where terms and variables from an infinite set of variables *Var* are allowed, also called *basic graph patterns (BGP)*. More complex graph patterns can be defined recursively, i.e., if P_1 and P_2 are graph patterns, $g \in I \cup Var$ and R is a filter expression, then P_1 `optional` P_2, P_1 `union` P_2, P_1 `filter` R, and `graph` g P_1 are graph patterns.

Graph Pattern Matching. Queries are evaluated by matching graph patterns against graphs in the dataset. In order to determine a query's solution, in the simplest case BGPs are matched against the *active graph* of the query, which is one particular graph in the dataset, identified as shown next.

Solutions of BGP matching consist of multisets of bindings for the variables mentioned in the pattern to terms in the active graph. Partial solutions of each subpattern are joined according to an algebra defining the `optional`, `union` and `filter` operators, cf. [10,18,12]. For what we are concerned with here, the most interesting operator though is the `graph` operator, since it changes the active graph. That is, the active graph

[7] Note that we allow *generalized RDF graphs* that may have blank nodes in property position.

is the default graph G_0 for any basic graph pattern not occurring within a **graph** sub pattern. However, in a subpattern **graph** $g\,P_1$, the pattern P_1 is matched against the named graph identified by g, if $g \in I$, and against any named graph u_i, if $g \in Var$, where the binding u_i is returned for variable g. According to [12], for a RDF dataset D and active graph G, we define $[\![P]\!]_G^D$ as the multiset of tuples constituting the *answer* to the graph pattern P. The solutions of a query $Q = (V, D, P)$ is the projection of $[\![P]\!]_G^D$ to the variables in V only.

3.1 SPARQL with Extended Datasets

What is important to note now is that, by the way how datasets are syntactically defined in SPARQL, the default graph G_0 can be obtained from merging a group of different source graphs, specified via several **from** clauses – as shown, e.g., in query (1) – whereas in each **from named** clause a single, separated, named graph is added to the dataset. That is, **graph** patterns will always be matched against a separate graph only. To generalize this towards dynamic construction of groups of merged named graphs, we introduce the notion of an *extended dataset*, which can be specified by enlarging the syntax of SPARQL with two additional dataset clauses:

- For $i, i_1, \ldots, i_m$ distinct IRIs ($m \geq 1$), the statement "**from named** $i(i_1 \ldots i_m)$" is called *extended dataset clause*. Intuitively, $i_1 \ldots i_m$ constitute a group of graphs to be merged: the merged graph is given i as identifying IRI.
- For $o \in I$ we call the statement "**using ontology** o" an *ontological dataset clause*. Intuitively, o stands for a graph that will merged with all graphs in a given query.

Extended RDF datasets are thus defined as follows. A *graph collection* $\mathcal{G}$ is a set of RDF graphs. An *extended RDF dataset* $\mathcal{D}$ is a pair $(\mathcal{G}_0, \{\langle u_1, \mathcal{G}_1\rangle, \ldots, \langle u_n, \mathcal{G}_n\rangle\})$ satisfying the following conditions: (i) each $\mathcal{G}_i$ is a nonempty graph collection (note that $\{\emptyset\}$ is a valid nonempty graph collection), (ii) each u_j is from I, and (iii) for all $i \neq j$, $\langle u_i, \mathcal{G}_i\rangle, \langle u_j, \mathcal{G}_j\rangle \in D$ implies $u_i \neq u_j$ and for $G \in \mathcal{G}_i$ and $H \in \mathcal{G}_j$, $blank(G) \cap blank(H) = \emptyset$. We denote $\mathcal{G}_0$ as $dg(\mathcal{D})$, the *default graph collection* of $\mathcal{D}$.

Let $\mathcal{D}$ and $\mathcal{O}$ be an extended dataset and a graph collection, resp. The ordinary RDF dataset obtained from $\mathcal{D}$ and $\mathcal{O}$, denoted $D(\mathcal{D}, \mathcal{O})$, is defined as

$$\left(\biguplus_{g \in dg(\mathcal{D})} g \uplus \biguplus_{o \in \mathcal{O}} o, \left\{\langle u, \biguplus_{g \in \mathcal{G}} g \uplus \biguplus_{o \in \mathcal{O}} o\rangle \mid \langle u, \mathcal{G}\rangle \in \mathcal{D}\right\}\right).$$

We can now define the semantics of extended and ontological dataset clauses as follows. Let F be a set of ordinary and extended dataset clauses, and O be a set of ontological dataset clauses. Let $graph(g)$ be the graph associated to the IRI g: the extended RDF dataset obtained from F, denoted $edataset(F)$, is composed of:

(1) $\mathcal{G}_0 = \{graph(g) \mid$ "**from** g" $\in F\}$. If there is no **from** clause, then $\mathcal{G}_0 = \emptyset$.
(2) A named graph collection $\langle u, \{graph(u)\}\rangle$ for each "**from named** u" in F.
(3) A named graph collection $\langle i, \{graph(i_1), \ldots, graph(i_m)\}\rangle$for each "**from named** $i(i_1 \ldots i_m)$" in F.

The graph collection obtained from O, denoted $ocollection(O)$, is the set $\{graph(o) \mid$ "**`using ontology`** o" $\in O\}$. The ordinary dataset of O and F, denoted $dataset(F, O)$, is the set $D(edataset(F), ocollection(O))$.

Let $\mathcal{D}$ and $\mathcal{O}$ be as above. The *evaluation* of a graph pattern P over $\mathcal{D}$ and $\mathcal{O}$ having active graph collection $\mathcal{G}$, denoted $[\![P]\!]_{\mathcal{G}}^{\mathcal{D},\mathcal{O}}$, is the evaluation of P over $D(\mathcal{D}, \mathcal{O})$ having active graph $G = \biguplus_{g \in \mathcal{G}} g$, that is, $[\![P]\!]_{\mathcal{G}}^{\mathcal{D},\mathcal{O}} = [\![P]\!]_{G}^{D(\mathcal{D},\mathcal{O})}$.

Note that the semantics of extended datasets is defined in terms of ordinary RDF datasets. This allows to define the semantics of SPARQL with extended and ontological dataset clauses by means of the standard SPARQL semantics. Also note that our extension is conservative, i.e., the semantics coincides with the standard SPARQL semantics whenever no ontological clauses and extended dataset clauses are specified.

3.2 SPARQL with Arbitrary Rulesets

Extended dataset clauses give the possibility of merging arbitrary ontologies into any graph in the dataset. The second extension herein presented enables the possibility of dynamically deploying and specifying rule-based entailments regimes on a per query basis. To this end, we define a generic $\mathcal{R}$-entailment, that is RDF entailment associated to a parametric ruleset $\mathcal{R}$ which is taken into account when evaluating queries. For each such $\mathcal{R}$-entailment regime we straightforwardly extend BGP matching, in accordance with the conditions for such extensions as defined in [10, Section 12.6].

We define an *RDF inference rule* r as the pair $(\mathcal{A}nte, \mathcal{C}on)$, where the *antecedent* $\mathcal{A}nte$ and the *consequent* $\mathcal{C}on$ are basic graph patterns such that $\mathcal{V}(\mathcal{C}on)$ and $\mathcal{V}(\mathcal{A}nte)$ are non-empty, $\mathcal{V}(\mathcal{C}on) \subseteq \mathcal{V}(\mathcal{A}nte)$ and $\mathcal{C}on$ does not contain blank nodes.[8] As in Example (5) above, we typically write RDF inference rules as

$$\mathcal{A}nte \rightarrow \mathcal{C}on \ . \tag{6}$$

We call sets of inference rules *RDF inference rulesets*, or *rulesets* for short.

Rule Application and Closure. We define *RDF rule application* in terms of the immediate consequences of a rule r or a ruleset $\mathcal{R}$ on a graph G. Given a BGP P, we denote as $\mu(P)$ a pattern obtained by substituting variables in P with elements of $I \cup B \cup L$. Let r be a rule of the form (6) and G be a set of RDF triples, then:

$$T_r(G) = \{\mu(\mathcal{C}on) \mid \exists \mu \text{ such that } \mu(\mathcal{A}nte) \subseteq G\}.$$

Accordingly, let $T_{\mathcal{R}}(G) = \bigcup_{r \in \mathcal{R}} T_r(G)$. Also, let $G_0 = G$ and $G_{i+1} = G_i \cup T_{\mathcal{R}}(G_i)$ for $i \geq 0$. It can be easily shown that there exists the smallest n such that $G_{n+1} = G_n$; we call then $Cl_{\mathcal{R}}(G) = G_n$ the *closure* of G with respect to ruleset $\mathcal{R}$.

We can now further define $\mathcal{R}$-*entailment* between two graphs G_1 and G_2, written $G_1 \models_{\mathcal{R}} G_2$, as $Cl_{\mathcal{R}}(G_1) \models G_2$. Obviously for any finite graph G, $Cl_{\mathcal{R}}(G)$ is finite. In order to define the semantics of a SPARQL query wrt. $\mathcal{R}$-*entailment* we now extend graph pattern matching in $[\![P]\!]_{G}^{D}$ towards respecting $\mathcal{R}$.

[8] Unlike some other rule languages for RDF, the most prominent of which being CONSTRUCT statements in SPARQL itself, we forbid blank nodes; i.e., existential variables in rule consequents which require the "invention" of new blank nodes, typically causing termination issues.

Definition 1 (Extended Basic Graph Pattern Matching for $\mathcal{R}$-Entailment). *Let D be a dataset and G be an active graph. The* solution of a BGP P wrt. $\mathcal{R}$-entailment, *denoted* $[\![P]\!]_G^{D,\mathcal{R}}$, *is* $[\![P]\!]_{Cl_{\mathcal{R}}(G)}^D$.

The solution $[\![P]\!]_G^{D,\mathcal{R}}$ naturally extends to more complex patterns according to the SPARQL algebra. In the following we will assume that $[\![P]\!]_G^{D,\mathcal{R}}$ is used for graph pattern matching. Our extension of basic graph pattern matching is in accordance with the conditions for extending BGP matching in [10, Section 12.6]. Basically, these conditions say that any extension needs to guarantee finiteness of the answers, and defines some conditions about a "*scoping graph*." Intuitively, for our extension, the scoping graph is just equivalent to $Cl_{\mathcal{R}}(G)$. We refer to [10, Section 12.6] for the details.

To account for this generic SPARQL BGP matching extension parameterized by an RDF inference ruleset $\mathcal{R}_Q$ per SPARQL query Q, we introduce another novel language construct for SPARQL:

– For $r \in I$ we call "**`using ruleset`** r" a *ruleset clause*.

Analogously to IRIs denoting graphs, we now assume that an IRI $r \in I$ may not only refer to graphs but also to rulesets, and denote the corresponding ruleset by $ruleset(r)$. Each query Q may contain zero or more ruleset clauses, and we define the query ruleset $\mathcal{R}_Q = \bigcup_{r \in R} ruleset(r)$, where R is the set of all ruleset clauses in Q.

The definitions of solutions of a query and the evaluation of a pattern in this query on active graph G is now defined just as above, with the only difference that answer to a pattern P are given by $[\![P]\!]_G^{D,\mathcal{R}_Q}$.

We observe that whenever $\mathcal{R} = \emptyset$, then $\mathcal{R}$-entailment boils down to simple RDF entailment. Thus, a query without ruleset clauses will just be evaluated using standard BGP matching. In general, our extension preserve full backward compatibility.

Proposition 1. *For $\mathcal{R} = \emptyset$ and RDF graph G,* $[\![P]\!]_G^{D,\mathcal{R}} = [\![P]\!]_G^D$.

Analogously, one might use *$\mathcal{R}$-entailment* as the basis for RDFS entailment as follows. We consider here the ρDF fragment of RDFS entailment [6]. Let $\mathcal{R}_{RDFS}$ denote the ruleset corresponding to the minimal set of entailment rules (2)–(4) from [6]:

```
?P rdfs:subPropertyOf ?Q . ?Q rdfs:subPropertyOf ?R .   → ?P rdfs:subPropertyOf ?R.
?P rdfs:subPropertyOf ?Q . ?S ?P ?O .                   → ?S ?Q ?O.
?C rdfs:subClassOf ?D . ?D rdfs:subClassOf ?E .         → ?C rdfs:subClassOf ?E.
?C rdfs:subClassOf ?D . ?S rdf:type ?C .                → ?S rdf:type ?D.
?P rdfs:domain ?C . ?S ?P ?O .                          → ?S rdf:type ?C.
?P rdfs:range ?C . ?S ?P ?O .                           → ?O rdf:type ?C.
```

Since obviously $G \models_{RDFS} Cl_{\mathcal{R}_{RDFS}}(\mathcal{G})$ and hence $Cl_{\mathcal{R}_{RDFS}}(\mathcal{G})$ may be viewed as a finite approximation of RDFS-entailment, we can obtain a reasonable definition for defining a BGP matching extension for RDFS by simply defining $[\![P]\!]_G^{D,RDFS} = [\![P]\!]_G^{D,\mathcal{R}_{RDFS}}$. We allow the special ruleset clause **`using ruleset`** `rdfs` to conveniently refer to this particular ruleset. Other rulesets may be published under a Web dereferenceable URI, e.g., using an appropriate RIF [23] syntax.

Note, eventually, that our rulesets consist of positive rules, and as such enjoy a natural monotonicity property.

Proposition 2. *For rulesets $\mathcal{R}$ and $\mathcal{R}'$, such that $\mathcal{R} \subseteq \mathcal{R}'$, and graph G_1 and G_2, if $G_1 \models_{\mathcal{R}} G_2$ then $G_1 \models_{\mathcal{R}'} G_2$.*

Entailment regimes modeled using rulesets can thus be enlarged without retracting former inferences. This for instance would allow to introduce tighter RDFS-entailment approximations by extending $\mathcal{R}_{RDFS}$ with further axioms, yet preserving inferred triples.

4 Translating SPARQL into Datalog and SQL

Our extensions have been implemented by reducing both queries, datasets and rulesets to a common ground which allows arbitrary interoperability between the three realms. This common ground is Datalog, wherein rulesets naturally fit and SPARQL queries can be reduced to. Subsequently, the resulting combined Datalog programs can be evaluated over an efficient SQL interface to an underlying relational DBMS that works as triple store.

From SPARQL to Datalog. A SPARQL query Q is transformed into a corresponding Datalog program D_Q. The principle is to break Q down to a series of Datalog rules, whose body is a conjunction of atoms encoding a graph pattern. D_Q is mostly a plain Datalog program in dlvhex [24] input format, i.e. Datalog with *external predicates* in the dlvhex language. These are explained along with a full account of the translation in [11,19]. Main challenges in the transformation from SPARQL to Datalog are (i) faithful treatment of the semantics of joins over possibly unbound variables [11], (ii) the multiset semantics of SPARQL [19], and also (iii) the necessity of Skolemization of blank nodes in **construct** queries [8]. Treatment of **optional** statements is carried out by means of an appropriate encoding which exploits negation as failure. Special external predicates of dlvhex are used for supporting some features of the SPARQL language: in particular, importing RDF data is achieved using the external *&rdf* predicate, which can be seen as a built-in referring to external data. Moreover, SPARQL **filter** expressions are implemented using the dlvhex external *&eval* predicate in D_Q.

Let us illustrate this transformation step by an example: the following query A asking for persons who are not named "Alice" and optionally their email addresses

```
select * from <http://alice.org/>
 where { ?X a foaf:Person. ?X foaf:name ?N.                              (7)
         filter ( ?N != "Alice") optional { ?X foaf:mbox ?M } }
```

is translated to the program D_A as follows:

```
(r1) "triple"(S,P,O,default)      :- &rdf[ "alice.org" ](S,P,O).
(r2) answer1(X_N,X_X,default)     :- "triple"(X_X,"rdf:type","foaf:Person",default),
                                     "triple"(X_X,"foaf:name",X_N,default),
                                     &eval[" ?N != 'Alice' ","N", X_N ](true).
(r3) answer2(X_M,X_X,default)     :- "triple"(X_X,"foaf:mbox",X_M,default).
(r4) answer_b_join_1(X_M,X_N,X_X,default)   :- answer1(X_N,X_X,default),
                                               answer2(X_M,X_X,default).
(r5) answer_b_join_1(null,X_N,X_X,default)  :- answer1(X_N,X_X,default),
                                               not answer2_prime(X_X,default).
(r6) answer2_prime(X_X,default)   :- answer1(X_N,X_X,default),
                                     answer2(X_M,X_X,default).
(r7) answer(X_M,X_N,X_X)          :- answer_b_join1(X_M,X_N,X_X,default).
```

where the first rule `(r1)` computes the predicate `"triple"` taking values from the built-in predicate *&rdf*. This latter is generally used to import RDF statements from the specified URI. The following rules `(r2)` and `(r3)` compute the solutions for the filtered basic graph patterns { `?X a foaf:Person. ?X foaf:name ?N.` **`filter`** `(?N != "Alice")` } and { `?X foaf:mbox ?M`}. In particular, note here that the evaluation of **`filter`** expressions is "outsourced" to the built-in predicate *&eval*, which takes a filter expression and an encoding of variable bindings as arguments, and returns the evaluation value (`true`, `false` or `error`, following the SPARQL semantics). In order to emulate SPARQL's **`optional`** patterns a combination of join and set difference operation is used, which is established by rules `(r4)–(r6)`. Set difference is simulated by using both *null* values and *negation as failure*. According to the semantics of SPARQL, one has to particularly take care of variables which are joined and possibly unbound (i.e., set to the `null` value) in the course of this translation for the general case. Finally, the dedicated predicate *answer* in rule `(r7)` collects the answer substitutions for Q. D_Q might then be merged with additional rulesets whenever Q contains **`using ruleset`** clauses.

From Datalog to SQL. For this step we rely on the system DLV^{DB} [25] that implements Datalog under stable model semantics on top of a DBMS of choice. DLV^{DB} is able to translate Datalog programs in a corresponding SQL query plan to be issued to the underlying DBMS. RDF Datasets are simply stored in a database D, but the native dlvhex $\&rdf$ and $\&eval$ predicates in D_Q cannot be processed by DLV^{DB}directly over D. So, D_Q needs to be post-processed before it can be converted into suitable SQL statements.

Rule `(r1)` corresponds to loading persistent data into D, instead of loading triples via the $\&rdf$ built-in predicate. In practice, the predicate `"triple"` occurring in program D_A is directly associated to a database table TRIPLE in D. This operation is done off-line by a loader module which populates the TRIPLE table accordingly, while `(r1)` is removed from the program. The $\&eval$ predicate calls are recursively broken down into `WHERE` conditions in SQL statements, as sketched below when we discuss the implementation of **`filter`** statements.

After post-processing, we obtain a program D'_Q, which DLV^{DB}allows to be executed on a DBMS by translating it to corresponding SQL statements. D'_Q is coupled with a mapping file which defines the correspondences between predicate names appearing in D'_Q and corresponding table and view names stored in the DBMS D.

For instance, the rule `(r4)` of D_A, results in the following SQL statement issued to the RDBMS by DLV^{DB}:

```
INSERT INTO answer_b_join_1
 SELECT DISTINCT answer2_p2.a1, answer1_p1.a1, answer1_p1.a2, 'default'
 FROM answer1 answer1_p1, answer2 answer2_p2
 WHERE (answer1_p1.a2=answer2_p2.a2)
 AND (answer1_p1.a3='default')
 AND (answer2_p2.a3='default')
 EXCEPT (SELECT * FROM answer_b_join_1)
```

Whenever possible, the predicates for computing intermediate results such as `answer1`, `answer2`, `answer_b_join_1`, ..., are mapped to SQL views rather than materialized tables, enabling dynamic evaluation of predicate contents on the DBMS side.[9]

Schema Rewriting. Our system allows for customizing schemes which triples are stored in. It is known and debated [26] that in choosing the data scheme of D several aspects have to be considered, which affect performance and scalability when handling large-scale RDF data. A widely adopted solution is to exploit a single table storing quadruples of form (s, p, o, c) where s, p, o and c are, respectively, the triple subject, predicate, object and context the triple belongs to. This straightforward representation is easily improved [27] by avoiding to store explicitly string values referring to URIs and literals. Instead, such values are replaced with a corresponding hash value.

Other approaches suggest alternative data structures, e.g., property tables [27,26]. These aim at denormalizing RDF graphs by storing them in a flattened representation, trying to encode triples according to the hidden "schema" of RDF data. Similarly to a traditional relational schema, in this approach D contains a table per each known property name (and often also per class, splitting up the `rdf:type` table).

Our system gives sufficient flexibility in order to program different storage schemes: while on higher levels of abstraction data are accessible via the 4-ary *triple* predicate, a schema rewriter module is introduced in order to match D'_Q to the current database scheme. This module currently adapts D'_Q by replacing constant IRIs and literals with their corresponding hash value, and introducing further rules which translate answers, converting hash values back to their original string representation.

Magic Sets. Notably, DLV^{DB} can post-process D'_Q using the magic sets technique, an optimization method well-known in the database field [28]. The optimized program mD'_Q tailors the data to be queried to an extent significantly smaller than the original D'_Q. The application of magic sets allows, e.g., to apply entailment rules $\mathcal{R}_{RDFS}$ only on triples which might affect the answer to Q, preventing thus the full computation and/or materialization of inferred data.

Implementation of `filter` Statements. Evaluation of SPARQL **`filter`** statements is pushed down to the underlying database D by translating filter expressions to appropriate SQL views. This allows to dynamically evaluate filter expressions on the DBMS side. For instance, given a rule $r \in D_Q$ of the form

```
h(X,Y) :- b(X,Y), &eval[f_Y](bool).
```

where the `&eval` atom encodes the **`filter`** statement (`f_Y` representing the filter expression), then r is translated to

```
h(X,Y) :- b'(X,Y).
```

where `b'` is a fresh predicate associated via the mapping file to a database view. Such a view defines the SQL code to be used for the computation of `f_Y`, like

```
CREATE VIEW B' AS ( SELECT X,Y FROM B WHERE F_Y )
```

[9] For instance, recursive predicates require to be associated with permanent tables, while remaining predicates are normally associated to views.

where F_Y is an appropriate translation of the SPARQL **filter** expression f_Y at hand to an SQL Boolean condition,[10] while B is the DBMS counterpart table of the predicate b.

5 Experiments

In order to illustrate that our approach is practically feasible, we present a quantitative performance comparison between our prototype system, GiaBATA, which implements the approach outlined before, and some state-of-the-art triple stores. The test were done on an Intel P4 3GHz machine with 1.5GB RAM under Linux 2.6.24. Let us briefly outline the main features and versions of the triple stores we used in our comparison.

AllegroGraph works as a database and application framework for building Semantic Web applications. The system assures persistent storage and RDFS++ reasoning, a semantic extension including the RDF and RDFS constructs and some OWL constructs (`owl:sameAs`, `owl:inverseOf`, `owl:TransitiveProperty`, `owl:hasValue`). We tested the free Java edition of AllegroGraph 3.2 with its native persistence mechanism.[11]

ARQ is a query engine implementing SPARQL under the Jena framework.[12] It can be deployed on several persistent storage layers, like filesystem or RDBMS, and it includes a rule-based inference engine. Being based on the Jena library, it provides inferencing models and enables (incomplete) OWL reasoning. Also, the system comes with support for custom rules. We used **ARQ** 2.6 with RDBMS backend connected to PostgreSQL 8.3.

GiaBATA [15] is our prototype system implementing the SPARQL extensions described above. GiaBATA is based on a combination of the DLV^{DB} [25] and dlvhex [24] systems, and caters for persistent storage of both data and ontology graphs. The former system is a variant of DLV [13] with built-in database support. The latter is a solver for HEX-programs [24], which features an extensible plugin system which we used for developing a rewriter-plugin able to translate SPARQL queries to HEX-programs. The tests were done using development versions of the above systems connected to PostgreSQL 8.3.

Sesame is an open source RDF database with support for querying and reasoning.[13] In addition to its in-memory database engine it can be coupled with relational databases or deployed on top of file systems. Sesame supports RDFS inference and other entailment regimes such as OWL-Horst [5] by coupling with external reasoners. Sesame provides an infrastructure for defining custom inference rules. Our tests have been done using Sesame 2.3 with persistence support given by the native store.

First of all, it is worth noting that all systems allow persistent storage on RDBMS. All systems, with the exception of ours, implement also direct filesystem storage. All

[10] A version of this translation can be found in [29].

[11] System available at `http://agraph.franz.com/allegrograph/`

[12] Distributed at `https://jena.svn.sourceforge.net/svnroot/jena/ARQ/`

[13] System available at `http://www.openrdf.org/`

cover RDFS (actually, disregarding axiomatic triples) and partial or non-standard OWL fragments. Although all the systems feature some form of persistence, both reasoning and query evaluation are usually performed in main memory. All the systems, except AllegroGraph and ours, adopt a persistent materialization approach for inferring data.

All systems – along with basic inference – support named graph querying, but, with the exception of GiaBATA, combining both features results in incomplete behavior as described in Section 2. Inference is properly handled as long as the query ranges over the whole dataset, whereas it fails in case of querying explicit default or named graphs. That makes querying of named graphs involving inference impossible with standard systems.

For performance comparison we rely on the LUBM benchmark suite [16]. Our tests involve the test datasets LUBMn for $n \in \{1, 5, 10, 30\}$ with LUBM30 having roughly four million triples (exact numbers are reported in [16]). In order to test the additional performance cost of our extensions, we opted for showing how the performance figures change when queries which require RDFS entailment rules (LUBM Q4-Q7) are considered, w.r.t. queries in which rules do not have an impact (LUBM Q1-Q3, see Appendix of [16] for the SPARQL encodings of $Q1$–$Q7$). Experiments are enough for comparing performance trends, so we didn't consider at this stage larger instances of LUBM. Note that evaluation times include the data loading times. Indeed, while former performance benchmarks do not take this aspect in account, from the semantic point of view, pre-materialization-at-loading computes inferences needed for complete query answering under the entailment of choice. On further reflection, dynamic querying of RDFS moves inference from this materialization to the query step, which would result in an apparent advantage for systems that rely on pre-materialization for RDFS data. Also, the setting of this paper assumes materialization cannot be performed *una tantum*, since inferred information depends on the entailment regime of choice, and on the dataset at hand, on a *per query* basis. We set a 120min query timeout limit to all test runs.

Our test runs include the following system setup: (i) "Allegro (native)" and "Allegro (ordered)" (ii) "ARQ"; (iii) "GiaBATA (native)" and "GiaBATA (ordered)"; and (iv) "Sesame". For (i) and (iii), which apply dynamic inference mechanisms, we use "(native)" and "(ordered)" to distinguish between executions of queries in LUBM's native ordering and in a optimized reordered version, respectively. The GiaBATA test runs both use Magic Sets optimization. To appreciate the cost of RDFS reasoning for queries $Q4$–$Q7$, the test runs for (i)–(iv) also include the loading time of the datasets, i.e., the time needed in order to perform RDFS data materialization or to simply store the raw RDF data.

The detailed outcome of the test results are summarized in Fig. 2. For the RDF test queries $Q1$–$Q3$, GiaBATA is able to compete for $Q1$ and $Q3$. The systems ARQ and Sesame turned out to be competitive for $Q2$ by having the best query response times, while Allegro (native) scored worst. For queries involving inference ($Q4$–$Q7$) Allegro shows better results. Interestingly, systems applying dynamic inference, namely Allegro and GiaBATA, query pattern reordering plays a crucial role in preserving performance and in assuring scalability; without reordering the queries simply timeout. In particular, Allegro is well-suited for queries ranging over several properties of a single class, whereas if the number of classes and properties increases ($Q7$), GiaBATA exhibits

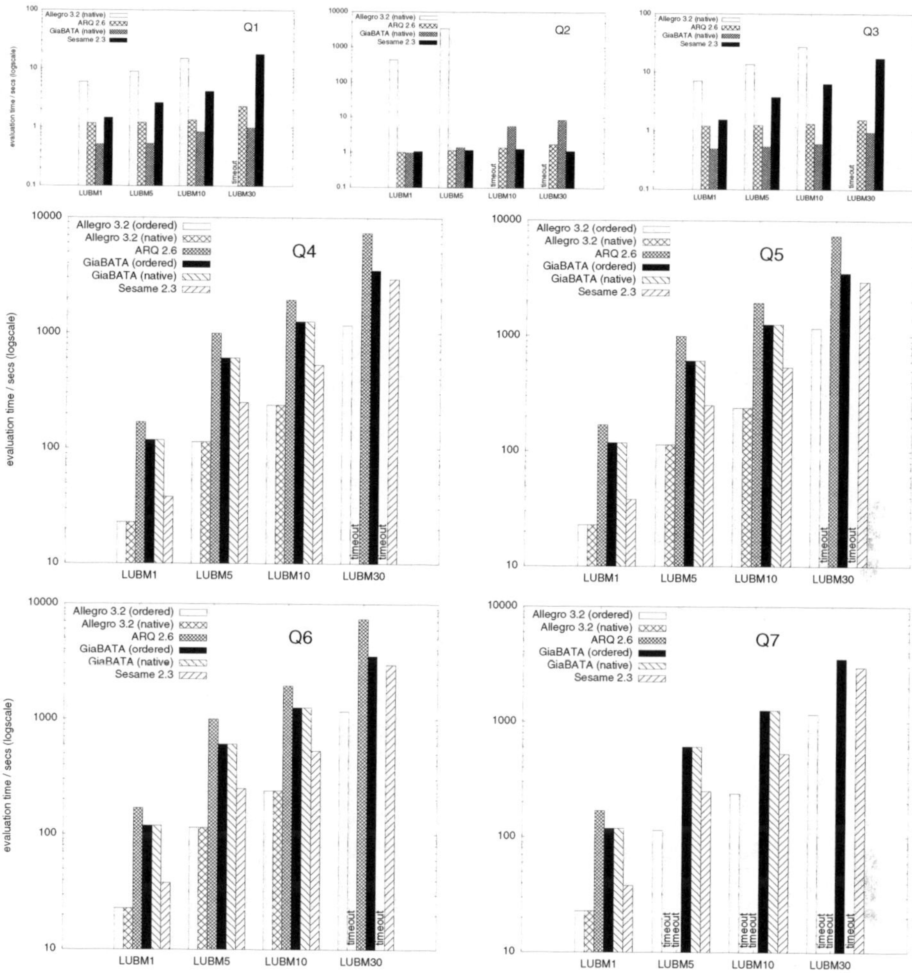

Fig. 2. Evaluation

better scalability. Finally, a further distinction between systems relying on DBMS support and systems using native structures is disregarded, and since figures (in logarithmic scale) depict overall loading and querying time, this penalizes in specific cases those systems that use a DBMS.

6 Future Work and Conclusion

We presented a framework for dynamic querying of RDFS data, which extends SPARQL by two language constructs: **`using ontology`** and **`using ruleset`**. The former is geared towards dynamically creating the dataset, whereas the latter adapts the entailment regime of the query. We have shown that our extension conservatively extends the standard SPARQL language and that by selecting appropriate rules in **`using ruleset`**, we may choose varying rule-based entailment regimes at query-time. We illustrated how such extended SPARQL queries can be translated to Datalog and SQL, thus providing entry points for implementation and well-known optimization techniques. Our initial experiments have shown that although dynamic querying does more computation at query-time, it is still competitive for use cases that need on-the-fly construction of datasets and entailment regimes. Especially here, query optimization techniques play a crucial role, and our results suggest to focus further research in this direction. Furthermore, we aim at conducting a proper computational analysis as it has been done for Hypothetical datalog [30], in which truth of atoms is conditioned by hypothetical additions to the dataset at hand. Likewise, our framework allows to add ontological knowledge and rules to datasets before querying: note however that, in the spirit of [31], our framework allows for hypotheses (also called "premises") on a per query basis rather than a per atom basis.

References

1. Klyne, G., Carroll, J.J. (eds.): Resource Description Framework (RDF): Concepts and Abstract Syntax. W3C Rec. (February 2004)
2. Brickley, D., Miller, L.: FOAF Vocabulary Specification 0.91 (2007), `http://xmlns.com/foaf/spec/`
3. Bojārs, U., Breslin, J.G., Berrueta, D., Brickley, D., Decker, S., Fernández, S., Görn, C., Harth, A., Heath, T., Idehen, K., Kjernsmo, K., Miles, A., Passant, A., Polleres, A., Polo, L., Sintek, M.: SIOC Core Ontology Specification. W3C member submission (June 2007)
4. Suchanek, F.M., Kasneci, G., Weikum, G.: Yago: A Core of Semantic Knowledge. In: WWW 2007. ACM, New York (2007)
5. ter Horst, H.J.: Completeness, decidability and complexity of entailment for RDF Schema and a semantic extension involving the OWL vocabulary. J. Web Semant. 3(2–3), 79–115 (2005)
6. Muñoz, S., Pérez, J., Gutierrez, C.: Minimal deductive systems for RDF. In: Franconi, E., Kifer, M., May, W. (eds.) ESWC 2007. LNCS, vol. 4519, pp. 53–67. Springer, Heidelberg (2007)
7. Hogan, A., Harth, A., Polleres, A.: Scalable authoritative owl reasoning for the web. Int. J. Semant. Web Inf. Syst. 5(2) (2009)
8. Polleres, A., Scharffe, F., Schindlauer, R.: SPARQL++ for mapping between RDF vocabularies. In: ODBASE 2007, pp. 878–896. Springer, Heidelberg (2007)
9. Harth, A., Umbrich, J., Hogan, A., Decker, S.: YARS2: A Federated Repository For Querying Graph Structured Data from the Web. In: Aberer, K., Choi, K.-S., Noy, N., Allemang, D., Lee, K.-I., Nixon, L.J.B., Golbeck, J., Mika, P., Maynard, D., Mizoguchi, R., Schreiber, G., Cudré-Mauroux, P. (eds.) ASWC 2007 and ISWC 2007. LNCS, vol. 4825, pp. 211–224. Springer, Heidelberg (2007)
10. Prud'hommeaux, E., Seaborne, A. (eds.): SPARQL Query Language for RDF. W3C Rec. (January 2008)
11. Polleres, A.: From SPARQL to rules (and back). In: WWW 2007, pp. 787–796. ACM, New York (2007)
12. Angles, R., Gutierrez, C.: The expressive power of SPARQL. In: Sheth, A.P., Staab, S., Dean, M., Paolucci, M., Maynard, D., Finin, T., Thirunarayan, K. (eds.) ISWC 2008. LNCS, vol. 5318, pp. 114–129. Springer, Heidelberg (2008)

13. Leone, N., Pfeifer, G., Faber, W., Eiter, T., Gottlob, G., Perri, S., Scarcello, F.: The DLV system for knowledge representation and reasoning. ACM Trans. Comput. Log. 7(3), 499–562 (2006)
14. Euzenat, J., Polleres, A., Scharffe, F.: Processing ontology alignments with SPARQL. In: OnAV 2008 Workshop, CISIS 2008, pp. 913–917. IEEE Computer Society, Los Alamitos (2008)
15. Ianni, G., Krennwallner, T., Martello, A., Polleres, A.: A Rule Systemfor Querying Persistent RDFS Data. In: Aroyo, L., Traverso, P., Ciravegna, F., Cimiano, P., Heath, T., Hyvönen, E., Mizoguchi, R., Oren, E., Sabou, M., Simperl, E. (eds.) ESWC 2009. LNCS, vol. 5554, pp. 857–862. Springer, Heidelberg (2009)
16. Guo, Y., Pan, Z., Heflin, J.: LUBM: A Benchmark for OWL Knowledge Base Systems. J. Web Semant. 3(2–3), 158–182 (2005)
17. Motik, B., Grau, B.C., Horrocks, I., Wu, Z., Fokoue, A., Lutz, C. (eds.): OWL 2 Web Ontology Language Profiles W3C Cand. Rec. (June 2009)
18. Pérez, J., Arenas, M., Gutierrez, C.: Semantics and complexity of SPARQL. In: Cruz, I., Decker, S., Allemang, D., Preist, C., Schwabe, D., Mika, P., Uschold, M., Aroyo, L.M. (eds.) ISWC 2006. LNCS, vol. 4273, pp. 30–43. Springer, Heidelberg (2006)
19. Polleres, A., Schindlauer, R.: dlvhex-sparql: A SPARQL-compliant query engine based on dlvhex. In: ALPSWS 2007. CEUR-WS, pp. 3–12 (2007)
20. Hayes, P.: RDF semantics. W3C Rec. (February 2004)
21. Ianni, G., Martello, A., Panetta, C., Terracina, G.: Efficiently querying RDF(S) ontologies with answer set programming. J. Logic Comput. 19(4), 671–695 (2009)
22. de Bruijn, J.: SemanticWeb Language Layering with Ontologies, Rules, and Meta-Modeling. PhD thesis, University of Innsbruck (2008)
23. Boley, H., Kifer, M.: RIF Basic Logic Dialect. W3C Working Draft (July 2009)
24. Eiter, T., Ianni, G., Schindlauer, R., Tompits, H.: Effective integration of declarative rules with external evaluations for semantic-web reasoning. In: Sure, Y., Domingue, J. (eds.) ESWC 2006. LNCS, vol. 4011, pp. 273–287. Springer, Heidelberg (2006)
25. Terracina, G., Leone, N., Lio, V., Panetta, C.: Experimenting with recursive queries in database and logic programming systems. Theory Pract. Log. Program. 8(2), 129–165 (2008)
26. Theoharis, Y., Christophides, V., Karvounarakis, G.: Benchmarking Database Representations of RDF/S Stores. In: Gil, Y., Motta, E., Benjamins, V.R., Musen, M.A. (eds.) ISWC 2005. LNCS, vol. 3729, pp. 685–701. Springer, Heidelberg (2005)
27. Abadi, D.J., Marcus, A., Madden, S., Hollenbach, K.J.: Scalable Semantic Web Data Management Using Vertical Partitioning. In: VLDB, pp. 411–422. ACM, New York (2007)
28. Beeri, C., Ramakrishnan, R.: On the power of magic. J. Log. Program. 10(3-4), 255–299 (1991)
29. Lu, J., Cao, F., Ma, L., Yu, Y., Pan, Y.: An Effective SPARQL Support over Relational Databases. In: SWDB-ODBIS, pp. 57–76 (2007)
30. Bonner, A.J.: Hypothetical datalog: complexity and expressibility. Theor. Comp. Sci. 76(1), 3–51 (1990)
31. Gutiérrez, C., Hurtado, C.A., Mendelzon, A.O.: Foundations of semantic web databases. In: PODS 2004, pp. 95–106. ACM, New York (2004)

Decidable Order-Sorted Logic Programming for Ontologies and Rules with Argument Restructuring

Ken Kaneiwa[1] and Philip H.P. Nguyen[2]

[1] National Institute of Information and Communications Technology, Japan
kaneiwa@nict.go.jp
[2] Department of Justice, Government of South Australia
philip.nguyen@sa.gov.au

Abstract. This paper presents a decidable fragment for combining ontologies and rules in order-sorted logic programming. We describe order-sorted logic programming with sort, predicate, and meta-predicate hierarchies for deriving predicate and meta-predicate assertions. Meta-level predicates (predicates of predicates) are useful for representing relationships between predicate formulas, and further, they conceptually yield a hierarchy similar to the hierarchies of sorts and predicates. By extending the order-sorted Horn-clause calculus, we develop a query-answering system that can answer queries such as atoms and meta-atoms generalized by containing predicate variables. We show that the expressive query-answering system computes every generalized query in single exponential time, i.e., the complexity of our query system is equal to that of DATALOG.

1 Introduction

In the Semantic Web context, conceptual knowledge representation and reasoning [23] have been studied for modeling ontologies in OWL (Web Ontology Language) [20]. In general, concepts are interpreted by sets of individuals, and concept hierarchies are constructed by subsumption (similar to IS-A relations). The formal semantics and reasoning of concept description languages are guaranteed by logical formalizations. Order-sorted logic [4,22,14] (as first-order logic with partially ordered sorts) provides sorts and sort hierarchy that represent concepts and their concept hierarchy, respectively. A predicate hierarchy, which is an extension of the sort hierarchy, consists of n-ary predicates that are conceptually related to each other. In [13], order-sorted logic programming was extended by introducing such a predicate hierarchy. Furthermore, the conceptual structure theory [19] was extended to include relation types and their type hierarchy for building complex ontologies.

Meta-level predicates (predicates of predicates) are expressions that can be employed for representing relationships between facts in knowledge bases. Similar to hierarchies of sorts and predicates, these meta-predicates can be used to conceptually construct a hierarchy, e.g., the meta-predicate *causes* implies the super

A. Bernstein et al. (Eds.): ISWC 2009, LNCS 5823, pp. 328–343, 2009.

meta-predicate *likelyCauses*. In the OWL family, meta-concepts are supported by OWL-Full (the most expressive language of OWL). The semantics of modeling for meta-concepts and the undecidability of meta-modeling in OWL-Full have been discussed in [18]. Further, higher-order expressions may conceptually yield a hierarchy when they are named using natural language words. However, order-sorted (or typed) logic programming lacks representation and reasoning for such meta-level predicates.

Alternatively, logic programming provides formal semantics and decidable reasoning services for RuleML (Rule Makeup Language) [1] in the Semantic Web. This language is a restricted fragment of first-order logic, and its complexities [5] have been studied in the area of automated deduction. It is known that full logic programming is undecidable, but function-free logic programming (i.e., DATALOG) is EXPTIME-complete with respect to the length of a program. In addition, non-recursive logic programming is NEXPTIME-complete, even if it includes functions.

However, SWRL (Semantic Web Rule Language) [11], a combination of OWL and RuleML, leads to undecidable reasoning between ontologies and rules (as shown in [10]). Several decidable fragments for combining ontologies and rules, such as DL-safe [9,21], DLP (Description Logic Programs) [7], and the rule-based language ELP [16] (related to the tractable language profile OWL 2 EL [2]), have been proposed by restricting the expressive power of rules. Similar to the approaches adopted in past studies, in order to make ontologies and rules in logic programing expressive and at the same time retain decidability, the logic programming language must be carefully extended for conceptual knowledge representation and reasoning. HILOG [3], which involves the second-order expression of meta-level predicates, has been developed as a decidable higher-order language for logic programming, and it may be more complex than the EXPTIME complexity of DATALOG. Unfortunately, in most cases, higher-order logic programming [12] makes reasoning increasingly difficult because complex structures of higher-order terms need to be treated.

To overcome the aforementioned difficulties related to expressiveness and complexity, we introduce meta-predicates and their hierarchy in a restricted and decidable fragment for combining ontologies and rules. In particular, we formalize an order-sorted logic programming language with a meta-predicate hierarchy. As a result, three kinds of hierarchies (sort hierarchy, predicate hierarchy, and meta-predicate hierarchy) are included in the syntax and semantics of the sorted logic programming. We develop the order-sorted Horn-clause calculus [8], which serves as a sorted deductive system, for reasoning on concept hierarchies where predicate assertions and relationships among the assertions are derived. This calculus terminates if the knowledge bases are function free (i.e., only constants such as 0-ary functions are allowed). Using this calculus, we develop a query-answering system that is extended by generalizing queries with predicate variables. Our result shows that the complexity of the expressive query system (even for meta-predicates and predicate variables) is single exponential time and equal to the complexity of DATALOG.

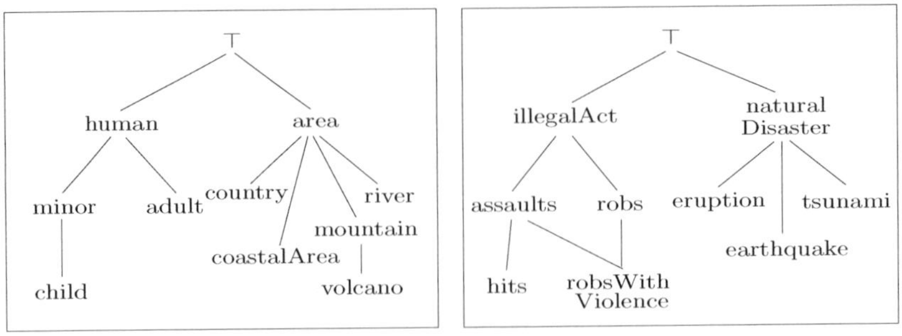

Fig. 1. Sort and predicate hierarchies

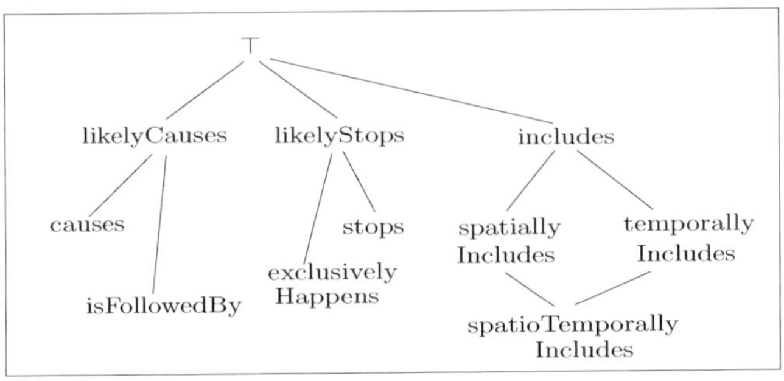

Fig. 2. A meta-predicate hierarchy

2 Motivating Examples

We now present some examples of hierarchies in a query-answering system. Given the sort, predicate, and meta-predicate hierarchies in Figs. 1 and 2, we consider logical reasoning using a knowledge base for the hierarchies. If the fact `hits(tom:minor,john:adult)` is valid, then the super predicate `illegalAct` can be derived in the predicate hierarchy (shown in Fig. 1).

```
hits(tom:minor,john:adult)
?-illegalAct(x:human)
yes
x=tom:minor
```

In this derivation, the second argument `john:adult` is deleted if the argument structure of the predicate `illegalAct` lacks the second argument of the predicate `hits`. Conceptually, both the name and argument structure of `illegalAct` are more abstract than `hits` in the predicate hierarchy.

Moreover, we employ meta-predicates (predicates of predicates) to express relationships among facts in the knowledge base. For example, the meta-predicate `isFollowedBy` is used to indicate that a tsunami in Phuket c_2 occurred after the earthquake in Indonesia c_1.

```
isFollowedBy(earthquake(c1:country),tsunami(c2:coastalArea))
?-likelyCauses(earthquake(c1:country),tsunami(c2:coastalArea))
yes
```

If the relationship between the two facts is valid, the super meta-predicate `likelyCauses` can be inferred in the meta-predicate hierarchy (shown in Fig. 2).

Additionally, the fact `earthquake(c1:country)` is derived from this relationship because it is the first argument of the meta-predicate `isFollowedBy`.

```
?-earthquake(c1:country)
yes
```

The assumption underlying the abovementioned derivation is that the meta-predicate points to the occurrence of facts in addition to indicating the existence of a relationship between them.

An expression with predicate variables X and Y is used to query the validity of a causal relationship between two natural disasters as follows.

```
?-likelyCauses(X:naturalDisaster(x:area),
                          Y:naturalDisaster(y:area))
yes
X=earthquake, x=c1:country, Y=tsunami, y=c2:coastalArea
```

Using the meta-predicate hierarchy, the reasoning engine should return the answer `yes` with a successful substitution of the variables, such as `X=earthquake`, `x=c1:country`, `Y=tsunami`, and `y=c2:coastalArea`.

In the Semantic Web context, the argument manipulation shown above is very useful when software agents derive semantically related terms and assertions using ontologies. This is because the differences between argument structures in predicates must facilitate such flexible reasoning for predicate assertions in the sorted logic programming.

3 Order-Sorted Logic with Meta-predicates

We introduce meta-predicates as new conceptual symbols in a sorted language. These meta-predicates represent n-ary relations among atomic predicate formulas and are used to construct a concept hierarchy.

Definition 1. *The alphabet of a sorted first-order language $\mathcal{L}$ with sort, predicate, and meta-predicate hierarchies contains the following symbols:*

1. *S: a countable set of sort symbols*
2. *F_n: a countable set of n-ary function symbols*
3. *P_n: a countable set of n-ary predicate symbols*
4. *Ψ_n: a countable set of n-ary meta-predicate symbols*

5. $\leftarrow, \{, \}$: *the connective and auxiliary symbols*
6. V_s: *an infinite set of variables* $x\colon s, y\colon s, z\colon s, \ldots$ *of sort* s

The set of all predicates is denoted by $P = \bigcup_{n\geq 1} P_n$, and the set of variables of all sorts is denoted by $V = \bigcup_{s\in S} V_s$.

Definition 2 (Sorted Signatures). *A signature of a sorted first-order language* $\mathcal{L}$ *with sort, predicate, and meta-predicate hierarchies (called a sorted signature) is a tuple* $\Sigma = (S, P, \Psi_n, \Omega, \leq)$ *such that:*

1. $(S, \leq)$ *is a partially ordered set of sorts (called a sort hierarchy);*
2. $(P, \leq)$ *is a partially ordered set of predicates (called a predicate hierarchy);*
3. $(\Psi_n, \leq)$ *is a partially ordered set of* n*-ary meta-predicates (called a meta-predicate hierarchy);*
4. Ω *is a set of function and predicate declarations such that*
   (a) *if* $f \in F_n$*, then there is a unique function declaration of the form* $f\colon s_1 \times \cdots \times s_n \to s \in \Omega$*, and*
   (b) *if* $p \in P_n$*, then there is a unique predicate declaration of the form* $p\colon s_1 \times \cdots \times s_n \in \Omega$*.*

The predicate hierarchy includes predicates with different argument structures, e.g., a binary predicate can be a subpredicate of a unary predicate. On the contrary, the meta-predicate hierarchy only contains meta-predicates with a fixed arity. In the sorted signature, Ω contains function and predicate declarations that determine the domains and ranges of functions f and predicates p. In particular, F_0 is the set of 0-ary functions (i.e., constants), and each constant $c \in F_0$ has a unique constant declaration of the form $c\colon \ \to s$.

We generally call sorts, predicates, and meta-predicates *concepts*. Let cp_1, cp_2, and cp_3 be three concepts. A concept cp_2 is called a upper bound for cp_1 if $cp_1 \leq cp_2$, and a concept cp_2 is called a lower bound for cp_1 if $cp_2 \leq cp_1$. The least upper bound $cp_1 \sqcup cp_2$ is a upper bound for cp_1 and cp_2 such that $cp_1 \sqcup cp_2 \leq cp_3$ holds for any other upper bound cp_3. The greatest lower bound $cp_1 \sqcap cp_2$ is a lower bound for cp_1 and cp_2 such that $cp_3 \leq cp_1 \sqcap cp_2$ holds for any other lower bound cp_3.

We define the following sorted expressions in the sorted signature Σ: terms, atoms (atomic formulas), meta-atoms (meta atomic formulas), goals, and clauses.

Definition 3 (Sorted Terms). *Let* $\Sigma = (S, P, \Psi_n, \Omega, \leq)$ *be a sorted signature. The set* $\mathcal{T}_s$ *of terms of sort* s *is defined by the following:*

1. *If* $x\colon s \in V_s$*, then* $x\colon s \in \mathcal{T}_s$*.*
2. *If* $t_1 \in \mathcal{T}_{s_1}, \ldots, t_n \in \mathcal{T}_{s_n}$*,* $f \in F_n$*, and* $f\colon s_1 \times \cdots \times s_n \to s \in \Omega$*, then* $f(t_1, \ldots, t_n)\colon s \in \mathcal{T}_s$*.*
3. *If* $t \in \mathcal{T}_{s'}$ *and* $s' \leq s$*, then* $t \in \mathcal{T}_s$*.*

Note that $\mathcal{T}_s$ contains not only terms of sort s but also terms of subsorts s' of sort s if $s' \leq s$. The set of terms of all sorts is denoted by $\mathcal{T} = \bigcup_{s\in S} \mathcal{T}_s$.

The function *sort* is a mapping from sorted terms to their sorts, defined by (i) $sort(x\colon s) = s$ and (ii) $sort(f(t_1, \ldots, t_n)\colon s) = s$. Let $Var(t)$ denote the set of variables occurring in a sorted term t. A sorted term t is called *ground* if $Var(t) = \emptyset$. $\mathcal{T}_0 = \{t \in \mathcal{T} \mid Var(t) = \emptyset\}$ is the set of sorted ground terms, and the set of ground terms of sort s is denoted by $\mathcal{T}_{0,s} = \mathcal{T}_0 \cap \mathcal{T}_s$. We write $\mathcal{T}_s^\Sigma$, $\mathcal{T}_0^\Sigma$, $\mathcal{T}_{s,0}^\Sigma$, and $\mathcal{T}^\Sigma$ for explicitly representing the sorted signature Σ.

In the following definition, sorted Horn clauses [17,6] are extended by meta-atoms $\psi(A_1, \ldots, A_n)$ that consist of meta-predicates ψ and atoms $A_1, \ldots, A_n$.

Definition 4 (Atoms, Meta-atoms, Goals, and Clauses) *Let $\Sigma = (S, P, \Psi_n, \Omega, \leq)$ be a sorted signature. The set $\mathcal{A}$ of atoms, the set $\mathcal{MA}$ of meta-atoms, the set $\mathcal{G}$ of goals, and the set $\mathcal{C}$ of clauses are defined by:*

1. *If $t_1 \in \mathcal{T}_{s_1}, \ldots, t_n \in \mathcal{T}_{s_n}$, $p \in P_n$, and $p\colon s_1 \times \cdots \times s_n \in \Omega$, then $p(t_1, \ldots, t_n) \in \mathcal{A}$.*
2. *If $A_1, \ldots, A_n \in \mathcal{A}$ and $\psi \in \Psi_n$, then $\psi(A_1, \ldots, A_n) \in \mathcal{MA}$.*
3. *If $L_1, \ldots, L_n \in \mathcal{A} \cup \mathcal{MA}$ $(n \geq 0)$, then $\{L_1, \ldots, L_n\} \in \mathcal{G}$.*
4. *If $G \in \mathcal{G}$ and $L \in \mathcal{A} \cup \mathcal{MA}$, then $L \leftarrow G \in \mathcal{C}$.*

Meta-atoms assert n-ary relations ψ over atoms $A_1, \ldots, A_n$ and can appear in the heads and bodies of extended Horn clauses. For example, the atoms $earthquake(c_1\colon country)$ and $tsunami(c_2\colon coastalArea)$ are used to assert the meta-atom $causes(earthquake(c_1\colon country), tsunami(c_2\colon coastalArea))$, where *causes* is a binary meta-predicate. A clause $L \leftarrow G$ is denoted by $L \leftarrow$ if $G = \emptyset$.

We define a sorted substitution such that each sorted variable $x\colon s$ is replaced with a sorted term in $\mathcal{T}_s$. Each sorted substitution is represented by $\{x_1\colon s_1/t_1, \ldots, x_n\colon s_n/t_n\}$. Let θ be a sorted substitution. Then, θ is said to be a sorted ground substitution if for every variable $x\colon s \in Dom(\theta)$, $\theta(x\colon s)$ is a sorted ground term. Let E be a sorted expression. The substitution θ is a sorted ground substitution for E if $E\theta$ is ground and $Dom(\theta) = Var(E)$. The composition $\theta_1\theta_2$ of sorted substitutions θ_1 and θ_2 is defined by $\theta_1\theta_2(x\colon s) = \theta_2(\theta_1(x\colon s))$.

In Σ, there are various argument structures in the predicate hierarchy $(P, \leq)$ because P contains predicates with various arities. Additionally, we declare the argument structure for each predicate $p \in P$ in Σ as follows.

Definition 5 (Argument Declaration). *Let $\Sigma = (S, P, \Psi_n, \Omega, \leq)$ be a sorted signature. An argument declaration Λ is a pair (AN, Π) of a set AN of argument names and a set Π of argument structures of the form $p\colon \langle a_1, \ldots, a_n\rangle$ where $p \in P_n$, $a_1, \ldots, a_n \in AN$, and for any $i \neq j$, $a_i \neq a_j$.*

Given an argument declaration $\Lambda = (AN, \Pi)$, we define an argument function $Arg\colon P \to 2^{AN}$ such that $Arg(p) = \{a_1, \ldots, a_n\}$ for each $p\colon \langle a_1, \ldots, a_n\rangle \in \Pi$. An argument declaration Λ is well arranged in the predicate hierarchy if $Arg(q) \subseteq Arg(p)$ for any $p, q \in P$ with $p \leq q$. Intuitively, the well-arranged argument declaration implies that the predicate q does not have any argument that its subpredicate p does not have.

Definition 6 (Argument Elimination). *Let $\Sigma = (S, P, \Psi_n, \Omega, \leq)$ be a sorted signature with an argument declaration $\Lambda = (AN, \Pi)$, let $\langle d_1, \ldots, d_n\rangle$ be an*

n-tuple, and let $p \in P_n$, $q \in P_m$ with $Arg(q) \subseteq Arg(p)$. An argument elimination from p to q is a function $\sigma^-_{p\to q}(\langle d_1,\ldots,d_n\rangle) = \langle d'_1,\ldots,d'_m\rangle$ such that

$$d'_i = d_j \textit{ if } a'_i = a_j \textit{ for each } 1 \leq i \leq m$$

where $p\colon \langle a_1,\ldots,a_n\rangle$ and $q\colon \langle a'_1,\ldots,a'_m\rangle$ in Π.

The argument eliminations will be used in the semantics and inference system of the order-sorted logic. An important property of argument eliminations that is used for the development of predicate-hierarchy reasoning is expressed as follows.

Proposition 1 (Transitivity of Argument Eliminations). *Let Σ be a sorted signature with an argument declaration Λ, let τ be an n-tuple, and let $p \in P_n$, $q \in P_m$, and $r \in P_k$. If $p \leq q$, $q \leq r$, and Λ is well arranged in Σ, then $\sigma^-_{q\to r}(\sigma^-_{p\to q}(\tau)) = \sigma^-_{p\to r}(\tau)$.*

This proposition guarantees that argument eliminations are safely embedded in predicate-hierarchy reasoning if the argument declaration is well arranged.

We define the semantics of the order-sorted logic with sort, predicate, and meta-predicate hierarchies as follows.

Definition 7 (Σ-Models). *Let Σ be a sorted signature with a well-arranged argument declaration Λ. A Σ-model M is a tuple $(U, U_{\mathcal{F}}, I)$ such that*

1. *U is a non-empty set of individuals;*
2. *$U_{\mathcal{F}}$ is a non-empty set of facts;*
3. *I is a function with the following conditions:*
   (a) *if $s \in S$, then $I(s) \subseteq U$ (in particular, $I(\top) = U$),*
   (b) *if $s_i \leq s_j$ for $s_i, s_j \in S$, then $I(s_i) \subseteq I(s_j)$,*
   (c) *if $f \in F_n$ and $f\colon s_1 \times \cdots \times s_n \to s \in \Omega$, then $I(f)\colon I(s_1) \times \cdots \times I(s_n) \to I(s)$,*
   (d) *if $p \in P_n$ and $p\colon s_1 \times \cdots \times s_n \in \Omega$, then $I(p)\colon I(s_1) \times \cdots \times I(s_n) \to 2^{U_{\mathcal{F}}}$,*
   (e) *if $p \leq q$ for $p \in P_n$ and $q \in P_m$, then $I(p)(\tau) \subseteq I(q)(\sigma^-_{p\to q}(\tau))$,*
   (f) *if $\psi \in \Psi_n$, then $I(\psi) \subseteq U^n_{\mathcal{F}}$,*
   (g) *if $\psi \leq \phi$ for $\psi, \phi \in \Psi_n$, then $I(\psi) \subseteq I(\phi)$.*

The class of Σ-models is a restricted class of standard models such that the domains and ranges of functions and predicates are constrained by sorts and the hierarchies of sorts, predicates, and meta-predicates are interpreted by subset relations over U, $U_{\mathcal{F}}$, and $U^n_{\mathcal{F}}$.

By the argument eliminations in the predicate hierarchy, the following two properties are derived in the class of Σ-models.

Proposition 2 (Conceptuality of Predicates). *Let $p \in P_n$, $q \in P_m$, and $r \in P_k$ and let $\tau_1 \in U^n$, $\tau_2 \in U^m$, and $\tau \in U^k$. Every Σ-model M has the following properties:*

1. *$p \sqcup q \leq r$ implies $I(p)(\tau_1) \cup I(q)(\tau_2) \subseteq I(r)(\tau)$ with $\tau = \sigma^-_{p\to r}(\tau_1) = \sigma^-_{q\to r}(\tau_2)$.*
2. *$r \leq p \sqcap q$ implies $I(r)(\tau) \subseteq I(p)(\sigma^-_{r\to p}(\tau)) \cap I(q)(\sigma^-_{r\to q}(\tau))$.*

This property is important for showing that predicates are consistently conceptualized in a hierarchy. However, this is not simple because predicates have their respective arguments that have different structures in the predicate hierarchy.

Even if predicates are conceptually interpreted as sets of tuples, it is necessary to define a model that can identify each fact expressed by predicate formulas.

Proposition 3 (Identifiability of Predicates). *Let τ be an n-tuple in U^n, and let $p \in P_n$, $q \in P_m$ $(p \neq q)$. Some Σ-models M have the following properties:*

1. *If $Arg(p) = Arg(q)$, then there are two facts $e_1 \in I(p)(\tau)$ and $e_2 \in I(q)(\tau)$.*
2. *If $Arg(p) \supsetneq Arg(q)$, then there are two facts $e_1 \in I(p)(\tau)$ and $e_2 \in I(q)(\sigma^-_{p \to q}(\tau))$.*

This proposition indicates that any two ground atoms with identical arguments $p(t_1, \dots, t_n)$ and $q(t_1, \dots, t_n)$ can be identified as distinct facts, if necessary. In the Σ-models, the set of facts $U_\mathcal{F}$ is used to identify ground atoms such that predicate assertions correspond to different elements in $U_\mathcal{F}$.

A variable assignment on a Σ-model $M = (U, U_\mathcal{F}, I)$ is a function $\alpha\colon V \to U$ where $\alpha(x\colon s) \in I(s)$. The variable assignment $\alpha[x\colon s/d]$ is defined by $(\alpha - \{(x\colon s, \alpha(x\colon s))\}) \cup \{(x\colon s, d)\}$. In other words, if $v = x\colon s$, then $\alpha[x\colon s/d](v) = d$, and otherwise $\alpha[x\colon s/d](v) = \alpha(v)$. Let $\Delta \subseteq U_\mathcal{F}$ be a valuation of facts on M. A Σ-interpretation $\mathcal{I}$ is a tuple (M, Δ, α) of a Σ-model M, a valuation of facts Δ on M, and a variable assignment α on M. The Σ-interpretation $(M, \Delta, \alpha[x\colon s/d])$ is simply denoted by $\mathcal{I}\alpha[x\colon s/d]$.

We define an interpretation of sorted terms and atoms as follows.

Definition 8. *Let $\mathcal{I} = (M, \Delta, \alpha)$ be a Σ-interpretation. The denotation function $[\![\ ]\!]_\alpha\colon \mathcal{T} \to U$ is defined by the following:*

1. $[\![x\colon s]\!]_\alpha = \alpha(x\colon s)$,
2. $[\![f(t_1, \dots, t_n)\colon s]\!]_\alpha = I(f)([\![t_1]\!]_\alpha, \dots, [\![t_n]\!]_\alpha)$ *with* $f\colon s_1 \times \cdots \times s_n \to s \in \Omega$,
3. $[\![p(t_1, \dots, t_n)]\!]_\alpha = I(p)([\![t_1]\!]_\alpha, \dots, [\![t_n]\!]_\alpha)$ *with* $p\colon s_1 \times \cdots \times s_n \in \Omega$.

The satisfiability of atoms, meta-atoms, goals, and clauses is defined by a Σ-interpretation $\mathcal{I}$.

Definition 9 (Σ-Satisfiability Relation). *Let $\mathcal{I} = (M, \Delta, \alpha)$ with $M = (U, U_\mathcal{F}, I)$ be a Σ-interpretation and let $F \in \mathcal{A} \cup \mathcal{MA} \cup \mathcal{G} \cup \mathcal{C}$. The Σ-satisfiability relation $\mathcal{I} \models F$ is defined inductively as follows:*

1. $\mathcal{I} \models A$ *iff* $[\![A]\!]_\alpha \cap \Delta \neq \emptyset$.
2. $\mathcal{I} \models \psi(A_1, \dots, A_n)$ *iff* $\mathcal{I} \models A_1, \dots, \mathcal{I} \models A_n$ *and* $([\![A_1]\!]_\alpha \times \cdots \times [\![A_n]\!]_\alpha) \cap I(\psi) \neq \emptyset$.
3. $\mathcal{I} \models \{L_1, \dots, L_n\}$ *iff* $\mathcal{I} \models L_1, \dots, \mathcal{I} \models L_n$.
4. $\mathcal{I} \models L \leftarrow G$ *iff for all* $d_1 \in I(s_1), \dots, d_n \in I(s_n)$, $\mathcal{I}\alpha[x_1\colon s_1/d_1, \dots, x_n\colon s_n/d_n] \models G$ *implies* $\mathcal{I}\alpha[x_1\colon s_1/d_1, \dots, x_n\colon s_n/d_n] \models L$ *where* $Var(L \leftarrow G) = \{x_1\colon s_1, \dots, x_n\colon s_n\}$.

Let $F \in \mathcal{A} \cup \mathcal{MA} \cup \mathcal{G} \cup \mathcal{C}$. An expression F is said to be Σ-satisfiable if for some Σ-interpretation $\mathcal{I}$, $\mathcal{I} \models F$. Otherwise, it is Σ-unsatisfiable. F is a consequence of a set of expressions $\mathcal{S}$ in the class of Σ-interpretations (denoted $\mathcal{S} \models F$) if for every Σ-interpretation $\mathcal{I}$, $\mathcal{I} \models \mathcal{S}$ implies $\mathcal{I} \models F$.

4 Horn-Clause Calculus for Predicate Hierarchies

In this section, we define the order-sorted Horn-clause calculus that is extended by adding inference rules for predicate and meta-predicate hierarchies. A knowledge base $\mathcal{K}$ is a finite set of sorted clauses in Σ where $\Sigma = (S, P, \Psi_n, \Omega, \leq)$ is a sorted signature with a well-arranged argument declaration Λ.

Definition 10 (Sorted Horn-Clause Calculus). *Let C be a ground clause, $\mathcal{K}$ be a knowledge base, and l be a label (non-negative integer). A derivation of C from $\mathcal{K}$ (denoted $\mathcal{K} \vdash l\colon C$) in the sorted Horn-clause calculus is defined as follows:*

- **Sorted substitution rule:** *Let $L \leftarrow G \in \mathcal{K}$ and θ be a sorted ground substitution for $L \leftarrow G$. Then, $\mathcal{K} \vdash l\colon (L \leftarrow G)\theta$ and l is incremented.*
- **Cut rule:** *Let $L \leftarrow G$ and $L' \leftarrow G' \cup \{L\}$ be ground clauses. If $\mathcal{K} \vdash l_1\colon L \leftarrow G$ and $\mathcal{K} \vdash l_2\colon L' \leftarrow G' \cup \{L\}$, then $\mathcal{K} \vdash l_2\colon L' \leftarrow G \cup G'$.*
- **Predicate hierarchy rule:** *Let $p(t_1, \ldots, t_n) \leftarrow G$ be a ground clause. If $\mathcal{K} \vdash l_1\colon p(t_1, \ldots, t_n) \leftarrow G$ and $p \leq q$, then $\mathcal{K} \vdash l_1\colon q(t'_1, \ldots, t'_m) \leftarrow G$ where $\sigma^-_{p \to q}(\langle t_1, \ldots, t_n \rangle) = \langle t'_1, \ldots, t'_m \rangle$.*
- **Meta-predicate hierarchy rule:** *Let $\psi(A_1, \ldots, A_n) \leftarrow G$ be a ground clause. If $\mathcal{K} \vdash l_1 : \psi(A_1, \ldots, A_n) \leftarrow G$ and $\psi \leq \phi$, then $\mathcal{K} \vdash l_1 : \phi(A_1, \ldots, A_n) \leftarrow G$.*
- **Fact derivation rule:** *Let $\psi(A_1, \ldots, A_n) \leftarrow G$ be a ground clause. If $\mathcal{K} \vdash l_1\colon \psi(A_1, \ldots, A_n) \leftarrow G$, then $\mathcal{K} \vdash l\colon A_i \leftarrow G$ with $1 \leq i \leq n$ and l is incremented.*

We simply write $\mathcal{K} \vdash l\colon L$ if $\mathcal{K} \vdash l\colon L \leftarrow$. The sorted substitution rule and the cut rule serve as sorted inference rules in ordinary order-sorted logic. The sorted substitution rule yields well-sorted ground clauses in the sort hierarchy. The predicate hierarchy rule and the meta-predicate hierarchy rule can be used to derive predicate and meta-predicate assertions in the predicate and meta-predicate hierarchies, respectively. The fact derivation rule derives atoms from meta-atoms, which was used in the third motivating example of Section 2.

To prove the completeness of the Horn-clause calculus, we construct extended Herbrand models for knowledge bases where positive atoms labeled by non-negative integers are used to identify different facts. We write $\mathcal{K} \vdash_{\psi(A_1,\ldots,A_n)} l\colon A_i$ if a labeled atom $l\colon A_i$ is directly derived from a labeled meta-atom $l_1\colon \psi(A_1, \ldots, A_n)$ using the fact derivation rule. Let $L \leftarrow G$ be a clause. We define $ground(L \leftarrow G)$ as the set of sorted ground clauses for $L \leftarrow G$. We define $ground(\mathcal{K}) = \bigcup_{L \leftarrow G \in \mathcal{K}} ground(L \leftarrow G)$ as the set of sorted ground clauses for all $L \leftarrow G$ in $\mathcal{K}$.

Definition 11 (Herbrand Models). *Let $\mathcal{K}$ be a knowledge base. A Herbrand model M_H for $\mathcal{K}$ is a tuple $(U_H, U_{\mathcal{F},H}, I_H)$ such that*

1. $U_H = \mathcal{T}_0$,
2. $U_{\mathcal{F},H} = \mathbb{N} - \{l \in \mathbb{N} \mid ground(\mathcal{K}) \vdash l\colon L \leftarrow G \;\&\; L \in \mathcal{MA}\}$,
3. *I_H is a function with the following conditions:*

(a) $I_H(s) = \mathcal{T}_{0,s}$ for each sort $s \in S$,
(b) if $f \in F_n$ and $f\colon s_1 \times \cdots \times s_n \to s \in \Omega$, then $I_H(f)(t_1, \ldots, t_n) = f(t_1, \ldots, t_n)\colon s$ where $t_1 \in I_H(s_1), \ldots, t_n \in I_H(s_n)$,
(c) if $p \in P_n$ and $p\colon s_1 \times \cdots \times s_n \in \Omega$, then $I_H(p)(\tau) = \bigcup_{q \leq p}\{l \in U_{\mathcal{F},H} \mid ground(\mathcal{K}) \vdash l\colon q(\tau')\}$ with $\sigma^-_{q \to p}(\tau') = \tau$,
(d) if $\psi \in \Psi_n$, then $I_H(\psi) = \bigcup_{\phi \leq \psi}\{(l_1, \ldots, l_n) \in U^n_{\mathcal{F},H} \mid$ for every $1 \leq i \leq n$, $ground(\mathcal{K}) \vdash_{\phi(A_1,\ldots,A_n)} l_i\colon A_i\}$.

A Herbrand interpretation $\mathcal{I}_H$ for $\mathcal{K}$ is a tuple (M_H, Δ_H, α) such that $M_H = (U_H, U_{\mathcal{F},H}, I_H)$ is a Herbrand model for $\mathcal{K}$, $\Delta_H = \bigcup_{\substack{p \in P \\ \tau \in \mathcal{T}_{0,s_1} \times \cdots \times \mathcal{T}_{0,s_n}}} I_H(p)(\tau)$ with $p\colon s_1 \times \cdots \times s_n \in \Omega$ is a valuation of facts on M_H, and α is a variable assignment on M_H.

We show that a Herbrand interpretation $\mathcal{I}_H$ is a Σ-interpretation that satisfies a knowledge base $\mathcal{K}$.

Lemma 1. *Let $\mathcal{K}$ be a knowledge base, let $\mathcal{I}_H$ be a Herbrand interpretation for $\mathcal{K}$, and let $L \leftarrow G$ be a clause. Then, the following statements hold:*

1. $\mathcal{I}_H \models L \leftarrow G$ *if and only if* $\mathcal{I}_H \models ground\ (L \leftarrow G)$.
2. $\mathcal{I}_H$ *is a Σ-interpretation of $\mathcal{K}$.*

We use the Herbrand model and the abovementioned lemma to prove the completeness of the Horn-clause calculus as follows.

Theorem 1 (Completeness of Horn-Clause Calculus). *Let $\mathcal{K}$ be a knowledge base in a sorted signature Σ and L be a ground atom or meta-atom. $\mathcal{K} \models L$ iff $\mathcal{K} \vdash l\colon L$.*

We show the termination of the Horn-clause calculus where a sorted signature is function-free.

Theorem 2 (Termination of Horn-Clause Calculus). *Let $\mathcal{K}$ be a knowledge base in a sorted signature Σ. Then, the Horn-clause calculus terminates if Σ is function-free.*

The termination of the calculus is proved by the fact that the set of derivable clauses $Con(\mathcal{K}) = \{L \leftarrow G \mid \mathcal{K} \vdash l\colon L \leftarrow G\}$ is finite. In other words, the calculus cannot generate terms and clauses infinitely because the cardinality of $Con(\mathcal{K})$ is bounded by finite constant, predicate, and meta-predicate symbols in $\mathcal{K}$.

We show the complexity of the derivation for atoms or meta-atoms L (not limited to ground) from a knowledge base where the set of ground atoms or meta-atoms $L\theta$ is computed using the Horn-clause calculus.

Corollary 1 (Complexity of Derivation for Atoms or Meta-atoms) *Let $\mathcal{K}$ be a knowledge base in a sorted signature Σ, L be an atom or meta-atom, and θ be a sorted ground substitution for L. If Σ is function-free, then deriving the set of ground atoms or meta-atoms $L\theta$ with $\mathcal{K} \vdash l\colon L\theta$ is (single) EXPTIME-complete (w.r.t. the length of $\mathcal{K}$).*

5 Query System

We describe a query-answering system for our order-sorted logic programming. In this system, query expressions are generalized by adding predicate variables in meta-atoms. The set of predicate variables is denoted by $\mathcal{V}$. The set of atoms with predicate variables is defined by $\mathcal{A}^{\mathcal{V}} = \{X\colon p(t_1,\ldots,t_n) \mid X \in \mathcal{V},\ p(t_1,\ldots,t_n) \in \mathcal{A}\}$. We call the form $X\colon p(t_1,\ldots,t_n)$ a predicate variable atom.

Definition 12 (Queries). *Let $\Sigma = (S, P, \Psi_n, \Omega, \leq)$ be a sorted signature with a well-arranged argument declaration Λ, and let $\mathcal{MA}^{\mathcal{V}} = \{\psi(A_1^+,\ldots,A_n^+) \mid \psi \in \Psi_n,\ A_1^+,\ldots,A_n^+ \in \mathcal{A} \cup \mathcal{A}^{\mathcal{V}}\}$ be the set of meta-atoms with predicate variables. The set $\mathcal{Q}$ of queries is defined by that if $L_1,\ldots,L_h \in \mathcal{A} \cup \mathcal{A}^{\mathcal{V}} \cup \mathcal{MA}^{\mathcal{V}}$, then $\{L_1,\ldots,L_h\} \in \mathcal{Q}$.*

We introduce substitutions for predicate variables $X \in \mathcal{V}$ such that each predicate variable atom $X\colon q(t'_1,\ldots,t'_m)$ is replaced with an atom $A \in \mathcal{A}$. We denote the set of atoms restricted to the subpredicates p of q by $\mathcal{A}_q = \{p(t_1,\ldots,t_n) \in \mathcal{A} \mid p \leq q \ \&\ \sigma^-_{p\to q}(\langle t_1,\ldots,t_n\rangle) = \langle t'_1,\ldots,t'_m\rangle\}$.

Definition 13 (Substitutions for Predicate Variables)
A substitution for predicate variables is a partial function $\delta\colon \mathcal{A}^{\mathcal{V}} \to \mathcal{A}$ such that $\delta(X\colon q(t'_1,\ldots,t'_m)) \in \mathcal{A}_q$ and the domain of δ (denoted $Dom(\delta)$) is finite.

The substitutions for predicate variables follow the predicate hierarchy, i.e., a subpredicate p of q is substituted for the predicate variable atom $X\colon q(\tau)$. A substitution δ is a most specific substitution for a predicate variable atom $X\colon q(\tau)$ if $\delta(X\colon q(\tau)) = p(\tau')$ with $\sigma^-_{p\to q}(\tau') = \tau$ and there is no other substitution δ' such that $\delta'(X\colon q(\tau)) = r(\tau'')$ with $\sigma^-_{r\to q}(\tau'') = \tau$ and $r \leq p$.

Definition 14 (Query System). *Let Q be a query in $\mathcal{Q}$, δ be a substitution for predicate variables in Q, and θ be a sorted substitution for $Q\delta$. Then, the query system $Query\colon \mathcal{Q} \to \{yes, no\}$ is defined by the following rule.*

(i) If there exists $\mathcal{K} \vdash l\colon Q\delta\theta$ such that $Var(Q\delta) \cap \mathcal{V} = \emptyset$ and $Var(Q\delta\theta) = \emptyset$, then $Query(Q) = yes$.
(ii) Otherwise, $Query(Q) = no$.

Without losing decidability, the query system is realized in the following two steps. First, atoms are substituted for predicate variable atoms in a query Q along with the predicate hierarchy. Second, predicate and meta-predicate assertions in the substituted query $Q\delta$ are derived using the Horn-clause calculus.

Theorem 3 (Termination of Query System). *Let $\mathcal{K}$ be a knowledge base in a sorted signature Σ. Then, the query system terminates if Σ is function-free.*

The termination leads to the following corollary that the complexity of the query-answering system is unaffected by the introduction of predicate variables in the queries.

Corollary 2 (Complexity of Query System). *Let $\mathcal{K}$ be a knowledge base in a sorted signature Σ and let Q be a query. If Σ is function-free, then deciding $Query(Q)$ is (single) EXPTIME-complete (w.r.t. the length of $\mathcal{K}$).*

6 Derivation Using Argument Restructuring

In the Horn-clause calculus (discussed in Section 4), redundant arguments in each predicate are deleted during the derivation of super predicates if the argument structures are well-arranged in a hierarchy. In this section, we generalize sorted signatures by removing the condition of their being well-arranged, i.e., some predicates may have an argument that their subpredicates do not have.

We give some examples of hierarchies in a query-answering system for the case where argument structures are not well-arranged in the sort, predicate, and meta-predicate hierarchies shown in Figs. 1 and 2. If the fact `assaults(tom:minor)` is valid, then the super predicate `illegalAct` can be derived in the predicate hierarchy as follows.

```
assaults(tom:minor)
?-illegalAct(x:human,mary:woman)
no
?-illegalAct(x:human,y:human)
yes
x=tom:minor, y=c:human
```

In the first case, there is no fact that indicates someone acts against the second argument `mary:woman` in the query. Thus, the answer to the first query is `no`. In the second case, we can obtain the answer `yes` to the second query from the fact `assaults(tom:minor)` and the predicate hierarchy. A new constant `c:human` is substituted for the variable y because the argument structure of the predicate `assaults` lacks the second argument of the predicate `illegalAct`.

For such argument structures in a predicate hierarchy (in a sorted signature), we perform the addition of missing arguments for the derivation of super predicates as follows.

Definition 15 (Naive Argument Restructuring). *Let $\Sigma = (S, P, \Psi_n, \Omega, \leq)$ be a sorted signature with an argument declaration $\Lambda = (AN, \Pi)$, let $\langle d_1, \dots, d_n\rangle$ be an n-tuple, and let $p \in P_n$ and $q \in P_m$. An argument restructuring from p to q is a function $\sigma^+_{p\to q}(\langle d_1, \dots, d_n\rangle) = \langle d'_1, \dots, d'_m\rangle$ such that*

$$d'_i = \begin{cases} d_j & \text{if } a'_i = a_j \\ c_i & \text{otherwise} \end{cases}$$

where p: $\langle a_1, \dots, a_n\rangle$ and q: $\langle a'_1, \dots, a'_m\rangle$ in Π and each c_i is a new element.

We refine the definition of Σ-models such a way that every argument elimination $\sigma^-_{p\to q}$ is replaced with an argument restructuring $\sigma^+_{p\to q}$. The satisfiability relation $\models$ is denoted by $\models_{\sigma^+}$ if an argument restructuring σ^+ is employed in each Σ-model. The conceptuality and identifiability of predicates in Propositions 2 and 3 hold for the case where the Σ-models are refined by replacement with an argument restructuring σ^+.

In order to embed an argument restructuring σ^+ in the Horn-clause calculus, we further extend the calculus as follows.

Definition 16 (Extended Sorted Horn-Clause Calculus) *Let C be a ground clause and $\mathcal{K}$ be a knowledge base. A derivation of C from $\mathcal{K}$ (denoted $\mathcal{K} \vdash_{\sigma^+} l\colon C$) in the sorted Horn-clause calculus is extended by replacing the predicate hierarchy rule with the following rule:*

– **Predicate hierarchy rule$^+$:** *Let $p(t_1, \ldots, t_n) \leftarrow G$ be a ground clause. If $\mathcal{K} \vdash l_1\colon p(t_1, \ldots, t_n) \leftarrow G$ and $p \leq q$, then $\mathcal{K} \vdash l_1\colon q(t'_1, \ldots, t'_m) \leftarrow G$ where $\sigma^+_{p \to q}(\langle t_1, \ldots, t_n \rangle) = \langle t'_1, \ldots, t'_m \rangle$.*

An atom A_1 is a parent of another atom A_2 if $\mathcal{K} \vdash_{\sigma^+} l\colon A_2 \leftarrow G$ is derived from $\mathcal{K} \vdash_{\sigma^+} l\colon A_1 \leftarrow G$ by an application of the predicate hierarchy rule. An atom A_1 is an ancestor of another atom A_2 if (i) A_1 is a parent of A_2 or (ii) A_1 is an ancestor of an atom A and A is a parent of A_2. Let A be an atom $p(t_1, \ldots, t_n)$ with $p\colon \langle a_1, \ldots, a_n \rangle \in \Pi$. We denote the occurrence of an argument name a_k and a term t_k in A by $A[a_k, t_k]$ if $1 \leq k \leq n$. The set of pairs of argument names and terms for a labeled atom $l\colon A$ is defined by $AL(l\colon A) = \{(a,t) \mid A[a,t]\} \cup \{(a,t) \mid A'[a,t] \text{ is an ancestor of } A\}$.

In the following definition, we introduce a label-based argument restructuring in order to solve the problem of incomplete derivation, i.e., the transitivity in Proposition 1 no longer holds if the argument structures are not well-arranged. Hence, it is necessary to solve the problem to prove the completeness of the extended sorted Horn-clause calculus.

Definition 17 (Label-Based Argument Restructuring in Derivation) *Let $\Sigma = (S, P, \Psi_n, \Omega, \leq)$ be a sorted signature with an argument declaration $\Lambda = (AN, \Pi)$, let $\langle d_1, \ldots, d_n \rangle$ be an n-tuple, let $p \in P_n$ and $q \in P_m$, and l be a label (non-negative integer). An argument restructuring from p to q is label-based if it is defined as a function $\sigma^*_{p \to q}(\langle t_1, \ldots, t_n \rangle) = \langle t'_1, \ldots, t'_m \rangle$ such that*

$$t'_i = \begin{cases} t_j & \text{if } a'_i = a_j \text{ with } (a_j, t_j) \in AL(l\colon p(t_1, \ldots, t_n)) \\ c_{l,a'_i} & \text{otherwise} \end{cases}$$

where $p\colon \langle a_1, \ldots, a_n \rangle$ and $q\colon \langle a'_1, \ldots, a'_m \rangle$ in Π and each c_{l,a'_i} is a new constant indexed by the pair of the label l and the argument name a'_i.

We denote the set of new constants that are used to add missing arguments in a label-based argument restructuring σ^* by $F_{0,new}$. The label-based argument restructuring σ^* can be applied to a tuple of terms $t_1, \ldots, t_n$ in a labeled atom $l\colon p(t_1, \ldots, t_n)$ in the derivation. This leads to the following transitivity, although the transitivity of naive argument restructurings σ^+ does not hold.

Proposition 4 (Transitivity of Label-Based Argument Restructurings) *Let Σ be a sorted signature with an argument declaration Λ, let τ be an n-tuple, and let $p \in P_n$, $q \in P_m$, and $r \in P_k$. If $p \leq q$ and $q \leq r$, then $\sigma^*_{q \to r}(\sigma^*_{p \to q}(\tau)) = \sigma^*_{p \to r}(\tau)$.*

The transitivity of label-based argument restructurings will be used to show the completeness of the extended sorted Horn-clause calculus.

Theorem 4 (Completeness of Extended Horn-Clause Calculus) *Let $\mathcal{K}$ be a knowledge base in a sorted signature Σ and L be a ground atom or meta-atom. $\mathcal{K} \models_{\sigma^+} L$ iff $\mathcal{K} \vdash_{\sigma^*} L$.*

Note that the consequence relation $\mathcal{K} \models_{\sigma^+} L$ is defined with a naive argument restructuring σ^+ but the derivation $\mathcal{K} \vdash_{\sigma^*} L$ is extended to contain a label-based argument restructuring σ^*. This is because $\mathcal{K} \vdash_{\sigma^+} L$ is incomplete for $\mathcal{K} \models_{\sigma^+} L$, i.e., the derivation is insufficient for the semantics.

However, the label-based argument restructurings σ^* lead to the undecidability of the extended sorted Horn-clause calculus as follows.

Theorem 5 (Undecidability of Extended Horn-Clause Calculus) *The extended Horn-clause calculus does not terminate for a knowledge base $\mathcal{K}$ in a function-free sorted signature Σ.*

Let p be an n-ary predicate and τ be an n-tuple of sorted terms. We denote an atom or meta-atom L by L_p if $L = p(\tau)$ or $L = \psi(A_1, \ldots, A_m)$ with $A_i = p(\tau)$ for some $1 \leq i \leq m$.

Definition 18 (Paths in a Knowledge Base). *Let $\mathcal{K}$ be a knowledge base in a sorted signature Σ, let L_p, L_q be atoms or meta-atoms, let a, a' be argument names, and let t be a sorted term. Then, there is a path from $L_p[a,t]$ to $L_q[a',t]$ in $\mathcal{K}$ if one of the following conditions holds:*

1. *$a = a'$, $p \leq q$, and $a \in \arg(p) \cap \arg(q)$,*
2. *$L_q[a', x\colon s] \leftarrow G$ where $L_p[a, x\colon s] \in G$ and $t \in \mathcal{T}_s$, and*
3. *there are two paths from $L_p[a,t]$ to $L_r[a'',t]$ and from $L_r[a'',t]$ to $L_q[a',t]$.*

In order to avoid the undecidability, we define a restricted set of knowledge bases (called safe knowledge bases).

Definition 19 (Safe Knowledge Bases). *A knowledge base $\mathcal{K}$ is safe if*

1. *$Var(L) \subseteq Var(G)$ for every clause $L \leftarrow G$ in $\mathcal{K}$, and*
2. *there is no path from $L_p[a,t]$ to $L_q[a',t]$ in $\mathcal{K}$ such that $q \leq p$, $a \neq a'$, $a \notin \arg(q)$, and $a \in \arg(p)$.*

Lemma 2. *Let $\mathcal{K}$ be a safe knowledge base in a sorted signature Σ. Then, the extended Horn-clause calculus with label-based argument restructuring does not generate new constants infinitely.*

Furthermore, we can show the complexity of the extended sorted Horn-clause calculus with label-based argument restructuring where Σ is function-free.

Theorem 6 (Complexity of Derivation for Atoms or Meta-atoms) *Let $\mathcal{K}$ be a safe knowledge base in a sorted signature Σ, L be an atom or meta-atom, and θ be a sorted ground substitution for L. If Σ is function-free, then deriving the set of ground atoms or meta-atoms $L\theta$ with $\mathcal{K} \vdash_{\sigma^*} l\colon L\theta$ is (single) EXPTIME-complete (w.r.t. the length of $\mathcal{K}$).*

Table 1. The complexities of Horn-clause calculus with argument manipulation

Horn-clause calculus	complexities
argument elimination	EXPTIME
naive argument restructuring	undecidable and incomplete
label-based argument restructuring	undecidable and complete
label-based argument restructuring for safe knowledge bases	EXPTIME

Table 1 lists the complexities of the Horn-clause calculus with argument elimination, naive argument restructuring, and label-based argument restructuring. We can extend the query system by using the Horn-clause calculus with label-based argument restructuring.

Theorem 7 (Complexity of Extended Query System)
Let $\mathcal{K}$ be a safe knowledge base in a sorted signature Σ and let Q be a query. If Σ is function-free, then deciding $Query(Q)$ is (single) EXPTIME-complete (w.r.t. the length of $\mathcal{K}$).

Due to spatial constraints, detailed proofs of the lemmas and theorems in this paper have been omitted (see http://kc.nict.go.jp/kaneiwa/).

7 Conclusions

We have developed an order-sorted logic programming language equipped with concept hierarchies of sorts, predicates, and meta-predicates. Predicates with differently structured arguments are conceptually interpreted in the semantics. According to the semantics, predicate-hierarchy reasoning is realized in the hierarchies of predicates and meta-predicates such that predicate assertions are used as arguments of meta-level predicates. To achieve such enhanced reasoning, we design inference rules for predicate and meta-predicate hierarchies in the order-sorted Horn-clause calculus. We employ the calculus to develop a query-answering system for generalized queries containing predicate variables. We show that the complexity of our expressive query-answering system is identical to that of DATALOG. We analyze several complexity results where argument restructuring gives rise to undecidable reasoning services in the derivation of super predicates in a predicate hierarchy, but a set of safe knowledge bases preserves the decidability of the derivation with argument restructuring.

References

1. `http://www.ruleml.org/`
2. `http://www.w3.org/tr/owl2-profiles/`
3. Chen, W., Kifer, M.: Sorted HiLog: Sorts in higher-order logic data languages. In: Y. Vardi, M., Gottlob, G. (eds.) ICDT 1995. LNCS, vol. 893, pp. 252–265. Springer, Heidelberg (1995)

4. Cohn, A.G.: Taxonomic reasoning with many sorted logics. Artificial Intelligence Review 3, 89–128 (1989)
5. Dantsin, E., Eiter, T., Gottlob, G., Voronkov, A.: Complexity and expressive power of logic programming. In: IEEE Conference on Computational Complexity, pp. 82–101 (1997)
6. Doets, K.: From Logic to Logic Programming. MIT Press, Cambridge (1994)
7. Grosof, B., Horrocks, I., Volz, R., Decker, S.: Description Logic Programs: Combining Logic Programs with Description Logics. In: Proc. of the Twelfth International World Wide Web Conference (WWW 2003), Budapest, Hungary (2003)
8. Hanus, M.: Logic programming with type specifications. In: Pfenning, F. (ed.) Types in Logic Programming. The MIT Press, Cambridge (1992)
9. Hitzler, P., Parsia, B.: Ontologies and rules. In: Staab, S., Studer, R. (eds.) Handbook on Ontologies, 2nd edn. (2009)
10. Horrocks, I., Patel-Schneider, P.F.: A proposal for an owl rules language. In: Proc. of the Thirteenth International World Wide Web Conference (WWW 2004), pp. 723–731. ACM, New York (2004)
11. Horrocks, I., Patel-Schneider, P.F., Boley, H., Tabet, S., Grosof, B., Dean, M.: SWRL: A Semantic Web Rule Language Combining OWL and RuleML. W3C Recommendation, `http://www.w3.org/submission/swrl/`
12. Jouannaud, J.-P., Okada, M.: Satisfiability of systems of ordinal notations with the subterm property is decidable. In: Leach Albert, J., Monien, B., Rodríguez-Artalejo, M. (eds.) ICALP 1991. LNCS, vol. 510, pp. 455–468. Springer, Heidelberg (1991)
13. Kaneiwa, K.: Order-sorted logic programming with predicate hierarchy. Artificial Intelligence 158(2), 155–188 (2004)
14. Kaneiwa, K., Mizoguchi, R.: An order-sorted quantified modal logic for meta-ontology. In: Beckert, B. (ed.) TABLEAUX 2005. LNCS (LNAI), vol. 3702, pp. 169–184. Springer, Heidelberg (2005)
15. Kaneiwa, K., Mizoguchi, R.: Distributed reasoning with ontologies and rules in order-sorted logic programming. In: Journal of Web Semantics (in press, 2009)
16. Krötzsch, M., Rudolph, S., Hitzler, P.: ELP: Tractable rules for OWL 2. In: Sheth, A.P., Staab, S., Dean, M., Paolucci, M., Maynard, D., Finin, T., Thirunarayan, K. (eds.) ISWC 2008. LNCS, vol. 5318, pp. 649–664. Springer, Heidelberg (2008)
17. Lloyd, J.W.: Foundations of Logic Programming. Springer, Heidelberg (1987)
18. Motik, B.: On the Properties of Metamodeling in OWL. Journal of Logic and Computation 17(4), 617–637 (2007)
19. Nguyen, P.H.P., Kaneiwa, K., Corbett, D.R., Nguyen, M.-Q.: An ontology formalization of relation type hierarchy in conceptual structure theory. In: Wobcke, W., Zhang, M. (eds.) AI 2008. LNCS (LNAI), vol. 5360, pp. 79–85. Springer, Heidelberg (2008)
20. Patel-Schneider, P.F., Hayes, P., Horrocks, I.: OWL Web Ontology Language Semantics and Abstract Syntax, W3C Recommendation, `http://www.w3.org/tr/2004/rec-owl-semantics-20040210/`
21. Rosati, R.: On the decidability and complexity of integrating ontologies and rules. Journal of Web Semantics 3(1), 41–60 (2005)
22. Socher-Ambrosius, R., Johann, P.: Deduction Systems. Springer, Heidelberg (1996)
23. Woods, W., Schmolze, J.: The KL-ONE family. Computers and Mathematics with Applications, Special Issue on Semantic Networks in Artificial Intelligence, Part 1 23(2–5), 133–178 (1992)

Semantic Web Service Composition in Social Environments

Ugur Kuter[1] and Jennifer Golbeck[2]

[1] Institute of Advanced Computer Studies,
University of Maryland,
College Park, Maryland 20742, USA
[2] College of Information Studies
University of Maryland,
College Park, Maryland 20742, USA

Abstract. This paper describes how to generate compositions of semantic Web services using social trust information from user ratings of the services. We present a taxonomy of features, such as interoperability, availability, privacy, security, and others. We describe a way to compute social trust in OWL-S style semantic Web services. Our formalism exploits the users' ratings of the services and execution characteristics of those services. We describe our service-composition algorithm, called Trusty, that is based on this formalism. We discuss the formal properties of Trusty and our implementation of the algorithm. We present our experiments in which we compared Trusty with SHOP2, a well-known AI planning algorithm that has been successfully used for OWL-S style service composition. Our results demonstrate that Trusty generates more trustworthy compositions than SHOP2.

1 Motivations

Web services are finding life in new forms, which is forecasting the future of what the web service environment will look like.

Apps for the iPhone and iPod touch are essentially web services with a wrapper to make them run on the iPhone. The Apple App Store provides a model of how web services are likely to evolve if they are to remain relevant and useful. There is a demand for these apps - there are roughly 20,000 apps in the store - and even simple services in app form receive thousands of downloads. As web services are integrated into forms like apps where there is user demand, we expect to see the same evolution with thousands of services available, many with the same functionality and semantic structure (inputs and outputs). As this becomes the web service environment, a mechanism for choosing from many similar services will be important for creating service compositions that meet the user's preferences and that the user trusts.

There are many apps offering classic web service functionality - finding weather conditions and forecasts, currency conversion, stock quotes, zip code lookup, and others. Consider weather forecasts; there are at least a dozen apps that provide current conditions for a given location. How does a user decide which app (or web service) to use if

A. Bernstein et al. (Eds.): ISWC 2009, LNCS 5823, pp. 344–358, 2009.

the functionality is identical? There are many attributes to consider. Cost is one; some apps are free, others charge a small fee. The quality of the information is another consideration; some weather services are more accurate or current than others. The service provider is also an option. A user may trust the National Weather Service as a provider because they have no commercial interest in the user's personal information while a corporate entity may store the user's personal data for other purposes. Other users may be more inclined to trust the way companies will handle their data more than the way the government would. Ratings of the app provided by other users are also very useful in making a decision about which to choose.

Artificial Intelligence (AI) planning techniques have been successfully used to decide which Web services to use and how to compose them in order to achieve a functionality on the Web. In this approach, a service composition is a ground sequence of service invocations that accomplishes a goal or task. A planning algorithm takes as an input a formal description of the Semantic Web service composition problem and generates a sequence of actions (a.k.a., a *plan* or a *composition*) that achieves the goals specified in the problem description.

Existing automated AI planning techniques have been focused on a purely *functional automation* of the service-composition process in great detail, and there have been great strides in this direction [1, 2, 3, 4, 5]. However, we believe that the next-generation of Web services will possibly have the following characteristics, which have not been entirely addressed in the existing Web Service Composition (WSC) research:

- One of the most significant aspects of planning and service composition on the Web is the "human factor;" i.e., the users who are exploiting the services and doing compositions on the Web to achieve some objectives. Although it makes sense to develop automated systems in order to perform service composition to achieve a functionality on the Web, in many situations it is important to appreciate the input of the human users who are initiating the composition process and generate reliable compositions that the user will be comfortable with.

  For example, the users may have concerns and preferences about the quality of a service and may not rate that service as reliable and trustworthy with respect to their objectives. The security and privacy policies of services may also be critically important to the users, even if they do not affect the direct outcome. The organization offering the service may also be of concern to users who want do not want to share their personal information with certain entities.

- Existing WSC planning algorithms usually make the assumption that the inputs/outputs/preconditions/results of a service process model are known; otherwise, the composition algorithm would not be able to do the chaining of services in order to generate a composition. In reality, however, it is not usually possible to know whether the execution of the service will follow exactly what is stated in the description/process model of that service. Web services do not always produce the outcomes as specified in their process models. This uncertainty makes the execution of the service unreliable for the user.

  For example, when a user is trying to finalize a travel itinerary something using a travel-agency Web service, the user may not get the desired itinerary due to a database problem at the end or the service may request the user to recreate it due to a communication problem between the provider of the service and the user. All these issues are natural, but makes the service unreliable in the eyes of the user. Then the question that a user will face is, given prior information about the Web services (e.g., his/her experience with it as well as other users' ratings of the service), should the user use that service?

2 Overview

In this paper, we describe a way to address the above aspects of Web service composition by using users' ratings about their experiences with the service and exploit this information to compute our "trust" in the service. A composition algorithm can then decide whether to execute the service based on that trust value.

Our contributions are as follows:

- We outline a taxonomy of features of Web services. This taxonomy specifies service characteristics such as privacy, security, usability/availability, reliability, and others. Most of the features in this taxonomy are already discussed in distributed systems and computing, multi-agent systems, and similar research areas, and our taxonomy is not complete by any means. But, to the best of our knowledge, these features have not been considered for the purposes of semantic Web service composition. While this feature taxonomy may change from application to application, it is meant to provide a general example here.
- We describe a formal treatment of how to take into account users' previous history of service ratings. This formalism is based on our taxonomy of service features mentioned above. The formalism allows us to develop ways of computing a user's trust in service composition process and we present such way of reasoning with trust in composing semantic Web services.
- We describe a new WSC planning algorithm, called Trusty. In Trusty, We focus on semantic Web services that are modeled in the prominent OWL-S service markup language [6]. Trusty generalizes the Hierarchical Task-Network (HTN) planner SHOP2, a well-known AI planning algorithm that has been successfully used for Web service composition [1], by incorporating reasoning mechanisms for social trust. This new WSC procedure uses the trust information on the services in order to infer the trustworthiness of a composition (or a partial composition), and returns the most trustworthy composition as a solution to a service-composition problem.
- We define three trust-computation strategies for Trusty; namely, overly-cautious, overly-optimistic, and average. These strategies differ in the way they compute the trust in a composite OWL-S service given the trust values of its sub-services. An overly-cautious trust computation produces compositions that maximize the minimum possible trust in the services. An overly-optimistic computation, on the other hand, maximizes the maximum possible trust in the services.
- We present theorems showing the correctness properties of Trusty.
- In our experimental evaluation, we compared Trusty using the three trust-computation strategies with SHOP2 in the well-known WSC domain of scheduling

doctor appointments for patients [1, 2]. Our statistical analyses of the results of the experiments, using ANOVA and t-tests as statistical methods, showed that overly-optimistic strategy enables Trusty to generate statistically more trustworthy compositions than SHOP2. The analyses also showed that overly-cautious strategy results in statistically worse trustworthy compositions compared to those generated by SHOP2. Although this might seem a bad result at first, note that it actually shows that the compositions generated by the overly-cautious strategy are the most trustworthy ones under the assumption that the worst will happen during the execution of those services. Consequently, both of these strategies, taken together, provide the most information for the human user doing the composition.

3 A Feature Taxonomy for the Behavioral Characteristics of Web Services

We compiled a feature taxonomy based on some known behavioral characteristics of Web Services. These characteristics describe the social aspects of a service that a user (human or machine alike) will consider when deciding whether to include the service in the composition being generated.While this feature taxonomy may change from application to application, it is meant to provide a general example here.

The following is a description of our feature taxonomy for WSC:

Interoperability. Web services provide a way for seamless communication, data and functional capability exchange between software applications and themselves. Service interoperability requires the following features to be implemented and handled properly by a service system: service management, data and meta-data handling, service descriptions, messaging, and service usability/availability.

Service management involves techniques for implementing how a service communicate and exchange data with other services on the Web.

Web services were originally designed to be *stateless*; i.e., a service is a functionality that performs a computation given some inputs and returns the outputs of that computation. However, many Web services use a database or a data storage/management backend in order to have a memory of their computations and operations and the results of those operations. *Data Handling* involves managing this backend system of a service by the service itself.

The *description* of a service also plays an important role in the interoperability of the service; the users of a service must know what the service does, what input/output requirements it has, and what conditions need to be satisfied in order to execute the service. If these are not known or the service does operate according to its description then the user may consider it as unreliable to include in a composition.

The *availability* of a service is also important for a user considering to use it. A service that publishes itself for the achievement of a functionality but that is never available (e.g., the server that runs the service is down or has limited processing power most of the time) is of no use to the user.

Privacy. Privacy concerns exist wherever personally identifiable information is collected and stored - in digital form or otherwise. In the context of Web services, the service is expected to implement a privacy policy in order to control the collection and dissemination of any personal information from the users. Thus, such privacy policies implemented by Web services are usually an important piece in the user's decision process of which service to use and whether he/she should execute it.

Security. The security of Web services have increasingly become an important issue. Security related features include the authentication of the service, its credentials, and whether a communication protocol (e.g., SSL) used by the service. *Authentication* and *credentials* describe how reliable the service is with respect to its provider and its description (see above for a discussion of service descriptions as an interoperability issue).

The communication protocol that a service uses is important for some users, depending on the objective of the service. For example, a user might look for secure connections from a Web service that performs banking transactions; on the other hand, a Web service that displays the current weather conditions at a particular location does not necessarily need a secure connection to its server.

Computation. Computational issues arise in Web services because a service usually operates on a remote server (or on a cluster of remote servers) on the Web and the user does not necessarily have access or control on that computational resource.

Thus, it is important for a service provider to ensure some measure of *computational reliability* of the service to the requesters of that service. Computational reliability of a service involves a good and honest description of what the service does (i.e., the functional definition of the service from its inputs to outputs), what it does not do, and what it may do when executed. This last issue arises if the description of the service and the outcome of its execution do not match to each other. In such cases, the user may not achieve his/her goals in using the service.

Another feature of a Web service is its *complexity* – i.e., its *functional granularity*. Web services are usually designed to be simple, achieving basic functionalities on the Web and designed to be composed into more complex functionalities via service-composition algorithms. A Web service that implements many functionalities in a very fine-grained way may or may not be attractive to the users depending their preferences and objectives.

The taxonomy and the features mentioned above are not complete by any means, but include the most typical examples of such behavioral features of Web services that influence users' decision on whether and how to use those services. Depending on the particular composition problem, this taxonomy can be expanded to include some domain-specific features as well. For example, a shopping service may require a licensing agreement with its provider and the user. Depending on the terms of that requirement, the user may or may not choose to use that service (e.g., if the terms are very restrictive, then the user may choose to use another service that has less restrictive requirements).

4 Social Trust

Users can determine the trust they have for a service by considering the features of the taxonomy above and their experience with a service. However, when a user has no experience with the service, it can be difficult to judge how trustworthy the service is. Information from other users who have interacted with the service can be used to help make these decisions. This is parallel to considering the ratings and reviews of apps for the iPhone. However, rating systems can be easily manipulated. For example, a malicious user can build a service that steals credit card numbers and then create hundreds of accounts to assign high ratings this service. To circumvent this, we consider *trust* between users along with ratings to help find trustworthy services for the composition.

4.1 Computing Trust between Users

There are several methods by which we can compute the trust that user c has in user u. These fall into two broad categories: trust can be computed using a social network or approximated through the ratings they have provided. Which approach to use, or if a combination of the two methods is preferable, will depend on the data available and the specific application.

When a social network is available, any trust-inference algorithm such as TidalTrust [7], advogato [8], Appleseed [9] , moletrust [10], and SUNNY [11] can be used to compute $\mathsf{trust}(c, u)$. In the Web service composition spaces, typically services are not associated with social networks but the data is available to approximate trust using user ratings. For example, in the Apple AppStore, each app/service is rated by many users and those ratings are publicly available for a user considering to use that app/service. Thus, we are going to focus here how to compute trust via user ratings.

Trust can also be computed from nuanced similarity measures over ratings users add into a system. In this context, the ratings would be on the web services. Research has shown that there is a strong correlation between trust and overall similarity [12, 13]. Further work indicates that trust is more complex than overall user-user similarity, and shows that trust can be estimated by a combination of more nuanced similarity measures [14]. That research identified four measures made over users' item ratings that relate to trust: overall similarity, similarity on extremes , the single largest difference, and the source's propensity to trust. We can compute similarity measures in two ways: as mean average error (MAE) and using the Pearson correlation. Thus, we had six total measures: the average difference (AD), overall correlation (COR), average difference on extremes (XD), correlation on extremes (XCOR), the single largest difference (MD), and the source's propensity to trust (PT). A linear combination of these values can predict trust is given in the following equation:

$$\begin{aligned}\mathsf{trust}(source, sink) = \theta_{AD} * AD + \theta_{COR} * COR + \theta_{XD} * XD + \\ \theta_{XCOR} * XCOR + \theta_{MD} * MD + \theta_{PT} * PT\end{aligned}$$

When the weights are estimated using multivariate linear regression, this equation generates a trust estimate accurate to within approximately 10%. This is as good or better than most trust inference algorithms that run on a social network as described above.

4.2 Semantic Web Services, User Ratings and Trust

We assume the existence of a finite set W of Web services that are given. Let F be a finite set of all of the available taxonomical features related to the services in W. A user's *service-composition rating* (or *rating*, for short) is partial function that models a user's rating of a service w:

$$\rho : W \times F \to [0, 1].$$

Intuitively, a user rating specifies a subjective value of a Web service w with respect to a behavioral feature f. For example, let f be the privacy policy of the service w. A low $\rho(w, f)$ value would mean that the user does not agree with the terms of the privacy requirements of the service, whereas a high $\rho(w, f)$ value would point to a more positive rating.

In most systems using social trust, a single value is computed that indicates the trust user c has for a user u. However, in the case of Web service composition, we have a set of features F of services and we can potentially compute the trust between c and u on feature f. Some methods for inferring trust between users will not have the data available to compute the values on each feature. In those cases, we will use the single trust value as the trust for all features. For example, if an algorithm computes that the user c should trust the user u at a level of 0.8 on a $[0, 1]$ scale, we would use 0.8 as the trust value for each feature $f \in F$.

As mentioned before, we focus on the well-known semantic Web services formalism OWL-S [6] to model Web services. OWL-S differentiates between *atomic* and *composite* services. The former can be directly executed in the world, whereas the latter must be decomposed into atomic ones for execution. Other formalisms model only atomic services. Below, we describe our definitions for trust to encapsulate both service formalisms.

Suppose $\mathcal{R}$ is a finite set of user ratings and let w be a Web service. Let U_ρ be the set of all users who rated the service w. In other words, for each user $u \in U_\rho$, there exists a user rating $\rho_u \in \mathcal{R}$ such that $\rho_u(w, f)$ is defined for some feature f. Finally, let c denote the user who is performing the Web service composition and $F_c(w)$ be the set of features the user c is interested in the service w.

We define the trust value $t_c(w, f)$ that the composer user c has in a Web service w based on the feature f as follows:

$$t_c(w, f) = \sum_{u \in U_\rho} \rho_u(w, f) \times \mathsf{trust}(c, u),$$

where $\mathsf{trust}(c, u)$ is the trust that c has in the user u, as described in the previous section.

Then, the expected trust that c has in the service w is:

$$t_c(w) = \frac{\sum_{f \in F_c(w)} t_c(w, f)}{|F_c(w)|}.$$

If there is a composite Web service w in W then the idea is to generate a composition (i.e., a sequence of atomic services) for the composite service and propagate the trust values of the atomic services in the composition upwards, starting from smaller services to the composed ones and eventually to w.

Let M be a set of process models for w and for each process model $m \in M$, let $\mathsf{sub}(m, w)$ denote the set of services that are sub-processes of w in m. We devise three different strategies that define the trust value $t(w)$ of w. These are as follows:

Overly-Cautious. The overly-cautious strategy for defining the trust in a service w aims to maximize the minimum expected trust value that the composer user has in the sub-processes of w. In other words, this strategy implements the well-known *minimax* game-tree evaluation criteria [15], adapted for social trust in Web service composition. Formally, the trust value in a service is

$$t(w) = max_{m \in M} Q(m)$$
$$Q(m) = \min_{w' \in \mathsf{sub}(m,w)} t(w')$$

Overly-Optimistic. This strategy can be seen as an opposite of the overly-cautious strategy above. The overly-cautious strategy assumes that if something bad could happen in the composed services (i.e., if the least trusted service fails our trust as expected), it would definitely happen. On the other hand, the overly-optimistic strategy assumes the opposite: nothing bad will happen (i.e., even if we have a low trust on a service, that service will not fail the expectations from it), and therefore, we can evaluate a composition based on the user's maximum trust on the sub-processes of a service. More formally,

$$t(w) = max_{m \in M} Q(m)$$
$$Q(m) = \max_{w' \in \mathsf{sub}(m,w)} t(w')$$

Average. Finally, the Average strategy computes the trust value in a process model m for w by taking the average of the trust computed for each of the sub-processes of w in m:

$$t(w) = max_{m \in M} Q(m)$$
$$Q(m) = \frac{\sum_{w' \in \mathsf{sub}(m,w)} t(w')}{|\mathsf{sub}(m,w)|}$$

Above $|\mathsf{sub}(m,w)|$ denotes the number of services in $\mathsf{sub}(m,w)$.

5 Web Service Composition with Social Trust

We start with our definitions for trust-based service composition. We define a *trust-based service composition problem* as a tuple $P = (S, c, G, W, F, \mathcal{R})$, where S is the initial state, c is the user that performs the composition, G is the OWL-S process model to be achieved, W is the available set of OWL-S services, F is the set of all features in the problem, and $\mathcal{R}$ is a finite set of user ratings.

A *solution* for P is a composition that maximizes the trust value of achieving G.

Figure 1 shows a high-level description of our Web service-composition algorithm, Trusty. Trusty is inspired by HTN planning algorithms for WSC, in particular, from the previous work on SHOP2 [1]. Unlike SHOP2, Trusty incorporates the notion of social trust and several trust computation strategies to reason with trust in order to generate trustworthy compositions.

Procedure Trusty$(s, c, G, W, F, \mathcal{R}, \pi)$
1. if $G = \emptyset$ then **return**(π)
2. nondeterministically choose a subgoal w from G that has no predecessors and remove w from G
3. if w can be achieved in s by an atomic service then
4. if there is no such process model in W for w then
5. **return**$(failure)$
6. insert the process model for w into π
7. $\tau \leftarrow 0$
8. for each feature $f \in F$ do
9. for each user u who has a rating for w do
10. $\tau \leftarrow \tau + \rho_u(w, f) \times$ trust(c, u)
11. $\Theta(w) \leftarrow \frac{t}{|\mathcal{R}||F|}$
12. $\pi \leftarrow$ Trusty$(s', c, G, W, F, \mathcal{R}, \pi)$
13. else
14. $\mathcal{D} \leftarrow \{d \mid m$ is a process model for w in s and d is a composition generated by applying m to w in $s\}$
15. $\pi_{max} \leftarrow \emptyset$
16. for each composition $d \in \mathcal{D}$ do
17. $G' \leftarrow$ UpdateComposition(G, s, t, m, d)
18. $\pi \leftarrow$ Trusty$(s, c, G', W, F, \mathcal{R}, \pi)$
19. if $\pi = failure$ then **return**$(failure)$
20. $Q(d) \leftarrow$ ComputeTrust(d, Θ)
21. if $Q(d) > \Theta(w)$ then $\pi_{max} \leftarrow \pi$; $\tau \leftarrow Q(d)$
22. $\Theta(w) \leftarrow \tau$
23. **return**(π)

Fig. 1. A high-level description of the Trusty procedure for Web service composition using trust. Above, s is the current state, c is the user who requested the composition, G are the goals of the user c, W is the set of available services, F is the set of features of those services, $\mathcal{R}$ is the set of user ratings, π is the (partial) solution composition. Initially s is the initial state, π is the empty composition, and β is 0. Θ is a global table that holds all of the trust information over the services that is computed by the algorithm.

The input to Trusty consists of a service composition problem with social trust $P = (s, c, G, W, F, \mathcal{R})$ and the empty plan π. We define Θ, a table that holds the trust-values on services. We assume that Θ is accessible globally by the Trusty procedure and initially, it specifies a trust value of 0 for each service in W.

With this input, Trusty starts from the goal G and recursively generates compositions for G and its subgoals, until a solution composition is generated or a $failure$ is returned during the problem-solving process.

At each invocation of the algorithm, Trusty first checks whether the goal functionality is achieved or not (i.e., whether $G = \emptyset$ or not). If the goal is empty, then Trusty returns the current composition π in Line 1. Otherwise, the algorithm nondeterministically chooses a subgoal w from G that has no predecessors in G. If w can be achieved by an atomic service, and there is a service process model defined for w

in W, then Trusty inserts that process model into π since when executed, that service process will achieve the goal w.

Next, Trusty computes the trust the current user c has that the service, when executed, will conform to c's composition profile (Lines 7–10). The algorithm then updates the trust-values table with the newly-computed trust value for the service for w (Line 11). It continues with the remainder of the goals to be accomplished in G via a recursive call to itself at Line 12.

If the subgoal w chosen at Line 2 is a nonprimitive functionality, then Trusty generates the set D of all possible compositions that might achieve w. At Line 14, the set D of compositions are generated via the available methods for w that are applicable in the current state of the world.

Then, for each possible composition for w, Trusty performs a series of steps as follows. Trusty first generates the next goal to achieve given the goal functionalities in the current composition for w and the remaining goals in G. The UpdateComposition function is responsible for this operation. UpdateComposition is a standard HTN task decomposition mechanism described in [1, 2], so we are not going into the details of this subroutine in this paper.

Then, Trusty calls itself recursively to achieve the new goal G'. When Trusty returns, there are two bits of knowledge that the algorithm needs to process. One is the partial composition π returned by the call. If π is a failure, than means that the current composition cannot generate a solution for w; in that case, Trusty ignores this composition and continues with the next one (if any).

If π is not failure when Trusty returned, then the trust table Θ must contain a value for the trust of c computed for each sub-service for w. The Q value of the current composition d is then computed by applying a trust computation strategy. As we described in the previous section, we developed three strategies for this purpose: namely, overly-cautious, overly-optimistic, and average. In Line 20, the ComputeTrust subroutine is responsible for implementing one of these strategies as specified by the user.

If the Q value of the decomposition d is greater than the current α value, then Trusty returns updates the trust value for the service w since it found a partial composition that has a higher value trustworthiness than the previously best one. It also keeps track of the current partial composition by marking as the current best solution candidate (see Line 21).

After Trusty processes and computes the best composition for w, it updates the trust table Θ with the trust value of that composition (Line 22) and returns that composition from the current invocation (Line 23).

The following theorems establish the soundness and completeness of Trusty:

Theorem 1. *Let $P = (s, c, G, W, F, \mathcal{R})$ be a social trust-based composition problem. Then, if Trusty returns a composition π for this problem P, then π is a solution of P.*

Note that there are two aspects of a solution π for P. First, π must achieve the functionality G specified in P. Second, if there is another composition π' then the user c's trust in π' must be lower or equal that in π. Trusty guarantees the first aspect given the set of services in the input W since we assume that for each goal that can be achieved by an atomic service, there is one corresponding process model in W, and for each goal that can be achieved by a composite service, there is one process model in W for that goal and for each of its subgoals. Then, the proof follows by induction on the number of decompositions of the G that needs to make until we generate a sequence of atomic services that, when executed, accomplishes G.

The second aspect follows from the bounding condition of Line 21 of the pseudo-code shown in Figure 1.

Theorem 2. *Let $P = (s, c, G, W, F, \mathcal{R})$ be a social trust-based composition problem. Then, if there exists at least one solution composition π for P then* Trusty *returns π.*

If there is no solution to P, then Trusty *returns $failure$.*

If there is a solution π for the service-composition problem P, then Trusty will eventually generate that solution since (1) it searches for every possible composition for the goal G, and (2) it prunes a composition only if that composition induces an inferior trust value compared to an alternative one. If there is no solution to P, then Trusty will try every possible composition (with possible prunings), and return failure in one of its recursive invocations.

The next section describes our implementation of Trusty and the experiments we ran with our implementation.

5.1 Discussion

Investigating the uncertainty due to the interactions between the users and their ratings is extremely interesting and realistic; however, reasoning about such uncertainty is hard and complex too. In fact, it indeed requires a different model to compute incremental inferences over social trust correctly, where a newer service with few or no ratings (and hence low trust) arrives and it is to end up in a solution composition, given the fact that there are many services that already have higher trust.

We to stay away from that uncertainty in our current model by assuming that each user and his/her rating is independent of another. This assumption allows us to adapt a typical assumption with most service-composition research that all the services (with their ratings in our case) are available up front and there are no changes to the service set during composition. Certainly, here are cases in which this assumption is not realistic; however this assumption simplified the initial model presented in this paper and allowed us to gain insights.

6 Experiments

We implemented a prototype of the Trusty algorithm using the code base of the SHOP2 planning system [16, 1], which was previously use for WSC quite successfully. We extended this planning system in order to implement our trust reasoning mechanisms and our optimizing search search algorithm to extend SHOP2's usual depth-first search implementation.[1] The particular implementation of optimization in Trusty depends on the strategies which one of the overly-cautious, overly-optimistic, or average trust computation is used.

[1] The SHOP2 planning system also implements a form of branch-and-bound search for simple optimization planning problems but it is designed for reasoning with action costs, not with trust values of both atomic and composite compositions.

Table 1. Trust values of the solution composition found by Trusty and SHOP2. In Trusty, we used three trust-inference strategies as described in text: Average, Overly-Cautious, and Overly-Optimistic. Each data point is an average of 50 random service-composition problems.

Algorithm	Trust-Inference Strategy	Average Trust
Trusty	Average	0.0043
Trusty	Overly-Cautious	0.0016
Trusty	Overly-Optimistic	0.0097
SHOP2	None (Depth-First Search)	0.0039

Our experimental scenario was based on the well-known WSC domain on *scheduling* of doctor-patient appointments. This scenario was first described in the Scientific American article about the Semantic Web [17]. In this domain, two people are trying to take their mother to a physician for a series of treatments and follow-up meetings. The service-composition problem is to come up with a sequence of appointments that will fit in to everyone's schedules, and at the same time, to satisfy everybody's preferences.

Since our system requires a mechanism for computing trust, we had to artificially generate realistic underlying data that could be used to compute trust. To do this, we used the social network and user ratings from FilmTrust [18], a social network where users rated their trust for one another and also rated movies. We kept the same social network and trust ratings between users, and randomly mapped users' ratings of movies to ratings of services for doing doctor-patient appointment assignments. This provided a realistic test dataset that represented user's distribution of preferences and their social relationships.

We generated 50 random mappings (i.e., random user ratings and trust values among users), and ran both SHOP2 and Trusty in this scenario with those ratings. We used the three trust-computation strategies in Trusty. In all of these runs, Trusty was able to generate compositions that have higher trust values than those generated by SHOP2, as summarized in Table 1.

Using the Overly-Optimistic trust-inference strategy, Trusty the average trust value of the solutions generated by SHOP2 was 0.0039, whereas that of the solutions generated by Trusty was 0.0097. The average trust value of the composition generated by Trusty's Average trust-inference strategy was 0.0016. An ANOVA shows statistically significant differences among the population, and pairwise t-tests indicate that the value of the Overly-Optimistic strategy statistically significantly higher than the Overly-Cautious strategy for $p < 0.01$.

In fact, our analyses showed that Trusty's Overly-Cautious strategy caused the algorithm to generate compositions whose average trust value is statistically worse than the average trust values generated by SHOP2. Although this result may suggest that this strategy is not good for service composition, this is not necessarily true. The objective of this strategy is to be conservative – i.e., the objective is to maximize the minimum possible trust in the services so that if a service fails fails to meet the composer's expectations, the ramifications of such a failure is minimal. The overly-cautious strategy just ensures this criterion while guaranteeing to generate service compositions whenever there are any for a composition problem.

Consequently, both of these strategies, taken together, provide the most information for the human user doing the composition.

We also ran a second set of experiments with Trusty using the average trust-computation strategy in order to determine the range of the trust values generated by the two planning systems. For these experiments, we used the same 50 randomly-generated mappings as above. In order to find the upper bound of the range of trust values, we set each trust value that appears in the user profiles to 1, the maximum possible trust value in our formalism. For the lower bound, we set each trust value that appears in the user profiles to 0.01, in order to estimate the minimum possible trust value in our formalism. We did not use 0's for the minimum trust values since then the compositions returned by both of the planners would always have 0 values.

When we set all of the trust values in the user profiles to 1, the average trust value of the solutions generated by SHOP2 in this setting was 0.0059, whereas that of the solutions generated by Trusty was 0.0065. When we set all of the trust values in the user profiles to 0.01, the average trust value of the solutions generated by SHOP2 in this setting was 0.000059, whereas that of the solutions generated by Trusty was 0.000065. In both cases, as above, Trusty's solution compositions had higher trust values than those of SHOP2.

The results with the maximum trust values in the profiles also show that SHOP2 cannot generate the best solution (i.e., the solution with maximum possible trust values) in our scenarios; the best composition it can generate has a trust value lower than that is generated by Trusty.

7 Related Work

Existing approaches for Web Service Composition formulate the problem in different ways, depending mainly on how the developers of those approaches perceive the problem. However, as we mentioned at the beginning of the paper, an important commonality among these approaches is that they focused on a purely *functional automation* of the service-composition process without any consideration for social trust, as we studied in this paper.

One of the first techniques developed for WSC is reported in [3]. Here, the states of the world and the world-altering actions are modeled as Golog programs, and the information-providing services are modeled as external functions calls made within those programs. The goal is stated as a Prolog-like query and the answer to that query is a sequence of world-altering actions that achieves the goal when executed in the initial state of the world. During the composition process, however, it is assumed that no world-altering services are executed. Instead, their effects are simulated in order to keep track of the state transitions that will occur when they are actually executed.

In [1], the WSC procedure is based on the relationship between the OWL-S process ontology [6] used for describing Web services and *Hierarchical Task Networks* as in HTN Planning [16]. OWL-S processes are translated into tasks to be achieved by the SHOP2 planner [16], and SHOP2 generates a collection of atomic process instances that achieves the desired functionality.

[19] extends the work in [1] to cope better with the fact that information-providing Web services may not return the needed information immediately when they are

executed, or at all. The ENQUIRER algorithm presented in this work does not cease the search process while waiting answers to some of its queries, but keeps searching for alternative compositions that do not depend on answering those specific queries.

[20] models Web services and information about the world using the "knowledge-level formulation" first introduced in the PKS planning system [21]. This formulation models Web services based not on what is actually true or false about them, but what the agent that performs the composition actually knows to be true or false about their operations and the results of those operations. A composition is formulated as a conditional plan, which allows for interleaving the executions of information-providing and world-altering services, unlike the work described above.

In [4] and [5], a planning technique based on the "Planning as Model Checking" paradigm is described for the automated composition of Web services. The BPEL4WS process models was first translated into state transition systems that describe the dynamic interactions with external services. Given these state-transition systems, the planning algorithm, using symbolic model checking techniques, returns an executable process rather than a linear sequence of actions.

8 Conclusions and Future Work

In this paper, we presented a new formalism for composing Web services when user ratings are present and Trusty, a new service-composition algorithm for creating trustworthy Web service compositions. This incorporates user preferences as considerations when generating compositions. The social trust component also allows users to benefit from the experiences and ratings of others, thus providing some information on the trustworthiness of unknown services. This, in turn, can be used as a mechanism for making choices when generating Web service compositions, resulting in selections and compositions that are personalized and more trustworthy from the user's perspective.

More extensive evaluation is the most important next step in this work. We provided some preliminary results in one sample domain. However, a more extensive study is needed to strongly demonstrate that Trusty will generate useful trust information about service compositions in a real domain and that users will significantly prefer these compositions. This will require recruiting users who have expertise and strong preferences regarding web services and who can understand the resulting compositions. Furthermore, we will require some method of extracting trust relationships between users. This can be done by implementing or recruiting subjects from an existing social network, or by computing trust from profile similarity, which will require a dense set of service ratings. With this data, we can illustrate that our algorithm effectively improves the quality of service compositions for users.

Acknowledgments. This work was supported in part by the DARPA/Air Force Integrated Learning Program, through the contract # FA8650-06-C-7606, and by the National Science Foundation. The opinions expressed in this paper are those of authors and do not necessarily reflect the opinions of the funders.

References

[1] Sirin, E., Parsia, B., Wu, D., Hendler, J., Nau, D.: HTN planning for web service composition using SHOP2. Journal of Web Semantics 1, 377–396 (2004)

[2] Kuter, U., Sirin, E., Parsia, B., Nau, D., Hendler, J.: Information gathering during planning for web service composition. Journal of Web Semantics (2005)
[3] McIlraith, S., Son, T.: Adapting Golog for composition of semantic web services. In: KR-2002, Toulouse, France (2002)
[4] Pistore, M., Barbon, F., Bertoli, P.G., Shaparau, D., Traverso, P.: Planning and monitoring web service composition. In: Bussler, C.J., Fensel, D. (eds.) AIMSA 2004. LNCS (LNAI), vol. 3192, pp. 106–115. Springer, Heidelberg (2004)
[5] Traverso, P., Pistore, M.: Automated composition of semantic web services into executable processes. In: McIlraith, S.A., Plexousakis, D., van Harmelen, F. (eds.) ISWC 2004. LNCS, vol. 3298, pp. 380–394. Springer, Heidelberg (2004)
[6] OWL Services Coalition: OWL-S: Semantic markup for web services, OWL-S White Paper (2004), http://www.daml.org/services/owl-s/1.1/owl-s.pdf
[7] Golbeck, J.: Computing and Applying Trust in Web-based Social Networks. PhD thesis, University of Maryland, College Park, MD, USA (2005)
[8] Levien, R., Aiken, A.: Attack-resistant trust metrics for public key certification. In: 7th USENIX Security Symposium, pp. 229–242 (1998)
[9] Ziegler, C.N., Lausen, G.: Spreading activation models for trust propagation. In: Proceedings of the IEEE International Conference on e-Technology, Taipei, Taiwan, IEEE Computer Society Press, Los Alamitos (2004)
[10] Avesani, P., Massa, P., Tiella, R.: Moleskiing.it: a trust-aware recommender system for ski mountaineering. International Journal for Infonomics (2005)
[11] Kuter, U., Golbeck, J.: Sunny: A new algorithm for trust inference in social networks, using probabilistic confidence models. In: Proceedings of the National Conference on Artificial Intelligence, AAAI (2007)
[12] Ziegler, C.N., Lausen, G.: Analyzing correlation between trust and user similarity in online communities. In: Proceedings of the Second International Conference on Trust Management (2004)
[13] Ziegler, C.N., Golbeck, J.: Investigating Correlations of Trust and Interest Similarity. Decision Support Services 1 (2006)
[14] Golbeck, J.: Trust and nuanced profile similarity in online social networks. In: Transactions on the Web (to appear, 2009)
[15] Chi, P., Nau, D.S.: Predicting the Performance of Minimax and Product in Game Tree Searching. In: Second Workshop on Uncertainty and Probability in Artificial Intelligence (1986)
[16] Nau, D., Au, T.C., Ilghami, O., Kuter, U., Murdock, W., Wu, D., Yaman, F.: SHOP2: An HTN planning system. JAIR 20, 379–404 (2003)
[17] Berners-Lee, T., Hendler, J., Lassila, O.: The Semantic Web. Scientific American 284, 34–43 (2001)
[18] Golbeck, J.: Generating predictive movie recommendations from trust in social networks. In: Proceedings of the Fourth International Conference on Trust Management, pp. 93–104 (2006)
[19] Kuter, U., Sirin, E., Nau, D.S., Parsia, B., Hendler, J.: Information gathering during planning for web service composition. In: McIlraith, S.A., Plexousakis, D., van Harmelen, F. (eds.) ISWC 2004. LNCS, vol. 3298, pp. 335–349. Springer, Heidelberg (2004)
[20] Martinez, E., Lespérance, Y.: Web service composition as a planning task: Experiments using knowledge-based planning. In: ICAPS-2004 Workshop on Planning and Scheduling for Web and Grid Services (2004)
[21] Petrick, R.P.A., Bacchus, F.: A knowledge-based approach to planning with incomplete information and sensing. In: AIPS (2002)

XLWrap – Querying and Integrating Arbitrary Spreadsheets with SPARQL

Andreas Langegger and Wolfram Wöß

Institute of Applied Knowledge Processing
Johannes Kepler University Linz
Altenberger Straße 69, 4040 Linz, Austria
{al,wolfram.woess}@jku.at

Abstract. In this paper a novel approach is presented for generating RDF graphs of arbitrary complexity from various spreadsheet layouts. Currently, none of the available spreadsheet-to-RDF wrappers supports cross tables and tables where data is not aligned in rows. Similar to RDF123, XLWrap is based on template graphs where fragments of triples can be mapped to specific cells of a spreadsheet. Additionally, it features a full expression algebra based on the syntax of OpenOffice Calc and various shift operations, which can be used to repeat similar mappings in order to wrap cross tables including multiple sheets and spreadsheet files. The set of available expression functions includes most of the native functions of OpenOffice Calc and can be easily extended by users of XLWrap.

Additionally, XLWrap is able to execute SPARQL queries, and since it is possible to define multiple virtual class extents in a mapping specification, it can be used to integrate information from multiple spreadsheets. XLWrap supports a special identity concept which allows to link anonymous resources (blank nodes) – which may originate from different spreadsheets – in the target graph.

1 Introduction

The translation of information stored in various legacy information systems and data formats to RDF is an important requirement of many Semantic Web applications. While for the Web of Data relational database management systems (RDBMS) are considered to be the most important legacy information systems, in case of corporate Semantic Web applications, spreadsheets play a similar important role. Spreadsheets are frequently used by people in companies, organizations, and research institutions to share, exchange, and store data. Whenever there is no database in place, spreadsheets are often the primary fall-back tool for maintaining structured information.

Compared to wrapping a relational database, which has a fixed schema, data types, and integrity constraints, wrapping spreadsheets is more difficult because the implicit schema has to be captured first when creating a formal mapping. Currently available spreadsheet wrappers treat spreadsheets as flat tables like single database relations or comma-separated value (CSV) files. In this paper we will present a novel mapping approach for spreadsheets which is based on template graphs similar to RDF123 [5], which we will explain in the related work part. However, the XLWrap mapping approach is not based on a simple row oriented iteration of tables. It allows to define

A. Bernstein et al. (Eds.): ISWC 2009, LNCS 5823, pp. 359–374, 2009.

template mappings as RDF graphs and to repeat them based on various shift and repeat operations in order to map arbitrary layouts including multi-dimensional cross tables and spreadsheets over multiple files. XLWrap supports expressions to reference cells and ranges from template graphs including sheet ranges and absolute references to other external spreadsheets such as supported by Microsoft Excel and OpenOffice Calc (the grammar is printed in Listing 1 on page 368).

Additionally, XLWrap features a server component called *XLWrap-Server*, which provides a SPARQL endpoint as well as a linked data browser similar to D2R-Server [3]. XLWrap-Server observes a configurable directory for mapping files and whenever a mapping file or one of the referred spreadsheet files of the mapping is added, modified, or removed, it automatically executes the translation process and caches the generated RDF data. XLWrap-Server integrates *Joseki* [6] to provide a SPARQL endpoint and *Snorql*, which has been adopted from D2R-Server and allows the user to explore the wrapped information on-the-fly. Together with the *Semantic Web Integrator and Query Engine* (SemWIQ) [7], XLWrap can be used to integrate various spreadsheets and other data sources such as relational databases and ODBC data sources. A simple setup procedure and ease of use have been two major requirements for the development of XLWrap-Server. The setup procedure is a matter of starting the server and putting mapping files into the observation folder. XLWrap-Server is also considered to become a practical tool for experimenting with linked data. It can be used to quickly expose information via SPARQL while editing it in a human-friendly way. XLWrap can be downloaded from its homepage at `http://www.langegger.at/xlwrap`

In the next section we will explore different spreadsheet layouts in order to create a sound mapping framework. In Section 3 related work is discussed. In Section 4 the XLWrap mapping formalism is presented and in Section 5 the transformation process is explained. In order to demonstrate XLWrap, an example is discussed as part of these sections. Section 6 concludes the contribution with some final remarks.

2 Background

In order to develop a generic mapping framework for spreadsheets, the spreadsheet paradigm has been analyzed and different ways of information representation have been examined. Some of these findings are presented as part of this section before related work will be discussed. In the following, we will use the terms *workbook* to refer to a file of a spreadsheet application, and *worksheet* (or just *sheet*) to denote a single two-dimensional sheet of a workbook.

2.1 Information Representation in Spreadsheets

Firstly, it is important to distinguish between the *information model* and the *representation model*, which is used to represent information within a spreadsheet. The *information model* is defined implicitly by the semantics of the entailed information. The resulting RDF graph should as closely as possible reflect the information model, as for example, expenditures by category, year, and sub-division or personal information about employees. The actual information representation as an RDF graph is a subject of

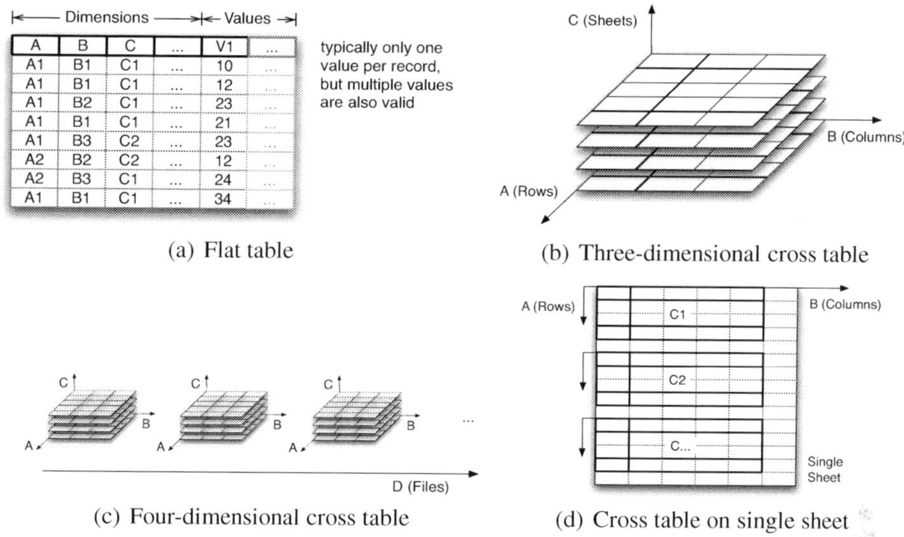

(a) Flat table (b) Three-dimensional cross table (c) Four-dimensional cross table (d) Cross table on single sheet

Fig. 1. Information representation in spreadsheets

data modeling and in fact, there are many ways how to represent expenditures in RDF. A generic wrapper should not enforce any fixed rules and structures that depend on the representation of the information model in the spreadsheet. Concerning the *representation model*, three different layouts could be identified, whereas the third one is a hybrid approach of the first two:

One-Dimensional Flat Table. In this layout (Figure 1(a)) information is represented, regardless of its dimensionality, in a flat table with a single column (*or* row) heading. It is used to represent information such as data lists, e.g. persons with fixed properties such as name, mailbox, age, etc. Except for RDF123, all existing wrappers create exactly one RDF resource per row as shown in Figure 2(a). Currently available spreadsheet wrappers, which will be discussed in Section 3, are restricted to this kind of representation.

However, flat tables can also be used to represent information with multiple dimensions (typically >2) in so-called de-normalized tables which are also the basis of pivot tables and OLAP programs. Some cells of the header represent dimensions of the information model, the other header cells represent values specific to each instance. The domain values of dimensions repeatedly occur in the instance rows as shown in Figure 1(a). Note that regarding the previous example of a person data list, no dimensions will be used and each value is regarded as specific to the person instance. Alternatively, each person property may be regarded as a dimension or *facet* (for instance, the dimension *first name* may have all occurring first names as its domain).

Cross Tables. In this layout, which is shown in Figure 1(b), the representation of information is organized in cross tables, which may span multiple columns, rows,

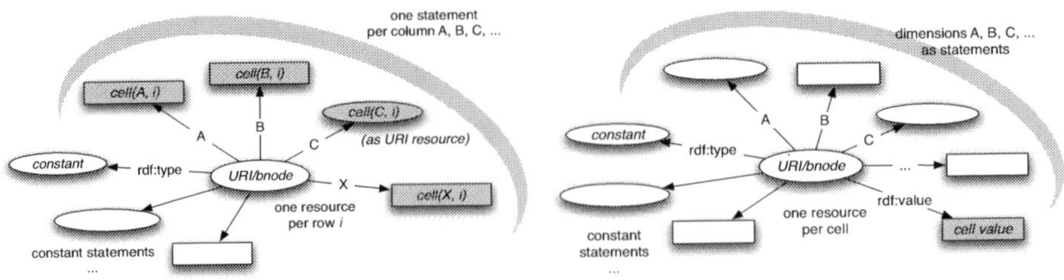

(a) One RDF resource per row (b) One RDF resource per cell

Fig. 2. Straight-forward translation of spreadsheet data to RDF resources

sheets, and even files and directories (Figure 1(c)). Each cell represents a single entity or instance. In a straight-forward approach, the translation could be done as shown in Figure 2(b). Instead of a single column/row header, cross tables have multiple headers, one for each dimension. Along the sheet axis, the header is defined either by the names of the sheets in the workbook or by text labels placed into the worksheet. Similarly, when using multiple files, the domain values of the corresponding axis are either defined by file names or text labels on worksheets. Because a single sheet is already restricted to columns and rows, cross tables are often represented by repeating similar table layouts on the same sheet as depicted in Figure 1(d). Similar to pivot tables or de-normalized tables as mentioned before, a higher dimensionality is broken down into two dimensions by repeating similar structures.

Hybrid Layouts. Finally, there is the possibility of combining flat tables with cross tables (e.g. de-normalized tables can be repeated across several sheets or files).

Independent of the representation model, a wrapper must be able to translate the information into an RDF graph by best reflecting the source information model. To give another example, revenues for different products, years, and countries can either be represented in a de-normalized flat table with the heading *(product, year, country, revenue)* or in a cross table with one sheet per *product*, each one representing *revenues* per *year* and *country*.

2.2 Definition of Spreadsheet Applications

The following definition for spreadsheet applications is the basis for the mapping framework presented in Section 4.

Definition 1. *A spreadsheet application A is defined as a set of named workbooks w_i, $i \in \mathbb{N}_0$, where* name(w_i) *is the canonical workbook filename denoted by its URI. A spreadsheet application has at least one workbook w_0, called the* base workbook *and optionally a set of* external workbooks w_i, $i \geq 1$, *which are referenced from w_0 by means of absolute cell references.*

Within companies it is very common that multiple spreadsheets are interlinked by external cell references. To be able to process external references in XLWrap mappings, this definition takes external workbooks into account. The URI scheme for local files is `file://`. Workbook files on the Web can be referenced by means of HTTP URIs.

Definition 2. *A workbook $w_i \in A$ is defined as a set of named worksheets s_j, $j \in \mathbb{N}$, where* name(s_j) *is the sheet name as specified in the application (a string).*

Beside single worksheets, it is also common to use multiple worksheets. As will be shown later, XLWrap supports a special sheet shift operation, which allows to repeatedly apply template mappings on multiple sheets.

Definition 3. *A worksheet $s_j \in w_i$ is defined as a matrix of cell values $V = [v_{c,r}]_{n \times m}$ where $c \in \mathbb{N}_0$, $c < m$, is the column index and $r \in \mathbb{N}_0$, $r < n$, is the row index of the corresponding worksheet table. m, n denote the total number of columns and rows used in worksheet s_j.*

Although there is a space limit for m and n in practice, there is no limitation enforced by this definition. The same applies to the number of worksheets in a workbook.

Definition 4. *A cell value $v_{c,r}$ has a* type annotation *denoted as* type($v_{c,r}$) $\rightarrow T$ *with* $T = \{t_{text}, t_{number}, t_{datetime}, t_{boolean}, t_{empty}\}$. *Additionally, a cell value may have a* formatting *(e.g. for number, currency, date/time formatting) and a* formula annotation *denoted as* formula($v_{c,r}$) $\rightarrow E$, *where $e \in E$ is a compositional expression according to a predefined grammar G_E.*

A cell formula, is not a specific type $\in T$. Instead, it is an annotation defining the expression which can be evaluated to reproduce or update $v_{c,r}$. Values with formula annotations are explicitly stored in the corresponding workbook files. In our definition of spreadsheet applications, the grammar G_E supports *range references* denoted as e_{ref}. Range references are used to refer to other cells of the same spreadsheet, other sheets of the same workbook, or external workbooks. In Section 4 it will be described, how expressions and range references are also used for XLWrap mappings to map spreadsheet cells to RDF graphs. The proper definition of a *range reference* is given below.

Definition 5. *A range reference sub-expression e_{ref} is generally defined as a set of partial 7-tuples of the form $(w_i, s_{j1}, c_1, r_1, s_{j2}, c_2, r_2)$. These components specify a workbook, a start cell consisting of a sheet, column, row, as well as a target cell specified by a sheet, column, and row. Some tuple components are optional and may be left empty. If $|e_{ref}| > 1$, the range reference is called* multi range reference, *because it contains multiple single range references. In case of OpenOffice Calc, the general lexical representation of a single range reference is:*

$$((w_i \text{ “\#\$” })?\ s_{j1} \text{ “.” })?\ c_1\, r_1\, (\text{ “:” } (s_{j2} \text{ “.” })?\ c_2\, r_2)?$$

For multi range references, single range references are separated with a semicolon. The following two sub-types are defined:

- Cell reference, *which has the optional components* w_i, s_{j1}, *the mandatory components* c_1, r_1, *and the component* s_{j2}, c_2, r_2 *left empty (e.g.* `A3`, `'Sheet 1'.A3`, *or* `file:foo.xls#$'Sheet 1'.A3`*)*
- Box reference, *which has the optional components* w_i, s_{j1}, s_{j2} *and the mandatory components* c_1, r_1, c_2, r_2 *(e.g.* `A3:B9`, *or* `file:foo.xls#$'Sheet 1'.A3:B28`.

Whenever the components w_i and s_{j1}, s_{j2} are omitted, the range reference is interpreted relative to the worksheet of its originating cell formula. Hence, a range reference can either be *absolute* or *relative* its base sheet or workbook file.

2.3 Dumping versus On-the-Fly Processing of SPARQL Queries

Concerning the wrapping approach, a distinction can be made based on the direction of the mapping formalism, which can either be source/spreadsheet-centric, or target/RDF-centric. In general, when mapping between two different data models, it is possible to define one of the data models as a view onto the other data model. Since in our case, the RDF model acts as the target model, the spreadsheet-centric variant is similar to the *Local-as-View* (LaV) approach and the RDF-centric variant is similar to the *Global-as-View* approach (GaV) in information integration [8]. Only XLWrap and RDF123, support the GaV-approach and allow the user to define RDF-centric mappings based on graphs. All other wrappers are based on a spreadsheet-centric view definition which map columns or cells to specific properties and do not allow the definition of custom target graphs.

Another distinction can be made concerning the actual transformation of spreadsheet data into RDF when executing SPARQL queries. For queries, the wrapping process can either be based on materialization involving a single data dump into a volatile or persistent RDF store, or it can be an on-the-fly query execution process (e.g. D2R-Server [3]). Actually, our initial motivation for developing XLWrap have been experiments towards a generic guideline for the generation of virtual RDF wrappers supporting SPARQL for the *Semantic Web Integrator and Query Engine* (SemWIQ). Based on experiences with D2R-Server while contributing some optimizations like push-down of filters into SQL, our assumption was, that simple RDF wrappers can be written based on Jena ARQ by supporting triple patterns and let ARQ process all higher level algebra operations. For instance, a wrapper providing SPARQL access to the directory structure of a file system or an FTP server can be written very easily this way. It only needs to correctly interpret subsequent triple patterns and generate the necessary variable bindings which correspond to a virtual RDF graph similar to a database access plan. However, for larger information sources, low-level index structures such as hash tables or B+ trees are necessary in order to achieve acceptable performance results. They may be managed by the wrapper in volatile memory caches and will have to be updated by means of a notification mechanism or pre-defined periods. Additionally, a wrapper for a large data source should provide cardinality estimations for triple patterns in order to support the re-ordering in a BGP by expected cardinalities. During the evaluation of a BGP, subsequent joins over triple patterns are executed by ARQ based on a substitute algorithm: already obtained variable bindings of triples produced by upper triple patterns are used as constraints for matching subsequent triple pattern. Thus, triple patterns with a lower cardinality should be executed first.

However, experiments have shown that in most situations it is still better to apply a materialized approach and dump data into an RDF store instead of maintaining separate index structures and insisting on the virtual approach. In particular, these findings apply to spreadsheets, because they are typically small in size (compared to databases) and there is actually no need to process queries on-the-fly. In fact, the dumping process is so fast, that even if the materialized cache has to be re-generated, it is only a matter of seconds or milliseconds. XLWrap tracks any changes in mapping files and referenced workbook files and re-generates caches in the background which is hardly noticed by end-users.

3 Related Work

Among the currently available spreadsheet wrappers there are two open source projects, *Excel2RDF* [4] and *RDF123* [5], and there is *TopBraid Composer* from TopQuadrant[1], which is a commercial ontology development environment integrating support for importing and exporting RDF from/to spreadsheet files. We were unable to find any further spreadsheet RDF wrapper in the literature and on the Web. There is another notable software called *Anzo for Excel* from Cambridge Semantics[2], which is actually not a dedicated RDF wrapper, but which can be used to map information from spreadsheets to vocabularies for distributed, collaborative work. Because Anzo is targeted to employees who typically do not know about RDF, it hides any details of the internals. Although it provides a graphical plugin for Microsoft Excel, which allows to map cells and ranges to vocabularies, it is unclear how it can be used to map spreadsheets to custom graphs. As a marginally related project the *Aperture*[3] framework developed for the *Nepomuk Semantic Desktop* is mentioned, which is extracting meta data from spreadsheet files. However, the content of spreadsheets is not processed or wrapped based on a formalized mapping. Instead, Aperture tries to extract significant features and terms from spreadsheets in order to support semantic search and interlinking of documents in the Nepomuk framework.

Unfortunately, nearly all existing dedicated wrappers are rather limited in practice. The simplest one is Excel2RDF and an extended version called *ConvertToRDF*, which supports basic mappings for column headers. As both tools follow the spreadsheet-centric mapping approach, the output graph is fixed and for each row of only a single spreadsheet table, one RDF resource is created with property/object pairs for all mapped columns. Resources as objects or typed literals are not supported. The spreadsheet import feature of TopBraid Composer is similarly basic. Actually, only CSV text files are supported and the mapping is similar to ConvertToRDF. Typed literals and resource objects are supported, but target types have to be specified manually.

The most relevant wrapper related to our work is RDF123 [5]. Beside XLWrap, it is the only wrapper that supports an RDF-centric mapping approach. However, although with RDF123 it is possible to define arbitrary target graphs, it is restricted to a specific spreadsheet layout, like all of the other existing wrappers. It only supports flat tables

[1] `http://www.topquadrant.com`
[2] `http://www.cambridgesemantics.com/`
[3] `http://aperture.sourceforge.net/`

as shown in Figure 1(a) of the previous section. The available prototype is capable of reading CSV text files but does not support Excel or OpenOffice spreadsheets. An RDF mapping in RDF123 is described as part of the spreadsheet itself in a special meta data section starting with the specific label `rdf123:metadata`. Among several Dublin Core metadata, the section specifies the start row, the end row, the start column, whether there is a row header which has to be skipped, and the URI of the template graph used for producing the results. When processing a spreadsheet, the execution engine scans the sheet for the metadata section and if it cannot be found, the complete sheet (only one worksheet is supported) is wrapped in a similar way as with Excel2RDF. If the metadata section is found, the execution engine retrieves the template graph from the specified location. The template graph is an RDF graph which may contain references to the columns of the current row being processed to generate nodes based on values obtained from the spreadsheet. Thus, the execution process is fixed to one row by row iteration. In order to refer to the columns of the active row being processed, expressions can be specified as special literals and faked URIs as for example `"Ex:$1"` and `<Ex:$1>`, which both refer to the first column. Unfortunately, this approach has the following severe consequences: all other literals starting with `Ex:` will be matched as expressions and cause errors. Furthermore, because expressions may contain spaces and special characters, encoding them as URIs (with `Ex:` as a pseudo protocol) will cause further troubles at runtime and lead to incompatibilities with other systems.

RDF123 provides a GUI which facilitates the process of creating mappings. However, the tool only provides a viewer for CSV files and a basic graph designer which both are not combined in any special way. The graph designer uses a proprietary format to store graphs specific to the Open JGraph library, which has been used. Developing a powerful graphical support tool is not easy and requires a lot of effort. It would be desirable for a graphical mapping tool to provide features specific to the mapping process such as drag and drop, highlighting of referenced cells, and a simulation feature for debugging the wrapping process.

Because XLWrap can be used to semantically integrate different spreadsheets and to query the entailed information based on logical inference, it can also be compared to the system proposed by [9] which is called LESS for *Logic Embedded in SpreadSheets*. Instead of writing typical numerical formulas into the cells of a spreadsheet, LESS allows to logically annotate facts and use logical functions for calculations.

4 XLWrap Mapping Formalism

XLWrap is based on an RDF-centric mapping approach which allows to map the information stored in a spreadsheet to arbitrary RDF graphs independent from the representation model as discussed in Section 2. Similar to RDF123, the output graph is defined by means of template graphs which are repeatedly applied during the wrapping process. With XLWrap expressions it is possible to refer to arbitrary cells and combine them with expressions like in spreadsheet applications.

4.1 XLWrap Mappings

In the following, XLWrap mappings are defined based on the definitions of Sect. 2.

Definition 6. *An XLWrap mapping M is defined as a set of* map templates $m_k, k \in \mathbb{N}$.

Definition 7. *A* map template $m_k = (w_k, s_k, C_k, G_k, F_k)$ *consists of the following components: a base workbook $w_k \in w_i$, a base worksheet $s_k \in s_j$, a constant RDF graph C_k, an RDF* template graph G_k, *and a sequence of* transform operations $F_k = (f_l)$, $l \in \mathbb{N}$.

Definition 8. *A* constant graph C_k *and a* template graph G_k *are valid RDF graphs, according to the W3C specification, which may contain literals of the custom data type* `xl:Expr` *called* XLWrap expressions[4].

Definition 9. *A transform operation f_l can modify the template graph G_k and change range references in expressions.*

During the wrapping process, each map template $m_k \in M$ contributes a sub-graph $[[G_k]]$ to the overall result similar to RDF123, but with the difference that a graph template is not moved from the first row to the last one in a fixed direction, instead, it is moved based on the transformation sequence defined by (f_l). The bracket notation $[[\ldots]]$ is used to denote the application of a template graph including the evaluation of all XLWrap expressions. While C_k is evaluated and merged into the target graph once, the template graph G_k is subsequently transformed by (f_l) and evaluated multiple times. If $|F_k| = 0$, no transformations are specified and G_k is applied once in its initial form.

Because by definition subjects and predicates cannot be literals, in order to specify XLWrap expressions for subjects or predicates, they have to be wrapped in blank nodes as part of the template graph. The special property `xl:uri` is then used to replace these blank nodes by URI resources in the target graph. Similarly, the property `xl:id` can be used to link blank nodes: XLWrap will produce blank nodes with equal local IDs in the target graph

Definition 10. *An XLWrap expression is defined as a compositional expression of basic expression symbols similar to spreadsheet formulas (see Definition 4). XLWrap expressions are parsed from the lexical representation by the custom data type implementation*[5] *according to the grammar defined in Listing 1.*

Range references, including cell, box, and multi range references are supported as specified in Definition 5. Additionally, XLWrap supports the special range types *null range*, *full sheet range*, and *any range*, which are lexically denoted as the empty string, (s_j ".*"), and "*.*". Depending on the semantics of operations and functions, only specific range sub-types are valid[6]. Optionally, a worksheet s_j can be prefixed with "#" and specified by the sheet number starting with 1 (e.g. "#1.A1", "#3.*").

XLWrap supports all standard arithmetic and also logical operators, string concatenation, and an extensible function library (additional implementations can be easily added at runtime). The most important functions for string manipulation, including SHA-1

[4] The full namespace for the prefix `xl:` is `http://langegger.at/xlwrap/vocab#`

[5] In Jena it is possible to register a custom data type handler by extending `BaseDatatype`.

[6] For example, while `SUM()` takes any range reference and also numbers as arguments (the number of arguments is not restricted), the expression `"A3:B5"` is invalid, since it is only possible to obtain values from a single cell.

```
XLExpression    =  "="? OrExpr <EOF> OrExpr          =  AndExpr (
"||" AndExpr )* AndExpr           =  Comparable ( "&&" Comparable )*
Comparable      =  Concatable ( CompOp Concatable )* Concatable =
Expr ( "&" Expr )* Expr           =  Term ( ("+"|"-") Term)* Term
=  Factor ( ("*"|"/") Factor)* Factor          = Atom ("^" Atom)*
Atom            =
    ("+"|"-"|"!")
    (
      <NUMBER> ("%")? |
      (<TRUE>|<FALSE>) |
      <STRING> |
      <CELLRANGE> |
      "(" Concatable ")" ("%")? |
      <FUNCIDENT> "(" ( Concatable ( (","|";") Concatable)* )? ")" ("%")?
    )
CompOp          =  "<=" | "<" | ">=" | ">" | ("!="|"<>") | ("=="|"=")
```

Listing 1. Grammar of XLWrap expressions

hashing (which, for instance, is required for `foaf:mbox_sha1sum` property values in FOAF applications), type casting (to enforce specific literal data types in the RDF output), aggregate functions such as `SUM()`, which takes cell, box, and multi ranges as arguments, have already been implemented.

The following transform operations are available in the current implementation:

- column shift: $f_{ColumnShift}(d, n, z, e_c)$
- row shift: $f_{RowShift}(d, n, z, e_c)$
- sheet shift: $f_{SheetShift}(d, n, z, e_c)$
- sheet repeat: $f_{SheetRepeat}((g_i), z, e_c)$
- file repeat: $f_{FileRepeat}((h_i), z, e_c)$

Common to all operations is z, a multi range reference, which can be used to restrict the transform operation on a set of ranges (default is `AnyRange`) and e_c can be a logical XLWrap expression, which is evaluated each time before the transformation is applied (default is *true*). For all the *shift* operations d is the amount of columns/rows/sheets to shift (defaults to 1), n is the number of times to apply the operation (defaults to the maximum integer value of the runtime system), and for the repeat operations, (g_i) and (h_i), respectively, specify the set of sheets or files to apply the template for. In order to dynamically wrap evolving spreadsheets, n can be omitted and the iteration can be controlled based on the condition e_c. As a consequence, the transform operation will be repeated until the condition evaluates to false. For convenience, the special function `EMPTY(`e_{ref}`)`, which takes a multi range argument and returns true if all cells in e_{ref} are empty, can be used to detect the end of a data range in the spreadsheet.

As mentioned along with Definition 5, a range reference can be *absolute* or *relative*. Relative range references are extended during the mapping process by the base workbook w_k and base worksheet s_k defined in the mapping M. For example, "A3" may refer to "file:foo.xls#$Sheet1.A3" at runtime. The sheet/file repeat transformations will override the sheet/file component as needed, but absolute range references are never modified by transform operations. There are special expression functions which can be used to track the origin of the generated RDF triple by obtaining the current filename, sheet name, sheet number, row and column of a cell at runtime: `FILENAME()`, `SHEETNAME()`, `SHEETNUM()`, `COLUMN()`, and `ROW()`. All these functions take a cell range as an argument.

4.2 Example Mapping

The source data for the example is printed in Table 1. The workbook w_0 used for this demonstration contains two sheets s_1, s_2 of information on revenues of a company. For each country the company operates in, revenues are organized in a cross table containing sold items and total revenue per product and year. As can be seen, data for 2008 is missing for Germany, and there is one more product for Germany.

Table 1. Source data for the discussed example ($s_1, s_2 \in w_0$)

Austria							
	2007		2008		2009		
product	items sold	revenue	items sold	revenue	items sold	revenue	
Product 1	342	7,866.00	376	8,648.00	490	11,760.00	
Product 2	4,333	1,005,256.00	5,655	1,328,925.00	3,493	838,320.00	
Product 3	3,312	1,136,016.00	4,566	1,598,100.00	5,993	1,917,760.00	
Product 4	45	19,350.00	56	24,304.00	54	23,328.00	
Totals	8,032	2,168,488.00	10,653	2,959,977.00	10,030	2,791,168.00	

Germany					
	2007		2009		
product	items sold	revenue	items sold	revenue	
Product1	2,431	55,913.00	3,419	82,056.00	
Product2	31,230	7,339,050.00	32,123	7,709,520.00	
Product3	23,121	8,092,350.00	31,039	9,932,480.00	
Product4	3,423	1,198,050.00	3,412	1,091,840.00	
Product5	121	52,514.00	312	134,784.00	
Totals	60,326	16,737,877.00	70,305	18,950,680.00	

Depending on the desired target graph, $C_k, G_k, and F_k$ can be specified differently. For instance, the target graph could be modeled as one resource per country having linked resources for all products, and years. A more direct representation of the multi-dimensional information is defined in the example mapping shown in Listing 2. We will describe the generation process for this mapping in the next section. A third representation using the *Statistical Core Vocabulary*[7] (SCOVO) is provided as part of the distribution (`mappings/iswc09-example.trig`). In order to be able to include the template graphs in the mapping specification, the TriG syntax[8], which allows to denote named graphs, is used. XLWrap searches for an instance of `xl:Mapping` in all graphs and starts parsing the specification. The XLWrap mapping vocabulary, which is published at `http://www.langegger.at/xlwrap/vocab#`, corresponds to the definitions provided in Section 4.1.

[7] `http://purl.org/NET/scovo`

[8] `http://www4.wiwiss.fu-berlin.de/bizer/TriG/`

```
@prefix rdf:  <http://www.w3.org/1999/02/22-rdf-syntax-ns#> .
@prefix xl:   <http://langegger.at/xlwrap/vocab#> .
@prefix ex:   <http://example.org/> .
@prefix :     <http://myApplication/mapping#> .

{ [] a xl:Mapping ;
    xl:template [
        xl:fileName "files/testing/iswc09-example.xls" ;
        xl:sheetNumber "0" ;
        xl:templateGraph :Revenues ;
        xl:transform [
            a rdf:Seq ;
            rdf:_1 [ a xl:RowShift ;
                xl:restriction "A4; B4:C4" ;
                xl:condition "LEFT(A4, 7) == 'Product'" ;
                xl:steps "1" ] ;
            rdf:_2 [ a xl:ColShift ;
                xl:restriction "B2; B4:C4"^^xl:Expr ;
                xl:condition "!EMPTY(B4:C4)" ;
                xl:steps "2" ] ;
            rdf:_3 [ a xl:SheetShift ;
                xl:restriction "#1.*"^^xl:Expr ;
                xl:repeat "2" ] ;
        ]
    ] .
}

:Revenues {
    [ xl:uri "'http://example.org/revenue_' & URLENCODE(SHEETNAME(A1) & '_' & B2 &
            '_' & A4)"^^xl:Expr ] a ex:Revenue ;
    ex:country      "DBP_COUNTRY(SHEETNAME(A1))"^^xl:Expr ;
    ex:year         "DBP_YEAR(B2)"^^xl:Expr ;
    ex:product      "A4"^^xl:Expr ;
    ex:itemsSold    "B4"^^xl:Expr ;
    ex:revenue      "C4"^^xl:Expr .
}
```

Listing 2. Example mapping specified in TriG syntax

In our approach, the mapping is stored separately from the actual spreadsheet files and not as part of them. It is assumed that the creator of the mapping file may not have the authorization or possibility to modify spreadsheet files to be wrapped. XLWrap is capable of loading local files and downloading spreadsheet files from the Web. Currently, Excel files, Open Document spreadsheets, and also CSV files are supported (although they could also be wrapped with other tools). The layout of CSV files (delimiters, separators, white spaces) can be specified. CSV files are streamed in order to support large data files. The implementation simulates a workbook and requests new rows from the input stream as needed by the execution engine. Because a single template graph typically refers to a small section of the whole spreadsheet, it is sufficient to keep the last n (where $n = 1000$ by default) rows in a cache.

5 Transformation Process

To give an overview of the transformation process, a flow chart is depicted in Figure 3. For each map template $m_k \in M$, C_k and all generated template graphs q_i are evaluated based on the current state of the execution context X, which contains a reference to the currently processed map template m_k (`activeTmpl`) in order to retrieve w_k and

s_k. The base workbook w_k and the base worksheet s_k are required for the evaluation of relative range references and obtaining values from the cells of the spreadsheets. The execution context also contains a reference to the target graph (`targetGraph`) where the generated statements are inserted. While in Section 4 we used $[[G_k]]$ to denote the evaluation of G_k including the application of transform operations F_k, the notation of $[[C_k]]_X$ and $[[q_i]]_X$ in the flow chart only represents the evaluation of XLWrap expressions for the given graphs C_k and q_i. The application of F_k is completely hidden by the `TemplateModelGenerator`, which subsequently applies the defined transform operations $f_l \in F_k$ against G_k and returns multiple template graphs q_i.

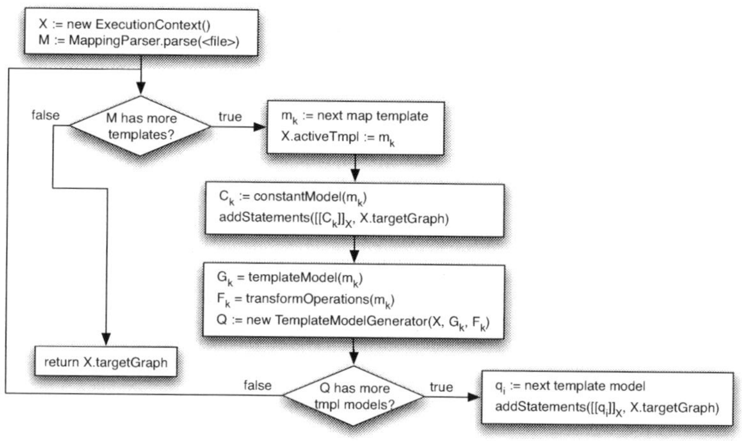

Fig. 3. Overview of the wrapping process

The `TemplateModelGenerator` is implemented as an iterator which uses a sequence of stacked instances of `TransformationStage` – one for each transform operation f_l. Each stage transforms its current *stage graph* $q^{f_l}_{i_1,\ldots,i_n}$ according to the corresponding transform operation as depicted in Figure 4. Initially, all stage graphs are equal to the template graph: $q^{f_l}_{0,\ldots,0} = G_k$. The blue nodes on the bottom represent final template graphs which are returned by `TemplateModelGenerator`. Each call to `TemplateModelGenerator.hasNext()` results in a transformation at the lowest stage that has more transformations to apply. For example, if there are no more rows to shift by f_1, the corresponding stage triggers its parent stage and tracks back its internal state. Likewise, if the condition defined for the transform operation does not hold, it is skipped and the parent stage is triggered.

When a template graph is applied, before its triples are added into the target graph, any blank node with a `xl:uri` property is replaced with a URI node, blank node labels with equal `xl:id` properties are aligned, and any `xl:Expr` literal is evaluated as an XLWrap expression e. The result of $[[e]]$ is an instance of `XLExprValue`, which can be a URI, blank node, string, long integer, double, boolean, or date value. When obtaining cell values, the type is automatically detected based on the type annotation (Definition 4). When creating literals for the target graph, long integers and floats are

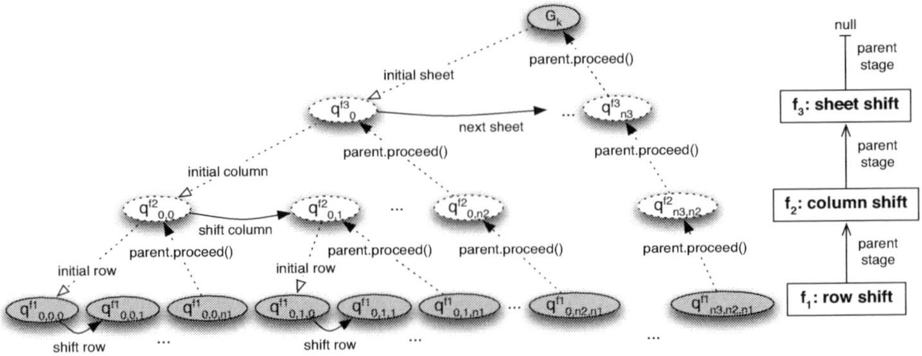

Fig. 4. Transform stages for the mapping specification of Listing 2

automatically reduced to the required size as long as they have not been explicitly casted with a type casting function before. Depending on the type, a matching typed literal is created.

For the example spreadsheet given in Table 1, after applying the mapping in Listing 2, the following triples are generated:

```
ex:revenue_Austria_2007_Product1 a ex:Revenue ;
      ex:country <http://dbpedia.org/resource/Austria> ;
      ex:itemsSold "342"^^<http://www.w3.org/2001/XMLSchema#short> ;
      ex:product "Product1" ;
      ex:revenue "7866"^^<http://www.w3.org/2001/XMLSchema#int> ;
      ex:year <http://dbpedia.org/resource/2007> .
ex:revenue_Austria_2007_Product2 a ex:Revenue ;
      ex:country <http://dbpedia.org/resource/Austria> ;
      ex:itemsSold "4333"^^<http://www.w3.org/2001/XMLSchema#short> ;
      ex:product "Product2" ;
      ex:revenue "1005256"^^<http://www.w3.org/2001/XMLSchema#int> ;
      ex:year <http://dbpedia.org/resource/2007> .
ex:revenue_Austria_2007_Product3 ...
ex:revenue_Austria_2007_Product4 ...
ex:revenue_Austria_2008_Product1 ...
...
ex:revenue_Austria_2009_Product1 ...
...
ex:revenue_Germany_2007_Product1 ...
...
ex:revenue_Germany_2009_Product1 ...
...
ex:revenue_Germany_2009_Product5 ...
```

Range reference sub-expressions of the stage template graph $q^{f1}_{0,0,0} = G_k$ are shifted down by one row first until the condition `LEFT(A4, 7) == 'Product'` is false, producing resources for all products sold in Austria in 2007. However, only those range references within the range restriction z_{f1} = `"A4; B4:C4"` are actually transformed. For instance, the expression `"A4"` (literal for `ex:product`) is subsumed by the restriction range and is therefore changed to `"A5"`, but `"DBP_YEAR(B2)"` remains unchanged. Next, $q^{f2}_{0,1}$ is calculated by a 2-step column shift of $q^{f2}_{0,0}$. The stage model of the sub-stage is initialized as $q^{f1}_{0,1,0} := q^{f2}_{0,1}$ for the next execution of f_1 (row shift). If both, f_1 and f_2

have no more transformations (or both stage conditions do not hold), the sheet is shifted according to f_3, producing similar RDF data for Germany.

Transform operations are not only applied to range references in `xl:Expr` literals of $q^{f_l}_{i_1,\ldots,i_n}$, they must be applied also to the range restrictions z^{f_l} and to the conditions $e^{f_l}_c$ of the corresponding transform operations. For instance, the range restriction on the row shift `"A4; B4:C4"` has to be shifted to `"A5; B5:C5"` in the first stage and then to `"A4; D4:E4"`, `"A5; D5:E5"`, and `"A4; F4:G4"`, `"A5; F5:G5"`, etc. in the second stage. When proceeding at the second stage, the transformation of the original f_1-restriction is itself restricted by the current range restriction of f_2, which is `"B2; B4:C4"`. As visualized in Figure 5, thus only a subset of `"A4; B4:C4"` is shifted leading to `"A4; D4:E4"`. Currently, XLWrap is not capable of automatically splitting arbitrary box ranges based on restrict ranges. This is why, "A4; B4:C4" was not specified as "A4:C4" in the mapping[9]. However, intersections of box ranges are detected during the initialization of a map template in order to be corrected.

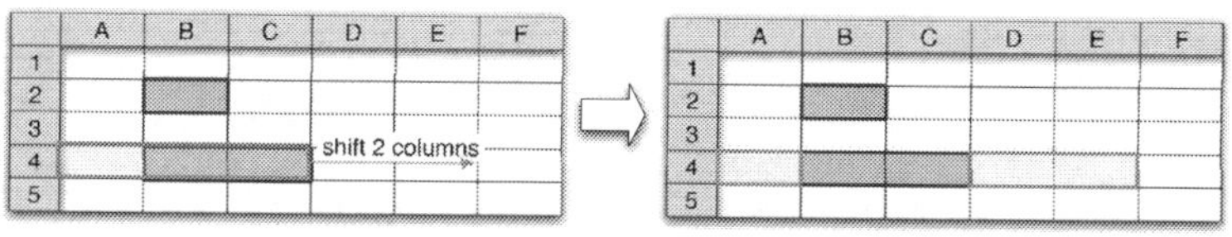

Fig. 5. Column shift of range restriction "A4; B4:C4" restricted by "B2; B4:C4"

6 Conclusion

In this contribution we have presented XLWrap, which is an RDF-centric mapping approach to support the transformation of spreadsheets with different representation models to arbitrary RDF graphs. The mapping concept has been formally defined and implemented based on the Jena Semantic Web framework. The server component called XLWrap-Server was not further discussed due to the page limit. It is a stand-alone Web application based on Joseki and Snorql from the D2R-Server project including a SPARQL endpoint and a linked data interface.

XLWrap is powerful enough to represent mappings for spreadsheets with different representation models and target graphs. Because it supports external references, HTTP URLs, and the wrapping of multiple spreadsheets into a combined cache including OWL inference, it can be used very easily to semantically integrate multiple spreadsheets locally or in intranets and extranets. The possibility of adding custom functions – which is a matter of extending `XLExprFunction` and providing an implementation for `eval()` – can be very practical for end-users. Beside adding custom mathematical and statistical functions, it is possible to access a database or Web resources by XLWrap functions. The future support for aggregate functions in SPARQL is a very important requirement in order to support typical operations on spreadsheet data.

[9] Especially in combination with the *multi sheet* and *any range*, ranges cannot be split straightforward and the implementation would additionally require support for exclusion ranges.

Future work will include the development of a graphical support tool including some kind of mapping debugger and auto-detection of cross-tables to facilitate the mapping specification task. Considerable work regarding auto-detection of headers and units has already been published [1,2].

Acknowledgements

This work is funded by the Austrian BMBWK (Federal Ministry for Education, Science and Culture), contract GZ BMWF-10.220/0002-II/10/2007. We would like to thank Bernhard Haslhofer, Bernhard Schandl, Thomas Leitner, and Berndt Zinnöcker for the helpful inputs and discussions held during the implementation of XLWrap.

References

1. Abraham, R., Erwig, M.: Header and unit inference for spreadsheets through spatial analyses. In: VLHCC 2004: Proceedings of the 2004 IEEE Symposium on Visual Languages - Human Centric Computing, Washington, DC, USA, pp. 165–172. IEEE Computer Society Press, Los Alamitos (2004)
2. Chambers, C., Erwig, M.: Dimension inference in spreadsheets. In: VLHCC 2008: Proceedings of the 2008 IEEE Symposium on Visual Languages and Human-Centric Computing, Washington, DC, USA, pp. 123–130. IEEE Computer Society Press, Los Alamitos (2008)
3. Cyganiak, R., Bizer, C.: D2R Server – Publishing Relational Databases on the Web as SPARQL Endpoints. In: Developers Track at the 15th International World Wide Web Conference (WWW2006), Edinburgh, Scotland (May 2006)
4. Group, M., Reck, R.P.: Excel2RDF, `http://www.mindswap.org/~rreck/excel2rdf.shtml` (Last visit, June 2009)
5. Han, L., Finin, T.W., Parr, C.S., Sachs, J., Joshi, A.: RDF123: From Spreadsheets to RDF. In: Sheth, A.P., Staab, S., Dean, M., Paolucci, M., Maynard, D., Finin, T., Thirunarayan, K. (eds.) ISWC 2008. LNCS, vol. 5318, pp. 451–466. Springer, Heidelberg (2008)
6. HP Labs, Bristol, UK: Joseki – A SPARQL Server for Jena, `http://www.joseki.org/` (Last visit, June 2009)
7. Langegger, A., Wöß, W.: SemWIQ – Semantic Web Integrator and Query Engine. In: Hegering, H.G., Lehmann, A., Ohlbach, H.J., Scheideler, C. (eds.) Beiträge der 38. Jahrestagung der Gesellschaft für Informatik e.V (GI), vol. 1. Bonner Köllen Verlag (2008)
8. Maurizio, L.: Data integration: a theoretical perspective. In: PODS 2002: Proceedings of the twenty-first ACM SIGMOD-SIGACT-SIGART symposium on Principles of database systems, pp. 233–246. ACM, New York (2002)
9. Valente, A., Van brackle, D., Chalupsky, H., Edwards, G.: Implementing logic spreadsheets in less. Knowl. Eng. Rev. 22(3), 237–253 (2007)

Optimizing QoS-Aware Semantic Web Service Composition*

Freddy Lécué

The University of Manchester
Booth Street East, Manchester, UK
firstname.lastname@manchester.ac.uk

Abstract. Ranking and optimization of web service compositions are some of the most interesting challenges at present. Since web services can be enhanced with formal semantic descriptions, forming the "semantic web services", it becomes conceivable to exploit the quality of semantic links between services (of any composition) as one of the optimization criteria. For this we propose to use the semantic similarities between output and input parameters of web services. Coupling this with other criteria such as quality of service (QoS) allow us to rank and optimize compositions achieving the same goal. Here we suggest an innovative and extensible optimization model designed to balance semantic fit (or functional quality) with non-functional QoS metrics. To allow the use of this model in the context of a large number of services as foreseen by the strategic EC-funded project SOA4All we propose and test the use of Genetic Algorithms.

Keywords: Semantic Web, Web service, Service composition, Quality of service and composition, Automated reasoning.

1 Introduction

The *Semantic Web* [1], where the semantic content of the information is tagged using machine-processable languages such as the *Web Ontology Language* (OWL) [2], is considered to provide many advantages over the current "formatting only" version of the World-Wide-Web. OWL is based on concepts from Description Logics [3] and ontologies, formal conceptualization of a particular domain. This allows us to describe the semantics of services, e.g., their functionality in terms of input and output parameters, preconditions, effects and invariants. Such descriptions can then be used for automatic reasoning about services and automating their use to accomplish "intelligent" tasks such as selection, discovery and composition.

Here we focus on web service composition and more specifically on its functional level, where a set of services is composed to achieve a goal on the basis of the semantic similarities between input and output parameters as indicators of service functionality. To measure semantic similarity, we use the concept of (functional) semantic link [4], defined as a semantic connection (i.e., part of data flow) between an output and an input parameter of two services. Web service compositions could thus be estimated

* Foundation Project: Supported by European Commission VII Framework IP Project Soa4All.

A. Bernstein et al. (Eds.): ISWC 2009, LNCS 5823, pp. 375–391, 2009.

and ranked not only along well known non functional parameters such as *Quality of Services* (QoS) [5] but also along the dimension of semantic similarity as indicator of functional fit [6]. Considering semantics on connections of services is useful in case the information required and provided by services does not match perfectly in every data flow. This is the case of semantic-based description of services. In this work we propose to unify both types of criteria in an innovative and extensible model allowing us to estimate and optimise the quality of service compositions.

Maximizing the quality of service composition using this model is essentially a multi-objective optimization problem with constraints on quality of services and semantic links, which is known to be NP-hard [7]. Most approaches in the literature addressing optimization in web service composition are based on *Integer linear Programming* (IP) e.g., [8]. However, IP approaches have been shown to have poor scalability [5] in terms of time taken to compute optimal compositions when the size of the initial set of services grows. Such a case can arise in the future semantic web, where a large number of semantic services will be accessible globally. This is the vision of SOA4All, a strategic EC-funded project. Rapid computation of optimal compositions is especially important for interactive systems providing service composition facilities for end users, where long delays may be unacceptable. Here we demonstrate that the optimisation problem can be automated in a more scalable manner using *Genetic Algorithms* (GAs), and propose an approach to tackle QoS-aware *semantic* web service composition.

The remainder of this paper is organised as follows. In the next section we briefly review i) semantic links, ii) their common descriptions and iii) the web service composition model. Section 3 introduces the quality criteria for QoS-aware semantic web service composition. Section 4 details the GA-based evolutionary approach, including the strategies of the crossover, mutation and fitness function. Section 5 reports and discusses results from the experimentations. Section 6 briefly comments on related work. Finally section 7 draws some conclusions and talks about possible future directions.

2 Background

In this section we describe how semantic links can be used to model web service composition. In addition we remind the definition of *Common Description* in semantic links.

2.1 Semantic Links between Web Services

In the semantic web, input and output parameters of services referred to concepts in a common ontology[1] or Terminology $\mathcal{T}$ (e.g., Fig.2), where the OWL-S profile [9] or SA-WSDL [10] can be used to describe them (through semantic annotations). At functional level web service composition consists in retrieving some semantic links [4] noted $sl_{i,j}$ (Fig.1) i.e.,

$$sl_{i,j} \doteq \langle s_i, Sim_{\mathcal{T}}(Out\_s_i, In\_s_j), s_j \rangle \quad (1)$$

[1] Distributed ontologies are not considered here but are largely independent of the problem addressed in this work.

between output parameters $Out\_s_i \in \mathcal{T}$ of services s_i and input parameters $In\_s_j \in \mathcal{T}$ of other services s_j. Thereby s_i and s_j are partially linked according to a matching function $Sim_{\mathcal{T}}$. Given a terminology $\mathcal{T}$, [11] and [12] value the range of the latter function along five matching types: i) *Exact* i.e., $Out\_s_i \equiv In\_s_j$, ii) *PlugIn* i.e., $Out\_s_i \sqsubseteq In\_s_j$, iii) *Subsume* i.e., $In\_s_j \sqsubseteq Out\_s_i$, iv) *Intersection* i.e., $\neg(Out\_s_i \sqcap In\_s_j \sqsubseteq \perp)$ and v) *Disjoint* i.e., $Out\_s_i \sqcap In\_s_j \sqsubseteq \perp$.

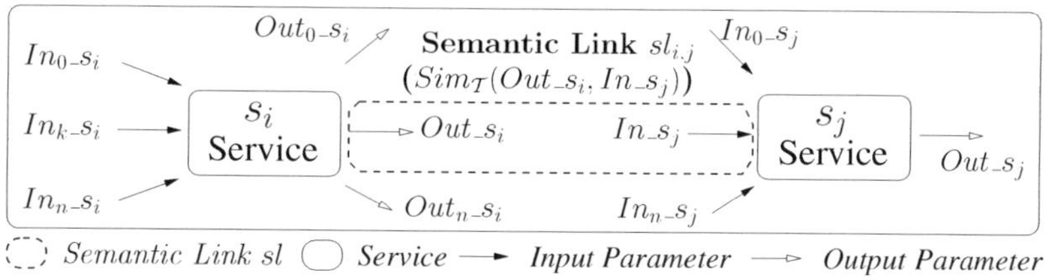

Fig. 1. A Semantic Link $sl_{i,j}$ between Services s_i and s_j

Example 1 ***(Matching Type)***
Suppose $sl_{1,2}$ (Fig.3) be a semantic link between two services s_1 and s_2 such that the output parameter `NetworkConnection` *of s_1 is (semantic) linked to the input parameter* `SlowNetworkConnection` *of s_2. According to Fig.2 this link is valued by a Subsume matching type since $NetworkConnection \sqsupseteq SlowNetworkConnection$.*

The matching function $Sim_{\mathcal{T}}$ enables, at design time, finding some types of semantic compatibilities (i.e., Exact, PlugIn, Subsume, Intersection) and incompatibilities (i.e., Disjoint) among independently defined service descriptions. In this direction the latter function can be used to define the quality of data flow in web service composition at semantic level.

2.2 Common Description of a Semantic Link

Besides computing the matching type of a semantic link, authors of [13] suggest computing a finer level of information i.e., the Extra and Common Descriptions between Out_s_i and In_s_j of a semantic link $sl_{i,j}$. They adapt the definition of syntactic difference [14] for comparing $\mathcal{ALE}$ DL descriptions and then obtaining a compact representation. The Extra Description $In\_s_j \backslash Out\_s_i$:

$$In\_s_j \backslash Out\_s_i \doteq \min_{\preceq_d}\{E | E \sqcap Out\_s_i \equiv In\_s_j \sqcap Out\_s_i\} \tag{2}$$

refers[2] to information required by In_s_j but not provided by Out_s_i to ensure a correct data flow between s_i and s_j. The Common Description of Out_s_i and In_s_j, defining as their Least Common Subsumer [15] lcs, refers to information required by In_s_j and provided by Out_s_i[3].

[2] With respect to the subdescription ordering $\preceq_d$.
[3] In case $Out\_s_i \sqcap \neg In\_s_j \sqsubseteq \perp$, $In\_s_j \backslash Out\_s_i$ is replaced by its more general form i.e., $In\_s_j \backslash lcs(In\_s_j, Out\_s_i)$.

$NetworkConnection \equiv \forall netPro.Provider \sqcap \forall netSpeed.Speed$
$SlowNetworkConnection \equiv NetworkConnection \sqcap$
$\forall netSpeed.Adsl1M$
$Adsl1M \equiv Speed \sqcap \forall mBytes.1M$

Fig. 2. Sample of an $\mathcal{ALE}$ Domain Ontology $\mathcal{T}$

Example 2 ***(Extra & Common Description)***
Suppose $sl_{1,2}$ in Example 1. On the one hand the Extra Description *missing in* `NetworkConnection` *to be used by the input parameter* `SlowNetworkConnection` *is defined by $SlowNetworkConnection \backslash NetworkConnection$ i.e., $\forall netSpeed.Adsl1M$. On the other hand the* Common Description *is defined by $lcs(SlowNetworkConnection, NetworkConnection)$ i.e., $NetworkConnection$.*

2.3 Modelling Web Service Composition

In this work, the process model of web service composition and its semantic links is specified by a statechart [16]. Its states refer to services whereas its transitions are labelled with semantic links. In addition some basic composition constructs such as sequence, conditional branching (i.e., OR-Branching), structured loops, concurrent threads (i.e., AND-Branching), and inter-thread synchronization can be found. To simplify the presentation, we initially assume that all considered statecharts are acyclic and consists of only sequences, OR-Branching and AND-Branching.

Example 3 ***(Process Model of a Web Service Composition)***
Suppose a composition extending Example 1 with six more services $s_{i,1\leq i\leq 8}$, eight more semantic links $sl_{i,j}$. Its process model is depicted in Fig.3.

The example 3 illustrates a composition wherein tasks T_i and abstract semantic link $sl^A_{i,j}$ have been respectively concretized by one of their n candidate services (e.g., s_i) and n^2 candidate links (e.g., $sl_{i,j}$). Indeed some services with common functionality,

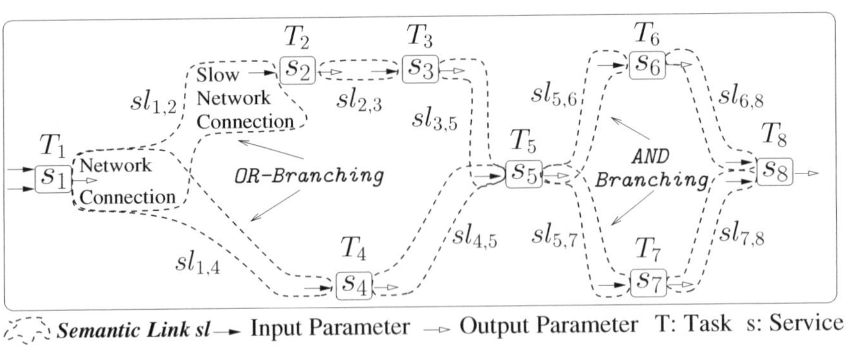

Fig. 3. A (Concrete) Web Service Composition

preconditions and effects although different input, output parameters and quality can be selected to perform a target task T_i and obtaining a concrete composition. Such a selection will have a direct impact on semantic links involved in the concrete composition.

In the following we assume that compositions of tasks (achieving a goal) have been pre-computed. This computation can be performed by *template-based* and *parametric-design-based* composition approaches [17]. So the control flow of the pre-computed compositions is fixed since it is pre-defined. The choice of the service (that fulfills a given task) will be done at composition time, based on both quality of i) services and ii) their semantic links (i.e., quality of data flow).

3 Quality Model

Here we present a quality criterion to value semantic links. Then we suggest to extend it with the non functional QoS to estimate both quality levels of any compositions.

3.1 Quality of Semantic Link

We consider two generic quality criteria for semantic links $sl_{i,j}$ defined by $\langle s_i, Sim_{\mathcal{T}}(Out\_s_i, In\_s_j), s_j\rangle$: its i) *Common Description* rate, and ii) *Matching Quality*.

Definition 1 ***(Common Description rate of a Semantic Link)***
Given a semantic link $sl_{i,j}$ between s_i and s_j, the Common Description rate $q_{cd} \in (0, 1]$ provides one possible measure for the degree of similarity between an output parameter of s_i and an input parameter of s_j. This rate is computed using the following expression:

$$q_{cd}(sl_{i,j}) = \frac{|lcs(Out\_s_i, In\_s_j)|}{|In\_s_j \backslash Out\_s_i| \;+\; |lcs(Out\_s_i, In\_s_j)|} \tag{3}$$

This criterion estimates the proportion of descriptions which is well specified for ensuring a correct data flow between s_i and s_j.

The expressions in between $|$ refer to the size of $\mathcal{ALE}$ concept descriptions ([18] p.17) i.e., $|\top|$, $|\bot|$, $|A|$, $|\neg A|$ and $|\exists r|$ is 1; $|C \sqcap D| \doteq |C| + |D|$; $|\forall r.C|$ and $|\exists r.C|$ is $1 + |C|$. For instance $|Adsl1M|$ is 3 in the ontology illustrated in Fig. 2.

Definition 2 ***(Matching Quality of a Semantic Link)***
The Matching Quality q_m of a semantic link $sl_{i,j}$ is a value in $(0, 1]$ defined by $Sim_{\mathcal{T}}(Out\_s_i, In\_s_j)$ i.e., either 1 (Exact), $\frac{3}{4}$ (PlugIn), $\frac{1}{2}$ (Subsume) or $\frac{1}{4}$ (Intersection).

The discretization of the matching types follows a partial ordering [19] where the assignment of values to matching types is driven by the data integration costs. Behind each matching type, tasks of XML (Extensible Markup Language) data type integration and manipulation are required. The PlugIn matching type is more penalized than the *Exact* matching type in this model. Indeed the data integration process is lower (in term of computation costs) for the *Exact* matching type than for *PlugIn* matching type.

Contrary to q_{cd}, q_m does not estimate similarity between the parameters of semantic links but gives a general overview (discretized values) of their semantic relationships. Given these quality criteria, the quality vector of a semantic link $sl_{i,j}$ is defined by:

$$q(sl_{i,j}) \doteq \left(q_{cd}(sl_{i,j}), q_m(sl_{i,j})\right) \quad (4)$$

The quality of semantic links can be compared by analysing their q_{cd} and q_m elements. For instance $q(sl_{i,j}) > q(sl'_{i,j})$ if $q_{cd}(sl_{i,j}) > q_{cd}(sl'_{i,j})$ and $q_m(sl_{i,j}) > q_m(sl'_{i,j})$. Alternatively we can compare a weighted average of their normalised components in case the value of the first element of $sl_{i,j}$ is better than the first element of $sl'_{i,j}$ but worse for the second element [20].

Example 4 ***(Quality of Semantic Links)***
Let s'_2 be another candidate service for T_2 in Fig.3 with `NetworkConnection` *as an input. The link $sl'_{1,2}$ between s_1 and s'_2 is better than $sl_{1,2}$ since $q(sl'_{1,2}) > q(sl_{1,2})$.*

In case s_i, s_j are related by more than one link, the value of each criterion is retrieved by computing their average. This average is computing by means of the And-Branching row of Table 1, independently along each dimension of the quality model.

3.2 QoS-Extended Quality of Semantic Link

We extend the latter quality model by exploiting the non functional properties of services (also known as QoS attributes [21] - given by service providers or third parties) involved in each semantic link. We simplify the presentation by considering only:

- **Execution Price** $q_{pr}(s_i) \in \Re^+$ of service s_i i.e., the fee requested by the service provider for invoking it.
- **Response Time** $q_t(s_i) \in \Re^+$ of service s_i i.e., the expected delay between the request and result moments.

A quality vector of a service s_i is then defined as follows:

$$q(s_i) \doteq (q_{pr}(s_i), q_t(s_i)) \quad (5)$$

Thus a QoS-extended quality vector of a semantic link $sl_{i,j}$:

$$\overset{*}{q}(sl_{i,j}) \doteq (q(s_i), q(sl_{i,j}), q(s_j)) \quad (6)$$

Given an abstract link between tasks T_i, T_j, one may select the link with the best functional quality (matching quality, common description rate), and non-functional (the cheapest and fastest services) quality values, or may be a compromise (depending on the enduser preferences) between the four by coupling (4) and (6) in (6). Moreover the selection could be influenced by predefining some constraints e.g., a service response time lower than a given value.

Example 5 ***(QoS-Extended Quality of Semantic Link)***
Suppose T_2 and its two candidate services s_2, s'_2 wherein $q(s'_2) < q(s_2)$. According to example 4, s'_2 should be preferred regarding the quality of its semantic link with s_1, whereas s_2 should be preferred regarding its QoS. So what about the best candidate for $sl^A_{1,2}$ regarding both criteria: $\overset{*}{q}$? Before addressing this question in Section 4 through equation (9)*, we first focus in quality of composition in Section 3.3.*

3.3 Quality of Composition

We present definitions for comparing and ranking different compositions along the common description rate and matching quality dimension. The rules for aggregating quality values (Table 1) for any concrete composition c are driven by them. In more details the approach for computing semantic quality of c is adapted from the application-driven heuristics of [6], while the computation of its non functional QoS is similar to [22].

Definition 3 ***(Common Description rate of a Composition)***
The Common Description rate of a composition measures the average degree of similarity between all corresponding parameters of services linked by a semantic link.

The Common Description rate Q_{cd} of both a sequential and AND-Branching composition is defined as the average of its semantic links' common description rate $q_{cd}(sl_{i,j})$. The common description rate of an OR-Branching composition is a sum of $q_{cd}(sl_{i,j})$ weighted by $p_{sl_{i.j}}$ i.e., the probability that semantic link $sl_{i,j}$ be chosen at run time. Such probabilities are initialized by the composition designer, and then eventually updated considering the information obtained by monitoring the workflow executions.

Definition 4 ***(Matching Quality of a Composition)***
The matching quality of a composition estimates the overall matching quality of its semantic links. Contrary to the common description rate, this criteron aims at easily distinguishing and identifying between very good and very bad matching quality.

The matching quality Q_m of a sequential and AND-Branching composition is defined as a product of $q_m(sl_{i,j})$. All different (non empty) matching qualities involved in such compositions require to be considered together in such a (non-linear) aggregation function to make sure that compositions that contains semantic links with low or high matching quality will be more easily identified, and then pruned for the set of potential solutions. The matching quality of an OR-Branching composition is defined as its common description rate by changing $q_{cd}(sl_{i,j})$ by $q_m(sl_{i,j})$.

Details for computing **Execution Price** Q_{pr} and **Response Time** Q_t can be found in Table 1, and further explained in [22].

Table 1. Quality Aggregation Rules for Semantic Web Service Composition

Composition Construct	Quality Criterion			
	Functional		Non Functional	
	Q_{cd}	Q_m	Q_t	Q_{pr}
Sequential/ AND- Branching	$\frac{1}{\lvert sl_{i.j}\rvert}\sum_{sl_{i.j}} q_{cd}(sl_{i,j})$	$\prod_{sl_{i.j}} q_m(sl_{i,j})$	$\frac{\sum_{s_i} q_t(s_i)}{\max_s q_t(s)}$	$\sum_{s_i} q_{pr}(s_i)$
OR-Branching	$\sum_{sl_{i.j}} q_{cd}(sl_{i,j}).p_{sl_{i.j}}$	$\sum_{sl_{i.j}} q_m(sl_{i,j}).p_{sl_{i.j}}$	$\sum_{s_i} q_t(s_i).p_{s_i}$	$\sum_{s_i} q_{pr}(s_i).p_{s_i}$

Using Table 1, the quality vector of any concrete composition can be defined by:

$$Q(c) \doteq (Q_{cd}(c), Q_m(c), Q_t(c), Q_{pr}(c)) \tag{7}$$

Although the adopted quality model has a limited number of criteria (for the sake of illustration), (4), (5), (6) as well as (7) are extensible: new functional criteria can be added without fundamentally altering the service selection techniques built on top of the model. In this direction the binary criterion of robustness [13] in semantic links can be considered[4]. In addition, other non-functional criteria such as reputation, availability, reliability, successful execution rate, etc., can also be considered in such an extension.

4 A Genetic Algorithm Based Optimization

The optimization problem (i.e., determining the best set of services of a composition with respect to some quality constraints) which can be formalized as a *Constraints Satisfaction Optimization Problem* (T, D, C, f) where T is the set of tasks (variables) in the composition, D is the set of services' domains for T (each D_i representing a set of possible concrete services that fulfil the task T_i), C is the set of constraints and f is an evaluation function that maps every solution tuple to a numerical value, is NP-hard. In case the number of tasks and candidate services are respectively n and m, the naive approach considers an exhaustive search of the optimal composition among all the m^n concrete compositions. Since such an approach is impractical for large-scale composition, we address this issue by presenting a GA-based approach [23] which i) supports constraints on QoS and also on quality of semantic links and ii) requires the set of selected services as a solution to maximize a given objective. Here compositions refer to their concrete form.

4.1 GA Parameters for Optimizing Composition

By applying a GA-based approach the optimal solution (represented by its *genotype*) is determined by simulating the evolution of an *initial population* (through generation) until survival of best *fitted* individuals (here compositions) satisfying some *constraints*. The survivors are obtained by *crossover*, *mutation*, *selection* of compositions from previous generations. Details of GA parameterization follow:

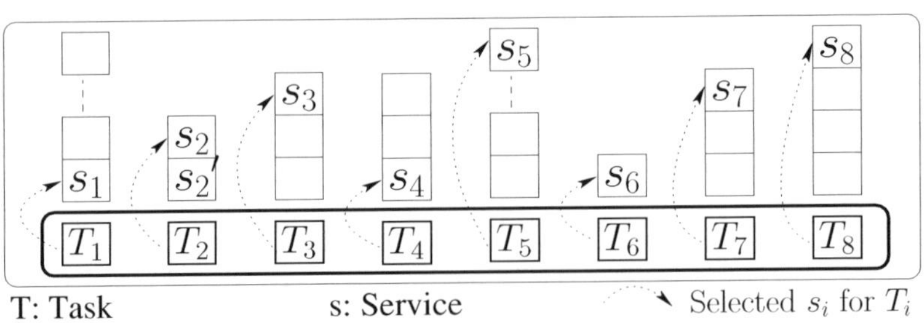

Fig. 4. Genotype Encoding for Service Composition

[4] Contrary to [6], we did not consider robustness because of its strong dependency with the matching quality criterion. Indeed they are not independent criteria since the robustness is 1 if the matching type is either Exact or PlugIn, and 0 otherwise.

• **Genotype:** It is defined by an array of integer. The number of items is equal to the number tasks involved in the composition. Each item, in turn, contains an index to an array of candidate services matching that task. Each composition, as a potential solution of the optimization problem, can be encoded using this genotype (e.g., Fig.4 is encoding the genotype of composition in Fig.3).

• **Initial Population:** It consists of an initial set of compositions (characterized by their genotypes) wherein services are randomly selected.

• **Global, Local Constraints** have to be met by compositions c e.g., $Q_{cd}(c) > 0.8$.

• **Fitness Function:** This function is required to quantify the "quality" of any composition c. Such a function f needs to maximize semantic quality attributes, while minimizing the QoS attributes of c:

$$f(c) = \frac{\omega_{cd}\hat{Q}_{cd}(c) + \omega_m \hat{Q}_m(c)}{\omega_{pr}\hat{Q}_{pr}(c) + \omega_t \hat{Q}_t(c)} \tag{8}$$

where $\hat{Q}_{l\in\{pr,t,cd,m\}}$ refer to Q_l normalized in the interval $[0,1]$. $\omega_l \in [0,1]$ is the weight assigned to the lth quality criterion and $\sum_{l\in\{pr,t,cd,m\}} \omega_l = 1$. In this way preferences on quality of the desired compositions can be done by simply adjusting ω_l e.g., the Common Description rate could be weighted higher.

In addition f must drive the evolution towards constraint satisfaction. To this end compositions that do not meet the constraints are penalized by extending (8) wrt. (9).

$$f'(c) = f(c) - \omega_{pe} \sum_{l\in \{pr,t, cd,m\}} \left(\frac{\Delta\hat{Q}_l}{\hat{Q}_l^{\max}(c) - \hat{Q}_l^{\min}(c)}\right)^2 \tag{9}$$

where $\hat{Q}_l^{\max}$, $\hat{Q}_l^{\min}$ are respectively the maximum and minimal value of the lth quality constraint, ω_{pe} weights the penalty factor and $\Delta\hat{Q}_{l\in\{pr,t,cd,m\}}$ is defined by:

$$\Delta\hat{Q}_l = \begin{cases} \hat{Q}_l - \hat{Q}_l^{\max} & \text{if } \hat{Q}_l > \hat{Q}_l^{\max} \\ 0 & \text{if } \hat{Q}_l^{\min} \le \hat{Q}_l \le \hat{Q}_l^{\max} \\ \hat{Q}_l^{\min} - \hat{Q}_l & \text{if } \hat{Q}_l < \hat{Q}_l^{\min} \end{cases} \tag{10}$$

Contrary to [5], compositions that violate constraints do not receive the same penalty. Indeed the factor ω_{pe} is further penalized in (9). This function avoids local optimal by considering also compositions that disobey constraints. Unfortunately, (9) contains a penalty for concrete compositions, which is the same at each generation. If, as usual, the weight ω_{pe} for this penalty factor is high, there is a risk that also concrete composition violating the constraints but "close" to a good solution could be discarded.

The alternative is to adopt a dynamic penalty, i.e., a penalty having a weight that increases with the number of generations. This allows, for the early generations, to also consider some individuals violating the constraints. After a number of generations, the population should be able to meet the constraints, and the evolution will try to improve only the rest of the fitness function. The dynamic fitness function (to be maximized) is:

$$f''(c, gen) = f(c) - \omega_{pe} \cdot \frac{gen}{maxgen} \cdot \sum_{l \in \substack{\{pr,t,\\ cd,m\}}} \left(\frac{\Delta \hat{Q}_l}{\hat{Q}_l^{\max}(c) - \hat{Q}_l^{\min}(c)}\right)^2 \quad (11)$$

gen is the current generation, while *maxgen* is the maximum number of generations.

• **Operators on Genotypes:** They define authorized alterations on genotypes not only to ensure evolution of compositions' population along generations but also to prevent convergence to local optimum. We use: i) *composition mutation* i.e., random selection of a task (i.e., a position in the genotype) in a concrete composition and replacing its service with another one among those available, ii) the standard two-points *crossover* i.e., randomly combination of two compositions and iii) *selection of compositions* which is fitness-based i.e., compositions disobeying the constraints are selected proportionally from previous generations.

• **Stopping Criterion:** It enables to stop the evolution of a population. First of all we iterate until the constraints are met (i.e., $\Delta Q_l = 0 \; \forall l \in \{pr, t, cd, m\}$) within a maximum number of generations. Once the latter constraints are satisfied we iterate until the best fitness composition remains unchanged for a given number of generations.

4.2 GA for Optimizing Composition in a Nutshell

The execution of the GA consists in i) defining the initial population (as a set of compositions), and computing the fitness function (evaluation criterion) of each composition, ii) evolving the population by applying mutation and crossover of compositions (Tasks with only one candidate service are disregarded), iii) selecting compositions, iv) evaluating compositions of the population, and v) back to step (ii) if the stopping criterion is not satisfied. Section 5 further details the parameters.

In case no solution exists, users may relax constraints of the optimization problem. Instead, fuzzy logic could be used to address the imprecision in specifying quality constraints, estimating quality values and expressing composition quality.

5 Experimental Results

We analyze the performances of our approach by i) discussing the benefits of combining QoS and functional criteria, ii) observing the evolution of the composition quality f'' in (11) (equal weights are assigned to the different quality criteria) over the GA generations by varying the number of tasks, iii) studying the behaviour of our approach regarding the optimisation of large scale compositions, iv) evaluating performance after decoupling the GA and the (on-line) DL reasoning processes, and v) comparing the convergence of our approach (11) with [5]. Before turning our attention to the latter five sets of experiments, we first draw the context of experimentation.

5.1 Context of Experimentation

Services, Semantic Links and their Qualities. Services[5] are defined by their semantic descriptions using an $\mathcal{ALE}$ ontology (formally defined by 1100 concepts and 390 properties, 1753 individuals, without data property), provided by a commercial partner. We have incorporated estimated values for Qos parameters (price and response time). Common description rate and matching quality of semantic links are computed according to an on-line DL reasoning process.

Implementation Details. The common description rate (3) is calculated by computing the *Extra Description* (2), the Least Common Subsumer [15], and the size ([18] p.17) of DL-based concepts. These DL inferences and the matching types have been achieved by a DL reasoning process i.e., an adaptation of Fact++ [24] for considering DL difference.

The aggregation rules of Table 1 are then used for computing each quality dimension of any composition. Finally the combination of QoS with semantic calculation is computed by means of (11), thus obtaining the final quality score for the composition.

Our GA is implemented in Java, extending a GPL library[6]. The optimal compositions are computed by using an elitist GA where the best 2 compositions were kept alive across generations, with a crossover probability of 0.7, a mutation probability of 0.1, a population of 200 compositions. The roulette wheel selection has been adopted as selection mechanism. We consider a simple stopping criterion i.e., up to 400 generations. We conducted experiments on Intel(R) Core(TM)2 CPU, 2.4GHz with 2GB RAM.

Compositions with up to 30 tasks and 35 candidates per task (35^2 candidate semantic links between 2 tasks) have been considered in Sections 5.2, 5.3, 5.5, 5.6, especially for obtaining convincing results towards their applicability in real (industrial) scenarios. Experiment results reported in Section 5.2 provide some benefits of combining QoS and functional criteria for the overall quality of composition, whereas those in Sections 5.3, 5.4, 5.5 and 5.6 are related to scalability.

5.2 Benefits of Combining QoS and Functional Criteria

Fig.5 reports the benefits of combining QoS and functional criteria. In more details, we studied the impact of functional quality on the costs of data integration, which enabling the end-to-end composition of services. The data integration process aligns the data flow specification of a composition by manipulating and transforming the semantic descriptions of contents of outgoing and incoming messages of annotated services.

Our approach and [5] are compared on ten compositions $c_{i,1\leq i\leq 10}$ with $Q_{cd}(c_i) = 10^{-1} \times i$ and $Q_m(c_i) = 10^{i-10}$ as functional quality to reflect gradually better quality.

[5] The choice of proprietary services has been motivated by the poor quality of existing benchmark services in terms of number of services or expressivity (limited functional specification, no binding, restricted RDF-based description). We plan further experimentations with OWL-S TC 3.0 (`http://www.semwebcentral.org/frs/?group_id=89`) and SA-WSDL TC1 (`http://projects.semwebcentral.org/projects/sawsdl-tc/`).

[6] http://jgap.sourceforge.net/

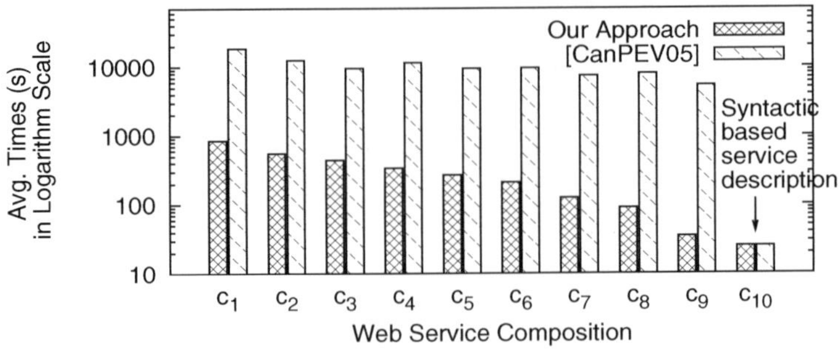

Fig. 5. Costs of Data Integration (through Data Flow Specification)

On the one hand, as expected, the costs of data integration is low (actually trivial) for both approaches, regarding the composition with the best quality (i.e., c_{10}). Indeed the parameters of services match exactly, hence no further specification is needed..

On the other hand, these costs decrease with the functional quality of compositions in our approach, whereas they are steady but very high for compositions computed by [5] (purely based on non functional quality of composition). This is due to i) the lack of specification of functional quality (hence a hard task to build semantic data flow from scratch), and ii) the manual approach used to link data in compositions.

Appropriate qualities of semantic links are very useful i) to discover data flow in composition, ii) to ease the specification (semi-automated process with Assign/Copy elements + XPath/XQuery processes a la BPEL4WS) of syntactic (and heterogeneous) data connections (for the persons who develop these mappings), so limiting the costs of data integration. Indeed the better the quality of semantic links the better the semantic mapping between the outgoing and incoming (SA-WSDL for instance) messages.

5.3 Evolution of the Composition Quality

Fig.6 reports the evolution of the composition quality over the GA generations, by varying the number of tasks. This illustrates different levels of convergence to a composition

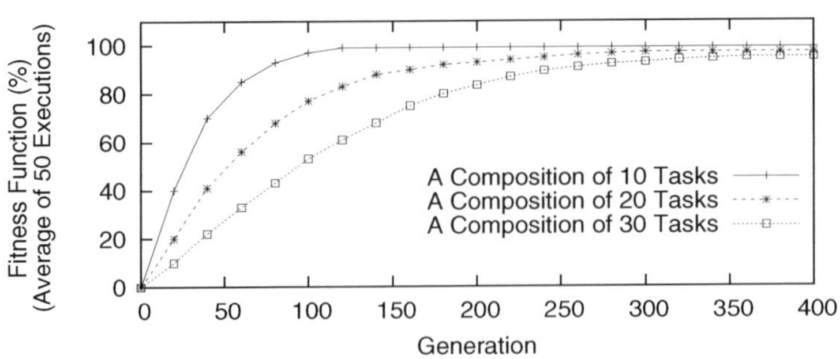

Fig. 6. Evolution of the Composition Quality

Table 2. Overview of Computation Costs

Tasks Num.	Max. Fitness (%)	Generation Num.	Time (ms)
10	99	120	1012
20	97	280	1650
30	95	360	3142

that meets some constraints and optimizes its different criteria by maximizing the common description and matching quality while minimizing price and response time.

Table 2 and Fig.6 present the computation costs and the number of generations required to obtain the maximal fitness value. The more tasks (and services) the more time consuming to converge to the optimum. Obviously, the population size and the number of generations should be extended to reach the optimum of more complex compositions.

5.4 Towards Large Scale Based Compositions

In this experiment we suggest to study the behaviour of our approach regarding the optimisation of compositions with a large number of tasks (up to 500 tasks) and candidate services (500). To this end we focus on its scalability and the impact of the number of generations as well as the population size on the GA success.

Table 3. Large Scale Compositions

Tasks Num.	Max. Fitness (%)	Generation Num./ Population Size	Time (ms)
100	85	400/200	4212
	96	700/400	9182
300	47	400/200	5520
	95	1500/500	19120
500	24	400/200	7023
	95	3000/1000	51540

As illustrated in Table 3, increasing both the number of generations and the population size does actually result in better fitness values for problems with a larger number of tasks and candidate services. For example, regarding the optimisation of a composition of 500 tasks with 500 candidate services, a number of generations of 400 and a population size of 200 do result in a low fitness value of 24% of the maximum, whereas considering a number of generations of 3000 and a population size of 1000 achieve 95% of the maximum. Note that better fitness values can be reached by further increasing the sizes of generations and populations. However doubling these sizes only improves the fitness value by 2%. This shows that each optimisation problem converges to a limit.

5.5 Decoupling GA Process and DL Reasoning

Since our approach is mainly depending on DL reasoning (i.e., Subsumption for q_m, Difference and lcs for q_{cd}) and the GA-based optimization process, we suggest to decouple and detail the computation costs of Table 2 in Fig.7.

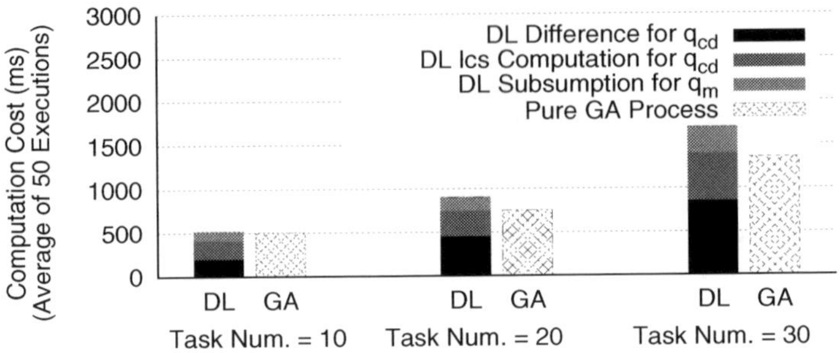

Fig. 7. DL and GA Processes in our Approach

DL reasoning is the most time consuming process in optimisation of QoS-aware semantic web service composition wherein the number of tasks and candidate services are greater than 10 and 35. This is caused by the critical complexity of q_{cd} computation through DL Difference (even in $\mathcal{ALE}$ DL).

5.6 Convergence of GA-Based Approaches

In this experiment, we compare the convergence of our approach (11) with the main alternative at present [5]. To this end the functional criteria of our approach are disregarded in order to focus only on the GA-driven aspects of the optimisation process.

According to Table 4, the advantage of our approach is twofold. Firstly we obtain better fitness values for the optimal composition than the approach of [5]. Secondly, our approach converges faster than the approach of [5]. In addition our function avoids getting trapped by local optimums by i) further penalizing compositions that disobey constraints (the factor of ω_{pe} in (9) and (11)) and ii) suggesting a dynamic penalty, i.e., a penalty having a weight that increases with the number of generations.

Table 4. Comparing GA-based Approaches (Population size of 200)

Tasks Num.	Approach	Max. Fitness (%)	Generation Num.	Time (ms)
10	Our Model (11)	99	120	1012
	[5]	97	156	1356
20	Our Model (11)	97	280	1650
	[5]	94	425	2896
30	Our Model (11)	95	360	3142
	[5]	85	596	6590

These results support the adoption of our model in the cases where a large number of tasks and services are considered.

6 Related Work

Review of existing approaches to optimising web service compositions reveals that no approach has specifically addressed optimisation of service composition using both *QoS* and *semantic similarities* dimensions in a context of *significant scale*.

Indeed main approaches focus on either QoS [5,8] or on functional criteria such as semantic similarities [6] between output and input parameters of web services for optimising web service composition. In contrast, we present an innovative model that addresses both types of quality criteria as a trade-off between data flow and non functional quality for optimizing web service composition.

Solving such a multi-criteria optimization problem can be approached using IP [8,6], GA [5], or Constraint Programming [25]. The results of [5] demonstrate that GAs are better at handling non-linearity of aggregation rules, and provide better scaling up to a large number of services per task. In addition they show that dynamic programming (such as IP-based approaches) is preferable for smaller compositions. We follow [5] and suggest the use of GAs to achieve optimization in web service composition, yet we also extend their model by i) using semantic links to consider data flow in composition, ii) considering not only QoS but also semantic quality (and contraints) of composition, iii) revisiting the fitness function to avoid local optimal solution (i.e., compositions disobeying constraints are considered).

The optimization problem can be also modelled as a knapsack problem [26], wherein [27] performed dynamic programming to solve it. Unfortunately the previous QoS-aware service composition approaches consider only links valued by Exact matching types, hence no semantic quality of compositions. Towards the latter issue [6] introduces a general and formal model to evaluate such a quality. From this they formulate an optimization problem which is solved by adapting the IP-based approach of [8]. All quality criteria are used for specifying both constraints and objective function.

7 Conclusion

We studied *QoS*-aware *semantic* web service composition in a context of significant scale i.e., how to effectively compute optimal compositions of QoS-aware web services by considering their semantic links. On the one hand the benefits of a significant domain such as the Web is clear e.g., supporting a large number of services providers, considering large number of services that have same goals. On the other hand, the benefits of combining semantic links between services and QoS are as following:

> *The computation of web services composition whilst optimising both the non functional qualities and the quality of semantic fit along non-trivial data flow, where the information required and provided by services does not match perfectly in every dataflow, using semantic-based description of services.*

By addressing non trivial data flow in composition , we aimed at limiting the costs of (semantic heterogeneity) data integration between services by considering appropriate quality of semantic links. To this end we have presented an innovative and extensible model to evaluate quality of i) web services (QoS), ii) their semantic links, and iii) their compositions. In regards to the latter criteria the problem is formalized as an optimization problem with multiple constraints. Since one of our main concerns is about optimization of large-scale web service compositions (i.e., many services can achieve a same functionality), we suggested to follow a GA-based approach, faster than applying IP. The experimental results have shown an acceptable computation costs of our GA-based approach despite the time consuming process of the on-line DL reasoning. In case of "semantic link"-less models, the benefits are mainly based on penalizing constraint violation (of the fitness function) which makes the approach faster than [5].

In future work we will consider a finer difference operator, which is also easy-to-compute in expressive DLs. Determining the most appropriate parameters for the GA phase requires further experimentations.

References

1. Berners-Lee, T., Hendler, J., Lassila, O.: The semantic web. Scientific American 284(5), 34–43 (2001)
2. Smith, M.K., Welty, C., McGuinness, D.L.: Owl web ontology language guide. W3c recommendation, W3C (2004)
3. Baader, F., Nutt, W.: The Description Logic Handbook: Theory, Implementation, and Applications (2003)
4. Lécué, F., Léger, A.: A formal model for semantic web service composition. In: Cruz, I., Decker, S., Allemang, D., Preist, C., Schwabe, D., Mika, P., Uschold, M., Aroyo, L.M. (eds.) ISWC 2006. LNCS, vol. 4273, pp. 385–398. Springer, Heidelberg (2006)
5. Canfora, G., Di Penta, M., Esposito, R., Villani, M.L.: An approach for qos-aware service composition based on genetic algorithms. In: GECCO, pp. 1069–1075 (2005)
6. Lécué, F., Delteil, A., Léger, A.: Optimizing causal link based web service composition. In: ECAI, pp. 45–49 (2008)
7. Papadimtriou, C.H., Steiglitz, K.: Combinatorial Optimization: Algorithms and Complexity. Prentice-Hall, Englewood Cliffs (1982)
8. Zeng, L., Benatallah, B., Dumas, M., Kalagnanam, J., Sheng, Q.Z.: Quality driven web services composition. In: WWW, pp. 411–421 (2003)
9. Ankolenkar, A., Paolucci, M., Srinivasan, N., Sycara, K.: The owl-s coalition, owl-s 1.1. Technical report (2004)
10. Kopecký, J., Vitvar, T., Bournez, C., Farrell, J.: Sawsdl: Semantic annotations for wsdl and xml schema. IEEE Internet Computing 11(6), 60–67 (2007)
11. Paolucci, M., Kawamura, T., Payne, T.R., Sycara, K.: Semantic matching of web services capabilities. In: Horrocks, I., Hendler, J. (eds.) ISWC 2002. LNCS, vol. 2342, pp. 333–347. Springer, Heidelberg (2002)
12. Li, L., Horrocks, I.: A software framework for matchmaking based on semantic web technology. In: WWW, pp. 331–339 (2003)
13. Lécué, F., Delteil, A.: Making the difference in semantic web service composition. In: AAAI, pp. 1383–1388 (2007)
14. Brandt, S., Kusters, R., Turhan, A.: Approximation and difference in description logics. In: KR, pp. 203–214 (2002)
15. Baader, F., Sertkaya, B., Turhan, A.Y.: Computing the least common subsumer w.r.t. a background terminology. In: DL (2004)

16. Harel, D., Naamad, A.: The statemate semantics of statecharts. ACM Trans. Softw. Eng. Methodol. 5(4), 293–333 (1996)
17. Motta, E.: Reusable Components For Knowledge Modelling Case Studies. In: Parametric Design Problem Solving. IOS Press, Netherlands (1999)
18. Küsters, R.: Non-Standard Inferences in Description Logics. LNCS (LNAI), vol. 2100. Springer, Heidelberg (2001)
19. Lécué, F., Boissier, O., Delteil, A., Léger, A.: Web service composition as a composition of valid and robust semantic links. IJCIS 18(1) (March 2009)
20. Hwang, C.-L., Yoon., K.: Multiple criteria decision making. Lecture Notes in Economics and Mathematical Systems (1981)
21. O'Sullivan, J., Edmond, D., ter Hofstede, A.H.M.: What's in a service? Distributed and Parallel Databases 12(2/3), 117–133 (2002)
22. Cardoso, J., Sheth, A.P., Miller, J.A., Arnold, J., Kochut, K.: Quality of service for workflows and web service processes. J. Web Sem. 1(3), 281–308 (2004)
23. Goldberg, D.E.: Genetic Algorithms in Search, Optimization, and Machine Learning. Addison-Wesley Publishing Company, Inc., Reading (1989)
24. Horrocks, I.: Using an expressive description logic: FaCT or fiction? In: KR, pp. 636–649 (1998)
25. Ben Hassine, A., Matsubara, S., Ishida, T.: A constraint-based approach to web service composition. In: Cruz, I., Decker, S., Allemang, D., Preist, C., Schwabe, D., Mika, P., Uschold, M., Aroyo, L.M. (eds.) ISWC 2006. LNCS, vol. 4273, pp. 130–143. Springer, Heidelberg (2006)
26. Yu, T., Lin, K.-J.: Service selection algorithms for composing complex services with multiple qos constraints. In: Benatallah, B., Casati, F., Traverso, P. (eds.) ICSOC 2005. LNCS, vol. 3826, pp. 130–143. Springer, Heidelberg (2005)
27. Arpinar, I.B., Zhang, R., Aleman-Meza, B., Maduko, A.: Ontology-driven web services composition platform. Inf. Syst. E-Business Management 3(2), 175–199 (2005)

Synthesizing Semantic Web Service Compositions with jMosel and Golog

Tiziana Margaria[1], Daniel Meyer[1], Christian Kubczak[2], Malte Isberner[2], and Bernhard Steffen[2]

[1] Chair Service and Software Engineering, Universität Potsdam, Germany
{margaria,meyerd}@cs.uni-potsdam.de
[2] Chair of Programming Systems, TU Dortmund, Germany
{christian.kubczak,malte.isberner,steffen}@cs.tu-dortmund.de

Abstract. In this paper we investigate different technologies to attack the automatic solution of orchestration problems based on synthesis from declarative specifications, a semantically enriched description of the services, and a collection of services available on a testbed. In addition to our previously presented tableaux-based synthesis technology, we consider two structurally rather different approaches here: using *jMosel*, our tool for *Monadic Second-Order Logic on Strings* and the high-level programming language *Golog*, that internally makes use of planning techniques. As a common case study we consider the Mediation Scenario of the Semantic Web Service Challenge, which is a benchmark for process orchestration. All three synthesis solutions have been embedded in the jABC/jETI modeling framework, and used to synthesize the abstract mediator processes as well as their concrete, running (Web) service counterpart. Using the jABC as a common frame helps highlighting the essential differences and similarities. It turns out, at least at the level of complication of the considered case study, all approaches behave quite similarly, both considering the performance as well as the modeling. We believe that turning the jABC framework into experimentation platform along the lines presented here, will help understanding the application profiles of the individual synthesis solutions and technologies, answering questing like when the overhead to achieve compositionality pays of and where (heuristic) search is the technology of choice.

1 Introduction

Dealing with (Web) services, semantics is gaining terrain as technology for application development and integration. However, semantics-based technology is still highly complicated in usage and implementation, which explains its relatively slow industrial adoption. It is the goal of the Semantic Web Service Challenge (SWSC) to overcome this situation by studying concrete case studies, with the goal of pinpointing the application profiles of the various approaches proposed so far. The corresponding leading case study is the Mediation Scenario [1], a business orchestration problem that requires adequate "translation" between different communication protocols and data models.

A. Bernstein et al. (Eds.): ISWC 2009, LNCS 5823, pp. 392–407, 2009.

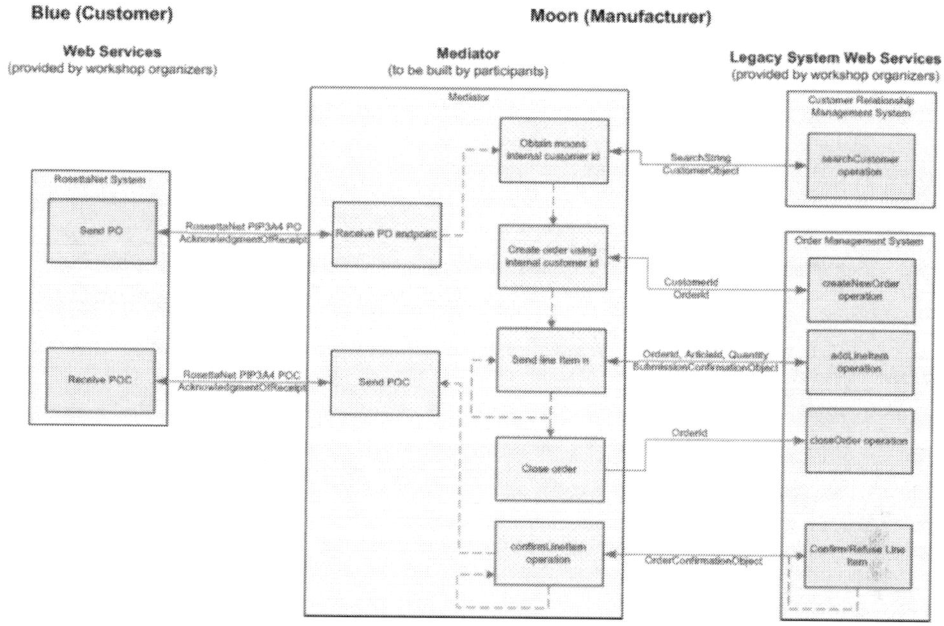

Fig. 1. The Mediation Scenario of the SWS Challenge

In this scenario, a customer named *Blue* (Fig. 1(left)) submits business orders in format and protocol compliant to the RosettaNet Pip3A4 protocol [2] to the backend system of*Moon* (Fig. 1(right)). Since *Moon* does not support RosettaNet the Pip3A4 communication standard, the task of the challenge is to automatically generate a mediation service that enables *Moon*'s legacy system to process the requests submitted by *Blue*.

This problem has been addressed by several research groups in the past: 5 solutions, 2 of which ours, have been presented and compared in [3]. We first directly modeled the mediator's orchestration in the jABC [4,5,6,7], our framework for service oriented development and model-driven orchestration, and generated and published the mediator using our jETI technology. In jABC, processes are orchestrations, they are modeled as directed flow graphs called Service Logic Graphs (SLG), where the nodes, which represent the services, are called *Service Independent Building Blocks* (SIBs).

Second, we enhanced the capabilities of the jABC by integrating a synthesis algorithm based on Semantic Linear Time Logic (SLTL) specifications [8]. Using this synthesis method, we were indeed able to automatically (re-)produce an equivalent orchestration [9]. This observation got us interested in using also other very different process/model synthesis techniques on the same problem, in order to compare (following the original purpose of the SWS Challenge) different methods and techniques, based on different semantic representations of declarative knowledge and goals.

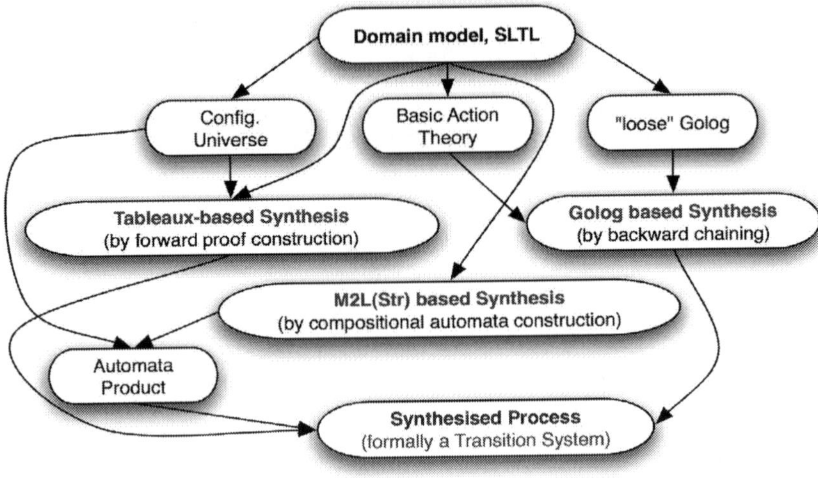

Fig. 2. The landscape of synthesis techniques in jABC

In this paper, we apply two alternative and conceptually very different approaches to planning to the Mediation Scenario:

- an approach based on monadic second order logic on strings M2L(Str) [10], which works by compositional automata construction in jMosel [11], and
- an approach based on Golog [12], a tool that internally uses backward chaining for solving planning/synthesis tasks. This work is based on the well known situation calculus planner by Reiter.

Fig. 2 provides a conceptual sketch covering the essence the three approaches we considered in details so far. Our study revealed that despite the huge conceptual difference these three approaches share quite some similarities. In particular, they were all able to quite naturally express the original problem in sufficient detail to automatically synthesize the desired solution to the Mediation problem (cf. Fig. 6). However there are also similarities at a different level, which became apparent when modeling the Golog-synthesis process itself as a process using jABC. To this aim, we implemented the steps described so far as a set of independent, reusable building blocks (SIBs). The resulting 7-step synthesis process is shown in Fig. 3. It is the same process we obtained with jMosel and with the previous (S)LTL based synthesis methods. Thus this process can be considered as a pattern for synthesis solutions, which allows one to integrate and/or combine various synthesis and transformation functionalities like the ones summarized in Fig. 2 to complex heterogeneous solutions. These solutions are then directly executable inside the jABC, or, as we did in the previous phases of the Challenge, they can be exported as a Web service.

After presenting our modeling framework jABC in Section 2, we show how we can easily plug in different algorithms and knowledge representation formalisms, namely *jMosel*, see Section 3 and *Golog*, Section 4. Subsequently we show how

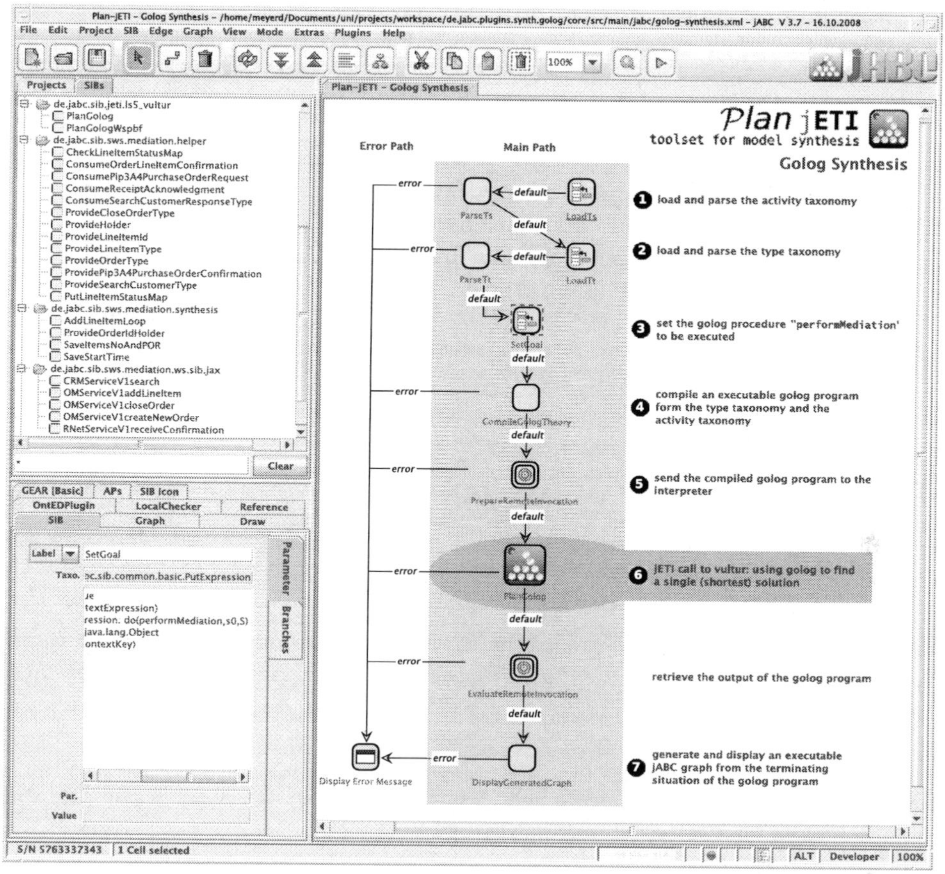

Fig. 3. The Golog-based synthesis process modeled in jABC

easily these at first sight completely different technologies can be operated on top of the same platform and can be used to obtain the same results, see Sect. 5.

2 Our Modeling Framework

Basic ingredient for the process mediation, e.g. for bridging the gap between Blue and Moon, are:

- a set of business objects, including the definition of their structure, properties, and data types,
- a set of services (SIBs) operating on the business objects, including knowledge about relevant properties, and
- domain knowledge, concerning semantic properties of the domain under consideration (typically, information on the data and on how the data can be manipulated by the services), and behavioral knowledge (often also called

procedural knowledge), that describes abstractly properties (restrictions, precedences, lose constraints) of the collaboration of such services.

Behavioral knowledge is often present in a domain description, and often the task to be performed by the orchestration is also already known in its behavioral traits. From the point of view of a user of semantic techniques, all the available knowledge should be optimally exploited by the techniques of a semantic web framework, in order to ensure optimal precision and efficiency. Thus, procedural knowledge is in our opinion much more than a way of formulating domain specific heuristics to make process synthesis computationally feasible: in real world applications, we need to be able to formulate constraints which must be satisfied by processes in order to be admissible. For example we may need to ensure

- general ordering properties (e.g., ensuring that some service is executed before another)
- abstract liveness properties (e.g., guaranteeing that a certain service is eventually executed)
- abstract safety properties (e.g., making sure that certain services are never executed simultaneously).

However, we do not want to specify the whole processes in a static, fixed way. On the contrary, we use *loose coordination specifications*, which leave room for variability inside the bounds of a declarative specification that includes behavioral constraints.

In jABC, we have implemented this adopting as semantic predicates abstraction concepts from *dataflow analysis* [13] in a *constructive* way, leading to the domain knowledge representation introduced in [8,14,15] and summarized in Sect. 2.1.

2.1 Modeling Domain Knowledge with Taxonomies

In jABC's modeling style, *business objects* are mapped to abstract semantic concepts that we call *Types*, and services, actually the collection of SIBs corresponding to single service operations, are mapped to semantic *activities*. Both types and activities are organized in *taxonomies*.

A taxonomy $\mathcal{T}ax = (T, CT, \rightarrow)$ is a directed acyclic graph (DAG) where CT is a set of concrete entities that are grounded to instances of the real world (as the elements of the A-box of Description Logics), and that are the sinks in the graph, T is a set of abstract concepts, and $\rightarrow$ relates concepts and concrete elements (T, CT) or pairs of concepts (T, T).

In a type taxonomy, as shown in Fig. 4 for the mediation problem, $\mathcal{T}_{ty} = (T_t, CT_t, is\_a)$, where CT_t is a set of semantic types that directly correspond to individual business objects, T_t is a set of abstract semantic types that represent groups of business objects, and edges reflect an *is_a* relationship. In our example, *OrderIDs* and *Confirmations* are in the group *Orders*. The concrete types

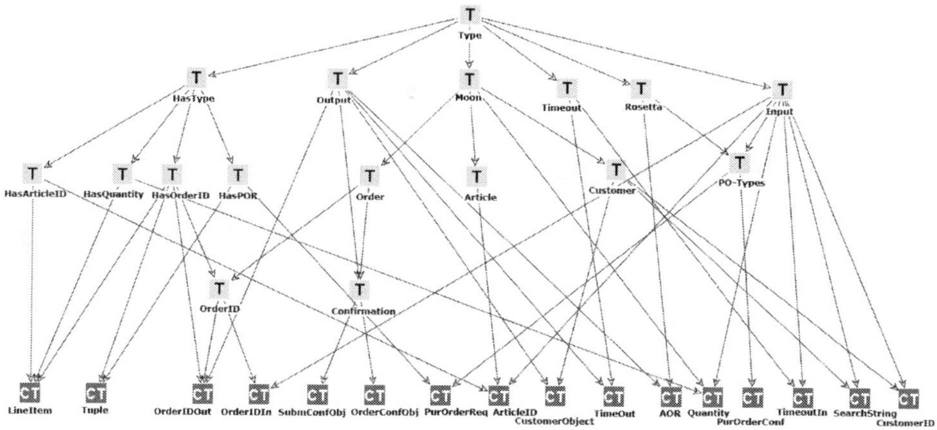

Fig. 4. The Type Taxonomy for the mediation scenario

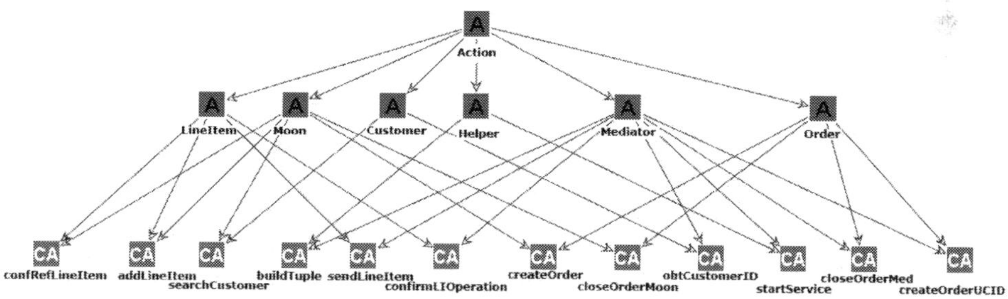

Fig. 5. The Service Taxonomy for the mediation scenario

(the leaves of the taxonomy)are the most concrete semantic entities we want to use in the synthesis problem - here directly the business objects used as input and output by the services.

The activity taxonomy $\mathcal{T}_a = (A, CA, is\_a)$ is defined in the same way, as shown Fig. 5: CA is a set of concrete semantic activities, representing individual SIBs, and abstract activities A represent groups of SIBs which share a common set of (non-)functional properties.

2.2 Dataflow Facts as Semantic Preconditions/Effects

We now need to formulate knowledge about how the activities operate on the types. At a technical level, Web services have very complex relations to business objects. At the semantic level, we abstract these complex relations into three basic functions, well known from dataflow analysis expressing the preconditions and effects of services, which are technically stored in the jABC (semantic) context:

Table 1. The SWS mediation Modules

name	input type (uses)	output type (gen)	description
`Mediator`			Maps RosettaNet messages to the backend
`startService`	$\{true\}$	*PurOrderReq*	Receives a purchase order request message
`obtCustomerID`	*PurOrderReq*	*SearchString*	Obtains a customer search string from the req. message
`createOrderUCID`	*CustomerObject*	*CustomerID*	Gets the customer id out of the customer object
`buildTuple`	*OrderID*	*Tuple*	Builds a tuple from the orderID and the POR
`sendLineItem`	*Tuple*	*LineItem*	Gets a LineItem incl. orderID, articleID and quantity
`closeOrderMed`	*SubmConfObj*	*OrderID*	Closes an order on the mediator side
`confirmLIOperation`	*OrderConfObj*	*PurOrderCon*	Receives a conf. or ref. of a LineItem and sends a conf.
`Moon`			The backend system
`searchCustomer`	*SearchString*	*CustomerObject*	Gets a customer object from the backend database
`createOrder`	*CustomerID*	*OrderID*	Creates an order
`addLineItem`	*LineItem*	*SubmConfObj*	Submits a line item to the backend database
`closeOrderMoon`	*OrderID*	*TimeoutOut*	Closes an order on the backend side
`confRefLineItem`	*Timeout*	*orderConfObj*	Sends a conf. or ref. of a prev. subm. LineItem

- $use(\cdot) : CA \rightarrow \mathcal{P}(CT)$
  in order for the activity a to be executable, values of the set of type $use(a)$ must be present in the execution context
- $gen(\cdot) : CA \rightarrow \mathcal{P}(CT)$, $gen(a)$ returns the set of types, values of which are added to the context after invocation of the activity a
- $kill(\cdot) : CA \rightarrow \mathcal{P}(CT)$, $kill(a)$ is the set of types, values of which are removed form the context if a is invoked

Table 1 lists in the first column a selection of activities from the mediation scenario and maps each activity to a set of input/use types (column 2) and a set of output/gen types (column 3). In this example there are no kill types.

The jMosel approach presented in Section 3 can directly work with this representation of the domain knowledge. In Sect. 4, we show how to translate it into situation calculus and Golog.

2.3 Expressing Behavioral Knowledge

The loose coordination specification language we use to express procedural knowledge is *Semantic Linear Time Logic (SLTL)* [8], a temporal (modal) logic that includes the taxonomic specifications of types and activities.

Definition 1 (SLTL)

The syntax of Semantic Linear Time Logic (SLTL) is given in BNF format by:

$$\Phi ::= \; tt \mid \mathsf{type}(t_c) \mid \neg\Phi \mid (\Phi \wedge \Phi) \mid \langle a_c\rangle\, \Phi \mid \mathbf{G}(\Phi) \mid (\Phi\, \mathbf{U}\, \Phi)$$

where t_c and a_c represent type and activity constraints, respectively, formulated as taxonomy expressions.

Taxonomy expressions are for this example propositional formulas that use as propositions the concepts in the taxonomies of Figs. 4 and 5 and the *use*, *gen*, *kill* predicates already introduced.

SLTL formulas are interpreted over the set of all *legal orchestrations*, which are coordination sequences, i.e. alternating type correct sequences of types and activities[1], which start and end with types. The semantics of SLTL formulas can now be intuitively defined as follows[2]:

- $\mathsf{type}(t_c)$ is satisfied by every coordination sequence whose first element (a type) satisfies the type constraint t_c.
- Negation $\neg$ and conjunction $\wedge$ are interpreted in the usual fashion.
- Next-time operator $\langle\rangle$:
  $\langle a_c\rangle\, \Phi$ is satisfied by coordination sequences whose second element (the first activity) satisfies a_c and whose *suffix*[3] satisfies Φ. In particular, $\langle tt\rangle\, \Phi$ is satisfied by every coordination sequence whose suffix satisfies Φ.
- Generally operator **G**:
  $\mathbf{G}(\Phi)$ requires that Φ is satisfied for every suffix satisfies Φ.
- Until operator **U**:
  $(\Phi\, \mathbf{U}\, \Psi)$ expresses that the property Φ holds at all type elements of the sequence, until a position is reached where the corresponding suffix satisfies the property Ψ. Note that $\Phi\, \mathbf{U}\, \Psi$ guarantees that the property Ψ holds eventually (strong until).
  The frequently occurring formula $(\mathit{true}\, \mathbf{U}\, \Psi)$ is called Eventually and is written $\mathbf{F}\, \Psi$.

The above definition of suffix may seem complicated at first. However, thinking in terms of path representations clarifies the situation: a sub path always starts with a node (type) again. However, users should not worry about these details: they may simply think in terms of pure activity compositions and should not care about the types (which are matched correctly by the synthesis algorithm), unless they explicitly want to specify type constraints.

[1] During the description of the semantics, types and activities will be called *elements* of the orchestration sequence.

[2] A formal definition of the semantics can be found online.

[3] According to the difference between activity and type components, a suffix of a coordination sequence is any subsequence which arises from deleting the first 2n elements (n any natural number).

The introduction of *derived operators*, like Eventually, supports a modular and intuitive formulation of complex properties. We support in the jABC meanwhile a rich collection of frequently occurring behavioral templates, which ease the declarative formulation of knowledge and goals.

SLTL is a specification language supported by jMosel. In Sect. 4, we will show how we link up to Golog in order to apply it to the mediation problem.

3 Solving the Mediation with M2L(Str) in jMosel

We first formally describe the semantics of service invocation with the associated *use*, *gen* and *kill* sets. At runtime, a service accepts certain inputs and produces certain outputs according to its WSDL description. Semantically, we describe a service (or concrete action) as a transformation on the power set of the set of (concrete) types in the orchestration's context:

$$\mathrm{eff}(\cdot)(\cdot)\colon CA \rightarrow (\mathcal{P}(CT) \rightarrow \mathcal{P}(CT)),$$

Accordingly, a service can only be invoked if all elements in $use(\cdot)$ are available (preconditions), and it produces elements the context according to $gen(\cdot)$, and invalidates the elements of the context specified in $kill(\cdot)$ (effects), thus

$$\mathrm{eff}(a)(T) = \begin{cases} (T \setminus kill(a)) \cup gen(a) & \text{if } use(a) \subseteq T \\ \text{undef} & \text{otherwise} \end{cases}.$$

Incidentally, this is the same abstraction used in Data Flow Analysis to describe the operational semantics of operations, like assignments, which have preconditions and (side) effects. This is the basis for constructing the structure the jMosel synthesis uses as the domain specification, the *configuration universe*, which is rather straightforward, and therefore omitted here due to lack of space.

3.1 jMosel, M2L and SLTL

The search for such a solution is done using a deterministic finite automaton semantically equivalent to the given formula. The input symbols the automaton is fed with are sets of atomic propositions (i.e., types) on one hand and actions (i.e., services) on the other. Since it is not feasible to consider alternating sequences of types and actions, some technical fine-tuning is needed: we now regard *steps*, consisting of an action (or *init*, which is not an action but is used for the head of the path) and a set of atomic propositions. Thus, instead of alternating sequences of sets of types and actions, we now consider strings over the *step alphabet* $\Sigma_{step} = \mathcal{P}(CT) \times (CA \uplus \{init\})$. For example, the path t_0, a_1, t_1, a_2, t_2 now is written as $(t_0, init)(t_1, a_1)(t_2, a_2)$.

There exist several methods to transform an LTL formula into a *Büchi automaton*. However, since we are interested in finite sequences of services rather than infinite words, a Büchi automaton is not quite what we want. One could modify the existing algorithms for generating Büchi automata in the way that NFAs or DFAs are constructed; we, however, want to describe a different approach, namely by using *monadic second-order logic on strings*, M2L(Str) [10], and our toolkit for this logic, jMosel [11,16].

jMosel calculates for a given input formula Φ a deterministic finite automaton A, with $L(\Phi) = L(A)$. A key characteristic of M2L(Str) is that it is *compositional*: atomic formulae are transformed into simple basic automata; for a compound formula, first the automata for its sub formulae are calculated, and then an automata synthesis operation is applied. For details refer to [17].

3.2 Mediator Synthesis

After the domain has been modeled by specifying the relevant modules and the corresponding type and action taxonomies, the mediator synthesis proceeds by the user

- entering a set of initial types T_0. In our example, we start with a *purchase order request*, POR,
- providing the temporal specification formula in SLTL or one of our dialects,
- specifying the desired kind of solution. For the case study, we selected minimality of the solution and the presentation in a format which is directly executable (rather than as a sequence of alternating types and activities), i.e. jABC's SLGs. This results in the same solution as provided via Golog in the next section (cf. Fig. 6.

Tab. 2 presents the corresponding SLTL formula already together with some intuitive explanations. The formula itself, which is just a combination of 4 next-time operators and one eventually operator would easily fit in one line.

Table 2. Explanation of the specification formula used for the jMosel synthesis

Formula element	Explanation
<`ConsumePip3A4POR`>	The synthesized sequence should start with this service.
<`SaveStartTime`>	A local service for time measurement. Since this is local to our requirements, the invocation of this service is required statically by the formula.
F	Find a path in the configuration universe, such that the following two services can be invoked (This is where the actual synthesis action happens).
<`SaveItemsNoAndPOR`>	Again, this is a service local to our requirements, and therefore statically requested.
<`OMServiceV1closeOrder`>	In the end, we want the order to be complete and therefore closed.
tt	Operand, required for the preceding unary `next` operator.

4 The Situation Calculus Solution with Golog

Golog is a high-level programming language based on the situation calculus [12] that was successfully used to solve Web service composition problems [18]. Here we show how to match the problem structure of the mediation scenario with situation calculus, how we generate Basic Action Theories from the domain knowledge representation of Sect. 2.1, how we model an abstract mediation process in Golog, and finally how we synthesize the mediator.

4.1 Intuitive Ontology of the Situation Calculus and Golog

The situation calculus is designed for representing and reasoning about stateful models, dynamically changing worlds, where changes to the world are due to performing named *actions*. S_0 is the initial state of the world, prior to any action. Changes are effected by executing actions in a situation: $s' = do(a, s)$ means that situation s' is obtained by executing the action a in situation s. For example, $do(store(A, B), do(ship(A, B), do(construct(A), S_0)))$ is a situation term denoting the sequence composition $construct(A) \circ ship(A, B) \circ store(A, B)$. Intuitively, situations are action execution paths, and denote indirectly states. Predicates and relations are true in some state and false in others, thus state variables are introduced by means of *relational fluents*. For example, the fluent $location(a, b, s)$ expresses that in the state of the world reached by performing the action sequence s, object a is located at location b.

Domain axiomatizations in the situation calculus are called **Basic Action Theories** and have the form

$$\mathcal{D} = \Sigma \cup \mathcal{D}_{ss} \cup \mathcal{D}_{ap} \cup \mathcal{D}_{una} \cup \mathcal{D}_{S_0}.$$

where Σ are the foundational axioms for situations, $\mathcal{D}_{ss}$ are the successor state axioms for fluents, $\mathcal{D}_{ap}$ is a set of action precondition axioms for situations, $\mathcal{D}_{una}$ is a set of unique names axioms for actions, $\mathcal{D}_{S_0}$ is a set of first order sentences, uniform in S_0. Basic action theories allow us to reason about action and change, but they offer no way to express a "story" about how certain effects can be achieved, thus no way of expressing procedural knowledge.

The high-level programming language Golog, built on top of Basic Action Theories, allows writing high-level non-deterministic procedures which model such a "story". Golog, in its essential form[4] provides the following language constructs, originally inspired by Algol and its operational semantics:

$Do(\phi?, s, s')$	Test actions
$Do(\delta_1; \delta_2, s, s')$	Sequential composition
$Do(\delta_1 \mid \delta_2, s, s')$	Non-deterministic choice of actions
$Do((\pi x)\delta(x), s, s')$	Non-deterministic choice of action arguments
$Do(\delta^*, s, s')$	Non-deterministic iteration.

[4] Golog has been extended in various ways to include e.g. concurrency, exogenous events, and sensing [12].

that "macro expand" into terms in the situation calculus. As usual, constructs like *if* and *while* can be defined in terms of these basic constructs according to the usual operational semantics. With these constructs, we can write Golog *procedures* specifying the story, or the behavioral knowledge, of a domain and of a solution.

Given a Golog procedure δ, Golog can prove constructively whether it is *executable* with respect to a given Basic Action Theory $\mathcal{D}$: $\mathcal{D} \models \exists s.Do(\delta, S_0, s)$. Since s is the terminating situation of the Golog program δ, the proof returns a sequence of *primitive actions* starting in the initial situation S_0 consistent with the procedure.

The resulting Basic Action Theory $\mathcal{D}_T$ allows us to determine for every situation whether a given activity(service) is executable, and what are the effects of its execution. We now concentrate on the Golog-based process synthesis.

4.2 'Loose' Golog: The :-Operator

In basic Golog, every action needs to be explicitly named in a procedure, and the only operator that allows chaining service executions is the (immediate) successor, corresponding to the SLTL operator Next. In order to specify loose coordinations, that include Eventually and Until, we need operators that allow replacement by an a priori undetermined number of steps. The underlying motivation is that those parts of the concrete processes are unspecified and any sequence that satisfies their boundary conditions is there admissible, building this way a (semantic) equivalence class.

When restricting ourselves to Eventually, which is sufficient for the considered case study, this can be achieved using the :-Operator introduced in [18]. This operator exploits planning/search to achieve the required preconditions of subsequent services by inserting an appropriate sequence of actions. With ':' we can easily specify the mediation task as the Golog procedure *performMediation*:

proc $performMediation$
$\quad consumePip3A4PurchaseOrderRequest(A, B, C, D, E, F, G)$;
$\quad saveStartTime$
:
$\quad saveItemsNoAndPOR(J, K)$;
$\quad omServiceV1closeOrderSIB(L, M, N)$
endProc.

Stating that the mediation process starts with the service $consumePip3A4$ $PurchaseOrderRequest$ and ends with the service $omServiceV1closeOrderSIB$. What happens in between depends on precision of modeling and, in this case, on Golog's search strategy for an action sequence, making the entire sequence executable. The result of the search is a suitable linear, deterministic sequence of actions/activities (a service composition). The services $saveStartTime$ and $saveItemsNoAndPOR$ are local services which need to be called in order for

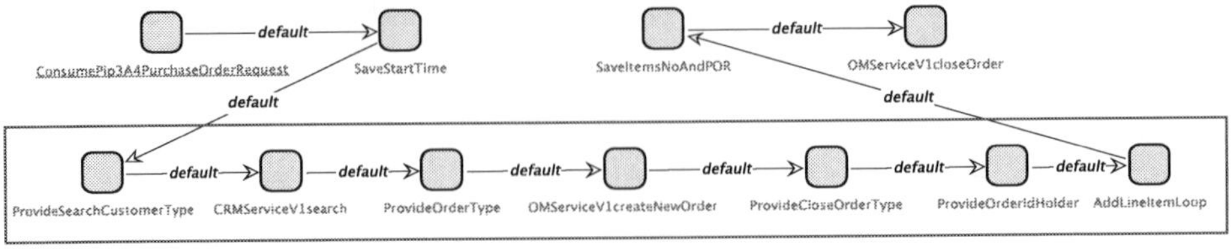

Fig. 6. The resulting Mediator process, visualized as a jABC orchestration. Highlighted in red is the synthesized sequence of actions.

data to be stored on disk for another process (the Mediator part 2, that we do not address here) to be able to retrieve it. The resulting resulting Mediator process is shown in Fig. 6.

5 Conclusion and Perspectives

We have presented two approaches to attack the automatic solution of orchestration problems based on synthesis from declarative specifications, a semantically enriched description of the services, and a collection of services available on a testbed. The first approach uses *jMosel* [11], our tool for solving *Monadic Second-Order Logic on Strings*[19], and the second approach is based on *Golog* [12], a programming language combining imperative features with a Prolog flavor. As a common case study we have considered the *Mediation Scenario* of the *Semantic Web Service Challenge* [1,3], with the goal to synthesize abstract mediator processes, and to describe how the concrete, running (Web) service composition is computed. As a result, together with the solution we had already described in [9,20,21], we have obtained three structurally rather different synthesis solutions (cf. Fig. 2), all integrated as running solutions in the jABC/jETI modeling framework [4,5,6,7].

This could be achieved despite their algorithmic and structural differences:

- our previously exploited method, originally presented in [8], is tableaux-based. It computes the service composition while constructing a proof tree in a fashion reminiscent of the approach presented in [22]
- the jMosel-approach uses a Monadic second-order (M2L) logic-based, compositional automata construction to infer an automaton description from the goal description in SLTL, and
- the Prolog flavored Golog approach synthesizes the mediation process via backward chaining.

There are also some strong similarities:

- All approaches are based on a domain modeling, which essentially consists of the specification of the available services in term of triples that specify the input types, an effect description, and an output type. For the Golog approach, this knowledge is specified within the Situation Calculus in terms

of Basic Action Theories, and in the other two approaches in the so-called Configuration Universe.
- All approaches synthesize an operational description of the mediator from a declarative specification of the available procedural knowledge. In Golog, the declarative description is given in 'loose' Golog, a variant of Golog, resembling an eventuality operator (cf. [18]). The synthesis transforms these loose specifications together with the Basic Action Theories into a concrete runnable Golog program. In the other two approaches the loose descriptions are given in SLTL, a logic specifically designed for temporally loose process specification. The synthesis then results in executable action/module sequences. As can be seen in Fig. 2, as for Golog, the tableaux-based approach directly exploits the domain model, while the M2L-based approach 'projects' the synthesized automaton onto the Configuration universe via simple product construction.
- From a semantical perspective, all approaches use a variant of (Kripke) Transitions Systems (KTS) [23] as their operational model, i.e. kinds of automata, where the edges are labelled with actions/activities, and where the nodes (implicitly) resemble type information/fluents. The fact that Golog is intuitively linked here to a tree structure rather than to a graph structure is technically of minor importance. However, here the fact transpires that Golog is intend to construct plans (essentially paths in a tree) essentially instance-driven, rather than to represent potentially all possible plans, as it is possible with the other approaches considered here.
- All approaches have a computational bottleneck: for the tableaux method it is the explosion of the proof tree, M2L-synthesis is known to be non-elementary (the automata construction may explode due to the required intermediate determination), and also backward chaining is classically known to be a hard problem. It is our goal to help understanding which bottleneck strikes when, and where to prefer which (combination of) which technologies.

This similarity/difference spectrum is ideal for an investigation of application profiles. We are therefore currently investigating *when which bottleneck strikes, and why.* Please note that there may be significant differences here between superficially similar approaches, with drastic effects on the performance depending on the considered situation. For comparison, consider the situation in Model Checking, where techniques like (BDD-based) symbolic model checking, bounded model checking, assume-guarantee techniques, partial order reduction, etc., are known to cover very different aspects of the so-called state explosion problem. Even in this well-studied field these effects are still not fully understood.

This is the motivation for us to work on a common platform for experimentation, where the different techniques can be evaluated, compared, modified and combined. Technological basis for this is the jABC/jETI modeling and experimentation framework [4,5,6,7], which has been initiated more than 10 years ago [14], and which has shown its power in other contexts, see e.g. our model learning environment [24,25]. At the moment four approaches have been integrated, and we plan to integrate more within the next year, in particular one exploiting traditional automata-theoretic methods for translation LTL to Büchi automata (cf. eg. [26]), one following the idea of Hierarchical Task Networks (cf. eg. [27]), which exploit given knowledge about task decomposition, and also the input/output function-focussed approach based on the Structural Synthesis Program described in [28]. This way we do not only want to be able to fairly compare the different scenarios in different contexts for their application profiles, but also to enlarge the landscape of Fig. 2 to a library of powerful synthesis and planning components which can be combined within jABC to new complex domain-specific planning or synthesis solutions. In order to make this possible at a larger scale, we plan to make the experimentation platform publicly available, allowing people not only to experiment with the integrated tools, but also to provide their own tools for others to experiment with. We hope that this will contribute to a better experimentation-based understanding of the various methods and techniques, and a culture of systematic application-specific construction of synthesis solutions.

References

1. Semantic Web Service Challenge (2009), `http://www.sws-challenge.org`
2. RosettaNet standard (2009), `http://www.rosettanet.org/`
3. Petrie, C., Margaria, T., Lausen, H., Zaremba, M. (eds.): Service-oriented Mediation with jABC/jETI. Springer, Heidelberg (2008)
4. Jörges, S., Kubczak, C., Nagel, R., Margaria, T., Steffen, B.: Model-driven development with the jABC. In: HVC - IBM Haifa Verification Conference, Haifa, Israel, IBM, October 23-26, 2006. LNCS. Springer, Heidelberg (2006)
5. Steffen, B., Margaria, T., Braun, V.: The electronic tool integration platform: Concepts and design. Int. Journal on Software Tools for Technology Transfer (STTT) 1(2), 9–30 (1997)
6. Margaria, T.: Web services-based tool-integration in the ETI platform. SoSyM, Int. Journal on Software and System Modelling 4(2), 141–156 (2005)
7. Steffen, B., Margaria, T., Nagel, R.: Remote Integration and Coordination of Verification Tools in jETI. In: Proc. of ECBS 2005, 12th IEEE Int. Conf. on the Engineering of Computer Based Systems, Greenbelt (USA), April 2005, pp. 431–436. IEEE Computer Society Press, Los Alamitos (2005)
8. Freitag, B., Steffen, B., Margaria, T., Zukowski, U.: An approach to intelligent software library management. In: Proc. 4th Int. Conf. on Database Systems for Advanced Applications (DASFAA 1995), National University of Singapore, Singapore (1995)
9. Margaria, T., Bakera, M., Kubczak, C., Naujokat, S., Steffen, B.: Automatic Generation of the SWS-Challenge Mediator with jABC/ABC. Springer, Heidelberg (2008)
10. Church, A.: Logic, arithmetic and automata. In: Proc. Int. Congr. Math.,Uppsala, Almqvist and Wiksells, vol. 1963, pp. 23–35 (1963)
11. Topnik, C., Wilhelm, E., Margaria, T., Steffen, B.: jMosel: A Stand-Alone Tool and jABC Plugin for M2L(str). In: Valmari, A. (ed.) SPIN 2006. LNCS, vol. 3925, pp. 293–298. Springer, Heidelberg (2006)
12. Reiter, R.: Knowledge in Action: Logical Foundations for Specifying and Implementing Dynamical Systems. MIT Press, Cambridge (2001)

13. Steffen, B.: Generating data flow analysis algorithms from modal specifications. Sci. Comput. Program. 21(2), 115–139 (1993)
14. Steffen, B., Margaria, T., Braun, V.: The electronic tool integration platform: Concepts and design. Int. Journal on Software Tools for Technology Transfer (STTT) 1(2), 9–30 (1997)
15. Margaria, T., Steffen, B.: LTL guided planning: Revisiting automatic tool composition in ETI. In: SEW 2007: Proceedings of the 31st IEEE Software Engineering Workshop, Washington, DC, USA, pp. 214–226. IEEE Computer Society Press, Los Alamitos (2007)
16. Wilhelm, C.T.E., Steffen, T.M.B.: jMosel: A stand-alone tool and jABC plugin for M2L(Str). In: Valmari, A. (ed.) SPIN 2006. LNCS, vol. 3925, pp. 293–298. Springer, Heidelberg (2006)
17. Margaria, T.: Fully automatic verification and error detection for parameterized iterative sequential circuits. In: Margaria, T., Steffen, B. (eds.) TACAS 1996. LNCS, vol. 1055, pp. 258–277. Springer, Heidelberg (1996)
18. McIlraith, S., Son, T.: Adapting golog for composition of semantic web services. In: Proceedings of the Eighth International Conference on Knowledge Representation and Reasoning (KR2002), Toulouse, France, April 22-25, 2002, pp. 482–493 (2002)
19. Kelb, P., Margaria, T., Mendler, M., Gsottberger, C.: MOSEL: A flexible toolset for monadic second-order logic. In: Brinksma, E. (ed.) TACAS 1997. LNCS, vol. 1217, pp. 183–202. Springer, Heidelberg (1997)
20. Kubczak, C., Margaria, T., Kaiser, M., Lemcke, J., Knuth, B.: Abductive synthesis of the mediator scenario with jABC and GEM. Technical Report LG-2009-01, Stanford University (2009), `http://logic.stanford.edu/reports/LG-2009-01.pdf`
21. Lemcke, J., Kaiser, M., Kubczak, C., Margaria, T., Knuth, B.: Advances in solving the mediator scenario with jABC and jABC/GEM. Technical Report LG-2009-01, Stanford University (2009), `http://logic.stanford.edu/reports/LG-2009-01.pdf`
22. Baier, J., McIlraith, S.: Planning with temporally extended goals using heuristic search. In: Proc. ICAPS 2006, Cumbria, UK. AAAI, Menlo Park (2006)
23. Müller-Olm, M., Schmidt, D.A., Steffen, B.: Model-checking: A tutorial introduction. In: Cortesi, A., Filé, G. (eds.) SAS 1999. LNCS, vol. 1694, pp. 330–354. Springer, Heidelberg (1999)
24. Raffelt, H., Steffen, B., Berg, T.: Learnlib: a library for automata learning and experimentation. In: Proc. of ACM SIGSOFT FMICS 2005, pp. 62–71. ACM Press, New York (2005)
25. Margaria, T., Raelt, H., Steen, B., Leucker, M.: The learnlib in FMICS-jETI. In: Proc. of ICECCS 2007, 12th IEEE Int. Conf. on Engineering of Complex Computer Systems, July 2007. IEEE Computer Soc. Press, Los Alamitos (2007)
26. Gastin, P., Oddoux, D.: Fast ltl to Büchi automata translation. In: Berry, G., Comon, H., Finkel, A. (eds.) CAV 2001. LNCS, vol. 2102, p. 53. Springer, Heidelberg (2001)
27. Sirin, E., Parsia, B., Wu, D., Hendler, J.A., Nau, D.S.: HTN planning for web service composition using shop2. In: ISWC 2003, vol. 1(4), pp. 377–396 (2003)
28. Matskin, M., Rao, J.: Value-added web services composition using automatic program synthesis. In: Bussler, C.J., McIlraith, S.A., Orlowska, M.E., Pernici, B., Yang, J. (eds.) CAiSE 2002 and WES 2002. LNCS, vol. 2512, pp. 213–224. Springer, Heidelberg (2002)

A Practical Approach for Scalable Conjunctive Query Answering on Acyclic $\mathcal{EL}^+$ Knowledge Base

Jing Mei[1], Shengping Liu[1], Guotong Xie[1], Aditya Kalyanpur[2], Achille Fokoue[2], Yuan Ni[1], Hanyu Li[1], and Yue Pan[1]

[1] IBM China Research Lab, Building 19 ZGC Software Park, Beijing 100193, China
{meijing,liusp,xieguot,niyuan,lihanyu,panyue}@cn.ibm.com
[2] IBM Watson Research Center, P.O. Box 704, Yorktown Heights, NY 10598, USA
{adityakal,achille}@us.ibm.com

Abstract. Conjunctive query answering for $\mathcal{EL}^{++}$ ontologies has recently drawn much attention, as the Description Logic $\mathcal{EL}^{++}$ captures the expressivity of many large ontologies in the biomedical domain and is the foundation for the OWL 2 EL profile. In this paper, we propose a practical approach for conjunctive query answering in a fragment of $\mathcal{EL}^{++}$, namely acyclic $\mathcal{EL}^+$, that supports role inclusions. This approach can be implemented with low cost by leveraging any existing relational database management system to do the ABox data completion and query answering. We conducted a preliminary experiment to evaluate our approach using a large clinical data set and show our approach is practical.

1 Introduction

The OWL 2 EL profile is a subset of OWL 2 that captures the expressive power of many large ontologies in the biomedical domain, such as the Gene Ontology [6], the Systematized Nomenclature of Medicine-Clinical Terms (SNOMED CT) [9], and large parts of the Galen Medical Knowledge Base ontology [13]. This profile is based on the Description Logic language $\mathcal{EL}^{++}$ [1,2] for which the concept subsumption and instance checking problem can be decided in polynomial time.

Meanwhile, conjunctive query answering, which originated from research in relational databases (RDB), is becoming a crucial requirement for ontology-based, data intensive applications. In biomedicine, there are several, very large $\mathcal{EL}^+$ ontological datasets, e.g., protein data annotated with the Gene Ontology and clinical data annotated using SNOMED CT. However, theoretical results have proved that conjunctive query answering is undecidable in both $\mathcal{EL}^+$ and $\mathcal{EL}^{++}$ [14]. To regain the decidability for addressing conjunctive queries, some fragments of $\mathcal{EL}^+$ and $\mathcal{EL}^{++}$ need to be identified. One approach is to exclude role inclusions, e.g., related work on $\mathcal{EL}$ [11] and its follow-up work on $\mathcal{ELH}^{dr}_{\bot}$ [12], while another approach is to impose restrictions on role inclusions, e.g., the so-called regular $\mathcal{EL}^{++}$ [10]. In either of the two approaches, less attention is paid to provide full support for role inclusions. Considering that role inclusion

A. Bernstein et al. (Eds.): ISWC 2009, LNCS 5823, pp. 408–423, 2009.

does play an important role in biomedical ontologies (such as transitivity for the *part-of* relationship in the Gene Ontology and the right identity axiom for *direct-substance* $\circ$ *active-ingredient* $\sqsubseteq$ *direct-substance* in SNOMED CT), we prefer to keep expressive roles in the logic and explore an alternate solution.

In this paper, we present a practical approach for scalable conjunctive query answering on acyclic $\mathcal{EL}^+$ KBs. The *acyclic* notion generalizes the classical one (as defined in the DL handbook [4]), by including general concept inclusions. Note that acyclic $\mathcal{EL}^+$ KBs can admit the irregularity in $\mathcal{EL}^+$ role inclusions. Also, acyclic $\mathcal{EL}^+$ is really useful in practice, because it is expressive enough for the commonly used biomedical ontologies, e.g., Gene Ontology, SNOMED CT.

At a glance, our approach is similar to the bottom-up rule evaluation approach which consists of: (1) doing an ABox completion to precompute all inferred assertions; (2) answering conjunctive queries on the completed ABox. In the first phase, we adopt the idea of "shared" individuals for existential restrictions, referred to as *canonical individuals* in this paper. For example, suppose $A \sqsubseteq \exists R.B$ in the TBox with $A(a)$ and $A(b)$ in the ABox, we will generate one and only one canonical individual u w.r.t. the existential restriction $\exists R.B$, and the completed ABox will have the assertions $B(u)$, $R(a, u)$, and $R(b, u)$. Thus, the cost of inference and the size of the completed dataset is greatly reduced, due to less fresh individuals. However, the canonical individual u is now "shared" by both a and b. If a query asks for $R(x_1, y)$ and $R(x_2, y)$, then the bindings $x_1 = a$ and $x_2 = b$ (with y bound to u) is not a correct answer. Therefore, in the second phase, we propose a *base path criterion* to decide the soundness of an answer on the completed ABox. Towards a practical implementation, we materialize the base path criterion by base path precomputation and query rewriting.

Accordingly, we implemented a prototype to evaluate our approach using a relational database (DB2). Initially, both TBox and ABox are stored in an RDB. The ABox completion is conducted by running an RDB-based Datalog engine with our ABox completion rules. Finally, queries are rewritten and transformed to SQL, making them executable on top of the RDB. The experimental results show that our approach is practical for a large ABox with millions of triples.

Our key contributions in this paper are as follows: (a) we propose a practical approach for conjunctive query answering for acyclic $\mathcal{EL}^+$ that supports role inclusions; (b) we describe a low-cost implementation for this approach on top of a standard relational database management system; (3) we demonstrate its effectiveness and efficiency on a large-scale TBox and ABox.

We direct readers to the Appendix[1] for all proofs.

2 Preliminaries

2.1 $\mathcal{EL}^+$ and $\mathcal{EL}$ Family

In $\mathcal{EL}$, concept descriptions are $C ::= A|\exists R.C|C_1 \sqcap C_2|\top$ where $\top$ is the top concept, A is a concept name and R is a role name. A TBox $\mathcal{T}$ is a finite set

[1] http://domino.research.ibm.com/comm/research_people.nsf/pages/jingmei.pubs.html/$FILE/iswc09_appendix.pdf

of general concept inclusions (GCIs) of the form $C_1 \sqsubseteq C_2$ where C_1 and C_2 are concept descriptions. An ABox $\mathcal{A}$ is a set of concept assertions $A(a)$ and role assertions $R(a_1, a_2)$ where A is a concept name, R is a role name, and a, a_1, a_2 are individual names. As usual in DLs, a knowledge base (KB) $\mathcal{K}$ is a pair $\langle \mathcal{T}, \mathcal{A} \rangle$ with $\mathcal{T}$ a TBox and $\mathcal{A}$ an ABox. We use NI (resp. NC and NR) to denote the set of named individuals (resp. concept names and role names).

$\mathcal{EL}^+$ [3] extends $\mathcal{EL}$ by also allowing in the TBox a finite set of role inclusions of the form $R_1 \circ \cdots \circ R_k \sqsubseteq R_{k+1}$ where each R_i is a role name and $1 \leqslant i \leqslant k+1$. Particularly important from the perspective of ontology applications, $\mathcal{EL}^+$ generalizes the following: role hierarchies $R \sqsubseteq S$ (aka. $\mathcal{ELH}$); transitive roles expressed by $R \circ R \sqsubseteq R$ (aka. $\mathcal{ELH}_{\mathcal{R}^+}$); and so-called left-identity rules $R \circ S \sqsubseteq S$ as well as right-identity rules $R \circ S \sqsubseteq R$. Finally, $\mathcal{EL}^{++}$ [1] is an extension of $\mathcal{EL}^+$ with the bottom concept $\bot$, the nominal $\{a\}$ and the concrete domain $p(f_1, \cdots, f_n)$.

As follows, we will focus on $\mathcal{EL}^+$ and its semantics is defined as usual. Moreover, $\mathcal{EL}^+$ TBoxes admit a normal form [3]. Any $\mathcal{EL}^+$ TBox in normal form consists of concept/role inclusions in one of the following forms: $A \sqsubseteq B, A_1 \sqcap A_2 \sqsubseteq B, A \sqsubseteq \exists R.B, \exists R.A \sqsubseteq B, R \sqsubseteq S$ and $R_1 \circ R_2 \sqsubseteq S$, where each A, B, A_1, A_2 is a concept name or the top concept $\top$, and each R, S, R_1, R_2 is a role name. Unless otherwise specified, this paper assumes that TBox $\mathcal{T}$ is in normal form.

A polynomial-time algorithm for TBox reasoning in $\mathcal{EL}^+$ has been proposed and implemented [3]. Using $C_1 \sqsubseteq_{\mathcal{T}} C_2$, we denote that C_1 is subsumed by C_2 w.r.t. $\mathcal{T}$. Given a TBox $\mathcal{T}$, we redefine cycles in $\mathcal{T}$ (to generalize the classical definition [4]). We say that A *directly uses* B in $\mathcal{T}$, if $A \sqsubseteq_{\mathcal{T}} \exists R.B$, and we call *uses* the transitive closure of the relation *directly uses*. Then $\mathcal{T}$ contains a cycle iff there exists a concept name in $\mathcal{T}$ that uses itself. Otherwise, $\mathcal{T}$ is called *acyclic*.

Finally, we also need to decide role subsumption in $\mathcal{EL}^+$. That is, given a TBox $\mathcal{T}$ and a finite sequence of role names $R_1, \cdots, R_k$ where $k \geqslant 3$, to find the set $\{R \mid R_1 \circ \cdots \circ R_k \sqsubseteq_{\mathcal{T}} R\}$. This problem can be computed in polynomial time in the size of $\mathcal{T}$. First, the role inclusions are expanded by exhaustively applying the following four rules: (1) if $R \sqsubseteq_{\mathcal{T}} S$ and $S \sqsubseteq_{\mathcal{T}} T$, then $R \sqsubseteq_{\mathcal{T}} T$; (2) if $R_1 \circ R_2 \sqsubseteq_{\mathcal{T}} S$ and $S \sqsubseteq_{\mathcal{T}} T$, then $R_1 \circ R_2 \sqsubseteq_{\mathcal{T}} T$; (3) if $R_1 \circ R_2 \sqsubseteq_{\mathcal{T}} S$ and $T \sqsubseteq_{\mathcal{T}} R_1$, then $T \circ R_2 \sqsubseteq_{\mathcal{T}} S$; (4) if $R_1 \circ R_2 \sqsubseteq_{\mathcal{T}} S$ and $T \sqsubseteq_{\mathcal{T}} R_2$, then $R_1 \circ T \sqsubseteq_{\mathcal{T}} S$. Then, given $R_1, \cdots, R_k$ where $k \geqslant 3$, we can recursively compute $\{S | R_1 \circ \cdots \circ R_i \sqsubseteq_{\mathcal{T}} S\}$ and $\{T | R_{i+1} \circ \cdots \circ R_k \sqsubseteq_{\mathcal{T}} T\}$, for $1 \leqslant i < k$, such that $\{R \mid S \circ T \sqsubseteq_{\mathcal{T}} R\}$.

2.2 Conjunctive Query Answering

Referring to [11,12], we introduce conjunctive queries and certain answers.

A conjunctive query q is an expression of the form $\exists \boldsymbol{y}.\varphi(\boldsymbol{x}, \boldsymbol{y})$ where $\boldsymbol{x} = \{x_1, \cdots, x_m\}$ and $\boldsymbol{y} = \{y_1, \cdots, y_n\}$ are vectors of variables, while φ is a conjunction of concept atoms $A(t)$ and role atoms $R(t_1, t_2)$. Variables in $\boldsymbol{x}$ are called distinguished variables (also called answer variables) and variables in $\boldsymbol{y}$ are non-distinguished (also called quantified variables). A term is a variable or a named individual, and we use $\mathrm{Term}(q)$ to denote the set of all terms in a query q.

Let $\mathcal{I}$ be an interpretation and $q = \exists \boldsymbol{y}.\varphi(\boldsymbol{x}, \boldsymbol{y})$ a conjunctive query. A match for $\mathcal{I}$ and q is a mapping $\pi : \text{Term}(q) \to \Delta^{\mathcal{I}}$ such that: (1) $\pi(a) = a^{\mathcal{I}}$ for all named individuals $a \in \text{Term}(q) \cap NI$; (2) $\pi(t) \in A^{\mathcal{I}}$ for all concept atoms $A(t) \in q$; (3) $(\pi(t_1), \pi(t_2)) \in R^{\mathcal{I}}$ for all role atoms $R(t_1, t_2) \in q$. If $\boldsymbol{x} = \{x_1, \cdots, x_m\}$ with $\pi(x_i) = a_i^{\mathcal{I}}$ for $1 \leqslant i \leqslant m$, then π is called an $(a_1, \cdots, a_m)$-match for $\mathcal{I}$ and q. If such a match exists, we write $\mathcal{I} \vDash q[a_1, \cdots, a_m]$.

A certain answer for a conjunctive query $q = \exists \boldsymbol{y}.\varphi(\boldsymbol{x}, \boldsymbol{y})$ and a knowledge base $\mathcal{K}$ is a tuple $[a_1, \cdots, a_m]$ such that, for any model $\mathcal{J}$ of $\mathcal{K}$, $\mathcal{J} \vDash q[a_1, \cdots, a_m]$. We use $cert(q, \mathcal{K})$ to denote the set of all certain answers for q w.r.t $\mathcal{K}$.

3 ABox Completion

The purpose of this phase is to build a completed ABox by enriching the original ABox with inferred assertions, so that conjunctive queries can be answered on the completed ABox, instead of doing ABox reasoning at runtime.

3.1 ABox Completion Rules

To complete the ABox, one of the main difficulties is handling the existential concept inclusion $A \sqsubseteq \exists R.B$, which is a generative axiom in that it introduces new individuals not occurring in the original ABox.

A naive way to do the ABox completion is to compute: (1) a mapping M_C from concepts to a subset of individuals; and (2) a mapping M_R from roles to a binary relation on individuals. Below is the list of ABox completion rules.

*AR*1. If $A \sqsubseteq B \in \mathcal{T}$, $x \in M_C(A)$ and $x \notin M_C(B)$,
then $M_C(B) := M_C(B) \cup \{x\}$

*AR*2. If $A_1 \sqcap A_2 \sqsubseteq B \in \mathcal{T}, x \in M_C(A_1)$, $x \in M_C(A_2)$ and $x \notin M_C(B)$,
then $M_C(B) := M_C(B) \cup \{x\}$

*AR*3. If $\exists R.A \sqsubseteq B \in \mathcal{T}, (x, y) \in M_R(R)$, $y \in M_C(A)$ and $x \notin M_C(B)$,
then $M_C(B) := M_C(B) \cup \{x\}$

*AR*4. If $R \sqsubseteq S \in \mathcal{T}$, $(x, y) \in M_R(R)$,
then $M_R(S) := M_R(S) \cup \{(x, y)\}$

*AR*5. If $R_1 \circ R_2 \sqsubseteq S \in \mathcal{T}, (x, y) \in M_R(R_1)$, $(y, z) \in M_R(R_2)$,
then $M_R(S) := M_R(S) \cup \{(x, z)\}$

*AR*6. If $A \sqsubseteq \exists R.B \in \mathcal{T}$, $x \in M_C(A)$, and there is no individual y', s.t. $y' \in M_C(B)$ and $(x, y') \in M_R(R)$, then generate a new individual y, and $M_C(B) := M_C(B) \cup \{y\}$ and $M_R(R) := M_R(R) \cup \{(x, y)\}$

However, there will be a large number of individuals generated. Suppose $A_i \sqsubseteq \exists R.A_j$ with $A_0(a)$, where $A_i \neq A_j$ and $0 \leqslant i < j \leqslant m$. By applying $AR6$ to the named individual a, there are $2^m - 1$ individuals generated. Let n be the number of A_0's instances. The size of the final generated individuals will blow up to $n \times (2^m - 1)$.

To address this problem, an idea of "shared" individuals is adopted in this paper. Specifically, for each existential restriction $\exists R.B$, we will generate a fresh new individual (viz. *canonical individual*) u with respect to R and B, denoted by

$CI(u, R, B)$. To elaborate, for each existential concept inclusion $A \sqsubseteq \exists R.B$ and all instances $a \in M_C(A)$, we ensure that $u \in M_C(B)$ and $(a, u) \in M_R(R)$. Below is an alternate ABox completion rule $AR6'$ that replaces the previous $AR6$.

$AR6'$. If $A \sqsubseteq \exists R.B \in \mathcal{T}$, $x \in M_C(A)$ and $CI(u, R, B)$,
then $M_C(B) := M_C(B) \cup \{u\}$ and $M_R(R) := M_R(R) \cup \{(x, u)\}$

It is not hard to show that, by repeatedly applying rules $AR1-5$ and $AR6'$, the rule application procedure will eventually terminate, when no more changes to M_C and M_R occur. In fact, canonical individuals are those newly introduced anonymous individuals, and we have generated all canonical individuals as the preprocessing step for ABox completion, so that M_C and M_R are computed using at most $n \cdot m$ rule applications, yielding the data set bounded by $\mathbf{O}(n \cdot m)$, where n and m are linear in the size of $\mathcal{T}$ and $\mathcal{A}$.

Now, applying $AR6'$ to the above example, we will only generate m individuals for each and every $\exists R.A_j$ where $1 \leqslant j \leqslant m$, even if there are n instances of A_0. Besides, our ABox completion can benefit a lot from canonical individuals. First, it dramatically reduces the number of anonymous individuals to be generated. Second, it can be pre-computed for a given TBox, so as to reduce the cost of doing the ABox completion.

Thus, similar to the theoretical result presented in [11], our ABox completion can be done in linear time and yield a completed ABox whose size is linear w.r.t. both the TBox and the ABox.

3.2 ABox Completion Implementation

Our approach is implemented using an RDB system. Actually, via introducing canonical individuals, our ABox completion rules $AR1-5$ and $AR6'$ can be transformed into Datalog rules, and the completed ABox can be computed by running an RDB-based Datalog engine via a bottom-up evaluation strategy.

First, we design a database schema s.t. the normalized TBox is stored in 6 tables, namely, `atomicSub`, `existsSub`, `gciInter`, `gciExists`, `subRole` and `roleChain`. These tables correspond to the TBox axioms in form of: $A \sqsubseteq B$, $A \sqsubseteq \exists R.B$, $A_1 \sqcap A_2 \sqsubseteq B$, $\exists R.A \sqsubseteq B$, $R \sqsubseteq S$ and $R_1 \circ R_2 \sqsubseteq S$, respectively. Similarly, the ABox is stored in two tables, `typeOf` and `roleStmt`, corresponding to assertions of $A(x)$ and $R(x, y)$, respectively. The pre-computed canonical individuals are stored in the table `canonInd`.

Second, we translate the ABox completion rules $AR1-5$ and $AR6'$ into Datalog rules, in addition to two initialization rules s.t., (1) if `typeOf`(x, y) then `infTypeOf`(x, y); (2) if `roleStmt`(R, x, y) then `infRoleStmt`$(R, x, y, 0)$. Now, running an RDB-based Datalog engine with the rules, and via a bottom-up evaluation strategy, all inferred assertions will be incrementally stored in tables `infTypeOf` and `infRoleStmt`. In particular, the last column in `infRoleStmt` is used to indicate the origin of inferred role assertions (the reason for doing that will be explained in the next section). Specifically, if $R(x, y)$ is inferred by $AR6'$, then we store it as `infRoleStmt`$(R, x, y, 1)$; and if $R(x, y)$ is inferred by $AR4-5$, then we store it as `infRoleStmt`$(R, x, y, 2)$. We remark that rules of

$AR4-5$ and $AR6'$ are allowed to fire, even if the role assertion $R(x,y)$ that they are attempting to infer has already been computed as $(x,y) \in M_R(R)$. For instance, suppose $\mathcal{T} = \{A \sqsubseteq \exists R_1.B, B \sqsubseteq \exists R_2.C, A \sqsubseteq \exists R_2.C, R_1 \circ R_2 \sqsubseteq R_2\}$ and $\mathcal{A} = \{A(a)\}$. After the rule execution, we have $\texttt{canonInd}(u_1, R_1, B)$ and $\texttt{canonInd}(u_2, R_2, C)$, in addition to $\texttt{infRoleStmt}(R_2, a, u_2, 1)$ because of $A \sqsubseteq \exists R_2.C$, and $\texttt{infRoleStmt}(R_2, a, u_2, 2)$ because of $R_1 \circ R_2 \sqsubseteq R_2$.

Thus, the ABox completion is done as described above, and we use $\mathcal{A}_\alpha$ to denote the completed ABox, where the set of canonical individuals in $\mathcal{A}_\alpha$ is denoted by CI.

As follows, we define an interpretation $\mathcal{I}_\alpha$ w.r.t. $\mathcal{A}_\alpha$ s.t.

- $\Delta^{\mathcal{I}_\alpha} := NI \cup CI$
- $a^{\mathcal{I}_\alpha} := a$ for all $a \in NI$
- $A^{\mathcal{I}_\alpha} := \{e \in \Delta^{\mathcal{I}_\alpha} \mid A(e) \in \mathcal{A}_\alpha\}$ for all $A \in NC$
- $R^{\mathcal{I}_\alpha} := \{(e_1, e_2) \in \Delta^{\mathcal{I}_\alpha} \times \Delta^{\mathcal{I}_\alpha} \mid R(e_1, e_2) \in \mathcal{A}_\alpha\}$ for all $R \in NR$

It is not hard to show that $\mathcal{I}_\alpha$ is a model of $\mathcal{K}$, and if $(a_1, \cdots, a_m) \in cert(q, \mathcal{K})$ then $\mathcal{I}_\alpha \models q[a_1, \cdots, a_m]$, but not vice versa.

4 Query Answering

After ABox completion, queries are ready for answering. However, problems arise due to the introduction of canonical individuals. In this section, we will illustrate a running example, followed by our solution and optimization.

4.1 The Running Example

We use a query $q = \exists y.R(x_1, y) \wedge R(x_2, y)$ and an $\mathcal{EL}^+$ KB $\mathcal{K} = \langle \mathcal{T}, \mathcal{A} \rangle$ as the running example, where $\mathcal{T} = \{C \sqsubseteq \exists R_2.B, B \sqsubseteq \exists R.D, R_1 \circ R \sqsubseteq R_3, R_2 \circ R \sqsubseteq R_4, R_3 \sqsubseteq R, R_4 \sqsubseteq R\}$ and $\mathcal{A} = \{B(b), C(c), R_1(a,b)\}$. Note that $\mathcal{T}$ is irregular. The query is shown below. By the ABox completion, there are two canonical individuals generated, i.e. u_1 for $\exists R_2.B$ and u_2 for $\exists R.D$, in addition, $\mathcal{A}_\alpha = \mathcal{A} \cup \{B(u_1), D(u_2), R(b, u_2), R_2(c, u_1), R(u_1, u_2), R_3(a, u_2), R(a, u_2), R_4(c, u_2), R(c, u_2)\}$, as sketched below. Now, matching the query q with the completed ABox $\mathcal{A}_\alpha$, we have the answer set, i.e., $\{(a,a), (b,b), (c,c), (a,b), (b,a), (b,c), (c,b), (a,c), (c,a)\}$. As shown below, the last four ones, $(b,c), (c,b), (a,c), (c,a)$, are incorrect answers.

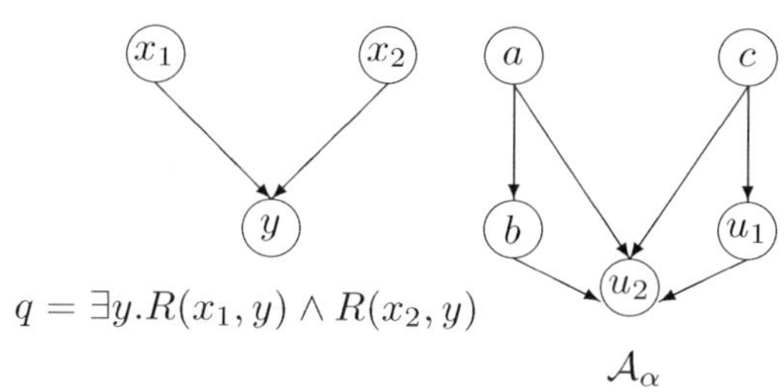

x_1	x_2	y	answer
a	a	u_2	correct
b	b	u_2	correct
c	c	u_2	correct
a	b	u_2	correct
b	a	u_2	correct
b	c	u_2	incorrect
c	b	u_2	incorrect
a	c	u_2	incorrect
c	a	u_2	incorrect

We observe that, first, there is a reverse tree in the query. Intuitively, a role atom $R(x, y)$ can be regarded as an edge with the direction pointing from x to y. Roughly, if there are multiple edges pointing to the same y, we call it a *reverse tree*. For q, role atoms $R(x_1, y)$ and $R(x_2, y)$ are regarded as two edges pointing to the same y, which makes this a reverse tree. Second, the completed data $R(a, u_2)$ and $R(c, u_2)$ matches the reverse tree incorrectly. This is due to the canonical individual u_2 that was generated for $\exists R.D$ and shared by b and u_1, due to $B \sqsubseteq \exists R.D$. Furthermore, the "sharing" is extended to a and c, due to $R_1 \circ R \sqsubseteq_{\mathcal{T}} R$ and $R_2 \circ R \sqsubseteq_{\mathcal{T}} R$, respectively, where $R_1(a, b)$ and $R_2(c, u_1)$. As proposed in [11,12], to guarantee sound answers, the query would be rewritten as $q^* = \exists y.R(x_1, y) \wedge R(x_2, y) \wedge (CI(y) \rightarrow x_1 = x_2)$ s.t., if y is matched with a canonical individual, then x_1 and x_2 need to be matched with the same individual. In this way, unsound answers like $(x_1/a, x_2/c)$ are filtered. However, answers are now incomplete, because some correct answers such as $(x_1/a, x_2/b)$ are also removed. The reason why the existing strategy of query rewriting fails in this case is due to the increased role expressivity. A big difference between $(x_1/a, x_2/c)$ and $(x_1/a, x_2/b)$ is that a and b share u_2 in a native way, since $R_1 \circ R \sqsubseteq_{\mathcal{T}} R$.

Therefore, we need to distinguish variable bindings considering reverse trees in the query. To this end, we first need to distinguish different role assertions in the completed ABox, some of which are generated by existential concept inclusions, and some of which are derived from role inclusions. The former is the origin of incorrect query answering, while the latter extends the incorrectness. Recalling our ABox completion implementation, we mark a tag in `infRoleStmt`, making $R(b, u_2), R_2(c, u_1), R(u_1, u_2)$ tagged by 1 and $R_3(a, u_2), R(a, u_2), R_4(c, u_2), R(c, u_2)$ tagged by 2. Thus, we distinguish role assertions tagged by 1 from those tagged by 2, which are "purely" derived by $AR6'$ and are the root cause of the problem. We call them *base triples*.

Furthermore, we build so-called *base paths* by traversing the graph of base triples. In our example, there are three base triples, i.e., $R(b, u_2), R_2(c, u_1), R(u_1, u_2)$, and three base paths, i.e., $bRu_2, cR_2u_1, cR_2u_1Ru_2$. Given a base path, the first individual is called the *head*, and the last individual is called the *tail*, while all individuals in the base path are said to be *directly located* in it. If there is a named individual reachable to the head of a base path, we refer to it as being *indirectly located* in the base path. For instance, a is reachable to b, making a indirectly located in bRu_2.

These base paths are now used for filtering incorrect answers. Informally, all terms occurring in a reverse tree of the query are required to be (in)directly located in a single base path, so called the *base path criterion*. In our example, a and b are, respectively, indirectly and directly located in bRu_2, so $(x_1/a, x_2/b)$ is a correct answer. Conversely, there is no base path making both a and c (in)directly located, so $(x_1/a, x_2/c)$ is not a correct answer.

Meanwhile, we realize that it is infeasible to check base paths at runtime during query answering. As illustrated above, there are at least 9 matches for $\mathcal{A}_\alpha$ and q in our example. To reduce the cost, we precompute and materialize all possible role assertions in the same base path in the table `BPath(path, tail, node1, node2, role)`, where the `path` has the `tail`, and `node1` is (in)directly located, while `node2` is directly located in the `path` via the `role` relationship. Below is the BPath storage of our example. For example, BPath(bRu_2, u_2, a, u_2, R) indicates the path bRu_2 has the tail u_2, and a is indirectly located while u_2 is directly located via the assertion $R(a, u_2)$.

Table 1. An example of the BPath storage

path	tail	node1	node2	role
cR_2u_1	u_1	c	u_1	R_2
bRu_2	u_2	b	u_2	R
bRu_2	u_2	a	u_2	R_3
bRu_2	u_2	a	u_2	R
$cR_2u_1Ru_2$	u_2	c	u_1	R_2
$cR_2u_1Ru_2$	u_2	u_1	u_2	R
$cR_2u_1Ru_2$	u_2	c	u_2	R_4
$cR_2u_1Ru_2$	u_2	c	u_2	R

Finally, our base path criterion is applicable for query rewriting. For q, we rewrite it as $q^* = \exists y.R(x_1, y) \wedge R(x_2, y) \wedge (CI(y) \rightarrow \exists p.BPath(p, y, x_1, y, R) \wedge BPath(p, y, x_2, y, R))$. Here, the satisfiability of $CI(y)$ means y's match is a canonical individual, while the satisfiability of both $BPath(p, y, x_1, y, R)$ and $BPath(p, y, x_2, y, R)$ means the matches of x_1 and x_2 are (in)directly located in a single base path p. As shown in Table 1, both BPath(bRu_2, u_2, a, u_2, R) and BPath(bRu_2, u_2, b, u_2, R) are satisfied, making $(x_1/a, x_2/b)$ a correct answer.

In the next three subsections, we will formally present our solution, including how to identify reverse tress in the query, how to compute base paths in the ABox completion, and how to rewrite the query according to our base path criterion.

4.2 Reverse Trees in Query

Before we provide a formal definition of reverse trees, we remark that little attention is paid to multiple edges pointing to the same y, where y is bound to a named individual. This is because only matches sharing some canonical individuals are problematic. Moreover, $\mathcal{EL}^+$ models are split [11], i.e., there is no role assertion pointing from an unnamed individual to a named one.

Formally, let $q = \exists \boldsymbol{y}.\varphi(\boldsymbol{x}, \boldsymbol{y})$ be a conjunctive query. We use $GT_q \subseteq \text{Term}(q)$ to denote the set of grounded terms in q, s.t. for each $t \in \text{Term}(q)$: (1) if $t \in NI \cup \boldsymbol{x}$ then $t \in GT_q$; (2) if $R(t, t') \in q$ and $t' \in GT_q$ then $t \in GT_q$.

Next, we use γ_q to denote the set of reverse tree roots in the query q, i.e., $\gamma_q := \{y \notin GT_q | \sharp\{R(x, y) \in q\} > 1\}$, where $\sharp$ denotes the cardinality.

Now, for each reverse tree root $y \in \gamma_q$, we compute $E_y^{(0)} := \{R(x, y) \in q\}$ and $E_y^{(i+1)} := E_y^{(i)} \cup \{R(s, t) \in q \mid t \in \{s' | R'(s', t') \in E_y^{(i)}\} \text{ and } t \notin GT_q\}$. The computation terminates, when $E_y^{(i)} = E_y^{(i+1)}$. $E_y^{(0)}$ refers to the set of multiple edges pointing to the same y. By incrementally expanding in a reverse direction, $E_y^{(i)}$ contains additional edges, all of which point to non-grounded terms, until we reach a fixedpoint. The resultant set $E_y := \bigcup_{i \geqslant 0} E_y^{(i)}$, and we use E_y to denote the set of all reverse tree edges rooted by $y \in \gamma_q$.

Finally, we use RT_q to denote the set of reverse trees, where each reverse tree is a pair (y, E_y) with $y \in \gamma_q$ and there is no pair $(y', E_{y'})$ in RT_q such that $E_y \subset E_{y'}$. We call a query q free of reverse trees, if RT_q is empty.

It is not hard to verify that GT_q and RT_q can be computed in time polynomial in the size of q. Moreover, the size of GT_q is linear in the size of q, and the size of RT_q is polynomial in the size of q.

As shown in the following theorem, for queries free of reverse trees, query answering over the completed ABox is sound and complete, even in cyclic $\mathcal{EL}^+$ KBs. Thus, if the application requires only queries free of reverse trees, the rest of this section can be skipped over.

Theorem 1. *$cert(q, \mathcal{K}) = ans(q, \mathcal{A}_\alpha)$, given that q is a conjunctive query free of reverse trees and $\mathcal{K}$ is an $\mathcal{EL}^+$ knowledge base.*

To see which queries are free of reverse trees, consider the following samples. First, $q_1 = \exists y_1, y_2.R_1(x_1, y_1) \wedge R_2(x_2, y_2)$ is forest-shaped (a notion well defined in DL literature [8]), and we claim that forest-shaped queries are all free of reverse trees. Next, neither $q_2 = \exists y_1, y_2, y_3.R(y_1, y_2) \wedge R(y_2, y_3) \wedge R(y_3, y_1)$ nor $q_3 = \exists y_1, y_2.R_1(x_1, y_1) \wedge R_2(x_2, y_1) \wedge R_3(y_1, y_2) \wedge R_4(y_2, x_3)$ is forest-shaped, while both q_2 and q_3 are free of reverse trees. In fact, there is no multiple edge in q_2, which makes it free of a reverse tree. In q_3, the multiple edges point to y_1 which is a grounded term due to the answer variable x_3, so there is no reverse tree root and q_3 is also free of a reverse tree.

4.3 Base Paths in ABox Completion

First, we need to identify base triples in the ABox completion. Taking advantage of the last column in `infRoleStmt`, BT is used to denote the set of base triples s.t. $BT := \{R(x, y) | \texttt{infRoleStmt}(R, x, y, 1)\} \setminus \{R(x, y) | \texttt{infRoleStmt}(R, x, y, 2)\}$. In other words, BT consists of triples which are "purely" derived by $AR6'$.

Next, we traverse the graph of BT to compute base paths. Given a canonical individual $u \in CI$, we use $BP(u)$ to denote the set of base paths ending up with u such that $BP(u) := \{u_0 R_1 u_1 \cdots R_k u_k | u_0 \in NI, u_k = u, u_i \in CI, R(u_{i-1}, u_i) \in BT, 1 \leqslant i \leqslant k\}$. The following figure illustrates the intuition of our definition.

There are two planes: the NI plane contains all triples between named individuals, and the CI plane contains only the base triples between canonical individuals. Meanwhile, base triples from named individuals to canonical individuals are those links from the NI plane to the CI plane. Here is the running example.

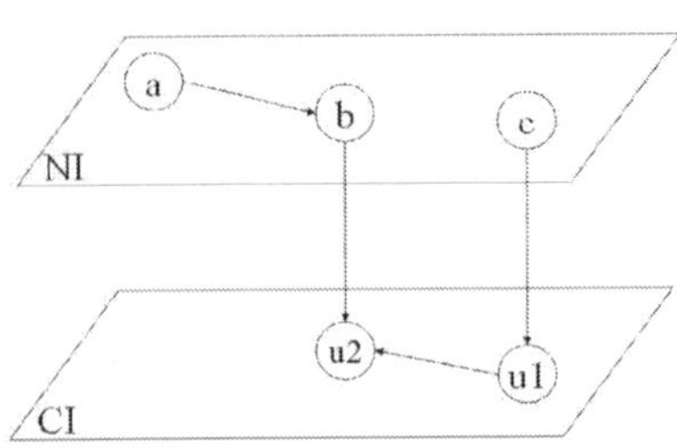

Intuitively, a base path starts from a named individual (i.e. the head), followed by a base triple jumping from the NI plane to the CI plane, and after traversal, it ends up with a canonical individual (i.e. the tail). Therefore, all individuals in a base path are canonical ones, except the head. Named individuals indirectly located in a base path are those reachable to its head in the NI plane.

It is not hard to show that, disregarding cyclic $\mathcal{EL}^+$ KBs, the base path computation will terminate, and using depth-first-search or breadth-first-search, its time complexity is proportional to the number of nodes plus the number of edges in the graph they traverse, i.e., $\mathbf{O}(|NI|+|CI|+|BT|)$. In other words, BP can be computed in time linear in the size of $\mathcal{A}_\alpha$. Unfortunately, in the worst-case, the size of base paths is exponential in the size of TBox and polynomial in the size of ABox. We believe this upper bound is unlikely to occur in practice, and Section 4.5 presents our optimization techniques to address this issue.

Finally, we precompute and materialize all possible role assertions in the same base path in the table `BPath(path, tail, node1, node2, role)`. For a base path p: $u_0R_1u_1\cdots R_ku_k$, we store it as $\mathrm{BPath}(p, u_k, u_i, u_j, R)$, where $0 \leqslant i < j \leqslant k$ and $R_{i+1}\circ\cdots\circ R_j \sqsubseteq_\mathcal{T} R$, as well, $\mathrm{BPath}(p, u_k, a_0, u_j, R)$ where $0 < j \leqslant k$ and $R_0 \circ R_1 \circ \cdots \circ R_j \sqsubseteq_\mathcal{T} R$ with $R_0(a_0, u_0) \in \mathcal{A}_\alpha$.

Here, we remark two key points about the `BPath` storage. One is how to compute the `role`, which denotes the super roles for the role chain from `node1` to `node2`. For $\mathrm{BPath}(p, u_k, u_i, u_j, R)$ where $0 \leqslant i < j \leqslant k$, we need to compute the set $\{R \mid R_{i+1}\circ\cdots\circ R_j \sqsubseteq_\mathcal{T} R\}$, as discussed in the Section 2 (Preliminaries). Similarly, for $\mathrm{BPath}(p, u_k, a_0, u_j, R)$, where $0 < j \leqslant k$, we compute the set $\{R \mid R_0\circ R_1\circ\cdots\circ R_j \sqsubseteq_\mathcal{T} R\}$. Another is how to compute all named individuals reachable to u_0. Suppose a_0 is reachable to u_0, via a sequence of named individuals $a_1, \cdots, a_n$ and a sequence of role names $S_1, \cdots, S_n$, where $a_n = u_0$ and $S_{i+1}(a_i, a_{i+1}) \in \mathcal{A}_\alpha$ for any $0 \leqslant i < n$, given $S_1\circ\cdots\circ S_n\circ R_1\circ\cdots\circ R_j \sqsubseteq_\mathcal{T} R$. Thanks to the TBox normalization [1], a role inclusion in form of $R_1\circ\cdots\circ R_k \sqsubseteq R_{k+1}$ will be normalized inductively by $R_1\circ\cdots\circ R_{k-1} \sqsubseteq R'_k$ and $R'_k\circ R_k \sqsubseteq R_{k+1}$, where R'_k is a newly introduced role name. In this respect, we are convinced that, given $S_1\circ\cdots\circ S_n\circ R_1\circ\cdots\circ R_j \sqsubseteq_\mathcal{T} R$, there is a role name R_0 (introduced by the normalization) s.t. $S_1\circ\cdots\circ S_n \sqsubseteq_\mathcal{T} R_0$ and $R_0\circ R_1\circ\cdots\circ R_j \sqsubseteq_\mathcal{T} R$. Thus, with u_0 pinpointed, the set of named individuals reachable to u_0 are $\{a_0 \mid R_0(a_0, u_0) \in \mathcal{A}_\alpha\}$.

4.4 Query Rewriting

Below, we define a base path criterion, then rewrite the query according to it.

Definition 1 (Base Path Criterion). *Let τ be a match for $\mathcal{I}_\alpha$ and q.*

We say that a reverse tree $(y, E_y) \in ST_q$ is satisfied by τ if, $\tau(y) \in CI$ implies there is a base path $u_0R_1u_1\cdots R_ku_k \in BP(\tau(y))$, and for any $R(s,t) \in E_y$, whenever $\tau(t) = u_j$ and $1 \leqslant j \leqslant k$,

- *$\tau(s) = u_i$ and $0 \leqslant i < j$, given $R_{i+1}\circ\cdots\circ R_j \sqsubseteq_\mathcal{T} R$, otherwise;*
- *$\tau(s) = a_0$ and $0 \leqslant i < n$, $(a_i, a_{i+1}) \in S_{i+1}^{\mathcal{I}_\alpha}$ with $a_n = u_0$, given $S_1\circ\cdots\circ S_n\circ R_1\circ\cdots\circ R_j \sqsubseteq_\mathcal{T} R$.*

The base path criterion is that all reverse trees of q are satisfied by the match τ for $\mathcal{I}_\alpha$ and q.

Since we have stored all base paths, the base path criterion can be directly adopted by query rewriting. The rewritten query q^* is defined as

$$q^* := q \wedge \bigwedge_{(y,E_y)\in RT_q} (CI(y) \rightarrow \exists p. \bigwedge_{R(s,t)\in E_y} (CI(t) \rightarrow BPath(p,y,s,t,R)))$$

and the size of rewritten queries is bounded by $\mathbf{O}(|q|^2)$.

By definition, an $(a_1,\cdots,a_m)$-match τ for $\mathcal{I}_\alpha$ and q satisfies a reverse tree $(y,E_y)\in ST_q$ if, $\tau(y)\in CI$ implies there is p such that, whenever $\tau(t)\in CI$, we have BPath$(\tau(p),\tau(y),\tau(s),\tau(t),R)$, for any $R(s,t)\in E_y$. If there exists such a match satisfying all reverse trees, we write $\mathcal{I}_\alpha \cup \text{BPath} \vDash q^*[a_1,\cdots,a_m]$.

Similar to [11,12], we use $ans(q,\mathcal{A}_\alpha)$ to denote the set of answers that a relational database system returns for q over $\mathcal{A}_\alpha$, while $ans(q^*,\mathcal{A}_\alpha\cup BPath)$ for q^* over $\mathcal{A}_\alpha$ with BPath. This paper contributes the following theorem.

Theorem 2. *$cert(q,\mathcal{K}) = ans(q^*,\mathcal{A}_\alpha \cup BPath)$, given that q is a conjunctive query and $\mathcal{K}$ is an acyclic $\mathcal{EL}^+$ knowledge base.*

Interestingly, if $\mathcal{K}$ is an $\mathcal{EL}$ knowledge base, then our base path criterion is reducible to the approach proposed for $\mathcal{EL}$ [11]. In fact, the Definition 1 will be simplified s.t., for any $R(s,t)\in E_y$, if $\tau(t)=u_j$ then $\tau(s)=u_{j-1}$, where $1\leqslant j\leqslant k$. Suppose $s_1,\cdots,s_l$ be a sequence of terms and all $R(s_i,t)\in E_y$ for $1\leqslant i\leqslant l$. Now, if $\tau(t)=u_j$ then $\tau(s_1)=\cdots=\tau(s_l)=u_{j-1}$, and in the rewritten query, it becomes $CI(t)\rightarrow(s_1=s_2\wedge\cdots\wedge s_{l-1}=s_l)$.

4.5 Optimization

Observing the possibility of using our approach for $\mathcal{EL}$, we consider removing some BPath records which contribute little to the *plus* part of $\mathcal{EL}^+$, towards an optimal storage. Formally, we define the set $NR^+ := \{R|R_1\circ R_2 \sqsubseteq_\mathcal{T} R_3, R\sqsubseteq_\mathcal{T} R_i, 1\leqslant i\leqslant 3\}$. Our strategy is to check the BPath storage, and for any BPath(p,z,x,y,R), if $R\notin NR^+$, then this BPath record can be removed.

Correspondingly, we also need to redefine the rewritten query $\bar{q}^*$ as follows.

$$\bar{q}^* := q \wedge \bigwedge_{(y,E_y)\in RT_q} (CI(y) \rightarrow \exists p. \bigwedge_{R(s,t)\in E_y, R\in NR^+} (CI(t) \rightarrow BPath(p,y,s,t,R))$$

$$\wedge \bigwedge_{(\{t_1,\cdots,t_l\},\zeta)\in \text{Fork}^y_=} (CI(t_\zeta) \rightarrow \bigwedge_{1\leqslant i<l} t_i = t_{i+1}) \wedge \bigwedge_{(I,\zeta)\in\text{Fork}^y_\mathcal{H}} (CI(t_\zeta) \rightarrow \bigvee_{R\in I} R(t_\zeta^{\text{pre}},t_\zeta)))$$

The definitions of $\text{Fork}^y_=$ and $\text{Fork}^y_\mathcal{H}$ are almost the same as those defined for $\mathcal{ELH}^{dr}_\perp$ [12]. The main differences are pre$(\zeta) := \{t|R(t,t')\in E_y$ for some $R\notin NR^+$ and $t'\in\zeta\}$ and in$(\zeta) = \{R|R(t,t')\in E_y$ for some $R\notin NR^+$ and $t'\in\zeta\}$. Here, $R(t,t')\in E_y$ and $R\notin NR^+$ are imposed in our context.

To see why our optimization retains the correctness, consider the query $q = \exists y_1, y_2, y.R_1(x_1, y_1) \wedge R_2(x_2, y_1) \wedge R(y_1, y) \wedge R(y_2, y)$ over an $\mathcal{EL}^+$ KB with $R_1 \sqsubseteq R_2$ and $R \circ R \sqsubseteq R$. There is one and only one reverse tree $(y, E_y) \in RT_q$, where E_y consists of all role atoms in q, and $NR^+ = \{R\}$. The non-optimized rewritten query q^* is $q \wedge (CI(y) \rightarrow \exists p.BPath(p, y, y_1, y, R) \wedge BPath(p, y, y_2, y, R) \wedge (CI(y_1) \rightarrow BPath(p, y, x_1, y_1, R_1) \wedge BPath(p, y, x_2, y_1, R_2)))$, and the optimized rewritten query $\bar{q}^*$ is $q \wedge (CI(y) \rightarrow \exists p.BPath(p, y, y_1, y, R) \wedge BPath(p, y, y_2, y, R) \wedge (CI(y_1) \rightarrow x_1 = x_2) \wedge (CI(y_1) \rightarrow R_1(x_2, y_1))$. Below shows the theorem for optimization, where $\overline{\text{BPath}}$ denotes the BPath storage after removal.

Theorem 3. *$cert(q, \mathcal{K}) = ans(\bar{q}^*, \mathcal{A}_\alpha \cup \overline{BPath})$, given that q is a conjunctive query and $\mathcal{K}$ is an acyclic $\mathcal{EL}^+$ knowledge base.*

This section is now concluded with the theorem for complexity and data size.

Theorem 4. *Conjunctive query answering on acyclic $\mathcal{EL}^+$ knowledge base can be done in time polynomial in the size of the query, the TBox and the ABox, with data of a polynomial-size blowup w.r.t. the query and the ABox while of an exponential-size blowup w.r.t. the TBox.*

5 Preliminary Experiments

We have implemented a prototype to evaluate our approach using Java SDK 1.5 and DB2 V9.3 Enterprise Edition. It performs three steps: (1) does ABox completion with canonical individuals and stores the completed ABox data in the RDB; (2)computes the base paths and stores them in the RDB as well; (3) rewrites the query according to the base path criteria, transforms the query into SQL and executes it on the completed ABox with base paths. The preliminary experiments are performed on an X-3650 server with 8G memory and two 3.0GHz Xeon CPUs.

TBox. The SNOMED CT ontology is generated by executing the perl script provided in the SNOMED CT 2009Jan release package, on the SNOMED CT 2007Jan release content files (because our data uses the 2007Jan version). It is formulated as an acyclic $\mathcal{EL}^+$ ontology that contains 308,832 concept names, 62 role names with 47,317 concept definitions and 261,514 primitive concept definitions. It also contains 11 role subsumption axioms and one right identity axiom: $caustiveAgent \circ hasActiveIngredient \sqsubseteq caustiveAgent$ [2]. After normalization, the TBox includes 463,971 concept names, 576,971 axioms of form $A \sqsubseteq B$, 428,905 axioms of form $A \sqsubseteq \exists R.B$, 49,454 axioms of form $\exists R.A \sqsubseteq B$ and 118,124 axioms of form $A_1 \sqcap A_2 \sqsubseteq B$.

[2] Actually, the stated right identity axiom in the SNOMED CT 2009Jan Release is $directSubstance \circ hasActiveIngredient \sqsubseteq directSubstance$. We do not use this axioms because there is no concept B appeared in both $A \sqsubseteq \exists directSubstance.B$ and $B \sqsubseteq \exists hasActiveIngredient.C$ in the 2007Jan Release.

ABox. The ABox of SNOMED CT ontology is constructed partly from a collection of 100,000 HL7 Clinical Document Architecture (CDA)[7] documents collected from a large hospital in southern China. CDA is a widely adopted standard to represent the electronic clinical documents, such as clinical notes and prescriptions. These CDA documents are CDA level 3 documents with clinical statements using codes from SNOMED CT 2007-Jan release.

We developed an XML-2-RDF transformer that can extract RDF triples from the structured body part of CDA documents. For each CDA document, the XML-2-RDF transformer outputs a single RDF document.In addition, because the collected CDA documents do not cover the concepts and roles involved in the right identity axiom, we generate additional instances randomly for the classes and roles involved in the right identity axiom, and then merge them into the ABox. In our experiment, we generate on average 30 instances per CDA document. The transformed RDF documents plus the randomly generated instances are partitioned into four data sets (ABoxes) according to the number of CDA documents, i.e, 0.1K, 1K, 10K and 100K.

ABox Data completion. After loading the normalized SNOMED CT ontology and the ABox into the RDB, we use the RDB-based datalog engine developed as part of the IBM SHER framework to perform ABox completion. To evaluate the benefits of canonical individual, we implemented both the intuitive approach and the canonical-individual-based approach.

In the intuitive approach, the algorithm first evaluates the ABox datalog rules ($AR1$-5) by a semi-naive approach, then iteratively does the following steps until no more new facts are generated: (1)For the TBox axiom: $A \sqsubseteq \exists R.B$, and ABox assertion $A(a)$, add new facts $R(a,u), B(u)$ into the ABox if there is no individual w satisfying both $R(a,w)$ and $B(w)$; (2) Incrementally evaluate the ABox completion rules with the new facts.

In the canonical-individual-based approach, the algorithm first generates canonical individuals for existential restrictions, then evaluates the ABox datalog rules $AR1$-5,$AR6'$ via a semi-naive evaluation. In this way, the overall number of generated anonymous (canonical) individuals is fixed – 76,614 in our case.

Table 2 summarizes the ABox completion results for both approaches. For each ABox, we report the number of individuals (IND), the numbers of triples (TRP) in the original ABox and in the completed ABox in both approaches, and the data completion time (excluding the ABox data loading time).

Table 2. ABox data completion results

Dataset	Original		Intuitive Approach			CI-based Approach		
	IND	TRP	IND	TRP	Time	IND	TRP	Time
0.1K	16K	22K	77K	1,018K	4.3h	16K	310K	53s
1K	169K	227K	793K	10,395K	58.4h	172K	4,099K	0.3h
10K	1,614K	2,155K	N/A	N/A	N/A	1,618K	39,923K	1.4h
100K	15,097K	22,157K	N/A	N/A	N/A	15,102K	387,316K	13.9h

As can be seen from the Table 2, in the intuitive approach, the completed ABox has a polynomial-size blowup. Also, it is very time-consuming to generate due to many iterative calls to the datalog engine. When using canonical individuals, the number of newly added individuals in the completed ABox is always less than a constant (76,614 in our case). The size of inferred triples is about 10-20 times that of the original triples because of the deep hierarchy of the normalized SNOMED CT ontology. For example, the concept "headache" has 15 super concepts in original ontology, but has 27 super concepts in the normalized ontology. Our experiment confirms that it is necessary to introduce canonical individual for ABox data completion in practice.

Base path generation and query performance. After performing the ABox completion, the base path can be generated and the query can be rewritten and executed on the completed ABox with base paths. We first compare the efficiency of base path generation of the initial approach and the optimized approach. Then we design three typical queries to test the query performances. These queries are designed according to the three typical query shape: point, tree and graph.

Q1(x)= $Headache(x)$

Q2(x) =$\exists y\ \ HearingNormal(x), findingSite(x,y), EntireLeftEar(y)$

Q3(x,y)=$\exists z\ \ causativeAgent(x,z), causativeAgent(y,z)$

In our implementation, the (rewritten) conjunctive query is represented using the RDF SPARQL query language. The SPARQL query is further transformed to SQL by a RDF-2-RDB mapping engine.

Table 3 summarizes the base path generation and query performance results for each ABox. We report the count of base path records (BPSize), the time to generate them (GenTime) and the query execution time for Q3 using the initial approach and the optimized approach. Since Q1 and Q2 are free of reverse trees, the base path optimization has no effect on their performances.

Table 3. Base path generation and query performance results

Dataset	without optimization			with optimization			Q1	Q2
	BPSize	GenTime	Q3	BPSize	GenTime	Q3		
0.1K	172K	480s	321ms	10K	36.6s	50ms	10ms	30ms
1K	1,917K	2,237s	3.5s	199K	228s	2.0s	61ms	500ms
10K	16,764K	4.2h	22.1s	1,812K	1.1h	12.7s	83ms	1.6s
100K	N/A	N/A	N/A	18,734K	7.8h	313s	712ms	14.6s

Our experiment is still at a preliminary stage due to the limitations of the ontology and data. However, from the results, we can observe that: (1) the base path optimization dramatically reduces the number of stored base paths in this case, because only five of SNOMED CT's 62 roles appear in the "real" role inclusions, and further improves the performance of the rewritten query Q3 because self-joins on the base path table are involved; (2) the query answering is scalable to a large data set because it can leverage the underlying relational database for scalability.

6 Related Work

In recent years, a rich literature on the subject of conjunctive query answering on the $\mathcal{EL}$ family has emerged, including both theoretical and practical work.

Theoretically, there are undecidable problems identified, such as answering conjunctive queries in both $\mathcal{EL}^+$ and $\mathcal{EL}^{++}$ [14]. Meanwhile, related work [10] presents that CQ answering on regular $\mathcal{EL}^{++}$ is decidable. However, we observed that the notion of regularity syntactically depends on the normal form of TBox. In other words, it can happen that a TBox becomes (ir)regular by normalizing it differently, even if this does not change the semantics. For example, suppose $R_1 \circ R_2 \circ R_3 \sqsubseteq R_3$ be a role inclusion originally asserted in the TBox. On the one hand, this TBox can be normalized as $R_1 \circ R_2 \sqsubseteq R'$ and $R' \circ R_3 \sqsubseteq R_3$, which appear as regular. On the other hand, this TBox can also be normalized as $R_2 \circ R_3 \sqsubseteq R''$ and $R_1 \circ R'' \sqsubseteq R_3$, which becomes irregular. Besides, our acyclic $\mathcal{EL}^+$ TBoxes admit the irregularity. Disregarding the rich expressivity of roles, CQ answering on $\mathcal{EL}$ and $\mathcal{ELH}$ has been proved decidable and PTIME-complete w.r.t. both data complexity and KB complexity [14].

As far as we know, the only practical approach for conjunctive query answering in $\mathcal{EL}$ has been proposed by Lutz et al. [11,12]. Our work can be regarded as an extension of their work by supporting role inclusions. A minor difference is that, in the ABox completion phase, we only generate one anonymous individual for each existential restriction, whereas their approach will generate one anonymous individual (called auxiliary object) for each subconcept of concepts used in $\mathcal{T}$. A major difference arises in the query rewriting phase, since our approach supports role inclusions. We firstly proposed a *base path criterion* and designed a base path storage schema to enable query rewriting based on this criterion. The query rewriting strategy in their approach is a special case of ours. Actually, as shown in the Section 4.5 for optimization, we can embed their query rewriting strategy into ours to handle the $\mathcal{ELH}$ part, while keeping the base path criterion dedicated for the "real" role inclusions in $\mathcal{EL}^+$.

Meanwhile, we can imagine a deterministic algorithm for query answering in $\mathcal{ELH}_{\mathcal{R}^+}$, which is a sub language of $\mathcal{SHIQ}$ [8]. Alternatively, DL-Lite [5] explores such a way that resorts simply to query rewriting without ABox completion. To employ either of them has to pay for an exponential blowup of the rewritten queries. Towards a practical implementation, query rewriting in our approach will only have a polynomial blowup, and we leverage RDB for query answering.

7 Conclusion

In this paper, we proposed a two-phase practical approach for query answering in $\mathcal{EL}^+$. In phase one, we do ABox completion using the notion of canonical individuals. For queries free of reverse trees, this phase is enough to provide sound and complete answers, w.r.t. arbitrary $\mathcal{EL}^+$ KBs. In phase two, we proposed the base path criterion and made the criterion applicable by the base path storage and query rewriting. A combination of the two phases guarantees sound and complete solutions to arbitrary queries on acyclic $\mathcal{EL}^+$ KBs.

As future work, we are planning to evolve the base path criterion, enabling queries on cyclic but regular $\mathcal{EL}^+$ ontologies. Development of more optimizations is also part of our ongoing work, especially reducing the blowup of base paths. Besides, for updating concerns, we will develop an incremental ABox completion, which is critical for the practical applications of our approach.

Acknowledgements

The authors wish to thank Carsten Lutz for useful discussions during the writing of this paper.

References

1. Baader, F., Brandt, S., Lutz, C.: Pushing the EL Envelope. In: Proc. of the 19th Int. Joint Conf. on Artificial Intelligence, pp. 364–369 (2005)
2. Baader, F., Brandt, S., Lutz, C.: Pushing the EL Envelope Further. In: Proc. of the Workshop on OWL: Experiences and Directions (2008)
3. Baader, F., Lutz, C., Suntisrivaraporn, B.: Efficient reasoning in $\mathcal{EL}^+$. In: Proc. of the Workshop on Description Logics (2006)
4. Baader, F., Nutt, W.: The Description Logic Handbook. In: Basic Description Logics, ch. 2. Cambridge University Press, Cambridge (2003)
5. Calvanese, D., De Giacomo, G., Lembo, D., Lenzerini, M., Rosati, R.: DL-Lite: Tractable Description Logics for Ontologies. In: Proc. of the 20th Nat. Conf. on Artificial Intelligence, pp. 602–607 (2005)
6. The Gene Ontology Consortium. Gene Ontology: Tool for the Unification of Biology. Journal of Nature Genetics 25, 25–29 (2000)
7. Dolin, R.H., Alschuler, L., Boyer, S., Beebe, C., Behlen, F.M., Biron, P.V., Shabo, A.: HL7 Clinical Document Architecture, Release 2.0. Journal of American Medical Informatics Association 13(1), 30–39 (2006)
8. Glimm, B., Horrocks, I., Lutz, C., Sattler, U.: Conjunctive Query Answering for the Description Logic $\mathcal{SHIQ}$. Journal of Artificial Intelligence Research 31, 157–204 (2008)
9. IHTSDO. Systematized Nomenclature of Medicine C Clinical Terms, `http://www.ihtsdo.org/snomed-ct/`
10. Krotzsch, M., Rudolph, S., Hitzler, P.: Conjunctive Queries for a Tractable Fragment of OWL 1.1. In: Proc. of Int. Semantic Web Conf. (2007)
11. Lutz, C., Toman, D., Wolter, F.: Conjunctive Query Answering in $\mathcal{EL}$ using a Database System. In: Proc. of the Workshop on OWL: Experiences and Directions (2008)
12. Lutz, C., Toman, D., Wolter, F.: Conjunctive Query Answering in $\mathcal{EL}$ using a Database System. In: Proc. of the 21st Int. Joint Conf. on Artificial Intelligence (2009)
13. Rector, A., Horrocks, I.: Experience building a large, reusable Medical Ontology using a Description Logic with Transitivity and Concept Inclusions. In: Proc. of the Workshop on Ontological Engineering, AAAI Spring Symposium, Menlo Park (1997)
14. Rosati, R.: On Conjunctive Query Answering in $\mathcal{EL}$. In: Proc. of the Workshop on Description Logics (2007)

Learning Semantic Query Suggestions

Edgar Meij[1], Marc Bron[1], Laura Hollink[2], Bouke Huurnink[1], and Maarten de Rijke[1]

[1] ISLA, University of Amsterdam, Science Park 107, 1098 XG Amsterdam
{edgar.meij,m.m.bron,bhuurnink}@uva.nl, mdr@science.uva.nl
[2] Dept. of Computer Science, VU University Amsterdam,
de Boelelaan 1081a, 1081 AH Amsterdam
hollink@cs.vu.nl

Abstract. An important application of semantic web technology is recognizing human-defined concepts in text. Query transformation is a strategy often used in search engines to derive queries that are able to return more useful search results than the original query and most popular search engines provide facilities that let users complete, specify, or reformulate their queries. We study the problem of *semantic query suggestion*, a special type of query transformation based on identifying semantic concepts contained in user queries. We use a feature-based approach in conjunction with supervised machine learning, augmenting term-based features with search history-based and concept-specific features. We apply our method to the task of linking queries from real-world query logs (the transaction logs of the Netherlands Institute for Sound and Vision) to the DBpedia knowledge base. We evaluate the utility of different machine learning algorithms, features, and feature types in identifying semantic concepts using a manually developed test bed and show significant improvements over an already high baseline. The resources developed for this paper, i.e., queries, human assessments, and extracted features, are available for download.

1 Introduction

Human-defined concepts are fundamental building blocks of the semantic web. When used as annotations for documents or text fragments they can provide explicit anchoring in background knowledge and enable intelligent search and browsing facilities. As such, an important application of ontological knowledge is augmenting unstructured text with links to relevant, human-defined concepts. For the author or reader of the text, this augmentation may supply useful pointers, for example to the concepts themselves or to other concepts related to the ones found. For ontology learning applications, such links may be used to learn new concepts or relations between them [34]. Recently, data-driven methods have been proposed to generate links between phrases appearing in full-text documents and a set of ontological concepts known a priori. [24] propose the use of several linguistic features in a machine learning framework to link phrases in full-text documents to Wikipedia articles and this approach is further improved upon by [25]. Because of the connection between Wikipedia and DBpedia, such data-driven linking methods help us establish links between textual documents and Linked Open Data [2, 3, 9, 33].

A. Bernstein et al. (Eds.): ISWC 2009, LNCS 5823, pp. 424–440, 2009.

Another, more challenging instantiation of linking text to human-defined concepts in a knowledge source is *semantic query suggestion*. Query suggestion is a strategy to derive terms that are able to return more relevant results than the initial query. Commonly used approaches to query suggestion (sometimes referred to as a form of query expansion) are highly data-driven and based mostly on term frequencies [21, Chapter 9]. *Semantic* query suggestion, in contrast, tries to understand (or learn) which concepts the user used in her query or, phrased alternatively, the concepts she is interested in and wants to find.[1] Moreover, the properties of each concept, and any other resources associated with it, could serve as additional, useful information for the user. In our current work, we use DBpedia as our target ontology. As an example of our task, consider the query "obama white house". A semantic query suggestion algorithm should return suggestions in the form of the (DBpedia) instances labeled "Barack Obama" and "White House". Identifying such semantic suggestions serves multiple purposes: it can (i) help the user acquire contextual information, (ii) suggest related concepts or associated terms that may be used for search, and (iii) provide valuable navigational suggestions.

In this paper we address the semantic query suggestion task and automatically link queries to DBpedia concepts. We propose a novel method that leverages the textual representation of each concept as well as query-based and concept-based features. Working with user queries, which are typically quite short [31], implies that we cannot use previously established approaches that rely on textual context such as shallow parsing or part-of-speech tagging [24]. One interesting aspect of working with user queries is that we have history-based information available (from the session or the user) that can potentially be used to help address the semantic query suggestion task.

Our approach proceeds as follows. First, we use language modeling for information retrieval (IR) techniques to retrieve the most relevant concepts for the full query and for each n-gram (i.e., contiguous sequence of n words) in the query. Second, we use supervised machine learning methods to decide which of the retrieved concepts should be kept and which should be discarded. In order to train the machine learner, we examined close to 1000 queries from a search engine for a digital multimedia archive and manually linked over 600 of these to relevant concepts in DBpedia.[2] The research questions we address are the following.

- Can we cast semantic query suggestion as a ranking problem?
- What is the best way of handling the terms in the input query?
- Can we use features pertaining to the query terms, concepts, and history to improve upon a basic retrieval based approach?
- Which type of the feature helps most? Which separate feature is most informative?

Our main contribution is threefold: we introduce the problem of semantic query suggestion, provide a very effective approach to the problem plus an analysis of the contributions of the features used, and we make available resources to enable future work on the problem. The remainder of this paper is structured as follows. In Section 2 we

[1] We use "ontology" to refer to the full spectrum of conceptualizations, ranging from glossaries to formal ontologies [22]. We refer to an instance in DBpedia as "concept" [33].

[2] The queries, human assessments, and extracted features are publicly available for download at `http://ilps.science.uva.nl/resources/iswc09_annotations`

discuss related work. Sections 3 and 4 detail the semantic query suggestion task and our approach. Our experimental setup is described in Section 5 and our results are presented in Section 6. We end with a concluding section.

2 Related Work

Linking terms or phrases to ontologies is related to several areas of research. These include semantic web areas such as ontology matching and semantic annotation, but also areas from language technology and information retrieval.

In *ontology matching* relations between concepts from different ontologies are found. These relations are based on a comparison of instances, concept labels, semantic structure, or ontological features such as constraints or properties, sometimes exploiting auxiliary external resources such as the lexical resource WordNet or the upper ontology DOLCE [30]. E.g., [36] develop a machine learning technique to learn the relationship between the similarity of instances and the validity of mappings between concepts. Other approaches are designed for lexical comparison of concept labels in the source and target ontology and do not use semantic structure nor instances (e.g. [32]). This type of matching is sometimes referred to as 'lexical matching' and is used in cases where the ontologies do not have any instances or structure; e.g., in [1] lexical comparison of labels is used to map both the source and the target ontology to a semantically rich external source of background knowledge.

Lexical matching is very similar to our task at hand, as we do not have any semantic structure in the queries. Since 2008, the Ontology Alignment Evaluation Initiative has contained a task similar to ours, where participants link a largely unstructured thesaurus, the GTAA, to DBpedia [10]. Our approach is different, however, in two ways. First, we use the interaction history with the system in the form of user sessions to obtain a certain amount of contextual information. Second, the fact that the source terms that we are trying to link are natural language captured in user queries instead of standardized concept labels makes the task intrinsically harder.

A lot of work has been done on semantic annotation: the process of relating documents to concepts from ontologies or other sources of structured knowledge, such as Wikipedia. Many restricted forms of the problem have been addressed. The detection of named entities in pieces of text is an important sub-problem of semantic annotation. For example, [20] create links between named entities in text and concepts from a light-weight ontology. [8] propose an interesting example of semantic document analysis, where named entities are related over time using Wikipedia. [11] do not restrict themselves to named entities, but instead link all words in a document to ontological concepts. They use latent topic models to learn links between words and concepts. Other sub-problems of semantic annotation include sense tagging and word sense disambiguation [13]. Some of the techniques developed there have fed into automatic link generation between full-text documents and Wikipedia. For example, [25] depend heavily on contextual information (terms and phrases) around the source text to determine the best Wikipedia articles to link to (like typical named entity recognition and sense tagging methods). This work was based on earlier work by [24]. Similarly, we also consider a two-step approach to linking text to Wikipedia/DBpedia; first keyword extraction is performedi, and next, each extracted keyword is linked to a Wikipedia article. The authors apply part-of-speech tagging and develop several ranking procedures for candidate Wikipedia articles. Our approach differs in that we do not limit ourselves to exact matches with the query terms. Another distinct difference between our approach and those mentioned above is that while they work with full-text documents, we utilize much sparser data in the form of user queries. As such we can not easily use techniques such as part-of-speech tagging or lean too heavily on context words for disambiguation. As will be detailed below, our approach uses session history and n-grams in the queries to obtain contextual information instead.

Turning to semantic query analysis (as opposed to semantic analysis of full documents), [14] perform named entity recognition in queries; they recognize a single entity in each query and subsequently classify it into one of a small set of predefined classes such as "movie" or "video game". We do not impose the restriction of having a single concept per query and, furthermore, our list of candidate concepts is much larger, i.e., all articles in Wikipedia. Several other approaches have been proposed that link queries to a limited set of categories. [26] use online product search engines to link queries to product categories; [5] link millions of queries to 17 topical categories based on a list of manually pre-categorized queries; [16] use commonly occurring multimedia terms to categorize audio, video, and image queries.

As to related work on query suggestions, some approaches use query logs to determine which queries or which rewrites occur frequently [18]. Others perform query analysis and try to identify the most relevant terms [6], to predict the query's performance a priori [40], or combine the two [7]. [6] use part-of-speech tagging and a supervised machine learning technique to identify the "key noun phrases" in natural language queries. Key noun phrases are phrases which convey the most information in a query and contribute most to the resulting retrieval performance; we evaluate several of the features proposed by these authors.

3 The Task

The semantic query suggestion task we address in this paper is the following. Given a query that is submitted to a search engine, identify the relevant concepts that the user entered in her query where the concepts are taken from an existing knowledge base or ontology. We address our task in the setting of a digital archive, specifically of the Netherlands Institute for Sound and Vision ("Sound and Vision"). Sound and Vision maintains a large digital audiovisual collection, currently containing over a million objects and daily updated with new broadcasts. Users of the archive consist primarily of media professionals who use an online search interface to locate audiovisual items to be used in new programs such as documentaries and news reviews.

Sound and Vision is in the process of linking all its digital resources to a variety of structured knowledge sources. Because of its central role in the Linked Open Data initiative, our knowledge source of choice for semantic query suggestion is DBpedia. Thus, in practical terms, the task we are facing is: given a query (within a session, for a given user), produce a ranked list of concepts from DBpedia that are mentioned or meant in the query. These concepts could then be used to suggest relevant multimedia items associated with each concept or to suggest contextual information, such as text snippets from the Wikipedia article.

4 Approach

Our approach to suggesting DBpedia concepts for user queries consists of two stages. In the first stage, a ranked list of possible concepts for the query is generated using a language modeling framework (cf. Section 4.1). To create this ranking we consider two approaches; one approach is to extract all n-grams in the query and generate a ranking for each. The other approach is to use the entire query to create a single concept ranking. We use various textual representations of each DBpedia concept, including the accompanying Wikipedia article text, its label, and the text used in the hyperlinks pointing to it. An example of the first stage of ranking concepts is provided in Table 1 for the query "obama white house". From the example it is clear why we consider these two approaches. Should we simply use the full query on its own (first row), we would miss the relevant concept "Barack Obama". However, as can be seen from the last two rows, considering all n-grams also introduces noise.

Table 1. An example of generating n-grams for the query "obama white house" and retrieved candidate concepts, ranked by retrieval score. Correct concepts in boldface.

N-gram (Q)	Candidate concepts
obama white house	**White House**; White House Station; President Coolidge; Sensation White
obama white	Michelle Obama; **Barack Obama**; Democratic Pre-elections 2008; January 17
white house	**White House**; White House Station; Sensation; President Coolidge
obama	**Barack Obama**; Michelle Obama; Presidential Elections 2008; Hillary Clinton
white	Colonel White; Edward White; White County; White Plains Road Line
house	House; Royal Opera House; Sydney Opera House; Full House

In the second stage supervised machine learning is used to decide which of the candidate concepts in the ranked lists should be kept as viable concepts for the query in question. In order to train these machine learning algorithms, we asked annotators to assess queries submitted to Sound and Vision and manually link them to relevant DBpedia concepts. More details with respect to the test collection and the manual annotations are listed in Section 5. The machine learning algorithms we consider are Naive Bayes, Decision Trees, and Support Vector Machines [35, 37] and are introduced in Section 4.2. As input to the machine learning algorithms we extract a set of features related to the query n-gram, concept, and the session in which the query appears as detailed in Section 4.3.

4.1 Ranking Concepts

We base our concept ranking framework within the language modeling paradigm, as it is a theoretically transparent retrieval approach that is competitive in terms of retrieval effectiveness [15, 28, 39]. Here, a query is viewed as having been generated from an underlying document language model, where some words are more probable to occur than others. At retrieval time each document is scored according to the estimated likelihood that the words in the query were generated by a random sample of the language model underlying the document. These word probabilities are generally estimated from the document itself (using maximum likelihood estimation) and combined with background collection statistics to overcome zero probability and data sparsity issues; a process known as *smoothing*.

For the n-gram based re-ranking, we extract all n-grams from each query $\mathbf{Q}$ (where $1 \leq n \leq |\mathbf{Q}|$) and create a ranked list of concepts for each individual n-gram, Q. For the full-query based reranking approach, we use the same method but add the additional constraint that $n = |\mathbf{Q}|$. The problem of ranking DBpedia concepts given an n-gram can then be formulated as follows. Each concept c should be ranked according to the probability that it was generated by the n-gram $P(c|Q)$, which can be rewritten using Bayes' rule as:

$$P(c|Q) = \frac{P(Q|c)P(c)}{P(Q)}.$$

Here, the term $P(Q)$ is the same for all concepts and be ignored for ranking purposes. The term $P(c)$ indicates the prior probability of selecting a concept, which we assume to be uniform. Assuming independence between the individual terms $q \in Q$, as is common in the field of information retrieval [4], we obtain

$$P(c|Q) \propto P(c) \prod_{q \in Q} P(q|c)^{n(q,Q)}, \tag{1}$$

where $n(q, Q)$ indicates the count of term q in n-gram Q. The probability $P(q|c)$ is smoothed using Bayes smoothing with a Dirichlet prior [39], which is formulated as:

$$P(q|c) = \frac{n(q,c) + \mu P(q)}{\sum_q n(q,c) + \mu}, \tag{2}$$

where $P(q)$ indicates the probability of observing q in a large background collection and μ is a hyperparameter that controls the influence of the background corpus.

Since each DBpedia concept is linked to its Wikipedia counterpart, we can use the textual representations of the associated wikipedia pages for retrieval. In particular, we perform retrieval using the title of the article, the content, and the text used for the hyperlinks pointing to it from other Wikipedia articles.

4.2 Learning to Rerank Concepts

Once we have obtained a ranked list of possible concepts for each n-gram in the query, we turn to concept selection. In this stage we need to decide which of the candidate concepts are most viable. We use a supervised machine learning approach, which takes as input a set of labeled examples (query to concept mappings) and several features of these examples (detailed below). We choose to compare a Naive Bayes (NB) classifier,

Table 2. Features used, grouped by type

N-gram features	
$LEN(Q) = \|Q\|$	Number of terms in the phrase Q
$IDF(Q)$	Inverse document frequency of Q
$WIG(Q)$	Weighted information gain using top-5 retrieved concepts
$QE(Q)$	Number of times Q appeared as *whole* query in query log
$QP(Q)$	Number of times Q appeared as *partial* query in query log
$QEQP(Q)$	Ratio between QE and QP
$SNIL(Q)$	Does a sub-n-gram of Q fully match with any concept label?
$SNCL(Q)$	Is a sub-n-gram of Q contained in any concept label?
Concept features	
$INLINKS(c)$	The number of concepts linking to c
$OUTLINKS(c)$	The number of concepts linking from c
$GEN(c)$	Function of depth of c in SKOS category hierarchy [25]
$CAT(c)$	Number of associated categories
$REDIRECT(c)$	Number of redirect pages linking to c
N-gram + concept features	
$TF(c, Q) = \frac{n(Q, c)}{\|c\|}$	Relative phrase frequency of Q in c, normalized by length of c
$TF_f(c, Q) = \frac{n(Q, c, f)}{\|f\|}$	Relative phrase frequency of Q in representation f of c, normalized by length of f
$POS_n(c, Q) = pos_n(Q)/\|c\|$	Position of nth occurrence of Q in c, normalized by length of c
$SPR(c, Q)$	Spread (distance between the last and first occurrences of Q in c)
$TF \cdot IDF(c, Q)$	The importance of Q for c
$RIDF(c, Q)$	Residual IDF (difference between expected and observed IDF)
$\chi^2(c, Q)$	χ^2 test of independence between Q in c and in collection $Coll$
$QCT(c, Q)$	Does q contain the label of c?
$TCQ(c, Q)$	Does label of c contain q?
$TEQ(c, Q)$	Does label of c equal q?
$SCORE(c, Q)$	Retrieval score of c w.r.t Q
$RANK(c, Q)$	Retrieval rank of c w.r.t Q
History features	
$CCIH(c)$	Number of occurrences of label of c appears as query in history
$CCCH(c)$	Number of occurrences of label of c appears in any query in history
$CIHH(c)$	Number of times c is retrieved as result for any query in history
$CCIHH(c)$	Number of times label of c equals title of any result for any query in history
$CCCHH(c)$	Number of times title of any result for any query in history contains label of c
$QCIHH(Q)$	Number of times title of any result for any query in history equals Q
$QCCHH(Q)$	Number of times title of any result for any query in history contains Q
$QCIH(Q)$	Number of times Q appears as query in history
$QCCH(Q)$	Number of times Q appears in any query in history

with a Support Vector Machine (SVM) classifier and a decision tree classifier (J48)—a set representative of the state-of-the-art in classification. We experiment with multiple classifiers in order to confirm that our results are generally valid, i.e., not dependent on any machine learning algorithm.

4.3 Features Used

We employ several *types* of features, each associated with either the current query n-gram, the current concept, their combination, or the current search history. Unless indicated otherwise, we consider Q to be a phrase when determining the features.

N-gram Features. These features are based on information from an n-gram and are listed in Table 2 (first group). $IDF(Q)$ indicates the relative number of concepts in which Q occurs, which is defined as $IDF(Q) = \log(|Coll|/df(Q))$, where $|Coll|$ indicates the number of documents in the collection and $df(Q)$ the number of documents in which Q occurs [4]. $WIG(Q)$ indicates the weighted information gain, that was proposed by [40] as a predictor of the retrieval performance of Q. It is the only feature which uses the set of all candidate concepts retrieved for this n-gram, C_Q, and determines the relative probability of Q occurring in these documents as compared to the collection. Formally:

$$WIG(Q) = \frac{\frac{1}{|C_Q|}\sum_{c \in C_Q} \log(P(Q|c)) - \log(P(Q))}{\log P(Q)}.$$

$QE(Q)$ and $QP(Q)$ indicate the number of times this n-gram appears in the entire query logs as a whole or partial query respectively.

Concept Features. Table 2 (second group) lists the features related to a DBpedia concept. This set of features is related to the knowledge we have of the current candidate concept, such as the number of other concepts linking to or from it, the number of associated categories (the count of the DBpedia property `skos:subject`), and the number of redirect pages pointing to it (the DBpedia property `dbpprop:redirect`).

N-gram + Concept Features. This set of features considers an n-gram and a concept (Table 2, third group). We consider the relative frequency of occurrence of the n-gram as a phrase in the corresponding Wikipedia article, in the separate document representations (title, content, anchor texts, first sentence, and first paragraph of the Wikipedia article), the position of the first occurrence of the n-gram, the distance between the first and last occurrence, and various IR-based measures. Of these, $RIDF$ [12] is the difference between expected and observed IDF for a concept, which is defined as $RIDF(c, Q) = \log(|Coll|/df(Q)) + \log(1 - \exp(-n(Q, Coll)/|Coll|))$. We also consider whether the label of the concept, i.e. the Wikipedia article title, matches Q and we include the current retrieval information.

History Features. Finally, we consider features based on the previous queries that were issued in the same session (Table 2, fourth group). These features look at either the current candidate concept or current n-gram and see whether they occur (partially) in the previous queries or in the retrieved candidate concepts.

Below, we compare the effectiveness of the features listed above for our semantic query suggestion task.

5 Experimental Setup

We introduce the experimental environment and the experiments that we perform to answer the research questions from Section 1. We also introduce our evaluation measures and describe our manual assessments, but start with detailing our data sets.

5.1 Data

Two main types of data are needed for our experiments: queries and DBpedia concepts. We have access to a set of 264,503 queries issued between 18 November 2008 to 15 May 2009 to Sound and Vision. Sound and Vision logs the actions of users on the site, generating session identifiers and time stamps. This allows a series of consecutive queries to be linked to a single search session, where a session is identified using a session cookie. A session is terminated once the user closes the browser. An example is given in Table 3. All queries were Dutch language queries; however, nothing in our semantic query suggestion approach is language dependent. As the "history" of a query, we took all queries previously issued in the same user session.

Table 3. An example of queries issued in a (partial) session, translated to English

Session ID	Query ID	Query (**Q**)
jyq4navmztg	715681456	santa claus canada
jyq4navmztg	715681569	santa claus emigrants
jyq4navmztg	715681598	santa claus australia
jyq4navmztg	715681633	christmas sun
jyq4navmztg	715681789	christmas australia
jyq4navmztg	715681896	christmas new zealand
jyq4navmztg	715681952	christmas overseas

The DBpedia version we use is the most recent Dutch version (3.2). We downloaded the Wikipedia dump from which this DBpedia version was created (dump date 20080609); this dump is used for all our text-based processing steps and features.

5.2 Training Data

For training and testing purposes, four assessors were asked to manually link 998 queries to DBpedia concepts. The assessors were presented with a list of sessions and the queries in them. Once a session had been selected, they were asked to find the most relevant DBpedia concepts given each query and its preceding session history if any. Our assessors were able to search through Wikipedia using the fields described in Section 4.1. Besides indicating relevant concepts, the assessors could also indicate whether a query was ambiguous, contained a typographical error, or whether they were unable to find any relevant concept. For our experiments, we removed all the assessed queries in these "anomalous" categories and were left with a total of 629 assessed queries in 193 sessions.[3] In this sample the average query length is 2.14 terms per query. Each query has 1.34 concepts annotated on average.

5.3 Parameters

As to retrieval, we use the entire Wikipedia document collection as background corpus and set μ to the average length of a Wikipedia article [39], i.e., $\mu = 315$ (cf. Eq. 2). For classification we use the following three machine learning algorithms in our experiments: J48, Naive Bayes and Support Vector Machines. The implementations are taken from the Weka machine learning toolkit [37]. J48 is a decision tree algorithm and the Weka implementation of C4.5 [29]. The Naive Bayes classifier uses the training data to estimate the probability that an instance belongs to the target class, given the presence of each feature. By assuming independence between the features these probabilities can be combined to calculate the probability of the target class given all features [19]. SVM uses a sequential minimal optimization algorithm for training with polynomial kernels as described in [27]. The training instances are represented as n-dimensional vectors and two decision hyperplanes are created that best separate the instances of the target classes. The distance between the hyperplanes is called the margin and by maximizing this distance the generalization error of the classifier is minimized. For all algorithms we do not perform extensive optimization of the parameter settings and use the default weka parameters.[4] Whether fine-grained parameter tuning is beneficial and, thus, whether our choice negatively influences the experimental outcomes is a topic for future work.

5.4 Testing and Evaluation

We define semantic query suggestion as a ranking problem. In this paper, the system has to return five concepts for a given input query; the assessments described above are used to determine the relevance of these five concepts. We employ several measures which are well-known in the field of information retrieval [4], namely precision@1 (P1; how many relevant concepts are retrieved at rank 1), r-precision (R-prec; precision@r where r equals the size of the set of known relevant concepts for this query), recall (which percentage of relevant concepts were retrieved?), mean reciprocal rank (MRR; the reciprocal of the rank of the first correct concept), and the success rate @5 (SR; a binary measure that indicates whether at least one correct concept has been returned in the top-5).

To verify the generality of the machine learning algorithms, we perform 10-fold cross validation [37], which reduces the possibility of errors being caused by artifacts in the data. The reported scores are averaged over all folds and all evaluation measures are averaged over the queries used for testing. For determining the statistical significance

[3] We focus on evaluating the actual semantic query suggestions and discard queries which the assessors deemed too anomalous to confidently link to any concept.

[4] See `http://weka.sourceforge.net/doc`

of the observed differences between the various runs we use one-way ANOVA to determine if there is a significant difference ($\alpha \leq 0.05$). We then use the Tukey-Kramer test to determine which of the individual pairs are significantly different. We designate the best result in each table in bold face.

6 Results

In this section we report on the experimental results which answer the research questions from Section 1. We compare three approaches to the semantic query suggestion task:

(i) a *baseline* which retrieves concepts based solely on their textual representation in the form of the associated Wikipedia article,
(ii) *n-gram based reranking* which extracts all n-grams from the query and uses machine learning to identify the best concepts, and
(iii) *full-query based reranking* which does not extract n-grams, but calculates feature values based on the full query.

After we have described the results for each approach, we zoom in on the most informative features and the specific feature types.

6.1 Baseline

As our baseline, we take the entire query as issued by the user and employ Eq. 1 to rank DBpedia concepts based solely on their textual representation (this techniques is similar to using a search engine to perform a search within the Dutch Wikipedia). We consider two forms of textual representation: "content" and "full text". The former consists of the textual content of a Wikipedia article corresponding to a DBpedia concept, the latter adds to this the title of the article and the anchor texts of hypertext links in Wikipedia that point to the article at hand.

Table 4. Results of ranking concepts based on using the entire query **Q** and either the content of the Wikipedia article or the full text associated with each DBpedia concept

	P1	R-prec	Recall	MRR	SR
full text	**0.5636**	**0.5216**	**0.6768**	**0.6400**	**0.7535**
content	0.5510	0.5134	0.6632	0.6252	0.7363

Table 4 shows the results of the baseline. From this table we observe that including the title and anchor texts of the incoming links results in improved retrieval performance overall. Notice that this is a strong baseline; on average, over 65% of the relevant concepts are correctly identified in the top-5 and, further, over 55% of the relevant concepts are retrieved at rank 1. The success rate indicates that for 75% of the queries at least one relevant concept is retrieved in the top-5.

6.2 N-gram Based Reranking

Table 5 shows the result for the baseline (last row) and the query "challenger wubbo ockels". As is clear from this example, the two relevant concepts are retrieved at ranks 1 and 4. When we look at the same results for the n-grams in the query, however, one of the relevant concepts is retrieved at the first position for each n-gram. This example and the one in Table 1 suggest that it will be beneficial to consider all possible n-grams in the query. In this section we report on the results of extracting n-grams from the query, generating features for each, and subsequently applying the machine learning algorithms to decide which of the suggested concepts to keep. The features used here are described in Section 4.2.

Table 5. An example of baseline results for the n-grams in the query "challenger wubbo ockels", ranked by retrieval score. Concepts labeled as correct in boldface.

N-gram	Candidate concepts
challenger	**Space Shuttle Challenger**; Challenger; Bombardier Challenger; STS-61-A
wubbo	**Wubbo Ockels**; Spacelab; Canon of Groningen; Superbus; André Kuipers
ockels	**Wubbo Ockels**; Spacelab; Superbus; Canon of Groningen; André Kuipers
challenger wubbo	**Wubbo Ockels**; STS-61-A; **Space Shuttle Challenger**; Spacelab; STS-9
wubbo ockels	**Wubbo Ockels**; Spacelab; Superbus; Canon of Groningen; Andr Kuipers
challenger wubbo ockels	**Wubbo Ockels**; STS-61-A; Spacelab; **Space Shuttle Challenger**

Table 6. Results for n-gram based reranking. $^{\blacktriangle}$ $^{\blacktriangledown}$ and $^{\circ}$ indicate that a score is significantly better, worse or statistically indistinguishable respectively. The leftmost symbol represents the difference with the baseline, the next with the J48 run, and the rightmost with the NB run.

Machine learner	P1	R-prec	Recall	MRR	SR
baseline	0.5636	0.5216	0.6768	0.6400	0.7535
J48	0.6510$^{\blacktriangle}$	0.5604°	0.7245°	0.7441$^{\blacktriangle}$	0.7958°
NB	0.4665$^{\blacktriangledown\circ}$	0.4344$^{\blacktriangledown\blacktriangledown}$	0.6984$^{\circ\circ}$	0.7100$^{\circ\circ}$	0.7614$^{\circ\circ}$
SVM	**0.8388**$^{\blacktriangle\blacktriangle\blacktriangle}$	**0.7170**$^{\circ\circ\blacktriangle}$	**0.7852**$^{\circ\circ\circ}$	**0.8500**$^{\blacktriangle\circ\blacktriangle}$	**0.8548**$^{\circ\circ\circ}$

Table 6 shows the results of applying the machine learning algorithms on the extracted n-gram features. We note that J48 and SVM are able to improve upon the baseline results from the previous section, according to all metrics. The Naive Bayes classifier performs worse than the baseline in terms of P1 and R-precision. SVM clearly outperforms the other algorithms and is able to obtain very high scores; significantly better than the baseline on all metrics. Interestingly, we see that the use of n-gram based reranking has both a precision enhancing effect for J48 and SVM (the P1 and MRR scores go up) and a recall-enhancing effect.

In Section 4.3 we identified several groups of features, relating to the n-gram, concept, their combination, or the session history. We will now zoom in on the performance of these groups. Table 7 shows the results when several of these groups are removed

Table 7. Results of removing specific feature types from the training data for the SVM classifier and n-gram based reranking. ▼ and ° indicate that a score is significantly worse or statistically indistinguishable. The leftmost symbol represents the difference with the all features run, the next with the without history features run, and the rightmost symbol the without concept features run.

	P1	R-prec	Recall	MRR	SR
All features	**0.8388**	**0.7170**	**0.7852**	**0.8500**	**0.8548**
Without history	0.7867 °	0.4687▼	0.6272°	0.8009°	0.8041°
Without concept	0.5826▼°	0.3282▼°	0.4554▼°	0.5826°°	0.5826▼°
Without history and concept	0.1929▼▼▼	0.1429▼▼°	0.1679▼▼▼	0.1929▼▼▼	0.1929▼▼▼

from the training data for the SVM classifier. It turns out that both the n-gram specific and n-gram + concept specific features are needed for classification. When these groups are removed, none of the relevant concepts are identified. From Table 7 we observe that removing the history features results in a drop in performance, albeit a small one. When the concept features are removed, the resulting performance drops even further and their combined removal yields very low scores. Classification without the concept based features results in performance that is statistically indistinguishable from the baseline.

Table 8. Results for full query-based reranking. ▲ ▼ and ° indicate that a score is significantly better, worse or statistically indistinguishable respectively. The leftmost symbol represents the difference with the baseline, the next with the J48 run, and the rightmost with the NB run.

Machine learner	P1	R-prec	Recall	MRR	SR
baseline	0.5636	0.5216	0.6768	0.6400	0.7535
J48	0.7055▲	0.5907°	0.6664°	0.6768°	0.7314°
NB	0.7110▲°	0.6004°°	0.7173°°	0.7121°°	0.7889°°
SVM	**0.8908▲▲▲**	**0.8604▲▲▲**	**0.8890▲▲▲**	**0.8173▲°▲**	**0.8963▲▲▲**

Next we turn to a comparison of n-gram based reranking and full query reranking. Table 8 shows the results when we omit the n-grams and only the full query is used to generate features. We again observe that SVM significantly outperforms J48, NB, as well as the baseline. We further note that these scores are the highest obtained so far and this approach is able to return almost 90% of all relevant concepts. This result is very encouraging and shows that the approach taken handles semantic query suggestions extremely well.

6.3 Feature Selection

Several methods exist with which to automatically determine the most informative features given training instances and their class labels (in our case the class label indicates whether the current concept is selected by the assessors). In this section we report on using the information gain based algorithm for feature selection [38].

Table 9. Results of calculating the information gain with respect to the class label for all features (truncated after 7 features). The higher this score, the more informative a feature is.

N-grams		Full-queries	
0.119	$RANK(c, Q)$	0.190	$RANK(c, Q)$
0.107	$DOCID$	0.108	$TEQ(c, Q)$
0.052	$INLINKS(c)$	0.080	$INLINKS(c)$
0.040	$TF_{anchor}(c, Q)$	0.056	$DOCID$
0.038	$OUTLINKS(c)$	0.041	$OUTLINKS(c)$
0.037	$TF_{title}(c, Q)$	0.033	$SCORE(c, Q)$
0.031	$TEQ(c, Q)$	0.025	$REDIRECT(c)$

Table 9 shows the features with the highest information gain scores for both n-gram and full-query based reranking. From this table we observe that the rank at which the retrieval framework puts a concept with respect to an n-gram is most informative. Also the number of in- and outlinks, and whether the n-gram matches the concept's label are informative. $DOCID$ is the internal identifier of each concept and not a feature that we explicitly implemented. However, it turns out that some DBpedia concepts have a higher a priori probability of getting selected. Indeed, in the assessments there were a total of 854 concepts identified, of which 505 are unique. Some of these repetitions are caused because of a coherent information need in the user sessions; when a user rewrites her query by adding or changing part of the query, the remaining concepts remain the same and were annotated as such. As to n-gram based reranking, the term frequency in the title and anchor texts are strong indicators of relevance for given phrase and concept.

7 Conclusion and Future Work

We have introduced the task of semantic query suggestion and presented a method that uses supervised machine learning methods to learn which concepts are used in a query. The concepts are obtained from an ontology and may be used to provide search or navigation suggestions to the user, or as an entry point into the Linked Open Data cloud. Our method extracts query, document, and history specific features from manual annotations and learns how to best rank candidate concepts given an input query.

Our results were obtained using the Dutch version of DBpedia and queries from a log of the Netherlands Institute for Sound and Vision. Although these resources are in Dutch, the framework we have presented is language-independent. Moreover, the approach is also generic in that several of the employed features could be used with ontologies other than DBpedia. However, as became clear from Table 7 and 9, DBpedia related features such as inlinks and redirects were especially helpful. Using Support Vector Machines and features extracted from the full input queries yields optimal results. The best performing run is able to locate almost 90% of the relevant concepts on average. Moreover, this particular run achieves a precision@1 of 89% which means that for this percentage of queries a relevant concept is returned as the first suggestion.[5]

In sum, we have shown that the semantic query suggestion problem can be successfully cast as a ranking problem. The best way of handling query terms is not as separate n-grams, but as a single unit—a finding also interesting from an efficiency viewpoint, since the number of n-grams is quadratic with respect to the length of the query. All types of feature were found to be helpful and, besides document and term features, we found that concept features were also important in achieving our best performance.

As to future work, we will look into expanding the list of features, for example by including more structural features such as ones pertaining to the structure of the ontology. Another question that should be answered is how much training data is needed in order to arrive at a reasonable level of performance. We also intend to go beyond suggesting concepts and look at which part of the query should be linked. Finally, we believe that there might be room for further improvement by using session history in other ways. One option would be a more fine-grained notion of session changes, for example using query overlap [17], or a wider one which considers user history over multiple sessions. Finally, our current approach is trained to find matches between (parts of) the user query and DBpedia concepts, comparable to finding `skos:exactMatch` or even `owl:equivalentClass` relations in an ontology matching task. However, semantic query suggestion can also be interpreted in a broader sense, where not only exact matches but also semantically related concepts are suggested [23]. We believe that our approach can be easily adapted to incorporate such semantically related suggestions.

Acknowledgments

We thank the anonymous reviewers for their constructive comments. This research was carried out in the context of the Virtual Laboratory for e-Science project and supported by the DuOMAn project carried out within the STEVIN programme which is funded by the Dutch and Flemish Governments under project number STE-09-12 and the Netherlands Organisation for Scientific Research (NWO) under project numbers 017.001.190, 640.001.501, 640.002.501, 612.066.512, 612.061.814, 612.061.815, 640.004.802.

References

[1] Aleksovski, Z., Klein, M., ten Kate, W., van Harmelen, F.: Matching unstructured vocabularies using a background ontology. In: Staab, S., Svátek, V. (eds.) EKAW 2006. LNCS (LNAI), vol. 4248, pp. 182–197. Springer, Heidelberg (2006)

[2] Auer, S., Bizer, C., Kobilarov, G., Lehmann, J., Cyganiak, R., Ives, Z.G.: Dbpedia: A nucleus for a web of open data. In: Aberer, K., Choi, K.-S., Noy, N., Allemang, D., Lee, K.-I., Nixon, L.J.B., Golbeck, J., Mika, P., Maynard, D., Mizoguchi, R., Schreiber, G., Cudré-Mauroux, P. (eds.) ASWC 2007 and ISWC 2007. LNCS, vol. 4825, pp. 722–735. Springer, Heidelberg (2007)

[5] Our high numbers can be partially explained by the fact that we have decided to focus on the quality of the suggested concepts and as such removed “anomalous” queries from the evaluation, i.e., queries with typos or that were too ambiguous for human assessors to be able to assign a concept to. The influence of this selection on the end-to-end results remains a topic for future work.

[3] Auer, S., Lehmann, J.: What have Innsbruck and Leipzig in common? Extracting semantics from wiki content. In: The Semantic Web: Research and Applications (2007)

[4] Baeza-Yates, R., Ribeiro-Neto, B.: Modern Information Retrieval. Addison Wesley, Reading (1999)

[5] Beitzel, S.M., Jensen, E.C., Chowdhury, A., Grossman, D., Frieder, O.: Hourly analysis of a very large topically categorized web query log. In: SIGIR 2004 (2004)

[6] Bendersky, M., Croft, W.B.: Discovering key concepts in verbose queries. In: SIGIR 2008 (2008)

[7] Bendersky, M., Croft, W.B.: Analysis of long queries in a large scale search log. In: WSCD 2009 (2009)

[8] Bhole, A., Fortuna, B., Grobelnik, M., Mladenic, D.: Extracting named entities and relating them over time based on wikipedia. Informatica 4(4), 463–468 (2007)

[9] Bizer, C., Cyganiak, R., Auer, S., Kobilarov, G.: DBpedia–querying Wikipedia like a database. In: WWW 2007 (2007)

[10] Caracciolo, C., Euzenat, J., Hollink, L., Ichise, R., Isaac, A., Malaisé, V., Meilicke, C., Pane, J., Shvaiko, P., Stuckenschmidt, H., Šváb, O., Svátek, V.: Results of the OAEI 2008. In: The Third International Workshop on Ontology Matching at ISWC (2008)

[11] Chemudugunta, C., Holloway, A., Smyth, P., Steyvers, M.: Modeling documents by combining semantic concepts with unsupervised statistical learning. In: Sheth, A.P., Staab, S., Dean, M., Paolucci, M., Maynard, D., Finin, T., Thirunarayan, K. (eds.) ISWC 2008. LNCS, vol. 5318, pp. 229–244. Springer, Heidelberg (2008)

[12] Church, K.W., Gale, W.A.: Inverse document frequency (IDF): A measure of deviations from poisson. In: Proc. Third Workshop on Very Large Corpora (1995)

[13] Fellbaum, C., Palmer, M., Dang, H.T., Delfs, L., Wolf, S.: Manual and automatic semantic annotation with wordnet. WordNet and Other Lexical Resources (2001)

[14] Guo, J., Xu, G., Cheng, X., Li, H.: Named entity recognition in query. In: SIGIR 2009 (2009)

[15] Hiemstra, D.: Using Language Models for Information Retrieval. PhD thesis, University of Twente (2001)

[16] Jansen, B.J., Goodrum, A., Spink, A.: Searching for multimedia: analysis of audio, video and image web queries. World Wide Web 3(4), 249–254 (2000)

[17] Jansen, B.J., Spink, A., Blakely, C., Koshman, S.: Defining a session on web search engines. J. Am. Soc. Inf. Sci. Technol. 58(6), 862–871 (2007)

[18] Jansen, B.J., Spink, A., Saracevic, T.: Real life, real users, and real needs: a study and analysis of user queries on the web. Information Processing and Management 36(2), 207–227 (2000)

[19] John, G.H., Langley, P.: Estimating continuous distributions in bayesian classifiers. In: UAI 1995 (1995)

[20] Kiryakov, A., Popov, B., Terziev, I., Manov, D., Ognyanoff, D.: Semantic annotation, indexing, and retrieval. J. Web Sem. 2(1), 49–79 (2004)

[21] Manning, C.D., Raghavan, P., Schütze, H.: Introduction to Information Retrieval. Cambridge University Press, Cambridge (2008)

[22] Mcguinness, D.L.: Ontologies come of age. In: Fensel, D., Hendler, J., Lieberman, H., Wahlster, W. (eds.) Spinning the Semantic Web: Bringing the World Wide Web to Its Full Potential. MIT Press, Cambridge (2003)

[23] Meij, E., Mika, P., Zaragoza, H.: Investigating the demand side of semantic search through query log analysis. In: SemSearch 2009 (2009)

[24] Mihalcea, R., Csomai, A.: Wikify!: Linking documents to encyclopedic knowledge. In: CIKM 2007 (2007)

[25] Milne, D., Witten, I.H.: Learning to link with wikipedia. In:CIKM 2008 (2008)

[26] Mishne, G., de Rijke, M.: A study of blog search. In: Lalmas, M., MacFarlane, A., Rüger, S.M., Tombros, A., Tsikrika, T., Yavlinsky, A. (eds.) ECIR 2006. LNCS, vol. 3936, pp. 289–301. Springer, Heidelberg (2006)

[27] Platt, J.C.: Fast training of support vector machines using sequential minimal optimization. In: Advances in kernel methods: support vector learning. MIT Press, Cambridge (1999)

[28] Ponte, J.M., Croft, W.B.: A language modeling approach to information retrieval. In: SIGIR 1998 (1998)

[29] Quinlan, R.J.: C4.5: Programs for Machine Learning. Morgan Kaufmann, San Francisco (1993)

[30] Shvaiko, P., Euzenat, J.: A survey of schema-based matching approaches. Journal on Data Semantics 4(3730), 146–171 (2005)

[31] Spink, A., Jansen, B.J., Wolfram, D., Saracevic, T.: From e-sex to e-commerce: Web search changes. IEEE Computer 35(3), 107–109 (2002)

[32] Stoilos, G., Stamou, G., Kollias, S.D.: A string metric for ontology alignment. In: Gil, Y., Motta, E., Benjamins, V.R., Musen, M.A. (eds.) ISWC 2005. LNCS, vol. 3729, pp. 624–637. Springer, Heidelberg (2005)

[33] Suchanek, F.M., Kasneci, G., Weikum, G.: Yago: a core of semantic knowledge. In: WWW 2007 (2007)

[34] van Hage, W.R., de Rijke, M., Marx, M.: Information retrieval support for ontology construction and use. In: McIlraith, S.A., Plexousakis, D., van Harmelen, F. (eds.) ISWC 2004. LNCS, vol. 3298, pp. 518–533. Springer, Heidelberg (2004)

[35] Vapnik, V.N.: The nature of statistical learning theory. Springer, Heidelberg (1995)

[36] Wang, S., Englebienne, G., Schlobach, S.: Learning concept mappings from instance similarity. In: Sheth, A.P., Staab, S., Dean, M., Paolucci, M., Maynard, D., Finin, T., Thirunarayan, K. (eds.) ISWC 2008. LNCS, vol. 5318, pp. 339–355. Springer, Heidelberg (2008)

[37] Witten, I.H., Frank, E.: Data Mining: Practical Machine Learning Tools and Techniques. Morgan Kaufmann, San Francisco (2005)

[38] Yang, Y., Pedersen, J.O.: A comparative study on feature selection in text categorization. In: ICML 1997 (1997)

[39] Zhai, C., Lafferty, J.: A study of smoothing methods for language models applied to information retrieval. ACM Trans. Inf. Syst. 22(2), 179–214 (2004)

[40] Zhou, Y., Croft, B.W.: Query performance prediction in web search environments. In: SIGIR 2007 (2007)

Investigating the Semantic Gap through Query Log Analysis

Peter Mika[1], Edgar Meij[2], and Hugo Zaragoza[1]

[1] Yahoo Research
Diagonal 177, 08018 Barcelona, Spain
{pmika,hugoz}@yahoo-inc.com
[2] ISLA, University of Amsterdam
Sciencepark 107, 1098 XG Amsterdam
edgar.meij@uva.nl

Abstract. Significant efforts have focused in the past years on bringing large amounts of metadata online and the success of these efforts can be seen by the impressive number of web sites exposing data in RDFa or RDF/XML. However, little is known about the extent to which this data fits the needs of ordinary web users with everyday information needs. In this paper we study what we perceive as the semantic gap between the supply of data on the Semantic Web and the needs of web users as expressed in the queries submitted to a major Web search engine. We perform our analysis on both the level of instances and ontologies. First, we first look at how much data is actually relevant to Web queries and what kind of data is it. Second, we provide a generic method to extract the attributes that Web users are searching for regarding particular classes of entities. This method allows to contrast class definitions found in Semantic Web vocabularies with the attributes of objects that users are interested in. Our findings are crucial to measuring the potential of semantic search, but also speak to the state of the Semantic Web in general.

1 Introduction

Semantic search is by its broadest definition a collection of approaches that aim at matching the Web's content with the information need of Web users at a semantic level. Most of the work in this area has focused on the *supply-side* of semantic search, in particular elevating Web content to the semantic level by relying on methods of information extraction [4] or working with explicit metadata embedded inside or linked to Web resources. With respect to explicit metadata, several studies have been done on the adoption of Semantic Web formats in the wild, mostly based on statistics from the crawls of Semantic Web search engines [8,7,6,14,10]. Much less effort has focused on the *demand-side* of semantic search, i.e. interpreting queries at the semantic level and studying the information needs of web users in terms of semantic categories. Conversely, little is known as to how much the supply of metadata actually matches the

A. Bernstein et al. (Eds.): ISWC 2009, LNCS 5823, pp. 441–455, 2009.

demand for information from ordinary web users, i.e. how large is the semantic gap between supply and demand on the Semantic Web. This question is central to the success of semantic search, but also to the success of the public Semantic Web in general.

In this paper, we divide our analysis in two parts and provide methods and tools for studying the semantic gap at both the level of instance data and vocabularies[1]. Section 3 covers our analysis of metadata on the Semantic Web. The question we seek to answer is to what extent data on the Semantic Web matches the information needs of the average Web search users as evidenced by search sessions sampled from the query log of a Web search engine. In addition, we look at how information extracted from individual sites with significant influence on the Web could be effective in filling in missing data. Last, we also investigate the particular categories of queries for which there is already metadata on the Web. These questions are pertinent because the success of semantic search hinges on the availability of data that covers user needs.

In Section 4, we address the problem of studying the information need of Web searchers at an ontological level, i.e., in terms of the particular attributes of objects they are interested in. We describe a set of methods for extracting the context words appearing in queries next to the instances of certain classes of objects. We implement these methods in an interactive tool called the Semantic Search Assist. The original purpose of this tool was to generate type-based query suggestions when there is not enough statistical evidence for entity-based query suggestions. However, from an ontology engineering perspective, this tool answers the question of what attributes a class of objects would have if the ontology for it was engineered purely based on the information needs of Web search users. As such it allows us to reflect on the gap between the properties defined in Semantic Web vocabularies and the attributes of objects that people are searching for on the Web. We evaluate our tool by measuring its predictive power on the query log itself.

Our main contribution is thus the usage-based perspective we take on analyzing metadata and vocabularies. We provide a set of methods and their implementation in tools for measuring the Semantic Web from this perspective. The results we provide can be independently validated and we plan to publish some of the detailed analysis in an online form. We conclude and summarize future work in Section 5.

2 Related Work

This work lies at the intersection of two separate streams of research on analyzing Semantic Web data and understanding user queries at a semantic level. In the first area, a number of studies have been done based on the crawls of Semantic Web search engines [8,7,6,14,10], although these studies have focused on

[1] In the following, we will use the term vocabulary instead of the term ontology when we want to put the emphasis on the surface forms of ontological elements. Otherwise we will use the two terms interchangeably.

data quality based on principles such as ontology reuse and interlinking, irrespective of particular applications of the data. These studies also have not touched upon embedded metadata (RDFa or microformat data), which are likely to have different characteristics, especially when it comes to user-generated content.

Analyzing query logs as a source of semantics bears many resemblances to mining semantics from folksonomies. Some of the related work use methods of networks analysis and unsupervised methods of data mining such as frequent itemset mining and hierarchical clustering among others. [13,11,15]. Krause et al. perform a network analysis of a 'logsonomy' which emerges by looking at queries as tags of clicked URLs and conclude that folksonomies and logsonomies share similar characteristics [12]. Francisco et al. generate a similar network and carry out clustering to mine semantically related queries [9]. Our analysis is different in that we are decomposing queries into entities and their context, and we use background knowledge in the form of entity to type mappings to associate queries.

3 The Data Gap

Discussions around the growth and adoption of the Semantic Web often revolve around the observable size of the Semantic Web, whether it's the number and size of datasets in the Linked Data cloud[2] or the number of pages annotated with microformats or using RDFa. As an example of this sort of analysis, Figure 1 shows the percentage of pages with certain types of microformats or RDFa data as observed in September, 2008 and March, 2009. We can read the growth rates from the chart, and observe, for example, that roughly 2% of webpages contained hCard data by the end of the observed period. But we have to ask ourselves the questions: how useful is this analysis? Just how big the Semantic Web should be?

One possible answer is that the Semantic Web should be *just big enough* to answer all the questions that we may want to ask.[3] The questions that we may want to ask to the data may depend on the particular application but considering semantic search on the public web, one valuable source of information are the logs collected by Web search engines. Due to the widespread, everyday use of search engines query logs provide an excellent record of the information needs of the collective of Web users.

Given a set of queries, the effectiveness of search still depends on the corpus as well as the search engine. Since our goal is not to evaluate semantic search engines, but to evaluate data, we fix the search engine in question by relying on Yahoo Search to retrieve web pages and look at the metadata associated with the results returned. Thus we assume that the current text search engine is a good approximation for a semantic search engine. We believe this is a reasonable assumption for embedded metadata (microformats, RDFa) where the metadata

[2] See `http://linkeddata.org`

[3] There is ample evidence that the Web is bigger than just enough: the three largest search engines crawl, index and query different parts of the Web and yet come up with qualitatively similar answers.

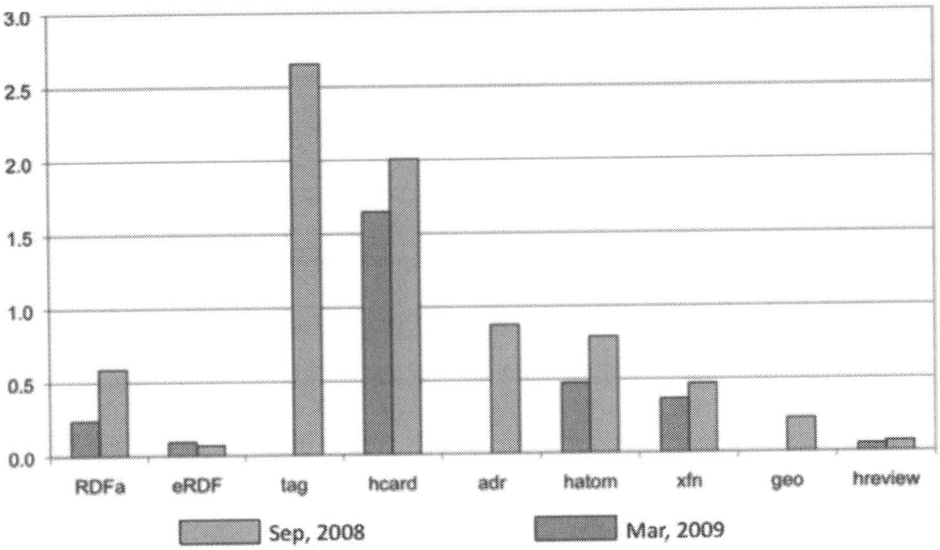

Fig. 1. Percentage of URLs with eRDF, RDFa data and certain popular microformats

is typically a structured representation of the main object presented in the page.[4] We also fix the corpus in the sense that we are naturally limited to the part of the Semantic Web that is crawled and indexed by Yahoo, which includes various microformats and RDFa data, but doesn't include Linked Data in general (e.g. RDF/XML documents).

In summary, we are interested in the forms and quantities of metadata that are returned with search results based on the behavior of the average Web user.

3.1 Methodology

For our analysis, we have taken a uniform sample of search sessions appearing in Yahoo's US query logs for the month of January, 2009.[5] This sample contained 10699 queries, with 7081 unique queries with at least one result. (We obtain the list of unique queries by taking each query once, no matter how often it occurred in the query log.) The distribution of queries follows the typical 'very long tailed' distribution observed in query logs [2]: 64% of the unique queries appear only once in the sample.[6] This is mostly a result of the fact that the same query

[4] However, we do not know of any studies that would have verified this commonly held assumption.

[5] The difference between sampling sessions and sampling queries directly is negligible for our analysis.

[6] In other words, this is the percentage of queries on the list of unique queries that appeared only once in the original sample. This is different from the ratio of unique queries, i.e. the percentage of queries that occur only once when counting with multiplicity.

can be written in multiple ways. Again, we rely on the search engine to return comparable results for equivalent queries.

We executed the queries in the log using the Yahoo BOSS search API, which has been recently extended to return embedded metadata with each search result.[7] The metadata is returned either as RDF/XML or in DataRSS format, where the RDF triples are grouped into 'adjuncts' based on the source of the metadata, i.e. RDFa or one of the recognized microformats.[8] DataRSS is a proprietary serialization format, but one that is fully compatible with RDFa, which means that the actual RDF triples can be extracted with any RDFa parser.

Table 1. The number of queries that return 1 to 10 results with metadata in particular formats, plus the total impressions for the entire set of queries and the average total impressions per query

format	1	2	3	4	5	6	7	8	9	10	TI	ATI
hcard	1457	370	93	11	3	0	0	0	0	0	2535	0.36
rel-tag	1317	350	95	44	14	8	6	3	1	1	2681	0.38
adr	456	77	21	6	1	0	0	0	0	0	702	0.10
hatom	450	52	8	1	0	0	0	0	0	0	582	0.08
license	359	21	1	1	0	0	0	0	0	0	408	0.06
xfn	339	26	1	1	0	0	0	1	0	0	406	0.06
RDFa	176	2	0	0	0	0	0	0	0	0	180	0.03
Any	2127	1164	492	244	85	24	10	5	3	1	7623	1.08

Using the adjunct ids returned for each query we can count the number of results with embedded metadata and the source of the information. The rows of Table 1 show the results for each format (RDFa or microformat) separately and for considering any format (Any). Each row shows the number of queries with 1 to 10 results with embedded metadata, the total impressions (TI) that is the total number of returned URLs that contained metadata, and the average total impressions (ATI), which is the average number of impressions per query.

As an example on how to read this table, the number 370 in the row labeled 'hcard' and the column labeled '2' shows that 370 queries returned two results with hCard data. In the same row, the column TI shows that 2535 of the returned results for all the queries contained hCard data, which makes an average of 0.36 results with hCard data per query. Note that the last row is not a total because it's not a simple sum of the rows above: a single page may contain multiple types of microformats, or a combination of microformats and RDFa.

Based on the results, we observe that 59% of the queries have at least one search result with metadata, with an average of about one search result with metadata. (Note that taking ten search results for each query, the ATI has a maximum value of 10.) hCard and rel-tag each appear on every third search result page on average, while other microformats appear a lot less frequently

[7] See `http://developer.yahoo.com/search/boss/structureddata.html`

[8] Yahoo Search converts microformats to RDF during indexing.

(the numbers in the last column are decreasing quickly). An RDFa enabled result would appear only for every 40th query at the time of the analysis (March, 2009).[9]

3.2 The Role of Popular Sites

It is a well-known phenomenon in Web search that the size of a web site doesn't necessarily correlate with its usefulness as determined by users. On the one hand, a web site doesn't have to be large to be popular with users: a well-known example is Wikipedia, which contains relatively a small amount, but diverse and high quality content, and as a result dominates search result pages beyond its size. At the other extreme, a large part of Web pages that are crawled are never returned by search engines. One can say that these pages are useful to search engine users only to the extent that they are linked or otherwise findable from pages that are being returned.

Table 2. Most popular hostnames in search results by total impression

host name	1	2	3	4	5	6	7	8	9	10	TI	ATI
en.wikipedia.org	1676	1	0	0	0	0	0	0	1	0	1687	0.24
www.youtube.com	475	1	0	0	0	0	0	2	0	0	493	0.07
www.amazon.com	345	3	0	0	0	0	1	0	0	0	358	0.05
www.answers.com	294	0	0	0	0	0	0	0	0	0	294	0.04
www.geocities.com	263	2	0	0	0	0	0	0	0	0	267	0.04
www.yellowpages.com	233	0	0	0	0	0	0	0	0	0	233	0.03
blog.360.yahoo.com	228	0	0	0	0	0	0	0	0	0	228	0.03
local.yahoo.com	220	1	0	0	0	0	0	0	0	0	222	0.03
www.imdb.com	197	0	0	0	0	0	0	0	0	0	197	0.03
www.myspace.com	163	0	0	2	0	0	0	0	0	0	171	0.02

We are interested in measuring the extent to which large sites dominate search results, and consequently the importance of the data they provide compared to the numerous but smaller contributions of average websites. To achieve this, we counted unique host names in search results exactly as we counted unique formats appearing. Table 2 shows the results in the same format as the previous table. Note that as a general rule Yahoo does not return more than two results from the same host except when the query is a URL or site query.

These results are illuminating in the sense that they show a surprisingly large influence of some websites. For example, if YouTube would introduce an entirely new microformat or one would extract information from this particular Web site, from the perspective of search users this data alone would be more significant

[9] Note that we don't count as RDFa triples in the XHTML namespace such as those generated by <link> elements with a rel attribute of icon or stylesheet. We choose to ignore these triples because they have no value for a semantic search engine. The only frequent-enough property in the XHTML namespace that does have a semantic value is xhtml:license, which we account for under rel-license.

than the total amount of XFN information on the Web contributed by millions of hosts. We also see that the most of the importance we can attribute to RDFa data comes from the adoption of RDFa by a single large site, myspace.com. We expect the relative importance of large sites to diminish over time, but it seems characteristic for the current early adoption phase of the Semantic Web.

3.3 The Influence of the Query Category

While query logs in general cover the breadth of information needs, we might be interested in measuring the potential of semantic search for particular categories of queries. The performance of current Web search technology in general strongly depends on the type of query (e.g. short queries vs. long queries, navigational vs. non-navigational) or domain of queries (e.g. person queries vs. product queries). Thus the potential for improvement using semantic technologies is consequently larger for certain kind of queries than others. Another reason to break down the results might be that certain kinds of queries are more important from the perspective of search advertising.

Given any classification of queries, the results of the analysis above can be easily broken down by category. The categories used to classify queries will depend on the type of application. For demonstration purposes, we show how the results break down for a small number of query categories defined by ourselves and used to categorize a set of 1000 queries.

Table 3. Average Total Impression (ATI) values for particular formats when restricting the query set by query category

Organization		Location		Person		Recent event	
hcard	0.40	hcard	0.7	rel-tag	0.65	hcard	0.65
rel-tag	0.35	adr	0.55	hcard	0.54	en.wikipedia.org	0.48
adr	0.23	local.yahoo.com	0.33	en.wikipedia.org	0.23	rel-tag	0.43
en.wikipedia.org	0.21	geo	0.31	hcalendar	0.16	hcalendar	0.40
geo	0.10	yelp.com	0.21	hatom	0.14	answers.com	0.15
local.yahoo.com	0.09	rel-tag	0.16	youtube.com	0.12	imdb.com	0.15
yelp.com	0.08	yellowpages.com	0.15	answers.com	0.10	myspace.com	0.15
hatom	0.08	en.wikipedia.org	0.11	facebook-video	0.07	hatom	0.12
Any	1.14	Any	1.31	Any	1.52	Any	1.66

Table 3 shows the ATI values for different formats and for the top four categories that surfaced most of the metadata: queries containing organizations, location names and person names, and recency sensitive queries, i.e. queries referring to news or events. There are again a number of noteworthy observations. As stipulated, and as shown by the last row, each of these restrictions of the data set resulted in returning more metadata per query than for the general case (where the ATI measure was 1.08), i.e. there is indeed more metadata to be exploited for particular classes of queries. We can also see significant changes in the relative importance of particular sites and different types of metadata. For example, Wikipedia's importance is significantly diminished for queries containing locations, which points to the fact that Wikipedia is rather incomplete when it comes to articles about places. Metadata in Facebook Share format[10] describing videos is not relevant for queries in general, but it has a relative importance to queries related to people (in particular, celebrities). Similarly, hCalendar did not appear in Table 1 because its significance was below that of the last entry (XFN). However, hCalendar data seems very significant to queries about events.

We will return to some of the limitations of this analysis in our conclusions in Section 5. In the second part of our paper we look at a parallel problem of measuring the semantic gap between the information needs (and corresponding vocabulary) of web searchers and the information captured in ontologies.

4 The Vocabulary Gap

We begin by observing that in Web search query logs and in particular for queries that contain a named entity, the *class* of the entity that the user is looking for often determines the query context, i.e., the terms written before (prefix) or after the name (suffix) of an entity, respectively. Put differently, entities of the same class often occur in the context of similar words, representing specific information users are interested in with respect to that particular class of entities. Table 4 shows some examples of queries with class-based contexts.

Table 4. Example queries, extracted entities, completions, and types

Query	Entity	Context	Class
aspirin side effects	ASPIRIN	+*side effects*	Anti-inflammatory drugs
how to take ibuprofen	IBUPROFEN	–*how to take*	Anti-inflammatory drugs
britney spears video	BRITNEY SPEARS	+*video*	American film actors
britney spears shaves her head	BRITNEY SPEARS	+*shaves her head*	American film actors

In this section, we look at how at a method to mine common attributes of classes of objects using query logs and class-membership information as background knowledge. The original use case of this analysis was to provide search suggestions based on the type of entity the user is looking, which is useful in situations where no good suggestions are available for the entity itself. However, the resulting structures are also interesting to compare to the explicit conceptualizations found in Web ontologies.

We start by selecting those queries from the query logs which have a named entity in them. Given this subset of the query logs, we assume queries can be

[10] `http://www.facebook.com/share_partners.php`

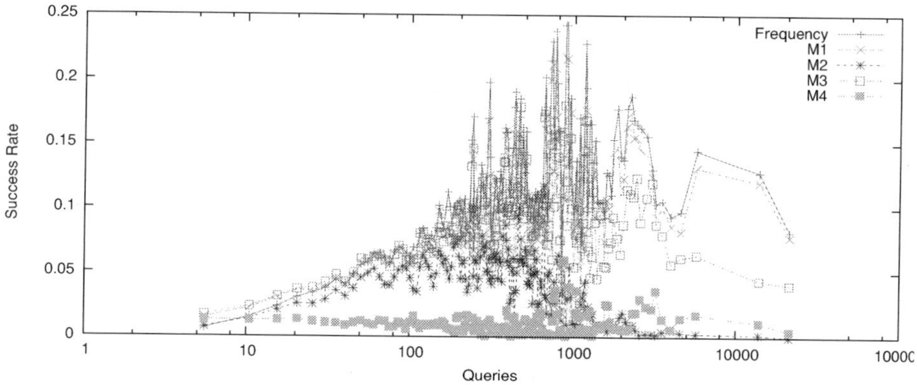

Fig. 2. Success rate for queries. The x-axis shows the queries which are averaged and binned by frequency of occurrence. On the left we see rare queries, on the right popular ones.

decomposed in an entity part e and a context part f and, further, that entities can be assigned a type T. In case the query contains a pre- and suffix, we treat it as two separate queries.

We then determine the matrix $N = (n_{ef})_{e,f}$, where n_{ef} is the number of times we see f with e. By grouping all entities of a certain type we can, for example, compute $n_{Tf} := \sum_{e \in T} n_{ef}$ which is the number of times we see completion f with an entity of type T. Using N, we can readily estimate probabilities such as $P(f)$, $P(f|e)$, $P(f|T)$, and $P(e|f,T)$ which we use to implement several intuitions regarding semantic query completion.

4.1 Extraction Methods

Imagine that a user is typing a query and we recognise what she has typed so far as an entity with a corresponding type. The most naive approach (and the one that is taken by most Web search engines) would be to suggest the most frequent completions for the current entity (**M0**): $score_{M0}(f, e) = P(f|e)$. Given an infinite amount of data this should suffice. However, it will probably fail for rare entities since we will have none or very few completions for them. For this reason we turn to to the entity type and smooth the entity distribution with the type distribution.

M1 aggregates completions over types and looks at the most likely completion for the current type:

$$score_{M1}(f, T) = P(f|T). \quad (1)$$

Another desirable property a completion should have, is being rare over all types. **M2** rewards such completions:

$$score_{M2}(f, T) = \frac{P(f|T)}{P(f)}. \quad (2)$$

Another intuition is that completions which are frequent as well as evenly distributed among the entities in the type should be rewarded (**M3**):

$$score_{M3}(f,T) = G(f|T) = \left(\prod_{e \in T} n_{ef}\right)^{1/|T|}. \tag{3}$$

The final method (**M4**) only considers the distribution of completions within the type:

$$score_{M4}(f,T) = H(\theta_{f|T}), \tag{4}$$

where H is the entropy of the multinomial $\theta_{f|T} = (P(e|f,T))_{e \in T}$.

To detect entities in queries, we require an ontology with a set of classes and a set of instances for each class. Consequently, we match the largest substring common to the query and the label of an instance.[11] In our experiments, we use DBpedia [4] considering both templates and categories as classes. Wikipedia entries can belong to many categories (e.g. 34 for Madonna) and reference a number of templates. We choose only one and try to select the *best* entity type. This is a challenging research question in itself; trivial methods such as choosing the most frequent or rarest type did not work well. Instead, we apply M1 on the training data and evaluate the performance of all possible types of each entity. We then choose the type that led to the best performance on the training set. For entities not present in the training set we select the type with the most entities.

4.2 Evaluation of Type-Based Context Prediction

We evaluate the success of our context extraction by measuring its predictive power. In particular, we compare the highest scoring completions of the various methods with the actual observed remainder of the queries in the test set. We use 6 consecutive days of query logs which we split equally into a training and a test set. We analyze each query and if it contains an entity we keep it. This results in 1,681,753 queries for training and 1,644,033 for testing. For each query, we compute the top $K = 10$ completions predicted by each method using postfixes only. The correct completion for that query is the one typed by the user. We are interested in two evaluation measures: (i) Success Rate @ K (SR), i.e. whether the completion is correctly predicted and (ii) Mean Reciprocal Rank @ K (MRR), i.e. the mean of the inverse of the ranks at which the completion was found, up to K.

Table 5 shows the results over all test queries. As is clear from the low absolute scores, the task of suggesting the correct completion is a difficult one. The highest obtained MRR lies around 0.18 for M0 with queries that occur around

[11] We remove any disambiguation part in the entry title. This has the adverse effect of introducing noise, e.g. collapsing Madonna (art) and Madonna (entertainer). Disambiguating such queries is beyond the scope of the current work but could, e.g., be achieved by leveraging a user's history [3].

Table 5. Aggregated results over all queries

	M0	M1	M2	M3	M4
MRR	0.081	0.068	0.014	0.046	0.006
SR	0.118	0.104	0.041	0.088	0.010

Table 6. Top ten prefixes and postfixes using our model M4 and Wikipedia templates as classes

infobox_settlement	infobox_musical_artist	drugbox	infobox_football_club
hotels	lyrics	buy	forum
map	buy	what is	news
map of	**pictures of**	tablets	website
weather	what is	what is	homepage
weather in	video	side effects of	tickets
flights to	download	hydrochloride	official website
weather	hotel	online	**badge**
hotel	dvd	overdose	fixtures
property in	mp3	capsules	free
cheap flights to	best	addiction	**logo**

1000 times. M0 outperforms the type-based methods on almost all queries and measures. However, as indicated by Figure 2, the type-based methods, in particular M1 and M3, perform slightly better than M0 for less frequent queries (occurring 40 times or less and making up 12.7% of the total query volume). For other queries, M0 outperforms all other methods although the difference with M1 is usually small. The reason for the lower scores at the most frequently occurring queries is that these mostly consist of entities such as "in", "to", and "uk" (which are actual Wikipedia entries).

In the future, we plan to complement this evaluation with a user study as we feel that some of the models might achieve a high prediction accuracy by overfitting popular entities. There are also many query contexts that are particular to the specific entity (e.g. *britney spears shaves her head*) but a user is likely to accept other reasonable suggestions based on the type (e.g. *britney spears videos*) when offered a choice.

4.3 Qualitative Analysis

We have implemented the above methods in a tool that can be used to dynamically query for the most common context words of an entity of a certain type. Shown in Figure 3, the tool allows to search for all entities using a text box that performs autocompletion. Once the user has selected an entity, the relevant types are retrieved, and the user can chose one of the available types. Based on the selected entity and the type, the tool shows both the entity-based and type-based context words. The tool relies on a number of indices built from the query

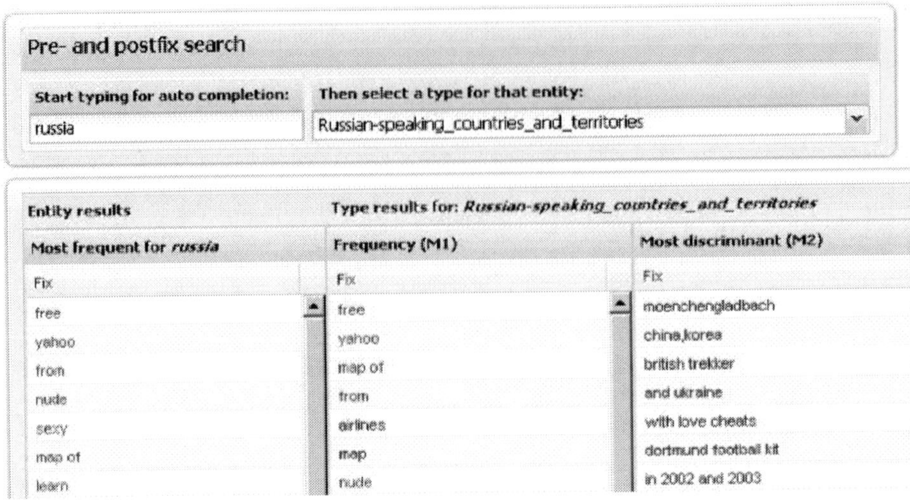

Fig. 3. Interactive search tool for the most common pre- and postfixes given an entity and a type

log using DBpedia as background knowledge. The tool could be used to perform the analysis for any other Semantic Web ontology by rebuilding the underlying indices.

In the following, we compare the results of context mining and the attributes found in DBpedia itself. This analysis is necessarily manual and qualitative because we would like to accept the situation where there is a semantic equivalence. For example, the users may be looking for 'pictures' while the ontology may contain a 'photo' property.

Table 6 shows the most common contexts (prefixes or suffixes) for five different Wikipedia templates, computed using method M4. We have chosen this particular model over our other models because it seems to give better type-specific results: even though our M1 has higher predictive power, at the same time it is over-fitting popular entities in the class. We have chosen these five templates because they vary in size from 43225 entities for *infobox_settlement* to 998 entities for *infobox_football_club*.

We show in **bold** the prefixes or suffixes that match an infobox property, i.e., where the user's query is likely to be satisfied by infobox data (assuming that the particular property is defined for the particular entity the user is searching for, i.e. that the infobox has been completed for this property). It is immediately obvious that there are very few of these. In fact, it seems the majority of these popular information needs cannot even be possibly satisfied by factual data. We leave it for further investigation to study whether it is the case that factual questions –which may be individually uncommon– would still make up a substantial portion of query volume.

It is interesting to note that there are also information needs where the answer could be relatively concise and expressed in a single sentence or paragraph. This is often reflected in the structure of articles, i.e. the division of information into sections. For example, articles on drugs often have sections titled 'Overdose' and 'Side Effects'. Even if the answer to a query such as *aspirin overdose* can not be answered by a single fact, the information the user is looking for may come from a single section or even a single paragraph within the Wikipedia article. This warrants further investigation of exploiting article structure when searching Wikipedia.

In summary, it is clear that if infobox data would be geared toward answering popular information needs as surfaced by our tool, the infoboxes would need to contain different information at different levels of granularity. This suggests that for answering these 'head' queries one may need to merge the methods of data retrieval with methods of structured retrieval and unstructured retrieval. Put differently, for using the output of our tool as an input for ontology engineering, the list of context words extracted will need to be filtered to those representing attributes of objects, i.e. properties that can be filled with simple values. Although this out of the scope for our current work, ontology learning in this context would be similar to ontology learning in folksonomies [5,13,1].

5 Conclusion and Future Work

Ultimately, the success of the Semantic Web depends not only on technology, but also how well the knowledge captured using microformats or Linked Data satisfies the needs of ordinary users. The two main factors in this respect are the coverage and quality of data and ontologies. In this paper, we have looked at the issue of coverage, in particular to what extent data on the Semantic Web is potentially useful in resolving queries and how well the vocabularies used match the implicit vocabularies of users as expressed by their queries.

We have presented methods of analysis and discussed the results of our evaluation. We plan to repeat these evaluations as the evolution of the semantic gap is just as interesting as a static picture of it. We have chosen Web search as our target domain, but the general ideas represented by these methods are equally applicable to vertical, enterprise or desktop search scenarios. The particular experiments we have performed can be reproduced using the BOSS API, which provides access to Web metadata crawled by Yahoo.

In terms of measuring the relevance of Semantic Web data to Web search, we have shown how we can measure the contributions of various forms of data by effectively replaying a large number of sessions sampled from query logs. We posit that just like in the case of the HTML web where often relatively small, but popular or qualitative websites serve a large number of user needs (such as Wikipedia), the Semantic Web also looks very different when looking at it from the perspective of user queries, instead of just to gauging the number of triples in public datasets. In fact, we find that popular sites have also a lot to contribute to the Semantic Web from this perspective, possibly just as much as the long tail of web sites. Last, we found useful breaking down the analysis into query categories, since such breakdown significantly influences the results and may point to query types where the Semantic Web has a particular potential.

We have also presented a number of methods and their implementation in an online tool for mining type-based query context information, i.e. query prefixes and postfixes that are common to a class of entities, while uncommon to other entities outside of their class. Postulating that these context words represent aspects of entities that search engine users are interested in, we proceeded to investigate on the case of Wikipedia the extent to which this schema of information needs matches the schema of available structured data. We find that at least for the most common context words the overlap is very low as the most common queries are not specific enough to be answered by factual data. We suggest that our tool could be used in the future to analyze, extend or create new ontologies based on the information needs extracted from query logs. In this case it is left to the ontology developer to consider which context words signify relevant attributes of objects to be included in an ontology.

The reader may note that throughout our analysis we attribute the same value to each query and to each piece of data. There might be very good reasons to attribute different value to different queries, for example, because the queries can be monetized to different extents and URLs may have different visibility in the search result page (e.g. top three positions vs. the rest). Certain data sets or combinations of data sets may provide extraordinary value to a small number of users. For example, a biomedical database may provide significant value to a researcher in biomedicine. This is not reflected in our average-value analysis. It is part of the future work to extend our analysis to weighted query sets.

Another limitation of our analysis is that we rely on existing query methods. One might argue that semantic search engines will allow the users to express different forms of queries (natural language queries, SPARQL queries, etc.) and the mere possibility to address information needs in a different form or the fact that semantic search engines will successfully answer new types of queries will change user behavior. Indeed, there are plenty of latent queries that users do not enter into Web search engines because they have learned they would not be answered. Often, these queries are turned into navigational queries, e.g. a user interested in flights from boston to san francisco would simply type in the name of an airline, knowing the search engine itself would not be able to return flight information directly. While such a transition toward rich, semantic queries may happen in the future, this change in user behavior will take some time. Similarly, as the Semantic Web grows and sees more usage in general, data may be more aligned with general information needs of Web users. In the meantime, semantic search engines will have to cope with the substantial gap in both data and vocabularies.

References

1. Angeletou, S., Sabou, M., Motta, E.: Folksonomy Enrichment and Search. In: Aroyo, L., Traverso, P., Ciravegna, F., Cimiano, P., Heath, T., Hyvonen, E., Mizoguchi, R., Oren, E., Sabou, M., Simperl, E.P.B. (eds.) ESWC. LNCS, vol. 5554, pp. 801–805. Springer, Heidelberg (2009)
2. Baeza-Yates, R., Gionis, A., Junqueira, F., Murdock, V., Plachouras, V., Silvestri, F.: The impact of caching on search engines. In: SIGIR 2007: Proceedings of the 30th annual international ACM SIGIR conference on Research and development in information retrieval, pp. 183–190. ACM, New York (2007)
3. Bai, J., Nie, J.-Y.: Adapting information retrieval to query contexts. IPM 44(6), 1901–1922 (2008)
4. Bizer, C.: DBPedia: Querying Wikipedia Like a Database. In: WWW 2007 (2007)
5. Brusilovsky, P., Davis, H.C. (eds.): HYPERTEXT 2008, Proceedings of the 19th ACM Conference on Hypertext and Hypermedia, Pittsburgh, PA, USA, June 19-21. ACM, New York (2008)
6. d'Aquin, M., Baldassarre, C., Gridinoc, L., Angeletou, S., Sabou, M., Motta, E.: Characterizing Knowledge on the Semantic Web with Watson. In: EON (2007)
7. Ding, L., Finin, T.: Characterizing the Semantic Web on the Web. In: Cruz, I., Decker, S., Allemang, D., Preist, C., Schwabe, D., Mika, P., Uschold, M., Aroyo, L.M. (eds.) ISWC 2006. LNCS, vol. 4273, pp. 242–257. Springer, Heidelberg (2006)
8. Ding, L., Zhou, L., Finin, T., Joshi, A.: How the Semantic Web is Being Used: An Analysis of FOAF Documents. In: HICSS 2005 (2005)
9. Francisco, A.P., Baeza-Yates, R.A., Oliveira, A.L.: Clique Analysis of Query Log Graphs. In: Amir, A., Turpin, A., Moffat, A. (eds.) SPIRE 2008. LNCS, vol. 5280, pp. 188–199. Springer, Heidelberg (2008)
10. Hausenblas, M., Halb, W., Raimond, Y., Heath, T.: What is the Size of the Semantic Web? In: I-Semantics 2008, Graz, Austria (2008)
11. Jäschke, R., Hotho, A., Schmitz, C., Ganter, B., Stumme, G.: Discovering shared conceptualizations in folksonomies. J. Web Sem. 6(1), 38–53 (2008)
12. Krause, B., Jäschke, R., Hotho, A., Stumme, G.: Logsonomy - social information retrieval with logdata. In: Brusilovsky, Davis (eds.) [5], pp. 157–166
13. Mika, P.: Ontologies Are Us: A Unified Model of Social Networks and Semantics. In: Gil, Y., Motta, E., Benjamins, V.R., Musen, M.A. (eds.) ISWC 2005. LNCS, vol. 3729, pp. 522–536. Springer, Heidelberg (2005)
14. Tummarello, G., Delbru, R., Oren, E.: Sindice.com: Weaving the Open Linked Data. The Semantic Web, 552–565 (2008)
15. Zhou, M., Bao, S., Wu, X., Yu, Y.: An Unsupervised Model for Exploring Hierarchical Semantics from Social Annotations. In: Aberer, K., Choi, K.-S., Noy, N., Allemang, D., Lee, K.-I., Nixon, L.J.B., Golbeck, J., Mika, P., Maynard, D., Mizoguchi, R., Schreiber, G., Cudré-Mauroux, P. (eds.) ASWC 2007 and ISWC 2007. LNCS, vol. 4825, pp. 680–693. Springer, Heidelberg (2007)

Towards Lightweight and Robust Large Scale Emergent Knowledge Processing

Vít Nováček and Stefan Decker

DERI, National University of Ireland, Galway
IDA Business Park, Galway, Ireland
vit.novacek@deri.org

Abstract. We present a lightweight framework for processing uncertain emergent knowledge that comes from multiple resources with varying relevance. The framework is essentially RDF-compatible, but allows also for direct representation of contextual features (e.g., provenance). We support soft integration and robust querying of the represented content based on well-founded notions of aggregation, similarity and ranking. A proof-of-concept implementation is presented and evaluated within large scale knowledge-based search in life science articles.

1 Introduction

On the Semantic Web, we often have to be able to represent and integrate statements coming from many resources with varying relevance in a bottom-up, emergent manner. Moreover, the statements themselves may be noisy and uncertain (e.g., inconsistent, potentially incorrect or having an explicit certainty degree). This is especially pertinent to a use case that has largely motivated our work – search for expressive statements instead of mere keywords in life science articles. More specifically, we want to allow life scientists to search for statements like *acute granulocytic leukemia : NOT is a : T-cell leukemia*, or *? : part of : immunization*. The former query is supposed to confirm whether acute granulocytic leukemia is different from T-cell leukemia by checking for similar statements in publications. Also, the result should provide articles supporting the query statement. The latter query is supposed to return everything that can be a part of the immunization process, plus any related statements and links to articles relevant to them.

Manual annotation of the publication knowledge to be exposed for such search is practically impossible in large scale. However, one can extract the knowledge from the article texts by ontology learning techniques [5] and link it to existing domain ontologies in order to increase the expressivity of the rather shallow extracted content. Such an approach still poses a couple of challenges, though: (i) The representation framework of choice should support uncertainty, as the extracted knowledge usually comes with explicit certainty degrees [5]. (ii) The representation should also straightforwardly support contextual features, namely at least provenance of statements (to link them to the respective source articles).

A. Bernstein et al. (Eds.): ISWC 2009, LNCS 5823, pp. 456–472, 2009.

(iii) Robust aggregation of the emergent statements based on relevance of respective resources is necessary, since we have to integrate noisy extracted knowledge with presumably more accurate manually designed domain ontologies. (iv) The processed knowledge has to be accessible by means of intuitive (i.e., nearly natural language) query answering, since we target users with no or little technical expertise. The query evaluation should also be approximate in order to provide useful answers even for queries partially evaluated on a lot of potentially noisy data.

Approaches like [13,11,4,7,8,15,1,12,2] provide particular solutions apt for coping with the challenges separately, however, to the best of our knowledge there is no off-the-shelf framework tackling all of them at once on a well-founded basis. The main contribution of this paper is two-fold. Firstly, we introduce a general notion of similarity-based lightweight semantics, integrally addressing all the above challenges (Section 2). Secondly, Section 3 presents a particular application of the general framework to knowledge-based search in life science articles. Promising results of an evaluation performed with domain experts are reported in Section 4. We discuss related approaches and conclude the paper in Sections 5 and 6, respectively.

2 General Framework

In the following, we first informally outline the essential notions of the proposed framework and briefly comment on their interplay. The outline is then expanded by more rigorous and explanatory subsections 2.1 (entities and their grounding) and 2.2 (knowledge bases, aggregation and query answering).

Central to our framework is a notion of *entities* that represent real and/or conceivable objects using unique identifiers and sets of positive or negative uncertain *relations* to other entities. To give an example, let us consider the d, a, c, t identifiers representing the `dog`, `animal`, `cat` concepts and `type` relationship, respectively. The `dog` entity can be further specified by binary relations $t(d, a)$ and $t(d, c)$ with a positive and negative certainty, respectively, meaning dogs are animals different from cats. To support contextual features of entity relationships (e.g., provenance or time-stamp), the relations may generally have arbitrary arities. A direct correspondence of sets of n-ary certainty-valued relations to n-dimensional tensors (generalisations of the scalar, vector and matrix notions) provides for a compact computational representation of entities. An entity E is then represented as $(e, \mathbf{E})$, i.e., its unique identifier and the respective compact representation of uncertain relations to other entities. To ensure accessibility for lay users, we link the somewhat abstract representation to corresponding natural language referents via a set of *grounding* functions. These may map, for instance, the `dog` entity to a preferred "dog" expression with a high certainty, but also to alternative synonyms like "doggy" or "hound", perhaps with a bit lower certainty. The other way around, a grounding would map the "mutt" word to the `dog` entity in the lexical domain of animals, but to a completely different entity in the domain of, say, humans. Thus the grounding provides a two-way bridge between the lexical (human-centric) and computational (machine-centric) aspects of the proposed lightweight semantics. The bridge is particularly important when answering user queries—formulated as mostly natural language statements—by means of a query answering service dealing with abstract entity representations.

Building on the compact computational representation of entities, we introduce the *aggregation* and *querying* services in order to tackle the remaining challenges specified in the introduction. Entity aggregation employs linear combinations that naturally model merging of possibly conflicting statements coming from sources with varying relevance. For instance, imagine a statement that dogs eat meat, coming from a highly relevant source, and an opposite, yet relatively irrelevant statement (vegetarian dogs actually exist, however, the respective rather exceptional sources are presumably less relevant). The sum of the corresponding representations, weighed by the relative source relevance, will result in a claim that dogs eat meat with a positive, but slightly lower certainty (as the knowledge from more relevant source prevails in the aggregation).

Query answering makes use of two notions of entity similarity. Let us imagine entities of `dog` and `cow`, eating and not eating meat, respectively. Evaluation of a query for meat-eating animals first checks for entities fitting to the context of the query, i.e., being animals and linked by an "eat" relation to meat. Both `dog` and `cow` entities fit the query within this coarse-grained approximation of similarity. A finer grained notion of similarity, taking the certainty degrees into account, can be naturally coined as dual to a distance defined on the set of entity representations. Utilising this type of similarity results into meat-eating `dog` being a much more certain answer to the query than `cow`, which is an animal, but does not eat meat. In more complex cases, we also sort the query results according to their relevance employing a generalised IR measure based on numbers of outgoing and incoming relations among stored entities.

2.1 Entities and Their Grounding

Entities. First we have to formalise certainty degrees, for which we use $\mathbb{R}$, i.e., real numbers. Positive and negative entity relationships are to be associated with positive and negative certainty values, respectively. 0 is of special importance, expressing absolute lack of certainty. We do not impose any restrictions on the range of certainty values, however, particular implementations may restrict the range to any set isomorphic with $\mathbb{R}$. A convenient variant (used throughout the paper) is $(-1, 1)$, which makes the certainty values compatible either with a recent approach to trust representation in RDF [7], or with general fuzzy and probabilistic formalisms (after transforming negative certainties into negative fuzzy/probabilistic statements). Openness of the certainty intervals reflects the fact that nothing is absolutely certain in emergent settings. Implementations may relax the assumption, though, and use a more traditional $[-1, 1]$ interval.

Moving on to defining entities themselves, let I be a non-empty countable set of unique entity identifiers (e.g., integer numbers or URIs) and $n \in \mathbb{N}_0$ a so called rank of an entity. Rank expresses the maximal arity of relations associated to an entity. An entity with an identifier $c \in I$ and rank $n = 0$ can be written down as a tuple (c, d), meaning that c merely exists (or does not exist) with the certainty d. In practice, more expressive entities with a rank $n > 0$ are required, though. An entity E with an identifier $c \in I$ and rank $n > 0$ can be written down as a set of tuples in the form $(c_1(c, c_2, \ldots, c_n), d)$, where $c_x \in I$ for $x \in \{1, \ldots, n\}$, $d \in \mathbb{R}$. The tuple elements encode a c_1 relation between c and other entities $c_2, \ldots, c_n$, and the relation's certainty, respectively. It is required that $\{(c_{i,1}, c_{i,2}, \ldots, c_{i,n}) | (c_{i,1}(c, c_{i,2}, \ldots, c_{i,n}), d_i) \in E\} = I^n$, meaning that the relations iterate through all possible identifier combinations. However, realistic entities are obviously associated only with a relatively small finite number of relations with a non-zero certainty. We distinguish a special zero-entity (denoted by $\mathcal{O}$ in the following text), which has all certainty degrees equal to 0, thus representing an absolutely uncertain object. $\mathcal{O}$ can be used namely to represent relations with an arity lower than n by n-ary ones (filling in the respective superfluous arguments as shown in Example 1).

Conceiving an entity as a set of relations associated with the entity's identifier is pretty intuitive. Such a form is, nonetheless, quite awkward for treating entities as compact objects. A more compact representation is possible using a direct correspondence between the sets of entity relations and the mathematical structure of tensor (multi-dimensional generalisation of the scalar, vector and matrix notions, which are tensors of ranks $0, 1, 2$, respectively). Using the tensor notation, an entity E with an identifier e and rank n can be represented as a tuple $E \equiv (e, \mathbf{E})$, where $e \in I$ and $\mathbf{E} \in T$ (a set of all tensors of rank n on the field $\mathbb{R}$). A tensor entity representation $(e, \mathbf{E})$ corresponds to a set of relation-degree tuples $\{(c_1(e, c_2, \ldots, c_n), \mathbf{E}_{c_1, c_2, \ldots, c_n}) | (c_1, c_2, \ldots, c_n) \in I^n\}$, where $\mathbf{E}_{c_1, c_2, \ldots, c_n}$ is the element of $\mathbf{E}$ with the respective indices.

Example 1. Here we illustrate the correspondence between the two entity notations (rank 2 is used to facilitate the presentation; higher ranks are direct generalisations of this case). Assuming `B` standing for `http://ex.org`, let $I = \{$ `B#null`, `B#type`, `B#cat`, `B#animal`, `B#eatsMeat` $\}$ abbreviated as $I = \{\perp, t, c, a, e\}$, respectively. `B#null` (or $\perp$) is an identifier of the zero entity $\mathcal{O}$. The `cat` entity E of rank 2 with an identifier c can be described by the following set of relation-degree tuples (omitting the ones with zero degrees): $\{(t(c, a), 0.99), (e(c, \perp), 0.99)\}$. The binary *type* relation says that cats are a type of animal, while the unary *eatsMeat* relation says that cats eat meat (both relations have a positive certainty). The respective tensor representation is $E = (c, \mathbf{E})$, where $\mathbf{E}$ is the following matrix:

	a	$\perp$
t	0.99	0
e	0	0.99

.

Note that we omit (also in the following examples) rows and columns with all degrees equal to zero when they are not required for facilitating the presentation.

Grounding. Let L be a non-empty countable set of language expressions (e.g., words upon an alphabet). Entities are grounded in a language via a so called entity grounding mapping $g_{ind} : I \to (L \to \mathbb{R})$. $g_{ind}(x)$ are total functions that assign certainty values to each element from L. The functions support the synonymy and antonymy lexical phenomenons via positive and negative certainty assignments, respectively. Going the other way around, language expressions are mapped to entity identifiers via a so called unique entity identifier assignment $g_{lan} : L \times I \rightarrow I$. The first argument of g_{lan} is an expression to be mapped to an entity identifier, while the second argument is a so called lexical domain – an entity providing a disambiguation context, catering for correct resolution of homonymous terms. Eventually, we need to ground the dimensions of the tensor representation (or argument positions in the relation-degree notation) to concept identifiers. This is done using a so called dimension grounding mapping $g_{dim} : \{1, \ldots, n\} \rightarrow I$, assigning an entity identifier to each entity index dimension.

Example 2. Assuming `B` standing for `http://ex.org`, let us extend the I set from Example 1 to $I = \{$ `B#null`, `B#type`, `B#cat`, `B#animal`, `B#eatsMeat`, `B#isVeggie`, `B#predicate`, `B#object`, `B#human`, `B#gld`, `B#sissy`$\}$. Furthermore, let $L = \{$*null entity, type, is a, cat, animal, pussycat, eatsMeat, isVeggie, meatEating, predicate, object, human, general lexical domain*$\}$. Let us consider functions $\mu_1, \ldots, \mu_{11}$ assigned by a sample entity grounding g_{ind} to the elements of I (in the order given in the beginning of the example). All the functions assign 0 to most elements of L, with the following exceptions: (i) $\mu_1(x) = 0.99$ for $x \in \{$*null entity*$\}$; (ii) $\mu_2(x) = 0.99$ for $x \in \{$*type, is a*$\}$; (iii) $\mu_3(x) = 0.99$ for $x \in \{$*cat*$\}$, $\mu_3(x) = 0.8$ for $x \in \{$*pussycat*$\}$; (iv) $\mu_4(x) = 0.99$ for $x \in \{$*animal*$\}$; (v) $\mu_5(x) = 0.99$ for $x \in \{$*eatsMeat, meatEating*$\}$, $\mu_5(x) = -0.99$ for $x \in \{$*isVeggie*$\}$; (vi) $\mu_6(x) = 0.99$ for $x \in \{$*isVeggie*$\}$, $\mu_6(x) = -0.99$ for $x \in \{$*eatsMeat, meatEating*$\}$; (vii) $\mu_7(x) = 0.99$ for $x \in \{$*predicate*$\}$; (viii) $\mu_8(x) = 0.99$ for $x \in \{$*object*$\}$; (ix) $\mu_9(x) = 0.99$ for $x \in \{$*human*$\}$; (x) $\mu_{10}(x) = 0.99$ for $x \in \{$*general lexical domain*$\}$; (xi) $\mu_{11}(x) = 0.99$ for $x \in \{$*pussycat*$\}$. Regarding the unique identifier assignment, the only ambiguous lexical expression is *pussycat*: $g_{lan}($*pussycat*, `B#human`$) =$ `B#sissy`, $g_{lan}($*pussycat*, `B#animal`$) =$ `B#cat`. All the other lexical expressions have obvious mappings to identifiers under the *general lexical domain*. Eventually, the dimension mapping g_{dim} can be defined as $g_{dim}(1) =$ `B#predicate`, $g_{dim}(2) =$ `B#object`. This roughly follows the RDF terminology in the sense that the first and second dimension of the tensor representation (i.e., the matrix row and column) correspond to predicate and object identifiers, respectively.

2.2 Knowledge Bases, Aggregation and Query Answering

Knowledge base of rank n is a tuple $(\mathcal{E}, n, I, L, \mathcal{G})$. I, L are the sets of entity identifiers and language expressions as introduced before. $\mathcal{E}$ is a set of entities $(e, \mathbf{E})$ such that $e \in I$ and $\mathbf{E} \in T$, where T is a set of all tensors of rank n defined on $\mathbb{R}$. $\mathcal{G}$ is a set of particular grounding mappings $g_{ind}, g_{lan}, g_{dim}$. Furthermore, a knowledge base must satisfy certain restrictions. Let $ind : \mathcal{E} \rightarrow I, ind((e, \mathbf{E})) = e$, $rep : \mathcal{E} \rightarrow T, rep((e, \mathbf{E})) = \mathbf{E}$ be projections mapping entities in a knowledge base to their identifiers and tensor representations, respectively. Then it is required that $ind(E) = ind(F)$ if and only if $rep(E) = rep(F)$ for every $E, F \in \mathcal{E}$ (consequently, $E = F$ iff $ind(E) = ind(F)$ or $rep(E) = rep(F)$). Also, the ind

projection has to be a bijection. Thus, every entity has a unique identifier and each identifier maps to an entity in a particular knowledge base.

As knowledge is often inherently context-dependent, we have to introduce an appropriate notion of context in our representation. We do so using so called contextual scopes, which are non-empty sets $S \subseteq I^n$ for a knowledge base $(\mathcal{E}, n, I, L, \mathcal{G})$. Briefly put, contextual scopes divide $\mathcal{E}$ into classes of entities associated with particular relations of non-zero certainty in direct correspondence to the elements of S. Each non-zero entity fits into at least one contextual scope. We refer to the minimal contextual scope fully covering a non-zero entity $\mathcal{E}$ by $scp : \mathcal{E} \setminus \mathcal{O} \to 2^{I^n}, scp(E) = \{(v_1, \ldots, v_n) | (v_1, \ldots, v_n) \in I^n \wedge rep(E)_{v_1,\ldots,v_n} \neq 0\}$. It is simply a set of indices of all non-zero elements in the respective entity representation. We define fitness of a non-zero entity E w.r.t. a general contextual scope S as $fit : \mathcal{E} \setminus \mathcal{O} \times 2^{I^n} \to [0,1], fit(E,S) = max(\frac{|scp(E) \cap S|}{|S|}, \frac{|scp(E) \cap S|}{|scp(E)|})$. Maximal fit of 1 is achieved if either all non-zero element indices of the entity are covered by the contextual scope, or if all elements of the contextual scope are covered by the entity's non-zero elements. Minimal fit of 0 is achieved if no index of any non-zero entity element appears in the contextual scope.

Example 3. In order to illustrate practical treatment of contextual scopes, let us extend the I set from previous examples to $I = \{$ `B#null`, `B#type`, `B#cat`, `B#animal`, `B#eatsMeat`, `B#dog`, `B#feline`, `B#canine`$\}$, abbreviated as $I = \{\bot, t, c, a, e, d, f, cn\}$, respectively. Let E and F be `cat` and `dog` entities, such that

$rep(E) =$

	a	f	cn	$\bot$
t	0.99	0.99	0	0
e	0	0	0	0.99

and $rep(F) =$

	a	f	cn	$\bot$
t	0.99	0	0.99	0
e	0	0	0	0.99

.

Contextual scopes corresponding to felines and canines can be defined as $S_1 = \{(t, f)\}, S_2 = \{(t, cn)\}$, respectively. Similarly, feline and canine animals correspond to contextual scopes $S_3 = \{(t, a), (t, f)\}, S_4 = \{(t, a), (t, cn)\}$. Consistently with common sense, $fit(E, S_1) = fit(F, S_2) = fit(E, S_3) = fit(F, S_4) = 1$, meaning that cats are in the context of felines and feline animals (similarly for canine dogs). Also, $fit(E, S_2) = fit(F, S_1) = 0$ meaning that cats do not fit in the context of canines and vice versa for dogs. However, $fit(E, S_4) = fit(F, S_3) = 0.5$, meaning that cats share certain properties (i.e., relations) with canine animals (i.e., being a type of animal), and vice versa for dogs.

In the following, we will need an auxiliary operator for entity trimming according to a contextual scope. It is defined as $\tau : T \times 2^{I^n} \to T, \tau(\mathbf{E}, S) = \mathbf{F}$, where $\mathbf{F}_{i_1,\ldots,i_n} = \mathbf{E}_{i_1,\ldots,i_n}$ for all $(i_1, \ldots, i_n) \in S$, otherwise $\mathbf{F}_{i_1,\ldots,i_n} = 0$. The trimming cuts all the relations not "belonging" to a contextual scope off an entity, rendering their certainty zero. Apparently, $\tau(rep(E), S) = rep(E)$ iff $scp(E) \subseteq S$. The operator is to be used when one needs to focus only on particular features of entities within their computational processing (e.g., aggregation or querying).

Entity Aggregation. Let $+, \cdot$ be operations of vector addition and scalar multiplication defined on T and $\mathbb{R}$ (e.g., element-wise tensor addition and scalar multiplication as a generalisation of the respective matrix operations). Then T forms a vector space and as such can provide a natural framework for weighed entity aggregation by means of linear combinations. An aggregation of entities $E_1, \ldots, E_k$ with rank n is a function $agg : 2^T \to T$ operating on the respective tensor representations:

$$agg(\{rep(E_1), \ldots, rep(E_k)\}) = \sum_{v \in V} \sum_{j \in J} r_{v,j} \tau(rep(E_j), \{v\}).$$

$V = \{(i_1, \ldots, i_n) | \exists x. x \in \{1, \ldots, k\} \wedge rep(E_x)_{i_1,\ldots,i_n} \neq 0\}$, $J = \{x | x \in \{1, \ldots, k\} \wedge rep(E_x)_{v_1,\ldots,v_n} \neq 0\}$, such that $(v_1, \ldots, v_n) = v$. $r_{v,j} \in \mathbb{R}_0^+$ are weights reflecting the relevance of the particular summation elements and τ is the entity trimming operator defined before. The generic aggregation definition flexibly covers intuitively applicable aggregation mechanisms, as shown in the following example.

Example 4. Assuming the I set from the previous examples, imagine two different dog entity representations $rep(E_1), rep(E_2)$, such that

$$rep(E_1) = \begin{array}{c||c|c} & a & \bot \\ \hline t & 0.99 & 0 \\ e & 0 & -0.5 \end{array} \text{ and } rep(E_2) = \begin{array}{c||c|c} & a & \bot \\ \hline t & 0.99 & 0 \\ e & 0 & 0.99 \end{array}.$$

Let the entity representations come from sources with relevance weights 0.2 and 1, respectively (the source conceiving a dog as a kind of vegetarian having much lower, although non-zero relevance). $agg(\{rep(E_1), rep(E_2)\})$ then expands as:

$$r_{(a,t),1} \begin{array}{c||c|c} & a & \bot \\ \hline t & 0.99 & 0 \\ e & 0 & 0 \end{array} + r_{(a,t),2} \begin{array}{c||c|c} & a & \bot \\ \hline t & 0.99 & 0 \\ e & 0 & 0 \end{array} + r_{(e,\bot),1} \begin{array}{c||c|c} & a & \bot \\ \hline t & 0 & 0 \\ e & 0 & -0.5 \end{array} + r_{(e,\bot),2} \begin{array}{c||c|c} & a & \bot \\ \hline t & 0 & 0 \\ e & 0 & 0.99 \end{array}.$$

Various mechanisms of aggregation can be achieved by setting the $r_{(a,t),1}, r_{(a,t),2}, r_{(e,\bot),1}, r_{(e,\bot),2}$ weights accordingly. E.g., $r_{(a,t),1} = r_{(a,t),2} = 0.5, r_{(e,\bot),1} = 0.2/1.2$, $r_{(e,\bot),2} = 1/1.2$ keeps equal elements unchanged, however, computes weighted mean for conflicting certainty values with the source relevances as particular weights, thus letting the statement from a more relevant source prevail.

Query Answering. We support soft anytime retrieval of entities from knowledge bases according to their *similarity* to so called primitive queries Q, with the results sorted by their *relevance*. Primitive queries are simply entities with an unknown identifier (i.e., variable). First approximation of the similarity is the fitness $fit(E, scp(Q))$. Assuming a knowledge base $(\mathcal{E}, n, I, L, \mathcal{G})$, entities $E \in \mathcal{E}$ with $fit(E, scp(Q)) > 0$ are plausible (possibly partial) answers for the query Q.

A more fine grained notion of similarity can be naturally defined using a metric $d : T^2 \to \mathbb{R}$ on the set T of entity representations. d can be any function satisfying the following properties for all $\mathbf{E}, \mathbf{F}, \mathbf{G} \in T$: (i) positive definiteness – $d(\mathbf{E}, \mathbf{F}) \geq 0$, $d(\mathbf{E}, \mathbf{F}) = 0$ if and only if $\mathbf{E} = \mathbf{F}$; (ii) symmetry – $d(\mathbf{E}, \mathbf{F}) = d(\mathbf{F}, \mathbf{E})$;

(iii) triangle inequality – $d(\mathbf{E},\mathbf{G}) \leq d(\mathbf{E},\mathbf{F}) + d(\mathbf{F},\mathbf{G})$. Similarity is conceptually dual to distance (i.e., metric). Therefore we can define similarity of entities $E, F \in \mathcal{E}$ as a a function $sim : \mathcal{E}^2 \to (0,1], sim(E,F) = \frac{1}{1+d(rep(E),rep(F))}$. The duality of sim and d is ensured by their apparent inverse proportionality. Moreover, sim has the following intuitively expected properties: $sim(E,E) = 1$ and $\lim_{x\to\infty} sim(E,F) = 0$, where $x = d(rep(E), rep(F))$.

Apart of similarity of candidate answers to queries, we establish the notion of entity relevance, which can be effectively used for ranking query results. Informally, relevance of an entity E in our framework is given by the number and certainty of relations that are associated to it, but also by the number and certainty of relations that reference it. Such a measure tells us how important E is w.r.t. determining the meaning of other entities. This is directly related to the hubs and authorities algorithm designed for ranking of web pages [9]. We only need to generalise it to support n-ary links with arbitrarily weighed relations and argument positions. The generalised hub measure of entities in a knowledge base $(\mathcal{E}, n, I, L, \mathcal{G})$ is recursively defined as $h : \mathcal{E} \to \mathbb{R}_0^+$ such that:

$$h(E) = \sum_{(u_1,\dots,u_n)\in scp(E)} |rep(E)_{u_1,\dots,u_n}| w_{arg}(1) w_{rel}(u_1) \sum_{k=2}^{n} w_{arg}(k) a(F),$$

where $F = ind^{-1}(u_k)$ is the entity referenced in the respective relation. Similarly, the generalised authority measure is defined as $a : \mathcal{E} \to \mathbb{R}_0^+$, such that:

$$a(E) = \sum_{F\in R} \sum_{(u_1,\dots,u_n)\in V} |rep(F)_{u_1,\dots,u_n}| w_{arg}(1) w_{rel}(u_1) h(F) \sum_{x\in Y} w_{arg}(x),$$

where $R = \{G | \exists G.rep(G)_{u_1,\dots,u_n} \neq 0 \wedge \bigvee_{i=1}^{n} ind(E) = u_i\}$ is a set of all entities referencing E, $V = \{(v_1,\dots,v_n) | rep(F)_{v_1,\dots,v_n} \neq 0 \wedge ind(E) \in \{v_1,\dots,v_n\}\}$ and $Y = \{y | y \in \{u_1,\dots,u_n\} \wedge y = ind(E)\}$. $w_{rel} : I \to \mathbb{R}_0^+$ and $w_{arg} : \{1,\dots,n\} \to \mathbb{R}_0^+$ are weights of particular relations and relation argument positions (generally including also the "zeroth" argument position, i.e., the relation identifier itself). Using the generalised measures, we can compute the hub and authority scores for entities $E \in \mathcal{E}$ with the iterative algorithm given in [9] (normalising the scores in each iteration to ensure convergence). The relevance of an entity E is then defined as $rel : \mathcal{E} \to \mathbb{R}_0^+, rel(E) = m(h(E), a(E))$, where $m : \mathbb{R}^2 \to \mathbb{R}$ is any aggregation function such that for all $x, y \in \mathbb{R}_0^+$, $min(x,y) \leq m(x,y) \leq max(x,y)$. Examples are min, max, or an arithmetic mean.

Having introduced all the necessary notions, we can finally specify the set of answers to a query $Q \in \mathcal{E}$ w.r.t. a knowledge base $(\mathcal{E}, n, I, L, \mathcal{G})$ as a function $ans : \mathcal{E} \to 2^{\mathcal{E}}$ such that $ans(Q) = \{s_1 A_1, \dots, s_k A_k\}$. $A_1, \dots, A_k \in \mathcal{E}$ and $s_i = sim(\tau(rep(A_i),\ scp(Q)), rep(Q))$ for $i \in \{1,\dots,k\}$. Note that to simplify the notation, we assume that $sA = s(a, \mathbf{A}) = (a, s\mathbf{A})$ for a multiplication of an entity (i.e., an identifier-tensor tuple) by a scalar value. It is required that $fit(A_1, scp(Q)) \geq \dots \geq fit(A_k, scp(Q)) > 0$. Moreover, every sequence $s_{i_1} A_{i_1}, \dots, s_{i_l} A_{i_l}$ such that $i_1, \dots,\ i_l \in \{1,\dots,k\}$, $i_1 \leq \dots \leq i_l$ and $fit(A_{i_1}, scp(Q)) = \dots = fit(A_{i_l}, scp(Q))$ must be lexicographically ordered according to the respective

$(s_x, rel(A_x))$ measures. Thus, $ans(Q)$ is a set of entities from $\mathcal{E}$ multiplied by their actual similarity to Q, taking only the minimal contextual scope covered by Q—i.e., $scp(Q)$—into account, though. The answers also must be ordered first regarding the fitness measure w.r.t. $scp(Q)$, then according to their similarity to the query (in the query's context), and finally according to their relevance.

Example 5. In the following, we employ similarity based on particular metric $d(\mathbf{E},\mathbf{F}) = \frac{1}{|V|}\sum_{(u_1,\dots,u_n)\in V} |\mathbf{E}_{u_1,\dots,u_n} - \mathbf{F}_{u_1,\dots,u_n}|$, where $V = scp(E) \cup scp(F)$. The metric simply sums up absolute values of differences across the representation indices referring to a non-zero value in $\mathbf{E}$ or in $\mathbf{F}$, normalising the result by the size of the summation range. The respective similarity $\frac{1}{1+d(rep(E),rep(F))}$ is essentially a very simple formalisation of the contrast model [14] (more sophisticated alternatives may, e.g., put specific weights on particular elements within the metric computation to reflect intensity and context in the sense of [14]).

Consider now the particular `cat` and `dog` entities E and F as given in Example 3 and a query Q asking for canine animals. The set of answers $ans(Q)$ then equals $\{A_1, A_2\}$, where

$$rep(Q) = \begin{array}{c||c|c} & a & cn \\ \hline \text{t} & 0.99 & 0.99 \end{array},\ A_1 = (d, \begin{array}{c||c|c|c} & a & cn & \perp \\ \hline \text{t} & 0.99 & 0.99 & 0 \\ \text{e} & 0 & 0 & 0.99 \end{array}), A_2 = (c, \begin{array}{c||c|c|c} & a & f & \perp \\ \hline \text{t} & 0.497 & 0.497 & 0 \\ \text{e} & 0 & 0 & 0.497 \end{array}).$$

When aggregating the hub and authority values using the max function and setting all weights to 1, except for the unary e relation weight set to 0, the relevance of the `cat, dog, animal, feline, canine` entities is $0.5, 0.5, 0.5, 0.25, 0.25$, respectively. However, apparently we do not need relevance in this simple example, as the fitness and similarity are enough to sort the results.

Raw sets of answers might not be particularly interesting for users in practical query-answering application scenarios. Therefore the implementations of the proposed framework may present just the corresponding ordered list of the answer entity identifiers (or their appropriate lexical labels). To provide additional information, such results may be associated with an aggregation of the respective fitness and similarity values, such as in the following: $\{dog : 1, cat : 0.5\}$ (using the corresponding lexical labels and min for the aggregation). Such an answer contains all the intuitively expected information – dogs are canine animals, while cats are animals, however, not canines. Therefore cats are present in the result, too, but with a lower explicit relevance.

3 Particular Implementation and Deployment

We have implemented a proof-of-concept prototype of the theoretical principles introduced so far, called EUREEKA (may be read as an acronym for *Efficient, Universal, Reasonable and Easy-to-use Emergent Knowledge Acquisition*). As mentioned in Section 1, the development and current deployment of the prototype has been motivated by the use case of knowledge-based search in life science articles. In order to realise this in an economically feasible way, we have to *extract* the respective knowledge from the texts, *represent* it in an appropriate manner, *integrate* it and *expose* it to the users in a robust and meaningful way. To address these tasks, we have recently delivered CORAAL (cf. `http://coraal.deri.ie:8080/coraal/`), which is a comprehensive life science publication search engine deployed on the data provided by Elsevier within their Grand Challenge contest (cf. `http://www.elseviergrandchallenge.com/`). EUREEKA forms the engine's crucial back-end part, catering for the *representation*, *integration* and *exposure* tasks, thus enabling the knowledge-based search functionalities.

For the initial knowledge extraction in CORAAL, we used a NLP-based heuristics stemming from [10,16] in order to process chunk-parsed texts into subject-predicate-object-score quads. The scores were derived from absolute and document frequencies of subject/object/predicate terms aggregated with subject/object co-occurrence measures. If a relation's score is not available for any reason (e.g., when importing legacy knowledge from crisp resources instead of extracting it from text), we simply set it to 1 (or -1) in the implementation. The extracted quads encoded three major types of ontological relations between concepts: (i) taxonomical—*type* or *same as*—relationships; (ii) concept difference (i.e., negative *type* relationships); and (iii) "facet" relations derived from verb frames in the input texts (e.g., *has part*, *involves* or *occurs in*). We imposed a taxonomy on the latter, considering the head verb of the respective phrase as a more generic relation (e.g., *involves expression of* was assumed to be a type of *involves*). Also, several artificial relation types were introduced to specify the semantics of some most frequent relations. Namely, (positive) *type* was considered transitive and anti-symmetric, and *same as* is set transitive and symmetric. Similarly, *part of* was assumed transitive and being inverse of *has part*.

After the initial knowledge extraction in CORAAL, EUREEKA comes into play in order to integrate the emergent statements, link them to precise domain thesauri and expose them to users via intuitive approximate querying. The remainder of this section outlines the most important features of the EUREEKA implementation that enabled its efficient deployment in CORAAL.

3.1 Relational Storage of Knowledge Bases

For low-level storage, we chose to employ a relational database, since it is a state of the art technology for scalable data management, which in addition allows for quite straightforward implementation of our framework. Considering a knowledge base $(\mathcal{E}, n, I, L, \mathcal{G})$, we can represent $\mathcal{G}, \mathcal{E}$ as two relational tables `grounding` and `entities`. The former serves for mapping of natural language inputs to unique internal identifiers and vice versa, while the latter supports the entity storage and operations according to Section 2.2.

The `grounding` table consists of the columns `lemma`, `identifier`, `scope`, `certainty` of `VARCHAR`, `INTEGER`, `INTEGER`, `FLOAT` types, and of indices `ls = (lemma,scope)`, `ic = (identifier,certainty)`. The sets I, L are given by the `identifier`, `lemma` columns, respectively (we store terms in their lemmatised, i.e., canonical lexical form). The table indices allow for a convenient and efficient implementation of the g_{ind} and g_{lan} mappings in $\mathcal{G}$ via the respective

SELECT operations. Note that inclusion of certainty into ic allows for direct access to, e.g., lexical expressions attached to an identifier with maximal positive or negative certainty. This retrieves an entity's preferred synonyms or antonyms, respectively. To save space, we use integer entity identifiers, however, these can be directly mapped to a respective URI scheme if required by an application. In the current deployment of EUREEKA, the grounding table is filled in according to the terms (and possibly their synonyms) coming from two sources: (i) EMTREE and NCI life science thesauri (cf. http://www.embase.com/emtree/, http://nciterms.nci.nih.gov, respectively); (ii) statements extracted from the Elsevier life science articles. The only lexical domains we currently distinguish are those corresponding to auxiliary *relation* and *generic* (i.e., non-relation) entities.

The entities table stores particular entities. These can be expressed as sets of relations associated with the respective certainty, as introduced in the beginning of Section 2. Such a notation can be directly transformed into a set of rows in a relational database table. However, a direct transformation of n-ary relations may be inadequate if n is not set firmly and/or if we have to store many relations with arities lower than n. These situations lead either to problems with maintenance, or to wasted space in the table.

Nevertheless, we process *subject-predicate-object* triples, all of which have a *provenance* (either an article, or a domain thesaurus), so we can explicitly represent the respective ternary relations in the entities table without wasting any space. For the representation of possible additional relation arities (such as location or other types of context), we associate each row in the entities table with a unique statement identifier stid. Then we can represent, e.g., quaternary relations in the form *bindsTo(drugX,proteinY,docID,bindingSiteZ)* as a ternary relation *at($stid_i$,bindingSiteZ,docID)*, assuming *at* grounding the fourth "location" argument. $stid_i$ is a statement identifier of *bindsTo(drugX,proteinY,docID)*. This procedure can be obviously generalised to arbitrary arities.

Following the design considerations, the entities table consists of columns stid, predicate, subject, object, provenance, certainty. All columns are INTEGER, except for the latter one, which is FLOAT. Provenance is modelled as a special entity linked to an article ID, title, text, etc. Besides the primary key (stid), indices on (subject,predicate,object), (subject,object), (object,predicate), (predicate,object) are defined. Explicit querying for provenance is not necessary in our use case (we only need to retrieve provenance as a function of particular statements), therefore we do not maintain respective indices. An entity $E \in \mathcal{E}$ directly corresponds to rows with subject equal to $ind(E)$ and to rows with subject referencing the respective stid values. The corresponding tensor entity representation $rep(E)$ can be directly constructed from the content of the rows as a multidimensional array of floats, with the necessary tensor-based operations implemented on the array data structure.

Regarding the particular implementation of entity ranking, we employ $w_{arg} = 1$ for predicate, object and $w_{arg} = 0$ for all other arguments (results in rather traditional binary hub and authority score computation). The relation weighing function makes use of the frequency of particular relation instances (i.e., number of statements having the relation identifier as a predicate): $w_{rel}(r) = \frac{1}{ln(e+f(r)-L)}$ if $f(r) \geq L$, $w_{rel}(r) = 0$ otherwise, where $f(r)$ is the absolute frequency of r. Relations with frequency below the limit are not taken into account at all. The heuristic weighing is designed to reduce the influence of very frequent, but rather generic relations (e.g., *type*), in favour of less frequent, but potentially significant ones (e.g., *involved in*). The L limit (set to 25 in the current implementation) serves for cutting accidental noise off the result. For the aggregation of the $h(E), a(E)$ hub and authority scores into $rel(E)$, we use the arithmetic mean.

3.2 Aggregating and Accessing the Emergent Knowledge

EUREEKA can smoothly merge facts extracted from different resources. This is done via decomposition of each entity into entities containing subject-predicate-object statements with equal provenance. The decomposed entities with same identifiers are merged using the *agg* operation into single entities with respective compound provenances. *agg* is implemented as weighted arithmetic mean (similarly to Example 4), with relevances 1, 0.2 for the thesauri and article provenance, respectively. This ensures significantly higher relevance of the manually designed thesauri in case of conflict with the automatically extracted knowledge.

In order to access the aggregated emergent knowledge, we implemented a service evaluating simple conjunctive queries with negation (for the query language specification, see `http://smile.deri.ie/projects/egc/quickstart`). The query evaluation and presentation of the answers is implemented essentially following Example 5[1]. In addition to the ranking of the answer entities, statements associated to an entity are sorted according to the relevance of their arguments in descending order. Example queries and selected top answer statements are (answer certainties in brackets): *Q: ? : type : breast cancer* ⇝ **cystosarcoma phylloides** *TYPE breast cancer (1); Q: rapid antigen testing : part of : ? AND ? : type : clinical study* ⇝ **dicom study** *USE protein info (0.8),* **initial study** *INVOLVED patients (0.9)*. The examples abstract from the result provenance, however, full-fledged presentation of answers to the above or any other queries can be tried live with CORAAL at `http://coraal.deri.ie:8080/coraal/`, using the *Knowledge* search tab or the guided query builder.

Currently the main means for accessing the EUREEKA deployment is the intuitive user-centric front-end in CORAAL. Applications may get RDF corresponding to the results presented in CORAAL from its Exhibit presentation layer, however, this is rather awkward. Therefore we are working on an API allowing for import and processing of arbitrary texts and RDF data in the N3 notation

[1] The translation from the query language into entity representations is quite straightforward – positive and negative crisp query statements form triple relations that are associated with maximal and minimal certainty values, respectively. Statements with variables in the "object" position are inverted, so that the query can be translated as a single entity. The answer candidates and their fitness measures are then computed on the top of (possibly nested for inverted statements) `SELECT` queries on the `entities` table, with `WHERE` conditions corresponding to the query statements.

(cf. http://www.w3.org/DesignIssues/Notation3). The processed data are to be exported as N3 RDF, with the certainties and provenance represented according to the W3C note at http://www.w3.org/TR/swbp-n-aryRelations/

4 Evaluation with Sample Users

In the CORAAL deployment, EUREEKA provides access to more than 15 million statements about ca. 350,000 unique entities that are referred to by about 620,000 natural language terms. The knowledge base is covering ca. 11,700 Elsevier articles mostly related to cancer research and treatment. With an assistance of a three-member domain expert evaluation committee, we assessed issues deemed to be most important by the committee regarding applicability of the framework: (i) ease of use, real-time response; (ii) quality of answers to queries (users want to have as many good results entailed by the articles and thesauri as possible); (iii) appropriateness of the result ranking (users want to find the relevant results on the top). Note that we do not discuss evaluation of the document retrieval here, since it is related to the CORAAL search engine as such, but not to the main contribution of this paper (presentation of the general emergent knowledge processing framework).

Ease of use was addressed by the simple queries close to natural language, guided query builder and faceted browsing (supported by Exhibit, cf. http://simile-widgets.org/exhibit/), all offered within the EUREEKA front-end in CORAAL. The response is actually not an issue – results are presented within units of seconds in CORAAL (up to 90% of the lag owing to the HTML rendering overhead, not to the query processing itself). The two remaining issues were mapped to these tasks: (i) assessing correctness (i.e., precision) and completeness (i.e., recall) of variable instances provided within answers to significant queries; (ii) assessing number of relevant statements as a function of their rank in answers. The latter task was evaluated using significant entities as queries (such results in effect provide statements assumed to be related to the query entities based on the fitness, similarity and relevance in direct correspondence to raw results in Example 5).The significance of queries and entities to be used for the evaluation was determined as follows. First we picked 100 random entity names and generated 100 random queries based on the extracted content. We let the evaluation committee assess the significance of respective concept and statement queries by 1-5 marks (best to worst). We used the following best-scoring queries—Q_1 : *? : type : breast cancer;* Q_2 : *? : part of : immunization;* Q_3 : *? : NOT type : chronic neutrophilic leukemia;* Q_4 : *rapid antigen testing : part of : ? AND ? : type : clinical study;* Q_5 : *? : as : complementary method AND ? : NOT type : polymerase chain reaction*—and entities—E_1 : *myelodysplastic syndrome;* E_2 : *p53;* E_3 : *BAC clones;* E_4 : *primary cilia;* E_5 : *colorectal cancer.*

For a base-line comparison, we employed the open source edition of OpenLink Virtuoso (cf. http://tinyurl.com/cf8ga2), a triple store with database

Table 1. Summary of the results

Approach	Correctness and completeness						Relevance per answer ranking				
	P	R	F	P_{nn}	R_{nn}	F_{nn}	1-10	11-50	51-100	101-200	201-...
EUREEKA	0.719	0.583	0.586	0.532	0.305	0.310	0.780	0.668	0.430	0.227	0.091
BASE	0.169	0.053	0.067	0.281	0.088	0.111	0.300	0.229	0.293	0.172	0.188

back-end supporting rule-based RDFS inference and querying[2]. The content fed to EUREEKA was transformed to crisp RDFS, omitting the unsupported negative statements and provenance arguments before import to the base-line. EUREEKA queries were mapped to statements with unique entity identifiers as per the `grounding` table and then translated to respective SPARQL equivalents to be executed using the base-line.

The evaluation results are summed up in Table 1. P, R, F columns contain precision, recall and F-measure ($\sim \frac{2(PR)}{P+R}$), respectively, averaged across the results of all evaluated queries. $X_{nn}, X \in \{P, R, F\}$ relate to average results of non-negative queries only (Q_1, Q_2, Q_4). Particular P, R values were computed as $P = \frac{c_r}{a_r}, R = \frac{c_r}{c_a}$, where c_r, a_r is a number of *relevant* and all answer entities returned, respectively. c_a is the number of all entities relevant to the query, as entailed by the documents in the CORAAL corpus (determined by the evaluation committee by means of manual analysis of full-text search results related to the entities occurring in the evaluated queries). The columns in the right hand part of Table 1 contain average values $\frac{s_r}{sz}$, where s_r, sz is the number of *relevant* and all statements in a given ranking range, respectively. The average goes across results corresponding to E_{1-5} query entities. The *relevance* was determined by unequivocal agreement of the evaluation committee. Results with certainty lower than 0.5 were disregarded (i.e., a statement was considered as a false positive iff it was deemed irrelevant and its absolute certainty value was 0.5 or more).

Regarding *correctness and completeness*, our approach offers almost three-times better results in terms of F-measure than the base-line. That holds for the negation-free queries supported by both frameworks. Obviously, the difference is even bigger for generic queries having no base-line results in two out of five cases. The increase in EUREEKA's precision was directly due to its two novel features unsupported by the base-line: (i) relevance-based aggregation of the initially extracted input; (ii) explicitly presented certainty of the results allowing for disregarding presumably uncertain ones. The increase in recall was caused by the approximate query evaluation that included also some correct results from answers with fitness lower than 1 (similar behaviour is not directly supported

[2] Alternatives [13,7] capable of either arbitrary meta-knowledge, or explicit trust representation in RDF were considered, too. However, the respective implementations allow neither for soft aggregation of emergent entities, nor for inherent exploitation of certainty in approximate answering of queries close to natural language. They can only expose the certainty and/or meta-knowledge via extended SPARQL queries. Therefore their capabilities are essentially equal to the "plain" Virtuoso RDF store base-line regarding our use case, while Virtuoso handles the relatively large amount of data more efficiently, presumably due to more mature data management engine.

in the base-line). The relevance of EUREEKA answers is a clearly decreasing function of the ranking. However, no similar pattern can be seen for the base-line.

The absolute EUREEKA results may still be considered rather poor (F-measure around 0.3), but the evaluation committee unequivocally considered the ability of EUREEKA to perform purely automatically as an acceptable trade-off for the presence of some noise in the not-entirely-complete results. In conclusion, the evaluation with sample users confirmed that the innovative principles of the proposed approach lead to a better applicability in the current use case, when compared to a base-line state of the art solution.

5 Related Work

An implemented approach [4] generalising Description Logics in order to support vagueness as one form of uncertainty exists, however, it does not allow for straightforward representation of contextual features. Moreover, logics-based approaches are usually not able to infer many meaningful conclusions from the rather sparse and noisy emergent inputs [3], which renders the querying in our use case practically unachievable if based on the logical inference.

The works [13,7] propose generic framework for representing contextual features like certainty or provenance in RDF. These features are considered rather as "annotations" of RDF triples and thus can be merely queried for. It is impossible to use the certainty as a first class citizen for robust entity integration and/or query answering, unless one builds an ad hoc application tackling that on the top of either [13], or [7]. Similarity-based query post-processing with imprecision support is tackled by [8], however, the suggested iSPARQL framework handles uncertainty merely concerning query result filtering, disregarding a priori imprecise knowledge. This makes it rather inapplicable both to partial query evaluation and processing of the emergent uncertain knowledge before the actual querying. The work [11] extends the crisp RDF semantics by fuzzy degrees, but supports neither robust querying nor integration capabilities, nor context representation. Integration of RDF ontologies based on graph theory is tackled in [15], but incorporation of certainty degrees and contextual features into the presented method is non-trivial, since [15] is based on crisp binary relations.

Papers [1,12] research techniques for ranking ontology concepts and anytime RDF query answering, respectively. The former approach is applicable for relevance-based sorting of query results, while the latter is apt for general robust, approximate and scalable query answering. However, both [1,12] lack explicit support for uncertainty and contextual features.

All the approaches discussed so far also neglect as clearly defined and universal interface between the lexical and computational aspects of semantics as proposed in our approach. The Textrunner framework [2] provides an expressive search service based on natural language, which is very similar to the deployment of our framework in CORAAL. However, the framework provides neither for extracted knowledge integration, nor for complex (i.e., conjunctive or negative) querying, lacking an appropriate underlying computational semantics model.

Conceptual spaces [6], a geometrical formalisation of meaning, shares some similarities with our approach, namely uncertainty-aware, non-logical nature of representation, and multi-dimensionality of concept features. However, exploitation of emergent relational statements is not particularly straightforward within the framework, since it is tailored primarily to non-symbolic connectionist input. Moreover, there is neither a standardised implementation, nor a universal and intuitively applicable querying mechanism available for conceptual spaces.

6 Conclusions and Future Work

We have introduced a framework that addresses all the challenges specified in Section 1 on a well-founded basis. The framework has been implemented in the form of a respective EUREEKA prototype. We applied and evaluated the prototype within a practical use case of knowledge-based life science publication search. Our approach is novel and promising regarding practical emergent knowledge processing, which has been proven not only by the results presented here, but also by our successful participation in the Elsevier Grand Challenge contest (cf. `http://www.elseviergrandchallenge.com/`).

In the near future, we are going to extend the user-centric query language by contexts and release the extended EUREEKA implementation as an open source module. In longer term, we have to investigate import of more complex ontologies into EUREEKA – so far we have covered only rather simple RDFS semantics of life science thesauri. Last but not least, we intend to provide means for distributed implementation of the principles introduced here in order to scale the framework up to arbitrarily large data.

Acknowledgments. We have been supported by the 'Líon II' project funded by SFI under Grant No. SFI/08/CE/I1380. Deployment of EUREEKA within CORAAL would not be possible without the great work of Tudor Groza and much appreciated senior support of Siegfried Handschuh. Finally, we are very grateful to our evaluators and testers: Doug Foxvog, Peter Gréll, MD, Miloš Holánek, MD, Matthias Samwald, Holger Stenzhorn and Jiří Vyskočil, MD.

References

1. Alani, H., Brewster, C., Shadbolt, N.: Ranking ontologies with AKTiveRank. In: Cruz, I., Decker, S., Allemang, D., Preist, C., Schwabe, D., Mika, P., Uschold, M., Aroyo, L.M. (eds.) ISWC 2006. LNCS, vol. 4273, pp. 1–15. Springer, Heidelberg (2006)
2. Banko, M., Etzioni, O.: The tradeoffs between open and traditional relation extraction. In: Proceedings of ACL 2008: HLT, pp. 28–36. ACL (2008)
3. Bechhofer, S., et al.: Tackling the ontology acquisition bottleneck: An experiment in ontology re-engineering (April 2003), `http://tinyurl.com/96w7ms`
4. Bobillo, F., Straccia, U.: FuzzyDL: An expressive fuzzy description logic reasoner. In: Proceedings of FUZZ 2008 (2008)

5. Buitelaar, P., Cimiano, P.: Ontology Learning and Population. IOS Press, Amsterdam (2008)
6. Gärdenfors, P.: Conceptual Spaces: The Geometry of Thought. MIT Press, Cambridge (2000)
7. Hartig, O.: Querying Trust in RDF Data with tSPARQL. In: Aroyo, L., Traverso, P., Ciravegna, F., Cimiano, P., Heath, T., Hyvönen, E., Mizoguchi, R., Oren, E., Sabou, M., Simperl, E. (eds.) ESWC 2009. LNCS, vol. 5554, pp. 5–20. Springer, Heidelberg (2009)
8. Kiefer, C., Bernstein, A., Stocker, M.: The fundamentals of isparql: A virtual triple approach for similarity-based semantic web tasks. In: ISWC/ASWC (2007)
9. Kleinberg, J.: Authoritative sources in a hyperlinked environment. Journal of the ACM 46(5) (1999)
10. Maedche, A., Staab, S.: Discovering conceptual relations from text. In: Proceedings of ECAI 2000. IOS Press, Amsterdam (2000)
11. Mazzieri, M.: A fuzzy RDF semantics to represent trust metadata. In: Proceedings of SWAP 2004 (2004)
12. Oren, E., Guéret, C., Schlobach, S.: Anytime query answering in RDF through evolutionary algorithms. In: Sheth, A.P., Staab, S., Dean, M., Paolucci, M., Maynard, D., Finin, T., Thirunarayan, K. (eds.) ISWC 2008. LNCS, vol. 5318, pp. 98–113. Springer, Heidelberg (2008)
13. Schueler, B., Sizov, S., Staab, S., Tran, D.T.: Querying for meta knowledge. In: Proceedings of WWW 2008. ACM Press, New York (2008)
14. Tversky, A.: Features of similarity. Psychological Review 84(2), 327–352 (1977)
15. Udrea, O., Deng, Y., Ruckhaus, E., Subrahmanian, V.S.: A graph theoretical foundation for integrating RDF ontologies. In: Proceedings of AAAI 2005 (2005)
16. Voelker, J., Vrandecic, D., Sure, Y., Hotho, A.: Learning disjointness. In: Franconi, E., Kifer, M., May, W. (eds.) ESWC 2007. LNCS, vol. 4519, pp. 175–189. Springer, Heidelberg (2007)

On Detecting High-Level Changes in RDF/S KBs

Vicky Papavassiliou[1,2], Giorgos Flouris[1], Irini Fundulaki[1], Dimitris Kotzinos[1,3], and Vassilis Christophides[1,2]

[1]FORTH-ICS, Greece
[2] University of Crete, Greece
[3] TEI of Serres, Greece
{papavas,fgeo,fundul,kotzino,christop}@ics.forth.gr

Abstract. An increasing number of scientific communities rely on Semantic Web ontologies to share and interpret data within and across research domains. These common knowledge representation resources are usually developed and maintained manually and essentially co-evolve along with experimental evidence produced by scientists worldwide. Detecting automatically the differences between (two) versions of the same ontology in order to store or visualize their deltas is a challenging task for e-science. In this paper, we focus on languages allowing the formulation of concise and intuitive deltas, which are expressive enough to describe unambiguously any possible change and that can be effectively and efficiently detected. We propose a specific language that provably exhibits those characteristics and provide a change detection algorithm which is sound and complete with respect to the proposed language. Finally, we provide a promising experimental evaluation of our framework using real ontologies from the cultural and bioinformatics domains.

1 Introduction

An increasing number of scientific communities rely on Semantic Web ontologies to share and interpret data within and across research domains (e.g., Bioinformatics or Cultural Informatics[1]). These community ontologies are usually developed and maintained manually while essentially co-evolve along with experimental evidence produced by scientists worldwide. Managing the differences (*deltas*) of ontology versions has been proved to be an effective and efficient method in order to synchronize them [5] or to explain the evolution history of a given ontology [13]. In this paper, we are interested in automatically detecting *both schema and data changes* occurring between asynchronously produced ontology versions.

Unless they are assisted by collaborative ontology development tools [8,9], ontology editors are rarely able or willing to systematically record the changes performed to obtain an ontology version. In particular, when there is no central authority responsible for ontology curation, manually created deltas are often absent, incomplete, or even erroneous [22]. Existing ontology diff tools, such as PromptDiff [14], SemVersion [23] and others [24] aim to satisfy this need. These tools are essentially based on a *language*

[1] www.geneontology.org, cidoc.ics.forth.gr

A. Bernstein et al. (Eds.): ISWC 2009, LNCS 5823, pp. 473–488, 2009.

of changes, which describes the semantics of the different change operations that the underlying algorithm understands and detects.

In its simplest form, a language of changes consists of only two *low-level* operations, *Add(x)* and *Delete(x)*, which determine individual constructs (e.g., triples) that were added or deleted [23,24]. In [10,14,15,17,19,22], *high-level* change operations are employed, which describe more complex updates, as for instance the insertion of an entire subsumption hierarchy. A high-level language is preferable than a low-level one, as it is more *intuitive*, *concise*, *closer to the intentions* of the ontology editors and *captures more accurately* the semantics of a change [10,21].

However, detecting high-level change operations introduces a number of issues. As the detectable changes get more complicated, so does the detection algorithm; complicated changes involve complicated detection procedures which may be inefficient, or based on matchers [6] and other heuristic-based techniques [10] that make it difficult to provide any formal guarantees on the detection properties. Another problem stems from the fact that it is impossible to define a complete list of high-level changes [10], so there is no agreed "standard" set of operations that one could be based on. Moreover, it is difficult to specify a language of changes that will be both high-level and able to handle all types of modifications (even fine-grained ones) upon an ontology.

The main contributions of our work are:

- the introduction of a *framework* for defining changes and of a *formal language of changes* for RDF/S ontologies [2,12] which considers operations in both data and schema and satisfies several desirable properties;
- the design of an *efficient* change detection algorithm which is *sound* and *complete* with respect to the proposed language;
- the experimental evaluation of our framework using *real ontologies* from the cultural (CIDOC [4]) and biological (GO [7]) domains.

The paper is organized as follows: Section 2 presents a motivating example that will be used for visualization purposes throughout the paper. In Section 3, we introduce the basic notions of RDF [12] and RDFS [2] as well as our language of high-level changes and show that the language and the proposed detection algorithm satisfy several desirable properties. Section 4 describes changes which require heuristics and matchers in order to be detected, thus extending our basic framework to include operations that are interesting in practice. Section 5 presents our experimental findings on real ontologies and section 6 discusses related work. We conclude in Section 7.

2 Motivating Example

In Figure 1 an example inspired from the CIDOC Conceptual Reference Model [4] is depicted; CIDOC is a core ontology intended to facilitate the integration, mediation and interchange of heterogeneous cultural heritage information. Table 1 shows the added and deleted triples (the low-level delta) as well as the high-level change operations that our approach will detect in this example. The table makes clear that even though the low-level delta contains all the changes that have been performed, it is not really useful as it captures the syntactical manipulations that led to the change, rather than

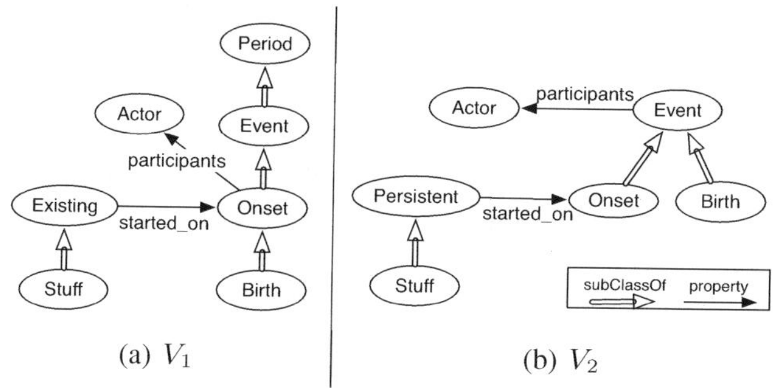

Fig. 1. Motivating Example

Table 1. Detected Low-level and High-level Changes (From Figure 1)

Added Triples (Low-Level Delta)	Deleted Triples (Low-Level Delta)	Detected Changes (High-Level Delta)
(participants, domain, Event)	(participants, domain, Onset)	Generalize_Domain(participants, Onset, Event)
(Birth, subClassOf, Event)	(Birth, subClassOf, Onset)	Pull_up_Class(Birth, Onset, Event)
–	(Period, type, class)	Delete_Class(Period, ∅, {Event}, ∅, ∅, ∅, ∅)
–	(Event, subClassOf, Period)	
(Stuff, subClassOf, Persistent)	(Stuff, subClassOf, Existing)	Rename_Class(Existing, Persistent)
(started_on, domain, Persistent)	(started_on, domain, Existing)	
(Persistent, type, class)	(Existing, type, class)	

the intentions of the editor. Our work is motivated by the belief that the "aggregation" of several low-level changes into more *coarse-grained*, *concise* and *intuitive* high-level changes (third column of Table 1) would lead to more useful deltas.

For instance, consider the change in the domain of property $participants$ from $Onset$ to $Event$ (Figure 1). The low-level delta reports two "changes", namely the deletion and the insertion of a domain for the property whereas the reported high-level operation *Generalize_Domain* combines them into one, capturing also the fact that the old domain is a *subclass of* the new domain (by exploiting semantical information in the two versions). A similar case appears in the change of the position of $Birth$ in the subsumption hierarchy which our framework reports as *Pull_Up_Class* and in the deletion of class $Period$ where the deletion of all edges originating from, or ending in, the deleted class are combined in a single operation. Regarding the latter, only a subclass relation is deleted ($Event$), whereas other relations, such as superclasses, supertypes, subtypes, comments and labels are absent (denoted by empty sets in Table 1). In total, only 4 high-level changes will be reported as opposed to 12 low-level ones.

Apart from being more concise, the reported high-level changes are also more intuitive. For example, the *Generalize_Domain* operation provides the additional information that the new domain is a superclass of the old. This may be useful for the evaluation and understanding of the change performed. For example, if we know only that a domain changed we cannot presume anything about the validity of the existing data, but if we know that the domain changed to a superclass we can assume, according to the RDF/S specification, that validity is not violated [10].

Another interesting example is the *Rename_Class* operation, which is reported instead of the deletion of class $Existing$ and the subsequent addition of $Persistent$. Unlike the changes discussed so far, the detection of *Rename* (as well as other operations, such as *Merge* and *Split*) requires the use of a matcher that would identify the two concepts ($Existing$ and $Persistent$) to be the same entity using heuristics.

Note also that all the triples of the low-level delta (in Table 1) are associated with one, and only one, high-level change. This allows the partition of low-level changes into well-defined high-level changes, in a unique way. This property guarantees that the detection algorithm will be able to handle all possible low-level deltas in a *deterministic* manner, i.e., that any set of low-level changes between two versions would be associated with *one, and only one*, set of high-level changes. The latter requirement calls for a careful definition of the change operations and is not directly related to the detection algorithm *per se*. Thus, we claim that the detection algorithm should be based on the defined language of changes, instead of the other way around.

Defining a language with the above properties is a challenging task, because it requires establishing a tradeoff between partly conflicting requirements. On the one hand, coarse-grained operations are necessary in order to achieve concise and intuitive deltas. On the other, fine-grained operations are necessary in order to capture subtle differences between a pair of versions. The existence of both fine-grained and coarse-grained operations in the language may allow the association of the same set of low-level changes with several different sets of high-level ones, thus jeopardizing determinism. In the next section, we will describe a language and a detection algorithm that avoids these problems and provably satisfies the above properties, while being efficient.

3 Change Detection Framework, Language and Algorithm

3.1 Formal Definitions

The representation of knowledge in RDF [12] is based on triples of the form (*subject*, *predicate*, *object*). Assuming two disjoint and infinite sets **U**, **L**, denoting the URIs and literals respectively, $\mathcal{T} = \mathbf{U} \times \mathbf{U} \times (\mathbf{U} \cup \mathbf{L})$ is the set of all triples. An *RDF Graph* V is defined as a set of triples, i.e., $V \subseteq \mathcal{T}$. RDFS [2] introduces some built-in classes (class, property) which are used to determine the *type* of each resource. Following the approach of [20], we assume that each resource is associated with one type determined by the triples that the resource participates in. The typing mechanism allows us to concentrate on nodes of RDF Graphs, rather than triples, which is closer to ontology curators' perception and useful for defining intuitive high-level changes. RDFS [2] provides also *inference semantics*, which is of two types, namely *structural inference* (provided mainly by the transitivity of subsumption relations) and *type inference* (provided by the typing system, e.g., if p is a property, the triple (p, type, property) can be inferred). The RDF Graph containing all triples that are either explicit or can be inferred from explicit triples in an RDF Graph V (using *both* types of inference), is called the *closure* of V and is denoted by $Cl(V)$. An *RDF/S Knowledge Base* (*RDF/S KB*) V is an RDF Graph which is closed with respect to *type inference*, i.e., it contains all the triples that can be inferred from V using type inference.

For a pair of RDF/S KBs ($V_1, V_2 \subseteq \mathcal{T}$), we define their *low-level delta* in a manner similar to symmetric difference:

Definition 1. *Let V_1, V_2 be two RDF/S KBs. The* low-level delta *between V_1, V_2, denoted by $\Delta(V_1, V_2)$ (or simply Δ) is a pair of sets of triples defined as: $\Delta(V_1, V_2) = \langle V_2 \setminus V_1, V_1 \setminus V_2 \rangle$. For brevity, we will use the notation Δ_1, Δ_2 for $V_2 \setminus V_1, V_1 \setminus V_2$ respectively ($\Delta_1 \subseteq \mathcal{T}$, $\Delta_2 \subseteq \mathcal{T}$).*

Note that Δ_1 corresponds to the triples *added* in V_1 to get V_2, and Δ_2 corresponds to the triples *deleted* from V_1 to get V_2. The low-level delta alone may not be enough to fully capture the intuition behind a change: sometimes we need to consider conceptual information that remained unchanged (see, e.g., the change of the domain of $participants$ in Figure 1 and the subsequent analysis in Section 2). Therefore, the definition of the detection semantics should consist of the triple(s) that must exist in the low-level delta, as well as of a set of conditions that must hold (in V_1 and/or V_2) in order for the detection to take place:

Definition 2. *A* change *c is defined as a triple $\langle \delta_1, \delta_2, \phi \rangle$, where:*

- $\delta_1 \subseteq \mathcal{T}$: required added triples. *Corresponds to the triples that should be in V_2 but not in V_1 (i.e., in Δ_1), in order for c to be detected.*
- $\delta_2 \subseteq \mathcal{T}$: required deleted triples. *Corresponds to the triples that should be in V_1 but not in V_2 (i.e., in Δ_2), in order for c to be detected.*
- ϕ: required conditions. *Corresponds to the conditions that should be true in order for c to be detected. A condition is a logical formula consisting of atoms of the form $t \in V$ or $t \notin V$, where $t \in \mathcal{T}$ and V is of the form V_i or $Cl(V_i)$ for $i \in \{1, 2\}$.*

For simplicity, we will denote by $\delta_1(c)$ ($\delta_2(c)$) the required added (deleted) triples of a change c, and by $\phi(c)$ the required conditions of c. Tables 2, 3 show the definition of some high-level changes. The complete list of defined changes can be found at [16]. We restrict our attention to changes for which $\delta_1 \cup \delta_2 \neq \emptyset$ and $\delta_1 \cap \delta_2 = \emptyset$. The first condition guarantees that at least *something* must be in Δ_1 or Δ_2 for a change to be detected. The second condition guarantees that no change would require the addition and deletion of the same triple to happen at the same time.

As discussed in Section 2, both fine-grained and coarse-grained high-level changes are necessary in order to support determinism, conciseness and intuitiveness. For this reason, we follow a common approach in the literature [10,17,21] and classify high-level changes into *basic* and *composite*. Basic changes are fine-grained and describe a change in one node or edge of the RDF/S KB taking into account RDF/S semantics. On the other hand, composite changes are coarse-grained and closer to the user's intuition, as they describe, in a concise way, changes affecting several nodes and/or edges of the RDF/S KB. The introduction of the two levels should be done carefully, as it may cause problems with determinism. For instance, in the motivating example (Figure 1 and Table 1), the deleted triple ($participants$, domain, $Onset$) could be associated with the basic change *Delete_Domain(participants,Onset)*, as well as with the composite change *Generalize_Domain(participants,Onset,Event)* (see also Tables 2, 3). This double association would jeopardize determinism, because the same low-level delta would correspond to two different high-level deltas. To avoid this problem, we define two different

Table 2. Formal Definition of some Basic Changes

Change	Delete_Superclass(x,y)	Add_Property_Instance(x_1, x_2,y)	Delete_Domain(x,y)
Intuition	IsA between x, y is deleted	Add property instance (x_1, y, x_2)	Domain y of property x is deleted
δ_1	$\emptyset$	$\{(x_1, y, x_2)\}$	$\emptyset$
δ_2	$\{(x, \text{subClassOf}, y)\}$	$\emptyset$	$\{(x, \text{domain}, y)\}$
ϕ	$(x, \text{type}, \text{class}) \in Cl(V_1)$	$(y, \text{type}, \text{property}) \in Cl(V_2)$	$(x, \text{type}, \text{property}) \in Cl(V_1)$

Table 3. Formal Definition of some Composite Changes

Change	Generalize_Domain(x,y,z)	Change_Domain(x,y,z)	Reclassify_Individual_Higher(x,Y,Z)
Intuition	Domain of property x changes from y to a superclass z	Domain of property x changes from y to a non-subclass/superclass z	Individual x is reclassified from class(es) Y to superclass(es) Z
δ_1	$\{(x, \text{domain}, z)\}$	$\{(x, \text{domain}, z)\}$	$\{(x, \text{type}, z) \| z \in Z\}$
δ_2	$\{(x, \text{domain}, y)\}$	$\{(x, \text{domain}, y)\}$	$\{(x, \text{type}, y) \| y \in Y\}$
ϕ	$(x, \text{type}, \text{property}) \in Cl(V_1) \wedge$ $(x, \text{type}, \text{property}) \in Cl(V_2) \wedge$ $(y, \text{subClassOf}, z) \in Cl(V_1) \wedge$ $(y, \text{subClassOf}, z) \in Cl(V_2)$	$(x, \text{type}, \text{property}) \in Cl(V_1) \wedge$ $(x, \text{type}, \text{property}) \in Cl(V_2) \wedge$ $((y, \text{subClassOf}, z) \notin Cl(V_1) \vee$ $(y, \text{subClassOf}, z) \notin Cl(V_2)) \wedge$ $((z, \text{subClassOf}, y) \notin Cl(V_1) \vee$ $(z, \text{subClassOf}, y) \notin Cl(V_2))$	$(x, \text{type}, \text{resource}) \in Cl(V_1) \wedge$ $(x, \text{type}, \text{resource}) \in Cl(V_2) \wedge$ $\forall y \in Y, \forall z \in Z :$ $(y, \text{subClassOf}, z) \in Cl(V_1) \wedge$ $(y, \text{subClassOf}, z) \in Cl(V_2)$

notions, *detectability* and *initial detectability*, and postulate that the detection of composite changes takes precedence over the detection of basic ones.

Definition 3. *Consider two RDF/S KBs V_1, V_2, their respective $\Delta(V_1, V_2)$ and a change c. Then, c is* initially detectable *iff $\delta_i(c) \subseteq \Delta_i$, $i \in \{1, 2\}$, and $\phi(c)$ is true.*
If c is a composite change, then c is detectable *iff it is initially detectable.*
If c is a basic change, then c is detectable *iff it is initially detectable and there is no initially detectable composite change (say c') for which $\delta_i(c) \subseteq \delta_i(c')$, $i \in \{1, 2\}$ and $\phi(c') \vdash \phi(c)$.*

In our running example, *Change_Domain(participants,Onset,Event)* is not initially detectable (thus, not detectable) because its conditions are not true (specifically, the part: $(Onset, \text{subClassOf}, Event) \notin Cl(V_1) \vee (Onset, \text{subClassOf}, Event) \notin Cl(V_2)$). On the other hand, *Generalize_Domain(participants,Onset,Event)* is initially detectable; given that it is a composite change, it is also detectable. Finally, the basic change *Delete_Domain(participants,Onset,Event)* is initially detectable, but not detectable (because *Generalize_Domain(participants,Onset,Event)* is initially detectable).

So far, we were only concerned with detection semantics of changes. However, changes can also be applied upon RDF/S KBs, where the application and detection semantics of a set of changes should be consistent. To be more precise, given two RDF/S KBs V_1, V_2, the application (upon V_1) of the delta computed between them should give V_2, irrespective of the order of application of the changes [24]. Therefore, we also need to define the application semantics of changes:

Definition 4. *Consider an RDF/S KB V and a change c. The* application *of c upon V, denoted by $V \bullet c$ is defined as: $V \bullet c = (V \cup \delta_1(c)) \setminus \delta_2(c)$.*

As an example, the application of *Generalize_Domain(participants,Onset,Event)* would lead to the addition of the triple $(participants, \text{domain}, Event)$ and the deletion of $(participants, \text{domain}, Onset)$.

3.2 Formal Results on the Proposed Language of Changes

Our framework was used to define $\mathcal{L}$, a specific language of changes (some of which are shown in Tables 2, 3) that satisfies several interesting properties. For a full definition of $\mathcal{L}$ and the proofs of the described properties, see [16].

First of all, $\mathcal{L}$ should conform to the property of *Completeness* by capturing any possible change, so that the detection algorithm can always process the input and return a delta. Moreover, the language should satisfy the property of *Non-ambiguity* by associating each low-level change with one, and only one, high-level change, and each set of low-level changes with one, and only one, set of high-level ones. These two properties are needed in order to guarantee that $\mathcal{L}$ supports a *deterministic detection process*.

Theorem 1. *Consider two RDF/S KBs V_1, V_2, their respective $\Delta(V_1, V_2) = \langle\Delta_1, \Delta_2\rangle$ and the set $C = \{c \in \mathcal{L} | c : detectable\}$. Then, for any $i \in \{1, 2\}$ and $t \in \Delta_i$, there is some $c \in C$ such that $t \in \delta_i(c)$.*

This theorem proves that $\mathcal{L}$ satisfies the property of *Completeness*. In order to prove that it also satisfies the property of *Non-ambiguity*, we must first show that each low-level change is associated with at most one detectable high-level change.

Theorem 2. *Consider two RDF/S KBs V_1, V_2, their respective $\Delta(V_1, V_2) = \langle\Delta_1, \Delta_2\rangle$ and two changes $c_1, c_2 \in \mathcal{L}$. Then one of the following is true:*

1. $\delta_i(c_1) \cap \delta_i(c_2) = \emptyset$ *for* $i \in \{1, 2\}$
2. $\delta_i(c_1) \nsubseteq \Delta_i$ *or* $\delta_i(c_2) \nsubseteq \Delta_i$ *for some* $i \in \{1, 2\}$
3. $\phi(c_j)$ *is not true for some* $j \in \{1, 2\}$
4. c_j *is a basic change,* c_k *is a composite change and* $\delta_1(c_j) \subseteq \delta_1(c_k)$, $\delta_2(c_j) \subseteq \delta_2(c_k)$ *and* $\phi(c_k) \vdash \phi(c_j)$ *for some* $j, k \in \{1, 2\}$, $j \neq k$

This theorem shows that the changes in $\mathcal{L}$ have been chosen in such a way that a change is either not detectable, or irrelevant to other detectable changes. In particular, if condition 1 is true then the required added and deleted triples of c_1 are disjoint from the ones of c_2. Hence, c_1, c_2 cannot be associated with the same low-level change. If conditions 2 or 3 are true then at least one of c_1, c_2 is not detectable (by Definition 3), so, again, a low-level change cannot be associated with both changes. Finally, if condition 4 is true, then change c_k is composite and more "general" than the basic change c_j. Therefore, by Definition 3 again, even if both of them are initially detectable, only c_k will be detectable. The usability of this theorem is to set the conditions that should hold for a change in order to allow us to add it to $\mathcal{L}$ without jeopardizing determinism.

Given this analysis, the following theorem is straightforward and proves that any two changes in $\mathcal{L}$ are non-ambiguous, ergo $\mathcal{L}$ satisfies the property of *Non-ambiguity*:

Theorem 3. *Consider two RDF/S KBs V_1, V_2, their respective $\Delta(V_1, V_2) = \langle\Delta_1, \Delta_2\rangle$ and the set $C = \{c \in \mathcal{L} | c : detectable\}$. Then, for any two changes $c_1, c_2 \in C$, it holds that $\delta_i(c_1) \cap \delta_i(c_2) = \emptyset$ for $i \in \{1, 2\}$.*

Theorems 1 and 3 guarantee a deterministic detection process. To see this, take any V_1, V_2, i.e., any Δ, and any triple $t \in \Delta$: by Theorem 1, t is associated with at least one detectable high-level change; moreover, by theorem 3, all detectable high-level changes have disjoint sets of required added (and deleted) triples; thus, t is associated with exactly one detectable high-level change. This means that any set of low-level changes can be fully partitioned into disjoint subsets, each subset being associated with a single detectable high-level change.

In the rest of this subsection, we will consider the application of changes and show that the detection and application semantics are such that, given V_1, V_2, the application of the set of detectable changes between them upon V_1 would give V_2. Before showing that, we must generalize Definition 4 to apply for sets of changes; given that elements in a set are unordered, before doing this generalization, we must first guarantee that the order of application does not matter.

Definition 5. *Two changes* c_1, c_2 *are called* conflicting *iff* $(\delta_1(c_1) \cap \delta_2(c_2)) \cup (\delta_1(c_2) \cap \delta_2(c_1)) \neq \emptyset$. *A set* C *of changes is called* conflicting *iff* C *contains at least one pair of conflicting changes.*

By definition, c_1, c_2 are conflicting iff the detection of c_1 requires the addition (or deletion) of a triple whose deletion (or addition) is required by c_2. For example, *Delete_Domain(participants,Event)* is conflicting with *Change_Domain(participants, Onset,Event)*, because the detection of the former requires the deletion of ($participants$, domain, $Event$) whereas the latter requires the same triple to be added. It is easy to see that when applying a conflicting set of changes upon a version, the order matters (e.g., in the above example, depending on the order, $Event$ would, or would not, be the domain of $participants$); however, for non-conflicting sets of changes, the order is irrelevant:

Theorem 4. *Consider an RDF/S KB* V *and a non-conflicting set of changes* $C = \{c_1, \ldots, c_n\}$. *Then, for any permutation* π *over the set of indices* $\{1, \ldots, n\}$ *it holds that:* $(\ldots((V \bullet c_1) \bullet c_2) \bullet \ldots) \bullet c_n = (\ldots((V \bullet c_{\pi(1)}) \bullet c_{\pi(2)}) \bullet \ldots) \bullet c_{\pi(n)}$.

Definition 6. *Consider an RDF/S KB* V *and a non-conflicting set of changes* $C = \{c_1, \ldots, c_n\}$. *The* application *of* C *upon* V, *denoted by* $V \bullet C$, *is defined as:* $V \bullet C = (\ldots((V \bullet c_1) \bullet c_2) \bullet \ldots) \bullet c_n$.

Theorem 5. *Consider two RDF/S KBs* V_1, V_2 *and the set* $C = \{c \in \mathcal{L} | c : detectable\}$. *Then* C *is non-conflicting and* $V_1 \bullet C = V_2$.

Definition 6 is the generalization of Definition 4 for non-conflicting sets. Given that we cannot define the application of sets of changes for conflicting sets, the result that C is non-conflicting is a critical part of Theorem 5. Theorem 5 shows that we can apply the detected delta upon one version in order to get the other.

An interesting corollary of Theorem 4 is that changes are *composable*, i.e., they can be applied either simultaneously or sequentially:

Theorem 6. *Consider an RDF/S KB* V *and two sets of changes* C_1, C_2 *such that* C_1, C_2, $C_1 \cup C_2$ *are non-conflicting. Then:* $(V \bullet C_1) \bullet C_2 = (V \bullet C_2) \bullet C_1 = V \bullet (C_1 \cup C_2)$.

Another useful property of $\mathcal{L}$ is *Reversibility*, i.e., for each change c, there is some change whose application cancels the effects of c. Thus, by keeping only the newest version of an RDF/S KB and the changes that led to it, previous versions can be restored.

Table 4. Look-up Table (Excerpt)

Low-Level Change Considered	Low-Level Change(s) Searched For	Potential High-Level Change
$(x, \text{domain}, z) \in \Delta_2$	–	*Delete_Domain(x,z)*
$(x, \text{domain}, z) \in \Delta_2$	$(x, \text{domain}, y) \in \Delta_1$	*Change_Domain(x,y,z)*
$(x, \text{domain}, z) \in \Delta_2$	$(x, \text{domain}, y) \in \Delta_1$	*Generalize_Domain(x,y,z)*

Definition 7. *A change* c_1 *is called the* reverse *of* c_2 *iff* $\delta_1(c_1)=\delta_2(c_2)$ *and* $\delta_2(c_1)=\delta_1(c_2)$.

Theorem 7. *Consider two changes* c_1, c_2 *such that* c_2 *is the reverse of* c_1. *Then,* c_1 *is the reverse of* c_2 *and* c_1, c_2 *are conflicting.*

Theorem 8. *Every change in* $\mathcal{L}$ *has a unique reverse.*

Theorem 8 shows that the reverse of a change always exists and is unique; we will denote by c^{-1} the reverse of $c \in \mathcal{L}$. For example, the reverse of *Change_Domain (participants,Onset,Event)* is *Change_Domain(participants,Event,Onset)*.

Theorem 9. *Consider two RDF/S KBs* V_1, V_2 *and the sets* $C = \{c \in \mathcal{L} | c : detectable\}$, $C^{-1} = \{c^{-1} | c \in C\}$. *Then,* C^{-1} *is non-conflicting and* $V_2 \bullet C^{-1} = V_1$.

Theorem 9 shows how a set of changes can be canceled by applying its reverse upon the result. This allows for both "undoing" an unwanted change, and reproducing older versions of an RDF/S KB.

3.3 Change Detection Algorithm

An essential part of our approach is the detection algorithm for $\mathcal{L}$ (Algorithm 1), which should be efficient, scalable and should correctly return the detectable changes. The first step of the algorithm is to pick a low-level change (i.e., a triple in Δ_1 or Δ_2), say $(participants, \text{domain}, Onset) \in \Delta_2$ (cf. Figure 1 and Table 1). Regardless of the particular input (V_1, V_2), there are certain high-level changes whose detection cannot be triggered by a given low-level change. For example, the deletion of triple $(participants$, domain, $Onset)$ cannot be related to the detection of *Delete_Superclass*, as no low-level change of this form appears in the required deleted triples of *Delete_Superclass* (see Table 2). On the other hand, it can potentially trigger the detection of a *Delete_Domain* or a *Change_Domain* operation if the latter is coupled with some other low-level change in Δ specifying the addition of a new domain for $participants$ (Table 3).

This kind of reasoning allows us to build a look-up table (excerpt shown in Table 4), which is used by $findPotentialChanges(t, \Delta)$ (line 3) to return the set of high-level changes $potC$, whose detection could, potentially, be triggered by the selected low-level change (t). $findPotentialChanges$ works as follows: if the selected t is in the left column of Table 4, then we check whether the low-level changes in the middle column appear in Δ; if so, then t *could* trigger the detection of the high-level change in the right column, so this high-level change is put in $potC$. In our example, $potC$ will contain *Delete_Domain(participants,Onset)*, *Change_Domain(participants,Onset,Event)* and *Generalize_Domain(participants,Onset, Event)*.

Algorithm 1. Change Detection Algorithm

1: $changes = \emptyset$
2: **for all** low-level changes t **do**
3:     potC := *findPotentialChanges(t, Δ)*
4:     **for all** $c \in potC$ **do**
5:         **if** $\phi(c)$ = true **then**
6:             $changes := changes \cup \{c\}$
7:             $\Delta_1 := (\Delta_1 \setminus \delta_1(c)), \Delta_2 := (\Delta_2 \setminus \delta_2(c))$
8:             break
9:         **end if**
10:     **end for**
11: **end for**
12: **return** $changes$

Method $findPotentialChanges$ performs a first "filtering", guaranteeing that the only (per Theorem 3) detectable high-level change associated with the low-level change under question is one of the changes in $potC$. To find it, we check the required conditions of each member of $potC$, and, once we find one whose conditions are true, we add it to the list of detectable changes (line 6). In our running example, *Generalize_Domain(participants,Onset,Event)* will be detected. Note that in order for the correct detection to take place, composite changes should be considered first in the for loop of line 4. Therefore, even though the conditions of the basic change *Delete_Domain(participants,Onset)* are also true, the algorithm will never reach that point (due to the "break" command in line 8). This is according to our definition that a basic change is detectable only if there is no composite change that is also detectable and more general. Note also that the elimination of the associated low-level changes from Δ (line 7) would not cause problems thanks to *Non-ambiguity* (done for performance purposes). The presented algorithm is sound and complete with respect to $\mathcal{L}$:

Theorem 10. *A change $c \in \mathcal{L}$ will be returned by Algorithm 1 iff c is detectable.*

Now suppose that the size of Δ is N. The look-up table used by $findPotentialChanges$ has a constant size, so it takes $O(1)$ time to search it. For each matching low-level change (left column in Table 4), a full search of the Δ is made for finding out the required low-level changes (middle column) by using a hash table, so it takes $O(N)$ time (worst-case). This determines the potential changes to be put in $potC$, per the right column of Table 4. Since the table is of constant size, the size of $potC$ will be $O(1)$ as well; therefore, computing $potC$ takes $O(N)$ in total.

For each change in $potC$, we need to determine whether its conditions are true. The time required for this depends on the change considered. For some changes (e.g., *Delete_Domain*), it takes $O(1)$ number of checks; for others, the cost is either $O(M)$ (e.g., *Delete_Class*) or $O(M^2)$ (e.g., *Reclassify_Individual_Higher*), where M is the number of triples in δ_1 and δ_2 of the respective high-level change. Note that each individual check can be done in $O(1)$, a result which can be achieved using sophisticated labeling algorithms, as described in [3].

To calculate the complexity of the algorithm, we will consider the worst-case scenario. The for loop (line 2) iterates over the low-level changes. Let us consider the i-th iteration: for the selected change, we need $O(N)$ time for $findPotentialChanges$, plus $O(1)$ iterations of $O(M_i^2)$ cost (lines 4-10), where M_i is the total size of δ_1 and δ_2 for the high-level change considered. Then, the total cost (for the entire algorithm) is: $O(\sum_{i=1,\ldots,N}(N + M_i^2))$. However, note that: $\sum_{i=1,\ldots,N}(N + M_i^2) = N^2 + \sum_{i=1,\ldots,N} M_i^2 \leq N^2 + (\sum_{i=1,\ldots,N} M_i)^2$. The sum in the last equation cannot exceed the size of Δ by more than a constant factor (i.e., it is $O(N)$). Therefore, the complexity of the algorithm is $O(N^2)$. As a final note, recall that the cost of computing $\Delta(V_1, V_2)$ is linear with respect to the larger of V_1, V_2. Thus:

Theorem 11. *The complexity of Algorithm 1 for input V_1, V_2 is $O(max\{N_1, N_2, N^2\})$, where N_i is the size (in triples) of V_i $(i = 1, 2)$ and N is the size of $\Delta(V_1, V_2)$.*

In practice, our algorithm will rarely exhibit the quadratic worst-case complexity described in Theorem 11. There are several reasons for that. First of all, the complexity of searching through Δ (in $findPotentialChanges$) was calculated to be $O(N)$; for most changes, this will be $O(1)$ on average, due to the use of hash tables. Secondly, evaluating the conditions (line 5) varies from constant to quadratic over M_i, depending on the type of changes in $potC$. Furthermore, even though M_i may, in the worst case, be comparable to N, this will rarely be the case; therefore, even operations that exhibit quadratic complexity over M_i, will rarely exhibit quadratic complexity over N. The above arguments appear more emphatically for basic changes, as the cost of evaluating the conditions of any basic change is $O(1)$. The above observations will be verified by the results of our experiments (Section 5).

4 Operations Based on Heuristics

The detection semantics of the changes described so far used no heuristics or other approximation techniques, and were based on the implicit assumption that no terminological changes occurred between the RDF/S KBs. However, as described in Section 2, this is not always true. In Figure 1 for example, a matcher could identify that classes $Existing$ and $Persistent$ correspond to the same entity, so a *Rename_Class* operation should be detected (rather than the addition of a class and the deletion of another). Operations like *Rename_Class* are different from the changes discussed so far, because they can only be detected using *matchers* [6], which employ various sophisticated, heuristic-based techniques for identifying elements with different names that correspond to the same real world entity.

For evaluation purposes, we implemented a simple matcher that associates elements based on the similarity of their "neighborhoods", i.e., the sets of nodes and links that are pointing from/to the elements under question. If the similarity exceeds a certain threshold, then a matching is reported. In particular, if an element in V_1 is matched with an element in V_2, we detect a *Rename* operation, whereas if it is matched with a set of elements in V_2, we detect a *Split* operation; on the other hand, if a set of elements in V_1 are matched with an element in V_2, a *Merge* operation is detected.

Another case where matchers are necessary appears when an object is associated with a different comment in V_1, V_2, which could be either because the old comment was deleted and a new one was added, or because the old comment was edited. In this case, we use the Levenshtein string distance metric [11] which compares the similarity of the respective comments and determines whether a pair of *Delete_Comment-Add_Comment*, or a single *Change_Comment*, should be returned (similarly for labels).

It should be noted that once the matchings are calculated and the corresponding detected operations are reported as above, we continue with the normal, non-approximate change detection process (as described in Section 3). This means in practice that the detection of heuristic changes takes precedence over composite ones, in the same way that the detection of composite changes takes precedence over basic changes. For example, in Figure 1, we would *not* report a *Change_Domain(started_on,Existing,Persistent)*, because $Existing$ and $Persistent$ are identified as the same class.

The focus of this paper is *not* on developing a sophisticated matcher, but on change detection. Our design was modular, so that any custom-made or off-the-shelf matcher could be used to calculate the required matchings; moreover, the user may choose to circumvent the matching process altogether. Thus, the matching process can be viewed as an optional, pre-processing phase to the actual change detection algorithm, and is an extension of our basic framework.

5 Experimental Evaluation

The evaluation of our approach was based on experiments performed on two well-established ontologies from the cultural (CIDOC [4]) and biological (GO [7]) domains. It aims at showing the intuitiveness and conciseness of the changes contained in $\mathcal{L}$ (Figure 2 and Table 5) as well as verifying that the performance of the implemented algorithm conforms to the average-case analysis of Section 3.3 (Table 6).

CIDOC consists of nearly 80 classes and 250 properties, but has no instances. For our experiments, we used versions v3.2.1 (dated 02.2002), v3.3.2 (dated 10.2002), v3.4.9 (dated 12.2003), v4.2 (dated 06.2005) and v5.0.1 (dated 04.2009), which are encoded in RDF and are available in [4]. The detected changes apply mostly on properties, and many involve the heuristic change *Rename* (see Figure 2). For the detection of the heuristic changes a special-purpose matcher was developed, that exploited CIDOC's

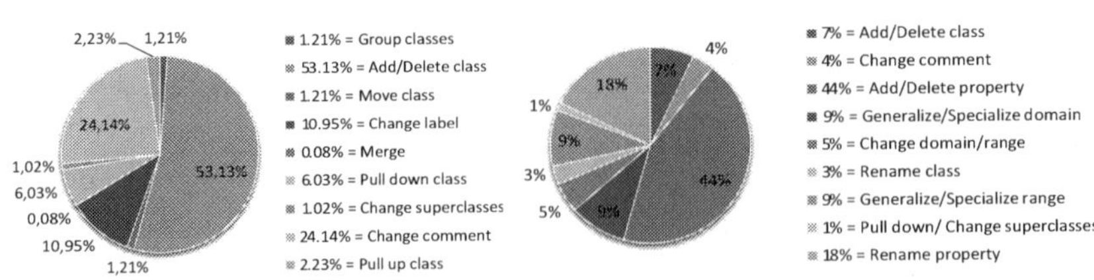

(a) GO: v25.11.08 - v26.05.09 (b) CIDOC: v3.2.1 - v5.0.1

Fig. 2. Overview of Composite and Heuristic Changes

naming policy which attaches a unique, change preserving ID in the names; this way, the precision and recall of the matchings for CIDOC was 100%. CIDOC versions are accompanied by release notes describing in natural language the differences with the previous version; our algorithm uncovered certain typos, omissions and other mistakes in these notes, which were verified by one of CIDOC's editors. This highlights the need for automated change detection algorithms, as even the most careful manual recording process may be inaccurate.

The Gene Ontology (GO) [7] describes gene products, and is one of the largest and most representative data sets for ontology evolution due to its size and update rate. GO is composed of circa 28000 classes (all instances of one meta-class), and 1350 property instances of *obsolete* which is, sometimes, used by the GO editors to mark classes as obsolete instead of deleting them. Although GO is encoded in RDF/XML format, the subsumption relationships are represented by user-defined properties instead of the standard subClassOf property, so we used versions of GO released by the UniProt Consortium [1], which use RDFS semantics. GO is updated on a daily basis, but UniProt releases a new version every month and only the latest version is available for download[2]. During the time of our experiments we were able to retrieve 5 versions of GO (dated 25.11.08, 16.12.08, 24.03.09, 05.05.09 and 26.05.09). The detected heuristic changes (*Merge*) were very few (0.08% of the total) as shown in Figure 2; the string matcher, on the other hand, detected several *Change_Comment* and *Change_Label* operations. The rest of the changes were mostly additions and deletions of classes, as well as changes in the hierarchy. The detected basic changes (not pictured) included, among others, additions of property instances. Even though we weren't able to find any recent official documentation regarding the changes on GO, the changes reported by certain studies (e.g., [25]) show that the detected operations capture the intuition of the editors.

Table 5. Evaluation Results

Versions	$V1$	$V2$	Δ	Basic	Basic + Composite + Heuristic
CIDOC					
v3.2.1 - v3.3.2	952	1081	870	834	202+120+39 = 361
v3.3.2 - v3.4.9	1081	1110	287	285	13+15+34 = 62
v3.4.9 - v4.2	1110	1254	571	538	287+6+10 = 303
v4.2 - v5.0.1	1254	1318	339	327	44+51+52=147
GO					
v25.11.08 - v16.12.08	183430	184704	2979	2260	326+296+307 = 929
v16.12.08 - v24.03.09	184704	188268	7312	5053	745+706+440 = 1891
v24.03.09 - v05.05.09	188268	190097	3108	2322	359+362+97 = 818
v05.05.09 - v26.05.09	190097	191417	2663	1983	265+312+147 = 724

Table 5 shows the number of detected changes between different pairs of CIDOC and GO versions. The columns report the compared versions and their sizes, the size of Δ and the number of detected basic (only) and high-level (in general, i.e., basic, composite and heuristic) changes. The number of detected basic changes is comparable

[2] ftp://ftp.uniprot.org/pub/databases/uniprot_datafiles_by_format/rdf/

to the size of Δ, showing that deltas consisting entirely of basic changes are not concise. On the other hand, the size of the delta is significantly reduced (44%-78% for CIDOC, 59%-74% for GO) when composite and heuristic changes are also considered.

Table 6 reports the running time of the detection algorithm, measured on a Linux machine equipped with a Pentium 4 processor running at 3.4GHz and 1.5GB of main memory. The times for the detection of basic changes were, in general, linear to the input verifying our average-case analysis in Section 3. With respect to composite changes, the execution time reveals some interesting anomalies. For example, comparing the results for versions v3.3.2-v3.4.9 and v3.4.9-v4.2 (for CIDOC) we see a reduction in the running time, despite the increase of the input size (cf. Table 5). This is due to the very small number of detected composite changes for the second input (see Table 5). Also, when comparing the results of v16.12.08-v24.03.09 to v25.11.08-v16.12.08 (GO) we see that the running time increases in a sub-linear fashion with respect to the input. This can be explained by considering the types of detected composite changes, which reveals that for versions v16.12.08-v24.03.09 the changes whose complexity for evaluating the conditions is quadratic are 4.5% of the total, whereas for v25.11.08-v16.12.08 such changes constitute 15% of the total. The slow execution times related to heuristic changes is due to the overhead caused by the employed matcher.

Table 6. Running Time

Versions	Δ	Basic Changes	Composite Changes	Heuristic Changes
		CIDOC		
v3.2.1 - v3.3.2	95.91 ms	13.53 ms	3.35 ms	26.19 ms
v3.3.2 - v3.4.9	91.45 ms	3.94 ms	1.01 ms	5.54 ms
v3.4.9 - v4.2	95.75 ms	8.05 ms	0.26 ms	9.68 ms
v4.2 - v5.0.1	120.58 ms	5.50 ms	2.12 ms	861.77 ms
		GO		
v25.11.08 - v16.12.08	35.214 s	133.79 ms	28.60 ms	45.195 s
v16.12.08 - v24.03.09	36.610 s	249.66 ms	39.65 ms	345.419 s
v24.03.09 - v05.05.09	36.684 s	146.20 ms	23.99 ms	38.006 s
v05.05.09 - v26.05.09	36.712 s	131.22 ms	24.45 ms	40.067 s

6 Related Work

Change detection algorithms in the literature report either low-level deltas ([23,24]), or high-level ones, which, like in our paper, are usually distinguished in basic and composite ([15,17,22]). In [10,14,15,18,19,21] authors describe several operations and the intuition behind them. However, a formal definition of the semantics of such operations ([10,14,15,19]), or of the corresponding detection process ([15]), is usually missing; thus, they cannot guarantee any useful formal properties.

Authors in [10,14] describe a fixed-point algorithm for detecting changes, which is implemented in PromptDiff, an extension of Protégé [8]. The algorithm incorporates heuristic-based matchers in order to detect the changes that occurred between two versions. Therefore, the entire detection process is heuristic-based, thereby introducing an uncertainty in the results: the evaluation reported by the authors showed that their algorithm had a recall of 96% and a precision of 93%. In our case, such metrics are not relevant, as our detection process does not use heuristics and any false positives or negatives will be artifacts of the matching process, not of the detection algorithm itself.

In [18] the Change Definition Language (CDL) is proposed as a means to define a language of high-level changes. In CDL, a change is defined and detected using temporal queries over a version log that contains recordings of the applied low-level changes. The version log is updated when a change occurs which overrules the use of this approach in non-curated or distributed environments. In our work, version logs are not necessary for the detection, as the low-level delta can be produced also *a posteriori*. Note also that, in [18] changes that require heuristics for their detection (such as *Rename*) are completely ignored. This reduces the usefulness of the proposed language.

In our framework, changes that require heuristics are considered separately. This way, we can support operations that require heuristics, while maintaining determinism for the operations that don't need them. In addition, we have the option to ignore such changes, which may be useful for applications that require perfect precision and recall.

7 Conclusion and Future Work

The need for dynamic ontologies makes the automatic identification of deltas between versions increasingly important for several reasons (storing and communication efficiency, visualization of differences etc). Unfortunately, it is often difficult or impossible for curators or editors to accurately record such deltas without the use of automated tools; this was also evidenced by the mistakes found in the release notes of CIDOC.

In this paper, we addressed the problem of automatically identifying deltas. We proposed a formal framework and used it for defining a language of high-level changes for both schema and data, $\mathcal{L}$, and an algorithm that correctly detects changes from $\mathcal{L}$. We proved that $\mathcal{L}$ satisfies several intuitive properties (*Completeness, Non-ambiguity, Reversibility*). Note that the existence of other languages satisfying these properties is not ruled out. However, if the intuitiveness of the changes is not taken into account, the languages will end up being artificial and without practical use in real-world scenarios. Thus, the intuitiveness of the changes that $\mathcal{L}$ contains was a critical factor in our design and experimental evidence on the usefulness of $\mathcal{L}$ was provided. The detection algorithm itself was shown to be quite efficient, namely of quadratic worst-case complexity (even though, in practice, it seems to exhibit linear average-case complexity). The approach can be extended to more expressive ontology languages but the details depend on the semantics of the language and must be determined. As future work, we plan to extend $\mathcal{L}$ by considering *complex changes*, which aggregate several composite changes together. Moreover, we plan to conduct empirical studies involving real users.

Acknowledgements

This work was partially supported by the EU project KP-Lab (FP6-2004-IST-4).

References

1. Bairoch, A., Apweiler, R., Wu, C.H., Barker, W.C., Boeckmann, B., Ferro, S., Gasteiger, E., Huang, H., et al.: The Universal Protein Resource (UniProt). Nucleic Acids Research (2005)
2. Brickley, D., Guha, R.V.: Rdf vocabulary description language 1.0: Rdf schema (2004), http://www.w3.org/TR/2004/REC-rdf-schema-20040210
3. Christophides, V., Karvounarakis, G., Plexousakis, D., Scholl, M., Tourtounis, S.: Optimizing taxonomic semantic web queries using labeling schemes. Web Semantics: Science, Services and Agents on the WWW (2004)
4. CIDOC-CRM, http://cidoc.ics.forth.gr/official_release_cidoc.html
5. Cloran, R., Irwin, B.: Transmitting rdf graph deltas for a cheaper semantic web. In: Proc. of SATNAC (2005)
6. Euzenat, J., Shvaiko, P.: Ontology Matching. Springer, Heidelberg (2007)
7. Hill, D.P., Smith, B., McAndrews-Hill, M.S., Blake, J.A.: Gene ontology annotations: What they mean and where they come from. BMC Bioinformatics (2008)
8. Protege Project (2002), http://protege.stanford.edu
9. Hozo, http://www.hozo.jp/
10. Klein, M.: Change Management for Distributed Ontologies. PhD thesis, Vrije Univ. (2004)
11. Levenshtein V I.: Binary Codes Capable of Correcting Deletions, Insertions, and Reversals. Soviet Physics-Doklady 10 (1966)
12. McBride, B., Manola, F., Miller, E.: Rdf primer (2004), http://www.w3.org/TR/rdf-primer
13. Noy, N.F., Chugh, A., Liu, W., Musen, M.A.: A Framework for Ontology Evolution in Collaborative Environments. In: Cruz, I., Decker, S., Allemang, D., Preist, C., Schwabe, D., Mika, P., Uschold, M., Aroyo, L.M. (eds.) ISWC 2006. LNCS, vol. 4273, pp. 544–558. Springer, Heidelberg (2006)
14. Noy, N.F., Musen, M.A.: PromptDiff: A Fixed-Point Algorithm for Comparing Ontology Versions. In: Proc. of AAAI (2002)
15. Palma, A., Haase, P., Wang, Y., dAquin, M.: D1.3.1 propagation models and strategies. Technical report, NeOn Deliverable D1.3.1 (2007)
16. Papavassiliou, V., Flouris, G., Fundulaki, I., Kotzinos, D., Christophides, V.: Formalizing high-level change detection for rdf/s kbs. Technical Report TR-398, FORTH-ICS (2009)
17. Plessers, P., De Troyer, O.: Ontology Change Detection Using a Version Log. In: Gil, Y., Motta, E., Benjamins, V.R., Musen, M.A. (eds.) ISWC 2005. LNCS, vol. 3729, pp. 578–592. Springer, Heidelberg (2005)
18. Plessers, P., De Troyer, O., Casteleyn, S.: Understanding Ontology Evolution: A Change Detection Approach. Web Semantics: Science, Services and Agents on the WWW (2007)
19. Rogozan, D., Paquette, G.: Managing Ontology Changes on the Semantic Web. In: Proc. of IEEE/WIC/ACM on Web Intelligence (2005)
20. Serfiotis, G., Koffina, I., Christophides, V., Tannen, V.: Containment and minimization of rdf/s query patterns. In: Proc. of ISWC (2005)
21. Stojanovic, L.: Methods and Tools for Ontology Evolution. PhD thesis, Univ. of Karlsruhe (2004)
22. Stojanovic, L., Maedche, A., Motik, B., Stojanovic, N.: User-Driven Ontology Evolution Management. In: Proc. of EKAW. Ontologies and the Semantic Web (2002)
23. Volkel, M., Winkler, W., Sure, Y., Ryszard Kruk, S., Synak, M.: Semversion: A versioning system for rdf and ontologies. In: Proc. of ESWC (2005)
24. Zeginis, D., Tzitzikas, Y., Christophides, V.: On the Foundations of Computing Deltas Between RDF Models. In: Proc. of ISWC+ ASWC (2007)
25. Zhdanova, A.V.: Community-Driven Ontology Evolution: Gene Ontology Case Study. In: Proc. of BIS (2008)

Efficient Query Answering for OWL 2

Héctor Pérez-Urbina, Ian Horrocks, and Boris Motik

Oxford University Computing Laboratory, Oxford, UK
{hector.perez-urbina,ian.horrocks,boris.motik}@comlab.ox.ac.uk

Abstract. The QL profile of OWL 2 has been designed so that it is possible to use database technology for query answering via query rewriting. We present a comparison of our resolution based rewriting algorithm with the standard algorithm proposed by Calvanese et al., implementing both and conducting an empirical evaluation using ontologies and queries derived from realistic applications. The results indicate that our algorithm produces significantly smaller rewritings in most cases, which could be important for practicality in realistic applications.

1 Introduction

Ontologies can be used as conceptual schemas to provide an intuitive and unified view over one or more data repositories, allowing queries to be independent of the structure and location of the data. The use of ontologies as conceptual schemas for data repositories has been extensively studied in a variety of contexts, such as information integration [4]. The use of data repositories to store instance data is becoming increasingly important, for instance in the semantic Web, due to the scalability requirements of many applications and the widespread use of ontologies.

In OWL 2—a new version of the OWL ontology language that is currently a W3C candidate recommendation—scalability requirements are addressed by *profiles*—subsets of the language that enjoy desirable computational properties. The OWL 2 QL profile was designed to allow query answering via *query rewriting*: a query over an OWL 2 QL ontology and a set of instance data stored in a data repository can be answered by *rewriting* the query w.r.t. the ontology and then answering the rewritten query in the data repository. In this paper we focus on the case where the data is stored in a relational database and accessed using SQL queries, but the same technique could be applied to data stored in a triple store and accessed via SPARQL queries.

OWL 2 QL is based on DL-Lite$_R$—one of a family of description logics developed by Calvanese et al. [3]. The DL-Lite$_R$ rewriting algorithm of Calvanese et al., which we will refer to as CGLLR, transforms a conjunctive query Q and a DL-Lite$_R$ ontology $\mathcal{O}$ into a *union of conjunctive queries* $Q_\mathcal{O}$ such that the answers to Q and any set of instance data $\mathcal{A}$ can be obtained by evaluating $Q_\mathcal{O}$ over $\mathcal{A}$ only. This technique has been implemented in reasoners such as QuOnto[1] and

[1] `http://www.dis.uniroma1.it/~quonto/`

A. Bernstein et al. (Eds.): ISWC 2009, LNCS 5823, pp. 489–504, 2009.

Owlgres[2]. Unfortunately, as shown by Calvanese et al., the size of $Q_{\mathcal{O}}$ is worst-case exponential w.r.t. the size of Q and $\mathcal{O}$ [3], and as we show in Section 4, realistic ontologies and queries can result in $Q_{\mathcal{O}}$ being extremely large (e.g., containing tens of thousands of conjunctive queries). Thus, on the one hand, $Q_{\mathcal{O}}$ may be costly to compute, and, on the other hand, evaluation by RDBMSs may be costly or even unfeasible. Trying to produce small rewritings is therefore of critical importance to the practical application of query rewriting in general, and of OWL 2 QL in particular.

Motivated by the prospect of applying deductive database techniques to improve the scalability of reasoners, in our previous work [9] we considered the problem of query rewriting for various logics of the DL-Lite and $\mathcal{EL}$ families, the latter being the basis for the OWL 2 EL profile. Our algorithm takes as input a conjunctive query Q and an ontology $\mathcal{O}$, and uses a resolution-based technique to produce a rewritten query $Q_{\mathcal{O}}$. Although $Q_{\mathcal{O}}$ will, in general, be a (possibly recursive) *datalog query*, and thus necessitate the use of a deductive database system, the algorithm exhibits "pay-as-you-go" behavior for various logics. In particular, if $\mathcal{O}$ is a DL-Lite$_R$ ontology, then $Q_{\mathcal{O}}$ is a union of conjunctive queries. Our algorithm can therefore be seen as a generalization and extension of CGLLR.

In this paper we describe a simplified version of our algorithm that we will refer to as RQR (Resolution-based Query Rewriting). Like CGLLR, RQR rewrites a query w.r.t. a DL-Lite$_R$ ontology to produce a union of conjunctive queries. RQR differs from CGLLR mainly in the way it handles existential restrictions in an ontology. First, RQR uses functional terms to keep track of successors whose existence is implied by such restrictions, while CGLLR relies on a so-called reduction step. Second, RQR directly handles qualified existential restrictions (i.e., those where the restriction class is not owl:Thing), whereas CGLLR requires the elimination of such restrictions using an encoding that introduces new "auxiliary" properties. We describe both algorithms in Section 3 and further discuss their differences in Section 3.3.

Both the reduction step and the introduction of auxiliary properties can increase the size of the rewriting. Therefore, we conjectured that RQR will often produce smaller rewritings than CGLLR. In order to test the practicality of query rewriting and the efficiency of the different rewriting techniques, we have implemented RQR in a query rewriting system that we call REQUIEM[3] (REsolution-based QUery rewrIting for Expressive Models), and compared its behavior with that of our implementation of CGLLR that we refer to as C. The comparison uses a benchmark suite containing realistic DL-Lite$_R$ ontologies and test queries as well as some artificial ontologies and queries designed to highlight the differences between the two algorithms. The benchmark suite also included versions of the ontologies in which qualified existential restrictions have been explicitly encoded using auxiliary properties, as this allowed us to compare the two

[2] `http://pellet.owldl.com/owlgres/`
[3] `http://www.comlab.ox.ac.uk/projects/requiem/`

implementations in cases where RQR's native handling of qualified existential restrictions is not advantageous.

Both algorithms are amenable to optimizations that can reduce the size of the rewritings. One obvious optimization would be to use query subsumption checks to eliminate redundant conjunctive queries from the rewriting; we discuss this and other optimization techniques in Section 3.4. In order to compare optimized versions of the two algorithms we additionally implemented REQUIEM-SC and C-SC, both of which first compute the rewriting as in the original version, and then apply the above-mentioned query subsumption optimization to the result.

Our evaluation showed that, even for ontologies in which qualified existential restrictions were already encoded, REQUIEM produced significantly smaller rewritings than C in most cases. In one case, for instance, C exceeded the maximum allowed run time (600 seconds) after producing more than 42,000 conjunctive queries; in contrast, REQUIEM completed the rewriting in less than half a second having produced only 624 queries. Moreover, the rewritings produced by REQUIEM were often similar or identical to those produced by REQUIEM-SC, something that was less often the case for C and C-SC; using RQR should, therefore, reduce the need for (potentially costly) query subsumption checking.

2 Ontology-Based Data Access via Query Rewriting

We next describe how to answer queries over an OWL 2 QL ontology and a database via query rewriting, illustrating the process by means of an example.

Suppose we have a relational database DB_0 containing a table `Professor` with attributes `name`, `department`, and `telephone`; and a table `Student` with attributes `name`, `major`, `address`, and `tutor`. We can use a suitable ontology as a conceptual schema that describes the structure of the data. For example, we might use the following ontology $\mathcal{O}_0$ to describe DB_0:[4]

$$\text{Teacher} \sqsubseteq \exists\text{teaches} \tag{1}$$

$$\text{Professor} \sqsubseteq \text{Teacher} \tag{2}$$

$$\exists\text{hasTutor}^- \sqsubseteq \text{Professor} \tag{3}$$

Axiom (1) states that teachers teach at least someone, axiom (2) states that professors are teachers, and axiom (3) states that the range of the property hasTutor is Professor.

Given suitable mappings from the classes and properties in the ontology to data in the database, queries posed in terms of the ontology can be answered using the database. This has several advantages: on the one hand, queries can be posed in terms of the conceptual structure of the data rather than its arrangement in the database, and on the other hand, the database provides data persistence and scalability.

Mappings from the ontology to the database are typically defined using expressions of the form $D \mapsto Q_D$, where D is a class or property occurring in

[4] We use the description logic syntax for the sake of compactness.

the ontology and Q_D is an SQL query over the database; Q_D could, however, equally well be a SPARQL query that accesses data in an RDF triple store. In our example, the mapping $\mathcal{M}_0$ between $\mathcal{O}_0$ and DB_0 is defined as follows:

Professor $\mapsto$ `SELECT Name FROM Professor`

hasTutor $\mapsto$ `SELECT Name, Tutor FROM Student`

Queries posed over the ontology are answered in two steps: first, the ontology is used to rewrite the query into a union of conjunctive queries; and second, the mappings are used to transform the rewritten query into an SQL query and evaluate it using an RDBMS. For example, consider the query

$$Q_0(x) \leftarrow \text{teaches}(x, y) \tag{4}$$

posed over $\mathcal{O}_0$. The rewriting $Q_{\mathcal{O}_0}$ of query (4) w.r.t. $\mathcal{O}_0$ contains (4) and the following queries:

$$Q_0(x) \leftarrow \text{Teacher}(x) \tag{5}$$

$$Q_0(x) \leftarrow \text{Professor}(x) \tag{6}$$

$$Q_0(x) \leftarrow \text{hasTutor}(y, x) \tag{7}$$

Transforming $Q_{\mathcal{O}_0}$ into an SQL query $\mathsf{sql}(Q_{\mathcal{O}_0})$ basically amounts to using $\mathcal{M}_0$ to replace each class or property D occurring in a query contained in $Q_{\mathcal{O}_0}$ with the corresponding SQL query Q_D, and forming the union of the resulting queries. Note that $\mathcal{M}_0$ does not contain a mapping for every class and property of $\mathcal{O}_0$. The answer to any query containing an atom for which there is no mapping will necessarily be empty, and we can therefore discard such queries. Consequently, queries (4) and (5) can be discarded in our example. As a result, we obtain the following rewritten SQL query:

$\mathsf{sql}(Q_{\mathcal{O}_0})$ = `SELECT Name FROM Professor UNION`
`SELECT Tutor FROM Student`

This query can now be directly evaluated in the RDBMS to compute the answers to the original query $Q_0(x)$.

In the rest of the paper we focus on the problem of computing the rewriting $Q_{\mathcal{O}}$ of a given query Q w.r.t. an OWL 2 QL ontology $\mathcal{O}$.

3 Query Rewriting Algorithms

In this section we describe the RQR and CGLLR query rewriting algorithms, discuss the differences between them, and present various optimizations that can help us reduce the size of the rewritings.

Before presenting the algorithms, we introduce some notation and definitions. We use the well-known notions of constants, variables, function symbols, terms, and atoms of first-order logic. A Horn clause C is an expression of the form $D_0 \leftarrow D_1 \wedge ... \wedge D_n$, where each D_i is an atom. The atom D_0 is called the *head*, and the set $\{D_1, ..., D_n\}$ is called the *body*. We require that all the variables occurring in the head of C occur at least in one of its body atoms. For instance, the expression teaches$(x, f(x)) \leftarrow$ Professor(x) is a Horn clause. The *depth* of a term t is defined as $\mathsf{depth}(t) = 0$ if t is a constant or a variable, and $\mathsf{depth}(f(s)) = 1 + \mathsf{depth}(s)$ if t is a functional term $f(s)$. The notion of depth is extended to an atom $R(t_1, \ldots, t_n)$ in the natural way: $\mathsf{depth}(R(t_1, \ldots, t_n)) = \max(\mathsf{depth}(t_i))$ for $1 \leq i \leq n$. An atom D occurring in a Horn clause C is said to be the *deepest* in C if $\mathsf{depth}(D) \geq \mathsf{depth}(D_i)$ for every atom D_i of C. For instance, the atom teaches$(x, f(x))$ is the deepest in the previously mentioned example clause.

A *conjunctive query* (CQ) Q posed over an ontology $\mathcal{O}$ is a Horn clause whose head predicate does not occur in $\mathcal{O}$, and whose body predicates are class and property names occurring in $\mathcal{O}$. For instance, (4) is a CQ over the ontology $\mathcal{O}_0$. A *union of conjunctive queries* (UCQ) over $\mathcal{O}$ is a set of conjunctive queries over $\mathcal{O}$ with the same head up to variable renaming [1]. For instance, the query $Q_{\mathcal{O}_0}$ composed of queries (4)–(7) is a UCQ over the ontology $\mathcal{O}_0$. A tuple of constants $\vec{a}$ is a *certain answer* to a UCQ Q over $\mathcal{O}$ and a set of instance data $\mathcal{A}$ iff $\mathcal{O} \cup \mathcal{A} \cup Q \models Q_P(\vec{a})$, where Q_P is the head predicate of Q, and Q is considered to be a set of universally quantified implications with the usual first-order semantics. The set of all answers to Q over $\mathcal{O}$ and $\mathcal{A}$ is denoted by $\mathsf{ans}(Q, \mathcal{O} \cup \mathcal{A})$. Given a conjunctive query Q and an ontology $\mathcal{O}$, a query $Q_{\mathcal{O}}$ is said to be a *rewriting* of Q w.r.t. $\mathcal{O}$ if $\mathsf{ans}(Q, \mathcal{O} \cup \mathcal{A}) = \mathsf{ans}(Q_{\mathcal{O}}, \mathcal{A})$ for every $\mathcal{A}$.

Both algorithms compute the rewriting $Q_{\mathcal{O}}$ of a given query Q w.r.t. a DL-Lite$_R$ ontology $\mathcal{O}$. DL-Lite$_R$ is the basis for OWL 2 QL. Extending the algorithms to handle the additional features of OWL 2 QL (e.g., datatypes, negative inclusions) is straightforward; we omit the details for the sake of simplicity.

3.1 CGLLR

The algorithm computes $Q_{\mathcal{O}}$ by using the axioms of $\mathcal{O}$ as rewrite rules and applying them to the body atoms of Q. The algorithm is shown in Algorithm 1. The partial function ref takes as input an axiom α and an atom D, and returns an atom $\mathsf{ref}(D, \alpha)$ as follows.

- If $D = A(x)$, then we have that (i) if $\alpha = B \sqsubseteq A$, then $\mathsf{ref}(D, \alpha) = B(x)$; (ii) if $\alpha = \exists P \sqsubseteq A$, then $\mathsf{ref}(D, \alpha) = P(x, \_)$; and (iii) if $\alpha = \exists P^- \sqsubseteq A$, then $\mathsf{ref}(D, \alpha) = P(\_, x)$.
- If $D = P(x, \_)$, then we have that (i) if $\alpha = A \sqsubseteq \exists P$, then $\mathsf{ref}(D, \alpha) = A(x)$; (ii) if $\alpha = \exists S \sqsubseteq \exists P$, then $\mathsf{ref}(D, \alpha) = S(x, \_)$; and (iii) if $\alpha = \exists S^- \sqsubseteq \exists P$, then $\mathsf{ref}(D, \alpha) = S(\_, x)$.
- If $D = P(\_, x)$, then we have that (i) if $\alpha = A \sqsubseteq \exists P^-$, then $\mathsf{ref}(D, \alpha) = A(x)$; (ii) if $\alpha = \exists S \sqsubseteq \exists P^-$, then $\mathsf{ref}(D, \alpha) = S(x, \_)$; and (iii) if $\alpha = \exists S^- \sqsubseteq \exists P^-$, then $\mathsf{ref}(D, \alpha) = S(\_, x)$.
- If $D = P(x, y)$, then we have that (i) if either $\alpha = S \sqsubseteq P$ or $\alpha = S^- \sqsubseteq P^-$, then $\mathsf{ref}(D, \alpha) = S(x, y)$; and (ii) if either $\alpha = S \sqsubseteq P^-$ or $\alpha = S^- \sqsubseteq P$, then $\mathsf{ref}(D, \alpha) = S(y, x)$.

Input: Conjunctive query Q, DL-Lite$_R$ ontology $\mathcal{O}$
$Q_\mathcal{O} = \{Q\}$;
repeat
  foreach *query* $Q' \in Q_\mathcal{O}$ **do**
    (reformulation) **foreach** *atom* D *in* Q' **do**
      foreach *axiom* $\alpha \in \mathcal{O}$ **do**
        if α *is applicable to* D **then**
          $Q_\mathcal{O} = Q_\mathcal{O} \cup \{Q'[D/\mathsf{ref}(D,\alpha)]\}$;
        end
      end
    end
    (reduction) **forall** *atoms* D_1, D_2 *in* Q' **do**
      if D_1 *and* D_2 *unify* **then**
        $\sigma = \mathsf{MGU}(D_1, D_2)$;
        $Q_\mathcal{O} = Q_\mathcal{O} \cup \{\lambda(Q'\sigma))\}$;
      end
    end
  end
until *no query unique up to variable renaming can be added to* $Q_\mathcal{O}$;
return $Q_\mathcal{O}$;

Algorithm 1. The CGLLR algorithm

Input: Conjunctive query Q, DL-Lite$_R$ ontology $\mathcal{O}$
$R = \Xi(\mathcal{O}) \cup \{Q\}$;
repeat
  (saturation) **forall** *clauses* C_1, C_2 *in* R **do**
    $R = R \cup \mathsf{resolve}(C_1, C_2)$;
  end
until *no query unique up to variable renaming can be added to* R ;
$Q_\mathcal{O} = \{C \mid C \in \mathsf{unfold}(\mathsf{ff}(R))$, and C has the same head predicate as $Q\}$;
return $Q_\mathcal{O}$;

Algorithm 2. Our resolution-based algorithm

If $\mathsf{ref}(D, \alpha)$ is defined for α and D, we say that α is *applicable* to D. The expression $Q[D/D']$ denotes the CQ obtained from Q by replacing the body atom D with a new atom D'. The function MGU takes as input two atoms and returns their most general unifier [1]. The function λ takes as input a CQ Q and returns a new CQ obtained by replacing each variable that occurs only once in Q with the symbol "_".

Starting with the original query Q, CGLLR continues to produce queries until no new queries can be produced. In the *reformulation* step the algorithm rewrites the body atoms of a given query Q' by using applicable ontology axioms as rewriting rules, generating a new query for every atom reformulation. Then, in the *reduction* step the algorithm produces a new query $\lambda(Q'\sigma)$ for each pair of body atoms of Q' that unify.

3.2 RQR

The algorithm first transforms Q and $\mathcal{O}$ into clauses and then computes $Q_{\mathcal{O}}$ by using a resolution-based calculus to derive new clauses from the initial set. The procedure is presented in Algorithm 2, where we show only those parts of the original algorithm that are relevant to DL-Lite$_R$. The expression $\Xi(\mathcal{O})$ denotes the set of clauses obtained from $\mathcal{O}$ according to Table 1. The function resolve takes two clauses C_1 and C_2, and it returns a set containing every clause C_R that can be obtained by combining the atoms of C_1 and C_2 according to the *inference templates* shown in Table 2. A template of the form $\frac{P_1 \quad P_2}{R}$ denotes that, if C_1 is a clause of the form of P_1 and C_2 is a clause of the form of P_2, then $\mathsf{resolve}(C_1, C_2)$ contains all clauses of the form of R that can be constructed from C_1 and C_2; otherwise, $\mathsf{resolve}(C_1, C_2) = \emptyset$. The function ff takes a set of clauses N and returns the subset of the function-free clauses in N. The function unfold takes a set of clauses N, and returns the set obtained by unfolding every clause in N; for example, if we have that $N = \{Q_P(x) \leftarrow A(x), A(x) \leftarrow B(x)\}$, then $\mathsf{unfold}(N) = N \cup \{Q_P(x) \leftarrow B(x)\}$, where $Q_P(x) \leftarrow B(x)$ is the result of unfolding $A(x) \leftarrow B(x)$ into $Q_P(x) \leftarrow A(x)$. A formal description of the unfolding step can be found in [9].

RQR computes $Q_{\mathcal{O}}$ in three steps: first, in the *clausification* step, the algorithm transforms Q and $\mathcal{O}$ into a set of clauses $\Xi(\mathcal{O}) \cup \{Q\}$; second, in the *saturation* step, the algorithm continues to produce clauses until no new clauses can be produced; third, in the *unfolding* and *pruning* step, clauses that are not function free are discarded, the remaining clauses are unfolded, and then clauses that do not have the same head predicate as Q are also discarded.

3.3 Differences

The algorithms mainly differ in the way they handle existential restrictions. This difference is twofold: first, while RQR deals with axioms containing existential quantifiers on the right-hand side by introducing functional terms, CGLLR does so by restricting the applicability of such axioms and relying on the reduction step; second, unlike RQR, CGLLR does not handle qualified existential restrictions natively—that is, there is no rewriting rule for axioms of the form $A \sqsubseteq \exists R.B$; instead, the algorithm requires a preliminary step in which each such axiom occurring in $\mathcal{O}$ is replaced with a set of axioms $\{A \sqsubseteq \exists P_1, \exists P_1^- \sqsubseteq B, P_1 \sqsubseteq R\}$, where P_1 is a new atomic property not occurring in $\mathcal{O}$. We explore these differences and their impact on the size of the rewritings by means of an example.

Consider an OWL 2 QL ontology $\mathcal{O}_1$ that consists of the following axiom

$$\text{Professor} \sqsubseteq \exists \text{teaches.Student}, \tag{8}$$

which states that a professor teaches at least some student, and the query

$$Q_1(x) \leftarrow \text{teaches}(x, y) \wedge \text{Student}(y). \tag{9}$$

Table 1. Translating $\mathcal{O}$ into a set of clauses $\Xi(\mathcal{O})$

DL-Lite$_R$ clause	DL-Lite$_R$ axiom
$B(x) \leftarrow A(x)$	$A \sqsubseteq B$
$P(x, f(x)) \leftarrow A(x)$	$A \sqsubseteq \exists P$
$P(x, f(x)) \leftarrow A(x)$	$A \sqsubseteq \exists P.B$
$B(f(x)) \leftarrow A(x)$	
$P(f(x), x) \leftarrow A(x)$	$A \sqsubseteq \exists P^-$
$P(f(x), x) \leftarrow A(x)$	$A \sqsubseteq \exists P^-.B$
$B(f(x)) \leftarrow A(x)$	
$A(x) \leftarrow P(x, y)$	$\exists P \sqsubseteq A$
$A(x) \leftarrow P(y, x)$	$\exists P^- \sqsubseteq A$
$S(x, y) \leftarrow P(x, y)$	$P \sqsubseteq S, P^- \sqsubseteq S^-$
$S(x, y) \leftarrow P(y, x)$	$P \sqsubseteq S^-, P^- \sqsubseteq S$

Note 1. Each axiom of the form $A \sqsubseteq \exists R.B$ is uniquely associated with a function symbol f.

Table 2. Inference templates for the function resolve

$$\frac{C(x) \leftarrow B(x) \quad B(f(x)) \leftarrow A(x)}{C(f(x)) \leftarrow A(x)}$$

$$\frac{B(x) \leftarrow P(x,y) \quad P(x, f(x)) \leftarrow A(x)}{B(x) \leftarrow A(x)} \qquad \frac{B(x) \leftarrow P(x,y) \quad P(f(x), x) \leftarrow A(x)}{B(f(x)) \leftarrow A(x)}$$

$$\frac{B(x) \leftarrow P(y,x) \quad P(x, f(x)) \leftarrow A(x)}{B(f(x)) \leftarrow A(x)} \qquad \frac{B(x) \leftarrow P(y,x) \quad P(f(x), x) \leftarrow A(x)}{B(x) \leftarrow A(x)}$$

$$\frac{S(x,y) \leftarrow P(x,y) \quad P(x, f(x)) \leftarrow A(x)}{S(x, f(x)) \leftarrow A(x)} \qquad \frac{S(x,y) \leftarrow P(x,y) \quad P(f(x), x) \leftarrow A(x)}{S(f(x), x) \leftarrow A(x)}$$

$$\frac{S(x,y) \leftarrow P(y,x) \quad P(x, f(x)) \leftarrow A(x)}{S(f(x), x) \leftarrow A(x)} \qquad \frac{S(x,y) \leftarrow P(y,x) \quad P(f(x), x) \leftarrow A(x)}{S(x, f(x)) \leftarrow A(x)}$$

$$\frac{Q_P(\vec{u}) \leftarrow B(t) \wedge \bigwedge D_i(\vec{t_i}) \quad B(f(x)) \leftarrow A(x)}{Q_P(\vec{u})\sigma \leftarrow A(x)\sigma \wedge \bigwedge D_i(\vec{t_i})\sigma}$$

where $\sigma = \mathsf{MGU}(B(t), B(f(x))$, and $B(t)$ is deepest in its clause.

$$\frac{Q_P(\vec{u}) \leftarrow P(s,t) \wedge \bigwedge D_i(\vec{t_i}) \quad P(x, f(x)) \leftarrow A(x)}{Q_P(\vec{u})\sigma \leftarrow A(x)\sigma \wedge \bigwedge D_i(\vec{t_i})\sigma}$$

where $\sigma = \mathsf{MGU}(P(s,t), P(x, f(x))$, and $P(s,t)$ is deepest in its clause.

$$\frac{Q_P(\vec{u}) \leftarrow P(s,t) \wedge \bigwedge D_i(\vec{t_i}) \quad P(f(x), x) \leftarrow A(x)}{Q_P(\vec{u})\sigma \leftarrow A(x)\sigma \wedge \bigwedge D_i(\vec{t_i})\sigma}$$

where $\sigma = \mathsf{MGU}(P(s,t), P(f(x), x)$, and $P(s,t)$ is deepest in its clause.

We first analyze the execution of CGLLR. Note that CGLLR cannot handle axiom (8) natively, and it must first be replaced with the following axioms:

$$\text{Professor} \sqsubseteq \exists R_{aux} \tag{10}$$
$$\exists R_{aux}^{-} \sqsubseteq \text{Student} \tag{11}$$
$$R_{aux} \sqsubseteq \text{teaches} \tag{12}$$

In the first iteration, axiom (12) is applicable to the atom teaches(x, y) in (9). Similarly, axiom (11) is applicable to Student(y) in (9). Therefore, we obtain the following queries in the reformulation step:

$$Q_1(x) \leftarrow R_{aux}(x, y) \wedge \text{Student}(y) \tag{13}$$
$$Q_1(x) \leftarrow \text{teaches}(x, y) \wedge R_{aux}(\_\,, y) \tag{14}$$

In this iteration no query can be obtained in the reduction step. In the next iteration, axiom (10) is not applicable to the atom $R_{aux}(x, y)$ in (13) because y occurs in (13) in more than one place. Axiom (10) cannot be applied to (13) because CGLLR does not keep track of information about role successors; furthermore, if we naively allowed existential quantification axioms to be applied, the resulting calculus would become unsound. To illustrate this point, suppose that (10) were applicable to $R_{aux}(x, y)$ in (13), and $\mathsf{ref}(R_{aux}(x, y), (10)) = \text{Professor}(x)$; we would then obtain the query

$$Q_1(x) \leftarrow \text{Professor}(x) \wedge \text{Student}(y). \tag{15}$$

Note that the relation between x and y is lost—that is, the fact that the individual represented by y must be a teaches-successor of the individual represented by x is not captured by query (15).

Although the applicability of (10) is restricted, axiom (11) is applicable to Student(y) in (13). Similarly, axiom (12) is applicable to teaches(x, y) in (14). Both reformulations produce the query

$$Q_1(x) \leftarrow R_{aux}(x, y) \wedge R_{aux}(\_\,, y). \tag{16}$$

In the next iteration, no axiom is applicable to any body atom of (16), so no query is added in the reformulation step. In the reduction step, however, the algorithm produces

$$Q_1(x) \leftarrow R_{aux}(x, \_\,) \tag{17}$$

by unifying the body atoms of (16). In the following iteration, axiom (10) is applicable to the only body atom of (17), producing

$$Q_1(x) \leftarrow \text{Professor}(x). \tag{18}$$

Note that without the reduction step, the algorithm would not have produced query (18). It can be easily verified that no more new queries can be produced; thus, CGLLR returns $\{(9), (13), (14), (16), (17), (18)\}$.

We now analyze the execution of RQR. According to Table 1, axiom (8) is translated into the following clauses:

$$\text{teaches}(x, f(x)) \leftarrow \text{Professor}(x) \tag{19}$$

$$\text{Student}(f(x)) \leftarrow \text{Professor}(x) \tag{20}$$

In the saturation step the algorithm produces

$$\mathsf{resolve}((9),(19)) = Q_1(x) \leftarrow \text{Professor}(x) \wedge \text{Student}(f(x)) \tag{21}$$

$$\mathsf{resolve}((9),(20)) = Q_1(x) \leftarrow \text{teaches}(x, f(x)) \wedge \text{Professor}(x) \tag{22}$$

$$\mathsf{resolve}((19),(22)) = Q_1(x) \leftarrow \text{Professor}(x) \tag{23}$$

Note the difference between queries (15) and (21). Since the function symbol f is uniquely associated with clause (19), unlike query (15), query (21) captures the fact that the individual represented by $f(x)$ must be a teaches-successor of the individual represented by x. It can easily be verified that no other clause is produced in the first step. Clearly, $\mathsf{ff}(R) = \{(9),(23)\}$. In this case, there is no unfolding to be done, so RQR returns $\{(9),(23)\}$.

As shown in the above example, the introduction of auxiliary properties can lead to an increase in the size of the rewritings. The reduction step alone, however, can also lead to larger rewritings. This situation arises especially in the case where part of the data of the database describes a graph. As a simple example, consider an OWL 2 QL ontology $\mathcal{O}_2$ that consists of the axiom

$$\text{Student} \sqsubseteq \exists \text{hasTutor}, \tag{24}$$

which states that a student has at least one tutor, and the query

$$Q_2(x) \leftarrow \text{hasTutor}(x, y) \wedge \text{hasTutor}(z, y) \wedge \text{hasTutor}(z, w) \wedge \text{hasTutor}(x, w). \tag{25}$$

When using CGLLR, axiom (24) is not applicable to query (25), so no query is produced in the reformulation step. However, every pair of body atoms in query (25) unify, and it is easy to see that for each query of this form with m body atoms, CGLLR produces $\binom{m}{2}$ new queries in the reduction step. Eventually, the reduction step produces the query

$$Q_2(x) \leftarrow \text{hasTutor}(x, \_\,). \tag{26}$$

Axiom (24) is now applicable to query (26), and the following query is produced in the reformulation step:

$$Q_2(x) \leftarrow \text{Student}(x) \tag{27}$$

Note, however, that several queries needed to be produced in the reduction step in order to produce query (27) in the reformulation step.

An important remark is in order. For every query Q' produced from a query Q in the reduction step, there is a substitution σ such that $Q\sigma \subseteq Q'$, in which case we say that Q *subsumes* Q'. It is well known that every query that is subsumed by another can be discarded after the rewriting has been computed without affecting completeness [5]; however, identifying such queries is not straightforward since CGLLR does not keep track of which queries were produced in the reduction step. In our example, query (25) subsumes query (26) by the substitution $\sigma = \{z \mapsto x, w \mapsto y\}$; therefore, query (26) can be safely discarded from the final rewriting. In this case, however, note that query (26) subsumes query (25) as well; therefore, it is sensible to eliminate query (25) instead since it is larger. Since both queries subsume each other, we say that they are *equivalent*. Moreover, since query (26) is the minimal equivalent subquery of query (25), we say that query (26) is a *condensation* of query (25) [2]. The potential generation of condensations by the reduction step plays an important role in the optimization of the rewritings. We discuss this aspect further in the following section.

The use of functional terms makes RQR more goal-oriented, in the sense that it does not need to derive the "irrelevant" queries produced by the reduction step of CGLLR in order to be complete. Moreover, RQR handles qualified existential restrictions natively, whereas CGLLR needs to encode them away by introducing new properties and axioms.

3.4 Optimizations

As discussed in the introduction, both algorithms are amenable to optimization. One obvious optimization technique is to check subsumption between pairs of conjunctive queries and eliminate any query that is subsumed by another. Such a procedure can be simply (albeit not necessarily optimally) applied a posteriori to the rewritings produced by RQR and CGLLR.

It is important to note that using the query subsumption optimization with RQR and CGLLR does not necessarily result in exactly the same rewritings. This is due to the fact that the CGLLR reduction step may produce *condensations*. We illustrate this point with an example. Consider an OWL 2 QL ontology $\mathcal{O}_3$ that consists of the following axiom

$$\exists \text{teaches}^- \sqsubseteq \text{Student}, \tag{28}$$

which states that someone that is taught is a student, and the query

$$Q_3(x) \leftarrow \text{teaches}(x, y) \wedge \text{Student}(y). \tag{29}$$

CGLLR produces a set containing (29) and the following queries:

$$Q_3(x) \leftarrow \text{teaches}(x, y) \wedge \text{teaches}(\_\,, y) \tag{30}$$

$$Q_3(x) \leftarrow \text{teaches}(x, \_\,) \tag{31}$$

Note that query (31) was produced in the reduction step from (30) and it is a condensation of (30). In the query subsumption check we have that query (31) subsumes query (29), so the latter is discarded. Note, however, that query (31) subsumes query (30) and *vice versa*. Therefore, since it is sensible to discard the larger query, the condensation (31) is kept and query (30) is discarded instead. It is easy to see that in the end we obtain $\{(31)\}$.

When using RQR, axiom (28) is translated into the following clause:

$$\text{Student}(x) \leftarrow \text{teaches}(y, x) \tag{32}$$

The algorithm then produces a set containing (29) and the following clause:

$$Q_3(x) \leftarrow \text{teaches}(x, y) \wedge \text{teaches}(z, y) \tag{33}$$

Since (33) subsumes (29), it is easy to see that after the query subsumption check we obtain $\{(33)\}$. As can be seen, (31) is slightly smaller than (33); it is also a condensation of (33). In our empirical evaluation (see Section 4), the optimized versions of RQR and CGLLR produced the same rewritings modulo the condensations produced by CGLLR. Modifying RQR to replace queries with their condensations before the query subsumption check would be straightforward.

Finally, we briefly describe two other well-known optimizations: forward and backward subsumption [2]. Both optimizations compare each new clause C produced in the saturation step with the set of previously generated clauses. In forward subsumption, C is discarded if the set of clauses already contains a clause C' that subsumes C; in backward subsumption, C' is removed from the set of clauses if it is subsumed by C.

Since RQR is based on a resolution calculus, both forward and backward subsumption can be straightforwardly applied without affecting completeness [2]. In the case of CGLLR, however, forward subsumption cannot be (straightforwardly) applied: every query produced in the reduction step is subsumed by another previously produced query; forward subsumption would thus effectively eliminate the reduction step, and so compromise completeness. For example, forward subsumption would remove query (26) in the above example, preventing the generation of query (27). It is not clear whether backward subsumption can be applied to CGLLR without affecting completeness.

4 Evaluation

In this section we present an empirical evaluation of our implementations of the RQR and CGLLR algorithms. RQR is implemented in a rewriting system that we call REQUIEM, while our CGLLR implementation is called C. We also implemented optimized versions of the two algorithms that try to reduce the size of the rewriting using an a posteriori query subsumption check—these are called REQUIEM-SC and C-SC, respectively. Note that C-SC eliminates queries that contain auxiliary properties (introduced by the encoding of qualified existential restrictions) *before* performing the query subsumption check. Both REQUIEM and C are available at REQUIEM's Web site.

The main goal of the evaluation is to compare the algorithms w.r.t. the *size of the rewritings* they produce. Simply counting the number of conjunctive queries in each rewriting might not provide a fair comparison as the queries themselves could differ in size; we therefore additionally measured the total number of symbols needed to represent the complete rewriting in the standard datalog notation. We also measured the time taken for each rewriting procedure. In view of our relatively naïve implementations, however, this may not provide a very meaningful measure of the likely cost of the rewriting process. We therefore also measured the number of *inferences* performed by each algorithm, where by an inference we mean the derivation of a query. Note that the number of inferences is not necessarily the same as the number of queries in the final rewriting since an algorithm may derive the same query more than once.

Tests were performed on a PC running Windows XP with a 2.59 GHz Intel processor and 1.87 GB of RAM. We used Java 1.6.0 Update 7 with a maximum heap size of 256 MB. Tests were halted if execution time exceeded 600 seconds.

4.1 Test Ontologies and Queries

The test set mainly consists of DL-Lite$_R$ ontologies that were developed in the context of real applications, along with test queries that are based on canonical examples of queries used in the corresponding application.

`V` is an ontology capturing information about European history, and developed in the EU-funded VICODI project.[5] `S` is an ontology capturing information about European Union financial institutions, and developed for ontology-based data access [10]. `U` is a DL-Lite$_R$ version of LUBM[6]—a benchmark ontology developed at Lehigh University for testing the performance of ontology management and reasoning systems—that describes the organizational structure of universities. `A` is an ontology capturing information about abilities, disabilities, and devices, and developed to allow ontology-based data access for the South African National Accessibility Portal [7].

We additionally included two synthetic ontologies in our tests in order to provide a controlled scenario to help us understand the impact of the reduction step. `P1` and `P5` model information about graphs: nodes are represented by individuals, and vertices are assertions of the form edge(a, b). The ontology `P5` contains classes representing paths of length 1–5, while `P1` contains a class representing paths of length 1 only.

Finally, for every ontology containing qualified existential restrictions, we created an ontology where the relevant axioms have been replaced by applying the encoding required in CGLLR. We included these ontologies in order to measure the impact of the encoding and the saturation step separately. These ontologies are identified with the name of the original ontology and the suffix `X`. In our discussion, we refer to these ontologies as the *AUX ontologies* and we refer to the others as the *original ontologies*. All the ontologies and queries are available at REQUIEM's Web site.

[5] `http://www.vicodi.org/`

[6] `http://swat.cse.lehigh.edu/projects/lubm/`

		Queries				Symbols				Inferences		Time			
O	Q	R	C	RSC	CSC	R	C	RSC	CSC	R/RSC	C/CSC	R	C	RSC	CSC
V	1	15	15	15	15	454	454	454	454	14	14	16	1	31	1
	2	10	11	10	10	762	812	762	762	9	10	16	1	47	15
	3	72	72	72	72	6,525	6,525	6,525	6,525	117	117	31	31	62	47
	4	185	185	185	185	11,911	11,911	11,911	11,911	328	328	62	47	141	110
	5	30	150	30	30	3,761	16,255	3,761	3,761	59	475	31	78	63	110
S	1	6	6	6	6	158	158	158	158	48	9	15	1	16	1
	2	160	204	2	2	11,422	13,680	154	66	689	876	78	78	109	141
	3	480	1,194	4	4	56,536	121,674	488	232	3,130	7,523	438	922	1,062	2,875
	4	960	1,632	4	4	111,092	173,088	468	224	5,841	10,270	828	1,109	2,171	4,156
	5	2,880	11,487	8	8	466,896	1,602,203	1,320	648	24,332	87,324	9,829	80,031	34,681	233,958
U	1	2	5	2	2	118	286	118	118	26	4	16	1	31	1
	2	148	287	1	1	10,378	18,496	68	30	307	589	47	62	93	125
	3	224	1,260	4	4	29,376	151,848	516	372	570	4,228	94	531	203	640
	4	1,628	5,364	2	2	113,270	348,782	124	56	5,028	18,541	797	4,610	4,093	8,390
	5	2,960	9,245	10	10	279,266	822,279	932	506	13,085	43,306	3,234	16,219	15,262	29,204
A	1	402	783	27	27	21,933	39,593	901	901	934	1,958	94	157	265	350
	2	103	1,812	50	50	7,122	116,137	3,783	3,783	308	4,986	47	687	78	719
	3	104	4,763	104	104	10,108	413,760	10,108	10,108	461	14,454	78	4,187	93	4,266
	4	492	7,251	224	224	33,454	461,549	16,069	16,069	1,405	21,377	156	8,000	422	8,250
	5	624	-	624	-	70,320	-	70,320	-	2,618	-	328	-	1,031	-
P1	1	2	2	2	2	42	42	42	42	1	1	1	1	1	1
	2	2	3	2	2	70	92	70	70	4	2	1	16	16	30
	3	2	7	2	2	98	262	98	98	9	9	1	16	16	30
	4	2	16	2	2	126	734	126	126	16	38	1	16	16	30
	5	2	33	2	2	154	1,824	154	154	25	129	15	31	16	47
P5	1	6	14	6	6	122	298	122	122	9	13	1	1	1	1
	2	10	86	10	10	286	2,814	286	286	75	141	15	15	15	31
	3	13	538	13	13	486	23,740	486	486	428	1,348	94	141	95	142
	4	15	3,620	15	15	708	200,696	708	708	2,238	12,471	922	3,546	1,045	3,607
	5	16	25,256	16	16	938	1,690,902	938	938	11,350	113,277	24,172	265,875	30,000	270,520
UX	1	5	5	5	5	286	286	286	286	39	4	16	1	31	1
	2	240	286	1	1	16,208	18,496	68	30	510	589	78	78	187	156
	3	1,008	1,248	12	12	125,304	151,848	1,452	1,060	3,240	4,228	641	532	2,093	2,484
	4	5,000	5,358	5	5	332,000	348,782	283	131	17,188	18,541	5,969	4,719	36,247	38,189
	5	8,000	9,220	25	25	737,000	822,279	2,240	1,220	37,635	43,306	19,141	15,844	106,055	108,034
AX	1	782	783	41	41	39,527	39,549	1,399	1,385	1,785	1,958	218	156	766	765
	2	1,781	1,812	1,431	1,431	114,537	116,137	91,576	91,145	5,073	4,986	1,016	688	5,547	5,282
	3	4,752	4,763	4,466	4,466	413,030	413,760	388,457	388,303	14,629	14,454	7,375	4,125	49,452	45,876
	4	7,100	7,251	3,159	3,159	454,807	461,549	199,604	197,955	21,137	21,377	13,032	8,047	72,781	79,377
	5	-	-	-	-	-	-	-	-	-	-	-	-	-	-
P5X	1	14	14	14	14	298	298	298	298	25	13	15	16	16	1
	2	77	86	25	25	2,616	2,814	784	728	182	141	31	15	63	47
	3	390	530	58	58	18,574	23,452	2,480	2,326	1,241	1,348	156	141	406	609
	4	1,953	3,476	179	179	120,232	193,608	10,010	9,520	7,832	12,471	2,375	3,469	9,610	26,438
	5	9,766	23,744	718	-	737,890	1,597,158	50,120	-	47,100	-	69,454	249,177	280,381	-

Fig. 1. Results

The number of classes, properties, and axioms are as follows:

	V	S	U	A	P1	P5	UX	AX	P5X
classes	194	18	34	74	2	6	35	74	6
properties	10	12	26	5	1	1	31	31	5
axioms	222	51	127	137	2	10	137	189	18

4.2 Results

Figure 1 shows the results of the empirical evaluation. For each ontology and query, the column "Queries" shows the number of conjunctive queries in the rewriting, the column "Symbols" shows the number of symbols needed to represent the rewriting in datalog notation, the column "Inferences" shows the number of inferences that were performed by each implementation to compute the rewritings (note that the number of inferences for REQUIEM and REQUIEM-SC, and for C and C-SC, is always the same), and the column "Time" shows the number of milliseconds taken to compute the rewritings.

Comparing REQUIEM to C w.r.t. the original ontologies, it can be seen that REQUIEM produced smaller rewritings than C in 25 out of 30 cases and equal sized rewritings in the remaining 5 cases. Moreover, REQUIEM was often faster and performed fewer inference steps, particularly in non-trivial cases (i.e., where both implementations took more than 1,000ms). The differences in the size of the rewritings are often significant: in the fifth queries over `U`, `S`, and `P5`, for example, the rewritings produced by C contain between two and four times as many queries (and contain correspondingly larger numbers of symbols). In the fifth query over `A`, C had already produced more than 42,000 queries when it exceeded the 600 second time limit; in contrast, REQUIEM completed the rewriting in less than 350ms and produced only 624 queries.

When we compare REQUIEM to REQUIEM-SC, we can see that they produced the same rewritings in 19 out of 30 cases. In some cases, however, the rewriting produced by REQUIEM was much larger: in the fifth query over `S`, for example, REQUIEM's rewriting contained 2,880 queries compared to only 8 for REQUIEM-SC. As we might expect given the larger rewritings produced by C, it produced the same rewritings as C-SC in only 6 out of 30 cases. The differences in size were also generally larger: in the fifth query over `S`, for example, C's rewriting contained 11,487 queries compared to only 8 for C-SC. The large size of the rewritings produced by C also mean that performing query subsumption tests over these rewritings can be costly. In the fifth query over `S`, for example, C-SC takes nearly three times as long as C.

Comparing REQUIEM-SC to C-SC, we can see that REQUIEM-SC produced the same rewritings as C-SC in 21 out of 30 cases, larger rewritings in 8 cases, and a smaller rewriting in the remaining case (due to the fact that C-SC exceeded the maximum time of 600 seconds). The differences in size are minimal, and the number of queries in the rewritings is always the same (with the exception of the case where C-SC exceeded the time limit). Note that the smaller rewritings produced by C-SC are due to its generation of condensations (see Section 3).

If we turn our attention to the AUX ontologies, we can see that REQUIEM still produced smaller rewritings than C in 12 out of 15 cases, although the differences were less marked. Moreover, REQUIEM still performed less inferences than C in 9 out of 15 cases. In contrast to the results with the original ontologies, REQUIEM was slower than C in 10 out of 15 cases; the differences, however, were generally small.

Our analysis suggests that REQUIEM will produce significantly smaller rewritings than C and will be significantly faster, particularly in cases where the queries are relatively complex and/or the ontologies contain a relatively large number of qualified existential restrictions. The size of the rewritings produced in some cases also means that a query subsumption check may be prohibitively costly in practice with CGLLR, even when queries containing auxiliary properties are removed before performing the check. Moreover, the results for the

AUX ontologies suggest that the reduction step alone has a negative impact on the size of the rewritings—that is, the introduction of auxiliary properties does contribute to producing large rewritings, but it is not the only cause.

5 Future Work

We plan to implement an ontology-based data access system using REQUIEM enhanced with various optimizations (i.e., forward/backward subsumption, query subsumption, and query condensation); we expect such a system to perform well both w.r.t. the size of the rewritings and the time needed to compute them. The practicality of such a system is, however, still open, as our results suggest that there are cases where the rewritings may be too large to evaluate. In such cases, we believe that a further optimization that uses the mappings to prune irrelevant queries (as described in Section 2) might produce rewritings of manageable proportions. We plan to test our system with actual data in order to discover if this is indeed the case. Finally, we plan to extend the system to support all of OWL 2 QL, which mainly involves adding support for datatypes.

References

1. Baader, F., Snyder, W.: Unification Theory. In: Robinson, A., Voronkov, A. (eds.) Handbook of Automated Reasoning, ch. 8, vol. I, pp. 445–532. Elsevier Science, Amsterdam (2001)
2. Bachmair, L., Ganzinger, H.: Resolution Theorem Proving. In: Robinson, A., Voronkov, A. (eds.) Handbook of Automated Reasoning, ch. 2, vol. 1, pp. 19–100. North Holland, Amsterdam (2001)
3. Calvanese, D., De Giacomo, G., Lembo, D., Lenzerini, M., Rosati, R.: Tractable Reasoning and Efficient Query Answering in Description Logics: The DL-Lite Family. J. of Automated Reasoning (2007)
4. Calvanese, D., Giacomo, G.D., Lenzerini, M., Nardi, D., Rosati, R.: Description Logic Framework for Information Integration. Principles of Knowledge Representation and Reasoning, 2–13 (1998)
5. Chang, C.-L., Lee, R.C.-T.: Symbolic Logic and Mechanical Theorem Proving. Academic Press, Inc., Orlando (1997)
6. Kazakov, Y.: Saturation-Based Decision Procedures for Extensions of the Guarded Fragment. PhD thesis, Universität des Saarlandes, Saarbrücken, Germany (2006)
7. Keet, C.M., Alberts, R., Gerber, A., Chimamiwa, G.: Enhancing web portals with ontology-based data access: The case study of south africa's accessibility portal for people with disabilities. In: OWLED (2008)
8. Motik, B.: Reasoning in Description Logics using Resolution and Deductive Databases. PhD thesis, Universität Karlsruhe (TH), Karlsruhe, Germany (2006)
9. Pérez-Urbina, H., Motik, B., Horrocks, I.: Tractable Query Answering and Rewriting under Description Logic Constraints. J. of Applied Logic (to appear, 2009), `http://web.comlab.ox.ac.uk/people/publications/date/Hector.Perez-Urbina.html`
10. Rodriguez-Muro, M., Lubyte, L., Calvanese, D.: Realizing ontology based data access: A plug-in for protégé. In: Proc. of the Workshop on Information Integration Methods, Architectures, and Systems (IIMAS 2008), pp. 286–289. IEEE Computer Society Press, Los Alamitos (2008)

Multi Visualization and Dynamic Query for Effective Exploration of Semantic Data

Daniela Petrelli[1], Suvodeep Mazumdar[2], Aba-Sah Dadzie[2], and Fabio Ciravegna[2]

[1] Department of Information Studies
[2] Department of Computer Science, University of Sheffield
Regent Court - 211 Portobello Street, S1 4DP, Sheffield, UK
{d.petrelli,s.mazumdar,a.dadzie,f.ciravegna}@shef.ac.uk

Abstract. Semantic formalisms represent content in a uniform way according to ontologies. This enables manipulation and reasoning via automated means (e.g. Semantic Web services), but limits the user's ability to explore the semantic data from a point of view that originates from knowledge representation motivations. We show how, for user consumption, a visualization of semantic data according to some easily graspable dimensions (e.g. space and time) provides effective sense-making of data. In this paper, we look holistically at the interaction between users and semantic data, and propose multiple visualization strategies and dynamic filters to support the exploration of semantic-rich data. We discuss a user evaluation and how interaction challenges could be overcome to create an effective user-centred framework for the visualization and manipulation of semantic data. The approach has been implemented and evaluated on a real company archive.

Keywords: Semantic Web, semantic multimedia data, graphical visualization, user interaction.

1 Introduction

Organisational memory, the ability of an organisation to record, retain and make use of information from the past to bear upon present activities [27], is a key issue for large organisations. The possibility of observing and reflecting on the past is particularly valuable in highly complex domains as it can inform and sustain decision-making. Civil aerospace engineering is one example: the life cycle of a gas turbine (commonly referred to as a 'jet engine') can last for 40-50 years from initial conception until the last engine is removed from service. During this long product lifetime a vast amount of heterogeneous data is created, i.e., text reports, numeric data, images, CAD (Computer Aided Design) drawings etc. [21]. Several everyday tasks require engineers to engage in sense-making activities, i.e., "a motivated, continuous effort to understand connections (which can be among people, places, and events) in order to anticipate their trajectories and act effectively" [16]. For example, issue identification and resolution for jet engines (a task performed when a generalised issue is suspected in a product, e.g. frequent excessive wear and tear of a set of components) can require access to information contained in tens of resources, including textual repositories

A. Bernstein et al. (Eds.): ISWC 2009, LNCS 5823, pp. 505–520, 2009.

(each containing several dozen of thousands of texts, spreadsheets, etc.), image repositories (same cardinality), raw data (a jet engine produces about 1G of vibration data per hour of flight) and some very large databases. Engineers must go through the different repositories searching and sieving for relevant information and any piece of evidence that can confirm or disprove their current hypothesis, or browse for patterns and trends that can spark an intuition. In a word, they use dispersed and diverse data and information to build up knowledge about a specific phenomenon. The task can last for several months [8].

Currently the work of evidence gathering and meaning structuring is done manually with the support of keyword search on textual documents and querying of unconnected distributed databases. Keyword based querying in this domain is rather ineffective due to low precision/recall [17]. Moreover the long time required to read each document and manually abstracting the data to identify trends and their possible causes implies only a limited number of hypotheses can be explored.

Semantic Web (SW) technologies can be used to semanticise such resources [8]. Semantic information enables to (i) formalise the unstructured information in texts, images and raw data, (ii) reconcile information contained in different information sources via a central ontology or a series of interconnected ontologies [3] and (iii) enable information integration across resources and formats. Semantic information is being generated over large scale using technologies for a) ontology-based knowledge capture using forms [3] and b) for information extraction from text that can be ported to new corpora by trained final users.

In this paper we explore the issue of browsing, querying and visualizing semantic information in such semantic repositories in a way that allows users to dynamically explore the data during a complex task such as issue identification and resolution. The solution provided is based on (i) the visual contextualisation of semantic information according to some easily graspable dimensions (e.g. space, time and topology) and (ii) the browsing of the displayed information by querying the knowledge base via dynamic filters that modify the visualization in order to focus on possible trends and patterns. This approach enables exploration of information and data currently highly challenging with existing technologies (especially commercial keyword-based systems) that could save thousands of hours a year of valuable resources to a company.

The paper is organized as follows: Section 2 overviews related work. Section 3 discusses the framework, section 4 the design rational, and section 5 provides some details on the implementation. Section 6 presents the user evaluation and our findings. An outline of the future work concludes the paper.

2 Related Work

Visualization is required whenever humans need to discover and reason about complex combinations of high volumes of data (e.g., [5]). Information visualization and visual data mining is not limited to the display, but aims at supporting human perceptual abilities during the data exploration process [15]. A vast literature exists on the topic. Cluster visualization has been used in such diverse fields as intelligence (i.e. to show correlation between people [36]) and image collection access (i.e. to show similarity in images [10]). Alternative visualizations have been used to make easy to identify patterns in homogeneous data (e.g. in geospatial data [1]); multiple visualizations, instead, map the strength of relationships between elements [4]. In text retrieval, much research has investigated the visualization of search results (see [12] and [13] for an overview), the visualization of the whole document collection (e.g., Treemaps [35]) or a large text corpus (e.g. Jigsaw [26]). Information exploration, an open-ended process that is iterative and multi-tactical [18, 33] is currently gaining interest and stimulating new user interactions beyond traditional text search [34, 24].

The issue of visualization of Semantic Web (SW) data has been recognized since the publication of the seminal book [11]. A tension exists: "the Semantic Web emphasises formal, machine readable [...] approaches. It focuses on the formal and even the meaning achieved through rigorously defined forms. In contrast, information visualization emphasizes the semantics and the meaning that can be conveyed by visual-spatial models to the users." [6]. Much research effort in semantic-based visualization has been spent on finding ways of visualizing complex graphs that derive from the interlinking of semantic data, the relation between different concepts [28], the different granularities [31], and (dis)connections [19]. The result is a large number of ontology-based visualization systems (some are reviewed in [9]).

More recent research has tried to make use of the special features of RDF to provide end-users with intuitive ways of accessing semantic data. BrowseRDF [20] uses the faceted browsing paradigm: facets are generated automatically from the data itself; the user can constrain one or more of the faceted provided to filter the data set.

Similarly, mSpace [23] sequences lists of facets, the item selected in a list constrains the following step. Users can combine facets in different ways: this allows an intuitive composition of complex filters for the purpose of exploration.

In IVEA [30, 31] the user creates their own view over a text collection by dragging and dropping ontology concepts onto a scatter plot panel. The filters provide a multi-dimensional view of the document collection as a matrix with colour coded values.

Exhibit, as part of the SIMILE Project[1], provides an interactive, web-based visualization widget developed to demonstrate the application of SW technologies to heterogeneous metadata. It interlinks geographical mapping and a timeline to display information about the USA (past) presidents, e.g., place of birth, term(s) in power, etc.

3 A User-Interaction Framework for Semantic Data Exploration

In a SW framework, information in text, images, tables and other forms of data can all be captured and mapped to ontology concepts, instances or relations and be represented as triples. SW technologies can pull together heterogeneous material in a single unified form and create a single organizational memory out of many different and scattered archives. However, SW-based organizational memory can be huge when derived from very large collections, encompassing dozens of repositories containing tens of thousands of documents which in turn produce millions or billions of triples. A real problem in knowledge discovery occurs when making use of such extremely large data set as no human could be expected to hold all the information in their mind. Specific tools that help users explore the knowledge and draw hypotheses

[1] The SIMILE Project: http://simile.mit.edu; Exhibit: http://simile.mit.edu/wiki/Exhibit

from it are essential for effective use of SW-based organizational memory. This requires the following fundamental steps:

1. The RDF repository has to be planned to support effective human interaction: Triples may not hold any context if it has not been captured, e.g. it is impossible to plot triples by time if the date is not there; A single ontology should map heterogeneous material into a single representation.
2. The visualization has to be intuitive to properly contextualize semantic data and the interaction tools have to be easy-to-use to support exploration and knowledge discovery, e.g. space and time contextualize semantic data in an intuitive, factual way.
3. Tools to facilitate data annotation should be smoothly integrated in the interaction flow to guarantee a sustained improvement of the quality of the repository along its use, especially when data is generated using automated means (e.g. by applying information extraction on legacy data).

In this paper we focus on the second point and propose the use of multiple visualizations as a way to help users explore, discover and reason; find confirmation of their intuition; and drill down to the data level when needed.

As discussed in the next section, the dimensions of visualization should come from the ontology and the values for each dimension from the semantic repository. Dimensions should be then mapped onto a structure that is appropriate for the final user. For example the dimension 'date' can be structured as a linear timeline (as done in this work) or as a calendar, e.g., to visualize publications. Both visualizations map the same semantic data but serve two very different user purposes.

Some dimensions are generic and likely to be valid across a wide range of applications. This is the case of time and space. Other dimensions are valid across a subset of domains, for example this paper uses topology, useful in engineering where a machine of some sort is the core of the domain ontology. Finally the user may define their own perspective, e.g., time and a continuous attribute could be plotted to facilitate the monitoring, for example, of financial markets.

4 Knowledge Visualization and Manipulation

For data and information visualization, Shneiderman advocated tools that provide the user with a progressive focus: "overview first, zoom and filter, then details-on-demand" [25]. The overview is to gain a sense of the whole data set; zoom and filters are used to focus the attention on (potentially) interesting patterns; details on demand to drill down to the level of single data and carefully inspect the content.

This section discusses our proposal for a concrete visualization of semantic data. This visualization complements ontology-based visualization, as it reduces the cognitive effort needed to understand the semantic data.

We adopted a user-centred iterative design approach [8]. The design rationale discussed in the next section emerged after workshops and observations with users aimed at collecting requirements, and a number of participatory design sessions with engineers in which layout and interaction were refined to maximise their usefulness. In opposition to the generic trend of using a single visualization to display semantic data and its connection, we contextualize data in multiple, complementary and inherently different visualizations where each "view" offers a perspective over the data: the user dynamically filters the data and moves from one view to another while following a personal investigation trial.

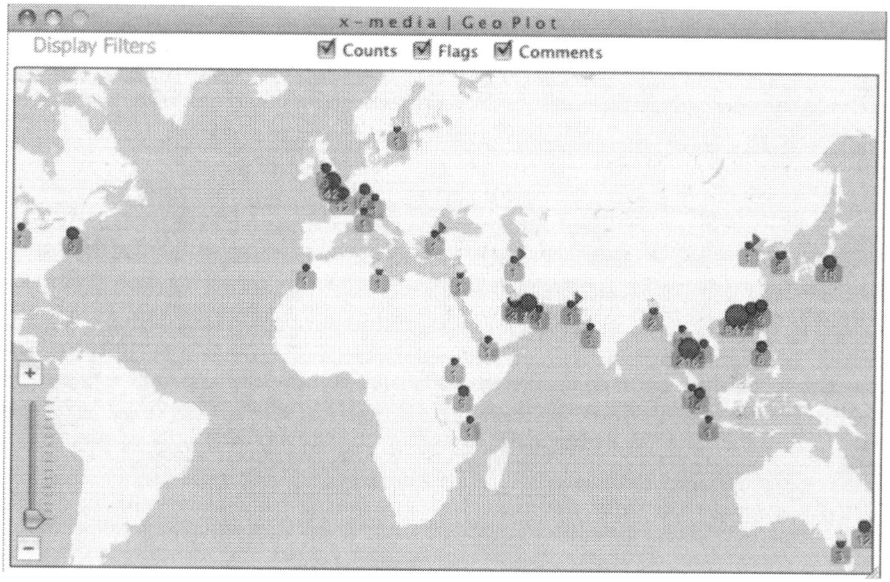

Fig. 1. The triple-store displayed on a world map - dots, flags and notes add meaning. The numbers in the map indicate the number of cases found per location. The filters used are identical to those in Figure 2.

Figure 1 shows the GeoPlot of 34,750 triples extracted from 4,958 event reports that are part of Rolls-Royce's organisational memory[2]. The extracted triples are displayed on a world map showing the distribution of events; the size of the dots codifies the number of events. Flags and comments can be added to keep track of personal intuition during the sense-making process.

Fundamental for effective exploration is the dynamic update of the display when the user manipulates the filters, called dynamic querying [2]. The filters (Fig. 2, left) allow the user to quickly set the parameters of interest and immediately see the effect on the display (Fig. 2, centre). The filters' interactive features depend on the data type: a slider to set a range for numeric data; a text field to enter codes; a single selection list items and group of check boxes for multiple selection. The result of the filtering is dynamically plotted: in Figure 2 the blue (darker) crosses match the query

[2] These are just a fraction of the triples that we generated from Rolls-Royce organisational memory. They were extracted by semi-automated information extraction and machine learning as part of a general effort to semanticise their legacy data. The final data set holds several hundreds of thousands of documents and covers several different types: one-page reports sent from all airports in the world covered by the Rolls-Royce service agreements, extended reports of workshop inspections, technical updates, workshop photos, tables, etc.

(multiple filtering), while the gray (lighter) ones are triples outside the result set. Filters on one visualization are applied to all the others simultaneously in order to maintain consistency. The geographical and the topological visualization needs two dimensions, but the TimeLine is uni-dimensional and therefore the Y axis can be dynamically changed using drag-and-drop with any concept from the ontology (Figure 2, right) onto the plot.

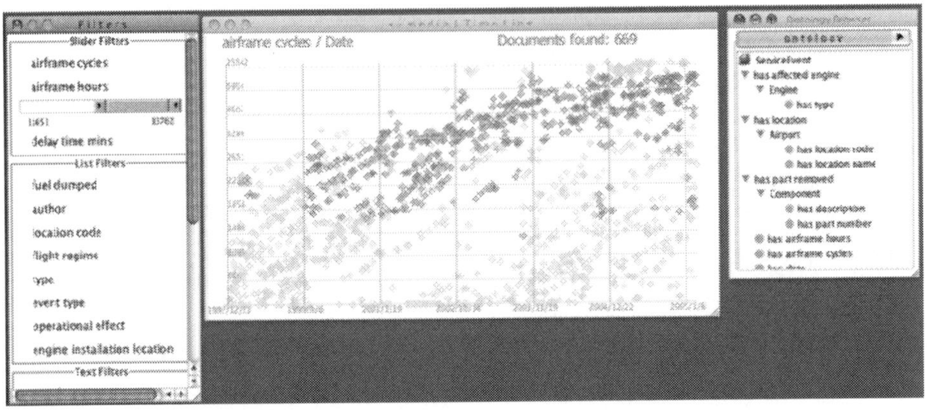

Fig. 2. The dynamic query filters (left) set the values for the TimeLine (centre). The X-Axis is time, the Y axis is the number of airframe cycles; the top right the number of matching documents.

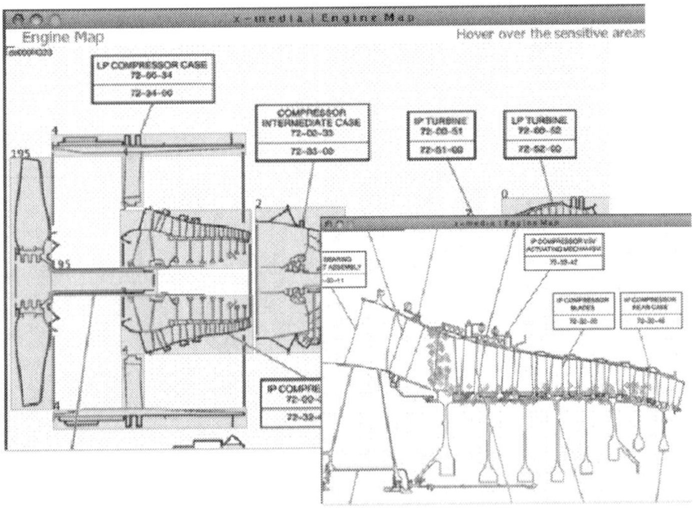

Fig. 3. The TopologicalPlot: each component in gray in the Engine Map (left) can be clicked on to zoom-in a more detailed view of the information associated to its subcomponents (right). Numbers is blue show the total count of issues identified per component. The filters are the same as in Figure 2.

A third visualization uses topological information, in our case a TopologicalPlot, as intuitive dimension to plot triples. Figure 3 left shows an engine overview: gray areas correspond to high-level ontology concepts. Hovering on an area shows a summary of the documents mapped to this engine part; clicking on the area opens a detailed map of that part where the documents are plotted, Figure 3 right, respect to finer grained concepts, the engine components. Their position is as faithful as the ontology allows, that is to say, the finer the grain of the ontology concept the more precise the position of the cross on the graph.

The three visualizations, TimeLine, GeoPlot and TopologicalPlot show the same data with respect to different dimensions that complement one another. The structure itself can be semantically enriched therefore adding new knowledge to the visualization that is not present in the data, i.e. providing semantic services attached to the world map. The visualization could be enriched father by coding properties the user wishes to monitor in a more salient and graphical way, e.g., suppose the user is interested in instances of the concept 'wear', an event often associated with sand friction, then these events could be highlighted in red and a GeoPlot can easily show their pattern in deserts regions.

5 Implementation

The starting point is an existing semantic repository (triple store) and its related ontology; interactive filters are automatically generated on the basis of the data found in the ontology. Data tables are created where each row is a document in the dataset, and each column is a concept annotated within the document. In Figure 4, the attributes *hasFormattedEventDate*, *hasLocation*, *hasComponent* are extracted to build TimeLine, GeoPlot and TopologicalPlot respectively.

```
<rdf:Description
rdf:about="http://kmi.open.ac.uk/projects/xmedia/RR1.owl#Event_Report.BKK.Event_Report_237">
    <rdf:type rdf:resource="http://kmi.open.ac.uk/projects/xmedia/RR1.owl#Event_Report"/>
    <j.0:has_file_location>BKK/Event_Report_237</j.0:has_file_location>
    <j.0:hasFormattedEventDate>26-Jul-1922</j.0:hasFormattedEventDate>
    <j.0:hasEventDate>26-Jul-22</j.0:hasEventDate>
    <j.0:hasAssociatedDate>28-Aug-22</j.0:hasAssociatedDate>
    <j.0:hasTSN>14613</j.0:hasTSN>
    <j.0:hasEngine_Serial_Number>ESN12345</j.0:hasEngine_Serial_Number>
    <j.0:hasLocation>BKK</j.0:hasLocation>
    <j.0:hasRegime>GROUND</j.0:hasRegime>
    <j.0:hasCSN>5362</j.0:hasCSN>
    <j.0:hasComponent>Fuel Metering Unit</j.0:hasComponent>
    </rdf:Description>
```

Fig. 4. An example of the annotations of a document in RDF format. (Values shown are realistic but fictitious and do not correspond to a real instance of an Event Report.)

While only some of the ontology concepts and relations become visualization structures, all of them become filters. The data table is read and each column is converted into a graphical widget, which one depends on its value range. A core set of filters, with different interaction affordances, is used to capture different aspects of the data set: sliders are used to define ranges of continuous data; text input is used to capture meaningful strings, e.g. engine serial number; check boxes and menus for selections in a closed set, e.g., *hasTSN* and *hasCSN* are mapped onto a slider; *hasEngine_Serial_Number* uses a text field; *hasRegime* uses a group of check box.

Different toolkits have been used to build the visualization modules. Prefuse [14], an interactive visualization toolkit with sophisticated visual features is used for the TimeLine. The X-Axis represents time, and the Y-Axis a continuous numeric value such as CSN ([flight] cycles since new). The user can dynamically change the Y-Axis concept using drag-and-drop from the ontology. This enables all the ontology concepts to be plotted against the timeline. A visibility filter controls the display of each visual item: the ones in the filtered set are highlighted, the others are greyed out.

The GeoPlot visualization is generated using the JXMapKit API that plots geographic coordinates on a world map downloaded from OpenStreetMaps.com. The geographic locations are airports, identified by their IATA (International Air Transport Association) codes, as extracted from the dataset. The IATA codes are used to automatically find the airport details such as geo-coordinates (used in the plot), airport name, city and country. The size and colour of the waypoints are calculated on the number of visual items associated with the airport.

The Topological Plot is composed of two interactive maps manually created using drawings from an engine user manual. For the top level, selected regions are annotated with high-level ontology concepts corresponding to the engine parts, each showing the number of visual items associated with the concept. A detailed view of the part is displayed on click; this drawing too has been manually annotated with finer grained concepts from the ontology that corresponds to engine components. Although the maps have been manually created it is easy to imagine a situation in which the CAD drawings of an engine has semantics associated and therefore the generation of the maps is automatic.

6 User Evaluation

The usability of the visualization and manipulation was carried out. Results are used to adjust and re-design the system before it is deployed for a monitored field trial at Rolls-Royce plc as an additional support to actual investigations. While the trial allowed us to measure the impact this technology has on real practice and observe its use in a naturalistic setting, the user evaluation reported here focuses on assessing its usability, that is to say to find out what works and what instead could be perfected. This 'evolutionary' approach to user evaluation, from lab to the field, has proved to be robust for the development of new technology for professional use [22].

6.1 Setup and Procedure

The user evaluation was set up to assess the usability of the visualization and dynamic querying of semantic data. 12 participants took part in the evaluation and were recruited by acquaintance, they were 4 women and 8 man, their age ranged from 25 and 45, they were PhD students and researchers; 3 participants were aware of information visualization tools but none had used any. As the time of professional engineers is a limited resource and the focus of this evaluation was the usability we considered the sample acceptable for the goal at hand. Participants carried out a number of small tasks to determine if:

1. The visualization mechanisms supported knowledge discovery: patterns were to be found in the data that could represent phenomena of interest;
2. The dynamic queries were intuitive to use: participants were required to manipulate different types of filters, slider, checkbox, text;
3. The overall visualization and dynamic queries strategy were usable.

The test was done individually. At arrival participants were introduced to the project and the purpose of the user evaluation. They were talked through the main features of the visualization and manipulation by an evaluator. Then participants familiarized themselves with the system using 7 simple, 1-step tasks covering the three visualizations and the filters. An indication of which visualization(s) was the most appropriate and how to use it (them) was given for each task. During the training participants could ask for explanation or support from the experimenter. Participants were then requested to carry out another 8 tasks on their own: each of these tasks asked the user to perform 2 or 3-steps, for a total of 18 steps. These tasks were slightly more complicated than the training to stimulate more articulated interactions and were designed to test different aspects of the system. Flexibility is a key point for user acceptance and we wanted to find out if our solution accommodated personal attitudes. Task distribution per visualization is reported in Table 1, T-TimeLine, G-GeoPlot, E-TopologicalPlot, and ALL the visualizations; C-F is for commenting and flagging in the GeoPlot.

The tasks were designed for users with no expertise on jet engines and did not require participants to understand the content of the documents that could be displayed on click. As users were not experts they were not required to identify trends directly, however the tasks used simulated the ones that an engineer will perform in order to identify trends. They generally required to identify (geographical, time or topological) areas where the count of events was either clearly above average or had specific characteristics. Most tasks were cumulative (i.e. they built on the results of the previous tasks) so to simulate a multi-step investigation.

Examples of cumulative tasks are[3]:

a) How many documents refer to the **registration number** *9V-SQD* with number of **airframe cycles** *>6500*?

b) Consider only what happened in ***SIN (Singapore)***, how many events occurred there?

c) Which component seems to have the highest number of cases associated with the **flight regime** *CLIMB*?

These tasks required the use of the different tools and strategies while performing each query:

a) The registration number requires entering text in a form field, setting an airframe cycle number requires manipulating a slide.

[3] Tasks are simplified here for the sake of clarity of exposition. Task b) was to be performed on the results of a); c) was to be performed on information retrieved in b).

b) Requires to interpret the previously retrieved data by focussing on a specific area of the GeoPlot
c) Requires now to move to the topological display and to drill down to the level of components. Flight regime is a checkbox.

The tasks enabled to measure all the available interactive features in a limited test time of 30-40 minutes. During the test the screen activity was recorded for further data analysis.

At the end of the evaluation participants were requested to fill in a user satisfaction questionnaire composed of 16 closed and open questions. Close questions were on a 5-points scale and addressed the system overall, its learnability, the task flow, the result display, the system's speed and reliability. Open questions asked about the most positive and negative aspects of the system. No questions focussed on comparison of visualizations as all three have been seen in use during the user requirements phase: timelines in presentations to summaries observed phenomena; geographical information was reported as one of the first inquiry done; and maps of the engine covered in post-it and annotations hung in meeting rooms. We are therefore confident on the usefulness of all three.

6.2 Analysis and Results

The results are analysed with respect to: efficiency, effectiveness, and user satisfaction (as in the ISO definition of usability [32]). Minor issues of interface inconsistency across the three visualizations were also identified, e.g. display of the number of results in the set displayed in different positions in the three panels. Both objective, numeric data, and subjective data, participant's opinion, have been analysed. Qualitative analysis, i.e. observation of participants' interaction, has been used to explain qualitative results, i.e. statistic on numeric values.

Efficiency has been calculated on the time participants needed to finish **a task** (that could include few different steps). Performance in both training and test varied greatly from task to task and from participant to participant, and from a min of 18 sec. to a max of 420 sec. Table 1 shows this variability by task, Table 2 by participants. The average time to complete a test task (including the time to think how to solve the task) was 87 sec. The average time for a simple 1-step training task was 67 sec. while the average time spent for complete 1-step during the test was 46 sec. showing an increase in efficiency after only an average of about 8 minutes training. The good efficiency is reflected on participants' opinion collected in the questionnaire: 82% rated the system speed very high.

On average, interactions with the GeoPlot lasted longer than other visualizations: T= 147 sec.; G=200 sec., E=138 sec. (cumulative values for participant, task only); however this is not statistically significant (one-way repeated measures ANOVA on time on T, G and E). Observation of the interaction shows that some users had difficulties in manipulating the GeoPlot: when zooming-in they were too fast and found themselves, for example, in Africa instead of Europe. Then, instead of zooming-out they preferred panning, an action that requires more time.

Table 2 shows the average time per participant. Variability among participants emerges during task with a polarization in two groups. Observations of the interaction

Table 1. Time on task in seconds. Each task was designed to test some features of the approach. The letters in parentheses describe the types of tools that the user was expected to use to perform the task in an appropriate way. All the users used at least the expected tools to carry out each task.

	Min.	Max.	Mean	Std.Dev.
Training 1 (T)	18	320	57	85
Training 2 (T)	20	60	32	12
Training 3 (G)	30	240	80	55
Training 4 (G)	15	200	84	50
Training 5 (C-F)	30	420	94	106
Training 6 (E)	20	120	55	30
Training 7 (E)	40	180	67	41
Task 1 (ALL)	20	240	125	81
Task 2 (T)	20	120	70	33
Task 3 (T)	25	150	77	36
Task 4 (C-F)	80	300	211	68
Task 5 (G)	50	300	122	81
Task 6 (G)	60	145	93	26
Task 7 (E)	15	100	52	24
Task 8 (E)	30	210	82	58

Table 2. Average time per task for each participant

	Mean (sec.) Training	Mean (sec.) Task	Mean (sec.) Training + Task
P1	88	127	110
P2	71	89	82
P3	170	119	141
P4	53	71	63
P5	33	78	59
P6	85	153	124
P7	52	136	100
P8	71	128	103
P9	68	66	67
P10	33	85	63
P11	35	103	73
P12	50	76	65

behaviour showed different strategies with faster participants setting all the filters at once and then exploring the visualization and the slower participants going step by step, i.e. setting the value of one filter then look at the result than set a second filter. Multi filter selection was not requested during training. Flexibility of use is one of the main requirements of any data exploration environment; therefore we consider this a positive result showing user could adopt personal strategies despite time differences.

Effectiveness is calculated on the accuracy of the answer provided as every task had a correct answer identified in advance. In 74% of tasks the exact answer was provided. The effectiveness rises to 82% when the simpler tasks of the training set are included. All wrong answers were 'near miss', i.e. participants incorrectly selected a value on the slide or selected the wrong value for a filter due to very similar spelling. Minor changes in the interface are needed to avoid unintended mistakes. Moreover we do not expect to see any spelling problems in field use as users knowledgeable in the domain would not mistake terms. Effectiveness was also affected by data density: some participants instead of zooming-in to gain a clearer view selected the wrong set in dense GeoPlot display.

Overall the approach emerges as effective in supporting the user browsing through a triple store, as the mistakes were due to interface issues that are easily fixable; this is confirmed by the fact that the errors were scattered and not linked to any specific condition, i.e. visualization or user.

User Satisfaction. Overall participants' opinion was positive; the system was judged easy (64%), satisfying (64%), stimulating (82%), fast (82%), and reliable (91%). While this result shows a very high level of engagement and trust (mainly in the stimulating judgement), it also points out to usability problems due in part to (i) some sub-optimal interaction strategy participants adopted to perform the tasks and (ii) interface limitations, that will be discussed below. Discussion with users showed that the latter accounted for the largest majority of the difficulty the user found (affecting the judgement on ease of use and their satisfaction). The dissatisfaction was not related to the general idea and users commented that – if the issues were fixed – their judgement would have changed.

Both issues above can be easily addressed: the first with a more extended training (which will be in any case given to the engineers); the second by re-designing the weak points in the interface.

Three questions addressed learnability: participants judged it easy to learn (82%), easy to explore (75%) and straightforward to use (63%). This last value was, again influenced by the same difficulties in manipulating some graphical elements. The task flow was considered easy to start (73%) and carry on (90%) while the manipulation of the results was problematic for some of the participants. 35% found the manipulation easy or very easy, 45% were neutral and 20% considered it difficult. Again, there is a dichotomy in judgement between the recognized value of the tool and the practical difficulties in manipulating it. Observations of the interaction and the comments left in the questionnaire "most negative aspects" explain this fact: as the values on the interface come from the RDF data, the values on the slide were not continuous nor the progression smooth. This was quite confusing for some participants who tried hard to set the slide to an inexistent value, e.g. the first value for 'airframe hours' (Figure 2) is 4350 but the slide starts from 0 so there is no change in the display until the user scrolls to 4350, that is halfway through the slide. This point can be fixed in the redesign by creating a tagged-slide that highlights the valid values (from the RDF) on a standard continuous slide.

The judgement on the browsing of the results was split: 45% judged it easy, 45% were neutral and 10% find it difficult. Observing the interactions we noticed that participants who lamented difficulties had problems in selecting the right graphical element: In the GeoPlot dots representing two different airports could overlap making it difficult to select one airport over the other. When the overlap occurs the correct interaction is to zoom-in but some participants did not use it despite having been demonstrated the feature before the test. Another point of difficulty occurred when documents were very dense, as for some engine components. I this case the zoom-in (enlarging the picture) is not effective in discriminating instances as the action does not add further details. A further level in the ontology would allow mapping to a more detailed drawing of the component and therefore a finer localization of the triple on the engine spatial representation. Alternatively instances can be listed: selecting an element highlights its position on the map and double clicking would open it.

Two open questions asked for the most negative and more positive aspects of the system. Besides the already mentioned problems with the slide, listed as negative, participants did not like to scroll up and down the filters (the list can indeed be very long if filters like the airport location is left open) and found some of the filters name cryptic. This last comment does not hold for professional engineers, as they are familiar with the data.

Appreciated across the whole sample was the tidy design of the interface, its intuitiveness and the instantaneous reaction and change of display after a new filter has been set, all features listed in the open question "most positive aspects". In addition participants commented positively the fact that the filter manipulation changes the three visualizations simultaneously therefore supporting an active engagement in the data exploration activity by simply swapping view.

6.3 Discussion

The user evaluation showed that the approach is efficient and effective and the interaction largely intuitive even with very limited training. Users found the approach stimulating and were able to identify trends in the data via interactive querying. The methodology used showed this in an indirect way: as participants were not experts in the domain, the tasks simulated the querying path that an expert would follow in order to identify trends. The simulated exploration paths have been observed and discussed with users, therefore the successful completion of the tasks would provide material to an expert for the identification of trends.

Some limited aspects of the interface needs some degree of re-design as the simple action of taking the data out of the RDF repository and into the user interface may produce a less-than optimal interaction, i.e. slides don't have a smooth progression but more of a 'jumpy' interaction style. Data-generated interactive filters and dense data display need careful considerations and specific interaction-design strategies particularly when scaled up to hundreds of thousands of triples displayed. Indeed what appeared to be critical is the combination of very dense semantic data onto small space and the tendency of participant to not zoom-into the detail to clarify the vision.

A research question concerns efficiency and effectiveness with respect to the technology currently available to our final users. We believe that the approach has the potential to save thousands of hours a year of search time (efficient) and to provide a way to more widely explore different hypotheses and therefore to discover more trends and patterns (more effective). We derive this by reflecting other evaluations where users performed similar tasks using other types of technologies, both semantic and more traditional. Performing the same tasks using traditional keyword-based searches would have required several weeks of searching and manually collating information[4]. Also, our approach is more effective because as side effect of efficiency users are enabled to explore different hypotheses and therefore to discover more trends and patterns. Such extensive exploration is currently largely impossible due to the scarce efficiency of the current methodologies (exploring more hypothesis means more time dedicated to the analysis, an often impossible task under the time pressure that some knowledge management tasks are worked under). Moreover traditional methods carry imprecision due to tiredness, which affects the quality of results on very long task.

Performing the same tasks with a semantic search-based system (e.g. [17]), would have required some days of work to extract different pieces of evidence and to group them manually around trends[5]. Questions like "which component reported most issues" would have required several dozens of queries.

---

[4] This estimate is based on discussion with real users and direct observation of working practices.

[5] This estimate is based on observation of search behaviour of users during the evaluation of the semantic search system.

7 Conclusions and Future Work

In this paper we proposed to complement the ontology-based, graph-based perspective with views that contextualise the concepts into vertical dimensions, like time, space and topology. The list of dimensions that could be used for this purpose is not exhaustive and others than those we used could be identified (mostly domain-specific). The different visualizations are created starting from the RDF data and, like a kaleidoscope, show different views on the same data set. Direct manipulation complements the display and engages users in the exploration: dynamic queries generated from the data are used to instantaneously change the visualizations.

Our aim was to provide a largely automatic way to visualize semantic data and support users in dynamic exploration and manipulation. We used the case of an existing organisational memory and complex knowledge management tasks observed in real work situations, i.e., issue identification and resolution in aerospace engineering. The user evaluation has demonstrated that this automatic mapping of multiple, context sensitive visualizations to ontology-based information stores provides an efficient way to display the result of complex queries that can combine several attributes. Moreover users can explore the result and effectively detect patterns and trends. The combination of powerful multiple, contextual visualizations and a highly dynamic interaction allows the exploration of semantic data to be carried out at scale. We have shown how our visualization approach improves in terms of efficiency and effectiveness with respect to technologies that are currently available to our users, i.e. keyword-based search and semantic search. To our knowledge this is the first study to show that using multiple visualizations is effective for document sense making in a complex organisational memory.

In our experience large organizations are willing to invest in semantic technologies for knowledge management, if they see a clear benefit and it is sustainable. The set of 4,958 documents used in this study correspond to a small chunk of the archives we are currently considering in the context of Rolls-Royce plc, but was instrumental to show the clear benefit of this innovative technology over the current practice. At the time of writing we are working in partnership with the company to extract information from large and heterogeneous archives and create a new semantic data set to support a field trial in the context of real practice. In the perspective of sustainability, we have already developed a technology to provide ontology-based knowledge capture using forms [3] and we are studying information extraction methodologies that can be ported to new corpora by a trained final users. In light of the experience above, the approach and tools proposed in this paper are deemed extremely useful as they allow engineers to rapidly make sense of the information and data.

The next prototype will incorporate the changes in the user interface pointed out in this study and will be applied to a larger and heterogeneous data set with the perspective on incrementally increase the size of the repository when new semantic data will be made available. Tests done on much larger document repositories show no particular strain on the technique adopted. A field trial at Rolls-Royce premises in Derby, UK, with the new prototype and new data is planned for the autumn and will last for a few months.

Acknowledgments. The work reported in this paper is part of the X-Media (www.x-media-project.org) project sponsored by the European Commission as part of the Information Society Technologies (IST) programme ISTFP6-026978. We are grateful to Mr. Ravish Bhagdev for providing the triples and the participants who took part in the evaluation. We are indebted to Mr. Andy Harrison, and Mr. Colin Cadas, Rolls-Royce plc., for their continuous support and fruitful discussions. Finally we thank the anonymous VC for their useful comments and their encouragement.

References

1. Aditya, T., Kraak, M.: A Search Interface for an SDI: Implementation and Evaluation of Metadata Visualization Strategies. Transactions in GIS 11(3), 413–435 (2007)
2. Ahlberg, C., Williamson, C., Shneiderman, B.: Dynamic Queries for Information Exploration: An Implementation and Evaluation. In: CHI 1992, pp. 619–626 (1992)
3. Bhagdev, R., Chakravarthy, A., Chapman, S., Ciravegna, F., Lanfranchi, V.: Creating and Using Organisational Semantic Webs in Large Networked Organisations. In: Sheth, A., Staab, S., Dean, M., Paolucci, M., Maynard, D., Finin, T., Thirunarayan, K. (eds.) ISWC 2008. LNCS, vol. 5318, pp. 723–736. Springer, Heidelberg (2008)
4. Brodbeck, D., Girardin, L.: Using multiple coordinated views to analyze geo-referenced high-dimensional datasets. In: Proceedings of InfoViz (2003)
5. Card, S., Mackinlay, J., Shneiderman, B.: Readings in Information Visualization: Using Vision To Think. Morgan Kaufmann Publishers Inc., San Francisco (1999)
6. Chen, C.: Information Visualization Versus the Semantic Web. In: Geroimenko, V., Chen, C. (eds.). Springer, Heidelberg (2003)
7. Dadzie, A.-S., Bhagdev, R., Chakravarthy, A., Chapman, S., Iria, J., Lanfranchi, V., Magalhães, J., Petrelli, D., Ciravegna, F.: Applying Semantic Web Technologies to Knowledge Sharing in Aerospace Engineering in Journal of Intelligent Manufacturing (June 2008) doi: 10.1007/s10845-008-0141-1
8. Dadzie, A.-S., Lanfranchi, V., Petrelli, D.: Seeing is Believing: Linking Data With Knowledge. Information Visualization Human-centered Information Visualization (September 2009)
9. Deligiannidis, L., Kochut, K., Sheth, A.: RDF Data Exploration and Visualization. In: CISM 2007 (2007)
10. Fan, J., Gao, Y., Luo, H., Keim, D., Li, Z.: A Novel Approach to Enable Semantic and Visual Image Summarization for Exploratory Image Search. In: Proc. of MIR (2008)
11. Geroimenko, V., Chen, C. (eds.): Visualizing the Semantic Web. Springer, Heidelberg (2003)
12. Hearst, M.: Search User Interfaces. Cambridge University Press, Cambridge (2009)
13. Hearst, M.: User Interfaces and Visualization. In: Baeza-Yates, R., Ribeiro-Neto, B. (eds.) Modern Information Retrieval. Addison-Wesley, Reading (1999)
14. Heer, J., Card, S., Landay, J.: Prefuse: A toolkit for interactive information visualization. In: Proc. CHI 2005, pp. 421–430 (2005)
15. Keim, D.: Information Visualization and Visual Data Mining. IEEE transactions on Visualization and Computer Graphics 8(1) (January-March 2002)
16. Klein, G., Moon, B., Hoffman, R.: Making Sense of Sensemaking 1: Alternative Perspectives. IEEE Intelligent Systems (July/August 2006)
17. Bhagdev, R., Chapman, S., Ciravegna, F., Lanfranchi, V., Petrelli, D.: Hybrid Search: Effectively Combining Keywords and Semantic Searches. In: Bechhofer, S., Hauswirth, M., Hoffmann, J., Koubarakis, M. (eds.) ESWC 2008. LNCS, vol. 5021, pp. 554–568. Springer, Heidelberg (2008)

18. Marchionini, G.: Exploratory Search: From finding to understanding. CAMC 49(4) (2006)
19. Mutton, P., Golbeck, J.: Visualization of Semantic Metadata and Ontologies. In: 7th International Conference on Information Visualization. IEEE Computer Society, Los Alamitos (2003)
20. Oren, E., Delbry, R., Decker, S.: Extending Faceted Navigation for RDF Data. In: Cruz, I., Decker, S., Allemang, D., Preist, C., Schwabe, D., Mika, P., Uschold, M., Aroyo, L.M. (eds.) ISWC 2006. LNCS, vol. 4273, pp. 559–572. Springer, Heidelberg (2006)
21. Petrelli, D., Lanfranchi, V., Moore, P., Ciravegna, F., Cadas, C.: Oh My, Where Is the End of the Context? Dealing with Information in a Highly Complex Environment. 1st IIiX (2006)
22. Petrelli, D.: On the Role of User-Centred Evaluations in the Advancement of Interactive Information Retrieval. In Information Processing and Management 44(1), 22–38 (2008)
23. Schraefel, M.C., Wilson, M., Russell, A., Smith, D.: mSpace: Improving Information Access to Multimedia Domains with MultiModal Exploratory Search. CACM 49(4), 47–49 (2006)
24. Schraefel, M.C.: Building Knowledge: What's beyond Keyword Search? IEEE Computer 42(3), 52–59 (2009)
25. Shneiderman, B.: The eyes have it: A task by data type taxonomy of information visualization. In: Bederson, B., Shneiderman, B. (eds.) The craft of information visualization. Morgan Kaufman, San Francisco (2003)
26. Stasko, G., Liu, Z.: Jigsaw: Supporting investigative analysis through interactive visualization. Information Visualization 7(2), 118–132 (2008)
27. Stern, E.W.: Organizational memory: Review of concepts and recommendations for management. International Journal of Information Management, 17–32 (1995)
28. Stuckenschmidt, H., et al.: Exploring Large Document Repositories with RDF Technology: The DOPE Project, May/June. IEEE Computer Society, Los Alamitos (2004)
29. Thai, V.T., Handschuh, S., Decker, S.: IVEA: An Information Visualization Tool for Personalized Exploratory Document Collection Analysis. In: Bechhofer, S., Hauswirth, M., Hoffmann, J., Koubarakis, M. (eds.) ESWC 2008. LNCS, vol. 5021, pp. 139–153. Springer, Heidelberg (2008)
30. Thai, V.T., Handschuh, S., Decker, S.: Tight Coupling of Personal Interests with Multidimensional Visualization for Exploration and Analysis of Text Collections. In: 12th International Conference Information Visualization IEEE, pp. 121–126 (2008)
31. Tu, K.W., Xiong, M., Zhang, L., Zhu, H.P., Zhang, J., Yu, Y.: Towards Imaging Large-Scale Ontologies for Quick Understanding and Analysis. In: Gil, Y., Motta, E., Benjamins, V.R., Musen, M.A. (eds.) ISWC 2005. LNCS, vol. 3729, pp. 702–715. Springer, Heidelberg (2005)
32. Van Welie, M., van der Veer, G.C., Eliens, A.: Breaking down Usability. In: Proc. INTERACT 1999, pp. 613–620 (1999)
33. White, R., Marchionini, G., Muresan, G.: Evaluating exploratory search systems. Information Processing Management 44(2), 433–436 (2008)
34. Wilson, M., Schraefel, M.C.: Improving Exploratory Search Interfaces: Adding Values or Information Overload?
35. Bederson, B.B., Shneiderman, B., Wattenberg, M.: Ordered and Quantum Treemaps: Making Effective Use of 2D Space to Display Hierarchies. ACM Transactions on Graphics (TOG) 21(4), 833–854 (2002)
36. Perer, A., Shneiderman, B.: Integrating statistics and visualization: case studies of gaining clarity during exploratory data analysis. In: CHI 2008 (2008)

A Conflict-Based Operator for Mapping Revision

Theory and Implementation

Guilin Qi[1,2], Qiu Ji[1], and Peter Haase[1]

[1] Institute AIFB, University of Karlsruhe, Germany
{gqi,qiji,pha}@aifb.uni-karlsruhe.de
[2] School of Computer Science and Engineering
Southeast University*, Nanjing 210096

Abstract. Ontology matching is one of the key research topics in the field of the Semantic Web. There are many matching systems that generate mappings between different ontologies either automatically or semi-automatically. However, the mappings generated by these systems may be inconsistent with the ontologies. Several approaches have been proposed to deal with the inconsistencies between mappings and ontologies. This problem is often called a mapping revision problem, as the ontologies are assumed to be correct, whereas the mappings are repaired when resolving the inconsistencies. In this paper, we first propose a conflict-based mapping revision operator and show that it can be characterized by two logical postulates adapted from some existing postulates for belief base revision. We then provide an algorithm for iterative mapping revision by using an ontology revision operator and show that this algorithm defines a conflict-based mapping revision operator. Three concrete ontology revision operators are given to instantiate the iterative algorithm, which result in three different mapping revision algorithms. We implement these algorithms and provide some preliminary but interesting evaluation results.

1 Introduction

Next generation semantic applications are employed by a large number of ontologies, some of them constantly evolving. As the complexity of semantic applications increases, more and more knowledge is embedded in ontologies, typically drawn from a wide variety of sources. This new generation of applications thus likely relies on a set of distributed ontologies, typically connected by mappings. One of the major challenges in managing these distributed and dynamic ontologies is to handle potential inconsistencies introduced by integrating multiple distributed ontologies.

For inconsistency handling in single, centralized ontologies, several approaches are known (see the survey in [7]). Recently, there are some works done on handling inconsistency in distributed ontologies connected by mappings, where a mapping between two ontologies is a set of correspondences between entities in the ontologies. In a distributed system consisting of two ontologies and a

* New affiliation since September 2009.

A. Bernstein et al. (Eds.): ISWC 2009, LNCS 5823, pp. 521–536, 2009.

mapping between them, correspondences in the mapping can have different interpretations. For example, in Distributed Description Logics (DDL) [3], a correspondence in a mapping is translated into two *bridge rules* that describe the "flow of information" from one ontology to another one. In [13], the authors deal with the problem of mapping revision in DDL by removing some bridge rules which are responsible for the inconsistency. The idea of their approach is similar to that of the approaches for debugging and repairing terminologies in a single ontology. Mappings can also be interpreted as sets of axioms in a description logic. A heuristic method for mapping revision is given in [12]. However, this method can only deal with inconsistency caused by disjointness axioms which state that two concepts are disjoint. Later on, Meilicke et al. proposed another algorithm to resolve the inconsistent mappings in [15]. The idea of their algorithm is similar to the linear base revision operator given in [16]. However, both methods given in [12] and [15] lack a rationality analysis w.r.t. logical properties.

In this paper, we first propose a conflict-based mapping revision operator based on the notion of a "conflict set", which is a subset of the mapping that is in conflict with ontologies in a distributed system. We then adapt two postulates from belief revision theory [8] and show that our mapping revision operator can be characterized by them (see Section 3). After that, in Section 4, we provide an iterative algorithm for mapping revision by using a revision operator in description logics and show that this algorithm results in a conflict-based mapping revision operator. We define a revision operator and show that the iterative algorithm based on it produces the same results as the algorithm given in [15]. This specific iterative algorithm has a polynomial time complexity if the satisfiability check of an ontology can be done in polynomial time in the size of the ontology. However, this algorithm may still be inefficient for large ontologies and mappings, because it requires a large number of satisfiability checks. Therefore, we provide an algorithm to implement an alternative revision operator based on the *relevance-based selection function* given in [11] which can be optimized by a module extraction technique given in [22]. Neither of the above proposed revision operators removes minimal number of correspondences to resolve inconsistencies. To better fulfil the principle of minimal change, we consider the revision operator given in [19] which utilizes a heuristics based on a *scoring function* which returns the number of *minimal incoherence-preserving sub-ontologies (MIPS)* that an axiom belongs to. Instantiating our iterative algorithm with this existing revision operator results in a new conflict-based mapping revision operator. Finally, we implement these algorithms and provide evaluation results for comparing their efficiency and effectiveness in Section 5.

Relationship with belief revision. This work is related to belief revision which has been widely discussed in the literature [5,10]. Our conflict-based mapping revision operator is inspired by the internal revision operator given in [8], and the postulate used to characterize our mapping revision operator is adapted from a postulate for internal revision operator given in [8]. The problem of mapping revision is not exactly the same as the problem of belief base revision because the mapping to be revised is dependent on ontologies in the distributed system and each correspondence in the mapping carries a confidence value which can be used to guide the revision. Our iterative algorithm is inspired by the iterative revision algorithm given in [18] and is tailored to produce a conflict-based revision operator.

2 Preliminaries

We assume that the reader is familiar with Description Logics (DL) and refer to Chapter 2 of the DL handbook [1] for a good introduction. Our method is independent of a specific DL language, and thus can be applied to any DL.

A DL-based ontology (or knowledge base) $O = (\mathcal{T}, \mathcal{R})$ consists of a set $\mathcal{T}$ of concept axioms (TBox) and a set $\mathcal{R}$ of role axioms (RBox). In this paper, we treat O as a set of axioms. Concept axioms (or terminology axioms) have the form $C \sqsubseteq D$, where C and D are (possibly complex) concept descriptions built from a set of concept names and some constructors, and role axioms are expressions of the form $R \sqsubseteq S$, where R and S are (possibly complex) role descriptions built from a set of role names and some constructors.

An interpretation $\mathcal{I} = (\triangle^{\mathcal{I}}, \cdot^{\mathcal{I}})$ consists of a non-empty domain set $\triangle^{\mathcal{I}}$ and an interpretation function $\cdot^{\mathcal{I}}$, which maps from concepts and roles to subsets of the domain and binary relations on the domain, respectively. Given an interpretation $\mathcal{I}$, we say that $\mathcal{I}$ satisfies a concept axiom $C \sqsubseteq D$ (resp., a role inclusion axiom $R \sqsubseteq S$) if $C^{\mathcal{I}} \subseteq D^{\mathcal{I}}$ (resp., $R^{\mathcal{I}} \subseteq S^{\mathcal{I}}$). An interpretation $\mathcal{I}$ is called a *model* of an ontology O, iff it satisfies each axiom in O. A concept C in an ontology O is unsatisfiable if for each model $\mathcal{I}$ of O, $C^{\mathcal{I}} = \emptyset$. An ontology O is incoherent if there exists an unsatisfiable concept in O.

Given two ontologies O_1 and O_2, describing the same or largely overlapping domains of interest, we can define correspondences between their elements.

Definition 1. *[4] Let O_1 and O_2 be two ontologies, Q be a function that defines sets of mappable elements $Q(O_1)$ and $Q(O_2)$. A correspondence is a 4-tuple $\langle e, e', r, \alpha \rangle$ such that $e \in Q(O_1)$ and $e' \in Q(O_2)$, r is a semantic relation, and α is a confidence value from a suitable structure $\langle D, \leq \rangle$, such as a lattice. A mapping $\mathcal{M}$ is a set of correspondences.*

In Definition 1, there is no restriction on function Q, semantic relation r and domain D. In the mapping revision scenario, we often consider correspondences between concepts and restrict r to be one of the semantic relations from the set $\{\equiv, \sqsubseteq, \sqsupseteq\}$, and let $D = [0, 1]$. A mapping is a set of correspondences whose elements are mappable. The following definition is adapted from the definition of a distributed system given in [23].

Definition 2. *A distributed system is a triple $\mathcal{D} = \langle O_1, O_2, \mathcal{M} \rangle$, where O_1 and O_2 are ontologies and $\mathcal{M}$ is a mapping between them. We call O_1 the source ontology and O_2 the target ontology.*

Example 1. Take the two ontologies CRS and EKAW in the domain of conference management systems as an example. They contain the following axioms:

crs : article $\sqsubseteq$ crs : document, crs : program $\sqsubseteq$ $\neg$crs : document,
ekaw : Paper $\sqsubseteq$ ekaw : Document, ekaw : Workshop_Paper $\sqsubseteq$ ekaw : Paper
ekaw : Conference_Paper $\sqsubseteq$ ekaw : Paper, ekaw : PC_Member $\sqsubseteq$ ekaw : Possible_Reviewer,

The correspondences in the mapping $\mathcal{M}$ between O_1 and O_2 which is obtained by the ontology matching system HMatch are listed as follows:

m$_1$: ⟨crs : article, ekaw : Conference_Paper, $\sqsubseteq$, 0.65⟩
m$_2$: ⟨ekaw : Workshop_Paper, crs : article, $\sqsubseteq$, 0.65⟩
m$_3$: ⟨ekaw : Document, crs : program, $\sqsubseteq$, 0.80⟩
m$_4$: ⟨crs : program, ekaw : Document, $\sqsubseteq$, 0.80⟩
m$_5$: ⟨crs : document, ekaw : Document, $\sqsubseteq$, 0.93⟩

Definition 3. *[13] Let $\mathcal{D} = \langle O_1, O_2, \mathcal{M}\rangle$ be a distributed system. The* union *$O_1 \cup_{\mathcal{M}} O_2$ of O_1 and O_2 connected by $\mathcal{M}$ is defined as $O_1 \cup_{\mathcal{M}} O_2 = O_1 \cup O_2 \cup \{t(m) : m \in \mathcal{M}\}$ with t being a translation function that converts a correspondence into an axiom in the following way: $t(\langle C, C', r, \alpha\rangle) = CrC'$.*

That is, we first translate all the correspondences in the mapping $\mathcal{M}$ into DL axioms, then the union of the two ontologies connected by the mapping is the set-union of the two ontologies and the translated axioms. Given $\mathcal{D} = \langle O_1, O_2, \mathcal{M}\rangle$, we use $Union(\mathcal{D})$ to denote $O_1 \cup_{\mathcal{M}} O_2$. Take a correspondence in Example 1 as an example, we have t(⟨crs:article, ekaw:Conference_Paper, $\sqsubseteq$, 0.65⟩) = crs:article $\sqsubseteq$ ekaw:Conference_Paper.

Definition 4. *[12] Given a mapping $\mathcal{M}$ between two ontologies O_1 and O_2, $\mathcal{M}$ is consistent with O_1 and O_2 iff there exists no concept C in O_i with $i \in \{1, 2\}$ such that C is satisfiable in O_i but unsatisfiable in $O_1 \cup_{\mathcal{M}} O_2$. Otherwise, $\mathcal{M}$ is inconsistent. A distributed system $\mathcal{D} = \langle O_1, O_2, \mathcal{M}\rangle$ is inconsistent if $\mathcal{M}$ is inconsistent with O_1 and O_2.*

An inconsistent mapping is a mapping such that there is a concept that is satisfiable in a mapped ontology but unsatisfiable in the union of the two ontologies together with the mapping. In Example 1, since ekaw:Workshop_Paper is satisfiable in both O_1 and O_2 but unsatisfiable in $O_1 \cup_{\mathcal{M}} O_2$, $\mathcal{M}$ is inconsistent. Note that $O_1 \cup O_2$ must be coherent if both O_1 and O_2 are coherent because they use different name spaces.

Definition 5. *A mapping revision operator $\circ$ is a function $\circ\langle O_1, O_2, \mathcal{M}\rangle = \langle O_1, O_2, \mathcal{M}'\rangle$ such that $\mathcal{M}' \subseteq \mathcal{M}$, where O_1 and O_2 are two ontologies and $\mathcal{M}$ is a mapping between them.*

Our definition of a mapping revision operator is similar to the definition of a revision function given in [14]. When repairing the mapping in a distributed system, we assume that ontologies are more reliable than the mapping and therefore only remove correspondences in the mapping to restore consistency. This makes the problem of mapping repair akin to the problem of belief revision. Thus we call the problem of repairing mappings *mapping revision*. However, this definition is very general and allows mapping revision operators that result in unintuitive results. That is, we can define two naive revision operators

$\circ_{Full}\langle O_1, O_2, \mathcal{M}\rangle = \langle O_1, O_2, \emptyset\rangle$ and $\circ_{Null}\langle O_1, O_2, \mathcal{M}\rangle = \langle O_1, O_2, \mathcal{M}\rangle$. In belief revision, the rationality of a revision operator is often evaluated by logical postulates. In this work, we will define a mapping revision operator and show that it can be characterized by an important logical postulate.

3 A Conflict-Based Mapping Revision Operator

In this section, we propose a method for mapping revision based on the idea of kernel contractions defined by Hansson in [9]. We adapt the notion of a minimal conflict set of a distributed system given in [13] as follows.

Definition 6. *Let $\langle O_1, O_2, \mathcal{M}\rangle$ be a distributed system. A subset $\mathcal{C}$ of $\mathcal{M}$ is a* conflict set *for a concept A in O_i $(i = 1, 2)$ if A is satisfiable in O_i but unsatisfiable in $O_1 \cup_{\mathcal{C}} O_2$. $\mathcal{C}$ is a* minimal conflict set *(MCS) for A in O_i if $\mathcal{C}$ is a conflict set for A and there exists no $\mathcal{C}' \subset \mathcal{C}$ which is also a conflict set for A in O_i.*

A minimal conflict set for a concept in one of the ontologies is a minimal subset of the mapping that, together with the ontologies, is responsible for the unsatisfiability of the concept in the distributed system. It is similar to the notion of a kernel in [9]. Note that if O_i $(i = 1, 2)$ is incoherent, then it is meaningless to define the notion of a MCS for an unsatisfiable concept. We use $MCS_{O_1,O_2}(\mathcal{M})$ to denote the set of all the minimal conflict sets for all unsatisfiable concepts in $O_1 \cup_{\mathcal{C}} O_2$. It corresponds to the notion of a *kernel set* in [9]. In Example 1, $MCS_{CRS,EKAW}(\mathcal{M}) = \{\{t(m_1), t(m_3)\}, \{t(m_2), t(m_3)\}, \{t(m_3), t(m_5)\}, \{t(m_1), t(m_2), t(m_3)\}, \{t(m_2), t(m_3), t(m_5)\}\}$.

Hansson's kernel contraction removes formulas in a knowledge base through an *incision function*, which is a function that selects formulas to be discarded. However, we cannot apply the notion of an incision function to mapping revision directly because the mapping to be revised is dependent on the ontologies in the distributed system. Therefore, the problem of mapping revision is not exactly the same as the problem of belief revision where the two knowledge bases may come from different sources. Furthermore, each correspondence in the mapping carries a confidence value which can be used to guide the revision. We use $\mathbf{D}$ and $\mathbf{M}$ to denote the set of all the distributed systems and the set of all the subsets of mappings in all distributed systems.

Definition 7. *An* incision function *σ: $\mathbf{D} \rightarrow \mathbf{M}$ is a function such that for any distributed system $\mathcal{D} = \langle O_1, O_2, \mathcal{M}\rangle$, we have*

(i) $\sigma(\mathcal{D}) \subseteq \bigcup(MCS_{O_1,O_2}(\mathcal{M}))$;
(ii) *if* $\mathcal{C} \neq \emptyset$ *and* $\mathcal{C} \in MCS_{O_1,O_2}(\mathcal{M})$, *then* $\mathcal{C} \cap \sigma(\mathcal{D}) \neq \emptyset$;
(iii) *if* $m = \langle C, C', r, \alpha\rangle \in \sigma(\mathcal{D})$, *then there exists* $\mathcal{C} \in MCS_{O_1,O_2}(\mathcal{M})$ *such that* $m \in \mathcal{C}$ *and* $\alpha = min\{\alpha_i : \langle C_i, {C'}_i, r_i, \alpha_i\rangle \in \mathcal{C}\}$.

The first two conditions say that an incision function selects from each kernel set at least one element. The third condition says that if a correspondence is selected by an incision function, then there must exist a MCS $\mathcal{C}$ such that its confidence value is the minimal confidence value of correspondences in $\mathcal{C}$. Going back to Example 1, the incision function σ may select m_1, m_2 and m_3 to resolve inconsistency.

We define our mapping revision operator based on an incision function.

Definition 8. *A mapping revision operator* $\circ$ *is called a* conflict-based mapping revision operator *if there exists an incision function* σ *such that:*

$$\circ\langle O_1, O_2, \mathcal{M}\rangle = \langle O_1, O_2, \mathcal{M} \setminus \sigma(MCS_{O_1,O_2}(\mathcal{M}))\rangle.$$

That is, we remove those correspondences in $\mathcal{M}$ that are selected by the incision function to restore consistency. We provide the representation theorem for conflict-based mapping revision. Before that, we need to define the notion of an inconsistency degree of a distributed system for a concept. Given a distributed system $\mathcal{D} = \langle O_1, O_2, \mathcal{M}\rangle$, a concept A in O_i $(i = 1, 2)$ is unsatisfiable in $\mathcal{D}$ if A is unsatisfiable in $O_1 \cup_{\mathcal{M}} O_2$.

Definition 9. *Given* $\mathcal{D} = \langle O_1, O_2, \mathcal{M}\rangle$*, the* β*-cut (resp. strict* β*-cut) set of* $\mathcal{D}$*, denoted as* $\mathcal{D}_{\geq\beta}$ *(resp.* $\mathcal{D}_{>\beta}$*), is defined as* $\mathcal{D}_{\geq\beta} = \langle O_1, O_2, \{\langle C, C', r, \alpha\rangle \in \mathcal{M} : \alpha \geq \beta\}\rangle$ *(resp.* $\mathcal{D}_{>\beta} = \langle O_1, O_2, \{\langle C, C', r, \alpha\rangle \in \mathcal{M} : \alpha > \beta\}\rangle$*).*

The β-cut set of $\mathcal{D}$ is a distributed system consisting of O_1, O_2 and correspondences in the mapping whose confidence values are greater than or equal to β. It is adapted from the notion of cut set in possibilistic DL in [20]. In Example 1, $\mathcal{D}_{>0.65} = \langle O_1, O_2, \{t(m_3), t(m_4), t(m_5)\}\rangle$.

Definition 10. *Given* $\mathcal{D} = \langle O_1, O_2, \mathcal{M}\rangle$*, the inconsistency degree of* $\mathcal{D}$ *for a concept* A *in* O_i $(i = 1, 2)$*, denoted by* $Inc(\mathcal{D})_A$*, is defined as* $Inc(\mathcal{D})_A = max\{\alpha : A$ *is unsatisfiable in* $\mathcal{D}_{\geq\alpha}\}$*. The inconsistency degree of* $\mathcal{D}$*, denoted as* $Inc(\mathcal{D})$*, is defined as* $Inc(\mathcal{D}) = max\{\alpha :$ *there exists an unsatisfiable concept in* $\mathcal{D}_{\geq\alpha}\}$*.*

It is easy to check that $Inc(\mathcal{D}) = max\{\alpha : \mathcal{D}_{\geq\alpha}$ is inconsistent$\}$. In Example 1, $\mathcal{D}_{\geq 0.93}$ is consistent but $\mathcal{D}_{\geq 0.8}$ is inconsistent since ekaw:Workshop_Paper is unsatisfiable. Thus, $Inc(\mathcal{D}) = 0.8$.

We give a postulate for mapping revision by generalizing the postulate (Relevance) for the internal partial meet revision operator given in [8]. It says that if a correspondence is removed from the mapping after revision, then it must be in a conflict set of the mapping for a concept and the confidence degree attached to it is minimal among all the confidence degrees in the conflict set.

Postulate (Relevance). Suppose $\circ\langle O_1, O_2, \mathcal{M}\rangle = \langle O_1, O_2, \mathcal{M}'\rangle$, if $m = \langle C, C', r, \alpha\rangle \in \mathcal{M}$ and $m \notin \mathcal{M}'$, then there exists a concept A in O_i $(i = 1, 2)$ and a subset $\mathcal{S}$ of $\mathcal{M}$ such that A is satisfiable in $\langle O_1, O_2, \mathcal{S}\rangle$ but is unsatisfiable in $\langle O_1, O_2, \mathcal{S} \cup \{m\}\rangle$ and $Inc(\langle O_1, O_2, \mathcal{S} \cup \{m\}\rangle)_A = \alpha$.

Relevance is an important postulate for minimal change. However, it does not constrain the number of correspondences to be removed. Therefore, it does not entail minimal change.

Algorithm 1. An iterative algorithm for mapping revision

Data: A distributed system $\mathcal{D} = \langle O_1, O_2, \mathcal{M} \rangle$ and a revision operator $\star$
Result: A repaired distributed system $\mathcal{D}_\star = \langle O_1, O_2, \mathcal{M}_\star \rangle$

begin
  if *either* O_1 *or* O_2 *is incoherent* **then**
    return $\mathcal{D}$
  Rearrange the weights in $\mathcal{M}$ such that $\beta_1 > \beta_2 > ... > \beta_l > 0$;
  $S_i := \{t(\langle C, C', r, \alpha \rangle) : \langle C, C', r, \alpha \rangle \in \mathcal{M}, \alpha = \beta_i\}, i = 1, ..., l$;
  while $\mathcal{M}$ *in* $\mathcal{D}$ *is inconsistent* **do**
    if $\beta_k = Inc(\mathcal{D})$ **then**
      $S_t := S_k \setminus (S_k \star (Union(\mathcal{D})_{>\beta_k}))$;
      $\mathcal{M} := \mathcal{M} \setminus \{\langle C, C', r, \alpha \rangle : t(\langle C, C', r, \alpha \rangle) \in S_t, \alpha = \beta_k\}$;
  return $\mathcal{D}$
end

We also need another postulate called *Consistency*.

Postulate (Consistency). For any $\mathcal{D} = \langle O_1, O_2, \mathcal{M} \rangle$ where O_i are coherent, $\circ\langle O_1, O_2, \mathcal{M} \rangle$ is a consistent distributed system.

The following theorem shows that our conflict-based mapping revision operator can be characterized by the postulates (Relevance) and (Consistency).

Theorem 1. *The operator $\circ$ is a conflict-based mapping revision operator if and only if it satisfies (Relevance) and (Consistency).*

To show the *if* direction of the theorem, we can construct $\sigma(\mathcal{D}) = \mathcal{M} \setminus \mathcal{M}'$ for $\mathcal{D} = \langle O_1, O_2, \mathcal{M} \rangle$ and $\circ(\mathcal{D}) = \langle O_1, O_2, \mathcal{M}' \rangle$, then show that σ is an incision function. Unlike revision operators given in [8], our conflict-based mapping revision operator is characterized by only two postulates. This is because the definition of a conflict already gives some constraints on how we can repair a mapping. According to Definition 5, ontologies in the distributed systems are not changed and revised mapping must be a subset of the original one. These two conditions correspond to (Success) and (Inclusion) for revision operators given in [8].

4 An Algorithm for Mapping Revision

In this section, we give an algorithm for mapping revision based on an ontology revision operator and then present some concrete ontology revision operators.

4.1 Algorithm

We describe the idea of our algorithm (Algorithm 1) as follows. Given a distributed system $\mathcal{D} = \langle O_1, O_2, \mathcal{M} \rangle$, if either O_1 or O_2 is incoherent, then we take $\mathcal{D}$ as the result of revision. That is, no change is needed. Suppose $\mathcal{M} = \{\langle C_i, C'_i, r_i, \alpha_i \rangle : i = 1, ..., n\}$ where n is the number of correspondences in $\mathcal{M}$. Let us rearrange the weights of axioms (i.e., α_i) in $\mathcal{M}$ such that $\beta_1 > \beta_2 > ... > \beta_l > 0$, where β_i $(i = 1, ..., l)$ are all the distinct weights appearing in $\mathcal{M}$. For each $i \in \{1, ..., l\}$, S_i consists of translated axioms of correspondences in $\mathcal{M}$ which have the confidence value β_i. Suppose $Inc(\mathcal{D}) = \beta_k$. We revise S_k by $Union(\mathcal{D}_{>\beta_k})$.

Suppose S_t is the set of axioms in S_k that are removed after revision of S_k by $Union(\mathcal{D}_{>\beta_k})$ using the operator $\star$. We then remove the correspondences in $\mathcal{M}$ that have confidence values β_k and are mapped to axioms in S_t by the translation function t. We iterate the revision process until the mapping becomes consistent.

In Algorithm 1, we need to compute the inconsistency degree of a distributed system. This can be easily done by adapting the algorithm for computing the inconsistency degree in [20] so we do not bother to provide it here.

We have not specified a revision operator in Algorithm 1. However, we require that the revision operator $\star$ used in the algorithm satisfy the following properties which are similar to the postulates *Inclusion*, *Success* and *Core-retainment* for kernel revision operator given in [9]:

- Inclusion: $O \star O' \subseteq O \cup O'$;
- Success: $O' \subseteq O \star O'$;
- Core-retainment: if $\phi \in O$ and $\phi \notin O \star O'$, then there exist a concept A in $O \cup O'$ and a subset O_s of O, such that A is satisfiable in $O_s \cup O'$ but is unsatisfiable in $O_s \cup O' \cup \{\phi\}$.

It is clear that Algorithm 1 generates a mapping revision operator. We show that this operator is a conflict-based mapping revision operator.

Theorem 2. *Suppose $\star$ satisfies Inclusion, Success and Core-retainment, and $\circ$ is a mapping revision operator such that, for any distributed system $\mathcal{D}$, $\circ(\mathcal{D})$ is the result of Algorithm 1 with $\star$ as an input parameter, then $\circ$ is a conflict-based mapping revision operator.*

4.2 Concrete Revision Operators

We first give a simple revision operator which is adapted from the linear base revision operator given in [16]. By SORT we denote a procedure that for each ontology $O = \{\phi_1, ..., \phi_n\}$, randomly ranks its elements as an ordered sequence $(\phi_1, ..., \phi_n)$. Let O and O' be two ontologies, and let $\text{SORT}(O) = \{\phi_1, ..., \phi_n\}$, the random linear base revision operator, denoted as $\circ_{linear}$, is defined inductively as follows $O \circ_{linear} O' = O' \cup S_1 \cup ... \cup S_n$, where S_i is defined by $S_i = \{\phi_i\}$ if $\{\phi_i\} \cup O' \cup \bigcup_{j=1}^{i-1} S_j$ is coherent, $\emptyset$ otherwise, for $i \geq 1$. It is easy to check that this revision operator satisfies conditions Inclusion, Success and Core-retainment. We show that the algorithm given in [15] is a special case of our iterative algorithm where the operator $\circ_{linear}$ is chosen.

Proposition 1. *For any distributed system $\mathcal{D} = \langle O_1, O_2, \mathcal{M}\rangle$ where O_1 and O_2 are coherent, suppose $\mathcal{D}_{\circ_{linear}}$ is the result of revision by Algorithm 1, then $\mathcal{D}_{\circ_{linear}}$ can be obtained by the algorithm given in [15] as well.*

As shown in [15], their algorithm only needs at most n satisfiability check, where n is the number of correspondences. Therefore, our iterative algorithm based on the revision operator $\circ_{linear}$ has a polynomial time complexity if the satisfiability check can be done in polynomial time in the size of union of ontologies and the mapping. However, this algorithm requires to rank correspondences with the same confidence value and there is no good principle to guide this ranking.

Table 1. Relevance-based mapping revision algorithm

```
REL_REVISION(O, O')
Input: Two ontologies O and O'
Output: A revised ontology O ⋆ O'
Global : 𝒥 ← ∅;
HS ← ∅;
for(C ∈ AllUnsatConcepts(O ∪ O')){
    k ← 1;
    O_t ← hs ← ∅;
    while(s_k(O ∪ O', C) ≠ ∅){
        O_t ← O_t ∪ s_k(O ∪ O', C);
        if(hs ≠ ∅){
            if((O \ hs) ∪ O' ⊭ C ⊑ ⊥)
                break;
            hs ← CONF(C, O_t ∩ (O \ hs), O_t ∩ O');
            HS ← HS ∪ hs;
        }else if(O_t ⊨ C ⊑ ⊥){
            hs ← CONF(C, O_t ∩ O, O_t ∩ O');
            HS ← HS ∪ hs;
        }
        k ← k + 1;
    } (end_while)
} (end_for)
return (O \ HS) ∪ O';
```

```
CONF(C, O, O')
Input: Two ontologies O and O', and an unsatisfiable concept C of O ∪ O'
Output: A hitting set hs in O for C w.r.t. O'
hs ← ∅ ;
while((O \ hs) ∪ O' ⊨ C ⊑ ⊥){
    J ← SINGLE_CONFLICT(C, O \ hs, O');
    𝒥 ← 𝒥 ∪ {J}
    hs = hs ∪ {ϕ} for some ϕ ∈ J;
}
return hs;
```

Furthermore, if the size of the union of ontologies and the mapping is big, then the algorithm may still be inefficient because it will need a large number of satisfiability checks over the union.

In the following, we present an algorithm REL_REVISION (see Table 1) to implement another concrete revision operator based on the relevance-based selection function. The motivation behind the algorithm is that when choosing between two correspondences to remove, we always remove the one which is more relevant to an unsatisfiable concept and thus is more likely to be problematic.

Given two axioms ϕ and ψ, ϕ is *directly relevant* to ψ iff there is an overlap between the signature of ϕ and the signature of ψ, where the signature of an axiom is the set of all concept names, role names and individual names appearing in it. Based on the notion of direct relevance, we can extend it to relevance relation between an axiom and an ontology. An axiom ϕ is relevant to an ontology O iff there exists an axiom ψ in O such that ϕ and ψ are directly relevant. We introduce a selection function defined in [11].

Definition 11. *[11] Let O be an ontology, ϕ be an axiom and k be an integer. The* relevance-based selection function, *written s_{rel}, is defined inductively as follows:*
$s_{rel}(O, \phi, 0) = \emptyset$
$s_{rel}(O, \phi, 1) = \{\psi \in O : \phi \text{ is directly relevant to } \psi\}$
$s_{rel}(O, \phi, k) = \{\psi \in O : \psi \text{ is directly relevant to } s_{rel}(O, \phi, k-1)\}$, *where* $k > 1$.

We call $s_{rel}(O, \phi, k)$ the k-relevant subset of O w.r.t. ϕ. For convenience, we define $s_k(O, \phi) = s_{rel}(O, \phi, k) \setminus s_{rel}(O, \phi, k-1)$ for $k \geq 1$.

Our algorithm REL_REVISION is based on Reiter's Hitting Set Tree (HST) algorithm [21]. Given a *universal set* U, and a set $K = \{s_1, ..., s_n\}$ of subsets of

U which are *conflict sets*, i.e. subsets of the system components responsible for the error, a *hitting set* T for K is a subset of U such that $s_i \cap T \neq \emptyset$ for all $1 \leq i \leq n$. To adapt HST algorithm to deal with revision of ontologies in DLs, we define the notion of a *minimal conflict set* of an ontology O for a concept C w.r.t. another ontology O'. A subset O_s of O is called a minimal conflict set of O for C w.r.t. O', if (1) C is unsatisfiable in $O_s \cup O'$ and (2) for any $O_t \subset O_s$, C is satisfiable in $O_t \cup O'$. A more general definition of a minimal conflict set is given in [2], where it is called a *minimal axiom set*.

In REL_REVISION, we handle unsatisfiable concepts in the union of the mapped ontologies and the ontology translated from the mapping one by one until we resolve the inconsistency. For each unsatisfiable concept to be handled, we first select axioms that are relevant to it iteratively by the relevance-based selection function until the concept is unsatisfiable in these axioms. $s_k(O, C)$ is the abbreviation of $s_k(O, C \sqsubseteq \bot)$. We find a hitting set for the selected sub-ontologies by calling the procedure CONF and update the existing incomplete hitting set HS. We then add to the selected sub-ontologies those axioms that are directly relevant to them and further expand the hitting set tree by calling to procedure CONF. We continue this process until the inconsistency is resolved. The procedure SINGLE_CONFLICT computes a minimal conflict set of O for C w.r.t. O'. This kind of procedure can be found in the literature, such as GETMUPS in [19]. It is possible that some axioms that are involved in a conflict set are not selected by the selection function. Therefore, when $s_k(O \cup O', C) = \emptyset$, we still have $(O \setminus HS) \cup O' \models C \sqsubseteq \bot$, then we set $s_k(O \cup O', C) = (O \cup O') \setminus s_{rel}(O \cup O', C \sqsubseteq \bot, k-1)$. Note that our algorithm may not remove minimal number of correspondences to resolve inconsistency because we only expand one branch of the hitting set tree in a depth-first manner. This is compensated by higher efficiency. Furthermore, although our algorithm does not remove minimal number of correspondences, the removals of correspondences are guided by a relevance-based selection function to improve the quality of removal. It is easy to see that the revision operator obtained by REL_REVISION satisfies conditions Inclusion, Success and Core-retainment.

In REL_REVISION, to resolve an unsatisfiable concept C in $O \cup O'$, we need to compute some minimal conflict sets of O for C w.r.t. O'. The time complexity of REL_REVISION depends on the DL under consideration. In the worst case, i.e., all the minimal conflict sets of all the unsatisfiable concepts are disjoint, our algorithm needs to compute all the minimal conflict sets for all the unsatisfiable concepts, which is a hard task [17]. For instance, the number of all the minimal conflict sets for an unsatisfiable concept is exponential in the worst case for lightweight ontology language EL^+ [2]. However, the average case complexity will be considerably lower: For many real ontologies, the number of all minimal conflict sets for an unsatisfiable concept is much less than the size of the ontology. Our algorithm usually does not compute all the minimal conflict sets for an unsatisfiable concept. Another complexity of our algorithm comes from the computation of a minimal conflict set, which is as hard as satisfiability checking of the underlying DL. Despite the high complexity of our algorithm, fortunately, there is an optimization technique to improve its efficiency. That is, for each unsatisfiable concept to be handled, we extract a so-called syntactic locality-based module [6] from $O \cup O'$ which contains all the *minimal conflict sets* of O for C w.r.t. O'. The module extraction step can be added between line 6 and line 7 in REL_REVISION. The correctness of our modified algorithm is ensured by the fact that the locality-based module contains all the minimal sub-ontologies of an ontology that are responsible for unsatisfiability of a concept shown in in [22].

Example 2. To illustrate our iterative algorithm (i.e. Algorithm 1) based on REL_REVISION, we follow Example 1. First of all, we need to reorder all distinct confidence values in a descending order $\beta_1 = 0.93 > \beta_2 = 0.8 > \beta_3 = 0.65$ and the corresponding layers of correspondence axioms are $S_1 = \{t(m_5)\}$, $S_2 = \{t(m_3), t(m_4)\}$ and $S_3 = \{t(m_1), t(m_2)\}$ respectively. Then, we go into line 6 since $\mathcal{M}$ is inconsistent. We obtain the inconsistency degree of $\mathcal{D}$ as 0.8. So $k = 2$. As we know that $\beta_2 = 0.8$, we use $Union(\mathcal{D}_{>0.8})$ to revise S_2 and the revision result is $(S_2 \setminus \{t(m_3)\}) \cup Union(\mathcal{D}_{>0.8})$ according to REL_REVISION (see "Illustration of REL_REVISION" below). Therefore, we remove m_3 from $\mathcal{M}$ (see line 9). Then we go to another iteration of the while loop. Since the modified $\mathcal{M}$ becomes consistent when m_3 is removed from it, the whole process of Algorithm 1 can be terminated and the result is $\mathcal{D}_\star = \langle O_1, O_2, \mathcal{M} \setminus \{m_3\} \rangle$.

Illustration of REL_REVISION: The input is $O = S_2$ and $O' = Union(\mathcal{D}_{>0.8})$. Suppose the first found unsatisfiable concept is *article*. We keep on selecting the k-relevant axioms in $O \cup O'$ w.r.t. the concept *article* until $O_t = O \cup O'$ (i.e. *article* becomes unsatisfiable in O_t). Then we go to line 14 and get the minimal conflict set $\{t(m_3)\}$ of O w.r.t. O' and a hitting set $hs = \{t(m_3)\}$ (see "Illustration of CONF" below). So $HS = \{t(m_3)\}$. After this, we go to another iteration of the while loop. Since all the axioms in O have been selected, we can terminate the process and return $(S_2 \setminus \{t(m_3)\}) \cup Union(\mathcal{D}_{>0.8})$.

Illustration of CONF: The input is $C = article$, $O = S_2$ and $O' = Union(\mathcal{D}_{>0.8})$ for CONF. First of all, we compute a MCS $J = \{t(m_3)\}$ in line 3. Since only one axiom in J, we get $hs = \{t(m_3)\}$ in line 5. We return $\{t(m_3)\}$ and update $\mathcal{J} = \{t(m_3)\}$.

Neither of the above proposed revision operators removes minimal number of correspondences to resolve inconsistencies. To better fulfil minimal change, we consider the revision operator given in Algorithm 1 in [19] which utilizes a heuristics based on a *scoring function* which computes the number of minimal incoherence-preserving sub-ontologies (MIPS) that an axiom belongs to. It is not difficult to check that this revision operator satisfies conditions Inclusion, Success and Core-retainment. A MIPS of ontology O *w.r.t.* another ontology O' is a minimal sub-ontology of O that is incoherent with O'. Instantiating our iterative algorithm with this operator results in a new conflict-based mapping revision operator. The disadvantage of this revision operator is that it needs to all the MIPSs obtained from all the minimal conflict sets of O for any concept w.r.t. O' by using a modified hitting set tree algorithm, thus its computational complexity is at least harder than those of previous revision operators.

5 Experimental Evaluation

In this section, we present the evaluation results of our algorithms by comparing them with existing algorithms. Our algorithms were implemented with the KAON2 API[1], using KAON2 as a black box reasoner. To fairly compare with the mapping repair algorithms in [13] and [15], we re-implemented them using the KAON2 API. More precisely, the following algorithms have been evaluated:

- **Weight-based-One:** Compute one minimal conflict subset each time and remove an element in it with lowest weights (see [13]).
- **Linear:** Our iterative algorithm based on the random linear base revision operator (it is equivalent to the algorithm given in [15]).
- **Weight-based-All:** Compute all minimal conflict subsets for an unsatisfiable concept and then resolve the unsatisfiability based on weights (see Algorithm 3 in [19]).
- **Relevance-based:** Our iterative algorithm based on the revision operator REL_REVISION defined in Table 1.
- **Score-based:** Our iterative algorithm based on the revision operator defined by Algorithm 1 in [19].

All of the experiments were performed on a Linux server with an Intel(R) CPU Xeon(TM) 3.2GHz running Sun's Java 1.5.0 with allotted 2GB heap space. Our system[2] including the implementation of the five algorithms can be downloaded, together with all the data sets and results.

5.1 Data Sets

We use the ontology matching data sets available in OAEI'08[3] (Ontology Alignment Evaluation Initiative), which provides a platform for evaluating ontology matching systems. For our experiments, the following individual ontologies in the domain of scientific conferences are used: confOf with 197 axioms, ekaw with 248 axioms and cmt with 246 axioms. The pairwise mappings have been generated by the matching systems participating in OAEI.[4] For simplicity, we use, for example, confOf-ekaw-DSSim to indicate a distributed system consisting of individual ontologies confOf and ekaw and a mapping between them which is generated by system DSSim.

5.2 Evaluation Results

We evaluated our algorithms with respect to the following measures: efficiency, scalability and correctness.

Efficiency and Scalability. To measure efficiency and scalability of the algorithms, we considered revision time, which includes the time used to check whether a

[1] `http://kaon2.semanticweb.org/`

[2] `http://radon.ontoware.org/downloads/mapping-revision-09.zip`

[3] `http://oaei.ontologymatching.org/2008/`

[4] In addition, we use FOAM (`http://ontoware.org/projects/map/`) to generate mappings by applying only some simple string-based algorithms.

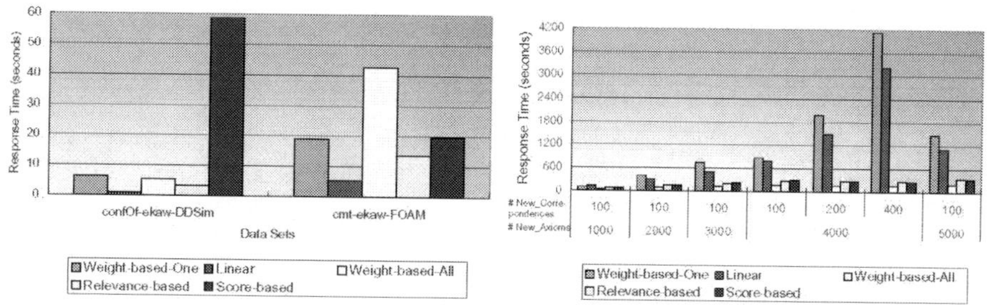

Fig. 1. The revision time of the algorithms

mapping is inconsistent and the time used to resolve inconsistencies. If a module extraction algorithm was applied, the time used to extract modules was also included. To test the efficiency of an algorithm, we continuously ran it for 30 times and took the average revision time. We have done the experiment based on the distributed systems $\mathcal{D}_1$:confOf-ekaw-DSSim and $\mathcal{D}_2$:cmt-ekaw-FOAM.

The left part of Figure 1 shows the average revision time over all runs for each algorithm. From this part we can see that for $\mathcal{D}_1$ and $\mathcal{D}_2$ which contain relatively small size of ontologies (i.e. no more than 250 axioms for each individual ontology) with mappings consisting of few correspondences (e.g. 19 and 55 correspondences for the mappings in $\mathcal{D}_1$ and $\mathcal{D}_2$ separately), Linear outperforms all the others. The second observation is that our Relevance-based outperforms all the others except Linear. It is because we expand only one branch of the hitting set tree in a depth-first manner and we apply the module extraction to optimize expanding the hitting set tree. Score-based has the worst performance for $\mathcal{D}_1$ but performs better for $\mathcal{D}_2$. This is because all correspondences in the mapping of $\mathcal{D}_1$ have the same weights so Score-based needs to compute all the minimal conflict sets for all unsatisfiable concepts in $\mathcal{D}_1$, whilst the correspondences in the mapping of $\mathcal{D}_2$ can be stratified by their weights so we only need to compute some minimal conflict sets. Another observation is that Weight-based-All does not perform well for $\mathcal{D}_2$ since it needs to compute all the minimal conflict sets for an unsatisfiable concept in each iteration.

In the right part, we show the scalability of the algorithms using the extended data sets based on $\mathcal{D}_3$:cmt-ekaw-Lily. This experiment is used to show that Linear may perform worse than some other algorithms if there are many axioms and correspondences that are not involved in the conflict sets. Here #New_Axioms means the number of axioms which are newly added to each individual ontology. #New_Correspondences indicates the number of newly added correspondences. Take the first column in the right part of Figure 1 as an example, we added 1000 dummy axioms to cmt and ekaw respectively and 100 dummy correspondences between the newly introduced concepts. Similarly we constructed other data sets by adding more axioms and correspondences. According to the right part of Figure 1, when adding more axioms and correspondences to $\mathcal{D}_3$, Weight-based-one and Linear perform worse and worse. In contrast, the other three algorithms are optimized by applying the module extraction technique and thus gain the advantage over Weight-based-one and Linear in the scalability test.

Table 2. Correctness of the algorithms

Distributed Systems	Algorithm	Repair precision			Repair recall		
		Max(Pr_i)	Avg(Pr_i)	Min(Pr_i)	Max(Rr_i)	Avg(Rr_i)	Min(Rr_i)
confOf–ekaw–DSSim	Weight-based-One	1.00	0.73	0.55	0.89	0.71	0.56
	Linear	1.00	0.75	0.50	0.78	0.67	0.56
	Weight-based-All	1.00	0.81	0.56	0.78	0.71	0.56
	Relevance-based	0.89	0.72	0.50	0.89	0.72	0.56
	Score-based	0.86	0.86	0.86	0.67	0.67	0.67
cmt–ekaw–FOAM	Weight-based-One	1.00	0.96	0.94	0.66	0.62	0.60
	Linear	1.00	0.97	0.93	0.56	0.56	0.56
	Weight-based-All	1.00	1.00	1.00	0.70	0.66	0.64
	Relevance-based	1.00	0.98	0.93	0.58	0.56	0.56
	Score-based	1.00	1.00	1.00	0.56	0.56	0.56

Correctness. In order to measure the correctness of the algorithms, we adopted the definitions of repair precision Pr and repair recall Rr in [13]. Assume $\mathcal{M}$ is a mapping between two ontologies and $\mathcal{G}$ is the reference mapping which is created manually by domain experts. Then $\mathcal{M}^- = \mathcal{M} - \mathcal{G}$ indicates those correspondences in $\mathcal{M}$ which are not correct. The repair precision and repair recall are defined as follows:

$$\text{Repair precision}: \mathsf{Pr} = \frac{\text{removed correspondences in } \mathcal{M}^-}{\text{all removed correspondences}}$$
$$\text{Repair recall}: \mathsf{Rr} = \frac{\text{removed correspondences in } \mathcal{M}^-}{|\mathcal{M}^-|}$$

This experiment is again based on $\mathcal{D}_1$ and $\mathcal{D}_2$, and we continuously ran each algorithm 30 times. For each run i (i=1,...,30), we compute the repair precision Pr_i and recall Rr_i. Table 2 shows the maximal, minimal and average repair precision and recall from all runs for each algorithm.

According to Table 2, we can see that Score-based has the highest repair precision and lowest repair recall in most cases. This shows that this algorithm best fulfils the principle of minimal change. On the other hand, since Score-based removes less correspondences, it may fail to remove some erroneous correspondences. We also noticed that Weight-based-All performs slightly better than all the others except Score-based w.r.t. both average repair precision and the average repair recall. For example, Weight-based-All reaches higher average precision (i.e. 0.81) and recall (i.e. 0.71) for $\mathcal{D}_1$ and the highest average repair precision (i.e. 1) and recall (i.e. 0.66) for $\mathcal{D}_2$. This shows that this algorithm removes more correspondences which are incorrect comparing with the results from other algorithms in most cases.

6 Conclusion and Discussion

In this paper, we discussed the problem of repairing inconsistent mappings in the distributed systems. We first defined a conflict-based mapping revision operator and provided a representation theorem for it. We then presented an iterative algorithm for mapping revision in a distributed system based on a revision operator in DLs and showed that this algorithm results in a conflict-based mapping revision operator. We showed that the algorithm given in [15], which we call Linear, can be encoded as a special iterative algorithm. We also provided an algorithm to implement an alternative revision operator based on the relevance-based selection function given in [11] which can be optimized by a module extraction technique and considered a revision operator based on a *scoring function* in [19]. All three specific iterative algorithms have been implemented. We compared these algorithms with two other existing algorithms for mapping revision in [13] and [19]. Although our experimental results are preliminary and do not tend to be conclusive, we can still make some interesting observations:

- For most of the tests, our iterative algorithms (where Linear is equivalent to the algorithm given in [15]) performed well compared with two existing mapping revision algorithms. It is interesting to see that Linear performed quite well for all the real life data, although it performed much worse than other iterative algorithms for the scalability test.
- The iterative algorithm Score-based showed its advantage over other algorithms w.r.t. minimal change and it produced the most stable results. However, it did not perform so well for the efficiency test.
- Our iterative algorithm Relevance-based was in a good position for the correctness test. It outperformed other algorithms except Linear w.r.t efficiency and had good performance on scalability test. Thus it is a good choice to revise inconsistent mappings in those distributed systems with large scale mappings and ontologies.
- Weight-based-One performed worst for the scalability test and it does not perform well for other tests, so we suggest that it can be replaced by Linear.
- We also noticed Weight-based-All had good performance for the correctness test, although it did not perform so well for the efficiency test. So it is a good alternative to the iterative algorithms.

In the future work, we could further optimize our iterative algorithms Relevance-based and Score-based. For example, since the module extraction algorithm used to optimize our algorithms is independent on the super-concept of a subsumption entailment and thus may result in a large module, we will consider a goal-directed module extraction method that can produce a smaller module than the syntactic locality-based module. For Relevance-based, we used the relevance-based selection function in this paper because it can be applied to any DL and is the basis of some other selection functions. This selection function may select too many axioms in each iteration. Other selection functions will be considered.

Acknowledgments

Research reported in this paper was partially supported by the EU in the IST project NeOn (IST-2006-027595, `http://www.neon-project.org/`). We would like to thank the anonymous reviewers as well as Richard Booth and Renata Wassermann for their helpful comments.

References

1. Baader, F., Calvanese, D., McGuinness, D.L., Nardi, D., Patel-Schneider, P.F.: The Description Logic Handbook: Theory, Implementation, and Applications. Cambridge University Press, New York (2003)

2. Baader, F., Peñaloza, R., Suntisrivaraporn, B.: Pinpointing in the description logic $\mathcal{EL}^+$. In: Proc. of KI, pp. 52–67 (2007)
3. Borgida, A., Serafini, L.: Distributed description logics: Assimilating information from peer sources. Journal of Data Semantics 1, 153–184 (2003)
4. Euzenat, J., Shvaiko, P.: Ontology Matching. Springer, Heidelberg (2007)
5. Gardenfors, P.: Knowledge in Flux-Modeling the Dynamic of Epistemic States. The MIT Press, Cambridge (1988)
6. Grau, B.C., Horrocks, I., Kazakov, Y., Sattler, U.: Just the right amount: extracting modules from ontologies. In: Proc. of WWW, pp. 717–726 (2007)
7. Haase, P., Qi, G.: An analysis of approaches to resolving inconsistencies in DL-based ontologies. In: Proc. of IWOD, pp. 97–109 (2007)
8. Ove Hansson, S.: Reversing the levi identity. Journal of Philosophical Logic 22(6), 637–669 (1993)
9. Ove Hansson, S.: Kernel contraction. Journal Symbolic Logic 59(3), 845–859 (1994)
10. Ove Hansson, S.: A Textbook of Belief Dynamics: Theory Change and Database Updating. Kluwer Academic Publishers, Dordrecht (1999)
11. Huang, Z., van Harmelen, F., ten Teije, A.: Reasoning with inconsistent ontologies. In: Proc. of IJCAI, pp. 454–459 (2005)
12. Meilicke, C., Stuckenschmidt, H.: Applying logical constraints to ontology matching. In: Proc. of KI, pp. 99–113 (2007)
13. Meilicke, C., Stuckenschmidt, H., Tamilin, A.: Repairing ontology mappings. In: Proc. of AAAI, pp. 1408–1413 (2007)
14. Meilicke, C., Stuckenschmidt, H., Tamilin, A.: Reasoning support for mapping revision. Journal of Logic and Computation (2008)
15. Meilicke, C., Völker, J., Stuckenschmidt, H.: Learning disjointness for debugging mappings between lightweight ontologies. In: Gangemi, A., Euzenat, J. (eds.) EKAW 2008. LNCS (LNAI), vol. 5268, pp. 93–108. Springer, Heidelberg (2008)
16. Nebel, B.: Base revision operations and schemes: Semantics, representation and complexity. In: Proc. of ECAI, pp. 341–345 (1994)
17. Peñaloza, R., Sertkaya, B.: Axiom pinpointing is hard. In: Proc. of DL (2009)
18. Qi, G.: A semantic approach for iterated revision in possibilistic logic. In: Proc. of AAAI, pp. 523–528 (2008)
19. Qi, G., Haase, P., Huang, Z., Ji, Q., Pan, J.Z., Völker, J.: A kernel revision operator for terminologies — algorithms and evaluation. In: Sheth, A.P., Staab, S., Dean, M., Paolucci, M., Maynard, D., Finin, T., Thirunarayan, K. (eds.) ISWC 2008. LNCS, vol. 5318, pp. 419–434. Springer, Heidelberg (2008)
20. Qi, G., Pan, J.Z., Ji, Q.: Extending description logics with uncertainty reasoning in possibilistic logic. In: Mellouli, K. (ed.) ECSQARU 2007. LNCS (LNAI), vol. 4724, pp. 828–839. Springer, Heidelberg (2007)
21. Reiter, R.: A theory of diagnosis from first principles. Artificial Intelligence 32(1), 57–95 (1987)
22. Suntisrivaraporn, B., Qi, G., Ji, Q., Haase, P.: A modularization-based approach to finding all justifications for OWL DL entailments. In: Domingue, J., Anutariya, C. (eds.) ASWC 2008. LNCS, vol. 5367, pp. 1–15. Springer, Heidelberg (2008)
23. Zimmermann, A., Euzenat, J.: Three semantics for distributed systems and their relations with alignment composition. In: Cruz, I., Decker, S., Allemang, D., Preist, C., Schwabe, D., Mika, P., Uschold, M., Aroyo, L.M. (eds.) ISWC 2006. LNCS, vol. 4273, pp. 16–29. Springer, Heidelberg (2006)

Functions over RDF Language Elements

Bernhard Schandl

University of Vienna
Department of Distributed and Multimedia Systems
bernhard.schandl@univie.ac.at

Abstract. RDF data are usually accessed using one of two methods: either, graphs are rendered in forms perceivable by human users (e.g., in tabular or in graphical form), which are difficult to handle for large data sets. Alternatively, query languages like SPARQL provide means to express information needs in structured form; hence they are targeted towards developers and experts. Inspired by the concept of spreadsheet tools, where users can perform relatively complex calculations by splitting formulas and values across multiple cells, we have investigated mechanisms that allow us to access RDF graphs in a more intuitive and manageable, yet formally grounded manner. In this paper, we make three contributions towards this direction. First, we present RDFunctions, an algebra that consists of mappings between sets of RDF language elements (URIs, blank nodes, and literals) under consideration of the triples contained in a background graph. Second, we define a syntax for expressing RDFunctions, which can be edited, parsed and evaluated. Third, we discuss Tripcel, an implementation of RDFunctions using a spreadsheet metaphor. Using this tool, users can easily edit and execute function expressions and perform analysis tasks on the data stored in an RDF graph.

1 Introduction

RDF is a highly generic model for data representation. Its fundamental information unit is the triple, which denotes a specific kind of relationship between two entities (resources), or between an entity and a literal value. As such, it can be used to represent arbitrary kinds of data, as shown by the Linked Data community project[1] and various applications, e.g., in the life sciences field [24] or the Semantic Desktop [25].

RDF data sets are currently accessed and used by applying one of the following two metaphors: either, users directly navigate and browse them using tools that display the graph in a human-perceivable manner (e.g., in tabular form of varying complexity [4,14]), or using graphical rendering (e.g., in the form of graphs[2] or using rendering template languages [7]). Such tools provide the user with direct, intuitive access to any resource or triple found in the graph, but

[1] Linked Data: http://linkeddata.org

[2] RDF-Gravity: http://semweb.salzburgresearch.at/apps/rdf-gravity

A. Bernstein et al. (Eds.): ISWC 2009, LNCS 5823, pp. 537–552, 2009.

cannot be reasonably applied to very large graphs. On the other hand, if RDF data are used within applications (i.e., graphs are not directly exposed to the end user), APIs provided by Semantic Web frameworks or query languages (of which SPARQL [23] is the most prominent one) are used. These allow developers to express information needs in a formalized and structured form. However, SPARQL queries can become complex to write and to evaluate [21]; additionally, SPARQL currently lacks certain features that are needed in several use cases (e.g., sub-queries and aggregates).

We are searching for metaphors to interact with Semantic Web data in a way that is intuitive for users, can be quickly authored and evaluated, but still is highly expressive and grounded in a formal model. We were inspired by the concept of *spreadsheets* [18], where data (usually, numbers and text) and calculations (formulas) are arranged in a grid structure of cells, and can be directly inspected and edited by the user. Formulas may refer to the contents of other cells by their coordinates, and evaluation results are displayed immediately after cells have been changed. Although the spreadsheet concept is not free of errors [22] it is a powerful tool and suitable for many application scenarios.

In this paper, we present our approach to merge the spreadsheet concept with the RDF data model. As its formal foundation we present *RDFunctions*, an algebra that consists of mappings between RDF language elements (URIs, blank nodes, and literals). These functions are evaluated not only using parameter values, but also under consideration of the triples stored in an additional *background graph*. Thus, RDFunctions can be used to perform data analysis and computation tasks over RDF data sources. In addition to this formal model, we define a concrete syntax for RDFunction expressions that can be edited by users, and parsed and evaluated by a corresponding engine. Finally, we introduce *Tripcel*, a spreadsheet application that implements the concepts described before, and present an evaluation of its applicability.

2 RDFunctions

As the conceptual basis of our approach we define *RDFunctions*, an algebra consisting of mappings between sets of RDF resources. An RDFunction takes a set of RDF elements (i.e., URIs, blank nodes, and literals) as input, and returns another set of RDF elements as result, whereas the evaluation of a function may consider the triples contained in a *background graph*.

A *background graph* G is an RDF graph as defined in [19] and hence consists of a set of triples $< s, p, o >$ which are constructed of elements of the set $\mathbb{U} \cup \mathbb{B} \cup \mathbb{L}$; i.e., URIs, blank nodes, and literals. Following [19] we denote with $universe(G)$ the set of elements that occur in the triples of G, and denote by U_G, B_G, and L_G the sets of URIs, blank nodes, and literals that are elements of this universe, i.e., $universe(G) = U_G \cup B_G \cup L_G$.

An RDFunction $f(\cdot)$ is a mapping $f : \mathcal{P}(\mathbb{U} \cup \mathbb{B} \cup \mathbb{L}) \mapsto \mathcal{P}(\mathbb{U} \cup \mathbb{B} \cup \mathbb{L})$; i.e., it takes a set of RDF elements as input and returns a set of RDF elements as output. These elements must not necessarily be contained in the associated background graph. However, an RDFunction can be defined under consideration of the background graph G; in this case we denote the function using an index $f_G(\cdot)$.

The design of this generic function signature is inspired by the fact that when querying data one is not necessarily interested in triples, but in resources or literals (*things*) that fulfill certain criteria: e.g., one might look for resources that fulfil certain criteria, or for literal values of certain properties. Hence we chose to put the RDF language elements (URIs, blank nodes, and literals) instead of triples into the focus of attention.

While an arbitrary number of concrete functions can be defined that fulfil this generic signature, we define in the following a core set of functions that are useful in a broad range of use cases.

Background Graph Access Functions. We define functions that return groups of elements contained in the background graph; *resources*(), *bnodes*(), *literals*(), and *properties*(), as follows:

$$resources_G(\cdot) := \{\ r \mid \forall s,p,o,r:\ (< r,p,o > \in G \vee < s,p,r > \in G) \\ \wedge\ r \in U_G \cup B_G\} \tag{1}$$

$$bnodes_G(\cdot) := B_G \tag{2}$$

$$literals_G(\cdot) := L_G \tag{3}$$

$$properties_G(\cdot) := \{\ p \mid \forall s,p,o :\ < s,p,o > \in G\} \tag{4}$$

As we can see from these definitions, all background graph access functions discard the input parameter set, i.e., their results are depending only on the triples contained in the background graph.

Construction Functions. In contrast to background graph access functions, the following group of functions construct RDF elements independent of whether they are contained in the background graph or not, hence the index $\cdot_G$ is not used in their definitions. For the three different forms of literals (plain, typed, and language-tagged) different construction functions are defined[3].

$$resource^{\texttt{uri}}(\cdot) := \texttt{<uri>} \tag{5}$$

$$bnode(\cdot) := \texttt{[]} \tag{6}$$

$$literal^{\texttt{lexicalform}}(\cdot) := \texttt{"lexicalform"} \tag{7}$$

$$literal^{\texttt{lexicalform, uri}}(\cdot) := \texttt{"lexicalform"\^{}\^{}uri} \tag{8}$$

$$literal^{\texttt{lexicalform, lang}}(\cdot) := \texttt{"lexicalform"@lang} \tag{9}$$

Unlike the background graph access functions described before, construction functions ignore the contents of the background graph G, as well as the provided input parameter.

[3] We use the triple notation [13] to serialize RDF elements.

Property Functions. The function $property^p$ returns all values (objects) for property p of the resources given as function parameters (I), based on the triples in the background graph G.

$$property_G^p(I) := \{\ o\ \mid\ \forall s,p,o : < s,p,o > \in G, s \in I\} \tag{10}$$

One concrete example of such a function is $property^{\texttt{rdfs:label}}$, which would return all labels of the resources given as parameters. Similarly, we define an inverse property function $invproperty^p$ that returns all subjects that have any of the resources given as function parameters as property values (objects) for property p:

$$invproperty_G^p(I) := \{\ s\ \mid\ \forall s,p,o : < s,p,o > \in G, o \in I\} \tag{11}$$

Examples. We illustrate applications of the functions we have defined so far by a number of concrete example. The function

$$invproperty_G^{\texttt{rdf:type}}(I)$$

returns all resources contained in the background graph whose `rdf:type` is one of the resources contained in the input set I. Since RDFunctions can be arbitrarily nested, we can use

$$property_G^{\texttt{rdfs:label}}(resources_G(\cdot))$$

to retrieve all `rdfs:labels` from all resources in the background graph (in this case, no input parameters are needed). The function

$$property_G^{\texttt{foaf:name}}(invproperty_G^{\texttt{foaf:knows}}(I))$$

returns the `foaf:name` values of all resources that `foaf:know` any of the resources contained in the input set I. Finally, the function

$$property_G^{\texttt{dc:creator}}(invproperty_G^{\texttt{rdf:type}}(resource^{\texttt{swrc:Publication}}(\cdot)))$$

returns the `dc:creators` of all resources that are typed as `swrc:Publication`.

Triple Functions. Property functions match a specific property to the predicate position of all triples in the background graph. These functions cover the cases where the property URI is known. To retrieve RDF elements that occur in conjunction with given input resources within a common triple in the background graph whereas the triple's predicate is not known, we define the following functions:

$$objects4subjects_G(I) := \{\ o \mid\ \forall s,p,o : < s,p,o > \in G, s \in I, s \in U_G \cup B_G\} \tag{12}$$

$$subjects4objects_G(I) := \{\ s \mid\ \forall s,p,o : < s,p,o > \in G, o \in I\} \tag{13}$$

$$predicates4subjects_G(I) := \{\ p \mid\ \forall s,p,o : < s,p,o > \in G, s \in I, s \in U_G \cup B_G\} \tag{14}$$

$$predicates4objects_G(I) := \{\ p \mid\ \forall s,p,o : < s,p,o > \in G, o \in I\} \tag{15}$$

Since literals are not allowed in the subject position of RDF triples, the functions $objects4subjects_G(\cdot)$ and $predicates4subjects_G(\cdot)$ consider only those elements of the input set I that are URIs or bnodes, while literals are discarded.

Aggregate Functions. An aggregate function returns a single value, which is computed from a set of input values. In the context of RDF, certain aggregate functions can be applied to all types of graph elements (e.g., `count()`), while others can be applied only to typed literal values, e.g., `avg()`, `min()`, or `sum()`. RDFunctions can be easily extended by aggregate functions; for the sake of brevity we give here as an example only the definition of the `count()` function that returns the input set's cardinality as typed RDF literal, where $\mid I \mid$ is the cardinality of I:

$$count(I) := \ literal^{|I|,\texttt{xsd:integer}}(\cdot) \tag{16}$$

Filter Functions. SPARQL provides a mechanism to test the values of RDF elements through the `FILTER` element (cf. [23], Section 11). The RDFunction framework provides the function `filter()` to incorporate SPARQL filter expressions; however the semantics of filter expression evaluation is different in RDFunctions. In contrast to SPARQL, RDFunctions are evaluated not over a graph pattern, but over the set of input elements I (i.e., URIs, blank nodes, and literals). Hence a filter evaluation cannot distinguish between bindings of different variables, as it is the case in graph patterns. Thus, when evaluating the filter expression, *all* variables are bound to the same element (taken from the input set I), which effectively implies that an RDFunctions filter function may contain only one variable. This restriction is indicated by the index of the SPARQL `FILTER` function in the definition of the filter function (17), which binds all expression variables to a single element e; this binding is indicated by the notion $?* = e$ in (17). All input elements for which the filter evaluates to `false` are discarded, and all other elements are added to the function's result set:

$$filter^{\texttt{expression}}(I) := \{e \mid e \in I \wedge \texttt{FILTER}_{?*=e}(\texttt{expression}) \neq \texttt{false}\} \tag{17}$$

Example. As an example that combines aggregate functions and filter functions, the following function returns the number of resources that have a `foaf:birthday` before September 18, 1979 as follows:

$$count(invproperty_G^{\texttt{foaf:birthday}}(\\ filter^{\texttt{?x < "1979-09-18T00:00:00"\^{}\^{}xsd:dateTime}}(literals_G(\cdot)))) \tag{18}$$

Discussion. RDFunctions are mappings between sets of RDF elements (URIs, bnodes, and literals) that consider the triples contained in a background graph for evaluation. RDFunctions are designed to be nested in order to formulate more complex mappings. Since they are evaluated against a background graph they can also be considered as query algebra over triples contained therein, and we have shown by a number of examples that they can be used to express complex information needs. Considered as a query language, RDFunctions differs from SPARQL because of the different underlying data model: SPARQL queries are evaluated against an RDF graph and return either a set of variable bindings (for `SELECT` queries) or an RDF graph (for `CONSTRUCT` and `DESCRIBE` queries). RDFunctions, on the other hand, are evaluated against a set of RDF language elements, and also return a set of RDF language elements. Consequently, several SPARQL features (e.g., joins or multi-variable `FILTER` expressions) cannot be represented in RDFunctions. On the other hand, RDFunctions provide several features that can currently only be found in proprietary SPARQL extensions; e.g., aggregate functions, arbitrary expression nesting, or sub-queries. Additionally, RDFunctions are easily extensible, since each function that can be reduced to a mapping between RDF elements can be integrated into the algebra.

The efficiency of expression processing (i.e., query execution) heavily depends on the order in which the elements of a formula are nested. For instance, in expression (18) literal elements are filtered according to their value before the RDF property is evaluated. This expression could be rewritten so that the selection based on the `foaf:birthday` property is conducted before the literal values are tested against the filter, which might lead to a more efficient evaluation depending on the structure of the background graph. However such an optimization depends on the actual implementation as well as knowledge about the underlying background graph and is out of the scope of this formal definition.

3 Tripcel: Applying RDFunctions in Spreadsheets

The spreadsheet concept is a powerful, widely-used metaphor for the analysis and processing of data. In essence, a spreadsheet is a collection of *cells* that are arranged in a 2-dimensional grid, the *sheet*. Each cell within a sheet may contain a value or a formula. Formulas are evaluated to return a single result value, which can be reused by other cells as input for evaluation. Cells are usually referred to using a coordinate system where columns are identified by letters, and rows are identified by numbers. The coordinate of a cell is obtained by concatenating its column and row identifiers (e.g., `C17` refers to the cell in column 3, row 17).

Spreadsheets are popular because of a number of reasons. First, they allow users to break down complex calculations into smaller units that are easier to understand. This decomposition is driven by the user, not the machine: it is up to the user to decide whether they prefer to write a single complex formula into one cell, or to split the formula into smaller parts and distribute them across multiple cells. Second, spreadsheets combine formal calculations with user-friendly presentation, since cells can be arranged and formatted according to the user's needs and taste. Finally, spreadsheets provide the possibility to explore and compare different scenarios in a simple manner: a user may change one single value in the sheet, and all other cells are immediately updated.

Based on the idea of spreadsheets, we propose *Tripcel*, a spreadsheet variant that considers not only cells and the values and formulas stored therein, but enriches them with the RDFunctions framework and with background information in the form of an RDF graph. In Tripcel each cell contains an RDFunctions expression, and as described in Section 2 the result of the evaluation of this formula depends not only on the results of other cells, but also on the information stored in the background graph. As a significant difference to traditional spreadsheets, Tripcel cells may evaluate to more than one result value; in fact cells may evaluate to sets of RDF language elements.

Tripcel strictly separates the contents of the spreadsheet (i.e., formulas and expressions) from the contents of the background graph. This means that the same Tripcel spreadsheet can be evaluated over different background graphs without any modification. The connection between the formulas in the spreadsheet and the triples in the background graph is established through the functions that consider the background graph G in their evaluation. All functions defined in Section 2 with an index $\cdot_G$ in their name are such functions.

3.1 The Tripcel Formula Syntax

In Section 2 an abstract algebra for functions over RDF language elements has been presented. For concrete applications it is however required to express ad serialize function expressions using a concrete syntax. To represent abstract function formulas we have developed an expression syntax under consideration of the following requirements:

- *Readability.* In spreadsheet-based applications, expressions and formulas are usually directly edited by the user. Hence it is necessary that the syntax of expressions is easily readable, understandable, and editable. This implies that all identifiers (like function names) carry meaningful names, and that the number of special language elements (symbols) is reduced to a minimum.
- *Accordance.* It is difficult for users to remember elements of different languages, which convey similar or equal semantics in different syntaxes. In our context, relevant languages include other spreadsheet expression languages, RDF query languages, as well as common mathematic symbols. Hence, identifiers and special symbols in the Tripcel expression language should be reused from these languages wherever possible.
- *Expressivity.* The language should cover all elements of the RDFunctions algebra, as specified in Section 2; this includes RDF elements (URIs, bnodes, literals) as well as functions. Additionally, the syntax must provide means to specify cell references as function parameters.
- *Unambiguousness.* The language should be easy to parse, and it should be possible to decide which model element a token belongs to.

The syntax for cell formulas is defined using EBNF and is reproduced in Figure 1. It defines five types of expressions; *literals*, *literal references*, *resources*, *functions*, and *cell references*. For each type, an example is given in Figure 2. A literal expression textually represents an RDF literal. Every cell resource that does

```
expression           = literal | literalref | resource | function |
                       cellreference ;
literal              = lexicalform ;
literalref           = ’"’ lexicalform ;
resource             = ’<’ { ’<’ uri ’>’ | curie } ’>’ ;
function             = ’=’ functionname [ ’[’ functionmodifier ’]’ ]
                       ’(’ functionparameter ’)’ ;
functionname         = propertyfunctionname | externalfunctionname ;
functionmodifier     = expression ;
functionparameter    = expression ;
propertyfunctionname = propertyfunction { ’/’ propertyfunction } ;
propertyfunction     = propertyname | invpropertyname ;
propertyname         = curie ;
invpropertyname      = ’~’ curie ;
cellreference        = ’=’ singlecellreference | enumcellreference |
                       areacellreference ;
singlecellreference  = { ’A’..’Z’ } { ’0’..’9’ } ;
enumcellreference    = singlecellreference { ’,’ singlecellreference } ;
rangecellreference   = singlecellreference ’..’ singlecellreference ;
externalfunctionname = alpha ;
```

Fig. 1. Tripcel formula syntax in EBNF. For the sake of brevity we omit the production rules for the symbols `lexicalform` (lexical form of literals), `uri` (URIs [3]), `curie` (compact URIs [5]), and `alpha` (alphabetic characters).

Type	EBNF Rule	Example
Literal	`literal`	`ISWC 2009`
Literal Reference	`literalref`	`"ISWC 2009`
Resource	`resource`	`<dogfood:conference/iswc/2009>` `<<http://www.semanticweb.org>>`
Function	`function`	`=filter[isLITERAL(?)](...)` `=filter[?>"M"](...)`

Cell Reference	`cellreference`	`=B4`
		`=A3,B6,F8`
		`=A5..C8`

Fig. 2. Examples of Tripcel expressions

not adhere to one of the special syntactic constructs described in the following is interpreted as literal. The entire content of the cell is used as the literal value, and the literal datatype is guessed by analyzing the textual representation (e.g., a representation consisting only of numeric characters is interpreted as `xsd:integer` literal). If the literal datatype cannot be guessed, `xsd:string` is used as default.

A literal reference is a cell expression that starts with a quotation mark ("). In contrast to a literal expression, which is directly translated into a literal value, literal references are interpreted with respect to the background graph: the interpretation of a literal reference is the set of RDF resources that have any property whose (literal) value is equal to the string representation of the literal reference. Hence, Tripcel literal references correspond to the $subjects4objects_G(\cdot)$ function defined in Section 2, where the input set I contains exactly one literal.

A resource can be explicitly instantiated by a cell expression that starts with an opening angle bracket (<), followed by a CURIE [5] that identifies the resource[4] and a closing angle bracket (>); alternatively, full URIs can be used by enclosing them into double angle brackets (<< and >>)[5]. The interpretation of a resource expression is a set that contains exactly one resource, which is identified by the specified URI (this corresponds to the $resource^{\mathtt{uri}}(\cdot)$ construction function).

Function expressions refer to other Tripcel functions (cf. Section 2). They consist of the function name, the function modifier expression, and the function parameter expression. The function name is a string that refers to the RDFunction to be used, while the function modifier and the function parameters can be any kind of cell expression, including functions; hence, nested and recursive expressions can be constructed. Function modifiers influence the behaviour of the respective function; for instance, the `filter` function interprets a provided modifier string as SPARQL `FILTER` expression and evaluates its input according to (17).

Because of their importance, a special syntax has been defined for property functions, which are identified by the property's abbreviated URI; an inverse property function is identified by a preceding tilde character. For example, the expression `=rdfs:label(...)` corresponds to the RDFunction $property_G^{\mathtt{rdfs:label}}(\cdot)$, while `=~foaf:knows(...)` corresponds to $invproperty_G^{\mathtt{foaf:knows}}(\cdot)$. Property functions can be concatenated using a slash character, whereas they are traversed from right to left: the expression `=foaf:name/~foaf:knows(...)` corresponds to $property_G^{\mathtt{foaf:name}}(invproperty_G^{\mathtt{foaf:knows}}(\cdot))$. This abbreviated syntax allows users to intuitively define property chains, which are often needed in analysis tasks.

Finally, cell references are used to link formulas across different cells. By using a cell reference, the results of the referenced cell(s) are inserted at the point of reference. Currently Tripcel supports three kinds of cell references; single, enumerated, and range. Single cell references are substituted by the referenced cell's result set; for enumerated and range cell references the union of all referenced

---

[4] We assume that in the context of the Tripcel sheet suitable URI prefixes are defined for all URIs under consideration. For a discussion on potential problems that may arise when URI prefixes are used in user interfaces we refer to [26].

[5] We are aware of the fact that the usage of CURIEs in combination with angle brackets does not correspond to typical RDF serialization formats. However we have chosen this syntax because we want to provide a possibility to enter CURIEs since they are easier to remember, but at the same time need a mechanism to unambiguously identify them as URIs.

cells' results is returned. For instance, the enumerated cell reference formula `=B3,B6,C6` will evaluate the formulas in the three specified cells and construct the union of all resulting RDF elements.

3.2 Implementation

We have implemented a prototypical spreadsheet application that uses the RDFunctions model and the Tripcel syntax, which have been presented in the previous sections[6]. This tool allows users to load background graphs; to edit, load, and safe Tripcel sheets; and to inspect the evaluation results of each cell in more detail.

The Tripcel application is divided into four layers, reflecting the conceptual components described so far. The basis of the Tripcel application is the *background graph layer*, which is implemented using the Jena Semantic Web framework[7]. All details of RDF storage are hidden by this framework, hence it is in principle possible to connect Tripcel to any RDF data source (e.g., in-memory, database-backed, or remote). However, Tripcel operates on both the Jena Model API and its SPARQL implementation ARQ[8] since several Tripcel functions cannot be efficiently implemented using pure SPARQL; consequently only such data sources can be connected that support both access methods.

On top of the background graph layer the *RDFunctions layer* is situated. This layer implements the semantics of RDFunctions as described in Section 2 in a flexible manner: functions are realized as Java classes that implement a specific interface, thus it is possible to extend this layer by new RDFunctions without modifications to existing code. The RDFunctions layer is responsible for the evaluation of Tripcel formulas; the RDF-specific parts of this layer are likewise implemented using Jena.

One level above, the *spreadsheet layer* implements the logic of Tripcel spreadsheets. Its responsibility is to manage the contents of cells and their interdependencies. It receives formulas (entered by the user) from the GUI layer (see below), passes them to the RDFunctions layer for evaluation, and buffers the returned evaluation results. It also maintains a cell dependency graph and, upon a cell change, propagates notifications to all depending cells.

Finally, the *GUI layer* provides a graphical representation of a Tripcel sheet (see Figure 3). It renders cells in a grid, provides an editing interface for formulas, and displays evaluation results. As Tripcel cells, in contrast to classical spreadsheets, may contain multiple values, the GUI additionally provides a separate detail window where all elements contained in the selected cell are displayed.

The Tripcel application provides interaction mechanisms similar to well-known spreadsheet applications. Normally cells are filled with their evaluation results. If the evaluation of a cell's formula results in more than one element, the number of results (e.g., *"(7 elements)"*) is displayed. When the user clicks on a cell,

---

[6] The prototype can be downloaded from `http://www.ifs.univie.ac.at/schandl/2009/06/tripcel`

[7] Jena Semantic Web Framework: `http://jena.sourceforge.net`

[8] ARQ: `http://jena.sourceforge.net/ARQ`

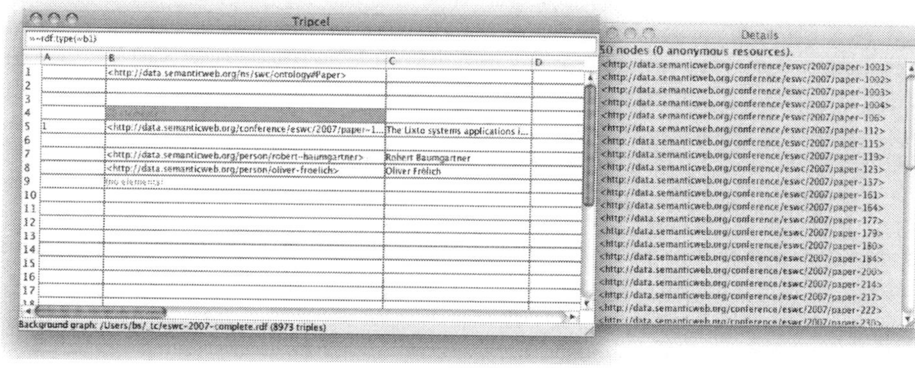

Fig. 3. Tripcel Screenshot: Spreadsheet Window (left) and Detail Window (right)

an editor line is provided where the user can inspect and modify the cell formula. When the user presses enter or selects a different cell, the edited formula is re-evaluated, and changes are propagated to all depending cells.

4 Evaluation

Qualitative Analysis. To estimate the usability and potential impact of our approach, we have performed a qualitative analysis on an initial group of 5 test persons, most of which are experts in the fields of Semantic Web, RDF, and query languages[9]. This analysis consisted of a think-aloud evaluation, followed by a structured questionnaire. In the course of the think-aloud evaluation, all candidates were asked to perform a tutorial of approx. 20 minutes length, during which they would learn the most important features of Tripcel and to get familiar with Tripcel formulas, their syntax, and the results.

During the think-aloud evaluation, the users individually performed the tasks described in a written tutorial, while they were observed by an interviewer. They were asked to immediately tell any thoughts they had during the task completion, regardless of whether they had to do with usability aspects, the entire Tripcel concept as such, or pure implementation issues and bugs. The goal of the think-aloud sessions was to estimate which associations and thoughts were triggered by Tripcel, and in which aspects the system could be improved. Most issues that were revealed during these sessions regarded the user interface (e.g., visual feedback or the application's general look and feel), missing features (which were structurally collected during the questionnaire, see below), or the syntax and semantics of formula expressions.

After the think-aloud session, each participant was asked to fill a questionnaire that was meant to reflect their impression on the concept and the tool, and

[9] The material that was used during the evaluation can be downloaded from `http://www.ifs.univie.ac.at/schandl/2009/06/tripcel`

to identify potential for improvements in a structured manner. The first part of the questionnaire consisted of questions that were to be answered on a 5-level Likert scale (1 = strongly disagree, 5 = strongly agree) and contained questions addressing Tripcel's general applicability and usability. The second part consisted of open questions addressing the user's impression on specific features as well as potential application fields. Finally, participants were asked to assess their familiarity with Semantic Web technologies and spreadsheets.

Participants rated the usefulness of the tool to get an overview on data with an average of 3.0 ($\sigma = 1.0$) for unknown data, and 3.6 ($\sigma = 1.5$) for known data. The syntax of Tripcel formula expressions was considered to be understandable (3.2, $\sigma = 1.3$) and even better memorizable (4.0, $\sigma = 1.0$). Participants agreed that RDF skills are required to use the tool; its usability for users without RDF skills was denied (1.8, $\sigma = 1.3$). However, for users with RDF skills the usability of the tool was rated very high (4.8, $\sigma = 0.4$). The participants considered themselves to be experts in Semantic Web technologies (RDF: 4.0, $\sigma = 1.4$; SPARQL: 3.8, $\sigma = 1.3$).

In a series of qualitative questions the participants were asked to judge the features of the application. Amongst the positively rated features were GUI aspects like the familiar interaction metaphors and their resemblance to spreadsheet applications, and the fast execution times of formula evaluation. The participants also liked the possibilities and expressivity of the formula language, especially the ability to formulate property paths and aggregate functions. Finally, the ability to apply the same formula sheet to different background graphs was appreciated.

The participants outlined several missing features, including the ability to load multiple RDF documents into one background graph, or to quickly switch between multiple background graphs. The ability to visualize the background graph or cell contents in the form of graphs or charts (as known from spreadsheet tools) was required by several participants. On the GUI level, features like auto-completion and syntax highlighting were mentioned, which would increase the application's usability and reduce the error rate. We are currently in the process of reviewing the detailed requirements for new features, which will be implemented subsequently.

Quantitative Analysis. As described in Section 3.2, Tripcel has been implemented on top of the Jena Semantic Web framework, and Tripcel functions are implemented using the Jena Model API or the ARQ SPARQL engine, depending on the function type. Consequently, the execution times of such calls and queries are not under the control of our implementation, and additionally depends on the size of the background graph. Here we refer to previous works on performance evaluation of different triple stores (e.g., [10,6]).

An essential feature of spreadsheets is the possibility to break down complex calculations into smaller units. By resolving a sheet's dependency graph, formulas that are distributed across multiple cells could be merged and optimized before they are translated into queries and executed against the background graph. However the user of a spreadsheet tool expects to be able to inspect intermediate results, which ultimately implies that each formula contained in a cell must be evaluated independently from other formulas. In our Tripcel implementation we follow the approach to buffer evaluation results in-memory for each cell as long as the cell formula (and the formula of any antecedent cell) is not modified. While this approach potentially requires more memory, it significantly reduces the time needed for formula evaluation.

5 Related Work

As mentioned before, SPARQL lacks a number of features that are needed in different application scenarios. These deficiencies have been acknowledged by previous works, which led to a number of proposed extensions. A number of these extensions are addressing similar issues as the RDFunctions framework does: Virtuoso SPARQL Extensions[10] or Jena ARQ[11] both provide aggregates and so-called pointer operators that reduce the number of variables needed in triple patterns. However we choose not rely on proprietary language extensions for our implementation. Currently, a number of proposed feature extensions for the next version of the SPARQL language are under review by the W3C SPARQL Working Group.

A number of languages have been proposed that allow to query for more complex triple patterns than it is currently possible using SPARQL. Many of these approaches provide mechanisms to navigate between nodes in an RDF graph, as can be done with the RDFunctions framework. This includes nSPARQL and rSPARQL [2], which provide means to express navigational expressions over RDF graphs, which can be evaluated under consideration of RDFS semantics. RDFunctions currently implements a subset of the features provided by nSPARQL for the sake of simplification. SPARQ2L [1], SPARQLeR [15], and ARQ extend SPARQL with functions for the analysis of path structures in an RDF graph, while the RDF path language of the SILK framework [27] and XsRQL [12] are independent languages. The ability to hierarchical nest property functions relates our work also to the family of RDF path query languages like Versa [20] or RPath [17], which could as well serve as the foundation for Tripcel.

Topic Map Query Language[12], although not designed for the RDF model, follows a similar conceptual model. All these approaches provide valuable input for further enhancement and extension of RDFunctions; however we want to ensure that the syntax for RDFunction expressions remains easily to remember.

Another approach comparable to Tripcel are Semantic Web Pipes [16] (which are inspired by Yahoo Pipes[13], an utility for meshing RSS feeds), where RDF data sources can be aggregated and manipulated through linked processing units. Our framework differs from Semantic Web Pipes in that we consider sets of RDF language elements as input and output of functions, rather than RDF graphs.

---

[10] Virtuoso SPARQL Extensions:
`http://docs.openlinksw.com/virtuoso/rdfsparql.html#sparqlextensions`

[11] ARQ Extensions: `http://jena.sourceforge.net/ARQ/documentation.html`

[12] Topic Map Query Language (TMQL): `http://www.isotopicmaps.org/tmql`

[13] Yahoo Pipes: `http://pipes.yahoo.com/pipes/`

The interrelationships between spreadsheets and semantic technologies have been studied in a number of works. Tools that are able to extract RDF data from spreadsheets include ConvertToRDF [9], which maps table column headings to ontological concepts, and RDF123 [11], which provides a special language to express the conversion parameters. Other approaches involve the use of GRDDL [8]; examples for this as described e.g., in the GRDDL Primer[14]. Vice versa, since SPARQL `SELECT` query results are tables they can be directly integrated into spreadsheets, as shown e.g., in [26]. However to the best of our knowledge there exists no approach so far that directly integrates processing of RDF language elements into the spreadsheet concept.

6 Conclusions and Further Research Directions

In this paper we have presented the concept of RDFunctions, which are mappings between sets of RDF elements under the consideration of background information expressed in an RDF graph. We have defined an extensible conceptual model for RDFunctions, as well as a number of basic functions for processing RDF language elements. These functions have been implemented in the form of Tripcel, a spreadsheet-based tool that allows users to use RDFunctions in order to analyse the contents of RDF graphs. To represent RDFunction expressions we have designed and implemented a formula language which is oriented towards the syntax of popular spreadsheet software. Our approach was evaluated in the course of a study among expert users, who judged Tripcel as being a useful tool for analyzing RDF data, and gave directions for further work.

In the future, we plan to improve the user interface of our implementation and extend it with features that improve usability (e.g., syntax and reference highlighting, auto completion, formula authoring assistants, and more efficient projection of three-dimensional results into the two-dimensional user interface) or that extend functionality (e.g., cell formatting, more advanced aggregate functions, etc.). We aim to extend the range of possible applications of Tripcel by integrating mechanisms that allow the software to connect to multiple remote data sources, which opens the door to evaluate spreadsheets against the Web of Data. Finally, we plan to integrate Tripcel with classical spreadsheet tools in order to facilitate data interoperability. In the first step, we will implement copy+paste functionality; in a second step we plan to implement direct data integration and live synchronization between Tripcel and other spreadsheet software.

Acknowledgements. Parts of this work have been funded by FIT-IT grants 812513 and 815133 from the Austrian Federal Ministry of Transport, Innovation, and Technology. The author thanks Leo Sauermann for initial inspiration, as well as the participants of the evaluation for their patience and support.

[14] GRDDL Primer: `http://www.w3.org/TR/grddl-primer`

References

1. Anyanwu, K., Maduko, A., Sheth, A.: SPARQ2L: Towards Support for Subgraph Extraction Queries in RDF Databases. In: WWW 2007: Proceedings of the 16th International Conference on World Wide Web, pp. 797–806. ACM Press, New York (2007)
2. Arenas, M., Gutierrez, C., Pérez, J.: An Extension of SPARQL for RDFS. In: Christophides, V., Collard, M., Gutierrez, C. (eds.) SWDB-ODBIS 2007. LNCS, vol. 5005, pp. 1–20. Springer, Heidelberg (2007)
3. Berners-Lee, T., Fielding, R., Masinter, L.: Uniform Resource Identifier (URI): Generic Syntax (RFC 3986). Network Working Group (January 2005)
4. Berners-Lee, T., Chen, Y., Chilton, L., Connolly, D., Dhanaraj, R., Hollenbach, J., Lerer, A., Sheets, D.: Tabulator: Exploring and Analyzing Linked Data on the Semantic Web. In: Proceedings of the 3rd International Semantic Web User Interaction Workshop (2006)
5. Birbeck, M., McCarron, S.: CURIE Syntax 1.0 — A Syntax for Expressing Compact URIs. World Wide Web Consortium (2009), http://www.w3.org/TR/curie/, http://www.w3.org/TR/curie/
6. Bizer, C., Schultz, A.: Benchmarking the Performance of Storage Systems that Expose SPARQL Endpoints. In: Proceedings of the 4th International Workshop on Scalable Semantic Web Knowledge Base Systems, SSWS 2008 (2008)
7. Champin, P.-A.: Tal4RDF: Lightweight Presentation for the Semantic Web. In: Proceedings of the 5th Workshop on Scripting and Development for the Semantic Web, SFSW 2009 (2009)
8. Connolly, D.: Gleaning Resource Descriptions from Dialects of Languages (GRDDL) (W3C) Recommendation World Wide Web Consortium (September 11, 2007)
9. Golbeck, J., Grove, M., Parsia, B., Kalyanpur, A., Hendler, J.: New Tools for the Semantic Web. In: Knowledge Engineering and Knowledge Management: Ontologies and the Semantic Web (2002)
10. Guo, Y., Pan, Z., Heflin, J.: LUBM: A Benchmark for OWL Knowledge Base Systems. Web Semantics: Science, Services and Agents on the World Wide Web 3(2-3), 158–182 (2005)
11. Han, L., Finin, T.W., Parr, C.S., Sachs, J., Joshi, A.: RDF123: From Spreadsheets to RDF. In: Sheth, A.P., Staab, S., Dean, M., Paolucci, M., Maynard, D., Finin, T., Thirunarayan, K. (eds.) ISWC 2008. LNCS, vol. 5318, pp. 451–466. Springer, Heidelberg (2008)
12. Katz, H.: XsRQL: an XQuery-style Query Language for RDF (RDF Data Access Working Group Submission) (2004), http://www.fatdog.com/xsrql.html (retrieved 09-June-2009)
13. Klyne, G., Carroll, J.J.: Resource Description Framework (RDF): Concepts and Abstract Syntax (W3C Recommendation) World Wide Web Consortium (February 10, 2004)
14. Kobilarov, G., Dickinson, I.: Humboldt: Exploring Linked Data. In: Proceedings of the Linked Data on the Web Workshop, LDOW 2008 (2008)
15. Kochut, K.J., Janik, M.: SPARQLeR: Extended SPARQL for Semantic Association Discovery. In: Franconi, E., Kifer, M., May, W. (eds.) ESWC 2007. LNCS, vol. 4519, pp. 145–159. Springer, Heidelberg (2007)

16. Le-Phuoc, D., Polleres, A., Hauswirth, M., Tummarello, G., Morbidoni, C.: Rapid Prototyping of Semantic Mash-ups through Semantic Web Pipes. In: WWW 2009: Proceedings of the 18th international conference on World wide web, pp. 581–590. ACM, New York (2009)
17. Matsuyama, K., Kraus, M., Kitagawa, K., Saito, N.: A Path-Based RDF Query Language for CC/PP and UAProf. In: IEEE International Conference on Pervasive Computing and Communications Workshops,l p. 3 (2004)
18. Mattessich, R.: Budgeting Models and System Simulation. The Accounting Review 36(3), 384–397 (1961)
19. Muñoz, S., Pérez, J., Gutierrez, C.: Minimal Deductive Systems for RDF. In: Franconi, E., Kifer, M., May, W. (eds.) ESWC 2007. LNCS, vol. 4519, pp. 53–67. Springer, Heidelberg (2007)
20. Ogbuji, C.: Versa: Path-Based RDF Query Language. XML.com (2005), `http://www.xml.com/pub/a/2005/07/20/versa.html`
21. Pérez, J., Arenas, M., Gutierrez, C.: Semantics and Complexity of SPARQL. In: Cruz, I., Decker, S., Allemang, D., Preist, C., Schwabe, D., Mika, P., Uschold, M., Aroyo, L.M. (eds.) ISWC 2006. LNCS, vol. 4273, pp. 30–43. Springer, Heidelberg (2006)
22. Powell, S.G., Baker, K.R., Lawson, B.: A Critical Review of the Literature on Spreadsheet Errors. Decision Support Systems 46(1), 128–138 (2008)
23. Prud'hommeaux, E., Seaborne, A.: SPARQL Query Language for RDF (W3C Recommendation), World Wide Web Consortium, (January 15, 2008)
24. Ruttenberg, A., Rees, J.A., Samwald, M., Marshall, M.S.: Life Sciences on the Semantic Web: the Neurocommons and Beyond. Briefings in Bioinformatics 10(2), 193–204 (2009)
25. Sauermann, L., Bernardi, A., Dengel, A.: Overview and Outlook on the Semantic Desktop. In: Decker, S., Park, J., Quan, D., Sauermann, L. (eds.) Proceedings of the 1st Semantic Desktop Workshop,CEUR Workshop Proceedings, Galway, Ireland, November 2005, vol. 175 (2005)
26. Schandl, B.: Representing Linked Data as Virtual File Systems. In: Proceedings of the 2nd International Workshop on Linked Data on the Web (LDOW), Madrid, Spain (2009)
27. Volz, J., Bizer, C., Gaedke, M., Kobilarov, G.: Silk – A Link Discovery Framework for the Web of Data. In: Proceedings of the 2nd International Workshop on Linked Data on the Web (LDOW), Madrid, Spain (2009)

Policy-Aware Content Reuse on the Web

Oshani Seneviratne, Lalana Kagal, and Tim Berners-Lee

MIT CSAIL, Cambridge
Massachusetts, USA
{oshani,lkagal,timbl}@csail.mit.edu

Abstract. The Web allows users to share their work very effectively leading to the rapid re-use and remixing of content on the Web including text, images, and videos. Scientific research data, social networks, blogs, photo sharing sites and other such applications known collectively as the Social Web have lots of increasingly complex information. Such information from several Web pages can be very easily aggregated, mashed up and presented in other Web pages. Content generation of this nature inevitably leads to many copyright and license violations, motivating research into effective methods to detect and prevent such violations.

This is supported by an experiment on Creative Commons (CC) attribution license violations from samples of Web pages that had at least one embedded Flickr image, which revealed that the attribution license violation rate of Flickr images on the Web is around 70-90%. Our primary objective is to enable users to do the right thing and comply with CC licenses associated with Web media, instead of preventing them from doing the wrong thing or detecting violations of these licenses. As a solution, we have implemented two applications: (1) Attribution License Violations Validator, which can be used to validate users' derived work against attribution licenses of reused media and, (2) Semantic Clipboard, which provides license awareness of Web media and enables users to copy them along with the appropriate license metadata.

1 Introduction

Content reuse, often called 'mash-ups', have existed for as long as content has existed. Musicians routinely use other songs and tunes in their compositions. Collage art is considered to be creative, and even original, although it is composed from many different sources. Scientists routinely utilize data from different sources to conduct their experiments. However, mash-ups are a peculiarly digital phenomenon of the Web age. They are entirely a product made possible by the portable, mixable and immediate nature of digital technology. A potential legal problem arises when more than one legally encumbered content or data stream is bound together with others in the form of a mash-up. The users of the original content should remain within the bounds of the permitted use of the components comprising the mash-up. They can choose to ignore these permissions, or follow them. Either way, this creates a burden on them. Ignoring the license terms puts them in peril of breaking the law, and following them slows the creative process.

A. Bernstein et al. (Eds.): ISWC 2009, LNCS 5823, pp. 553–568, 2009.

Licenses or policies in general are pervasive in Web applications. They play a crucial role in enhancing security, privacy and usability of the services offered on the Web [3]. In this paper we limit the 'policy awareness' to scenarios that involve content reuse. We expect the policies in this context to comprise of licenses that can be expressed semantically, that are widely deployed on a range of media, and that have a large community base. CC licenses fit this description perfectly, as they provide a very clear and a widely accepted rights expression language implemented using Semantic Web technologies [11]. These licenses are both machine readable and human readable, and clearly indicate to a person, who wishes to reuse content, exactly how it should be used by expressing the accepted use, permissions, and restrictions of the content.

Popular search engines, including Google, Yahoo, and even sites such as Flickr, blip.tv, OWL Music Search and SpinXpress, have advanced search options to find CC licensed content on the Web [5,32,9,2,20,26]. However, even with these human-friendly licenses and the tools to support license discovery, license violations occur due to many reasons; Users may be ignorant as to what each of the licenses mean, or forget or be too lazy to check the license terms, or give an incorrect license which violates the original content creator's intention, or intentionally ignore the CC-license given to an original work in their own interests. Therefore, it is important that we have tools and techniques to make users aware of policies that they must follow while making the process of being license-compliant as painless as possible for the user, and make it difficult for someone to become license in-compliant either deliberately or by mistake.

In essence, the work described in this paper supports the principles of information accountability [29], policy-awareness and after-the-fact violations detection instead of relying on strict up-front enforcement. This paper is organized as follows: Section 2 gives the background and an overview of the technologies used. Section 3 outlines an experiment conducted to assess the level of CC attribution license violations on the Web using Flickr images. This experiment provided the motivation to develop tools for policy aware content reuse as described later in the paper in Section 4. Section 5 discusses the related work in this area and Section 6 discusses some future work on the tools we have developed. Finally, we conclude the paper with a summary of the contributions in Section 7.

2 Background

To be useful, metadata needs to have three important characteristics: it has to be easy to produce, be easily embeddable within the data they describe, and be easily readable [18]. The easiest way to produce metadata is to have it be produced automatically. Any metadata that has to be produced manually by the user usually doesn't get produced at all. The easiest way to ensure that the link between metadata and the data it describes is not broken is by embedding the former inside the latter. This way, the two travel together inseparably as a package. Finally, metadata has to be accessible easily, readable both manually as well as programmatically. At best, the metadata should be readable by crawlers of various search engines. Since metadata and data are traveling together, if popular search engines such as Google and Yahoo can read the metadata, by default the data become available to anyone who searches for it. RDF [23] satisfies all these criteria, has lot of community support, has been adopted widely and is a W3C recommendation.

2.1 Inline Provenance Using Metadata

The Extensible Metadata Platform (XMP) [31] is a technology that allows one to transfer metadata along with the content by embedding the metadata in machine readable RDF using a pre-defined set of classes and properties. This technology is widely deployed to embed licenses in free-floating multimedia content such as images, audio and video on the Web. Another format which is nearly universal when it comes to images is the Exchangeable Image File format (EXIF) [7]. The International Press Telecommunications Council (IPTC) photo metadata standard [14] is another well known standard. The metadata tags defined in these standards cover a broad spectrum of properties including date & time information, camera settings, thumbnail for previews and more importantly, the description of the photos including the copyright information. However both EXIF and IPTC formats do not store metadata in RDF. Also, one major drawback of inline metadata formats such as XMP, EXIF and IPTC is that it is embedded in a binary file, completely opaque to nearly all users, whereas metadata expressed in RDFa [24] will require colocation of metadata with human visible HTML. In addition to that, these metadata formats can only handle limited number of properties and lack the rich expressivity required for many content reuse policies.

2.2 Policies for Rights Enforcement on the Web

Policies governing the reuse of digital content on the Web can take several forms. It can be upfront enforcement mechanisms such as *Digital Rights Management* (DRM) approaches, or rights expression mechanisms such as *Creative Commons* licenses where users are given the freedom to reuse content, subject to several restrictions and conditions.

When it comes to DRM, distribution and usage of copyrighted content is often controlled by up-front policy enforcement. These systems usually restrict access to the content, or prevent the content from being used within certain applications. The core concept in DRM is the use of digital licenses, which grant certain rights to the user. These rights are mainly usage rules that are defined by a range of criteria, such as frequency of access, expiration date, restriction to transfer to another playback device, etc. An example of a DRM enforcement would be a DRM enabled playback device not playing a DRM controlled media transferred from another playback device, or not playing the media after the rental period has ended. The use of DRM to express and enforce rights on content on the Web raises several concerns. First, consumer privacy and anonymity are compromised. Second, the authentication process in the DRM system usually requires the user to reveal her identity to access the protected content, leading to profiling of user preferences and monitoring of user activity at large [8]. Third, the usability of the content is questionable since the user is limited to using proprietary applications to view or play the digital content, producing vendor lock-in. Similarly, 'copyright notices' or 'end-user license agreements' describe the conditions of usage of copyrighted material. A user of that particular material should abide by the license that covers the usage, and if any of the conditions of usage described in that license are violated, then the original content creator has the right to take legal action against the violator.

In contrast, CC has been striving to provide a simple, uniform, and understandable set of licenses that content creators can issue their content under to enable reuse with much less restrictions. Often, people tend to post their content with the understanding that it will be quoted, copied, and reused. Further, they may wish that their work only be used with attribution, used only for non-commercial use, distributed with a similar license or will be allowed in other free culture media. To allow these use restrictions CC has composed four distinct license types: *BY* (attribution), *NC* (non-commercial), *ND* (no-derivatives) and *SA* (share-alike) that can be used in combinations that best reflect the content creator's rights. In order to generate the license in XHTML easily, CC offers a license chooser that is hosted at `http://creativecommons.org/license`. With some user input about the work that is being licensed, the license chooser generates a snippet of XHTML that contains the RDFa [24] to be included when the content is published on the Web. Content creators have the flexibility to express their licensing requirements using the Creative Commons Rights Expression Language (*ccREL*)[1] [11] and are not forced into choosing a pre-defined license for their works. Also, they are free to extend licenses to meet their own requirements. ccREL allows a publisher of a work to give additional permissions beyond those specified in the CC license with the use of the *cc:morePermissions* property to reference commercial licensing brokers or any other license deed, and *dc:source* to reference parent works.

3 Motivation

Unless a particular piece of content on the Web has some strict access control policies, most users do not feel the need to check for the license it is under and be license compliant. To verify this hypothesis we conducted an experiment to assess the level of license violations on the Web. Specifically, the goal of the experiment was to obtain an estimation for the level of CC *attribution license violations* on the Web using Flickr images[2].

[1] *ccREL* is the standard recommended by the Creative Commons for machine readable expressions of the meaning of a particular license.

[2] As of April 2009, Flickr has over 100 million Creative Commons Licensed images. Thus it provided a large sample base for our experiment.

3.1 Experiment Setup

The sampling method used for the experiment was simple random sampling on clusters of Web pages gathered during a particular time frame. To ensure a fair sample we used the Technorati blog indexer[3] without hand-picking Web pages to compose the sample to check for attribution license violations. We limited the number of Web pages to around 70, and the number of images to less than 500, so that we could do a manual inspection to see if there are any false positives, false negatives and/or any other errors. This also enabled us to check if the different samples contained the same Web pages. We found that the correlation among the samples was minimal.

Sample Collection using the Technorati API: The Technorati blog indexer crawls and indexes blog-style Web pages and keeps track of what pages link to them, what pages they link to, how popular they are, how popular the pages that link to them are, and so on. Technorati data are time dependent, and therefore the *Technorati Authority Rank*, a measurement that determines the top n results from any query to the Technorati API, is based on the most recent activity of a particular Web page[4]. The *Technorati Cosmos* querying functions allow the retrieval of results for blogs linking to a given base URI based on the *authority rank*. Therefore to generate the samples, we used the *Technorati Cosmos* functions by retrieving results for Web pages linking to Flickr server farm URIs that have this particular format: `http://farm<farm-id>.static.flickr.com/<server-id>/<id>_<secret>.(jpg|gif|png)`[5]. Since the Flickr site has several server farms, each time the experiment was run, the base URIs were randomly generated by altering the Flickr server *farm-ids*. In addition to that, we made sure that the samples were independent of each other and the correlation among the samples were low by running the experiment three times with two weeks between each trial. This is because the *Authority Rank* given to a Web page by Technorati, and hence the results returned from the *Cosmos* query functions dynamically changes as new content gets created.

Criteria for Checking Attribution: Flickr is still using the older CC 2.0 recommendation. Therefore, Flickr users do not have that much flexibility in specifying their own *attributionURL* or the *attributionName* as specified in ccREL. However, it is considered good practice to give attribution by linking to the Flickr user profile or by giving the Flickr user name (which could be interpreted as the *attributionURL* and the *attributionName* respectively), or at least, point

---

[3] Implemented using the Technorati API detailed at `http://technorati.com/developers/api`

[4] We expected that the use of the *Technorati Authority Rank* would introduce a bias in our sample. This is because the top Web pages from the Technorati blog indexer are probably well visited, hence more pressure on the Web page owners to fix errors in attribution. However, the results proved otherwise.

[5] According to `http://www.flickr.com/services/api/misc.urls.html`, all Flickr images have that particular URI pattern.

to the original source of the image [12]. Therefore, the criteria for checking attribution consist of looking for the *attributionURL* or the *attributionName* or any *source citations* within a reasonable level of scoping from where the image is embedded in the Document Object Model (DOM) of the corresponding Web page.

3.2 Results

The results from the 3 trials are given in Fig 1. These results have misattribution and non-attribution rates ranging from 78% to 94% signaling that there is a strong need to promote license or policy awareness among reusers of content. The entire result set includes the total number of Web pages tested, number of images in all of those Web pages, number of properly attributed images, number of misattributed or non-attributed images, and the number of instances that led to an error due to parsing errors resulting from bad HTML. Using these values, the percentages of misattribution and non-attribution for each sample were calculated.

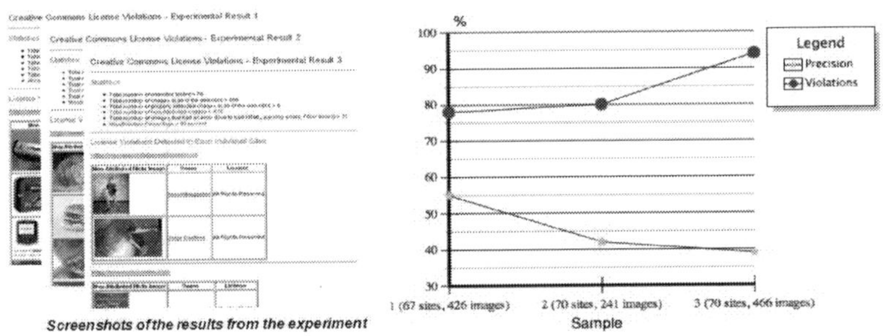

Fig. 1. **Left:** Screenshots of the results from the experiment (Refer to *http://dig.csail.mit.edu/2008/WSRI-Exchange/results* for more information). **Right:** Attribution violations rate and Precision obtained after correcting for self-attribution.

3.3 Issues and Limitations of the Experiment

As in any experiment, there are several issues and limitations in this experiment.

Results include cases where the users have not attributed themselves: For example, consider the case where a user uploads her photos on Flickr, and uses those photos in one of her own Web pages without attributing herself. As the copyright holder of the work, she can do pretty much whatever she wants with those, even though the CC BY license deed states: *"If You Distribute you must keep intact all copyright notices for the Work and provide (i) the name of the Original Author..."* [4]. However if she fails to include the license notice and the attribution to herself, she may be setting a precedent for the violation of her own rights in the long run. From the point of view of the experiment, it was difficult to infer the page owner from the data presented in the page.

Even if that was possible, it is hard to make a correlation between the Flickr photo owner and the page owner. However, we manually inspected the samples to see whether the misattributed images were actually from the user or not, and flagged the ones which are definitely from the original user as *false positives* in the results set to obtain the precision rate. After this correction, we found the precision rate of the experiment to be between 55% to 40%.

Low adoption of ccREL and Attribution Scoping: We found out that a majority of the Web pages examined in this experiment have not used ccREL in marking up attribution. Therefore, we used a heuristic to check for the existence of attribution in the pages used in the trials. This heuristic includes the *attributionName* constructed from the Flickr user name, or the *attributionURL* constructed from the Flickr user profile URI, or the original source document's URI. We expected to find the attribution information in the parent of the DOM element or in one of the neighboring DOM elements. This can be visually correlated to finding the attribution information immediately after the content that is being reused. However, since there is no strict definition from CC as to how attribution should be scoped, someone could also attribute the original content creator somewhere else in the document. However, considering that it is possible the user intended to include more than one image from the same original content creator, and by mistake failed to attribute some images, while correctly attributing all the others, we only checked attribution information within the neighboring DOM elements, and not at the document level.

Blog Aggregators such as Tumble-logs cutting down the text and favoring short form, mixed media posts over long editorial posts: Use of such blog aggregators (for example `tumblr.com`) is another problem in getting an accurate assessment of attribution license violations. For example, in a blog post where a photo was reused, the original owner of the photograph may have been duly attributed. But when the tumble-log pulls in the feed from that post in the original Web page and presents the aggregated content, the attribution details may be left out. This problem is difficult to circumvent because there is no standard that specifies how aggregation should take license and attribution details into consideration.

4 Tools to Enable Policy Awareness

As a proof of concept we developed two tools that can be used to enable policy awareness when reusing *images* on the Web. Even though both tools are currently limited to image reuse, it can be easily extended to support other types of media.

4.1 Attribution License Violations Validator for Flickr Images

When someone aggregates content from many different sources, it is inevitable that some attribution details may be accidentally forgotten. The Attribution License Violations Validator is designed to check whether the user has properly cited the source by giving the due attribution to the original content creator. In order to make sure that no CC license terms of the user are violated, the author can run the CC License Violations Validator and see if some sources have been left out or whether some have been misattributed.

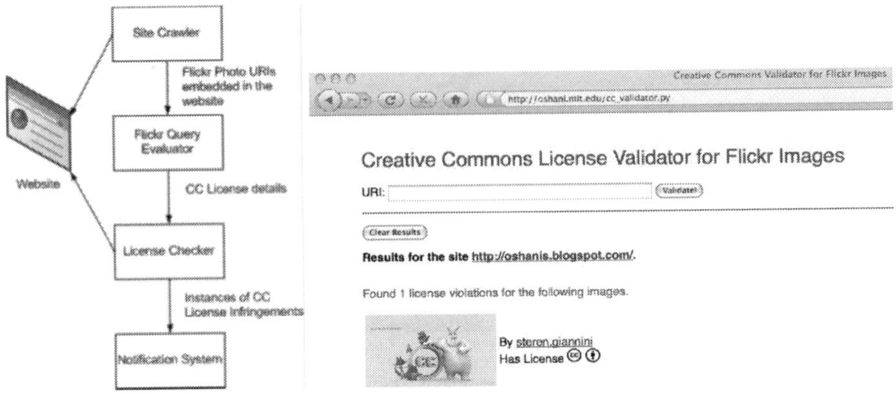

Fig. 2. Left: The Design of the Validator. **Right:** Output from the Validator showing the image that was not attributed properly, who the image belongs to and what license it is under.

Design and Implementation: The tool has four major components as shown in the left half of Fig 2. Once the user gives the URI where the composite work can be found, the site crawler will search for all the links embedded in the given Web page and extract any embedded Flickr photos. From each of these Flickr photo URIs, it is possible to glean the Flickr photo id. Using this photo id, all the information related to the photo is obtained by calling several methods in the Flickr API. This information includes the original creator's Flickr user account id, name, and CC license information pertaining to the photo, etc. Based on the license information of the Flickr photo, the tool checks for the attribution information that can be either the *attributionName*, *attributionURL*, source URI or any combination of those within a reasonable scoping in the containing DOM element in which the image was embedded. The 'reasonable scoping' in this case, is taken to be within the parent or the sibling nodes of the element that has the embedded image. If such information is missing, the user is presented with the details of the original content creator's name, the image along with its URI, and the license it is under, enabling the user to compose the XHTML required to properly attribute the sources used.

Challenges and Limitations: The license violations detection can only work if the image URI is actually linked from the Flickr site. Therefore if a user wants to cheat, she can easily do so by changing the image URI by uploading it to another Web space. Another complication is that a Flickr user can upload and assign CC licenses regardless of that user having the actual rights to do so. In other words, if someone uploads a copyrighted photo from *Getty Images* and assigns a CC license on Flickr, and an innocent user downloads and uses this photo, then that user will be violating the copyright law inadvertently. Therefore, we need to have some capability to track provenance of image data, and be able to identify whether a particular image has been used elsewhere in a manner that violates the original license terms. One of the major assumptions we have made in developing this tool is that attribution is specified within the parent node or the sibling nodes of the containing image element. Otherwise we classify it an instance of non-attribution. This assumption works in practice and appears to be the most logical thing to do. However, since there is no standard agreement as to what the correct scoping for attribution is, this assumption can lead to a wrong validation result. The solution to this problem is two-fold. (1) CC should give a guideline as to what the correct scoping of attribution should be relative to the content that is attributed. (2) Flickr (or any other such service) should expose the license metadata as RDF, instead of providing an API to query with. Exposing license metadata as RDF is preferred as it enables data interoperability and relieves the tool authors from having to write data wrappers for each service.

4.2 Semantic Clipboard

The Semantic Clipboard is a Firefox Web browser based tool integrated as part of the Tabulator, a linked data browser that can be installed as a Firefox extension [28]. The primary goal of this tool is to let users reuse content with minimal effort.

Design and Implementation: The design of the Semantic Clipboard is given in Fig 3. The tool uses the 'RDFa Extractor' to glean the Creative Commons license information expressed in RDFa from the HTML page the user browses. The 'UI Enhancer' implements several menu options in the Firefox browser to select licensed images with the proper intention. The available options are given in Fig 3. For example, if a user want to see images that can be used for 'Commercial Purposes', she can select the corresponding menu item. Then the images that do not have the CC-NC clause (Creative Commons Non Commercial use) will be highlighted with an overlay on the image. The 'Attribution XHTML Constructor' is called when the user issues a copy instruction on a particular image by right-clicking on the image and selecting the context menu option 'Copy Image with License' as shown in the right half of Fig 3. Based on the license information for that particular image, the attribution XHTML snippet is constructed as specified by Creative Commons, and copied to the system clipboard. Currently two data flavors are supported: ASCII text and HTML. Therefore if the target application accepts HTML such as in a rich text editor, the source text (with the angle brackets) will not be displayed.

Challenges and Limitations: One of the hazards of combining multiple data sources is that incompatible licenses can get mixed up creating a license that basically freezes the creative process. Take for example a Non-Commercial (NC)

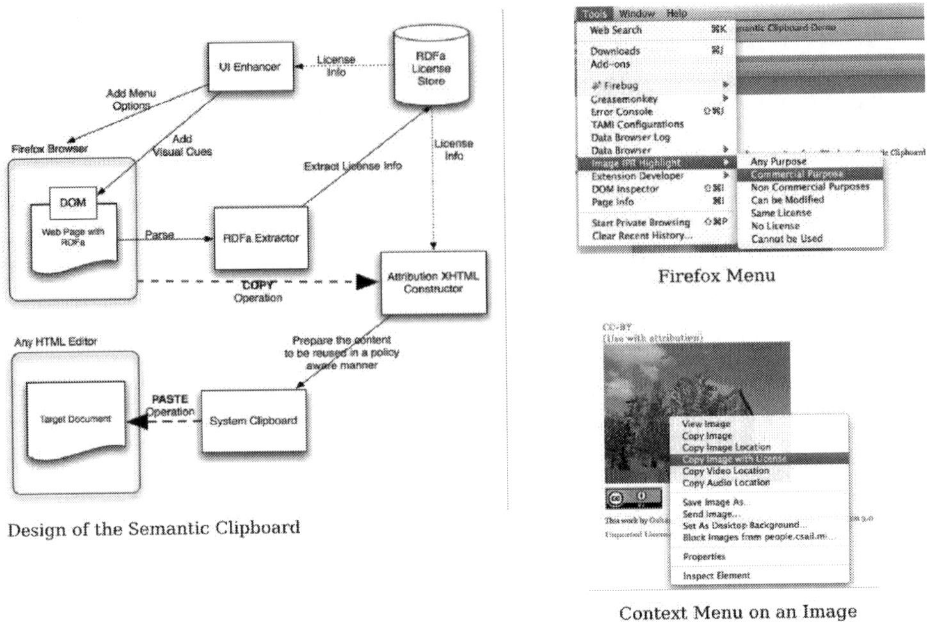

Fig. 3. Left: Semantic Clipboard Architecture. **Right:** Semantic Clipboard User Interface.

license that gets mixed with a Share-Alike (SA) license. An SA license requires that the resulting product be shared under exactly the same conditions as the component product under SA. The resulting license in our scenario becomes NC-SA. But while the result satisfies the first license by also being NC, it fails the second license by not being *only* SA. We cannot simply ignore the NC clause and give the resulting work only the SA license because somebody else might use the resulting derivative work which does not have the NC clause for some commercial use violating the rights of the original creator who composed the NC component. The Semantic Clipboard does not handle such license conflicts.

5 Related Work

Reuse detection is important in domains such as plagiarism detection and even in biological sequence mining. Significant research has been carried out to detect reuse of text. This includes information retrieval techniques as mentioned in [17,25], where the document is treated as a sequence of symbols and substring based fingerprints are extracted from the document to determine repetitive patterns.

The CC License ***Syntax*** Validation Service [13] can be used to parse documents for embedded licenses in RDFa. After parsing the document, this service gives a list of licensed objects and each of their license authorship, version, jurisdiction, whether the license has been superseded or deprecated and whether the work is allowed in free cultural works, etc. However, it does not give the information as to whom the attribution should be given when reusing these license objects, like in the attribution license violations validator we have developed. In addition to that, CC has put much focus on coming up with ways to enable tool builders to use the CC licenses very effectively. For example, the *live box* on the *License Deed Page* as shown in Fig 4 suggests how to attribute a particular work. This is created when a CC license hyperlink that has the *attributionName* and the *attributionURL* properties to the *License Deed Page* is dereferenced. There are also several license aware Mozilla Firefox extensions developed by the CC. MozCC [19] is one such tool. It provides a specialized interface for displaying CC licenses, where the user receives visual cues when a page with RDFa metadata is encountered. This includes the display of specific CC branded icons in the browser status bar when the metadata indicates the presence of a CC license. However, this software does not offer the capability to copy the license attribution XHTML as in the Semantic Clipboard that we have developed.

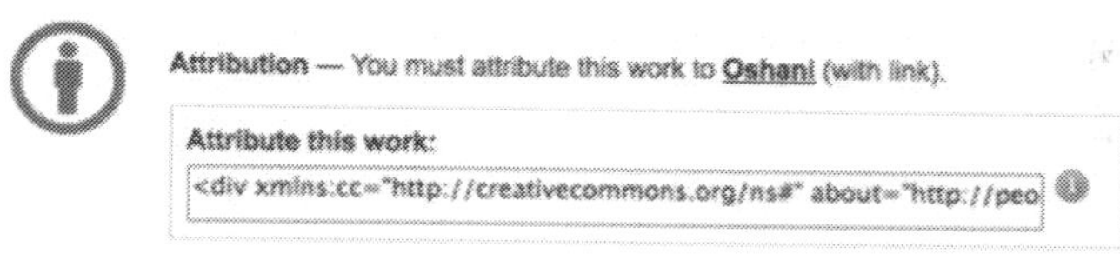

Fig. 4. CC Deed Page Displaying the Attribution XHTML

The Semantic Clipboard was actually inspired from the work done on 'XHTML Documents with Inline, Policy-Aware Provenance' [15] by Harvey Jones. Jones developed a document format that can represent information about the sources of its content, a method of excerpting from these documents that allows programs to trace the excerpt back to the source, a CC reasoning engine which calculates the appropriate license for the composite document, and a bookmarklet that uses all these components to recommend permissible licenses. But this tool requires all the source documents to be annotated with a special document fragment ontology, and the *Paste Operation* is limited to inserting copied XHTML in the top level of the document only, i.e. it does not allow copying inside existing document fragments. The Semantic Clipboard addresses these issues by eliminating the reliance of an external document fragment ontology and utilizing the operating system's clipboard to paste the image with the associated license metadata in XHTML. The only requirement for the Semantic Clipboard to work is that the license information about the work must be expressed in RDFa in the source documents.

There are several tools that can be used to automatically embed the license metadata from Flickr. Applications such as ThinkFree, a Web based commercial office suite [27], and the open source counterpart of it, the "Flickr image reuse for OpenOffice.org" [10] are examples of such applications. These applications allow the user to directly pick an image from the Flickr Web site and automatically inject the license metadata with it into a document in the corresponding *office suite.* A severe limitation of this approach is that they only support Flickr images. The Semantic Clipboard can be used to copy any image to any target document along with the license as long as the license metadata is expressed in RDFa.

Attributor [1], a commercial application, claims to continuously monitor the Web for its customers' photos, videos, documents and let them know when their content has been used elsewhere on the Web. It then offers to send notices to the offending Web sites notifying link requests, offers for license, requests for removal or shares of the advertisement revenue from the offending pages. Another commercial application called PicScout [21] claims that it is currently responsible for detecting over 90% of all online image infringement detections. They also claim to provide the subscribers of their service with a view into where and how their images are being used online. The problem with these services is that they penalize the infringers, rather than encouraging them to do the right thing upfront [16]. In addition to that, since their implementations are based on bots that crawl the Web in search of infringements, these services take up valuable Internet bandwidth [30]. Also, these services are not free, which bars many content creators who wish to use such services to find license violations of their content from using the service.

6 Future Work

Currently, the images that are copied with their metadata to the Semantic Clipboard are overwritten when some new content is copied to the clipboard. In other words, the tool only supports copying of one image at a time. But it would be useful to have a persistent data storage to register images or any other Web media along with their license metadata, index them, make persistent across browser sessions, and use the copied content whenever the user needs it in a license-compliant manner. One other main drawback of the Semantic Clipboard is that it is Firefox browser-dependent. Developing an Opera Widget, a Chrome Extension, a Safari Plugin, an Internet Explorer Content Extension or completely making this tool browser independent seems to be a viable future direction of the project.

Also, the tools we have developed only works if every image found on the page has it's own license. Possible extensions would be to have higher level of granularity to determine the license of an image when it does not have a license of it's own, but is contained within a page that has a license or is a member of a set of images (e.g. a photo album) that has a license. The Protocol for Web Description Resources (POWDER) [22], a mechanism that allows the provision of descriptions for groups of online resources, seems like a viable method to making the license descriptions about the resources explicit. Tool builders can then rely on the POWDER descriptions to help users to make appropriate content reuse decisions.

It would be interesting to measure how user behavior changes with the introduction of tools such as the License Violations Validator and the Semantic Clipboard. A measurement of the change in the level of license awareness would be an important metric in determining the success of these tools. Therefore, we plan to perform a controlled user study in the future.

We have only explored one domain of content, specifically image reuse on the Web. However, there are billions of videos uploaded on YouTube, and potentially countless number of documents on the Web, which have various types of licenses applied. While organizations such as Mobile Picture Association of America (MPAA), Recording Industry Association of America (RIAA) and other such organizations are working towards preserving the rights of the works of their artists on YouTube, other video and audio sharing sites and peer-to-peer file sharing networks, there are no viable alternatives for ordinary users who intend to protect their rights using CC. Thus a solution of this nature which detects CC license violations based on the metadata of other types of free-floating Web media will be very useful.

The requirement for attribution is a form of social compensation for reusing one's work. While mentioning one's name, homepage or the WebID when attributing draws attention to an individual, other forms of *'attention mechanisms'* can also be implemented. For example, a content creator can obligate the users of her works to give monetary compensation or require that they include certain ad links in the attribution XHTML or give attribution in an entirely arbitrary manner. These extra license conditions can be specified using the *cc:morePermissions* property. Tools can be built to interpret these conditions and give credit to the original creator as requested.

We envision that the same principle used for checking *attribution* license violations could be used for checking other types of license violations. Detecting whether an image has been used for any commercial use would be of much interest to content creators, especially if the second use of the image decreases the monetary value of the original image. The CC deed for Non Commercial (NC) use specifies that a license including the NC term may be used by anyone for any purpose that is not *"primarily intended for or directed towards commercial advantage or private monetary compensation"*. However, this definition can be vague in certain circumstances. Take, for example, the case where someone uses a CC-BY-NC licensed image in her personal blog properly attributing the original content creator. The blog is presumably for non commercial use, and since she has given proper attribution, it appears that no license violation has occurred. However there might be advertisements in the page that are generated as a direct result of the embedded image. Our user might or might not actually generate revenue out of these advertisements. But if she does, it could be interpreted as a 'private monetary compensation'. Hence we believe that the perception as to what constitutes a 'Commercial Use' is very subjective.

CC recently conducted an online user survey to gather general opinions as to what people perceive a 'Commercial Use' is [6]. An important finding from this survey is that 37% of the users who make money from their works do so indirectly through advertisements on their Web pages. Therefore, it seems that there aren't any clear cut definitions of a 'Non Commercial Use' yet to find out violations and gather experimental results. But, if the definition of *non commercial use* becomes clearer and much more objective, a validator can be implemented to check for such violations as well. It would also be interesting to check for share-alike license violations. These violations happen when a conflicting license is given when the content is reused. The solution, therefore, is to check the RDFa in both the original page and in the page where the image was embedded to see if the latter is the same as the original CC license.

7 Conclusion

As our license violations experiment indicated, there is a strong lack of awareness of licensing terms among content reusers. This raises the question as to whether machine readable licenses are actually working. Perhaps more effort is needed to bring these technologies to the masses, and more tools are needed to bridge the gap between the license-aware and the license-unaware. There should also be methods to find out license violations when users are not cooperative.

An important research question that stems from this work is the method of provenance preservation of content on the Web. We have trivially assumed URIs to be the provenance preservation mechanism when developing the tools described. However, it would be an interesting challenge to track provenance based on the content itself, without having to rely on a unique identifier such as a URI. This would enable us to find out license violators, in addition to validating one's own work for any violations. Also, programmatically determining whether a particular reuse of material is allowable or not is subjective, especially since some of the laws and standards have been quite ambiguous in defining these terms.

In general, social constraints are functions of any part of the blossoming Social Web we are experiencing today. As we are living in an era of increasing user generated content, these constraints can be used to communicate the acceptable uses of such content. We need tools, techniques and standards that strike an appropriate balance between the rights of the originator and the power of reuse. The rights of the originator can be preserved by expressing what constitutes appropriate uses and restrictions using a rights expression language. These rights will be both machine and human readable. Reuse can be simplified by providing the necessary tools that leverage these machine readable rights to make the users more aware of the license options available and ensure that the user be license or policy compliant. Such techniques can be incorporated in existing content publishing platforms or validators or even Web servers to make the process seamless. This paper has demonstrated several tools that enable the development of such policy-aware systems, and we hope that these will stimulate research in this area in the future.

Acknowledgements

Parts of this work was done while the first author was undergoing the 'Networks for Web Science Research Exchange' program under the guidance of Nigel

Shadbolt at University of Southampton, UK. Special thanks for his advice all throughout the project. In addition, the authors wish to thank their colleagues, Danny Weitzner, Hal Abelson, Gerry Sussman, and other members at DIG for their contribution to the ideas expressed in this paper.

This work was supported by NSF Cybertrust award CT-M: 0831442, IARPA award FA8750-07-2-0031, and UK Engineering and Physical Sciences Research Council (EPSRC) grant EP/F013604/.

References

1. Attributor - Subscription based web monitoring platform for content reuse detection, http://www.attributor.com
2. blip.tv - Hosting, distribution and advertising platform for creators of web shows
3. Bonatti, P.A., Duma, C., Fuchs, N., Nejdl, W., Olmedilla, D., Peer, J., Shahmehri, N.: Semantic web policies - a discussion of requirements and research issues. In: Sure, Y., Domingue, J. (eds.) ESWC 2006. LNCS, vol. 4011, pp. 712–724. Springer, Heidelberg (2006)
4. Creative Commons BY 3.0 Unported Legal Code, http://creativecommons.org/licenses/by/3.0/legalcode
5. Creative Commons Customized Search in Google, http://creativecommons.org/press-releases/entry/5692
6. Creative Commons Noncommercial study interim report,http://mirrors.creativecommons.org/nc-study/NC-Use-Study-Interim-Report-20090501.pdf
7. Exchangeable Image File Format, http://www.exif.org/specifications.html
8. Feigenbaum, J., Freedman, M.J., Sander, T., Shostack, A.: Privacy engineering for digital rights management systems. In: Sander, T. (ed.) DRM 2001. LNCS, vol. 2320, pp. 76–105. Springer, Heidelberg (2002)
9. Flickr API, http://www.flickr.com/services/api
10. Flickr image reuse for openoffice.org, http://wiki.creativecommons.org/Flickr-Image-Re-Use-for-OpenOffice.org
11. Abelson, H., Adida, B., Linksvayer, M., Yergler, N.: ccREL: The Creative Commons Rights Expression Language. Creative Commons Wiki (2008)
12. How to attribute Flickr images, http://www.squidoo.com/cc-flickr/#module12311035
13. Dworak, H.: Creative Commons License Validation Service, http://validator.creativecommons.org
14. International Press Telecommunications Council Photo Metadata Format, http://www.iptc.org/IPTC4XMP
15. Jones, H.C.: Xhtml documents with inline, policy-aware provenance. Master's thesis, Massachusetts Institute of Technology (May 2007)
16. Doctor, K.: Blog Entry on Attributor Fair Syndication Consortium Completes Newspaper Trifecta, http://www.contentbridges.com/2009/04/attributor-ad-push-on-piracy-completes-newspaper-trifecta.html
17. Kim, J.W., Candan, K.S., Tatemura, J.: Efficient overlap and content reuse detection in blogs and online news articles. In: 18th International World Wide Web Conference WWW 2009 (April 2009)
18. Kishor, P., Seneviratne, O.: Public policy: Mashing-up technology and law. In: Mashing-up Culture: The Rise of User-generated Content, COUNTER workshop, Uppsala University (May 2009)

19. MozCC - Firefox extension to discover Creative Commons licenses, http://wiki.creativecommons.org/MozCC
20. OWL Music Search, http://www.owlmusicsearch.com
21. picScout - Image tracker for stock photography agencies and professional photographers, http://www.picscout.com
22. Protocol for Web Description Resources (POWDER), http://www.w3.org/2007/powder
23. RDF, Resource Description Framework, http://www.w3.org/TR/rdf-syntax-grammar
24. RDFa, Resource Description Framework in Attributes, http://www.w3.org/2006/07/SWD/RDFa/syntax
25. Shivakumar, N., Garcia-Molina, H.: Scam: A copy detection mechanism for digital documents. In: Second Annual Conference on the Theory and Practice of Digital Libraries (1995)
26. SpinXpress - Collaborative media production platform, http://spinxpress.com
27. Think Free - Java based web office suite, http://www.thinkfree.com
28. Berners-Lee, T., Hollenbach, J., Lu, K., Presbrey, J., Prud'ommeaux, E., Schraefel, M.C.: Tabulator Redux: Browing and Writing Linked Data. In: Linked Data on the Web Workshop at WWW (2008)
29. Weitzner, D.J., Abelson, H., Berners-Lee, T., Feigenbaum, J., Hendler, J., Sussman, G.J.: Information accountability. Communications of the ACM (June 2008)
30. Faulkner, W.: Tales From The IT Side: PicScout, Getty Images and Goodbye iStockPhoto..!, http://williamfaulkner.co.uk/wordpress/2007/09/picscout-getty-images-and-goodbye-istockphoto
31. XMP - Extensible Metadata Platform, http://www.adobe.com/products/xmp/index.html
32. Yahoo Creative Commons Search, http://search.yahoo.com/cc

Exploiting Partial Information in Taxonomy Construction

Rob Shearer and Ian Horrocks

Oxford University Computing Laboratory, Oxford, UK

Abstract. One of the core services provided by OWL reasoners is *classification*: the discovery of all subclass relationships between class names occurring in an ontology. Discovering these relations can be computationally expensive, particularly if individual subsumption tests are costly or if the number of class names is large. We present a classification algorithm which exploits partial information about subclass relationships to reduce both the number of individual tests and the cost of working with large ontologies. We also describe techniques for extracting such partial information from existing reasoners. Empirical results from a prototypical implementation demonstrate substantial performance improvements compared to existing algorithms and implementations.

1 Introduction

Among the most important relationships between the class names occurring in an ontology is *subclassing*: a class B is a subclass of a class A if and only if every member of B must also be a member of A. One of the core services provided by reasoners for ontology languages is *classification*, the discovery of all such subclass relationships. In fact, classification is the primary (and often the only) reasoning service exposed by ontology engineering tools. The Protégé-OWL editor, for example, includes a "Reasoning" button which performs classification. The resulting hierarchy of subclass relationships is used to organize classes within all aspects of Protégé's interface, and the subclass relationships which arise as implicit consequences of an ontology are the primary mechanism authors use to check that the axioms they write are consistent with their intuitions about the structure of the domain. Finally, other reasoning services, such as explanation and query answering, typically exploit a cached version of the classification results; classification is thus usually the first task performed by a reasoner.

For some less expressive ontology languages, such as the OWL 2 EL profile,[1] it may be possible to derive all subclass relationships in a single computation [Baader *et al.*, 2005]. For OWL (2) DL, however, it is in general necessary to "deduce" the set of all such relationships by performing a number of *subsumption tests*; each such test checks whether or not there is a subclass relationship between two particular ontology classes.

In this paper, we present a novel algorithm which can greatly reduce the number of subsumption tests needed to classify a set of ontology classes. Our algorithm is also able to exploit *partial information* about subclass relationships—for example, the set of

[1] `http://www.w3.org/TR/owl2-profiles/`

A. Bernstein et al. (Eds.): ISWC 2009, LNCS 5823, pp. 569–584, 2009.

subclass relationships which are explicitly stated in the ontology—to further reduce the number of tests; our approach thus generalizes a wide range of commonly-implemented optimizations. We further describe how a great deal of such partial information can be gathered from the data generated by reasoning systems when performing individual subsumption or consistency tests. Finally, we present an empirical evaluation that demonstrates the significant advantages of our techniques over the approaches implemented in existing reasoners, such as Pellet and FaCT++.

2 Overview

For a set of n classes, there are a total of n^2 possible subclass relationships, thus a naïve representation of classification results can require large amounts of storage. Most systems exploit the transitivity of subclassing and store only the "most specific" superclasses and "most general" subclasses of each class; we call such a compressed representation a *taxonomy*. A taxonomy lends itself to representation as a graph or tree, often called a *class hierarchy*.

It is clear that a set of n class names can be classified by simply performing all n^2 individual subsumption tests, but for the tree-shaped class hierarchies typically found in realistic ontologies much better results can be achieved using algorithms that construct the taxonomy incrementally. Class names are inserted into the taxonomy one by one, and the correct location for each is found by traversing the partially-constructed hierarchy, performing subsumption tests as each node of the graph is visited.

This kind of algorithm suffers from two main difficulties. First, individual subsumption tests can be computationally expensive—for some complex ontologies, even state-of-the-art reasoners may take a long time to perform a single test. Second, even when subsumption tests themselves are cheap, an ontology containing a very large number of class names will obviously result in a very large taxonomy, and repeatedly traversing this structure can be costly. This latter problem is particularly acute for the relatively flat (i.e., broad and shallow) tree-shaped hierarchies often found in large biomedical ontologies. In general, subsumption tests must be performed between every pair of class names with the same direct superclass(es), thus broad hierarchies require many such tests. These two difficulties clearly interact: large numbers of class names require large numbers of subsumption tests, each of which can be expensive.

The first difficulty is usually addressed by using an optimized construction that tries to minimize the number of subsumption tests performed in order to construct the taxonomy. Most implemented systems use an "enhanced traversal" algorithm due to Ellis [1991] and to Baader *et al.* [1994] which adds class names to the taxonomy one at a time using a two-phase strategy. In the first phase, the most specific superclasses of a class C are found using a top-down breadth-first search of the partially-constructed taxonomy. In this phase, subtrees of non-superclasses of C are not traversed, which significantly reduces the number of tests performed. The second phase finds the most general subclasses of C using a bottom-up search in a similar way. The algorithm exploits the structure of the ontology to identify "obvious" superclasses (so-called *told-subsumers*) of each class, and uses this information in a heuristic that chooses the order in which classes are added, the goal being to construct the taxonomy top-down; it also exploits information from the top-down search in order to prune the bottom-up search.[2]

The second difficulty can be addressed by optimizations that try to identify a subset of the class names for which complete information about the subclass relationships can be deduced without performing any individual subsumption tests. This can be achieved, e.g., by identifying *completely-defined* classes [Tsarkov *et al.*, 2007]—those between which only structurally-obvious subclass relationships hold. Having constructed part of the taxonomy using such a technique, the remaining class names can be added using the standard enhanced traversal algorithm.

In Section 3 we present a new classification algorithm that generalizes and refines the above techniques. Our approach is based on maintaining, and incrementally extending, two sorts of information: a set of pairs of classes A and B such that we know that A is a subclass of B (the *known subsumptions*), and a set of pairs of classes such that we know that A is not a subclass of B (the *known non-subsumptions*). The key insight is that information from these two sets can often be combined to derive new information. For example, if we know that A is a subclass of B, and that A is *not* a subclass of C, then we can conclude that B is not a subclass of C. In section 3.1 we show how to derive the largest possible set of such inferences.

Our classification algorithm is straightforward: at each stage we pick a pair of classes A and B which does not appear in either the set of known subsumptions or the set of known non-subsumptions, perform a subsumption test between the two, and use the result of the test to further extend the sets of known subsumptions and non-subsumptions. Each such test adds at least one pair to one of the sets (i.e. A and B themselves). Eventually, every possible pair of classes is present in one set or the other, and the set of known subsumptions thus contains all subsumptions between class names.

An important advantage of our algorithm is that the initial sets of known (non-)subsumptions may be empty, may include partial information about all classes, and/or may include complete information about some classes; in all cases the information is maximally exploited. Such information can be derived from a variety of sources, ranging from syntactic analysis of the ontology to existing classification results for a subset of class names from the ontology (e.g. independent modules or classified sub- or super-sets of the ontology) to data derived in the course of reasoning.

In Section 4 we show how the models constructed by (hyper)tableau-based reasoners in the course of subsumption and satisfiability testing can be used as sources of partial information about subclass relationships. For example, if a reasoner produces a model containing an individual which is a member of both the class A and the complement (negation) of the class B, then A is clearly not a subclass of B; more sophisticated analysis based on the *dependency tracking* structures typically maintained by tableau reasoners also allows detection of subclass relationships. The models generated by tableau reasoners are typically very rich sources for this type of information; in fact, for ontologies which do not result in nondeterminism, including all OWL 2 EL ontologies, the model constructed by a hypertableau-based reasoner when checking the satisfiability of

[2] Other optimizations can be used to decrease the cost of individual subsumption tests (see, e.g., [Tsarkov *et al.*, 2007]), but these techniques are largely orthogonal to classification optimizations.

a class A will contain sufficient information to determine if A is (not) a subclass of B for all class names B occurring in the ontology.

Our approach to partial-information derivation provides an efficient generalization of the told-subsumer and completely-defined optimizations, both of which derive partial information from structural analysis of the ontology. When the known (non-) subsumption information is incomplete, our algorithm incrementally computes additional (non-)subsumption relationships, and maximally exploits the resulting information to refine the sets of known and possible subsumers; this can be seen as a generalization of the search-pruning optimizations introduced by Baader *et al.*.

We have used a prototypical implementation of our new algorithm to compare its behavior with that of the classification algorithms implemented in state-of-the-art OWL reasoners. The comparison shows that our algorithm can dramatically reduce the number of subsumption tests performed when classifying an ontology. Moreover, in contrast to the completely-defined optimization, the behavior of our algorithm degrades gracefully as the gap between the sets of initially-known and possible subsumption relationships increases.

3 Deducing a Quasi-Ordering

We first introduce some notation and definitions that will be useful in what follows.

Given a set of elements $U = \{a, b, c, ...\}$, let R be a binary relation over U, i.e., a subset of $U \times U$. We say that there is a *path* from a to b in R if there exist elements $c_0, ..., c_n \in U$ such that $c_0 = a$, $c_n = b$, and $\langle c_i, c_{i+1}\rangle \in R$ for all $0 \leq i < n$. The *transitive closure* of R is the relation R^+ such that $\langle a, b\rangle \in R^+$ iff there is a path from a to b in R. The *transitive-reflexive closure* R^* of R is the transitive closure of the reflexive extension of R, i.e. $R^+ \cup \{\langle a, a\rangle \mid a \in U\}$.

A binary relation is a *quasi-ordering* if it is both reflexive and transitive. Clearly, the subsumption relation on a set of classes is a quasi-ordering. Note, however, that it is not a *partial-ordering*, because it is not antisymmetric: $C \sqsubseteq D$ and $D \sqsubseteq C$ does not imply that $C = D$.

The *restriction* of a relation R to a subset D of U is the relation $R[D] = R \cap (D \times D)$. All restrictions of a reflexive relation are reflexive, and all restrictions of a transitive relation are transitive; thus, a restriction of a quasi-ordering is itself a quasi-ordering. Further, if $R \subseteq S$ for relations R and S, then $R[D] \subseteq S[D]$ for all $D \subseteq U$.

Given a universe U, a quasi-ordering R over U and a finite set of elements $D \subseteq U$, we consider the problem of computing the restriction $R[D]$ via tests of the form $\langle a, b\rangle \in^? R$. If U is the set of (arbitrary) class expressions which can be represented in ontology language $\mathcal{L}$, R is the subsumption relation over U, and D is the set of class names occurring in an ontology $\mathcal{O}$ written in language $\mathcal{L}$, then computing $R[D]$ is equivalent to classifying $\mathcal{O}$, and the relevant tests are subsumption tests.

We assume that we begin with partial information about R: we are provided with a set $K = \{\langle a_0, b_0\rangle, ..., \langle a_m, b_m\rangle\}$ where $\langle a_i, b_i\rangle \in R$ for $0 \leq i \leq m$, and also with a set $K_{neg} = \{\langle c_0, d_0\rangle, ..., \langle c_n, d_n\rangle\}$ where $\langle c_i, d_i\rangle \notin R$ for $0 \leq i \leq n$. We call the set K the *known* portion of R. In this paper we do not operate on the set K_{neg} directly; our presentation instead refers to its complement $U \times U \setminus K_{neg}$, which we denote by

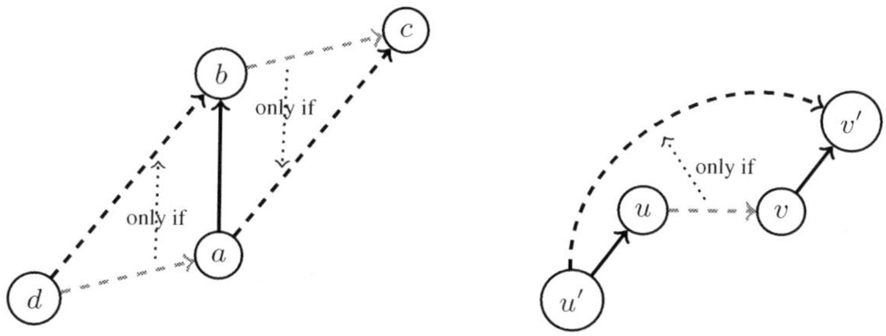

(a) Simple cases **(b)** General case

Fig. 1. Eliminating possible edges: if the solid edges are known to be in quasi-ordering R, then the gray edges can be in R only if the indicated dashed edges are in R

P and call the *possible* portion of R. It is thus the case that $K \subseteq R \subseteq P$. If no partial information is available, then $K = \emptyset$ and $P = U \times U$.

We can use the result of each test $\langle a, b \rangle \in^? R$ to further refine the bounds on R by either adding $\langle a, b \rangle$ to K or removing it from P; eventually $K[D] = R[D] = P[D]$. We next show, however, that the bounds on R can sometimes be refined without performing additional tests by combining information from K and P.

3.1 Maximizing Partial Information

The key to minimizing the number of explicit tests required to discover $R[D]$ is maximizing the information gained from K and P. To do so, we exploit the knowledge that R is a quasi-ordering. In this case, $K \subseteq R$ obviously implies that $K^* \subseteq R$, so we can use K^* to obtain a tighter lower bound on R. Less obvious is the fact that we can also obtain a tighter upper bound on R by identifying tuples in P which are not consistent with K and the transitivity of R.

For example, consider the case shown in Figure 1(a). If we know that b is a successor of a in R (i.e., $\langle a, b \rangle \in K$), then an element c can be a successor of b only if it is also a successor of a (if $\langle a, c \rangle \notin P$ then $\langle b, c \rangle \notin R$). Further, a can be a successor of an element d only if b is also a successor of d.

Both of these examples are special cases of the structure shown in Figure 1(b): if u is a successor of u' and v' is a successor of v, then an edge from u to v would form a path all the way from u' to v', requiring v' to be a successor of u'. Since R is reflexive we can choose $u' = u$ or $v = v'$ to see that v can be a successor of u only if v is a successor of u' and v' is also a successor of u. We use this to formalize a subset $\lfloor P \rfloor_K$ of P, and show that $\lfloor P \rfloor_K$ is the tightest possible upper bound on R.

Definition 1. *Let K and P denote two relations such that $K^* \subseteq P$. We define the* reduction $\lfloor P \rfloor_K$ of P due to K *as follows:*

$$\lfloor P \rfloor_K = P \cap \{\langle u, v \rangle \mid \forall u', v' : \{\langle u', u \rangle, \langle v, v' \rangle\} \subseteq K^* \rightarrow \langle u', v' \rangle \in P\}$$

Lemma 1. *Let K and P denote two relations such that $K^* \subseteq P$. (i) For all quasi-orders R such that $K \subseteq R \subseteq P$, it is the case that $R \subseteq \lfloor P \rfloor_K$. (ii) Let S be a proper subrelation of $\lfloor P \rfloor_K$. Then there exists a quasi-ordering R such that $K \subseteq R \subseteq P$ and $R \nsubseteq S$; i.e. $\lfloor P \rfloor_K$ is minimal.*

Proof. *(i)* Let $\langle u, v \rangle$ be a tuple in R. For every u', v' such that $\{\langle u', u \rangle, \langle v, v' \rangle\} \subseteq K^*$, $K^* \subseteq R$ implies that $\{\langle u', u \rangle, \langle v, v' \rangle\} \subseteq R$. Because R is transitive and $\langle u, v \rangle \in R$, it must also be the case that $\langle u', v' \rangle \in R$ and thus that $\langle u', v' \rangle \in P$. Consequently, $\langle u, v \rangle \in \lfloor P \rfloor_K$, so $R \subseteq \lfloor P \rfloor_K$.

(ii) Choose elements a and b such that $\langle a, b \rangle \in \lfloor P \rfloor_K$ but $\langle a, b \rangle \notin S$. Let R be the transitive-reflexive closure of the relation $K \cup \{\langle a, b \rangle\}$. Clearly $K \subseteq R$ and $R \nsubseteq S$. Let $\langle u, v \rangle$ be any tuple in R. There are three cases:

1. $\langle u, v \rangle = \langle a, b \rangle$. Then $\langle u, v \rangle \in P$ since $\langle a, b \rangle \in \lfloor P \rfloor_K$ and $\lfloor P \rfloor_K \in P$.
2. $\langle u, v \rangle \in K^+$. Then $\langle u, v \rangle \in P$ since $K^* \subseteq P$.
3. $\langle u, a \rangle \in K^*$ and $\langle b, v \rangle \in K^+$. Then $\langle u, v \rangle \in P$ since $\langle a, b \rangle \in \lfloor P \rfloor_K$.

For any tuple $\langle u, v \rangle \in R$, it is the case that $\langle u, v \rangle \in P$, thus $K \subseteq R \subseteq P$ and $R \nsubseteq S$. □

Note that $\lfloor P \rfloor_K$ itself is not necessarily transitive: given three elements a, b, and c and the relation $P = \{\langle a, a \rangle, \langle b, b \rangle, \langle c, c \rangle, \langle a, b \rangle, \langle b, c \rangle\}$, it is the case that $\lfloor P \rfloor_\emptyset = P$. Of course no transitive subrelation R of P contains both $\langle a, b \rangle$ and $\langle b, c \rangle$.

3.2 Taxonomy Construction and Searching

As described in Section 3.1, given relations K and P such that $K \subseteq R \subseteq P$ for some unknown quasi-ordering R, a tuple $\langle a, b \rangle$ is an element of R if $\langle a, b \rangle \in K^*$, and $\langle a, b \rangle$ is not an element of R if $\langle a, b \rangle \notin \lfloor P \rfloor_K$; the only "unknown" elements of R are the tuples in $\lfloor P \rfloor_K \setminus K^*$. Further, if $\langle a, b \rangle \in \lfloor P \rfloor_K \setminus K^*$, then a test of the form $\langle a, b \rangle \in^? R$ provides additional information which can be used to extend K or restrict P. This suggests the following simple procedure for deducing the restriction $R[D]$ of a quasi-ordering R to domain D:

COMPUTE-ORDERING(K, P, D)

while $K^*[D] \neq \lfloor P \rfloor_K[D]$
  do choose some $a, b \in D$ such that $\langle a, b \rangle \in \lfloor P \rfloor_K \setminus K^*$
    if $\langle a, b \rangle \in^? R$ **then** add $\langle a, b \rangle$ to K
      else remove $\langle a, b \rangle$ from P
return $K[D]$

Completely recomputing K^* and $\lfloor P \rfloor_K$ in each iteration of the above loop is clearly inefficient. Practical implementations would instead maintain both relations and update them as P and K change. Techniques for updating the transitive-reflexive closure of a relation are well-known; we provide below a naïve algorithm that, given $\lfloor P \rfloor_K$, $K' \supseteq K$ and $P' \subseteq P$, computes an updated relation $\lfloor P' \rfloor_{K'}$.

The algorithm exploits the technique for eliminating edges that was described in Section 3.1 and Figure 1(b): it removes a tuple $\langle u, v \rangle$ from the set of possible tuples

$\lfloor P \rfloor_K$ when adding it to the set of known tuples K' would imply, due to the transitivity of R, that some other tuple would be at the same time both known (i.e., in K') and not possible (i.e., not in $\lfloor P' \rfloor_{K'}$). This is done incrementally by considering tuples that have either become known (i.e., are in $K' \setminus K$) or been shown to be impossible (i.e., are in $\lfloor P \rfloor_K \setminus P'$).

First, $\lfloor P \rfloor_K$ is copied to $\lfloor P' \rfloor_{K'}$. Then, in lines 2–4, each tuple $\langle u', v' \rangle$ that has been shown to be impossible is considered and, if there are tuples $\langle u', u \rangle$ and $\langle v, v' \rangle$ in K', then $\langle u, v \rangle$ is clearly not possible either (it would imply that $\langle u', v' \rangle$ is not only possible but known) and so is removed from $\lfloor P \rfloor_K$. Next, in lines 5–13, each tuple $\langle x, y \rangle$ that has become known is considered. There are two possible cases: one where x, y correspond to u', u in Figure 1(b), and one where they correspond to v, v'. Lines 6–9 deal with the first case: if there are tuples $\langle u, v \rangle$ in $\lfloor P \rfloor_K$ and $\langle v, v' \rangle$ in K', but $\langle u', v' \rangle$ is not in P', then $\langle u, v \rangle$ is clearly not possible (it would imply that $\langle u', v' \rangle$ is not only possible but known) and so is removed from $\lfloor P \rfloor_K$. Lines 10–13 deal similarly with the second case: if there are tuples $\langle u, v \rangle$ in $\lfloor P \rfloor_K$ and $\langle u', u \rangle$ in K', but $\langle u', v' \rangle$ is not in P', then $\langle u, v \rangle$ is removed from $\lfloor P \rfloor_K$.

PRUNE-POSSIBLES($\lfloor P \rfloor_K, K, P', K'$)
1 $\lfloor P' \rfloor_{K'} \leftarrow \lfloor P \rfloor_K$
2 **for** each $\langle u', v' \rangle \in \lfloor P \rfloor_K \setminus P'$
3   **do for** each u, v such that $\langle u', u \rangle \in K'$ and $\langle v, v' \rangle \in K'$
4   **do** remove $\langle u, v \rangle$ from $\lfloor P' \rfloor_{K'}$
5 **for** each $\langle x, y \rangle \in K' \setminus K$
6   **do let** $u' \leftarrow x$ and $u \leftarrow y$
7   **for** each v such that $\langle u, v \rangle \in \lfloor P \rfloor_K$
8   **do if** there exists v' such that $\langle v, v' \rangle \in K'$ and $\langle u', v' \rangle \notin P'$
9   **then** remove $\langle u, v \rangle$ from $\lfloor P' \rfloor_{K'}$
10   **let** $v \leftarrow x$ and $v' \leftarrow y$
11   **do for** each u such that $\langle u, v \rangle \in \lfloor P \rfloor_K$
12   **do if** there exists u' such that $\langle u', u \rangle \in K'$ and $\langle u', v' \rangle \notin P'$
13   **then** remove $\langle u, v \rangle$ from $\lfloor P' \rfloor_{K'}$
14   **return** $\lfloor P' \rfloor_{K'}$

In the case where no information about the quasi-ordering $R[D]$ is available other than K and P, the COMPUTE-ORDERING procedure performs well. In many cases, however, some general properties of $R[D]$ can be assumed. In the case where R represents subsumption relationships between class expressions, for example, $R[D]$ is typically much smaller than $D \times D$ (i.e., relatively few pairs of class names are in a subsumption relationship). In such cases, it is beneficial to use heuristics that exploit the (assumed) properties of $R[D]$ when choosing a and b in line 2 of the above procedure.

We summarize below a variant of COMPUTE-ORDERING which performs well when the restriction to be computed is treelike in structure and little information about the ordering is available in advance. This procedure is designed to perform individual tests in an order similar to the enhanced traversal algorithm; however, it minimizes the number of individual tests performed by maximally exploiting partial information.

The algorithm chooses an element of $a \in D$ for which complete information about $R[D]$ is not yet known. It identifies the subset $V^{\uparrow} \subseteq D$ of elements b for which $\langle a, b\rangle \in R$, and the subset $V^{\downarrow} \subseteq D$ of elements b for which $\langle b, a\rangle \in R$, updating K and P accordingly. In order to compute these sets efficiently, we make use of the subroutines SUCCESSORS and PREDECESSORS, which perform the actual tests. The SUCCESSORS and PREDECESSORS functions are derived from the enhanced traversal algorithm: they perform a breadth-first search of the *transitive reduction* $K^{\diamond}$ of the known subsumptions K—the smallest relation whose transitive closure is K^*. In order to avoid the cost of repeated traversals of $K^{\diamond}$, we restrict the searches to, respectively, the possible successors and predecessors of a. We omit the details of these search routines for the sake of brevity.

COMPUTE-ORDERING-2(K, P, D)

1 **while** $K^*[D] \neq \lfloor P \rfloor_K[D]$
2   **do** choose some $a, x \in D$ s.t. $\langle a, x\rangle \in \lfloor P \rfloor_K \setminus K^*$ or $\langle x, a\rangle \in \lfloor P \rfloor_K \setminus K^*$
3   let B be the possible successors of a, i.e. $D \cap \{b \mid \langle a, b\rangle \in \lfloor P \rfloor_K \setminus K^*\}$
4   **if** $B \neq \emptyset$ **then** $V^{\uparrow} \leftarrow$ SUCCESSORS$(a, K^{\diamond}[B])$
5   add $\langle a, b\rangle$ to K for every element b of $V^{\uparrow}$
6   remove $\langle a, b\rangle$ from P for every element b of $B \setminus V^{\uparrow}$
7   let B be the possible predecessors of a, i.e. $D \cap \{b \mid \langle b, a\rangle \in \lfloor P \rfloor_K \setminus K^*\}$
8   **if** $B \neq \emptyset$ **then** $V^{\downarrow} \leftarrow$ PREDECESSORS$(a, K^{\diamond}[B])$
9   add $\langle b, a\rangle$ to K for every element b of $V^{\downarrow}$
10   remove $\langle b, a\rangle$ from P for every element b of $B \setminus V^{\downarrow}$
11 **return** $K[D]$

3.3 Example

Consider the process of using COMPUTE-ORDERING-2 to discover the subsumption relation $\{\langle a, a\rangle, \langle b, a\rangle, \langle b, b\rangle, \langle c, a\rangle, \langle c, c\rangle, \langle d, a\rangle, \langle d, b\rangle, \langle d, d\rangle\}$ with no initial partial information available. We initialize K to $\{\langle a, a\rangle\langle b, b\rangle, \langle c, c\rangle, \langle d, d\rangle\}$ and P to $\{a, b, c, d\} \times \{a, b, c, d\}$; this situation is shown in the left diagram of Figure 2. Each element appears in a tuple occurring in $K^* \setminus \lfloor P \rfloor_K$, so on the first execution of line 2 of COMPUTE-ORDERING-2 we are free to choose any element; assume that we choose d. Then SUCCESSORS performs the tests $d \sqsubseteq^? a$, $d \sqsubseteq^? b$, and $d \sqsubseteq^? c$ (discovering that a and b are the only successors of d), we add $\langle a, d\rangle$ and $\langle b, d\rangle$ to K, and we remove $\langle c, d\rangle$ from P. PREDECESSORS performs the tests $a \sqsubseteq^? d$, $b \sqsubseteq^? d$, and $c \sqsubseteq^? d$ (discovering that d has no predecessors), and we remove each of $\langle a, d\rangle$, $\langle b, d\rangle$, and $\langle c, d\rangle$ from P. Further, because $d \sqsubseteq a$ but $d \not\sqsubseteq c$, we can conclude that $a \not\sqsubseteq c$; similar reasoning shows that $b \not\sqsubseteq d$; $\lfloor P \rfloor_K$ is thus restricted to the set $\{\langle a, a\rangle, \langle a, b\rangle, \langle b, a\rangle, \langle b, b\rangle, \langle c, a\rangle, \langle c, b\rangle, \langle c, c\rangle, \langle d, a\rangle, \langle d, b\rangle, \langle d, d\rangle\}$. The states of K^* and $\lfloor P \rfloor_K$ at this point are shown in the left-center diagram of Figure 2.

In the next iteration through the COMPUTE-ORDERING-2 loop we cannot choose d since it does not occur in any tuple of $K^* \setminus \lfloor P \rfloor_K$; assume that we instead choose b. The only possible successor of b is a, so SUCCESSORS searches this one-element subgraph by performing the single test $b \sqsubseteq^? a$ (which returns TRUE), and we add $\langle b, a\rangle$ to K. The element d is already known to be a predecessor of b, so PREDECESSORS searches the

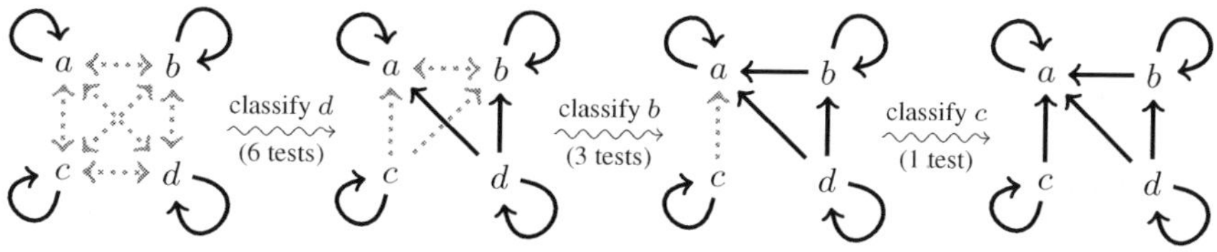

Fig. 2. As classification proceeds, known edges (denoted by solid black arrows) are discovered, and possible edges (denoted by dotted gray arrows) are eliminated

subgraph $K^\diamond[\{a, c\}]$ by performing the tests $a \sqsubseteq^? b$ and $c \sqsubseteq^? b$, finding no predecessors, and the two corresponding tuples are removed from P. The states of K^* and $\lfloor P \rfloor_K$ are shown in the right-center diagram of Figure 2.

After two iterations, the set $K^* \setminus \lfloor P \rfloor_K$ contains only the pair $\langle c, a\rangle$; if we choose c in the next iteration then SUCCESSORS tests $c \sqsubseteq^? a$, we add $\langle c, a\rangle$ to K, and $K^* = \lfloor P \rfloor_K$. The final subsumption relation is given by K^* and is shown in the right-hand diagram of Figure 2.

COMPUTE-ORDERING-2 thus classifies this ontology using 10 subsumption tests instead of the 16 required by a naïve brute-force approach. Note, however, that the results of the seven tests $a \sqsubseteq^? c$, $a \sqsubseteq^? d$, $b \sqsubseteq^? a$, $b \sqsubseteq^? d$, $c \sqsubseteq^? a$, $c \sqsubseteq^? b$, and $d \sqsubseteq^? b$ are sufficient to extend K and restrict P such that $K^* = \lfloor P \rfloor_K$, providing a full classification for this ontology. Identifying such a minimal set of tests is, however, extremely difficult without prior knowledge of the final taxonomy.

4 Extracting Subsumption Information from Models

We next turn our attention to the specific case of identifying all subsumption relationships between the class names occurring in an ontology $\mathcal{O}$. Instead of treating a reasoning service as an oracle that answers boolean queries of the form "is A subsumed by B w.r.t. $\mathcal{O}$?" (which we will write $\mathcal{O} \models^? A \sqsubseteq B$), we consider how information generated internally by common reasoning algorithms can be exploited to discover information about the subsumption quasi-ordering.

4.1 Identifying Non-subsumptions

Most modern reasoners for Description Logics, including HermiT, Pellet, and FaCT++, transpose subsumption queries into satisfiability problems; in particular, to determine if $\mathcal{O} \models A \sqsubseteq \bot$, these reasoners test whether the class A is satisfiable w.r.t. $\mathcal{O}$. They do this by trying to construct (an abstraction of) a Tarski-style *model* of $\mathcal{O}$ in which the extension of A is nonempty. We begin by providing an abbreviated formalization of such models (see Baader *et al.* [2003] for more details):

Definition 2. *Given sets of class names N_C, property names N_R and individual names N_I, an* interpretation $\mathcal{I} = (\Delta^\mathcal{I}, \cdot^\mathcal{I})$ *consists of a nonempty set $\Delta^\mathcal{I}$ and an* interpretation function $\cdot^\mathcal{I}$ *which maps every element of N_C to a subset of $\Delta^\mathcal{I}$, every element of N_R to* *a subset of* $\Delta^{\mathcal{I}} \times \Delta^{\mathcal{I}}$ *and every element of* N_I *to an element of* $\Delta^{\mathcal{I}}$*. An interpretation* $\mathcal{I}$ *is a model of an axiom* $A \sqsubseteq B$ *if* $A^{\mathcal{I}} \subseteq B^{\mathcal{I}}$ *(similar definitions hold for other kinds of statement); it is a model of an ontology* $\mathcal{O}$ *if it models every statement in* $\mathcal{O}$*.*

Let A *and* B *be classes. A model* $\mathcal{I}$ *of* $\mathcal{O}$ *is a* witness for the satisfiability of A w.r.t. $\mathcal{O}$ *if* $A^{\mathcal{I}}$ *is nonempty; it is a* witness for the non-subsumption $A \not\sqsubseteq B$ w.r.t. $\mathcal{O}$ *if* $A^{\mathcal{I}} \not\subseteq B^{\mathcal{I}}$*, i.e., if there exists* $i \in \Delta^{\mathcal{I}}$ *s.t.* $i \in A^{\mathcal{I}}$ *and* $i \notin B^{\mathcal{I}}$*.*

The algorithms in question typically represent the model being constructed as an ABox, i.e., as a set of *assertions* of the form $x : C$ and $\langle x, y \rangle : R$ for individuals x, y, classes C, and properties R [Baader *et al.*, 2003]. An ABox including the assertion $x : C$ represents a model in which $x^{\mathcal{I}} \in C^{\mathcal{I}}$. To construct a witness for the satisfiability of a class A, the ABox is initialised with an assertion $x : A$ and the construction proceeds in a goal-directed manner by adding further assertions only as necessary in order to ensure that the ABox represents a model of $\mathcal{O}$.

Assuming that the construction is successful, the resulting ABox/model provides a rich source of information. For example, for any class B such that $x : (\neg B)$ is in the ABox, it is the case that $x^{\mathcal{I}} \notin B^{\mathcal{I}}$; thus the model is a witness for the non-subsumption $\mathcal{O} \models A \not\sqsubseteq B$ for all such classes B. In many cases, the non-presence of $x : B$ in the ABox is sufficient to conclude the relevant non-subsumption; in fact, when using a hypertableau algorithm, this is *always* the case. When using a tableau algorithm, the non-subsumption conclusion can be drawn if B is a so-called primitive concept, i.e., if the unfoldable part of the TBox includes an axiom $B \sqsubseteq C$ for some concept C; in this case, adding $x : (\neg B)$ to the ABox trivially results in a witness for the non-subsumption (this is closely related to the widely employed model merging optimization [Tsarkov *et al.*, 2007]).

The goal-directed nature of the ABox construction means that the models constructed are typically quite small. As a result, these models tend to be extremely rich in non-subsumption information: in a typical witness for the satisfiability of A, i.e., a model $\mathcal{I}$ of $\mathcal{O}$ with $i \in A^{\mathcal{I}}$, there will be relatively few other class names B such that $i \in B^{\mathcal{I}}$, and thus $\mathcal{I}$ will identify the vast majority of class names in $\mathcal{K}$ as non-subsumers of A. For this reason, it is almost always more efficient to record the set $P_A = \{B \mid i \in A^{\mathcal{I}} \text{ and } i \in B^{\mathcal{I}} \text{ for some } i\}$ of "possible subsumers" of A.

4.2 Identifying Subsumptions

While single models allow us to detect non-subsumptions, additional information about the space of possible models is required in order to identify subsumption relationships. Sound and complete tableau reasoning algorithms systematically explore the space of all "canonical" models (typically tree- or forest-shaped models), on the basis that, if any model exists, then one of these canonical models also exists. In particular, when $\mathcal{O}$ includes disjunctions or other sources of nondeterminism, it may be necessary to choose between several possible ways of modeling such statements, and to backtrack and try other possible choices if the construction fails.

For such algorithms, it is usually easy to show that, if the ABox was initialized with $x : A$, the construction did not involve any nondeterministic choices, and the resulting ABox includes the assertion $x : B$, then it is the case that in any model $\mathcal{I}$ of $\mathcal{K}$, $i \in A^{\mathcal{I}}$

implies $i \in B^{\mathcal{I}}$, i.e., that $\mathcal{K} \models A \sqsubseteq B$. Moreover, as we have already seen in Section 4.1, such an ABox is (at least in the hypertableau case) a witness to the non-subsumption $\mathcal{K} \models A \not\sqsubseteq B$ for all class names B such that $x : B$ is not in the ABox. Thus, when testing the satisfiability of a class A, it may be possible to derive *complete information* about the subsumers of A.

The hypertableau-based HermiT reasoner is designed to reduce nondeterminism, and avoids it completely when dealing with Horn-$\mathcal{SHIQ}$ (and OWL 2 EL) ontologies; for such ontologies it is thus able to derive complete information about the subsumers of a class A using a single satisfiability test. This allows HermiT to derive all relevant subsumption relationships in a Horn-$\mathcal{SHIQ}$ ontology as a side effect of performing satisfiability tests on each of the class names [Motik *et al.*, 2007].

This idea can be extended so as to also derive useful information from nondeterministic constructions by exploiting the dependency labeling typically used to enable "dependency-directed backtracking"—an optimization which reduces the effects of nondeterminism in reasoning [Horrocks, 1997]. In the resulting ABoxes, each assertion is labelled with the set of choice points on which it depends. An empty label indicates that the relevant assertion will always be present in the ABox, regardless of any choices made during the construction process. Thus, if the ABox is initialized with $x : A$, an empty-labelled assertion $x : B$ in the resulting ABox can be treated in the same way as if the construction had been completely deterministic. Performing a satisfiability test on A may, therefore, allow some subsumers of A to be identified even when nondeterministic choices are made during reasoning. In practice, almost all of the actual subsumers of A can usually be identified in this way.

It is easy to see that this idea is closely related to, and largely generalizes, the told subsumer and completely-defined optimizations. For a completely defined class name A, a satisfiability test on A will be deterministic (and typically rather trivial), and so will provide complete information about the subsumers of A. Similarly, if B is a told subsumer of A, then an ABox initialized with $x : A$ will always produce $x : B$, and almost always deterministically. (It is theoretically possible that $x : B$ will be added first due to some nondeterministic axiom in the ontology).

5 Related Work

Computing a quasi- (or partial-) ordering for a set of n incomparable elements clearly requires n^2 individual tests—naïvely comparing all pairs is thus "optimal" by the simplest standard. The literature therefore focuses on a slightly more sophisticated metric which considers both the number of elements in the ordering as well as the *width* of the ordering—the maximum size of a set of mutually incomparable elements. Faigle and Turán [1985] have shown that the number of comparisons needed to deduce an ordering of n elements with width w is at most $O(wn \log(n/w))$ and Daskalakis *et al.* provide an algorithm which approaches this bound by executing $O(n(w + \log n))$ comparisons [2007]. Taxonomies, however, tend to resemble trees in structure, and the width of a subsumption ordering of n elements is generally close to $n/2$. Further, the algorithms of Faigle and Turán as well as Daskalakis *et al.* rely on data structures which require $O(nw)$ storage space even in the best case, and thus exhibit quadratic performance when constructing a taxonomy.

A taxonomy-construction strategy which performs well for tree-like relations is described by Ellis [1991]: elements are inserted into the taxonomy one at a time by finding, for each element, its subsumers using a breadth-first search of all previously-inserted elements top-down, and then its subsumees using a breadth-first search bottom-up. Baader *et al.* further refine this technique to avoid redundant subsumption tests during each search phase: during the top search phrase, a test $\mathcal{K} \models^? A \sqsubseteq B$ is performed only if $\mathcal{K} \models A \sqsubseteq C$ for all subsumers C of B [1994]. This can be seen as a special case of our $\lfloor P \rfloor_K$ pruning of possible subsumers, with the restriction that it only applies to subsumption tests performed in a prescribed order.

The traversal algorithms described by Ellis and by Baader *et al.* perform subsumption tests between every pair of siblings in the final taxonomy. Such algorithms are thus very inefficient for taxonomies containing nodes with large numbers of children; completely flat taxonomies result in a quadratic number of subsumption tests. Haarslev and Möller propose a way to avoid the inefficiencies of these traversals by *clustering* a group of siblings $A_1, ..., A_n$ by adding the class expression $\bigsqcap_{1 \leq i \leq n} A_i$ to the taxonomy [2001]. Our approach can easily incorporate this technique by including partial or complete information about such new class expressions in K and P. In practice, however, the partial information extracted from tableau models almost always includes non-subsumptions between siblings in such flat hierarchies, so these refinements are unnecessary.

Baader *et al.* also describe techniques for identifying subsumers without the need for multiple subsumption tests by analyzing the syntax of class definitions in an ontology: if an ontology contains an axiom of the form $A \sqsubseteq B \sqcap C$ where A and B are class names, then B is a "told subsumer" of A, as are all the told subsumers of B. The various simplification and absorption techniques described by Horrocks [1997] increase the applicability of such analysis. Haarslev *et al.* further extend this analysis to detect non-subsumption: an axiom of the form $A \sqsubseteq \neg B \sqcap C$ implies that A and B are disjoint, thus neither class name subsumes the other (unless both are unsatisfiable) [2001]. Tsarkov *et al.* describe a technique for precisely determining the subsumption relationships between "completely defined classes"—class names whose definitions contain only conjunctions of other completely defined classes [2007]. All these optimizations can be seen as special cases of (non-)subsumption information being derived from (possibly incomplete) calculi as described in Section 4.

6 Empirical Evaluation

In order to determine if our new algorithm is likely to improve classification performance in practice we conducted two experiments using large ontologies derived from life-science applications.

First, we compared the performance of our new algorithm with the enhanced traversal algorithm. In order to analyze how much improvement is due to the information extracted directly from models and how much is due to our new approach to taxonomy construction, we extend the enhanced traversal algorithm such that it first performs a satisfiability test on every class name and constructs a cache of information derived from the resulting models using the techniques described in Section 4. During the subsequent taxonomy construction, subsumption tests are performed only if the relevant

Table 1. Algorithm Comparison

Relation Size		ET		New	
Known	Possible	Tests	Seconds	Tests	Seconds
335 476	335 476	0	190	0	17
335 476	2 244 050	152 362	246	24 796	22
335 476	4 147 689	303 045	257	49 308	31
335 476	6 046 804	455 054	292	73 945	33
335 476	7 940 847	606 205	305	98 613	34
251 880	335 476	80 878	634	19 773	28
251 880	2 244 050	439 002	740	50 143	32
251 880	4 147 689	794 513	809	79 038	40
251 880	6 046 804	1 151 134	836	107 416	46
251 880	7 940 847	1 506 752	919	136 190	50
168 052	335 476	143 913	1079	62 153	62
168 052	2 244 050	673 768	1267	146 823	91
168 052	4 147 689	1 201 904	1320	226 670	93
168 052	6 046 804	1 729 553	1414	304 784	98
168 052	7 940 847	-	-	381 330	130

subsumption relationship cannot be determined by consulting the cache. Note that this caching technique strictly subsumes the "told subsumer" and "primitive component" optimizations described by Baader *et al.*.

We implemented both algorithms within the HermiT reasoner [Motik *et al.*, 2007] and performed testing using an OWL version of the well-known US National Cancer Institute thesaurus (NCI), a large but simple ontology containing 27,653 classes.[3] The models constructed by HermiT during satisfiability testing of these classes provide complete information about the subsumption ordering for this ontology, so both algorithms are able to classify it without performing any additional tests. To study how the algorithms compare when less-than-complete information is available, we limited the amount of information extracted from HermiT's models, reducing the number of known subsumptions and increasing the number of possible subsumptions to varying degrees. The number of full subsumption tests required for classification as well as the total running times (including both classification and satisfiability testing) for each implementation are given in Table 1.

As the table shows, our simple implementation of the enhanced traversal algorithm (ET) is substantially slower than the new algorithm even when complete information is available; this is the result of the "insertion sort" behavior of ET described in Section 5.

When complete information is not available, our algorithm consistently reduces the number of subsumption tests needed to fully classify the ontology by an order of magnitude.

In a second experiment, we compared the implementation of our new algorithm in HermiT with the widely-used Description Logic classifiers FaCT++ and Pellet. Both of

[3] The latest version of the NCI ontology contains over 34,000 classes; the older version used here was, however, sufficient for our purposes.

Table 2. System Comparison

		FaCT++		Pellet		HermiT	
Ontology	Classes	Tests	Seconds	Tests	Seconds	Tests	Seconds
NCI	27 653	4 506 097	2.3	-	16.1	27 653	22
$\text{NCI}^{\exists}$	27 654	8 658 610	4.4	-	16.7	27 654	21.0
$\text{NCI}^{\sqcup}$	27 655	8 687 327	5.1	10 659 876	95.4	48 389	37.0
$\text{NCI}^{\exists\forall}$	27 656	18 198 060	473.9	10 746 921	1098.3	27 656	20.8
GO	19 529	26 322 937	8.6	-	6.0	19 529	9.2
$\text{GO}^{\exists}$	19 530	26 904 495	12.7	-	6.9	19 530	9.7
$\text{GO}^{\sqcup}$	19 531	26 926 653	15.5	21 280 377	170.0	32 614	15.2
GALEN	2749	313 627	11.1	131 125	8.4	2749	3.3
$\text{GALEN}^{\exists}$	2750	327 756	473.5	170 244	9.7	2750	3.5
$\text{GALEN}^{\sqcup}$	2751	329 394	450.5	175 859	9.8	4657	40.5

these systems are quite mature and implement a wide range of optimizations to both taxonomy construction and subsumption reasoning; we were thus able to compare our new algorithm with existing state-of-the-art implementations.

In this case, in addition to NCI we used the Gene Ontology (GO), and the well-known GALEN ontology of medical terminology[4]. Both NCI and GO have been specifically constructed to fall within the language fragment which existing reasoners are able to classify quickly; GALEN, in contrast, necessitates substantially more difficult subsumption testing but contains an order of magnitude fewer class names. In order to estimate how the different algorithms would behave with more expressive ontologies, for each ontology $\mathcal{O}$ we constructed two extensions: $\mathcal{O}^{\exists}$, which adds the single axiom $\top \sqsubseteq \exists R.A$ for a fresh property name R and fresh class name A, and $\mathcal{O}^{\sqcup}$ which adds the axiom $\top \sqsubseteq A \sqcup B$ for fresh class names A and B. For NCI we constructed a further extension $\text{NCI}^{\exists\forall}$ by adding the axioms $\top \sqsubseteq \exists R.A$ and $C \sqsubseteq \forall R.B$ for each of the 17 most general class names C occurring in the ontology. Each of these extensions increases the complexity of individual subsumption tests and reduces the effectiveness of optimizations that try to avoid performing some or all of the tests that would otherwise be needed during classification.

The number of class names occurring in each ontology as well as the number of tests performed (including all class satisfiability and subsumption tests) and the total time taken by each reasoner to fully classify each ontology are shown in Table 2. The Pellet system makes use of a special-purpose reasoning procedure for ontologies that fall within the $\mathcal{EL}$ fragment [Baader *et al.*, 2005]; for such ontologies we do not, therefore, list the number of subsumption tests performed by Pellet.

As Table 2 shows, HermiT's new classification algorithm dramatically reduces the number of subsumption tests performed when classifying these ontologies. This does not, however, always result in faster performance. This is largely due to optimizations used by the other reasoners which greatly reduce the cost of subsumption testing

[4] All test data is available from `http://www.comlab.ox.ac.uk/rob.shearer/2009/iswc-classification-ontologies.tgz`

for simple ontologies: the overwhelming majority of subsumption tests performed by FaCT++, for example, can be answered using the pseudo-model merging technique described by Horrocks [1997].

Most of these optimizations could equally well be used in HermiT, but in the existing implementation each subsumption test performed by HermiT is far more costly. The number of subsumption tests performed by HermiT is, however, far smaller than for the other reasoners, and its performance also degrades far more gracefully as the complexity of an ontology increases: adding a single GCI or disjunction to an ontology can prevent the application of special-case optimizations in Pellet and FaCT++, greatly increasing the cost of subsumption testing and, due to the very large number of tests being performed, vastly increasing the time required for classification. The $\text{NCI}^{\exists\forall}$ ontology, for example, eliminates any benefit from the pseudo-model merging optimization (since no two pseudo-models can be trivially merged), and this causes the classification time to increase by roughly two orders of magnitude for both Pellet and FaCT++. In contrast, HermiT's classification time is unaffected. The relatively poor performance of HermiT on the $\text{GALEN}^{\sqcup}$ ontology is due to the fact that the underlying satisfiability testing procedure is particularly costly when there are large numbers of branching points, even if no backtracking is actually required.

7 Discussion and Future Work

We have described a new algorithm for taxonomy construction that effectively exploits partial information derived from structural analysis and/or reasoning procedures, and we have shown that, when compared to the widely-used enhanced traversal algorithm, it can dramatically reduce both the number of individual comparisons and the total processing time for realistic data sets. For simple ontologies, our prototype implementation makes the HermiT reasoner competitive with state-of-the-art reasoners which implement special-purpose optimizations of the subsumption testing procedure for such cases; on more expressive ontologies our new system substantially outperforms existing systems.

Future work will include extending HermiT to incorporate some of the subsumption testing optimizations used in other systems, in particular reducing the overhead cost of individual subsumption tests. We believe that this will greatly improve HermiT's performance on simple ontologies; as we have seen, it is already highly competitive on more complex ontologies.

The procedure we describe in Section 4 extracts subsumption relationships involving only the class used to initialize a model. This is because the dependency labeling implemented in tableau reasoners is currently designed only to allow the application of dependency-directed backtracking, and discards a great deal of dependency information. We intend to explore more sophisticated dependency labeling strategies which allow the extraction of additional subsumption information.

We also want to investigate meaningful complexity bounds for taxonomy searching and construction tasks. As we have seen, a completely naïve search routine is optimal if only the number of elements is considered. We will attempt to obtain tighter bounds for certain classes of relation: relations with linear taxonomy graphs, for example, can be deduced with only $n \log n$ comparisons. Bounds based on more sophisticated metrics may also be possible; e.g., bounds based on the size of the subsumption relation instead of the number of elements.

Finally, preliminary testing demonstrates that when significant partial information is available, the COMPUTE-ORDERING-2 procedure, based on the breadth-first search of the enhanced traversal algorithm, offers little advantage over COMPUTE-ORDERING, which performs tests in an arbitrary order; in many cases the performance of COMPUTE-ORDERING-2 is actually worse. Investigating other heuristics for choosing the order in which to perform tests will also be part of our future work.

References

[Baader *et al.*, 1994] Baader, F., Hollunder, B., Nebel, B., Pro Tlich, H.-J., Franconi, E.: An empirical analysis of optimization techniques for terminological representation systems or: Making KRIS get a move on. Applied Artificial Intelligence. Special Issue on Knowledge Base Management 4, 270–281 (1994)

[Baader *et al.*, 2003] Baader, F., Calvanese, D., McGuinness, D., Nardi, D., Patel-Schneider, P.F. (eds.): The Description Logic Handbook: Theory, Implementation and Applications. Cambridge University Press, Cambridge (2003)

[Baader *et al.*, 2005] Baader, F., Brandt, S., Lutz, C.: Pushing the $\mathcal{EL}$ envelope. In: Proc. of the 19th Int. Joint Conf. on Artificial Intelligence (IJCAI 2005), pp. 364–369 (2005)

[Daskalakis *et al.*, 2007] Daskalakis, C., Karp, R.M., Mossel, E., Riesenfeld, S., Verbin, E.: Sorting and selection in posets. CoRR, abs/0707.1532 (2007)

[Ellis, 1991] Ellis, G.: Compiled hierarchical retrieval. In: 6th Annual Conceptual Graphs Workshop, pp. 285–310 (1991)

[Faigle and Turán, 1985] Faigle, U., Turán, G.: Sorting and recognition problems for ordered sets. In: Mehlhorn, K. (ed.) STACS 1985. LNCS, vol. 182, pp. 109–118. Springer, Heidelberg (1984)

[Haarslev and Möller, 2001] Haarslev, V., Möller, R.: High performance reasoning with very large knowledge bases: A practical case study. In: Nebel, B. (ed.) Proceedings of Seventeenth International JointŁConference on Artificial Intelligence, IJCAI 2001, pp. 161–166 (2001)

[Haarslev *et al.*, 2001] Haarslev, V., Möller, R., Turhan, A.-Y.: Exploiting pseudo models for tBox and aBox reasoning in expressive description logics. In: Goré, R.P., Leitsch, A., Nipkow, T. (eds.) IJCAR 2001. LNCS (LNAI), vol. 2083, pp. 61–75. Springer, Heidelberg (2001)

[Horrocks, 1997] Horrocks, I.: Optimising Tableaux Decision Procedures for Description Logics. PhD thesis, University of Manchester (1997)

[Motik *et al.*, 2007] Motik, B., Shearer, R., Horrocks, I.: Optimized Reasoning in Description Logics Using Hypertableaux. In: Pfenning, F. (ed.) CADE 2007. LNCS (LNAI), vol. 4603, pp. 67–83. Springer, Heidelberg (2007)

[Tsarkov *et al.*, 2007] Tsarkov, D., Horrocks, I., Patel-Schneider, P.F.: Optimising terminological reasoning for expressive description logics. J. of Automated Reasoning (2007)

Actively Learning Ontology Matching via User Interaction*

Feng Shi[1], Juanzi Li[1], Jie Tang[1], Guotong Xie[2], and Hanyu Li[2]

[1] Department of Computer Science and Technology
Tsinghua National Laboratory for Information Science and Technology
Tsinghua University, Beijing, 100084, China
{shifeng,ljz,tangjie}@keg.cs.tsinghua.edu.cn
[2] IBM China Research Laboratory, Beijing 100094, China
{xieguot,lihanyu}@cn.ibm.com

Abstract. Ontology matching plays a key role for semantic interoperability. Many methods have been proposed for automatically finding the alignment between heterogeneous ontologies. However, in many real-world applications, finding the alignment in a completely automatic way is *highly infeasible*. Ideally, an ontology matching system would have an interactive interface to allow users to provide feedbacks to guide the automatic algorithm. Fundamentally, we need answer the following questions: How can a system perform an efficiently interactive process with the user? How many interactions are sufficient for finding a more accurate matching? To address these questions, we propose an active learning framework for ontology matching, which tries to find the *most informative* candidate matches to query the user. The user's feedbacks are used to: 1) correct the mistake matching and 2) propagate the supervise information to help the entire matching process. Three measures are proposed to estimate the confidence of each matching candidate. A correct propagation algorithm is further proposed to maximize the spread of the user's "guidance". Experimental results on several public data sets show that the proposed approach can significantly improve the matching accuracy (+8.0% better than the baseline methods).

1 Introduction

The growing need of information sharing poses many challenges for semantic integration. Ontology matching, aiming to obtain semantic correspondences between two ontologies, is the key to realize ontology interoperability [10]. Recently, with the success of many online social networks, such as Facebook, MySpace, and Twitter, a large amount of user-defined ontologies are created and published on the Social Web, which makes it much more challenging for the ontology matching problem. At the same time, the Social Web also provides some opportunities (e.g., rich user interactions) to solve the matching problem.

* The work is supported by the Natural Science Foundation of China (No. 60703059), Chinese National Key Foundation Research (No. 2007CB310803), National High-tech R&D Program (No. 2009AA01Z138), and Chinese Young Faculty Research Fund (No. 20070003093).

A. Bernstein et al. (Eds.): ISWC 2009, LNCS 5823, pp. 585–600, 2009.

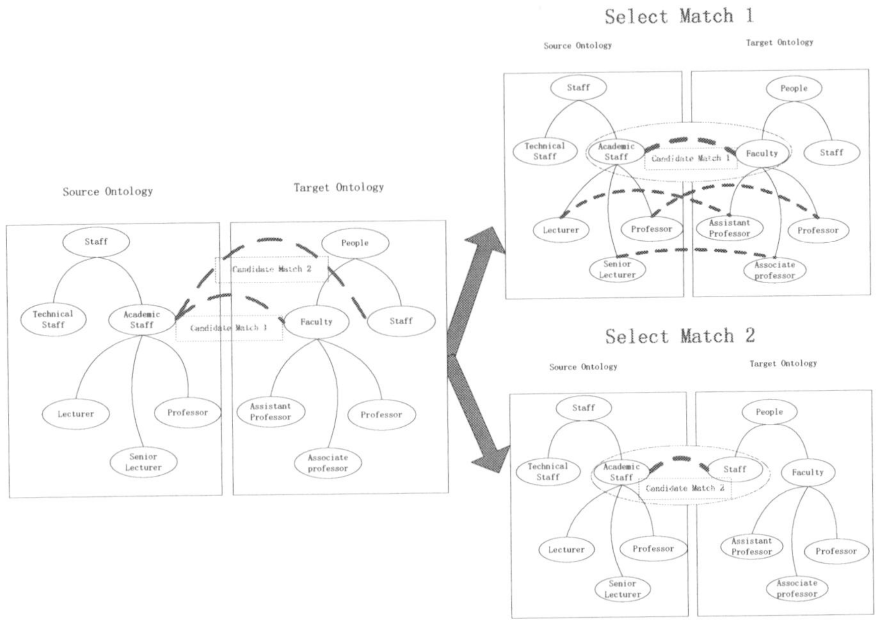

Fig. 1. Example of the candidate match selection. Ideally, we hope to query candidate match 1, instead of match 2. With the user correction, we can propagate the user correction to guide the matching process of other elements, as illustrated in the top-right.

Much efforts has been made for ontology matching. Methods such as Edit Distance [13], KNN [3] and Bayesian similarity [26], have been proposed to calculate the similarity between elements (e.g., concepts). However, most of existing works aim to find the ontology matching in a completely automatic way, although the complete automation is infeasible in many real cases [24]. One promising solution is to involve user interactions into the matching process to improve the quality of matching results [24]. However, regarding the large size of ontologies, random user interactions are really limited for helping ontology matching. Now, the question is: is there any way to minimize the amount of user interactions, while maximize the effect (accuracy improvement) of interactive efforts? Or a more specific question: given a fixed number of user interactions, can a system "actively" query the user so as to maximize spread of the interactions?

It is non-trivial to address the problem. A simple way is to let the user select candidate matches or to select matches with a low confidence (similarity) to query. Such queries can benefit the queried matches, however, it may not be helpful to the other non-queried candidate matches. Our goal is not only to correct the possibly wrong matches through the user interactions, but also to maximize the correction via spread (propagation) of the interactions. Thus, how to design an algorithm to actively select candidate matches to query is a challenging issue.

Motivating Example. We use an example to demonstrate the motivation of the work. Figure 1 shows an example of the candidate match selection. The source and target ontologies are both about academic staff. If we select the candidate match 1 ("Academic Staff", "Faculty") to query, with the user correction, we can further infer and correct another potential error match ("Academic Staff", "Staff"). Moreover, the sub matches of ("Lecturer", "Assistant Professor") and ("Senior Lecturer", "Associate Professor") would also gain a higher confidence. While if we just select the candidate match 2 ("Academic Staff", "Staff"), it is very possible to be no improvement.

Our Solution. In this paper, we propose a novel problem of *ontology matching with active user interaction*. In particular, we propose an active learning framework for ontology matching, which tries to find the most informative candidate matches to query, and propagate the user correction according to the ontology structure to improve the matching accuracy. We present a simple but effective algorithm to select the threshold with user feedbacks. Three measures to evaluate the confidence of each match. A *correct propagation* algorithm is proposed to spread the user corrections. Experimental results demonstrate that the proposed approach can significantly improve (+8.0%) the matching performance with only a few queries (< 10).

The rest of this paper is organized as follows. Section 2 formalizes the problem. Section 3 describes our active learning framework for ontology matching. Section 4 explains the method of threshold selection, three measures for error matches detection, and the correct propagation algorithm. Section 5 presents the experimental results. Finally, we discuss related work in Section 6 and conclude in Section 7.

2 Problem Formulation

In this section, we first give the definition of ontology and then formalize the problem of ontology matching with active user interaction.

An ontology usually provides a set of vocabularies to describe the information of interest. The major components of an ontology are concepts, relations, instances and axioms [26].

In this paper, our ontology matching mainly focuses on concepts and relations. According to the relations between concepts, an ontology can be viewed as a directed graph, in which vertexes represent concepts and edges represent relations.

Given a source ontology O_S, a target ontology O_D, and an element (a concept or a relation) e_i in O_S, the procedure to find the semantically equivalent element e_j in O_D to e_i is called ontology matching, denoted as M. Formally, for each element (e_i in O_S), the ontology matching M can be represented [26] as:

$$M(e_i, O_S, O_D) = \{e_j\} \quad (1)$$

Furthermore, M could be generalized to elements set matching, that is for a set of elements $\{e_i\}$ in O_S, the ontology matching is defined as:

$$M(\{e_i\}, O_S, O_D) = \{e_j\} \quad (2)$$

If $\{e_i\}$ contains all the elements of O_S, the matching can be predigested as

$$M(O_S, O_D) = \{e_j\} \quad (3)$$

3 An Active Learning Framework for Ontology Matching

We propose an active learning framework for ontology matching. Algorithm 1 gives an overview of the active learning framework. Assume that O_S is the source ontology, O_D is the target ontology. M is a traditional method of ontology matching, L is the set of confirmed matches submitted by users, and N is the iteration number, which is also the number of candidate matches to query.

Algorithm 1. The Active Learning Framework for Ontology Matching

Input:

- the source ontology O_S, the target ontology O_D,
- A traditional method of ontology matching M,
- the confirmed match set L,
- number of matches need to be confirmed N

Initialization:

- apply M to map O_S to O_D, and get the match result R
- initialize L with Ø

Loop for N iterations:

- let $<(e_S, e_D), ?>$ = SelectQueryMatch();
- query users to confirm the match $<(e_S, e_D), ?>$
- add $<(e_S, e_D), l>$ to L
- improve the matching result R with $<(e_S, e_D), l>$

The basic process of the framework is as follows: first, it applies the traditional method of ontology matching M to map O_S to O_D, and gets the match result R, where multi-method results of different types are usually more useful for the next step. Second, it selects an informative candidate match $<(e_S, e_D), ?>$ to query users for confirmation with the result of the first step and the structure information of the two ontologies O_S and O_D. After the user confirmation, it adds the match $<(e_S, e_D), l>$ to the confirmed match set L, and improves the match result R with the confirmed matches. Then it repeats the second step for N iterations, or until it gets a result good enough.

Algorithm 1 is merely a shell, serving as a framework to many possible instantiations. What separates a successful instantiation from a poor one is the follow two problems:

1. First, how to select the most informative candidate match to query.
2. Second, how to improve the matching result with the confirmed matches.

In the next section, we will give the solutions to these two problems.

4 Match Selection and Correct Propagation

This part introduces our solution to the two core problems of ontology matching with active learning: the candidate match selection and the matching result improvement. We first present a simple but effective algorithm to select the threshold for ontology matching with user feedback, and then give several measures to detect informative candidate matches to query. In the end of this section, we describe our correct propagation algorithm to improve the matching result with the user confirmed matches.

4.1 Threshold Selection with User Feedback

Most methods of ontology matching find matches through calculating the similarities between elements of source and target ontologies. The similarity can be string similarity, structure similarity, semantic similarity and so on [26]. No matter what kind of similarity is chosen, it needs a threshold to estimate which matches are correct. So it is very important to select a suitable threshold. However, the threshold selection is very difficult, especially when there is no any supervised information.

Through analysis we find the relationship of thresholds and matching results in most cases, as shown in Figure 2 (precisions, recalls and F1-Measures whose definitions are introduced in section 5).

From Figure 2, we can find that the precision curve is an increasing one, while the recall curve is a decreasing one, and the magnitude of change is getting smaller as the threshold getting bigger. So the F1-Measure curve has a maximum value on some threshold, which is our aim.

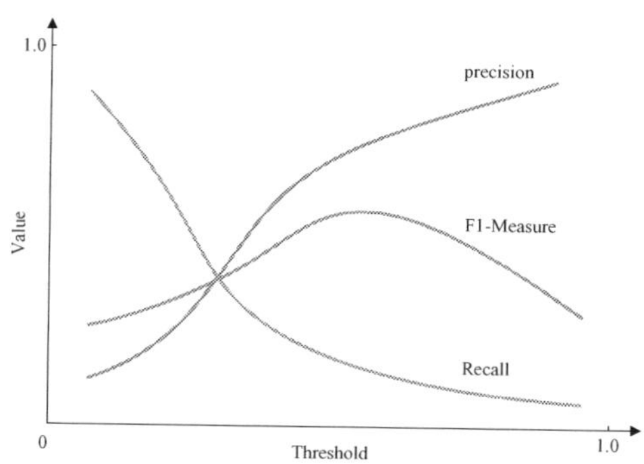

Fig. 2. Relationship of thresholds and matching performance (precision, recall and f1-Measure)

Algorithm 2 shows our algorithm of threshold selection. The input of the algorithm consists of the similarity set S, which contains all the matches and their similarity degree, an update step st for the threshold's updating, and an attenuation factor λ for st, and an initial threshold θ_0. First, the similarity set S needs to be normalized, and all the similarity degrees should be normalized into $[0, 1]$, and then let the threshold θ be the initial value θ_0. Second, it finds the match (e_S, e_D) whose similarity degree is the closest to θ, and let a user check whether the match is correct. If it is correct, the threshold θ

increases by st, otherwise θ decreases by st. The second step is an iterative process, and st updates according to the correctness of the selected match each iteration. If the correctness of the selected match is different from last one, the update step st will multiply the attenuation factor λ. Because the attenuation factor is a decimal in range $(0, 1)$, so after sufficient iterations, the update step st will be small enough so that the threshold θ will stabilize at some value, which is our final threshold.

The algorithm cannot always achieve a good result, but if the value of F1-Measure with the threshold increases first and then decreases, which is typically the case, our algorithm can usually achieve a good value. Moreover, when the data is huge, our algorithm usually can get the result after a few iterations. That is to say the number of iteration will not increase much as the data becomes huge.

Algorithm 2. Threshold Selection

Input: The similarity set: S, an initial threshold: θ_0, an update step: st, an attenuation factor: λ.
Output: The threshold of the matching result: θ.
Normalize the similarity set S
Let θ be θ_0
While st is big enough
  let $(e_S, e_D) = \min\{|similarity(e_S, e_D) - \theta|\}$
  ask users to comfirm the match (e_S, e_D)
  if (e_S, e_D) is correct
    if last match is not correct
      $st = st * \lambda$
    end if
    $\theta = \theta - st$
  else
    if last match is correct
      $st = st * \lambda$
    end if
    $\theta = \theta + st$
  end if
end while

4.2 Candidate Match Selection

One of the key points of ontology matching with active learning is to select informative matches to query. The most informative match means the match that can maximize the improvement of the matching performance. If the correctness of a match is found different from the matching result after the user confirmation, we call this match an *error match*. An error match is considered to be informative, because the result can be improved as long as the error is corrected. If the size of the data is small, or the original match result is already very good, this kind of improvement will be limited. If the data size is not small, or the original result is not good, we can also use the information of the *error match* to find other errors to significantly improve the matching result, which will be introduced in the next subsection.

The probability that a match is an error match is measured with *error rate*, and we propose three measures to estimate the error rate of a match as follows. Finally we combine these three measures to find the error matches.

Confidence. The first measure we define is called *confidence*. Assume e_S and e_D are elements of the source ontology O_S and the target ontology O_D respectively. f is a similarity computing function of some ontology matching method M, and θ is its threshold. The confidence of M on a match (e_S, e_D) can be defined as follows:

$$Confidence(f(e_S, e_D)) = |\theta - f(e_S, e_D)| \tag{4}$$

The confidence can measure how sure the method M is about the correctness of the match. So the match with least confidence is most possible to be an error match, which is called least confidence selection.

If there are k ontology matching methods of different types: $\{M_1, M_2, ..., M_k\}$, we can extend the least confidence selection as follows:

$$Q = \min\{\sum_{f_i \in \{M_1, M_2, ..., M_k\}} w_i * |\theta_i - f_i(e_S, e_D)|\} \tag{5}$$

In the formula, f_i is one of the similarity computing functions of different ontology matching methods $\{M_1, M_2, ..., M_k\}$, θ_i and w_i are its threshold and weight respectively. Q is the match selecting function. It means that the similarity of a match is closer to the threshold, it is more possible to be an error match.

Similarity Distance. The second measure is named *similarity distance*. Assume e_S is an element from the source ontology O_S, and method M maps the element e_S to e_D and $e_D^{'}$, which are two elements from the target ontology O_D. If the difference of $f(e_S, e_D)$ and $f(e_S, e_D^{'})$ is very small, there is very likely to be *error matches* in these two matches. Finally, we select the match that has the minimum difference. Formally, similarity distance is defined as:

$$SD(e_S, e_D) = \min\{|f(e_S, e_D) - f(e_S^{'}, e_D^{'})|\}; \quad (e_S = e_S^{'} \; or \; e_D = e_D^{'}) \tag{6}$$

The similarity distance is very efficient in a one-to-one matching, in which most methods only select the best one from the similar matches.

Contention Point. We define another measure, named, *contention point*, to find mistakes from the contention of different methods. This measure is defined based on the results of the matching results of some other algorithms such as edit distance or vector-space based similarity. The contention point is defined as:

$$ContentionPoint = \{< (e_s, e_D), ? > \in U | \exists i, j \; st. \; R_i(e_S, e_D) \neq R_j(e_S, e_D)\} \tag{7}$$

For a match (e_s, e_D), some of the k methods $\{M_1, M_2, ..., M_k\}$ consider it as matched, while the others consider not. Thus there must be mistakes among these methods, that is to say it is likely to be an error match. The contentious degree of a contention point can be further defined as:

$$Q = \min_{(e_S,e_D)\in ContentionPoint} \{ \max_{f_i\in\{M_1,M_2,\ldots,M_k\}} Confidence(f_i(e_S,e_D)) - \max_{f_j\in\{M_1,M_2,\ldots,M_k\}} Confidence(f_j(e_S,e_D))\}; \quad f_i(e_S,e_D) \neq f_j(e_S,e_D) \tag{8}$$

Intuitively, a contention one indicates that for the given match some methods consider it correct and some others consider not. That is to say that the matching algorithms have a maximal disagreement with the similarity confidences. In this case, the final matching result is very likely to be an error match.

4.3 Correct Propagation

When the size of the ontology is huge, correcting the selected error match only is far from sufficient. It is desirable that an algorithm can propagate the supervised information to help to correct the other potential error matches between the two ontologies.

Based on this consideration, we propose a *correct propagation* algorithm, which aims at detecting related error matches to the selected one. Thus when selecting a match to query, we need consider not only the error rate but also the effect of the error match on others, where the effect on others is called *propagation rate*.

Firstly, we introduce the concept of similarity propagation graph, which comes from the algorithm of similarity flooding [18]. A similarity propagation graph is an auxiliary data structure derived from ontologies O_S and O_D. The construction of propagation(PG) abides the principle as follows:

$$((a,b),p,(a_1,b_1)) \in PG(O_S,O_D) \Longleftrightarrow (a,p,a_1) \in O_S \text{ and } (b,p,b_1) \in O_D \tag{9}$$

Each node in the propagation graph is an element from $O_S \times O_D$. Such nodes are called map pairs. The intuition behind arcs that connect map pairs is the following. For map pairs (a,b) and (a_1,b_1), if a is similar to b, then probably a_1 is somewhat similar to b_1. Figure 3 gives an example of the propagation graph.

For every edge in the propagation graph, it adds an additional edge going in the opposite direction against the original one. The weights placed on the edges of the propagation graph indicate how well the similarity of a given map pair propagates to its neighbors and back. These so-called propagation coefficients range from 0 to 1 inclusively and can be computed in many different ways.

Our algorithm of *correct propagation* is also based on the propagation graph, but we both consider the negative and active effects of the propagation arcs. According to the character of the propagation graph, for a map pair (a,b) and (a_1,b_1), if a is not matched with b, then probably a_1 is not matched with b_1. With the measurement of error rate, error matches are easier to be detected, and we can correct more error matches related to the confirmed match according to the propagation graph, which is called correct propagation.

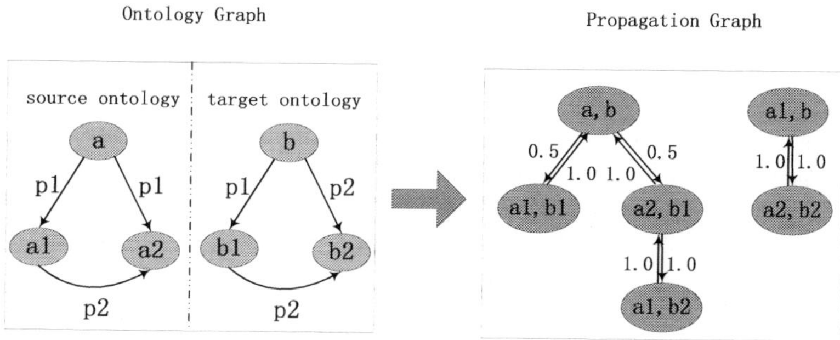

Fig. 3. Example of the similarity propagation graph

Before introducing the propagation, we consider the match selection again. To correct more error matches, we should not only consider the error rate, but also the propagation rate which measures the influence ability of a match. It mainly includes two factors: first, the number of matches that a match can influence. The bigger of the number, the range that the match can influence is wider, accordingly it is possible to correct more error matches. Second, the similarity differences between the match and its related matches. If the similarity difference is big, it is very possible to be error match among the match and its related ones.

Now, our match selection is according to the calculation of both error rate and propagation rate. When the correction (or confirmation) of the selected matches is provided by users, we conduct the correct propagation to update all the matches. Taking Figure 3 as an example, assume that we select the match (a_2, b_1) to query, and it is proved to be an error match. If the match (a_2, b_1) is confirmed to be an error by the user, then the similarities of the matches (a, b) and (a_1, b_2) which are related to the match (a_2, b_1) would be decreased. On the contrary, if the match (a_2, b_1) is confirmed to be correct, then the similarities of (a, b) and (a_1, b_2) should be increased. The update (decrease or increase) should be related to the similarities of the selected match, the error rates of related matches, and the weight of the arcs. Therefore, we can define the following update rules:

$$\begin{aligned} sim(a_i, b_i) = sim(a_i, b_i) + \alpha * w((x, y), (a_i, b_i)) \\ *(1 - sim(x, y)) * (1 - er(a_i, b_i)); \\ (x, p, a_i) \in O_S, (y, p, b_i) \in O_D \end{aligned} \tag{10}$$

$$\begin{aligned} sim(a_i, b_i) = sim(a_i, b_i) - \alpha * w((x, y), (a_i, b_i)) * sim(x, y) * er(a_i, b_i); \\ (x, p, a_i) \in O_S, (y, p, b_i) \in O_D \end{aligned} \tag{11}$$

In the formula, the match (x, y) is the selected error match, and $sim(x, y)$ is its similarity degree. The match (a_i, b_i) is one of the matches related to the match (x, y), and $w((x, y), (a_i, b_i))$ is the weight of their relation, and $er(a_i, b_i)$ stands for the error rate of the match (a_i, b_i), and α is an effect factor which is used to control the rate of the propagation. If the match (x, y) is correct (by the user), the update function uses Formula 10, else it uses Formula 11.

The correct propagation runs in an iterative process. In each iteration, it selects the match for user feedback with the error rate and the propagation rate, and then let users to confirm the selected match. After the confirmation, it updates the similarity degree, error rate and the propagation rate of related matches. Then it repeats this process until convergence (e.g., no any change) or the number of query times reaches a predefined threshold.

5 Experiments

We present details of the experiments in this section.

5.1 Experiment Setup, Data, and Evaluation Methodology

We implement all the algorithms using Java 2 JDK version 1.6.0 environment. All experiments are performed on a PC with AMD Athlon 4000+ dual core CPU (2.10GHz) and 2GB RAM Windows XP Professional edition OS.

Data sets. For our experiments of the first two groups, we use the OAEI 2008 30x benchmark [2]. There are four data sets in the group of benchmark 30x, in which each size is no more than 100 concepts and relations. The traditional matching results on these data sets is very high, hence it is very suitable for the first two experiments. For the experiment of the correct propagation, we use part of the OAEI 2005 Directory benchmark [1], which consists of aligning web sites directory (like open directory or Yahoo's) with more than two thousand elementary tests. The reason we select this data set lies in its available ground truth and its low matching accuracy by the traditional methods [16].

Platform. We conduct all the experiments on the ontology matching platform RiMOM [26], which is a dynamic multi-strategy ontology alignment framework. With RiMOM, we participated into the campaigns of the Ontology Alignment Evaluation Initiative (OAEI) from 2006 to 2008, and our system is among the top three performers on the benchmark data sets.

Performance Metrics. We use precision, recall, F1-Measure to measure the performance of the matching result. They are defined next.

Precision: It is the percentage of the correct discovered matches in all discovered matches.

Recall: It is the percentage of the correct discovered matches in all correct matches.

F1-Measure: F1-Measure considers the overall result of precision and recall.

$$F1 - Measure = 2(Precison * Recall)/(Precision + Recall) \quad (12)$$

5.2 Threshold Selection

We first analyze the performance of our approach for threshold selection. Figure 4 shows the results on the OAEI 2008 benchmark 301 [2], and the matching method is a combination of KNN [3], Edit Distance [13] and the method using the thesaurus WordNet [4].

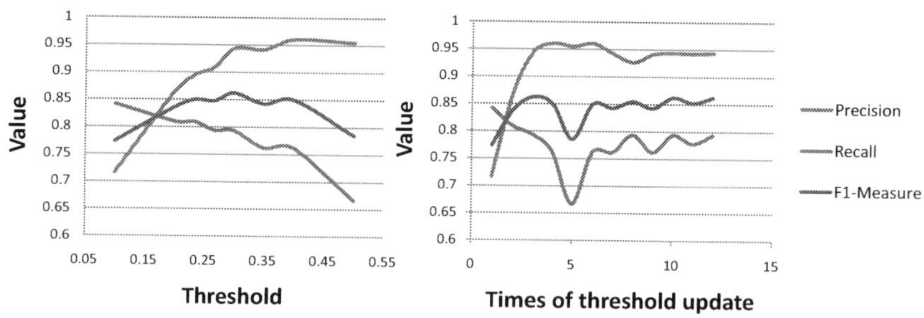

Fig. 4. Performance of threshold selection on OAEI 2008 benchmark 301

The left one in Figure 4 shows the relationship between thresholds and performances of matching results (precision, recall and F1-Measure), and we can see it is consistent with our point introduced in section 4 except a few dithering points. The right one in Figure 4 present the result of our approach.

5.3 Measurements of Error Match Selection

In this subsection, we evaluate the effectiveness of the different strategies for error match selection: confidence, similarity distance and contention point.

Figure 5 is an experiment on the OAEI 2008 benchmark 304. From the precision figure (left), we note that the measurement combined least confidence and similarity distance performs much better than others. But after about 10 matches confirmed, it is hard to keep improving the matching accuracy. The reason is that the size of the ontology is small, and the original matching accuracy is already high.

Figure 6 is another experiment on the OAEI 2008 benchmark 301. The results are very similar to Figure 5. From the recall figure (right) we note that it improves the recall slightly. While the recall figure (right) of Figure 5 has no improvement. The reason why the recall has little improvement is that the thresholds chosen for the original matching results are very low, and almost all the matches with similarity lower than the threshold are not correct ones. Our approach can only correct the errors. Thus a draft conclusion is that if there are no error matches below the threshold, the approach cannot improve the recall value.

Figure 7 is an experimental result on the OAEI 2008 benchmark 302, which is the best result of all the four benchmarks. From the figure we note that the measurement combined with least confidence, similarity distance and contention point improves fastest, but these measurements themselves improve slightly. This confirms us that combining these three measurements is useful for ontology matching.

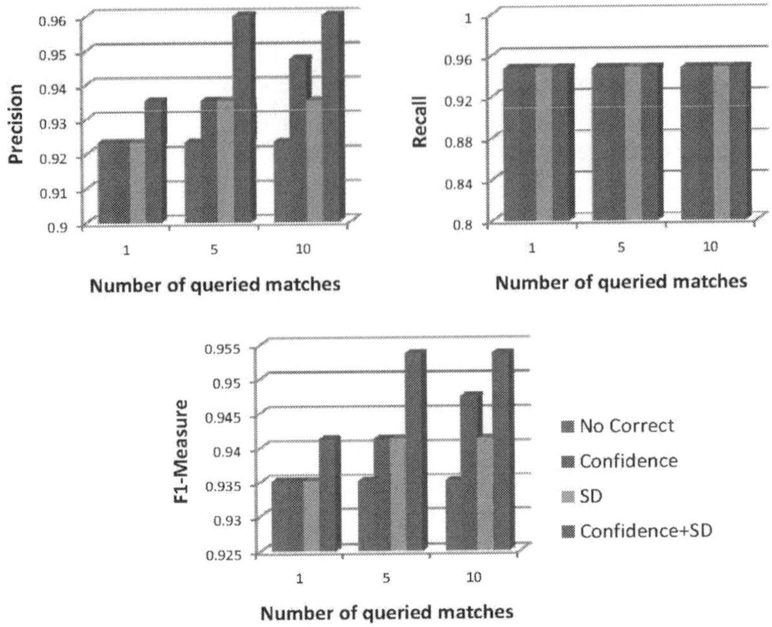

Fig. 5. Performance of matching after correcting error matches on OAEI 2008 benchmark 304

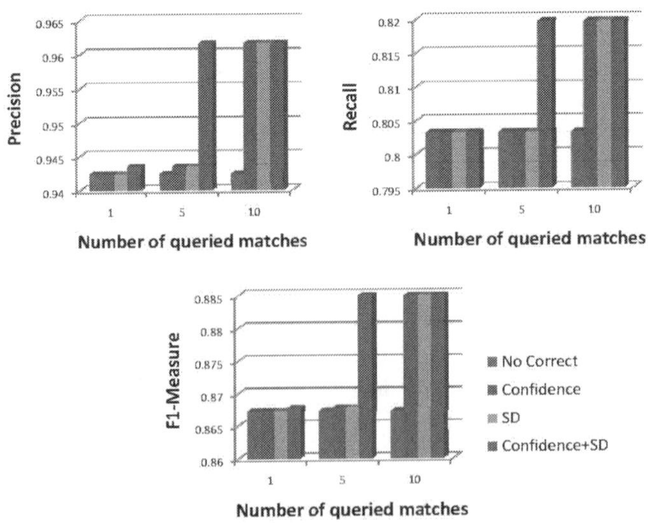

Fig. 6. Performance of matching after correcting error matches on OAEI 2008 benchmark 301

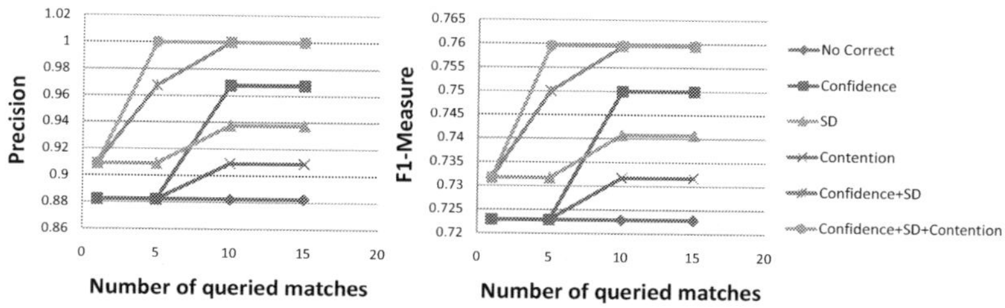

Fig. 7. Performance of matching after correcting error matches on OAEI 2008 benchmark 302

5.4 Correct Propagation

Figure 8 is an experiment on the approach of correct propagation with the OAEI 2005 Directory benchmark [1]. From the precision figure (left) we note that the result of correct propagation is much better than the approach of just correcting error matches. This implies that after propagation, more error matches are corrected with the selected one. Sometimes, the selected match is not an error match, so the approach of correcting error matches has no improvement, but the approach of correct propagation has. From the F1-Measure figure (below), it is not surprising that the approach of correct propagation grows faster than the others. Moreover, we find that the curve is steeper at the beginning. The reason is that the first few matches have bigger propagation rate, which means it can help to find more error matches.

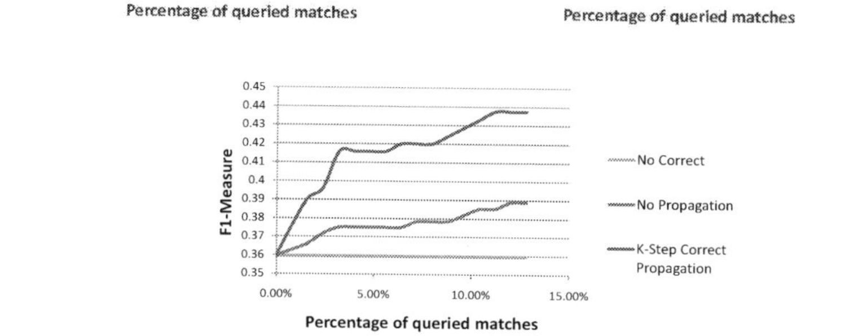

Fig. 8. Performance of matching after correct propagation on OAEI 2005 Directory

5.5 Summary

We summarize the experimental results as follows.

- First, in most cases our method of threshold selection can efficiently find a good threshold after a few queries.
- Second, all the three measures for match selection can help find the error matches, which are helpful to improve the matching result.
- Third, our approach of correct propagation can further improve the matching result. The improvement is more significant at the beginning than later. This also satisfies the limit of user feedback, that is also the reason we can improve the matching result greatly via only a few queries.

6 Related Work

6.1 Ontology Matching

Many works have addressed ontology matching in the context of ontology design and integration [6][17][19][21]. Some of them use the names, labels or comments of elements in the ontologies to suggest the semantic correspondences. [7] gives a detailed compare of various string-based matching techniques, including edit-distance [13] and token-based functions, e.g., Jaccard similarity [25] and TF/IDF [22]. Many works do not deal with explicit notions of similarity. They use a variety of heuristics to match ontology elements [17][19].

Some other works consider the structure information of ontologies. [15] uses the cardinalities of properties to match concepts. The method of similarity flooding is also an example using structure information [18]. Another type of method utilizes the background knowledge to improve the performance of ontology matching. For example, [4] proposes a similarity calculation method by using thesaurus WordNet. [12] presents a novel approximate method to discover the matches between concepts in directory ontology hierarchies. It utilizes information from Google search engine to define the approximate matches between concepts. [5] makes semantic mappings more amenable to matching through revising the mediated schema. Other methods based on instances of ontologies [28] or reasoning [27] also achieve good results.

6.2 Active Learning

Active learning can be viewed as a natural development from the earlier work on optimum experimental design [11]. This method is widely used in the domain of machine learning. [23] introduces an active-learning based approach to entity resolution that requests user feedback to help train classifiers. Selective supervision [14] combines decision theory with active learning. It uses a value of information approach for selecting unclassified cases for labeling. Co-Testing [20] is an active learning technique for multi-view learning tasks.

There are many works addressing ontology matching with user interaction, like GLUE [8], APFELi [9], [28], etc. Nevertheless, the annotation step is time-consuming and expensive, and users are usually not patient enough to label thousands of concept pairs for the relevance feedback. So our approach takes the concept of active learning to alleviate the burden of confirming large amounts of candidate matches, and the measurements we propose are based on the features of ontology. Our approach of correct propagation uses the propagation graph as the approach of similarity flooding [18]. The difference is that we propagate the similarity partly and purposely, and do not do it up and down. Our approach is more focused and more efficient than similarity flooding.

7 Conclusion and Future Work

In this paper we propose an active learning framework for ontology matching. But the framework is just a shell, and what separates a successful instantiation from a poor one is the selection of matches to query and the approach to improve the traditional matching result with the confirmed matches by users. We present a series of measurements to detect the error match. Furthermore we propose an approach named correct propagation to improve the matching result with the confirmed error matches. We also present a simple but effective method of selecting the threshold with user feedback, which is also helpful for the error match selection. Experimental results clearly demonstrate the effectiveness of the approaches.

As the future work, one interesting direction is to explore other types of user feedbacks. In our experiments, we take the standard boolean answer as the user feedback. However, in some cases users cannot give a simple correct or not answer for the queried match, especially when the ontologies are defined for some special domain. One solution is to select matches that the users are familiar with for confirmation, or translate the match into a question that the users can easily answer. Another interesting topic is how to reduce the negative effect of user mistakes.

References

1. Oaei directory download site (2005), `http://oaei.ontologymatching.org/2005/`
2. Ontology alignment evaluation initiative, `http://oaei.ontologymatching.org/`
3. Baily, T., Jain, A.K.: A note on distance-weighted k-nearest neighbor rules. IEEE Transaction System Man Cybern 8(4), 311–313 (1978)
4. Budanitsky, A., Hirst, G.: Evaluating wordnet based measures of lexical semantic relatedness. Computational Linguistics 32(1), 13–47 (2006)
5. Chai, X., Sayyadian, M., Doan, A., Rosenthal, A., Seligman, L.: Analyzing and revising mediated schemas to improve their matchability. In: Proceedings of the VLDB Conference 2008 (2008)
6. Chalupsky, H.: Ontomorph: A translation system for symbolic knowledge. In: Principles of Knowledge Representation and Reasoning, pp. 471–482 (2000)
7. Cohen, W., Ravikumar, P., Fienberg, S.: A comparison of string metrics for matching names and records. In: Proceedings of the IJCAI 2003 Workshop on Information Integration on the Web, pp. 73–78 (2003)
8. Doan, A., Madhavan, J., Domingos, P., Halevy, A.: Ontology matching: A machine learning approach. Springer, Heidelberg (2003)

9. Ehrig, M., Staab, S., Sure, Y.: Bootstrapping ontology alignment methods with APFEL. In: Gil, Y., Motta, E., Benjamins, V.R., Musen, M.A. (eds.) ISWC 2005. LNCS, vol. 3729, pp. 186–200. Springer, Heidelberg (2005)
10. Euzenat, J., Shvaiko, P.: Ontology Matching. Springer, Heidelberg (2007)
11. Fedorov, V.: Theory of Optimal Experiments. Academic Press, London (1972)
12. Gligorov, R., Aleksovski, Z., Kate, W., Harmelen, F.: Using google distance to weight approximate ontology matches. In: Proceedings of the 16th International World Wide Web Conference (WWW), pp. 767–776 (2007)
13. Gusfield, D.: Algorithms on strings, trees, and sequences. Cambridge University Press, Cambridge (1997)
14. Kapoor, A., Horvitz, E., Basu, S.: Selective supervision: Guiding supervised learning with decision-theoretic active learning. In: Proceedings of the IJCAI Conference 2007, pp. 877–882 (2007)
15. Lee, M., Yang, L., Hsu, W., Yang, X.: Xclust: Clustering xml schemas for effective integration. In: Proceedings of the 11th International Conference on Information and Knowledge Management (CIKM), pp. 292–299 (2002)
16. Li, Y., Zhong, Q., Li, J., Tang, J.: Result of ontology alignment with rimom at oaei07. OM, 1 (2007)
17. McGuinness, D., Fikes, R., Rice, J., Wilder, S.: The chimaera ontology environment. In: Proceedings of the 17th National Conference on Articial Intelligence (AAAI), pp. 1123–1124 (2000)
18. Melnik, S., Garcia-Molina, H., Rahm, E.: Similarity flooding: A versatile graph matching algorithm and its application to schema matching. In: Proceedings of 18th International Conference of Data Engineering (ICDE), pp. 117–128 (2002)
19. Mitra, P., Wiederhold, G., Jannink, J.: Semi-automatic integration of knowledge sources. In: Proceedings of the 2nd International Conference On Information FUSION (1999)
20. Muslea, I.: Active learning with multiple views. PhD thesis, Department of Computer Science, University of Southern California (2002)
21. Noy, N., Musen, M.: Prompt: Algorithm and tool for automated ontology merging and alignment. In: Proceedings of the National Conference on Articial Intelligence (AAAI), pp. 450–455 (2000)
22. Salton, G., Buckley, C.: Term-weighting approaches in automatic text retrieval. In: Information Processing and Management, pp. 513–523 (1988)
23. Sarawagi, S., Bhamidipaty, A.: Interactive deduplication using active learning. In: Proceedings of the 8th ACM SIGKDD international conference on Knowledge discovery and data mining, pp. 269–278 (2002)
24. Shvaiko, P., Euzenat, J.: Ten challenges for ontology matching. In: On the Move to Meaningful Internet Systems, OTM (2008)
25. Tan, P., Steinbach, M., Kumar, V.: Introduction to Data Mining (2005)
26. Tang, J., Li, J., Liang, B., Huang, X., Li, Y., Wang, K.: Using bayesian decision for ontology mapping. Web Semantics 4(4), 243–262 (2006)
27. Udrea, O., Getoor, L., Miller, R.J.: Leveraging data and structure in ontology integration. In: Proceedings of the 26th International Conference on Management of Data (SIGMOD 2007), pp. 449–460 (2007)
28. Wang, S., Englebienne, G., Schlobach, S.: Learning concept mappings from instance similarity. In: Sheth, A.P., Staab, S., Dean, M., Paolucci, M., Maynard, D., Finin, T., Thirunarayan, K. (eds.) ISWC 2008. LNCS, vol. 5318, pp. 339–355. Springer, Heidelberg (2008)

Optimizing Web Service Composition While Enforcing Regulations

Shirin Sohrabi and Sheila A. McIlraith

Department of Computer Science, University of Toronto, Toronto, Canada
{shirin,sheila}@cs.toronto.edu

Abstract. To direct automated Web service composition, it is compelling to provide a template, workflow or scaffolding that dictates the ways in which services can be composed. In this paper we present an approach to Web service composition that builds on work using AI planning, and more specifically Hierarchical Task Networks (HTNs), for Web service composition. A significant advantage of our approach is that it provides much of the how-to knowledge of a choreography while enabling customization and optimization of integrated Web service selection and composition based upon the needs of the specific problem, the preferences of the customer, and the available services. Many customers must also be concerned with enforcement of regulations, perhaps in the form of corporate policies and/or government regulations. Regulations are traditionally enforced at design time by verifying that a workflow or composition adheres to regulations. Our approach supports customization, optimization and regulation enforcement all at composition construction time. To maximize efficiency, we have developed novel search heuristics together with a branch and bound search algorithm that enable the generation of high quality compositions with the performance of state-of-the-art planning systems.

1 Introduction

Increasingly, corporations are providing services within and between organizations by publishing programs on corporate intranets or on the World Wide Web. Many of these programs represent component software that can be composed together either manually or automatically to provide value-added service. To direct automated Web Service Composition (WSC), it is compelling to provide some sort of template, workflow or scaffolding that dictates the ways in which services can be composed while leaving enough flexibility for different possible realizations of the programs within the template. A template-based composition is compelling for many applications in domains including e-science (e.g., [1]), e-government (e.g., [2]), and Grid computing (e.g., [3]).

A WSC template is designed with respect to a particular task to be performed. It provides high-level guidance on how to perform a task, but leaves many of the details to run-time synthesis. For many WSC problems, the task can be realized by a diversity of different services, offering comparable, but not identical services. Also unknown at the outset is the data that serves as choice points

A. Bernstein et al. (Eds.): ISWC 2009, LNCS 5823, pp. 601–617, 2009.

in a WSC – the availability of goods, their properties and pricing, etc. A composition template streamlines the generation of a problem, and customer-specific WSC, while enabling the individual customer to customize the composition with respect to their preferences and constraints and/or those of the corporation they work for, the laws of the countries in which they are doing business, etc.

A composition template can be represented in a variety of different ways. One way to represent a template is to use a workflow or a flowchart. This can be expressed pictorially as a schematic or alternatively in a form akin to a procedural programming language. The Algol-inspired Golog agent programming language provides one such procedural language (e.g., [4]). Indeed, the first template-based approach to WSC exploited Golog to provide a so-called generic procedure that provided a template specification of the composition [5,6]. The Golog procedures were combined with individual user constraints (e.g., *"I want to fly with a star alliance carrier"*) at run time, resulting in *dynamic binding of Web services.* However, the user constraints considered were hard constraints, i.e., realizations that did not satisfy those constraints were eliminated. In [7], we extended this framework to be able to deal with *soft* user constraints (i.e., preferences). The proposed preference language handled a wide variety of user constraints. It enabled the synthesis of a composition of services, where the selection of services and service groundings (e.g., in the case of travel, the selection of the specific flight) was customized to individual users at run time. Unfortunately, the implementation of the system, GologPref was not optimized.

Another type of composition template that can be used is based on Hierarchical Task Networks (HTNs) [8]. Like Golog, HTNs provide useful control knowledge — advice on how to perform a composition. However, this how-to knowledge is specified as a *task network.* The task network provides a way of hierarchically abstracting the composition into a set of tasks that need to be performed and that decompose in various ways into leaf nodes realized by programs. Sirin et al. [9] used SHOP2, a highly-optimized HTN planner for the task of WSC. The HTN induces a family of compositions and the if-then-else ordering of SHOP2 provided a means of reflecting a preference for achieving a task one way over another. However this limited form of preference was hard-coded into the SHOP2 domain description (i.e., the method description) and could not be customized by an individual user without recoding the HTN. In [10], an HTN-DL formalization was proposed in which they combined reasoning about Web service ontologies using a DL reasoner with HTN planning. Like their predecessor, they exploited SHOP2 domain ordering to reflect preferences, but these were again not easily customizable to an individual user. They further provided a means of preferring services according to their class descriptions, but did not optimize the selection of service groundings.

Most recently, Lin et al. [11] proposed an algorithm for HTN planning with preferences described in the Planning Domain Definition Language PDDL3 [12] that did allow for preferences over service groundings. They implemented a prototype of the algorithm in a planner, **scup**, tailored to the task of WSC. A merit of this work over previous HTN work is that it is not restricted to SHOP2 syntax and as such provides the nondeterminism (flexibility) necessary for preference-based planning. Unfortunately, the ability of the planner to deal with preferences was somewhat limited. In particular, it appears to be unable to handle conflicting user preferences. The authors indicate that conflicting preferences are removed (rather than resolved) during a pre-processing step prior to run time.

In this paper, we build on our previous work on GologPref, our previous work on HTN planning with rich user preferences, and on the previous work of others on HTN WSC to propose another template-based WSC system, based on HTN planning, **HTNWSC-P**. Our work advances the state of the art by providing an HTN-based WSC system that:

1. synthesizes compositions that adhere to policies and regulations expressed as a subset of linear temporal logic (LTL);
2. exploits a preference language that is truly tailored to WSC with HTNs and that can express preferences over how a task is to be decomposed, as well as preferences over service and data selection;
3. imports and exploits OWL-S profiles for Web service selection;
4. synthesizes a composition that simultaneously optimizes, at run time, the selection of services based on functional and non-functional properties and their groundings, while enforcing stated regulations; and that
5. provides an implementation that combines HTN templates, the optimization of rich user preferences, and adherence to LTL regulations within one system, that reflects and exploits state-of-the-art techniques for planning with preferences. In particular, we exploit our own recent work on HTN planning with preferences [13] as the computational foundation for **HTNWSC-P**.

Elaborating on the first point, many customers must be concerned with enforcement of regulations, perhaps in the form of corporate policies and/or government regulations. Software that is developed for use by a particular corporation or jurisdiction will have the enforcement of such regulations built in. For Web services that are published for use by the masses this is not the case, and the onus is often on the customer to ensure that regulations are enforced when a workflow is constructed from multiple service providers. For inter-jurisdictional or international business, different regulations may apply to different aspects of the composition. In this paper we provide a mechanism for generating compositions from templates that adhere to such regulations.

Figure 1 provides a high-level depiction of our WSC framework. We assume that Web services are described in OWL-S, a leading ontology for describing Web services [14]. We also assume that the composition template (e.g., for trip planning, commodity purchasing, etc.) is described in OWL-S, though it need not be. The user's task (e.g., the specifics of the trip) are specializations of the composition template. We provide a translation from OWL-S to HTN that not only translates OWL-S process models, but also translates service profiles (see Section 2). A user's task, is translated to an initial task network in the HTN framework. User preferences and regulations are also important elements of the structure and could be described within an OWL ontology, though likely not in a

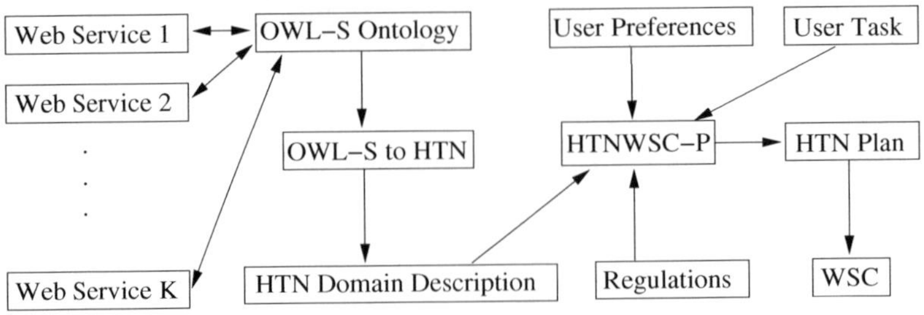

Fig. 1. The overall structure of our composition framework

way that preserves their semantics given their description in LTL. We do not address this issue in the paper. Our HTN WSC planner, **HTNWSC-P**, then takes user preferences specified in PDDL3 syntax (see Section 3) along with a user's task specified as an initial task network and computes a preferred plan, pruning plans that do not meet the imposed regulations specified by LTL formulae. The final HTN plan is then converted to a composition of Web services. In the sections that follow, we elaborate on individual components of our framework.

2 Preliminaries

In this section, we first overview OWL-S [14], a Web ontology for Web services. Next, we describe HTN planning [8] and show how OWL-S can be translated into HTN. The translation is similar to [9], but has some key differences that make service selection based on the non-functional properties possible.

2.1 OWL-S

OWL-S [14] is a Web ontology [15] for Web services with a view to supporting automated discovery, enactment and composition of Web services. The OWL-S ontology has three major components: service profile, process model, service grounding. The service profile is used to advertise the service by describing its functional properties (i.e., input, output, precondition, and effects) and non-functional properties (e.g., service trust, reliability, subject, cost, etc.). The process model describes how the service works, similar to workflow. Finally, service grounding explains how to interact with the service.

OWL-S defines three classes of processes: atomic, composite and simple processes. Each process has input, output, precondition and effects. Atomic processes have no subprocesses and can be executed in a single step. Simple processes provide an abstract view for an existing process. However, unlike atomic processes a simple process is not associated with a grounding. A composite process is composed of other processes via control constructs such as Sequence, Split, Any-Order, Choice, If-Then-Else, Repeat-While, Repeat-Until, and Iterate.

Web service composition systems generally translate OWL-S process models into internal representations such as HTN, PDDL, or Golog that are more amenable to AI planning [7,9,10,11]. Our approach to WSC also translates OWL-S description into an HTN planning problem. In the next section, we briefly describe HTN planning and then describe this translation.

2.2 HTN Planning

Hierarchical Task Network (HTN) planning [8] is a popular and widely used planning paradigm that has been shown to be promising for the task of Web service composition (e.g., [9,11]). In HTN planning, the planner is provided with a set of tasks to be performed, together with a set of so-called *methods* that tell how to decompose tasks into subtasks. Given an initial task network, a plan is formulated by repeatedly decomposing tasks into smaller and smaller subtasks until a primitive decomposition of the initial task network is found. Most of the basic definitions that follow originate in [8].

Definition 1 (HTN Planning Problem). *An HTN planning problem is a 3-tuple* $\mathcal{P} = (s_0, w_0, D)$ *where* s_0 *is the initial state,* w_0 *is a task network called the initial task network, and* D *is the HTN planning domain which consists of a set of operators and methods.*

A domain is a pair $D = (O, M)$ where O is a set of operators and M is a set of methods. An operator is a primitive action, described by a triple o =*(name(o), pre(o), eff(o))*, corresponding to the operator's name, preconditions and effects.

A *task* consists of a task symbol and a list of arguments. A task is primitive if its task symbol is an operator name and its parameters match, otherwise it is *nonprimitive.* A method, m, is a 4-tuple *(name(m), task(m),subtasks(m), constr(m))* corresponding to the method's name, a nonprimitive task and the method's task network, comprising subtasks and constraints. Method m is relevant for a task t if there is a substitution σ such that $\sigma(t)$ =*task*(m). Several methods can be relevant to a particular nonprimitive task t, leading to different decompositions of t. An operator o may also accomplish a ground primitive task t if their names match.

Definition 2 (Task Network). *A task network is a pair w=(U, C) where U is a set of task nodes and C is a set of constraints. The constraints normally considered are of type precedence constraint, before-constraint, after-constraint or between-constraint.*

Definition 3 (Plan). $\pi = o_1 o_2 \ldots o_k$ *is a plan for HTN planning program* $\mathcal{P}$ *if there is a primitive decomposition,* w*, of* w_0 *of which* π *is an instance.*

2.3 From OWL-S to HTN

In this section, we describe how to translate an OWL-S description into an HTN planning domain and problem. We first describe how to encode an OWL-S process model as elements of HTN planning (i.e., operators and methods). Then we describe how to encode the service profile. Encoding the service profile as a component of HTN planning will enable users to specify preferences over how to select services based on their non-functional properties (i.e., those specified in the service profile).

Our translation is similar to that in [9]. In particular, we encode each atomic process as an HTN operator just as in [9]. We also encode each composite and simple process as an HTN method. Where our translation differs is that we associate each method with a unique name. Having a name for a method allows preferences to refer to methods by their name. This is particularly important in preferences that describe how to decompose a particular task. Since a task can be realized by more than one method, being able to distinguish each method by its name allows the user to express preferences over which methods they prefer, or in other words, how they prefer the task to be realized. In the next section, we will give examples of such preferences. Below we show how to translate the Sequence construct. The translations for the rest of the constructs is similar.

Translate-Sequence(Q)
Input: a OWL-S definition of a composite process Q in the form $Q_1;Q_2;...;Q_k$ with Sequence control construct.
Output: an HTN method M.
Procedure:
(1) let $\boldsymbol{v}$ = the list of input parameters defined for Q
(2) let Pre = conjunct of all preconditions of Q
(3) **for all** $i : 1 \leq i \leq k$: let n_i be a task node for Q_i
(4) let $C = \{before(n_1, Pre), (n_i, n_{i+1}) | 1 \leq i < k\}$
(5) **Return** $M = (N_m, Q(\boldsymbol{v}), \{n_1, n_2, ..., n_k\}, C)$, where N_m is a unique method name.

In addition, for every process and subprocesses in the process model that is associated with a service (i.e., is executable on the Web), we compile its service profile as extra properties of their corresponding HTN element. Hence, if an atomic/composite process is associated with a service, its corresponding HTN operator/method will be associated with that service profile. We capture this extra property using a predicate *isAssociatedWith*.

For example, let us assume that the Air Canada service can be described by an atomic process *AP* and service profile *SP*. In addition, assume that the service profile *SP hasName* AirCanada, *has-url* www.aircanada.com, *has-Language* English, *has-trust* high, *has-reliability* high. Then we will encode the atomic process *AP* into an HTN operator with the same name as described above. Next, we would capture the service profile of the service Air Canada associated with the atomic process *AP* by the binary predicate *isAssociatedWith(AP, SP)*. Note *AP* is the name of the encoded HTN operator. In the case of composite process we would have the name of the corresponding HTN method. The profile information of the service profile *SP* would now be described by predicates *has-language(SP, English)*, *has-trust(SP, High)*, and *has-reliability(SP, High)*.

3 WSC with Preferences

In this section, we describe the syntax of the preference language we use for specifying user preferences. The preference language is an extension of the Planning Domain Definition Language, PDDL3 [12] that we proposed in [13]. The preference language supports specification of preferences over how tasks are decomposed analogous to how the process model is realized. It also allows users to specify preferences over the non-functional properties of services as well as their preferred parameterizations of tasks analogous to processes. The semantics of the preference language is defined using the situation calculus [4]. We will not discuss the semantics here and direct readers to [13].

Illustrative Example (Travel Example). To help illustrate our preference language, consider the problem of arranging travel for a conference. The problem can be viewed as having a top-level composite process *bookTravelforConference* that is composed of several other composite processes via the Choice construct. One of the composite processes among them can be viewed as a composite process that is constructed via an Any-Order construct into registering for a conference, arranging transportation, accommodations, local transportation, and getting insurance for the trip. Each of these processes can be constructed via the Choice construct to consider several different Web services that offer flights, hotels, cars, trains, buses, insurance, etc.

3.1 Specifying Preferences in Our PDDL3 Extension

The Planning Domain Definition Language (PDDL) is a standard input language for many planning systems. PDDL3 extends PDDL2.2 to support the specification of preferences and hard constraints over *state* properties of a trajectory. In [13], we extended PDDL3 to support preferences that are over how to decompose tasks as well as expressing preferred parameterization of a task (i.e., constraints over *action* properties). In the context of WSC and OWL-S, following the translation from OWL-S to HTN, how to decompose a tasks is analogous on how to realize a service using its process model. This is particularly important when the process model is constructed using the Choice construct and users may prefer one choice over another. Each preference formula is given a name and a metric value (i.e., penalty if the preference formula is not satisfied). The quality of a plan is defined using a *metric function*. A lower metric value indicates higher satisfaction of the preferences, and vice versa.

PDDL3 supports specification of preferences that are temporally extended in a subset of Linear Temporal Logic (LTL). *always*, *sometime*, *at-most-once*, *sometime-after*, *sometime-before* are among the constructs allowed in PDDL3.

We extended PDDL3 to give users the ability to express preferences over how to decompose tasks as well as expressing preferences over the preferred parameterization of a task. We added three new constructs to PDDL3: **occ**(a), **initiate**(x) and **terminate**(x), where a is a primitive task (i.e., an operator or an atomic process), and x is either a task (i.e., a composite process' name and its input parameters) or a method name (i.e., the unique method name assigned for each method during the translation). **occc**(a) states that the primitive task a occurs in the present state. On the other hand, **initiate**(t) and **terminate**(t) state, respectively, that the task t is initiated or terminated in the current state. Similarly, **initiate**(n) (resp. **terminate**(n)) states that the application of method named n is initiated (resp. terminated) in the current state. Below are some examples from our travel domain given a particular origin Origin and destination Dest[1] that use the above extension.

```
(preference p1 (sometime-after (terminate arrange-trans)(initiate arrange-acc)))
(preference p2 (sometime-after (terminate arrange-acc)(initiate get-insurance)))
(preference p3 (always (not (occ (pay MasterCard)))))
(preference p4 (sometime (initiate (book-flight SA Eco Direct WindowSeat))))
(preference p5
       (imply (different Origin Dest) (sometime (initiate by-flight-trans))))
(preference p6 (imply (and (hasBookedFlight ?Y)(hasAirline ?Y ?X)(member ?X SA))
       (sometime (occ (pay ?Y CIBC)))))
(preference p7 (imply (hasBookedCar ?Z) (sometime (occ (pay ?Z AE)))))
```

p1 states that the task associated with the *arrange-trans* process is terminated before the task associated with the *arrange-acc* process begins (for example: finish arranging your transportation before booking a hotel). Similarly, **p2** states that the task associated with the *arrange-acc* process is terminated before the task associated with the *get-insurance* process begins. The **p3** preference states that the user never pays by Mastercard. Note here that payment with MasterCard is thought of as an atomic process. The **p4** preference states that at some point the user books a direct economy window-seated flight with a Star Alliance (SA) carrier. Here, booking a flight is believed to be a composite process. The **p5** preference states that if Origin and Dest are different, the user prefers that at some point a method named *by-flight-trans* is chosen for decomposition of a task (i.e., the arrange transportation process). The **p6** preference states that if a flight is booked with a Star Alliance (SA) carrier, pay using the user's CIBC credit card. Finally **p7** preference states that if a car is booked, the user prefers to pay with their American Express (AE) credit card.

The **metric function** defines the quality of a plan, generally depending on the preferences that have been achieved by the plan. PDDL3 defines an `is-violated` function, that takes as input a preference name and returns the *number of times* the corresponding preference is violated. It is also possible to define whether we want to maximize or minimize the metric, and how we want to weight its different components. For example, the PDDL3 metric function:

```
(:metric minimize (+ (* 40 (is-violated p1)) (* 20 (is-violated p2))))
```

specifies that it is twice as important to satisfy preference `p1` as to satisfy preference `p2`. Note that it is always possible to transform a metric that requires maximization into one that requires minimization, henceforth, we will assume that the metric is always being *minimized*.

[1] For simplicity, many parameters have been suppressed. Variables start with ?

Further note that inconsistent preferences are handled automatically using the PDDL metric function as discussed above. The metric function is a weighted sum of individual preference formulae. This function is then minimized by our planning approach. In doing so, it makes an appropriate trade off between inconsistent preferences so that it can optimize the metric function.

3.2 Service Selection Preferences

Service selection or discovery is a key component of WSC. However, the only other approaches, to our knowledge, that treat this as a preference optimization task *integrated with* actual composition are [10] and our previous Golog work [7]. In [10], they rely on extending an OWL-S ontology to include abstract processes that refer to service profiles. These descriptions also need to be represented as assertions in an OWL ontology, and an OWL-DL reasoner needs to undertake the task of matching and ranking services based on their service selection preferences. Unfortunately, combining OWL-DL reasoning with planning can create significant performance challenges since one needs to call the OWL-DL reasoner many times during the planning phase, leading to very expensive computations.

Our approach is different. Following discussion in Section 2, during the translation phase we compile each service profile as an extra property of its corresponding HTN element. Note that not all processes will be associated to a service since a process can correspond to an internal subprocess of the service. We only associate profiles with Web-accessible processes. We capture the profile property using a binary predicate *isAssociatedWith(process, service-profile)*. The service-profile serves as an index for the profile information and is encoded as additional predicates (e.g., *has-trust(service-profile, trust)*, *has-reliability(service-profile, reliability)*, etc). Below are some service selection preferences for our travel domain.

```
(preference p8 (always
   (imply (and (initiate ?X)(isAssociatedWith ?X ?Y))(has-trust ?Y High)))
(preference p9 (sometime
   (and (initiate ?Z)(isAssociatedWith ?Z ?Y)(has-name ?Y AirCanada))))
(preference p10 (never
   (and (initiate ?Z)(isAssociatedWith ?Z ?Y)(has-reliability ?Y Low))))
```

p8 states that the user prefers selecting services that have high trust values. **p9** states that a user prefers to invoke the AirCanada service. Lastly, **p10** states that the user prefers to never select low reliability services.

4 Regulation-Based Composition

Policies and regulations are an important aspect of semantic Web services. A number of researchers have proposed approaches to both regulation representation and regulation enforcement as part of semantic Web service tasks (e.g., [16]). Kolovski et al. [17] proposed a formal semantics for the WS-policy [18] language by providing a mapping to a Web ontology language OWL [15] and describing how an OWL-DL reasoner could be used to enforce policies. They provided two translations of WS-Policy to OWL-DL by treating policies as instances and classes in the DL framework. Chun et al. [19] considered policies imposed on both service selection and on the entire composition, expressed using RuleML [20]. In their work, policies take the form of condition-action pairs providing an action-centric approach to policy enforcement.

Regulations are traditionally enforced at design time by verifying that a workflow or composition adheres to the regulations. In our approach, we enforce regulations during composition construction. In particular, during the planning phase we consider only those partial plans that adhere to the regulations while pruning those that do not. In the next section, we provide an algorithm that specifies exactly how this pruning occurs within the HTN algorithm.

In this paper, we focus on regulations that are more geared towards the verification community, particularly those that can be specified as safety constraints. During our regulation enforcement phase, we ensure that the computed composition preserves certain properties of the world. These types of regulations can be specified potentially by state conditions that must hold during the composition. Hence, rather than having action-centric rules in the form of RuleML or rule-based languages, we are interested in assertions that must be enforced during the composition. Classically this form of verification has been represented in Linear Temporal Logic (LTL) [21] or some combination of first-order logic with temporal logic (e.g., [22]). Here, we are not concerned with the representation of regulations within an ontology but rather with how we enforce them within our framework. Hence, for the purpose of this paper we represent regulations in a subset of LTL considering for the most part the *never* and *always* constructs. Below are some example regulations that corporations might impose on their employees when traveling: (1) Always book flights with US-carriers. (2) Never book business or first-class flights. (3) Get pre-approval for travel outside the US. (4) Always pay for flights and hotels with your corporate credit card. As an example, the first regulation above can be written in LTL as follows[2]:

$$\Box\ [(\texttt{(hasBookedFlight ?Y)} \wedge \texttt{(hasAirline ?Y ?X)}) \Rightarrow \texttt{(USCarrier ?X)}]$$

5 Computing Preferred WSC Adhering to Regulations

In this section we address the problem of how to compute a preferred Web service composition while enforcing regulations. Having the HTN encoding of the problem in hand, we turn to planning techniques to help guide construction of the composition. In particular, we exploit our developed heuristics for HTN planning and augment our algorithm [13] to enforce regulations.

Our algorithm is outlined in Figure 2. Our HTNWSC planner performs best-first, incremental search (i.e., always improves on the quality of the plans returned). It takes as input a planning problem (s_0, w_0, D), a metric function MetricFn, a heuristic function HeuristicFn, and regulations Regulations.

[2] $\Box$ is a symbol for *always*.

```
function HTNWSC(s0, w0, D, MetricFn, HeuristicFn, Regulations)
  frontier ← ⟨s0, w0, ∅⟩                                   ▷ initialize frontier
  bestMetric ← worst case upper bound
  while frontier is not empty do
    current ← Extract best element from frontier
    ⟨s, w, partialP⟩ ← current
    if SatisfiesRegulations(s) then              ▷ pruning to enforce regulations
      lbound ← MetricBoundFn(s)
      if lbound < bestMetric then              ▷ pruning suboptimal partial plans
        if w = ∅ and current's metric < bestMetric then
          Output plan partialP
          bestMetric ← MetricFn(s)
        succ ← successors of current
        frontier ← merge succ into frontier
```

Fig. 2. A sketch of our HTNWSC algorithm

frontier contains the nodes in the search frontier. Each of these nodes is of the form $\langle s, w, partialP\rangle$, where s is a plan state, w is a task network, and $partialP$ is a partial plan. *frontier* is initialized with a single node $\langle s_0, w_0, \emptyset\rangle$, where $\emptyset$ represents the empty plan. Its elements are always sorted according to the function HeuristicFn. *bestMetric* is a variable that stores the metric value of the best plan found so far initialized to a high value representing a worst case upper bound. In each iteration of the while loop, the algorithm extracts the best element from the *frontier* and places it in *current*. If the state violates the regulations (i.e., SatisfiesRegulations(s) returns false), this node will be pruned from the search space. LTL regulations are enforced by progression of the formula as the plan is constructed (e.g., [23]). The LTL formulation is more expressive than HTN-type constraints and thus the enforcement is different that e.g., Redux [24]. Using the function MetricBoundFn a lowerbound estimation of the metric value is computed. If *lbound* is greater than or equal to *bestMetric* this node would again be pruned. If *current* corresponds to a plan, *bestMetric* is updated, and the plan is returned. All successors to *current* are computed using the Partial-order Forward Decomposition procedure (PFD) [8], and merged into the *frontier*. The algorithm terminates when *frontier* is empty.

Although templates specified in HTN greatly reduce the search space, a task can be decomposed by a fairly large number of methods corresponding to a large number of services that can carry out the same task. Hence, we use the heuristics proposed in [13] to guide the search towards finding a high-quality composition quickly. We will use four heuristic functions as follows: Optimistic Metric Function (OM), Pessimistic Metric Function (PM), Lookahead Metric Function (LA), and Depth (D). The OM function estimates optimistically the metric value resulting from the current task network w. Recall that in PDDL3 the metric function defines the quality of a plan. The PM function is the dual of OM. LA function estimate the metric of the *best successor* to the current node. It first solves the current node up to a certain depth, then it computes a single primitive decomposition for each of the resulting nodes. In the end, it returns the best metric value among all the fully decomposed nodes. D is another heuristic to guide the search. This heuristic encourages the planner to find a decomposition soon. The HeuristicFn function we use in our algorithm is a *prioritized sequence* of the above heuristics. However, as shown in [13] the best combination is to use D, LA, OM, and PM, in that order, when comparing two nodes. Hence, if the depths are equal, we use the other heuristics in sequence to break ties. We will use this prioritized sequence in our evaluations.

The search space for a WSC is reduced by imposing the template, imposing the regulations, and by further sound pruning that results from the incremental search. In particular, the OM function provides sound pruning if the metric function is non-decreasing in the number of satisfied preferences, non-decreasing in plan length, and independent of other state properties. A metric is non-decreasing in plan length if one cannot make a plan better by increasing its length only (without satisfying additional preferences).

Using inadmissible heuristics does not guarantee generation of an optimal plan. However, we have shown in [13] that in the case the search is exhausted, the last plan returned is guaranteed to be optimal. In our algorithm we are pruning those states that violate the regulations, so optimality is with respect to the subset of plans that adhere to the regulations.

Proposition 1. *If the algorithm performs sound pruning, then the last plan returned, if any, is optimal.*

6 Implementation and Evaluation

We implemented our Web service composition engine using templates specified in HTN, user preferences specified in PDDL3 syntax, regulations specified as LTL formulae, and the user's initial task specified as HTN's initial task network. Our implementation, **HTNWSC-P**, builds on our earlier work **HTNPlan-P** [13] that itself is a modification of the LISP version of **SHOP2** [25]. It implements the algorithm and heuristic described above. We used a 15 minute time out and a limit of 1 GB per process in all our experiments.

HTNWSC-P builds on the effective search techniques for **HTNPlan-P**, which was shown to generate better quality plans faster that the leading planners from the IPC-5 planning competition. We do not repeat these experimental results here. However, as such, we had three main objectives in performing our experimental evaluation. We wanted to evaluate the performance of our implementation as we increased the number of preferences and the number of services. We also wanted to compare our work with other WSC preference-based planners that use HTNs. Unfortunately, we were unable to achieve our third objective, since we could not obtain a copy of **scup**[11], the only other HTN preference-based planner (for WSC) we know of (See Section 7 for a qualitative comparison).

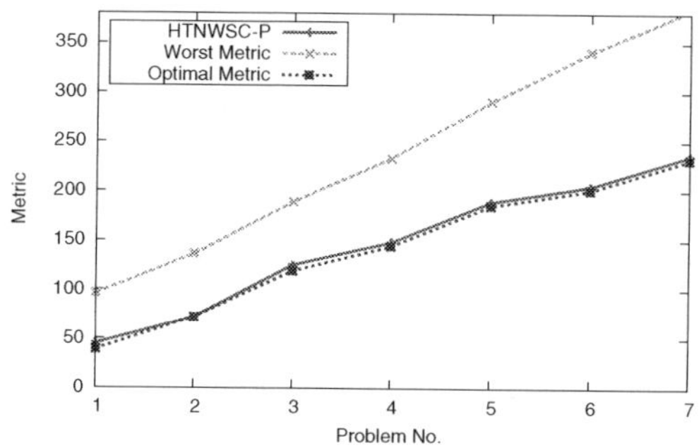

Fig. 3. Evaluating the quality of the last plan as the number of preferences increases. A low metric value means higher quality plan. Worst Metric is a metric value if none of the preferences are satisfied.

Prb #	Ser #	FirstPlan Time(s)	LastPlan Time(s)
1	10	0.22	580.00
2	30	0.23	610.00
3	50	0.21	636.00
4	70	0.22	640.00
5	110	0.23	643.00
6	130	0.24	656.00
7	150	0.24	660.00
8	170	0.26	668.00
9	190	0.24	671.00
10	210	0.25	673.00

Fig. 4. Time comparison between the first and last plan returned as we increase the number of services in the problem

We used the Travel domain described in this paper as our benchmark. (Note that **HTNPlan-P** was additionally evaluated with IPC-5 planning domains.) The problem sets we used were designed to help us achieve our first and second objectives. We achieved this by adding more preferences some of which could potentially be conflicting with each other, and by increasing the number of services, achieved by increasing the branching factor and grounding options of the domain. To this end, we automatically generated 7 problems where the number of services were kept constant and the number of preferences were increased. We similarly generated 10 problems with increasing number of services, keeping the number of preferences constant. The preferences were rich, temporally extended preferences over task groundings and task decompositions. Note that we used a constant number of policies in each problem.

Figure 3 shows the last metric value returned by **HTNWSC-P** for the 7 problems with increasing number of preferences and constant number of services. It also shows the Worst and Optimal Metric value for these problems. Worst Metric is the metric value of the problem if none of the preferences are satisfied while Optimal Metric is the best possible metric value achievable. The result shows that **HTNWSC-P** finds a very close to optimal solution within the time limit. Furthermore, similar to our work in [13], we observe a rapid improvement during the first seconds of search, followed by a marginal one after that.

Next, we evaluated the performance of **HTNWSC-P** by increasing the number of available services. This results in having more methods and operators in the HTN description, hence, the number of possible ways to decompose a single task increases. This causes the number of nodes in the *frontier* to blow up according to the algorithm described in Section 5, and the planner to run out of stack. There are two common ways HTN planners solve this problem. Combining the advantages of both, we propose a middle-ground solution to the problem.

One way to avoid the problem is to have a limit on the size of the *frontier* as in [11]. However, this approach only works if the size is relatively small. Moreover, many possible decompositions and high-quality solutions could potentially be removed from the search space. Another approach is to use the if-then-else non-determinism semantics taken for example by **SHOP2**. In this semantics, if there are several methods m_1 to m_k that can be used to decompose a task t, method $m1$ should be used if it is applicable, else method m_2, else method m_3, and so forth. Hence, the order in which the methods are written in the domain description can influence the quality of the results. This simple ordering is considered a form of user preferences in [25]. Hence, users must write different versions of a domain description to specify their preferences. However, this form of preferences is very limited and is analogous to writing different templates for different users as opposed to customizing one fixed template to meet users' differing needs.

In this experiment, we employed a combination of the above two approaches, modifying our algorithm to place a limit on the number of applicable methods for a task. Our search considered *all* tasks by considering all of their corresponding nodes in the *frontier* but we limited the number of applicable methods for each task. Note that with this approach we might also potentially prune good-quality plans but the likelihood of this is small compared to limiting the size of the frontier. Nevertheless, our optimality result does not hold for this experiment. Our results are summarized in Figure 4.

Figure 4 shows the time to find the first and the last plan within the time-out. The experiments are run on the 10 problem sets with constant preferences and increasing service numbers. Note that the metric value of all the first and last plans is equal since all 10 problems use the same sets of preferences. The result shows that as the number of services increases, the time to solve the problem increases only slightly.

Finally, recall that our implementation is incremental, performing search in a series, each one returning a better-quality plan. To see how effective this approach is, we calculated the *percent metric improvement* (PMI), i.e., the percent difference between the metric of the first and the last plan returned by our planner (relative to the first plan). The average PMI for the problems used in our experiments is 23%.

7 Summary and Related Work

A number of researchers have advocated using AI planning techniques to address the task of Web service composition including planners that are based on model checking (e.g., [26]) and planners that use a regression-based approach [27]. Previous work has also considered using a template or workflow to ease the task of composition including the work using Golog [5,6,13] and HTNs [9,10,11]. Work by Calvanese, de Giacomo and colleagues on the so-called Roman model is another example of a template-like approach in that they provide a desired behaviour to be synthesized (e.g., [28]) by a set of services. This desired behaviour plays a similar role to that of a template however the synthesis itself is performed using techniques from finite state controller synthesis. Following in this tradition, we also take a template-based approach to WSC. Our templates are specified using HTN domain descriptions and can be customized by the specification of rich user preferences and by the specification of hard regulations. Users specify their preferences in our PDDL3 extension that supports conditional, temporally extended, service selection preferences as well as preferences over how to parameterize and how to decompose a task. Regulations are specified at LTL formulae. We provide translation from OWL-S to HTN that not only translates OWL-S process models, but also translates service profiles into our HTN framework. Our composition engine, **HTNWSC-P**, then takes user preferences and computes a preferred composition while pruning those that do not meet the imposed regulations. Our algorithm is based on our previous work on HTN preference-based planning that has been demonstrated to outperform leading planners. Experimental evaluation shows that our approach can be scaled as we increase the number of preferences and the number of services.

Most of the related work with respect to specifying and imposing regulations has already been discussed in Section 5. There has also been work on compliance checking using a constraint-based approach that is similar in spirit to regulation enforcement (e.g., [29]). Also, recent work [30] has considered integrity constraints, and proposed various ways to solve the ramification problem. Solving the ramification problem is not a focus of this paper.

The most notable and closest work to ours that uses both HTNs and preferences developed for IPC-5 is [11]. Unfortunately, the **scup** prototype planner is not available for experimental comparison. There are several differences among our works. In particular, they translate user preferences into HTN constraints and preprocess the preferences to check if additional tasks need to be added. They also have an interesting approach to the problem by combining HTN planning with DL, and by using a DL reasoner. However, their preferences are specified in PDDL3, while our preferences can be expressed in the PDDL3 extension that uses HTN-specific preference constructs. Moreover, they do not translate service profiles; hence, they are unable to specify preferences over service selections. Additionally, they do not consider handling regulations, a hallmark of our work. Further, their algorithm cannot handle conflicting user preferences at run-time, and so conflicts need to be detected as a pre-processing step.

Acknowledgements. We thank our colleague Jorge Baier for helpful discussion. We gratefully acknowledge funding from the Natural Sciences and Engineering Research Council of Canada (NSERC) and the Ontario Ministry of Innovations Early Researcher Award (ERA).

References

1. Cheung, W.K.W., Gil, Y.: Privacy enforcement through workflow systems in e-science and beyond. In: Proceedings of the ISWC 2007 Workshop on Privacy Enforcement and Accountability with Semantics (PEAS) (2007)
2. Chun, S.A., Atluri, V., Adam, N.R.: Policy-based Web service composition. In: Proceedings of the 14th International Workshop on Research Issues on Data Engineering: Web Services for E-Commerce and E-Government ApplicationsRIDE, pp. 85–92. IEEE Computer Society, Los Alamitos (2004)

3. Gil, Y., Deelman, E., Blythe, J., Kesselman, C., Tangmunarunkit, H.: Artificial intelligence and grids: Workflow planning and beyond. IEEE Intelligent Systems 19(1), 26–33 (2004)
4. Reiter, R.: Knowledge in Action: Logical Foundations for Specifying and Implementing Dynamical Systems. MIT Press, Cambridge (2001)
5. McIlraith, S., Son, T., Zeng, H.: Semantic Web services. IEEE Intelligent Systems. Special Issue on the Semantic Web 16(2), 46–53 (2001)
6. McIlraith, S., Son, T.: Adapting Golog for composition of semantic Web services. In: Proceedings of the 8th International Conference on Knowledge Representation and Reasoning (KR), pp. 482–493 (2002)
7. Sohrabi, S., Prokoshyna, N., McIlraith, S.A.: Web service composition via generic procedures and customizing user preferences. In: Cruz, I., Decker, S., Allemang, D., Preist, C., Schwabe, D., Mika, P., Uschold, M., Aroyo, L.M. (eds.) ISWC 2006. LNCS, vol. 4273, pp. 597–611. Springer, Heidelberg (2006)
8. Ghallab, M., Nau, D., Traverso, P.: Hierarchical Task Network Planning. In: Automated Planning: Theory and Practice. Morgan Kaufmann, San Francisco (2004)
9. Sirin, E., Parsia, B., Wu, D., Hendler, J., Nau, D.: HTN planning for Web service composition using SHOP2. Journal of Web Semantics 1(4), 377–396 (2005)
10. Sirin, E., Parsia, B., Hendler, J.: Template-based composition of semantic Web services. In: AAAI 2005 Fall Symposium on Agents and the Semantic Web (2005)
11. Lin, N., Kuter, U., Sirin, E.: Web service composition with user preferences. In: Bechhofer, S., Hauswirth, M., Hoffmann, J., Koubarakis, M. (eds.) ESWC 2008. LNCS, vol. 5021, pp. 629–643. Springer, Heidelberg (2008)
12. Gerevini, A., Long, D.: Plan constraints and preferences for PDDL3. Technical Report 2005-08-07, Department of Electronics for Automation, University of Brescia, Brescia, Italy (2005)
13. Sohrabi, S., Baier, J.A., McIlraith, S.A.: HTN planning with preferences. In: Proceedings of the 21st International Joint Conference on Artificial Intelligence(IJCAI), pp. 1790–1797 (2009)
14. Martin, D., Burstein, M., McDermott, D., McIlraith, S., Paolucci, M., Sycara, K., McGuinness, D., Sirin, E., Srinivasan, N.: Bringing semantics to Web services with OWL-S. World Wide Web Journal 10(3), 243–277 (2007)
15. Horrocks, I., Patel-Schneider, P., van Harmelen, F.: From $\mathcal{SHIQ}$ and RDF to OWL: The making of a Web ontology language. Journal of Web Semantics 1(1), 7–26 (2003)
16. Tonti, G., Bradshaw, J.M., Jeffers, R., Montanari, R., Suri, N., Uszok, A.: Semantic Web languages for policy representation and reasoning: A comparison of KAoS, Rei, and Ponder. In: Fensel, D., Sycara, K., Mylopoulos, J. (eds.) ISWC 2003. LNCS, vol. 2870, pp. 419–437. Springer, Heidelberg (2003)
17. Kolovski, V., Parsia, B., Katz, Y., Hendler, J.A.: Representing Web service policies in OWL-DL. In: Gil, Y., Motta, E., Benjamins, V.R., Musen, M.A. (eds.) ISWC 2005. LNCS, vol. 3729, pp. 461–475. Springer, Heidelberg (2005)
18. WS-Policy: Web service policy framework (WS-policy), `http://www.w3.org/Submission/WS-Policy/`
19. Chun, S.A., Atluri, V., Adam, N.R.: Using semantics for policy-based Web service composition. Distrib. Parallel Databases 18(1), 37–64 (2005)
20. RuleML: Rule markup language (RuleML), `http://ruleml.org/`
21. Emerson, E.A.: Temporal and modal logic. In: Handbook of theoretical computer science: formal models and semantics **B**, pp. 995–1072 (1990)

22. Gerth, R., Peled, D., Vardi, M.Y., Wolper, P.: Simple on-the-fly automatic verification of linear temporal logic. In: Proceedings of the 15th International Symposium on Protocol Specification, Testing and Verification (PSTV), pp. 3–18 (1995)
23. Bacchus, F., Kabanza, F.: Using temporal logics to express search control knowledge for planning. AI Magazine 16, 123–191 (2000)
24. Petrie, C.J.: The Redux Server. In: Proc. Intl. Conf. on Intelligent and Cooperative Information Systems (ICICIS), pp. 134–143 (1993)
25. Nau, D.S., Au, T.C., Ilghami, O., Kuter, U., Murdock, J.W., Wu, D., Yaman, F.: SHOP2: An HTN planning system. Journal of Artificial Intelligence Research (JAIR) 20, 379–404 (2003)
26. Traverso, P., Pistore, M.: Automatic composition of semantic Web services into executable processes. In: McIlraith, S.A., Plexousakis, D., van Harmelen, F. (eds.) ISWC 2004. LNCS, vol. 3298, pp. 380–394. Springer, Heidelberg (2004)
27. McDermott, D.V.: Estimated-regression planning for interactions with Web services. In: Proceedings of the 6th International Conference on Artificial Intelligence Planning and Scheduling (AIPS), pp. 204–211 (2002)
28. Calvanese, D., Giacomo, G.D., Lenzerini, M., Mecella, M., Patrizi, F.: Automatic service composition and synthesis: the Roman Model. IEEE Data Eng. Bull. 31(3), 18–22 (2008)
29. Hoffmann, J., Weber, I., Governatori, G.: On compliance checking for clausal constraints in annotated process models. In: Journal Information Systems Frontiers (2009)
30. Hoffmann, J., Bertoli, P., Helmert, M., Pistore, M.: Message-based Web service composition, integrity constraints, and planning under uncertainty: A new connection. Journal of Artificial Intelligence Research (JAIR) 35, 49–117 (2009)

A Weighted Approach to Partial Matching for Mobile Reasoning

Luke Albert Steller, Shonali Krishnaswamy, and Mohamed Methat Gaber

Faculty of Information Technology, Monash University, Melbourne, Australia
{Luke.Steller,Shonali.Krishnaswamy,Mohamed.Gaber}@infotech.monash.edu

Abstract. Due to significant improvements in the capabilities of small devices such as PDAs and smart phones, these devices can not only consume but also provide Web Services. The dynamic nature of mobile environment means that users need accurate and fast approaches for service discovery. In order achieve high accuracy semantic languages can be used in conjunction with logic reasoners. Since powerful broker nodes are not always available (due to lack of long range connectivity), create a bottleneck (since mobile devices are all trying to access the same server) and single point of failure (in the case that a central server fails), on-board mobile reasoning must be supported. However, reasoners are notoriously resource intensive and do not scale to small devices. Therefore, in this paper we provide an efficient mobile reasoner which relaxes the current strict and complete matching approaches to support anytime reasoning. Our approach matches the most important request conditions (deemed by the user) first and provides a degree of match and confidence result to the user. We provide a prototype implementation and performance evaluation of our work.

1 Introduction

The number of mobile subscribers is reaching the 3 billion mark, world wide [20]. This provides significant opportunities for new mobile applications which meet mobile user demands for improved access to information in their environment. The real usefulness of these small devices is in their interaction [23], that is, information sharing. [24] advocates the usefulness of service oriented architectures and accurate service discovery for mobile services, because of the need to support user mobility in dynamic environments. In addition, advances in device capability mean that mobile devices can act not only as service consumers but also as service providers or will do so in the near future [19,16]. Most current service discovery architectures perform matching on a high-end broker server. However, "for dynamic ad-hoc mobile environments, assuming the existence of devices that are stable and powerful enough to play the role of the central service registries [or brokers] is inappropriate" [16]. In addition, with the growth of increasing services, it is seen that such a centralised approach could become a bottle neck for the whole system [22]. Cost is also a significant factor which determines whether users are likely to use a mobile service. There are clear evidence in studies that the benefits must exceed the cost [9]. It is also noteworthy,

A. Bernstein et al. (Eds.): ISWC 2009, LNCS 5823, pp. 618–633, 2009.

that in terms of monitoring costs as well as cost energy efficiency, communication from mobile devices to a remote location has been established as more expensive than performing computation on the mobile device.

Employing a partially or completely decentralised approach in which the matching occurs on the mobile device itself overcomes all of these problems. Onboard reasoning would remove a central point of failure, deployment is cheap and scalable because adding more users does not involve any additional network provision and elevates privacy concerns. The ontologies used in the reasoning process could be collected by devices as the user walks past other devices, short-range download points (such as a Kiosk or shopping centre entrances) or downloaded previously (eg at home or work) from the Internet. However, mobile service discovery and matching must meet two important user requirements. Matching must be accurate / useful [9] and fast [14].

Accuracy can be achieved by adopting semantic web ontology languages (OWL) and using logic based reasoners to infer relationships to model requirements and using logic reasoners to infer new information from these ontologies. Most current semantic reasoners utilise description logic (DL) which provides the most expressive, decidable OWL logic. However, these logic reasoners are notoriously slow and resource intensive [25]. As such current reasoners cannot be ported to resource constrained mobile devices such as smart phones and PDAs, in their current form, especially since mobile users generally require a result within 10-15 seconds [14]. Current reasoners give only a $true$ or $false$ result, or no result at all, if the inference task is not completed in full. [4] advocates the need to relax these traditional notions of absolute accuracy and completeness for increased performance without reduced expressiveness. Therefore, in this paper, we provide such an approach, a prototype and an evaluation, which has the following features:

1. match the most important request conditions first (as deemed by the user explicitly or implicitly by preferences);
2. handle heterogeneous inaccurate data by continuing to reason after a term failed (if there is sufficient time);
3. handle time/resource pressures by supporting early stopping of the reasoner and providing a current degree of inference to the user.

The remainder of the paper is structured as follows. In section 2 we describe related work. In section 3 we discuss the current approach to reasoning using Tableaux. In section 4, we provides an adaptive reasoning approach to meet the three goals above. In section 5 we discuss our implementation and provide evaluations of our work. Finally in section 6 we conclude the paper.

2 Related Work

While current service discovery architectures such as Jini [1] do not make use of semantic languages, there is a growing emergence of OWL-S semantic matchmakers such as CMU Matchmaker [15] which requires a centralised high-end node to perform reasoning. Architectures such as DIANE [11] and Gaia [13] provide semantically driven context middleware. EASY [12] takes context and QoS into consideration and performs indexed classification of the ontology hierarchy offline such that subsequent lookup is much faster. However all of these architectures require the existence of a high-end central node, due to their reliance on current reasoners.

Gu et al. [6] an RDF reasoner which runs on mobile devices on J2ME with acceptable performance, but it only supports a subset of semantic technologies and forward chaining rule inference. Kleeman et. al. [10] have developed KRHyper, a first order logic (FOL) reasoner for deployment on resource constrained devices. However, performance is comparable with RacerPro. [5] compares ontology terms by Google distance, but this requires Internet access to Google. [7] disregards non-horn clauses, to provide a faster but less accurate result by reducing expressiveness. These works do not implement logical level optimisations to enable mobile reasoning, other than by reducing the language expressively.

The work provided in [18] approximates class membership inference and iteratively matches all inference conditions. The reasoner can be stopped maturely, and a *true* result is given if the current approximation / iteration holds. [21] provides conjunctive query answering and instance retrieval. The main limitation of these approaches is that they rely on the Tableaux algorithm. It is evident in [2] that DL Tableaux checks conditions in depth first order. Current approaches do not take level of importance of each condition into consideration and do not provide degree of match / confidence metrics. Therefore, we provide an approach to address this need. In the next section we describe current Tableaux reasoning then introduce our novel architecture in section 4.

3 Tableaux Reasoning

The power of semantics means that in addition to explicit definitions and assertions, implicit knowledge can be made explicit using inference. The most expressive OWL language which retains decidability [3] is known as OWL-DL which is based on description logic (DL) [2]. OWL-DL inference proves include Pellet[1], FaCT++[2], RacerPro[3] and KAON2[4]. Most of these reasoners use the Tableaux algorithm which has "dominated recent DL research" [2].

DL comprises a TBox $\mathcal{T}$ containing terminological class concept definitions and an ABox $\mathcal{A}$ comprising assertion definitions. The ABox assertions are called nodes or individuals x which are defined by their membership to a class concept C and by its relations R to other objects to form a graph. These two definitions are given in equation 1.

$$C(x) \qquad or \qquad R(x_1, x_2) \tag{1}$$

[1] http://clarkparsia.com/pellet
[2] http://owl.man.ac.uk/factplusplus
[3] http://www.racer-systems.com
[4] http://kaon2.semanticweb.org

Tableaux [8] is an unsatisfiability algorithm which attempts to prove an inference by refuting it. For instance, in order to prove an inference of the form $C(x)$, its negation is added the ABox $\mathcal{A}$ such that $\mathcal{A}_0 = \mathcal{A} \cup \neg C(x)$. The inference is proven if any attempt to extend the asserted negation into a complete interpretation will lead to a logical contradiction (a concept and its negation are present for the same individual). Alternatively, if a sound and non-contradictory interpretation is found, then this represents a counter example that disproves the conjectured inference.

Tableaux minimises the amount of space used by utilising a Tableaux expansion tree $\mathcal{T}$. This tree imposes an ordering on the application of expansion rules. Disjunctive concepts give rise to expansion and every possible expansion is searched in turn until a fully expanded and clash free tree is found, or all possibilities have been shown to lead to a contradiction. When $\mathcal{A}$ contains the disjunction $(C_1 \sqcup C_2)(x)$, disjunction transformation rule replaces $\mathcal{A}$ with $\mathcal{A}' = \mathcal{A} \cup \{C_1(x)\}, \mathcal{A}'' = \mathcal{A} \cup \{C_2(x)\}$. As such $(C_1 \sqcup C_2)(x)$ represents a branching point node n in the expansion tree $\mathcal{T}$ which contains new ABox states, and the concepts $C_1(x)$ and $C_2(x)$ each represent possible future branching point nodes n. In this paper, we distinguish between branching point nodes $n \in \mathcal{T}$ and individuals $x \in \mathcal{A}$, by referring to n as a node and x as an individual.

Tableaux also labels each concept in an individual and each role in an edge with a dependency set, indicating the branching points on which it depends. A concept $C \in \mathcal{L}(x)$ depends on branching point n if C was added to $\mathcal{L}(x)$ at the branching point n or if C depends on another concept D (or role R), and D (or R) depends on the branching point n. A role $R = \mathcal{L}(\langle x, y \rangle)$ depends on concept D when $\langle x, y \rangle$ was labelled R by the application of an expansion rule that used D. When a clash occurs, the reasoner state is restored to the most recent branch point where exploring another branch might alleviate the cause of the clash.

It is evident in [2] that DL Tableaux employs depth first expansion, which we illustrate in algorithm 1. Initially, this algorithm is started using the call $TabTreeTraverse(f, H, \mathcal{A}, \mathcal{T})$ where f is the top branch node in the expansion tree, H is the assertion $\neg C(x)$ for the inference check $C(x)$, $\mathcal{A}$ is the ABox, which is initialised with all the explicit assertions and relations in the semantic graph (knowledge base) and $\mathcal{T}$ is the expansion tree. The algorithm attempts to expand the tree and apply Tableaux transformation rules in order to generate a clash to prove all alternatives of the the tree, which are all of the conditions of the inference. If it returns *true* then the inference is proven. The algorithm utilises $ApplyTransRules(\mathcal{A})$ which applies the normal Tableaux DL transformation rules as defined in [2], on ABox $\mathcal{A}$ until a clash is detected (return *true*) or there are no more rules to apply (return $false$). The backjumping functionality is provided in algorithm 2. We note that we define a branch point node n identifier as $branchID(n)$ and we define a concept C or role R dependency set using $depBranchIDs(C)$. Let $currBranchNode(\mathcal{A})$ contain the currently active branch point node in the expansion tree for the ABox $\mathcal{A}$.

An example of an dependency directed, expansion tree is shown in figure 1. This figure represents the inference check $C(x)$ where C is a conjunction of the

Algorithm 1. TabTreeTraverse$(n, W, \mathcal{A}, \mathcal{T})$

Inputs: BranchNode n, Assertion W, ABox $\mathcal{A}$, ExpansionTree $\mathcal{T}$ where W is of the form $C(x)$
Outputs: Boolean $clashFound$
Let $clash$ be a clashing concept if one exists
$clash \leftarrow ApplyTransRules(\mathcal{A})$ * runs the standard Tableaux rules *\
if $clash \neq null$ **then**
    return true
end if
Let $clashFound \leftarrow$ **false**
Let $branchID(n)$ denote the branch point node identifier for n
if $W = (D_1 \sqcup D_2 \sqcup ... \sqcup D_m)(x)$ * has children *\ **then**
    for all $D_j \in W$ **do**
        Let c be a new BranchNode in the expansion tree $\mathcal{T}$ for $\mathcal{A}$, $\mathcal{T} \leftarrow \mathcal{T} \cup \{c\}$
        $branchID(c) \leftarrow branchID(c) + 1$
        $currBranchNode(\mathcal{A}) \leftarrow c$
        $\mathcal{L}(x) \leftarrow \mathcal{L}(x) \cup \{D_j(x)\}$ where x is an individual in $\mathcal{A}$
        $clashFound \leftarrow TabTreeTraverse(c, D_i(x), \mathcal{A}, \mathcal{T})$
        remove c from $\mathcal{T}$
        if $clashFound =$ **false then**
            return false* failed to prove inference for disjunct c *\
        else
            $restoreTo(n)$ * backjump to n *\
        end if
    end for
end if
return $clashFound$

Algorithm 2. RestoreTo$(u, \mathcal{A})$

Inputs: int u and ABox where u is the branch node identifier to restore $\mathcal{A}$ to.
Outputs: ABox $\mathcal{A}$
for all $x_i \in \mathcal{A}$ where x_i is an individual * all individuals in ABox $\mathcal{A}$ *\ **do**
    for all $A_j \in \mathcal{L}(x_i)$ where A_j is a class concept **do**
        Let t be the minimum value in $depBranchIDs(A_j)$ where $depBranchIDs(A_j)$ contains the branch dependency identifiers $branchID$ for class concept A_j
        if $t > u$ * A_j added after u *\ **then**
            remove A_j from $\mathcal{L}(x_i)$
        end if
    end for
end for
return $\mathcal{A}$

form $C \equiv ((C_1 \sqcap C_2) \sqcap C_3)$. Also assume $\{C_1, C_2, C_3\} \subseteq \mathcal{L}(x)$ where $x \subseteq \mathcal{A}$. Since Tableaux proves inference by refutation C is transformed into a disjunction of the form $\neg C \equiv ((\neg C_1 \sqcup \neg C_2) \sqcup \neg C_3)$. In figure 1a, a clash is detected caused by C_1 which is part of the disjunction $(\neg C_1 \sqcup \neg C_2)$, added to $\mathcal{L}(x)$ at branch node 2. Therefore, the Tableaux is restored to its earlier state, by a backjump to branch 2. Then in 1b, the second element $\neg C_2$ of $(\neg C_1 \sqcup \neg C_2)$ is applied, giving rise to another clash. The reasoner backjumps to expansion tree node 1 (1c), after which the second disjunct $\neg C_3$ of $((\neg C_1 \sqcup \neg C_2) \sqcup \neg C_3)$ is applied giving rise to a third clash. All branch nodes clashed, so the inference $C(x)$ is proven. Note that grey nodes indicate those which have already been evaluated in a previous iteration. In this example, only one individual x was used for briefness, however in practise the expansion tree may apply disjuncts to several different individuals.

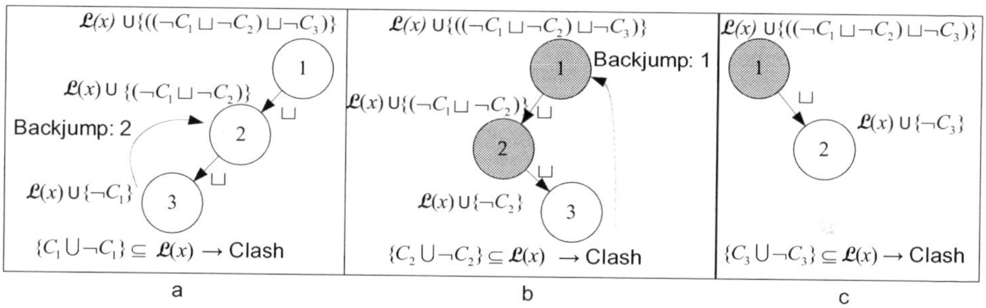

Fig. 1. Standard Tableaux Expansion Search Tree

Current Tableaux has the following shortcomings:

1. Only a true/false answer is provided. A weighted degree of match and a level of confidence in this match based on how much of the inference was checked, should be provided;
2. The order of condition and subcondition evaluation is depth-first. These should be applied in weighted order of importance to the user, so that the highest possible degree of match can be found in the time available;
3. Tableaux returns $false$ as soon as any condition or subcondition in the inference check fails. Since pervasive environments inherently contain dynamic heterogeneous data the reasoner should continue to check subsequent inference conditions, even if some fail (time/resource permitting).

Our approach, which addresses these shortcomings, is introduced in the next section.

4 Weighted Adaptive Reasoning

We propose an approach to Tableaux inference proof, which supports a reduction in result accuracy if greater efficiency is required. Our approach associates a level of importance with each branch leaf and evaluates these branches first. The reasoner will continue even if some conditions fail, and may be stopped at any point in the reasoning process and a degree of match result is provided to the user.

4.1 Weighted Expansion Tree Traversal

Certain inference conditions may have a different level of importance to the user. Therefore, we associate weights with each condition and sub-condition. These weights may be entered explicitly by the user at the time that the user specifies the request as a rating from 1 to 10, high/medium/low priority, or mandatory/non-mandatory. These values are converted to relative weights (see below). Alternatively, weights may be gathered implicitly using historical user preference data. For instance, if for half his or her requests, the user has invoked services which are close in proximity, this characteristic may be inserted to future requests with a weight of 0.5. In summary, weights may be assigned by different means, however, the assignment process is not the principle focus on this paper, therefore, we do not discuss this in further detail.

Let rw denote a relative weight value for each a condition or sub-condition derived from each conjunct or subconjunct in the class concept C, for the inference check $C(x)$. Since Tableaux proves the inference using negation, C is converted into a disjunction where $\neg C \equiv (\neg D_1 \sqcup (\neg D_2 \sqcup \neg D_3) \sqcup ... \neg D_m), 1 \leq j \leq m$. Each disjunct and sub-disjunct are organised as pairs containing the disjunct $\neg D_j$ and relative weight $rw(D_j)$ pairs and stored in descending $rw(D_i)$ order in a queue, such that $\mathcal{Q} = \{\langle \neg D_1, rw(\neg D_1)\rangle, \langle \neg D_2, rw(\neg D_2)\rangle, \langle \neg D_3, rw(\neg D_3)\rangle, ..., \langle \neg D_m, rw(\neg D_m)\rangle\}$, as shown in algorithm 3 where $\neg C$ is passed as input from the inference $C(x)$.

Weights are said to be relative, because the sum of the $rw(D_j)$ values for all children nodes much equal the $rw(p_e)$ of the parent node. Let p_e denote a node, and let $\mathcal{C}$ denote a set of child nodes for p_e such that $\mathcal{C} = \{c_1, c_2, ..., c_t \mid (1 \leq s \leq t)\}$. rw values must be defined such that $rw(p_e) = \sum_{c_s \in \mathcal{C}} rw(c_s)$ where $\langle p_e, rw(p_e)\rangle$. Relative weights ensure that a disjunct D_1 in $P = (D_1 \sqcup D_2)$, cannot be applied before its parent disjunction P is itself applied.

Algorithm 3. CreateQueue(A)

1: **Inputs:** ClassConcept A
2: **Outputs:** Queue $\mathcal{Q}$
3: **if** $A = (D_1 \sqcup ... \sqcup D_m)$ **then**
4:     **for all** $D_j \in D$ **do**
5:         $\mathcal{Q} \leftarrow \mathcal{Q} \cup \{\langle D_j, rw(D_j)\rangle\}$
6:     **end for**
7: **end if**

Algorithm 4 presents a general illustration of our adaptive reasoning search tree expansion process, which replaces depth first with condition relative weight order. This algorithm is invoked by the call $AdapTreeTraverse(\mathcal{Q}, x, \mathcal{A}, \mathcal{T})$ where $\mathcal{Q}$ is the pre-initialised queue as specified above, of disjuncts and sub-disjuncts which have been expressed in $\neg C$ for the inference check $C(x)$, and their relative weights. x is the individual in the inference check $C(x)$. $\mathcal{A}$ is the ABox, which contains the explicit assertions and relations in the semantic graph (knowledge base) and $\mathcal{T}$ is the expansion tree. Unlike Algorithm 1, AdapTreeTraverse (algorithm 4) is not a recursive algorithm, rather it iterates through all the concepts in the queue $\mathcal{Q}$, as specified above, in descending relative weight order. It attempts to expand the tree and apply Tableaux transformation rules to generate clashes to prove each condition. The algorithm continues until all conditions are checked, or the algorithm is stopped prematurely when $KeepMatching()$ results a $false$.

Algorithm 4 uses the functions defined as follows. Let $currBranchNode(\mathcal{A})$ be a holder containing the currently active branch point node in the expansion tree for ABox $\mathcal{A}$. Let $ApplyTransRules(\mathcal{A})$ denote a function which applies the normal Tableaux DL transformation expansion rules on ABox $\mathcal{A}$ until a clash is detected (return $true$) or there are no more rules to apply (return $false$). Functions $GetDegMatch()$ and $GetConf()$ return the degree of match and the confidence values (see section 4.3), which also make use of the $clashingConcept(W)$ and $checkedConcept(W)$ return value.

The function, $KeepMatching()$ is used to determine whether the reasoner should stop execution prematurely, based on constraints. There may be several constraint parameters P, such that $P = \{p_1, p_2, ..., p_n \mid (1 \leq i \leq n)\}$ and threshold values T for parameters, such that $T = \{t_1, t_2, ..., t_n \mid (1 \leq i \leq n)\}$. Each p_i corresponds to a threshold value t_i. A p_i represents a ratio $(0 \leq p_i \leq 1)$, such as memory used over total memory, current confidence over total total possible confidence, time elapsed over total time available (user specified), and battery life remaining over total battery life. Each t_i is specified by application defaults or explicit user requirements, such that $0 \leq t_i \leq 1$. $KeepMatching()$ returns $false$ iff any $t_i \geq p_i$. In the current implementation of the algorithm (see section 5), we utilise time as our only resources parameter, such that $p_1 = timeElapsed/totalTime$, where $totalAvailTime$ is user specified, eg "I want a result in 10 seconds".

Figure 2 illustrates an example utilising our adaptive reasoning approach, for the inference check $C(x)$ where $\neg C \equiv ((\neg C_1 \sqcup \neg C_2) \sqcup \neg C_3)$ is given the relative weights, such that $\{\langle \neg C, 1.0\rangle, \langle(\neg C_1 \sqcup \neg C_2), 0.7\rangle, \langle \neg C_1, 0.5\rangle, \langle \neg C_3, 0.3\rangle, \langle \neg C_1, 0.2\rangle\}$. Assume also that $\{C_1, C_2, C_3\} \subseteq \mathcal{L}(x)$ where $x \in \mathcal{A}$. The nodes are executed in relative weight rw order (weights and node identifiers are displayed on the nodes in figure 2). Notice that in contrast to depth-first Tableaux (see figure 1), concept $\neg C_3$ is applied in step b, before the concept $\neg C_2$ from $(\neg C_1 \sqcup \neg C_2)$ is applied in step c. This is because $\neg C_3$ had a relative weight of 0.3, which was higher than 0.2 for $\neg C_2$. The grey nodes denote those nodes which have already been expanded in a previous step and do not need to be re-applied. In this example, only one individual x was used for briefness, however in practise the expansion tree may apply disjuncts to several different individuals.

4.2 Branch Identifiers and State Management

The, relative weight values control execution, as opposed to the traditional depth or breath-first ordering. Therefore, it is likely that there may be multiple unfinished expansion branches. As shown in section 3, in depth-first reasoning,

Algorithm 4. AdapTreeTraverse($\mathcal{Q}, x, \mathcal{A}, \mathcal{T}$)

Inputs: Queue $\mathcal{Q}$, Individual x, ABox $\mathcal{A}$, ExpansionTree $\mathcal{T}$.
Outputs: $\langle$Decimal $degMatch$, Decimal $conf\rangle$
* $\mathcal{Q}$ is in descending order by $rw(n_j)$ where $\mathcal{Q} = \{\langle D_1, rw(D_1)\rangle, \langle D_2, rw(D_2)\rangle, ..., \langle D_m, rw(D_m)\rangle \mid (1 \leq j \leq m)\}$ and D is a class concept *\
for all $D_j \in \mathcal{Q}$ **do**
  if $KeepMatching() =$ **false then**
    return $\langle GetDegMatch(\mathcal{Q}), GetConf(\mathcal{Q})\rangle$
  end if
  Let h be a new BranchNode in the expansion tree $\mathcal{T}$ for $\mathcal{A}$ such that $\mathcal{T} \leftarrow \mathcal{T} \cup \{h\}$
  $currBranchNode(\mathcal{A}) \leftarrow h$
  $\mathcal{L}(x) \leftarrow \mathcal{L}(x) \cup \{D_j\}$ * where x is an individual in $\mathcal{A}$ *\
  $checkedConcept(D_j) \leftarrow$ **true*** for confidence *\
  Let $clash$ be a clashing concept, if one exists
  $clash \leftarrow ApplyTransRules(\mathcal{A})$ * runs the standard Tableaux rules *\
  if $clash \neq null$ **then**
    $clashingConcept(clash) \leftarrow$ **true*** for degree of match & confidence *\
  end if
end for
return $\langle GetDegMatch(\mathcal{Q}), GetConf(\mathcal{Q})\rangle$

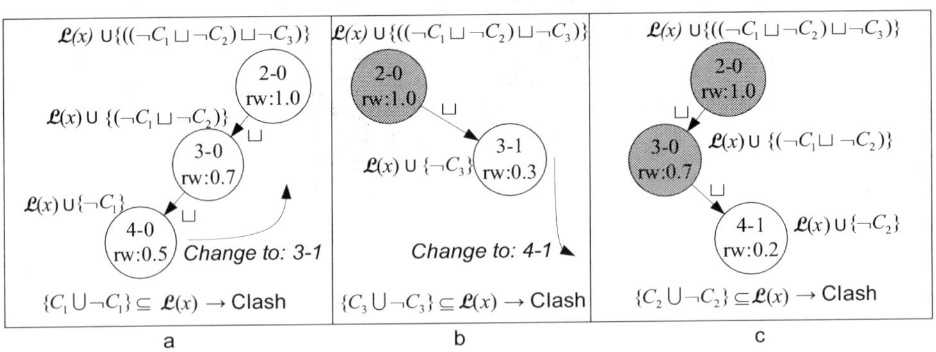

Fig. 2. Tableaux Adaptive Expansion Search Tree

a branch is continually expanded using transformation rules such that $\mathcal{A}' = \mathcal{A} \cup \{...\}$. This depth first expansion occurs until either a clash is found or there are no more branches to apply. Where a clash is found the ABox $\mathcal{A}_n$ is restored to an earlier branch node n_i state $\mathcal{A}_i$, and the current state $\mathcal{A}_n$ is discarded. This means that all ABox assertions added after $branchID(n_i)$ are removed. However, under our adaptive reasoning strategy, the expansion tree is expanded in branch node weight order, rather than depth first. This implies that there may be several unfinished open branches at one time, for instance in figure 2, branch point 3-0 from step a, is preserved for step c, while a new branch is opened in step b. This was because concept C_3 was considered more important than C_2.

Support for this functionality requires modifications to:

- Branch point node identifiers *branchID*
- State of node labels $\mathcal{L}(x)$ and $\mathcal{L}(\langle x, y\rangle)$ where x and y are individuals

In terms of branch point node identifiers, a $branchID(n)$ must be unique. However, since a traditional $branchID(n)$ is only a depth count, it is no longer unique if multiple branches are present in the tree $\mathcal{T}$. Therefore, we specify new branch identifiers $adapBranchID(n)$, of the form given in expression 2, to take breath into account. Let $adapBranchID(n)$ denote the branch identifier for a node n in a tree formation where $depth(n)$ denotes the distance from the top most object in the tree down to node n and $breath(n)$ is a count which increases for every node in the tree which has the same $depth(n)$, such that $\{2 \leq depth(n)$, $0 \leq breath(n)$. Note that *adapBranchID* 0-0 and 1-0 are reserved for explicit assertions and pre-processing. Let $adapDepBranchIDs(n)$ contain a dependency list of $adapBranchID(n)$ for the node n. $adapDepBranchIDs(n)$ contains the branch identifier for node n and for all of its parents in the expansion tree $\mathcal{T}$, such that $adapDepBranchIDs(n) = adapBranchID(n) \cup \{\bigcup_{p\in P} adapBranchID(p)\}$, where P contains all parent nodes of n in $\mathcal{T}$.

$$adapBranchID(n) = depth(n) \ominus breath(n) \tag{2}$$

In terms of type label state management standard Tableaux does not support multiple unfinished open branches. The state of type labels $\mathcal{L}(x)$ and $\mathcal{L}(\langle x, y\rangle)$ where x and y are individuals in ABox $\mathcal{A}$ is determined by parent branch point nodes. Class concepts and role assertions are progressively added to type labels by tree branch expansions and by other transformation rules as they occur, and removed when a backjump (restore) occurs. However, in order to support multiple open branches, previously applied branch nodes much be retained. Type labels support multiple states, and their contents (state) at any one time must reflect the current branch point node without requiring re-application of all parent node expansions and transformations, when moving back to a previously partially expanded branch. Therefore, we redefine type labels, such that concepts C and role assertions $R(x, y)$ added to type labels $\mathcal{L}(x)$ and $\mathcal{L}(\langle x, y\rangle)$, respectively, are indexed by branch identifier.

In this section we refer to both $\mathcal{L}(x)$ and $\mathcal{L}(\langle x, y\rangle)$, simply as $\mathcal{L}$, such that $\mathcal{L} = \{\langle adapBranchID(\alpha_1), \alpha_1\rangle, ..., \langle adapBranchID(\alpha_p), \alpha_p\rangle \mid (1 \leq k \leq p)\}$. Let $V(w)$ be a subset of the elements α_k (concepts and relations) in $\mathcal{L}$, where w is the currently active branch node in the expansion tree (see algorithm 4) and where w has a set of branch node dependencies and all the elements in $V(w)$ must have a branch identifier which is in this set of dependencies. This is defined in algorithm 5, such that $V(w) = \{\alpha_1, ..., \alpha_v \mid (1 \leq u \leq v)\}$ where $adapBranchID(\alpha_u) \in adapDepBranchIDs(w)$. Therefore, $V(w)$ can be considered as the state of $\mathcal{L}$ when the currently active expansion node is w, to support swapping between simultaneous unfinished branches. For example, if $\mathcal{L} = \{\langle 2\text{-}1, \alpha_1\rangle, \langle 2\text{-}3, \alpha_2\rangle, \langle 3\text{-}1, \alpha_3\rangle, \langle 4\text{-}0, \alpha_4\rangle\}$ and $adapDepBranchIDs(w) = \{2\text{-}1, 3\text{-}1\}$ where $w = currBranchNode(\mathcal{A})$, then $GetTypeLabel$ returns $\{\alpha_1, \alpha_3\}$

Algorithm 5. GetTypeLabel($\mathcal{L}$, ABox $\mathcal{A}$)

```
Inputs: TypeLabel L, ABox A
Outputs: Set V
Let w = currBranchNode(A) \* currently active branch point node for A *\
for all β_i ∈ L where β_i = ⟨adapBranchID(α_i), α_i⟩ do
    if adapBranchID(α_i) ∈ adapDepBranchIDs(w) then
        V ← V ∪ {α_i}
    end if
end for
return V
```

4.3 Degree of Match and Match Confidence

In our adaptive reasoning strategy we provide a degree of inference match and a level of confidence in this result. Degree of match is the known, weighted degree to which a particular inference $C(x)$, holds. It is a normalised value based on those conditions actually checked in the processing time available, where 1 denotes a complete match and 0 denotes no match. A degree of match is the sum of all relative weight rw values for all the concepts C where a clash was detected. Let Q denote a set containing all conditions and sub-conditions to be checked, as defined in defined in section 4.1, $\mathcal{Q} = \{\langle D_1, rw(D_1)\rangle, \langle D_2, rw(D_2)\rangle, ..., \langle D_p, rw(D_p)\rangle \mid (1 \leq k \leq p)\}$ as defined in section 4.1 where $rw(D_k)$ is the relative weights for each condition specified in $\neg C$ for the inference check $C(x)$. Let $getDegMatch(T)$, given in formula 3, return the known degree of match for a particular inference. Let $clashingConcept(W)$ contain $true$ if the concept W caused a clash, set by algorithm 4.

$$GetDegMatch(\mathcal{Q}) = \sum_{e_j \in \mathcal{Q}} rw(D_k) \text{ iff } clashingConcept(D_k) = true \quad (3)$$
$$\text{where } e_j = \langle D_k, rw(D_k)\rangle$$

Confidence is the normalised ratio providing a measure indicating the proportion of the total reasoning process, which has been completed. Confidence is the ratio of worst-case weighted number of conditions to execute over the number actually executed. Let $GetConf(\mathcal{Q})$ denote the confidence value, given in formula 4. Let $checkedConcept(W)$ return $true$ if the class concept W was actually applied by the algorithm 4.

$$GetConf(\mathcal{Q}) = \frac{\sum_{e_j \in \mathcal{Q}} rw(D_k) \text{ iff } checkedConcept(D_k) = true}{\sum_{e_j \in \mathcal{Q}} rw(D_k)} \quad (4)$$
$$\text{where } e_j = \langle D_k, rw(D_k)\rangle$$

In this section we provided our adaptive reasoning approach. Both our adaptive algorithm (see algorithm 4) and standard Tableaux (see algorithm 1) have a worst case complexity of $O(n)$ where n is the total number of branch point nodes which can be expanded in the tree $\mathcal{T}$. However, if the level of confidence required is reduced, the adaptive reasoning strategy has a best case complexity of $O(1)$ while Tableaux has a best case complexity of $O(m)$ where m is the average number of nodes in a branch in the expansion tree $\mathcal{T}$.

In the next section we discuss the implementation of our adaptive reasoning approach and provide an evaluation to illustrate the effect on performance and result accuracy.

5 Implementation and Evaluation

In this section we discuss the implementation of our adaptive reasoning strategy provide a performance evaluation using a case study.

5.1 Case Study

To demonstrate our adaptive reasoning we approach we developed a case study in which Bob is searching for particular attributes in a foreign city centre, such as a movie cinema, printing service, Internet cafe etc. The case study comprises several request queries as follows (relative weight in brackets):

1. RetailOutlet (0.33), MovieCinema (0.3), Internet (0.23), Cafe (0.14)
2. RetailOutlet (0.40), MovieCinema (0.32), Internet (0.2), Cafe (0.04), Colour Fax WiFi Printer (0.04)
3. Internet (0.5), Colour Fax WiFi Printer (0.3), Cafe (0.2)

Our scenario also has 8 different service advertisements which we label A, B, C, D, E, F, G, H, where A matches all requests, B-G only partially match each request, while H does not match any attribute in any request. The ontologies containing these service offerings comprise 204 classes and 241 individuals.

5.2 Implementation

We have implemented our adaptive strategies as an extension to the Pellet[5] 1.5 reasoner which supports OWL-DL with $\mathcal{SHOIN}$ expressiveness [2]. Pellet is open source, allowing us to provide a proof of concept. We selected Pellet over FaCT++ because it is written in Java, making it easily portable to small devices such as PDAs and mobile phones, while FaCT++[6] is written in C++. All our evaluations were conducted on a HP iPAQ hx2700 PDA, with Intel PXA270 624Mhz processor, 64MB RAM, running Windows Mobile 5.0 with Mysaifu Java J2SE Virtual Machine (JVM)[7], allocated 15MB of memory. Note, that in the next section, results for both standard reasoning and adaptive reasoning utilise optimisations from previous work which enable mobile reasoning [17].

[5] http://clarkparsia.com/pellet
[6] http://owl.man.ac.uk/factplusplus
[7] http://www2s.biglobe.ne.jp/ dat/java/project/jvm/index_en.html

5.3 Evaluation

Figure 3 shows the degree of match obtained by comparing request 1 with service advertisements A-H (see section 5.1), for standard reasoning and adaptive reasoning. Both standard and adaptive reasoning stop on failure (stop as soon as any request condition fails to match). Standard reasoning provides a meaningless *false* result when stopped prematurely, while our adaptive reasoning strategy provides a degree of match result for those request conditions, which were successfully matched in the available time. This demonstrates a far better reporting of inference results to the user. The "Check All" results, figure 3, demonstrate our support for heterogeneous data (continue matching when some conditions fail). Figure 4 demonstrates that our adaptive reasoning strategy matches the most important conditions first, for the services A-H, as well as supporting premature stopping and a degree of match result to the user. For most service advertisements, most of the degree of match result is obtained in the first 10 seconds of processing.

Figure 5 compares request 2 and 3 against service advertisement A, showing similar results. For instance in 5(left), a degree of match of almost 80% was reached in the first 10s of operation, while the remaining 20% required 40 seconds to establish. Figure 5(right) shows that when condition weights are similar, the correspondence between degree of match and processing time is more uniform. When standard reasoning was used, no result was provided until the reasoning was completed in full, requiring 40 seconds. Figure 6 further highlights the way in which our adaptive reasoning strategy matches the most important attributes first to provide a higher degree or match more quickly, in order to make the best use of processing time available. This figure compares adaptive reasoning against the result which would be obtained if standard Tableaux reasoners provided a degree of match. Where request 2 is matched with service A (left), after 10 seconds the adaptive reasoner found a degree of match of almost 80% while the standard reasoner had successfully matched conditions with weights that added up to under 20%. This is because standard reasoners match attributes in depth-first, arbitrary order. Therefore, it is often the case that attributes which contribute greatly to the degree of match, are checked late in the process. Also note while a cumulative result is shown for standard reasoners for illustrative purposes only, standard reasoners cannot be stopped early.

The strategies presented in this paper are shown to effectively meet the following challenges:

1. match the most important request conditions first
2. handle heterogeneous inaccurate data by continuing to reason after a term failed (if sufficient time).
3. handle time/resource pressures by supporting early stopping of the reasoner and providing a current degree of inference to the user.

In this section we have clearly demonstrated that our adaptive reasoning approach provides a significant advantage over current reasoning strategies for mobile users. Our approach effectively provides flexible reasoning to meet the resource and time challenges which characterise mobile environments.

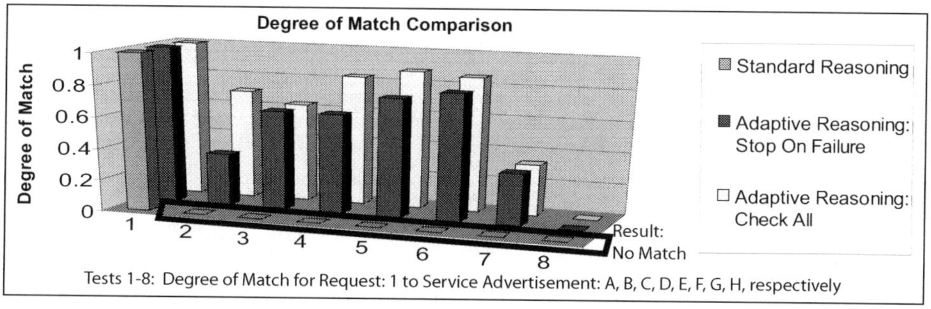

Fig. 3. Comparison of degree of match for standard (stop after any request condition fails) and adaptive reasoning (check all conditions), where Request 1 is compared against 8 different potential services A-H see section 5.1

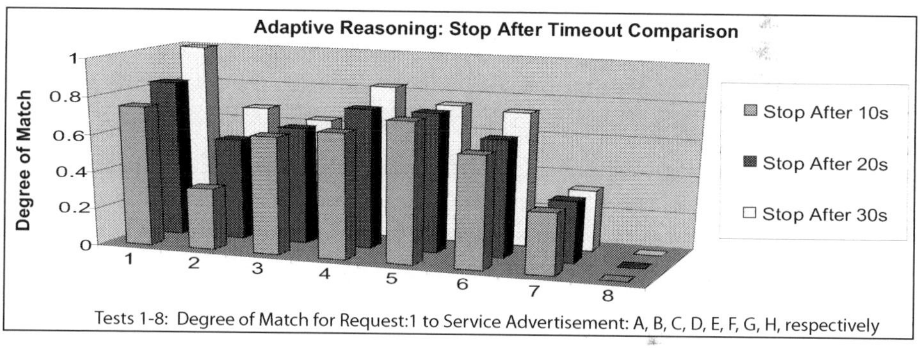

Fig. 4. Comparison of degree of match for adaptive reasoning after a timeout of 10, 20 and 30 seconds, where Request 1 is compared against 8 different potential services: A-H see section 5.1

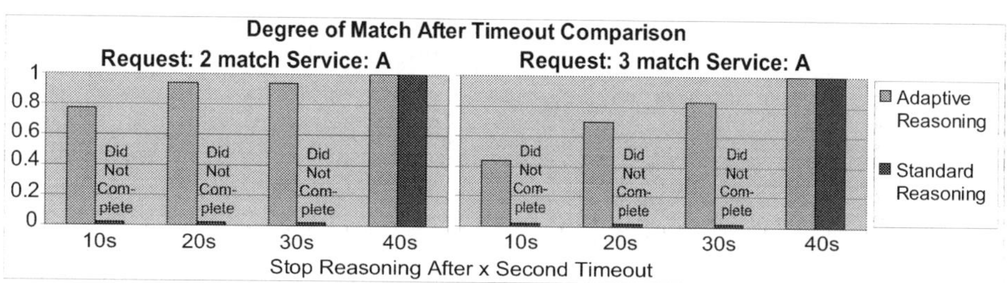

Fig. 5. Comparison of degree of match for adaptive reasoning after a timeout of 10, 20, 30 and 40 seconds, for requests 2 (left) and 3 (right), matched against service advertisement A

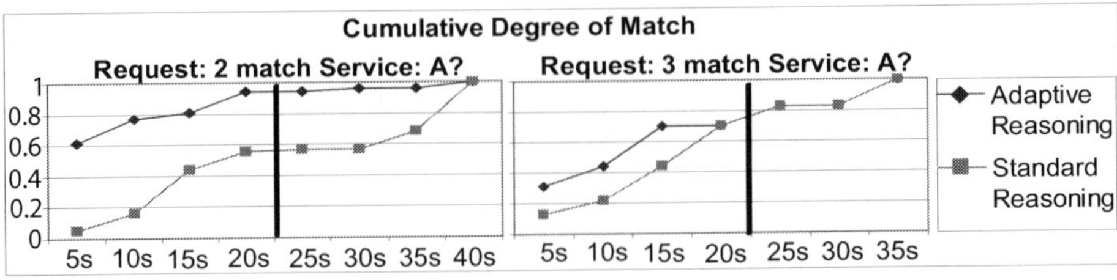

Fig. 6. Illustration of the cumulative degree of match for adaptive reasoning and standard reasoning, where requests 2 (left) and 3 (right) are matched against service advertisement A

6 Conclusion

We have presented a novel adaptive reasoning strategy which supports partial matching, premature stopping and provides a degree of inference match, based on those request conditions already checked. Our evaluations demonstrate that our strategies effectively meet these goals, by matching the most important request conditions first, to achieve the highest possible degree of match result within the time available. This is a significantly more effective approach than current approximate reasoning techniques which match request conditions arbitrarily and completely before providing any results.

References

1. Arnold, K., O'Sullivan, B., Scheifler, R.W., Waldo, J., Woolrath, A.: The Jini Specification. Addison-Wesley, Reading (1999)
2. Baader, F., Calvanese, D., McGuinness, D.L., Nardi, D., Patel-Schneider, P.F.: The Description Logic Handbook: Theory, Implementation, and Applications. Cambridge University Press, Cambridge (2003)
3. Donini, F.M., Lenzerini, M., Nardi, D., Schaerf, A.: Reasoning in description logics (1997)
4. Fensel, D., van Harmelen, F.: Unifying reasoning and search to web scale. Internet Computing, IEEE 11(2), 96 (2007)
5. Gligorov, R., ten Kate, W., Aleksovski, Z., van Harmelen, F.: Using google distance to weight approximate ontology matches. In: 16th international Conference on World Wide Web. ACM, New York (2007)
6. Gu, T., Kwok, Z., Koh, K.K., Pung, H.K.: A mobile framework supporting ontology processing and reasoning. In: 2nd Workshop on Requirements and Solutions for Pervasive Software Infrastructure (RSPS) in conjunction with the 9th International Conference on Ubiquitous Computing (Ubicomp 2007), Austria (2007)
7. Hitzler, P., Vrandecic, D.: Resolution-Based Approximate Reasoning for OWL DL. In: Semantic Web - ISWC. Springer, Heidelberg (2005)
8. Horrocks, I., Sattler, U.: A tableaux decision procedure for shoiq. In: 19th International Conference on Artificial Intelligence, IJCAI 2005 (2005)

9. Kargin, B., Basoglu, N.: Factors affecting the adoption of mobile services. In: Portland International Center for Management of Engineering and Technology, pp. 2993–3001. IEEE, Los Alamitos (2007)
10. Kleemann, T.: Towards mobile reasoning. In: International Workshop on Description Logics (DL 2006), Windermere, Lake District, UK (2006)
11. Kuster, U., Konig-Ries, B., Klein, M.: Discovery and mediation using diane service descriptions. In: Second Semantic Web Service Challenge 2006 Workshop, Budva, Montenegro (2006)
12. Mokhtar, S.B., Preuveneers, D., Georgantas, N., Issarny, V.: Easy: Efficient semantic service discovery in pervasive computing environments with qos and context support. Journal Of System and Software 81(5) (2008)
13. Ranganathan, A., Campbell, R.H.: A middleware for context-aware agents in ubiquitous computing environments. In: ACM/IFIP/USENIX International Middleware Conference, Rio de Janeiro, Brazil, p. 143–161 (2003)
14. Roto, V., Oulasvirta, A.: Need for non-visual feedback with long response times in mobile hci. In: International World Wide Web Conference Committee (IW3C2), Chiba, Japan (2005)
15. Srinivasan, N., Paolucci, M., Sycara, K.: Semantic web service discovery in the owl-s ide. In: 39th International Conference on System Sciences, IEEE, Hawaii (2005)
16. Srirama, S.N., Jarke, M., Parinz, W.: Mobile web service provisioning. In: Advanced International Conference on Telecommunications and International Conference on Internet and Web Applications and Services, AICT/ICIW (2006)
17. Steller, L., Krishnaswamy, S., Gaber, M.M.: Enabling scalable semantic reasoning for mobile services. International Journal of Semantic Web Information Systems - Special Issue on Scalability and Performance of Semantic Systems (2009)
18. Stuckenschmidt, H., Kolb, M.: Partial matchmaking for complex product and service descriptions. Multikonferenz Wirtschaftsinformatik (2008)
19. Tergujeff, R., Haajanen, J., Leppanen, J., Toivonen, S.: Mobile soa: Service orientation on lightweight mobile devices. In: International Conference on Web Services (ICWS), pp. 1224–1225. IEEE, Salt Lake City (2007)
20. Veijalainen, J.: Mobile ontologies: Concept, development, usage, and business potential. International Journal on Semantic Web & Information Systems 4(1), 20–34 (2008)
21. Wache, H., Groot, P., Stuckenschmidt, H.: Scalable instance retrieval for the semantic web by approximation. In: WISE 2005. Springer, Heidelberg (2005)
22. Wang, Z., Hu, Y.: An approach for semantic web service discovery based on p2p network. In: 4th International Conference on Wireless Communications, Networking and Mobile Computing (WiCOM 2008), pp. 1–4. IEEE, Los Alamitos (2008)
23. Weiser, M.: The computer of the 21st century. Scientific American 3(265), 66–75 (1991)
24. Xiaosu, C., Jian, L.: Build mobile services on service oriented structure. In: International Conference on Wireless Communications, Networking and Mobile Computing, vol. 2, pp. 1472–1476. IEEE, Los Alamitos (2005)
25. Zacharias, V., Abecker, A., Vrandecic, D., Borgi, I., Braun, S., Schmidt, A.: Mind the web. In: ASWC 2007 and ISWC 2007, vol. 291. IEEE, Los Alamitos (2007)

Scalable Distributed Reasoning Using MapReduce

Jacopo Urbani, Spyros Kotoulas, Eyal Oren, and Frank van Harmelen

Department of Computer Science,
Vrije Universiteit Amsterdam,
The Netherlands

Abstract. We address the problem of scalable distributed reasoning, proposing a technique for materialising the closure of an RDF graph based on MapReduce. We have implemented our approach on top of Hadoop and deployed it on a compute cluster of up to 64 commodity machines. We show that a naive implementation on top of MapReduce is straightforward but performs badly and we present several non-trivial optimisations. Our algorithm is scalable and allows us to compute the RDFS closure of 865M triples from the Web (producing 30B triples) in less than two hours, faster than any other published approach.

1 Introduction

In this paper, we address the problem of *scalable distributed reasoning*. Most existing reasoning approaches are centralised, exploiting recent hardware improvements and dedicated data structures to reason over large-scale data [8, 11, 17]. However, centralised approaches typically scale in only one dimension: they become faster with more powerful hardware.

Therefore, we are interested in parallel, distributed solutions that partition the problem across many compute nodes. Parallel implementations can scale in two dimensions, namely hardware performance of each node and the number of nodes in the system. Some techniques have been proposed for distributed reasoning, but, as far as we are aware, they do not scale to orders of 10^8 triples.

We present a technique for materialising the closure of an RDF graph in a distributed manner, on a cluster of commodity machines. Our approach is based on MapReduce [3] and it efficiently computes the closure under the RDFS semantics [6]. We have also extended it considering the OWL Horst semantics [9] but the implementation is not yet competitive and it is should be considered as future work. This paper can be seen as a response to the challenge posed in [12] to exploit the MapReduce framework for efficient large-scale Semantic Web reasoning.

This paper is structured as follows: we start, in Section 2, with a discussion of the current state-of-the-art, and position ourselves in relation to these approaches. We summarise the basics of MapReduce with some examples in Section 3. In Section 4 we provide an initial implementation of forward-chaining

A. Bernstein et al. (Eds.): ISWC 2009, LNCS 5823, pp. 634–649, 2009.

RDFS materialisation with MapReduce. We call this implementation "naive" because it directly translates known distributed reasoning approaches into MapReduce. This implementation is easy to understand but performs poorly because of load-balancing problems and because of the need for fixpoint iteration. Therefore, in Section 5, an improved implementation is presented using several intermediate MapReduce functions. Finally, we evaluate our approach in Section 6, showing runtime and scalability over various datasets of increasing size, and speedup over increasing amounts of compute nodes.

2 Related Work

Hogan *et al.* [7] compute the closure of an RDF graph using two passes over the data on a single machine. They implement only a fragment of the OWL Horst semantics, to allow efficient materialisation, and to prevent "ontology hijacking". Our approach borrows from their ideas, but by using well-defined MapReduce functions our approach allows straightforward distribution over many nodes, leading to improved results.

Mika and Tummarello [12] use MapReduce to answer SPARQL queries over large RDF graphs, and mention closure computation, but do not provide any details or results. In comparison, we provide algorithm details, make the code available open-source, and report on experiments of up to 865M triples.

MacCartney *et al.* [13] show that graph-partitioning techniques improve reasoning over first-order logic knowledge bases, but do not apply this in a distributed or large-scale context. Soma and Prasanna [15] present a technique for parallel OWL inferencing through data partitioning. Experimental results show good speedup but on relatively small datasets (1M triples) and runtime is not reported.In contrast, our approach needs no explicit partitioning phase and we show that it is scalable over increasing dataset size.

In previous work [14] we have presented a technique based on data-partitioning in a self-organising P2P network. A load-balanced auto-partitioning approach was used without upfront partitioning cost. Conventional reasoners are locally executed and the data is intelligently exchanged between the nodes. The basic principle is substantially different from the work here presented and experimental results were only reported for relatively small datasets of up to 15M triples.

Several techniques have been proposed based on deterministic rendezvous-peers on top of distributed hashtables [1, 2, 4, 10]. However, these approaches suffer of load-balancing problems due to the data distributions [14].

3 What Is the MapReduce Framework?

MapReduce is a framework for parallel and distributed processing of batch jobs [3] on a large number of compute nodes. Each job consists of two phases: a map and a reduce. The mapping phase partitions the input data by associating each element with a key. The reduce phase processes each partition independently. All data is processed based on key/value pairs: the map function

Algorithm 1. Counting term occurrences in RDF NTriples files

```
map(key, value):
  // key: line number
  // value: triple
  emit(value.subject, blank); // emit a blank value, since
  emit(value.predicate, blank); // only amount of terms matters
  emit(value.object, blank);

reduce(key, iterator values):
  // key: triple term (URI or literal)
  // values: list of irrelevant values for each term
  int count=0;
  for (value in values)
    count++; // count number of values, equalling occurrences
  emit(key, count);
```

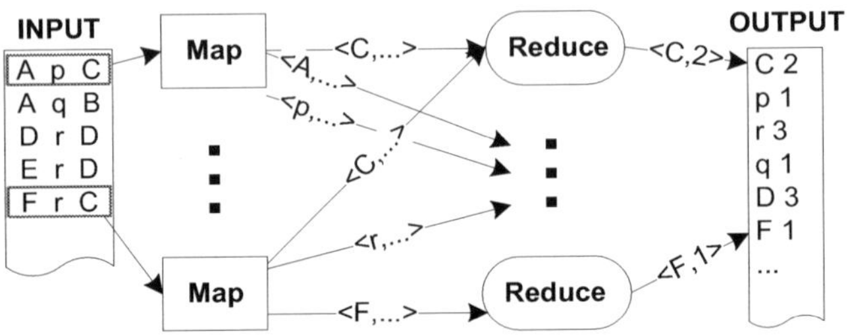

Fig. 1. MapReduce processing

processes a key/value pair and produces a set of new key/value pairs; the reduce merges all intermediate values with the same key into final results.

We illustrate the use of MapReduce through an example application that counts the occurrences of each term in a collection of triples. As shown in Algorithm 1, the *map* function partitions these triples based on each term. Thus, it emits intermediate key/value pairs, using the triple terms (s,p,o) as keys and blank, irrelevant, value. The framework will group all intermediate pairs with the same key, and invoke the *reduce* function with the corresponding list of values, summing these the number of values into an aggregate term count (one value was emitted for each term occurrence).

This job could be executed as shown in Figure 1. The input data is split in several blocks. Each computation node operates on one or more blocks, and performs the map function on that block. All intermediate values with the same key are sent to one node, where the reduce is applied.

This simple example illustrates some important elements of the MapReduce programming model:

- since the *map* operates on single pieces of data without dependencies, partitions can be created arbitrarily and can be scheduled in parallel across many nodes. In this example, the input triples can be split across nodes arbitrarily, since the computations on these triples (emitting the key/value pairs), are independent of each other.

Table 1. RDFS rules [6]

1:	$s\ p\ o$ (if o is a literal)		⇒ _:n rdf:type rdfs:Literal
2:	p rdfs:domain x	& $s\ p\ o$	⇒ s rdf:type x
3:	p rdfs:range x	& $s\ p\ o$	⇒ o rdf:type x
4a:	$s\ p\ o$		⇒ s rdf:type rdfs:Resource
4b:	$s\ p\ o$		⇒ o rdf:type rdfs:Resource
5:	p rdfs:subPropertyOf q	& q rdfs:subPropertyOf r	⇒ p rdfs:subPropertyOf r
6:	p rdf:type rdf:Property		⇒ p rdfs:subPropertyOf p
7:	$s\ p\ o$	& p rdfs:subPropertyOf q	⇒ $s\ q\ o$
8:	s rdf:type rdfs:Class		⇒ s rdfs:subClassOf rdfs:Resource
9:	s rdf:type x	& x rdfs:subClassOf y	⇒ s rdf:type y
10:	s rdf:type rdfs:Class		⇒ s rdfs:subClassOf s
11:	x rdfs:subClassOf y	& y rdfs:subClassof z	⇒ x rdfs:subClassOf z
12:	p rdf:type rdfs:ContainerMembershipProperty		⇒ p rdfs:subPropertyOf rdfs:member
13:	o rdf:type rdfs:Datatype		⇒ o rdfs:subClassOf rdfs:Literal

- the *reduce* operates on an iterator of values because the set of values is typically far too large to fit in memory. This means that the reducer can only partially use correlations between these items while processing: it receives them as a stream instead of a set. In this example, operating on the stream is trivial, since the reducer simply increments the counter for each item.
- the *reduce* operates on all pieces of data that share some key, assigned in a *map*. A skewed partitioning (i.e. skewed key distribution) will lead to imbalances in the load of the compute nodes. If term x is relatively popular the node performing the *reduce* for term x will be slower than others. To use MapReduce efficiently, we must find balanced partitions of the data.

4 Naive RDFS Reasoning with MapReduce

The closure of an RDF input graph under the RDFS semantics [6] can be computed by applying all RDFS rules iteratively on the input until no new data is derived (fixpoint). The RDFS rules, shown in Table 1, have one or two antecedents. For brevity, we ignore the former (rules 1, 4a, 4b, 6, 8, 10, 12 and 13) since these can be evaluated at any point in time without a join. Rules with two antecedents are more challenging to implement since they require a join over two parts of the data.

4.1 Encoding an Example RDFS Rule in MapReduce

Applying the RDFS rules means performing a join over some terms in the input triples. Let us consider for example rule 9 from Table 1, which derives `rdf:type` based on the sub-class hierarchy. We can implement this join with a *map* and *reduce* function, as shown in Figure 2 and Algorithm 2:

In the *map*, we process each triple and output a key/value pair, using as value the original triple, and as key the triple's term (`s,p,o`) on which the join should

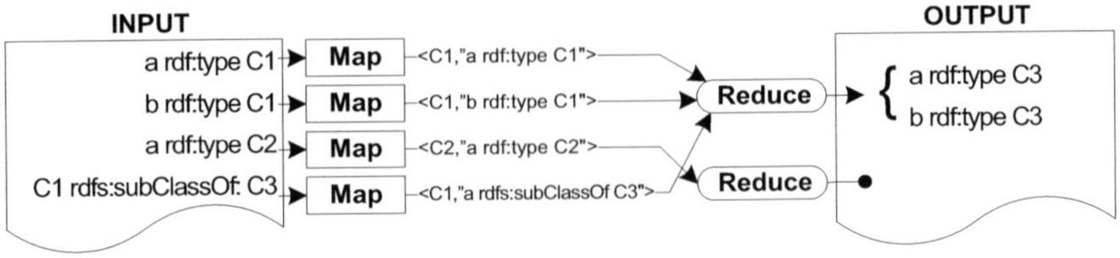

Fig. 2. Encoding RDFS rule 9 in MapReduce

Algorithm 2. Naive sub-class reasoning (RDFS rule 9)

```
map(key, value):
  // key: linenumber (irrelevant)
  // value: triple
  switch triple.predicate
  case "rdf:type":
    emit(triple.object, triple); // group (s rdf:type x) on x
  case "rdfs:subClassOf":
    emit(triple.subject, triple); // group (x rdfs:subClassOf y) on x

reduce(key, iterator values):
  // key: triple term, eg x
  // values: triples, eg (s type x), (x subClassOf y)
  superclasses=empty;
  types=empty;

  // we iterate over triples
  // if we find subClass statement, we remember the super-classes
  // if we find a type statement, we remember the type
  for (triple in values):
    switch triple.predicate
    case "rdfs:subClassOf":
      superclasses.add(triple.object) // store y
    case "rdf:type":
      types.add(triple.subject) // store s

   for (s in types):
    for (y in classes):
     emit(null, triple(s, "rdf:type", y));
```

be performed. To perform the sub-class join, triples with `rdf:type` should be grouped on their object (eg. "x"), while triples with `rdfs:subClassOf` should be grouped on their subject (also "x"). When all emitted tuples are grouped for the reduce phase, these two will group on "x" and the reducer will be able to perform the join.

4.2 Complete RDFS Reasoning: The Need for Fixpoint Iteration

If we perform this *map* once (over all input data), and then the *reduce* once, we will not find *all* corresponding conclusions. For example, to compute the transitive closure of a chain of n `rdfs:subClassOf`-inclusions, we would need to iterate the above *map/reduce* steps n times.

Obviously, the above *map* and *reduce* functions encode only rule 9 of the RDFS rules. We would need to add other, similar, *map* and *reduce* functions to implement each of the other rules. These other rules are interrelated: one rule can derive triples that can serve as input for another rule. For example, rule 2 derives `rdf:type` information from *rdfs:domain* statements. After applying that rule, we would need to re-apply our earlier rule 9 to derive possible superclasses.

Thus, to produce the complete RDFS closure of the input data using this technique we need to add more *map/reduce* functions, chain these functions to each other, and iterate these until we reach some fixpoint.

5 Efficient RDFS Reasoning with MapReduce

The previously presented implementation is straightforward, but is inefficient because it produces duplicate triples (several rules generate the same conclusions) and because it requires fixpoint iteration. We encoded, as example, only rule 9 and we launched a simulation over the Falcon dataset, which contains 35 million triples. After 40 minutes the program had not yet terminated, but had already generated more than 50 billion triples. Considering that the unique derived triples from Falcon are no more than 1 billion, the ratio of unique derived triples to duplicates is at least 1:50. Though the amount of duplicate triples depends on the specific data set, a valid approach should be able to efficiently deal with real world example like Falcon.

In the following subsections, we introduce three optimisations to greatly decrease the number of jobs and time required for closure computation:

5.1 Loading Schema Triples in Memory

Typically, schema triples are far less numerous than instance triples [7]; As also shown in Table 2, our experimental data[1] indeed exhibit a low ratio between schema and instance triples. In combination with the fact that RDFS rules with two antecedents include at least one schema triple, we can infer that joins are made between a large set of instance triples and a small set of schema triples. For example, in rule 9 of Table 1 the set of `rdf:type` triples is typically far larger than the set of `rdfs:subClassOf` triples. As our first optimisation, we can load the small set of `rdfs:subClassOf` triples in memory and launch a MapReduce job that streams the instance triples and performs joins with the in-memory schema triples.

5.2 Data Grouping to Avoid Duplicates

The join with the schema triples can be physically executed either during the map or during the reduce phase of the job. After initial experiments, we have concluded that it is faster to perform the join in the reduce, since doing so in the map results in producing large numbers of duplicate triples.

[1] from the Billion Triple challenge 2008, `http://www.cs.vu.nl/~pmika/swc/btc.html`

Table 2. Schema triples (amount and fraction of total triples) in datasets

schema type	amount	fraction
domain, range (p rdfs:domain D, p rdfs:range R)	30.000	0.004%
sub-property (a rdfs:subPropertyOf b)	70.000	0.009%
sub-class (a rdfs:subClassOf b)	2.000.000	0.2%

Let us illustrate our case with an example based on rule 2 (`rdfs:domain`). Assume an input with ten different triples that share the same subject and predicate but have a different object. If the predicate has a domain associated with it and we execute the join in the mappers, the framework will output a copy of the new triple for each of the ten triples in the input. These triples can be correctly filtered out by the reducer, but they will cause significant overhead since they will need to be stored locally and be transfered over the network.

We can avoid the generation of duplicates if we first group the triples by subject and then we execute the join over the single group. We can do it by designing a mapper that outputs an intermediate tuple that has as key the triple's subject and as value the predicate. In this way the triples will be grouped together and we will execute the join only once, avoiding generating duplicates.

In general, we set as key those parts of the input triples that are also used in the derived triple. The parts depend on the applied rule. In the example above, the only part of the input that is also used in the output is the subject. Since the key is used to partition the data, for a given rule, all triples that produce some new triple will be sent to the same reducer. It is then trivial to output that triple only once in the reducer. As value, we emit those elements of the triple that will be matched against the schema.

5.3 Ordering the Application of the RDFS Rules

We analyse the RDFS ruleset with regard to input and output of each rule, to understand which rule may be triggered by which other rule. By ordering the execution of rules we can limit the number of iterations needed for full closure. As explained before, we ignore some of the rules with a single antecedent (1, 4, 6, 8, 10) without loss of generality: these can be implemented at any point in time without a join, using a single pass over the data. We first categorise the rules based on their output:

- rules 5 and 12 produce schema triples with *rdfs:subPropertyOf* as predicate,
- rules 11 and 13 produce schema triples with *rdfs:subClassOf* as predicate,
- rules 2, 3,and 9 produce instance triples with *rdf:type* as predicate,
- rule 7 may produce arbitrary triples.

We also categorise the rules based on the predicates in their antecedents:

- rules 5 and 10 operate only on triples with sub-class or sub-property triples,
- rules 9, 12 and 13 operate on triples with type, sub-class, and sub-property,
- rule 2, 3 and 7 can operate on arbitrary triples.

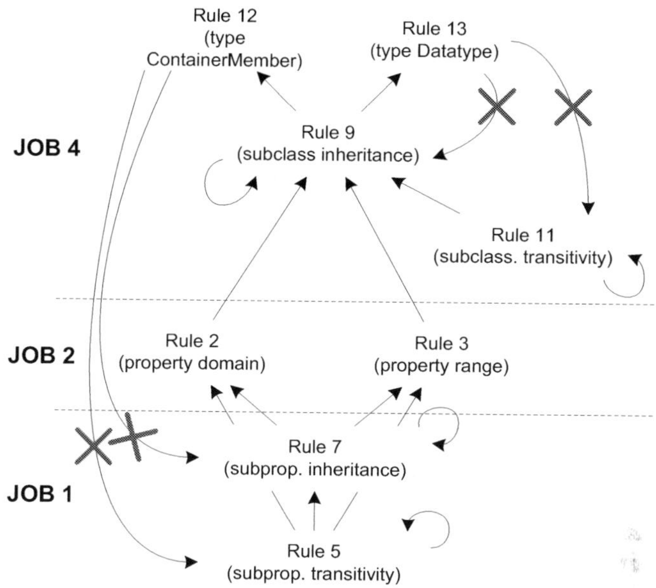

Fig. 3. Relation between the various RDFS rules

Figure 3 displays the relation between the RDFS rules, connecting rules based on their input and output (antecedents and consequents). An ideal execution should proceed from the bottom of the picture to the top: first apply the transitivity rules (rule 5 and 11), then apply rule 7, then rule 2 and 3, then rule 9 and finally rules 12 and 13.

It may seem that rule 12 and 13 could produce triples that would serve as input to rules 5 and 11; however, looking carefully we see that this is not the case: Rule 12 outputs `(?s rdfs:subPropertyOf rdfs:member)`, rule 13 outputs `(?s rdfs:subClassOf rdfs:Literal)`. For rules 5 and 11 to fire on these, *rdfs:member* and *rdfs:Literal* must have been be defined as sub-classes or sub-properties of something else. However, in RDFS none of these is a sub-class or sub-property of anything. They could of course be super-classed by arbitrary users on the Web. However, such "unauthorised" statements are dangerous because they can cause ontology hijacking and therefore we ignore them following the advice of [7]. Hence, the output of rules 12 and 13 cannot serve as input to rules 5 and 11. Similarly, rules 2 and 3 cannot fire.

Furthermore, rule 9 cannot fire after rule 13, since this would require using literals as subjects, which we ignore as being non-standard RDF. The only rules that could fire after rule 12 are rules 5 and 7. For complete RDFS inferencing, we would need to evaluate these rules for each container-membership property found in the data, but as we will show, in typical datasets these properties occur very rarely.

As our third optimisation, we conclude that instead of having to iterate over all RDFS rules until fixpoint, it is sufficient to process them only once, in the order indicated in Figure 3.

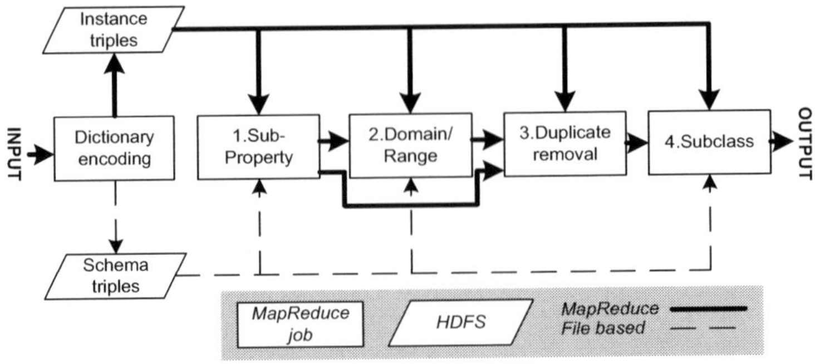

Fig. 4. Data flow. The solid lines refer to data split partitioned using MapReduce while the dashed lines refer to shared data.

5.4 The Complete Picture

In this section, we present an updated algorithm implementing the above optimisations. The complete algorithm consists of five sequential MapReduce jobs, as shown in Figure 4. First, we perform dictionary encoding and extract the schema triples to a shared distributed file system. Then, we launch the RDFS reasoner that consists in a sequence of four MapReduce jobs.

The first job applies the rules that involve the sub-property relations. The second applies the rules concerning domain and range. The third cleans up the duplicated statements produced in the first step and the last applies the rules that use the sub-class relations. In the following subsections, each of these jobs is explained in detail.

Distributed Dictionary Encoding in MapReduce. To reduce the physical size of the input data, we perform a dictionary encoding, in which each triple term is rewritten into a unique and small identifier. We have developed a novel technique for distributed dictionary encoding using MapReduce, rewriting each term into an 8-byte identifier; the encoding scales linearly with the input data. Due to space limitations, we refer the reader to [16]. Encoding all 865M triples takes about 1 hour on 32 nodes. Note that schema triples are extracted here.

First Job: Apply Rules on Sub-Properties. The first job applies rules 5 and 7, which concern sub-properties, as shown in Algorithm 3. Since the schema triples are loaded in memory, these rules can be applied simultaneously.

To avoid generation of duplicates, we follow the principle of setting as the tuple's key the triple's parts that are used in the derivation. This is possible because all inferences are drawn on an instance triple and a schema triple and we load all schema triples in memory. That means that for rule 5 we output as key the triple's subject while for rule 7 we output a key consisting of subject and object. We add an initial flag to keep the groups separated since later we have to apply a different logic that depends on the rule. In case we apply rule 5, we output the triple's object as value, otherwise we output the predicate.

Algorithm 3. RDFS sub-property reasoning

```
map(key, value):
  // key: null
  // value: triple
  if (subproperties.contains(value.predicate)) // for rule 7
    key = "1" + value.subject + "-" + value.object
    emit(key, value.predicate)
  if (subproperties.contains(value.object) &&
      value.predicate == "rdfs:subPropertyOf") // for rule 5
    key = "2" + value.subject
    emit(key, value.object)

reduce(key, iterator values):
  // key: flag + some triples terms (depends on the flag)
  // values: triples to be matched with the schema
  values = values.unique // filter duplicate values

  switch (key[0])
   case 1: // we are doing rule 7: subproperty inheritance
     for (predicate in values)
        // iterate over the predicates emitted in the map and collect superproperties
       superproperties.add(subproperties.recursive_get(value))
     for (superproperty in superproperties)
       // iterate over superproperties and emit instance triples
       emit(null, triple(key.subject, superproperty, key.object)
   case 2: // we are doing rule 5: subproperty transitivity
     for (predicate in values)
       // iterate over the predicates emitted in the map, and collect superproperties
       superproperties.add(subproperties.recursive_get(value))
     for (superproperty in superproperties)
       // emit transitive subproperties
       emit(null, triple(key.subject, "rdfs:subPropertyOf", superproperty))
```

The reducer reads the flag of the group's key and applies to corresponding rule. In both cases, it first filters out duplicates in the values. Then it recursively matches the tuple's values against the schema and saves the output in a set. Once the reducer has finished with this operation, it outputs the new triples using the information in the key and in the derivation output set.

This algorithm will not derive a triple more than once, but duplicates may still occur between the derived triples and the input triples. Thus, at a later stage, we will perform a separate duplicate removal job.

Second Job: Apply Rules on Domain and Range. The second job applies rules 2 and 3, as shown in Algorithm 4. Again, we use a similar technique to avoid generating duplicates. In this case, we emit as key the triple's subject and as value the predicate. We also add a flag so that the reducers know if they have to match it against the domain or against the range schema. Tuples about domain and range will be grouped together if they share the same subject since the two rules might derive the same triple.

Third Job: Delete Duplicate Triples. The third job is simpler and eliminates duplicates between the previous two jobs and the input data. Due to space limitations, we refer the reader to [16].

Algorithm 4. RDFS domain and range reasoning

```
map(key, value):
  // key: null
  // value: triple
  if (domains.contains(value.predicate)) then // for rule 2
    key = value.subject
    emit(key, value.predicate + "d")
  if (ranges.contains(value.predicate)) then // for rule 3
    key = value.object
    emit(key, value.predicate +''r'')

reduce(key, iterator values):
  // key: subject of the input triples
  // values: predicates to be matched with the schema
  values = values.unique // filter duplicate values
  for (predicate in values)
    switch (predicate.flag)
     case "r": // rule 3: find the range for this predicate
       types.add(ranges.get(predicate))
     case "d": // rule 2: find the domain for this predicate
       types.add(domains.get(predicate))
  for (type in types)
     emit(null, triple(key, "rdf:type", type))
```

Algorithm 5. RDFS sub-class reasoning

```
map(key, value):
  // key: source of the triple (irrelevant)
  // value: triple
  if (value.predicate = "rdf:type")
    key = "0" + value.predicate
    emit(key, value.object)
  if (value.predicate = "rdfs:subClassOf")
    key = "1" + value.predicate
    emit(key, value.object)

reduce(key, iterator values):
  //key: flag + triple.subject
  //iterator: list of classes
  values = values.unique // filter duplicate values

  for (class in values)
    superclasses.add(subclasses.get_recursively(class))

  switch (key[0])
   case 0: // we're doing rdf:type
    for (class in superclasses)
      if !values.contains(class)
        emit(null, triple(key.subject, "rdf:type", class))
   case 1: // we're doing subClassOf
    for (class in superclasses)
      if !values.contains(class)
        emit(null, triple(key.subject, "rdfs:subClassOf", class))
```

Fourth Job: Apply Rules on Sub-Classes. The last job applies rules 9, 11, 12, and 13, which are concerned with sub-class relations. The procedure, shown in Algorithm 5, is similar to the previous job with the following difference: during the map phase we do not filter the triples but forward everything to the reducers instead. In doing so, we are able to also eliminate the duplicates against the input.

6 Experimental Results

We use the Hadoop[2] framework, an open-source Java implementation of MapReduce. Hadoop is designed to efficiently run and monitor MapReduce applications on clusters of commodity machines. It uses a distributed file system and manages execution details such as data transfer, job scheduling, and error management.

Our experiments were performed on the DAS-3 distributed supercomputer[3] using up to 64 compute nodes with 4 cores and 4GB of main memory each, using Gigabit Ethernet as an interconnect. We have experimented on real-world data from the Billion Triple Challenge 2008[4]. An overview of these datasets is shown in Table 3, where dataset *all* refers to all the challenge datasets combined except for Webscope, whose access is limited under a license. All the code used for our experiments is publicly available[5].

6.1 Results for RDFS Reasoning

We evaluate our system in terms of time required to calculate the full closure. We report the average and the relative deviation σ (the standard deviation divided by the average) of three runs. The results, along with the number of output triples, are presented in Table 3. Figure 6 shows the time needed for each reasoning phase. Our RDFS implementation shows very high performance: for the combined dataset of 865M triples, it produced 30B triples in less than one hour. This amounts to a total throughput of 8.77 million triples/sec. for the output and 252.000 triples/sec. for the input. These results do not include dictionary encoding, which took, as mentioned, one hour for all datasets combined. Including this time, the throughput becomes 4.27 million triples/sec. and 123.000 triples/sec. respectively, which to the best of our knowledge, still outperforms any results reported both in the literature [11] and on the Web[6].

Besides absolute performance, an important metric in parallel algorithms is how performance scales with additional compute nodes. Table 4 shows the

Table 3. Closure computation using datasets of increasing size on 32 nodes

dataset	input	output	time	σ
Wordnet	1.9M	4.9M	3'39"	9.1%
Falcon	32.5M	863.7M	4'19"	3.8%
Swoogle	78.8M	1.50B	7'15"	8.2%
DBpedia	150.1M	172.0M	5'20"	8.6%
others	601.5M			
all	864.8M	30.0B	56'57"	1.2%

[2] http://hadoop.apache.org
[3] http://www.cs.vu.nl/das3
[4] http://www.cs.vu.nl/~pmika/swc/btc.html
[5] https://code.launchpad.net/~jrbn/+junk/reasoning-hadoop
[6] E.g. at esw.w3.org/topic/LargeTripleStores

Table 4. Speedup with increasing number of nodes

(a) Falcon

nodes	runtime (s)	speedup	efficiency
1	3120	1	1
2	1704	1.83	0.92
4	873	3.57	0.89
8	510	6.12	0.76
16	323	9.65	0.60
32	229	13.61	0.43
64	216	14.45	0.23

(b) DBpedia

nodes	runtime (s)	speedup	efficiency
1	1639	1	1
2	772	2.12	1.06
4	420	3.9	0.98
8	285	5.76	0.72
16	203	8.07	0.5
32	189	8.69	0.27
64	156	10.53	0.16

speedup gained with increasing number of nodes and the resulting efficiency, on the Falcon and DBpedia datasets. Similar results hold for the other datasets. To the best of our knowledge, the only published speedup results for distributed reasoning on a dataset of this size can be found in [14]; for both datasets, and all numbers of nodes, our implementation outperforms this approach.

The speedup results are also shown in Figure 5. They show that our high throughput rates are already obtained when utilising only 16 compute nodes. We attribute the decreasing efficiency on larger numbers of nodes to the fixed Hadoop overhead for starting jobs on nodes: on 64 nodes, our computation per node is not big enough to compensate platform overhead.

Figure 6 shows the division of runtime over the computation phase from Figure 4, and confirms the widely-held intuition that subclass-reasoning is the most expensive part of RDFS inference on real-world datasets.

We have verified the correctness of our implementation on the (small) Wordnet dataset. We have not stored the output of our algorithm: 30B triples (each of them occupying 25 bytes using our dictionary encoding) produce 750GB of data. Mapping these triples back to the original terms would require approx. 500 bytes per triple, amounting to some 15TB of disk space.

In a distributed setting load balancing is an important issue. The Hadoop framework dynamically schedules tasks to optimize the node workload. Furthermore, our algorithms are designed to prevent load balancing problems by intelligently grouping triples (see sections 5.1 and 5.2). During experimentation, we did not encounter any load balancing issues.

6.2 Results for OWL Reasoning

We have also encoded the OWL Horst rules [9] to investigate whether our approach can be extended for efficient OWL reasoning. The OWL Horst rules are more complex than the RDFS rules, and we need to launch more jobs to compute the full closure. Due to space restrictions, we refer to [16], for the algorithms and the implementation.

On the LUBM(50) benchmark dataset [5], containing 7M triples, we compute the OWL Horst closure on 32 nodes in about 3 hours, resulting in about 13M triples. In comparison, the RDFS closure on the same dataset is computed

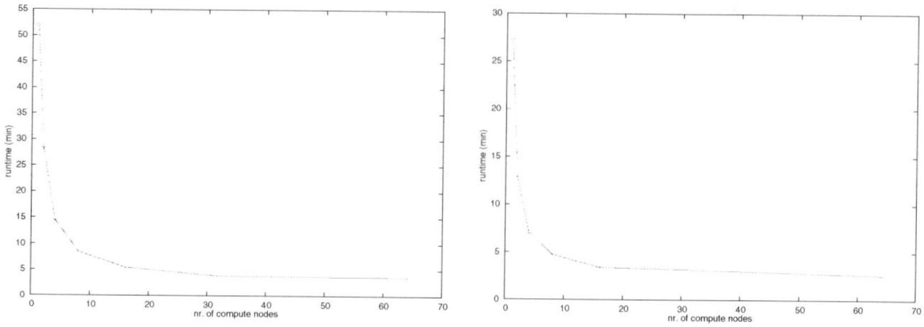

(a) Falcon (b) DBpedia

Fig. 5. Speedup with increasing number of nodes

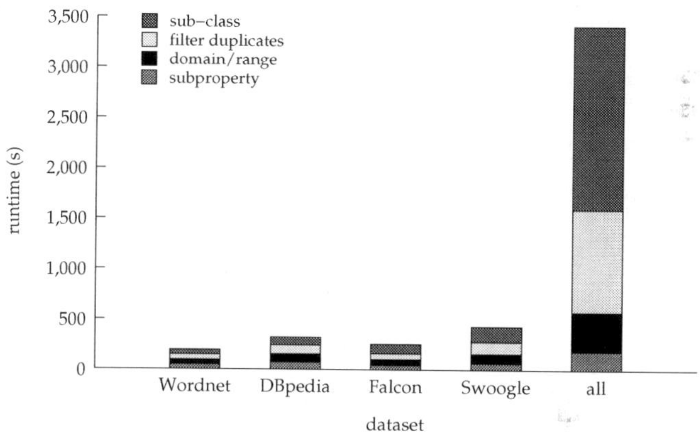

Fig. 6. Detailed reasoning time (per phase) for various datasets (32 nodes)

in about 10 minutes, resulting in about 8.6M triples. On a real-world dataset (Falcon, 35M triples) we stopped OWL Horst inference after 12 hours, at which point more than 130 MapReduce jobs had been launched, and some 3.8B triples had been derived. Clearly, our implementation on OWL Horst has room for optimisation; on RDFS, our optimisations drastically reduced computation time.

6.3 Discussion

Some datasets produce very large amounts of output. For example, the closure of the Swoogle and Falcon datasets is around 20× the original data. We attribute these differences to the content and quality of these datasets: data on the Web contains cycles, override definitions of standard vocabularies, etc. Instead of applying the standard RDFS and OWL rules, Hogan *et al.* [7] propose to only consider "authoritative" statements to prevent this data explosion during reasoning.

In this paper, we did not focus on data quality. To avoid the observed inference explosion, the approach from [7] can be added to our algorithms.

As explained, the presented algorithm performs incomplete RDFS reasoning. We ignore RDF axiomatic triples because this is widely accepted practice and in line with most of the existing reasoners. We omit the rules with one antecedent since parallelizing their application is trivial and they are commonly ignored by reasoners as being uninteresting. If standard compliance is sought, these rules can be implemented with a single *map* over the final data, which very easy to parallelise and should not take more than some minutes. Similarly, we have ignored the rule concerning container-membership properties since these occur very rarely: in all 865M triples, there are only 10 container-membership properties, of which one is in the `example.org` namespace and two override standard RDFS. If needed, membership properties can be implemented in the same way as the subproperty-phase (albeit on much less data), which takes approximately 3 minutes to execute on the complete dataset, as seen in Figure 6.

7 Conclusion

MapReduce is a widely used programming model for data processing on large clusters, and it is used in different contexts to process large collections of data. Our purpose was to exploit the advantages of this programming model for Semantic Web reasoning; a non-trivial task given the high data correlation. We have shown a scalable implementation of RDFS reasoning based on MapReduce which can infer 30 billion triples from a real-world dataset in less than two hours, yielding an input and output throughput of 123.000 triples/second and 4.27 million triples/second respectively. To the best of our knowledge, our system outperforms any other published approach. To achieve this, we have presented some non-trivial optimisations for encoding the RDFS ruleset in MapReduce. We have evaluated the scalability of our implementation on a cluster of 64 compute nodes using several real-world datasets.

A remaining challenge is to apply the same techniques successfully to OWL-Horst reasoning. Our first experiments have shown this to be more challenging.

We would like to thank Christian Rossow for reviewing our work. This work was supported by the LarKC project (EU FP7-215535).

References

[1] Battré, D., Heine, F., Höing, A., Kao, O.: On triple dissemination, forward-chaining, and load balancing in DHT based RDF stores. In: Moro, G., Bergamaschi, S., Joseph, S., Morin, J.-H., Ouksel, A.M. (eds.) DBISP2P 2005 and DBISP2P 2006. LNCS, vol. 4125, pp. 343–354. Springer, Heidelberg (2007)

[2] Cai, M., Frank, M.: RDFPeers: A scalable distributed RDF repository based on a structured peer-to-peer network. In: WWW Conference (2004)

[3] Dean, J., Ghemawat, S.: Mapreduce: Simplified data processing on large clusters. In: Proceedings of the USENIX Symposium on Operating Systems Design & Implementation (OSDI), pp. 137–147 (2004)

[4] Fang, Q., Zhao, Y., Yang, G.-W., Zheng, W.-M.: Scalable distributed ontology reasoning using DHT-based partitioning. In: Domingue, J., Anutariya, C. (eds.) ASWC 2008. LNCS, vol. 5367, pp. 91–105. Springer, Heidelberg (2008)
[5] Guo, Y., Pan, Z., Heflin, J.: LUBM: A benchmark for OWL knowledge base systems. Journal of Web Semantics 3, 158–182 (2005)
[6] Hayes, P. (ed.): RDF Semantics. W3C Recommendation (2004)
[7] Hogan, A., Harth, A., Polleres, A.: Scalable authoritative OWL reasoning for the web. Int. J. on Semantic Web and Information Systems 5(2) (2009)
[8] Hogan, A., Harth, A., Polleres, A.: Saor: Authoritative reasoning for the web. In: Domingue, J., Anutariya, C. (eds.) ASWC 2008. LNCS, vol. 5367, pp. 76–90. Springer, Heidelberg (2008)
[9] ter Horst, H.J.: Completeness, decidability and complexity of entailment for RDF schema and a semantic extension involving the OWL vocabulary. Journal of Web Semantics 3(2–3), 79–115 (2005)
[10] Kaoudi, Z., Miliaraki, I., Koubarakis, M.: RDFS reasoning and query answering on top of DHTs. In: Sheth, A.P., Staab, S., Dean, M., Paolucci, M., Maynard, D., Finin, T., Thirunarayan, K. (eds.) ISWC 2008. LNCS, vol. 5318, pp. 499–516. Springer, Heidelberg (2008)
[11] Kiryakov, A., Ognyanov, D., Manov, D.: OWLIM – a pragmatic semantic repository for OWL. In: Web Information Systems Engineering (WISE) Workshops, pp. 182–192 (2005)
[12] Mika, P., Tummarello, G.: Web semantics in the clouds. IEEE Intelligent Systems 23(5), 82–87 (2008)
[13] MacCartney, B., McIlraith, S.A., Amir, E., Uribe, T.: Practical partition-based theorem proving for large knowledge bases. In: IJCAI (2003)
[14] Oren, E., Kotoulas, S., et al.: Marvin: A platform for large-scale analysis of Semantic Web data. In: Int. Web Science conference (2009)
[15] Soma, R., Prasanna, V.: Parallel inferencing for OWL knowledge bases. In: Int. Conf. on Parallel Processing, pp. 75–82 (2008)
[16] Urbani, J.: Scalable Distributed RDFS/OWL Reasoning using MapReduce. Master's thesis, Vrije Universiteit Amsterdam (2009), `http://www.few.vu.nl/~jui200/thesis.pdf`
[17] Zhou, J., Ma, L., Liu, Q., Zhang, L., Yu, Y., Pan, Y.: Minerva: A scalable OWL ontology storage and inference system. In: Mizoguchi, R., Shi, Z.-Z., Giunchiglia, F. (eds.) ASWC 2006. LNCS, vol. 4185, pp. 429–443. Springer, Heidelberg (2006)

Discovering and Maintaining Links on the Web of Data

Julius Volz[1], Christian Bizer[2], Martin Gaedke[1], and Georgi Kobilarov[2]

[1] Chemnitz University of Technology
Distributed and Self-Organizing Systems Group
Straße der Nationen 62, 09107 Chemnitz, Germany
volz@hrz.tu-chemnitz.de,
gaedke@cs.tu-chemnitz.de
[2] Freie Universität Berlin
Web-based Systems Group
Garystr. 21, 14195 Berlin, Germany
chris@bizer.de,
georgi.kobilarov@fu-berlin.de

Abstract. The Web of Data is built upon two simple ideas: Employ the RDF data model to publish structured data on the Web and to create explicit data links between entities within different data sources. This paper presents the Silk – Linking Framework, a toolkit for discovering and maintaining data links between Web data sources. Silk consists of three components: 1. A link discovery engine, which computes links between data sources based on a declarative specification of the conditions that entities must fulfill in order to be interlinked; 2. A tool for evaluating the generated data links in order to fine-tune the linking specification; 3. A protocol for maintaining data links between continuously changing data sources. The protocol allows data sources to exchange both linksets as well as detailed change information and enables continuous link recomputation. The interplay of all the components is demonstrated within a life science use case.

Keywords: Linked data, web of data, link discovery, link maintenance, record linkage, duplicate detection.

1 Introduction

The central idea of Linked Data is to extend the Web with a data commons by creating typed links between data from different data sources [1,2]. Technically, the term Linked Data refers to a set of best practices for publishing and connecting structured data on the Web in a way that data is machine-readable, its meaning is explicitly defined, it is linked to other external datasets, and can in turn be linked to from external datasets. The data links that connect data sources take the form of RDF triples, where the subject of the triple is a URI reference in the namespace of one dataset, while the object is a URI reference in the other [2,3].

A. Bernstein et al. (Eds.): ISWC 2009, LNCS 5823, pp. 650–665, 2009.

The most visible example of adoption and application of Linked Data has been the Linking Open Data (LOD) project[1], a grassroots community effort to bootstrap the Web of Data by interlinking open-license datasets. Out of the 6.7 billion RDF triples that are served as of July 2009 by participants of the project, approximately 148 million are RDF links between datasets[2].

As Linked Data sources often provide information about large numbers of entities, it is common practice to use automated or semi-automated methods to generate RDF links. In various domains, there are generally accepted naming schemata, such as ISBN and ISSN numbers, ISIN identifiers, EAN and EPC product codes. If both datasets already support one of these identification schemata, the implicit relationship between entities in the datasets can easily be made explicit as RDF links. This approach has been used to generate links between various data sources in the LOD cloud. If no shared naming schema exists, RDF links are often generated by computing the similarity of entities within both datasets using a combination of different property-level similarity metrics.

While there are more and more tools available for publishing Linked Data on the Web [3], there is still a lack of tools that support data publishers in setting RDF links to other data sources, as well as tools that help data publishers to maintain RDF links over time as data sources change. The Silk – Linking Framework contributes to filling this gap. Silk consists of three components: 1. A link discovery engine, which computes links between data sources based on shared identifiers and/or object similarity; 2. A tool for evaluating the generated RDF links in order to fine-tune the linking specification; 3. A protocol for maintaining RDF links between continuously changing data sources.

Using the declarative *Silk - Link Specification Language (Silk-LSL)*, data publishers can specify which types of RDF links should be discovered between data sources as well as which conditions data items must fulfill in order to be interlinked. These link conditions can apply different similarity metrics to multiple properties of an entity or related entities which are addressed using a path-based selector language. The resulting similarity scores can be weighted and combined using various similarity aggregation functions. Silk accesses data sources via the SPARQL protocol and can thus be used to discover links between local or remote data sources.

The main features of the Silk link discovery engine are:

- It supports the generation of owl:sameAs links as well as other types of RDF links.
- It provides a flexible, declarative language for specifying link conditions.
- It can be employed in distributed environments without having to replicate datasets locally.

[1] http://esw.w3.org/topic/SweoIG/TaskForces/CommunityProjects/LinkingOpenData

[2] http://esw.w3.org/topic/TaskForces/CommunityProjects/LinkingOpenData/DataSets/LinkStatistics

- It can be used in situations where terms from different vocabularies are mixed and where no consistent RDFS or OWL schemata exist.
- It implements various caching, indexing and entity preselection methods to increase performance and reduce network load.

Datasets change and are extended over time. In order to keep links between two data sources current and to avoid dead links, new RDF links should be continuously generated as entities are added to the target dataset and invalidated RDF links should be removed. For this task, we propose the *Web of Data – Link Maintenance Protocol (WOD-LMP)*. The protocol automates the communication between two cooperating Web data sources: The *link source* and the *link target*, where the link source is a Web data source that publishes RDF links pointing at data published by the target data source. The protocol supports:

- Notifying the target data source that the link source has published a set of links pointing at the target source. This allows the target to track incoming links and decide whether it wants to set back-links.
- The link source to request a list of changes from the target source. Based on the changes, the link source can recompute existing links and generate additional links pointing at new resources.
- The link source to monitor resources in the target dataset by subscribing to be informed about changes that occur to these resources.

This paper is structured as follows: Section 2 gives an overview of the Silk – Link Specification Language along a concrete usage example. In Section 3, we present the Silk user interface to evaluate generated links. We describe the Web of Data – Link Maintenance Protocol in Section 4 and give an overview of the implementation of the Silk framework in Section 5. Section 6 reviews related work.

2 Link Specification Language

The *Silk – Link Specification Language (Silk-LSL)* is used to express heuristics for deciding whether a semantic relationship exists between two entities. The language is also used to specify the access parameters for the involved data sources, and to configure the caching, indexing and preselection features of the framework. Link conditions can use different aggregation functions to combine similarity scores. These aggregation functions as well as the implemented similarity metrics and value transformation functions were chosen by abstracting from the link heuristics that were used to establish links between data sources in the LOD cloud.

Figure 1 contains a complete Silk-LSL link specification. In this particular use case, we want to discover `owl:sameAs` links between the URIs that are used by DBpedia [4], a data source publishing information extracted from Wikipedia, and by the Linked Data version of DrugBank [5], a pharmaceutical database, to identify medical drugs. In line 22 of the link specification, we thus configure the `<LinkType>` to be `owl:sameAs`.

```
<Silk>
    <DataSource id="dbpedia">
        <EndpointURI>http://dbpedia.org/sparql</EndpointURI>
        <Graph>http://dbpedia.org</Graph>
        <DoCache>1</DoCache>
        <PageSize>1000</PageSize>
    </DataSource>
    <DataSource id="drugbank">
        <EndpointURI>http://www4.wiwiss.fu-berlin.de/drugbank/sparql</EndpointURI>
    </DataSource>
    <Metric id="jaroSets">
        <Param name="item1" />
        <Param name="item2" />
        <AVG>
            <Compare metric="jaroWinklerSimilarity">
                <Param name="str1" path="?item1" />
                <Param name="str2" path="?item2" />
            </Compare>
        </AVG>
    </Metric>
    <Interlink id="drugs">
        <LinkType>owl:sameAs</LinkType>
        <SourceDataset dataSource="dbpedia" var="a">
            <RestrictTo>
                ?a rdf:type dbpedia:Drug
            </RestrictTo>
        </SourceDataset>
        <TargetDataset dataSource="drugbank" var="b">
            <RestrictTo>
                ?b rdf:type drugbank:drugs
            </RestrictTo>
        </TargetDataset>
        <LinkCondition>
            <AVG>
                <MAX weight="1">
                    <Compare metric="maxSimilarityInSets">
                        <Param name="set1" path="?a/rdfs:label" />
                        <Param name="set2" path="?b/rdfs:label" />
                        <Param name="submetric" value="jaroSets" />
                    </Compare>
                    <Compare metric="maxSimilarityInSets" optional="1">
                        <Param name="set1" path="?a/rdfs:label" />
                        <Param name="set2" path="?b/drugbank:synonym" />
                        <Param name="submetric" value="jaroSets" />
                    </Compare>
                    <Compare metric="maxSimilarityInSets" optional="1">
                        <Param name="set1" path="?a/rdfs:label" />
                        <Param name="set2" path="?b/drugbank:genericName" />
                        <Param name="submetric" value="jaroSets" />
                    </Compare>
                </MAX>
                <MAX optional="1" weight="5">
                    <Compare metric="stringEquality" optional="1">
                        <Param name="str1">
                            <Transform function="concat">
                                <Param name="str1" path="?a/dbpedia:atcprefix" />
                                <Param name="str2" path="?a/dbpedia:atcsuffix" />
                            </Transform>
                        </Param>
                        <Param name="str2" path="?b/drugbank:atcCode" />
                    </Compare>
                    <Compare metric="stringEquality" optional="1">
                        <Param name="str1" path="?a/dbpedia:casNumberLink" />
                        <Param name="str2" path="?b/drugbank:casRegistryNumber" />
                    </Compare>
                    <Compare metric="stringEquality" optional="1">
                        <Param name="str1" path="?a/dbpedia:pubchem" />
                        <Param name="str2" path="?b/drugbank:pubchemCompoundId" />
                    </Compare>
                </MAX>
                <Compare metric="numSimilarity" optional="1" weight="2">
                    <Param name="num1" path="?a/dbpedia:molecularweight" />
                    <Param name="num2" path="?b/drugbank:molecularWeightAverage" />
                </Compare>
            </AVG>
        </LinkCondition>
        <Thresholds accept="0.9" verify="0.7" />
        <Limit max="1" method="metric_value" />
        <Output acceptedLinks="drug_accepted_links.n3"
         verifyLinks="drug_verify_links.n3" format="n3" mode="truncate" />
    </Interlink>
</Silk>
```

Fig. 1. Example: Interlinking drugs in DBpedia and DrugBank

2.1 Data Access

For accessing the source and target data sources, we first specify access parameters for the DBpedia and DrugBank SPARQL endpoints using the `<DataSource>` directive. The only mandatory data source parameter is the SPARQL endpoint URI. Besides this, it is possible to define other data source access options, such as the graph name and to enable in-memory caching of SPARQL query results. In order to restrict the query load on remote SPARQL endpoints, it is possible to set a delay between subsequent queries using the `<Pause>` parameter, specifying the delay time in milliseconds. For working against SPARQL endpoints that restrict result sets to a certain size, Silk uses a paging mechanism. The maximal result size is configured using the `<PageSize>` parameter. The paging mechanism is implemented using SPARQL `LIMIT` and `OFFSET` queries. Lines 2 to 7 in Figure 1 show how the access parameters for the DBpedia data source are set to select only resources from the named graph `http://dbpedia.org`, enable caching and limit the page size to 1,000 results per query.

The configured data sources are later referenced in the `<SourceDataset>` and `<TargetDataset>` clauses of the link specification. Since we only want to match drugs, we restrict the sets of examined resources to instances of the classes `dbpedia:Drug` and `drugbank:drugs` in the respective datasets by supplying SPARQL conditions within the `<RestrictTo>` directives in lines 25 and 30. These statements may contain any valid SPARQL expressions that would usually be found in the `WHERE` clause of a SPARQL query.

2.2 Link Conditions

The `<LinkCondition>` section is the heart of a Silk link specification and defines how similarity metrics are combined in order to calculate a total similarity value for a pair of entities. For comparing property values or sets of entities, Silk provides a number of built-in similarity metrics. Table 1 gives an overview of these metrics. The implemented metrics include string, numeric, date, URI, and set comparison methods as well as a taxonomic matcher that calculates the semantic distance between two concepts within a concept hierarchy using the distance metric proposed by Zhong et al. in [8]. Each metric in Silk evaluates to a similarity value between 0 or 1, with higher values indicating a greater similarity.

These similarity metrics may be combined using the following aggregation functions:

- AVG – weighted average
- MAX – choose the highest value
- MIN – choose the lowest value
- EUCLID – Euclidian distance metric
- PRODUCT – weighted product

Table 1. Available similarity metrics in Silk

jaroSimilarity	String similarity based on Jaro distance metric[6]
jaroWinklerSimilarity	String similarity based on Jaro-Winkler metric[7]
qGramSimilarity	String similarity based on q-grams
stringEquality	Returns 1 when strings are equal, 0 otherwise
numSimilarity	Percentual numeric similarity
dateSimilarity	Similarity between two date values
uriEquality	Returns 1 if two URIs are equal, 0 otherwise
taxonomicSimilarity	Metric based on the taxonomic distance of two concepts
maxSimilarityInSet	Returns the highest encountered similarity of comparing a single item to all items in a set
setSimilarity	Similarity between two sets of items

To take into account the varying importance of different properties, the metrics grouped inside the AVG, EUCLID and PRODUCT operators may be weighted individually, with higher weighted metrics having a greater influence on the aggregated result.

In the `<LinkCondition>` section of the example (lines 33 to 76), we compute similarity values for the the labels, PubChem IDs[3], CAS registry numbers[4], ATC codes[5] and molecular weights between datasets and calculate a weighted average of these values.

Most metrics are configured to be optional since the presence of the respective RDF property values they refer to is not always guaranteed. In cases where alternating properties refer to an equivalent feature (such as `rdfs:label`, `drugbank:synonym` and `drugbank:genericName`), we choose to perform comparisons for both properties and select the best evaluation by using the `<MAX>` aggregation operator. The `<MAX>` operator is also used to choose the maximum value of the comparisons between any of the exact drug identifiers. Weighting of results is used within the metrics comparing these exact values (line 52), with the metric weight raised to 5, as well as within the molecular weight comparison using a weighting factor of 2.

Lines 11 to 20 demonstrate how a user-defined metric is specified. User-defined metrics may be used like built-in metrics. In the example, the defined `jaroSets` metric is used as a submetric for the `maxSimilarityInSets` evaluations in lines 36-50 for the pairwise comparison of elements of the compared sets. In this case, the user-defined metric is mainly a wrapper around a `jaroWinklerSimilarity` call to achieve type-compatibility with the set comparison interface.

Property values are often represented differently across datasets and thus need to be normalized before being compared. For handling this task, it is possible to apply data transformation functions to parameter values before passing them to a similarity metric. The available transformation functions are shown in Table 2. In

[3] `http://pubchem.ncbi.nlm.nih.gov/`

[4] `http://www.cas.org/expertise/cascontent/registry/regsys.html`

[5] `http://www.who.int/classifications/atcddd/en/`

Table 2. Available transformation functions in Silk

removeBlanks	Remove whitespace from string
removeSpecialChars	Remove special characters from string
lowerCase	Convert a string to lower case
upperCase	Convert a string to upper case
concat	Concatenate two strings
stem	Apply word stemming to a string
alphaReduce	Strip all non-alphabetic characters from a string
numReduce	Strip all non-numeric characters from a string
replace	Replace all occurrences of a string with a replacement
regexReplace	Replace all occurences of a regex with a replacement
stripURIPrefix	Strip the URI prefix from a string
translateWithDictionary	Translate string using a provided CSV dictionary file

the drug linking example, a drug's ATC code in the DBpedia dataset is split into a prefix and a suffix part, while it is stored in a single property on the DrugBank side. Hence, we use the `concat` transformation function to concatenate the code's prefix and suffix parts on the DBpedia side before comparing it to the single-property code in DrugBank (lines 55 to 58).

After specifying the link condition, we finally specify within the `<Thresholds>` clause that resource pairs with a similarity score above 0.9 are to be interlinked, whereas pairs between 0.7 and 0.9 should be written to a separate output file in order to be reviewed by an expert. The `<Limit>` clause is used to limit the number of outgoing links from a particular entity within the source dataset. If several candidate links exist, only the highest evaluated one is chosen and written to the output files as specified by the `<Output>` directive. In this example, we permit only one outgoing `owl:sameAs` link from each resource.

Discovered links can be outputted either as simple RDF triples and/or in reified form together with their creation date, confidence score and the URI identifying the employed interlinking heuristic.

2.3 Silk Selector Language

Especially for discovering semantic relationships other than entity equality, a flexible way for selecting sets of resources or literals in the RDF graph around a particular resource is needed. Silk addresses this requirement by offering a simple RDF path selector language for providing parameter values to similarity metrics and transformation functions. A Silk selector language path starts with a variable referring to an RDF resource and may then use several path operators to navigate the graph surrounding this resource. To simply access a particular property of a resource, the forward operator (/) may be used. For example, the path "`?drug/rdfs:label`" would select the set of label values associated with a drug referred to by the `?drug` variable.

Sometimes, we need to navigate backwards along a property edge. For example, drugs in DrugBank contain a `drugbank:target` property pointing to the drug's target molecule. However, there exists no explicit reverse property like `drugbank:drug` in the drug target's resource. So if a path begins with a drug target and we need to select all of the drugs that apply to it, we may use the backward operator (\ ) to navigate property edges in reverse. Navigating backwards along the property `drugbank:target` would select the applicable drugs.

The filter operator ([ ]) can be used to restrict selected resources to match a certain predicate. To select only drugs amongst the ones applicable to a target molecule which have been marked as *approved*, we could for instance use the RDF path "`?target\drugbank:target[drugbank:drugType drugType:approved]`". The filter operator also supports numeric comparisons. For example, to select drugs with a molecular weight above 200, the path "`?target\drugbank:target [drugbank:molecularWeightAverage > 200]`" can be used.

2.4 Pre-matching

To compare all pairs of entities of a source dataset S and a target dataset T would result in an unsatisfactory runtime complexity of $O(|S| \cdot |T|)$. Even after using SPARQL restrictions to select suitable subsets of each dataset, the required time and network load to perform all pair comparisons might prove to be impractical in many cases. To avoid this problem, we need a way to quickly find a limited set of target entities that are likely to match a given source entity. Silk provides this by offering rough index pre-matching.

When using pre-matching, all target resources are indexed by the values of one or more specified properties (most commonly, their labels) before any detailed comparisons are performed. During the subsequent resource comparison phase, the previously generated index is used to look up potential matches for a given source resource. This lookup uses the BM25[6] weighting scheme for the ranking of search results and additionally supports spelling corrections of individual words of a query. Only a limited number of target resources found in this lookup is then considered as candidates for a detailed comparison.

An example of such a pre-matching specification that could be applied to our drug linking example is presented in Figure 2. This directive instructs Silk to index the drugs in the target dataset by their `rdfs:label` and `drugbank:synonym` property values. When performing comparisons, the `rdfs:label` of a source resource is used as a search term into the generated indexes and only the first ten target hits found in each index are considered as link candidates for detailed comparisons.

If we neglect a slight index insertion and search time dependency on the target dataset size, we now achieve a runtime complexity of $O(|S| + |T|)$, making it feasible to interlink even large datasets under practical time constraints. Note however that this pre-matching may come at the cost of missing some links during discovery, since it is not guaranteed that a pre-matching lookup will always find all matching target resources.

---

[6] `http://xapian.org/docs/bm25.html`

```
<PreMatchingDefinition sourcePath="?a/rdfs:label" hitLimit="10">
    <Index targetPath="?b/rdfs:label" />
    <Index targetPath="?b/drugbank:synonym" />
</PreMatchingDefinition>
```

Fig. 2. Pre-matching

3 Evaluating Links

In real-world settings, data is often not as clean and complete as we would wish it to be. For instance, two data sources might both support the same identification schema, like EAN, ISBN or ISIN numbers, but due to a large number of missing values, it is nevertheless necessary to use similarity computations in addition to identifier matching to generate links. Such data quality problems are usually not known in advance but discovered when a data publisher tries to compute links pointing to a target data source. Therefore, finding a good linking heuristic is usually an iterative process. In order to support users with this task, Silk provides a Web interface for evaluating the correctness and completeness of generated links. Based on this evaluation, users can fine-tune their linking specification, for example by changing weights or applying different metrics or aggregation functions.

3.1 Resource Comparison

The resource comparison component of the Silk web interface allows the user to quickly evaluate links according to the currently loaded linking specification. A screenshot of this interface is shown in Figure 3.

The user first enters a set of RDF links into the box at the top of the screen. Silk then recomputes these links and displays the resulting similarity scores for each link in an overview table. For further examination, a drill-down view of a specific pair comparison can be shown by clicking on one of the table rows. This drill-down shows in a tree-like fashion the exact parameterizations and evaluations of all submetrics and aggregations employed. This information allows the user to spot parts of the similarity evaluation which did not behave as expected.

An example drill-down of a comparison between the DrugBank and DBpedia resources describing the drug Lorazepam is shown in Figure 4. As evident from the illustration, the two drug entries are matched successfully with a high total similarity score although several subcomparisons return infavorable results. For example, the comparison of the DBpedia resource's label with the synonyms on the DrugBank side finds only a similarity of 0.867. However, since perfectly matching labels exist on both sides, the `<MAX>` operator grouping these name-related property comparisons evaluates to a total similarity value of 1. Similarly, due to a dataset error, the section aggregating exact numeric drug identifiers contains a similarity value of 0 for the CAS registry numbers. This erroneously low value is corrected by the availability of other exactly matching identifiers in a `<MAX>` aggregation.

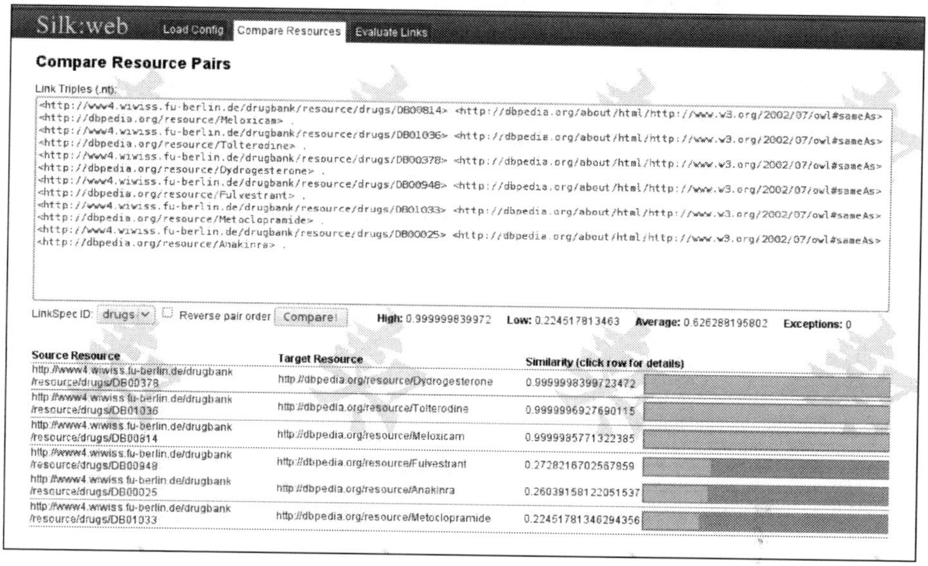

Fig. 3. Comparing resource pairs with the Silk web interface

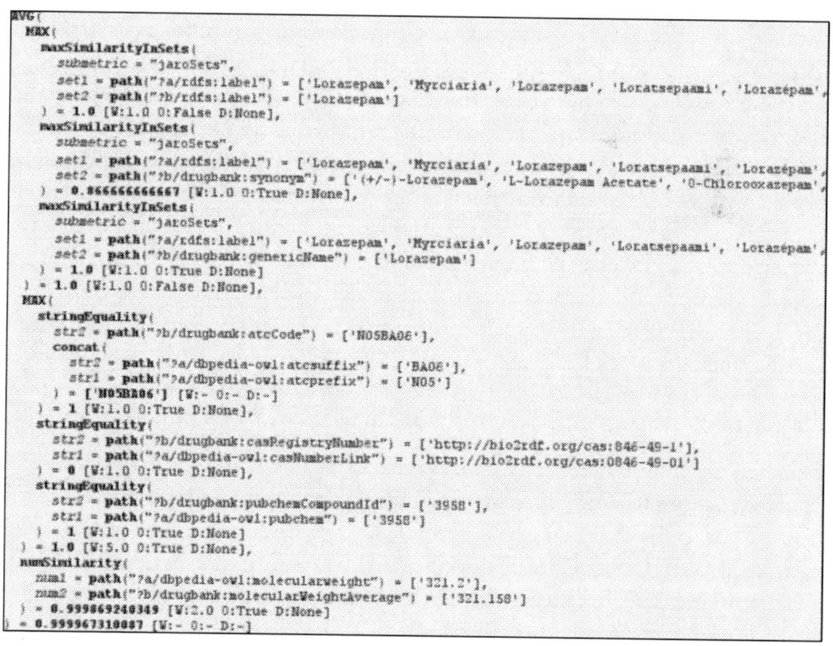

Fig. 4. Detailed drill-down into a resource pair comparison

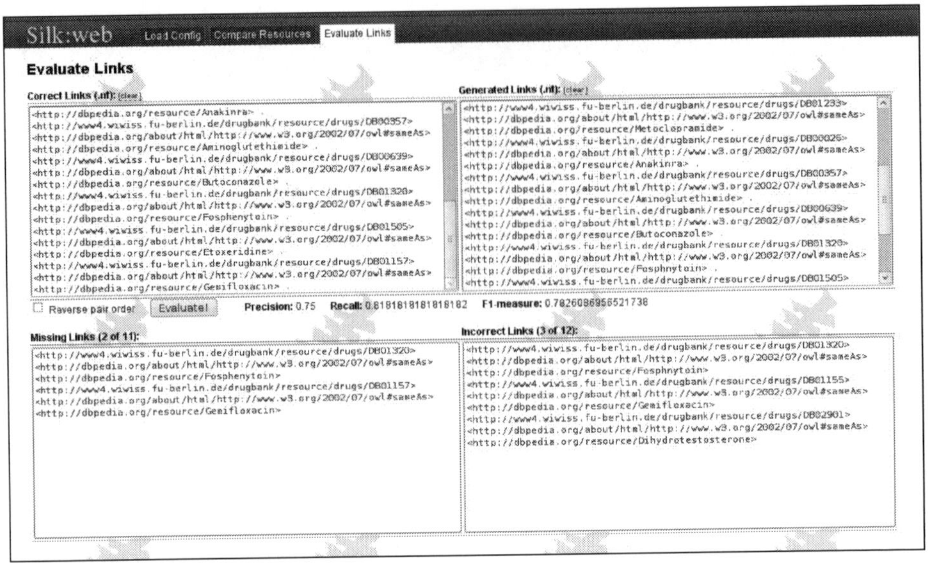

Fig. 5. Evaluating linksets with the Silk web interface

3.2 Evaluation against a Reference Linkset

A methodology that proved useful for optimizing link specifications is to manually create a small reference linkset and then optimize the Silk linking specification to produce these reference links, before Silk is run against the complete target data source. Once such a reference linkset is available, the Silk web interface provides a linkset evaluation component which allows the comparison of generated linksets to the reference set. This component is shown in Figure 5.

Silk displays which links are missing from the generated set as well as which resource pairs were interlinked erroneously. To give an overall indication about the linkset quality, Silk also computes statistical measures pertaining to completeness and correctness of the generated links. A *Precision* value indicates the correctness of generated links, while a *Recall* value measures the completeness of discovered links. Finally, the F_1*-measure* calculates the weighted harmonic mean of both, providing an overall-quality measure of the linkset.

3.3 Improving the DBpedia/DrugBank Link Specification

We compared 3,134 drugs in DBpedia with 4,772 drugs in DrugBank. As a result of applying the linking specification shown in Figure 1, Silk discovered 1,227 confident links above the threshold of 0.9 and found 32 more links above the threshold of 0.7. To evaluate the quality of the retrieved links, we created a reference linkset pertaining to 50 drugs selected randomly from DrugBank and found 38 manually researched links to DBpedia. We then ran Silk a second time to find only links from these 50 selected DrugBank resources to DBpedia and compared both the generated and the reference linkset.

The evaluation revealed 4 missing links and one incorrectly discovered link. This corresponded to a Precision of 0.97, a Recall of 0.89 and an F_1-measure of 0.93. To better understand why certain links are missing and why one link was incorrect, we then compared their source and target resources via the resource comparison web interface. One link was missed because of radically differing molecular weights in both datasets. Three other missing links were not discovered due to the fact that their CAS registry numbers did not match while at the same time no other exact identifiers were present. Finally, one link was discovered incorrectly since the resource labels were very similar and no other relevant property values were present in the datasets. In a subsequent tuning of the link specification, we mitigated the effect of a single mismatching exact identifier by lowering the weight for the surrounding aggregation to 3 and setting a default value of 0.85 for the IDs in the same `<MAX>` aggregation in case the corresponding RDF properties were not available. This lowered the negative effect of a single incorrect identifier while preserving a high rating in this `<MAX>` aggregation whenever a matching value is found. After this improvement, only 2 links were missing, which means that we now reached a Recall value of 0.95 and an F_1-measure of 0.96.

4 Web of Data – Link Maintenance Protocol

Changes or additions in either of the interlinked datasets can invalidate existing links or imply the need to generate new ones. With the *Web of Data – Link Maintenance Protocol (WOD-LMP)*, we propose a solution to this problem.

The WOD-LMP protocol automates the communication between two cooperating Web data sources. It assumes two basic roles: *Link source* and *link target*, where the link source is a Web data source that publishes RDF links pointing at data published by the target data source. The protocol covers the following three use cases:

4.1 Link Transfer to Target

In the simplest use case, a link source wants to send a set of RDF links to the target data source so that the target may keep track of incoming links and can decide whether it wants to set back-links. Afterwards, the source wants to keep the target informed about subsequent updates (i.e. additions and deletions) to the transferred links. To achieve the transfer of the initial set of links and of subsequently generated ones, a *Link Notification* message is sent to the target data source. This notification includes the generated links along with the URL of the WOD-LMP protocol endpoint at the source side. Single deletion of links by the source is communicated to the target in a *Link Deletion Notification* message, which in turn contains the link triples to be deleted.

4.2 Request of Target Change List

In this use case, the source data source asks the target to supply a list of all changes that have occurred to RDF resources in a target dataset within a specific time period. The source may then use the provided change information for periodic link recomputation. The protocol also provides requesting only additions, updates or deletions of resources. WOD-LMP uses incremental sequence numbers to identify resource changes. The changes are requested by the remote data source by sending a *Get Changes* message, which contains both the desired change sequence number range as well as the desired change type filter options. The target replies to this with a *Change Notification*, which lists the requested changes together with their corresponding sequence numbers and change types. If no upper sequence number is supplied, the target sends all changes to the latest change.

This case of selective link recomputation requires periodic polling of the remote data source by the source but has the advantage of working without maintaining a persistent relationship between the linked data sources.

4.3 Subscription of Target Changes

The protocol also supports fine-grained link recomputation by monitoring the resources in the target dataset that were used to compute links. As illustrated in Figure 6, the source informs the target dataset via a *Link Notification* message about a group of generated links and for each transferred link, supplies the URIs of the resources in the target dataset that were used to compute the link. The target saves this information and monitors the resources. If one of them changes or is deleted, the target notifies the source about these changes by sending a *Change Notification* message. The source may then use this information to recompute affected links and possibly delete invalidated ones. In this case, it notifies the target about deleted links with a *Link Deletion Notification*, which cancels the subscription of resources relevant to these links.

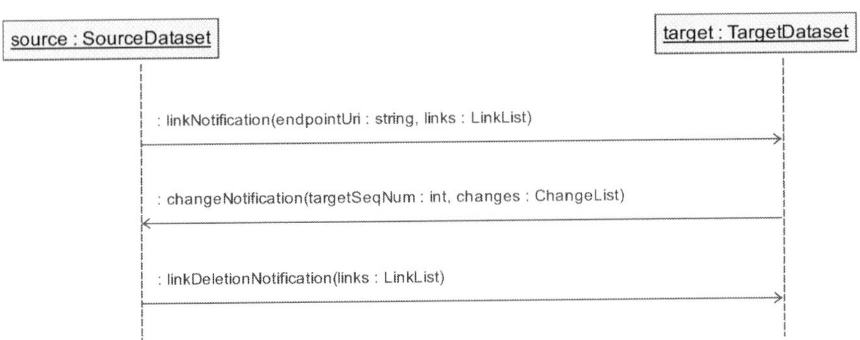

Fig. 6. Subscribing to resource changes in the target data source

The implementation of the WOD-LMP protocol is based on SOAP. The complete specification of the protocol is available at `http://www4.wiwiss.fu-berlin.de/bizer/silk/wodlmp/`

The WOD-LMP protocol is used to maintain the links between DBpedia and DrugBank. Links generated on the DrugBank side are sent and integrated into DBpedia, while DBpedia notifies the DrugBank Silk instance about changes to subscribed resources. This synchronization will become especially important as DBpedia will start to utilize the Wikipedia live update stream to continuously extract data from changed Wikipedia pages. Thus, DBpedia resources will be continuously updated to match Wikipedia, while at the same time the DrugBank Silk instance will be able to maintain and recompute links to DBpedia.

5 Implementation

Silk is written in Python and is run from the command line. When generating linksets, Silk is started as a batch process. It runs as a daemon when serving the web interface or WOD-LMP protocol endpoints. The framework may be downloaded from Google Code[7] under the terms of the BSD license. For calculating string similarities, a library from Febrl[8], the Freely Extensible Biomedical Record Linkage Toolkit, is used, while Silk's pre-matching features are achieved using the search engine library Xapian[9]. The web interface was realized with the Werkzeug[10] toolkit, while the link maintenance protocol endpoints use the free soaplib[11] library for the exchange of SOAP messages.

6 Related Work

There is a large body of related work on record linkage [7] and duplicate detection [9] within the database community as well as on ontology matching [10] in the knowledge representation community. Silk builds on this work by implementing similarity metrics and aggregation functions that proved successful within other scenarios. What distinguishes Silk from this work is its focus on the Linked Data scenario where different types of semantic links should be discovered between Web data sources that often mix terms from different vocabularies and where no consistent RDFS or OWL schemata spanning the data sources exist.

Related work that also focuses on Linked Data includes Raimond et al. [11] who propose a link discovery algorithm that takes into account both the similarities of web resources and of their neighbors. The algorithm is implemented within the GNAT tool and has been evaluated for interlinking music-related datasets. In [12], Hassanzadeh et al. describe a framework for the discovery of

[7] `http://silk.googlecode.com`
[8] `http://sourceforge.net/projects/febrl`
[9] `http://xapian.org`
[10] `http://werkzeug.pocoo.org`
[11] `http://trac.optio.webfactional.com/`

semantic links over relational data which also introduces a declarative language for specifying link conditions. Their framework is meant to be used together with relational database to RDF wrappers like D2R Server or Virtuoso RDF Views. Differences between LinQL and Silk-LSL are the underlying data model and Silk's ability to more flexibly combine metrics through aggregation functions. A framework that deals with instance coreferencing as part of the larger process of fusing Web data is the KnoFuss Architecture proposed in [13]. In contrast to Silk, KnoFuss assumes that instance data is represented according to consistent OWL ontologies.

Furthermore, approaches to track changes and updates in Linked Data sources include PingtheSemanticWeb[12], a central registry for Web of Data documents which offers XML-RPC and REST APIs to notify the service about new or changed documents. A further approach to making change information available is proposed by Auer et al. and implemented in Triplify[14]. Similar to the second WOD-LMP use case, change information is requested on a peer-to-peer basis instead of being aggregated into a central registry, such as PingtheSemanticWeb. This approach is also implemented by DSNotify[15], which runs as an add-on to a local data source and uses indexes to track resource changes. DSNotify supports the active notification of subscribers as well as providing change data on demand. It further uses heuristics to determine the cause of a resource change and whether a deleted link target has become available under a different URI.

7 Conclusion

We presented the Silk framework, a flexible tool for discovering links between entities within different Web data sources. The Silk-LSL link specification language was introduced and its applicability was demonstrated within a life science use case. We then proposed the WOD-LMP protocol for synchronizing and maintaining links between continuously changing Linked Data sources.

Future work on Silk will focus on the following areas: We will implement further similarity metrics to support a broader range of linking use cases. To assist users in writing Silk-LSL specifications, machine learning techniques could be employed to adjust weightings or optimize the structure of the matching specification. Finally, we will evaluate the suitability of Silk for detecting duplicate entities within local datasets instead of using it to discover links between disparate RDF data sources.

The value of the Web of Data rises and falls with the amount and the quality of links between data sources. We hope that Silk and other similar tools will help to strengthen the linkage between data sources and therefore contribute to the overall utility of the network.

The complete Silk – LSL language specification, WoD Link Maintenance Protocol specification and further Silk usage examples are found on the Silk project website at `http://www4.wiwiss.fu-berlin.de/bizer/silk/`

[12] `http://pingthesemanticweb.com`

References

1. Berners-Lee, T.: Linked Data - Design Issues, `http://www.w3.org/DesignIssues/LinkedData.html`
2. Bizer, C., Heath, T., Berners-Lee, T.: Linked Data - The Story So Far. Journal on Semantic Web and Information Systems (in press, 2009)
3. Bizer, C., Cyganiak, R., Heath, T.: How to publish Linked Data on the Web, `http://www4.wiwiss.fu-berlin.de/bizer/pub/LinkedDataTutorial/`
4. Bizer, C., et al.: DBpedia - A Crystallization Point for the Web of Data. Journal of Web Semantics: Sci. Serv. Agents World Wide Web (2009), doi:10.1016/j.websem.2009.07.002
5. Jentzsch, A., et al.: Enabling Tailored Therapeutics with Linked Data. In: Proceedings of the 2nd Workshop about Linked Data on the Web (2009)
6. Jaro, M.: Advances in Record-linkage Methodology as Applied to the 1985 Census of Tampa, Florida. Journal of the American Statistical Society 84(406), 414–420 (1989)
7. Winkler, W.: Overview of Record Linkage and Current Research Directions. Bureau of the Census - Research Report Series (2006)
8. Zhong, J., et al.: Conceptual Graph Matching for Semantic Search. The 2002 International Conference on Computational Science (2002)
9. Elmagarmid, A.K., Ipeirotis, P.G., Verykios, V.S.: Duplicate Record Detection: A Survey. IEEE Transactions on Knowledge and Data Engineering 19(1), 1–16 (2007)
10. Euzenat, J., Shvaiko, P.: Ontology Matching. Springer, Heidelberg (2007)
11. Raimond, Y., Sutton, C., Sandler, M.: Automatic Interlinking of Music Datasets on the Semantic Web. In: Proceedings of the 1st Workshop about Linked Data on the Web (2008)
12. Hassanzadeh, O., et al.: Semantic Link Discovery Over Relational Data. In: Proceedings of the 18th ACM Conference on Information and Knowledge Management (2009)
13. Nikolov, A., et al.: Integration of Semantically Annotated Data by the KnoFuss Architecture. In: 16th International Conference on Knowledge Engineering and Knowledge Management, pp. 265–274 (2008)
14. Auer, S., et al.: Triplify – Light-Weight Linked Data Publication from Relational Databases. In: Proceedings of the 18th International World Wide Web Conference (2009)
15. Haslhofer, B., Popitsch, N.: DSNotify – Detecting and Fixing Broken Links in Linked Data Sets. In: Proceedings of 8th International Workshop on Web Semantics (2009)

Concept and Role Forgetting in $\mathcal{ALC}$ Ontologies*

Kewen Wang[1], Zhe Wang[1], Rodney Topor[1], Jeff Z. Pan[2], and Grigoris Antoniou[3]

[1] Griffith University, Australia
{k.wang,jack.wang,r.topor}@griffith.edu.au
[2] University of Aberdeen, UK
jeff.z.pan@abdn.ac.uk
[3] University of Crete, Greece
antoniou@ics.forth.gr

Abstract. Forgetting is an important tool for reducing ontologies by eliminating some concepts and roles while preserving sound and complete reasoning. Attempts have previously been made to address the problem of forgetting in relatively simple description logics (DLs) such as DL-Lite and extended $\mathcal{EL}$. The ontologies used in these attempts were mostly restricted to TBoxes rather than general knowledge bases (KBs). However, the issue of forgetting for general KBs in more expressive description logics, such as $\mathcal{ALC}$ and OWL DL, is largely unexplored. In particular, the problem of characterizing and computing forgetting for such logics is still open.

In this paper, we first define semantic forgetting about concepts and roles in $\mathcal{ALC}$ ontologies and state several important properties of forgetting in this setting. We then define the result of forgetting for concept descriptions in $\mathcal{ALC}$, state the properties of forgetting for concept descriptions, and present algorithms for computing the result of forgetting for concept descriptions. Unlike the case of DL-Lite, the result of forgetting for an $\mathcal{ALC}$ ontology does not exist in general, even for the special case of concept forgetting. This makes the problem of how to compute forgetting in $\mathcal{ALC}$ more challenging. We address this problem by defining a series of approximations to the result of forgetting for $\mathcal{ALC}$ ontologies and studying their properties and their application to reasoning tasks. We use the algorithms for computing forgetting for concept descriptions to compute these approximations. Our algorithms for computing approximations can be embedded into an ontology editor to enhance its ability to manage and reason in (large) ontologies.

1 Introduction

The amount of semantically annotated data available on the Web is growing rapidly. Often, the formal model used for representing such information is an ontology in some description logic. As more ontologies are used for annotating data on the Web, and as the populated ontologies become larger and more comprehensive, it becomes increasingly important for the Semantic Web [4] to be able to construct and manage such ontologies. Examples of large ontologies currently in use include the Systematised

* This work was partially supported by the Australia Research Council (ARC) Discovery Project 0666107.

A. Bernstein et al. (Eds.): ISWC 2009, LNCS 5823, pp. 666–681, 2009.

Nomenclature of Medicine Clinical Terms (SNOMED CT) containing 380K concepts, GALEN, the Foundational Model of Anatomy (FMA), the National Cancer Institute (NCI) Thesaurus containing over 60K axioms, and the OBO Foundry containing about 80 biomedical ontologies.

While it is expensive to construct large ontologies, it is even more expensive to host, manage and use a large, comprehensive ontology when a smaller ontology would suffice. Therefore, tools to reduce large ontologies to smaller ontologies that meet the needs of specific applications aid and encourage the use of existing ontologies. However, as the tool evaluation study in [5] shows, existing tools, such as Protégé [28], NeOn [29] and TopBraid [30], are far from satisfactory for this purpose.

Ontology engineers thus face the task of reducing existing, large ontologies to smaller, better focussed ontologies by hiding or forgetting irrelevant concepts and roles while preserving required reasoning capabilities. Such ontology reductions can be applied to ontology extraction, ontology summary, ontology integration, and ontology evolution.

We consider two typical scenarios, in ontology extraction and ontology summary, respectively.

Ontology Extraction. To avoid building new ontologies from scratch, it is preferable to reuse existing ontologies whenever possible. But often, only part of an existing ontology is required. For exxample, if we had an ontology of medical terms, such as SNOMED, but were only interested in infectious diseases, it would be desirable to forget all other terms in the ontology before starting to reason about infectious diseases.

Ontology summary: As argued in [1,18], a key problem faced by ontology engineers is the task of constructing a summary of an existing ontology so that other users can decide whether or not to use it. This task involves first identifying the key concepts (and roles) in the ontology and then hiding or forgetting all other concepts (and roles). For example, an astronomical ontology of the solar system might have planets as key concepts and asteroids and comets as less important concepts.

However, an ontology is often represented as a logical theory, and the removal of one term may influence other terms in the ontology. Thus, more advanced methods for dealing with large ontologies and reusing existing ontologies are desired.

Forgetting (or uniform interpolation) has previously been studied for propositional and first-order logic and logic programming[15,16,6], where it has proved a useful technique for reducing the size of a logical theory while preserving sound and complete reasoning in the resulting smaller theory.

However, description logic (DL) [3] is a different, important. knowledge representation framework, which is the basis for ontology languages such as OWL, that are widely used in the Semantic Web. In this context, an ontology is a knowledge base (KB) in a particular description logic, where a knowledge base consists of a terminology box (TBox) and an assertion box (ABox).

Although most description logics are equivalent to fragments of first-order logic (FOL), the forgetting for first-order logic introduced in [16] is not directly applicable to description logics for at least two reasons. First, the correspondence between DLs and FOL is useless in investigating forgetting for DLs because the result of forgetting in a theory of the first-order logic (FOL) may only be expressible in second-order logic.

Second, it is preferable to perform forgetting in description logics directly rather than transforming an ontology into a first-order theory, forgetting and then transforming back to an ontology.

Attempts have previously been made to address the problem of forgetting in relatively simple description logics (DLs) such as DL-Lite [21,14] and extended $\mathcal{EL}$ [12]. The ontologies used in these attempts were mostly restricted to KBs with empty ABoxes. Forgetting also generalizes previous work on *conservative extensions* [9,7,17] and the *modularity* defined in [8,10,13]. A definition of forgetting for TBoxes in the more expressive DL $\mathcal{ALC}$ was given in [7].

However, the issue of forgetting for general KBs, with nonempty ABoxes, in more expressive DLs, such as $\mathcal{ALC}$ and OWL DL, is largely unexplored. In particular, the problem of characterizing and computing forgetting for such logics is still open.

In this paper we first give a semantic definition of forgetting for ontologies in the description logic $\mathcal{ALC}$ and state several important properties of forgetting. We choose $\mathcal{ALC}$ to study in this paper because it allows all boolean operations and most expressive DLs are based on it. Others have argued [19] that practical ontologies such as SNOMED would benefit from a more expressive DL. We then define the result of forgetting for concept descriptions in $\mathcal{ALC}$, state the properties of forgetting for concept descriptions, and present algorithms for computing the result of forgetting for concept descriptions. (Forgetting for concept descriptions in $\mathcal{ALC}$ has previously been investigated under the name of *uniform interpoloation* in [20].) Unlike the case of DL-Lite, the result of forgetting for an $\mathcal{ALC}$ ontology does not exist in general, even for the special case of concept forgetting. This makes the problem of how to compute forgetting in $\mathcal{ALC}$ more challenging. We address this problem by defining a series of approximations to the result of forgetting for $\mathcal{ALC}$ ontologies and studying their properties and their application to reasoning tasks. We use the algorithms for computing forgetting for concept descriptions to compute these approximations. Our algorithms for computing approximations can be embedded into an ontology editor to enhance its ability to manage and reason in (large) ontologies.

Our work significantly extends previous work in at least two ways: (1) We make the first attempt to study forgetting for an expressive description logic, instead of DL-Lite and variants of $\mathcal{EL}$. (2) Ontologies in this paper are KBs with nonepty ABoxes, rather than TBoxes only in previous work. In addition, our definitions and results hold for forgetting about both concepts and roles.

Due to space limitation, proofs are omitted in this paper but can be found at `http://www.cit.gu.edu.au/~kewen/Papers/alc_forget_long.pdf`

2 Description Logic $\mathcal{ALC}$

In this section, we briefly recall some preliminaries of $\mathcal{ALC}$, the basic description logic which contains all boolean operators. Further details of $\mathcal{ALC}$ and other DLs can be found in [3].

First, we introduce the syntax of *concept descriptions* for $\mathcal{ALC}$. To this end, we assume that N_C is a set of *concept names* (or *concept*), N_R is a set of *role names* (or *roles*) and N_I is a set of individuals.

Elementary concept descriptions consist of both *concept names* and *role names*. So a concept name is also called *atomic concept* while a role name is also called *atomic role*. Complex concept descriptions are built inductively as follows: A (atomic concept); $\top$ (universal concept); $\bot$ (empty concept); $\neg C$ (negation); $C \sqcap D$ (conjunction); $C \sqcup D$ (disjunction); $\forall R.C$ (universal quantification) and $\exists R.C$ (existential quantification). Here, A is an (atomic) concept, C and D are concept descriptions, and R is a role.

An interpretation $\mathcal{I}$ of $\mathcal{ALC}$ is a pair $(\Delta^{\mathcal{I}}, \cdot^{\mathcal{I}})$ where $\Delta^{\mathcal{I}}$ is a non-empty set called the *domain* and $\cdot^{\mathcal{I}}$ is an interpretation function which associates each (atomic) concept A with a subset $A^{\mathcal{I}}$ of $\Delta^{\mathcal{I}}$ and each atomic role R with a binary relation $R^{\mathcal{I}} \subseteq \Delta^{\mathcal{I}} \times \Delta^{\mathcal{I}}$. The function $\cdot^{\mathcal{I}}$ can be naturally extended to complex descriptions:

$$\top^{\mathcal{I}} = \Delta^{\mathcal{I}} \qquad \bot^{\mathcal{I}} = \emptyset$$
$$(\neg C)^{\mathcal{I}} = \Delta^{\mathcal{I}} - C^{\mathcal{I}} \qquad (C \sqcap D)^{\mathcal{I}} = C^{\mathcal{I}} \cap D^{\mathcal{I}}$$
$$(C \sqcup D)^{\mathcal{I}} = C^{\mathcal{I}} \cup D^{\mathcal{I}}$$
$$(\forall R.C)^{\mathcal{I}} = \{a \in \Delta^{\mathcal{I}} \ : \ \forall b.(a,b) \in R^{\mathcal{I}} \text{ implies } b \in C^{\mathcal{I}}\}$$
$$(\exists R.C)^{\mathcal{I}} = \{a \in \Delta^{\mathcal{I}} \ : \ \exists b.(a,b) \in R^{\mathcal{I}} \text{ and } b \in C^{\mathcal{I}}\}$$

An $\mathcal{ALC}$ *assertion box* (or *ABox*) is a finite set of *assertions*. An assertion is a *concept assertion* of the form $C(a)$ or a *role assertion* of the form $R(a, b)$, where a and b are individuals, C is a concept name, R is a role name.

An interpretation $\mathcal{I}$ *satisfies* a concept assertion $C(a)$ if $a^{\mathcal{I}} \in C^{\mathcal{I}}$, a role assertion $R(a, b)$ if $(a^{\mathcal{I}}, b^{\mathcal{I}}) \in R^{\mathcal{I}}$. If an assertion α is satisfied by $\mathcal{I}$, it is denoted $\mathcal{I} \models \alpha$. An interpretation $\mathcal{I}$ is a *model* of an ABox $\mathcal{A}$, written $\mathcal{I} \models \mathcal{A}$, if it satisfies all assertions in $\mathcal{A}$.

A *inclusion axiom* (simply *inclusion*, or *axiom*) is of the form $C \sqsubseteq D$ (C is *subsumed* by D), where C and D are concept descriptions. The inclusion $C \equiv D$ (C is *equivalent* to D) is an abbreviation of two inclusions $C \sqsubseteq D$ and $D \sqsubseteq C$. A *terminology box*, or *TBox*, is a finite set of inclusions. An interpretation $\mathcal{I}$ satisfies an inclusion $C \sqsubseteq D$ if $C^{\mathcal{I}} \subseteq D^{\mathcal{I}}$. $\mathcal{I}$ is a model of a TBox $\mathcal{T}$, denoted $\mathcal{I} \models \mathcal{T}$, if $\mathcal{I}$ satisfies every inclusion of $\mathcal{T}$. $\mathcal{T} \models C \sqsubseteq D$ if for any $\mathcal{I}$, $\mathcal{I} \models \mathcal{T}$ implies $\mathcal{I} \models C \sqsubseteq D$.

Formally, a knowledge base (KB) is a pair $(\mathcal{T}, \mathcal{A})$ of a TBox $\mathcal{T}$ and an ABox $\mathcal{A}$. An interpretation $\mathcal{I}$ is a model of $\mathcal{K}$ if $\mathcal{I}$ is a model of both $\mathcal{T}$ and $\mathcal{A}$, denoted $\mathcal{I} \models \mathcal{K}$. If α is an inclusion or an assertion, $\mathcal{K} \models \alpha$ if every model of $\mathcal{K}$ is also a model of α. Two KBs $\mathcal{K}$ and $\mathcal{K}'$ are equivalent, written $\mathcal{K} \equiv \mathcal{K}'$, if they have the same models. "$\equiv$" can be similarly defined for ABoxes and TBoxes.

The signature of a concept description C, written $\mathsf{sig}(C)$, is the set of all concept and role names in C. Similarly, we can define $\mathsf{sig}(\mathcal{A})$ for an ABox $\mathcal{A}$, $\mathsf{sig}(\mathcal{T})$ for a TBox $\mathcal{T}$, and $\mathsf{sig}(\mathcal{K})$ for a KB $\mathcal{K}$.

3 Forgetting in $\mathcal{ALC}$ Ontologies

In this section, we will first give a semantic definition of what it means to forget about a set of variables in an $\mathcal{ALC}$ KB, and then state and discuss several important properties of forgetting that justify the definition chosen. This is the first study of forgetting both concepts and roles for arbitrary knowledge bases in $\mathcal{ALC}$.

As explained earlier, given an ontology $\mathcal{K}$ on signature $\mathcal{S}$ and $\mathcal{V} \subset \mathcal{S}$, in ontology engineering it is often desirable to obtain a new ontology $\mathcal{K}'$ on $\mathcal{S} - \mathcal{V}$ such that reasoning tasks on $\mathcal{S} - \mathcal{V}$ are still preserved in $\mathcal{K}'$. As a result, $\mathcal{K}'$ is weaker than $\mathcal{K}$ in general. This intuition is formalized in the following definition.

Definition 3.1 (KB-forgetting). *Let $\mathcal{K}$ be a KB in $\mathcal{ALC}$ and $\mathcal{V}$ be a set of variables. A KB $\mathcal{K}'$ over the signature* $\mathsf{sig}(\mathcal{K}) - \mathcal{V}$ *is a* result of forgetting *about $\mathcal{V}$ in $\mathcal{K}$ if*

(KF1) $\mathcal{K} \models \mathcal{K}'$;
(KF2) *for each concept inclusion $C \sqsubseteq D$ in $\mathcal{ALC}$ not containing any variables in $\mathcal{V}$, $\mathcal{K} \models C \sqsubseteq D$ implies $\mathcal{K}' \models C \sqsubseteq D$;*
(KF3) *for each membership assertion $C(a)$ or $R(a,b)$ in $\mathcal{ALC}$ not containing any variables in $\mathcal{V}$, $\mathcal{K} \models C(a)$ (resp., $\mathcal{K} \models R(a,b)$) implies $\mathcal{K}' \models C(a)$ (resp., $\mathcal{K}' \models R(a,b)$);*

Condition **(KF3)** extends previous definitions [7] to allow for nonempty ABoxes in KBs.

To illustrate the above definition of semantic forgetting and how forgetting can be used in ontology extraction, consider the following example of designing an $\mathcal{ALC}$ ontology about flu.

Example 3.1. Suppose we have searched the Web and found an ontology about human diseases (such a practical ontology could be very large):

$Disease \sqsubseteq \forall attacks.Human$,
$Human \equiv Male \sqcup Female$,
$Human \sqcap Infected \sqsubseteq \exists shows.Symptom$,
$Disease \equiv Infectious \sqcup Noninfectious$,
$Influenza \sqcup HIV \sqcup TB \sqsubseteq Infectious$.

We want to construct a (smaller) ontology only about flu by reusing the above ontology. This is done by forgetting about the undesired concepts $\{Disease, Noninfectious, HIV, TB\}$. As a result, the following ontology is obtained:

$Influenza \sqsubseteq Infectious$,
$Infectious \sqsubseteq \forall attacks.Human$,
$Human \equiv Male \sqcup Female$,
$Human \sqcap Infected \sqsubseteq \exists shows.Symptom$,

The next example shows that the result of forgetting in an $\mathcal{ALC}$ ontology may not exist in some cases.

Example 3.2. Let $\mathcal{K} = (\mathcal{T}, \mathcal{A})$ be an $\mathcal{ALC}$ KB where $\mathcal{T} = \{ A \sqsubseteq B,\ B \sqsubseteq C,\ C \sqsubseteq \forall R.C,\ C \sqsubseteq D \}$, and $\mathcal{A} = \{ B(a),\ R(a,b) \}$.

Take $\mathcal{K}_1 = (\mathcal{T}_1, \mathcal{A}_1)$ where $\mathcal{T}_1 = \{ A \sqsubseteq C,\ C \sqsubseteq \forall R.C,\ C \sqsubseteq D \}$ and $\mathcal{A}_1 = \{ C(a),\ R(a,b) \}$. Then $\mathcal{K}_1$ is a result of forgetting about concept B in $\mathcal{K}$.

However, there does not exist a result of forgetting about $\{B, C\}$ in $\mathcal{K}$. To understand this, we note that the result of forgetting about $\{B, C\}$ in $\mathcal{K}$ should include the following inclusions:

$A \sqsubseteq D$, $A \sqsubseteq \forall R.D$, $A \sqsubseteq \forall R.\forall R.D$, $\ldots$, and
$D(a)$, $(\forall R.D)(a)$, $(\forall R.\forall R.D)(a)$, $\ldots$, and
$R(a,b)$, $D(b)$, $(\forall R.D)(b)$, $(\forall R.\forall R.D)(b)$, $\ldots$

In fact, there is no finite $\mathcal{ALC}$ KB which is equivalent to the above infinite set of inclusions.

If the result of forgetting about $\mathcal{V}$ in $\mathcal{K}$ is expressible as an $\mathcal{ALC}$ KB, we say $\mathcal{V}$ is *forgettable* from $\mathcal{K}$.

In the rest of this section, we state and discuss some important consequences of this definition of forgetting for KBs in $\mathcal{ALC}$. These properties provide evidence that the defintion is appropriate.

Proposition 3.1. *Let $\mathcal{K}$ be a KB in $\mathcal{ALC}$ and $\mathcal{V}$ a set of variables. If both $\mathcal{K}'$ and $\mathcal{K}''$ in $\mathcal{ALC}$ are results of forgetting about $\mathcal{V}$ in $\mathcal{K}$, then $\mathcal{K}' \equiv \mathcal{K}''$.*

This proposition says that the result of forgetting in $\mathcal{ALC}$ is unique up to KB equivalence. Given this result, we write $\mathsf{forget}(\mathcal{K}, \mathcal{V})$ to denote any result of forgetting about $\mathcal{V}$ in $\mathcal{K}$ in $\mathcal{ALC}$. In particular, $\mathsf{forget}(\mathcal{K}, \mathcal{V}) = \mathcal{K}'$ means that $\mathcal{K}'$ is a result of forgetting about $\mathcal{V}$ in $\mathcal{K}$.

In fact, forgetting in TBoxes is independent of ABoxes as the next result shows.

Proposition 3.2. *Let $\mathcal{T}$ be an $\mathcal{ALC}$ TBox and $\mathcal{V}$ a set of variables. Then, for any $\mathcal{ALC}$ ABox $\mathcal{A}$, $\mathcal{T}'$ is the TBox of $\mathsf{forget}((\mathcal{T}, \mathcal{A}), \mathcal{V})$ iff $\mathcal{T}'$ is the TBox of $\mathsf{forget}((\mathcal{T}, \emptyset), \mathcal{V})$.*

For simplicity, we write $\mathsf{forget}(\mathcal{T}, \mathcal{V})$ for $\mathsf{forget}((\mathcal{T}, \emptyset), \mathcal{V})$ and call it the result of TBox-forgetting about $\mathcal{V}$ in $\mathcal{T}$.

The following result, which generalizes Proposition 3.1, shows that forgetting preserves implication and equivalence relations between KBs.

Proposition 3.3. *Let $\mathcal{K}_1, \mathcal{K}_2$ be two KBs in $\mathcal{ALC}$ and $\mathcal{V}$ a set of variables. Then*

- $\mathcal{K}_1 \models \mathcal{K}_2$ *implies* $\mathsf{forget}(\mathcal{K}_1, \mathcal{V}) \models \mathsf{forget}(\mathcal{K}_2, \mathcal{V})$;
- $\mathcal{K}_1 \equiv \mathcal{K}_2$ *implies* $\mathsf{forget}(\mathcal{K}_1, \mathcal{V}) \equiv \mathsf{forget}(\mathcal{K}_2, \mathcal{V})$.

However, the converse of Proposition 3.3 is not true in general. Consider $\mathcal{K}$ and $\mathcal{K}_1$ in Example 3.2, it is obvious that $\mathsf{forget}(\mathcal{K}, \{B\}) \equiv \mathsf{forget}(\mathcal{K}_1, \{B\})$. However, $\mathcal{K}$ and $\mathcal{K}_1$ are not equivalent.

Consistency and query answering are two major reasoning tasks in description logics. It is a key requirement for a reasonable definition of forgetting to preserve these two reasoning forms.

Proposition 3.4. *Let $\mathcal{K}$ be a KB in $\mathcal{ALC}$ and $\mathcal{V}$ a set of variables. Then*

1. *$\mathcal{K}$ is consistent iff $\mathsf{forget}(\mathcal{K}, \mathcal{V})$ is consistent;*
2. *for any inclusion or assertion α not containing variables in $\mathcal{V}$, $\mathcal{K} \models \alpha$ iff $\mathsf{forget}(\mathcal{K}, \mathcal{V}) \models \alpha$.*

The next result shows that the forgetting operation can be divided into steps, with a part of the signature forgotten in each step.

Proposition 3.5. *Let $\mathcal{K}$ be a KB in $\mathcal{ALC}$ and $\mathcal{V}_1, \mathcal{V}_2$ two sets of variables. Then we have*

$$\mathsf{forget}(\mathcal{K}, \mathcal{V}_1 \cup \mathcal{V}_2) \equiv \mathsf{forget}(\mathsf{forget}(\mathcal{K}, \mathcal{V}_1), \mathcal{V}_2).$$

To compute the result of forgetting about $\mathcal{V}$ in $\mathcal{K}$, it is equivalent to forget the variables in $\mathcal{V}$ one by one, *i.e.*, forgetting can be computed incrementally.

4 Forgetting in $\mathcal{ALC}$ Concept Descriptions

Forgetting in a concept description has been investigated under the name of *uniform interpolation* in [20]. In this section, we reformulate the definition of the forgetting about concept and role names in $\mathcal{ALC}$ concept descriptions (briefly, c-forgetting) and introduce some results that will be used in the next section. From the view point of ontology management, the issue of forgetting in concept descriptions is less important than that for KBs and TBoxes. However, we will show later that c-forgetting can be used to provide an approximation algorithm for KB-forgetting in $\mathcal{ALC}$, as well as its theoretical importance.

Intuitively, the result C' of forgetting about a set of variables from a concept description C should be weaker than C but as close to C as possible. For example, after the concept $Male$ is forgotten from a concept description for "Male Australian student" $Australians \sqcap Students \sqcap Male$, then we should obtain a concept description $Australians \sqcap Students$ for "Australian student". More specifically, C' should be a concept description that defines a minimal concept description among all concept descriptions that subsume C and are syntactically irrelevant to $\mathcal{V}$ (*i.e.*, variables in $\mathcal{V}$ do not appear in the concept description).

Definition 4.1 (c-forgetting). *Let C be a concept description in $\mathcal{ALC}$ and $\mathcal{V}$ a set of variables. A concept description C' on the signature* $\mathsf{sig}(C) - \mathcal{V}$ *is a* result of c-forgetting *about $\mathcal{V}$ in C if the following conditions are satisfied:*

(CF1) $\models C \sqsubseteq C'$.
(CF2) *For every $\mathcal{ALC}$ concept description C'' with* $\mathsf{sig}(C'') \cap \mathcal{V} = \emptyset$, $\models C \sqsubseteq C''$ *implies* $\models C' \sqsubseteq C''$.

The above (CF1) and (CF2) correspond to the conditions (2) and (3) of Theorem 8 in [20]. A fundamental property of c-forgetting in $\mathcal{ALC}$ concept descriptions is that the result of c-forgetting is unique under concept description equivalence.

Proposition 4.1. *Let C be a concept description in $\mathcal{ALC}$ and $\mathcal{V}$ a set of variables. If two concept descriptions C' and C'' in $\mathcal{ALC}$ are results of c-forgetting about $\mathcal{V}$ in C, then* $\models C' \equiv C''$.

As all results of c-forgetting are equivalent, we write $\mathsf{forget}(C, \mathcal{V})$ to denote an arbitrary result of c-forgetting about $\mathcal{V}$ in C.

Example 4.1. Suppose the concept "Research Student" is defined by $C = Student \sqcap (Master \sqcup PhD) \sqcap \exists supervised.Professor$ where "Master", "PhD" and "Professor" are all concepts; "supervised" is a role and $supervised(x, y)$ means that x is supervised by y. If the concept description C is used only for students, we may wish to forget about $Student$: $\mathsf{forget}(C, Student) = (Master \sqcup PhD) \sqcap \exists supervised.Professor$. If we do not require that a supervisor for a research student must be a professor, then the filter "Professor" can be forgotten: $\mathsf{forget}(C, Professor) = Student \sqcap (Master \sqcup PhD) \sqcap \exists supervised.\top$.

A concept description C is *satisfiable* if $C^{\mathcal{I}} \neq \emptyset$ for some interpretation $\mathcal{I}$ on $\mathsf{sig}(C)$. C is unsatisfiable if $\models C \equiv \bot$. By Definition 4.1, c-forgetting also preserves satisfiability of concept descriptions.

Proposition 4.2. *Let C be a concept description in $\mathcal{ALC}$, and $\mathcal{V}$ be a set of variables. Then C is satisfiable iff* $\mathsf{forget}(C, \mathcal{V})$ *is satisfiable.*

Similar to forgetting in KB, the c-forgetting operation can be divided into steps.

Proposition 4.3. *Let C be a concept description in $\mathcal{ALC}$ and $\mathcal{V}_1, \mathcal{V}_2$ two sets of variables. Then we have*

$$\models \mathsf{forget}(C, \mathcal{V}_1 \cup \mathcal{V}_2) \equiv \mathsf{forget}(\mathsf{forget}(C, \mathcal{V}_1), \mathcal{V}_2).$$

Given the above result, when we want to forget about a set of variables, they can be forgotten one by one. Also, the ordering of c-forgetting operation is irrelevant to the result.

Corollary 4.1. *Let C be a concept description in $\mathcal{ALC}$ and let $\mathcal{V} = \{V_1, \ldots, V_n\}$ be a set of variables. Then, for any permutation $(i_1, i_2, \ldots, i_n)$ of $\{1, 2, \ldots, n\}$,*

$$\models \mathsf{forget}(\mathsf{forget}(\mathsf{forget}(C, V_{i_1}), V_{i_2}), \ldots), V_{i_n}) \equiv \\ \mathsf{forget}(\mathsf{forget}(\mathsf{forget}(C, V_1), V_2), \ldots), V_n).$$

The following result, which is not obvious, shows that c-forgetting distributes over union $\sqcup$.

Proposition 4.4. *Let $C_1, \ldots, C_n$ be concept descriptions in $\mathcal{ALC}$. For any set $\mathcal{V}$ of variables, we have*

$$\models \mathsf{forget}(C_1 \sqcup \cdots \sqcup C_n, \mathcal{V}) \equiv \mathsf{forget}(C_1, \mathcal{V}) \sqcup \cdots \sqcup \mathsf{forget}(C_n, \mathcal{V}).$$

However, c-forgetting for $\mathcal{ALC}$ does not distribute over intersection $\sqcap$. For example, if the concept description $C = A \sqcap \neg A$, then $\mathsf{forget}(C, A) = \bot$, since $\models C \equiv \bot$. But $\mathsf{forget}(A, A) \sqcap \mathsf{forget}(\neg A, A) \equiv \top$.

An important reason for this is that c-forgetting does not distribute over negation. Actually, we have

$$\models \neg\mathsf{forget}(C, \mathcal{V}) \sqsubseteq \neg C \sqsubseteq \mathsf{forget}(\neg C, \mathcal{V}).$$

These subsumptions may be strict, *e.g.*, if C is $A \sqcap B$ and $\mathcal{V}$ is $\{A\}$, then $\neg\mathsf{forget}(C, \mathcal{V})$ is $\neg B$, but $\mathsf{forget}(\neg C, \mathcal{V})$ is $\top$.

The next result shows that c-forgetting distributes over quantifiers. Since c-forgetting does not distribute over negation, the two statements in the following proposition do not necessarily imply each other. The proof uses tableau reasoning for $\mathcal{ALC}$ and is surprisingly complex.

Proposition 4.5. *Let C be a concept description in $\mathcal{ALC}$, R be a role name and $\mathcal{V}$ be a set of variables. Then*

- $\mathsf{forget}(\forall R.C, \mathcal{V}) = \top$ *for $R \in \mathcal{V}$, and* $\mathsf{forget}(\forall R.C, \mathcal{V}) = \forall R.\mathsf{forget}(C, \mathcal{V})$ *for $R \notin \mathcal{V}$;*
- $\mathsf{forget}(\exists R.C, \mathcal{V}) = \top$ *for $R \in \mathcal{V}$, and* $\mathsf{forget}(\exists R.C, \mathcal{V}) = \exists R.\mathsf{forget}(C, \mathcal{V})$. *for $R \notin \mathcal{V}$;*

These results suggest a way of computing c-forgetting about set $\mathcal{V}$ of variables in a complex $\mathcal{ALC}$ concept description C. That is, to forget about each variable V in $\mathcal{V}$ one after another, and to distribute the c-forgetting computation to subconcepts of C.

In what follows, we introduce an algorithm for computing the result of c-forgetting through rewriting of concept descriptions (syntactic concept transformations) [20]. This algorithm consists of two stages: (1) C is first transformed into an equivalent disjunctive normal form (DNF), which is a disjunction of conjunctions of simple concept descriptions; (2) the result of c-forgetting about $\mathcal{V}$ in each such simple concept description is obtained by removing some parts of the conjunct.

Before we introduce disjunctive normal form (DNF), some notation and definitions are in order. We call an (atomic) concept A or its negation $\neg A$ a *literal concept* or simply a *literal*. A *pseudo-literal* with role R is a concept description of the form $\exists R.F$ or $\forall R.F$, where R is a role name and F is an arbitrary concept. A *generalized literal* is either a literal or a pseudo-literal.

Definition 4.2. *A concept description D is in* disjunctive normal form *(DNF) if $D = \bot$ or $D = \top$ or D is a disjunction of conjunctions of generalized literals $D = D_1 \sqcup \cdots \sqcup D_n$, where each $D_i \not\equiv \bot$ $(1 \leq i \leq n)$ is a conjunction $\bigsqcap L$ of literals, or of the form*

$$\bigsqcap L \sqcap \bigsqcap_{R \in \mathcal{R}} \left[\forall R.U_R \sqcap \bigsqcap_k \exists R.(E_R^{(k)} \sqcap U_R) \right]$$

where $\mathcal{R}$ is the set of role names that occur in D_i, and each U_R and each $E_R^{(k)} \sqcap U_R$ is a concept description in DNF.

We note that, to guarantee the correctness of the algorithm, the above DNF for $\mathcal{ALC}$ is more complex than we have in classical logic and DL-Lite. See Example 4.2 for an example of a concept description in DNF.

Each concept description in $\mathcal{ALC}$ can be transformed into an equivalent one in DNF by the following two steps: (1) first transform the given concept description into a disjunction of conjunctions of pseudo-literals using De Morgan's laws, distributive laws and necessary simplifications, and then (2) for each conjunction in the resulting concept description, perform the following three transformations in order:

$$\begin{aligned} C &\rightsquigarrow \forall R.\top \sqcap C, \quad \text{for } C = \exists R.C_1 \sqcap \cdots \sqcap \exists R.C_m, m > 0 \\ \forall R.C_1 \sqcap \exists R.C_2 &\rightsquigarrow \forall R.C_1 \sqcap \exists R.(C_1 \sqcap C_2) \\ \forall R.C_1 \sqcap \cdots \sqcap \forall R.C_n &\rightsquigarrow \forall R.(C_1 \sqcap \cdots \sqcap C_n). \end{aligned}$$

The first transformation above is to transform a concept description containing only existential quantifier into the normal form. For example, if C is concept name, $\exists R.C$, which is not in normal form, can be transformed into the normal form $\forall R.\top \sqcap \exists R.C$. While the second is to assemble several quantifications with the same role name into a single one, the third is crucial for guaranteeing the correctness of our algorithm.

Once an $\mathcal{ALC}$ concept description D is in the normal form, the result of c-forgetting about a set $\mathcal{V}$ of variables in D can be obtained from D by simple symbolic manipulations (ref. Algorithm 1).

According to Algorithm 1, an input concept description must first be transformed into the normal form before the steps for forgetting are applied. For instance, if we

---

Algorithm 1. (Compute c-forgetting)
Input: An $\mathcal{ALC}$ concept description C and a set $\mathcal{V}$ of variables in C.
Output: $\mathsf{forget}(C, \mathcal{V})$.
Method:
Step 1. Transform C into its DNF D. If D is $\top$ or $\bot$, return D; otherwise, let $D = D_1 \sqcup \cdots \sqcup D_n$ as in Definition 4.2.
Step 2. For each conjunct E in each D_i, perform the following transformations:

- if E is a literal of the form A or $\neg A$ with $A \in \mathcal{V}$, replace E with $\top$;
- if E is a pseudo-literal in the form of $\forall R.F$ or $\exists R.F$ with $R \in \mathcal{V}$, replace E with $\top$;
- if E is a pseudo-literal in the form of $\forall R.F$ or $\exists R.F$ where $R \notin \mathcal{V}$, replace F with $\mathsf{forget}(F, \mathcal{V})$, and replace each resulting $\forall R.(\top \sqcup F)$ with $\top$.

Step 3. Return the resulting concept description as $\mathsf{forget}(C, \mathcal{V})$.

Fig. 1. Forgetting in concept descriptions

want to forget A in the concept description $D = A \sqcap \neg A \sqcap B$, D is transformed into the normal form, which is $\bot$, and then obtain $\mathsf{forget}(D, A) = \bot$. We note that B is not a result of forgetting about A in D.

Example 4.2. Given a concept $D = (A \sqcup \exists R.\neg B) \sqcap \forall R.(B \sqcup C)$, we want to forget about concept name B in D. In Step 1 of Algorithm 1, D is firstly transformed into its DNF $D' = [A \sqcap \forall R.(B \sqcup C)] \sqcup [\forall R.(B \sqcup C) \sqcap \exists R.(\neg B \sqcap C)]$. Note that $\exists R.(\neg B \sqcap C)$ is transformed from $\exists R.[\neg B \sqcap (B \sqcup C)]$. Then in Step 2, each occurrence of B in D' is replaced by $\top$, and $\forall R.(\top \sqcup F)$ is replaced with $\top$. We obtain $\mathsf{forget}(D, \{B\}) = A \sqcup \exists R.C$. To forget about role R in D, Algorithm 1 replaces each pseudo-literals in D' of the form $\forall R.F$ or $\exists R.F$ with $\top$, and returns $\mathsf{forget}(D, \{R\}) = \top$.

Obviously, the major cost of Algorithm 1 is from transforming the given concept description into its DNF. For this reason, the algorithm is exponential time in the worst case. However, if the concept description C is in DNF, Algorithm 1 takes only linear time (w.r.t. the size of C) to compute the result of c-forgetting about $\mathcal{V}$ in C. And the result of c-forgetting is always in DNF.

Theorem 4.1. *Let $\mathcal{V}$ be a set of concept and role names and C a concept description in $\mathcal{ALC}$. Then Algorithm 1 always returns* $\mathsf{forget}(C, \mathcal{V})$.

The proof of this theorem uses the tableau for $\mathcal{ALC}$.

5 Approximate Forgetting in $\mathcal{ALC}$ Ontologies

As we showed in Section 3, the result of forgetting for an $\mathcal{ALC}$ KB might not exist. However, if our goal is to determine whether a given set Σ of inclusions and assertions are logical consequences of the result of forgetting for a KB $\mathcal{K}$, we show in this section how to determine this *without* computing the result of forgetting at all. Instead, given an upper bound on the size of the inclusions and assertions in Σ, we show how to compute a finite KB $\mathcal{K}'$ such that $\mathcal{K}' \models \Sigma$ if and only if $\mathcal{K} \models \Sigma$. Here, we assume Σ does not contain any of the variables that are being forgotten. The finite KB $\mathcal{K}'$ is called an approximation to the result of forgetting for $\mathcal{K}$. In fact, we show how to compute a sequence of approximations that can determine whether inclusions and assertions of increasingly large size are logical consequences of the result of forgetting. These approximations are computed using Algorithm 1 for c-forgetting. We can thus obtain the benefits of forgetting, by computing a smaller KB, and performing reasoning, without having to actually compute the (non-existent) result of forgetting.

As a special case, we first introduce an approximation to TBox-forgetting. Example 3.2 shows that, for some TBox $\mathcal{T}$, $\mathsf{forget}(\mathcal{T}, \mathcal{V})$ may not be expressible as a finite $\mathcal{ALC}$ TBox. Thus, it is natural to consider a sequence of (finite) TBoxes that approximate the result of forgetting in $\mathcal{T}$ in the sense that the sequence is non-decreasing in terms of logical implication and the limit of the sequence is the result of forgetting. Such a consequence is constructed by using results developed for c-forgetting in Section 4.

We note that, for an inclusion $C \sqsubseteq D$ in $\mathcal{T}$, $\mathsf{forget}(C, \mathcal{V}) \sqsubseteq \mathsf{forget}(D, \mathcal{V})$ may not be a logical consequence of $\mathcal{T}$ and thus may not be in $\mathsf{forget}(\mathcal{T}, \mathcal{V})$. However, if we transform $\mathcal{T}$ into an equivalent singleton TBox $\{\top \sqsubseteq C_{\mathcal{T}}\}$, where $C_{\mathcal{T}} = \bigsqcap_{C \sqsubseteq D \in \mathcal{T}} (\neg C \sqcup D)$, then inclusion α_0 of the form $\top \sqsubseteq \mathsf{forget}(C_{\mathcal{T}}, \mathcal{V})$ is a logical consequence of $\mathcal{T}$. In general, the singleton TBox $\{\alpha_0\}$ is not necessarily equivalent to $\mathsf{forget}(\mathcal{T}, \mathcal{V})$. However, it can be a starting point of a sequence whose limit is $\mathsf{forget}(\mathcal{T}, \mathcal{V})$. Note that $\mathcal{T}$ is also equivalent to $\{\top \sqsubseteq C_{\mathcal{T}} \sqcap \forall R.C_{\mathcal{T}}\}$ for an arbitrary role name R in $\mathcal{T}$. Hence, inclusion α_1 of the form $\top \sqsubseteq \mathsf{forget}(C_{\mathcal{T}} \sqcap \forall R.C_{\mathcal{T}}, \mathcal{V})$ is a logical consequence of $\mathcal{T}$, and it can be shown that TBox $\{\alpha_1\}$ is logically stronger then $\{\alpha_0\}$. That is, $\mathsf{forget}(\mathcal{T}, \mathcal{V}) \models \{\alpha_1\} \models \{\alpha_0\}$. Let α_2 be $\top \sqsubseteq \mathsf{forget}(C_{\mathcal{T}} \sqcap \forall R.C_{\mathcal{T}} \sqcap \forall R.\forall R.C_{\mathcal{T}}, \mathcal{V})$, then we have $\mathsf{forget}(\mathcal{T}, \mathcal{V}) \models \{\alpha_2\} \models \{\alpha_1\} \models \{\alpha_0\}$. In this way, we can construct a sequence of TBoxes with increasing logical strength, whose limit is $\mathsf{forget}(\mathcal{T}, \mathcal{V})$.

For $n \geq 0$, define

$$C_{\mathcal{T}}^{(n)} = \bigsqcap_{k=0}^{n} \bigsqcap_{R_1,\ldots,R_k \in \mathcal{R}} \forall R_1 \cdots \forall R_k.C_{\mathcal{T}}$$

where $C_{\mathcal{T}} = \bigsqcap_{C \sqsubseteq D \in \mathcal{T}} (\neg C \sqcup D)$ and $\mathcal{R}$ is the set of role names in $\mathcal{K}$.

We now define a sequence of TBoxes, which essentially provides an approximation to the result of TBox-forgetting.

Definition 5.1. *Let $\mathcal{T}$ be an $\mathcal{ALC}$ TBox and $\mathcal{V}$ be a set of variables. For each $n \geq 0$, the TBox*

$$\mathsf{forget}^n(\mathcal{T}, \mathcal{V}) = \{\, \top \sqsubseteq \mathsf{forget}(C_{\mathcal{T}}^{(n)}, \mathcal{V}) \,\}$$

*is called the n-*forgetting *about $\mathcal{V}$ in $\mathcal{T}$.*

Note that the above n-forgetting for TBoxes is defined in terms of forgetting in concept descriptions (c-forgetting).

Example 5.1. Consider the TBox $\mathcal{T}$ in Example 3.2, we have $C_{\mathcal{T}} = (\neg A \sqcup B) \sqcap (\neg B \sqcup C) \sqcap (\neg C \sqcup \forall R.C) \sqcap (\neg C \sqcup D)$, and $C_{\mathcal{T}}^{(0)} = C_{\mathcal{T}}$, $C_{\mathcal{T}}^{(1)} = C_{\mathcal{T}} \sqcap \forall R.C_{\mathcal{T}}, \ldots,$ $C_{\mathcal{T}}^{(n)} = C_{\mathcal{T}} \sqcap \forall R.C_{\mathcal{T}}^{(n-1)}$ $(n \geq 2)$.

Let $\mathcal{V} = \{B, C\}$. For each $n \geq 0$, the forget$^n(\mathcal{T}, \mathcal{V})$ can be computed as follows.
forget$^0(\mathcal{T}, \mathcal{V}) = \{\, \top \sqsubseteq \neg A \sqcup D \,\}$, which is equivalent to $\{\, A \sqsubseteq D \,\}$.
forget$^1(\mathcal{T}, \mathcal{V}) = \{\, \top \sqsubseteq \neg A \sqcup (D \sqcap \forall R.D) \,\}$, which is $\{\, A \sqsubseteq D,\ A \sqsubseteq \forall R.D \,\}$.
......

$$\text{forget}^n(\mathcal{T}, \mathcal{V}) = \{\, A \sqsubseteq D,\ A \sqsubseteq \forall R.D,\ \ldots,\ A \sqsubseteq \underbrace{\forall R.\forall R \cdots \forall R}_{n\ Rs}.D \,\}.$$

We call $(\bigsqcap_{i=1}^{n} C_i)(a)$ the *conjunction* of assertions $C_1(a), \ldots, C_n(a)$ while $(\bigsqcup_{i=1}^{n} C_i)(a)$ is called the *disjunction* of these assertions.

Before we can perform forgetting on a given ABox, we need to preprocess it and thus transform it into a normal form. To this end, we give the following definition.

Definition 5.2. *An ABox $\mathcal{A}$ in $\mathcal{ALC}$ is* complete *if for any individual name a in $\mathcal{A}$ and assertion $C(a)$ with $\mathcal{A} \models C(a)$, we have $\models C' \sqsubseteq C$, where $C'(a)$ is the conjunction of all the concept assertions about a in $\mathcal{A}$.*

For example, ABox $\mathcal{A} = \{\forall R.A(a), R(a,b)\}$ is incomplete, because $\mathcal{A} \models A(b)$ whereas no assertion $C(b)$ exists in $\mathcal{A}$ such that $\models C \sqsubseteq A$. After adding assertions $A(b)$, $\exists R.A(a)$ and $\top(a)$ into $\mathcal{A}$, the resulting ABox is complete.

In complete ABoxes, concept assertion entailment can be reduced to concept subsumption, and is independent of role assertions.

However, there exist incomplete ABoxes that are not equivalent any (finite) complete ABox. For example, the ABox $\{R(a,b), R'(b,a)\}$ has infinitely many logical consequences of the form $(\exists R.\exists R'.C \sqcup \neg C)(a)$ where C is an arbitrary concept description. This kind of situations are caused by certain cycles in ABoxes. We say an ABox is *acyclic* if there exists no cycle of the form $R_1(a, a_1), R_2(a_1, a_2), \ldots, R_i(a_i, a)$ in $\mathcal{A}$ in the ABox. Many practical ABoxes are acyclic or can be transformed to equivalent acyclic ABoxes.

Note all the role assertions in an acyclic ABox form tree-shape relations between individuals. We call an individual without any predecessor a *root* individual, and that without any successor a *leaf* individual.

Algorithm 2 is developed to transform a given acyclic $\mathcal{ALC}$ ABox into an equivalent complete ABox. The correctness of the algorithm shows that any acyclic ABox can be transformed to an equivalent complete ABox in $\mathcal{ALC}$.

Note that Algorithm 2 always terminates.

Lemma 5.1. *Given an acyclic $\mathcal{ALC}$ ABox $\mathcal{A}$, Algorithm 2 always returns a complete ABox $\mathcal{A}'$ that is equivalent to $\mathcal{A}$.*

With the notion of complete ABox, we can extend n-forgetting in TBoxes and define n-forgetting for an $\mathcal{ALC}$ KB as follows.

For the remainder of this section, we assume that the ABox of every KB $\mathcal{K}$ is acyclic.

Definition 5.3. *Let $\mathcal{K} = (\mathcal{T}, \mathcal{A})$ be an $\mathcal{ALC}$ KB and $\mathcal{V}$ a set of variables. For each $n \geq 0$, the KB* forget$^n(\mathcal{K}, \mathcal{V}) = (\mathcal{T}', \mathcal{A}')$ *is called the result of n-*forgetting *about $\mathcal{V}$ in $\mathcal{K}$, where $\mathcal{T}' =$* forget$^n(\mathcal{T}, \mathcal{V}) = \{\, \top \sqsubseteq$ forget$(C_{\mathcal{T}}^{(n)}, \mathcal{V}) \,\}$ *and $\mathcal{A}'$ is obtained from $\mathcal{A}$ through the following steps:*

Algorithm 2. (Complete an acyclic ABox)
Input: An acyclic $\mathcal{ALC}$ ABox $\mathcal{A}$.
Output: An equivalent complete $\mathcal{ALC}$ ABox $\mathcal{A}'$.
Method:
Step 1. Starting from root individuals, for each individual a and each role assertion $R(a,b)$ in $\mathcal{A}$, let $C(a)$ be the conjunction of all the concept assertions about a in $\mathcal{A}$.
Transform C into its DNF $C = \bigsqcup_{i=1}^{n} D_i$ as in Definition 4.2. Let $\forall R.U_i$ be the universal quantified conjunct of R in D_i. Add $(\bigsqcup_{i=1}^{n} U_i)(b)$ to $\mathcal{A}$.
Step 2. For each individual a in $\mathcal{A}$, add $\top(a)$ to $\mathcal{A}$.
Step 3. Starting from leaf individuals, for each individual b and each role assertion $R(a,b)$ in $\mathcal{A}$, add assertion $(\exists R.E)(a)$ to $\mathcal{A}$, where $E(b)$ is the conjunction of all the concept assertions about b in $\mathcal{A}$.
Step 4. Return the resulting ABox.

Fig. 2. Transform an acyclic ABox into a complete ABox

1. *For each individual name a in $\mathcal{A}$, add $C_{\mathcal{T}}^{(n)}(a)$ to $\mathcal{A}$.*
2. *Apply Algorithm 2 to obtain a complete ABox, still denoted $\mathcal{A}$.*
3. *For each individual name a in $\mathcal{A}$, replace $C(a)$ with $(\mathsf{forget}(C,\mathcal{V}))(a)$, where $C(a)$ is the conjunction of all the concept assertions about a in $\mathcal{A}$.*
4. *Remove each $R(a,b)$ from $\mathcal{A}$ where $R \in \mathcal{V}$.*

The basic idea behind Definition 5.3 is to transform the given KB into a new KB such that forgetting can be done in its ABox and TBox, separately, in terms of c-forgetting for individual assertions and inclusions.

Example 5.2. Consider the KB $\mathcal{K} = (\mathcal{T}, \mathcal{A})$ in Example 3.2 and let $\mathcal{V} = \{B, C\}$.

For each $n \geq 0$, let $\mathcal{A}_n$ be the ABox of $\mathsf{forget}^n(\mathcal{K}, \mathcal{V})$. We will elaborate the computation of $\mathcal{A}_0$ as follows: Note that $\mathcal{A}$ is acyclic. First $C_{\mathcal{T}}^{(0)}(a)$ and $C_{\mathcal{T}}^{(0)}(b)$ are added into $\mathcal{A}$, where $C_{\mathcal{T}}^{(n)}$ is the same as in Example 5.1. After applying Algorithm 2 to $\mathcal{A}$, the resulting ABox is equivalent to

$$\{\ (B \sqcap C \sqcap \forall R.C \sqcap D)(a),\ R(a,b),\ ((\neg A \sqcup B) \sqcap C \sqcap \forall R.C \sqcap D)(b),$$
$$\exists R.((\neg A \sqcup B) \sqcap C \sqcap \forall R.C \sqcap D)(a)\ \}$$

By applying c-forgetting to the conjunctions of concept assertions about, respectively, a and b, we obtain $\mathcal{A}_0 = \{\ D(a),\ R(a,b),\ D(b)\ \}$.

Similarly, we can compute $\mathcal{A}_1, \ldots, \mathcal{A}_n$ as:

$$\mathcal{A}_1 = \{\ D(a),\ (\forall R.D)(a),\ R(a,b),\ D(b),\ (\forall R.D)(b)\ \}.$$
$$\cdots\cdots$$
$$\mathcal{A}_n = \{\ D(a),\ (\forall R.D)(a),\ \ldots,\ (\underbrace{\forall R.\forall R \cdots \forall R}_{n\ Rs}.D)(a),\ R(a,b),$$
$$D(b),\ (\forall R.D)(b),\ \ldots,\ (\underbrace{\forall R.\forall R \cdots \forall R}_{n\ Rs}.D)(b)\ \}.$$

The following result shows that n-forgetting preserves logical consequences of the original KB.

Given a concept description C, let $|C|$ be the number of all different subconcepts of C. For a TBox $\mathcal{T}$, define $|\mathcal{T}| = \sum_{C \sqsubseteq D \in \mathcal{T}}(|C| + |D|)$. Similarly, for an ABox $\mathcal{A}$, define $|\mathcal{A}| = \sum_{C(a) \in \mathcal{A}} |C|$. Then for a KB $\mathcal{K} = (\mathcal{T}, \mathcal{A})$, $|\mathcal{K}| = |\mathcal{T}| + |\mathcal{A}|$.

Proposition 5.1. *Let $\mathcal{K}$ be an $\mathcal{ALC}$ KB and $\mathcal{V}$ be a set of variables. Then* $\mathsf{forget}^n(\mathcal{K}, \mathcal{V})$ *satisfies the following conditions:*

1. $\mathcal{K} \models \mathsf{forget}^n(\mathcal{K}, \mathcal{V})$.
2. *Let C and D be two concept descriptions containing no variable in $\mathcal{V}$. If $n \geq 2^{|C|+|D|+|\mathcal{K}|}$, then $\mathcal{K} \models C \sqsubseteq D$ iff* $\mathsf{forget}^n(\mathcal{K}, \mathcal{V}) \models C \sqsubseteq D$.
3. *Let C be a concept description containing no variable in $\mathcal{V}$, and a an individual name in $\mathcal{K}$. If $n \geq 2^{|C|+|\mathcal{K}|}$, then $\mathcal{K} \models C(a)$ iff* $\mathsf{forget}^n(\mathcal{K}, \mathcal{V}) \models C(a)$.
4. *Let R be a role name not in $\mathcal{V}$, and a, b two individual names in $\mathcal{K}$. Then $\mathcal{K} \models R(a, b)$ iff* $\mathsf{forget}^n(\mathcal{K}, \mathcal{V}) \models R(a, b)$.

Recall from the definition of KB-forgetting that, with respect to inclusions and assertions not containing variables in $\mathcal{V}$, $\mathcal{K}$ is logically equivalent to $\mathsf{forget}(\mathcal{K}, \mathcal{V})$. Proposition 5.1 tells us that if we know *which* inclusions and assertions not containing variables in $\mathcal{V}$ we wish to reason about in advance, then we can derive a value for n, compute $\mathsf{forget}^n(\mathcal{K}, \mathcal{V})$, and use the fact that, with respect to these inclusions and assertions, $\mathsf{forget}^n(\mathcal{K}, \mathcal{V})$ is logically equivalent to $\mathcal{K}$ and hence to $\mathsf{forget}(\mathcal{K}, \mathcal{V})$. In this way, we can use $\mathsf{forget}^n(\mathcal{K}, \mathcal{V})$ as a practical approximation to $\mathsf{forget}(\mathcal{K}, \mathcal{V})$.

The above proposition shows that, for any $n \geq 0$, each $\mathsf{forget}^n(\mathcal{K}, \mathcal{V})$ is logically weaker than $\mathsf{forget}(\mathcal{K}, \mathcal{V})$. Also, as the number n is sufficiently large, $\mathsf{forget}^n(\mathcal{K}, \mathcal{V})$ preserves more and more consequences of $\mathcal{K}$. Therefore, the sequence of KBs $\{\mathsf{forget}^n(\mathcal{K}, \mathcal{V})\}_{n \geq 0}$ is non-decreasing w.r.t. semantic consequence as the next proposition shows.

Proposition 5.2. *Let $\mathcal{K}$ be an $\mathcal{ALC}$ KB and $\mathcal{V}$ a set of variables. Then, for any $n \geq 0$, we have* $\mathsf{forget}^{n+1}(\mathcal{K}, \mathcal{V}) \models \mathsf{forget}^n(\mathcal{K}, \mathcal{V})$.

Based on the above two results, we can show the main theorem of this section as follows, which states that the limit of the sequence of n-forgettings captures the result of forgetting.

Theorem 5.1. *Let $\mathcal{K}$ be an $\mathcal{ALC}$ KB and $\mathcal{V}$ a set of variables. Then*

$$\mathsf{forget}(\mathcal{K}, \mathcal{V}) = \bigcup_{n=0}^{\infty} \mathsf{forget}^n(\mathcal{K}, \mathcal{V}).$$

So, by Theorem 5.1, we can compute $\mathsf{forget}(\mathcal{K}, \mathcal{V})$, if it exists, using algorithms introduced in the paper.

Corollary 5.1. *Let $\mathcal{K}$ be an $\mathcal{ALC}$ KB and $\mathcal{V}$ be a set of variables. $\mathcal{V}$ is forgettable from $\mathcal{K}$ if and only if there exists $N \geq 0$ such that* $\mathsf{forget}^n(\mathcal{K}, \mathcal{V}) \equiv \mathsf{forget}^N(\mathcal{K}, \mathcal{V})$ *for all $n \geq N$. In this case,* $\mathsf{forget}(\mathcal{K}, \mathcal{V}) = \mathsf{forget}^N(\mathcal{K}, \mathcal{V})$.

As we can see from Example 3.2, the sizes of consequences (assertions and inclusions) of $\mathcal{K}$ not containing variables in $\mathcal{V}$ do not have an upper bound. If it does not exist, we can always choose n large enough to "approximate" $\mathsf{forget}(\mathcal{K}, \mathcal{V})$. However, two issues are still unclear to us: First, the computation of $\mathsf{forget}^{n+1}(\mathcal{K}, \mathcal{V})$ is not based on $\mathsf{forget}^{n}(\mathcal{K}, \mathcal{V})$. Next, it would be interesting to find a way to measure how close $\mathsf{forget}^{n}(\mathcal{K}, \mathcal{V})$ is from $\mathsf{forget}(\mathcal{K}, \mathcal{V})$.

6 Conclusion

We have presented a theory and methods for forgetting for knowledge bases in the expressive description logic $\mathcal{ALC}$. This is the first work that deals with the combination or concepts and roles, nonempty ABoxes, and $\mathcal{ALC}$. Because the result of forgetting may not exist for $\mathcal{ALC}$ knowledge bases, we define a sequence of finite knowledge bases that approximate the result of forgetting and serve as a basis for query answering over the ontology with forgotten concepts and roles. We provide algorithms for computing these approximations, using algorithms for forgetting for concept descriptions, and note their correctness. Proofs of our results may be found in an online report.

Many interesting problems remain unsolved. It would be useful to clarify the decision problem of whether the result of forgetting for an $\mathcal{ALC}$ knowledge base exists. It would be useful to extend the results of this paper to even more expressive description logics. It would be interesting to find lower bounds on the complexity of forgetting for description logics. It would be useful to find an incremental algorithm for computing approximations. It would be useful implement our algorithms and incorporate them into ontology editors such as Protégé [28]. See `http://www.cit.gu.edu.au/~kewen/DLForget` for our progress on this task.

References

1. Alani, H., Harris, S., O'Neil, B.: Winnowing ontologies based on application use. In: Sure, Y., Domingue, J. (eds.) ESWC 2006. LNCS, vol. 4011, pp. 185–199. Springer, Heidelberg (2006)
2. Antoniou, G., Harmelen, F.: A Semantic Web Primer, 2nd edn. MIT Press, Cambridge (2008)
3. Baader, F., Calvanese, D., McGuinness, D., Nardi, D., Patel-Schneider, P.: The Description Logic Handbook. Cambridge University Press, Cambridge (2002)
4. Berners-Lee, T., Hendler, J., Lassila, O.: The semantic web. Scientific American, 29–37 (May 2001)
5. Dzbor, M., Motta, E., Buil, C., Gomez, J.M., Görlitz, O., Lewen, H.: Developing ontologies in owl: an observational study. In: Proc. Workshop on OWL: Experiences and Directions (2006)
6. Eiter, T., Wang, K.: Semantic forgetting in answer set programming. Artificial Intelligence 172(14), 1644–1672 (2008)
7. Ghilardi, S., Lutz, C., Wolter, F.: Did i damage my ontology? a case for conservative extensions in description logics. In: Proc. KR 2006, pp. 187–197 (2006)
8. Grau, B., Kazakov, Y., Horrocks, I., Sattler, U.: A logical framework for modular integration of ontologies. In: Proc. IJCAI 2007, pp. 298–303 (2007)
9. Grau, B., Parsia, B., Sirin, E.: Combining OWL ontologies using e-connections. Journal of Web Semantics 4(1), 40–59 (2006)

10. Cuenca Grau, B., Kazakov, Y., Horrocks, I., Sattler, U.: Just the right amount: Extracting modules from ontologies. In: Proc. WWW 2007, pp. 717–726 (2007)
11. Konev, B., Walther, D., Wolter, F.: The logical difference problem for description logic terminologies. In: Armando, A., Baumgartner, P., Dowek, G. (eds.) IJCAR 2008. LNCS (LNAI), vol. 5195, pp. 259–274. Springer, Heidelberg (2008)
12. Konev, B., Walther, D., Wolter, F.: Forgetting and uniform interpolation in large-scale description logic terminologies. In: Proc. IJCAI 2009 (2009)
13. Kontchakov, R., Wolter, F., Zakharyaschev, M.: Modularity in DL-Lite. In: Proc. DL 2007 (2007)
14. Kontchakov, R., Wolter, F., Zakharyaschev, M.: Can you tell the difference between DL-Lite ontologies? In: Proc. KR 2008, pp. 285–295 (2008)
15. Lang, J., Liberatore, P., Marquis, P.: Propositional independence: Formula-variable independence and forgetting. J. Artif. Intell. Res. 18, 391–443 (2003)
16. Lin, F., Reiter, R.: Forget it. In: Proc. AAAI Fall Symposium on Relevance, pp. 154–159 (1994)
17. Lutz, C., Walther, D., Wolter, F.: Conservative extensions in expressive description logics. In: Proc. IJCAI 2007, pp. 453–458 (2007)
18. Peroni, S., Motta, E., d'Aquin, M.: Identifying key concepts in an ontology, through the integration of cognitive principles with statistical and topological measures. In: Domingue, J., Anutariya, C. (eds.) ASWC 2008. LNCS, vol. 5367, pp. 242–256. Springer, Heidelberg (2008)
19. Rector, A., Brandt, S.: Why do it the hard way? The Case for an Expressive Description Logic for SNOMED. In: Proc.of the 3rd Int Conf. on Knowledge Representation in Medicine (KR-MED 2008), p. 16 (2008)
20. ten Cate, B., Conradie, W., Marx, M., Venema, Y.: Definitorially complete description logics. In: Proc. KR 2006, pp. 79–89 (2006)
21. Wang, Z., Wang, K., Topor, R., Pan, J.Z.: Forgetting concepts in DL-Lite. In: Bechhofer, S., Hauswirth, M., Hoffmann, J., Koubarakis, M. (eds.) ESWC 2008. LNCS, vol. 5021, pp. 245–257. Springer, Heidelberg (2008)
22. Wang, Z., Wang, K., Topor, R.: Forgetting for Knowledge Bases in DL-Lite$_{bool}$. In: Proc. ARCOE 2009 (IJCAI 2009 Workshop) (2009)
23. `http://www.fmrc.org.au/snomed/`
24. `http://www.openclinical.org/prj_galen.html`
25. `http://fma.biostr.washington.edu/`
26. `http://ncit.nci.nih.gov/`
27. `http://www.obofoundry.org/`
28. `http://protege.stanford.edu`
29. `http://www.neon-toolkit.org`
30. `http://www.topquadrant.com/products/TB_Composer.html`

Parallel Materialization of the Finite RDFS Closure for Hundreds of Millions of Triples

Jesse Weaver and James A. Hendler

Tetherless World Constellation, Rensselaer Polytechnic Institute, Troy, NY, USA
{weavej3,hendler}@cs.rpi.edu

Abstract. In this paper, we consider the problem of materializing the complete finite RDFS closure in a scalable manner; this includes those parts of the RDFS closure that are often ignored such as literal generalization and container membership properties. We point out characteristics of RDFS that allow us to derive an embarrassingly parallel algorithm for producing said closure, and we evaluate our C/MPI implementation of the algorithm on a cluster with 128 cores using different-size subsets of the LUBM 10,000-university data set. We show that the time to produce inferences scales linearly with the number of processes, evaluating this behavior on up to hundreds of millions of triples. We also show the number of inferences produced for different subsets of LUBM10k. To the best of our knowledge, our work is the first to provide RDFS inferencing on such large data sets in such low times. Finally, we discuss future work in terms of promising applications of this approach including OWL2RL rules, MapReduce implementations, and massive scaling on supercomputers.

1 Introduction

At present, the semantic web consists of ever-increasing Resource Description Framework[1] (RDF) data. In RDF, the fundamental unit of information is a triple; a triple describes a relationship between two things. Some triples in conjunction with each other can give rise to new knowledge. Consider rule *rdfs2* from [1]:

$$(\ ?a \text{ rdfs:domain } ?x) \wedge (\ ?u\ ?a\ ?y\ ) \rightarrow (?u \text{ rdf:type } ?x)$$

Deriving such inferences in the semantic web poses several challenges. First, data on the web is distributed making it difficult to ensure that the appropriate triples (e.g., rdfs:domain) are discovered together to derive the appropriate inferences (e.g., rdf:type). Second, the semantic web continues to grow creating vast amounts of information. Computation capable of scaling to large data sets is necessary. Third, the amount of time required to derive inferences should be reasonable (which depends on a use case).

[1] `http://www.w3.org/RDF/`

A. Bernstein et al. (Eds.): ISWC 2009, LNCS 5823, pp. 682–697, 2009.

These three issues of dependency, scalability, and time must be addressed to make inferencing on the semantic web practical and useful. We look at how to resolve these issues for computing the finite RDF Schema (RDFS) closure of large data sets. We find that the RDFS rules have certain properties which allow us to solve the issue of dependency rather easily. Scalability and time are addressed by forming an embarrassingly parallel[2] algorithm for such rules that allows for strong scaling.

We consider the problem of materializing the complete finite RDFS closure in a scalable manner. We define the finite RDFS closure as follows.

Definition 1. *We define the* ***finite axiomatic triples***—*denoted* T_{rdfs}—*as the RDF and RDFS axiomatic triples [1] minus any triples that describe resources of the form rdf:_#.*

Definition 2. *We define the* ***finite RDFS rules*** *as the set of rules concerning literal generalization (lg, gl) and the RDF and RDFS entailment rules (rdf1-2, rdfs1-13)[1], and also the following rule which we call cmp:*

$$(?n\ rdf{:}type\ rdfs{:}Resource) \wedge ?n \text{ is of the form } rdf{:}\_\# \longrightarrow (?n\ rdf{:}type\ rdfs{:}ContainerMembershipProperty)$$

Definition 3. *The* ***finite RDFS closure*** *is defined the same way as the RDFS closure [1] with the following exceptions: (1) the finite axiomatic triples are used instead of all of the RDF and RDFS axiomatic triples; and (2) rule cmp is included in the last step of rule applications.*

To our knowledge, only a few systems exist that produce the complete RDFS closure, none of which scale to large data sets. We address this problem by defining a partitioning scheme and showing that the finite RDFS rules can be applied to such a partitioning to produce the finite RDFS closure. In addition to deriving an embarrassingly parallel algorithm, we discuss other challenges such as parallel file I/O for RDF data and handling blank nodes. We present and evaluate an implementation of the algorithm written in C using the Message Passing Interface[3] (MPI) on a cluster of large memory Opteron machines for large subsets of the LUBM 10,000-university data set, scaling up to 128 processes. We also give an evaluation for the amount of duplicate inferences in the output, and promising applications are discussed as future work.

2 Related Work

The work most related to ours is MaRVIN [3,4] and the parallel Web Ontology Language[4] (OWL) inferencing work by [5,6].

[2] An embarrassingly parallel computation is "the 'ideal' computation from a parallel computing standpoint—a computation that can be divided into a number of completely independent parts, each of which can be executed by a separate processor." [2]

[3] `http://www.mpi-forum.org/`

[4] `http://www.w3.org/2004/OWL/`

MaRVIN provides sound, anytime, and eventually complete RDFS reasoning using a "divide-conquer-swap" strategy. Every process uses a reasoner and processes a fraction of the data to locally produce any possible inferences. A scoring mechanism is used to determine which triples are most useful for further inferencing, and these triples are exchanged between processes in an attempt to mix up the triples and find more inferences. Their evaluation in [4] shows presumably nearly complete reasoning on 14.9 million SwetoDBLP[5] triples in roughly 23 minutes.

[6] presents an approach to rule-based reasoning for OWL ontologies with so-called "OWL Horst" (perhaps better known as pD*) semantics [7]. Before parallel inferencing occurs, a fair amount of preprocessing is required. OWL ontologies are compiled into rules and a partitioning is determined ahead of time. Four partitioning approaches are presented, three of which are classified as data partitioning and one of which is classified as rule partitioning. Data partitioning gives each process a fraction of the data and all of the rules, while in rule partitioning, each process receives all of the data and a fraction of the rules. While it is uncertain what execution times were actually achieved, they indicate that they were able to achieve a speedup of 18 for 16 processes on the Lehigh University Benchmark[6] (LUBM) [8] 10-university data set (approximately one million triples). Other data sets evaluated, however, did not see as much speedup.

These two projects have not shown scaling to the extent that we demonstrate for our system, and they also handle only limited subsets of existing reasoning standards or perform approximation. We show scaling that has not been previously demonstrated in other systems (to the best of our knowledge) for complete finite RDFS reasoning. Furthermore, our inferencing times are smaller than those reported for the two previously mentioned works.

Also, BigOWLIM[7] reports reasoning times on large data sets, but it is uncertain how to compare the performance of BigOWLIM with our system. BigOWLIM is a semantic web repository that reportedly can perform some level of reasoning on 8 billion statements. Loading, inferencing, and query evaluation of the LUBM benchmark on LUBM8k took 15.2 hours. Our work focuses on inferencing as a large computation without concerns of storage or indexing, so the two systems are not quite comparable.

Less related works include reasoning over distributed hash tables [9,10]; potentially parallel evolutionary query answering [11]; composition of approximate, anytime reasoning algorithms executed in parallel mentioned at the end of [12]; proposals of parallel computation techniques for ontology reasoning [13]; and approaches for exploiting vertical parallelism in tableaux-based description logic reasoning [14].

The SOAR work [15,16,17] uses assumptions and observations similar to some of those used in this paper. While SOAR focuses on a scalable, disk-based, reasoning system using some RDFS and OWL-based rules on a single machine,

---

[5] `http://lsdis.cs.uga.edu/projects/semdis/swetodblp/`

[6] `http://swat.cse.lehigh.edu/projects/lubm/`

[7] `http://ontotext.com/owlim/benchmarking/lubm.html`

our system differs greatly from SOAR in that it is not disk-based, it is parallel, and it provides complete (finite) RDFS reasoning.

3 Our Approach to Parallel RDFS Reasoning

Our approach takes advantage of modern parallel computation techniques to compute the finite RDFS closure of large data sets. Previous work has used approximation to achieve higher scalability while other work focuses on minimizing dependencies in partitioning the work load. The former has the disadvantage of sacrificing soundness and/or completeness while the latter often requires time-intensive sequential computation to prepare the data for parallel reasoning. We focus on finding properties of RDFS reasoning that allow for natural parallelization in all parts of the computation. Therefore, we maintain soundness and completeness without requiring any cumbersome preparation of the data. We discuss challenges and their solutions in the following subsections.

3.1 Workload Partitioning

Workload partitioning involves breaking down the problem in such a way that it can be executed in parallel. Ideally, each process should take the same amount of time for the entire computation. In [6], the distinction between data partitioning and rule partitioning is made. For RDFS reasoning, we consider it self-evident that data partitioning is the appropriate approach considering that RDFS reasoning has fewer than 20 rules while data sets could scale into billions of triples and beyond.

Before continuing, we define a few terms used throughout this paper:

Definition 4. *An* ***ontological triple*** *is a triple used in describing an ontology and from which significant inferences can be derived. For RDFS, these are triples with predicate rdfs:domain, rdfs:range, rdfs:subClassOf, or rdfs:subPropertyOf and also triples with predicate rdf:type with object rdfs:Datatype, rdfs:Class, or rdfs:ContainerMembershipProperty.*

For clarification, we emphasize that not all triples with predicate *rdf:type* are ontological triples in the context of RDFS. Only *rdf:type*-triples for which the object is *rdfs:Datatype*, *rdfs:Class*, or *rdfs:ContainerMembershipProperty* are considered ontological.

Definition 5. *An* ***assertional triple*** *is any triple that is not an ontological triple.*

Considering the finite RDFS rules, a particular property becomes apparent. We find that each rule has at most one triple pattern in its body that can match an assertional triple. For example, consider rule *rdfs3*:

$$(?a \text{ rdfs:range } ?x) \wedge (?u\ ?a\ ?v) \rightarrow (?v \text{ rdf:type } ?x)$$

The first triple pattern of the rule body will not match assertional triples but only ontological triples. The second triple pattern, however, could match any triple.

Definition 6. *An M1 rule is a rule whose body has at most one triple pattern which can match an assertional triple.*

Proposition 1. *By inspection, the finite RDFS rules are all M1.*

Proposition 1 gives way to our partitioning scheme. By allowing each process to have all ontological triples but only a fraction of the assertional triples, we can distribute the data in such a way that the workload can be executed in parallel. We consider this approach to be reasonable since ontologies tend to be fixed data sets that are relatively very small compared to the potentially ever-increasing assertional data.

Definition 7. *An* ***abox partitioning*** *is a data partitioning scheme in which each process/partition gets all ontological triples and a fraction of the assertional triples.*

Theorem 1. *A single application of M1 rules to the partitions in an abox partitioning produces the same inferences as in a single partition with all triples. More formally, define a single application of a rule r to a set of triples G, denoted r(G), as the triples resulting from satisfying the antecedent of r without adding such triples back into G. Then, given an M1 rule* r_m *and an abox partitioning* $(G_1, G_2, \ldots, G_n)$ *of a set of triples G,* $r_m(G) = \bigcup_{i=1}^{n} r_m(G_i)$.

Proof. Since an M1 rule needs at most one assertional triple to be satisfied, it can be satisfied on the partition that has such a triple in the same way it is satisfied on a single partition with all triples since all partitions have all ontological triples. □

Therefore, our approach is to give each process a partition of an abox partitioning, and that process will apply all finite RDFS rules until no more inferences can be found. This is shown in Algorithm 1. Line 1 is to be interpreted as parallelism in which i denotes the rank of the process. Line 4 iterates over the finite RDFS rules in the appropriate order to satisfy data dependencies between consequents and antecedents of rules such that only a single pass over the rules is required.

It remains to be shown that this algorithm correctly produces the RDFS closure. Although M1 rules produce all the appropriate inferences when applied once, it must be shown that placing the inferences in the partition from which they were derived sufficiently maintains the abox partitioning so that subsequent applications of the rules will also produce the correct inferences.

Lemma 1. *If an M1 rule sufficiently maintains an abox partitioning after adding its inferences back to the partition from which they were derived, then the M1 rule can be applied multiple times and produce sound and complete inferences.*

Proof. The proof is intuitive. Since M1 rules produce sound and complete inferences in a single application to an abox partitioning, if the result after adding the

Algorithm 1. Parallel RDFS Inferencing Algorithm

Input: A set of assertional triples T_A, a set of ontological triples T_O, and a number of processes p . T_{rdfs} is the set of finite axiomatic triples.
Output: All triples together with all inferences from the computation of finite RDFS closure.

// outer loop denotes parallelism where i is rank of process
for $i = 0$ *to* $p - 1$ **do**
    $T_{A_i} = \{ t \mid t \in T_A \land t \notin T_{A_j}, \forall j \neq i \}$
    // T_{A_i} contains roughly $|T_A|/p$ triples from T_A
    // that are given only to process i
    $T_i = T_{A_i} \cup T_O \cup T_{rdfs}$
    foreach *rule* $\in$ *finite RDFS rules* **do**
        repeat
            apply *rule* to T_i to get inferences
            add inferences to T_i
        until *no new inferences*
    end
end
return $\bigcup_{i=0}^{p-1} T_i$

inferences is an abox partitioning, the M1 rules can be applied again to produce sound and complete inferences. □

Definition 8. *Say that a rule that fits the description in Lemma 1 is* ***abox partitioning safe (APS)****.*

Proving that a rule (or set of rules) is APS consists of proving it is M1 and that it sufficiently maintains the abox partitioning. (We return to the issue of sufficiency later in the paper.) We define several classes of M1 rules that sufficiently maintain an abox partitioning, and thus, such rules are APS. Before doing so, we make the following assumption.

Axiom 1. *Other than those mentioned in the axiomatic triples or those produced from the entailment rules, the resources in the RDF and RDFS vocabulary have no superclasses, subclasses, superproperties, subproperties, domains, or ranges.*

This axiom allows us to disregard odd cases that might occur if such triples were included (like if rdfs:Class has a superclass). Such triples could modify the semantics of RDFS, and so we simply disallow them. We now go on to prove that the finite RDFS rules are APS using sketch proofs to be concise and excluding proofs of propositions that are straightforward.

Definition 9. *Say that a rule is* ***abox partitioning easy (APE)*** *if it is M1 and produces only assertional triples.*

Theorem 2. *APE rules are APS.*

Proof. Adding the produced assertional triples from APE rules to any partition results in an abox partitioning since the ontological triples are unaffected, so all partitions still have all ontological triples. □

Proposition 2. *By inspection, rules lg, gl, rdf1, rdf2, rdfs1, rdfs4a, rdfs4b, and rdfs7 are APE.*

Definition 10. *Say that a rule is* ***abox partitioning ontological (APO)*** *if the triple patterns in its body can match only ontological triples. (Note that this implies they are M1.)*

Theorem 3. *APO rules are APS.*

Proof. Since all partitions have all ontological triples, such rules will produce all of their inferences on all partitions. If an APO rule produces ontological triples, they will be produced on all partitions, and thus all partitions still have all ontological triples. □

Proposition 3. *By inspection, rules rdfs5, rdfs8, rdfs10, rdfs11, rdfs12, and rdfs13 are APO.*

Definition 11. *Say that a rule is* ***abox partitioning friendly (APF)*** *if it is M1 and it produces ontological triples only when the body is satisfied by only ontological and/or axiomatic triples.*

Theorem 4. *APF rules are APS.*

Proof. Since all partitions have all ontological triples and axiomatic triples (every partition adds the axiomatic triples at the beginning), when APF rules produce ontological triples, they are produced on all partitions, and thus, all partitions still have all ontological triples. □

Proposition 4. *By inspection, rules rdfs2, rdfs3, and rdfs9 are APF.*

Definition 12. *Say that a rule is* ***abox partitioning trivial (APT)*** *if it is M1 and the ontological triples it produces do not contribute to the inferencing of new triples that would not otherwise be inferred by other rules.*

Theorem 5. *APT rules are APS.*

Proof. This is where "**sufficiently** maintains an abox partitioning" from Lemma 1 is important. APT rules do not necessarily ensure that all partitions will have all ontological triples. Instead, they ensure that the ontological triples that they produce are insignificant for further inferencing. Therefore, even though such triples may have the form of ontological triples, they fail to meet the part of the definition of ontological triple which states that significant inferences are derived from them. Therefore, we can disregard the ontological triples produced by APT rules, including them only for completeness. □

Proposition 5. *Rule rdfs6 is APT.*

Proof. We include a brief proof for this proposition since it is a little less intuitive than the other propositions. *rdfs6* produces triples of the form (?u rdfs:subPropertyOf ?u). Such triples can help to satisfy rules *rdf1*, *rdfs2*, *rdfs3*, *rdfs4a*, *rdfs4b*, *rdfs5*, and *rdfs7*. The first five rules are intuitive and are not elaborated upon here. Using triples produced by *rdfs6*, rules *rdfs5* and *rdfs7* merely produce triples that already exist. □

Definition 13. *Say that a rule is* ***abox partitioning dynamic (APD)*** *if it is M1 and it produces parts of the ontology (ontological triples) only if the partition needs them to produce inferences that otherwise would not be produced.*

Theorem 6. *APD rules are APS.*

Proof. The concern is that APD rules may create some ontological triples on some partitions and not on others. By definition, though, if the other partitions need these ontological triples to produce sound and complete inferences, then they would have been produced on that partition also by the APD rules. Therefore, all partitions can be considered to have all ontological triples produced by APD rules in the sense that all partitions have the APD rules and would produce the triples if needed. □

Proposition 6. *Rule cmp is APD.*

Proof. If a *rdf:_#* resource is mentioned in the triples of a partition, then the triple (*rdf:_#* rdf:type rdfs:Resource) will eventually be produced. (If used as a property, by way of *rdf1* and *rdfs4a*; if used as a subject, by way of *rdfs4a*; and if used as an object, by way of *rdfs4b*.) Then, *cmp* is satisfied by that triple, and the appropriate ontological triples for *rdf:_#* are produced. □

Corollary 1. *By Theorems 1-6 and Propositions 1-6, all of the finite RDFS rules are APS. Therefore, the finite RDFS closure can be computed in parallel using abox partitioning.*

3.2 Parallel File I/O for RDF Data

As mentioned, we desire that our approach require no preprocessing of the data; everything should be performed in parallel. Therefore, we must determine a way to read and write RDF data in parallel. Most parallel I/O approaches use a "chunking" method in which each process gets a fairly even number of bytes from the data file. However, in RDF, the fundamental unit of data is the triple, not the byte. RDF syntaxes are string-based and therefore cannot be divided into mere bytes assuring that each process will get a set of *complete* triples. Some RDF syntaxes such as RDF/XML [18] make this particularly difficult. If an RDF/XML file is divided into portions of bytes, it becomes extremely difficult to determine which triples are complete and which triples are incomplete in that portion. We take advantage of the simple RDF syntax N-triples[8].

[8] `http://www.w3.org/TR/2004/REC-rdf-testcases-20040210/#ntriples`

In N-triples syntax, every triple occupies a single line. When a process reads its portion of bytes, it can determine where the first complete triple begins by locating the first end-line character, and it can determine where the last complete triple ends by locating the last end-line character. Each process i sends the triple fragment at the beginning of its bytes to process $i-1$ (process 0 exempted), and each process uses this fragment to complete the last (partial) triple. In this way, each process has a set of complete triples, and we assume that reading a fairly equal number of bytes will generally result in a fairly even number of triples.

This method of reading triple sets from an N-triples file on disk is used to partition the assertional triples as described in line 2 of Algorithm 1. Thus, we require no preprocessing of data, although we do require that it is in N-triples format.

3.3 Distributed Blank Nodes

After partitioning the assertional triples, each process essentially has its own RDF graph. Thus, some meaning is lost if blank nodes are distributed. In the original graph, we know that two blank nodes are the same because they have the same label within the same graph. Now, however, we have partitioned the graph into smaller graphs, and there is no guarantee that two blank nodes with the same label in different graphs are actually the same node. (Note that since we require data in N-triples format, all blank nodes have labels.) To resolve this problem, each blank node is replaced by a special URI with scheme "b" and with URI body equal to the blank node identifier. This is done while reading the data to ensure that we will always be able to refer to blank nodes uniquely. For example, blank node _:bnode123 would become <b:bnode123>. Then, as the results are written to file, the special URIs are turned back into blank nodes.

This handles the case of already-existing blank nodes, but it becomes more difficult when applying rules *lg*, *rdf2*, and *rdfs1* in which blank nodes are uniquely allocated to literals. We use a similar approach as with the already-existing blank nodes. We turn the literals into URIs by encoding the literal into an appropriate blank node identifier and then using that identifier as the URI body of a special URI with scheme "l." We use a simple encoding that we call a z-encoding.

Definition 14. *A **z-encoding** of a literal is generated as follows. The literal is first represented in N-triples syntax (including unicode escapes). Then, each character in the string that is not 0-9, A-Z, or a-y is replaced by "zHH" where HH is the hexadecimal representation of the character.*

Using the special l-scheme URIs with z-encoded labels ensures that each process produces the same blank node identifier for each blank node allocated to a literal, and the blank nodes are referred to in a consistent way across processes. These URIs are also converted into blank nodes during output.

4 Implementation

Our implementation is written in C using Redland[9] for in-memory RDF storage and query processing. We use Redland's tree storage structure for efficient loading and querying of RDF data. We implement each finite RDFS rule as a SPARQL Protocol and RDF Query Language[10] (SPARQL) query followed by a function call. The SPARQL query serves to find all potential matches for the rule and the function serves to further restrict the results if needed (e.g., the well-formed requirement in *rdf2*) and produce the appropriate triples. Most rules are implemented as a simple CONSTRUCT query followed by a function that simply adds the resulting triples. For example, *rdfs3*'s query (prefix declarations omitted) is:

CONSTRUCT { ?v rdf:type ?x } { ?a rdfs:range ?x . ?u ?a ?v }

This rule is simply executed and the results added to the process' triples. A more complicated case would be rule *lg*:

CONSTRUCT { ?u ?a ?l } { ?u ?a ?l . FILTER(isLITERAL(?l)) }

In this case, the query is used to find all triples with literal objects, but an additional function is needed to actually allocate a blank node to ?l.

We use MPI for parallel I/O and interprocess communication (only necessary when reading assertional triples). However, unlike the set-union operator on line 11 of Algorithm 1, we simply write the results of each partition/process to a file without eliminating any duplicate triples between partitions. Ensuring that only one process has any given triple (i.e., removing duplicates) would take away from the embarrassingly parallel nature of the algorithm, and so to emphasize the scalability of the algorithm, we do not remove duplicates. Writing to different files allows for better parallel I/O performance since processes will not have to compete for file locks, and all the files will be "self-describing" in the sense that they each have the ontology. The downside, however, is that the overall resulting data set will be larger than necessary, so we present an evaluation in the following section for how much duplication of inferences actually occurs.

The overall process involves initialization of the environment, loading ontological triples from an N-triples file, loading assertional triples from a different N-triples file, performing inferencing, writing results to separate files (one per process), and finalizing the environment.

5 Evaluation

For a data set, we generated the LUBM 10,000-university data set (LUBM10k) which, when the generated OWL files are translated directly to N-triples files,

[9] `http://librdf.org/`
[10] `http://www.w3.org/TR/rdf-sparql-query/`

contains 1,382,072,494 triples. We broke LUBM10k down into subsets by continually halving the data set, generating data sets that we denote as LUBM10k/2, LUBM10k/4, ..., LUBM10k/1024. LUBM10k/1024 is the smallest data set for which we perform an evaluation, and it contains 1,349,680 triples. LUBM10k/4 is the largest data set for which we perform an evaluation, and it contains 345,518,123 triples.

We perform our evaluation on the Opteron blade cluster at Rensselaer Polytechnic Institute's (RPI) Computational Center for Nanotechnology Innovations[11] (CCNI) using only the large memory machines. Each machine is an IBM LS21 blade server running RedHat Workstation 4 Update 5 with two dual-core 2.6 GHz AMD Opteron processors with gigabit ethernet and infiniband interconnects and system memory of 16 GB. We read and write files to/from the large General Parallel File System[12] (GPFS) which has a block size of 1024 KB, scatter block allocation policy, and 256 KB RAID device segment size using a RAID5 storage system.

We ran each job with an estimated time limit of 30 minutes which—due to the nature of the job queuing system—lessened the waiting time for execution. After 30 minutes, the scheduler terminates the job if it is not finished. We found this to be reasonable since in our experience, if it took longer than 30 minutes, it was usually because memory usage was at maximum capacity and the time to finish (if possible) would far exceed 30 minutes due to swapping. When the job is run, it has full control over its nodes. Four processes are on one machine since each machine has four cores, and this causes contention for memory between processes. The main source of contention, though, is among our processes and among external processes in the CCNI that may create a high demand of service on the disk, thus slowing disk I/O. We attempted to perform our evaluation at times when the CCNI was least used to try and reduce competition for disk service. Timings (wall clock, not just CPU) were measured using RDTSC (ReaD Time Stamp Counter).

5.1 Performance

Since the algorithm is embarrassingly parallel, it is no surprise to see in Figure 1 that the time to inference halves as the number of processes doubles. Similarly, as the size of the data set doubles, the time to inference doubles.[13] On 128 processes, LUBM10k/1024 takes 1.10 seconds, and LUBM10k/4 takes 291.46 seconds.

Figure 2 shows that the overall time of the computation is generally linear, but as the smallest data set is run on a larger number of processes, the speedup

[11] `http://www.rpi.edu/research/ccni/`

[12] `http://www-03.ibm.com/systems/clusters/software/gpfs/index.html`

[13] While LUBM data is well-structured and evenly distributed, data that is more uneven (high skew) would likely exhibit similar linear scalability, especially as the number of processes increases. Further performance evaluation using different data sets is part of future work.

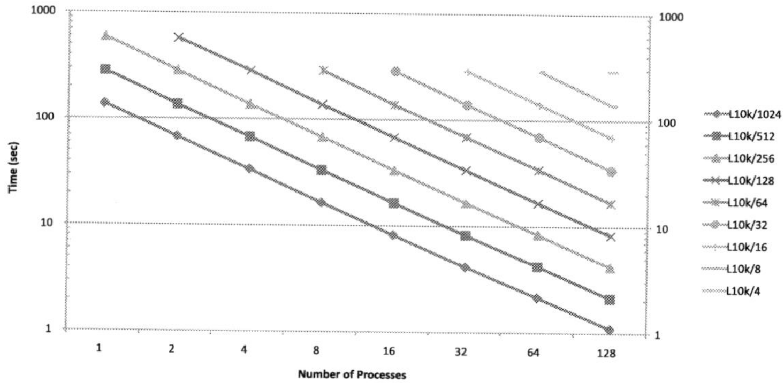

Fig. 1. Time for inferencing only, averaged across processes, for different-size data sets and varying number of processes

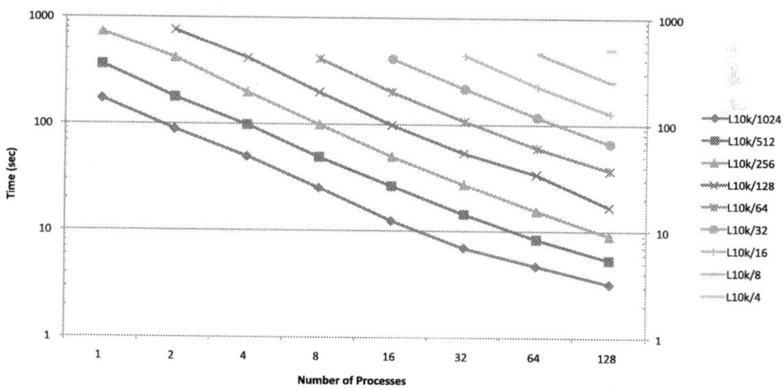

Fig. 2. Overall time averaged across processes, for different-size data sets and varying number of processes

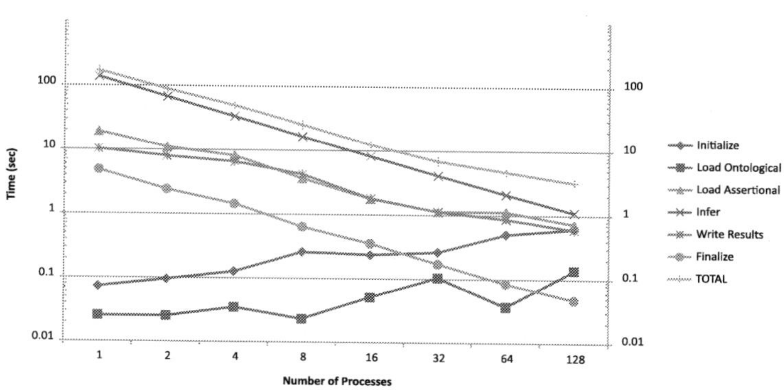

Fig. 3. Breakdown of times for computing finite RDFS closure on LUBM10k/1024 for varying number of processes

decreases. This behavior is likely caused by the small amount of work per process when a larger number of processes are employed. Figure 3 shows this more clearly for LUBM10k/1024 on varying numbers of processes. Time to infer tends to dominate the computation time and decreases as the number of processes increases. The time to initialize is generally very small, but increases with the number of processes. For 128 processes, it took roughly the same amount of time to infer as it did to initialize the environment. Therefore, as the amount of work per process gets smaller, the overall time reaches the overhead cost of computing in parallel.

5.2 Inferences Produced

As mentioned in Section 5, our implementation does not eliminate duplicate triples in different partitions; instead, for performance, we simply have each process write all of its triples to separate files. Therefore, we present an evaluation on how many duplicate inferences are produced by this approach as we scale across number of processes and data set size. When run with one process, no duplicates are created, and so we use the number of inferences from one process as our standard for comparison. However, only the three smallest data sets could be run on one process, so we provide evaluation only for those three data sets. For evaluation purposes, we count the axiomatic triples as inferences since they are added during the inferencing process, and the axiomatic triples are duplicated on all partitions.

Table 1. Information about data sets and inferences produced. Note that inferences are unique only for a single process.

Data sets	# Assertions	Inferences for Varying Number of Processes							
		1	2	4	8	16	32	64	128
10k/1024	1,349,680	1,200,186	1,205,309	1,215,685	1,235,831	1,274,320	1,339,677	1,446,088	1,621,892
10k/512	2,699,360	2,397,066	2,402,267	2,412,473	2,433,087	2,473,273	2,549,186	2,678,194	2,888,254
10k/256	5,398,720	4,785,083	4,790,285	4,800,510	4,821,327	4,861,330	4,940,744	5,091,493	5,348,844
10k/128	10,797,441		9,572,280	9,582,529	9,603,158	9,644,372	9,725,863	9,885,081	10,184,984
10k/64	21,594,882				19,169,424	19,210,312	19,291,708	19,453,885	19,774,980
10k/32	43,189,765					38,337,401	38,420,096	38,583,284	38,906,839
10k/16	86,379,530						76,671,931	76,835,141	77,163,076
10k/8	172,759,061							153,348,394	153,676,871
10k/4	345,518,123								306,700,784

Table 1 shows the number of inferences produced for different data sets and number of processes. Note that only the numbers of inferences for one process are unique inferences only (no duplicates). In Figure 4, the percentage of duplicate inferences grows super-linearly at first (across processes) but soon begins to taper off. As the data set doubles in size, the percentage of duplicates seems to halve; this indicates that duplicates result from fairly static information that somehow ends up in all partitions. This would include ontological data, axiomatic triples, and inferences that proliferate fairly easily. The rules that created the most duplicate inferences are those that most users would probably choose to

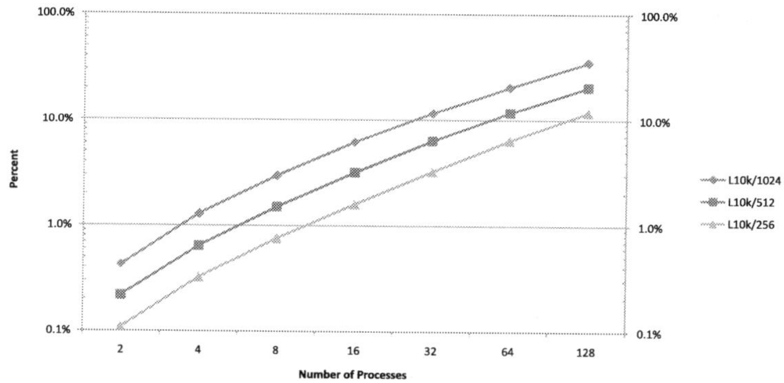

Fig. 4. The percentage of inferences duplicated for the three smallest data sets for varying number of processes

exclude such as *rdfs4a/b* (everything is a rdfs:Resource) and *rdfs1* (all literals are rdfs:Literals). The exception, though, is *rdfs9* which infers rdf:type triples from subclass hierarchy. Interestingly, the only rule for which there were any inferences and no duplicate inferences was *rdfs7*, inferring statements from subproperty hierarchy.

6 Future Work

We see two general directions for future work: scalability and expressivity.

Scalability can be improved by using more efficient in-memory representations of RDF data and by scaling to more processes. In the former case, we hope to employ BitMat [19], a very compact in-memory representation of RDF that also allows for efficient, basic graph pattern querying. In the latter case, we would like to move our code to a more scalable environment like the Blue Gene/L (BG/L) at RPI's CCNI. Our system could also be extended to use a pipelining approach allowing us to scale to data sets which are limited only by disk space. We could simply read in as much of the assertional triples as will fit in a single process, perform inferencing, write results, and then request more assertional triples from disk.

Expressivity can be improved by simply adding more rules that fit the M1 classes described thus far and also by discovering more classes of rules that are APS. In the former case, we have already identified a handful of OWL2RL[14] rules that can be supported in this paradigm (e.g., symmetric properties, equivalent classes, has-value restrictions, etc.), although a multiple-pass approach may be needed instead of the single pass approach in Algorithm 1. We plan to add support for these features in the near future. In the latter case, we are investigating with other colleagues how to handle joins among triple patterns in such a cluster environment so that we can handle non-M1 rules.

[14] `http://www.w3.org/TR/2009/WD-owl2-profiles-20090421/#OWL_2_RL`

We also believe that this approach my be useful in a MapReduce implementation of an RDFS reasoner. In the map phase, ontological triples could be mapped to all reducers and assertional triples to only one reducer each; then, in the reduce phase, the rules can be applied to all the triples mapped to that partition. Of course, considerations may have to be made to ensure that the partitions are not overloaded with too many triples.

7 Conclusion

We have defined a partitioning scheme—abox partitioning—and five classes of rules which can be used to perform complete parallel inferencing on abox partitions. We showed that all of the finite RDFS rules are abox partitioning safe and derived an embarrassingly parallel algorithm for producing the finite RDFS closure. We implemented a C/MPI version of the algorithm and performed an evaluation on a cluster of large memory Opteron machines that showed linear scaling for the inferencing time. We also showed that although some inferences are duplicated, the percentage is generally small for larger data sets. Our results exceed the results reported by related work in that we scale to 128 processes, producing a closure of roughly 650 million triples from an initial data set of roughly 345 million triples in only 8 minutes and 25 seconds, for which the actual inferencing time was only 4 minutes and 51 seconds. To our knowledge, no other system exists that can produce the (finite) RDFS closure on such large data sets in so little time.

Acknowledgements. We thank Medha Atre and Gregory Todd Williams for their insightful comments in reviewing this paper.

References

1. Hayes, P.: RDF Semantics (2004), `http://www.w3.org/TR/2004/REC-rdf-mt-20040210/`
2. Wilkinson, B., Allen, M.: Parallel Programming: Techniques and Applications Using Networked Workstations and Parallel Computers, 2nd edn. Prentice-Hall, Englewood Cliffs (2005)
3. Anadiotis, G., Kotoulas, S., Oren, E., Siebes, R., van Harmelen, F., Drost, N., Kemp, R., Maassen, J., Seinstra, F.J., Bal, H.E.: MaRVIN: a distributed platform for massive RDF inference (2008), `http://www.larkc.eu/marvin/btc2008.pdf`
4. Oren, E., Kotoulas, S., Anadiotis, G., Siebes, R., ten Teije, A., van Harmelen, F.: MaRVIN: A platform for large-scale analysis of Semantic Web data. In: Proceeding of the WebSci 2009: Society On-Line (March 2009)
5. Soma, R., Prasanna, V.K.: A Data Partitioning Approach for Parallelizing Rule Based Inferencing for Materialized OWL Knowledge Bases. Technical report, University of Southern California (2008)
6. Soma, R., Prasanna, V.K.: Parallel Inferencing for OWL Knowledge Bases. In: ICPP 2008: Proceedings of the 2008 37th International Conference on Parallel Processing, Washington DC, USA, pp. 75–82. IEEE Computer Society Press, Los Alamitos (2008)

7. ter Horst, H.J.: Combining RDF and part of OWL with rules: Semantics, decidability, complexity. In: International Semantic Web Conference, pp. 668–684 (2005)
8. Guo, Y., Pan, Z., Heflin, J.: LUBM: A benchmark for OWL knowledge base systems. Web Semantics: Science, Services and Agents on the World Wide Web 3(2-3), 158–182 (2005)
9. Fang, Q., Zhao, Y., Yang, G., Zheng, W.: Scalable Distributed Ontology Reasoning Using DHT-Based Partitioning. In: Domingue, J., Anutariya, C. (eds.) ASWC 2008. LNCS, vol. 5367, pp. 91–105. Springer, Heidelberg (2008)
10. Kaoudi, Z., Miliaraki, I., Koubarakis, M.: RDFS Reasoning and Query Answering on Top of DHTs. In: Sheth, A.P., Staab, S., Dean, M., Paolucci, M., Maynard, D., Finin, T., Thirunarayan, K. (eds.) ISWC 2008. LNCS, vol. 5318, pp. 499–516. Springer, Heidelberg (2008)
11. Oren, E., Gueret, C., Schlobach, S.: Anytime Query Answering in RDF through Evolutionary Algorithms. In: Sheth, A.P., Staab, S., Dean, M., Paolucci, M., Maynard, D., Finin, T., Thirunarayan, K. (eds.) ISWC 2008. LNCS, vol. 5318, pp. 98–113. Springer, Heidelberg (2008)
12. Rudolph, S., Tserendorj, T., Hitzler, P.: What is Approximate Reasoning? In: Calvanese, D., Lausen, G. (eds.) RR 2008. LNCS, vol. 5341, pp. 150–164. Springer, Heidelberg (2008)
13. Bock, J.: Parallel Computation Techniques for Ontology Reasoning. In: Sheth, A.P., Staab, S., Dean, M., Paolucci, M., Maynard, D., Finin, T., Thirunarayan, K. (eds.) ISWC 2008. LNCS, vol. 5318, pp. 901–906. Springer, Heidelberg (2008)
14. Liebig, T., Muller, F.: Parallelizing Tableaux-Based Description Logic Reasoning. In: Meersman, R., Tari, Z., Herrero, P. (eds.) OTM-WS 2007, Part II. LNCS, vol. 4806, pp. 1135–1144. Springer, Heidelberg (2007)
15. Hogan, A., Harth, A., Polleres, A.: Scalable Authoritative OWL Reasoning on a Billion Triples. In: Proceedings of Billion Triple Semantic Web Challenge (2008)
16. Hogan, A., Harth, A., Polleres, A.: SAOR: Authoritative Reasoning for the Web. In: Domingue, J., Anutariya, C. (eds.) ASWC 2008. LNCS, vol. 5367, pp. 76–90. Springer, Heidelberg (2008)
17. Hogan, A., Harth, A., Polleres, A.: Scalable Authoritative OWL Reasoning for the Web. International Journal on Semantic Web and Information Systems (2009)
18. Beckett, D.: RDF/XML Syntax Specification, Revised (2004), `http://www.w3.org/TR/2004/REC-rdf-syntax-grammar-20040210/`
19. Atre, M., Srinivasan, J., Hendler, J.A.: BitMat: A Main Memory RDF Triple Store. Technical report, Rensselaer Polytechnic Institute (January 2009)

Live Social Semantics

Harith Alani[1], Martin Szomszor[1], Ciro Cattuto[2], Wouter Van den Broeck[2], Gianluca Correndo[1], and Alain Barrat[3,2]

[1] Intelligence, Agents, Multimedia
School of Electronics and Computer Science
University of Southampton, Southampton, UK
{h.alani,mns2,gc3}@ecs.soton.ac.uk
[2] Complex Networks and Systems Group
Institute for Scientific Interchange (ISI) Foundation, Turin, Italy
ciro.cattuto@isi.it, wouter.vandenbroeck@isi.it
[3] Centre de Physique Théorique (CNRS UMR 6207), Marseille, France
alain.barrat@cpt.univ-mrs.fr

Abstract. Social interactions are one of the key factors to the success of conferences and similar community gatherings. This paper describes a novel application that integrates data from the semantic web, online social networks, and a real-world contact sensing platform. This application was successfully deployed at ESWC09, and actively used by 139 people. Personal profiles of the participants were automatically generated using several Web 2.0 systems and semantic academic data sources, and integrated in real-time with face-to-face contact networks derived from wearable sensors. Integration of all these heterogeneous data layers made it possible to offer various services to conference attendees to enhance their social experience such as visualisation of contact data, and a site to explore and connect with other participants. This paper describes the architecture of the application, the services we provided, and the results we achieved in this deployment.

1 Introduction

Most conference attendees would agree that *networking* is a crucial component of their conference activities. Conferences, and similar events, are indeed often judged not only by their popularity or scientific qualities, but also by the social experiences they provide. Consequently, conference organisers are keen to enhance the social experience by offering activities or technologies that encourage and support social interactions.

We strove to significantly further the state of the art by developing a Semantic Web application that integrates (a) the available wealth of linked semantic data, (b) the rich social data from *existing* major social networking systems, and (c) a physical-presence awareness infrastructure based on active radio-frequency identification (RFID). The resulting prototypical application was deployed at the 2009 European Semantic Web Conference (ESWC09).

Making Use of Linked Data. The amount and variety of semantic data available on the web is continuously growing. The Linked Data initiative has been instrumental in this.

A. Bernstein et al. (Eds.): ISWC 2009, LNCS 5823, pp. 698–714, 2009.

Data from various conferences (e.g. ESWC, ISWC, WWW) has been consistently collected and published in recent years [12], and can be retrieved from sites such as `data.semanticweb.org`. This data has been merged with data from several publication databases (e.g. CiteSeer, DBLP) by the RKBExplorer system [5]. From this data we inferred Communities of practice (COP), which offer a first insight into the scientific networks of the participants. We used this to provide awareness of the presence of their COP members at the conference, and of any talks they might be giving there.

Mining folksonomies. The tags that people use on various Web 2.0 sites tend to represent their personal interests [10]. Avid users are often active across several such sites, each of which solicits different aspects of a user's interests. If these are brought together, a far richer understanding of a user's interests can be obtained and subsequently used for superior personalisation, recommendation or awareness services [17]. Users often carry some of their tagging selections and patterns across different folksonomies [18]. Retrieving interests of conference attendees from multiple social sites, interests that might transcend the academic or scientific domain, could lead to more interesting matchmaking services. To this end, we generated Profiles of Interests (POI) for participants to allow people to explore each others' interests.

Meshing online and real-life social networks. Social relationship data from online social networks could provide a useful substrate for constructing social services. However, since such networks generally capture only part of the actual social network, meshing this data with knowledge of real-life social activities would greatly improve this potential. To this end, we deployed a novel active-RFID based sensor platform [1] that is capable of detecting real-life social interactions in terms of sustained face-to-face proximity events. This not only enabled us to provide participants with various novel services, such as logs and summaries of their social interactions, but also to integrate this with information from people's social profiles of interest, scientific communities of practice, and their online social contacts. This meshing not only leads to superior services, but can also facilitate the extension of both networks, online as well as offline [14].

The following Section sheds some light on related work. A full description of the Live Social Semantics application is given in section 3. Section 4 covers various aspects of the results of the application deployment at ESWC. Discussion and future work are given in Section 5, followed by conclusions in Section 6.

2 Related Work

The interplay of networking and social contact at a conference gathering was initially investigated in the context of opportunistic networking for mobile devices [9] by using wearable Bluetooth-enabled devices. Subsequent work focused on sensing organisational aspects [4] by using Bluetooth-enabled mobile phones, and on characterising some statistical properties of human mobility and contact [20,15]. All of these early experiments involved a small number of participant, and could not assess face-to-face human contact in a large-scale setting, as they mostly relied on Bluetooth communication.

Recently, the SocioPatterns project[1] investigated patterns of human contact at large-scale social gatherings by deploying a distributed RFID platform that is scalable and attains reliable detection of face-to-face interactions as a proxy of social contact [1]. The application presented here leveraged that platform to mine real-time social contacts.

IBM used RFIDs to track attendees of a conference in Las Vegas in 2007. The devices were used to track session and meal attendance [20]. The information they collected were limited to the name, title and company of attendees. No social or semantics data were collected nor used. Fire Eagle[2] by Yahoo! is a service that detects the geographical location of users (e.g. based on wifi points), and allows them to share it with their online friends. To the best of our knowledge, our application is the first where real-world face-to-face contacts are mashed up in real time with semantic data from on-line social networking systems.

The novelty of the user profiling work reported here is in the amalgamation of multiple Web 2.0 user-tagging histories to build up personal semantically-enriched models of interest. This process involves dealing with several problems, such as filtering of tags, disambiguating them, associating tags with semantics, and identifying interests.

The free nature of tagging generates various vocabulary problems: tags can be too personalised; made of compound words; mix plural and singular terms; they can be meaningless; they can be synonymous, etc. [11,6,7]. This total lack of control obstructs its analysis [10]. In our work, we follow the approach of *cleaning* existing tags using a number of term filtering processes, similar in spirit to those used in [8]. Our filtering process is fully described in [3,18] and produces a *cleaned* tag cloud for each user.

Tag ambiguity is a well recognised problem, yet still under researched. Clustering of tags was investigated for tag disambiguation [2], where similar tags were grouped together to facilitate distinguishing between their different meanings when searching. Similar clustering techniques were investigated to automatically identify emergent communities that correspond to a tag's different interpretations [21]. While such techniques have demonstrated that the underlying folksonomy structure does contain information that can enable automatic disambiguation, they are too computationally expensive and lack any semantic grounding. The latter has been investigated in [16] where clusters of related tags are grounded to Semantic Web concepts.

The Meaning Of A Tag (MOAT) framework is a system in which users can manually select appropriate semantic web URIs for their tags from existing ontologies [13]. In contrast, the work reported in this paper explores a strategy for the automatic selection of URIs to maintain the essential simplicity of tagging, an approach also followed in [19] where DBPedia[3] concept URIs are automatically suggested for Delicious[4] tags. We make use of our own tagging and disambiguation ontologies since the ones provided by MOAT do not maintain tag ordering - an important feature when automatically determining tag semantics.

[1] http://www.sociopatterns.org
[2] http://fireeagle.yahoo.net/
[3] http://dbpedia.org/
[4] http://delicious.com/

3 Live Social Semantics Application

At ESWC09, the *semantic web*, the *social web*, and the *physical world* were brought together to create a rich and integrated network of information. Acquiring and integrating these heterogeneous, but overlapping, data sources enabled us to provide a new experience and services to conference attendees. The main goal was to encourage conference participants to network, to find people with similar interests, to locate their current friends, and to make new ones.

The Live Social Semantics application was deployed for 4 days (1-4 June 2009) during the European Semantic Web Conference (ESWC), which was located in Crete. More than 300 people attended the conference, out of which 187 accepted to participate in using our application. Each participant was issued with a uniquely numbered RFID badge. Users were asked to enter their RFID ID number on a website dedicated to this social application. On this website, users were also able to provide their Delicious, Flickr, and lastFM[5] account names, as well as activating a Facebook application that collected their social contacts. Out of the 187 who collected an RFID badges, 139 of them also created accounts in our application site (see Section 4).

3.1 General Architecture

Data from various Web 2.0 sources were imported using APIs or screen scraping, and subsequently converted RDF. The aim was to provide a service endpoint that supports the collection and reasoning over the data. Figure 1 provides a global picture of the Live Social Semantics framework. The vertical axis partitions the diagram according to two spaces: the virtual world (i.e. data about individuals held in the web), and the real world (i.e. RFID contact data). Data in the virtual world is sourced from social networking sites, to obtain social tagging data and contact networks, as well as the Semantic Web (SW), to obtain information about publications, projects, and the individuals COP (via RKBExplorer and semanticweb.org). All data is sourced directly from linked data sites, or converted to a linked data representation via the Extractor Daemon, and stored in a triple store (center, right of diagram). The Profile Builder (center, top of diagram) processes an individual's tagging activities and links them to DBpedia[6] URIs using the TAGora Sense Repository (sec. 3.4). Similarly, their favourite music artists from LastFM are linked to DBpedia URIs using DBTune.[7] In turn, the Profile Builder automatically suggests to users a list of interests that they can edit, and elect to expose to other participants. DBPedia was our choice *lingua franca* for representing participant's interests.

Data from the real world, i.e. that representing the social interactions of the conference participants, is collected and processed by a local server that communicates via RDF / HTTP with the triples store. A custom *Contact* ontology[8] was used to represent social interactions between individuals, recording the total contact time on a daily basis.

[5] `http://last.fm/`

[6] `http://dbpedia.org`

[7] `http://dbtune.org/`

[8] `http://tagora.ecs.soton.ac.uk/schemas/LiveSocialSemantics`

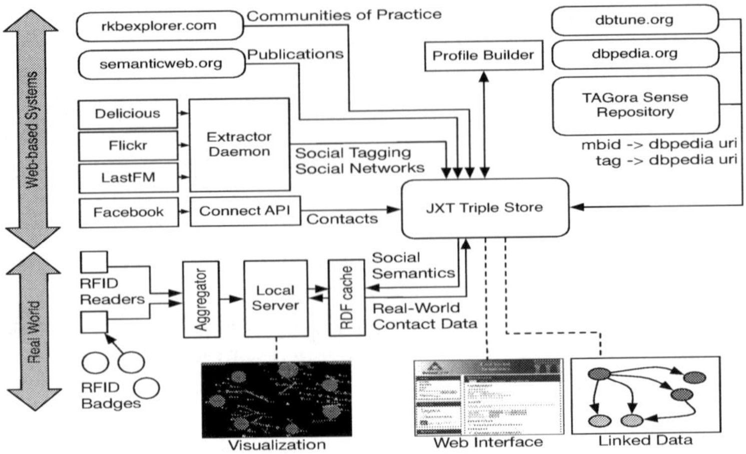

Fig. 1. Live Social Semantics Architecture

3.2 Semantically Interlinked Personal Data

To provide a practical framework that supports the integration of personal data, we employed a technique we refer to as Distinct Separated Identity Management (DSIM). DSIM provides each participant with a foaf:Person URI, that can be linked with other Semantic Web URIs that expose different metadata about the individual, whether that be an external linked data source such as data.semanticweb.org, or internally created data such as the contact data derived from RFID badges. This means that individuals contact data is stored in a separate graph to that of their Facebook friends, Delicious tags, COP, etc. The advantage of this approach is that it closely approximates a distributed linked data scenario (i.e queries must be expanded and run over multiple SPARQL endpoints, contact networks must be flattened), as well as allowing different processes to update the data model asynchronously. This asynchronous nature proved particularly useful when managing the real-time contact data since it enabled separate systems to simply push/pull data whenever needed, whether that be the local RFID server updating the central Triple Store with participants contact data, or the inclusion of Social Semantics (i.e. social networking contacts and profiles) in the visualisation client.

3.3 Real-Time Social Contacts

In order to mine the real-world interactions of conference attendees, we deployed the hardware and software infrastructure developed by the SocioPatterns project [1]. The name badges of those attendees who volunteered to become users of the application were equipped with active RFID badges.[9] The RFID badges engage in multi-channel

[9] Each RFID is equipped with a common button-size battery cell that can last for one week on average.

bi-directional radio communication, and by exchanging low-power signals which are shielded by the human body, they can reliably assess the continued face-to-face proximity of two individuals.

We assume continued face-to-face proximity, within a distance of approximately 1 meter, to be a good proxy for a social interaction between individuals. Contacts are not recorded if people are facing the same direction (e.g. listening to a speaker), unless they turn and face each other for around 10 seconds or more. This kind of resolution is a result of the particular distributed sensing technology we use here, which pushes the state of the art of RFID platforms.

The real-world proximity relations are relayed from RFID badges to RFID readers installed in the conference venue. The readers encapsulate the RFID packets into UDP packets and forward them over a local Ethernet network to a central server. There, the UDP packets from RFID badges are aggregated and fed to a post-processing server that builds and maintains a real-time graph representation of the proximity relations among the tagged attendees. This instantaneous contact graph is represented as a time-dependent adjacency matrix A^t_{ij}, such that $A^t_{ij} = 1$ if individuals i and j are in contact at discrete time t, and $A^t_{ij} = 0$ otherwise. The adjacency matrix was updated every 5 seconds.

The post-processing server also maintains a weighted graph representation of their cumulative proximity relations of the tagged attendees over time. The (normalized) adjacency matrix of the cumulative graph during the time interval $[t_1, t_n]$ is defined as $C_{ij}(t_1, t_n) = (1/n) \sum_{k=0}^{n} A^{t_k}_{ij}$. The matrix element $C_{ij}(t_1, t_n) \in [0, 1]$ is the fraction of application time that individuals i and j spent together. Periodically, the cumulated proximity graph is thresholded, and those relations for which $C_{ij} > C_0$ are represented as a set of RDF triples describing the cumulated real-world proximity of attendees, and periodically uploaded to the triple store via RDF / HTTP.

The real-world proximity relations of the instantaneous proximity graph are mashed up by the server with the web-based attendee relations that it periodically pulls from the triple store. This allows the visualisation clients to display real-world relations in the context of their on-line counterparts.

Moreover, the post-processing server uses the real-world and web-based relations to compute simple recommendation schemes. For example, if two attendees are in contact at a given time, the server provides access to those attendees who may not be present at the same time, but are nevertheless connected to the two users in one of the web-based social networks covered in the application. The visualisation clients (specifically, the user-centerer views) can then use this information to enhance the presented information and support browsing of the social network. More precisely, when creating a personalised view for attendee u_0, the system considers the set $\mathcal{V}$ of attendees who are currently in contact with u_0, or who have been (significantly) in contact with u_0, i.e.,

$$\mathcal{V}(u_0) = \{v \in \mathcal{U} \,|\, A_{u_0,v} \neq 0 \vee C_{u_0,v} > C_0\} ,$$

where $\mathcal{U}$ is the set of all attendees. Subsequently, the system considers the neighbors of attendees $\mathcal{V}$ along the web-based social networks obtained from the triple store, and builds a new set of attendees $\mathcal{W}$ such that members of $\mathcal{W}$ are connected (along web-based systems) to both a member of $\mathcal{V}$ *and* to the focused attendee u_0, closing triangles that have one edge grounded in (current or cumulated) physical proximity, and two edges grounded in on-line relationships. The instantaneous contact graph, the cumulated contact graph, and the web-based graphs are then restricted to the set $u_0 \cup \mathcal{V} \cup \mathcal{W}$ and sent to the visualisation clients, that lay them out for the final user.

Proximity data from RFID devices were taken in the conference area only, covering conference sessions and coffee breaks, but excluding breakfasts, lunches, and evenings.

3.4 Profiles of Interest

In previous work [17], we devised an architecture to automatically generate a list of DBpedia URIs to represent interests a person might have by processing their social tagging activity. Under the assumption that the tags used most often by an individual correspond to the topics, places, events and people they are interested in, we sought to provide a novel dimension to the social interaction at the conference by providing people with a basis to expose their interests, both professional and personal, and see those of others at the conference. Central to this idea is that these profiles can be built automatically, only requiring a short verification phase from the user.

Within the Live Social Semantics architecture, any social tagging information from Delicious and Flickr is collected and converted to an RDF representation (according to the TAGora tagging ontology[10]) by the Extractor Daemon (Figure 1). For each URI that represents a user's tag (for example a Delicious tag *ontologymapping*), a property is created that links it to the Global Tag in the TAGora Sense Repository (TSR).[11] When queried with a tag, the TSR will attempt to find DBpedia.org URIs and Wordnet Synsets that correspond to the possible meanings of the tag. This linked data resource provides information about the possible senses of a tag with mappings to DBpedia resources. Figure 2 contains example URIs that show how the FOAF file produced by our system for Martin (`http://tagora/eswc2009/foaf/4`) is linked to the interest *Semantic integration* via the delicious tag *ontologymapping*.

Profile Building Algorithm. To build a Profile of Interests (POI), we first check to see if the user has a LastFM account. Using DBTune, a linked data site providing metadata about music, we can map the MusicBrainz[12] ID associated to their top artists in LastFM to a resource in DBpedia. The top 5 artists with a DBpedia mapping are added to the user's POI. The second phase of the profile generation procedure is to map the user's tags to DBpedia resources that represent their topics of interest. This is achieved with the following steps:

1. **Disambiguate Tags.** When tags are associated to multiple senses (i.e. more than 1 DBpedia resource), we compare the similarity (using a cosine measure) of the user's cooccurrence vector for that tag (i.e. all other tags that occur in the same post, and their frequencies) against the term frequencies associated with the possible DBpedia senses. If one of the similarity scores is above a threshold value,

[10] `http://tagora.ecs.soton.ac.uk/schemas/tagging`
[11] `http://tagora.ecs.soton.ac.uk/`
[12] `http://musicbrainz.org/`

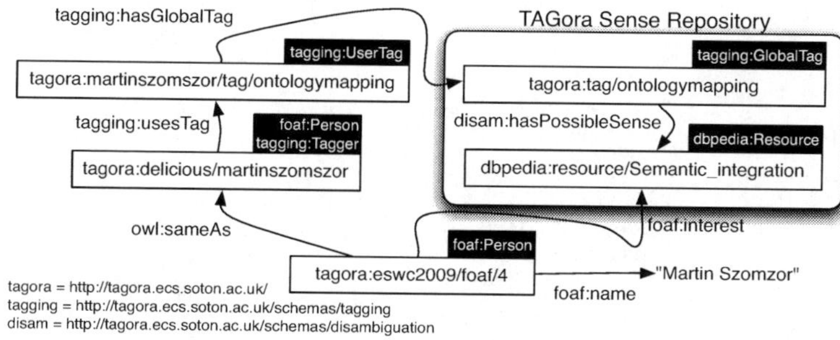

Fig. 2. Linking participants to their interests. The boxes in the diagram represent linked data URIs that provide metadata about various aspects of a participant's social networking data. They link participants to each other through the contact data exposed in various social networking sites, as well as associating them with interests that have been mined from their tagging activity.

(0.3 in this case), we conclude that this is the correct sense for that tag. If more than one (or zero) senses score above the threshold, we do not associate a meaning to the tag. By iterating through all tags associated to a user (i.e. through Delicious or Flickr), we are able to build a candidate resource list C.

2. **Calculate Interest Weights.** For each DPpedia resource $r \in C$, we calculate a weight $w = f_r * u_r$, where f_r is the total frequency of all tags disambiguated to sense r, and u_r is a a time decay factor. This factor $u_r = \lceil days(r)/90 \rceil$. Hence tags used within the last 3 months are given their total frequency, tags used between 3 and 6 months ago are given $1/2$ their frequency, 6 - 9 months a third, etc.
3. **Create Interest List.** If more than 50 candidate resources have been found, we rank them by weight and suggest the top 50. Since users are required to edit and verify this list, we believe it important to keep the number of suggestions to a reasonable amount.

Such semantic POIs could be used to find users with similar interests.

3.5 Visualisation

Two kinds of real-time visualisations were provided. The first, the *spatial view*, was publicly displayed on large screens in the main lobby area. The second, the *user focus view*, was accessible by means of a web browser on the conference LAN, and is linked to from each user's account page on the application site. Both are dynamic visualisations driven by regular updates received through a TCP socket connection with the local post-processing server.

Spatial view. This view provides an overview of the real-time contact graph. It represents the RFID-badge wearing participants within range of the RFID readers, as well as ongoing social contacts (see section 3.3). Each participant is represented by a labelled yellow disc or, when available, by the Facebook profile picture. The contacts are represented by yellow edges, whose thickness and opacity reflects the weight of the contact. The edges are decorated, where applicable, with small Facebook, Flickr, Delicious,

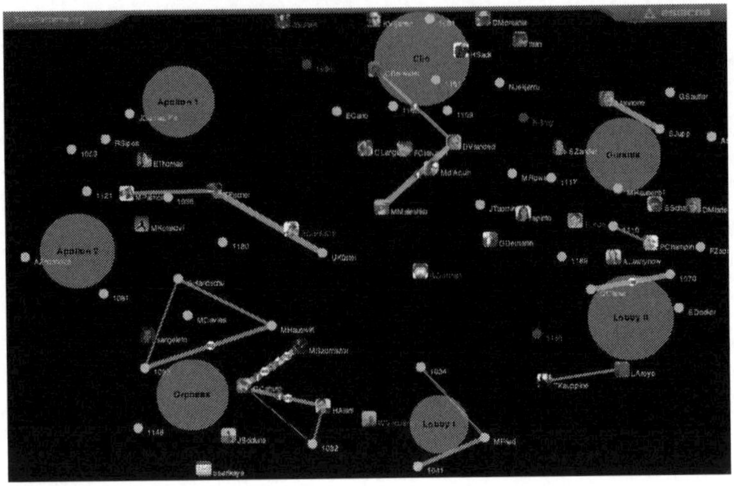

Fig. 3. Screenshot of the spatial view grabbed during a session

lastFM or COP icons, marking the occurrence of that relationship in the respective network. This approach constitutes a projection of said networks onto the real-time contact network.

The SocioPatterns project is primarily concerned with the real-time detection of the contact topology. The precise localisation of the participants in the physical space is of lesser concern. However, a coarse-grained localisation of the participants with respect to the RFID readers is possible. This enabled us to not only represent the contact topology, but also give an indication of which area the participants are in. To this end, the RFID readers were represented by labelled grey shapes, equiangularly laid out on a circumcentric oval, and the participants' shapes are positioned near or in between the readers' marks they are close to. This approach adds spatial structure to the contact graph representation.

User-focus view. This view displays the social neighbourhood of the focussed upon participant. It represents all participants with whom this user has ongoing contact or had significant (cumulative) contact with so far. All physical interactions between these participants are shown as edges, the current ones in yellow, the historical in grey.

This view furthermore attempts to *close relevant triangles*, by which we mean that all participants that are in some way linked to both the focus participant and any of the initially included participants (i.e. those with whom the focus participant has or had contact), are also included, as well as the concerned links, decorated with the relevant icons like in the spatial view. The objective was to provide the users, after focussing upon themselves, with an overview of that subsection of their social neighbourhood that is relevant for their networking activities at that moment.

3.6 Privacy

Permission was sought from all participants for collecting and using their data. A form was prepared which explained what the data is, how it was going to be used, and for how

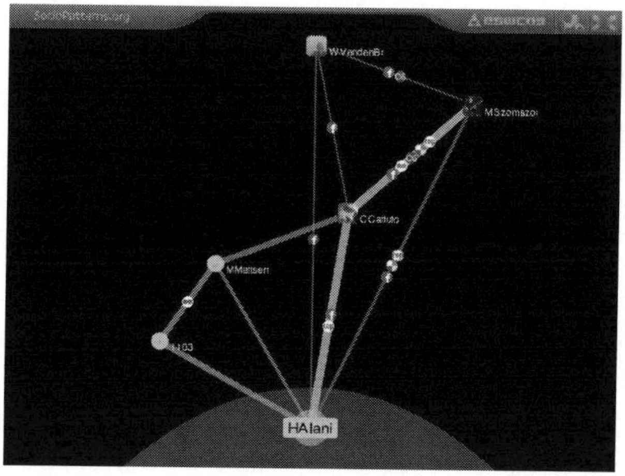

Fig. 4. User-focus visualisation in which *HAlani* has the focus. He has ongoing contacts with *MMattsen* and an anonymous user with badge id *1103*, as indicated by the yellow edges. These two users are also in contact, and they are Flickr friends as indicated by the yellow edge and the Flickr icon that decorates it. There has been significant contact between HAlani and *CCattuto*, as indicated by the thick grey line. They are also Facebook friends and share a COP. Both WVandenBr and MSzomszor were included in order to close relevant triangles. The cyan coloured edges indicate that the users are (only) linked in one or more of the social or COP networks.

long. Users were shown how the RFID badges are used, and the geographical limits of where their face-to-face contacts can be detected (conference building). When creating an account on the application site, each user was given the option of destroying their data after the end of the event.

As explained in Section 3.1, a POI was generated for each user who declared an account in any of the tagging systems we supported (Delicious, Flickr, and lastFM). To ensure that the users are happy with those interests to be viewed by others, each user was asked to verify and edit their list of interests. These profiles only become visible to other users once their owners activate them.

As an extra security, all data from the RFID devices were encrypted to ensure that could only be processed by our systems. All the data gathered by this application were stored in private triplestores, only accessible to the developers of this application.

4 Results

In this section we will report on various results of the application launch at ESWC09; numbers of participants and their shared networking accounts, interest profile generation, RFID usage, and privacy outcomes.

4.1 Participation

Out of the 305 conference attendees, 187 of them took part in Live Social Semantics. Out of these 187 users, 139 of them managed to create an account on the application

Table 1. Number of social networking accounts entered by users into the application site

Account	Facebook	Delicious	lastFM	Flickr	**Total**
Quantity	78	59	57	52	**246**

site, Hence about 26% of the users who collected an RFID badge did not submit any information about themselves (e.g. name, email, social network accounts). Face-to-face contacts of such users were captured, but were not associated with any personal profiles.

4.2 Social Networking Accounts

The application site allowed users to declare their accounts on Delicious, Flickr, lastFM, and Facebook. Table 1 shows how many social networking accounts were entered into our system by all our 139 registered participants.

Table 2 shows that about 35% of our registered users did not declare any social networking accounts (49 users). It also shows that over 61% of our 139 users had more than one social networking account.

Table 2. Number of users who entered 0,1,2,3 or 4 social networking accounts into the Live Social Semantics site

Number of Social Networking Accounts	0	1	2	3	4	**Total**
Number of Users	49	36	28	13	13	**139**

After the conference, we emailed all 49 users who did register on our site, but did not enter any social networking accounts. Aim was to understand the reasons behind that. Table 3 lists the 22 responses we received so far. Out of those 22 participants, 9 (41%) of them simply did not have *any* social networking accounts, and only 1 of these 9 indicated that s/he have an almost empty Facebook account. Four participants (18%) indicated that they use *other* networking accounts, (LinkedIn was named twice). Only 2 (9%) of the 22 replies we received cited privacy reasons for not sharing their social networking accounts. Six replies (27%) picked answer *d*, and four of them blamed the slow internet connection at the conference venue. One participant (5%) picked *e* for being "too busy' during the conference'.

Table 3. Reasons why some users didn't enter any social network accounts to our application site

Option	**Reason**	**No. Users**	**%**
a	don't have those accounts (or rarely use them)	9	41%
b	use different networking sites	4	18%
c	don't like to share them	2	9%
d	didn't get a chance to share them (eg no computer, slow internet)	6	27%
e	other	1	5%
	Total	22	100%

Table 4. Statistics of the profile generation, editing, and saving

	Global	Delicious	Flickr
Concepts Generated	1210	922	288
Concepts Removed	247	156	91
Concepts Added	19		
Concepts Saved	982	766	197

4.3 Social Profiles-of-Interest Results

Out of the 90 people who entered at least one social networking account (Table 2), 59 of them entered at least one account from Delicious, Flickr, or lastFM (remaining 31 only entered Facebook accounts, which we do not use when generating POIs). Although our profile building framework had the potential to utilise all three of these accounts, the linked data site DBTune was offline for the duration of the conference, and hence, we were unable to associate a user's favourite lastFM artists to a DBPedia concept. 41 individuals viewed and saved their POI, of which 31 had a non-empty profile generated. Empty profiles were generated for a number of users who registered that had a very small or empty tag-cloud. Table 4 summarises the results in terms of the number of concepts automatically generated, the number that were removed manually by users, the number that were added manually, and the size of the final profile they saved.

A total of 1210 DBPedia concepts were proposed (an average of 39 per person across the 31 non-empty profiles), out of which 247 were deleted. While it would be useful to know exactly why users deleted a concept, whether it be simply inaccurate (i.e. incorrect disambiguation), it didn't reflect an actual interest (i.e. a very general concept), or it was something they wished to keep private, we considered it too much of a burden to ask users this question when editing their profiles. The total number of concepts deleted was 20% of those suggested. Although a facility was included on the website for users to add their own interests, few did - only 19 new concepts were added. When comparing the results from Delicious and Flickr, we see that 17% of concepts proposed from Delicious Tags were deleted, and 32% respectively for Flickr tags. This suggests that the accuracy of topics harvested from Delicious tags was more accurate than those from Flickr. Inspection of the concepts removed shows that Flickr was likely to suggest concepts referring to years and names.

4.4 RFID Results

Data from RFID badges were taken for a continuous interval of about 80 hours, fully covering the three days of the ESWC09 main track. During that interval, one snapshot of the instantaneous contact graph was recorded every 5 seconds, for a total of $57,240$ snapshots covering the approximate location and the proximity relations of 174 RFID devices[13].

The first column of Table 5 reports the fraction of possible pair-wise contacts that involve face-to-face proximity for a time interval longer than a given threshold. As

[13] Out of the 187 RFIDs we gave out, 13 were used during workshops only.

Table 5. Properties of the cumulative contact graph, as a function of the contact duration threshold. The edge fraction is the percentage of possible edges that are present. The average degree is the average number of contacts to distinct attendees. Clustering is the average node clustering of the graph. The number of connected components and their average size is also reported, together with the fraction of isolated nodes.

threshold	edge fraction	avg. degree	clustering	# conn. comp.	avg. comp. size	isolated nodes
1 min	17.1%	14.9	0.36	1	173.0	0.5%
2 min	11.4%	10.2	0.34	2	84.5	2.9%
5 min	5.5%	5.4	0.20	1	153.0	12.1%
15 min	1.7%	2.6	0.21	7	14.6	41.4%
30 min	0.5%	1.4	0.08	18	3.1	68.4%
60 min	0.1%	1.3	0.17	6	2.3	92.0%

Table 6. Number of individuals met by the 5 most social attendees at the conference, as a function of the contact duration threshold

threshold	#1	#2	#2	#4	#5
1 min	61	59	40	35	27
2 min	49	44	24	21	18
5 min	22	17	15	11	10
15 min	7	7	5	4	4
30 min	2	2	2	2	2
60 min	2	2	1	1	1

expected, the cumulative contact graph is dominated by contacts of short duration, and the introduction of a threshold on contact significance makes the graph more and more sparse as the threshold increases. Table 5 also reports standard network metrics for the cumulative contact graphs over the entire conference duration, for different values of the contact duration threshold. It is apparent that the heterogeneity of the contact graph makes it impossible to choose a single threshold for the significance of social contacts.

Table 6 reports the number of distinct attendees met by the 5 most social attendees of the conference, for different values of the contact duration threshold. As the RFID contact data were taken during the conferences sessions and coffee breaks (and not during lunchtime, for example), only few contacts of long duration were observed.

4.5 Privacy Results

Naturally, privacy is always a concern in such contexts, where personal data is being collected and processed in various ways. As explain in section 3.6, we took various measures to secure the data and protect privacy, even though most of the data we were gathering was actually in the public domain (e.g. shared tags).

Some participants asked for the data to be kept without any anonymisation, to be stored for reuse in coming events, and even to be published so they can link to their profiles and contacts logs from their websites and blogs. On the other hand, some participants were only prepared to take part if the data is anonymised. Table 7 shows the

Table 7. Numbers of participants who chose for their anonymised data to be kept, or destroyed

Option	No. Users	%
I agree for my data to be used for research purposes after the end of this experiment if properly anonymised	85	61%
I do not agree for my data to be kept after the end of this experiment	54	39%
Total	139	100%

two options given to the users in the Terms & Conditions form (section 3.6) when they come to register on the application website. The table shows that 61% of the participants were happy for their data to be kept, while 39% requested the destruction of their data.

The numbers in section 4.2 indicate that the majority were happy to share their social networking accounts. However, we cannot extend this observation to the other 48 who collected and RFID but never registered any information into our site.

5 Discussion and Future Work

The deployment of the application at ESWC2009 was the first where all components were put together and a good number of participants got to use it. We observed quite a few technical and sociological issues, which we discuss in the following.

There are many social networking sites, but in this first version we only supported four currently popular ones. We are working on a open plug-in architecture that allows external parties to develop the functionality needed to connect to, and crawl data from, other networking systems. We also plan to let users submit their FOAF files.

The number of available social networking sites on the web is always on the increase, and the popularity of such sites is never constant. In our application, only four of such networking systems were taken into account. Although the ones we selected are currently amongst the most popular ones, several users wished to add other accounts, such as FOAF files and LinkedIn. One approach to increase extendibility and increase coverage is to use an open architecture to allow external parties to develop and plug applications and services to connect to, and crawl data from, other networking systems, or sources such as FOAF files.

We had devised a privacy and data retention policy in which we pledged to anonymise the resulting data set, and allowed participants to request the complete removal of their data after the end of the event. This approach introduced a number of inconsistencies and ambiguities. As highlighted in section 4.5, many users expressed their interest in acquiring their data after the conference. However, the anonymisation actually precludes that. Other issues, such as whether a participant holds the right to access information on recorded contacts with participants that choose to have their data fully removed, points to the need to reconsider our privacy and data retention policies for future deployments. We believe it would be important to allow people to retain all their data, including user accounts, profiles, contact logs, etc. This will not only enable them to access their activity log, but it also allow them to carry their accounts across conferences where this application is deployed.

The visualisation displays of live contacts were popular points of attraction during the conference. They gave accurate reflections of real-time social interactions during the conference. People were often coming to those displays, searching for their colleagues, session chairs, organisers, etc. We plan to extend these visualisations to highlight conference organisers, session chairs, authors. Furthermore, we plan to introduce support for Twitter, both as another source of on-line social links and as a way of providing additional conference-related content in the visualisations. For example, the node corresponding to a person in the visualisation could be highlighted when s/he sends a message on Twitter that relates to the conference.

The results in table 3 show that 27% of our users could not log into our system to enter further data (social networking accounts, edit POI, etc) because of bandwidth issues at the conference venue. To avoid rushing everyone to enter their data while at the conference, we plan to make the application site available well ahead of the starting date of conferences where this application will be deployed next.

Extractions of POIs has so far been limited to users' online tagging activities. However, many of the participants have authored papers which can be used to determine their research interests, and some of these interests are already available on `semanticweb.org` in the form of paper keywords. Acquiring such interests can be added to the system and used to improve recommendations on talks or sessions to attend, or people to meet. Also, information from social networking accounts can be used to avoid recommending existing friends. We furthermore believe it will be advantageous to organise the interests URIs into hierarchies, to support inference and fuzzy matching.

The use of Flickr tags to identify interests seemed to be less accurate than when using Delicious tags. This indicates the need for alternative approaches when dealing with Flickr tags.

In the application deployment reported here, recommendation of attendees based on physical proximity and on links in social networking systems was performed by means of the simple scheme of section 3.3. One could consider ranking schemes for suggested attendees, to make the recommendation more useful and serendipitous. Specifically, a representation of the context of the conference and of the attendees' interests can help in ranking suggested social connections in terms of their predictability based on the conference context.

More services will be provided in future application runs, such as a 'search for person', 'I want to meet', and 'find people with similar interests'. Data from RFIDs can be used to identify 'best attended session or talk'. Social contacts from social networking systems and COPs could be used to find out who has made new contacts, especially if we can compare data over several application launches.

6 Conclusions

The Live Social Semantics application was a demonstration of how semantics from several different sources can be harnessed and used to enhance the real-world interactions of people at a social gathering. In particular, the combination of semantic data from social media with the real-world encounters of attendees provides a new way of connecting to people, both in the real world and on-line.

Exposing real-world encounters in digital form facilitates mining interesting and serendipitous social connections, and greatly facilitates the process of establishing new on-line connections to encountered people. On the other hand, connections in social networking systems such as Facebook can be used to stimulate real-world encounters on the basis of shared acquaintances and interests. All of these opportunities were explored by the participants of ESWC09, and their reactions, observations, and responses provided valuable input on the future evolution of the platform.

In general, this application goes in the direction of making the co-evolution of real-world social networks and on-line social networks more transparent to users, more lightweight, and more usable. The Live Social Semantics application also provided a first opportunity to expose the semantics of social encounters, and investigate recommendation schemes in bodies of data that mix links from social media with links from real-world encounters.

Overall, this application has great potential for further growth over future deployments at conferences and similar event in a wide variety of domains.

Acknowledgement

This research has been supported by the TAGora project funded by the Future and Emerging Technologies program (IST-FET) of the European Commission under the contract IST-34721. The information provided is the sole responsibility of the authors and does not reflect the Commission's opinion. The Commission is not responsible for any use that may be made of data appearing in this publication. This research was supported by the SocioPatterns.org project and its partners. The authors thank Milosch Meriac from Bitmanufaktur, and also Alexander Wahler from STI Int. for supporting our launch at ESWC09. The authors also thank Dr Michael Hausenblas for making ESWC data available in RDF.

References

1. Barrat, A., Cattuto, C., Colizza, V., Pinton, J.-F., den Broeck, W.V., Vespignani, A.: High resolution dynamical mapping of social interactions with active rfid (2008), http://arxiv.org/abs/0811.4170
2. Begelman, G., Keller, P., Smadja, F.: Automated tag clustering: Improving search and exploration in the tag space. In: Proc. 17th Int. World Wide Web Conf., Edinburgh, UK (2006)
3. Cantador, I., Szomszor, M., Alani, H., Fernändez, M., Castells, P.: Enriching ontological user profiles with tagging history for multi-domain recommendations. In: Proc. Workshop on Collective Semantics: Collective Intelligence and the Semantic Web (CISWeb 2008), in 5th ESWC, Tenerife, Spain (2008)
4. Eagle, N., (Sandy) Pentland, A.: Reality mining: sensing complex social systems. Personal Ubiquitous Comput. 10(4), 255–268 (2006)
5. Glaser, H., Millard, I., Jaffri, A.: Rkbexplorer.com:a knowledge driven infrastructure for linked data providers. In: Proc. European Semantic Web Conference, Tenerife, Spain (2008)
6. Golder, S.A., Huberman, B.A.: Usage patterns of collaborative tagging systems. Journal of Information Science 32, 198–208 (2006)
7. Guy, M., Tonkin, E.: Tidying up tags? D-Lib Magazine 12(1) (2006)

8. Hayes, C., Avesani, P., Veeramachaneni, S.: An analysis of the use of tags in a log recommender system. In: Int. Joint Conf. Artificial Intelligence (IJCAI), Hyderabad, India (2007)
9. Hui, P., Chaintreau, A., Scott, J., Gass, R., Crowcroft, J., Diot, C.: Pocket switched networks and human mobility in conference environments. In: WDTN 2005: Proc. 2005 ACM SIGCOMM workshop on Delay-tolerant networking. ACM, New York (2005)
10. Li, X., Guo, L., Zhao, Y.E.: Tag-based social interest discovery. In: Proc. 19th Int. World Wide Web Conf (WWW), Beijing, China (2008)
11. Mathes, A.: Folksonomies - cooperative classification and communication through shared metadata. Computer Mediated Communication - LIS590CMC (December 2004), `http://www.adammathes.com/academic/computer-mediated-communication/folksonomies.html`
12. Mller, K., Heath, T., Handschuh, S., Domingue, J.: Recipes for semantic web dog food - the eswc and iswc metadata projects. In: Recipes for Semantic Web Dog Food - The ESWC and ISWC Metadata Projects, Busan, Korea (2007)
13. Passant, A., Laublet, P.: Meaning of a tag: A collaborative approach to bridge the gap between tagging and linked data. In: Workshop on Linked Data on the Web (LDOW), Int. Word Wide Web Conference, Beijing, China (2008)
14. Passant, A., Mulvany, I., Mika, P., Maisonneauve, N., Löser, A., Cattuto, C., Bizer, C., Bauckhage, C., Alani, H.: Mining for social serendipity. In: Dagstuhl Seminar on Social Web Communities (2008)
15. Scherrer, A., Borgnat, P., Fleury, E., Guillaume, J.-L., Robardet, C.: Description and simulation of dynamic mobility networks. Comput. Netw. 52(15), 2842–2858 (2008)
16. Specia, L., Motta, E.: Integrating folksonomies with the semantic web. In: Franconi, E., Kifer, M., May, W. (eds.) ESWC 2007. LNCS, vol. 4519, pp. 624–639. Springer, Heidelberg (2007)
17. Szomszor, M., Alani, H., Cantador, I., O'Hara, K., Shadbolt, N.: Semantic modelling of user interests based on cross-folksonomy analysis. In: Sheth, A.P., Staab, S., Dean, M., Paolucci, M., Maynard, D., Finin, T., Thirunarayan, K. (eds.) ISWC 2008. LNCS, vol. 5318, pp. 632–648. Springer, Heidelberg (2008)
18. Szomszor, M., Cantador, I., Alani, H.: Correlating user profiles from multiple folksonomies. In: Proc. Int. Conf. Hypertext (HT 2008), Pittsburgh, PA, USA (2008)
19. Tesconi, M., Ronzano, F., Marchetti, A., Minutoli, S.: Semantigy delicious: automatically turn your tags into senses. In: Social Data on the Web, Workshop at the 7th ISWC (2008)
20. Thibodeau, P.: IBM uses RFID to track conference attendees (2007), `http://pcworld.about.com/od/businesscenter/IBM-uses-RFID-to-track-confere.htm`
21. Yeung, C.-M.A., Gibbins, N., Shadbolt, N.: Tag meaning disambiguation through analysis of tripartite structure of folksonomies. In: IEEE/WIC/ACM International Conferences on Web Intelligence and Intelligent Agent Technology Workshops, pp. 3–6 (2007)

RAPID: Enabling Scalable Ad-Hoc Analytics on the Semantic Web

Radhika Sridhar, Padmashree Ravindra, and Kemafor Anyanwu

North Carolina State University
{rsridha,pravind2,kogan}@ncsu.edu

Abstract. As the amount of available RDF data continues to increase steadily, there is growing interest in developing efficient methods for analyzing such data. While recent efforts have focused on developing efficient methods for traditional data processing, analytical processing which typically involves more complex queries has received much less attention. The use of cost effective parallelization techniques such as Google's Map-Reduce offer significant promise for achieving Web scale analytics. However, currently available implementations are designed for simple data processing on structured data.

In this paper, we present a language, RAPID, for scalable ad-hoc analytical processing of RDF data on Map-Reduce frameworks. It builds on Yahoo's Pig Latin by introducing primitives based on a specialized join operator, the MD-join, for expressing analytical tasks in a manner that is more amenable to parallel processing, as well as primitives for coping with semi-structured nature of RDF data. Experimental evaluation results demonstrate significant performance improvements for analytical processing of RDF data over existing Map-Reduce based techniques.

Keywords: RDF, Scalable Analytical Processing, Map-Reduce, Pig Latin.

1 Introduction

The broadening adoption of Semantic Web tenets is giving rise to a growing amount of data represented using the foundational metadata representation language, Resource Description Framework (RDF) [19] In order to provide adequate support for knowledge discovery tasks such as exists in scientific research communities, many of which have adopted Semantic Web technologies, it is important to consider how more complex data analysis can be enabled efficiently at Semantic Web scale. Analytical processing involves more complex queries than traditional data processing often requiring multiple aggregations over multiple groupings of data. These queries are often difficult to express and optimize using traditional join, grouping and aggregation operators and algorithms. The following two examples based on the simple Sales data in Figure 1 illustrate the challenges with analytical queries. Assume we would like to find "*for each customer, their total sales amounts for Jan, Jun and Nov for purchases made in the state NC*", i.e., to compute the relation (*cust, jansales, junsales, novsales*).

A. Bernstein et al. (Eds.): ISWC 2009, LNCS 5823, pp. 715–730, 2009.

Using traditional query operators this query will be expressed as a union query, resulting in three sub queries (each computing the aggregates for each of the months specified) and then an outer join for merging the related tuples for each customer. Each of these sub queries will need a separate scan of the same typically large table. A slightly more demanding example would be to find *"for each product and month of 2000, the number of sales that were between the previous and following months' average sales"*. Computing the answer to this query requires that for each product and month, we compute aggregates from tuples outside the group (the next and previous month's average sales). After these values are computed, we have enough information to compute the output aggregate (count). This query also requires multiple pass aggregation with a lot of repeated processing of the same set of tuples such as repeated scans on relations just to compute slightly different values e.g. previous month aggregate vs. next month aggregates. High-end database systems and OLAP servers such as Teradata, Tandem, NCR, Oracle-n CUBE, and Microsoft and SAS OLAP servers with specialized parallel architectures and sophisticated indexing schemes employ techniques to mitigate this inefficiency. However, such systems are very expensive and are targeted at enterprise scale processing making it difficult to scale them to the Web in a straightforward and cost effective way.

Sub	Prop	Obj
C1	TYPE	CUSTOMER
C1	BOUGHT	P1
P1	LOC	NC
P1	PRICE	100
P1	MONTH	JAN
C1	BOUGHT	P2
P2	LOC	CA
P2	PRICE	50
P2	MONTH	JUNE
C1	BOUGHT	P4
P4	LOC	NC
P4	PRICE	50
P4	MONTH	JUNE
C2	TYPE	CUSTOMER
C2	BOUGHT	P3
P3	LOC	MC
P3	PRICE	100
P3	MONTH	NOV

(a)

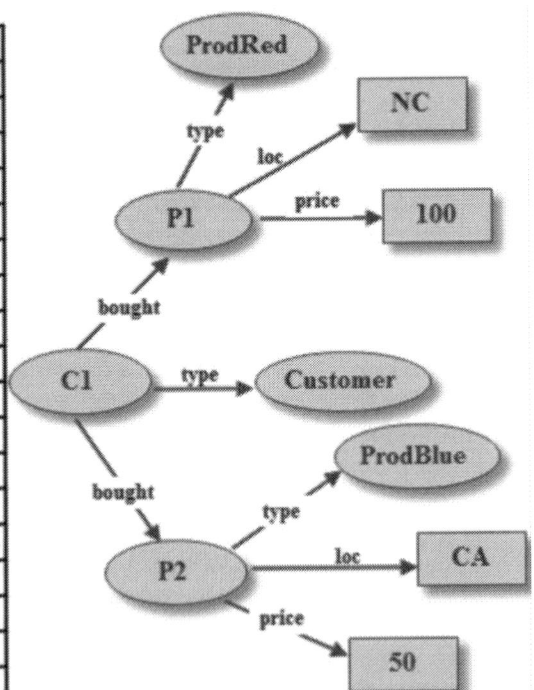

(b)

Fig. 1. RDF representation of Sales relation

One promising direction is to leverage the recent advancements made by search engine companies that utilize parallel data processing frameworks based on clusters of commodity grade machines that can be scaled easily and cost effectively for processing at Web scale. Examples are Google's *Map-Reduce [6]* framework and an open source version developed by Yahoo called *Hadoop [18]*. These frameworks also offer the benefit of allowing users abstract from low level parallelization issues, greatly simplifying the process of writing parallel programs for their tasks. In furtherance of the ease-of-use goal, a high level dataflow language called Pig Latin [12] in the spirit of SQL has been developed for Map-Reduce data processing tasks. Such a high level language offers clear advantages like automatic optimization and code reuse over the traditional Map-Reduce framework that relies on a black box approach in which users writing their own implementations for their tasks making it difficult to automatically optimize and reuse them. However, the current Pig Latin language is targeted at processing structured data, limiting its use for semi-structured data such as RDF. Further, it currently provides only a limited set of primitives that is insufficient for efficient expression of complex analytical queries. This limitation leads to the earlier mentioned problem of avoidable multi-pass aggregations. As an example, encoding our first example in a Pig Latin program yields a 10-step whereas the approach that we propose here results in a 3 step program. This not only improves usability of such systems but also leads to a significant performance savings.

Related Work. The earlier generation RDF storage engines were architected either as main memory stores [3][5] or layered on top of relational databases [2][4][15][16]. These enabled enterprise-scale performance for traditional data processing queries but would be challenged by Web scale processing and analytics-style workloads. More recently, native stores have been developed with a focus on achieving efficient Web scale performance using specialized storage schemes. Some notable ones include (i) vertically partitioned column stores [1] which combine the advantages of a vertically partitioned scheme for property bound queries with the compressibility advantages of column-oriented stores; (ii) multi-indexing techniques [14] that use multiple indexes on triples to produce different sort orders on the different components of a triple, thereby trading off space for the possibility of utilizing only fast merge joins for join processing. In [11], the multi-indexing approach was combined with cost-based query optimization that targets optimizing join-order using statistical synopses for entire join paths. Multi-indexing techniques have also been combined with distributed query processing techniques in [9] for processing queries in federated RDF databases. However, these techniques may pose limitations for ad-hoc processing on the Web, because they require that RDF document content be exported into a database systems for preprocessing (index construction) before data processing can occur. Further, as analytical queries involve filter, join as well as multiple aggregations over different grouping, the indexes will only confer an advantage on the filter and join operations, while the aggregations over groupings would need to occur on the intermediate results that are not indexed. Finally, as we shall see later, the nature of query operators used in the query expression plays a significant role in optimizability of such complex queries.

The second line of work that is relevant is that based on extending Google's Map-Reduce parallel processing framework with support for data processing tasks. The original Map-Reduce framework is designed to work with a single dataset which creates a problem when we would need to perform binary operations like `JOIN` operations. There are methods for overcoming this limitation by doing what are known as *Map-Side* or a *Reduce-Side Joins* respectively. However, the rigid Map-Reduce execution cycle of a Map phase followed by a Reduce, forces such join operations to only be realizable with additional Map-Reduce cycles in which some phases perform no-op. Such wasted cycles lead to inefficiencies and motivated an extension of the Map-Reduce framework called *Map-Reduce-Merge* [17] framework. In this programming model, an additional *merge* phase is added to the typical 2-phase cycle, where the `JOIN` operation is performed between the two sets of data that are partitioned and grouped together in the *map* and the *reduce* phases. However, in these approaches, users bear the responsibility of implementing their tasks in terms of Map and Reduce functions which makes them less amenable to automatic query optimization techniques and offers limited opportunity for code reuse. Pig Latin[12] is a high level dataflow language that overcomes this limitation by provide dataflow primitives with clear semantics similar to query primitives in declarative query languages. However, the set of primitives provided are still limited to basic data processing of structured relational data. Some techniques [8][10] have been developed for processing RDF data at Web scale based on the Map-Reduce platform. However, these efforts have primarily focused on graph pattern matching queries and queries involving inferencing and not analytical queries requiring complex grouping and aggregation.

Contributions and Outline. The goal of the work presented here is to offer a framework for scalable ad-hoc analytical processing of RDF data that would be easy to use, cost effective and directly applicable to RDF document sources without requiring expensive preprocessing. The approach is based on integrating RDF-sensitive and advanced analytical query operators into Map-Reduce frameworks. Specifically, we make the following contributions:

- We propose a dataflow language, RAPID, which extends Yahoo's Pig Latin language with query primitives for dealing with the graph structured nature of RDF.
- We propose a technique for efficient expression of complex analytical queries that offers better opportunities for optimization and parallelization on Map-Reduce platforms. The technique is based on integrating into Pig Latin, primitives for implementing an efficient query operator for analytical processing called MD-join [6].
- We present evaluation results on both synthetically generated benchmark datasets as well as real life dataset.

The rest of the paper is organized as follows: Section 2 gives an overview of analytical queries using a sophisticated operator called MD-join and introduces the Map-Reduce framework and the data flow language, Pig Latin. Section 3 presents the RAPID approach and section 4 shows its experimental evaluation. Section 5 concludes the paper.

2 Preliminaries

2.1 MD-Join and Analytical Querying

Earlier examples showed that expressing complex analytical queries using traditional query operators can lead to inefficiencies due to redundant table scans. One cause of this problem identified in [6] is the tight coupling of the GROUPBY and *aggregation clauses* in query expressions that makes it difficult for the optimizer to identify and coalesce related computations. This led to the proposal of a hybrid join and aggregation operator called the `MD-Join` [6] which decouples the *grouping* and the *aggregation clauses* in query expressions. The MD-Join operator `MD(B, R, l, Θ)` defines a relation with schema B (*base table*), R is the *fact table*, *l* is a list of aggregation functions that need to be computed over attributes and Θ is a set of conditions involving the attributes of B and R. The *fact table* contains all the data that needs to be aggregated and the *base table* contains all combinations of key values representing each group on which aggregations have to be computed. Using this operator, we can express the second example query in the following way:

```
MD( MD( MD( B, R, avg(sales), θ1), R, avg(sales), θ2), R, count(sales), θ3)
where B  – Base relation (the list of Product-Month combinations)
      R  – Fact table (Sales relation)
      θ  – the list of conditions
      θ1 – R.ProdId = ProdId and R.Month = Month+1 and R.year = "2000"
      θ2 – R.ProdId = ProdId and R.Month = Month-1 and R.year = "2000"
      θ3 – R.ProdId = ProdId and R.sales >= prev_month_avg and
           R.sales <= next_month_avg and R.year = "2000"
```

Fig. 2. Expressing second example query using the MD-join

This example also shows the MD-joins can be nested where the result of each MD-Join forms the base relation for the outer MD-Join. The details of the algorithm we omit for brevity but shown by way of an example.

Table 1. (a) Fact table representing Sales data (b) Base table being updated

Cust	Prod	Month	Year
C1	P1	JAN	2008
C1	P2	JUNE	2007
C1	P4	JUNE	2009
C2	P2	JUNE	2007
C2	P3	NOV	2006

(a)

Prod	Month	Sum(Price)
P1	JAN	25
P2	JUNE	~~35~~ 70
P2	DEC	NULL
P3	NOV	NULL
P4	JUNE	50

(b)

Figure 1 is the fact table consisting of the *Sales* data. Each tuple in the Base table is one of the desired groups consisting of a *Product-Month* combination. The algorithm scans the fact table R and loops over all tuples of Base table *B* to identify matches based on a condition θ. If a match is found, the appropriate aggregate columns are updated. Table 1 (a) shows an example fact table and Table 1 (b) shows the base table for *Product*, *Month* groups and a placeholder for the Sum(*Price*) aggregate initially set to NULL. The arrows show the tuples that have just been processed by the MD-join algorithm. The group (*P2*, *FEB*) which had an original sum of *35* based on the second tuple is now updated to *70* after the price of the next tuple (being pointed to) in that group is processed in the fact table. One attractive property of the MD-join operator is that is easily amenable to parallelization. This paper primarily focuses on how to achieve that in the context of Map-Reduce frameworks.

2.2 Data Processing in Map-Reduce Framework

In the Map-Reduce programming model, users encode their tasks in terms of two functions: the *Map* and the *Reduce*. These functions are independent and the execution of each can be parallelized on a cluster of machines. The *Map* function groups together related data items e.g. records with same key values, and the *Reduce* function focuses on performing aggregation operations on the grouped data output by the Map function. The (key, value) mappings can be represented as:

Map (**k1, v1**) → list (**k2, v2**) **and** Reduce (**k2,** list (**v2**)) → list (**v2**)

Recently, Pig Latin, a high level dataflow language was proposed to overcome these limitations. It achieves a sweet spot between the declarative style of the languages like SQL and the low level procedural style of the Map-Reduce programming. It provides a set of predefined functions and query expressions that can be used to describe the data processing tasks. Its data model consists of an atom that holds a single atomic value, a *tuple* that holds a series of related values, a *bag* that forms a collection of tuples and a *map* that contains collection of key value pairs. A tuple can be nested to an arbitrary depth. Table 2 provides an example of a nested tuple *t* with fields' *f1*, *f2* and *f3* with *f2* containing tuples. It also shows data expressions for retrieving components of the tuple *t*.

Table 2. Expressions in Pig Latin

$$t = \left(C1, \begin{Bmatrix} NC,25 \\ CA,35 \end{Bmatrix}, P1\right)$$

Expression type	Example	Value for t
Field by position	$3	P1
Field by name	f1	C1
Projection	f2.$1	{(25), (35)}
Function evaluation	SUM(f2.$1)	25 + 35 = 60

The other functions offered by Pig Latin language are LOAD that implements the reading an input file, FOREACH which is used as an iterator to loop through a collection, FILTER for filtering out tuples, the JOIN function, grouping functions like GROUP and COGROUP, and other common commands reminiscent of SQL. In addition to these primitives, the language also allows supports User Defined Functions (UDFs).

3 The RAPID Language

This section introduces a dataflow language RAPID that enables complex manipulation of RDF graphs and builds on Pig Latin and the MD-Join operator. It includes primitives for describing nodes, their properties and associated paths, as well as primitives for expressing queries in terms of MD-joins. The model of RAPID is shown in Figure 3.

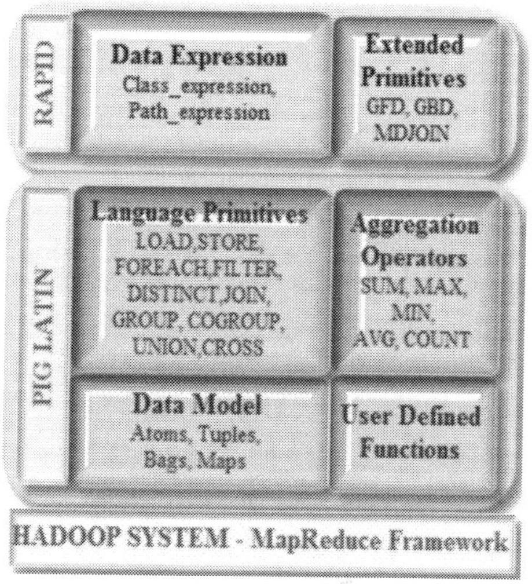

Fig. 3. Architecture of RAPID

3.1 Primitives for Basic Manipulation of RDF Graphs in RAPID

RAPID includes *Class* and *Property* expression types that are very natural for querying RDF data. It also supports *Path expressions* – defining resources in terms of their class or property types or the navigational patterns that they are reachable by.

Table 3. Expressions for manipulating RDF data

Expression	Example	Answer
Class expressions	type: Customer	{C1}
Property expressions	bought	{P1, P2 ..}
Path expressions	C1.bought.price	{25,35}

Class expressions consist of the list of classes that are in desired groups e.g. *Customer*, *Product* denoted as *type:Class*. These expressions evaluate to the resources of type Class. Property expressions list the properties that are used in the query and evaluate to the list of objects of that property. Finally, path expressions specify navigational patterns required to access the resources participating in desired aggregations. Table 3 summarizes these expressions and shows examples based on the example data in Figure 1.

3.2 Primitives for Analytical Processing

In order to implement the MD-Join operation, we introduce three functions `GFD`, `GBD` and `MDJ` for computing the fact table, base table and MD-Join respectively.

3.2.1 Generating Fact Table (GFD)

The role of the GFD function is to reassemble related triples that represent the n-ary relationships typically stored in an OLAP fact table. However, since a query may only involve some parts of the relationship, the GFD function allows users to specify, using path expressions, the specific parts (eq. to columns) of the relationships that are required for grouping or aggregation. Desired filtering conditions (eq. rows) are specified in a similar way. The function assumes that the input RDF document is in the N3-like format where triples of the form *<Subject, Property, Object>* are separated by a special character E.g. ':'. The `GFD` function is passed as a parameter to the Pig Latin's `LOAD` command which uses it as a file handler to reads an RDF file and generate fact table tuples. An example command is shown below:

fact_dataset = LOAD 'input.rdf' USING GFD (class_expression; property_expression; aggregation_pathExpression;filter_pathExpression);

where "input.rdf" is the RDF file to be loaded, *Class_Expression* lists the classes for elements of each group, property_expressions indicate the properties whose range values need to be aggregated.

Pseudo code for GFD Function

```
GFD( class_expression c_e; property_expression p_e;
  aggregation_pathExpression ag_e; filter_pathExpression f_e)
FOREACH <sub,prop,obj> in the input file
  | Group by sub as key
  | Insert <sub, bag of grouped tuples <prop, obj> into
  | Hash Map DataMap
END FOR
FOREACH <key s,value valBag> in DataMap
  | IF any tuple in valBag matches c_e
  |   | Insert <s, valBag> into a Hash Map BaseMap
  | END IF
END FOR
FOREACH property p in p_e  // repeat with ag_e and f_e
  | Loop through BaseMap and DataMap and
  | Generate < s1_s2, p2, o2 >
  | where s1 = instance of c_e
  |       p2 = last property in p_e
  |       s2 = subject corresponding to p2
  |       o2 = object corresponding to p2
END FOR
```

(a)

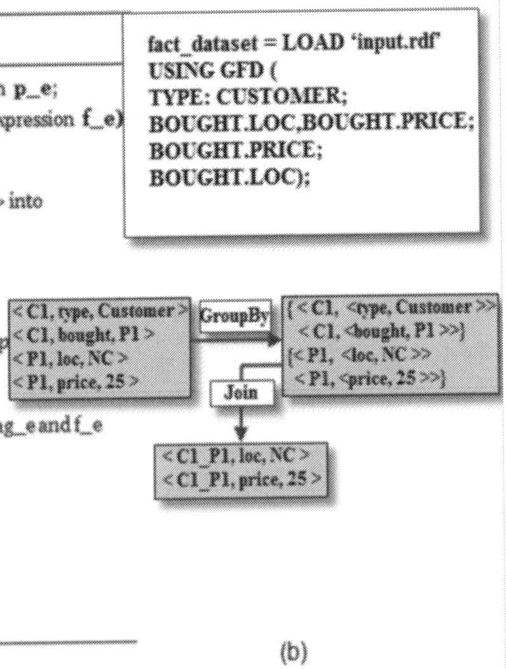

(b)

Fig. 4. GFD Function (a) Pseudo code (b) Execution

The *filter_pathExpression* and *aggregation_pathExpression* describe the navigational paths leading to the nodes that need to be filtered / aggregated. The GFD loads those triples involved in the expressions in the query and groups them using the subject values as keys. Then it performs the necessary join operations using the Pig

Latin JOIN operator (indicated by the path expressions) to reassemble the connections between values. Figure 4 (a) shows the pseudo code for the GFD function. Figure 4 (b) shows the steps in executing GFD.The output consists of triples of the form <*Subject, Property, Object*> where *Subject* is a canonical representation of key value generated using the path expression. In the output, the key *C1_P1* represent the nodes *C1* and *P1* leading to pairs (*loc, NC*) and (*price, 25*). The output of this function is the fact table and is stored using Pig Latin's STORE command and subsequently retrieved as an input to the MDJ function. The following section discusses this in more detail.

Map-Reduce Workflow for GFD Operator. In this section, we discuss the details of the workflow of GFD operator in the context of the Hadoop environment. The Hadoop system consists of one *Job Tracker* which acts as a master process, reads input data, divides this dataset into chunks of equal size and assigns these chunks of data to each of the *Task Trackers*. *Task Trackers* are processors that are designed to perform the *map* or the *reduce* functions called the *Mapper* and *Reducer* respectively.

Mapper Design. The *Job Tracker* assigns each Mapper process with a chunk of the dataset, using which, the Mapper generates <*key, value*> pairs as output. Figure 5 (a) shows the pseudo code for the *map* function and Figure 5 (b) shows the output generated by the *map* function.

Pseudo code for Map Function

```
Input: Key: File Name
       Value: Chunk of data
Map (String Map, String Value)
    For Each line in the value
        // line is of the form <Subject, Property, Object>
        Output (<Subject, <Property, Object>>)
    End of For
End of Map
```

(a)

Subject	<Property, Object>
C1	TYPE, CUSTOMER
C1	BOUGHT,P1
P1	LOC, NC
...	

(b)

Fig. 5. Map function (a) Pseudo code (b) Result

A Map function contains a subroutine called the *Combiner.* All the *objects* having the same *key* are grouped together into a collection in the Combiner function. If the query contains any user specified filter conditions, then the corresponding <*key, Collection of values*> are filtered out based on the given condition in the Combiner function. In [6], the authors show that the tuples for which the filter condition is not true will never be considered by the `MD-Join` so these tuples can be eliminated from the dataset. This reduces the number of records that need to be processed in the *reducer* function, thus increasing the efficiency of processing.

Given the data in Figure 1, suppose the user wants to compute the total price for each customer and the product. Figure 6 (b) shows the result after the execution of the Combiner code. The subject column in the Figure 6 (b) shows how the Combiner function combines multiple properties to generate the composite key when the user query involves multi-dimensional groups. Figure 6 (a) shows the pseudo code for the reducer implementation.

```
Pseudo code for Combiner Function

Input: MapOutput – Which has the output of the Map function in
the form <subject, <Property, Object>>

Combiner (Collection MapOutput)
  FOREACH Subject in the Output
    Get all the records with the same Subject
    IF any filter condition
      FOR EACH record
        validate the filter condition
      END FOR
    END IF
    IF Aggregation is on Multidimensional Properties
      Key: Generate composite key based on the
           Multi-dimensional properties
      Value: <Property, Object>
    ELSE
      Key: Subject
      Value: <Property, Object>
    END IF
    Output (<Key, [Value]>)
  END OF FOR
END OF Combiner
```

(a)

Subject	Value
C1_P1	[<Price, 100>, <Price,105>]
C1_P2	[<Price, 50>]
C1_P4	[<Price,50>]
C2_P3	[<Price,100>]
....	

(b)

Fig. 6. Combiner function (a) Pseudo code (b) Result

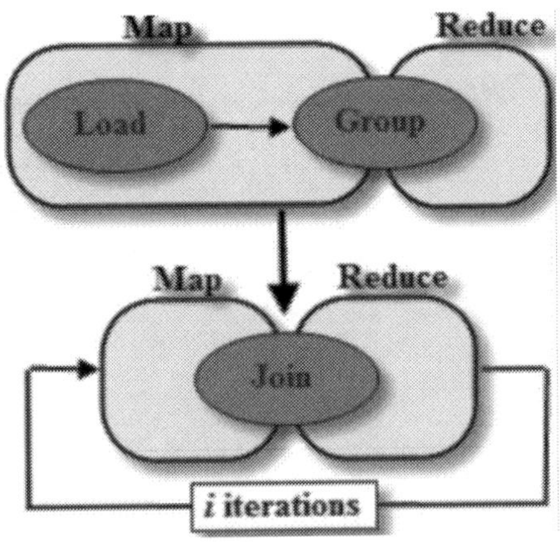

Fig. 7. Compilation of GFD on Map-Reduce framework

Reducer Design. The set of keys and the collection of values obtained from the Combiner function is the input for the reducer function. Each reducer will have a set of tuples. Same keys for the tuples are grouped together and all the corresponding properties of these keys are collected together as part of a reducer function. Algorithm for the reducer function is similar to that of the Combiner function, except that the data for the Combiner function is limited from the corresponding Mapper process, while the data for the reducer function could be from the output of various Mapper/Combiner functions. Figure 7 shows the compilation of the GFD operator on Map-Reduce framework.

3.2.2 Generating Base Table (GBD)

The role of the base table in the MD-Join algorithm is to maintain container tuples representing each group for which aggregations are desired. For every tuple in the fact dataset, the corresponding combination in the base dataset is obtained and the aggregation results are updated in the container tuples. Similar to GFD, the GBD operator is executed with the LOAD function and the class_expression and property_expression play similar roles.

base_dataset = LOAD 'input.rdf' USING GBD (class_expression; property_expression; FLAG);

The FLAG holds either the value "NULL" or "BOTH" indicating whether the aggregation is to be computed on properties got from the property_expression or a combination of the property value and the type class. The tuples generated by the GBD are of the type <Subject, Base, NULL> where "Base" is a keyword that indicates that the tuple belongs to the base table. The NULL value acts as placeholder for aggregates that will be computed by the MD-join step. Figure 8 shows the steps in executing GFD for the given example

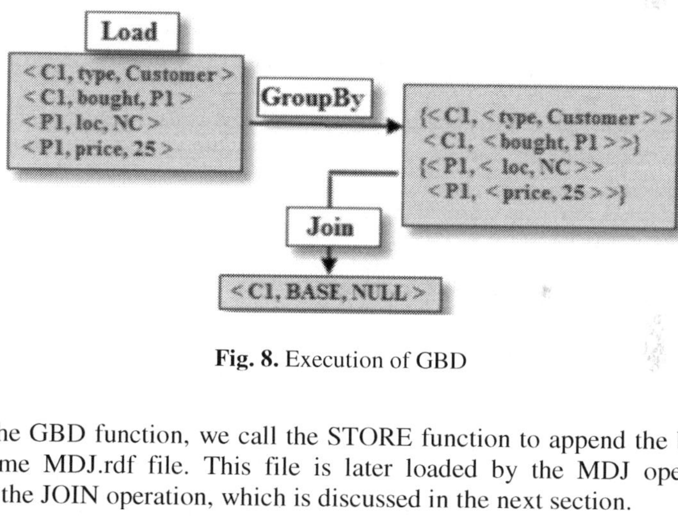

Fig. 8. Execution of GBD

Within the GBD function, we call the STORE function to append the base dataset into the same MDJ.rdf file. This file is later loaded by the MDJ operator while performing the JOIN operation, which is discussed in the next section.

base_dataset = LOAD 'input.rdf' USING GBD (TYPE: CUSTOMER; BOUGHT.LOC, BOUGHT.PRICE; NULL);

Map-Reduce workflow for GBD operator. The Map-Reduce function for the GBD operator is very similar to the GFD operator. The difference is in the Reduce function for these two operators. In the Reduce function of the GBD operator, after the <Key, <Collection>> of values is grouped together. For each key and its group; a corresponding <Key, <Base, Null>> record is created. Thus the output from the reduce function is the set of base tuples that are required for the MDJ operation. The compilation of GBD operator is similar to GFD operator as show in Figure 7.

3.2.3 Multi-dimensional Join in RAPID

In [6], it was shown that if you partition the base table, execute the MD-Join on each partition and union the results, the result is equivalent to executing the MD-Join on the non-partitioned base table. This result leads to natural technique for parallelizing the MD-join computation. It is possible to derive an analogous result to the partitioning of the fact table as well. Due to this limited set of fact and base data records in each partition, the execution of the MD-Join algorithm is much faster because for each fact record we will only iterate through all the base records having the same Key as the fact record. More formally, assume that we can partition the base table B into $B_1 \cup B_2 \ldots \cup B_n$ where $B_i = \sigma_i(B)$, i.e. σ_i is the range condition that tuples in B_i satisfy. Then, $MD(\sigma_i(B), R, l, \Theta) = MD(\sigma_i(B), \sigma'_i(R), l, \Theta)$. In other words, the result of an MD-join of a member of the partition of the base table say B_i, and the entire fact table is equivalent to B_i and the corresponding partition of the fact table - $\sigma'_i(R)$. The result of the GFD and GBD operators seen in the above sections generates the input dataset for the MDJ operator in such a way that efficient partitioning of the records is possible to perform MDJ operation in parallel. Hence each Map processor will receive one set of records containing both the fact and the base dataset with the same keys set. The "MDJ.rdf" file created by the GFD and GBD operators as mentioned in 3.2.1and 3.2.2, contains fact and base tuple sets. The MDJ operator executes on these datasets and also takes as input the filter condition on which the aggregation needs to be computed and the aggregation function such as the SUM, COUNT, MAX, MIN, AVG. As is shown in the pseudo code for MDJ operator in Figure 9 (a), when there is a match between the base and fact tuple, the aggregation is computed and the base dataset is updated. The grouping of related datasets is

Pseudo code for MDJ Function

```
MDJ (class_name class; aggregation_function :
aggregation_property ag; filter_property: filter_condition f_c)
FOREACH t<key, prop, obj> in output of GFD, GBD functions
    IF prop == BASE
        Insert t into a Hash Map BaseTableMap
    ELSE
        Insert t into a Hash Map FactTableMap
    END IF
END FOR
FOREACH <key, valBag> in FactTableMap
    Filter based on f_c
END FOR
FOREACH entry in FactTableMap and BaseTableMap
    IF FactTableMap.Key == BaseTableMap.Key
        Compute the Aggregation Function on property ag
        Update the BaseTableMap with the new value
    END IF
END FOR
```

(a)

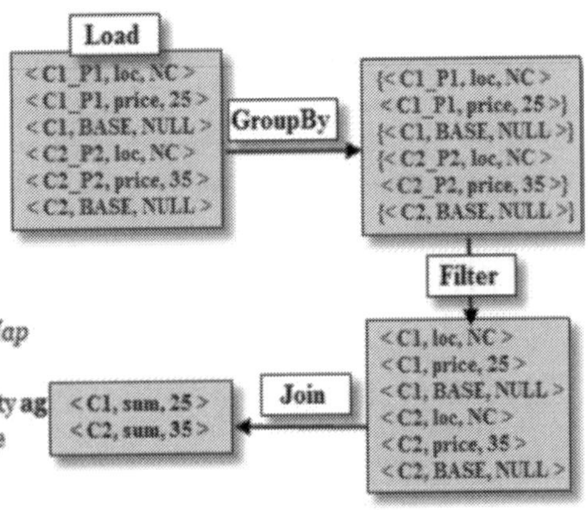

(b)

Fig. 9. MDJ Function (a) Pseudo code (b) Execution

separated into the map function while the aggregation on this grouped data is computed within the reducer function. The syntax for MDJ operator is as follows:

output_dataset = LOAD 'MDJ.rdf' USING MDJ (class_name; aggregation_func: aggregation_property; filter_property: filter_condition);

Figure 9 (b) shows the steps in executing MDJ for the given example. The output generated is stored in an output file using the Pig Latin primitive the STORE function

output_dataset = LOAD 'MDJ.rdf' using MDJ(CUST; SUM:PRICE; LOC:NC);

Map-Reduce workflow for MDJ operator. When performing the LOAD operation for MDJ we load both sets of data (the fact and base dataset) in the *map* function. The reduce function will execute the pseudo code as shown in Figure 9 (a). The compilation of MDJ on Map-Reduce is show in Figure 10.

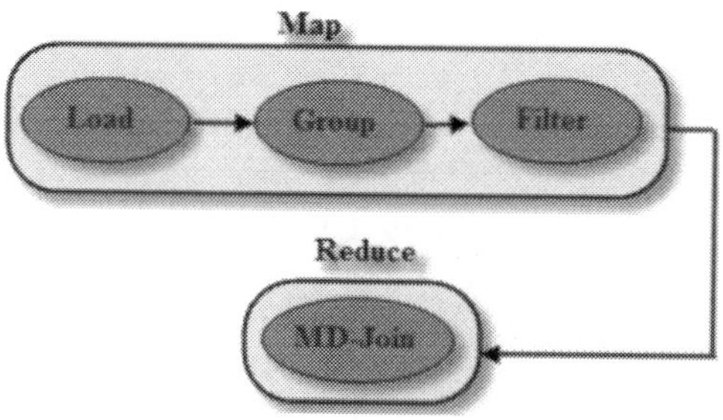

Fig. 10. Compilation of MDJ on Map-Reduce framework

4 Experiments and Results

Our experimental setup involved clusters of 5 nodes where each node is either a single or duo core Intel X 86 machines with 2.33 GHz processor speed and 4G memory and running Red Hat Linux. Our test bed includes a real life dataset, SwetoDBLP [20] and a synthetic dataset generated by the benchmark dataset generator called BSBM [21]. The queries used for the evaluation consists of 5 queries for each dataset.

4.1 Datasets and Results

SWETO dataset contains an RDF representation of the bibliographic data in DBLP such as information about Authors, Books and other types of Publications, Year of Publication, etc. Table 4 shows the set of queries for the DBLP dataset.

Table 4. DBLP Query table

Query 1: Compute the total number of books for all combinations of author-year-publication
Query 2: Compute the total number of book for all combinations of author-publication in the year "2000"
Query 3: Compute the number of books for each author for the year "1999", "2003" and "2007"
Query 4: Compute the total number of books published where the count is greater than the average count for all combinations of author-publisher-month during the year "2008"
Query 5: Compute the total number of books published where the number of books published for all combinations of author-year-publication is greater than 50

The Berlin SPARQL Benchmark (BSBM) is a synthetic benchmark dataset for evaluating the performance of SPARQL queries. The dataset contains information about the Vendor, the Offers and the Product types and the relationships between them. Table 5 shows the set of queries executed on the BSBM dataset were used for evaluation of the two approaches.

Table 6 compares the reduction in the table scans when using RAPID as the number of GROUPBY and JOIN operation reduces. This results in more scalable and efficient approach for processing analytical queries. Figure 11 and Figure 12 (b), show a comparison of the execution times for the two approaches on DBLP and BSBM dataset respectively. Figure 12 (a) shows the number of sub-queries required for each query. The figure clearly shows that the RAPID approach offers better usability. The reduced number of steps in the RAPID approach is due to the coalescing of JOINS and

Table 5. BSBM Query table

Query 6: Compute the number of offers present for all combinations of country-vendor- product-productType
Query 7: Compute the sume of price for each offer for all combinations of country-product for the month "Jan"
Query 8: Compute the number of offers made for each product in the country "US","India","China" and "Japan"
Query 9: Compute the number of offers present where the total price of the offers for all combinations of country-vendor-product-productType is less than "$1000"
Query 10: Compute the number of offers made by vendors which were above the average offer, when the data is viewed from all combinations of productType-validTo-country

Table 6. Number of full table scans for Pig and RAPID operations

Query	Pig	RAPID	Query	Pig	RAPID
Query 1	16	9	Query 6	32	13
Query 2	16	9	Query 7	16	9
Query 3	17	11	Query 8	17	11
Query 4	27	10	Query 9	52	12
Query 5	27	10	Query 10	27	10

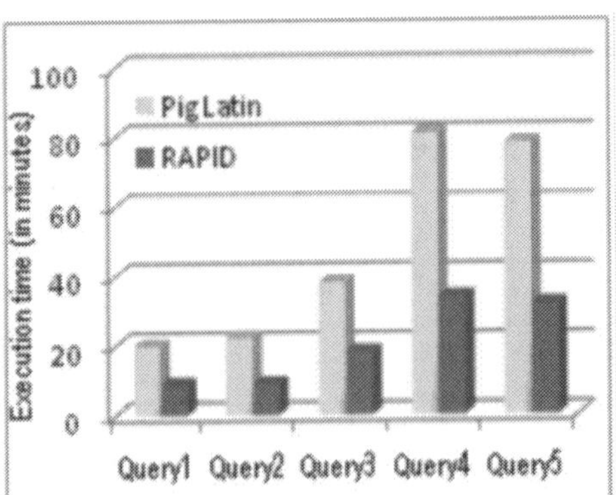

Fig. 11. Cost analysis on DBLP dataset based on execution time (In minutes)

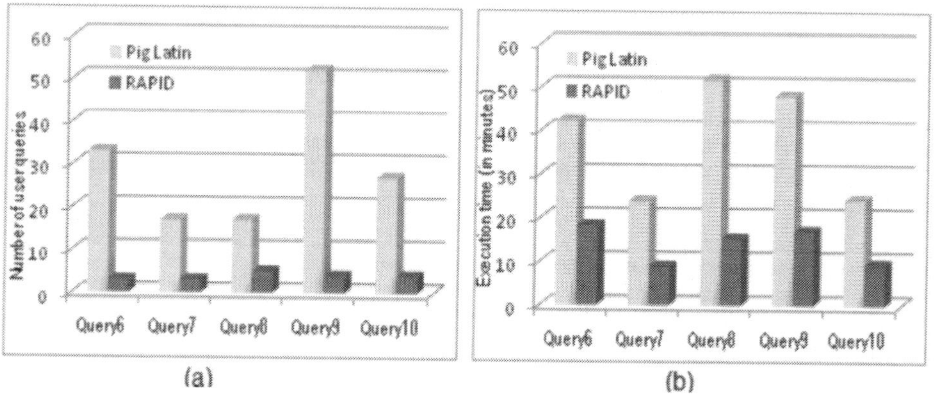

Fig. 12. Cost analysis on BSBM dataset based on (a) Number of user queries (b) Execution time (In minutes)

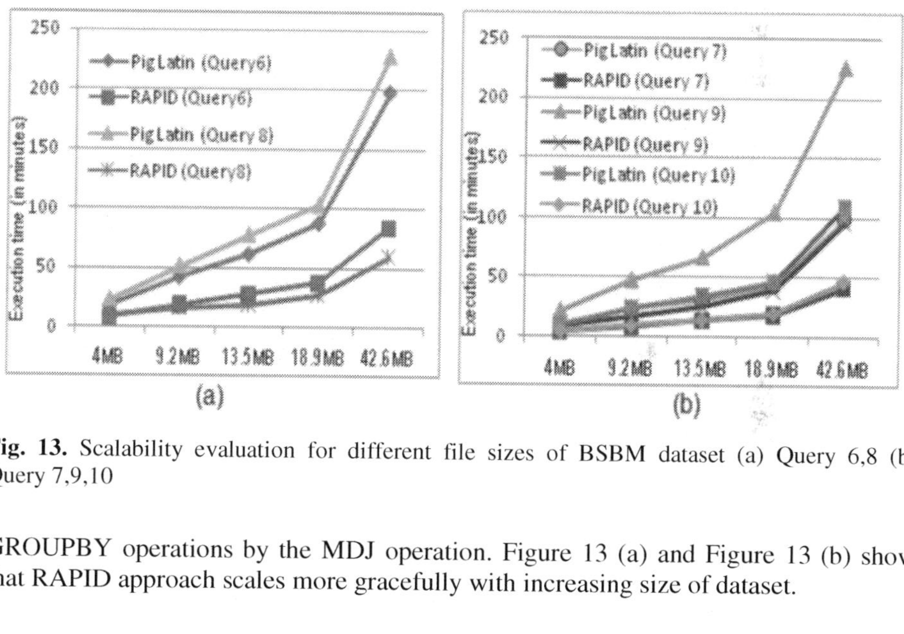

Fig. 13. Scalability evaluation for different file sizes of BSBM dataset (a) Query 6,8 (b) Query 7,9,10

GROUPBY operations by the MDJ operation. Figure 13 (a) and Figure 13 (b) show that RAPID approach scales more gracefully with increasing size of dataset.

5 Conclusion

In this paper, we present an approach for scalable analytics of RDF data. It is based on extending easy-to-use and cost-effective parallel data processing frameworks based on Map-Reduce such as Pig Latin with dataflow/query operators that support efficient expression and parallelization of complex analytical queries as well as for easier handling of graph structured data such as RDF. The results show significant query performance improvements using this approach. A worthwhile future direction to pursue is the integration of emerging scalable indexing techniques for further performance gains.

References

[1] Abadi, D.J., Marcus, A., Madden, S.R., Hollenbach, K.: Scalable Semantic Web Data Management Using Vertical Partitioning. In: Proc. of VLDB 2007, pp. 411–422 (2007)
[2] Alexaki, S., Christophides, V., Karvounarakis, G., Plexousakis, D., Tolle, K.: The ICS-FORTH RDFSuite: Managing voluminous RDF description bases. In: SemWeb (2001)
[3] Beckett, D.: The design and implementation of the Redland RDF application framework. In: WWW (2001)
[4] Broekstra, J., Kampman, A., van Harmelen, F.: Sesame: A generic architecture for storing and querying RDF and RDF Schema. In: Horrocks, I., Hendler, J. (eds.) ISWC 2002. LNCS, vol. 2342, p. 54. Springer, Heidelberg (2002)
[5] Carroll, J.J., Dickinson, I., Dollin, C., Reynolds, D., Seaborne, A., Wilkinson, K.: Jena: implementing the Semantic Web recommendations. In: WWW (2004)
[6] Chatziantoniou, D., Akinde, M., Johnson, T., Kim, S.: The MD-join: an operator for Complex OLAP. In: ICDE 2001, pp. 108–121 (2001)
[7] Dean, J., Ghemawat, S.: MapReduce: Simplified Data Processing on Large Clusters. In: Proc. of OSDI 2004 (2004)
[8] Erling, O., Mikhailov, I.: Towards Web Scale RDF. In: 4th International Workshop on Scalable Semantic Web Knowledge Base Systems, SSWS 2008 (2008)
[9] Harth, A., Umbrich, J., Hogan, A., Decker, S.: YARS2: A Federated Repository for Querying Graph Structured Data from the Web. In: Aberer, K., Choi, K.-S., Noy, N., Allemang, D., Lee, K.-I., Nixon, L.J.B., Golbeck, J., Mika, P., Maynard, D., Mizoguchi, R., Schreiber, G., Cudré-Mauroux, P. (eds.) ASWC 2007 and ISWC 2007. LNCS, vol. 4825, pp. 211–224. Springer, Heidelberg (2007)
[10] Newman, A., Li, Y., Hunter, J.: Scalable Semantics – The Silver Lining of Cloud Computing. eScience, 2008. In: IEEE Fourth International Conference on eScience 2008 (2008)
[11] Neumann, T., Weikum, G.: RDF-3X: a RISC-style engine for RDF. PVLDB 1(1), 647–659 (2008)
[12] Olston, C., Reed, B., Srivastava, U., Kumar, R., Tomkins, A.: Pig Latin: a not-so-foreign language for data processing. In: Proc. of ACM SIGMOD 2008, pp. 1099–1110 (2008)
[13] Pavlo, A., Paulson, E., Rasin, A., Abadi, D.J., DeWitt, D.J., Madden, S., Stonebraker, M.: A Comparison of Approaches to Large-Scale Data Analysis. In: Proc. of SIGMOD 2009 (2009)
[14] Weiss, C., Karras, P., Bernstein, A.: Hexastore: Sextuple Indexing for Semantic Web Data Management. In: Proc. of VLDB (2008)
[15] Wilkinson, K.: Jena property table implementation. In: SSWS (2006)
[16] Wilkinson, K., Sayers, C., Kuno, H.A., Reynolds, D.: Efficient RDF storage and retrieval in Jena2. In: SWDB (2003)
[17] Yang, H., Dasdan, A., Hsias, R.-L., Parket, D.S.: Map-Reduce-Merge: Simplified Relational Data Processing on Large Clusters. In: Proc. SIGMOD 2007, pp. 1029–1040 (2007)
[18] Apache Projects Proceedings, http://hadoop.apache.org/core/
[19] W3C Semantic Web Activity Proceedings, http://www.w3.org/RDF/
[20] Swetodblp, http://lsdis.cs.uga.edu/projects/semdis/swetodblp/
[21] BSBM, http://www4.wiwiss.fu-berlin.de/bizer/BerlinSPARQLBenchmark/spec/index.html#dataschema

LinkedGeoData: Adding a Spatial Dimension to the Web of Data

Sören Auer, Jens Lehmann, and Sebastian Hellmann

Universität Leipzig, Institute of Computer Science,
Johannisgasse 26, 04103 Leipzig, Germany
lastname@informatik.uni-leipzig.de
http://aksw.org

Abstract. In order to employ the Web as a medium for data and information integration, comprehensive datasets and vocabularies are required as they enable the disambiguation and alignment of other data and information. Many real-life information integration and aggregation tasks are impossible without comprehensive background knowledge related to spatial features of the ways, structures and landscapes surrounding us. In this paper we contribute to the generation of a spatial dimension for the Data Web by elaborating on how the collaboratively collected OpenStreetMap data can be transformed and represented adhering to the RDF data model. We describe how this data can be interlinked with other spatial data sets, how it can be made accessible for machines according to the linked data paradigm and for humans by means of a faceted geo-data browser.

1 Introduction

It is meanwhile widely acknowledged that the Data Web will be an intermediate step on the way to the Semantic Web. The Data Web paradigm combines lightweight knowledge representation techniques (such as RDF, RDF-Schema and simple ontologies) with traditional Web technologies (such as HTTP and REST) for publishing and interlinking data and information.

In order to employ the Web as a medium for data and information integration, comprehensive datasets and vocabularies are required as they enable the disambiguation and alignment of other data and information. With DBpedia [1], a large reference dataset providing encyclopedic knowledge about a multitude of different domains is already available. A number of other datasets tackling domains such as entertainment, bio-medicine or bibliographic data are available in the emerging linked Data Web[1].

Many real-life information integration and aggregation tasks are, however, impossible without comprehensive background knowledge related to spatial features of the ways, structures and landscapes surrounding us. Such tasks include,

[1] See, for example, the listing at: http://esw.w3.org/topic/TaskForces/CommunityProjects/LinkingOpenData/DataSets

A. Bernstein et al. (Eds.): ISWC 2009, LNCS 5823, pp. 731–746, 2009.

for example, to depict locally the offerings of the bakery shop next door, to map distributed branches of a company or to integrate information about historical sights along a bicycle track.

With the OpenStreetMap (OSM)[2] project, a rich source of spatial data is freely available. It is currently used primarily for rendering various map visualizations, but has the potential to evolve into a crystallization point for spatial Web data integration. In this paper we contribute to the generation of an additional spatial dimension for the Data Web by elaborating on:

- how the OpenStreetMap data can be transformed and represented adhering to the RDF data model,
- how this data can be interlinked with other spatial data sets,
- how it can be made accessible for machines according to the linked data paradigm and for humans by means of a faceted geo-data browser.

The resulting RDF data comprises approximately 2 billion triples. In order to achieve satisfactory querying performance, we have developed a number of optimizations. These include a one-dimensional geo-spatial indexing as well as summary tables for property and property value counts. As a result, querying and analyzing LinkedGeoData is possible in real-time; thus enabling completely new spatial Data Web applications.

The paper is structured as follows: after introducing the OpenStreetMap project in Section 2, we describe how the OSM data can be transformed into the RDF data model in Section 3 and be published as Linked Data in Section 4. We present a mapping to existing data sources on the Data Web in Section 5. In Section 6 we showcase a faceted geo-data browser and editor and conclude in Section 7 with an outlook to future work.

2 The OpenStreetMap Project

OpenStreetMap is a collaborative project to create a free editable map of the world. The maps are created by using data from portable GPS devices, aerial photography and other free sources. Registered users can upload GPS track logs and edit the vector data by using a number of editing tools developed by the OSM community. Both rendered images and the vector dataset are available for downloading under a Creative Commons Attribution-ShareAlike 2.0 license. OpenStreetMap was inspired by the Wiki idea - the map display features a prominent 'Edit' tab and a full revision history is maintained.

Until now the OpenStreetMap project has succeeded in collecting a vast amount of geographical data (cf. Figure 1), which in many regions already surpasses by far the quality of commercial geo-data providers[3]. In other regions, where currently only few volunteers contribute, data is still sparse. The project,

[2] `http://openstreetmap.org`

[3] Data about the Leipzig Zoo, for example, includes the location and size of different animals' vivariums.

however, enjoys a significant growth in both active contributors and daily contributed data so that uncharted territory vanishes gradually. For some regions the project also integrates publicly available data (as with the TIGER data in the U.S.) or data donated by cooperations (as in The Netherlands).

Table 1. OSM statistics as of June 2009

Category	**Overall Amount**	**Daily Additions** (avg.)	**Monthly Growth** in the last year
Users	127,543	200	11%
Uploaded GPS points	915,392,139	1,600,000	10%
Nodes	374,507,436	400,000	5%
Ways	29,533,841	30,000	7%
Relations	136,245	300	6%

The OSM data is represented by adhering to a relatively simple data model. It comprises three basic types - *nodes*, *ways* and *relations* - each of which are uniquely identified by a numeric id. Nodes basically represent points on earth and have longitude and latitude values. Ways are ordered sequences of nodes. Relations are, finally, groupings of multiple nodes and/or ways. Each individual element can have a number of arbitrary key-value pairs (tags in the OSM terminology). Ways with identical start and end nodes are called closed and are used to represent buildings or land use areas, for example.

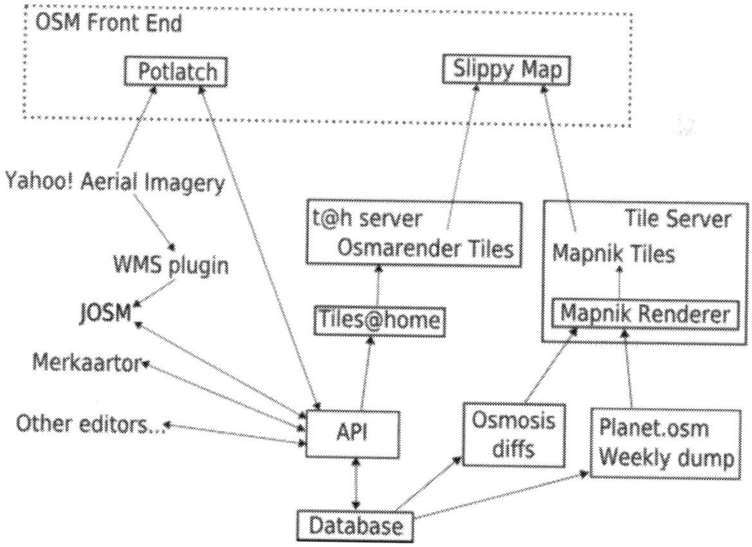

Fig. 1. Technical OpenStreetMap architecture and components. Source: `http://wiki.openstreetmap.org/wiki/Image:OSM_Components.png`

The various OSM components are depicted in Figure 1. The data is stored in a relational database. The data can be accessed, queried and edited by using a REST API, which basically uses HTTP GET, PUT and DELETE requests with XML payload (as shown in Figure 2). The data is also published as complete dumps of the database in this XML format on a weekly basis. It currently accounts for more than 6GB of Bzip2 compressed data. In minutely, hourly and daily intervals the project additionally publishes changesets, which can be used to synchronize a local deployment of the data with the OSM database.

Different authoring interfaces, accessing the API, are provided by the OSM community. These include the online editor *Potlatch*, which is implemented in Flash and accessible directly via the edit tab at the OSM map view, as well as the desktop applications *JOSM* and *Merkaartor*. Two different rendering services are offered for the rendering of raster maps on different zoom levels. With *Tiles@home*, the performance-intense rendering tasks are dispatched to idle machines of community members; thus achieving timeliness. The *Mapnik* renderer, in turn, operates on a central tile server and re-renders tiles only in certain intervals.

```xml
<node id="26890002" lat="51.051934" lon="13.7415877" version="10"
    changeset="766465" user="saftl" uid="7989" visible="true"
    timestamp="2009-03-09T08:49:48Z">
  <tag k="name" v="Frauenkirche" />
  <tag k="created_by" v="Potlatch 0.10e" />
  <tag k="tourism" v="viewpoint" />
  <tag k="url" v="http://www.frauenkirche-dresden.de/" />
  <tag k="denomination" v="lutheran" />
  <tag k="wikipedia:en" v="Frauenkirche_Dresden" />
  <tag k="religion" v="christian" />
  <tag k="amenity" v="place_of_worship" />
  <tag k="wikipedia:de" v="Frauenkirche_(Dresden)" />
</node>
```

Fig. 2. OSM XML excerpt representing a node

Apart from geographical features, the key-value pairs associated with OSM elements are a rich source of information. Such annotations are, for example, used to distinguish different types of roads, to annotate points-of-interest or to influence the map rendering. While initially intended primarily to guide the map rendering, the key-value annotations now already contain a multiplicity of information, which is actually not rendered on the map. This includes, for example, opening hours, links to Web sites or speed limits. An overview over the community-agreed annotations can be found at: `http://wiki.openstreetmap.org/wiki/Map_Features`

3 Transforming OSM into RDF Data Model

A straightforward transformation of OSM data into RDF is not practical, since the resulting 2 billion triples are difficult to handle by existing triple stores. Current triple stores might generally be able to load and query this amount of data; however, response times are according to our experiments not sufficient for practical applications. A particular issue is the storage of the longitude/latitude information, which can currently by far be more efficiently handled by relational database indexing techniques.

As a result of these considerations, we chose to follow a mixed approach in which part of the data is stored in relations and another part is stored according to the RDF data model. But even with regard to the latter part some additional assumptions can considerably reduce the amount of data and increase the querying performance. For example, OSM element ids (used to identify nodes, ways and relations) are always positive integer values. Taking this into account, the space allocated for storing subjects in RDF triples can be significantly reduced and the indexing can be performed more efficiently. Another optimization we performed is to store 'interesting' nodes, ways and relations (i.e. those tagged with certain tags) together with their coordinates in a summary table named `elements`. The resulting database schema is visualized in Figure 3.

The database is populated by importing the XML files which are published in regular intervals by the OpenStreetMap project[4]. In order to be able to import

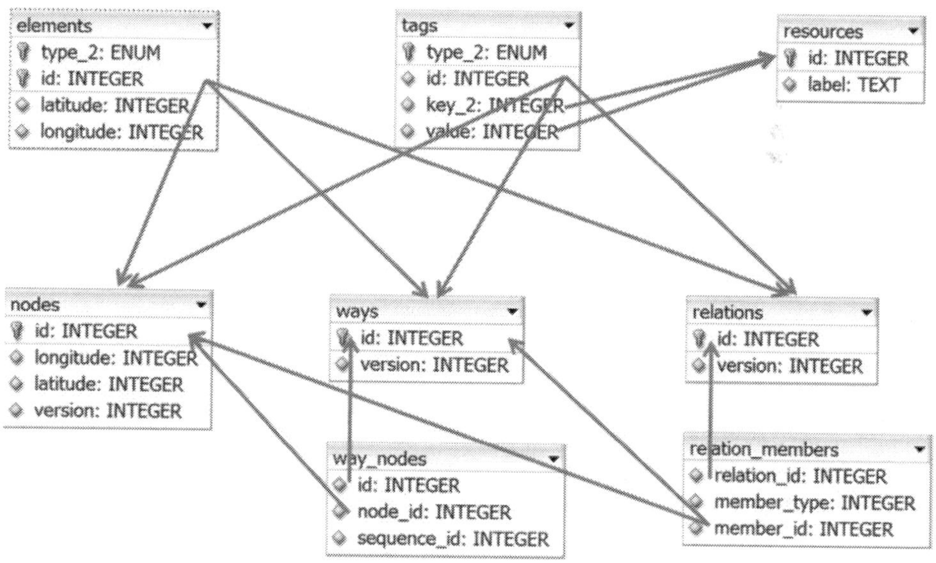

Fig. 3. LinkedGeoData database schema

[4] `http://planet.openstreetmap.org/`

the gigantic XML exports more quickly, we developed our own optimized import script, which is by a factor 3-5 faster than the Osmosis tool. It also handles incremental updates, which are published by OSM on a minutely basis and allow to syncronize a local with the OSM database.

The LinkedGeoData Ontology

A part of the LGD ontology[5] is derived from the relational representation as shown in Figure 3. It includes (or reuses) classes like `geo-wgs84:SpatialThing` with subclasses `node`, `way`, `relation` and properties such as `geo-wgs84:lat`, `geo-wgs84:lon`, `locatedNear`, `rdfs:label`. A major source of structure, however, are the OSM tags, i.e. attribute-value annotations to nodes, ways and relations.

There are no restrictions whatsoever regarding the use of attributes and attribute values to annotate elements in OSM. Users can create arbitrary attributes and attribute values. This proceeding is deliberate in order to allow new uses as well as to accommodate unforeseen ones. There is, however, a procedure in place to recommend and standardize properties and property values for common uses. This procedure involves a discussion on the OSM mailinglist and after acceptance by the community the documentation of the attribute on the OSM wiki[6].

When we examined the commonly used attributes we noticed that they fall into three categories:

- *classification attributes*, which induce some kind of a class membership for the element they are applied to. Example include: `highway` with values `motorway`, `secondary`, `path` etc. or `barrier` with values `hedge`, `fence`, `wall` etc.
- *description attributes*, which describe the element by attaching to it a value from a predefined set of allowed values. Examples include: `lit` (indicating street lightning) with values `yes`/`no` or `internet_access` with values `wired`, `wlan`, `terminal` etc.
- *data attributes*, which annotate the element with a free text or data values. Examples include: `opening_hours` or `maxwidth` (indicating the maximal allowed width for vehicles on a certain road).

We employ this distinction to obtain an extensive class hierarchy as well as a large number of object and datatype properties. The class hierarchy is derived from OSM classification attributes. All classification attributes are interpreted as classes and their values are represented as subclasses. Thus `secondary`, `motorway` and `path`, for example, become subclasses of the class `highway`. OSM elements tagged with classification attributes are represented in RDF as instances of the respective attribute value. In some cases the value of classification attributes is just `yes` - indicating that an OSM element is of a certain type, but no sub-type is

[5] The LGD ontology is available at: `http://linkedgeodata.org/vocabulary`

[6] `http://wiki.openstreetmap.org/wiki/Map_Features`

known. In this case we, assign the element to be an instance of the class derived from the classification attribute. Consequently, a way tagged with `highway=yes` would become an instance of the class `highway`. Description attributes are converted into object properties, the respective values into resources. Data attributes are represented as datatype properties and their values are represented as RDF literals.

The resulting ontology contains roughly 500 classes, 50 object properties and ca. 15,000 datatype properties. Only half of the datatype properties, however, are used more than once and only 15% are used more than 10 times. We aim at making this information timely available to the OSM community so that the coherence and integration of OSM information can be increased.

4 Publishing LinkedGeoData

For publishing the derived geo data, we use Triplify [2]. Triplify is a simplistic but effective approach to publish Linked Data from relational databases. Triplify is based on mapping HTTP-URI requests onto relational database queries. Triplify transforms the resulting relations into RDF statements and publishes the data on the Web in various RDF serializations, in particular as Linked Data.

The database schema we developed for representing the OSM data can be easily published using Triplify. Figure 4 shows an example of a generated RDF for an OSM node. However, in order to retrieve information, the point or way

```
lgd-node:26890002      rdfs:comment            "Generated by Triplify V0.5" .
lgd-node:26890002      cc:license                    cc:by-sa/2.0 .
lgd-node:26890002      lgd-vocabulary:attribution "This data is derived" .
lgd-node:26890002#id   rdf:type                      lgd-vocabulary:node .
lgd-node:26890002#id   geo-wgs84:long                "13.7416"^^xsd:decimal .
lgd-node:26890002#id   geo-wgs84:lat                 "51.0519"^^xsd:decimal .
lgd-node:26890002#id   lgd-vocabulary:created_by  lgd:Potlatch+0.10e .
lgd-node:26890002#id   lgd-vocabulary:religion    lgd:christian .
lgd-node:26890002#id   lgd-vocabulary:name        "Frauenkirche" .
lgd-node:26890002#id   lgd-vocabulary:tourism     lgd:viewpoint .
lgd-node:26890002#id   lgd-vocabulary:amenity     lgd:place_of_worship .
lgd-node:26890002#id   lgd-vocabulary:wikipedia%2525de
                    "http://de.wikipedia.org/wiki/Frauenkirche_(Dresden)" .
lgd-node:26890002#id   lgd-vocabulary:wikipedia%2525en
                     "http://en.wikipedia.org/wiki/Frauenkirche_Dresden" .
lgd-node:26890002#id   lgd-vocabulary:denomination   lgd:lutheran .
lgd-node:26890002#id   lgd-vocabulary:url
                                 "http://www.frauenkirche-dresden.de/" .
lgd-node:26890002#id   lgd-vocabulary:locatedNear   lgd-way:23040893> .
lgd-node:26890002#id   lgd-vocabulary:locatedNear   lgd-way:23040894> .
```

Fig. 4. RDF/N3 representation of OSM node with id 26890002 (Dresdner Frauenkirche)

identifiers (i.e. primary keys from the respective columns) have to be known, which is usually not the case. A natural entry point for retrieving geo data, however, is the neighborhood around a particular point, possibly filtered by points holding certain attributes or being of a certain type. To support this usage scenario, we have developed a spatial Linked Data extension, which allows to retrieve geo data of a particular circular region. The structure of the URIs used looks as follows:

Longitude Latitude Radius Property

http://LinkedGeoData.org/near/48.213,16.359/1000/amenity=pub

The linked geo data extension is implemented in Triplify by using a configuration with regular expression URL patterns which extract the geo coordinates, radius and optionally a property with associated value and inject this information into an SQL query for retrieving corresponding points of interest. The following represents an excerpt of the LinkedGeoData Triplify configuration:

```
1  /^near\/(-?[0-9\.]+),(-?[0-9\.]+)\/([0-9]+)\/?$/=>
2  SELECT CONCAT("base:",n.type,"/",n.id,"#id") AS id,
3    CONCAT("vocabulary:",n.type) AS "rdf:type",
4    longitude AS "wgs84_pos:long^^xsd:decimal",
5    latitude AS "wgs84_pos:lat^^xsd:decimal",
6    rv.label AS "t:unc", REPLACE(rk.label,":","%25"),
7    HAVERSINE(latitude,longitude) AS "distance^^xsd:decimal"
8  FROM  elements n INNER JOIN tags t USING(type,id)
9   INNER JOIN resources rk ON(rk.id=t.k)
10  INNER JOIN resources rv ON(rv.id=t.v)
11 WHERE longitude BETWEEN CEIL($2-($3/1000)/abs(cos(radians($1))*111))
12     AND CEIL($2+($3/1000)/abs(cos(radians($1))*111))
13  AND latitude BETWEEN CEIL($1-($3/1000/111)) AND CEIL($1+($3/1000/111))
14 HAVING distance < $3 LIMIT 1000'
```

The first line contains the regular expression, which is evaluated against HTTP request URIs. If the expression matches, the references to parenthesized subpatterns in the SQL query (lines 2-14) will be replaced accordingly. In this particular case $1 in the SQL query will be replaced with the longitude, $2 with the latitude and $3 with the radius. The SQL query is optimized so as to retrieve first points in the smallest rectangle covering the requested circular area (lines 11-13) and then cutting the result set into a circular area by using the Haversine formula (line 14), which is, for the purpose of brevity, in the example called as a stored procedure.

Triplify requires the results of the SQL queries to adhere to a certain structure: The first column (line 2) must contain identifiers, which are used as subjects in the resulting triples, while the column names are converted into property identifiers (i.e. triple predicates) and the individual cells of the result into property values (i.e. triple objects). In our example, we reuse established vocabularies for typing the elements (line 3) and associating longitude and latitude values with them (lines 4 and 5). The datatype for literal values can be encoded by appending two carets and the datatype to the column (prospective property) name (as can be seen in lines 4, 5 and 7). For transforming the tags, which are already stored in a key-attribute-value notation in the `tags` table, into RDF, we use the special "t:unc" column name, which instructs Triplify to derive the property URIs from the next column in the result set, instead of using the current column name (line 6).

Table 2. Performance results for retrieving points-of-interest in different areas

Location	Radius	Property	Results	Time
Leipzig	1km	-	291	0.05s
Leipzig	5km	amenity=pub	41	0.54s
London	1km	-	259	0.28s
London	5km	amenity=pub	495	0.74s
Amsterdam	1km	-	1811	0.31s
Amsterdam	5km	amenity=pub	64	1.25s

Table 3. LinkedGeoData services provided using Triplify

Description	URL
Points of interest in a **circular area**	`lgd:near/%lat%,%lon%/%radius%`
Example: Points of interest in a 1000m radius around the center of Dresden	`lgd:near/51.033333,13.733333/1000`
Points of interest in a **circular area having a certain property**	`lgd:near/%lat%,%lon%/%radius%/%category%`
Example: Amenities in a 1000m radius around the center of Dresden	`lgd:near/51.033333,13.733333/1000/amenity`
Points of interest in a **circular area having a certain property value**	`lgd:near/%lat%,%lon%/%radius%/%property%=%value%`
Example: Pubs in a 1000m radius around the center of Dresden	`lgd:near/51.033333,13.733333/1000/amenity=pub`
A **particular point** of interest (identified by its OSM id)	`lgd:node/%OSMid%`
Example: The Cafe B'liebig in Dresden	`lgd:node/264695865`
A **particular way** (identified by its OSM id)	`lgd:way/%OSMid%`
Example: Alte Mensa at TU Dresden	`lgd:way/27743320`

The Triplify configuration can be also used to create a complete RDF/N3 export of the LinkedGeoData database. The dump amounts to 16.3 GB file size and 122M RDF triples. The different REST services provided by LinkedGeoData project by means of Triplify are summarized in Table 3. Some performance results for retrieving points-of-interest in different areas are summarized in Table 2.

5 Establishing Mappings with Existing Datasources

Interlinking a knowledge base with other data sources is one of the four key principles for publishing Linked Data according to Tim Berners-Lee[7]. Within the Linking Open Data effort, dozens of data sets have already been connected to each other via `owl:sameAs` links. A central interlinking hub is DBpedia, i.e. if we are able to build links to DBpedia, then we are also connected to data sources such as Geonames, the World Factbook, UMBEL, EuroStat, and YAGO. For this reason, our initial effort consists of matching DBpedia resources with LinkedGeoData. In future work, we may extend this further.

When matching two large data sets such as DBpedia and LinkedGeoData, it is not feasible to compare all entities in both knowledge bases. Therefore, we first restricted ourselves to those entities in DBpedia, which have latitude and longitude properties. We then experimented with different matching approaches and discovered that in order to achieve high accuracy, we had to take type information into account. To detect interesting entity types, we queried DBpedia for those classes in the DBpedia ontology which have instances with latitude and longitude properties.

To proceed, we had to discover how those classes are represented in OSM. For this task, we built a test set, which we later also used for evaluating the matching quality. The set consisted of those entity pairs, where a link from a place in OSM (and therefore a LinkedGeoData entity) to a Wikipedia page (and therefore a DBpedia resource) already exists. Such links were set by OSM contributors manually. This resulted in pairs of user-created `owl:sameAs` links between LinkedGeoData and DBpedia. For each of the common DBpedia ontology classes, we picked their instances within this test set. The required schema matching between DBpedia and LGD could then be understood as a supervised machine learning problem, where those instances were positive examples. This problem was solved by using DL-Learner [7,8] and the result can be found in Table 4. For the cases where we did not have instances in the test set, we consulted the OSM wiki pages. Most results were straightforward, but we discovered that suburbs are often typed as cities in DBpedia and that it is often useful to combine the DBpedia ontology with the UMBEL class hierarchy.

The matching heuristic was then defined as a combination of three criteria: type information, spatial distance, and name similarity. Given a DBpedia entity, we proceeded as follows: 1.) Determine the type in LGD according to Table 4. 2.) Query all LGD points, which are within a certain maximum distance from the DBpedia point. The maximum distance depends on the type, e.g. it is higher

[7] `http://www.w3.org/DesignIssues/LinkedData.html`

Table 4. Schema Level Matching between DBpedia and LGD. In doubt, we preferred a more general expression in LGD, since it does not effect matching quality negatively.

Type	DBpedia	LinkedGeoData
city	`dbpedia-owl:City` or `umbel-sc:City`	`lgd:city` or `lgd:town` or `lgd:village` or `lgd:suburb`
railway station	`dbpedia-owl:Station`	`lgd:station`
university	`dbpedia-owl:University`	`lgd:university`
school	`dbpedia-owl:School` or `umbel-sc:SchoolInstitution`	`lgd:school`
airport	`dbpedia-owl:Airport` or `umbel-sc:AirField`	`lgd:aerodrome`
lake	`dbpedia-owl:Lake` or `umbel-sc:Lake`	`lgd:water`
country	`dbpedia-owl:Country` or `umbel-sc:Country`	`lgd:country`
island	`dbpedia-owl:Island` or `umbel-sc:Island`	`lgd:island`
mountain	`dbpedia-owl:Mountain` or `umbel-sc:Mountain`	`lgd:peak`
river	`dbpedia-owl:River` or `umbel-sc:River`	`lgd:waterway`
lighthouse	`dbpedia-owl:LightHouse` or `umbel-sc:Lighthouse`	`lgd:lighthouse`
stadium	`dbpedia-owl:Stadium` or `umbel-sc:Stadium`	`lgd:stadium`

for a country than for a university. This restricts the set of points to consider and improves performance. 3.) Compute a spatial score for each LGD point depending on its distance. 4.) Compute a name similarity score for each LGD point. 5.) Pick the LGD point with the highest combined spatial and name similarity score, if this score exceeds a certain threshold. The outcome is either "no match", i.e. the highest score is below the threshold, or an LGD entity, which is matched via `owl:sameAs` to the given DBpedia entity.

For computing the name similarity, we used `rdfs:label` in DBpedia and additionally a shortened version of the label without disambiguation information, e.g. "Berlin" instead of "Berlin, Connecticut". Within LGD, we used the properties `name`, `name%25en`, and `name_int` (not all of those are defined for each LGD point). For comparing the name strings, we used a Jaro distance metric. The name similarity was then defined as the maximum of the six comparisons between the 2 DBpedia and 3 LGD names.

For the spatial distance metric, we used a quadratic function, which had value 1 if two points coincide exactly and value 0 at the mentioned maximum distance of the given type. It should be noted that the coordinates in DBpedia and LinkedGeoData were often not exactly the same, since for larger entities, e.g. cities, both Wikipedia and OSM choose a reference point, which has to be

Table 5. Evaluation Results

Type	Entities of this type	Matches found	Correct matches	Precision	Recall
city	275	239	235	98.3%	85.5%
railway station	56	38	38	100.0%	67.9%

a representative, but there are no strict guidelines as to how exactly this point is chosen. For brevity, we omit a detailed discussion of the parameters of the matching heuristic and threshold values.

We evaluated the heuristic on the above described test set. Only cities and railway stations were contained in this set more than 20 times, so we had to limit our evaluation to those two types. Table 5 summarizes the results. We defined precision as the number of correct matches divided by the number of reported matches and recall as the number of correct matches divided by the number of entities of this type in the test set. As intended, the heuristic has a high precision with a lower recall. This was desired since not setting an `owl:sameAs` link is much less severe than setting a wrong `owl:sameAs` link. Upon manual inspection, the incorrect matches turned out to be errors in the test set (places within a city linking to the Wikipedia article about the city). Missed matches were usually due to missing names or missing classification information.

Finally, Table 6 presents the overall matching results. More than 50.000 matchings could be found by the script, which required a total runtime of 47 hours. Most of the time was spend for SPARQL queries to our local DBpedia and LGD endpoints. Despite our strict matching heuristic, the matches cover 53.8% of all DBpedia entities of the given types and can therefore be considered a valuable addition to the Linking Open Data effort. Most of the 53.010 matches found are cities, since they are common in Wikipedia and well tagged in OSM. Many DBpedia entities, which cannot be matched, do either not exist in LGD, are not classified, or misclassified in DBpedia (e.g. the German city Aachen is typed as `dbpedia-owl:Country` since it used to be a country in the Middle Ages).

Table 6. Matching Results: The second column is the total number of matches found for this type. The third column is the percentage of DBpedia entities of this type, which now have links to LGD.

Type	#Matches	Rate
city	45729	70.9%
railway station	929	24.8%
university	210	13.3%
school	1483	38.4%
airport	649	8.4%
lake	1014	22.1%

Type	#Matches	Rate
country	160	20.1%
island	313	29.8%
mountain	1475	24.5%
river	677	32.0%
lighthouse	25	4.3%
stadium	346	17.0%

6 Faceted LinkedGeoData Browser and Editor

In order to showcase the benefits of revealing the structured information in OSM, we developed a facet-based browser and editor for the linked geo data (cf. Figure 5)[8]. It allows to browse the world by using a slippy map. Once a region is selected, the browser analyzes the descriptions of nodes and ways in that region and generates facets for filtering. Once a facet or a specific facet value has been selected, matching elements are displayed as markers on the map and in a list. If the selected region is changed, these are updated accordingly.

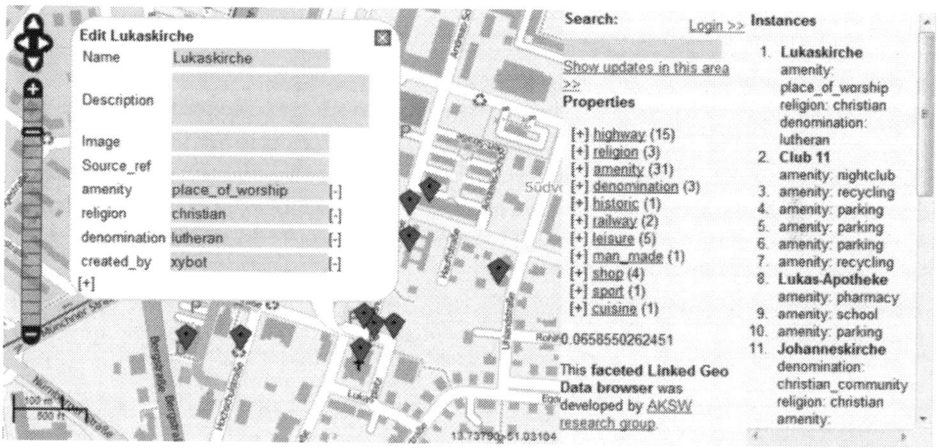

Fig. 5. Faceted Linked Geo Data Browser and Editor

If a user logs into the application by using her OSM credentials, the displayed elements can directly be edited in the map view. For this, the browser generates a dynamic form based on existing properties. The form also allows to add arbitrary additional properties. In order to encourage reuse of both properties and property values, the editor performs a type-ahead search for existing properties and property values and ranks them according to the usage frequency. When changes are made, these are stored locally and propagated to the main OSM database by using the OSM API.

Performing the facet analysis naively, i.e. counting properties and property values for a certain region based on longitude and latitude, is extremely slow. This is due to the fact that the database can only use either the longitude or the latitude index. Combining both - longitude and latitude - in one index is also impossible, since, given a certain latitude region, only elements in a relatively small longitude region are sought for.

A possible solution for this indexing problem is to combine longitude and latitude into one binary value, which can be efficiently indexed. The challenge is

[8] Available online at: `http://linkedgeodata.org/browser`

to find a compound of longitude and latitude, which preserves closeness. This is possible by segmenting the world into a raster of, for example, 2^{32} tiles, whose x/y coordinates can be interleaved into a 32-bit binary value[9]. The resulting tiles are squares with an edge length of about 600m, which is sufficient for most use cases[10].

The 32-bit tile address for a given longitude and latitude value can be efficiently computed by the DBMS using the following formula:

```
(CONV(BIN(FLOOR(0.5 + 65535*(longitude+180)/360)), 4, 10)<<1)
    | CONV(BIN(FLOOR(0.5 + 65535*(latitude+90)/180)), 4, 10)
```

In this formula "`<<1`" symbolizes a bit-shift by one digit to the left, "`|`" is the bitwise "OR" and `CONV` converts the first argument from number base given as second argument to the number base given as third argument.

After elements are associated with the tiles they are located on and after tiles are indexed by the DBMS, elements located on a certain tile can be fairly efficiently retrieved. If the user browses to a certain area, the application has to determine all the tiles encircled by that area. Since co-located tiles are assigned to adjacent tile numbers, a certain area usually consists of a small number of tile ranges, which can be efficiently processed by the DBMS.

Even these indexing optimizations were not yet sufficient to obtain acceptable response times for the faceted browser. In order to further increase the querying performance, we precomputed the counts for all properties on all tiles, as well as the counts of all property values for a set of predefined properties of which we know that they have only a limited number of values. We did that not only for the highest zoom level, but for each zoom level which users are able to select. The lower the zoom level, the more the number of tiles reduces and the faster corresponding property and property value count aggregates can be computed.

7 Conclusions, Related and Future Work

7.1 Conclusions

The transformation and publication of the OpenStreetMap data according to the Linked Data principles adds a new dimension to the Data Web: spatial data can be retrieved and interlinked on an unprecedented level of granularity. This enhancement enables a variety of new Linked Data applications such as geo-data syndication or semantic-spatial searches. The dynamic of the OpenStreetMap project will ensure a steady growth of the dataset. Furthermore, we established mappings with DBpedia as the central interlinking hub in the Web of Data. We also presented an efficient browser and editor for semantically enriched geo-data.

---

[9] This is also discussed on `http://wiki.openstreetmap.org/wiki/Quadtile`

[10] In fact, the precision can be increased arbitrarily by using simply a larger number of tiles.

7.2 Related Work

The two main areas of relevant related work concern 1.) techniques for converting relational databases to Semantic Web standard formats and 2.) ontology matching.

There is a large body of work dedicated to converting relational databases to RDF and OWL. The W3C RDB2RDF incubator group, which we are participating in, has the aim to classify and standardize such approaches. For a comprehensive overview, we refer to `http://esw.w3.org/topic/Rdb2RdfXG/StateOfTheArt` or the latest survey of the incubator group. For the article, we restrict ourselves to naming a few relevant tools in this area: Tirmizi [11] has a formal system to capture all information contained in a database based on the idea that all domain semantics are already contained in it. i [9], DB2OWL [6], and RDBToOnto [4] use less complete extraction rules and partially allow to refine the resulting knowledge base. In particular, DB2OWL allows to align the knowledge base to a reference ontology. For LinkedGeoData, we decided to use Triplify. Its use requires manual effort to write SQL mapping queries, but it is very light-weidth, easy to use, and most importantly sufficiently efficient to handle the large volumes of data in OpenStreetMap.

Regarding ontology mapping, there have been several decades of research starting with the integration of different database schemata. Tools like COMA [5] provide rich support for various matching operations between data bases as well as between RDF knowledge bases. For this article, we can limit ourselves to instance matching, since our main goal is to match specific points of interests in different knowledge bases. While there has been work on spatial matching methods, our experiments indicated that it is difficult to apply them automatically due to efficiency issues and the specifics of the involved knowledge bases. SILK [12] is a framework aiming to overcome this problem, but currently lacks support for spatial matching features.

[3] describes a semantic approach for matching export schemas of geographical database Web services, based on the use of a small set of typical instances. The paper also contains an extensive experiment, carried out within the context of two gazetteers, Geonames and the ADL gazetteer, to illustrate the approach. [10] describes an approach integrating geo data from multiple sources, which also incorporates a temporal dimension.

7.3 Future Work

Regarding the mapping approach described in Section 5, we aim to extend it in three different directions: 1.) We intend to interlink LinkedGeoData with further geographic knowledge bases. For instance for Geonames[11], one of the benefits will be that the tagging structures in OpenStreetMap will be complemented by the hierarchical structural features in Geonames. 2.) We may integrate the efficient matching methods we have used in ontology matching tools like SILK.

[11] `http://geonames.org`

3.) We intend to use machine learning techniques to facilitate proper choices of parameters and threshold values in the matching method.

In general, we plan to build a community around LinkedGeoData and encourage people to use the data provided by OpenStreetMap in novel ways through our interfaces, SPARQL endpoint, and Linked Data.

References

1. Auer, S., Bizer, C., Kobilarov, G., Lehmann, J., Ives, Z.: Dbpedia: A nucleus for a web of open data. In: Aberer, K., Choi, K.-S., Noy, N., Allemang, D., Lee, K.-I., Nixon, L.J.B., Golbeck, J., Mika, P., Maynard, D., Mizoguchi, R., Schreiber, G., Cudré-Mauroux, P. (eds.) ASWC 2007 and ISWC 2007. LNCS, vol. 4825, pp. 722–735. Springer, Heidelberg (2007)
2. Auer, S., Dietzold, S., Lehmann, J., Hellmann, S., Aumueller, D.: Triplify - lightweight linked data publication from relational databases. In: Proceedings of the 17th International Conference on World Wide Web, WWW 2009, Madrid, Spain, April 20-24, pp. 621–630 (2009)
3. Brauner, D.F., Intrator, C., Freitas, J.C., Casanova, M.A.: An instance-based approach for matching export schemas of geographical database Web services. In: Proc. of the IX Brazilian Symp. on GeoInformatics (GEOINFO), pp. 109–120 (2007)
4. Cerbah, F.: Learning highly structured semantic repositories from relational databases. In: Bechhofer, S., Hauswirth, M., Hoffmann, J., Koubarakis, M. (eds.) ESWC 2008. LNCS, vol. 5021, pp. 777–781. Springer, Heidelberg (2008)
5. Do, H.H., Rahm, E.: COMA - A system for flexible combination of schema matching approaches. In: VLDB, pp. 610–621. Morgan Kaufmann, San Francisco (2002)
6. Ghawi, R., Cullot, N.: Database-to-ontology mapping generation for semantic interoperability. In: Third International Workshop on Database Interoperability (InterDB 2007), held in conjunction with VLDB 2007 (2007)
7. Lehmann, J.: DL-Learner: Learning concepts in description logics. Journal of Machine Learning Research (to appear 2009)
8. Lehmann, J., Hitzler, P.: A refinement operator based learning algorithm for the ALC description logic. In: Blockeel, H., Ramon, J., Shavlik, J., Tadepalli, P. (eds.) ILP 2007. LNCS (LNAI), vol. 4894, pp. 147–160. Springer, Heidelberg (2008)
9. Li, M., Du, X., Wang, S.: A semi-automatic ontology acquisition method for the semantic web. In: Fan, W., Wu, Z., Yang, J. (eds.) WAIM 2005. LNCS, vol. 3739, pp. 209–220. Springer, Heidelberg (2005)
10. Manguinhas, H., Martins, B., Borbinha, J.L.: A geo-temporal web gazetteer integrating data from multiple sources. In: ICDIM, pp. 146–153. IEEE, Los Alamitos (2008)
11. Tirmizi, S.H., Sequeda, J., Miranker, D.P.: Translating SQL applications to the semantic web. In: Bhowmick, S.S., Küng, J., Wagner, R. (eds.) DEXA 2008. LNCS, vol. 5181, pp. 450–464. Springer, Heidelberg (2008)
12. Volz, J., Bizer, C., Gaedke, M., Kobilarov, G.: Silk–a link discovery framework for the web of data. In: Proceedings of the 2nd Workshop about Linked Data on the Web, LDOW 2009 (2009)

Enrichment and Ranking of the YouTube Tag Space and Integration with the Linked Data Cloud

Smitashree Choudhury[1], John G. Breslin[1,2], and Alexandre Passant[1]

[1] DERI, National University of Ireland, Galway, Ireland
[2] School of Engineering and Informatics, National University of Ireland, Galway, Ireland
{smitashree.choudhury,john.breslin,alexandre.passant}@deri.org

Abstract. The increase of personal digital cameras with video functionality and video-enabled camera phones has increased the amount of user-generated videos on the Web. People are spending more and more time viewing online videos as a major source of entertainment and "infotainment". Social websites allow users to assign shared free-form tags to user-generated multimedia resources, thus generating annotations for objects with a minimum amount of effort. Tagging allows communities to organise their multimedia items into browseable sets, but these tags may be poorly chosen and related tags may be omitted. Current techniques to retrieve, integrate and present this media to users are deficient and could do with improvement. In this paper, we describe a framework for semantic enrichment, ranking and integration of web video tags using Semantic Web technologies. Semantic enrichment of folksonomies can bridge the gap between the uncontrolled and flat structures typically found in user-generated content and structures provided by the Semantic Web. The enhancement of tag spaces with semantics has been accomplished through two major tasks: (1) a tag space expansion and ranking step; and (2) through concept matching and integration with the Linked Data cloud. We have explored social, temporal and spatial contexts to enrich and extend the existing tag space. The resulting semantic tag space is modelled via a local graph based on co-occurrence distances for ranking. A ranked tag list is mapped and integrated with the Linked Data cloud through the DBpedia resource repository. Multi-dimensional context filtering for tag expansion means that tag ranking is much easier and it provides less ambiguous tag to concept matching.

1 Introduction

A key feature of the Social Web is the change in the role of a user from simply being a consumer of media: they are now content creators. It is not just textual content that can be shared, annotated or discussed, but any multimedia content such as pictures, videos, or even presentation slides. With tools like iMovie for video creation and digital cameras with built-in WiFi for instant uploads, web users can easily add their multimedia content to social media websites. With this ease of creation, there is an ever increasing amount of multimedia in various formats becoming available on the

A. Bernstein et al. (Eds.): ISWC 2009, LNCS 5823, pp. 747–762, 2009.

Social Web. Recently YouTube[1] reported that 20 hours of video were being uploaded per minute, which amounts to 28,800 hours of video uploaded in one day to that site.

All of these videos are being annotated by users with free unstructured keywords. Some video sharing sites also permit sharing and collaboration in the tagging process by allowing other users to tag a video, thereby giving a sense of collective intelligence. Current techniques to retrieve, integrate and present this tagged media to users are deficient and could certainly benefit from improvement. Semantic technologies make it possible to give richer descriptions to media, facilitating the process of locating, combining diverse media from various sources and personalising content recommendation.

A major problem is that the textual annotations vary in terms of quality and their ability to describe the video content. The tags include not only the content but also information about the user, their subjective opinion of the content, misspelling, and emerging co-joined tags. Given the ambiguity, subjectivity and noise in tags, one of the fundamental problems is to learn the relevance of the tag corresponding to the content. Unstructured and informal descriptions rule out any kind of interoperability of the resources across similar and related content. An attempt to give a well-defined structure and to formalise the tag space for user-generated videos will be the first step towards a desired solution. Moreover, we believe that this could be an efficient way to add relevant semantics to videos on the Web, in combination with existing initiatives, such as the MPEG7 [10] standard and its associated RDF(S)/OWL mappings (that can be used to represent image regions and add particular annotations about them) and the current tasks of the W3C Media Annotation Working Group, as defined in their document on "web video"[2].

In this study, we have designed a framework to explore the contribution of various types of contextual data to the tag space and their relevance in ranking. Information embedded in video contexts such as social, spatial and temporal contexts are a good source for video tag suggestions. Enriching tags, though helpful for more reliable descriptions of content, can at the same time add noise to the resulting video metadata. In order to attenuate the noise from the tag space, we need to rank the tags. Studies have been carried out recently on the relevance of ranking tags for documents and images, but to our knowledge there is no study yet to rank tags for user videos on the Web. After tag ranking, we consider linking this enriched data to the open Web following the principles envisioned in the Linking Open Data (LOD) initiative [15]. This rich data cloud gives each object and concept a unique identifier (URI) which is referenceable and linkable on the Web, such that they can make reference to each other irrespective of the vocabulary used. Three video resources may be described with three different tags "new york city", "nyc" and "big apple" by three different users, but the intended meaning is the same, i.e. the city of New York. When we look for "new york city", we may not find the other two even though both of them are describing the same content. If we can disambiguate these three and link to one identifier, this makes retrieval much easier. To address this problem a solution is to disambiguate each tag to an ontological concept identified by its own URI [27]. Since tags are simple uncontrolled keywords, they inherit the same IR-related

---

[1] http://www.youtube.com/

[2] http://www.w3.org/2008/WebVideo/Annotations/

problems of synonyms and polysemy, as described in [26] and [27]. A robust disambiguation method is needed for direct tag-to -concept matching. In the present study, we have not described the tag-to-concept matching module in much detail, but rather we have described the applicability of tag-to-concept matching and the benefits of interlinking to the structured world. The final output of the framework is a set of RDF triples describing the video and its contextual metadata with the support of a video model and various existing lightweight ontologies such as Dublin Core, SIOC, MOAT, FOAF, etc.

The rest of the paper is organised as follows. Section 2 describes various related studies in tag suggestion ranking and semantic integration. Section 3 describes the system architecture and its modules. Section 4 describes the integration of the enriched video tag space and metadata into the Linked Data Cloud. This is followed by experiments and evaluation in section 5, after which we will conclude with some remarks and future directions in the final section.

2 Related Tag Studies

Strongly descriptive and unambiguous tags are the first step toward more effective retrieval and interoperability across Social Web data sources. Much research has been carried out recently in refining user-generated tags to make them more semantically interoperable. Numerous studies [1], [8] have been carried out to suggest relevant tags for media documents based on supervised learning techniques, where the models are built for specific domains and co-relations of low-level features to tags are learned. However, due to the numerous amounts of visual variations, many efforts are far from satisfactory and moreover are restricted to a small domain of applications. Manual and collaborative tagging is one of the alternatives adopted by most popular media sharing sites such as Flickr and YouTube. This of course adds other problems as described in the first section of this paper. These problems have led to many studies in the field of folksonomies, user-tagging behaviours, semantic tagging and tag refinement. We will describe some of the studies relevant to the present study and outline how they differ from the present study. These studies mainly come under three different groupings: tag suggestion, tag ranking and tag semantics.

2.1 Tag Suggestion

In the field of tag suggestion, different but simultaneous approaches have been pursued by researchers to improve both automatic annotations and multimedia annotation quality. Researchers from the machine vision community are now focusing on gathering contextual data together with content processing to bridge the semantic gap [19], while other researchers [17] are purely focusing on social data combined with a knowledge base to augment media with social annotations[3]. Though both approaches have their valid points, harvesting social data is not only inexpensive but can contribute significantly to bootstrapping the content understanding process.

The informal nature of tagging means that semantic information cannot be directly inferred from an annotation, as any user can tag any resource with whatever strings

[3] http://acronym.deri.org/

they wish. However, studying the collective tagging behaviour of a large number of users allows emergent semantics to be derived [14]. Through a combination of such mass collaborative 'structural' semantics (via tags, geo-temporal information, ratings, etc.) and extracted multimedia 'content' semantics (which can be used for clustering purposes, e.g. image similarities or musical patterns), relevant annotations can be suggested to users when they contribute multimedia content to a community site by comparing new items with related semantic items in one's implicit and explicit networks.

2.2 Tag Ranking

Research into tag ranking began with studies [1] and [8] where ranks were assigned with respect to visual content as the result of supervised machine learning approaches, where models map relationships between visual features and semantic concepts. Uncontrolled visual content where there are a vast number of concepts involved makes the above approach less effective, and led to another approach for tag ranking which followed usage statistics by studying tag co-occurrence over a large corpus. Sigurbjörnsson et al. ranked Flickr tags [3] by means of co-occurring tags. Hotho et al. [11] suggested Folkrank for community detection in Delicious tags. Relevance ranking by means of frequency counting for "neighbouring" images (in terms of visual similarity) was conducted by Li et al. [4], where they selected common tags from neighbouring images for higher ranking. A recent study on tag ranking by Liu et al [9] proposed a tag rank for Flickr images by means of a random walk. Our ranking module is in the same general domain with the exception that we enriched our tag space before ranking to tackle the problem of tag sparsity in YouTube videos.

2.3 Tag Semantics

Studies in tag semantics fall into two broad categories: a corpus-based or statistical approach and a knowledge-based approach. Initial studies [2] on folksonomies explored means of leveraging the statistical co-occurrence relations between tags to define their semantics, and knowledge-based approaches refer to external knowledge sources such as thesaurus and ontologies to define the tag meaning [20]. Rattenbury et al. [5] explored tag-usage statistics to determine the events and place semantics from Flickr tags using burst detection analysis. Research in [6] used online ontologies and WordNet [17] to map tags for Flickr tags to concepts. Simon et al. [7] used Wikipedia categories and template structures to classify Flickr tags and these were mapped to WordNet concepts.

Other works on the topic include studies regarding the emergent semantics of tagging systems. Among others, [22] used an approach based on related co-occurrences of tags to extract hierarchical relationships between concepts, modeled in RDFS, while [23] defined a socially-aware approach for building ontologies by combining social network analysis and clustering algorithms based on folksonomies. More recently, FolksOntology [29] and FLOR [28] also provide frameworks for automated semantic enrichment of tagged data.

Finally, various models have been developed to capture the semantics of tagging systems using lightweight ontologies, such as SCOT [25], MOAT [24] or CommonTag[4].

3 System Architecture

In this section, we will give a detailed description of the tag expansion and ranking system that we have built. We begin with a general overview of the different modules, followed by an explanation of the tag filtering and expansion step, then we will describe the tag graph creation process, and finally we will detail the tag ranking methods by means of spreading activation over the tag graph.

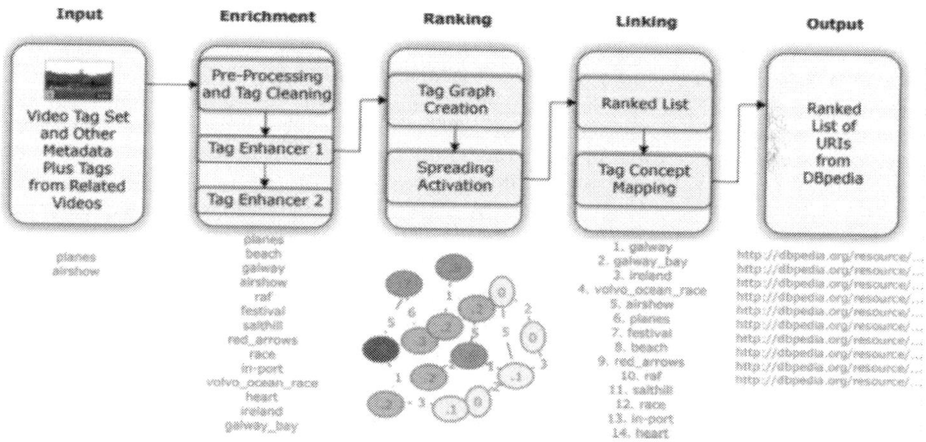

Fig. 1. Work flow of the tag enrichment, ranking and linking processes

The goal of this work is to enrich the user-generated tag space, and to rank and interlink the tags to DBpedia concepts for greater integration with other datasets. DBpedia is considered as a central node in the LOD cloud (the DBpedia nucleus), and linking to DBpedia also allows one to reach other datasets, thanks to the network effect of this project. There are three main modules in the system, each of which consists of many sub-modules. Figure 1 shows the normal work flow of the system: (1) context analysis and tag expansion; (2) tag ranking, and (3) concept mapping and linking to the Semantic Web.

3.1 Context Analysis and Tag Expansion

In this section, we describe our first module that implements the tag expansion strategy. Because of the sparseness of video tags, we need to expand the tag base with various other contextual sources such as social, temporal and geographical contexts.

---

[4] http://commontag.org/

We will begin with a description of our pre-processing step. User-generated tags consist of three broad categories of tags: functional tags (meaningful and mostly single keywords), noisy tags, and compound or emerging tags. Compound or emerging tags are those tags consisting of two or more keywords without any white space such as "friendsoftheearth", "iswc09" (used for friends_of_the_earth, ISWC_2009 respectively). There are other categories of tags which are subjective or judgmental tags, as studied by [21] and these reflect a user's view point rather than the video content, for example, "funny", "wonderful", "watch this", etc. In our system, we excluded tags with less than three characters, subjective tags, non-English tags, and tags describing usernames for the purpose of this study. However, there can be some difficulty with compound tags as these tags are not common words, and further work must be performed to identify meaningful tags from these composite sets. The textual content from the video title and descriptions are subject to the same kind of pre-processing described above, including stopword removal.

3.2 Semantic Tag Space Enrichment

In this module, we work on the tag space enrichment process where the sparse video tag space is enriched with multiple contextual sources. The sources are of various natures and exist in the context of the video in question:

1. Other textual contexts such as title and description of the video
2. Geospatial contexts, such as the place where the video has been recorded (latitude and longitude coordinates available through the YouTube API)
3. Temporal contexts, e.g. recording time
4. Social contexts, e.g. groups or playlists that include the tagged video as an item
5. Related videos, i.e. videos sharing some specific characteristics such as tags or time and space
6. User contexts, such as the type of user that includes the video in their bookmarks or favorites list
7. Context from the Web itself, i.e. other websites delivering information about these tags

We have considered the first five contextual sources to increase the tag space, and omitted the last two, which may be the subject of another study. Textual contexts such as video titles, descriptions and categories are used to rank the tag weights and sometimes add extra tags that are missing in the tag space itself. To avoid noise propagation, weights are added to different sources.

A *playlist* "extreme sailing" can include videos whose tag space is more compact and clustered from the general "sailing" tag space. Playlist and group structures where videos and users are members can propagate tags to the individual video items [18].

"Related Videos" in YouTube are those videos that are considered similar to the original video in some aspects. YouTube provides a related video feed for each video. It is not known on what basis YouTube ranks the relatedness of a video, and sometimes the results are unexpected. Moreover, YouTube feeds cannot be filtered with complex queries such as "*give me the videos related to the query where relatedness is based on a shared tag space, should be from the same place, and must be within a*

time range, but not from the same user" without a lot of work, so we decided to generate a list of related videos for each video from our own data set. The related videos are judged based on mutual content information in tag space. We adopted a space and time normalisation criteria in selecting the related videos. To explain this, if videos share a time and place value with the original video, they are ranked higher in relatedness. The intuitive explanation for this is that videos from the same place and same time are more likely to capture the same events and content [5]. Videos from Galway (a geographical area) from 30-05-2009 to 01-06-2009 are more likely to contain the events "Salthill air show" and "Volvo Ocean Race", so accordingly there is a definitive pattern of high-frequency tags such as "Salthill", "air show", "red arrows", "beach", "Volvo Ocean Race", etc.

Spatial context is information regarding the geolocation where the video has been recorded or the place that the content describes, which can be extracted from the geo coordinates. *Temporal context* is the time of video recording (not publishing). Table 1 shows the comparative tag spaces of related videos with and without time and space filters. This contextual information not only expands the initial tag space, but it also adds weights to the tags. The intermediate list of tags is the input for the final phase of tag expansion and recommendation based on tag co-occurrence. Table 2 shows the first phase of tag expansion.

Table 1. Comparative tag spaces of related videos with and without filters

Original Video Tags	*Related Video Tags Without Filter*
Planes, Air show, Galway	Planes, Air show, red_arrows, Volvo Ocean Race, Galway, Ireland, Panasonic, NV-GS330, NV, GS, 330, NV-GS
Original Video Tags	*Related Video Tags With Time and Space Filters*
"MOV04687", "Galway", "Ireland", "air show"	"heart", "festival", "raf", "in-port", "race", "beach", "galway_bay", "salthill", "Volvo Ocean Race", "red arrows", "beach"

Table 2. Multi-contextual tag expansion

Title	Planes Salthill Galway	"Planes", "beach", "Galway", "Air show", "raf", "festival", "salthill", "red arrows", "race", "in-port", "volvo_ocean_race", "heart", "Ireland", "galway_bay"
Description	same planes!!	
Tags	Planes, air show	
Related Videos	"heart", "festival", "raf", "in-port", "race", "beach", " galway_bay," "salthill", "Volvo Ocean Race", "red arrows", "beach"	
Geolocation	Galway, Ireland	

Tag co-occurrence is one of the key enablers towards creating a more comprehensive semantically-related tag space for the video. Co-occurrence between two tags occurs when both the tags are used to label the same resource. We opted for a second phase of tag expansion based on tag co-occurrence if the tag set (N < 5) is less than five after the first stage of expansion. Raw co-occurrence gives a weak relationship as there may be many occurrences of less descriptive tags such as "news" for all news category videos. Therefore, it is natural to normalise the count in order to reduce the bias.

Intuitively, one resource will not be tagged with equivalent tags but rather with related tags in which case the distance between them will not be symmetrical: $d(t_1, t_2) \neq d(t_2, t_1)$. We have adopted an asymmetric approach of measuring co-occurrence using the equation:

$$cd(t_1, t_2) = |t_1 \cap t_2| / |t_1| \quad (1)$$

It captures how often tag 2 (t_2) occurs with tag 1 (t_1) given the total number of occurrences. It gives a more diverse tag space when compared to a symmetric co-occurrence coefficient.

When we get a list of co-occurring tags for each of the tags from the list above we need a mechanism to aggregate them so that we can prepare the final list. Aggregation can be a simple voting mechanism where frequent candidate tags are ranked higher.

3.3 Tag Ranking

In this section we describe the detection of ranked nodes in the graph, beginning with an overview of the tag graph creation, and then describing spreading activation over the graph to rank the nodes.

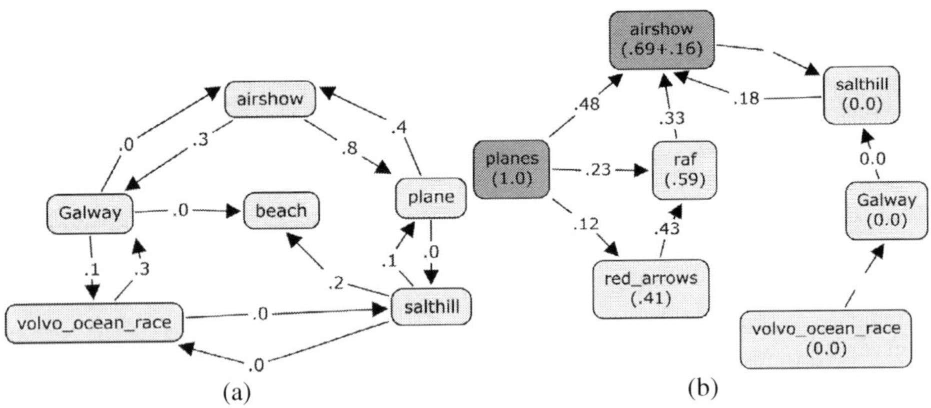

Fig. 2. (a) Tag graph for a video and (b) spreading activation from the node "planes"

Given a video $v \in V$ and an extended tag set $ET = \{t_1, t_2, \ldots t_n\}$, we create a local graph of tags. The tag graph is a directed weighted graph with tags as nodes and the links between nodes are weighted edges. The edge weight is an asymmetric correlation based on their co-occurrence. If the correlation value is less than a threshold (τ) the tags are not connected. The co-occurrence relation is calculated as per equation 1. Figure 2 (a) shows a tag graph for the video.

We have used the tag list as a loose semantic network based on their correlations, and processed the network using spreading activation. Spreading activation as an information processing algorithm is based on the theory of cognitive science [16] and human memory. It works on a semantic network of nodes and links where links are connections between nodes based on certain relationship. Information processing starts when the activated node starts spreading its energy towards the neighboring nodes. At the end of processing, all nodes will have some activation value which was contributed through its relation with neighboring nodes.

In our work, we have used the video tags as nodes of the network and their co-occurring relations as the weighted link between nodes. The relationship between tags can be semantic as in WordNet and other ontologies, or it can be based on co-occurrence patterns as observed in web data. We have opted for co-occurrence relations to connect the tags as a network. The activation process includes the following steps:

1. The graph nodes are assigned an initial value of 0 except for the firing node which has an activation value of 1.0.
2. It spreads its activation to all nodes in its immediate neighborhood that are connected to the source node.
3. Output activation is a function of (*initial activation* + (*initial activation* * *edge strength*) * *d*), where d is a decay factor set up experimentally. In our case it is .85.
4. If the node value exceeds a threshold, we fire the node again.
5. The node activation value is the weighted sum of its contributing nodes.
6. Each node activates once in the process.

Following the above steps, we rank our extended tag list turned local graph. Experimentally we set up the decay factor to be .85 and the iteration was 1 for all the nodes as this graph is not a complex nested graph. The activation starts with the top node in the tag list. All the nodes except the firing nodes are assigned a value of 0. Once the energy propagation starts, the node spreads its energy to the connected nodes and the receiving amount of energy is a function of relationship strength and decaying energy factor. Figure 2 (b) shows the activation process starting with the node "plane" and spreading to three nodes "air show", "raf", "red arrows" (this node again spreads and contributes to the "air show" node).

3.4 Linked Data Creation

Once the tags are cleaned and ranked, they can then be connected to other similar and related resources. For example, there are videos of the "Volvo Ocean Race" on YouTube as well as on other media sharing sites such as Vimeo or Joost. To discover the entire spectrum we need to create a mapping mechanism from user tags to ontological resources so that they can be more connected and discoverable. This is where the Linked Open Data initiative fits in. As part of its principles, one needs to identify every entity (object, concept, event, people) with a unique web identifier or URI on the Web. Following one URI will lead to the discovery of some more related information. This simple yet powerful idea has rapidly gained momentum recently. The Linked Open Data cloud now consists of more than a hundred datasets and billions of interlinked facts as entities.

There is the question of how best to integrate user-generated videos with the existing structured LOD cloud. Automatic mapping from tags to concepts is desirable, but challenging due to multiple contexts of the concept. Studies in MOAT[5] showed that users are willing to do this manually when they realise the benefits of such an effort, for instance, if they get advanced browsing or querying features.

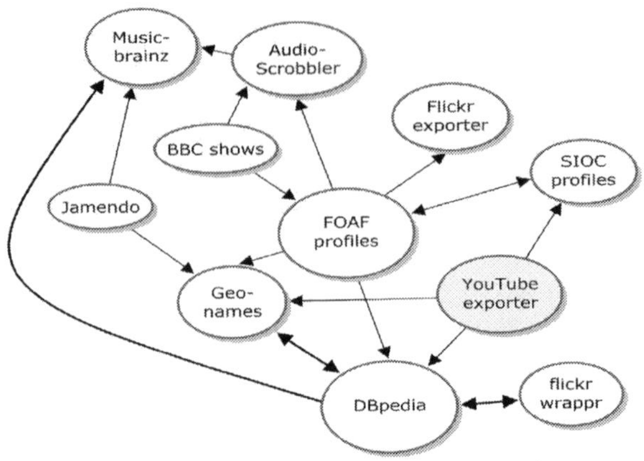

Fig. 3. Connecting to multimedia datasets already interlinked within the Linked Data cloud

In the present study, since we have a limited domain, the number of mappings is quite small so we used the semantic indexing engine Sindice[6] to query DBpedia and select the most appropriate URIs. Automatic mapping of tags to concepts is ongoing work to be reported on later. Links can also be added to user accounts (SIOC) and locations (GeoNames) obtained from the YouTube API (Figure 3).

3.5 Tags-to-Concept Mapping

Tag-to-concept matching is not yet fully implemented and will be part of our future work. As part of the experiment we have used DBpedia resources. We presume that cross-resource mapping to other sources from the LOD initiative (such as Freebase) can easily be adapted. Depending on the context, some particular datasets may also be considered, e.g. a genes database when dealing with medical videos. Here, we will briefly describe our approach for tag-to-concept matching. Once the tags are finalised, we use a two-step process for assigning concept identifiers. The tags are fed into a local WordNet module and some simple heuristics are followed:

1. If the tag matches with a WordNet noun, and if there is only one matching synset, we select the corresponding WordNet URI in DBpedia (Figure 4).

---

[5] http://www.moat-project.org/
[6] http://sindice.com/

2. If there are more than one WordNet synset, we send the tag and its context tags to a similarity module to compute the cosine similarity between the current tag context and already-existing tag URIs. (The similarity module is based on the Lucene[7] text retrieval Java library and on other work in progress).
3. For those tags that are not part of WordNet, we send them to the semantic indexing engine Sindice to look for resources. Once we get the top k URIs for the query, the user can select the URI manually or else it is fed into another disambiguation module where URIs can be contextually disambiguated (not implemented yet).

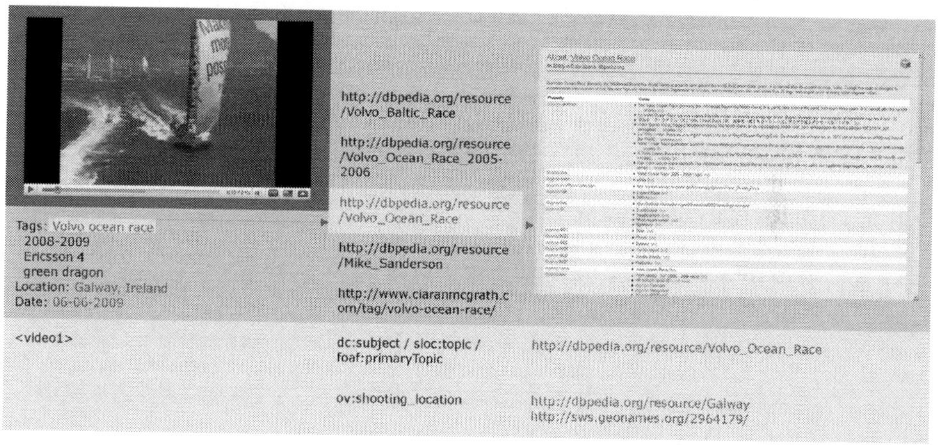

Fig. 4. Matching YouTube tags to DBpedia concepts

4 Experiments and Evaluation

4.1 Results

We have collected 3,990 YouTube videos. All video metadata including the metadata of related videos was collected through the YouTube API. We collected videos of specific categories such as "skiing", "sailing" and "cricket". The data includes video tags, dates, places (if available), titles, descriptions and group tags (if available). The total number of unique tags is more than 11,900 which includes many misspellings, number tags, co-joined tags, and subjective as well as meaningless tags. There are 2,261 distinct users in the data set. On average, one user has less than two videos. Since users tag differently depending on their background and expertise, we can assume a relatively heterogeneous tag source. We did a preliminary filtering of tags by removing stop words, tags with two characters and number tags. Though the tag list is far from clean, this reduces a lot of noise. This tag set is used for extracting the co-occurrence statistics.

[7] http://lucene.apache.org/java/docs/

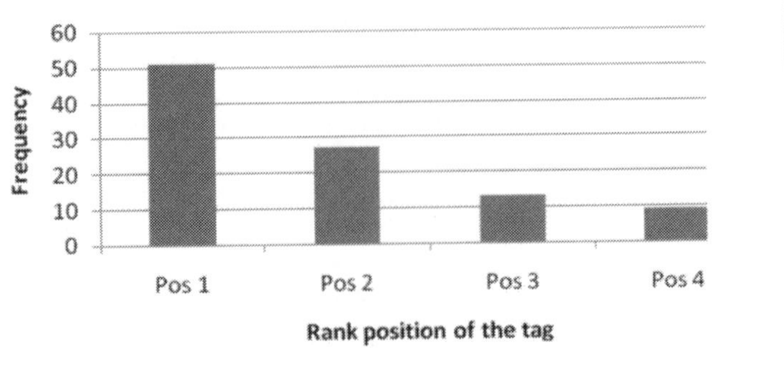

Fig. 5. The number of times the most relevant tag was suggested at different positions

We conducted a preliminary evaluation to explore the quality of our ranking method and tag enhancement. We randomly selected 100 videos from the larger set to explore potential benefits and problems. Three users familiar with the topics were asked to rank the tag lists of these videos on a scale of 1 to 4: "most relevant", "relevant", "partially relevant" and "irrelevant", according to their depicted content. We computed the amount of times the most relevant tag was ranked as "most relevant" by users, and we will now discuss the quality of our tag enhancement.

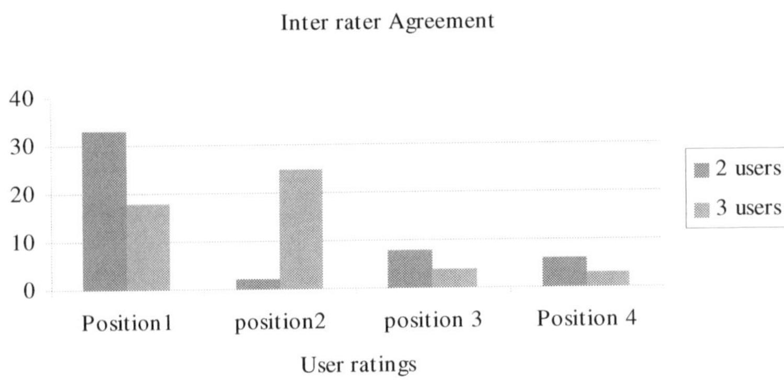

Fig. 6. Inter-rater agreement over the most relevant tag in the first four positions

Though relatively small, the user feedback gave some interesting results. Regarding inter-rater variation, we considered a final rank when a minimum of two users agreed on the same rank. For the 100 videos, Figure 5 shows that the most relevant tag came in at the top position 51 times (where a minimum of two users have agreed), whereas the top tag came second 27 times. Therefore, for almost 80% of the time, the top-ranked tag was in either of the first two positions. Figure 6 shows the inter-rater agreement over the tag relevance in the first four positions.

Similarly, we tried to explore the tag enrichment task and its effectiveness in describing the content of the video. This evaluation was conducted at two stages: before

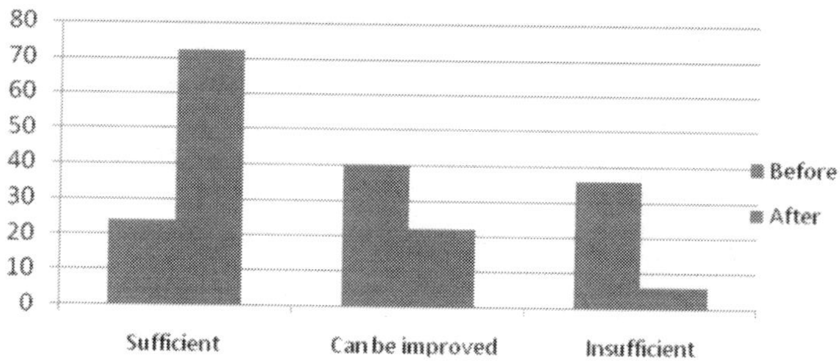

Fig. 7. Comparative evaluation of the quality of the tag space

the tag expansion and after the tag expansion. The users were asked to rank the original tag list on a three-point scale of (1) sufficient for the content, (2) okay, but can be improved, and (3) insufficient. The comparative results are in Figure 7.

The result showed that the tag enrichment process increased the content understanding considerably. Still, 28% of the videos need improvement. There may be many reasons for this such as more specific tag cleaning, insufficiency of the tag list, or perhaps due to noise propagation from the different contextual sources which were used.

In the present study, inter-rater agreement evaluation is only focused on the top-ranked tag. Out of the 51 times where the system-suggested top tag was considered most relevant by users, two users agreed 33 times and three users agreed 18 times. However, in position two, out of 27 total times, all three users agreed that the tag was most relevant 25 times. An exhaustive analysis of user agreement over the relevance of suggested tags is desirable to explore possible room for improvements.

We will conclude this section by discussing some potential benefits in applications such as (1) semantic tag-based search and retrieval, and (2) improved video categorisation.

4.2 Semantic Tag-Based Search and Retrieval

We evaluated our work in a retrieval framework and describe here two use cases of tag-based retrieval. Given a query q, the system will retrieve all the videos tagged with q but they will be ranked according to the ranked position of q in the tag space. Thus, if two videos tagged with q are retrieved, the video having q in a higher position will be ranked higher. Identifying tags with DBpedia URIs opens up many possibilities of knowledge discovery. A resource tagged with "Volvo Ocean Race" and identified with a URI such as *"http://dbpedia.org/resource/Volvo_ocean_race"* will lead us to discover information about "St._Petersburg, Russia" as it is related to the query by means of destination port. A video tag identified with *"http://dbpedia/resource/Galway"* will lead us to discover more about the culture, events and history of Galway.

4.3 Improved Video Categorisation

Categories in YouTube are selected by users when uploading videos. Sometimes, a user selects a less-relevant category for their content as it is a flat single category system and the choices are also quite limited. In practice, video content may belong to more than one category, and moreover, it may follow a hierarchal structure. Based on our enriched tag space, we can suggest a hierarchical categorisation for a video by exploiting the relations between tags. A video tagged with "news, Japan, earthquake, building, tsunami" can be categorized under News >> Natural Hazard >> Earthquake conforming to the hierarchical structure of existing ontologies such as the Large Scale Concept Ontology for Multimedia (LSCOM) [13]. In LSCOM, "earthquake is a subclass of natural hazard".

5 Conclusions

In this paper, we propose a roundtrip semantic framework which provides some steps towards solving the above problem. The key modules that have been implemented are for tag enrichment, tag ranking, concept mapping and semantic linking. Tag enrichment involves various contextual analyses of the video and the contribution of these contexts towards content understanding. Context not only includes textual descriptions but also the temporal, spatial and social contexts in which the video is being used and shared. Interplaying a combination of contexts provides for improved enrichment. The advantages of the proposed algorithm for tag enrichment are: (1) multiple sources can make the tags more reliable for content description; (2) the subjective ambiguities are reduced; (3) the method is scalable since it does not require any domain specific model training; and (4) it can evolve with tag usage.

We have also proposed a tag-ranking algorithm performed over a local tag graph by means of spreading activation. The tag graph is based on the enriched tag set and is connected by means of co-occurrence strength. Spreading activation helps to activate the focused nodes and reduces the strength of noisy nodes. We also described the tag-to-resource mapping (with DBpedia as our semantic repository) and outlined how Linked Data principles can aid with linking to user-generated video content. Successful linking to DBpedia web identifiers can harness other related data sources from the LOD cloud and then enrich information discovery from videos.

Our future work includes a complete concept-to-URI matching mechanism and an explorative evaluation of the approach with more users.

Acknowledgements

This work was supported by Science Foundation Ireland under grant number SFI/08/CE/I1380 (Líon 2).

References

1. Barnard, K., Duygulu, P., Forsyth, D., Freitas, D.N., Blei, D.M., Jordan, M.I.: Matching words and pictures. JMLR, 1107–1135 (2003)
2. Begelman, G., Keller, P., Smadja, F.: Automated tag clustering: Improving search and exploration in the tag space. In: WWW Collaborative Web Tagging Workshop (2006)
3. Sigurbjörnsson, B., van Zwol, R.: Flickr tag recommendation based on collective knowledge. In: Proc. of the International World Wide Web Conference (2008)

4. Li, X.R., Snoek, C.G.M., Worring, M.: Learning tag relevance by neighbour voting for social image retrieval. In: Proc. of the ACM International Conference on Multimedia Information Retrieval (2008)
5. Rattenbury, T., Good, N., Naaman, M.: Towards automatic extraction of event and place semantics from Flickr tags. In: Proc. of SIGIR, pp. 103–110 (2007)
6. Angeletou, S., Sabou, M., Motta, E.: Semantically enriching folksonomies with FLOR. In: Proc. of the 5th ESWC: CISWeb, Tenerife, Spain (2008)
7. Overell, S., Sigurbjörnsson, B., van Zwol, R.: Classifying tags using open content resources. In: Proceedings of the Second ACM International Conference on Web Search and Data Mining (WSDM 2009), ACM, Barcelona, Spain (2009)
8. Li, J., Wang, J.Z.: Real-time computerized annotation of pictures. IEEE Transactions on Pattern Analysis and Machine Intelligence (2008)
9. Dong, L., Xian-Seng, H., Yang, L.: Tag ranking. In: Proc. of the International World Wide Web Conference (2009)
10. Hunter, J.: Adding multimedia to the Semantic Web – Building an MPEG-7 ontology. In: 1st International Semantic Web Working Symposium (SWWS 2001), California, USA, pp. 261–281 (2001)
11. Hotho, A., Jäschke, R., Schmitz, C., Stumme, G.: Information retrieval in folksonomies: search and ranking. In: Sure, Y., Domingue, J. (eds.) ESWC 2006. LNCS, vol. 4011, pp. 411–426. Springer, Heidelberg (2006)
12. Wu, X., Zhang, L., Yu, Y.: Exploring social annotations for the Semantic Web. In: Proc. of the International World Wide Web Conference (2006)
13. Naphade, M.: Large-scale concept ontology for multimedia. IEEE Multimedia 13(3), 86–91 (2006)
14. Mika, P.: Ontologies are us: A unified model of social networks and semantics. In: Gil, Y., Motta, E., Benjamins, V.R., Musen, M.A. (eds.) ISWC 2005. LNCS, vol. 3729, pp. 522–536. Springer, Heidelberg (2005)
15. Bizer, C., Cyganiak, R., Heath, T.: How to publish Linked Data on the Web (2007), http://www4.wiwiss.fu-berlin.de/ bizer/pub/ LinkedDataTutorial/
16. Quillian, M.R.: Semantic memory. In: Minsky, M. (ed.) Semantic Information Processing, pp. 227–270. MIT Press, Cambridge (1968)
17. WordNet, http://wordnet.princeton.edu/ (last accessed March 12, 2009)
18. Celma, O., Ramírez, M., Herrera, P.: Foafing the music: A music recommendation system based on RSS feeds and user preferences. In: Proc. of the 6th International Conference on Music Information Retrieval (ISMIR 2005), London, UK, pp. 464–457 (2005)
19. Oge, M., Lux, M.: An exploratory study on joint analysis of visual classification in narrow domains and the discriminative power of tags. In: Proc. of the 2nd ACM workshop on Multimedia semantics, Vancouver, British Columbia, Canada (2008)
20. Schmitz, P.: Inducing ontology from Flickr tags. In: Proc. of the Collaborative Web Tagging Workshop at the International World Wide Web Conference, Edinburgh, UK (2006)
21. Golder, S., Huberman, B.A.: The Structure of Collaborative Tagging Systems. Journal of Information Sciences 32(2), 198–208 (2006)
22. Halpin, H., Robu, V., Shepard, H.: The Dynamics and Semantics of Collaborative Tagging. In: Proceedings of the 1st Semantic Authoring and Annotation Workshop (SAAW 2006), The 5th International Semantic Web Conference (ISWC 2006), Athens, Georgia, USA (2006)

23. Mika, P.: Ontologies are us: A unified model of social networks and semantics. In: Gil, Y., Motta, E., Benjamins, V.R., Musen, M.A. (eds.) ISWC 2005. LNCS, vol. 3729, pp. 522–536. Springer, Heidelberg (2005)
24. Passant, A., Laublet, P.: Meaning Of A Tag: A Collaborative Approach to Bridge the Gap Between Tagging and Linked Data. In: Proceedings of the Linked Data on the Web Workshop (LDOW 2008) at the 17th International Conference on the World Wide Web (WWW 2008), Beijing, China (April 2008)
25. Kim, H.L., Yang, S.K., Breslin, J.G., Kim, H.G.: Simple Algorithms for Representing Tag Frequencies in the SCOT Exporter. In: Proceedings of the IEEE/WIC/ACM International Conference on Intelligent Agent Technology, Fremont, California, USA, pp. 536–539 (2007)
26. Mathes, A.: Folksonomies: Cooperative Classification and Communication Through Shared Metadata. Computer Mediated Communication, LIS590CMC, Graduate School of Library and Information Science, University of Illinois Urbana-Champaign (2004)
27. Passant, A.: Using Ontologies to Strengthen Folksonomies and Enrich Information Retrieval in Weblogs: Theoretical background and corporate use-case. In: ICWSM 2007 (2007)
28. Angeletou, S.: Semantic Enrichment of Folksonomy Tagspaces. In: Sheth, A.P., Staab, S., Dean, M., Paolucci, M., Maynard, D., Finin, T., Thirunarayan, K. (eds.) ISWC 2008. LNCS, vol. 5318, pp. 889–894. Springer, Heidelberg (2008)
29. Van Damme, C., Hepp, M., Siorpaes, K.: Folksontology: An integrated approach for turning folksonomies into ontologies. In: Bridging the Gap between Semantic Web and Web 2.0 (SemNet 2007), pp. 57–70 (2007)

Produce and Consume Linked Data with Drupal!*

Stéphane Corlosquet[1,2], Renaud Delbru[1], Tim Clark[2,3],
Axel Polleres[1], and Stefan Decker[1]

[1] Digital Enterprise Research Institute, National University of Ireland, Galway, Ireland
[2] Massachusetts General Hospital, Department of Neurology, Boston, MA, USA
[3] Harvard Medical School, Department of Neurology, Boston, MA, USA

Abstract. Currently a large number of Web sites are driven by Content Management Systems (CMS) which manage textual and multimedia content but also - inherently - carry valuable information about a site's structure and content model. Exposing this structured information to the Web of Data has so far required considerable expertise in RDF and OWL modelling and additional programming effort. In this paper we tackle one of the most popular CMS: Drupal. We enable site administrators to export their site content model and data to the Web of Data without requiring extensive knowledge on Semantic Web technologies. Our modules create RDFa annotations and – optionally – a SPARQL endpoint for any Drupal site out of the box. Likewise, we add the means to map the site data to existing ontologies on the Web with a search interface to find commonly used ontology terms. We also allow a Drupal site administrator to include existing RDF data from remote SPARQL endpoints on the Web in the site. When brought together, these features allow networked RDF Drupal sites that reuse and enrich Linked Data. We finally discuss the adoption of our modules and report on a use case in the biomedical field and the current status of its deployment.

1 Introduction

Since the late 90ies and early 2000s a paradigm shift has taken place in Web publishing towards a separation of data (content) and structure (mainly layout). The first ideas to have the data represented in a way which allows its reuse in various ways and not only HTML, emerged in parallel in various systems such as Typo3 (1997), Plone (1999), WebML (2000) [6]. These open source systems and their commercial counterparts are typically subsumed under the term *Content Management Systems* (CMS).

While it is worthwhile to mention that the first of these systems appeared at around the same time as Semantic Web technologies emerged, with RDF [14] being standardized in 1999, the development of CMSs and Semantic Web technologies have gone largely separate paths. Semantic Web technologies have matured to the point where they are increasingly being deployed on the Web. But the HTML Web still dwarfs this emerging Web of Data and – boosted by technologies such as CMSs – is still growing at much faster pace than the Semantic Web.

* This work has been supported by Science Foundation Ireland under Grant No. SFI/08/CE/I1380 (Lion-2). Preliminary results of this work were presented at the SFSW2009 workshop. We thank the reviewers and workshop audience for their valuable feedback.

A. Bernstein et al. (Eds.): ISWC 2009, LNCS 5823, pp. 763–778, 2009.

This is a pity, since actually both worlds could significantly benefit from each other. Particularly, large amounts of RDF data can now be accessed over the Web as *Linked Data* and built into Web applications [11]. Yet, widely available support for exporting and consuming Linked Data in CMSs – especially for non-experts – is still lacking, despite the fact that the need for dynamic inclusion and export of structured data in Web sites is widely recognized: at present, RSS feeds are the only agreed way to share and syndicate data across Web sites. However, the RSS format is quite limited when it comes to semantics. It was designed to carry news information, but today it is employed for carrying other types of information likely due to unawareness of compelling alternatives such as Linked (RDF) Data. We believe RDF is a good candidate for improving interoperability between sites because RSS can only carry a flat list of news item, while RDF can express any structured data and – by RDFS and OWL – even describe its own structure. Moreover, it offers a structured query language and protocol in order to extract and retrieve data matching a set of criteria, whereas with RSS, one first needs to fetch an entire feed and then process it locally to extract the desired information.

It is remarkable therefore, that although most common CMSs support RSS, RDF is still being largely ignored as a much richer potential syndication format. This is especially true since CMSs are typically built on a structured model of the domain of the site, which is reflected in both the underlying database, but – also and often more accurately – in the content types defined in the CMS itself by the site administrator. It is reasonable to assume that such structured models map naturally to RDFS and OWL. Additionally, the recently finished RDFa [1] standard could support the exposure of structured data on CMSs by allowing RDF to be embedded directly into the HTML pages, as opposed to a separate document (like in RSS). Hence RDFS, OWL and RDFa seem to be more adequate means to expose machine-readable Linked Data on Web sites. Likewise, the consumption and aggregation of Linked Data on a CMS site offer new possibilities further described in this paper. Approaching site administrators of widely used CMSs with easy-to-use tools to enhance their site with Linked Data will not only be to their benefit, but also significantly boost the Web of Data.

To this end, we extend Drupal – a state-of-the art CMS which has been gaining popularity recently by its rapidly growing user community, openness and modular design – with a number of features:

Firstly, we make it possible to expose the content types and fields typically defined by a site administrator using Drupal's Content Construction Kit (CCK) automatically in terms of classes and properties of a canonical OWL ontology, the *site vocabulary*. With our RDF CCK module, the site data itself become available as RDFa embedded in the live Web site following Linked Data principles. While this off-the-shelf solution already makes any Drupal site which installs our module amenable to Semantic Web tools such as RDF crawlers, search engines, and SPARQL engines to search and query the site, this is only a first step.

Secondly, for better linkage to the Web of Data, we enable mappings of the site vocabulary to existing ontologies. To this end, we extend the RDF CCK module to allow importing existing RDF vocabularies and map the site model to their terms.

Thirdly, to keep the learning curve low for site administrators, we support them in finding existing ontologies to reuse by a search facility. Unintended modifications of existing ontologies as well as the introduction of potential inconsistencies on the Semantic Web are avoided by our user interface as far as possible.

Fourthly, upon installation of an additional module, the site data can – on top of the RDFa data embedded in the HTML pages – be accessible via a standard SPARQL query interface, following the SPARQL protocol.

Finally, we allow administrators to dynamically integrate data from other RDF enhanced Drupal sites or Linked Data producers.

In the rest of this paper we briefly outline how our approach differs from earlier work in Section 2. Then we move on to describing specific details on Drupal in Section 3 and outline the goals our implementation should achieve in Section 4, before we look closer into each of our modules in Section 5. Adoption and deployment of our work are discussed in Section 6, conclusions are drawn in Section 7.

2 Related Works

Let us explain how our approach differs from previous attempts to link CMSs with the Semantic Web and why we believe that it is more adequate.

Although some earlier approaches proposed ontology based CMSs running natively on ontology management systems with RDF stores as back-ends [22,13], current CMSs run on very traditional Web application servers with relational databases back-ends. We do not aim to replace established infrastructures, but to build on top of them, with minimal intrusion and maximal reuse of the existing CMS infrastructure. We aim to extract and link ontologies from the content models and within the tools that site administrators are familiar with nowadays. We believe that taking users from where they are will enable more rapid adoption of Semantic Web technologies and lower entry barriers.

Somewhat closer to our approach is the idea to map the relational database schema underlying a CMS to RDF/RDFS [23]. Triplify[2] also follows this path, providing a generic mapping tool from relational databases to Linked Data (e.g. providing some predefined mappings to wrap Drupal sites' back-end databases into RDF). Our approach is again significantly different, as we want to capture the site content model and its constraints rather than the underlying database structure. Actually, a good part of the success of CMSs is grounded precisely in the fact that site administrators do not need to delve into the details of the underlying database system or schema – which may vary between different versions of Drupal and be affected by changes of the content model in non-obvious ways. Our approach works on a more abstract level (Drupal's CCK) directly in the API and user interface the site administrator is used to. We consider this model of the information structure more adequate than the underlying database schema.

Finally, none of the above approaches provide an all-in-one solution for exporting, aggregating and mapping Linked Data from within a commonly used CMS. This is what distinguishes our work, which is tailored for easy of use.

As for pre-existing work specifically in Drupal, a recent paper [8] describes the related SCF (Science Collaboration Framework) Node Proxy architecture. This module, developed specifically for biomedical applications, enables RDF from defined SPARQL queries to be mapped to specific Drupal content types. These mappings must be generated individually per content type - Node Proxy is not a general RDF-to-Drupal mapping facility, it does not support CCK, and it is read-only (into Drupal from Linked Data). Nonetheless, it was a significant first step along the path we develop further and more generally here. The Node Proxy architecture is currently used to retrieve Gene information from a common RDF store for StemBook[1], an open access review of stem cell biology. It will be superseded in the SCF Drupal distribution by the much more general and fully-featured modules we describe here. We will provide more details on the SCF use case in Section 6.

3 Drupal: A Popular CMS

Drupal (`http://drupal.org/`) is among the top three open-source CMS products in terms of market share [21] and accounts for more than 175 000 installations on the Web[2]. The system facilitates the creation of Web sites by handling many aspects of site maintenance, such as data workflow, access control, user accounts, and storage of data in the database.

As is typical for CMSs, a *site administrator* initially sets up a site by installing the core Drupal Web application and choosing from a large collection of *modules*. Site administrators do not write code; this is done by module developers instead. After the site has been set up, Drupal allows non-technical users to add content and handle routine maintenance tasks.

Each item of content in Drupal is called a *node*. Nodes usually correspond more or less directly to the pages of a site. Nodes can be created, edited and deleted by content authors. Some modules extend the nodes, for example a taxonomy module allows assignment of nodes to categories.

The *Content Construction Kit (CCK)* is one of the most popular and powerful modules used on Drupal sites. It allows site administrators to define types of nodes, called *content types*, and to define *fields* for each content type. Fields can be plain text fields, dates, file uploads, or references to other nodes, etc. Additional field types can be added via modules. When defining content types and fields, the site administrator has to provide *ID*, *label*, and *description* for content types and fields. Additionally, CCK allows to specify the following constraints on fields: (i) *Cardinality:* fields can be optional or required, and may have a maximum cardinality (ii) *Domain:* fields can be shared among one or more content types; (iii) *Range:* fields can be of type text, integer, decimal, float, date, file attachment, or node reference; node reference fields can be restricted to nodes of specific content types; text fields can be restricted to a fixed list of text values.

3.1 Motivating Example: The Project Blogs Site

Throughout this paper we will use a running example to illustrate our approach: a project blogs Web site[3] contains various information about the researchers at DERI and their collaborators, including their publications, blog posts and projects they work for. Our goal is to expose the site data and structure in a machine-readable form as well

[1] http://www.stembook.org/

[2] `http://drupal.org/project/usage/drupal`

[3] Demo-site available at `http://drupal.deri.ie/projectblogs/`

as pull in data available from the Linked Data cloud or other Drupal sites in order to enrich the information displayed on the site.

Fig. 1 shows the typical look and feel of a Drupal page and administrative interface for the *Person* content type, without our extensions installed. This content type offers fields such as *name*, *homepage*, *email*, *colleagues*, *blog url*, *current project*, *past projects*, *publications*. Particularly, we will illustrate in the further sections how to extend the *publications* field to automatically display a list of publications pulled from various data endpoints.

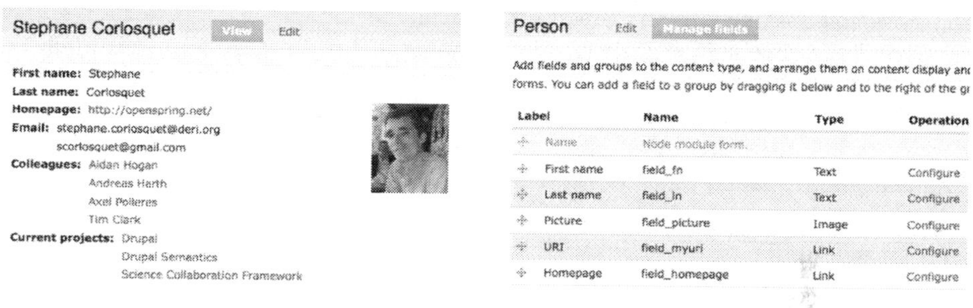

Fig. 1. A user profile page (left), editing a content type in Drupal's CCK (right)

4 Publishing and Consuming Linked Data with a CMS

Given a Drupal CCK content model consisting of content types, fields, and nodes that instantiate the content types, what would be a good way of representing it in RDF? We consider the following features desirable for the RDF output which are in line with the Linked Data principles and best practices [3]:

(i) Resolvable HTTP URIs for all resources, to take advantage of existing tools that can consume Linked Data style RDF content. That is, when resolving URIs, one should find machine-readable information describing the URI. On the one hand in Drupal, typically URIs of the running site are simply URLs pointing to Web pages, but on the other hand, each of these pages also represents a node of a certain content type in the CCK content model. Thus, in our model, each node becomes an RDF resource, and the HTML Web page describing the node is enriched with RDFa [1] that reflects the links in the content model. That is, for each node URI

- we add an `rdf:type` triple asserting the membership of the node to a class corresponding to the content type of the node,
- we add a triple for each *field* displayed on the page where the predicate is a property representing the field itself and the field value is either a datatype literal (for text, integer, decimal, float, or date fields) or the URI of the respective node reference.

(ii) Expressing Drupal CCK constraints in OWL. Constraints that are defined on the types and fields (domains, ranges, cardinalities, disjointness) should be automatically published as RDF Schema [5] or OWL [9] expressions. We will enable this by an

auto-generated *site vocabulary* that is linked from the site and which describes all content type and field URIs as classes and properties in an ontology that reflects exactly the constraints expressible in CCK. We will explain this mapping in detail in Section 5.1.
(iii) Re-use of published ontology terms. To support sites talking about arbitrary domains, pre-defined/auto-generated RDF classes and properties are most likely insufficient. In fact, the default site vocabulary only comprises an isolated ontology not related to the rest of the Semantic Web. In order to link content to existing ontologies, we have to provide means to the site administrator to select terms from existing ontologies when setting up the content model. This requires that sites may reuse/import vocabulary terms from common existing ontologies. We will explain this in more detail in Section 5.1.
(iv) Safe vocabulary re-use. Mixing the content model constraints with constraints of a published ontology might have unintended semantic effects, especially since most site administrators will not be familiar with the details of the OWL semantics. The system must prevent such effects as far as possible. Practical examples are given in Section 5.1.
(v) Exposing a query interface. We rely on the SPARQL protocol [19], i.e. the site should expose its data in a SPARQL endpoint associated with the site. It should be easy to set up and should not be a burden for the site administrator.
(vi) Reuse of Linked Data. Where possible, linkage should be defined to other instances of the Linked Data cloud.

5 Implementation

We will now present our extensions of Drupal, which are designed to fulfill the goals outlined in the previous section, in more detail.

5.1 RDF CCK: From Content Models to Site Vocabularies

Administrators use CCK to define a site-specific content model, which is then used by content authors to populate the site. The focus of our work is to expose (i) such a CCK site content model as an OWL ontology that reflects the site structure which the administrator had in mind and (ii) the site content as RDF data using this ontology. We have implemented a Drupal module that enhances Drupal's CCK with the ability to auto-generate RDF classes and properties for all content types and fields. We build a so-called *site vocabulary*, i.e. an RDFS/OWL ontology which describes the content types and fields used in the data model as classes and properties. The field and type names are extracted from field and type *ID*s from CCK, such that – following common conventions – fields are assigned a property name, and content types are assigned a class name. Field and content type labels and descriptions are likewise exported as `rdfs:labels` and `rdfs:comments`. Here goes a typical content type and field definition extracted from CCK into RDFS:

```
site : Person a rdfs : Class; rdfs : label "Person";
  rdfs : comment "Researchers in DERI and their collaborators".
site : fn a rdf : Property; rdfs : label "First name";
  rdfs : comment "First name of a Person";
```

Likewise, field constraints from CCK are reflected in the site vocabulary: *Cardinality* is mapped to cardinality restrictions in OWL, i.e. *required* fields are restricted to `owl:cardinality 1`. whereas fields with a maximum cardinality n are restricted to `owl:maxCardinality` n. For instance, if we assume that each *Person* is required to have a *name* and works in at most 5 *projects*, these constraints in CCK would be exported in OWL as follows.

```
site:Person a rdfs:Class; rdfs:subclassof
  [a owl:Restriction; owl:onProperty site:name; owl:cardinality 1],
  [a owl:Restriction; owl:onProperty site:project; owl:maxCardinality 5].
```

Domains are reflected by the `rdfs:domain` constraints. Here, fields used by a single type can be modeled by a simple `rdfs:domain` triple. For instance, assuming that the *colleagues* field for *Persons* is not shared with any other content type in the current content model, we can simply write:

```
site:colleagues rdfs:domain site:Person.
```

CCK fields shared among several types have the union of all types sharing the field as domain. E.g., as *Publication* and *Blog post* share the *title* field, the site vocabulary contains

```
site:title rdfs:domain [owl:unionOf (site:Publication site:Blog_post) ].
```

Ranges of fields are analogously encoded by the `rdfs:range` property. Additionally, we distinguish between fields of range text, float, integer, decimal, or date, and those referring to file attachments or node references. We declare the former as `owl:Datatype-Property` and assign the datatypes supported in Drupal with their respective XSD datatypes, i.e. $text \rightarrow$ `xs : string`, $float \rightarrow$ `xs : float`, $integer \rightarrow$ `xs : integer`, $decimal \rightarrow$ `xs : decimal`, or $date \rightarrow$ `xs : date`. For instance, the text field *name* is reflected in the site vocabulary as:

```
site:name rdfs:range xs:string; a owl:DatatypeProperty.
```

Fields that range over texts restricted to a fixed list of text values are assigned an enumerated class of values using `owl:DataRanges`, e.g. *gender* is modeled as

```
site:gender a owl:DatatypeProperty; rdfs:range
  [ a owl:DataRange; owl:oneOf ("male" "female") ].
```

Adhering to Linked Data principles. Following the conventions mentioned in the previous section, the site vocabulary is generated and published automatically at the site URL under the default namespace `http://`*`siteurl`*`/ns#`, which we denoted by the namespace prefix `site:` in the examples before. Likewise, any Drupal page on a site will be annotated with RDFa triples that dereference terms of this site vocabulary as classes and properties linking Drupal content nodes as subjects and objects. We are in line with the Linked Data principles and best practices [3] as we provide resolvable HTTP URIs for all resources: Each of the pages also represents a node of a certain content type in the CCK content model. That is, as mentioned before, each node becomes an RDF resource, and the HTML Web page describing the node has describing embedded RDFa [1] using the site vocabulary. By this design, any Drupal site using our module is off-the-shelf amenable to existing tools that can consume Linked Data.

Mapping to Existing Ontologies. While the functionality we have described previously fits Drupal sites well into the Linked Data world, so far, we have created nothing more than an isolated ontology based on the existing site content model. However, the benefits of this exercise remain limited, unless we additionally allow linking the site vocabulary to existing vocabularies and ontologies populating the Semantic Web. For instance, instead of just exposing the *Person* type as a class in the site vocabulary, we might want to reuse a class in an existing ontology, such as `foaf:Person` from the FOAF[4] ontology which some other publishers on the Web already used. Likewise, we may wish to state that a *Publication* is actually a `foaf:Document`, or that a Publications are linked to their Publications by Dublin Core's[5] `dc:creator` property, etc.

To this end, our module adds a new tab "Manage RDF mappings" to the content type administration panel of CCK for managing such mappings cf. Fig. 2. An auto-complete list of suggested terms is shown, based on the keyword entered by the user. The terms are coming from two different sources, which are detailed below.

External vocabulary importer service. The module RDF external vocabulary importer (evoc)[6] has been created to allow the import of vocabularies available on the Web and make the imported terms available in the mapping interface. The site administrator simply needs to fill in a form with the vocabulary URI and the prefix to be used in the system to refer to the vocabulary term when using the CURIE [4] format. A set of SPARQL queries are processed against the vocabulary to extract its classes and properties and some information about them like label, comment, superclass, domain, and range. These are then cached locally to provide a smoother user experience. Given their popularity, the Dublin Core, FOAF and SIOC vocabularies are imported automatically upon installation of the evoc module.

External ontology search service. We have also developed an ontology search service to help users to find ontologies published on the Web of Data. The search engine is entity-centric, i.e. instead of returning a list of relevant ontologies, it returns a list of relevant ontology entities to the user request. The service is currently covering cleaned up Web crawls of DERI's SWSE.org and Sindice.com search engines comprising Web data documents that define properties and classes (+100.000 documents).

Data Pre-Processing. Before being indexed, we perform a sequence of pre-processing tasks on the ontology data. Among them, the most important ones are reasoning and splitting. Reasoning is applied on each ontology in order to infer useful information for a search engine, such as the hierarchy of classes and properties, the domains and ranges of properties, etc. Reasoning over semantically structured documents enable to make explicit what would otherwise be implicit knowledge: it adds value to the information

[4] `http://xmlns.com/foaf/0.1/`

[5] `http://purl.org/dc/elements/1.1/`

[6] `http://drupal.org/project/evoc`

and enables an entity-centric search engine to ultimately be much more competitive in terms of precision and recall [16]. Ontologies are then split into smaller pieces on a per-entity basis. For each authoritative URI[7] found in the ontology, we simply extract all its outgoing links.

Indexing Model. The simplest semi-structured indexing method is to represent an ontology entity as a set of attribute-value pairs using a field-based approach [15]. For example, the ontology class `foaf:Person` will have fields *label*, *comment* and *subClassOf*; index terms are constructed by concatenating the field names with values of this field, for example as *subClassOf:Agent*.

Objects of type *literals* and *URI* are normalised (tokenised) before being concatenated with their field name. It is thus possible to use full-text search not only on literals, but also on URIs identifying ontology terms. For example one could search for "`Agent`" to match `foaf:Agent`, ignoring the namespace.

We allow search for plain keywords, combinations of keywords, or structured queries (e.g. `student AND subClassOf:Person` or `name AND domain:Person`), search examples are shown in Fig. 2; find more details on the user interface below.

Mapping process. The terms suggested by both of the import service and the ontology search service can be mapped to each content type and their fields. For mapping content types, one can choose among the classes of the imported ontologies and for fields, one can choose among the properties. The local terms will be linked with `rdfs:subClassOf` and `rdfs:subPropertyOf` statements, e.g. `site:Person rdfs:subClassOf foaf:Person` to the mapped terms in the site vocabulary; wherever a mapping is defined, extra triples using the mapped terms are exposed in the RDFa of the page.

The use of subclasses and subproperties for mapping to existing external ontologies – instead of reusing the imported terms directly in the definitions of the site vocabulary – is a simple way of minimizing unintended conflicts between the semantics of local vocabulary and public terms. This way we ensure that, e.g. constraints on local terms such as the cardinality restrictions which we derive from CCK do not propagate to the imported ontology. This ensures *safe vocabulary re-use*, i.e. avoids what is sometimes referred to as "Ontology Hijacking" [12].

Intuitively, safe reuse means that a vocabulary importing another one does not modify the meaning of the imported vocabulary or "hijack" instance data of the imported vocabulary.

We remark that this measure alone is not sufficient to guarantee consistency of the site vocabulary. Contradicting cardinalities in the site vocabulary and imported properties could make site instance data inconsistent or imply unwanted equalities. Our mapping tool makes several restrictions in this respect such as disallowing properties for mapping with cardinality restrictions that do not comply with those specified in CCK for the respective field.

---

[7] The Linked Data principles suggest that URIs for named entities should be dereferenceable and should link directly to the data describing the entity itself. Following this recommendation, we define as an *authoritative URI*, a URI that is dereferenceable and is linked to the ontology.

We emphasize that we cannot detect all inconsistencies possibly introduced by ontology reuse in our system. We cannot afford full reasoning services as we want our tool to fit in the Drupal ecosystem as a normal module that is installable on a site without the need to install separate external software components such as a DL reasoner. Our current approach is a best-effort approach to avoid "misuse" of imported ontologies as far as possible while deliberately keeping the interface simple. We refer the interested reader to [7] for further details.

We emphasize that the whole mapping step is optional, and the main benefit of the Web of Data – exposing site data for re-use by third parties – is essentially realized by the default mapping already.

User experience. The first example illustrated on the left of Fig. 2 is the mapping of the Person content type to an RDF class. When typing `person` in the text field, a list of suggestions is pulled from both the local set of terms and from the external search service. The local terms appear first: `foaf:Person` is a CURIE and reuses the prefix definition of the site which can be defined during the import step. Then the list is completed with the terms from the external service, which includes a much greater variety of terms. This might help the user in discovering new terms or ontologies which she may not previously have encountered. Note that the external terms are displayed as full URI as we want to avoid imposing any prefix on them. We are currently evaluating the best way to display these.

Fig. 2. RDF mappings management through the Drupal interface: RDF class mapping (left) and RDF property mapping (right)

The second example on the right of Fig. 2 illustrates the case where the user wants to reuse the Relationship Ontology[8] to express relationships between colleagues who work with each other. Despite the fact that the Relationship Ontology was not imported locally, the external ontology search Web service (5.1) was able to suggest the right term URI.

[8] `http://purl.org/vocab/relationship/`

5.2 Exposing and Consuming Linked Data in Drupal with SPARQL

We extend our running use case to the Linked Data cloud environment where we can demonstrate the interoperability of multiple Web sites, as illustrated on Fig. 3. Our goal is to use our project blogs Web site as a hub containing information federated from various remote locations:

- DBLP is a public SPARQL endpoint containing metadata on scientific publications. It is part of the Linking Open Data cloud and runs on a D2R server[9].
- The Science Collaboration Framework Web site which contains information about the SCF team and their scientific publications. It runs Drupal and the modules described in this paper.

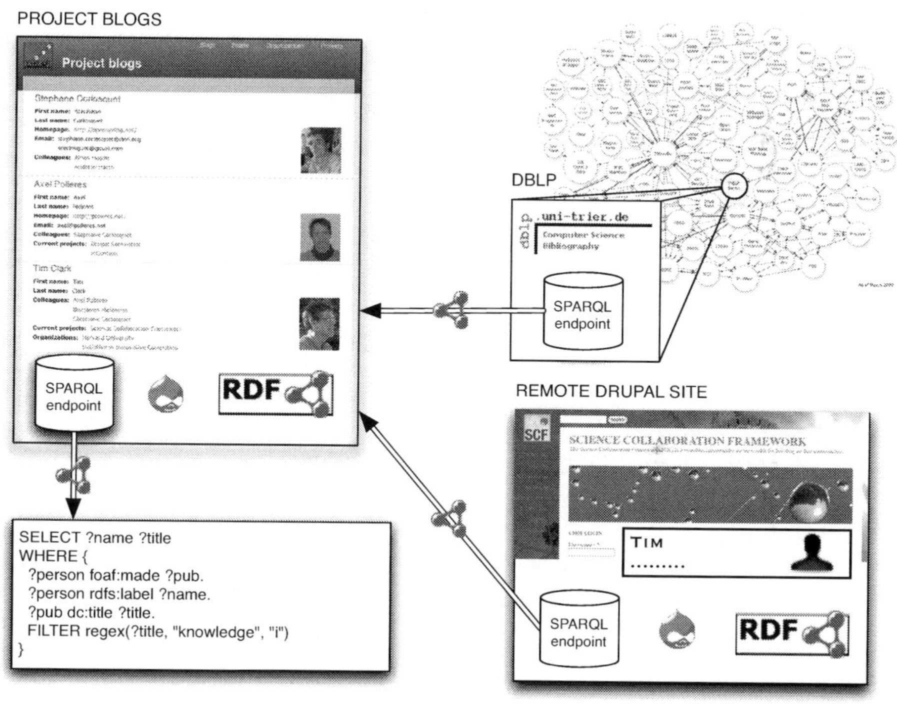

Fig. 3. Extended example in a typical Linked Data eco-system

Exposing RDF data with a SPARQL endpoint. The first step to ensure interoperability on the Web of Data is to provide an endpoint which exposes RDF data. The *RDF SPARQL endpoint* module uses the PHP ARC2 library[10]. Upon installation, the module will create a local RDF repository which will host all the RDF data generated by the RDF CCK module (see Section 5.1). The site can then be indexed with a simple click.

[9] `http://www4.wiwiss.fu-berlin.de/bizer/d2r-server/`
[10] `http://arc.semsol.org/`

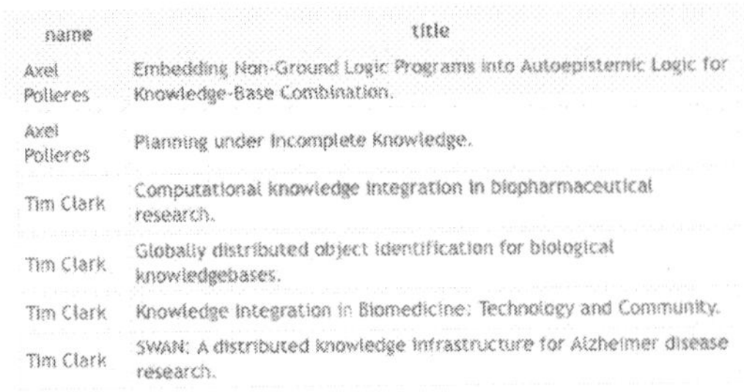

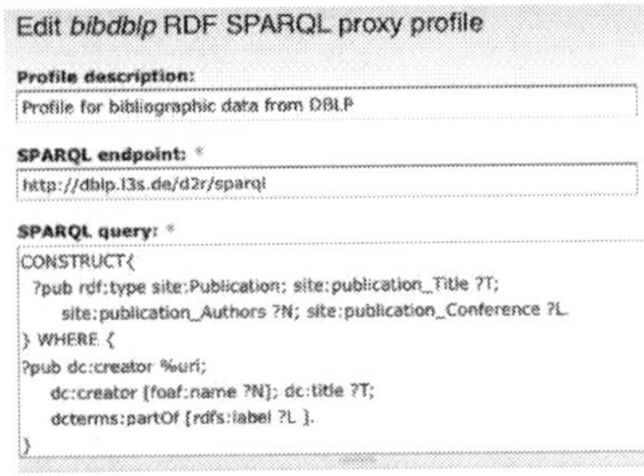

Fig. 4. A list of SPARQL results (left) and an RDF SPARQL Proxy profile form (right)

The RDF data of each node is stored in a graph which can be kept up to date easily when the node is updated or deleted. Fig. 4 (left) depicts a list of publications whose title contains the keyword "knowledge".

Consuming Linked Data by lazy loading of distant RDF resources. With an ever growing amount of data available on the Semantic Web, one site cannot technically afford to host all the data available on the Web, even if the scope was restricted to a specific domain. Instead, each piece of data can be retrieved only when needed. This design pattern is known as lazy loading [10]. Information on the Web of Data is made available in endpoints which can be queried and from where information can be retrieved according to the specific needs of an application. The SPARQL query language [18] allows complex WHERE patterns able to extract the pieces of information desired, but another feature of SPARQL is the ability to specify a schema in which the RDF data should be returned. These CONSTRUCT queries are useful in the case where the information retrieved should retain a particular structure which would not fit in flat array of results such as a SELECT SPARQL query would provide.

Building atop the existing RDF schema provided by the RDF CCK module presented in Section 5.1, we developed *RDF SPARQL Proxy*, a module which allows to import RDF instances on demand, via a CONSTRUCT query in which the WHERE clause corresponds to the schema of the distant data on the SPARQL endpoint, and the CONSTRUCT clause corresponds to the local site schema defined by RDF CCK. As depicted on Fig. 4 (right), site administrators can define profiles, which specify the rules for creating or updating the local RDF instances based on the schema of the distant RDF data. In order to keep these profiles generic, we allow input parameters such as URIs. In this example, we map the publications by an author represented by her URI `%uri` along with the information about each publication (title, name of authors, conference) to our local schema. The value of the `%uri` parameter will be replaced for the value given as input, either in the address bar of the browser or by the API calling the RDF SPARQL Proxy module. For our use case, we have setup two such profiles: one for bridging the DBLP SPARQL endpoint to the project blogs Web site, and a second for bridging the Science Collaboration Framework Web site. When visiting Tim's profile page, the relevant publication information will be fetched from both DBLP and SCF Web sites, and either new nodes will be created on the site or older ones will be updated if necessary.

6 Adoption and Deployment

Our hypothesis and general rationale is that ease-of-use and a one-click solution to export Linked Data from CMSs will enable many concrete applications of the Semantic Web, and create a bridge between the CMS and Semantic Web technology ecosystems to the benefit of both. Particular use cases and applications of this approach are discussed below.

6.1 Usability

We did a limited-scale user evaluation aimed at showing that linking a site to existing vocabularies with our Drupal module does not impose a significant burden on site administrators. We argue that the benefits of exposing Semantic Web data such as greatly improved searchability, will typically outweigh this extra effort.

Our evaluation was carried out on a group of 10 users, moderately familiar with Drupal and more or less familiar with the Semantic Web. We found that on average, the RDF mapping process took about 50% of the time required to setup the content model. For more detailed results, we refer the reader to [7]. While linking to external vocabularies was subjectively experienced as easy by all users, a significant time was actually spent deciding to which properties and classes to link with the CCK fields. Inspired by this finding we put on our agenda to further investigate how we can better assist non-Semantic-Web-savvy users in finding the "right" classes and properties for their needs.

6.2 Adoption

Besides our closed evaluation, we have also released the RDF CCK module on drupal.org. Since its release Nov. 2008, the RDF CCK module has – steadily increasing – reached a number of 63 deployed installations[11] as shown in Fig. 5. This is encouraging. Our module is currently being tested and will be deployed in the next version of the Science Collaboration Framework (SCF) platform, a special Drupal distribution developed at the Massachusetts General Hospital and Harvard University in collaboration with DERI and other participating institutions [8].

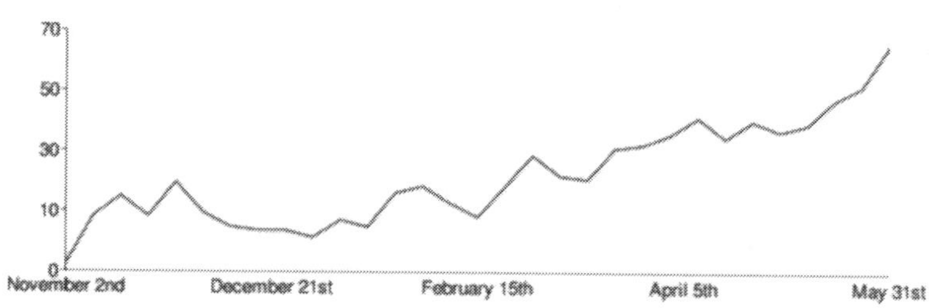

Fig. 5. Evolution of the number of installations of RDF CCK

[11] According to `http://drupal.org/project/usage/rdfcck`

6.3 Motivation and Benefits - The SCF Use Case

Harvard Medical School and DERI are collaborating on a larger use case in the course of which the technologies mentioned in this paper were developed. The Science Collaboration Framework (SCF) [8] is a distributed Drupal installation launched in Beta version at various institutions working in the Biomedical domain.

Biomedical informatics provide one particularly cogent and well-researched set of use-cases for the facility we have built and there are big expectations for the use of Linked Data in this domain, especially in the SCF Project. SCF is building Drupal based sites and tools that will enable scientific collaboration and Semantic search in this area.

Mapping of graph-based metadata embodying controlled terminologies and relationships (ontologies) to CMS-managed content promises to be exceptionally useful in biomedical informatics, and more broadly in scientific communications on the Web. Biomedicine, a highly descriptive, inductive and experimentally based discipline, is rife with complex terminologies. Synonym, subsumption, and other semantic relationships in such terminologies are natural and necessary. But currently we are still limited in the power of text searching across documents and sites if the relationships and properties in the text are not computable across the elements of these terminologies (or ontologies). This requires that certain elements in the text be assigned a semantic context which is computable in the CMS. This is a use case for semantic tagging of documents, which can leverage the well-defined ontologies in this domain.

For example, from scientific papers in this domain we may extract text strings such as "nf-κB", "nuclear factor kappa B", or "nf-kappa-B". By adequate thesauri, or user tagging using CommonTag[12] all of these could actually be matched to the query string "NFKB1", which the HUGO official gene names[13] and potentially other synonyms all resolve to a common URI represented in the Neurocommons [20] triple store, and the synonymy relationship is represented in RDF available at the Neurocommons SPARQL endpoint. Such extended search facilities, are next on our agenda, once the simple annotation of publications authors like presented in a simplified form in this paper is realised. Here, mapping RDF to an associated CCK generated type in Drupal will import the synonymy relationships and enable term expansion to increase search power.

Existing biomedical ontologies and database records which represent information about genes and other biomedical terms represent structured relationships all of which can be found in RDF and drawn into our site.

This use case becomes particularly compelling when one considers that biomedical research consists of myriad sub-specialities ranging across from basic research to clinical practice, as well as incorporating divisions by biological process, organ, species, cell type, molecule, protein family, technological approach, clinical orientation, disorder, and so forth. Each of these areas can and often does have its own slightly different semantic universe and forms of discourse. The ability to intersect documents and information from and about researchers across these domains of discourse, at scale, with assistance from computers, is dependant upon our ability to leverage formal

[12] `http://commontag.org`

[13] HUGO Gene Nomenclature Committee `http://www.genenames.org/`

terminologies and ontologies by linking them to text in scientific communications. That is precisely the purpose of the module described in this paper. The experts in these domains are hardly IT or Semantic Web experts, though they are able to use easy-configurable tools for aggregating and setting up content like Drupal, setting up the required modules via SCF on their site, and enter relevant data.

At the moment, we deploy RDF CCK in the SCF Beta version and the other modules mentioned in this paper are shortly before deployment and several of them have been co-developed or inspired by existing SCF modules such as SCF Node Proxy module, which we mentioned in the Introduction.

7 Conclusions and Outlook

We have presented a number of extensions to Drupal that enable the exposure of site content as Linked Data and likewise allow to aggregate and reuse existing RDF data from the Web in your Drupal site. A list of the modules used throughout this paper is available at `http://drupal.deri.ie/projectblogs/about`. Our most widely deployed module RDF CCK – available at the official Drupal site `http://drupal.org/project/rdfcck`) – allows to link existing and newly deployed Drupal sites to the Web of Data by a few clicks. It auto-exports the content model of a Drupal site to an ontology that is published following common best practices for ontology publication and enables the exposure of Drupal site content as RDFa. With the Evoc module, we link to existing properties and classes from other Semantic Web vocabularies by subclass/subproperty relations and offer a search facility for commonly used vocabulary terms on the Web of Data. A third module – RDF SPARQL Endpoint – exposes upon installation a SPARQL endpoint on the site data without additional configuration steps for site administrators who wish this feature. Finally, a fourth module – RDF SPARQL Proxy – allows to dynamically load data into the site, and displays this data using a lazy loading strategy for minimizing delays in the user experience.

In combination, all four modules offer new possibilities to create networked Web applications and pushing further population of the Web of Data by significantly lowering entry barriers for a large user community – CMS administrators.

Next steps include the extension of RDF/OWL export to other Drupal modules such as the taxonomy module which allows to define tag hierarchies usable in Drupal, which we plan to expose as RDF using SKOS [17], and content type hierarchies, to be reflected by subclass relationships. Moreover, we shall also lower the burden for users of the RDF SPARQL Proxy – which is at the moment only accessible to users knowledgeable in SPARQL – to really reveal the full potential of our approach to a wider audience.

We shall further develop and deploy our approach in the Science Collaboration Framework which we have presented as a promising early adopter use case.

The "infection" of emerging power-user communities such as the rapidly growing Drupal site administrator and developer groups is in our opinion a key in boosting Semantic Web technologies. We shall provide easy-to-use, unobtrusive RDF exposure in a way general enough for a variety of sites, thus potentially contributing significantly to populating the Web with high-quality RDF data.

References

1. Adida, B., Birbeck, M., McCarron, S., Pemberton, S. (eds.): RDFa in XHTML: Syntax and Processing, W3C Rec (2008)
2. Auer, S., Dietzold, S., Lehmann, J., Hellmann, S., Aumüller, D.: Triplify - Lightweight Linked Data Publication from Relational Databases. In: WWW (2009)
3. Berrueta, D., Phipps, J. (eds.): Best Practice Recipes for Publishing RDF Vocabularies, W3C Working Group Note (August 2008)
4. Birbeck, M., McCarron, S. (eds.): CURIE Syntax 1.0: A syntax for expressing Compact URIs, W3C Cand. Rec (January 2009)
5. Brickley, D., Guha, R. (eds.): RDF Vocabulary Description Language 1.0: RDF Schema, W3C Rec (2004)
6. Ceri, S., Fraternali, P., Bongio, A.: Web Modeling Language (WebML): A Modeling Language for Designing Web Sites. In: WWW 2000, Amsterdam, The Netherlands, pp. 137–157 (2000)
7. Corlosquet, S., Cyganiak, R., Decker, S., Polleres, A.: Semantic Web Publishing with Drupal. Tech. Report DERI-TR-2009-04-30, DERI (April 2009)
8. Das, S., Girard, L., Green, T., Weitzman, L., Lewis-Bowen, A., Clark, T.: Building biomedical Web communities using a semantically aware content management system. Briefings in Bioinformatics 10(2), 129–138 (2009)
9. Dean, M., et al. (eds.): OWL Web Ontology Language Reference, W3C Rec (2004)
10. Fowler, M.: Patterns of Enterprise Application Architecture. Addison-Wesley, Reading (2002)
11. Hausenblas, M.: Exploiting Linked Data For Building Web Applications. IEEE Internet Computing 13(4), 80–85 (2009)
12. Hogan, A., Harth, A., Polleres, A.: Scalable Authoritative OWL Reasoning for the Web. Int'l Journal on Semantic Web and Information Systems 5(2) (2009)
13. Jin, Y., Decker, S., Wiederhold, G.: OntoWebber: Model-Driven Ontology-Based Web Site Management. In: Cruz, I.F., Decker, S., Euzenat, J., McGuinness, D.L. (eds.) SWWS (2001)
14. Lassila, O., Swick, R. (eds.): Resource Description Framework (RDF) Model and Syntax Specification, W3C Rec (1999)
15. Luk, R.W., Leong, H.V., Dillon, T.S., Chan, A.T., Croft, W.B., Allan, J.: A survey in indexing and searching XML documents. Journal of the American Society for Information Science and Technology 53(6), 415–437 (2002)
16. Mayfield, J., Finin, T.: Information retrieval on the Semantic Web: Integrating inference and retrieval. In: SIGIR Workshop on the Semantic Web (August 2003)
17. Miles, A., Bechhofer, S. (eds.): SKOS Simple Knowledge Organization System Reference, W3C Cand. Rec (March 2009)
18. Prud'hommeaux, E., Seaborne, A. (eds.): SPARQL query language for RDF, W3C Rec (2008)
19. Clark, K.G., Feigenbaum, L., Torres, E. (eds.): SPARQL protocol for RDF, W3C Rec (2008)
20. Ruttenberg, A., Rees, J., Samwald, M., Marshall, M.S.: Life sciences on the Semantic Web: the Neurocommons and beyond. Briefings in Bioinformatics 10(2), 193–204 (2009)
21. Shreves, R.: Open Source CMS Market Share. White paper, Water & Stone, `http://waterandstone.com/downloads/2008OpenSourceCMSMarketSurvey.pdf`
22. Staab, S., Angele, J., Decker, S., Erdmann, M., Hotho, A., Maedche, A., Schnurr, H.P., Studer, R., Sure, Y.: Semantic community Web portals. Computer Networks 33(1-6), 473–491 (2000)
23. Stojanovic, L., Stojanovic, N., Volz, R.: Migrating data-intensive Web sites into the Semantic Web. In: SAC 2002: Proceedings of the 2002 ACM symposium on Applied computing. ACM, New York (2002)

Extracting Enterprise Vocabularies Using Linked Open Data

Julian Dolby, Achille Fokoue, Aditya Kalyanpur, Edith Schonberg, and Kavitha Srinivas

IBM Watson Research Center, P.O. Box 704, Yorktown Heights, NY 10598, USA
{dolby,achille,adityakal,ediths,ksrinivs}@us.ibm.com

Abstract. A common vocabulary is vital to smooth business operation, yet codifying and maintaining an enterprise vocabulary is an arduous, manual task. We describe a process to automatically extract a domain specific vocabulary (terms and types) from unstructured data in the enterprise guided by term definitions in Linked Open Data (LOD). We validate our techniques by applying them to the IT (Information Technology) domain, taking 58 Gartner analyst reports and using two specific LOD sources – DBpedia and Freebase. We show initial findings that address the generalizability of these techniques for vocabulary extraction in new domains, such as the energy industry.

Keywords: Linked Data, Vocabulary Extraction.

1 Introduction

Most enterprises operate with their own domain-specific vocabularies. A vocabulary can be anything from a set of semantic definitions to a formal ontology. Vocabularies are necessary to work effectively in a global environment and to interact with customers. They facilitate common tasks, such as searching a product catalog or understanding general trends. However, building and maintaining a vocabulary manually is both time-consuming and error-prone.

This paper presents a process to fully automate the construction of a domain-specific vocabulary from unstructured documents in an enterprise. The constructed vocabulary is a set of terms and types, where we label each term by its type(s). We applied this process to the IT (information technology) domain, and automatically built an IT vocabulary taking 58 Gartner analyst reports as our input corpus. For this domain, we capture types such as *Distributed Computing Technology*, *Application Server*, and *Programming Language*, and use them to label terms like “Cloud Computing”, “IBM WebSphere”, and “SmallTalk”.

Our approach is based on a simple observation: people searching for term definitions on the Web usually find answers in either a glossary or Wikipedia. We use LOD (Linked Open Data) as our source for domain-specific types. We decided to focus on two specific subsets of LOD as our reference data – DBpedia and Freebase. Both datasets derive from Wikipedia and thus have broad domain

A. Bernstein et al. (Eds.): ISWC 2009, LNCS 5823, pp. 779–794, 2009.

coverage. Also, both have type information, e.g., DBpedia associates entities with types from the YAGO and Wikipedia type hierarchies [1].

To perform vocabulary extraction, a key step is to determine what the relevant domain-specific types are for a particular corpus. Simply looking up corpus terms in LOD is not adequate since the terms may have different senses in LOD. Many or all of these senses may be unrelated to the domain. Therefore, we use statistical techniques to automatically isolate domain-specific types found in LOD. These types are then used to label corpus terms. This core vocabulary extraction process based on LOD is discussed in Section 2. For the Gartner case, this technique produced results with high precision (80%) but poor recall (23.8%).

To address this, we present two techniques to boost recall for vocabulary extraction. The first technique, presented in Section 3, directly improves recall by increasing the coverage in LOD. Numerous instances in DBpedia or Freebase have incomplete or no type information at all, which explains some of the poor recall. The second technique, presented in Section 4, improves recall by automated machine learning.

Our technique to improve coverage in LOD is a form of type inference based on the attributes of a particular instance, and based on fuzzy definitions of domains/ranges for these attributes. As an example, we can infer that an instance has a type *Company* if it has an attribute *revenue*, because statistically, across the DBpedia dataset, instances with the attribute *revenue* tend to have *Company* as a type. We performed this type inference on the entire LOD dataset, and then re-applied our core vocabulary extraction algorithm. Type inference improved recall to 37.6% without altering the precision.

Our technique to boost recall using machine learning relies on building statistical named entity recognition (NER) models automatically. We accomplished this by using seeds generated directly from LOD and exploiting structural information in both Wikipedia and DBpedia to generate high quality contextual patterns (features) for the model. The end result was an effective, general-purpose NER model that worked well across different corpora, i.e., it was trained on Wikipedia but applied to the domain-specific corpus, the IT analyst reports. Adding results from NER to our previous output gave us a net recall of 46% and precision of 78%.

We discuss strengths and limitations of our overall approach in Section 5. In particular, we describe initial findings that show the generalizability of the vocabulary extraction techniques in new domains.Finally, we discuss related work and conclusions in Section 6.

In summary, the main contributions of this paper are:

- We describe a process to automatically extract a domain specific vocabulary from unstructured data in the enterprise using information in LOD. We validate our techniques by applying them to a specific domain, the IT industry, with a precision of 78% and recall of 46%.
- We describe a set of techniques to boost recall in vocabulary extraction. The first improves the coverage of structured information in DBpedia and Freebase (two core pieces of LOD). By making LOD more robust, it becomes

more useful for a range of applications, including our current task of extracting vocabularies. The improved versions of LOD improved the recall of our vocabulary extraction process by 14% without affecting precision. The second technique improves recall by relying on techniques for automated named entity recognition, and this improves recall by an additional 8%.

2 Vocabulary Extraction

Our process extracts domain specific terms from an unstructured text corpus, and labels these terms with appropriate domain-specific types. This is a different problem than traditional NER. Off-the-shelf NERs are typically trained to recognize a fixed set of high-level types such as *Person*, *Organization*, *Location* etc. In our case, we need to discover the types for a specific domain, and use these discovered types to label terms in the corpus. We rely on sources outside the corpus, in particular LOD, since appropriate types may not even appear in the corpus.

We also note that, typically, domain-specific terms in a corpus are found using *tf-idf* scores. To the best of our knowledge, we have not seen type information being used as an additional dimension to filter domain-specific terms, and this is one of the differentiators of our approach.

Briefly, our process performs the following steps:

1. Extract a population of terms (noun phrases) from the domain corpus.
2. Extract a seed set of domain-specific terms from the corpus using traditional *tf-idf* scores.
3. Extract domain-specific types from LOD, using the seed terms from step 2.
4. Filter the set of all terms from step 1, based on the domain-specific types from step 3. The result is a set of domain-specific terms, labeled by their respective type(s). This allows certain relevant corpus terms that have low *tf-idf* scores to be selected based on their relevant domain-specific type.

The following subsections provide the details of each of these steps, and results for the IT domain.

2.1 Term Population

We extract the noun phrases from a domain corpus, using a standard NLP part-of-speech tagger (from OpenNLP[1]). These noun phrases comprise the population of all terms. For our case study, the domain corpus was 58 IT analyst reports from Gartner. The resulting term population extracted consisted of approximately 30,000 terms. To measure the precision and recall of our process, we created a gold standard from the term population. We randomly selected 1000 terms from the population, and four people judged whether each term in this sample was relevant to the domain. A term was considered relevant only if all four judges agreed. In the end, 10% of the sample terms were considered relevant to the IT domain. Therefore, we expect 3,000 terms in the population to be relevant.

[1] http://opennlp.sourceforge.net/

2.2 Domain-Specific Seed Terms

The next step is to extract an initial set of domain-specific terms from the corpus based on traditional *tf-idf* metrics. These terms are subsequently used to look up types in LOD, and form the basis for domain-specific type selection. For this task, it is more important to find a precise set of domain-specific terms, than to find all of the domain-specific terms. We use an off-the-shelf tool, GlossEx [2], that extracts words and noun phrases along with their frequency and domain-specificity. This associated data allows low-frequency, low-specificity terms to be filtered out. For our analyst report corpus, we selected proper noun phrases and common noun phrases with frequency ≥ 2 and with an appropriate tool-specific domain-specificity threshold. We obtained 1137 domain-specific terms from the Gartner IT reports.

2.3 Domain-Specific Types

Using the domain-specific seed terms, we discover a set of relevant interesting types from LOD. Our algorithm to discover domain-specific types is outlined in Table 1. The first step is to find a corresponding LOD entity for a domain-specific term. The most precise and direct way to do this is to encode the term directly as an LOD URI (i.e., by adding the DBpedia URL prefix) and check if it exists. This produced matching entities for 588 of the terms.

Ideally it should be possible to simply look up the type(s) of each of these entities and mark them as interesting. However, this does not work since a term can ultimately map to different types with different senses. For example, "Java" is a programming language and an island, and we may select the incorrect LOD entity and hence sense. Even if there is a single LOD entity for a term, it may not be the sense that is relevant for the domain. For example, the term "Fair Warning" is a software product, but there is only one type sense in LOD, which is `Category:MusicAlbum` (pointing to an album with the same name released by Van Halen).

Table 1. Extraction of Domain-Specific Types

Input: S_{dt} : set of domain-specific seed terms, threshold parameters $\alpha_Y, \alpha_F, \alpha_C$
Output: τ set of domain-specific types from LOD

(1) initialize potential type list $\tau_p \leftarrow \emptyset$
(2) for each domain-specific seed $T \in S_{dt}$
(3) encode T as a DBpedia URI U
(4) $\tau_p \leftarrow \tau_p \cup$ $types(U)$
where $types(U)$ = values of prop. *rdf:type* for U $\cup$
values of prop. *skos:subject* for U $\cup$
'equivalent' types obtained via mappings to Freebase
(5) $\tau \leftarrow T \in \tau_p$ if
(freq(T) $\geq \alpha_Y$ and T is a *Yago* Type) or
(freq(T) $\geq \alpha_F$ and T is a *Freebase* Type) or
(freq(T) $\geq \alpha_C$ and T is a Wiki *Category*)
(6) for each type $X \in (\tau_p - \tau)$
(7) if there exists a type $Y \in \tau$ s.t. LOD $\models X \sqsubseteq Y$
(8) $\tau \leftarrow \tau \cup \{X\}$

To address this problem, we filter out uninteresting types using simple statistical information. We score LOD types based on the number of terms they match across all the documents in our corpus and filter out infrequent types (those whose frequency is below a pre-determined threshold). One issue here is the different kinds of type information in LOD – YAGO types from DBpedia, Freebase types and Categories that come from Wikipedia. We found that having separate frequency thresholds ($\alpha_Y, \alpha_F, \alpha_C$ resp. in Table 1) for each of the types produced better results.

Another issue we noticed with DBpedia is that several entities do not have any values in their *rdf:type* field, but had interesting type information in the *skos:subject* field. For example, the term "NetBIOS" has no *rdf:type*, though its *skos:subject* is mentioned as `Category:Middleware`. Hence, we take SKOS subject values into account as well when computing types for a seed term (step 4 of the algorithm). Also, DBpedia and Freebase come with different types for the same instance, and we exploit these in step 4 to obtain additional related type information from Freebase.

Filtering also removes several low frequency types that are interesting, e.g., `yago:XMLParsers`. To address this issue, we consider low frequency types as interesting if they are subsumed by any of the high frequency types. For example, `yago:XMLParsers` is subsumed by `yago:Software` in the Yago type hierarchy, and is thus considered relevant as well (steps 6-8 of the algorithm).

We ran the type-discovery algorithm with 1137 seed terms and setting appropriate type-frequency thresholds ($\alpha_Y = 4$, $\alpha_F = 4$, $\alpha_C = 4$) based on manual inspection of highly frequent types in τ_p. This produced 170 interesting types. Upon manual inspection, we found the output to have extremely high precision (98%). As expected, there is a tradeoff between precision and recall – reducing the type-frequency thresholds increases the size of the output types but decreases its precision.

2.4 Filtered Terms and Types and Discussion of Initial Results

Once we have the set of domain-specific types, we revisit the entire population of corpus terms from Section 2.1. We find terms in this population that belong to one of the domain-specific types. This set is more comprehensive than the initial seed set in 2.2. Specifically, we select the terms from the population that are 'closely related' to entities in LOD which belong to at least one domain-specific type. In order to find 'closely related' entities, we perform a keyword search for each term over a database, populated with data from DBpedia and FreeBase, and indexed using Lucene. We select matches with a relevance score greater than 0.6. For example, searching for the term "WebSphere" over the Lucene index produces entity matches such as "IBM_WebSphere", whose corresponding *rdf:type/skos:subject* value belonged to one of our domain-specific types. Therefore, 'WebSphere' is selected as a domain-specific term.

This process resulted in 896 type-labeled terms using the 170 interesting types found in the previous step. We evaluated the precision of both the terms and their types by manually evaluating a 200 term sample. This gave us a precision of

80%. We also computed recall by taking into account our gold standard estimate and the precision, and found that recall was 23.8%.

Although the precision of our initial results was reasonable, our recall was poor. Our first approach to improving recall was to directly improve the coverage of types in LOD. The techniques developed are outlined in the next section.

3 Improving Coverage of LOD

In this section, we discuss techniques to add new knowledge to our reference LOD datasets – DBpedia and Freebase. For DBpedia, we used data dumps for DBpedia 3.1. For Freebase, we used the WEX data dumps from July [3].

Sometimes there are no types in this combined dataset, which is a problem when using the dataset to perform vocabulary extraction for any domain. We therefore enhanced the type information by linking instances and types across datasets, and by inferring new types for instances.

3.1 Adding Type Information from Linking

An obvious step to improve type coverage in Linked Open Data is to leverage the fact that DBpedia and Freebase might have different types for the same instance. We linked DBpedia instances to Freebase instances using their shared Wikipedia name. This technique allowed us match all 2.2 million Freebase instances except 4,946, mainly because of differences in Wikipedia versions between the two datasets.

Although linking instances achieves the aggregation of types across Freebase and DBpedia, the two type systems are still disconnected. It would be useful to know mappings between the two sets of types for vocabulary extraction. For instance, if we knew both that a type in DBpedia such as `yago:Byway102930645` maps to the corresponding Freebase type `freebase:/transportation/road`, and also knew that `yago:Byway102930645` is as an interesting domain specific type, then `freebase:/transportation/road` is likely an interesting type as well. DBpedia has 159,379 types and Freebase has a much smaller set of 4,158 types that are specified at a coarser level of granularity. We therefore used the relative frequency with which a given DBpedia type A co-occurs with a Freebase type B to drive the mapping. We considered a mapping valid if the conditional probability $p(FreebaseType|DBpediaType)$ was greater than .80. Manual inspection of a random sample of 110 pairings revealed that 88% mappings were correct. With this technique, we were able to map 91,558 DBpedia out of 152,696 DBpedia types to Freebase types (we excluded mappings which mapped to the freebase type /common/topic because its a top level type like `owl:Thing` or `yago:Entity`). In all, because a single DBpedia type can map to multiple Freebase types (e.g., `yago:InternetCompaniesEstablishedIn1996` is mapped to `freebase:/business/company` and `freebase:/business/employer`), we had 140,063 mappings with the 80% threshold.

3.2 Type Inference

We propose a simple statistical technique to extract fuzzy domain and range restrictions for properties of an individual, and use these restrictions to perform type inference, as we describe below.

To illustrate how our technique works in DBpedia, take as an example the instance `yago:Ligier`, which is a French automobile maker that makes race cars. DBpedia has the types `yago:FormulaOneEntrants` and `yago:ReliantVehicles` as types, neither of which is a company. Yet, `dbpedia:Ligier` has properties specific to companies, such as `dbpedia-owl:Company#industry`, `dbpedia-owl:Company#parentCompany`, etc. If we had a predefined ontology, where `dbpedia-owl:Company#industry` had a domain of the type `yago:Company`, we could have used RDFS or OWL reasoning to infer that `yago:Ligier` is really a Company. However, manually defining domain and range restrictions is not an option, because DBpedia has 39,345 properties. We therefore used statistical techniques to define fuzzy notions of domains and ranges to perform type inference.

Our approach to type inference is based on correlating what properties an entity has with its explicit types. The idea is that if many instances of a particular type have a certain set of edges, then other entities with that same set of edges probably are instances of that type too. In performing this type of inference, we relied on the DBpedia to Freebase mappings we established in 3.1 to infer Freebase types. As discussed earlier, because Freebase types are specified at a coarser level of granularity, type inference is more robust for these types because of the larger sample size of instances.

More formally, we define the notion of a property implying a type based on the fraction of the given edge that pertain to instances with that property being greater than some threshold τ. Note that this notion applies to both subjects and objects of edges.

$$I_{subj}(p,t) \equiv \frac{|\{p(x,y)\,|x:t\}|}{|\{p(x,y)\,|\exists t_1 x:t_1\}|} > \tau \qquad I_{obj}(p,t) \equiv \frac{|\{p(x,y)\,|y:t\}|}{|\{p(x,y)\,|\exists t_1 y:t_1\}|} > \tau$$

where $i:t$ is an `rdf:type` assertion between an instance i and a type t, and $p(i,x)$ is a role assertion which links instance i to instance x on a property p. This step can be thought of as inferring domains and ranges for properties, as was done in [4]; however, rather than use these types directly as such constraints, we use them in a voting scheme to infer types for subject and object instances.

Given the notion of a property implying a type, we define the notion of properties voting for a type, by which we simply mean how many of a given instance's properties imply a given type:

$$V(i,t) \equiv \left|\{p\,|\,(\exists x\; p(i,x) \wedge I_{subj}(p,t)) \vee (\exists x\; p(x,i) \wedge I_{obj}(p,t))\,\}\right|$$

We additionally define the notion of all edges that take part in voting, i.e. the number of edges that pertain to a given instance that imply any type:

$$V_{any}(i) \equiv \left|\{p\,|\exists t\,(\,(\exists x\; p(i,x) \wedge I_{subj}(p,t)) \vee (\exists x\; p(x,i) \wedge I_{obj}(p,t))\,)\,\}\right|$$

Finally, given the notion of voting, we define the implied types of an instance simply as those types that receive the greatest number of votes from properties of that instance compared to the total number of properties of that instance that could vote for *any* type:

$$T(i) \equiv \left\{ t \middle| (\forall_{t_1} V(i,t) \geq V(i,t_1)) \wedge \frac{V(i,t)}{V_{any}(i)} \geq \lambda \right\}$$

We applied this technique to our data setting τ and λ to .5. We inferred types for 1.1M instances of the 2.1M instances in Linked Open Data. To help evaluate these types, we introduce a technique next to automatically detect when we might have inferred invalid types.

3.3 Evaluating Type Inferences

At its core, the detection of invalid type inference is based on the observation that if two types are known to be logically disjoint, such as *Person* and *Place*, and we infer a type that is disjoint with any of the explicitly asserted types of an instance, this constitutes an error in our type inference. In prior work, ontology reasoning has been used to automatically detect invalid type inference in text extraction [5]. However, extending this approach to Linked Open Data is not easy. Although YAGO types are organized in a hierarchy, there are no obvious levels in the hierarchy to insert disjoints. Freebase has no hierarchy at all: the type structure is completely flat.

Our approach therefore was to first statistically define a type hierarchy for Freebase, and then use that hierarchy to define disjoint classes to detect invalid type inferences.

Because Freebase has no type structure, each instance in Freebase is annotated with a flat set of types, in which more-general types occur along with less-general types. We use this fact to approximate the usual notion of supertype: a supertype Y by definition contains all the instances of its subtype X, and, in normal circumstances, thus we would expect to see the following, where P denotes probability: $P(i \in Y | i \in X) == 1$

Because the Freebase data is noisy, this probability will most likely be less than 1; to account for this, we can recast the above constraint as follows: $P(i \in Y | i \in X) > \tau$

Thus, $X \subset Y$ if instances of X are almost always instances of Y. This allows us to define groups of types that are related as subtypes; in particular, related groups are maximal groups that are closed under subtype, i.e. G is a group if and only if $\forall_{x,y} x \in G \wedge (y \subset x \vee x \subset y) \Rightarrow y \in G$

This definition gave us 78 groups, with τ set to .65. Of these, 26 groups were singletons, and the largest group had 409 types. We further manually grouped the 78 groups by those that appeared to belong to the same domain; this gave us 35 larger groups of types, which covered 1,281 types out of 4,158 types. In practice, these groups were overwhelmingly disjoint, and we dropped the few types that occurred in multiple groups. The 35 groups of types were declared as pairwise disjoint (i.e., each type T in group A was declared disjoint from each type Q in group B).

To evaluate the 1.1M inferred types better, we divided them into 3 categories: (i) *Verified* for the entities for which at least one of the inferred types is the same as an explicitly declared one (this category had 808,849 instances), (ii) *Additional inferred types* for the entities for which the inferred types were not disjoint with any existing type assertion, i.e. these denote additional inferred type assertions that helps improve coverage in DBpedia (this category had 279,407 instances), and finally (iii) *Invalid* for the entities for which at least one of the inferred types conflicted with an explicitly asserted type (this category had 6,874 instances).

To determine the accuracy of our inferred types, we took samples of the invalid and additional inferred types, and evaluated the precision for these categories with two random samples of 200 instances each. An instance was considered to be typed correctly if all the inferred types for an instance were correct. In the additional inferred types category, we typed 177 instances correctly, and 23 incorrectly. In the invalid category, we typed 21 instances correctly, and 179 wrong. Taking the overall results for all the categories into account, we achieved a net recall of 49.1% and an estimated precision of 95.8% accuracy. Note that we have the usual trade-off between precision and recall based on values for the parameters τ, λ, however we achieved a reasonably high F-score of 64.9. We added only the *Additional inferred types* into our version of Linked Open Data, and re-ran our vocabulary extraction. The results are described in the next section.

3.4 Results with Improved LOD

Note that the previous execution of the vocabulary extraction algorithm (Table 1) using off-the-shelf LOD datasets produced 170 interesting types and 896 type-labeled terms in the output, with a precision of 80% and recall of 23.8%. We re-applied the algorithm with the newly discovered type assertions added to DBpedia and the new type mappings from DBpedia to Freebase (and the same input parameters as earlier) and discovered 188 interesting types and a net output of 1403 type-labeled terms. Manual inspection of a 200 term sample revealed that the precision was unaltered, and recall had increased to 37.6%, which validated our coverage enhancement techniques.

4 Improving Coverage Using Statistical NER

Since LOD coverage is incomplete, the techniques described above do not produce a complete domain-specific vocabulary. From our gold standard, described in Section 2.1, we expect to find around 3K domain-specific terms, and the output of the previous step is still quite short. In order to improve the coverage of our solution, we automatically build an NER model for domain-specific types.

The previous step produced a large number of interesting types (>170). These include YAGO types, Freebase types and Wikipedia Categories, which have related groups, such as, `yago:ComputerCompanies`, `freebase:venture_funded_company` and `Category:CompaniesEstablishedIn1888`, all conceptually subclasses of *Company*. Given the large number of closely related types, it does not make sense to build an NER model for each of the types. Instead we decided to look only at top-level types in the output (types that were not subsumed by any other). Furthermore, given the noise in the type-instance information in LOD, we decided to restrict ourselves to Yago types that have Wordnet sense ID's attached to them (e.g. `yago:Company108058098`), since they have a precise unambiguous meaning and their instances are more accurately represented in LOD. In our case, this yields five YAGO/Wordnet types: `yago:Company108058098`, `yago:Software106566077`, `yago:ProgrammingLanguage106898352`, `yago:Format106636806` and `yago:WebSite106359193`.

The process of building a statistical model to do NER is inspired by techniques described in systems such as Snowball [6] and PORE [7]. The basic methodology is the following – start with a set of training seed tuples, where each tuple is an *<instance, type>* pair; generate a set of 'textual patterns' (or features) from the context surrounding the instance in a text corpus; and build a model to learn the correlation between contextual patterns for an *instance* and its corresponding *type*. We combine the best ideas from both Snowball and PORE and make significant new additions (see the Related Work (Section 6) for a detailed comparison).

We could not afford to train the model on the IT corpus itself for two reasons: (i) lack of sufficient contextual data (we only had 58 reports), (ii) lack of adequate training seed tuples (even if we took the most precise term-type pairs generated in the previous section, it was not enough data to build a robust model). However, Wikipedia, combined with LOD, provides an excellent and viable alternative. This is because we can automatically obtain the training seed tuples from DBpedia, without being restricted to our domain-specific terms. We look for instances of the YAGO/Wordnet types in DBpedia, and find the context for these instances from the corresponding Wikipedia page.For our learning phase, we took either 1000 training seed instances per type or as many instances as were present in LOD (e.g., `yago:ProgrammingLanguage106898352` had only 206 instances). This gave us a total of 4679 seed instances across all five types.

A key differentiator in our solution is the kind of text patterns we generate (by patterns here, we mean a sequence of strings). For example, suppose we want to detect the type *Company*. The following text pattern [*X, acquired, Y*], where X, Y are proper nouns, serves as a potentially interesting pattern to infer that X is of type *Company*. However, a more selective pattern is the following: [*X acquired* <Company>]. Knowing that Y is of type *Company*, makes a stronger case for X to be a *Company*. Adding type-information to patterns produces more selective patterns.

Table 2. Pattern Generation Algorithm

Input: Sentence S containing training instance I, entities with Wikipedia URLs $WN_1..WN_k$; and complete Type-Outcome set for model OT
Output: Set CP of patterns (string sequences)
(1) Run S through OpenNLP POS tagger to get token sequence $TK : [.., < word, POS >, ..]$
(2) Remove tokens in TK where the word is an adverb , modifier or determiner
(3) Replace common nouns/verbs in TK with respective word *stems* using WordNet
(4) for each occurrence of pair $< I, POS(I) >$ in TK,
  (where pos_i is position index of pair)
(5) $SPANS \leftarrow$ ExtractSpans(TK, pos_i)
(6)   for each [start-pos, end-pos] $\in SPANS$
(7) $TK_{span} \leftarrow$ subsequence TK(start-pos, end-pos)
(8) $CP \leftarrow CP \cup$ word sequence in TK_{span}
(9) $CP \leftarrow CP \cup$ word sequence in TK_{span} replacing
      proper nouns/pronouns with resp. POS tag
(10) $CP \leftarrow CP \cup$ word sequence in TK_{span} replacing
      $WN_1..WN_k$ with corresponding types and their resp.
      super-types from LOD (provided that type is in OT)
(11)   Remove adjectives (JJ) from TK
      repeat (8)-(10) once
(Note: When generating patterns in steps (8)-(10), we replace training instance I by tagged variable 'X:POS(I)'

Subroutine: ExtractSpans(TK, pos_i)
(1) $SPANS \leftarrow \emptyset$
(2) for each j, $0 \leq j \leq (pos_i - 1)$
(3)   if $TK(j)$.POS $=$ *verb* or *noun*
(4) $SPANS \leftarrow SPANS \cup [j, pos_i]$
(5) for each j, $(pos_i + 1) \leq j \leq length(TK)$
(6)   if $TK(j)$.POS $=$ *verb* or *noun*
(7) $SPANS \leftarrow SPANS \cup [pos_i, j]$
(8) return SPANS

To add precise type information, we exploit the structure of Wikipedia and DBpedia. In Wikipedia, each entity-sense has a specific page, other Wikipedia entities mentioned on a page are typically hyperlinked to pages with the correct sense. E.g., the "Oracle_Corporation" page on Wikipedia has the sentence *"Oracle announces bid to buy BEA"*. In this sentence, the word *BEA* is hyperlinked to the 'BEA Systems' page on Wikipedia (as opposed to Bea, a village in Spain). Thus, using the hyperlinked Wikipedia URL as the key identifier for a particular entity sense, and obtaining type-information for the corresponding DBpedia URL, enables us to add precise type information to patterns. For the example sentence above, and the seed <*Oracle*, `yago:Company108058098`>, we generate the pattern [*X:NNP, announces, bid, to, buy,* <`yago:Company108058098`>] because "BEA_Systems" has the type <`yago:Company108058098`> (among others) in LOD, while X here is a variable representing the seed instance *Oracle*, and is tagged as a proper noun.

Moreover, not only do we substitute a named entity in a pattern by all its corresponding types in LOD, we add in super-type information, based on the type-hierarchy in LOD. We only focus on the YAGO/Wordnet types in the hierarchy which are the most precise. This generalization of patterns further helps improve recall of the model. Besides using type information, we also use a stemmer/lemmatizer (using a Java WordNet API[2]) to generalize patterns, and

[2] http://sourceforge.net/projects/jwordnet

Table 3. Sample High Scoring Patterns

Type	Pattern
Company	[<NNP>, *be, acquire, by, X:NNP*]
Software	[<Company>, *release, version, of, X:NNP*]
ProgLang	[<Software>, *write, in, X:NNP*]
Format	[*encode, X:NNP*]
Website	[*X:NNP, forum*]

Table 4. Evaluating our NER model

Domain	No Feedback			With Feedback		
	Prec.	Rec.	F	Prec.	Rec.	F
Wikipedia	71.1	41.5	52.4	69.5	42.5	52.8
IT Corpus	76.4	38.6	51.3	76.5	52.3	62.1

a part-of-speech tagger to eliminate redundant words (e.g., determiners) and add POS information for proper nouns and pronouns in patterns. Details of our pattern generation is described in the algorithm in Table 2.

We use the text patterns as features to train a Naive Bayesian classifier that recognizes the concerned types. Finally, we repeat the recognition phase. Newly recognized term-type tuples are fed back into the system and used to rescore patterns taking in the new contexts, and also to generate new contexts for the remaining unrecognized terms by adding in type information. This process repeats until nothing changes. This feedback loop is effective, since we produce several patterns with type information in them, and these patterns are not applicable unless at least some terms in the context already have types assigned. For example, the sentence fragment *"IBM acquired Telelogic"* appears in our text corpus; initially, we detect that the term "IBM" has type *Company*, and feeding this information back to the system helps the machine recognize "Telelogic" is a *Company* as well (based on the pattern [*<Company>, acquired, X*] for Type(X):*Company*). We have to be careful during the feedback process since terms that have incorrectly recognized types, when fed back to the system, may propagate additional errors. To prevent this, we only feedback terms whose types have been recognized with a high degree of confidence ($Pr(T_i) > 0.81$). Some sample high scoring patterns captured by our model are shown in Table 3.

4.1 Evaluation of Our NER Model

We evaluated our NER separately on Wikipedia data and the IT corpus. For the Wikipedia evaluation, we took our initial set of 4679 seed instances from LOD, and randomly selected 4179 instances for training and set aside the remaining 500 instances for evaluation. For evaluation on the IT corpus, we manually generated a gold-standard of 159 *<term, type>* pairs, by randomly selecting a

Table 5. Sample IT Vocabulary Extracted

Software Developer
BMC Software, Fujitsu, IBM Software Group, Automattic, Apache Software Foundation,...
Telecommunications equipment vendors
Alcatel-Lucent, Avaya, SonicWALL, Lucent Technologies ...
Service-oriented business computing
Multitenancy, B2B Gateway, SaaS, SOA Governace, Cloud Computing
Content Management Systems
PHP-Nuke, OsCommerce, Enterprise Content Management, WordPress, Drupal
Software
Corep, Lotus Sametime Advanced, rhype, AIM Pro Business Edition, BPMT, Agilense ...
Programming language
Joomia, ABAP, COBOL, BASIC, ruby, Java
Java Plaform
NetBeans, JDevelopper, Java Software, Java ME, Oracle JDevelopper, ZAAP
Website
Twitter, Microsoft Live, Office Online, GMail, Second Life

sample of 200 pairs from the output of Section 2.4, and then manually fixing erroneous pairs. The results are shown in Table 4.

The table shows precision, recall and F-scores for our model over each of the domains, with and without the feedback loop implemented. The results are encouraging. While not near the performance of state-of-the-art NER's (which achieve F-scores in 90% range, e.g., [8]), there are several key points to keep in mind.

First, typical NERs detect a pre-defined set of types and are specially optimized for the types using a combination of hand-crafted patterns/rules and/or a large amount of manually annotated training data. We have taken a completely automated approach for both recognizing domain-specific types and generating training data and patterns, and our scores are comparable to, and in some cases even better than, similar approaches such as [6], [7]. The quality issues in LOD adversely affects our results, and thus the more we can improve the quality of LOD the better our results should be. Second, there is scope for improvement using a more robust classifier based on Support Vector Machines (SVMs).

Finally, the performance of our model across domain corpora is significant. The model, which is trained on Wikipedia and LOD and applied to IT corpus, performs comparably well without feedback, and substantially better with feedback (esp. recall). This indicates that the kind of patterns we learn on Wikipedia, using information from LOD, can be interesting and generic enough to be applicable across different domains. The performance improvement with feedback on the IT corpus was due to better uniformity in the writing style, and thus incorporating text-patterns for recognized terms in the feedback loop helped generate additional interesting domain-specific patterns.

4.2 Results with NER Model

As a result of applying our NER model to the IT reports generated 381 new term-type pairs, which was added to the output of Section 3.4 to give us 1784 terms in all in our domain-specific vocabulary. Using the same evaluation process as earlier, we found that precision dropped a bit to 78%, but recall increased to 46%. Table 5 shows a sample of the extracted domain-specific vocabulary.

5 Discussion

Benefits of Using LOD: There are some direct benefits of using LOD that we should mention. First, having labeled domain-specific terms with appropriate types, a next logical step is to arrange the type labels in a hierarchy for classification purposes. Here, we can leverage the YAGO Wordnet hierarchy in DBpedia in addition to using our inferred type hierarchy from Freebase. Second, a way to enrich the vocabulary is to capture relations between terms, e.g., the *developerOf* relation between the company *IBM* and the software *WebSphere.* Such relation information exists in LOD, making it possible to extract relevant relations between vocabulary terms. Third, because Wikipedia does cover a broad set of topics, our techniques can generalize to a new domain. For example, we conducted a preliminary experiment which suggests that our vocabulary extraction techniques can be generalized to the energy domain.

In our experiment, the input was a corpus of 102 news articles for the energy sector, drawn from various websites that specialize in news for the energy industry. The 102 news articles matched the 58 IT reports in terms of length (i.e., number of words). We started with a set of 1469 domain-specific terms drawn from GlossEx [2], and drew a sample of 500 terms. Of this sample, 204 terms were evaluated by four people to be relevant to the energy domain, leaving us with an estimate of 599 relevant terms in the overall sample. Our vocabulary extraction technique extracted 260 terms and types, of which 190 were correct terms and types (73% precision and recall of 32%). Sample terms and types we found in the energy sector included companies such as *Gazprom*, *Petrobras*, important people in energy such as *Kevin Walsh*, *Chris Skrebowski*, countries or regions relevant to energy such as *Sakhalin*, and *South Ossetia*, and terms such as *Tar sands*, *Bitumen*, *LNG*, and *Methane Hydrates.*

Limitations: Our current recall score is still less than 50% inspite of improving the coverage of information in our two LOD sources and using automated NER, both of which independently had reasonably good performance. Obviously, we could improve coverage by considering additional LOD sources or building more NER models. Yet another way to improve recall is to look at the general Web for additional type information. The idea, explored in previous work such as [9], is to capture *is-a* relations in text using Hearst Patterns encoded as queries to a Web search engine. For example, given the entity *IBM Websphere*, we can issue the following phrase query *"IBM Websphere is a"* to Google, parse the outputs looking for a noun phrase (NP) following the input query and use the NP as a basis for the type. There are two challenges here – recognizing type-phrases that are conceptually similar (e.g., *'Application Server'*, *"Web Server"*, *'Software Platform"*), which is typically done using clustering algorithms based on WordNet etc., and dealing with multiple type senses and figuring out the relevant type for a particular domain (e.g. *Java* could be a *Programming Language* or the *Island*). For the latter, we use our automatically discovered domain-specific types as filters. Initial results in using our type-discovery algorithm as an output filter to approaches such as [9] have been very promising – boosting recall to 75% for the Gartner case without altering precision, and we plan to further investigate this approach in the future.

6 Related Work and Conclusions

There are a number of attempts to define taxonomies from categories in Wikipedia, and map the classes to Wordnet (see [1], [10]), which address a different problem from inferring hierarchies from a relatively flat type structure like Freebase. Wu et al. [4] address the problem of creating a class hierarchy from Wikipedia infoboxes. Although their major focus is on detecting subsumption amongst these infobox classes, one aspect of their work is to infer ranges for infobox properties. For this, they examine what types of instances are referenced by these properties. This is related to what we do for type inference; however, they focus on inferring ranges for individual properties, whereas we use the domain and range information of all incident edges to infer types for instances themselves.

Research in NER has mainly focused on recognizing a fixed set of generic types, and little or no work has been done on recognizing a larger set of domain-specific types (as is our scenario). Alternately, there has been a lot of recent interest on relationship detection (e.g. Snowball [6], PORE [7]), and type detection can be seen as a special case of it (*is-a* relation). However, we differentiate ourselves in several ways. Like Snowball, we build text-patterns to represent the context. However, we obtain the appropriate training seeds automatically from LOD. Our patterns capture long-distance dependencies by not being limited to a fixed size context as in Snowball, and we add part-of-speech information to improve pattern quality. Also, both Snowball and PORE add type information to patterns, however Snowball uses an off-the-shelf NER, which suffers from granularity and PORE adds type information by looking at textual information on the Wikipedia page (e.g., Categories), which can be noisy and non-normative. As described in Section 4, our patterns contain precise type information (i.e. Yago Wordnet senses) from LOD for the precise-entity sense obtained by looking at Wikipedia URIs, and we generalize types by looking at the Yago-Wordnet type hierarchy. Finally, we have demonstrated that our patterns are generalizable across domains, a point not addressed in previous solutions.

On the broader problem of extracting semantic relationships from Wikipedia, Kylin's self-supervised learning techniques to extract a large number of attribute/value pairs from Wikipedia[11] has recently demonstrated very good results. Similar to our approach, the learning is performed on the structured information available in Wikipedia in the form of infoboxes. Our system differs from Kylin in two important ways. First, our goal is to identify and extract only domain specific types, not all the values of the type attribute. Second, Kylin's technical approach assumes that the value of the attribute appears in the text used in the evaluation phase. In our two use cases, this assumption clearly does not hold for the type attribute. In fact, many of the discovered domain types were not mentioned in either the analyst reports or the energy articles.

In conclusion, we have shown that general-purpose structured information in Linked Open Data (LOD) combined with statistical analysis can be used for automated extraction of enterprise-specific vocabularies from text. We applied this idea to the IT domain looking at Gartner analyst reports, and automatically generated a vocabulary with 78% precision and 46% recall. As part of our solution, we have developed an algorithm to automatically detect domain-specific types for a corpus; a set of techniques to improve coverage of information in LOD. We have shown initial promising results for the Energy industy and are working on improving the solution coverage by leveraging the general Web.

References

1. Suchanek, F.M., Kasneci, G., Weikum, G.: Yago: A large ontology from wikipedia and wordnet. Web Semant 6(3), 203–217 (2008)
2. Park, Y., Byrd, R.J., Boguraev, B.K.: Automatic glossary extraction: beyond terminology identification. In: Proceedings of the 19th international conference on Computational linguistics, pp. 1–7. Association for Computational Linguistics, Morristown (2002)
3. Metaweb Technologies: Freebase data dumps (2008), `http://download.freebase.com/datadumps/`
4. Wu, F., Weld, D.S.: Automatically refining the wikipedia infobox ontology. In: Proc. of 17th international conference on World Wide Web (WWW), pp. 635–644. ACM, New York (2008)
5. Welty, C., Murdock, J.W.: Towards knowledge acquisition from information extraction. In: Cruz, I., Decker, S., Allemang, D., Preist, C., Schwabe, D., Mika, P., Uschold, M., Aroyo, L.M. (eds.) ISWC 2006. LNCS, vol. 4273, pp. 709–722. Springer, Heidelberg (2006)
6. Agichtein, E., Gravano, L.: Snowball: Extracting relations from large plain-text collections. In: Proceedings of the Fifth ACM International Conference on Digital Libraries (2000)
7. Wang, G., Yu, Y., Zhu, H.: Pore: Positive-only relation extraction from wikipedia text. In: Aberer, K., Choi, K.-S., Noy, N., Allemang, D., Lee, K.-I., Nixon, L.J.B., Golbeck, J., Mika, P., Maynard, D., Mizoguchi, R., Schreiber, G., Cudré-Mauroux, P. (eds.) ASWC 2007 and ISWC 2007. LNCS, vol. 4825, pp. 580–594. Springer, Heidelberg (2007)
8. Chieu, H.L., Ng, H.T.: Named entity recognition with a maximum entropy approach. In: Proceedings of the seventh conference on Natural language learning at HLT-NAACL 2003, pp. 160–163. Association for Computational Linguistics, Morristown (2003)
9. Cimiano, P., Staab, S.: Learning by googling. SIGKDD Explorations 6(2), 24–34 (2004)
10. Ponzetto, S., Strube, M.: Deriving a large scale taxonomy from wikipedia. In: Proceedings of the 22nd National Conference on Artificial Intelligence (AAAI 2007), Vancouver, B.C, July, pp. 1440–1447 (2007)
11. Wu, F., Hoffmann, R., Weld, D.S.: Information extraction from wikipedia: moving down the long tail. In: KDD, pp. 731–739 (2008)

Reasoning about Resources and Hierarchical Tasks Using OWL and SWRL

Daniel Elenius, David Martin, Reginald Ford, and Grit Denker

SRI International, Menlo Park, California, USA
firstname.lastname@sri.com

Abstract. Military training and testing events are highly complex affairs, potentially involving dozens of legacy systems that need to interoperate in a meaningful way. There are superficial interoperability concerns (such as two systems not sharing the same messaging formats), but also substantive problems such as different systems not sharing the same understanding of the terrain, positions of entities, and so forth. We describe our approach to facilitating such events: describe the systems and requirements in great detail using ontologies, and use automated reasoning to automatically find and help resolve problems. The complexity of our problem took us to the limits of what one can do with OWL, and we needed to introduce some innovative techniques of using and extending it. We describe our novel ways of using SWRL and discuss its limitations as well as extensions to it that we found necessary or desirable. Another innovation is our representation of hierarchical tasks in OWL, and an engine that reasons about them. Our task ontology has proved to be a very flexible and expressive framework to describe requirements on resources and their capabilities in order to achieve some purpose.

1 Introduction

In military training and testing events, many heterogeneous systems and resources, such as simulation programs, virtual trainers, and live training instrumentation, are used. Often, the systems were not designed to be used together. Therefore, many interoperability problems arise. These problems range from the superficial – such as two systems not sharing the same messaging formats – to more substantive problems such as different systems not sharing the same understanding of the terrain, positions of entities, and so forth.

In [1], we described an approach for automated analysis of military training events. The approach used OWL ontologies describing the systems and the purpose for which they were to interoperate. A custom Prolog program was used to produce warnings concerning potential interoperability problems, as well as "configuration artifacts" such as a chain of mediation components that could be used to connect two systems that otherwise would not be able to communicate.

A. Bernstein et al. (Eds.): ISWC 2009, LNCS 5823, pp. 795–810, 2009.

This paper describes our further work in this area[1], which overcomes many of the limitations of our previous work. We provide two main contributions. First, we have extended OWL with a representation of hierarchical *tasks* in OWL. Tasks describe the structure of events and the requirements of resources that perform them. This has proved to be a flexible and expressive framework to describe military training and testing events in a way that allows automated reasoning in order to find problems and their solutions. Second, where we previously used hard-coded Prolog rules to detect interoperability problems, we are now using SWRL[2] rules and constraints. This approach is more declarative and extensible. We discuss shortcomings of SWRL as applied to our problem domain, and propose solutions to overcoming some of these. We have implemented general-purpose task processing tools to manage and reason about resources and hierarchical tasks. While our work has been driven by military training and testing domain, most of the problems and solutions discussed in this paper are applicable to other domains.

The paper is organized as follows. Section 2 describes some of the core ontologies underlying our approach. Section 3 discusses our task and task plan concepts. Section 4 discusses some specific uses of SWRL, and the benefits and limitations derived from its use. Section 5 describes our implementation of task processing tools. Section 6 discusses related work. Section 7 summarizes the lessons that we have learned and our conclusions.

2 Ontologies

Our approach to interoperability analysis depends on good-quality, authoritative ontologies. Since our intent is to find very subtle problems, many details must be ontologized regarding simulators, training instrumentation, vehicles, communication architectures, terrain, training and testing events, and so forth. It is also important that the task processing software does not depend on specific domain ontologies. Most of these will not be under the control of ONISTT developers, and we also want the software to be reusable for other domains. To that end, we have defined a very small set of core ontologies. Different organizations can create and maintain their own domain-specific ontologies that import the core ontologies and create subclasses of classes therein.

We investigate system interoperability in the context of a specific *purpose*. The key concepts to capture purpose are as follows. A `Task` is an intended action that requires some resource(s), and potentially has some additional constraints associated with it. A `Role` is a "slot" of a task that needs to be filled with some resource. A `TaskPlan` is a plan for how to perform a task, including assignment of resources to roles. The task and task plan ontologies are described in more detail in Section 3.

---

[1] This work was performed under the ONISTT project, sponsored by DUSD/R-RTPP (Training Transformation) and the ANSC project, sponsored by USD/AT&L-TRMC (S&T Portfolio).

[2] `http://www.w3.org/Submission/swrl/swrl.owl`

The key concepts to capture details about resources are as follows. A `Resource` is a thing that can *do* something, through the use of capabilities. Examples of resources are a tank, a human, a simulator, or training instrumentation. We also allow for intangible resources such as software systems. A resource can have subresources. A `Capability` is a discrete piece of functionality that belongs to a resource. Examples include the capability to send a "weapon fired" message using some message format, the capability to physically move (at some maximum speed), or the capability to detect things in the infrared spectrum. A `Confederation` is a set of resources.

A `Deployment` connects the purpose and resource descriptions. It describes an event (such as a training or testing event) in terms of one `TaskPlan` and one `Confederation`.

As an example of extending these core ontologies, we have an ontology of capability types for military training and testing resources containing subclasses of the `Capability` class, such as `MovementCapability`, `DetectabilityCapability` and `DirectFireCapability`, and an ontology of military `Task` types such as `TST` (time-sensitive targeting) and `JCAS` (joint close air support).

3 Tasks and Task Plans

A task is a description of the structure and requirements of some set of intended actions. The structural aspects of tasks are based on two fundamental ideas: *composition* and *refinement*. It is useful to express tasks in a compositional way, grouping tasks into composite tasks. This allows reuse of larger units, and makes it easier to understand a complex task at a high level. Furthermore, there are often different ways to perform a task, imposing different requirements on the participating resources. We model this using abstract tasks that can be refined into several different more concrete tasks.

In the following, we define a task ontology. One could view this ontology as an encoding of the *syntax* of a *task language*. This is analogous to the OWL encoding of SWRL and OWL-S[3]. In all three cases, an OWL encoding is useful in order to achieve a tight integration with OWL. However, as in the case of OWL-S process models and SWRL rules, OWL cannot express the intended semantics of our task concepts. Therefore, we provide a dedicated semantics for the new concepts. We have in effect extended the OWL language. We have also defined new reasoning problems (task analysis and task synthesis) that could not practically be reduced to the standard OWL reasoning problems (such as class subsumption checking).

3.1 Task Ontology

The task ontology is shown in Figure 1. Tasks are structured in a hierarchical way. There are three kinds of tasks: abstract, primitive, and composite. All tasks have *formal arguments*. The arguments are instances of the `Role` class.

[3] `http://www.daml.org/services/owl-s/`

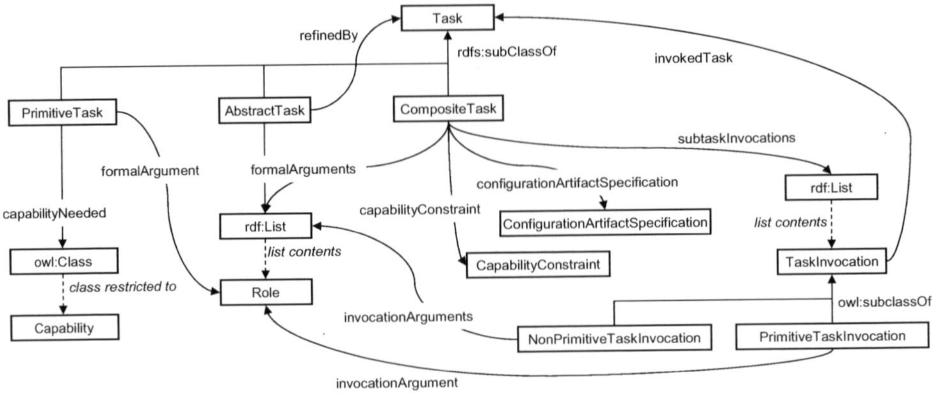

Fig. 1. Task ontology

Arguments are variables that can be assigned to resources in order to say what resources are involved in performing a task. Role-resource assignments are discussed below.

Abstract tasks can be performed, or *refined*, in several alternative ways. Each refinement is itself a task. Composite tasks have a list of *subtasks*, all of which must be performed for the composite task to succeed. *Task invocations* are used to bind the arguments of the composite task to those of its subtasks. Composite tasks can also have *constraints* on their subtasks, and *configuration artifacts* derived from their subtasks. Constraints and configuration artifacts are explained below. Primitive tasks have only one formal argument and thus are performed by one resource. Primitive tasks have an associated *capability needed*. The capability needed is a subclass of the `Capability` class[4]. A resource can be assigned to a primitive task only if it has a capability individual that is an instance of the capability needed class.

3.2 Semantics of Tasks

The following semantics is intended to facilitate the understanding of the meaning of tasks, and to provide an exact criterion for whether or not a task can be performed.

The intended meaning of a task is a set of Horn clauses. The *task atom* of a task T with formal arguments $\bar{\varphi}$ is defined as the atomic formula $\alpha(T) = T(\bar{\varphi})$. Similarly, a *task invocation atom* of a subtask invocation for task S with invocation arguments $\bar{\varphi}'$ is defined as the atomic formula $\iota(S) = S(\bar{\varphi'})$[5]. We define the function $\mathcal{H}$, denoting the Horn clauses of a task, in the following way (keeping the universal quantification of the variables in the clauses implicit). For an abstract task

[4] This is a use of classes-as-instances, and puts the ontology in OWL Full. This does not cause us problems, because we do not perform DL reasoning on task structures.

[5] From the perspective of the task ontology, the arguments are `Roles`. Logically, they are variables ranging over `Resources`.

A with refining tasks $R_1 \ldots R_n$, $\mathcal{H}(A) = \{\alpha(A) \Leftarrow \alpha(R_1), \ldots, \alpha(A) \Leftarrow \alpha(R_n)\}$. For a composite task C with subtask invocations $S_1 \ldots S_n$, $\mathcal{H}(C) = \{\alpha(C) \Leftarrow \iota(S_1) \wedge \ldots \wedge \iota(S_n)\}$. For a primitive task P with $\alpha(P) = P(x)$ and capability needed `Cap`, $\mathcal{H}(P) = \{\alpha(P) \Leftarrow \texttt{Resource}(x) \wedge \texttt{capability}(x, y) \wedge \texttt{Cap}(y)\}$. A *task library* L is a set of tasks $T_1 \ldots T_n$, and we extend the translation so that $\mathcal{H}(L) = \mathcal{H}(T_1) \cup \ldots \cup \mathcal{H}(T_n)$. Let KB be an OWL knowledge base, defined in the usual way, possibly containing SWRL rules, and $\mathcal{T}$ a function that translates such knowledge bases to their first-order logic equivalent [2]. A task T in a task library L is *performable*, given KB, iff $\alpha(T)$ is satisfiable by $\mathcal{T}(KB) \cup \mathcal{H}(L)$.

The mathematical descriptions account for the satisfiability of tasks. However, tasks have other characteristics that do not affect satisfiability but can be used by a task processing engine to produce useful information for the user, or to guide its processing. These features include constraints and configuration artifacts, both of which are described below.

It may reasonably be asked why we do not use SWRL rules (being a representation of Horn clauses) directly to represent the tasks. One reason was mentioned above: the "additional features" of tasks that are not formalized in the Horn clause semantics. The other reason is that we want to constrain the users somewhat, and not allow arbitrary rules as task descriptions.

3.3 Task Plans

The task plan ontology (see Figure 2) provides a layer on top of the task ontology, whereby one can describe *how* to perform a task. Task plans add five types of information to task descriptions:

- Role-resource assignments. A determination of which resource is to fill a particular role of the task.
- Capability assignments. A determination of which capability of the resource assigned to a primitive task is to be used for the task.
- Choices of which refining tasks to use for abstract tasks. For each abstract task, at most *one* refining task is chosen, by creating a refining task plan.
- Information about which constraints failed.
- Values for configuration artifacts.

Task plans can be partial, i.e. they do not have to have resource or capability assignments, refining plans, or subtask invocation plans.

Figure 3 shows an example of how the task and task plan ontologies can be instantiated. The task instances are in the left half of the figure, and the task plan instances in the right half. A `RoleAssignment` is used on the primitive task plan `SendPlan_1` to assign the resource `UAV1` to the `Sender` role. A capability of `UAV1` is also assigned to the task plan via the `assignedCapability` property. Note that the task instances are reusable, whereas the task plan instances are not reusable to the same extent – they show a particular way of performing the task.

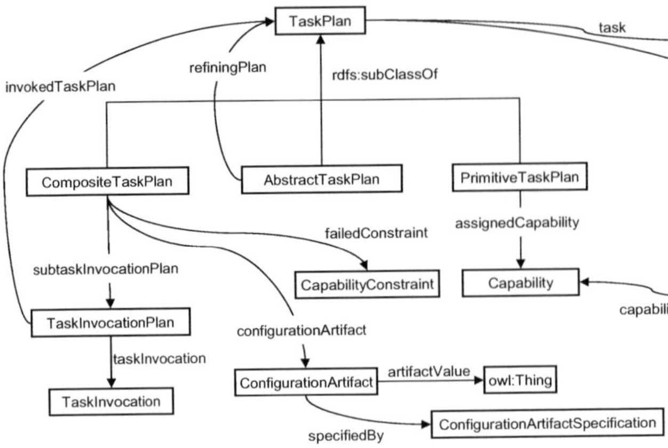

Fig. 2. Task plan ontology

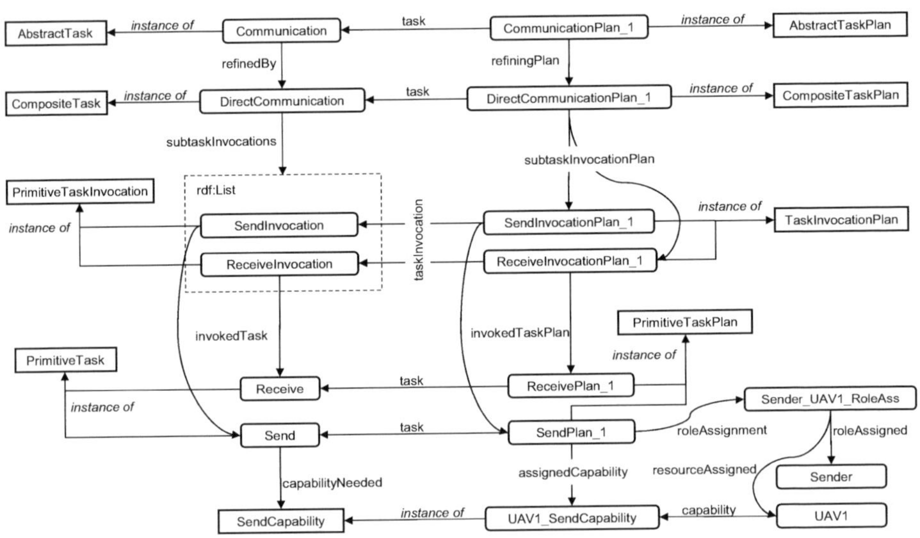

Fig. 3. Example of task and task plan instances

3.4 Semantics of Task Plans

The following is a sketch of the semantics of task plans. We leave out the precise details because of space considerations.

The meaning of a task plan P is a set of Horn clauses $\mathcal{H}_p(P)$. A role assignment is a substitution of a resource for a variable (role) that can be applied to a formula. Similarly, a capability assignment is a substitution of a capability for the variable corresponding to the capability needed in a primitive task. Intuitively, the Horn clauses for a task plan are the Horn clauses for the corresponding task, with the role and capability assignments applied as variable substitutions where appropriate, and the choices of refining tasks for abstract tasks narrowed down to what is selected in the task plan. We define the *task plan atom* α_p of a task plan P for task T and with role assignments R as the result of applying R to $\alpha(T)$. A task plan P is *valid*, given an OWL knowledge base KB, iff $\alpha_p(P)$ is satisfiable by $\mathcal{T}(KB) \cup \mathcal{H}_p(P)$. Note that if a task plan is valid, the corresponding task is performable, but the opposite is not true in general. We call the determination of task plan validity the *Task Plan Analysis problem.*

A task plan is *complete* if it is fully specified to the primitive level, and assigns all roles and capabilities. Note that a complete task plan is a set of *ground* Horn clauses. A task plan P_c is a *completion* of a task plan P if $\mathcal{H}_p(P_c)$ can be produced by instantiating all the variables (roles), and removing alternative clauses for abstract tasks, in $\mathcal{H}_p(P)$. A completion is always complete. Note that there are no valid completions of an invalid task plan. The *Task Plan Synthesis problem* is to generate all *valid completions* from a task plan.

The constraints and artifacts part of task plans does not affect validity, and is not part of the semantics, but task plan processors can use these fields to return useful information. This is discussed in more detail below. In Section 5 we describe our implementation of an engine that performs both task plan analysis and task plan synthesis.

3.5 Constraints

As mentioned above, a composite task can have additional *constraints* on its subtasks. While primitive tasks place constraints on individual resources by forcing them to have a capability of a certain type, the constraints on a composite task are usually used to specify requirements on the interaction between several resources that perform subtasks of the composite task. For example, suppose we have a `TransferVideo` composite task with primitive subtasks `ProvideVideo` and `ConsumeVideo`. A constraint could be used to say that all the `supportedResolutions` of the provider have to be supported by the consumer.

A `CapabilityConstraint` of a composite task has an associated message, a severity, and a constraint atom (see Figure 4). The constraint atom is a SWRL atom. Typically, the predicate of the atom is defined using a SWRL rule. A task processor should try to prove the constraint atom in the context of the surrounding Horn clause. If it fails to do so, the constraint failure is reported in the result of the analysis, using the `failedConstraint` property on the generated task plan. The message associated with a constraint is a natural language description of the problem, which can be shown to the user. The task processor should also generate an overall score for each solution, by adding up the weights of the severities of all the constraints that failed. In other words, a lower score is better, and a score of zero signifies the absence of any known problems. All constraints are *soft*, and as mentioned previously do not affect the performability of the task.

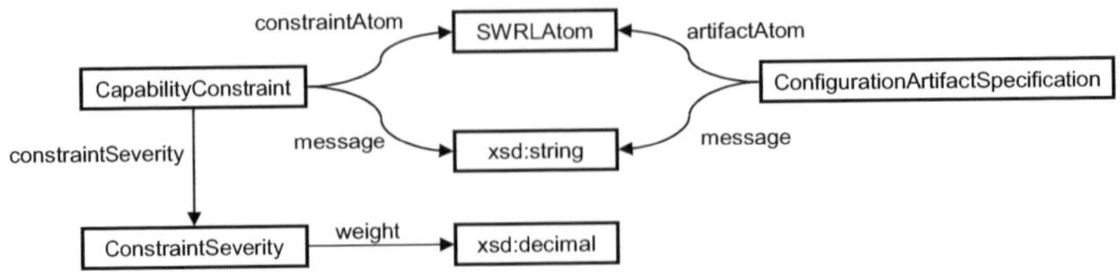

Fig. 4. Ontology elements for constraints and configuration artifacts

3.6 Configuration Artifacts

Configuration artifacts are similar to constraints (see Figure 4), except that they also have a return value. The return value is captured by letting the second argument of the artifact atom be a variable. This variable is bound when the task processor evaluates the artifact atom. The return value could be an `rdf:List` (created using the SWRL list built-ins) or any OWL individual or data value.

A typical use of configuration artifacts is to explain why a constraint failed. Continuing on the example above with supported resolutions, we could have a configuration artifact returning all the resolutions supported by the provider but not by the consumer. If the constraint succeeded, the list would be empty, but if it failed, the configuration artifact would show why it failed.

4 Benefits and Limitations of SWRL

We discuss our various uses of SWRL, the benefits we derived from its use, and the limitations that we have run into. We have put SWRL to use in many areas, such as defining constraints and configuration artifacts, reasoning about units, and ontology mapping. While SWRL has allowed us to go far beyond OWL, we have identified several limitations that appeared in different application areas: the limitation to unary and binary predicates, the lack of negation-as-failure and other nonmonotonic operations, and the inability to produce new individuals as a result of evaluating a rule.

We recognize that SWRL is not a standard, but most of what is said here also applies to other prospective rule languages such as RIF[6], and is of general concern.

4.1 Defining Constraints

SWRL rules provide a rather flexible way to evaluate capabilities of resources and perform various operations on them. Indeed, with the SWRL builtins[7], we can even do limited forms of "programming" with rules. There are some serious limitations to the usefulness of SWRL, however. First, SWRL "predicates" are OWL

[6] `http://www.w3.org/2005/rules/wiki/RIF_Working_Group`

[7] `http://www.w3.org/Submission/SWRL/#8`

classes or properties, and can therefore take only one or two arguments. SWRL builtin atoms solve this by using an `rdf:List` to contain the arguments. The same approach can be used with the other types of SWRL atoms in order to get an arbitrary number of arguments, and we use this approach when necessary. However, it is an awkward solution. First, we have to create a list to put arguments into, in the constraint atom. Then, the SWRL rule has to "unpack" the arguments from the list, using the list operations `swrlb:first` and `swrlb:rest`. Predicates with an arbitrary number of arguments is therefore high on our wish list for a future Semantic Web rule language.

The lack of negation in SWRL rules may be an even more serious limitation. In principle, SWRL allows any class expression, which means that one can use complement classes, but this allows us to negate only class expressions, not arbitrary SWRL formulas. In addition, this is classical negation, whereas we often need negation-as-failure. Going back to an example from Section 3, we may want to check that a video consumer can handle all the resolutions that a video provider can provide. Some capability of the provider and some capability of the consumer have a property `supportedResolution`. We need to check that all the provider's property values are also property values of the consumer. However, without negation-as-failure or a "closed world assumption" we cannot express this in OWL or SWRL. OWL's open world assumption means that there could always be more values of a property than what has been asserted. One could introduce several additional axioms to express that the property values are exactly the asserted property values:

```
Individual(ProvideVideoCapability
  type(restriction(supportedResolution cardinality(3)))
  value(supportedResolution 640x480)
  value(supportedResolution 800x600)
  value(supportedResolution 1024x768))

DifferentIndividuals(640x480 800x600 1024x768)
```

However, this quickly becomes unwieldy if one has to do this on all capabilities and properties in order to evaluate constraints on them like the one discussed here. Furthermore, we do not *want* to limit the capability to specific values in general – we want to retain the ability to define a capability across different ontologies in an open-ended manner. A better solution would be to make a "local closed world assumption" inside the rule. Of course, SWRL offers no such capability. For the time being, we decided on the following solution. We introduce a new SWRL builtin, called `allKnown`. This works similarly to the `setof` predicate in Prolog – it returns a list of all the "known" values of some property, for some individual. Once we have lists of property values, we can use SWRL's list built-ins to check whether one list is contained in another and so forth. The `allKnown` operation is nonmonotonic, and does not fit neatly into OWL's semantic framework. A more principled approach would be desirable, and is something we would like to see in a future Semantic Web rule language. A promising starting point is presented in [3], where a subset of the description logic underlying OWL is augmented with an auto-epistemic operator **K** that can be used with class and role expressions. More general approaches of combining description logic with "logic programming" are presented in [4].

4.2 Defining Configuration Artifacts

The representation of configuration artifacts using SWRL shares all the issues discussed above, and introduces an additional problem. As we mentioned in Section 3, configuration artifacts are defined using a predicate (object property) where the second argument is a "return value". The value can be any OWL individual. However, there is no way to produce a *new* individual using a SWRL rule (with the exception that some SWRL built-ins for lists can produce new lists). We will see in the following sections how the same problem reappears in different contexts.

4.3 Ontology Mapping

Addressing military training and testing problems on a large scale using our ontology-based approach requires many detailed domain ontologies spanning a wide range of subjects. The development of such ontologies will be distributed among different stakeholders. However, not everyone will adhere to the same ways of describing resources and capabilities. Therefore, ontology mapping will be necessary. Ontology mapping is a wide topic, studied by many researchers using different approaches [5]. For our purposes, what is needed is to bring in knowledge from outside ontologies under our own upper ontologies (of resources, capabilities and so forth).

A mapping from one ontology to another can be defined in some special format, but a more flexible approach is to use a well-known logical formalism to describe the relationships between the two ontologies. When there is a relatively simple correspondence between entities in the two ontologies, one can define class and property equivalence or subsumption between the existing classes and properties, using OWL axioms. With SWRL rules, we can define more complex mappings.

As an example, suppose we have a "normative" `video` ontology to describe video provider resources, and an "external" `camera` ontology that we want to map to the `video` ontology (see Figure 5).

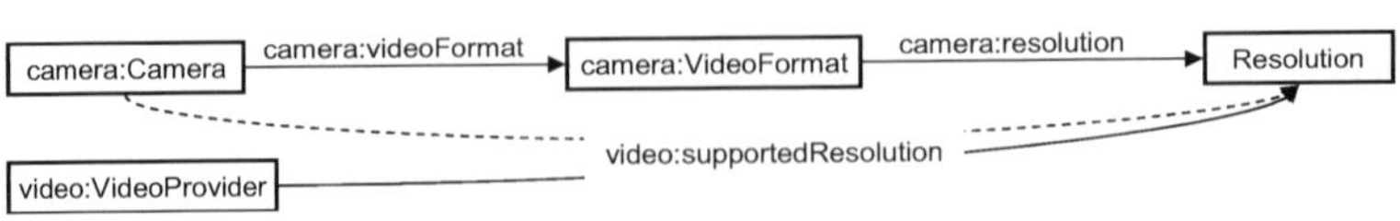

Fig. 5. Different representations of supported resolutions of a video provider resource. Dashed line shows relationship that needs to be generated.

We can define `camera:Camera` to be a subclass of `video:VideoProvider`, but the `camera` ontology describes the supported resolutions of the video provider in a different way than the `video` ontology. This difference in representation can be mapped using the SWRL rule.

$$\texttt{camera:Camera}(?x) \wedge \texttt{camera:videoFormat}(?x, ?y) \wedge \texttt{camera:resolution}(?y, ?z) \Rightarrow \texttt{video:supportedResolution}(?x, ?z)$$

However, consider what happens if the "intermediate" instance is in the target ontology. This is a typical case in our real ontologies: Resources have *capabilities* (the intermediate individuals), which in turn have properties. For example, a resource might have a `video:ProvideVideoCapability` that has the `video:supportedResolution` property. Figure 6 shows the desired mapping.

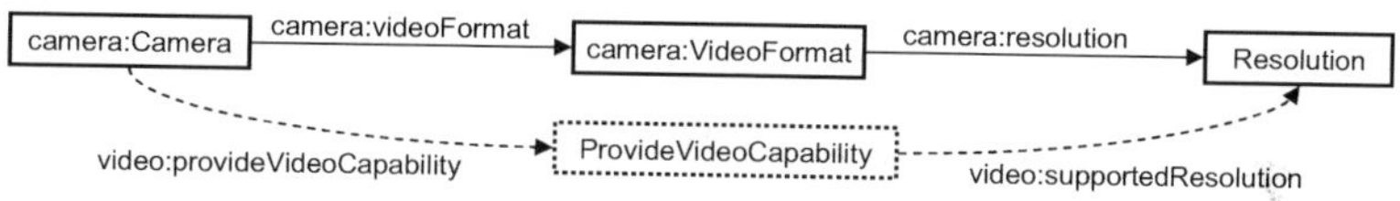

Fig. 6. Ontology mapping example. New property values (dashed lines) as well as a new instance (dashed box) need to be generated.

The mapping can be expressed by the rule

$$\texttt{camera:Camera}(?x) \wedge \texttt{camera:videoFormat}(?x, ?y) \wedge \texttt{camera:resolution}(?y, ?z) \Rightarrow \exists ?c\texttt{:}\ \texttt{capability}(?x, ?c) \wedge \texttt{video:ProvideVideoCapability}(?c) \wedge \texttt{video:supportedResolution}(?c, ?z)$$

Here we have an existentially quantified variable in the rule head, representing a "new instance" that needs to be created, viz. the `ProvideVideoCapability` of the resource. This cannot be expressed in Horn logic or SWRL rules. A limited form of existential quantification is possible by using `someValuesFrom` class expressions in a rule. However, the example above can still not be encoded in this way. The pattern discussed here is very common in OWL, since all structured data is described using property-value chains. Thus, SWRL can be used only for relatively simple types of mapping, where there is little structure in the target ontology.

4.4 Reasoning about Units

A common problem in ontologies is how to represent and reason about units of measure. Units are ubiquitous in our military training and testing domain, for example, in describing formats representing time and space positions, or the speed of vehicles. OWL by itself can be used to represent information about units, but is not adequate for making the right inferences from the information. However, this is an area where SWRL can be used to great advantage.

Figure 7 shows the essence of our "quantity" ontology. A quantity is an entity that has a unit and a magnitude, for example 5 kg or 10 lbs. A unit has a primary unit and a conversion factor to its primary unit. For example, the unit lbs has primary unit kg and conversion factor 0.4535924. Based on this quantity ontology, we have developed ontologies for engineering values (e.g., accelerations, areas, frequencies) and computation values (e.g., bits per second, megabytes, mebibytes).

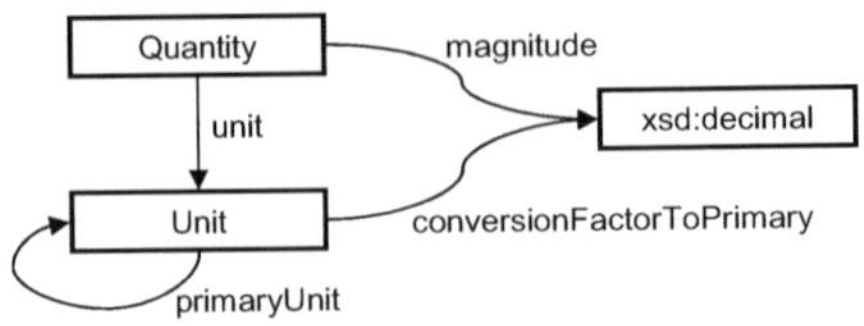

Fig. 7. Quantity ontology

Being able to describe quantities is only the first step, however. We also want to do things with them. For example, we want to compare quantities in different units. The quantity ontology defines a number of operations on quantities, using SWRL rules. First, we define a "helper" predicate `primaryMagnitude`. This is the magnitude of a quantity in its primary unit. `swrlb:multiply` is a SWRL built-in, where the first argument is the result of multiplying the rest of the arguments.

$$\texttt{magnitude}(?q, ?mag) \wedge \texttt{unit}(?q, ?u) \wedge \texttt{conversionFactorToPrimary}(?u, ?convf) \wedge \\ \texttt{swrlb:multiply}(?pmag, ?mag, ?convf) \Rightarrow \texttt{primaryMagnitude}(?q, ?pmag)$$

Next, we can define equals, less than, and so on, using this helper predicate:

$$\texttt{primaryMagnitude}(?q1, ?pmag1) \wedge \texttt{primaryMagnitude}(?q2, ?pmag2) \wedge \\ \texttt{swrlb:lessThan}(?pmag1, ?pmag2) \Rightarrow \texttt{qLessThan}(?q1, ?q2)$$

These operations can be used to determine for example that 10 lbs is less than 5 kg. This shows a common use of SWRL in our domain: We define some "abstract data type" [6] along with some operations on it. Another example (not shown here), which builds on the previously defined operations (e.g., less than) is quantity intervals, with operations such as checking whether two quantity intervals overlap.

One limitation is that we cannot use SWRL to define operations that *produce* new entities. This cannot be expressed using SWRL rules because of the limitation regarding existential variables illustrated in the discussion about ontology mapping above. For example, adding two quantities produces a new quantity that is the sum of the two. For some operations, like division, the result could even have a different unit than the inputs.

5 Implementation

We have implemented tools that realize the task planning framework described in Section 3. The two main components are a Protégé plug-in and a task engine, described below. The overall architecture is shown in Figure 8.

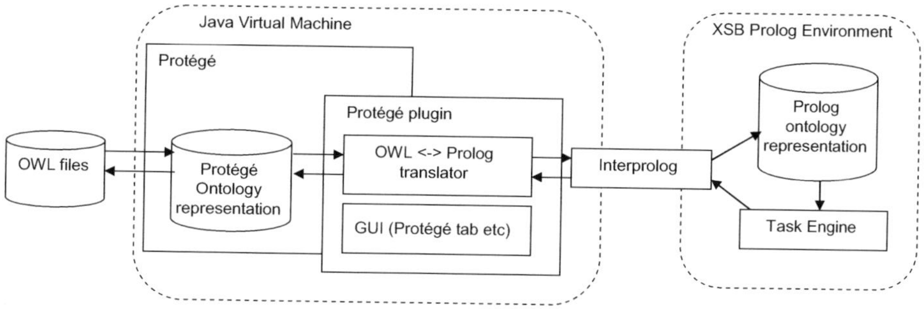

Fig. 8. Implementation architecture. The new components are the Protégé plug-in and task engine. As shown by arrows, results from the task engine can be translated all the way back to OWL files.

5.1 Task Engine

Our task engine is a program that solves the task plan analysis and synthesis problems discussed in Section 3.3. The engine is implemented in XSB Prolog[8]. Prolog was a natural choice because it provides built-in backtracking, which we use to generate all solutions during task plan synthesis.

Given that the meaning of a task or task plan is a set of Horn clauses, for task plan *analysis* we could just translate task plans directly into Prolog rules, query the task plan atom, and see if it succeeded or not. However, for the more interesting task plan *synthesis* problem, we also want to know *how* the task succeeded. This means that the engine has to "interpret" the task descriptions and construct structured result terms. This is similar to evaluating rules and returning the call tree and all variable assignments generated along the way.

As mentioned in Section 3, the engine depends on the entire OWL KB (i.e., knowledge beyond the task structure) for two purposes:

- When a primitive task is evaluated, the engine must perform a KB query of the form $\texttt{Resource}(x) \wedge \texttt{capability}(x, y) \wedge \texttt{Cap}(y)$.
- When a constraint atom or artifact atom is evaluated, the engine must perform a KB query given by atom.

To perform these queries, the OWL KB is translated to Prolog (by the OWL $\leftrightarrow$ Prolog translator component in Figure 8). The translation uses the well-known correspondence of a large subset of OWL, called DLP [7], to Horn clauses

[8] `http://xsb.sourceforge.net/`

(the translation is the same as in our previous work [1]). This means that not all of OWL's semantics is covered (i.e., the query answering is not complete), but in practice we have not found this to be a limitation for the ontologies that we work with, as we do not tend to rely on the more complex OWL axioms and the inferences that they would enable.

5.2 Protégé Plug-In

We developed a plug-in for Protégé [8] to help in creating tasks and task plans, invoking the task engine, and navigating results from the task engine.

Figure 9 shows one of several views of the results of running task synthesis for a given deployment. Results can be fairly complex. Our plug-in helps the user explore the result in terms of which resources were used, what warnings were generated, and the structure of the generated task plan.

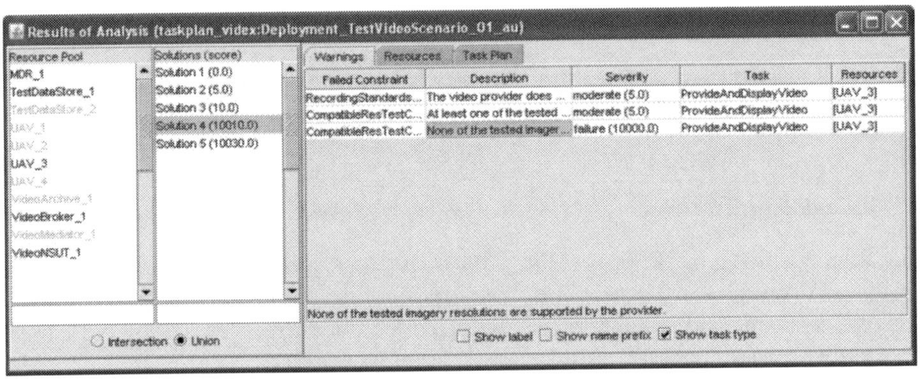

Fig. 9. Results of running the task engine from the Protégé plug-in. This view shows which resources were used to generate the selected solution. In the view shown, the user can explore warnings due to failed constraints.

6 Related Work

A related paradigm for task planning is *Hierarchical Task Network (HTN) Planning* [9]. HTNs have *tasks* (corresponding to our abstract tasks), *methods* (corresponding to our composite tasks), and *primitive tasks*. The goal is to decompose tasks down to primitive tasks, and assign *operations* to the primitive tasks. This is constrained by preconditions and effects on the tasks. Thus, the HTN planning engine has to keep track of the state of the world, and changes to it that tasks achieve. This makes HTN planning a harder problem than ours, since we do not care about the *order* in which tasks are performed. On the other hand, HTN planning is also *easier* than our problem, because the arguments to tasks are known in advance, whereas our task planning paradigm allows us to run the planning with variable "roles", which then need to be assigned by the planning engine.

Another related paradigm is Semantic Web Services, and OWL-S in particular. OWL-S is used to describe *services* based primarily on their inputs, outputs, preconditions, and effects (IOPEs). This is useful in order to infer which combination of services can be used to achieve a particular goal. Services can be thought of as tasks, and indeed HTN planning has been used with OWL-S service descriptions [10]. However, OWL-S is an open-ended representation scheme without any particular mandated computational paradigm. Our task ontology focuses on requirements on resources, whereas OWL-S focuses on IOPEs and sequencing of services, using control constructs such as *sequence* and *split-join*.

Finally, there is a wealth of work generally referred to as "scheduling" or "planning with resources" [11]. This focuses on deciding which resources can be used for multiple tasks at the same time and how tasks should be scheduled onto resources in an optimal way. In contrast, we assume that all resources can be used for any number of tasks.

7 Conclusions

Military training and testing events are highly complex affairs, potentially involving dozens of legacy systems that need to interoperate in a meaningful way. Our approach to facilitating such events is ambitious: describe the systems and requirements in great detail using ontologies, and use automated reasoning to automatically find and fix problems.

Our approach relies on ontologies, and it will be infeasible for us (the authors) to create all the domain ontologies. Therefore, a standard ontology language on which everyone can agree is critical, and OWL is the de facto standard. However, the complexity of our problem took us to the limits of what one can do with OWL, and we needed to introduce some innovative techniques of using and extending it.

One of our main contributions is our task and task plan concepts, which can be viewed as extensions to the OWL language. These concepts allow us to represent events and their requirements in a structured way, and break down an overwhelming amount of detail into manageable and reusable chunks. The second main contribution is a discussion of the benefits and limitations of SWRL. We put SWRL to use in many areas, such as defining constraints and configuration artifacts, reasoning about units, and ontology mapping. Among the limitations are the restriction to unary and binary predicates, the lack of negation-as-failure and other nonmonotonic operations, and the inability to produce new individuals as a result of evaluating a rule.

Our next task will be to define and evaluate a large real-world event using the techniques described in this paper. The long-term goal is to provide a complete system that is usable by military training and testing experts who are not necessarily knowledgeable in Semantic Web technologies. For such a transition to be successful, several different Semantic Web technologies and research areas need to progress further. The scale and distributed nature of the necessary ontology development will require significant improvement in ontology engineering approaches and tools.

References

1. Elenius, D., Ford, R., Denker, G., Martin, D., Johnson, M.: Purpose-aware reasoning about interoperability of heterogeneous training systems. In: Aberer, K., Choi, K.-S., Noy, N., Allemang, D., Lee, K.-I., Nixon, L.J.B., Golbeck, J., Mika, P., Maynard, D., Mizoguchi, R., Schreiber, G., Cudré-Mauroux, P. (eds.) ASWC 2007 and ISWC 2007. LNCS, vol. 4825, pp. 750–763. Springer, Heidelberg (2007)
2. Tsarkov, D., Riazanov, A., Bechhofer, S., Horrocks, I.: Using Vampire to reason with OWL. In: McIlraith, S.A., Plexousakis, D., van Harmelen, F. (eds.) ISWC 2004. LNCS, vol. 3298, pp. 471–485. Springer, Heidelberg (2004)
3. Grimm, S., Motik, B.: Closed world reasoning in the Semantic Web through epistemic operators. In: Grau, B.C., Horrocks, I., Parsia, B., Patel-Schneider, P. (eds.) Second International Workshop on OWL: Experiences and Directions (OWLED 2006), Galway, Ireland (2005)
4. Motik, B., Rosati, R.: A faithful integration of description logics with logic programming. In: Veloso, M.M. (ed.) Proc. 20th Int. Joint Conference on Artificial Intelligence (IJCAI 2007), Hyderabad, India, pp. 477–482. Morgan Kaufmann Publishers, San Francisco (2007)
5. Kalfoglou, Y., Schorlemmer, M.: Ontology mapping: The state of the art. In: Semantic Interoperability and Integration. Number 04391 in Dagstuhl Seminar Proc., Internationales Begegnungs- und Forschungszentrum fuer Informatik (IBFI), Schloss Dagstuhl, Germany (2005)
6. Ehrig, H., Mahr, B.: Fundamentals of Algebraic Specification 1. Springer, Heidelberg (1985)
7. Grosof, B.N., Horrocks, I., Volz, R., Decker, S.: Description logic programs: Combining logic programs with description logic. In: Proc. of the 2nd International Semantic Web Conference, ISWC 2003 (2003)
8. Knublauch, H., Fergerson, R.W., Noy, N.F., Musen, M.A.: The Protégé OWL plugin: An open development environment for Semantic Web applications. In: McIlraith, S.A., Plexousakis, D., van Harmelen, F. (eds.) ISWC 2004. LNCS, vol. 3298, pp. 229–243. Springer, Heidelberg (2004)
9. Ghallab, M., Nau, D., Traverso, P.: Hierarchical task network planning. In: Automated Planning: Theory and Practice, ch.11. Morgan Kaufmann Publishers, San Francisco (2004)
10. Sirin, E., Parsia, B., Wu, D., Hendler, J., Nau, D.: HTN planning for web service composition using SHOP2. Web Semantics: Science, Services and Agents on the World Wide Web 1, 377–396 (2004)
11. Ghallab, M., Nau, D., Traverso, P.: Planning and resource scheduling. In: Automated Planning: Theory and Practice, ch.15. Morgan Kaufmann Publishers, San Francisco (2004)

Using Hybrid Search and Query for E-discovery Identification

Dave Grosvenor and Andy Seaborne

Hewlett-Packard Laboratories, Bristol
{dave.grosvenor,andy.seaborne}@hp.com

Abstract. We investigated the use of a hybrid search and query for locating enterprise data relevant to a requesting party's legal case (e-discovery identification). We extended the query capabilities of SPARQL with search capabilities to provide integrated access to structured, semi-structured and unstructured data sources. Every data source in the enterprise is potentially within the scope of e-discovery identification. So we use some common enterprise structured data sources that provide product and organizational information to guide the search and restrict it to a manageable scale. We use hybrid search and query to conduct a rich high-level search, which identifies the key people and products to coarsely locate relevant data-sources. Furthermore the product and organizational data sources are also used to increase recall which is a key requirement for e-discovery Identification.

Keywords: SPARQL, e-discovery, identification, hybrid search and query.

1 Introduction

E-discovery is the process of collecting, preparing, reviewing, and producing electronic documents in a variety of criminal and civil actions and proceedings [1]. In this paper we address the problems of scale and recall in the *identification* stage of e-discovery which is responsible for learning a coarse location of data relevant to a legal case. There are two components to our approach to these problems. The first component was to add search directives to SPARQL [2] to give two different information retrieval models in a hybrid search and query. This gives integrated access to both structured and unstructured data sources in the enterprise. However the second component was to exploit some common product and organizational data sources to both guide the searches to cope with scale, and to increase recall.

This paper is organized into five sections. Firstly we extend the introduction with an overview of e-discovery, identification and related work. Secondly we motivate our approach to hybrid search and query. Thirdly we discuss our use of some common data sources to both guide the search, and restrict it to a manageable scale. Fourthly we examine a hypothetical patent violation e-discovery case and give examples on the use of hybrid search and query. Finally we give our conclusions on the investigation.

A. Bernstein et al. (Eds.): ISWC 2009, LNCS 5823, pp. 811–826, 2009.

1.1 E-discovery

E-discovery is a new issue for enterprises created in 2006 by the Federal Rules for Civil Procedure (FRCP) [3] in the US which legally require enterprises:

- To disclose the identities of all individuals likely to have discoverable information relevant to a legal case.
- To either provide a copy, or the location and description of all Electronically Stored Information (ESI) relevant to a legal case.

The courts can potentially impose punitive damages on an enterprise for a failure to comply.

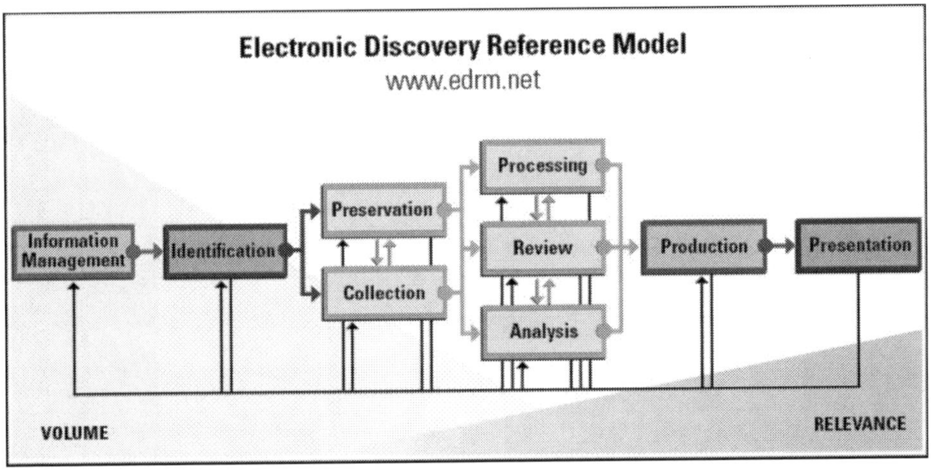

Fig. 1. Reference Model for E-discovery

The ESI includes both structured and unstructured information within the enterprise which specifically includes e-mails, web-pages, word processing files, and databases stored in the memory of computers, magnetic disks (such as computer hard drives and floppy disks), optical disks (such as DVDs and CDs), and flash memory (such as "thumb" or "flash" drives). Currently email is the most important source of discoverable information (80-90% according to a Magistrate Judge Survey [4]). For example email analysis was used extensively in the ENRON case and is a subject of the TREC legal track [5].

E-discovery requests can be initiated by arbitrary legal cases which makes it difficult to prepare for in advance. According to the "Magistrate Judge Survey" [4] only a few types of case are responsible for most of the e-discovery issues. These types of cases were: Individual plaintiffs in Employment cases; General Commercial cases, Patent or Copyright cases, Class Employment actions, and Product Liability cases. The Federal Judicial Centre provides examples of e-discovery requests [6], preservation orders [7], and 'meet and confer' forms [8]. The Sedona conference [9] has been influential in the development of approaches to e-discovery. In this paper we examine a fictitious patent violation case.

E-discovery is a complex guided search task conducted over a long duration (perhaps several months) by an expert team consisting of legal and IT experts acting on behalf of the disclosing party. The *Electronic Discovery Reference Model* (EDRM) [10] has been developed to explain the different stages occurring in the e-discovery process. Figure 1 shows the sequence of stages and indicates that the earlier stages must process a greater volume of data with each unit of data being unlikely to be relevant to a request, whereas the later stages process less data but with each data unit is being more likely to be relevant.

The reference model decomposes the e-discovery process into:

- An *information management* stage responsible for all preparation prior to an e-discovery request, including records management and policies.
- An *identification* stage responsible for providing a broad characterization of the data relevant to the discovery request.
- A *preservation* stage responsible for ensuring the ESI identified in the previous step is protected from inappropriate alteration or destruction.
- A *collection* step responsible acquiring the identified ESI.
- A *processing* stage where the volume of data is further reduced and converted into a suitable format for *Review* and *Analysis*.
- A *review* stage where a disclosing party's legal team sorts out both the responsive documents to produce, and the privileged documents to withhold.
- An *analysis* stage where the collection of materials is evaluated to determine relevant summary information, such as key topics of the case, important people, specific vocabulary and jargon, and important individual documents.
- A *production* stage where the ESI is delivered to the requesting party in both an appropriate form and by an appropriate delivery mechanism.
- A *presentation* stage where the ESI produced is displayed before legal audiences (dispositions, hearings, trials, etc..).

Search technology is used for many stages of the reference model [11]. However e-discovery is not just an enterprise search application for two reasons. Firstly e-discovery is not just concerned with the retrieval of documents. The FRCP requires the identity of every person likely to hold relevant information. Furthermore during the identification stage (which we address) is concerned with finding not only specific documents, but also relevant people, projects, organizations and data repositories. Secondly the emphasis in e-discovery is on returning all potentially responsive data whereas enterprise search returns a selection of the documents most likely to be relevant to a request. Precision and recall are two measures of information retrieval performance [12] that are commonly used.

- Precision is the fraction of retrieved documents which are relevant.
- Recall is the fraction of the relevant documents which have been retrieved.

Enterprise search is primarily concerned with precision and typically focuses on finding the few documents most likely to answer a user's question. Whereas e-discovery is concerned with recall and all responsive ESI must be located.

1.2 Identification

Identification is responsible for the initial preparation that allows later automated searches to be performed. It is usual for identification to have a strong manual element and it can be completely manual. It starts from the subpoena and interviews of potential custodians which are performed by the e-discovery team (referred to as custodian-led identification). Identification begins with the key players because the data of individuals who played a central role in the dispute are likely to contain the majority of information relevant to the dispute. An aim of the interviews is a basic framing of the request such as the relevant time frame, the impacted products and organizations, the key witnesses and custodians of the relevant data sources needed. But they also collect information needed for driving later automated searches, such as keyword lists, special language, jargon, acronyms, and technical terms. It will also include creating a data map showing the type and location of the disclosing company's data sources relevant to the request.

Our aim was to provide automated support for identification. We introduced a hybrid of search and query to support a rich model of the identification search task which not only retrieves documents, but people, organizations and products. However identification poses two problems to the use of retrieval technologies:

- The scale of the data which is potentially retrievable during identification.
- The requirement for high recall.

During identification any person, product, organization and data-source within the organization is potentially within its scope. This scope makes the brute force use of search technology more difficult during identification. An output of the identification stage is a manageable selection of the enterprise data that is potentially relevant to the e-discovery request. Only the data selected in identification is subjected to the more detailed review and analysis during of the later e-discovery stages. The recall obtained during identification provides an upper bound for the overall recall during e-discovery and so high recall during identification is very important. As a result identification casts a broad net in characterizing what constitutes relevant data.

1.3 Related Work

There are many approaches to hybrid search and query to which we provide some brief pointers. Research has aimed at search-like retrieval to access structured data, providing keyword search, imprecise queries, top-k selection and ranking [13]. Similarly other research has aimed at more query-like retrieval from unstructured data, using information extraction to provide: querying of unstructured text [14] and searches returning entities rather than just documents [15]. This research simply underlines the value of both models of information retrieval and our approach has been to make use of both. This pragmatic approach has been followed previously, such as in the WSQ/DSQ work [16] which combines the query facilities of traditional databases with existing search engines on the Web. WSQ/DSQ leverages both results from Web searches to enhance SQL queries over a relational database, and uses information stored in the database to enhance and explain Web searches. Parametric search (such as supported by Lucene [17] and most enterprise search engines) provides a similar capability to hybrid search and query by both associating metadata fields with each document that is indexed, and allowing queries to retrieve their value and to restrict searches to particular values.

There is closely related work [18][19][20][21] which adds full text search capability to SPARQL. They all are concerned with searching the literal strings in RDF datasets. The Sesame like-operator [18] simply filters the results using regular expression matches on the literals in the result sets. Virtuoso [21] provides a system of rules for selecting which RDF triples are indexed. We use an ARQ mechanism for extending SPARQL that allows us to access arbitrary indexes and are not restricted to RDF datasets. In addition ARQ provides support for both full text search of RDF datasets, and selectively indexing RDF triples using Lucene [17].

There are standard approaches to increasing recall using domain knowledge, such as query expansion [22] and spreading activation [23]. Both are automatic means of obtaining more responses to an original query using domain knowledge. This is important for e-discovery identification. Query expansion operates in the query space and transforms the query using the domain knowledge and co-occurring terms to find related or more general search terms and constraints. Spread activation operates in the result space and uses the initial results as seeds that are used to activate other related concepts during a propagation phase. We use neither technique, but we use the product catalog and the organizational data to identify related people and products which are used both in additional queries, and to generate further results.

2 Our Approach to Hybrid Search and Query

In this section we describe our approach to accessing structured and unstructured data using a hybrid of search and query. We give our motivation for using SPARQL extended with search directives, and explain how the search directives are evaluated within a hybrid query.

2.1 Motivation

To exploit both structured and unstructured data sources for e-discovery requires some form of information integration. The semantic web provides useful technology for such integration. We use RDF as common data model to integrate some diverse enterprise data including organizational and product information. SPARQL is used as the common query language. This approach provides a low cost of entry allowing you to query and navigate RDF instance data without the need for semantic integration. This is important as e-discovery potentially requires ad hoc integration to bring together data sources for the particular legal request that would not be used together during the normal operation of the business. Pragmatically we chose to extend SPARQL with search directives to retrieve unstructured documents because search is the predominant means of retrieving unstructured documents.

2.2 SPARQL

SPARQL [2] is a standard query language, protocol and XML result set format as defined by a W3C working group. It became a W3C recommendation in January

2008. A SPARQL query consists of a graph pattern and a query form (one of `SELECT`, `CONSTRUCT`, `DESCRIBE`, `ASK`). The simplest graph pattern is the basic graph pattern (BGP), a set of triple patterns that must all be matched. This is an extension point of the language and we utilize it to add semantic relationships which are not directly present in the data. In particular, we use other indexing technologies, such as free-text indexes to relate text query strings with document URIs.

2.3 Property Functions

ARQ [24] is a query engine for Jena [25] that implements SPARQL. ARQ provides *property functions* as a way to extending ARQ while remaining within the SPARQL syntax, and new capabilities can be added by applications for local needs without needing to modify the ARQ query engine code base. A property function causes some custom code to be executed instead of the usual matching of a triple pattern against the graph data. ARQ executes any property functions in a way that respects the position of the property function in the containing BGP so which variables are already bound at that point in the query does not change.

2.4 Free Text Searches

The property function mechanism has been used to provide access to different indexing technologies, including Lucene, Autonomy Enterprise Search, Google, and Wikipedia. ARQ itself does not provide the free text indexing but provides the bridge between a SPARQL query and the index. The property function implementing the search directives takes a search string and accepts other parameters for controlling the search and return values that are RDF terms. This is usually the URI of the document. In this case the indexing technologies use the body of some arbitrary document as the indexing text which is not part of the knowledge base itself.

3 Data Sources

We address the problems of scale and recall posed by identification by using some common structured and semi-structured data sources both to guide the searches to restrict the scale, and to increase the recall. This use of particular data sources contrasts with the generic but document centric approach followed by the EDRM reference model which is suitable for arbitrary e-discovery requests on arbitrary data sources.

We use some common structured data sources giving the organizational hierarchy and product catalog:

- The organizational hierarchy provides personal contact information for all people working within HP together with information about the reporting structure and a high-level business area names of the organizational structure.
- The product catalog is used by different content management systems within HP to provide different kinds of product related information ranging from product specification to collections of unstructured documents about products intended for use by sales or marketing.

- The product catalog and organizational hierarchy have common fields allowing connections between people and products to be made. For example, the products business area can be used to identify the high-level organization responsible for a product line.

The semi-structured data sources are important because they provide connections between the structured and the unstructured data sources. They connect structured entities to unstructured text which can be used to characterize topics for the search of other unstructured data sources. For example we can retrieve the technical reports written by a particular author to provide document text which can be used to characterize topics. Semi-structured data sources also connect unstructured text to structured entities which can be joined with other structured data sources. So unstructured text in the semi-structured data source can be searched and the entities of the responsive documents retrieved. For example we can find documents responsive to a topic and return the authors of these documents.

- There are many different content management systems within HP which are used to generate the external HP web site, organize unstructured sales brochures and support information. They associate the product catalog with many different forms of structured and unstructured data. They are an important data source for e-discovery.
- There are several repositories of technical reports for which author, creation date, abstract structured fields are maintained. Some of these technical reports are grouped by business area and maintain a record of a report's reviewers.
- The email repositories are very important semi-structured data sources for most e-discovery cases. But for an organization the size of HP they are costly to search without narrowing the search down to particular people and time intervals. They associate people and time intervals to unstructured text and titles which can be searched.
- Patent repositories are semi-structured data sources with structured fields linking people, publications and other patents.

4 An E-discovery Example

In a fictitious case HP is alleged to have infringed a patent on the use of *impressive print technology* assigned to *Another Photo Print Company*. HP is required to disclose information relevant to: the development and use of this technology in its products; estimating the likely profit associated with the use of this technology; showing how sales and marketing made use of the technology. HP is the disclosing party in this e-discovery case, and Another Photo Print Company is the requesting party who initiated the subpoena. HP has their own research and development program for print technology and so whilst complying with the e-discovery request HP is also keen to establish any prior art on the development of such a technology.

The subpoena triggers a duty to disclose and preserve all information relevant to the patent violation case. A team is assembled which is responsible for satisfying the legal request and applying legal holds to preserve relevant data. An identification process is initiated to identify the key witnesses, custodians of data sources, and finding the location of data relating to individuals and organizations. The initial problem is to get a better characterization of the topic, the people, organizations and products relevant to the case.

The patent alleged to have been infringed will be cited in the subpoena together with some related patents and cited publications. These cited documents can be used to provide an initial characterization of the topics relevant to the case (e.g. as sets of keywords and phrases). This obtains an initial characterization of the technical areas related to the impressive print technology. Similarly the authors of the cited documents are identified and provide some an initial set of people to seed our searches.

Our approach is to use the structured and semi-structured data sources to expand and corroborate the people, products and topics related to the case. We show examples of simple tactics for deriving a set of entities from other entities. These simple tactics can be composed with others to obtain more complex tactics. There is a need for an e-discovery environment to: manage the entities retrieved by such tactics; support the composition of complex tactics; and record the evidence of how they relate to the legal case.

4.1 Finding Relevant Products

The alleged patent infringement is concerned with printing, but the subpoena did not identify the products that may have used the impressive print technology. The e-discovery process must identify these products because the potential value attributed to the impressive print technology needs to be assessed. The subpoena did cite some patents and publications which can be used to characterize some relevant topics. So we use a tactic to find products using these initial topics. The tactic searches for web pages on the HP site for product names and numbers co-occurring with terms related to one of these topics.

This tactic uses the Google search engine to perform a search of the HP public web sites dedicated to product sales and support. Alternatively we could perform a search of the internal content management system which generates the content for this web site. Throughout this paper we will use both parameterized queries to represent such tactics, and the convention that the variable for the parameter is prefixed with a dollar sign and those prefixed with a question mark are bound by the query evaluation.

```
select ?product ?doc{
     ?product product_catalog:product_name ?name.
     ?product product_catalog:product_number ?number .
     ?query ext:printf(
          "site:http://h10010.www1.hp.com %s %s %s"
          $topic ?name ?number).
     ?doc ext:GoogleSearch(?query $n)
   }
```

The example uses a property function performing `printf` function to assemble a query string composed from the topic string parameter together with the product names and product numbers which are retrieved from the product catalog during evaluation of the query. The `GoogleSearch` property function calling the Google search engine takes a query string for the search and a parameter controlling the maximum number of results to be returned. The Google site query requires all terms to be present for a web page to be returned. So not all searches will return any results and so the search will select products for which there are documents responsive to the topic string. This tactic provides a means of obtaining products related to the topic.

Documents matching the `GoogleSearch` are returned in the (ranked) order that Google returns them. But this is not used to rank the products returned as this would require merging the relevance of scores from distinct searches. Although in this example it would be plausible to do so because each search shares the same terms contained in the topic. It also has some different terms because of the product name and product number. For example, we might want to return the products mentioned in documents that are most responsive to the topic search. But we might also take into account the number and relevance of the documents which are matched.

The ranking problem is better illustrated with a similar tactic which retrieves the products mentioned in documents created by a given person in the technical reports database.

```
select ?product ?doc ?score{
      ?product product_catalog:product_name ?name.
      ?product product_catalog:product_number ?number .
      ?query ext:printf(
            "%s %s"
            ?name ?number).
      (?doc ?score)
            ext:techreportsTextSearch(?query $n $rel).
      ?doc techreports:author $author .
   }
```

This tactic takes three arguments: the author, the maximum number of results, and the minimum relevance of a document. It uses the product catalog to generate query strings using just the product name and number. A search directive uses the generated query string to search the technical reports database. The responsive documents retrieved by the search directive are only returned by the tactic when they were created by the given author. The documents retrieved by the search directive are returned in ranked order for each query string. But for different products the query string is different and so it is not meaningful to use the relevance score of a document to rank the products. We do not address the ranking problem in this tactic, but return additional information with the product result that can be used to discriminate the returned products. We return the document URI and its relevance score. This would then allow someone to discriminate the products returned by arbitrary means, such as the number of responsive documents and their relevance information, or use other attributes of the documents such as date and topic, or whether other products were mentioned in the same documents. We did not address the ranking issue because discrimination of the results seemed sufficient for identification where we were more concerned with recall than obtaining the top-k most relevant results. However the example does illustrate the general problem of ranking the results of such hybrid search and queries.

4.2 Expanding the Set of Products

Once we have identified an initial set of products we can find related products using both the product catalog and the organizational hierarchy.

The product catalog indicates products that are related through being part of series of products addressing a market. For example there will be a printer targeting the consumer market and others targeting the small business office. Over time there will be a series of products addressing this market. Even within these markets products will address different price points and have different specifications. So we can expand from an initial seed product to retrieve all the other printers that are part of the same series or which address the same market.

The product catalog also gives information about how the product is made and where it is supported. For example, every product has a product line which is the responsibility of some organization. This allows us to group products using the organizational structure as well as the market structure. So we can expand from an initial product seed to retrieve all the products that are produced by the same product line and organization. Furthermore we can use the organizational hierarchy for further expansion. For example in the tactic below a product line is followed to its organization which in turn is followed to the business unit, then the direction along the hierarchy is reversed to find all the product lines and products produced by this business unit.

```
select ?product {
  $seed_product product_catalog:product_line ?seed_pl.
  ?pl_org hp_ba_hierarchy:product_line ?seed_pl.
  ?pl_org hp_ba_hierarchy:business_unit ?unit.
  ?org hp_ba_hierarchy:partof ?unit.
  ?org hp_ba_hierarchy:product_line ?pl.
  ?product product_catalog:product_line ?pl }
```

4.3 Finding Relevant People

The subpoena provides an initial characterization of the topics related to the impressive print technology. We now examine a tactic for using a topic to obtain a set of relevant people using a semi-structured data source. We use the HP Labs technical reports repository which we have indexed using both the Autonomy Enterprise search engine and Lucene. Semi-structured data sources, such as the technical reports repository and the content management systems, are very important because they allow the structured and unstructured data sources to be used together.

```
select ?person ?doc ?score {
  (?doc ?score) ext:TechReportsTextSearch($topic
                                          $n $relevance).

  ?doc TechReports:author ?person
}
```

The text search property function used for searching the technical reports repository again takes three arguments (the search string, the maximum number of results and the minimum relevance score). The search returns the document URI and its relevance score which are both returned by the SPARQL select query because they provides the evidence for the relevance of the person to the topic. The results of this query will be in the ranked order returned by the single text search which gives a meaningful relevance score and ordering because a single search was used.

For example, this tactic for finding relevant people to a topic returned the groups of (fictious) authors of documents relevant to one of the seed topics.

- David Shaken, Neil Arrested, Ant Fame, Iris Retinex
- Daniel Lyon, Ron Glass, Gary Circle
- Neil Arrested, David Shaken, Ace Beach
- Daniel Lyon, Ron Glass
- Ron Glass
- David Shaken, Neil Arrested, Ant Fame, Iris Retinex
- Matt Goat, Kelvin Chemistry
- Peter Wilder
- Ernest House

Two of the authors (Ron Glass and Peter Wilder) were also authors of cited papers and/or related patents cited in the subpoena. This provides further corroboration of the relevance of these people to the case. The relevance of an entity to the case is corroborated when the same entities are returned by distinct search paths. Several people who were authors of these cited papers or patents did not show up in the search because they did not write any technical reports, but they did write lots of patents and so a similar query on a patents database would also find patents related to the topic.

So we can derive a larger set of people who are potentially related to some of the seed topics without using any of the people seeds given by the subpoena. Not all of these are as good as each other, but the emphasis in e-discovery is on performing a search with high recall and will not be discarded at this stage. Some of these seeds were directly derived from the patent and some were corroborated by the searches. i.e. when the same people were derived using different search strategies.

4.4 Expanding the Set of People

There are several tactics with which we can expand the set of people considered relevant to the case using the structured and unstructured data sources. These are simple tactics that take a person and retrieves a larger set of people using only the structured data sources and not used the hybrid search capability. However these simple tactics can be combined with some other tactics deriving a set of people relevant to a topic which can be used to corroborate or rank the expanded set of people. For example, a simple tactic uses the organizational hierarchy to expand the potential set of people. This exploits the heuristic that people in the same group have similar skills or work on similar products (which is not always true but which is still useful during identification), and so would be likely to possess information relevant to the case.

```
select ?person {
 ?manager hp_org:manages $seed_person.
 ?manager hp_org:manages ?person
}
```

Similarly other tactics use the semi-structured data sources. A simple tactic expands the set of people by retrieving the co-inventors of the patents written by the seed inventor.

```
select ?person {
    ?patent Patents:inventor $inventor.
    ?patent Patents:inventor ?person
}
```

A similar tactic uses the technical report repository to retrieve the co-authors of documents patents written by a given person.

Such tactics can be combined with other tactics which derive people relevant to a topic to obtain stricter queries. For example, the following tactic makes the expansion to all organizational peers of the seed person conditional on the manager having written a relevant patent.

```
select ?person ?doc ?score {
 ?manager hp_org:manages $seed_person.
 ?manager hp_org:manages ?person.
  (?patent ?score) ext:PatentTextSearch($topic
                                          $n $relevance).
  ?report Patents:inventor ?manager
}
```

4.5 Expanding a Topic to Generate Related Topics

The technical reports repository can also be used to derive other related topics. The repository has title and abstract fields for each document that are good sources of the keywords used to characterize a topic. For example the following tactic retrieves the titles of relevant technical reports to create an expanded set of topics.

```
select ?title {
  ?doc ext:TechReportsTextSearch($topic,
                                 $n,$relevance).
  ?doc TechReports:title ?title
}
```

We obtained an expanded set of topics (fictious) with one of the initial seed topics.

- "Ink location for Color Halftones"
- "Geometric Screening"
- "Fundamental Characteristics of Halftone Textures"
- "Curved dithering"
- "Multi-pass Error-Diffusion for Color Halftones"
- "Lossless Compression of half-toned Images"
- "Anti-aliasing Methods for Computer Graphics Hardware"
- "Inverse Half-toning for Error Diffusion"
- "n-Simplex gamut mapping"

Some of these topics are inappropriate. Instead of the topics being concerned with printing the topics are concerned with computer graphics and image compression. Some topics whilst concerned with printing are concerned with color gamut mapping and are not relevant to this particular request. Such related topics might still be useful indicators of people with related skills and interests.

4.6 Corroboration

We can corroborate the current set of relevant entities or concepts whenever an existing entity or concept can be retrieved independently by one of these tactics. This was encountered earlier in this paper, when we found that "Ron Glass" was also: an inventor of a related patent, a publisher of a cited paper, and the author of a technical report responsive to one of the seed topics. We can strengthen the constraints on some of our tactics by introducing a requirement for such corroboration.

Similarly if we suspect two different entities or concepts have a relevant relation, we can corroborate this relation by deriving some other common entities starting from either. For example we can corroborate the relation between a topic and product by both using a tactic to find people related to topic, and using another tactic to find people within organizations responsible for this product. The directness of the relationship of people to the product will corroborate the relationship between the topic and the product.

Unfortunately if we perform a topic search of the technical reports repository we find that every author of a responsive document will occur in Labs which is not responsible for any product line. Either we need a different repository, or we need to use a weaker relation between people and products. For a weaker relation we use the existence of an email message relevant to a topic between the person from labs and someone in the business.

The tactic below takes a topic and a product as arguments, plus some controls for the two searches that are used. One performs a search of the technical reports, but the other performs a parameteric search capability on the email repository which restricts the search to emails between two people. The tactic returns the evidence that the topics are related to the product. This evidence is the document and email that are relevant to the topic, and the people who communicated about the topic – one of whom is in an organization responsible for the product.

```
select ?personA ?personB ?doc ?dscore ?email ?escore{
(?doc ?dscore) ext:techreportsTextSearch($topic,$n1,
                                   $relevance1).
 ?doc techreports:author ?personA.
 Sproduct productmaster:product_line ?product_line.
 ?product_line hp_org:is_in_org ?org.
  ?personB hp_org:is_in_org ?org.
 (?email ?escore) email:EmailTextSearch(?personA,?personB,
                              $topic,$n2, $relevance2).
}
```

This evidence could then be analyzed to provide a more sophisticated scoring of the quality of the corroborating link between the topic and the product. We need some means of scoring the corroboration that would take into account either the quality of the relevance of the technical report or the email message to the topic, or the number of relevant communications between the two people. At the moment we just return the evidence.

5 Conclusions

We implemented a hybrid search and query by extending SPARQL using property functions which returned ranked search results. This gives a rich retrieval model allowing text search and query of structured and semi-structured data to be used together. This was used to exploit product and organizational structure to increase recall by finding potentially related people and products. Furthermore problems with scale are avoided because the structured data sources are used to guide the search and so avoid searching everything in the enterprise.

Unfortunately both product and organizational data sources are constantly changing. For our approach to be most effective there is a need to find the organization and product structure for the particular periods of time relevant to the e-discovery request. This need not be information that is kept during the normal operation of a business. For example, it is common to keep only the current organizational information.

E-discovery legislation only requires an enterprise to disclose information that is held by the enterprise for the normal operation of its business. It does not require enterprises to store additional information. In fact, it is for this reason that proactive records management systems are proposed for e-discovery, as they enforce policies which stipulate that only data with a business purpose should be kept. However maintaining historical organizational and product information to help e-discovery is optional because this is not part of the normal operation of a business. Interestingly the product and organizational structure themselves may not be relevant ESI. They are merely a means of finding relevant ESI and perhaps of understanding its significance.

There are further opportunities for the application of semantic technologies in e-discovery:

- Litigation readiness – where semantic technologies can be used to represent knowledge which helps the finding of ESI without keeping additional ESI. In our investigation we assumed that structured information about people and products was available, which was not in itself ESI but which could help to find ESI which already exists. An important business issue is deciding what information should be kept.
- An information map of the data sources available in the company.
- Representing the semantics of data selected during e-discovery.
- Describing the provenance of data through the different stages of e-discovery.

Acknowledgements

Brian McBride, Dave Reynolds, Mark Butler, and Ian Dickinson were instrumental in developing the approach to e-discovery followed in this paper.

References

1. Rothstein, B.J., Hedges, R.J., Wiggins, E.C.: Managing Discovery of Electronic Information: A Pocket Guide for Judges. Federal Judicial Center (2007), http://www.fjc.gov/public/pdf.nsf/lookup/eldscpkt.pdf/file/eldscpkt.pdf
2. Prud'hommeaux, E., Seaborne, A. (eds.): SPARQL Query Language for RDF. W3C Recommendation (2008)
3. Federal Rules for Civil Procedure (2008), http://www.uscourts.gov/rules/CV2008.pdf
4. Francis, J.C., Schenkier, S.I.: Surviving E-discovery. In: Magistrates Workshop (2006), http://www.fjc.gov/public/pdf.nsf/lookup/MagJ0608.ppt/$file/MagJ0608.ppt
5. TREC Legal Track, http://trec-legal.umiacs.umd.edu/
6. Example Electronic Discovery Request, http://www.fjc.gov/public/pdf.nsf/lookup/ElecDi13.pdf/$file/ElecDi13.pdf
7. Example Preservation Order, http://www.fjc.gov/public/pdf.nsf/lookup/ElecDi21.pdf/$file/ElecDi21.pdf
8. Example Meet and Confer Form, http://www.fjc.gov/public/pdf.nsf/lookup/ElecDi22.rtf/$file/ElecDi22.rtf
9. The Sedona Conference, http://www.thesedonaconference.org
10. Electronic Discovery Reference Model, http://www.edrm.net
11. EDRM Search Guide. EDRM (2009), http://www.edrm.net/files/EDRM-Search-Guide%20v1.14.pdf
12. Baeza-Yates, R., Ribeiro-Neto, B.: Modern Information Retrieval. Addison-Wesley, ACM Press (1999)
13. Chaudhuri, S., Ramakrishnan, R., Weikum, G.: Integrating DB and IR Technologies: What is the Sound of One Hand Clapping? In: Proceedings of the 2005 CIDR Conference (2005)
14. Cafarella, M.J., Re, C., Suiciu, D., Etzioni, O., Banko, M.: Structured Querying of Web Text. In: 3rd Biennial Conference on Innovative Data Systems Research (CIDR), Asilomar, California, USA (2007)
15. Cheng, T., Yan, X., Chang, K.C.: Supporting Entity Search: A Large-Scale Prototype Search Engine. In: ACM SIGMOD 2007, Beijing, China (2007)
16. Goldman, R., Widom, J.: WSQ/DSQ: A Practical Approach for Combined Querying of Databases and the Web. In: ACM SIGMOD International Conference on Management of Data (2000), http://www-db.stanford.edu/~royg/wsqdsq.pdf
17. Hatcher, E., Gospodnetić, O., McCandless, M.: Lucene in Action. Manning Publications (2004) ISBN 1932394281
18. Sesame - the like operator, http://www.openrdf.org/doc/sesame/users/ch06.html#section-like
19. Virtuoso - bif:contains full text search, http://docs.openlinksw.com/virtuoso/rdfsparqlrulefulltext.html

20. Glitter - textlike and textmatch operators, http://www.openanzo.org/projects/openanzo/wiki/SPARQLExtensions
21. Minack, E., Sauermann, L., Grimnes, G., Fluit, C., Broekstra, J.: The Sesame LuceneSail: RDF Queries with Full-text Search. NEPOMUK Technical Report (2008), http://www.dfki.uni-kl.de/~sauermann/papers/Minack+2008.pdf
22. Andreou, A.: Ontologies and Query Expansion. Master of Science, School of Informatics, University of Edinburgh (2005), http://www.inf.ed.ac.uk/publications/thesis/online/IM050335.pdf
23. Crestani, F.: Application of Spreading Activation Techniques in Information Retrieval. Artificial Intelligence Review 11(6), 453–482 (1997)
24. ARQ home page, http://jena.sf.net/ARQ
25. Carroll, J.J., Dickinson, I., Dollin, C., Reynolds, D., Seaborne, A., Wilkinson, K.: Jena: Implementing the Semantic Web Recommendations. In: Proceedings of the 13th International World Wide Web Conference (2004)

Bridging the Gap between Linked Data and the Semantic Desktop

Tudor Groza, Laura Drăgan, Siegfried Handschuh, and Stefan Decker

DERI, National University of Ireland, Galway
IDA Business Park, Lower Dangan, Galway, Ireland
{tudor.groza,laura.dragan,siegfried.handschuh,stefan.decker}@deri.org
http://www.deri.ie/

Abstract. The exponential growth of the World Wide Web in the last decade brought an explosion in the information space, which has important consequences also in the area of scientific research. Finding relevant work in a particular field and exploring the links between publications is currently a cumbersome task. Similarly, on the desktop, managing the publications acquired over time can represent a real challenge. Extracting semantic metadata, exploring the linked data cloud and using the semantic desktop for managing personal information represent, in part, solutions for different aspects of the above mentioned issues. In this paper, we propose an innovative approach for bridging these three directions with the overall goal of alleviating the information overload problem burdening early stage researchers. Our application combines harmoniously document engineering-oriented automatic metadata extraction with information expansion and visualization based on linked data, while the resulting documents can be seamlessly integrated into the semantic desktop.

1 Introduction

The World Wide Web represents an essential factor in the dissemination of scientific work in many fields. At the same time, its exponential growth is reflected in the substantial increase of the amount of scientific research being published. As an example, in the biomedical domain, the well-known MedLine [1] now hosts over 18 million articles, having a growth rate of 0.5 million articles / year, which represents around 1300 articles / day [1]. In addition, we can also mention the lack of uniformity and integration of access to information. Each event has its own online publishing means, and there is no central hub for such information, even within communities in the same domain. Consequently, this makes the process of finding and linking relevant work in a particular field a cumbersome task.

On the desktop, we can find a somewhat similar problem, though on a smaller scale. A typical researcher acquires (and stores) an significant number of publications over time. Generally, the files representing these publications have a non-intuitive name (often the same cryptic name assigned by the system publishing

[1] http://medline.cos.com/

A. Bernstein et al. (Eds.): ISWC 2009, LNCS 5823, pp. 827–842, 2009.

them), and may, in the best case scenario, be structured in intuitive folder hierarchies. Thus, finding a particular publication or links between the existing ones represents quite a challenge, even with the help of tools like Google Desktop[2].

Semantic Web technologies have been proved to help at alleviating, at least partially, the above mentioned issues. And at the foundation of the Semantic Web we find semantic metadata. Used in particular contexts, semantic metadata enables a more fertile search experience, complementing full text search with search based on different facets (e.g., one can search for a publication by a specific author and with some specific keywords in its title). In addition, subject to its richness, it can also leverage links between publications, e.g. citation networks.

Looking at the status of semantic metadata for scientific publications in the two directions, i.e. the Web and the Desktop, we observe the following. With the emergence of the Linked Open Data (LOD)[3] initiative, an increasing number of data sets were published as linked metadata. Regarding scientific publications, efforts like the Semantic Web Dog Food Server started by Möller et al. [2] represent pioneering examples. The repository they initiated acts as a linked data hub, for metadata extracted from different sources, such as the International or European Semantic Web conferences, and now hosts metadata describing over 1000 publications and over 2800 people. The manual creation of metadata is their main drawback, as well as of the other similar approaches. Within the second direction, i.e. on the desktop, different Semantic Desktop efforts improve the situation, by extracting shallow metadata, either file-related (e.g. creator, date of creation), or even publication-related, such as title or authors. In conclusion, we currently have two directions targeting similar goals and having the same foundation: (i) the <LOD — semantic metadata> bridge, linking publications on the web, and (ii) the <Semantic Desktop — semantic metadata> bridge, linking publications and personal information, on the desktop.

In this paper, we propose a solution for bridging the two directions, with the goal of enabling a more meaningful searching and linking experience on the desktop, having the linked data cloud as the primary source. Our method consists of a three step process and starts from a publication with no metadata, each step carried out incrementally to enrich the semantic metadata describing the publication. The three steps are: (i) *extraction* – we extract automatically metadata from the publication based on a document-engineering oriented approach; (ii) *expansion* – we use the extracted raw metadata to search the linked data cloud, the result being a set of clean and linked metadata; and (iii) *integration* – the metadata is further enriched by embedding it within the semantic desktop environment, where it is automatically linked with the already existing personal metadata. The main results of our process are: a simple and straightforward way of finding related publications based on the metadata extracted and linked automatically, and the opportunity of weaving the linked publication data on the desktop, by means of usual desktop applications (e.g. file and Web browser).

---

[2] http://desktop.google.com/

[3] http://linkeddata.org/

The remainder of the paper is structured as follows: Sect. 2 introduces the scenario used for exemplifying our approach, while in Sect. 3 we detail the technical elements of each of the three steps in the process. Sect. 4 describes the preliminary evaluation we have performed, and before concluding in Sect. 6, we have a look at related efforts in Sect. 5.

2 Scenario

To illustrate the problem mentioned in the previous section in a particular context, in the following, we will consider a typical scenario for an early stage researcher (or any kind of researcher) that steps into a new field. The amount of existing publications, and their current growth rate, makes the task of getting familiarized with the relevant literature in a specific domain highly challenging. From an abstract perspective, the familiarization process consists of several stages, as follows: (i) the researcher starts from a publication provided by the supervisor; (ii) she reads the publication, thus grasping its claims and associated argumentation; (iii) reaching a decision point, she either decides to look for another publication, or she follows backwards the chain of references, possibly including also publications by the same authors. This last step usually involves accessing a search engine (or a publication repository) and typing the names of the authors, or the title of the publication, to be able to retrieve similar results.

Each of the above activities has an associated corresponding time component: (i) the time assigned to reading the entire publication (or most of it) — needed to decide whether the publication is of interest or not, (ii) the time associated with finding appropriate links and references between publications,[4] and (iii) the time associated with searching additional publications, based on different metadata elements, and (manually) filtering the search results. This time increases substantially when an individual is interested in all the publications of a particular author.

Finally, analyzing the pairs <activity, time component> from the metadata perspective, and what it can do to improve the overall process, we can conclude that:

- searching and linking related publications as mentioned above, entails the (manual) extraction and use of shallow metadata. Thus, performing automatic extraction of shallow metadata and using it within the linked data cloud, will significantly reduce the time component associated with this activity.
- reading the publication to a large extent corresponds to mining for the discourse knowledge items, i.e. for the rhetorical and argumentation elements of the publication, representing its deep metadata. Consequently, extracting automatically such discourse knowledge items from a publication, will provide the user with the opportunity of having a quick glance over the publication's main claims and arguments and thus decrease the time spent on deciding whether the publication is relevant or not.

[4] Following all the references of a publication is obviously not a feasible option. Thus, the decision is usually done based on the citation contexts mentioning the references in the originating publication.

Transposing these elements into engineering goals led us to a three step process, detailed in the following section: *extraction* – automatic extraction of shallow and deep metadata; *expansion* – using the extracted metadata within the linked data cloud for cleaning and enriching purposes; *integration* – embedding the resulted linked metadata within the personal desktop to ensure a smooth search and browse experience, by using the ordinary desktop applications.

As a side remark to the proposed scenario, an interesting parallel can be made with the music domain. Similarly to publications, music items (e.g. music files, tracks, etc) are also acquired and stored by people on their personal desktops, in numbers usually increasing with time. And as well as publications, these can embed (depending on the format) shallow metadata describing them, such as, band, song title, album or genre. Thus, conceptually, the *extraction* — *expansion* — *integration* process we propose can be applied also in this domain. In practice, there already exist tools that deal with parts of this process. For example, on the extraction side, there are tools that help users to create or extract ID3 tags embedded into MP3 files or on the expansion side, there exist tools, such as Picard,[5] that clean the metadata based on specialized music metadata repositories (e.g. MusicBrainz). As we shall see in the next section, the result of our work is quite similar to these, but applied on scientific publications.

3 Implementation

One of our main goals was reducing as much as possible the overhead imposed by collateral activities that need to be performed while researching a new field, in parallel with the actual reading of publications. And at the same time, we targeted an increase of the user's reward, by ensuring a long-term effect of some of the achieved results. An overall figure of the three step process we propose, is depicted in Fig. 1. The first step, *extraction*, has as input a publication with no metadata and it outputs two types of metadata:

(i) shallow metadata, i.e. title, authors, abstract, and (ii) deep metadata, i.e. discourse knowledge items like claims, positions or arguments. It represents the only step that appears to have no direct reward (or value) for the user (except for

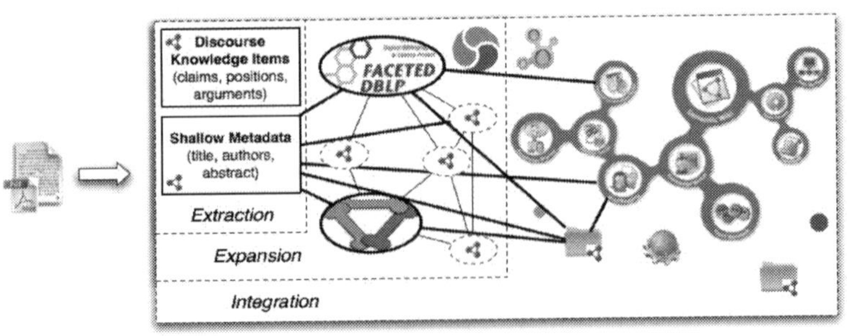

Fig. 1. Incremental metadata enrichment process

[5] http://musicbrainz.org/doc/PicardTagger

the discourse knowledge items). Nevertheless, it is compulsory in order to start the process, each subsequent step building on its results, and thus enabling an incremental approach to the enrichment of the semantic metadata describing the publication. Since the extraction process is based on a hybrid 'document engineering – computational linguistic' approach, the resulting metadata may contain errors. These errors can be corrected in the *expansion* step, in addition to enriching the basic set of metadata with linked data, coming from different sources. As we shall see, we opted for a clear distinction of the semantics of the `owl:sameAs` and `rdfs:seeAlso` relations. Finally, the *integration* step embeds the linked metadata into the semantic desktop environment, thus connecting it deeper within the personal information space, and fostering long-term effects of the overall process.

In terms of implementation, the first two steps are developed as part of a standalone application[6]. The extraction currently targets publications encoded as PDF documents and preferably using the ACM and LNCS styles, while the expansion is achieved via the Semantic Web Dog Food Server and the Faceted DBLP[7] linked data repositories. The integration of the metadata is done using the services provided by the KDE NEPOMUK Server,[8] while the searching and browsing experience is enabled via the usual KDE Desktop applications, such as Dolphin (the equivalent of Windows Explorer) and Konqueror (a KDE Web browser). The application we have developed is highly customizable, each step being represented by a module. Therefore, adding more functionality is equivalent to implementing additional modules, for example, an extraction module for MS Word documents, or an expansion module for DBpedia.

To have a better understanding of the result of each step, we will use as a running example throughout the following sections, the metadata extracted from a publication entitled *Recipes for Semantic Web Dog Food – The ESWC and ISWC Metadata Projects.*[9] More precisely, we will assume that a user would start her quest from this publication, and show the incremental effect of using our application on the created metadata.

3.1 Extraction

The extraction of shallow metadata was developed as a set of algorithms that follow a low-level document engineering approach, by combining mining and analysis of the publication's text based on its formatting style and font information. The algorithms currently work only on PDF documents, with a preference for the ones formatted with the LNCS and ACM styles. Each algorithm in the set deals with one aspect of the shallow metadata. Thus, there are individual algorithms for extracting the title, authors, references and the linear structure.

A complete description of the algorithms can be found in [3]. Nevertheless, to provide the basic idea of how they work, we will describe shortly the authors extraction algorithm. There are four main processing steps: (i) We first merge the

[6] Demo at http://sclippy.semanticauthoring.org/movie/sclippy.htm
[7] http://dblp.l3s.de/
[8] http://nepomuk.kde.org/
[9] http://iswc2007.semanticweb.org/papers/795.eps

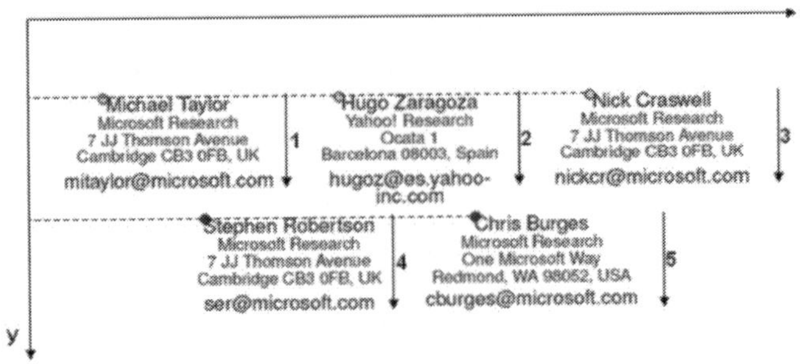

Fig. 2. Authors extraction algorithm example

consecutive text chunks on the first page that have the same font information and are on the same line (i.e. the Y coordinate is the same); (ii) then, we select the text chunks between the title and the abstract and consider them author candidates; (iii) the next step is the linearization of the author candidates based on the variations of the Y axis; (iv) finally, we split the author candidates based on the variations of the X axis.

Fig. 2 depicts an example of a publication that has the authors structured on several columns. The figure shows the way in which the authors' columns containing the names and affiliations are linearized, based on the variation of the Y coordinate. The arrows in the figure show the exact linearization order. The variations on the X axis can be represented in a similar manner.

The extraction of deep metadata, i.e. discourse knowledge items (claims, positions, arguments), was performed based on a completely different approach. Having as foundational background the Rhetorical Structure of Text Theory (RST) [4], we have developed a linguistic parser that mines the presence of rhetorical relations within the publication's content. In order to automatically identify text spans and the rhetorical relations that hold among them, we relied on the discourse function of cue phrases, i.e. words such as *however*, *although* and *but*. An exploratory study of such cue phrases provided us with an empirical grounding for the development of an extraction algorithm. The next phase consisted of an experiment for determining the initial probabilities for text spans to represent knowledge items, based on the participation in a rhetorical relation of a certain type and its block placement in the publication (i.e. abstract, introduction, conclusion or related work). The parser was implemented as a GATE[10] plugin. Detailing the actual extraction mechanism is out of the scope of this paper, as our focus is on the incremental process that realizes the bridging the Linked Web of Data and the Semantic Desktop. Nevertheless, it is worth mentioning that we do extract also deep metadata, as it brings added value to the user, and as we shall see later in this section, enables meaningful queries in the bigger context of the full set of extracted metadata.

As mentioned, the first two steps of our process are implemented as a standalone application. The left side of Fig. 3 depicts the main interface of this

[10] http://gate.ac.uk/

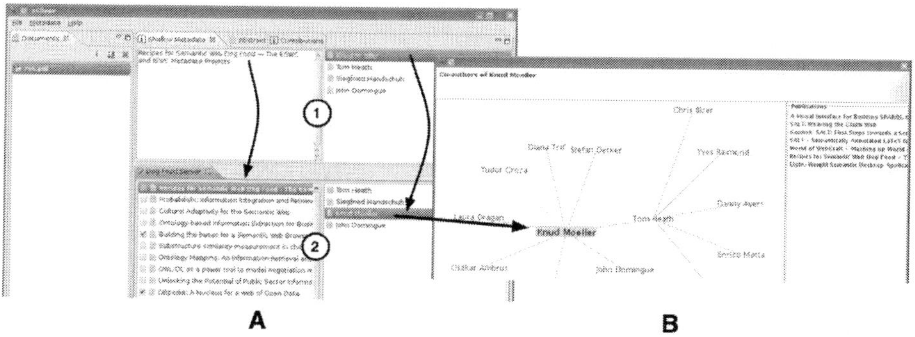

Fig. 3. Screenshot of the application's interface: [A] – The main window; [B] – Co-authors graph visualization

application, while with <1> we indicated the place where the result of the *extraction* is displayed. At the same time, the listing below summarizes elements of the metadata extracted after this first step, i.e. title, authors, the text of the abstract, and a list of claims (i.e. the most important contributions statements of the publication). For shaping the metadata, we used a mixture of ontologies and vocabularies, such as SALT (Semantically Annotated LaTeX) framework [5], DublinCore and FOAF. One particular element that should be noted here, is that this metadata may contain errors. As it can be seen in the listing below the name of the first author is incorrect: *Mo "ller* instead of *Möller*. The user has the chance to correct such mistakes manually, or advance to the next step, where the correction can be done automatically — if the publication under scrutiny is found in the linked data repository. In any case, already at this point, the user can decide to jump to the *integration* step, simply just export this metadata as an individual file, or embed it directly into the originating PDF. From the scenario's point-of-view, the researcher already gains value, as she can quickly grasp the main claims of the publication, by inspecting the extracted deep metadata.

```
<pub> a sdo:Publication .
<pub> dc:title ''Recipes for Semantic Web ...'' .
<pub> dc:creator knud .
<pub> dc:creator tom .
<pub> dc:creator siegfried .
...
<knud> a foaf:Person .
<knud> foaf:name ''Knud Mo"ller'' .
<abs> a sro:Abstract .
<abs> konnex:hasText ''Semantic Web ...'' .
<pub> dcterms:abstract abs .
<claim> a sro:Claim .
<claim> konnex:hasText ''This paper ...'' .
<pub> konnex:hasClaim claim .
```

3.2 Expansion

The *expansion* step takes the metadata extracted previously and, under the user's guidance, corrects existing errors and enriches it, by using Linked Data repositories. We have currently implemented expansion modules for the Semantic Web

Dog Food Sever and Faceted DBLP. The actual expansion is done based on the extracted title and authors. On demand, these are individually used for querying the SPARQL endpoints of the Linked Data repositories. As a prerequisite step, both the title and the authors (one-by-one) are cleaned of any non-letter characters, and transformed into regular expressions. The title is also chunked into multiple variations based on the detected nouns, while each author name is chunked based on the individual parts of the full name, discarding the parts that are just one letter long. Consequently, each element will have an associated array of sub-strings used for querying.

In the case of the title, the query result will be a list of resources that may contain duplicates, and among which there might also be the publication given as input. In order to detect this particular publication, we perform a shallow entity identification. First, to mask possibly existing discrepancies in the title, we use string similarity measures. An empirical analysis led us to using a combination of the Monge-Elkan and Soundex algorithms, with fixed thresholds. The first one analyzes fine-grained sub-string details, while the second looks at coarse-grained phonetic aspects. The titles that pass the imposed thresholds (0.75 and 0.9) advance to the next step. Secondly, we consider the initially extracted authors and compare them with the ones associated with the publications that pass over the above mentioned thresholds. The comparison is done using the same similarity measures, but with different thresholds (0.85 and 0.95). The publications satisfying both conditions have their models retrieved and presented to the user as candidates. A similar approach is also followed on the authors' side.

The outcome of the expansion features three elements: (i) a list of candidates, to be used for cleaning and linking the initially extracted metadata (with their linked model and authors' models), (ii) a list of similar publications, mainly the ones that did not satisfy the two conditions of the shallow entity resolution (again with their linked model and authors' models), and (iii) for each author of the given publication found, the full linked model and the complete list of publications existing in the respective repository. From the scenario perspective, this outcome provides the researcher with the chance of analyzing both publications that might have similar approaches and inspect all the publications of a particular author.

At this stage, there are three options that can be followed. The first option is to use the best retrieved candidate to correct and link the initial metadata. Both the publication and the authors will inherit the discovered `owl:sameAs` links, that will later provide the opportunity to browse different instances of the same entity in different environments. The second option is to link other publications that she considers relevant to the one under scrutiny. While at the interface level this is done based on the user's selection (see pointer 2 in Fig. 3), at the model level we use the `rdfs:seeAlso` relation. We thus make a clear distinction in semantics between `owl:sameAs` and `rdfs:seeAlso`. The former represents a strong link between different representations of the same entity, while the latter acts as a weak informative link, that will later help the user in re-discovering similarities between several publications. The third and last option is specific for authors, and allows the user to navigate through the co-authors networks of a particular author (part

B of Fig. 3). An interesting remark here, is that the visualization we have developed can act as a uniform graph visualization tool for any co-author networks emerging from a linked dataset.

Returning to our running example, the output of this step is an added set of metadata, presented briefly in the listing below. Thus, in addition to the already existing metadata, we can now find the above mentioned `owl:sameAs` and `rdfs:seeAlso` relations, and the incorrectly extracted name *Mo "ller*, now corrected to *Moeller*, based on the `foaf:name` found in the linked data.

```
<knud> foaf:name ''Knud Moeller'' .
<knud> owl:sameAs http://data.semanticweb.org/person/knud-moeller .
<knud> owl:sameAs http://dblp.l3s.de/d2r/resource/authors/Knud_M\%C3\%B6ller .
...
<pub> owl:sameAs http://data.semanticweb.org/conference/iswc-aswc/2007/.../papers/795 .
<pub> owl:sameAs http://dblp.l3s.de/d2r/resource/publications/conf/semweb/MollerHHD07 .
<pub> rdfs:seeAlso <pub2> .
...
<pub2> a sdo:Publication .
<pub2> dc:title ''DBPedia: A Nucleus for a Web of Open Data ... '' .
<pub2> dc:creator <richard> .
<pub2> dc:creator <georgi> .
<pub2> owl:sameAs http://dblp.l3s.de/d2r/resource/publications/conf/semweb/AuerBKLCI07 .
<pub2> owl:sameAs http://data.semanticweb.org/conference/iswc-aswc/2007/.../papers/715 .
```

3.3 Integration

The last step of our process is the *integration*, which embeds the extracted and linked metadata into the personal information space, managed by the Semantic Desktop, and thus realizing the actual bridge between the Linked Data and the Semantic Desktop. To achieve this, we have used the NEPOMUK–KDE implementation of the Semantic Desktop. This provides a central local repository for storing structured data and it is well integrated with the common desktop applications, such as the file and Web browsers. In terms of foundational models, it uses the NEPOMUK Ontologies[11] suite. In our case, the actual integration was done at the metadata level, where we had to align the instances previously extracted with the ones already existing in the local repository.

Currently, we deal with two types of alignments: person alignment and publication alignment, that are described within the Semantic Desktop context by means of the NCO (NEPOMUK Contact Ontology), NFO (NEPOMUK File Ontology) and PIMO (Personal Information Model Ontology) ontologies.

The person alignment resumes to checking whether an extracted author is already present in the personal information model, and in a positive case, merging the two models in an appropriate manner. This is done based on a two step mechanism, similar to finding authors in a Linked Data repository. We first query the local repository for the name of the author and the associated substrings resulted

[11] http://www.semanticdesktop.org/ontologies/

from chunking the name into several parts. Using the same similarity measures, we filter out only the realistic candidates. These candidates are then analyzed based on the existing `owl:sameAs` links and their linked publications. If a candidate is found to have one identical `owl:sameAs` link and one identical publication with the initial author, we consider it a match and perform the merging of the two models. In a negative case, the author's model is stored as it is and we advance to the publication alignment. The result of this alignment is exemplified in the listing below. In addition to the already existing metadata, the author now has attached an email address and the birth date, both found within the user's personal information space.

The publication alignment is straightforward, as from a local and physical perspective, the publication is represented by a file. Thus, considering that the user started the process from such a file (which is always the case), we query the repository for the information element corresponding to that file, having the `fileUrl` (or path) as the main indicator. The conceptual model found to be associated with the file is then merged with the extracted publication model. The listing below shows this alignment as part of our running example, the last statement creating the actual grounding of the publication onto a physical file.

The *integration* enables, in particular, two important elements: (i) firstly, more meaningful ways of finding and linking publications on the desktop, and (ii) secondly, an opportunity of weaving the linked data present on the desktop, using ordinary desktop applications. Fig. 4 depicts the first aspect, using Dolphin, the

```
<knud> nco:birthDate ''1980-11-01'' .
<knud> nco:emailAddress knud.moeller@deri.org .
...
<pubFile> a nfo:FileDataObject .
<pubFile> nfo:fileSize 1353543 .
<pubFile> nfo:fileUrl file:///home/user/research/papers/p215.eps .
<pub> pimo:groundingOccurence <pubFile> .
```

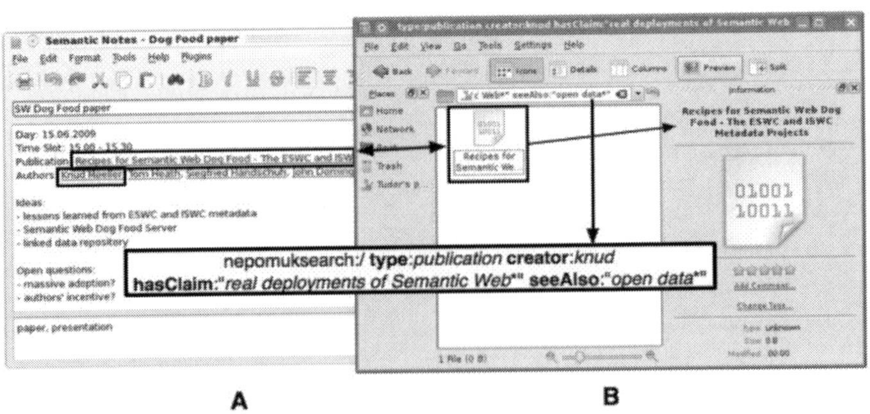

Fig. 4. Deep metadata integration in KDE applications: [A] SemNotes; [B] Dolphin

KDE file browser (part B), and SemNotes,[12] a KDE semantic note-taking application (part A). As shown in the figure, to retrieve all `publications` having `knud` among the authors, `claim`-ing that they deal with *"real deployments of Semantic Web..."* and being related to (`seeAlso`) publications that speak about *"open data"*, resolves to using the following query in Dolphin:

```
nepomuksearch:/ type:publication creator:knud hasClaim:''real
    deployments of Semantic Web*'' seeAlso:''open data*''
```

The result of the query will be a virtual folder that is automatically updated in time (enabling the long-term effect), thus showing also the publications that are added at a later stage and that satisfy the query, independently of the name of the physical file or its location. Similarly, while taking notes during the presentation of this particular paper, SemNotes will automatically link both the publication and the author mentioned in the note, therefore providing added information about the publication and its authors.

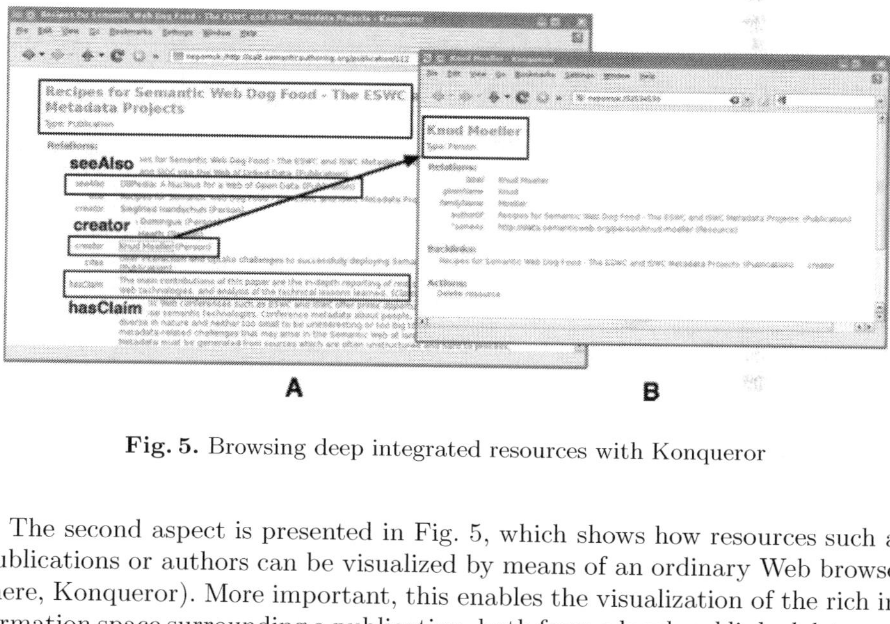

Fig. 5. Browsing deep integrated resources with Konqueror

The second aspect is presented in Fig. 5, which shows how resources such as publications or authors can be visualized by means of an ordinary Web browser (here, Konqueror). More important, this enables the visualization of the rich information space surrounding a publication, both from a local and linked data perspective. Without even opening the actual publications, the user can: (i) quickly grasp the main ideas of the publications, via the presented claims, (ii) see related publications, via the `rdfs:seeAlso` links, or (iii) inspect the publications' authors, either via their personal contact information, or via their different instances on the Web (`owl:sameAs` links). We believe that this approach combines harmoniously the Linked Data perspective with the Semantic Desktop perspective, thus enabling the weaving of Linked Data on the desktop.

[12] http://smile.deri.ie/projects/semn

4 Preliminary Evaluation

We evaluated the extraction of semantic metadata and performed a short-term usability study of the overall approach. The shallow metadata extraction achieved high accuracies for the title (95%) and abstract (96%) extraction, and a lower accuracy for authors extraction (90%). The evaluation method and complete results can be found in [3]. In this section, we focus on the usability study, as we believe that the developed application has to be easy to learn and use, and to provide the most appropriate information.

The study was conducted together with 16 evaluators, a mixture of PhD students and Post Doctorands from our institute, that were asked to perform a series of tasks covering all the application's functionalities. Example of such tasks included: extraction and manual correction of metadata from publications, expansion of information based on the same publications or exploration of the co-authors graph. At the end, the evaluators filled in a questionnaire, comprising of 18 questions, with Likert scale-based or free form answers, concentrating on two main aspects: (i) suitability and ease of use, and (ii) design, layout and conformity to expectancies. The complete results of the questionnaire can be found at http://smile.deri.ie/sclippy-usabilitystudy

Overall, the application scored very well in both categories we have targeted. The vast majority of the evaluators (on average more than 90%) found the tool well suited for the extraction and exploration of shallow and deep metadata. The same result was achieved also for the exploration of the information space surrounding the chosen publication, based on the extracted and linked metadata. In addition, the information presented by the application, both for publications and authors, was found helpful (100% for publications and 72.8% for authors), while 93.8% of the evaluators found an added value in our tool when compared to the original expansion environment.

In the other category, all evaluators considered the application easy to learn and use (100%) while having the design and layout both appealing (87.5%) and suited for the task (93.6%). Issues were discovered in two cases: (i) the self-descriptiveness of the application's interface (only 68.8% found it self-descriptive), mainly due to the lack of visual indicators and tooltips, and (ii) the suggested list of similar publications (again only 68.8% found it relevant). Although the application always found the exact publication selected for expansion in the repository, the proposed list of similar publications created some confusion.

Apart from these findings, directly taken from the questionnaires, we observed that even without any training and documentation, the evaluators experienced a very smooth learning curve. Additionally, most of them enjoyed our exercise, while some were interested in using the application on a daily basis. On the other hand, the study pointed out a number of issues and led us to a series of directions for improvement. First of all, the need to make use of a more complex mechanism for suggesting similar publications. As we expected, the shallow similarity-based heuristics we used for building the list of suggested publications left plenty of space for improvement. Unfortunately, its improvement is directly dependent on the quantity and quality of information provided by the linked data repository. As an example, while we could use the abstract provided by the Semantic Web Dog Food Server to extract discourse knowledge items, and then perform similarity measures at this level, this would not be possible when using the Faceted DBLP, where such information does not exist. For this case, a possible solution, would be to drill deeper into the linked web of data. Secondly, augmenting the expanded information with additional elements (e.g. abstract, references, citation contexts), thus providing a deeper insight into the publications and a richer experience for the users.

5 Related Work

To our knowledge, until now, there was no attempt to combine in such a direct manner automatic metadata extraction from scientific publications, linked open data and the semantic desktop. Nevertheless, there are efforts that deal with parts of our overall approach, and, in this section, we will focus on them. Hence, we will cover: (i) automatic extraction of shallow metadata, including the context of the semantic desktop, and (ii) information visualization for scientific publications.

Before detailing the two above-mentioned directions, we would like to discuss the position of the alignments described in the *expansion* and *integration* steps to the current state of the art. To a certain extent, these person and publication alignments are similar to performing coreference resolution. While in the person case the resolution is solved directly via string similarity measures, in the publication case we add the authors list as an extra condition. This makes our approach more simple and straightforward than the more accurate algorithms existing in the literature. Examples of such techniques include: naive Bayes probability models and Support Vector Machines [6], K-means clustering [7] or complex coreference based on conditionally trained uni-directed graph models using attributes [8].

Extensive research has been performed in the area of the Semantic Desktop, with a high emphasis on integration aspects within personal information management. Systems like IRIS [9] or Haystack [10] deal with bridging the different isolated data silos existing on the desktop, by means of semantic metadata. They extract shallow metadata from the desktop files and integrate it into a central desktop repository. Compared to our approach, the metadata extraction is file-oriented and shallow, whereas we extract specific publication metadata and integrate it within the already existing semantic desktop data. The closest effort to ours was the one of Brunkhorst et al. [11]. In their Beagle++ search engine, developed in the larger context of the NEPOMUK Semantic Desktop [12], the authors also perform metadata extraction from scientific publications, but limited to title and authors.

Regarding the general context of automatic extraction of metadata from publications, there have been several methods used, like regular expressions, rule-based parsers or machine learning. Regular expressions and rule-based systems have the advantage that they do not require any training and are straightforward to implement. Successful work has been reported in this direction, with emphasis on PostScript documents in [13], or considering HTML documents and use of natural language processing methods in [14]. Our approach is directly comparable with these, even though the target document format is different. In terms of accuracy, we surpass them with around 5% on title and authors extraction, and with around 15% on linear structure extraction, while providing additional metadata (i.e. abstract or references).

Although more expensive, due to the need of training data, machine learning methods are more efficient. Hidden Markov models (HMMs) are the most widely used among these techniques. However, HMMs are based on the assumption that features of the model they represent are not independent from each other. Thus, HMMs have difficulty exploiting regularities of a semi-structured real system. Maximum entropy based Markov models [15] and conditional random fields [16] have been introduced to deal with the problem of independent features. In the same category, but following a different approach, is the work performed by Han et al. [17], who uses Support Vector Machines (SVMs) for metadata extraction.

With respect to information visualization of scientific publications, a number of methods and tools have been reported in the literature. The 2004 InfoVis challenge had motivated the introduction of a number of visualization tools highlighting different aspects of a selected set of publications in the Information Visualization domain. Faisal et. al. [18] reported on using the InfoVis 2004 contest dataset to visualize citation networks via multiple coordinated views. Unlike our work, these tools were based on the contents of a single file, which contained manually extracted and cleaned metadata. As noted by the challenge chairs, it was a difficult task to produce the metadata file [19] and hence the considerable efforts required made it challenging for wide-spread use. In [20], a small scale research management tool was built to help visualizing various relationships between lab members and their respective publications. A co-authorship network visualization was built from data entered by users in which nodes represented researchers together with their publications, and links showed their collaborations. A similar effort to visual domain knowledge was reported by [21], with their data source being bibliographic files obtained from distinguished researchers in the "network science" area. While this work was also concerned with cleansing data from noisy sources, the metadata in use was not extracted from publications themselves and no further information available from external sources such as Faceted DBLP was utilized. Another tool targeting the exploration of the co-authorship network is CiteSpace [22]. CiteSpace tries to identify trends or salient patterns in scientific publications. The source of information for CiteSpace is also from bibliographic records crawled from different publishers on the web, rather than extracted metadata.

6 Conclusion and Future Developments

In this paper we presented an approach for dealing, at least to some extent, with the information overload issue both on the Web and on the Desktop, and having as target early stage researchers. Our solution, inspired from the typical process of getting familiarized with a particular domain, combines elements from the Linked Web of Data and the Semantic Desktop, using semantic metadata as a common denominator. The result consists of three steps (*extraction – expansion – integration*) that incrementally enrich the semantic metadata describing a publication, from no metadata to a comprehensive model, linked and embedded within the personal information space.

Each step has associated a series of open challenges that we intend to address as part of our future work. As currently the *extraction* works only on publications published as PDF documents, and formatted preferably with the LNCS and ACM styles, we plan to improve extraction algorithms to accommodate any formatting style, as well as develop new extraction modules for other document formats, such as Open Document formats. At the moment, the *expansion* uses only two Linked Data repositories, i.e. the Semantic Web Dog Food Server and the Faceted DBLP. Future developments will include also other repositories, in addition to means for creating ad-hoc mash-ups between them, thus allowing the user to see data coming from different sources in an integrated and uniform view. Last, but not least, we plan an even tighter *integration* within the Semantic Desktop, therefore enabling more meaningful queries and a richer browsing experience, and ultimately a complete automatization of the process, thus reducing the overhead to the minimum possible.

Acknowledgments

The work presented in this paper has been funded by Science Foundation Ireland under Grant No. SFI/08/CE/I1380 (Lion-2).

References

1. Tsujii, J.: Refine and pathtext, which combines text mining with pathways. In: Keynote at Semantic Enrichment of the Scientific Literature 2009, Cambridge, UK (2009)
2. Möller, K., Heath, T., Handschuh, S., Domingue, J.: Recipes for Semantic Web Dog Food — The ESWC and ISWC Metadata Projects. In: Aberer, K., Choi, K.-S., Noy, N., Allemang, D., Lee, K.-I., Nixon, L.J.B., Golbeck, J., Mika, P., Maynard, D., Mizoguchi, R., Schreiber, G., Cudré-Mauroux, P. (eds.) ASWC 2007 and ISWC 2007. LNCS, vol. 4825, pp. 802–815. Springer, Heidelberg (2007)
3. Groza, T., Handschuh, S., Hulpus, I.: A document engineering approach to automatic extraction of shallow metadata from scientific publications. Technical Report 2009–06–01, Digital Enterprise Research Institute (2009)
4. Mann, W.C., Thompson, S.A.: Rhetorical Structure Theory: A theory of text organization. Technical Report RS-87-190, Information Science Institute (1987)
5. Groza, T., Handschuh, S., Möller, K., Decker, S.: SALT - Semantically Annotated LaTeX for Scientific Publications. In: Franconi, E., Kifer, M., May, W. (eds.) ESWC 2007. LNCS, vol. 4519, pp. 518–532. Springer, Heidelberg (2007)
6. Han, H., Zha, H., Giles, C.: A Model-based K-means Algorithm for Name Disambiguation. In: Proc. of the Workshop on Semantic Web Technologies for Searching and Retrieving Scientific Data @ ISWC 2003, Sanibel Island, Florida, USA (2003)
7. Han, H., Giles, L., Zha, H., Li, C., Tsioutsiouliklis, K.: Two supervised learning approaches for name disambiguation in author citations. In: Proc. of the 4th ACM/IEEE-CS joint conference on Digital Libraries, Tuscon, AZ, USA (2004)

8. Wellner, B., McCallum, A., Peng, F., Hay, M.: An integrated, conditional model of information extraction and coreference with application to citation matching. In: Proc. of the 20th Conf. on Uncertainty in Artificial Intelligence, Banff, Canada (2004)
9. Cheyer, A., Park, J., Giuli, R.: IRIS: Integrate. Relate. Infer. Share. In: Proc. of the Semantic Desktop and Social Semantic Collaboration Workshop (2006)
10. Quan, D., Huynh, D., Karger, D.R.: Haystack: A Platform for Authoring End User Semantic Web Applications. In: Proc. of the 2nd International Semantic Web Conference, pp. 738–753 (2003)
11. Brunkhorst, I., Chirita, P.A., Costache, S., Gaugaz, J., Ioannou, E., Iofciu, T., Minack, E., Nejdl, W., Paiu, R.: The beagle++ toolbox: Towards an extendable desktop search architecture. Technical report, L3S Research Centre, Hannover, Germany (2006)
12. Bernardi, A., Decker, S., van Elst, L., Grimnes, G., Groza, T., Handschuh, S., Jazayeri, M., Mesnage, C., Möller, K., Reif, G., Sintek, M.: The Social Semantic Desktop: A New Paradigm Towards Deploying the Semantic Web on the Desktop. In: Semantic Web Engineering in the Knowledge Society. IGI Global (2008)
13. Shek, E.C., Yang, J.: Knowledge-Based Metadata Extraction from PostScript Files. In: Proc. of the 5th ACM Conf. on Digital Libraries, pp. 77–84 (2000)
14. Yilmazel, O., Finneran, C.M., Liddy, E.D.: Metaextract: an nlp system to automatically assign metadata. In: JCDL 2004: Proc. of the 4th ACM/IEEE-CS joint conference on Digital libraries, pp. 241–242 (2004)
15. McCallum, A., Freitag, D., Pereira, F.: Maximum entropy markov models for information extraction and segmentation. In: Proc. of the 17th Int. Conf. on Machine Learning, pp. 591–598 (2000)
16. Lafferty, J., McCallum, A., Pereira, F.: Conditional random fields: Probabilistic models for segmenting and labeling sequence data. In: Proc. of the 18th Int. Conf. on Machine Learning, pp. 282–289 (2001)
17. Han, H., Giles, C.L., Manavoglu, E., Zha, H., Zhang, Z., Fox, E.A.: Automatic document metadata extraction using support vector machines. In: Proc. of the 3rd ACM/IEEE-CS Joint Conf. on Digital libraries, pp. 37–48 (2003)
18. Faisal, S., Cairns, P.A., Blandford, A.: Building for Users not for Experts: Designing a Visualization of the Literature Domain. In: Information Visualisation 2007, pp. 707–712. IEEE Computer Society Press, Los Alamitos (2007)
19. Plaisant, C., Fekete, J.-D., Grinstein, G.: Promoting Insight-Based Evaluation of Visualizations: From Contest to Benchmark Repository. IEEE Transactions on Visualization and Computer Graphics 14(1), 120–134 (2008)
20. Neirynck, T., Borner, K.: Representing, analyzing, and visualizing scholarly data in support of research management. In: Information Visualisation 2007, pp. 124–129. IEEE Computer Society Press, Los Alamitos (2007)
21. Murray, C., Ke, W., Borner, K.: Mapping scientific disciplines and author expertise based on personal bibliography files. In: Information Visualisation 2006, pp. 258–263. IEEE Computer Society Press, Los Alamitos (2006)
22. Chen, C.: CiteSpace II: Detecting and visualizing emerging trends and transient patterns in scientific literature. J. of the American Society for Information Science and Technology 57(3), 359–377 (2006)

Vocabulary Matching for Book Indexing Suggestion in Linked Libraries – A Prototype Implementation and Evaluation

Antoine Isaac[1,2], Dirk Kramer[2], Lourens van der Meij[1,2], Shenghui Wang[1,2], Stefan Schlobach[1], and Johan Stapel[2]

[1] Vrije Universiteit Amsterdam
[2] Koninklijke Bibliotheek, Den Haag
{aisaac,lourens,swang,schlobac}@few.vu.nl,
{Dirk.Kramer,Johan.Stapel}@kb.nl

Abstract. In this paper, we report on a technology-transfer effort on using the Semantic Web (SW) technologies, esp. ontology matching, for solving a real-life library problem: book subject indexing. Our purpose is to streamline one library's book description process by suggesting new subjects based on descriptions created by other institutions, even when the vocabularies used are different. The case at hand concerns the National Library of the Netherlands (KB) and the network of Dutch local public libraries. We present a prototype subject suggestion tool, which is directly connected to the KB production cataloguing environment. We also report on the results of a user study and evaluation to assess the feasibility of exploiting state-of-the art techniques in such a real-life application. Our prototype demonstrates that SW components can be seamlessly plugged into the KB production environment, which potentially brings a higher level of flexibility and openness to networked Cultural Heritage (CH) institutions. Technical hurdles can be tackled and the suggested subjects are often relevant, opening up exciting new perspectives on the daily work of the KB. However, the general performance level should be made higher to warrant seamless embedding in the production environment—notably by considering more contextual metadata for the suggestion process.

1 Introduction

Motivation Cultural Heritage (CH) institutions usually own well-described collections of objects, and publishing (descriptions of) their assets to unknown users is an inherent part of their mission. Of course, both aspects are equally crucial for the Web of Data: data with structured meta-data, and the drive to publish data for unforeseen reuse.

This has lead to many CH institutions being at the forefront of developing and applying Semantic Web (SW) technology. Important initiatives for representing knowledge on the SW, such as Dublin Core [1] or SKOS [2], have been driven by expertise and requirements from the CH domain. In turn, many recent projects

A. Bernstein et al. (Eds.): ISWC 2009, LNCS 5823, pp. 843–859, 2009.

have largely benefited from using SW techniques to make CH material accessible more easily, *e.g.*, in [3,4].

Given these success stories, it is no surprise to see yet another fruitful cross-fertilisation opportunity between the two fields: *semantic interoperability*. In the SW community, recognition of this problem has lead to exhaustive research in ontology matching [5], the goal of which is to make interoperable the data expressed in different ontologies. Again, CH institutions have strong interest for adapting this technology in their daily routine and for their future vision. Collections from different institutes are indeed described with different knowledge organisation systems (KOS), such as thesauri. In a more and more inter-linked CH world requiring cross-collection applications, alignments between different KOSs become crucial.

Linking vocabularies across Dutch libraries. The National Library of the Netherlands (KB) holds numerous books also described by other libraries, many of them having their own way of semantically describing (indexing) the content of books. This leads to human indexers at the KB still going through each book to re-index it in "the KB way," *i.e.*, using their own thesauri, although that book might have already been described by some other institutions.

Such re-indexing process can be streamlined by suggesting indexers relevant concepts from KB thesauri, based on other institutions' descriptions. In the context of the STITCH project, we implemented a prototype that provides such a functionality. For each incoming book already indexed with concepts from the Biblion thesaurus (used by Dutch public libraries), it suggests a new indexing with the Brinkman subject thesaurus used in the KB. The suggestion is based on semantic links between these two thesauri which are derived from generic ontology matching methods [6,7].

This prototype has been directly connected to the KB production cataloguing environment, used daily by indexers. To the best of our knowledge, KB is one of the first non-academic institutions to use and thoroughly evaluate generic ontology matching technology for a real-life application. This paper describes the index suggestion prototype, an exhaustive evaluation, and some lessons learned, particularly from the perspective of the SW technology that was used.

Evaluating the SW approach in practice. In order to assess the usefulness in practice, we thoroughly evaluated our proposed methodology and prototype implementation. We had two goals, for which involvement of KB staff at all levels (management, technical and user—indexers) was critical:

- technology transfer. STITCH gathers researchers and practitioners to investigate how SW technology can be used to solve CH interoperability issues. In previous experiments, researchers have played a leading role. Here, KB is sought to actively participate in all steps of the development of a SW-based tool, from data preparation to design and testing.
- feasibility study. This prototyping experiment had to determine whether—and to which extent—KB could benefit in the medium term from the techniques employed. The tool must deliver appropriate suggestion performance,

and its design has to fit existing processes seamlessly, both at the conceptual and technical levels. Evaluation is thus key, as is the realistic embedding of the developed tool into the production process.

Findings Our suggestion prototype, on one hand, shows the limitation of the used matching methods to provide high quality suggestions. However, the close collaboration with KB staff gives us much deeper insight into the problem and, importantly, points out potential improvements. In particular, exploiting existing book descriptions for matching thesauri is a valid approach. Only, it may require more thesauri to be linked, in order to make the re-indexing process more efficient.

From that perspective, it is crucial that the prototype demonstrated that SW components can be successfully plugged into the KB production environment. This brings the latter a higher level of flexibility and openness, making it better interoperable with other providers (and consumers) of semantic data.

In Sec. 2 we present the specific case we addressed in our experiment. Sec. 3 then gives a general technical overview of our prototype. Sec. 4 describes our user study and its results, before we conclude.

2 Book Re-indexing at the KB

To manage and allow access to its collections, KB relies on a careful description of its books. This includes *subject indexing*, a concise and controlled reformulation of book subjects, typically done by assigning concepts from a KOS. Indexing requires trained employees to analyze the content of books and carefully pick the most appropriate concept(s) for describing them. This is labor-intensive, and KB is aiming to assist indexers by using (semi-)automatic techniques.

2.1 Existing Work on Assisting Document Description

The first approach to automated document description is to apply natural language processing tools to the textual content (or summary) of documents. In SW research, text analyzers have been used to produce (structured) semantic annotations along formal ontologies that are either pre-defined or learned on-the-fly [8]. Similar techniques have also been deployed to produce document annotations with KOSs that are closer to the ones used in libraries—see [9,10].

However, these techniques require a sufficient amount of textual data, which is not always available, especially for KB whose books have mostly not been digitized. As a result, one has to find sets of related documents (as in the CHOICE project [10]) or to exploit the limited textual information present in the metadata record (the title, sometimes a summary). KB already experimented with the latter approach, without obtaining convincing results.

We have opted for a different approach, exploiting descriptions resulting from a principled interpretation. Our problem is therefore to bridge across different interpretations, using ontology matching, rather than bridge the semantic gap between uninterpreted text content and controlled indexing.

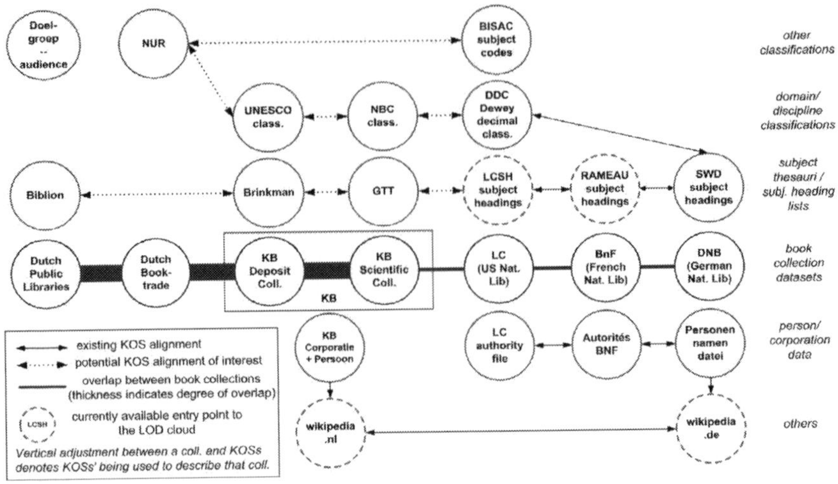

Fig. 1. (Partial) cloud of related collections and KOSs in KB's environment

2.2 The Need for Re-indexing at KB

Our proposal is to assist indexers by re-using existing subject descriptions for books, as contributed by other institutions. Quite often, indeed, a same book can be of interest—and thus described—by several actors. KB thus holds numerous books that are also held by other Dutch libraries, by publishers, or even by foreign institutions. For instance, the collections of KB and the Dutch (local) public libraries contain around a quarter of million books in common. Figure 1 shows some collections that are related to the KB ones. As the goals of the holding institutions often overlap, there is a certain amount of redundancy. Sharing and connecting descriptions in such a network would thus generate interesting synergies and productivity gains.

In fact, many libraries already share their descriptions. For instance, book metadata in the KB is stored in a database—OCLC-Pica's *GGC*—which is also used by other libraries for shared cataloguing, in particular, by the Dutch academic libraries and the Dutch public libraries. However, if integration of descriptions at a basic level (authors, publication date, etc.) could be done, these libraries still perform subject indexing with their own KOSs, as depicted in Fig. 1. Indexing is indeed largely application-specific: different collections are gathered for specific information needs, and require descriptions with different granularity or expertise levels, or even in different languages. There can be significant semantic overlap between the KOSs used, but it is not possible to benefit directly from the work done by other institutions. To re-use existing descriptions in the KB application context, they have first to be fit into KB needs.

In this paper, we focus on the specific KB–public libraries case. KB, for indexing the subject of books in its *legal deposit* collection, uses the *Brinkman subject thesaurus* (hereafter called "Brinkman"). This thesaurus has a generic scope and contains more than 5,000 topical concepts. The main subject thesaurus used in the public libraries is the *Biblion general keyword thesaurus* (hereafter called "Biblion") which contains approximately 18,000 concepts. KB indexers quite often have to *re-index*, with Brinkman concepts, books that have already been indexed with Biblion, while the scope and structure of the two general thesauri overlap quite much. The question is whether this re-indexing can be streamlined by an automatic process providing indexers with candidate subjects.

2.3 Re-indexing Requirements

One solution is to create semantic correspondences between the elements of the KOSs at hand, allowing to convert descriptions from one system to the other. This scenario, described in [11], can be regarded as a problem of ontology matching.

In our case, re-indexing requires to implement a translation process from concepts of a Biblion-indexed book into Brinkman ones, so as to yield a Brinkman indexing of that book. Such a functionality can be formalized by the function:

$$f_r : 2^{\mathcal{BI}} \rightarrow 2^{\mathcal{BR}},$$

where $2^{\mathcal{BI}}$ and $2^{\mathcal{BR}}$ denote the powersets of Biblion and Brinkman concepts.

At first sight, one expects such function to relate Biblion concepts to Brinkman concepts that have semantically equivalent or close meaning. *Variability* issues can however make such simple equivalence associations unfit. First, one same book may be described by two semantically different concepts, even though each of these concepts has an equivalent concept in the other KOS. This may for instance stem from institutions' indexing policies aiming at different description granularity levels. Second, related to the way KOSs are designed, a concept from one vocabulary may be expressed using several concepts from the other KOS, which will be combined together to describe the subject of a book.

To the best of our knowledge, there is no similar work deploying ontology matching for migrating subject indexes from a KOS to another. Ontology matching has been identified as a solution for the more general data migration/translation problem, but it is difficult to find examples of concrete deployments [5]. Also, in more industry-oriented database migration efforts, the focus is rather on translating data from one schema to another—that is, focusing on the relation between (metadata) fields rather than the values that populate them. Finally, in the CH domain, most projects exploiting KOS alignments focus on query reformulation or cross-collection browsing (as in HILT[1] or the SW-based Europeana Thought Lab[2]), which are related but different application issues.

3 Prototype Design

3.1 General Description

The core functionality of our Brinkman subject suggestion tool (SST) is to use available Biblion indexing and other metadata to suggest new Brinkman

[1] `http://hilt.cdlr.strath.ac.uk/`

[2] `http://www.europeana.eu/portal/thought-lab.html`

indexing. This functionality is coupled to the *WinIBW* software used to describe books in the production process of the KB acquisition and cataloguing department. WinIBW is connected to the aforementioned GGC cataloguing system, which is also used by the public libraries. The metadata including Biblion indexing can therefore be accessed and exploited seamlessly.

The SST can be activated when an indexer describes a book already indexed with a Biblion subject. A pop-up window presents then the new Brinkman suggestions. The correct suggestions can be selected and automatically inserted within the description of the book currently being edited in WinIBW. After this, the description is saved and stored in GGC.

In the following, we describe further the use and functioning of the SST, as well as some aspects of the suggestion process itself.

3.2 User Interface

The SST can be launched for a given book being described in the WinIBW system. It comes as a pop-up window split in two panels, as shown in Fig. 2. This window gathers the existing metadata of the book and suggests new subjects.

The left-hand side displays the existing metadata, using a layout similar to WinIBW. We reproduced this basic layout so as to integrate the prototype as seamlessly as possible in the current process. In particular, our panel keeps to the

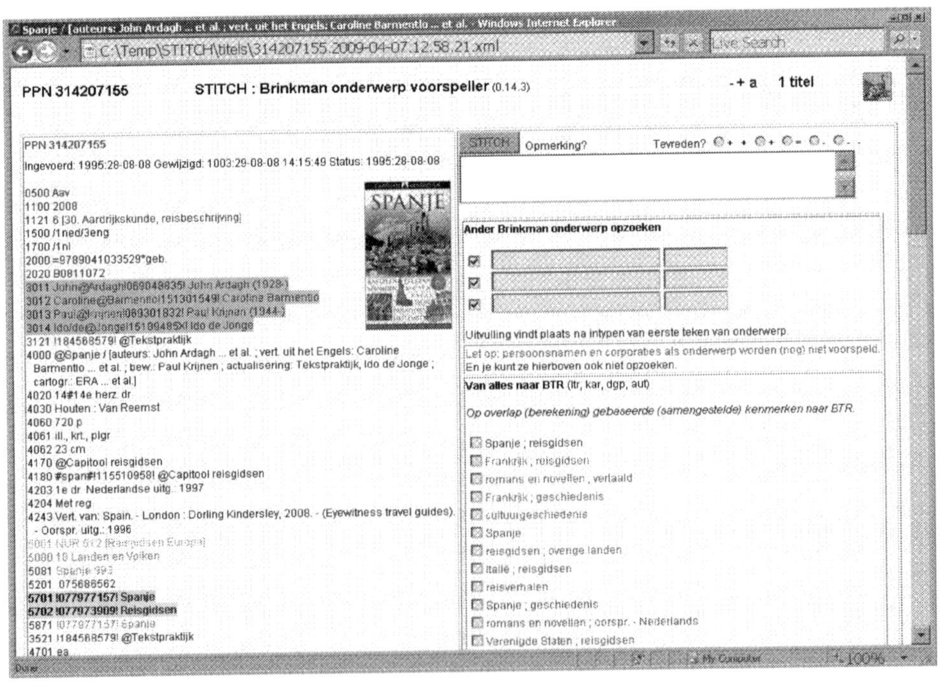

Fig. 2. The Suggestion Tool's user interface

original description codes—"kenmerkcodes," or KMCs [12]. The only additions are the use of color for spotting different elements, and the access to basic additional information services. For instance, 112X corresponds to the "UNESCO" classification (very general, mostly subject-based classification of books, *e.g.*, "linguistics"), 111X to "KAR" (the "characteristic", roughly corresponding to the type of publication, *e.g.*, "thesis"), 1401 to DGP (intellectual level/target group, *e.g.*, "youth fiction, 7-8 years") and 5061 to NUR (Dutch book trade (topical) classification focused on genres, *e.g.*, "national history"). Further descriptions of these elements can be accessed via online help resources.

The right-hand side of the screen presents the suggestions (bottom). Three colors (blue, purple, red) are used to mirror the confidence level of the suggestions (see Sec. 3.4 for more details). Indexers can select the suggestion(s) they agree with by ticking them. There are also empty boxes for searching new concepts to add to the description ("Ander Brinkman onderwerp opzoeken").

For the user study, indexers are offered the possibility to give a general quality assessment of the suggestions for the book ("tevreden" radio buttons) as well as a comment ("opmerking"). When the screen has been filled, users can save the new indexing and go back to WinIBW by pressing the "STITCH" button.

There is also the possibility to ask for more information on suggestions (the '+' top-right), such as the concepts from the record that triggered them, and their numerical confidence value. This shows more suggestions (with lower confidence). Previously established alignments based on equivalence of concept labels (see Sec. 3.4) can also be requested to provide more suggestions (the 'a' top-right).

3.3 Architecture

Our SST prototype consists of various distributed components, as illustrated in Fig. 3.

Indexers start from the WinIBW cataloguing interface, a Windows client. Simply clicking a button, they launch a VisualBasic script that converts the current book data in XML format and displays it into an Internet Explorer pop-up window.

This window uses an XSLT stylesheet to display the book metadata, as well as querying various services and displaying the information got from them as XML data. To enhance information presentation, the stylesheet triggers access to two business web services—*Bol.com* for book images and *Picarta* for obtaining additional book information. To fulfill its main concept suggestion mission, it accesses two "alignment web services" hosted on machines at the Free University of Amsterdam. Primary suggestions are provided by a SWI-Prolog server that exploits the statistical rules described in Sec. 3.4. Additional (lexical) alignments (for the 'a' option) are accessed via a SPARQL endpoint [13] on top of a Sesame[3] RDF store.

At this stage, indexers can add one or more subjects, which they find with the help of a vocabulary service (JavaScript/Ajax). When they are finished,

[3] `http://openrdf.org`

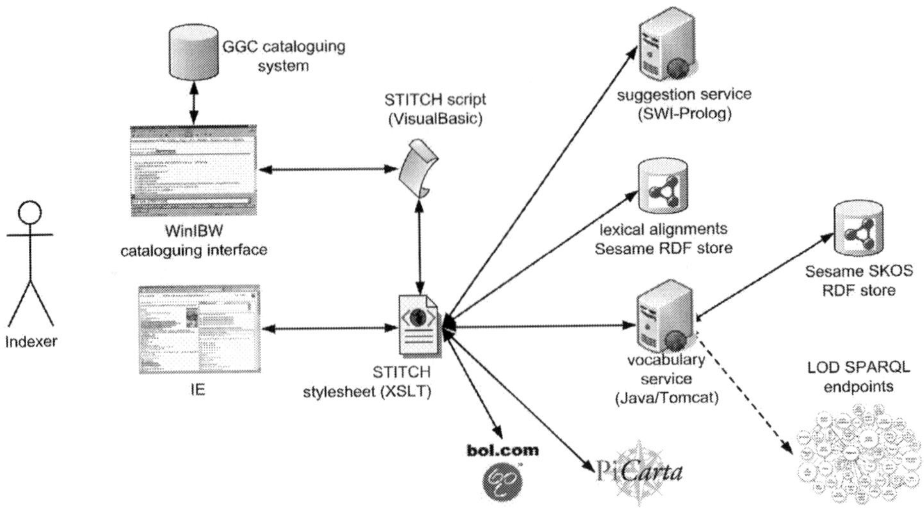

Fig. 3. The Suggestion Tool's architecture

clicking on the STITCH button sends back the created metadata into WinIBW (JavaScript/VBScript).

For the purpose of evaluation, existing book metadata, suggestions and user input (including the evaluation-specific satisfaction level and comments) are also stored in XML (VBScript). A dedicated XLST stylesheet enables the re-displaying of that information for each considered book, as well as the creation of tables with all gathered data.

Access to concept information is done via a generic SKOS vocabulary service[4] we implemented in STITCH. To streamline management of KOSs in our architecture, we have opted for converting them to the SKOS standard to represent KOSs in RDF [2]. The service allows to retrieve SKOS data for concepts (labels, documentation, semantic relationships to other concepts) and enables searching for concepts in a given KOS, *e.g.*, by means of autocompletion. It thus provides the basic means to explore vocabularies.

Crucially, the service can be used to access distributed RDF KOS sources, published via either a SPARQL endpoint or other instances of the service that are loaded with the corresponding SKOS files. This allows to seamlessly extend our prototype to deal with unforeseen KOSs, when these are published in SKOS on the Web of Data. Our SST could then be articulated with more collections, even ones already linked at the semantic level—see Fig. 1.

3.4 Brinkman Subject Suggestion Rules

The suggestion process relies on statistical techniques to find association between the elements found in the available book metadata and the Brinkman concepts to

[4] `http://stitch.cs.vu.nl/repository`

be suggested for indexing. For the sake of brevity, we limit ourselves here to the description of what the suggestion tool outputs and which information sources it uses, and only give an overall description of the suggestion rules creation process.

Metadata Used for Creating and Applying Suggestion Rules. In a first version of the suggestion tool, we tried to build and exploit an alignment between Biblion (LTR field) alone and Brinkman. The idea was to provide suggestions based on simple, ready-to-use individual "translation rules" from Biblion to Brinkman, as obtained from lexical comparison of their labels.

However, as previously mentioned, our case comprises books which are shared between the KB and the public library collections. Many of these books (nearly 238K) are thus indexed with both Biblion and Brinkman. Following the re-indexing evaluation approach of [11], we used the existing Brinkman subjects of these books as a gold standard with which we could automatically compare new suggested subjects. Based on the low performance of this preliminary automatic evaluation, we decided to drop these first suggestion rules. They are only accessible via the advanced 'a' option mentioned above, as a complementary source of suggestions.

Matching techniques using the *extension* of concepts, namely the way they co-occur in book descriptions, seemed much more promising. From a theoretical perspective, they fit well the application scenario, which is about reproducing indexing practices instead of trying to grasp the "intrinsic meaning" of concepts. Practical evaluations in the case of other KOSs—*e.g.*, for the Library track of the Ontology Alignment Evaluation Initiative [14] and for STITCH research [15]—have further demonstrated that these techniques indeed perform well for re-indexing.

As a matter of fact, the dually indexed corpus that we used for automatic evaluation above could also be used as a learning set for deriving statistical associations between Biblion and Brinkman concepts. In order to fit the application scenario better and to take into account potential indexing- and vocabulary-related granularity mismatches, we sought to find suggestions based on *combinations* of one, two or three Biblion concepts. We also extended the matching process by considering the values found in *several* metadata fields. This provides more information to elicit finer-grained suggestion rules. Our prototype suggestion tool uses as input the following four metadata fields (and their associated vocabularies of controlled values) used in books with Biblion indexing:

– LTR – Biblion concepts,
– AUT – main authors of books,
– KAR – (coded) "characteristic" and
– DGP – (coded) intellectual level/target group.

Establishment of Suggestion Rules. The most important source of information for the SST is the set of books which are described by both Brinkman and Biblion. In order to analyse the reliability of the rules, we extracted a third of this set as a new test set for automatic evaluation.

Suggestion rules are established by looking for statistical associations between a source combination of metadata values (including at least one Biblion concept, and optionally, one or two values for the AUT, KAR, DGP fields mentioned above) and target Brinkman concepts. We adapted previous work on instance-based thesaurus matching [6,7] to fit the re-indexing scenario. Given a source combination of metadata values, noted as C_i, the probability that a Brinkman concept B_j should be suggested is calculated as

$$p(C_i \rightarrow B_j) = \frac{|C_i \cap B_j|}{|C_i|} \tag{1}$$

where $|C_i \cap B_j|$ is the number of books which are described by both C_i and B_j and $|C_i|$ is the number of books which are described by C_i. In order to reduce the bias from the sparseness of the data, we took a variation of (1):

$$p(C_i \rightarrow B_j) = \frac{|C_i \cap B_j|}{|C_i|} - 0.46 \times \frac{2}{|C_i| + 1}, \tag{2}$$

which gives the best performance in practice—as observed by automatic evaluation over the third of the dataset which we extracted for testing.

In this way, we established a set of 1.5 million suggestion rules, each with a "confidence level"—in fact, the probability for a suggestion to be correct—determined by formula (2). Table 1 presents some of them. The left-hand side of each rule corresponds to a combination of metadata values found in the training set; the right-hand side corresponds to the Brinkman suggestion. The rightmost column gives "correct books," the number of books in our test set that had both

Table 1. Subject suggestion rules

Source combination → target concept	Confidence level	Correct books / Total
DGP:Jeugd fictie; vanaf 13 jaar (youth fiction; from 13 year) + KAR:Stripverhaal (comics) → BTR:stripverhalen (comics)	0.995	182/182
LTR:Reisgidsen (travel guides) + LTR:Spanje (Spain) → BTR:Spanje ; reisgidsen (Spain; travel guides)	0.982	50/50
...		
LTR:Brabantse dialecten (Brabant dialects) → BTR:Nederlandse dialecten (Dutch dialects)	0.967	27/27
...		
LTR:Liefde (love) + AUT:Jeanette Winterson → BTR:Romans en novellen ; vertaald (novels and novellas; translated)	0.540	1/1
...		
LTR:Bouwkunde (building engineering) → BTR:leermiddelen ; bouwtechniek (learning material ; building engineering)	0.196	25/123

the source combination and the target concept, and "total," the number of books that matched the source combination.

Working the Suggestion Rules. When a book's metadata contains a suitable source combination (including Biblion subject indexing, and possibly, AUT, KAR and DGP annotations), the SST suggests a list of Brinkman subjects, each with a confidence value. This allows to rank the suggested concepts, as well as to classify them in broader confidence categories indicated by specific colors, as mentioned in Sec 3.2. Thus, blue suggestions correspond to a confidence level higher than 0.54—all suggestions based on concept combination (co-)occurring in a single book have this measure. Purple suggestions have a confidence level between 0.1 and 0.54, and the red suggestions are between 0.02 and 0.1.

4 User Study

4.1 Aim and Settings

Our user study aims to determine the extent to which the SST could be used at the KB. We are also interested in users' suggestions for improving the tool, or applying it in alternative ways.

We gathered six indexers of the *deposit* cataloguing team (out of a total 16). They were all trained indexers, with experience ranging between 10 and 30 years—five of them having at least 20 years experience.

The tests were originally planned for a period of three weeks, but this was lengthened to 6 weeks after it appeared that too little suitable material could be gathered in that time span.[5] In total, 284 books were evaluated.

The evaluation task was seamlessly integrated in the daily work of the evaluators. As, when describing books with WinIBW, they recognized a book for which Brinkman suggestions could be done, they activated the tool, and then resumed to their "normal" tasks after that book had been indexed.

As already described, the SST interface required the evaluators to select which were the correct suggestions, or complete these with non-suggested concepts when required. It was also equipped with radio buttons with which evaluators could indicate their level of satisfaction, and a free-text field where they could enter their comments on the suggestions for the book at hand.

Before the test, the evaluators were given in a group briefing on the task and SST's features. We also gave each of them a questionnaire to fill at the end of the test period—see Appendix. After the test period, we organized a (group) evaluation interview. There, the evaluators where given an extra question list we devised after analyzing the input from the first questionnaire, so as to get

[5] Currently, KB indexes books very shortly after their publication; the number of books indexed by Biblion *before* being indexed by KB turned thus to be smaller than we hoped. Luckily, our committed evaluators pretty soon decided themselves to collect all suitable books from their entire department, beyond the ones attributed to them *via* the normal KB process.

more precise (and quantitative) feedback on specific points—see Appendix. The list also provided a sound structure to guide the interview, during which the evaluators had to answer it.

4.2 Suggestion Performance

Using the correct indexing selected or added by the evaluators, we could measure the performance of our concept suggestions in terms of precision and recall. The 284 books evaluated were together given 4,021 Brinkman subject suggestions, for a total 468 correct concepts. Table 2 gives the results for each of the three suggestion classes mentioned in Sec. 3, where the "non suggested" category lists the number of individual concept indices that could not be found by the system and had to be added by indexers.

Table 2. Suggestion performance

Suggestion class	# suggestions	# correct	precision	recall
blue	308	224	72.7%	47.9%
purple	1,188	127	10.7%	27.1%
red	2,525	28	1.11%	5.98%
non suggested		89		19.0%

Looking at the "blue" row, we achieve a recall of 47.9% with a precision of 72.7%. Taking the blue, purple and red suggestion results together, we can achieve a recall of 81% if we accept a much lower precision of one in ten correct suggestions (9.4%).

These results confirm our expectations regarding the categorization of suggestions in terms of precision classes. Interestingly enough, they are also consistent with the results of the automatic evaluation over the testing set, mentioned in Sec. 3.4—see Appendix.

4.3 User Feedback

User Satisfaction. We evaluated the more subjective aspect of user satisfaction based on the following sources:

- the general satisfaction level given to each book's suggestions;
- the comments given for each book;
- the answers given to the first and second questionnaires;
- the informal feedback obtained during the group interview.

At the level of books, the satisfaction level seemed first relatively high. Out of the 264 books for which this information was filled, 193 were given a "++" or a "+", the two best marks available out of five. However, this positive appreciation does not correlate with the perceived global usefulness of the tool, as it emerged from latter comments and questionnaires. Indeed, when asked if they would continue using the tool *as it is* , only two evaluators answered positively.

A first source of complaints is the robustness of the tool: the vocabulary service hosted in Amsterdam crashed a couple of times. But the evaluators could abstract from this, and they generally appreciated the functionality and the design by the prototype.

Rather, the most important issue was clearly the perceived quality of individual concept suggestions: the tool did not hinder evaluators' work, yet they expected more accurate suggestions. The current prototype displays quite many concepts, many of which were considered useless. At low confidence levels, suggestions can be based on just a insignificant overlap that is not meaningful from an indexing perspective, as for "Cross-culturele psychologie" (cross-cultural psychology) and " groepsdynamica" (group dynamics). While including them still brings some correct suggestions, the "serendipity" effect we wanted to test was not considered as an advantage, in such a controlled indexing process. The basic ranking of suggestions, based on their confidence level and expressed by the color distinctions, was judged a very positive feature. Yet evaluators still felt they were compelled to examine all suggestions, even the least certain ones. This suggests finer strategies have to be devised concerning the way suggestions are presented to the user, depending on their confidence level. A higher confidence threshold might therefore be required for a suggestion to be displayed among the first choices.

Our prototype also suggests that better suggestions could (and should) be obtained, in absolute. The way suggestions are obtained is indeed judged very promising, as testified by reactions collected on individual suggestions. Suggesting new indexing based on previous ones often successfully managed to bridge the "indexing gap" between the two collections at hand, *e.g.*, for "Perzische taal" (Persian language) and "Iraanse taal- en letterkunde" (Iranian language and literature). In fact in some cases it provided with suggestions which evaluators would not have anticipated, but that turned out to be correct.

Feedback for Improvement. As a token of interest, the evaluators strongly suggested to *have the same approach applied to other situations*, *i.e.*, not only books indexed with Biblion. They hinted that the tool could also exploit metadata fields like the ones using the aforementioned NUR or other elements of Fig. 1, such as the two lists of Persons as subjects and Corporations, which have a strong connection with topical concepts. As a matter of fact, these extra sources may be used in the first place to improve the performance of the Biblion-based process, they confirmed us. Exploiting co-occurrences of a larger set of concepts would allow to grasp some interesting patterns. But they might also be sufficient to do useful suggestions when a Biblion concept is missing. Finally, showing their awareness of the semantic connections between the different indexing systems in their environment—and their potential value—they suggested that concepts from other vocabularies than Brinkman could be suggested. In particular, the UNESCO classification, when it is not yet given for a book, would be a useful and easy target to aim at. Other elements of Fig. 1 such as the two lists of Persons as subjects and Corporations, would be valuable too.

In fact, as they felt that suggestion process was similar to the way they perform indexing themselves, our evaluators also brought ideas to investigate in the longer term, to make that similarity even higher. For instance, using "negative rules" such as "LTR:Zussen (sisters) → BTR:Zussen, `unless` KAR:roman is present"; or setting up a more interactive suggestion process that actualises the list of suggested Brinkman concepts once a first one has been selected.

A last issue, which evaluators felt was clearly more urgent to tackle, is the incompleteness of the data that is used for the suggestions and of the set of concepts that can be suggested. In many cases, they expected more suggestions to be made, considering the available metadata that could be used as a learning set—which they have rather extensive knowledge on. This can be explained by our selecting only two thirds of the dually indexed books as a learning set. But the original dump of records itself dates back to 2006, and is therefore relatively obsolete. Similarly, the SKOS versions of the vocabularies which we exploit both for the suggestion and the vocabulary services (esp. to allow users to search for additional concepts using auto-completion) are mere snapshots. They were already slightly outdated at the time of our experiment, which was detected by indexers. Update mechanisms should be set up to exploit the most recent data.

5 Conclusion

As regards our initial aims, the main results are:

- SW technology uptake: despite a few robustness issues that should be solved in a next implementation phase, the main technical goals have been met. Exploiting Semantic Web technologies, we could develop a prototype with features both innovative and relevant from a domain perspective, and which *can be integrated into the KB production process.*
- feasibility of the re-indexing approach: our evaluation showed strong interest from the book indexers. The quality of the suggestions must clearly be improved before they start using such a tool. However, the prototype has validated the principle of using previous indexing to suggest new one, as well as most of our design decisions. Further, the user study provided the development team with clear and feasible solutions to start improving on the current prototype, wrt. both interface and suggestion quality.

An important point that motivates the adoption of SW solutions for the case at hand is their genericity and flexibility. We could well have developed a prototype with similar functionality, using more traditional techniques, on top of the same collections and KOSs. In fact, some of the key benefits of SW solutions, such as relying on unambiguous identifiers for resources, are already available for use in the current application context. However, our technological choice guarantees high flexibility while not requiring much additional implementation effort—off-the-shelf RDF stores can be deployed easily and accessed by simple scripts using SPARQL. Especially, the prototype can be adapted so as to use other metadata (and their associated KOSs) to derive more precise suggestion rules, which is one of the most important directions for future improvement.

As a matter of fact, our tool could be seamlessly deployed with completely different collections and vocabularies, provided the associated metadata is also made available in the appropriate RDF format—which was one of the bottlenecks we identified as we proceeded with our own data conversion effort. Considering the growing interest of CH institutions for Open Linked Data[6] developments, *e.g.*, [16,17,18], one can be optimistic about it. For instance, as shown in Fig 1, collections that are indexed by the subject heading lists of the American Library of Congress and the French National Library (resp., LCSH and RAMEAU) overlap with the KB one. Our prototype could be extended to have KB benefit from the work that is being done in these libraries. In that specific case we could even seek to exploit existing semantic mappings between the two aforementioned KOSs, which SW techniques made much easier to publish and access.[7]

Indeed, we believe that an important step has been made at the KB towards embracing the Open Linked Data vision. In the library domain, as in many others, institutions will not adopt new technologies unless they have been demonstrated with scenarios bringing clear value to them. Here, we have been able to show that a library can benefit from having other libraries' data being accessible, even for a back-office, mission-specific process like indexing.

In turn, our prototype confirmed that there is value for KB making accessible its own data and vocabularies. If other institutes could benefit from KB descriptions in their own metadata creation process as well, this would boost the productivity of the entire network. In fact, as reflected in Fig. 1, this experiment was also the opportunity to identify in a clearer way:

- from a technical perspective, the various vocabularies that can be connected at the semantic level to enhance the indexing process;
- from a more institutional perspective, the collection holders that form together a neighbourhood of peers in the network, which can benefit from the virtuous circle mentioned in the previous paragraph.

This may drastically enhance synergies, and make common adoption of these new technologies an even more realistic perspective. Other institutes in the KB neighbourhood, such as the Dutch association of public libraries, are actually already considering moving to RDF and SKOS [19]. The time is now relatively near, when a library cloud such as the one in Fig. 1 will lend itself to treatments that allow the participating institutions to benefit from each other's efforts.

Acknowledgements. This work is funded by the NWO CATCH and EU eContentPlus programmes (STITCH and TELPlus projects). We are especially grateful to the team of KB indexers who participated the evaluation and provided crucial feedback.

---

[6] `http://linkeddata.org/`

[7] See `http://id.loc.gov` and `http://stitch.cs.vu.nl/rameau`

References

1. Dublin Core Metadata Initiative: Dublin Core Metadata Element Set. DCMI Recommendation (2008), `http://dublincore.org/documents/dces/`
2. Miles, A., Bechhofer, S.: SKOS Reference. W3C Proposed Recommendation (2009), `http://www.w3.org/TR/skos-reference/`
3. Hyvönen, E., Mäkelä, E., Salminen, M., Valo, A., Viljanen, K., Saarela, S., Junnila, M., Kettula, S.: MuseumFinland – Finnish Museums on the Semantic Web. Journal of Web Semantics 3(2) (2005)
4. Wielemaker, J., Hildebrand, M., van Ossenbruggen, J., Schreiber, G.: Thesaurus-based search in large heterogeneous collections. In: Sheth, A.P., Staab, S., Dean, M., Paolucci, M., Maynard, D., Finin, T., Thirunarayan, K. (eds.) ISWC 2008. LNCS, vol. 5318, pp. 695–708. Springer, Heidelberg (2008)
5. Euzenat, J., Shvaiko, P.: Ontology Matching. Springer, Heidelberg (2007)
6. Isaac, A., van der Meij, L., Schlobach, S., Wang, S.: An empirical study of instance-based ontology matching. In: Aberer, K., Choi, K.-S., Noy, N., Allemang, D., Lee, K.-I., Nixon, L.J.B., Golbeck, J., Mika, P., Maynard, D., Mizoguchi, R., Schreiber, G., Cudré-Mauroux, P. (eds.) ASWC 2007 and ISWC 2007. LNCS, vol. 4825, pp. 253–266. Springer, Heidelberg (2007)
7. Wang, S., Isaac, A., van der Meij, L., Schlobach, S.: Multi-concept alignment and evaluation. In: Proc. 2nd ISWC Workshop on Ontology Matching (OM 2007), Busan, Korea (2007)
8. Buitelaar, P., Cimiano, P. (eds.): Ontology Learning and Population: Bridging the Gap between Text and Knowledge. IOS Press, Amsterdam (2008)
9. Golub, K.: Automated subject classification of textual web documents. Journal of Documentation 62(3), 350–371 (2006)
10. Gazendam, L., Malaisé, V., de Jong, A., Wartena, C., Brugman, H., Schreiber, G.: Automatic annotation suggestions for audiovisual archives: Evaluation aspects. J. Interdisciplinary Science Reviews 34(2-3), 172–188 (2009)
11. Isaac, A., Wang, S., Zinn, C., Matthezing, H., van der Meij, L., Schlobach, S.: Evaluating thesaurus alignments for semantic interoperability in the library domain. IEEE Intelligent Systems 24(2), 76–86 (2009)
12. OCLC: Kenmerkcodes in titelrecords, `http://oclcpica.org/?id=13&ln=nl&par=p-kmc` (in Dutch, accessed in August 2009)
13. Clark, K.G., Feigenbaum, L., Torres, E.: SPARQL Protocol for RDF. W3C Recommendation (2008), `http://www.w3.org/TR/rdf-sparql-protocol/`
14. Euzenat, J., Isaac, A., Meilicke, C., Shvaiko, P., Stuckenschmidt, H., Svab, O., Svatek, V., van Hage, W.R., Yatskevich, M.: Results of the Ontology Alignment Evaluation Initiative 2007. In: Proc. 2nd ISWC Workshop on Ontology Matching (OM 2007), Busan, Korea (2007)
15. Schopman, B.A.C., Wang, S., Schlobach, S.: Deriving concept mappings through instance mappings. In: Domingue, J., Anutariya, C. (eds.) ASWC 2008. LNCS, vol. 5367, pp. 122–136. Springer, Heidelberg (2008)
16. Malmsten, M.: Making a Library Catalogue Part of the Semantic Web. In: Proc. International Conference on Dublin Core and Metadata Applications (DC 2008), Berlin, Germany (2008)
17. Summers, E., Isaac, A., Redding, C., Krech, D.: LCSH, SKOS and Linked Data. In: Proceedings of the International Conference on Dublin Core and Metadata Applications, Berlin, Germany (2008)

18. Neubert, J.: Bringing the "Thesaurus for Economics" on to the Web of Linked Data. In: Proc. WWW Workshop on Linked Data on the Web (LDOW 2009), Madrid, Spain (2009)
19. Vereniging Openbare Bibliotheken: Informatiearchitectuur Openbare Bibliotheken, `http://www.debibliotheken.nl/content.jsp?objectid=22642` (in Dutch, 2008)

Appendix

An online annex to this paper (`http://www.few.vu.nl/~aisaac/iswc2009/`) features the 1st and 2nd user study questionnaires and the results of the preliminary automatic evaluation of suggestions using existing dually indexed books.

Semantic Web Technologies for the Integration of Learning Tools and Context-Aware Educational Services

Zoran Jeremić[1], Jelena Jovanović[1], and Dragan Gašević[2]

[1] FON-School of Business Administration, University of Belgrade, Serbia
[2] School of Computing and Information Systems, Athabasca University, Canada
jeremycod@yahoo.com, jeljov@gmail.com, dgasevic@acm.org

Abstract. One of the main software engineers' competencies, solving software problems, is most effectively acquired through an active examination of learning resources and work on real-world examples in small development teams. This obviously indicates a need for an integration of several existing learning tools and systems in a common collaborative learning environment, as well as advanced educational services that provide students with right in time advice about learning resources and possible collaboration partners. In this paper, we present how we developed and applied a common ontological foundation for the integration of different existing learning tools and systems in a common learning environment called DEPTHS (Design Patterns Teaching Help System). In addition, we present a set of educational services that leverages semantic rich representation of learning resources and students' interaction data to recommend resource relevant for students' current learning context.

Keywords: Semantic web, ontologies, collaborative learning, project-based learning, software patterns, context-awareness.

1 Introduction

The major concern of today's software engineering education is to provide students with the skills necessary to integrate theory and practice; to have them recognize the importance of modeling and appreciate the value of a good software design; and to provide them with ability to acquire special domain knowledge beyond the computing discipline for the purposes of supporting software development in specific domain areas. In addition, it is essential that students learn how to exploit previous successful experiences and knowledge of other people in solving similar problems. This knowledge about successful solutions to recurring problems in software design is also known as software design patterns (DPs) [1].

Apart from learning individual DPs and the principles behind them, students should learn how to understand and apply patterns they have not seen before, how to integrate different DPs, and how to use this knowledge in real-life situations.

This indicates a rising need for the social constructivist approach in software engineering education. In particular, an active learning paradigm is needed which recognizes that student activity is critical to the learning process. The basic philosophy of active learning paradigm [2] is to foster deep understanding of subject matter by

A. Bernstein et al. (Eds.): ISWC 2009, LNCS 5823, pp. 860–875, 2009.

engaging students in learning activities, not letting them be passive recipients of knowledge. Moreover, the students are involved in the knowledge construction and sharing through social interactions in the given learning context.

Following this paradigm, we identified several requirements that a learning environment for software DPs needs to meet [3]:

1. Enable students to learn at the pace and in a place that best suits them.
2. Integrate software development tools that would enable students to experience patterns-based software development in the context of real-world problems.
3. Include collaborative tools such as discussion forums, chat, and tools for software artifacts exchange.
4. Enable seamless access to online repositories of software DPs and communities of practice that will provide students with right-in-time access to the relevant online resources, that is, to the resources on software DPs relevant for the problem at hand.
5. Provide tools for informing teachers about students learning activities, their usage of learning content and other valuable information that could help them improve the learning content and/or the chosen teaching approach.

Even though the above mentioned kinds of tools do exist today, they are not used in an integrated way [4]. Instead, current approaches to learning software patterns are based on individual use of these tools. The major problem with this 'fragmented' approach is in its lack of means for enabling exchange of data about the activities that students performed within individual learning tools and learning artifacts they have produced during those activities. Besides, with such an approach it is very hard to provide support for context-aware learning services and offer personalized learning experience to students.

In this paper we describe an integrated learning environment for software DPs called DEPTHS (Design Patterns Teaching Help System) [3]. DEPTHS integrates an existing Learning Management System (LMS), a software modeling tool, diverse collaboration tools and relevant online repositories of software DPs. The integration of these different learning systems and tools into a common learning environment is achieved by using the Semantic Web technologies, ontologies in particular. Specifically, ontologies enabled us to formally represent and merge data about students interactions with the systems and tools integrated in DEPTHS. On top of that data, we have built context-aware educational services that are available throughout the DEPTHS environment. These services enrich and foster learning processes in DEPTHS in two main ways:

- recommendation of appropriate learning content (i.e., Web page(s), lessons or discussion forum threads describing software DP). We can recommend fine-grained learning resources, and make the recommendations aware of the recognized learning needs.
- fostering informal learning activities by bringing together students and experts that are dealing with the same software problem or have experience in solving similar problems.

2 Scenario of Use

Effective learning of software DPs requires a constructive approach to be applied in the teaching process. It is very important that students experience software development and the application of DPs on real-world examples, in order to develop a deep understanding of the basic principles behind them and to learn how to apply them in different situations. Having this in mind, we have explored a number of theories and research fields in the area of project-based and computer supported collaborative learning. Based on the guidelines for teaching software engineering to students [5] [6] and our own teaching experience, we have identified the following three as the most important for teaching/learning software DPs: Learning through Design, Project-based learning (PBL) and Engagement theory. More details on pedagogical aspects of this work could be found at [7]. In what follows, we present a usage scenario which illustrates how these theories are applied for learning software DPs in DEPTHS.

A typical scenario for learning software DPs with DEPTHS assumes a project-based learning approach with collaborative learning support (Fig. 1). In particular, a teacher defines a specific software design problem that has to be solved in a workshop-like manner by performing several predefined tasks: brainstorming, creating and submitting solutions, evaluating solutions etc.

First, a student is asked to present his ideas about possible ways for solving the problem under study and to discuss and rate his peers' ideas. In order to get the required information for performing this task, he searches online repositories about software DPs and the related course content. DEPTHS makes this search more effective by providing semantically-enabled context-aware learning services for finding related online and internally produced resources (Fig. 1B). Moreover, to get some initial directions on the performing task, the student uses semantically-enabled peers finding service (Fig. 1A) to find people who have shared interests and are engaged in similar problems. As we explain in the following section, both kinds of services are enabled by leveraging formally represented semantics of the learning context and learning resources (both online resources and those internally produced). Afterwards, the student has to find associations between the gained knowledge and the problem that has to be solved and to propose potential solution strategies. Later, consultations are directed at confirming the idea and refining it to accommodate criticisms.

Previous works on similar problems could be useful for students as they give them opportunities to learn from positive examples; and provide them with new facts, information, and an idea how to apply the same approach (design patterns) in a similar situation. Moreover, exploring previous works provokes critical thinking as it helps the student think about alternatives along with their advantages and disadvantages. DEPTHS context-aware learning services for discovery of relevant learning resources (both external and internal) greatly facilitate this task. These services are powered by semantic annotations of learning resources: ontologies enable capturing and formal representation of the semantics of the content of those resources, as well as the context of their creation and usage (see Section 3).

Having acquired the required knowledge, students should complete the deliverable using the software modeling tool. This learning activity requires students to externalize their knowledge, to analyze possible solutions and to provide a design rationale.

After completing the project, students are asked to evaluate their own project, as well as to perform evaluation of each other's work. Students reflect critically on their own and others' contributions, and acquire knowledge about other possible solutions. This helps them recognize possible improvements in their own solutions. DEPTHS uses ontologies to capture the semantic of the students' evaluations, so that they can be used for recommendations as well as feedback provisioning.

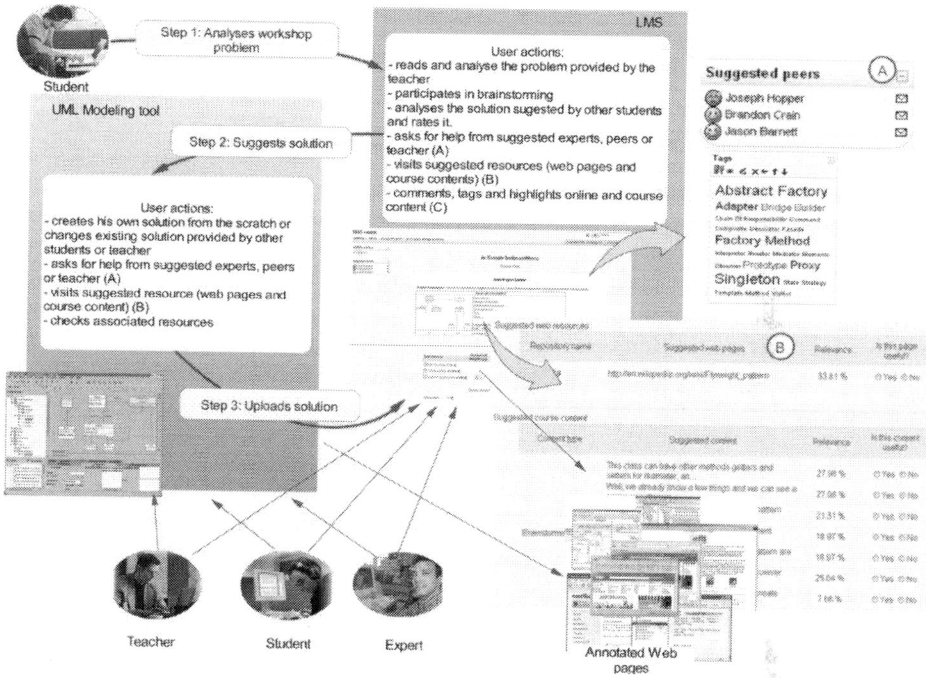

Fig. 1. An example learning scenario with DEPTHS: problem-based learning with collaborative learning support

All students' projects are published and publicly available; they are stored together with contextual semantic-rich metadata which facilitates their discovery and reuse. Students may be anxious that their work will be so visible, but it does seem to push them along to polish their projects. Moreover, students can learn from each other as portions of their projects became available before the final due date.

3 The DEPTHS Learning Environment

DEPTHS integrates existing, proven learning systems, tools and services in order to provide an effective collaborative environment for teaching and learning software DPs (Fig. 2). In particular, DEPTHS currently integrates an LMS (Fig. 2A), a software modeling tool (Fig. 2B), a feedback provisioning tool for teachers (Fig. 2C), a collaborative annotation tool (Fig. 2D), and online repositories of software patterns

(Fig. 2E). This integration is achieved through a flexible underlying ontology-based framework called LOCO (Fig. 2F).

3.1 The Ontological Foundation of DEPTHS

LOCO (Learning Object Context Ontologies) [4] is a generic framework capable of formally representing all particularities of the given learning context: the learning activity, the learning content that was used or produced, and the student(s) involved. Accordingly, the framework integrates a number of learning-related ontologies, such as learning context ontology, a user model ontology, and domain ontologies. These ontologies allow one to formally represent all the details of any given learning context, thus preserving its semantics in machine interpretable format and allowing for development of context-aware learning services. The LOCO ontologies are developed by following the Linked Data best practices (http://www4.wiwiss.fu-berlin.de/bizer/pub/LinkedDataTutorial/). In particular, linkages were established with well-known Web ontologies such as the Dublin Core vocabulary, FOAF (Friend-Of-A-Friend, http://xmlns.com/foaf/0.1), and SIOC (Semantically Interlinked Online Communities, http://sioc-project.org). All the ontologies of this framework are publicly accessible (http://iis.fon.rs/LOCO-Analyst/loco.html).

DEPTHS currently makes use of two ontologies of this framework Learning Context ontology and domain ontology, that is used for representing the domain of software patterns. The Learning Context ontology allows for semantic representation of the data about a student's overall interactions with learning content and other students during different learning activities. Based on this data, DEPTHS can perform context-aware retrieval of resources on software DPs from online repositories and its own repository of software artifacts (which may also contain artifacts produced by other students and shared by the community). It can also identify and draw students' attention to the related threads in discussion forums, as well as identify peers that could help in the given learning situation (i.e., learning context). This ontology was extended to allow for an unambiguous representation of learning contexts specific to the systems, tools and services that DEPTHS integrates (the ontologies are available at the project's website: www.learningdesignpatterns.org).

Activities are very important part of the learning process in DEPTHS as they lead to realization of learning objectives. Examples of such activities are reading lessons, visiting online pages, participating in a workshop or doing an assignment, solving design problems, quizzing and collaborating with other participants. In the Learning Context ontology, some of these activities are recognized and grouped as three basic types of activities: reading, doing an assessment and collaborating. However, some specific types of activities and events typically occurring within software modeling tools were not supported by the formalisms of the LOCO's Learning Context ontology. Accordingly, we extended this ontology with a set of classes and properties that enable collecting data about user's activities in software development tools, namely Brainstorming, Submitting and Assessing.

Another important concept determining any particular learning context is content item, which represents any kind of content available in a learning environment (e.g. a lesson, course, question, discussion forum post). We have extended the *ContentItem* class with three concepts relevant for learning software DPs, namely, *DesignProblem*,

Task and *Diagram* (Fig. 3). The *DesignProblem* class identifies a software problem defined by a teacher and presents the basis for project-based learning in the DEPTHS environment. Students are expected to suggest their own solutions of this problem by producing a UML diagram(s) as a graphical representation of the solution. To allow for formally representing students software models, we have introduced the Diagram class as a subclass of *ContentItem*. The *Task* class represents a task that a teacher can define in the context of solving a specific software design problem. We have modeled them as subclasses of the *Task* class (Brainstorm, Submission and Assessment).

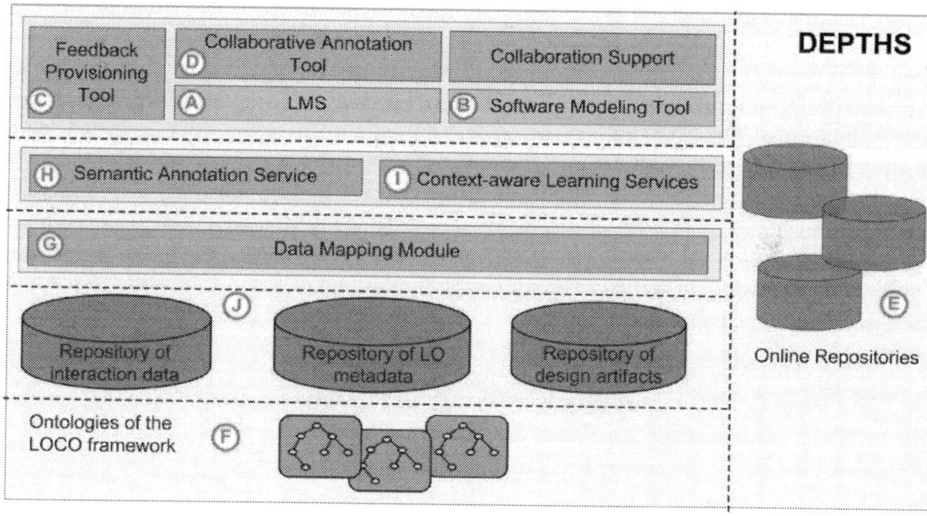

Fig. 2. DEPTHS architecture

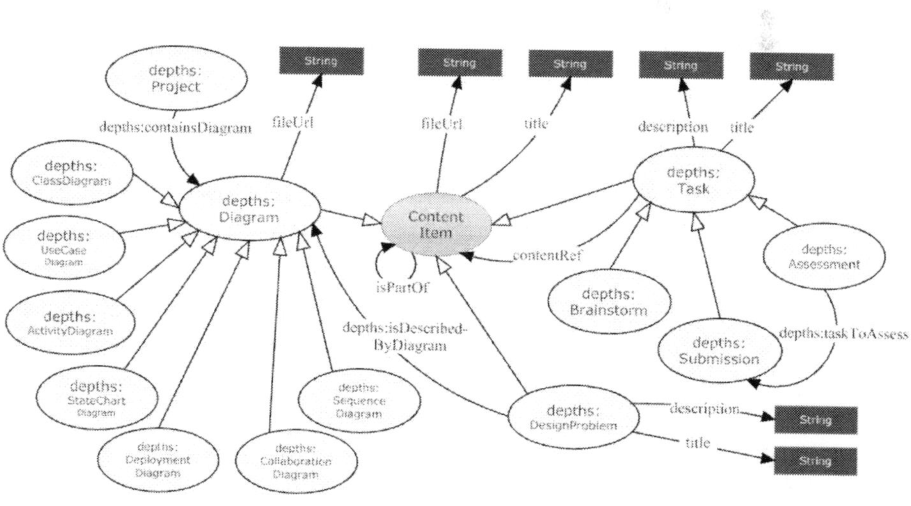

Fig. 3. The extension of the Learning Context ontology for the DEPTHS environment

Since DEPTHS is devised as an environment for teaching/learning software patterns, it leverages an ontology of software patterns as its domain ontology. DEPTHS uses this ontology to annotate semantically relevant online resources and extract metadata that is subsequently used for finding resources appropriate for a student's current learning context. It also annotates semantically the products of learning activities, such as chat messages or discussion posts. By leveraging the semantic annotations, DEPTHS can easily connect the products of learning activities with online learning resources, and use this information to further improve its context-aware support by being able to mash-up knowledge scattered in different activities.

Rather than developing new design pattern ontology from scratch, we decided to (re)use an existing ontology. Among the available ontologies for the design patterns domain [8][9][10][11], we have chosen the set of ontologies suggested in [11] to serve as the domain ontology of the DEPTHS framework. Comparing these with the other ontologies, we found that they provide a very intuitive and concise way to describe DPs and patterns collections, and to offer more information on usability knowledge and the contextual factors that impact this knowledge domain. This approach to pattern representation has the ability to federate distributed pattern collections. These ontologies include a set of pattern forms (e.g., Coplien Form [12], Gang of Four Form [1]) arranged in an inheritance hierarchy which can be easy extended with additional pattern forms (Fig. 4).

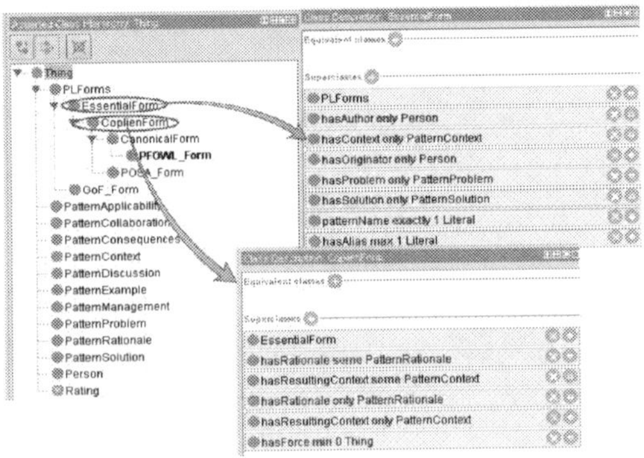

Fig. 4. Domain ontology for describing software design patterns

3.2 The Architecture of DEPTHS

The LOCO ontologies are used as the basis for the storage and exchange of data among DEPTHS components. In particular, these ontologies underlie two DEPTHS repositories (Fig. 2J):

Repository of interaction data stores data about students' interaction with learning content as well as their mutual interactions in the DEPTHS learning environment. The interaction data are stored in the form of RDF triples compliant with the extended Learning Context ontology of the LOCO framework (e.g., {ContentItem} loco:hasUserEvaluation {UserNote}).

Repository of LO metadata stores semantic metadata about all kinds of learning objects (LO) used in the courses under study. This metadata formally define the semantics of the learning content the metadata is attached to. They are stored as RDF triples compliant with the extended Learning Context ontology and the domain ontology of software DPs (e.g., {ContentItem} loco:hasDomainTopic {dp:DesignPattern}).

DEPTHS also includes the *Repository of design artifacts* which uses open standard formats to store software artifacts created by students. The students' artifacts are stored in two formats: XML Metadata Interchange (XMI) and Scalable Vector Graphic (SVG). The former facilitates storing UML diagrams in the format suitable for later reuse in any software modeling tool, whereas the latter is suitable for content presentation in a Web browser.

Since different learning systems, tools and services use different formats for representing and storing interaction data, DEPTHS integrates *Data Mapping Module* (Fig. 2G) which performs the mapping of those native data formats (e.g., the exchanged chat messages stored within the LMS's database, using a proprietary database schema) into RDF triples compliant with the LOCO's Learning Context ontology (e.g., {ChatMessage} loco:sentBy {User}). The resulting (RDF) data is stored in the *Repository of interaction data*. Data mapping is performed as a two step process: the initial mapping, that is performed during the initialization of the system; and the real-time mapping that is performed throughout each learning session in order to keep the semantic repository updated (with data about the interactions occurring during that session). The initial mapping is performed using D2RQ (http://www4.wiwiss.fu-berlin.de/bizer/D2RQ/spec/) – an open source platform that facilitates the mapping of relational databases into ontological models. This way a lot of valuable data that resided in the LMS' database prior to the DEPTHS initialization are made accessible to DEPTHS in the form of RDF statements. Real-time mapping is based on events in the DEPTHS environment (e.g., posting a message in a discussion forum thread, reading chat message). DEPTHS uses Sesame API (http://www.openrdf.org/) to create RDF statements (compliant with the LOCO's Learning Context ontology) for each event and stores them in the *Repository of interaction data*.

3.3 Educational Services in DEPTHS

The next layer in the DEPTHS architecture consists of learning support services, namely Semantic Annotation and Indexing Service and Context-aware Learning Services.

Semantic Annotation and Indexing Service (Fig. 2H) is used for the semantic annotation and indexing of documents from publicly accessible repositories of DPs, as well as internal content created by students and teachers (e.g., exchanged messages, and assignments). Since all these resources are used by Context-aware Learning Services to help students in specific learning contexts, DEPTHS needs additional information about them in order to use them in an appropriate way. Specifically, the most important information for DEPTHS regarding these resources is what each specific resource is about and how relevant it is for a specific DP. However, this information is not directly available. Having analyzed many online repositories of DPs (e.g., Portland Pattern Repository, http://c2.com/ppr/; Hilside patterns, http://hillside.net/patterns/; etc.), we found that the majority of them contains mass of documents, but only a few of those documents describe a specific DP. In addition, most of relevant documents describe in detail exactly one DP (we refer to it as *dominant pattern* throughout the section), but also mention many other DPs (e.g., in the form of links to other documents and related DPs). Moreover, one particular DP is often discussed in many different documents. In order to make this crowd of documents useful for learning purposes, DEPTHS has to analyze how relevant each document is for learning a specific DP, that is, it has to perform semantic annotation and indexing of these documents.

Having tested many of the available tools for semantic annotation, we decided to use the KIM framework (http://www.ontotext.com/kim/index.html), that provides APIs for automatic semantic annotation of documents. Semantic annotation in KIM is based on the PROTON (http://proton.semanticweb.org/) ontology which we have extended with the ontology of software patterns (see Section 3.1) in order to make KIM aware of the concepts from the domain of software DPs. In that way, we were able to use KIM annotation facilities to automatically annotate documents from an online repository. To make documents from an online repository available to KIM for processing, DEPTHS uses a Web crawler that traverses through the structure of the online repository, collects URLs of all its documents and sends them to the KIM annotation facilities.

Semantic annotation of the internally produced content (e.g., chat messages, forum messages and ideas) is performed in the similar way, immediately after the user creates a content unit (i.e., following the event of submitting a new content unit to the system). Additionally, each recognized term (DP label) is changed to the hyperlink, used for launching web-recourse and internal content finding services.

Based on the semantic (meta)data generated using KIM, DEPTHS is aware of the DPs mentioned in each document. However, in order to find out what the most relevant DP for a given document is, and what the most relevant document is for a specific DP in the whole corpus of documents, additional indexing is performed.

The index data for each document contains two values: the dominant DP (i.e., concept from the DP ontology) and its accompanying relevance value. To find the dominant software DP in each document (i.e., what the document is about), DEPTHS uses the term frequency-inverse document frequency (TF-IDF) [13], which is a proven statistical measure used in information retrieval to evaluate how important a word is to a document in a collection. DEPTHS uses a document's semantic annotations to create a collection of all software DPs mentioned in that document. Subsequently, it checks how relevant the document is for each DP found in it in order to find the dominant DP. If the TF-IDF value for the dominant DP is lower than the predefined threshold, DEPTHS eliminates this document as not enough relevant for any specific DP. As the result of this process, we got a set of documents, each of which describes one DP. Additionally, for each of the indexed documents (i.e., those whose dominant DP pass the threshold), we also capture the relevance value. The relevance value is computed as a cosine similarity between the TF-IDF value of a document's dominant DP (concept) and the vector of the TF-IDF values of the DPs (concepts) discovered while the document was semantically annotated. Thus, we can use this relevance value in the process of ranking. That is, the data about what each document is about (i.e., concept of the dominant pattern) and how relevant it is for a specific DP (i.e., relevance value), are stored in the *Repository of LO metadata*, and used later by the Web resource finding service for the recommendation of Web resources.

Context-aware learning services. (Fig. 2I) are accessible to all systems and tools integrated in the DEPTHS environment and are exposed to end users (students) as context-aware learning features. They include (but are not limited to):

Web resource finding. Based on the metadata provided by the Semantic Annotation and Indexing Service, this service generates a list of recommended Web resources from publicly accessible repositories of software DPs. The service gets activated when a student selects a hyperlink to a domain topic (i.e., a topic related to software DPs) available within different kinds of learning content (e.g., lessons, chat messages and project descriptions). These hyperlinks result from the semantic annotation of the learning content done by the Semantic Annotation and Indexing Service. Afterward, the service sends a query to the *Repository of LO metadata* to extracts information about all Web resources that are relevant for the desired domain topic (i.e., DP), as well as their relevance values. Additionally, this service makes use of students estimations of the resource's relevance for the current DP (each time a student visits a suggested Web resource he is asked to rate its relevance for the given DP). Students' positive and negative ratings affect the resource's overall rating according to the influence factor (value between 0 and 1) defined by the teacher or the system administrator.

Discovery of relevant internally produced resources. This service suggests internally created resources (e.g., discussion threads, brainstorming notes, and project description) that could be useful for a student to solve a problem at hand in the given learning context. The computation of relevance is done in a similar manner to the one above described for external, Web resources.

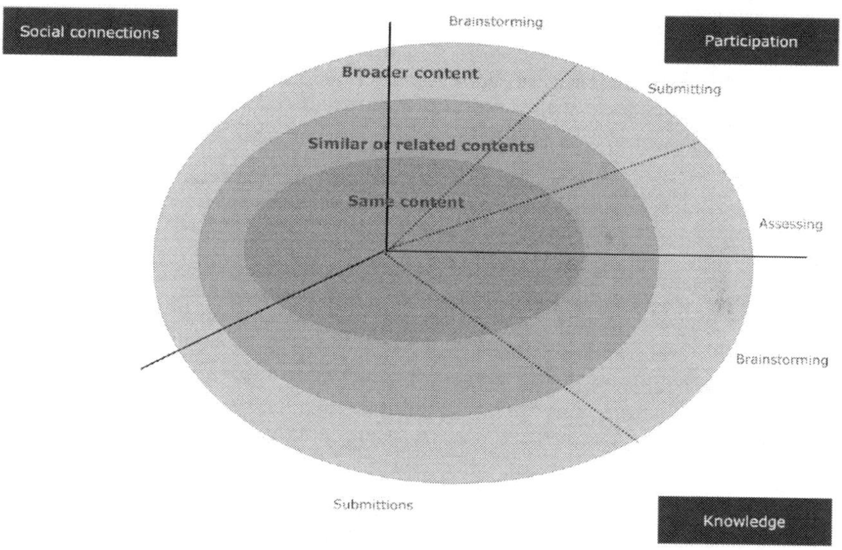

Fig. 5. Factors affecting estimation of potential collaborators' competences in DEPTHS

Experts, teachers and peers discovery. Based on the current learning context, this service suggests other students or experts as possible collaborators. Collaborators are selected and sorted using an algorithm which considers their competences on three different levels (Fig. 5): same content (i.e., current software problem), similar or related learning content (i.e., similar software problem) and broader content (i.e., software problem in the same course). Estimation of a peer's competence (C) on each level is performed through assessing three types of competence indicators:

- participation (C_p) in learning activities (e.g., brainstorming, submitting or assessing peers' works). Each activity has different impact factor on the system's estimation of the student's competences. This factor is defined in the system itself, but could be changed by the teacher.
- knowledge level (C_k) estimated by the teacher and obtained through the peers' evaluations, including projects evaluations and idea ratings. However, not all ratings have the same influence on the knowledge level estimation. For example, a high mark given by a student with high competences on the given topic has more impact on the final knowledge level appraisal than a high mark given by a student with average or low competences.
- social connections (C_{sc}) with the peer asking for help - the stronger the social connection with a specific person, the more suitable that person is for the help provision [14]. We believe that an already appointed social connection could be much more successful and effective than new connections with people one does not know. Social connections among students are mined from their interaction data (represented in accordance with the Learning Context ontology and stored in the DEPTHS's *Repository of interaction data*). For newcomers we are using their FOAF profiles, if available. Otherwise, this indicator is not considered.

All this data is collected through the queries performed on both DEPTHS' semantic repositories. Each type of competence indicator has different influence on the overall competence which depends on the predefined weight factor (WF) assigned to it (default systems' values could be changed by the teacher). To compute the overall competence of a peer, we use the following formula:

$$C_{total} = \frac{C_p \cdot WF_p + C_k \cdot WF_k + C_{sc} \cdot WF_{sc}}{3}$$

where $WF_p + WF_k + WF_{sc} = 1$, and $0 \le WF_p, WF_k, WF_{sc} \le 1$.

Context-based semantic relatedness. This service is used by all other services, as it allows for: i) computing context-based semantic relatedness between tags that students define and/or use in the given learning contexts [15]; ii) connecting students' tags with appropriate concepts of the domain ontology (i.e. disambiguation of the tags with the domain concepts); iii) resolution of students' queries containing both, tags and domain concepts relevant for the given learning context. This service connects tags with the concepts of the domain ontology as well as resources annotated with these concepts and stored in the DEPTHS semantic repositories (Repository of LO metadata and Repository of interaction data). That way the system selects only appropriate tags to show in the given moment based on the current learning context, and connects these tags with appropriate domain concepts related to it. This service is currently in its development stage and uses context-based semantic relatedness measure described in our previous work [15].

3.4 DEPTHS Implementation

The DEPTHS framework proposed in this work is a generic one. It is not dependent on any specific learning system/tool. However, for the purpose of our research we have implemented DEPTHS by leveraging open-source solutions and extending them with Semantic web technologies. Specifically, we have integrated Moodle LMS (http://moodle.org), ArgoUML software modeling tool (http://argouml.tigris.org), OATS (Open Annotation and Tagging System) tool (http://ihelp.usask.ca/OATS) for collaborative tagging and highlighting and LOCO-Analyst tool to provide teachers with feedback regarding students' activities [15]. Moreover, we use semantic annotation services of the KIM framework (http://www.ontotext.com/kim) and the Sesame server (http://www.openrdf.org) for semantic repositories. In order to provide students with context-aware educational services of the DEPTHS framework, we have extended both Moodle and ArgoUML so that they can make use of these services. Moreover, we have developed a Moodle module that supports project-based collaborative learning, that is, it supports and integrates in DEPTHS several kinds of collaborative activities such as brainstorming, submitting and assessing projects. Coupled with ArgoUML and educational services in DEPTHS it provides effective learning of software DPs, as described in Section 3 (screenshots used in Fig 1 are taken from this implementation).

4 Evaluation

The evaluation of DEPTHS was conducted in February 2009, in the context of a course that the first author of this paper taught at the Department of Computer Science of Military Academy in Belgrade, Serbia. DEPTHS was evaluated with a group of 13 students of the fifth year of the computer science program who took part in our course on software development. The students already had some elementary knowledge in the domain of software DPs, but they were not familiar with the particular software DPs used in this experiment (Facade, Adapter, Strategy, Composite, Visitor and Decorator).

The students were divided into 4 groups (3 groups with 3 students and 1 group with 4 students), based on the teacher's subjective opinion about their knowledge in the domain of software development and their previous results in the courses related to software engineering. The size of the groups is based on our belief and teaching experience that work in small size groups (3 or 4 students) is a necessity for effective education of software engineers.

The aim of the evaluation was to determine how effective DEPTHS is for learning DPs. Specifically, we wanted to evaluate the perceived usefulness of the use of engagement theory in software engineering education. Moreover, we wanted to check if active students' involvement in real world problems and the employment of context-aware educational services could ensure a more effective way of imparting knowledge in the domain of software development.

Before the experiment started, a demonstration of DEPTHS functionalities along with a training using a task similar to the one used in the experiment, were performed with students. Each group was assigned a different task (i.e., a software design problem). Students were asked to suggest solutions and evaluate each others' solutions within one week period of time. Actually, project organization used in the experiment was based on the learning workflow described in Section 3.

We used an interview to collect data about the students' satisfaction with and attitudes towards learning with the DEPTHS system. The interview was also supposed to reveal the students' perceptions regarding the effectiveness of learning with DEPTHS.

The questions were divided into three sections based on the type of information we were interested in. The first section (14 questions) gathered data regarding the students' previous experience with computer-assisted learning. The questions of the second section (15 questions) were related to the DEPTHS system and the third section (11 questions) was aimed at evaluating the learning program on software DPs offered by DEPTHS system. Most of the questions (33) were multiple-choice questions with 5 possible answers, ranging from 1 (most negative) to 5 (most positive). There were 6 open-ended questions and 1 combined (multiple choice and open-ended).

We used three methodologies to analyze the results gained in this experiment. First, we analyzed the results of the overall corpus of the students using standard descriptive statistic instruments such as frequency, mean, median, and average. The second kind of analysis consisted of comparing the groups of students that were derived by splitting the results data based on the students' answers on the questions from the first section of the interview. Finally, we used Pearson's chi-square test to find if there is significant association between different variables. We used SPSS tool (www.spss.com) to process data and analyze the results.

Having analyzed the results, we found that the majority of students (84.62%) have experience in using Internet to find relevant information, collaborate with colleagues on solving common tasks (53.85%) and use tools for message exchange and discussion (84.62%). However, they have far less experience with online learning tools (only 23.07% are familiar with e-learning tools) and using the Internet to find peers for solving problems (only 38.46% answered positively).

The DEPTHS system received high marks from the students. Majority of them (53.85%) reported they have learned as effectively as in traditional way, and 30.77% reported that they have learned more than in traditional way. The students reported it was intuitive and very easy to use (76.92%), but they also have reported some technical issues. These issues were caused by the software bug that caused problems in uploading UML diagrams to the repository, and have been resolved one day after the beginning of the evaluation. We believe that this issue could have affected students' confidence in the system. The students felt educational services provided in DEPTHS are very helpful: Web resource recommending service - 92,30%; course content recommending service – 84,61%; and peers recommending service – 76,92%. They also thought that the activities provided within the tasks considerably contribute to the learning process (brainstorming – 76.92%, and evaluating each other's works – 100%).

Having analyzed the trends of the different groups of students based on their answers on the first group of questions, we found that all students that have taken a course offered through an LMS consider educational services provided in DEPTHS as very useful. Students that have used computers to find relevant collaborator for domain that they are currently working on, have much more positive attitude for brainstorming tool in DEPTHS and learning from other students ideas. There is no significant difference between those students who are familiar with e-learning and those who are not with respect to their experience of learning in DEPTHS environment. However, it is interesting to notice that all students who are not familiar with e-learning gave the highest mark for educational service for Web pages recommending.

We have identified 27 variables' pairs that have significant association through the use of Pearson's chi-square (we used standard level of significance $p<0.05$). For example, there is a significant association between the students' answers on the question if they used tools for sending messages and discussions, and the question if they think that the other students' ideas were useful. Many useful conclusions could be drawn from these associations about how close various variable impacts might be on student achievement. For example, we found that students' satisfaction with the educational services for recommending relevant Web pages and course content affected their latter satisfaction with their learning results using this program. We believe that this association is strongly related with the students' level of adoption of these services as very important and useful part of a learning system. However, the deeper analysis of these results is out of scope of this paper.

5 Related Work

The framework proposed in this paper is related to two favored research fields: collaborative learning in the domain of software engineering and context-aware learning. Even though extensive work has been done in both research fields, to the best of our knowledge there were very few attempts in developing collaborative learning environments that support knowledge creation and sharing through the collaborative learning process based on the active learning principles.

In [16], the authors suggested an approach similar to the one presented in this work. They have developed MICE – a learner-centered platform for regulating learners' programming styles when studying a programming language using an integrated development environment. It also integrates an LMS and a set of tools for communication and collaboration among users. Even though MICE follows a similar approach to the integration of existing tools, it still lacks context-aware educational support (e.g., recommending online resources relevant for the given learning context) that is available in our framework. Besides our framework promises additional support for collaborative learning as it offers social tagging support.

One of the main objectives of the EU project APOSDLE is to develop a system that would be able to provide knowledge workers with learning resources relevant for their present work context [17]. In particular, based on the immediate work context of a user, the system should identify his/her missing competencies and learning needs and suggest appropriate learning resources. These learning resources are created on-the-fly from a variety of resources (documents, videos, expert profiles, and so on) already stored in the workplace and may be in the form of learning material or suggestions to contact experts and /or colleagues. The system's functionality is based primarily on its knowledge base that stores an integrated representation of various kinds of knowledge (e.g., domain, task, and instructional). Knowledge integration and advanced search and retrieval capabilities (associative information retrieval) are enabled by the Semantic Web technologies. Obviously, this approach has a lot of commonalities with the one we suggested in this paper. Nonetheless, having grounding our approach in pedagogical theories and best practices of collaborative learning, we can expect to provide students with better learning experience.

An e-learning framework proposed in [18] supports a peers' recommendation based on the student's context. Student's context is defined as the result of the interaction of three key elements: the knowledge potential, the social proximity and the technical access. Comparing to DEPTHS approach of peers' recommendation, this approach is advantageous as it considers technical context that includes factors that may influence e-learning such as technical media or time proximity. However, unlike the approach proposed in DEPTHS, this approach does not consider the influence of student's participation in the learning activities on his competences to help other students. Another approach [19] suggests use of ontologies to support online professional communities. By constructing a semantic model of the content, the interactions and the structure of the community, the activities of an online professional community can be supported. However, our work goes a step further and uses ontologies in concrete learning contexts.

Another related research work is going on within the recently started ENSEMBLE project (www.ensemble.ac.uk). The project aims at exploring the potential of the emerging Semantic Web to support teaching in complex and rapidly evolving fields where case-based learning is the pedagogical approach of choice. The main idea is to combine key elements of digital repositories, semantic web technologies, and features of 'social software' in order to allow for reuse through reconfiguration, adaptation, and collective action. However, this project is still in its early research stage.

6 Conclusions

Collaborative learning through project-based work helps students reflect on their learning experiences in ways that promote abstraction from experience, explanation of results, and understanding of conditions of DPs applicability in real world situations; it also provides the experience of working in software development teams. Following this paradigm, we have developed a learning environment for software DPs which leverages semantic technologies to integrate several existing learning systems and tools, and provide context-aware educational services that together allow for effective learning of software DPs. Our present implementation and first evaluation results convince us that this environment could significantly contribute to effective teaching and learning of DPs. The use of Semantic Web technologies greatly facilitated the development process. In particular, by using a common RDF-based data model as an exchange mechanism among content and (interaction) data providers (i.e., learning systems/tools integrated within DEPTHS), we enabled seamless integration of heterogeneous data and content formats different providers rely upon. Moreover, by leveraging the integrated data and the semantics of the interaction data model (i.e., the Learning Context ontology) we developed educational services that enable fast and effective search for relevant resources and possible peers.

We are encouraged with the results of the initial evaluation study that show very positive students' attitude toward learning in DEPTHS. Students' perception of system's usefulness is valuable and encouraging for our further research. However, the results we got still do not have a statistical power, as the participants' sample was too small. Further research is required that would include sufficient participants to ensure the general applicability of the findings. In addition, in our future work we intend to do a more precise evaluation of each specific educational service as well as a quantitative evaluation of the students' learning effectiveness. Moreover, beside the support for working with text-based and UML based learning objects, our intention for future work is to extend the system to support other non text-based learning objects (e.g. Flash animations, videos, graphics). We hope this will make the DEPTHS framework useful in broad range of domains.

References

[1] Gamma, E., Helm, R., Johnson, R., Vlissides, J.: Design Patterns: Elements of Reusable Object-Oriented Software. Addison-Wesley, Reading (1995)

[2] Warren, I.: Migrating to a Teaching Style that Facilitates Active Learning. CiLTHE Stage 1 Dissertation, Lancaster University (2002)

[3] Jeremić, Z., Jovanović, J., Gasević, D.: Towards a Semantic-rich Collaborative Environment for Learning Software Patterns. In: Proc. of the 3rd European Conference on Technology Enhanced Learning, pp. 155–166 (2008)

[4] Jovanović, J., et al.: Using Semantic Web Technologies for the Analysis of Learning Cotent. IEEE Internet Computing 11(5) (2007)

[5] Jazayeri, M.: The Education of a Software Engineer. In: Proceedings of the 19th IEEE International Conference on Automated Software Engineering, pp. xviii–xxvii (2004)

[6] Bagert, D., Hilbum, T., Hislop, G., Lutz, M., McCracken, M., Mengel, S.: Guidelines for Software Engineering Education, Version 1.0. Technical Report CMU/SEI-99-TR-032, Software Engineering Institute, Carnegie Mellon University, Pittsburgh PA (1999)

[7] Jeremić, Z., Jovanović, J., Gašević, D.: Project-based Collaborative Learning Environment with Context-aware Educational Services. In: Proceedings of 4th European Conference on Technology Enhanced Learning (ECTEL 2009), Nice, France (accepted, 2009)

[8] Dietrich, J., Elgar, C.: A Formal Description of Design Patterns using OWL. In: Proceedings of ASWEC. IEEE Comp. Soc., Los Alamitos (2005)

[9] Montero, S., Diaz, P., Aedo, I.: Formalization of web design patterns using ontologies. In: Proceedings of the 1st International Atlantic Web Intelligence. Conf. (AWIC), Spain, pp. 179–188 (2003)

[10] Kampffmeyer, H., Zschaler, S.: Finding the Pattern You Need: The Design Pattern Intent Ontology. In: Engels, G., Opdyke, B., Schmidt, D.C., Weil, F. (eds.) MODELS 2007. LNCS, vol. 4735, pp. 211–225. Springer, Heidelberg (2007)

[11] Henninger, S.: A Framework for Flexible and Executable Usability Patterns Standards. In: 31st IEEE Software Engineering Workshop (SEW-31), USA, pp. 23–34 (2007)

[12] Coplien, J.O.: Software Patterns. SIGS Books, New York (1996)

[13] Manning, C.D., Raghavan, P., Schütze, H.: Introduction to Information Retrieval. Cambridge University Press, Cambridge (2008)

[14] Bull, S., Greer, J.: Peer Help for Problem-Based Learning. In: Proceedings of ICCE/ICAI, Taiwan, pp. 1007–1015 (2000)

[15] Torniai, C., Jovanovic, J., Gasevic, D., Batemen, S., Hatala, M.: E-Learning Meets the Social Semantic Web. In: Proc. of 8th IEEE Int'l Conf. on Advanced Learning Technologies, pp. 389–393 (2008)

[16] Jovanović, J., Rao, S., Gašević, D., Devedžić, V., Hatala, M.: An Ontological Framework for Educational Feedback. In: Proceedings of the 5th International Workshop on Ontologies and Semantic Web for Intelligent Distributed Educational Systems, USA, pp. 54–64 (2007)

[17] Ghidini, C., Pammer, V., Scheir, P., Serafini, L., Lindstaedt, S.: APOSDLE: Learn@work with semantic web technology. In: I-Know 2007, Graz, Austria (2007)

[18] Yanlin, Z., Yoneo, Y.: A Framework of Context Awareness support for peer recommendation in the e-learning context. British Journal of Educational Technology 38(2), 197–210 (2007)

[19] Ankolekar, A., Sycara, K., Herbsleb, J., Kraut, R., Welty, C.: Supporting online problem-solving communities with the semantic web. In: Proc. of the 15th International WWW Conference, Scotland, pp. 575–584 (2006)

Semantic Enhancement for Enterprise Data Management

Li Ma[1], Xingzhi Sun[1], Feng Cao[1], Chen Wang[1], Xiaoyuan Wang[1], Nick Kanellos[2], Dan Wolfson[2], and Yue Pan[1]

[1] IBM China Research Laboratory
ZhongGuanCun Software Park #19, Beijing 100094, China
{malli,sunxingz,caofeng,chwang,wangxyxy,panyue}@cn.ibm.com
[2] IBM Software Group, 11400 Burnet Rd. Austin, TX 78758-3415, USA
kanellos@ca.ibm.com, dwolfson@us.ibm.com

Abstract. Taking customer data as an example, the paper presents an approach to enhance the management of enterprise data by using Semantic Web technologies. Customer data is the most important kind of core business entity a company uses repeatedly across many business processes and systems, and customer data management (CDM) is becoming critical for enterprises because it keeps a single, complete and accurate record of customers across the enterprise. Existing CDM systems focus on integrating customer data from all customer-facing channels and front and back office systems through multiple interfaces, as well as publishing customer data to different applications. To make the effective use of the CDM system, this paper investigates semantic query and analysis over the integrated and centralized customer data, enabling automatic classification and relationship discovery. We have implemented these features over IBM Websphere Customer Center, and shown the prototype to our clients. We believe that our study and experiences are valuable for both Semantic Web community and data management community.

Keywords: Master Data Management, Semantic Query, Mapping, Reasoning.

1 Introduction

With the increase of market competition, modern companies become more and more seriously dependent on their information, such as customer and product data. A typical situation that companies are facing is information incomplete and inconsistent across many systems, such as one product having different codes and descriptions in various markets, and one customer having different IDs in various systems. This situation mainly results from the lack of global standards (or insufficient application of standards), and the fact that data is captured many times, in many different systems. Therefore, it is critically important to keep a single truth of core entities across various systems within an enterprise to improve business efficiency.

Core business entities that a company uses repeatedly across many business processes and systems are called master data [7], such as lists or hierarchies of customers, accounts, products. Customer data is the most important kind of master data, and customer data management is becoming critical for modern enterprises because it maintains a single, complete and accurate record of customers across the enterprise

A. Bernstein et al. (Eds.): ISWC 2009, LNCS 5823, pp. 876–892, 2009.

[6]. Recently, major data management solution providers, such as IBM, Oracle and SAP, have released their master data management (MDM) solutions [2][3][4], especially on customer data management (CDM, also called customer data integration, CDI) and product information management (PIM). But it does not mean that all problems in MDM have been well solved and these commercial systems meet all customers' requirements. In fact, it is reported in [6][7] that there are still some technical challenges in MDM, for instance, enterprise-wide master data models, federation and identity management. In our previous work on PIM [14], we proposed to build an expressive and flexible ontology to represent the semantics of product information, which dynamically varies with business changes. The proposed approach enables automatic categorization and relationship discovery for product data.

Different from PIM, mature industry models for customer data have been developed [8], such as the customer model for insurance and the patient model for healthcare. The CDM system can directly make use of these industry models without changes or with few changes. Also, there are no strong business needs to change the customer data model after it is deployed. So, based on the industry models, the CDM system can develop well-optimized database schema and thus run in an operational way. IBM Websphere Customer Center (WCC) V6.5 [4] is such an example. It adopts different data models for different industries, uses an object-oriented schema for storage and provides more than 500 business services to manage customer data. Integrating customer data from heterogeneous data sources for a single, complete and accurate record of customers has been widely studied for decades, such as the ETL approach supported by commercial MDM solutions. In this paper, we investigate how to search and analyze the integrated customer data stored in an operational store by semantic Web technologies, instead of discussing data cleansing and integration issues. Besides customer data, the proposed method could be easily adapted to various enterprise data management solutions.

The rest of this paper is organized as follows. Section 2 introduces ontology enabled classification and relationship query. Our prototype over IBM WCC [31] is presented in Section 3, including ontology representation, mapping, SPARQL query evaluation, and reasoning. Section 4 presents experiment results and Section 5 concludes this paper.

2 Semantic Query and Analysis for Customer Data Management

Based on W3C's specifications on semantic Web (such as RDFS[1], OWL[10]), we develop semantic technologies for consumption by CDM, which enables:

- Exposing customer data as RDF views and linking them with the open ontology/data for effective collaboration.
- Declarative definition of customer entities and relationships using logic languages supported by Semantic Web.
- Analytics to discover valuable but implicit knowledge hidden in the rich relationships among customer data.

Domain Ontology Based Classification. In reality, lots of shallow ontologies [19] (like taxonomy) for classification purpose are developed, such as profession ontology in Figure 1(a). These ontologies encoding common knowledge can be leveraged by existing CDM systems to categorize customer data. By simply associating customer data with such domain ontologies, users can classify and retrieve their customers by profession, sector etc.. For instance, we can say "Customer A is a Cosmologist". Then, when querying all "Scientist", we will obtain Customer A. This sort of concept query is based on class inheritance in RDFS/OWL.

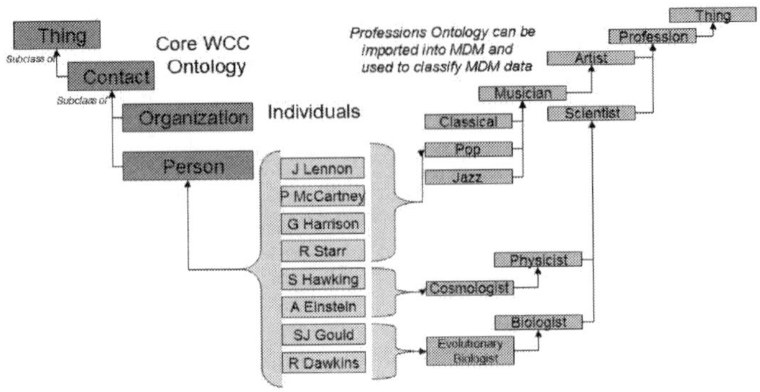

a) Domain ontology based classification

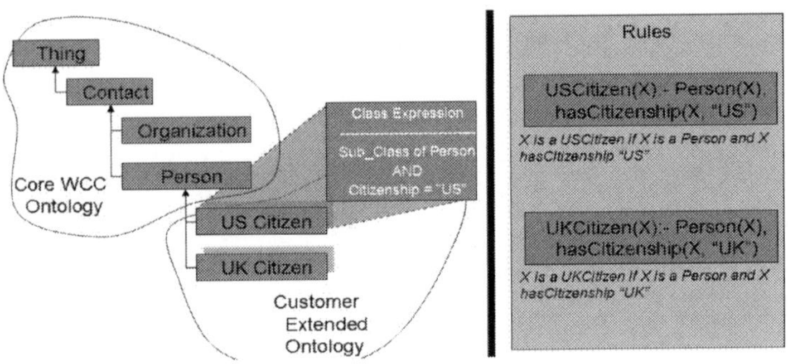

b) Classification based on Formal Vocabularies

Fig. 1. Semantic Query and Analysis for CDM

Classification based on Formal Vocabularies. The expressivity of OWL allows the definition of logical classes (using intersection, union and complement operators). This enables automatic classification for customer data. Also, OWL restrictions support the creation of new category of customer data. As the example in Figure 1(b) shows, we can use OWL intersection and hasValue restrictions to define a new category "US Citizen" which is a subclass of the pre-built category "Person". Using OWL reasoning, instances of this newly-defined category can be automatically retrieved. Besides OWL restrictions, we can also apply rules to formally describe new categories. An example is to define VIP customers as those whose minimum payment is larger than \$10K. That is, users can extend the core CDM vocabulary using OWL expressions or business rules, and freely define new categories at query time and classify customers using reasoning, rather than changing the deployed CDM data model and storage model to represent new categories. This definitely helps to improve the flexibility of CDM systems. Domain ontology based classification allows users to leverage open ontologies to classify and query customers using relatively simple reasoning, whereas OWL restrictions or rules allow users to define new categories based on existing classes and properties of the core CDM ontology and use expressive ontology reasoning or rule reasoning for classification.

Relationship Discovery and Query. OWL allows the definition of richer relationships. In OWL, it is possible to define an Object Property as symmetric, functional, or transitive. OWL Object Properties are suitable to describe the complex relationships among customers, as well as those between customers and other entities like contracts. Followings are some examples of relationships in CDM: *Partner, Sibling, Trustee-Trustee, Owner between party and contract, Part-of among contracts*. Besides using OWL to define the type of relationships, we can use rules to define more general relationships in CDM. Using OWL and Datalog rule inference, we can effectively discover implicit relationships from explicit assertions. For instance, the query "Find all Insurance Contracts held by Persons having the same address as an insured property" needs to compute two implicit rule paths: *Contract->Holding->Property->Address->Person* and *Contract->Person->Address*.

Building an ontology representation for customer data (named core WCC ontology in Figure 1) will facilitate us to implement the above attractive features over existing CDM systems. Moreover, end users can query customer data via classes and relationships defined in ontologies and rules, for example "Find all Scientists living in UK" to send mailing regarding sale of lab coats. In this paper, we refer to the above features as semantic query and analysis over CDM systems, which can be realized by SPARQL queries [12]. In addition, an advantage provided by ontology and RDF is to give each CDM entity a Universal Resources Identifier (URI). That is, we can expose customer data as linked RDF data and allow effective synchronization of CDM utilities to other core business entities, such as those in PIM, or to open data like DBpedia [30].

3 System Implementation

In this section, we take IBM WCC [4] as an example to detail our implementation of semantic query and analysis for CDM [31]. But technologies developed here can be used in other CDM systems, and easily adapted to various data management solutions.

3.1 Architecture

A CDM system provides a unified customer view and an update environment to multiple channels. So, it is impractical to export all customer data into an RDF store for semantic query and analysis. Figure 2 shows an overview of the proposed system, which adds a semantic data access engine over a CDM system through a RDB-to-ontology mapping. The engine exposes a virtual RDF view for customer data.

Firstly, we create a mapping to link customer data with the OWL ontology generated and enriched from the CDM logical data model. That is, we construct a formal ontology representation for customer data stored in the CDM system. Then, users can define business rules for knowledge discovery, and can introduce domain ontologies or OWL restrictions for classification. At query time, users can issue SPARQL queries to the semantic data access engine for data retrieval. The query can include classes and properties defined in both core and user-extended ontologies, and those in domain ontologies. A query will be parsed into a SPARQL query pattern tree. Each node in the tree will be analyzed and checked whether it implies ontology reasoning or rule reasoning. When the rule reasoning is needed, the Datalog rule evaluation will be conducted. If ontology reasoning for instance classification is required, the ontology reasoner will be called. During the reasoning process, the SPARQL query pattern tree will be modified according to the reasoning result. Finally, a SQL generator will generate a single SQL statement, which is evaluated by the database engine where the CDM store is built. In our current implementation, SOR ontology repository [17] is used to store ontologies and rules, and SHER engine [15] for ontology classification reasoning.

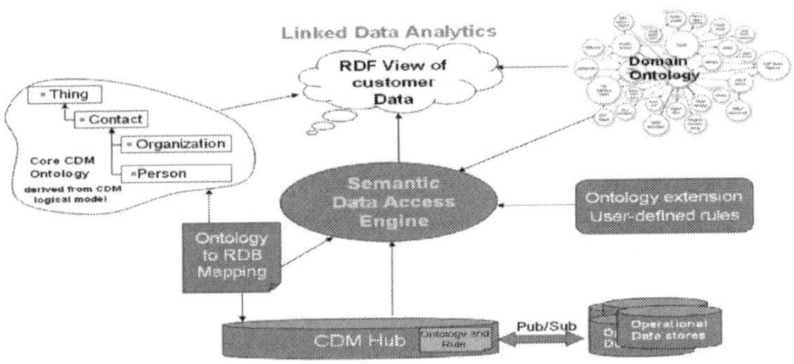

Fig. 2. The overview of Semantic CDM

Business users are able to directly use semantic query to retrieve customer data and analyze the relationships among entities involved in CDM. Another valuable use scenario of semantic query is to enable application developers to develop general analytic services over CDM systems. Now, most existing CDM systems provide only built-in services and do not allow direct SQL queries. For example, WCC provides more than 500 services for various operations and may use different optimization techniques for different services, such as materialized views. When users want additional analytic services that can be realized by the semantic queries, developers just need to write SPARQL to implement them and do not have to consider complex SQL statements and reasoning. That is, we provide a general way to develop new analytic services by SPARQL queries.

3.2 Ontology and Mapping Building

The physical storage model of CDM is similar to its logical data model pre-built for a particular industry. As introduced in the Section 1, the data model of CDM is relatively stable. The CDM's object-oriented storage model makes it possible to take full advantage of mature relational database technologies. The stable data model sacrifices flexibility but contributes to the performance. Therefore, we should construct a relatively stable, lightweight, and coarse-grained ontology model for CDM, over which the business rules and class expressions still can be defined for expressivity. Meanwhile, we need a RDB-to-ontology mapping between the CDM storage model and the constructed ontology model so that an ontology query can be executed smoothly over the CDM system. Note that there are two types of ontologies in our system. The first type of ontology is a semantic representation of underlying customer data and derived from the logical model of the CDM system. The second type of ontology is the imported domain ontology, which captures the domain knowledge and enables users to view customer data based on the domain vocabulary.

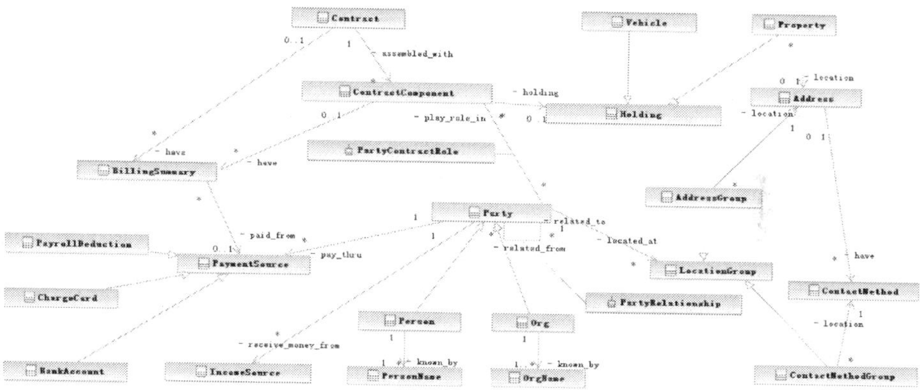

Fig. 3. Part of High Level Industry Model for CDM

CDM Data and Storage Model. Figure 3 shows part of the WCC's data model, in which the attributes of entities are omitted. Within the model, *Party* is the most important concept, with *Person* and *Organization* as its two subconcepts. Other entities are associated with *Party* via various types of relationships. *Contract* is another important concept in CDM, which represents the purchasing agreements of customers. In this model, most relationships are represented as entities, e.g., *Party_Relationship, Contract_Relationship, Party_Contract_Role* (linking *Party* with *Contract*). Besides the entities and relationships, there are some simple hierarchies among them. For example, *Bank_Account_Record, Charge_Card_Record,* and *Payroll_Deduction* are disjoint sub-classes of *Payment_Source*. The WCC's storage model is very close to this logical data model, by considering entities as the tables, their attributes as the columns in the corresponding tables, and their relations between entities as foreign keys or tables.

Ontology Representation of CDM. In our study, we found that RDFS can capture the semantics of the CDM data model. We construct the ontology representation of the CDM model in a relatively straightforward way. For example, *Party, Person,* and *Organization* are defined as classes, and their hierarchy is modeled by *subclassOf* semantics. All objects stored in *Party, Person* and *Organization* tables are their instances. Although the CDM ontology's expressivity is RDFS, users can define very expressive OWL classes over it.

In CDM, many N-Ary relationships are used to describe the variability of customer data. For instance, relationships can be time-dependant, such as *Party_Relationship* having attributes *Start_Datetime* and *End_Datetime*. Moreover, CDM has a set of predefined types for *Party_Relationship*, such as *AncestorOf*. OWL only allows the definition of binary relationships, but we need to represent N-Ary relationships in CDM. The best practices documented at [23] for N-Ary relationship representation in OWL provide some choices as a starting point. In our case, we model N-Ary relationships as classes. For example, *Party_Relationship* is an OWL class, and the specific relationship between two parties is an OWL individual and has *Start_Datetime* and *End_Datetime* as properties. The characteristics of relationships (such as symmetric, transitive) can be enforced by rules. We can use a rule engine for relationships inference, which is discussed in more detail in next sub-sections.

RDB-to-Ontology Mapping. Given the designed CDM ontology, we have to map customer data to it for semantic queries. The mapping is critical to translate a SPARQL query on ontologies into an executable SQL on customer data. In 2008, W3C has formed an incubator group to investigate the standardization of a mapping language between ontology and relational database. Here, we use D2RQ mapping language [9], which is a declarative language for defining the mapping between an ontology and a relational data model. To generate a complete mapping file, we adopt a semi-automatic method. We firstly generate an initial D2RQ mapping based on WCC's physical schema using the D2RQ mapping tool, and then revise it manually according to our ontology.

3.3 SPARQL Query Evaluation

SPARQL queries allow to access customer data, user-defined categories and imported domain ontologies. Figure 4 shows the high level workflow of our query processing. A SPARQL query is firstly parsed into a query pattern tree (shown in Figure 5), and then the Datalog rule engine checks if there are rule heads in query nodes and expands rule heads accordingly. If there is a recursive loop in expanding the rule heads, the Datalog engine will evaluate the corresponding query nodes directly and generate a temporary table as evaluation results. In this process, the Datalog engine will call the SQL generator, which is responsible to generate optimized SQL queries, to get SQL statements for rule reasoning. Accordingly, the Datalog engine updates the SPARQL pattern tree by modifying the query nodes that need reasoning. At last, the SQL generator translates the updated SPARQL pattern tree into a single SQL statement. As mentioned in Section 2, we use an ontology reasoner for instance classification. If a SPARQL query needs classification reasoning (i.e. the triple pattern using rdf:type as predicate), the ontology reasoner will be invoked and the results are temporarily stored in a *TypeOf* table, which contains user-defined classification information as well (namely user-specified *TypeOf* information through domain ontologies). That is, the query nodes with rdf:type as predicate are translated into a sub-query over a *TypeOf* table.

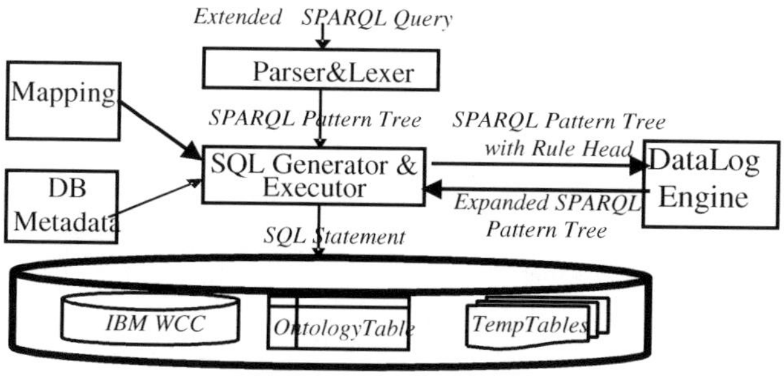

Fig. 4. The Workflow of Query Processing

Some translation methods have been proposed to translate a SPARQL query into SQL queries for evaluation based on the RDB-to-ontology mapping, such as D2R [20], Virtuoso's SPARQL to SQL translation [28], etc.. But these approaches may generate inefficient SQL queries in some cases. Our method makes use of mapping and all other relevant information (such as database catalog) to optimize query generation. To fully utilize the well-developed SQL optimization engine, our method translates a SPARQL query into a single SQL statement. The generated SQL query can be directly embedded into other CDM SQL queries as a sub-query, which provides a way to integrate normal CDM queries with semantic queries.

As introduced earlier, there are some N-Ary relationships in CDM. Unfortunately, the current SPARQL specification does not directly support N-Ary relationships. Although SPARQL may express N-Ary relationships by combining many triple patterns, it is very inconvenient. Therefore, we extend the SPARQL query with a new pattern, called N-ARY pattern, denoting N-Ary relationships and the head predicates in the user-defined rules. The following example shows a SPARQL query with two N-ARY patterns *VIP_Organization* and *Legal_Relationship*, which finds all persons with their addresses (if available) who became the legal persons of VIP organizations that have Party relationship with IBM or HP after year 2000. Also, we show rules used to answer this query.

```
SELECT ?person ?address
WHERE  {
        ?org1 wcc:PartyRelationship ?org2 .
        ?org2 rdf:type wcc:Organization .
        VIP_Organization(?org1).
        Legal_Relationship (?person, ?org1, ?year).
        OPTIONAL {?person wcc:address ?address} .
        {
          {?org2 wcc:Org_Name "IBM"} UNION
          {?org2 wcc:Org_Name "Hp"}
        } .
        FILTER (?year > "2000-1-1") }
```

```
r1. Subs_Org_Relationship(x, y) :-
wcc:Organization(x), wcc:Organization(y), wcc:PartyRelationship(z),
wcc:Related_From(z, x), wcc: Related_To(z, y),
wcc:PartyRelationshipType(z, "Subsidiary") ;;.

r2. Sub_Org_Relationship(x, y) :-
Subs_Org_Relationship(x, z), Subs_Org_Relationship(z, y);;.

r3. Legal_Relationship(x, y, z) :-
wcc:Person(x), wcc:Organization(y), wcc:PartyRelationship(u),
wcc:Related_From(u, x), wcc:Related_To(u, y),
wcc:Start_Time(u, z), PartyRelationshipType(u, "Legal Person");;.

r4. Legal_Relationship(x, y, z) :-
Legal_Relationship(x, u, z), Sub_Org_Relationship(u, y);;.

r5.VIP_Organization(x):-
wcc:Organization(x), wcc:Client_Importance(x, "High");;.
```

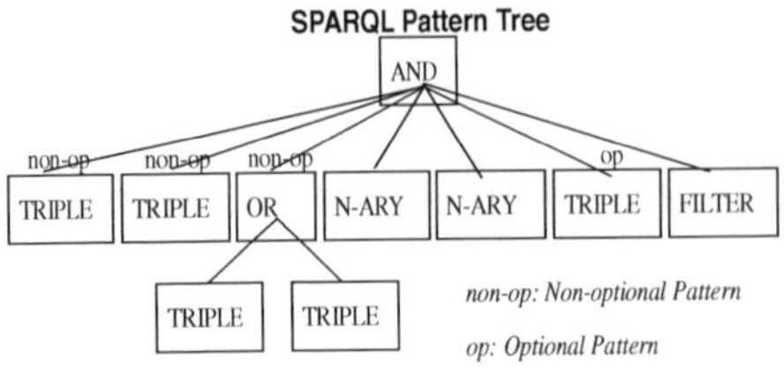

Fig. 5. A SPARQL Pattern Tree

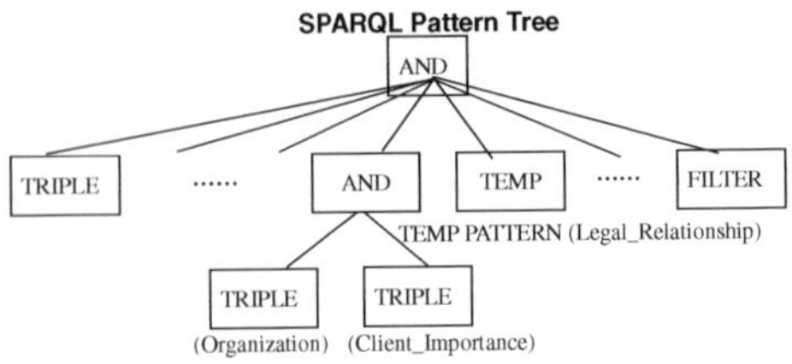

Fig. 6. Rule Expansion on A Pattern Tree

Generally, the graph pattern of a SPARQL query can be expressed as a SPARQL pattern tree, which is introduced in our previous work [16]. The pattern tree shows the backbone of the SPARQL query and is very useful for the SPARQL-to-SQL translation. Figure 5 shows the SPARQL pattern tree of the above example query. There are five types of nodes in the pattern tree:

AND node: It corresponds to the conjunction of graph patterns in a SPARQL query. Each child node has a flag to indicate whether the node is optional or not. By this way, optional patterns in SPARQL are also covered by AND nodes.
OR node: It corresponds to a UNION pattern in a SPARQL query.
TRIPLE node: It represents a graph pattern with a single triple. The subject, predicate and object in the triple could be constants or variables.
FILTER node: It represents a filter expression in a SPARQL query.
N-ARY node: It represents an N-ARY pattern and often directly corresponds to a rule head. Each element in the N-ARY node could be a constant or a variable.

Due to space limitations, we highlight new structures and optimization schemes used in our mapping based translation method.

(1) *Facet of IRI/Literal pattern to translate filter expressions.* Extended from the facet of literal defined in [16], a facet of an IRI/Literal pattern is a complex SQL expression.
(2) *Semi-SQL structure to bridge the columns in SQL and the variables in SPARQL.* Semi-SQL structure consists of several sections. These sections are connected by UNION in semantics. Each section includes two fields: a) SQL field, to keep a SQL statement as the intermediate translation result. b) Mapping field, to record the mapping between each variable and IRI/Literal pattern binding.
(3) *Integrity constraint based simplification for IRI/Literal comparison.* The basic ideas are: a) to compare the IRI/Literal pattern on the raw columns instead of the generated IRIs or literals; b) to compare the constants in IRI/Literal pattern level instead of instance level. TRIPLE nodes often include some constants. Taking the constants as a special IRI/Literal pattern, the prefixes and the data types of the constants are adopted to filter out unmatched property bridges and class maps, which can effectively reduce the number of sub-queries.
(4) *Flattened SQL to detect unnecessary JOINs.* When an AND-pattern contains more than one sub-pattern, the SQL statements of the sub-patterns should be merged for one SQL. An alternative way of merging sub-SQL statements is to take one sub-SQL as a table and embed the table in the FROM clause of the other sub-SQL. However,

the unnecessary JOINs in embedded IRI/Literal patterns may lose the opportunity of being detected and removed. Therefore, we adopt a flattened SQL strategy to merge the SELECT, FROM and WHERE clauses of the sub-SQL statements, respectively. By this way, the IRI/Literal patterns can be directly generated using the columns in the new SELECT clause. Once the sub-queries from child pattern nodes are flattened, it is easy to find the unnecessary self-joins on primary key. In that case, two tables in the FROM clause can be reduced into one, and the costly join operation is avoided.

3.4 Reasoning in CDM

In this section, we first define the semantics of the reasoning system and then introduce the reasoning approach.

3.4.1 Semantics of CDM reasoning

For the better understanding of reasoning semantics, we first categorize the reasoning in CDM according to the reasoning scope as follows.

1) Reasoning for ontology: The reasoning introduced in this category is performed based on the CDM ontology. Specifically, it is either to classify data into newly-defined classes or to find implicit assertions from the CDM ontology. Under this category, we further classify the reasoning as the following two types:
Reasoning based on Description Logic (DL): We apply IBM SHER engine [15] to support all concept assertions. SHER reasoner uses a novel method that allows for efficient querying of SHIN ontologies with large ABoxes in databases. Currently, this method focuses on instance retrieval that queries all individuals of a given class in the ABox. It is well known that all queries over DL ontologies can be reduced to consistency check, which is usually checked by a tableau algorithm. Motivated by the fact that in most real ontologies: 1) individuals of the same class tend to have the same assertions with other individuals; 2) most assertions are in fact irrelevant for purposes of consistency check, we proposed to group individuals which are instances of the same class into a single individual to generate a summary ABox of a small size. Then, consistency check can be done on the dramatically simplified summary ABox, instead of the original ABox. The SHER reasoner implements this reasoning approach on top of SOR's storage component. It is reported in [15] that SHER can process ABox queries with up to 7.4 million assertions efficiently.

Reasoning based on Description Logic Programs (DLP): DLP reasoning is applied to discover the implicit role assertions (relationships) in the CDM ontology. Description logic programs (DLP) is an intersection of description logic (DL) and logic programs (LP). As a subset of LP, DLP can be fulfilled by horn rule reasoning. An example rule in DLP could be *wcc:p(x,z):-wcc:p(x,y) . wcc:p(y,z)*, where *p* is a transitive property in the CDM ontology. Compared with DL reasoning, the complexity of which is NEXPTIME-complete, the LP can be computed in P-time. It has been proved that DLP reasoning is sufficient for RDFS ontology. For the ontology with more expressivity, such as OWL-DL, DLP reasoning is sound but incomplete. In our application, DLP is applied in the part of ontology reasoning. The program, which consists of a fixed set of horn rules for DLP reasoning, is denoted as P_{DLP}.

2) Reasoning for user-defined rules: In our application, users can define a set of business rules (denoted as program P_{user}) on top of ontology to enrich the semantics. The rules are in the form of Datalog-safe rules [27], with the extension of aggregation functions. For example, the rules r1 to r5 in Section 3.3 are user-defined rules. Specifically, r1 and r2 define the sub-organization relationship. R3 and r4 define the legal-reprehensive relationship between a person and an origination. Also, the rule r5 defines the concept of VIP organization. An important constraint for user-defined rule is that the rule head could not be the DL predicate.

For clarity, we refer to DL reasoning as ontological reasoning and logic programs as rule reasoning. In general, the combination of ontological reasoning and rule reasoning causes the undecidability problem. In [24], Motik et al. proposed a decidable combination of OWL-DL with rules by restricting the rules to so-called DL-safe ones. In our work, we separated these two types of reasoning to gain the decidability. The semantics of our knowledge base system is as follows. First, let *KB* be the knowledge base corresponding to the CDM ontology. In addition, let *KB'* denote the extension of *KB* by importing domain ontology and/or defining new OWL restrictions. We apply the ontological reasoning $\prod$ on *KB'* to find all concept assertions, and the result model is represented as $\prod(KB')$. Given the program $P = P_{DLP} \cup P_{user}$ (where P_{DLP} and P_{user} are the program for DLP rules and user-defined rules respectively), the fixpoint [27] of $\prod(KB')$ can be computed as $P(\prod(KB'))$. Any assertion α is called entailed by our reasoning system iff it is contained in $P(\prod(KB'))$.

Note that the example in [25] shows that separating ontological reasoning and rule reasoning may cause incompleteness problem. However, due to the simplicity of the CDM ontology (RDFS), our reasoning scheme is practical for customer data management in terms of decidability and soundness, with the minor drawback on completeness.

3.4.2 Proposed solution

After defining the semantics, we introduce the high-level design of our reasoning system. In terms of the reasoning scheme, we apply a runtime-based reasoning, i.e., the reasoning is performed at query time. Compared with materialize-based reasoning, in which all the inference results are pre-computed and stored in database, despite having longer response time, the runtime reasoning occupies much less disk space and does not have the update problem. Given the large size customer data, it is obviously not practical to materialize all inference results. We have two reasoning components. For ontological reasoning, SHER engine is deployed to do domain ontology based classification and classification based on formal vocabularies and to discover all concept assertions for the CDM ontology. As another key component, a Datalog engine is designed and implemented for supporting the reasoning for all the rules.

Tree expansion approach. Recall that the functionality of the CDM reasoning system is to rewrite original SPARQL query tree, such that every node on the tree can be directly interpreted and processed by the SPARQL query engine. In other words, after processed by the reasoning system, the tree becomes ready to be translated into one SQL statement by the SQL generator component.

Technically, SHER engine will be invoked for every node corresponding concept assertion in the SPARQL query tree. For each node related to the rule head, the rule

```
Algorithm: TreeExpansion
Input:
node: the root of the SPARQL query tree
P: the program for rule reasoning (the set of rules)
dGraph: dependency graph of program P
hashMap: the mapping between rule-head predicate and temporary table
Output:
The root of the expanded SPARQL query tree
Method:
if (node is not a leaf)
        for each child node chNode
                chNode = TreeExpansion(chNode, hashMap, dGraph);
        return node;
/* cases for leaf node */
if (node does not requires reasoning) return node;
if (node corresponds to a type-of assertion) /* require ontological reasoning */
        Set node based on SHER reasoning result;
        return node;
/* node is related to rule head */
if (node can be found in hashMap) /* i.e., has been evaluated before */
        node = corresponding temporary pattern in hashMap;
        return node;
else /* rule reasoning */
        if (node is not in a strong connected subgraph of dGraph) /* not recursive */
                Build sub-tree sRoot by expanding node based on related rules;
                return TreeExpansion(sRoot, hashMap, dGraph);
        else /* recursive rule evaluation */
                Find the strong connected sub-graph scGraph that contains node;
                for each node dNode that is depended by a node in scGraph
                  and dNode✂scGraph
                        dNode = TreeExpansion(dNode, hashMap, dGraph);
                        Call SQLGenerator(dNode) to get SQL statement;
                        Create temporary table for dNode based the SQL;
                        Update hashMap;
                /* call datalog evaluation for a set of rules */
                Find the set of rules ruleSet that corresponds to all nodes in scGraph;
                DatalogEvaluatior(ruleSet, hashMap);// hashMap is updated
                node = corresponding temporary pattern in hashMap;
                return node;
```

Fig. 7. TreeExpansion Algorithm

reasoning component will process it based on the program *P* (i.e., the set of all rules). According to the characteristics of related rule(s), each rule-head node will either be expanded into a sub pattern tree or be replaced by a temporary pattern (corresponding to a temporary table). Specifically, if the node corresponds to one or multiple non-recursive rules[1], it can be expanded as a sub pattern tree. For example, suppose that in Figure 5, the N-Ary pattern on the left relates to a non-recursive rule r5. Then, we can replace the N-Ary pattern by a sub pattern tree that is built based on r5, as shown in Figure 6. For the node corresponding to recursive rules, we have to trigger the Datalog engine to compute the fixpoint for all related rules and append a temporary pattern to replace the node. Suppose in Figure 5 the N-Ary pattern on the right corresponds to the rule head of r4. Because r4 is a self-recursive rule, the N-Ary pattern can not be

[1] A rule *r* is called recursive if it depends on itself or mutually depends on other rules. Otherwise, *r* is non-recursive rule.

expanded as a sub pattern tree. Instead, we do Datalog evaluation and create the temporary pattern for it, as can be seen in Figure 6.

Importantly, the principle of SPARQL tree rewriting is to maintain the tree structure (i.e., expand the node) as much as possible and generate the temporary pattern (i.e., trigger the Datalog evaluation) only when necessary. The aim is to give the SQL generator more information for optimization. Intuitively, the more information the SQL generator can have, the better chance exists to get the final generated SQL optimized.

TreeExpansion algorithm. To follow the above principle, we propose a key algorithm to rewrite the SPARQL query tree. Figure 7 shows the pseudo code of the algorithm. The algorithm recursively expand the query tree based on the ontological reasoning and rule reasoning. Since the SHER engine has been introduced in [27], for simplicity, in this section, we focus on the rule reasoning part. If a node in the SPARQL query corresponds to a rule-head predicate, the rule reasoning engine is triggered. First, as in traditional Datalog approach, by checking dependency graph[2] of program *P*, we can determine if the related rule(s) is recursive or not. For non-recursive rule (i.e., the rule head is not in a strong connected sub-graph[3] of the dependency graph), we expand the node as a sub-tree and then recursively call *TreeExpansion* method on the sub-tree. If the rule is recursive, we compute the strong connected sub-graph *scGraph* it belongs to. Then, we search on the dependency graph to find any *dNode* such that 1) *dNode* is not contained in *scGraph* and 2) at least one node in *scGraph* depends on *dNode*. If there does not exist such *dNode*, it means that the all recursive rules corresponding to the nodes in *scGraph* do not depends on other rules. In this case, we can call Datalog evaluation for computing the fixpoint for these recursive rules. Otherwise (*dNode* exists), before the Datalog evaluation for recursive rules related to *scGraph*, we have to get the temporary table for each *dNode*. So, we treat each *dNode* as an N-Ary pattern. After recursively call *TreeExpansion* to expand *dNode*, the SQL generator is called to get the SQL statement, and the data for *dNode* is retrieved and stored in a temporary table. From the algorithm, we can see that our Datalog evaluation will call *DatalogEvaluatior* only when we need to compute the fixpoint for a set of recursive rules. *DatalogEvaluatior* is implemented based on the semi-naïve evaluation approach with Magic Sets optimization [27]. After the evaluation, *DatalogEvaluatior* will update the temporary table *hashMap*, which is designed to avoid evaluating the same set of rules multiple times.

4 Experimental Results

We implemented our approach to semantic query and analysis over IBM Websphere Customer Center [31]. In this section, we report the experiment results to show the

---

2 A dependency graph of a program *P* is defined as a pair <*V*, *E*>, where *V* is the set of the rule-head predicates in *P*, and *E* is the set of direct edges that shows the dependent relationship among elements in *V*.

3 A graph G is called a strong connected graph if for every node pair *(n1, n2)* in *G*, there exist a path from *n1* to *n2*. Given a program *P*, a rule in *P* is recursive if its head predicate is in a strong connected sub-graph of *DG*, where *DG* is the dependency graph of *P*.

effectiveness and efficiency of the proposed approach. All experiments are conducted on a 2.8GHz Intel Pentium-D PC with 2GB RAM, running Windows XP Professional. Through a 1.0 Gbps intranet, a 2.66GHz Intel Core2 PC with 2GB RAM serves as the backend WCC server, with IBM DB2 9.1 Enterprise edition as the database.

In the experiments, we used a data set from a real WCC customer. The main entities in the data set include 1.9M contracts, 1.0M claims, 2.0M contacts (parties) and 3.2M location groups. The data set contains 1.9M records for *contact-contract* relationship. To estimate the performance of semantic queries that are of interest to WCC customers, additional synthetic data is generated to enrich the data set. We generate 1.0M records for *contact-contact* relationship, *contract-contract* relationship, *contact-claim* relationship and *contract-claim* relationship, respectively.

Since our approach translates a SPARQL query into a single SQL statement, its effectiveness can be evaluated by the correctness of the resulting SQL statements. The system performance is demonstrated in terms of query translation time, response time, and retrieval time, i.e., the duration from the time that a query is issued to the time that a SQL statement is generated, to the time that first answer is retrieved, and to the time that the last record in the result is retrieved, respectively.

Table 1 gives the sample rules used in the experiments. Note that in our test cases, all reasoning tasks are covered by the rule inference. R1 and R2 define the relationships that will be frequently used. R3 to R7 are designed for the scenario below: an insurance company wants to analyze each contract to see how many large claims (claims with amount greater than $5000) are made from it. Table 2 gives the semantics of queries used in the experiments. Indeed, these types of queries are interested and suggested by customers. Not all sample queries are shown in Table 3 due to space limitations. Except for MQ4, all queries can be translated into a SQL statement without any inference result materialized in temporary tables. MQ2-1 and MQ2-2 have the same query semantics but MQ2-2 utilizes rule predicate (N-Ary pattern) to specify *partner-trustee* relationships. The same situation happens for MQ3-1 and MQ3-2.

Figure 8 shows the translation time, response time and retrieval time for queries MQ0, MQ1, MQ2 and MQ3. We can see that the translation time is relatively little compared with retrieval time. Comparing MQ2-1 and MQ2-2, the expansion of the rule predicate in the SPARQL introduces a marginal additional cost. However, rules can be re-used and can make SPARQL queries concise. Note that the retrieval time is less than 3 seconds for these quite complex semantic queries. This is mainly because that our query translation approach outputs near optimal SQL statements.

To answer MQ4, rules from R3 to R7 will be triggered. Note that R4 and R6 are recursive rules. So, we have to call the Datalog engine to do evaluation twice

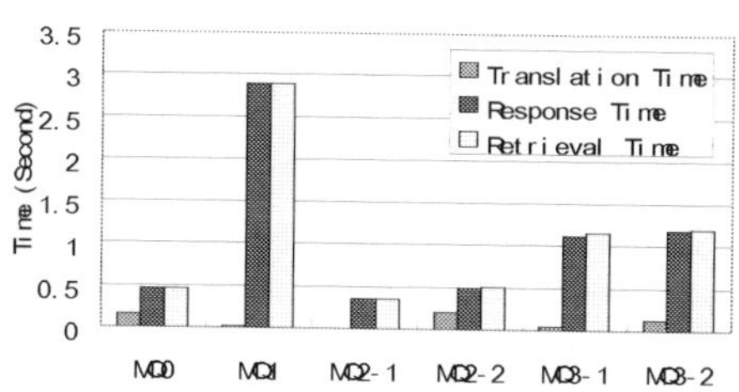

Fig. 8. Query Performance

Table 1. Sample Rules

ID	Rule
R1	Partner_Relationship(?x, ?y) :- wcc:CONTACT_Relationship_FROM(?r, ?x), wcc:CONTACT_Relationship_TO(?r, ?y), wcc:hasCONTACT_Relationship_Type(?r, ?v), wcc:Contact_Relationship_FROM_TO_NAME(?v, 'Partner');;. ***Explanation*** *: Define the partner relationship between two parties.*
R2	Trustee_Relationship(?x, ?y) :- wcc:CONTACT_Relationship_FROM(?w, ?x), wcc:CONTACT_Relationship_TO(?w, ?y), wcc:hasCONTACT_Relationship_Type(?w, ?z1), wcc:Contact_Relationship_FROM_TO_NAME(?z1, 'Trustee');;. ***Explanation*** *: Define the trustee relationship between two parties.*
R3	Sub_Contract(?x,?y):- wcc:CONTRACT_Relationship_FROM(?u, ?x), wcc:CONTRACT_Relationship_TO(?u, ?y), wcc:hasContract_Relationship_Type(?u, ?z), wcc:Contract_Relationship_FROM_TO_NAME(?z, 'Sub Agreement');;. ***Explanation*** *: Define the sub-contract relationship between two contracts*
R4	Sub_Contract(?x,?y):-Sub_Contract(?x,?z),Sub_Contract(?z,?y);;. ***Explanation*** : Define the sub-contract relationship is transitive
R5	Large_Claim_Agreement(?clm, ?contract, ?amount):- wcc:CLAIM_PAID_Amount(?clm,?amount), wcc:CLAIM_CONTRACT_Relationship_To(?u, ?clm), wcc:CLAIM_CONTRACT_Relationship_From(?u, ?contract); ?amount > "5000"^^xsd:double;. ***Explanation*** *: Define the relationship Large_Claim_Agreement, indicating that a contract has a claim with the amount more than $5000*
R6	Large_Claim_Agreement(?clm, ?contract, ?amount):- Large_Claim_Agreement(?clm, ?subcontract, ?amount), Sub_Contract(?subcontract,?contract);;. ***Explanation*** *: Define that a contract is involved in the large_claim_Agreement relationship if its sub-contract has large claims*
R7	Contract_NumOfLClms(?contract, ?Agg_numclm):-Large_Claim_Agreement(?clm, ?contract, ?amount);; Group by <?contract> Count(?clm) as ?Agg_numclm. ***Explanation*** *: Compute the contract and the total number of large claims of the contract*

Table 2. Meaning of Queries

ID	Query Explanation
MQ0	Find the claims that are associated to a given contract (1162960045467) and have the amount greater than $5000
MQ1	Find the claimants of the claims that are associated to the contract 1162960045467 and have the amount greater than $5500
MQ2	Find the person who has the partner 1000020 and is the trustee of the contact 1000045.
MQ3	Find the claimant (with their guardians) who are related to the replacement of contract with id 1000
MQ4	Find the number of large claims (amount greater than 5000) made on contract 1162960045467

(for R4 and R6 respectively) and store the inference results in temporary tables. A naïve way to answer MQ4 is to firstly to compute the number of large claims for all contracts, and then to select the record for the given contract 1162960045467. In the experiment, this naïve approach takes 40.109 seconds as the retrieval time. A smarter approach is to apply the filter condition at early stage of the query so that the inference is performed only for the relevant data. The classical Magic Sets algorithm [29] helps to realize the later approach by rewriting the rules based on the filter condition. To improve the query performance, we implemented the Magic Sets algorithm in our system. As a result, the retrieval time for MQ4 is improved to 6.453 seconds.

We also tried to run MQ0, MQ1, MQ2-1 and MQ3-1 on D2R sever [9] (V0.3.2 in our tests), an open source engine for RDF access to relational databases. However,

Table 3. Sample Queries

ID	SPARQL query
MQ0	Select ?clm ?amount where { ?u wcc:CLAIM_CONTRACT_Relationship_To ?clm. ?u wcc:CLAIM_CONTRACT_Relationship_From <WCC65DB.CONTRACT/1162960045467>. ?clm wcc:CLAIM_PAID_Amount ?amount. Filter(?amount>5000)}
MQ1	Select ?contact ?clm ?amount where {?u wcc:CLAIM_CONTRACT_Relationship_To ?clm. ?u wcc:CLAIM_CONTRACT_Relationship_From <wcc65db.contract/3000>. ?clm wcc:CLAIM_PAID_Amount ?amount. ?v wcc:CLAIM_ROLE_Relationship_From ?clm. ?v wcc:CLAIM_ROLE_Relationship_To ?contact. ?v wcc:hasCLAIM_ROLE_Relationship_Type ?type. ?type wcc:CLAIM_ROLE_NAME 'Claimant'. Filter (?amount >5500)}
MQ2-1	Select ?y where { ?r wcc:CONTACT_Relationship_FROM <WCC65DB.CONTACT/1000020>. ?r wcc:CONTACT_Relationship_TO ?y. ?r wcc:hasCONTACT_Relationship_Type ?v. ?v wcc:Contact_Relationship_FROM_TO_NAME 'Partner'. ?r1 wcc:CONTACT_Relationship_FROM ?y. ?r1 wcc:hasCONTACT_Relationship_Type ?v1. ?r1 wcc:CONTACT_Relationship_TO <WCC65DB.CONTACT/1000045>. ?v1 wcc:Contact_Relationship_FROM_TO_NAME 'Trustee'};
MQ2-2	Select ?y where {Partner_Relationship(<WCC65DB.CONTACT/1000020> ?y). Trustee_Relationship(?y <WCC65DB.CONTACT/1000045>)};
MQ4	Select ?contract ?numofClaims where { Contract_NumOfLClms (?contract ?numofClaims). FILTER (?contract = < WCC65DB.CONTRACT/1162960045467>)}

D2R cannot well support these queries especially when the queries may introduce a number of JOINs. This is because that D2R server provides a memory based solution and does not use enough optimizations (not optimized for graph traversals). As D2R does not support N-Ary patterns and reasoning, it cannot execute MQ2-2, MQ3-2 and MQ4.

5 Conclusions

In this paper, we presented the use of semantic Web technologies for enterprise customer data management. Taking WCC as an example, we introduced an ontology representation for industry customer data model. An effective and practical approach to SPARQL query processing was proposed, with the ontology and rule reasoning incorporated. We believe that the presented approach could be easily adapted to various enterprise data management solutions for semantic query and analysis.

References

[1] Customer Data Integration: Market Review & Forecast for 2005-2006, A CDI Institute MarketPulse™ In-Depth Report

[2] Oracle Data Hub (2005), `http://www.oracle.com/data_hub/index.html`

[3] SAP Master Data Management (2005), `https://www.sdn.sap.com/irj/sdn/developerareas/mdm`

[4] IBM Websphere Customer Center, `http://www-306.ibm.com/software/data/masterdata/customer/`

[5] IBM Websphere Product Center, `http://www-306.ibm.com/software/data/masterdata/product-info/`

[6] Gartner Reports, Magic Quadrant for Product Information Management (2006)

[7] Morris Henry, D., Dan, V.: Managing Master Data for Business Performance Management: The Issues and Hyperion' s Solution. IDC white paper (2005)

[8] IBM Industry Models,
http://www-306.ibm.com/software/data/ips/products/industrymodels/
[9] The D2RQ Platform v0.5.1 - Treating Non-RDF Relational Databases as Virtual RDF Graphs, http://sites.wiwiss.fu-berlin.de/suhl/bizer/D2RQ/spec/
[10] Smith, M.K., Welty, C., McGuinness, D.: OWL web ontology language guide. W3C recommendation (Febuary 2004)
[11] Brickley, D., Guha, R.V.: RDF vocabulary description language 1.0: RDF schema. W3C recommendation (February 2004)
[12] Prud'hommeaux, E., Seaborne, A.: SPARQL query language for RDF W3C recommendation (January 2008), http://www.w3.org/TR/rdf-sparql-query/
[13] Hayes, P., Welty, C.: Defining N-ary Relations on the Semantic Web, W3C Working Group Note (April 12, 2006)
[14] Brunner, J., Ma, L., Wang, C., Zhang, L.: Explorations in the Use of Semantic Web Technologies for Product Information Management. In: WWW 2007, pp. 747–756 (2007)
[15] Dolby, J., Fokoue, A., Kalyanpur, A., Kershenbaum, A., Ma, L., Schonberg, E., Srinivas, K.: Scalable Semantic Retrieval Through Summarization and Refinement. In: Proc. of AAAI 2007, pp. 299–304 (2007)
[16] Lu, R., Cao, F., Ma, L., Yu, Y.: An Effective SPARQL Support over Relational Databases. In: Joint ODBIS & SWDB workshop on Semantic Web, Ontologies, Databases, VLDB 2007 (2007)
[17] Ma, L., Wang, C., Lu, J., Cao, F., Pan, Y., Yu, Y.: Effective and Efficient Semantic Web Data Management over DB2. SIGMOD, 1183–1194 (2008)
[18] Motik, B.: On the properties of meta-modeling in OWL. In: Gil, Y., Motta, E., Benjamins, V.R., Musen, M.A. (eds.) ISWC 2005. LNCS, vol. 3729, pp. 548–562. Springer, Heidelberg (2005)
[19] Shadbolt, N., Berners-Lee, T., Hall, W.: Nigel Shadbolt, Tim Berners-Lee and Wendy Hall, The Semantic Web Revisited. IEEE Intelligent Systems 21(3), 96–101 (2006)
[20] Bizer, C., Cyganiak, R.: D2R Server – Publishing Relational Databases on the Semantic Web. In: Proc. of ISWC (2006)
[21] Chen, H., Wu, Z., Wang, H., Mao, Y.: RDF/RDFS-based Relational Database Integration. In: Proc. of ICDE (2006)
[22] Cyganiak, R.: A relational algebra for SPARQL. HP-Labs Technical Report (September 2005)
[23] The best practices of N-Ary relationships representation in Semantic web,
http://www.w3.org/TR/swbp-n-aryRelations/
[24] Motik, B., Sattler, U., Studer, R.: Query Answering for OWL-DL with Rules. In: International Semantic Web Conference (2004)
[25] Levy, A.Y., Rousset, M.-C.: Combining Horn Rules and Description Logics in CARIN. Artifical Intelligence 104(1-2) (1998)
[26] Donini, F.M., Lenzerini, M., Nardi, D., Schaerf, A.: AL-log: Integrating Datalog and Description Logics. J. Intell. Inf. Syst. 10(3) (1998)
[27] Ceri, S., Gottlob, G., Tanca, L.: Logic Programming and Databases. Springer, Heidelberg (1990)
[28] Mapping relational data to rdf in virtuoso,
http://virtuoso.openlinksw.com/wiki/main/Main/VOSSQLRDF
[29] Bancilhon, F., Maier, D., Sagiv, Y., Ullman, J.: Magic sets and other strange ways to implement logic programs. In: Proc. of the Fifth ACM Symposium on Principles of Database Systems (1986)
[30] DBpedia (2009), http://dppedia.org/About
[31] Wang, X., Sun, X., Cao, F., Ma, L., et al.: SMDM: Enhancing Enterprise-Wide Master Data Management Using Semantic Web Technologies. In: VLDB, pp. 1594–1597 (2009)

Lifting Events in RDF from Interactions with Annotated Web Pages

Roland Stühmer[1], Darko Anicic[1], Sinan Sen[1], Jun Ma[1], Kay-Uwe Schmidt[2], and Nenad Stojanovic[1]

[1] FZI Research Center for Information Technology
Haid-und-Neu-Straße 10-14, 76131 Karlsruhe, Germany
{roland.stuehmer,darko.anicic,sinan.sen,jun.ma,nenad.stojanovic}@fzi.de
http://www.fzi.de/
[2] SAP AG, Research
Vincenz-Prießnitz-Straße 1, 76131 Karlsruhe
kay-uwe.schmidt@sap.com
http://www.sap.com

Abstract. In this paper we present a method and an implementation for creating and processing semantic events from interaction with Web pages which opens possibilities to build event-driven applications for the (Semantic) Web. Events, simple or complex, are models for things that happen e.g., when a user interacts with a Web page. Events are consumed in some meaningful way e.g., for monitoring reasons or to trigger actions such as responses. In order for receiving parties to understand events e.g., comprehend what has led to an event, we propose a general event schema using RDFS. In this schema we cover the composition of complex events and event-to-event relationships. These events can then be used to route semantic information about an occurrence to different recipients helping in making the Semantic Web active. Additionally, we present an architecture for detecting and composing events in Web clients. For the contents of events we show a way of how they are enriched with semantic information about the context in which they occurred. The paper is presented in conjunction with the use case of Semantic Advertising, which extends traditional clickstream analysis by introducing semantic short-term profiling, enabling discovery of the current interest of a Web user and therefore supporting advertisement providers in responding with more relevant advertisements.

Keywords: Complex Event Processing, Event Representation, Semantic Advertising, User Profiling, RDFa, event-condition-action, ECA.

1 Introduction

Currently the Semantic Web largely obeys a request/response style of communication. This is owed to the predominating way in which the traditional Web is used. However, the Active Web is making progress demonstrated by applications such as Twitter which are able to *push* information to recipients at the

A. Bernstein et al. (Eds.): ISWC 2009, LNCS 5823, pp. 893–908, 2009.

time when the information is created. Event-driven ways of disseminating data were identified as being efficient as well as intuitive for many business applications e.g., in [1]. The (Semantic) Web of the future will also benefit from active technologies which help deliver information on time and in a resource saving manner i.e., without having to poll for new or changed information.

In this paper we propose an architecture for capturing and processing complex events on the Web, based on the use of Semantic Web technologies, making it suitable for the future (Active) Semantic Web. As part of this we produce an event representation which facilitates mutual understanding of events on the Semantic Web. Additionally, we demonstrate the benefits of event processing and evaluate them in the field of Semantic Advertising as a use case scenario.

To achieve this, we assess Semantic Web technologies to enable reactivity and design and implement a client-side framework to generate and process events. Furthermore, we demonstrate this framework in the aforementioned advertising use case. The main contributions of this work will be an extensible representation for events on the (Semantic) Web, as well as an implementation of a client-side framework to create these events based on user interactions with Web documents as a source of events. These contributions serve to advance the state of reactivity on the Web and promote new ways of efficiently communicating Web-based information, which we see as a necessary factor for future Semantic Web applications.

This paper is structured as follows: In Section 2 we will describe the requirements for generating meaningful events from Web documents, followed by means of processing these events to detect complex situations. In subsequent sections we will name key parts and technologies for our architecture such as a client-side reactive rule language in Section 3, an RDFS ontology for representing complex events in Section 4 and a way of obtaining meaningful context for events from annotations in Section 5. We will then present implementation details of our architecture in Section 6. The whole approach will be evaluated for performance and usefulness in the advertising scenario in Section 7 and we will discuss related work and conclude the paper in the last remaining sections.

2 Event Generation and Processing from Semantic Web Pages

The main issue in making the Web active is to enable capturing of actions or changes in Web documents. These can be treated as events, which an event-driven system will react to. For our use case of advertising we will focus on events created from a user's interaction with Web documents. After having extracted events from a Web document, they must be processed in order to interpret them semantically, to be able to react on them appropriately. The following two subsections describe our approach for these two issues: generation and processing of Web events.

2.1 Event Generation

A simple event in Web clients is characterized by two dimensions; the type of event (e.g. click, mouseover) and the part of the Web page, where the event occurred (e.g. a node of the Document Object Model of the Web document). This node is, however, just a syntactical artifact of the document as it is presented in a Web browser. Adding this node or parts of it to the event body will not significantly add meaning to the event and not ease the understanding of the event for the recipient of the event.

We therefore propose to add semantic information to the event which pertains to the actual domain knowledge that the Web page is about. In order to enable this, the first step is to represent the content of a Web page in a form that can be used for generating meaningful events. To do so without having to manually annotate every Web document, we envision a mechanism, which ensures the relevance of the annotations. This can be done in many (semi-) automatic ways, e.g. by providing Web forms (page templates), which for a given user's input, automatically adds the proper semantic relationships between the form fields. In this way all user generated content will be annotated. The Web forms are created based on supported vocabularies for a particular Web site. Our particular focus is on widely spread vocabularies such as Dublin Core[1], Creative Commons [2], FOAF[3], GeoRSS[4] and OpenCalais[5]. Regarding the format of structured data, RDFa [2], eRDF[6] and Microformats[7] are all good candidates for this purpose. They support semantics embedded within actual Web page data and allow reusable semantic markup inside of Web pages. In our implementation we use RDFa, since in comparison to eRDF it is a more encouraged candidate by the W3C. Comparing it further to Microformats, RDFa is more flexible in mixing different existing vocabularies.

In the remaining part of this section we give an example demonstrating the generation of events in the context of a Semantic Advertising scenario. The ad space is a part of the Web page which can be dynamically filled by an ad provider as a response to an event the client sends. In our approach ad content is created based on a current user's attention. In order to accomplish this we need as much (meta-) information as possible about the content of the Web page. Therefore, we assume semantically enriched Web content such that context extraction is easier and more precise. Additionally, every page is split up in a number of *Semantic Web Widgets* (SWW). We introduce *Semantic Web Widgets* as self-contained components annotated with semantic data and displayed in a Web page. Semantic Web Widgets give a high-level description of the content, and provide the basic context of data contained in the widgets. For instance on a

[1] Dublin Core: `http://dublincore.org`
[2] Creative Commons: `http://creativecommons.org`
[3] FOAF: `http://foaf-project.org`
[4] GeoRSS: `http://georss.org`
[5] OpenCalais: `http://opencalais.com`
[6] eRDF: `http://research.talis.com/2005/erdf`
[7] Microformats: `http://microformats.org`

```
<div xmlns:rdf = "http://www.w3.org/1999/02/22-rdf-syntax-ns#"
     xmlns:dc = "http://purl.org/dc/elements/1.1/"
     xmlns:vCard = "http://www.w3.org/2001/vcard-rdf/3.0#"
     xmlns:iCal = "http://www.w3.org/2002/12/cal/ical#">
    <ul about="events/Mary_Poppins_Show">
      <li typeof="cal:Vevent">
        <a property="ical:categories">Classic, Comedy, Kid Friendly, Musical<
            /a>
        <a property="cal:dtstart" content="20081008T180000Z">October 8th at
            18am</a>
        <a property="cal:duration" content="PT2H">2 hour</a>
        <vCard:TEL rdf:parseType="Resource">
         <rdf:value>(212) 307-4100</rdf:value>
         <rdf:type rdf:resource="http://www.w3.org/2001/vcard-rdf/3.0#work"/>
        </vCard:TEL>
        <vCard:ADR rdf:parseType="Resource">
         <vCard:Street>   214 West 42nd Street </vCard:Street>
         <vCard:Locality> New York City </vCard:Locality>
         <vCard:Pcode>    NY 10036 </vCard:Pcode>
         <vCard:Country>  USA </vCard:Country>
        </vCard:ADR>
        <a property="cal:description">Mary Poppins takes up residence at
            magnificent New Amsterdam Theater.
        </a>
      </li>
    </ul>
</div>
```

Listing 1. An example for a musical listed in a Semantic Web Widget

news portal incorporating semantic advertising one widget could be used for listing all news belonging to one subcategory, e.g., politics, another one for arts, etc..

In Figure 1 we show an RDFa example of the semantic description for an arts event[8] listed in a widget related to musicals. The code snippet presents an event named "Mary_Poppins_Show" described using RDF Schemata for Dublin Core, vCard and iCal vocabularies. Information such as categories, start and duration of the musical are provided together with contact information, location and so on.

2.2 Complex Event Processing

Simple events extracted from Web documents can be combined in order to detect complex situations. This is the task of Complex Event Processing, that we describe in the context of the Semantic Advertising use case. Detecting the behavior of Web users according to our proposal is divided into design time and run time. The design time consists of (i) semantically enhancing the Web page and then (ii) recording average viewing statistics of the annotated elements, e.g. from log files. From the statistical data we generate client-side rules. Once these rules are created they are pulled by the next client request and loaded into the rule engine for the run time.

For the run time we have developed a client-side event-condition-action (ECA) rule engine. It uses a lightweight rule language which supports ECA rules

[8] An *event* in the sense of a gathering of people.

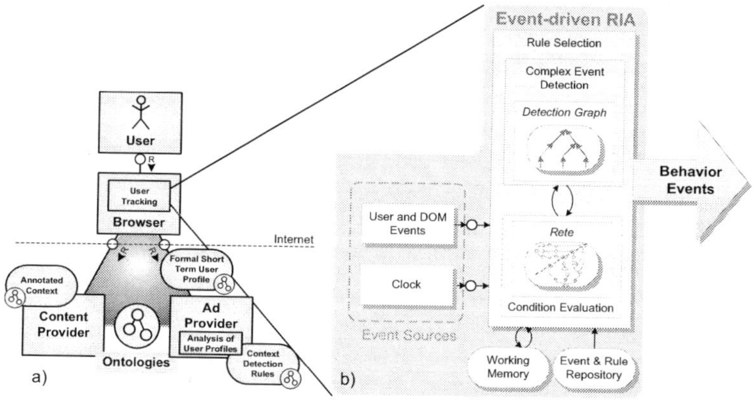

Fig. 1. Architecture: a) Logical Architecture b) Client-side User Behavior Analysis

described in more detail in Section 3. The rules on the client serve to detect users exhibiting interesting behavior as learned from the average usage patterns. The user causes events to occur by interacting with the Web page, detected by the event processor and rule engine. Rules are triggered which create intermediate events in a hierarchy of event abstraction. These events are subsequently accumulated until sufficient interest according to the ad provider is recorded (threshold achieved) and actions can be taken by further rules.

The distinction between run time and design time in this section is not a strict temporal distinction as the names would suggest. Rather, because new users will inevitably alter our knowledge of what is interesting there is a loop in the process, feeding back from the run time into the design time to evolve new rules for future users.

Figure 1 shows a rough architecture of our approach: Part b) on the right hand side of the figure depicts the components of our client-side rule engine. Multiple event sources provide input for the event detection, creating complex events. Also, a working memory submits its changes to a Rete network, evaluating rule conditions. The logic for both the event detection and condition evaluation is supplied by rules from a repository, generated from past user activities. Part a) on the left hand side places the client-side components above the protocol boundary dividing client and server. Below on the server or several distributed servers hold the Web content as well as the advertising content. The Web content is annotated, providing semantical relations to the advertisements. Short-term user models provide a temporal model of how a user interacts with the Web content. The ad provider analyses user models to provide up-to-date and personalized advertisements.

In the next three sections we present in detail three mechanisms which enable the realization of the approach we present in Section 6.

3 JSON-Rules: A Client-Side Rule Language

To facilitate client-side advertisement we use JSON-Rules, our client-side rule language. It resembles a lightweight reaction rule language tailored to the needs of Rich Internet Applications, specifically applications that profit from or require Complex Event Processing, condition evaluation on a working memory, and running rule actions written in JavaScript. As a representation for our rules we use JSON[9], because it is natively usable within JavaScript. JSON can specify objects, arrays and primitives. Rule objects in our JSON-Rules language contain the three attributes `event`, `condition` and `action`. The event part consists of patterns in the event pattern language Snoop [3]. The condition part consists of conjunctive predicates over variables from a working memory. The action part in turn contains one or more JavaScript code blocks to gain a maximum degree of versatility for the rule author. Alternatively for rule actions we offer to trigger certain desired events as well as manipulations of the working memory. The latter types of action offer greater declarativity while formulating rules. This increase is, however, bought at the price of some flexibility. Thus, we still offer all three kinds of rule actions which can be freely mixed.

For the event part of each rule the usual Snoop operators are available: $Or(E_1, E_2)$, $And(E_1, E_2)$, $Any(m, E_1, E_2, \ldots)$, $Seq(E_1, E_2)$, $A(E_1, E_2, E_3)$, $A^*(E_1, E_2, E_3)$, $P(E_1, TI[\text{:parameters}], E_3)$, $P^*(E_1, TI\text{:parameters}, E_3)$, $Not(E_1, E_2, E_3)$, and $Plus(E_1, TI)$. We only briefly list them here, their semantics are documented in [3]. Additionally we define further event operators $Mask(E_1, \text{condition})$ and $Thres(E_1, \text{threshold})$ as follows. The operator $Filter$ enforces a condition on each occurrence of events E_1. This allows e.g. for fine-grained content-based filtering/masking of events. The operator $Thres$ is another content-based operator which we need to extend the Snoop algebra with. $Thres(E_1, \text{threshold})$ accumulates the events of type E_1 until the boolean function `threshold` returns true, releasing all accumulated events as a complex event and starting accumulation anew.

A condition in our language may use comparison operators on facts from the working memory and literal values. The condition part is a conjunction of predicates. Comparison operators are $<$, $>$, $=$, $<=$ and $>=$. Variables specify items from the working memory. For the advertising use case conditions are not needed, the current rules react purely to events.

Rule actions are JavaScript code blocks or new events or modifications to the working memory to be triggered on rule execution. A code block has access to the set of events and facts that has led to the firing of the rule. Thus, rule authors may create applications that do calculations on the parameters of the collected events and matched condition variables.

Listing 2 shows an example rule. It can be automatically created from analyzing histories of interesting behavior. The only requirement is knowledge, that e.g. states that only two percent of users look at a politics item followed by a science item. The actual rule consists of an event part starting at line 5 and an

[9] JavaScript Object Notation: `http://www.json.org/`

```
1 {
2   "meta": {
3     "rule": "Politics->Science=>2%"
4   },
5   "event": {
6     "type": "SEQ",
7     "children": [
8       {
9         "type": "DOM",
10        "selector": "div[property=dc:keywords][content?=politics]",
11        "event": "click"
12      },
13      {
14        "type": "DOM",
15        "selector": "div[property=dc:keywords][content?=science]",
16        "event": "click"
17      }
18    ]
19  },
20  "action": [
21    {
22      "type": "EVENT",
23      "trigger": "unusual",
24      "parameters": {"probability": 0.02}
25    }
26  ]
27 }
```

Listing 2. Example of a single Rule

action part starting at line 20. The rule resembles an event-condition-action rule where the condition is left blank, i.e. is always true.

The event part in this example describes a sequence of two sub-events. Both sub-events are of type "DOM" which means they are adding handlers to the Web page. In this case each one listens to clicks on DIV elements in the document object model (DOM) where the keywords politics and science are annotated. The rule action is of type "EVENT" which means the rule raises another event. The event to be created is called "unusual" and carries a parameter containing a probability. This event can be subscribed to by further rules. In our case there is a rule aggregating all events of this type until enough unusualness (in terms of aggregated probability) is observed. This example rule is a small part of our whole architecture [4,5] which detects, aggregates and finally submits the user profile in form of one or more complex events.

4 RDFS

In this chapter we present an RDFS ontology for events. It covers a schema for simple events and for modeling derived events such as events composed of one or more simpler event occurrences. The basic classes of the ontology are shown in Figure 2.

An event in terms of our ontology is either a simple event or a complex event. All events have a type, some timestamps and may contain a body (also called payload). A simple event is an instantaneous occurrence where start and end times are equal. A complex event is composed of one or more events which means that

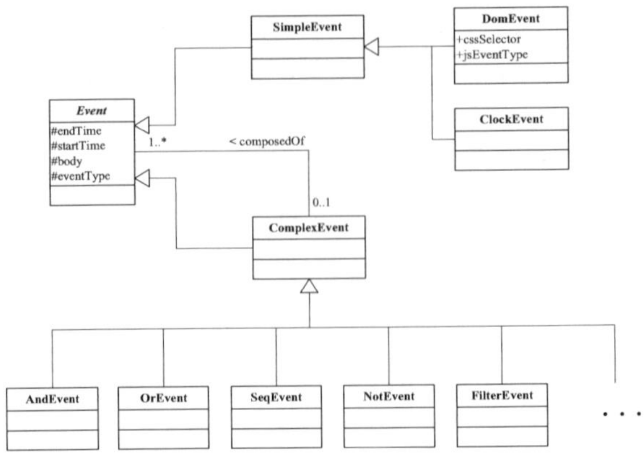

Fig. 2. RDFS ontology for events. The composition of complex events is conceptualized via classes with the names of the operators used in their detection. For simple events we currently only require the subclasses `DomEvent` for user interaction with a Web page and `ClockEvent` for purely temporal events such as timers, reminders, etc..

a complex event can occur over an interval of time, where start and end time are different. A complex event instance is usually detected by applying an event pattern to a stream of events. A complex event may contain its contributing events which led to the detection of the complex event. So, for example for a pattern of $And(E_1, E_2)$ an instance contains one event of each type E_1 and E_2. To distinguish such a complex event from other complex event which are incidentally composed of two contributing events, we create a type for the complex event pertaining to the event operator which was used in the detection.

Therefore, we obtain an ontology of subtypes of `ComplexEvent` which represent event operators. To come up with a list of operators we started with one of the oldest general purpose event languages named Snoop [3]. Snoop contains a fairly comprehensive list of boolean and temporal operators. They are modeled in our ontology. What is missing in Snoop are operators which inspect the contents of input events such as attributes other than timestamps and type. Therefore, we added a `FilterEvent` as an example of what is needed to filter events by their content.

For simple events we identify two significant subtypes which are relevant to the client-side of the Web. These are the actual events created in the browser after a user interaction is detected and events form the client's clock. They are represented as `DomEvent` and `ClockEvent` respectively. These two types of events make up all of the client-side event load in a Web browser.

5 RDFa

In the previous section we described our schema for events, simple or complex. In this section we focus on enriching simple events with semantics from the context of the Web page in which the event occurred.

A simple event in Web clients is characterized by two dimensions; the type of event (e.g. click, mouseover) and the part of the Web page, where the event occurred (e.g. a node in the Document Object Model of the Web document). Subscribing to simple events of these types therefore requires the specification of type and the specification of the node or nodes where the events may originate. Both dimensions are retained in an event instance by using the attributes `jsEventType` and `cssSelector` (cf. Figure 2).

In order to better understand these events and make sense of what happened we must enrich the content of events when they are produced. The `jsEventType` tells us what a user has done and the `cssSelector` tells us where on the Web page the user did it. However, the latter is a purely presentation-dependent measure. There is no semantics which has any meaning beyond the context of a specific Web page structure. We propose to extract presentation-independent semantic information from the Web page if present. Instead of creating events from interaction with purely syntactic items of a Web document, we create events about interaction with semantic concepts which the document stands for. As an example, an event should not represent e.g., a click on a certain headline element of a Web document but rather a user's interaction with an article talking about politics and certain persons mentioned within.

To annotate a Web page with semantic data such as the topics of an article, we use RDFa. Defined in [2] RDFa is a means of adding RDF data to existing Web pages by using inline XHTML attributes.

After detecting an event which happened in the context of a certain DOM node of a Web document, we collect all semantic information in the Web page about the thing that is reported in that given DOM node. We currently achieve this by employing the client-side RDFa library ubiquity[10]. The lifting of context is achieved in a two-phase process. In the first phase we collect the list of RDF subjects of possible triples. This is done close to where the event happened in the document to provide *accurate* context. In the second phase we collect every triple with these subjects from the overall document in order to provide a very *rich* context.

To find valid subjects the first phase traverses the node where the event happened and its complete subtree[11]. If the given main node does not contain a subject, the immediate dominator node containing a subject is added to the list. This serves two purposes, guaranteeing a single root subject for orphan properties and objects in the subtree and guaranteeing a non-empty result set.

In the second phase all triples with the given subjects are collected from the entire document tree[12]. The gathered triples are then reified and appended as a bag to the event payload.

[10] The Ubiquity RDFa parser project: `http://ubiquity-rdfa.googlecode.com/`

[11] In the use case example this could include a news article about a politician and all the contained paragraphs.

[12] In the use case example this includes all triples about the politician from the article as well as possible extra information such as scattered information areas on the remaining Web page.

Even if the event itself becomes part of more complex events during the process of correlating and aggregating events, this basic data is retained as part of the simple event.

6 Implementation: Client-Side Event-Enabled Rule Engine

For our implementation we chose JavaScript from the available Web programming languages, for reasons of widespread availability. The data structures and program logic we implemented are roughly divided into the following areas: adapters for the rule language and remote event sources, the working memory, condition representation and evaluation as well as complex event detection.

For Complex Event Processing we are using a graph based approach as proposed in [3]. Initially the graph is a tree with nested complex events being parents of their less deeply nested sub-events, down to the leaves being simple events. However, common subtrees may be shared by more than one parent. This saves space and time compared to detecting the same sub-events multiple times, and renders the former tree a directed acyclic graph.

When using the term *event*, the distinction must be drawn between event occurrences (i.e. instances) and event types, usually done implicitly. In the detection graph the nodes are event types, they exist before there are any instances. Event instances exist after simple instances arrive and are fed into the graph at the leaves. Complex instances are then formed at the parent nodes, which in turn propagate their results upwards. Every complex event occurrence carries pointers to the set of its constituent event occurrences, so that the events and their parameters can be accessed later. Once an occurrence is computed at a node which is attached to a rule, the state of the associated Rete node is started and actions are triggered.

7 Evaluation

In this section we first evaluate the performance of our client side complex event processor. After that we present a first discussion of a running demo application collecting and aggregating events from test users we invited to a demo "news portal".

7.1 Performance

In this section we show the results from a performance test demonstrating that CEP and a client-side rule engine for the Web are indeed feasible. It is a technical evaluation of the event-processing capabilities of our ECA-rule-based framework. We will show that an event rate of about 64 events per second is possible with a given rule set on our test machine. Our test machine is a 2.4 GHz Intel Core2 CPU with four cores. Since JavaScript execution is inherently single-threaded it

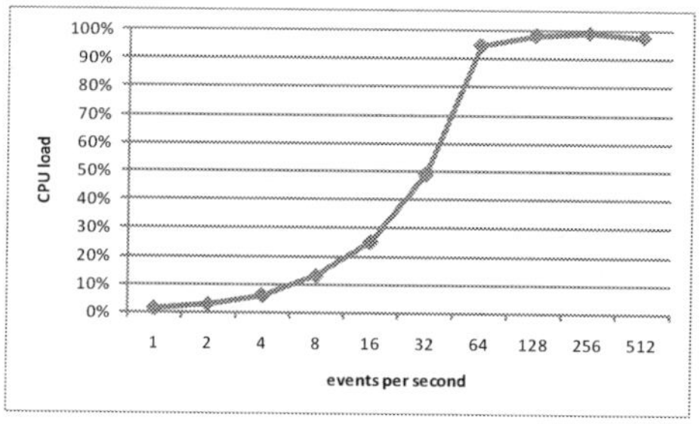

Fig. 3. CPU load by increasing event frequency

profits only from one CPU core. Having spare cores for other tasks combined with a generally low operating system load provides results which are uninfluenced by other running tasks. The chosen JavaScript engine is Mozilla Firefox 3.0.3 for Windows using the Firebug[13] profiler. The browser was installed freshly with no extra plug-ins.

We start out with the BEAST benchmark [6]. BEAST is an attempt at measuring CEP performance in early CEP applications, which use a similar event algebra than Snoop. We borrow some of the rules which are applicable to our CEP engine. Some of the event operators were not applicable to Snoop, like the count based window operator (cf. Figure 3 on page 8 of [6]). The event expressions from BEAST which are tested are described as follows, using event operators from the Snoop algebra:

$$\texttt{SEQ}(\mathrm{E}_1, \mathrm{E}_2) \tag{1}$$
$$\texttt{NOT}(\mathrm{E}_3, \mathrm{E}_4, \mathrm{E}_5) \tag{2}$$
$$\texttt{SEQ}(\mathrm{E}_6, \texttt{SEQ}(\texttt{OR}(\mathrm{E}_7, \mathrm{E}_8), \mathrm{E}_9)) \tag{3}$$

Event expression 1 represents a sequence of two events. Event expression 2 represents the non-occurrence of E_4 in the interval of two other specified events and finally: event expression 3 represents the sequence of one event followed by a disjunction and followed by another event.

The tested rules contain empty rule actions, so only the CPU load for Complex Event Processing is measured. We run each test for 30 seconds at various frequencies of simple events per second. The simple events in the mentioned patterns are entered into the detection system in a round robin manner. We then measure the load percentage caused by the detection system matching the incoming events and producing complex events.

The results are shown in Figure 3. The chart shows that our event detector can handle a maximum of about 64 events per second in real time. After that the JavaScript engine is used up to capacity and further incoming events are

[13] Firebug Web site: `http://getfirebug.com/`

queued up. Rule conditions, unlike event expressions, are not strictly needed for our approach to online advertising. Their evaluation can be found in [5].

The general use case for our framework is to monitor the user's event-based interaction with a Web page. This means reacting mainly to human actions. Since events from the human user are not occurring at millisecond rates, our framework should be fast enough to handle events at near real-time. Although expensive rule actions may lessen these results, upcoming new browsers promise a significant increase in general JavaScript performance due to newer compiling techniques. Currently, we expect the results to be sufficient for most client-side applications including our application of event-driven online advertising.

7.2 Ad Quality

To evaluate the return of targeted advertisements we created a demo Web page with some news articles. Each news article is contained in a separate part of the page, termed Semantic Web Widget (cf. Section 2.1). Each widget is annotated using RDFa using basic keywords and concepts pertaining to the article. For a user entering our demo, each widget is at first partially concealed. This is done to solicit an action from the user when "unfolding" the widget. Thereby the user expresses interest. This creates explicit events which can then be processed by our engine. Our initial evaluation of the ad quality was performed as follows:

1. We selected three different news domains (politics, culture, sports) in order to prove the domain-independence of the approach and pull into the demo Web page, as separate evaluation sessions.
2. We selected five users (PhD students from the Institute) with different cultural backgrounds.
3. The users should browse the demo Web page and judge about the relevance of generated ad-keywords in the case of a) the keywords generated statistically from the Web page (Google approach) and b) keywords generated by using the event-driven approach described in this paper. In order to ensure a fair comparison, the users did not know which list of ad-keywords was produced by which method.

We ask the users to rate the gathered keywords in terms of relevance to what they had been doing in the news portal and to compare this with a static list of keywords extracted from the overall page. The results are very encouraging: in the average 85% of keywords generated in our approach were described as "very relevant" and 98% as "relevant" (very similar results across all three domains). The traditional approach achieved 65% success for "very relevant" and 85% sucess for "relevant" ad-keywords. This result demonstrates the advantages of our approach for generating very relevant ads.

In comparison, Web Usage Mining (e.g., [7]) is used on log files which are analyzed on the server side at certain intervals or possibly in a continuous fashion. It is important, however, to stress that our approach detected all events on the client. Events occurred purely by folding and unfolding widgets as parts of the page. No communication with the server took place and hence no artifacts are visible in server log files. Thus, our approach extends clickstream analysis to regions which were previously invisible to server-based mining techniques.

Moreover, our approach is a truly event-driven application, meaning that we detect events in real-time, as soon as they happen. In contrast, traditional mining techniques function in a query-driven manner where results are only created at intervals, such as daily analyses of the log files.

8 Related Work

In this section we discuss related work from several fields of research relevant for this paper, namely reactivity for the Web, online advertising related to our use case and Complex Event Processing in general and specifically for the Web.

There exist several approaches to reactivity for the Web. The approach from [8] describes a rule-based event processing language for XML events. The language design is described as focusing on aspects of data extraction, event composition (operators), temporal relationships and event accumulation. The approach is based on logic programming. Some drawbacks are inherited from this. The most striking fact is that events (simple and complex) are detected and reacted to in a query-driven fashion. This means that event patterns are only fulfilled when the query engine asks for the patterns. There is no data-driven way of fulfilling patterns in the moment each event arrives. This behavior is based on the fact that logic programming systems such as Prolog operate in a backward-chaining way, fulfilling queries only when they are posed. There is no first class notion of continuous queries. This means that the approach from [8] as well as others such as [9] are not truly event-driven, because events are not handled when they occur but are stored until the query is posed for the next time. Furthermore, it is unclear where the events come from and how they are entered into the logic programming system; There is no notion of subscribing to input streams or similar ways of accessing event sources. Consumption of events is also not defined; Events seem to have indefinite life-time and be reused in new patterns over and over. In comparison to our work there is no focus on client-side events which occur in a browser, e.g. from humans interacting with Semantic Web documents.

Another event processing language for the Web is presented in [10]. It is likewise an event-condition-action (ECA) rule-based approach, with pluggable language dialects for each of the E, C and A parts of a rule. An ontology of the compositional approach is presented. The question of connecting event sources is addressed in this work, but requires a degree of cooperation of nodes on the Web which is currently not practical. For example, a possible source of events is said to be the changes to XML data. However, such events are only created if change is monitored, e.g. with the help of an *active* XML database. As a workaround, so-called *ECA services* are proposed which provide active notifications from passive nodes. However, as this requires polling/querying, it is again not strictly event-driven. Our solution actively publishes events when they occur and as such is fully event-driven. In a federated setup of the mentioned related work, our solution could possibly be used as a source of events.

In Web advertising there are essentially two main approaches, *contextual advertising* and *behavioral advertising*. Contextual advertising [11] is driven by the user's context, represented usually in the form of keywords that are extracted from the Web page content, are related to the user's geographical location, time and other contextual factors. An ad provider (ad serving service) utilizes these meta data to deliver relevant ads. Similarly, a users' search words can also be used to deliver related advertisement in search engine results page, Google's second pillar in online advertising. However, contextual advertising, although exploited today by major advertising players (e.g., GoogleAdsense[14], Yahoo! Publisher Network[15], Microsoft adCenter[16], Ad-in-Motion[17] etc.), shows serious weaknesses. Very often the automatically detected context is wrong, and hence ads delivered within that context are irrelevant[18]. For instance, a banner ad offering a travel deal to Florida can possibly be seen side-by-side to a story of a tornado tearing through Florida. This is happening because the context was determined using purely keywords such as "Florida, "shore" etc (i.e., without taking keyword semantics into account). While there are improvements in contextual advertising (e.g., language-independent proximity pattern matching algorithm [12]), this approach still often leads companies to investments that are wasting their advertising budgets, brand promotion and sentiment. In contrast, our approach utilizes semantics to cure major drawbacks of today's contextual advertising. Semantic Web technologies can be used to improve analysis of the meaning of a Web page, and accordingly to ensure that the Web page contains the most appropriate advertising.

The second approach to Web advertising is based on the user's behavior, collected through the user's Web browsing history (i.e., *behavioral targeted advertising*). The behavior model for each user is established by a persistent cookie. For example, Web sites for online shopping utilize cookies to record the user's past activities and thereby gain knowledge about the user or a cluster of users. There are several reasons why behavioral targeted advertisement via cookies is not a definitive answer to all advertisement problems. First, if a user, after browsing the information about an item purchases that item, he or she will not be interested in that particular good afterwards. Therefore, all ads and "special deals" offered to the user later while browsing that Web site are useless. Also, the short-term user interest should be detected more quickly (i.e., during the current user session). Displayed ads need to reflect current moods or transient user interest. For example, a user looking hastily to buy a gift of flowers is not interested in ads related to his/her long-term profile, created during previous

[14] GoogleAdsense: `http://google.com/adsense`

[15] Yahoo! Publisher Network: `http://publisher.yahoo.com`

[16] Microsoft adCenter: `http://adcenter.microsoft.com`

[17] Ad-in-Motion: `http://ad-in-motion.com`

[18] Adam Ostrow, When Contextual Advertising Goes Horribly Wrong - Mashable: `http://mashable.com/2008/06/19/contextual-advertising`

purchases unrelated good or services. Further on, there are problems with cookies. Computers are sometimes shared and users get to see ads governed by other user's cookies. Finally, given the European Union's Directive and US legislation concerned with restricted use of cookies, behavioral targeted advertisement based on cookies is not a promising direction for Web advertising.

We believe that *short-term profiling* (in contrast to long-term profiles created by cookies) is a valid and possibly augmenting approach in terms of personalization and identification of the user's interest. We realize a short-term profiling using client-side Complex Event Processing techniques (cf. Section 2.2), and background semantics (cf. Section 2). Such profiles are automatically detected, are always up-to-date and fully personalized.

The work from [13] describes event processing for Web clients. Events are observed on the client, however, complex events are not detected in the client. All simple events are propagated to the server for detection of patterns. This incurs latency and reduced locality for the processing of events, so the advantages of client-side event processing are lost.

9 Conclusion

In this paper we present a novel approach for generating and processing complex events form Web pages which opens possibilities to build event-driven applications for the (Semantic) Web. We envision the future of the (Semantic) Web as a huge, decentralized event repository (the so-called Event Cloud in CEP terminology), which will contain information about the real-time activities of different Web users. Such an event repository will enable different kinds of processing of the real-time information, making the Semantic Web really active, i.e. the environment can react and adapt itself on the signals sensed from the environment, connecting the Internet of Things with the Internet of Services, two basic elements of the Future Internet. As a result this paper makes contributions to the integration of the Semantic Web in the Future Internet. The presented Semantic Advertising use case shows clearly how efficient the contribution of active components and semantic technologies in future Internet applications can be.

Acknowledgments. We would like to thank Weiping Qu for his contribution to the RDF schema and to the implementation.

References

1. Luckham, D.C.: The Power of Events: An Introduction to Complex Event Processing in Distributed Enterprise Systems. Addison-Wesley Longman Publishing Co., Inc., Boston (2001)
2. Adida, B., Birbeck, M., McCarron, S., Pemberton, S.: Rdfa in xhtml: Syntax and processing (October 2008), `http://www.w3.org/TR/rdfa-syntax/`

3. Chakravarthy, S., Krishnaprasad, V., Anwar, E., Kim, S.K.: Composite events for active databases: Semantics, contexts and detection. In: Bocca, J.B., Jarke, M., Zaniolo, C. (eds.) 20th International Conference on Very Large Data Bases, Santiago, Chile proceedings, Los Altos, CA 94022, USA, September 12–15, pp. 606–617. Morgan Kaufmann Publishers, San Francisco (1994)
4. Schmidt, K.U., Stühmer, R., Stojanovic, L.: From business rules to application rules in rich internet applications. Scalable Computing: Practice and Experience 9(4), 329–340 (2008)
5. Schmidt, K.U., Stühmer, R., Stojanovic, L.: Gaining reactivity for rich internet applications by introducing client-side complex event processing and declarative rules. In: Stojanovic, N., Abecker, A., Etzion, O., Paschke, A. (eds.) The 2009 AAAI Spring Symposium on Intelligent Event Processing, Association for the Advancement of Artificial Intelligence, March 2009, pp. 67–72 (2009)
6. Geppert, A., Berndtsson, M., Lieuwen, D., Roncancio, C.: Performance evaluation of object-oriented active database systems using the beast benchmark. Theor. Pract. Object Syst. 4(3), 135–149 (1998)
7. Liu, B.: Web Data Mining. Data-Centric Systems and Applications. Springer, Heidelberg (2007)
8. Bry, F., Eckert, M.: Rule-based composite event queries: The language xchangeeq and its semantics. In: Marchiori, M., Pan, J.Z., de Sainte Marie, C. (eds.) RR 2007. LNCS, vol. 4524, pp. 16–30. Springer, Heidelberg (2007)
9. Paschke, A., Kozlenkov, A., Boley, H.: A homogenous reaction rules language for complex event processing. In: International Workshop on Event Drive Architecture for Complex Event Process (2007)
10. May, W., Alferes, J.J., Amador, R.: An ontology- and resources-based approach to evolution and reactivity in the semantic web. In: Meersman, R., Tari, Z. (eds.) OTM 2005. LNCS, vol. 3761, pp. 1553–1570. Springer, Heidelberg (2005)
11. Kenny, D., Marshall, J.: Contextual marketing–the real business of the Internet. Harvard Business Review 78(6), 119–125 (2000)
12. Schonfeld, E.: Proximic signs deals with yahoo and ebay to turn product listings into contextual ads; taking on adsense. Online Article (January 2008), `http://www.techcrunch.com/2008/01/15/proximic-signs-deals-with-yahoo-and-ebay-to-turn-product-listings-into-contextual-ads-taking-on-adsense/` (Last visited August 2009)
13. Carughi, G.T., Comai, S., Bozzon, A., Fraternali, P.: Modeling distributed events in data-intensive rich internet applications. In: Benatallah, B., Casati, F., Georgakopoulos, D., Bartolini, C., Sadiq, W., Godart, C. (eds.) WISE 2007. LNCS, vol. 4831, pp. 593–602. Springer, Heidelberg (2007)

A Case Study in Integrating Multiple E-commerce Standards via Semantic Web Technology

Yang Yu, Donald Hillman, Basuki Setio, and Jeff Heflin

Department of Computer Science and Engineering. Lehigh University
19 Memorial Drive West, Bethlehem, PA 18015
{yay208,djh3,bas207,heflin}@cse.lehigh.edu

Abstract. Internet business-to-business transactions present great challenges in merging information from different sources. In this paper we describe a project to integrate four representative commercial classification systems with the Federal Cataloging System (FCS). The FCS is used by the US Defense Logistics Agency to name, describe and classify all items under inventory control by the DoD. Our approach uses the ECCMA Open Technical Dictionary (eOTD) as a common vocabulary to accommodate all different classifications. We create a semantic bridging ontology between each classification and the eOTD to describe their logical relationships in OWL DL. The essential idea is that since each classification has formal definitions in a common vocabulary, we can use subsumption to automatically integrate them, thus mitigating the need for pairwise mappings. Furthermore our system provides an interactive interface to let users choose and browse the results and more importantly it can translate catalogs that commit to these classifications using compiled mapping results.

Keywords: e-commerce, ontology integration.

1 Introduction

Internet business-to-business transactions afford both dramatically increased flexibility and present great challenges in merging information coming from so many sources. To provide B2B services, suppliers must deal with the problem of heterogeneity underlying their customers' product, catalog, and document descriptions. Due to the lack of global common standards to classify products, a key task for these marketplaces is to effectively and efficiently manage different description styles. In real-world marketplaces, developing a scalable approach for information integration has become essential to expanding business.

The Federal Cataloging System (FCS) is an extensive taxonomy for naming, classifying, and describing items of supply. The Defense Logistics Information Service (DLIS) created the FCS and uses it to catalog and differentiate items of supply. DLIS routinely catalogs a varied population of items. The conventional coding of products for internal operations has resulted in product identifiers that

A. Bernstein et al. (Eds.): ISWC 2009, LNCS 5823, pp. 909–924, 2009.

are not easily exchangeable across taxonomies, causing significant duplication and inefficiencies in business processes. For DLIS to have full access to suppliers and their products everywhere and at all times, the FCS taxonomy must be aligned with supplier taxonomies.

There are many taxonomies and classification schemes and more than 40 of them have been publicly identified. Most of them organize data into domains viewed from the perspective of a specific use case. ECl@ss is a Standardized Material and Service Classification, while Product Service Classification (PSC), United Nations Standardized Products and Services Code (UNSPSC), Common Procurement Vocabulary (CPV), Common Procurement Code (CPC), RosettaNet Technical Dictionary (RTD), and the Customs Harmonized Tariff Code (HS) are commodity classifications. It is not only products that appear in taxonomies. For example, the Standard Industrial Classification (SIC) code and the North American Industrial Classification Code (NAIC) organize companies by principal occupation, such as Manufacturer, Wholesaler, Retailer, etc.

In this project we integrated four representative commercial classification systems with the FCS by focusing on their underlying knowledge structures rather than on their surface characteristics. Since these four systems adopt significantly differing classification strategies and cover most leading classification styles, we believe that the experience of this development is sufficiently general and capable of being used in most other popular classifications. The four classifications we used were the UNSPSC, CPV, eCl@ss, and the ISO 13584-511 Fasteners Dictionary. (The official acronym for ISO 13584 is PLIB, so we use PLIB-511 as its short name in this paper.)

The paper is organized as following. Section two introduces the necessary background knowledge for better understanding. In section three we briefly overview the approach we devised. The process of constructing an ontology for the taxonomies is described in section four. Section five discusses how we built mappings among the ontologies. Section six introduces prototype tools. The last section describes the conclusion and future work.

2 Background

It is necessary to first make clear the definitions of the terms "taxonomy" and "ontology" because they are distinct concepts that we frequently misused as synonyms. A taxonomy is a particular classification arranged in a hierarchical structure organised by subtype-supertype relationships. Guarino defines that "an ontology is a logical theory accounting for the intended meaning of a formal vocabulary [7]." Simply speaking, a taxonomy only contains the corresponding vocabulary with a hierarchical structure in the ontology; however an ontology can include axioms to define and restrict the semantics and relationships in the taxonomy, which is more important. Most ontologies have a taxonomy, but not all taxonomies can be easily converted to ontologies, due to ill-defined parent-child relationships. In a formal ontology, strict is-a interpretation is used for the taxonomy, i.e. every instance of a child is an instance of the parent. But in informal taxonomies parent-child relationships are often topic-oriented. Therefore an important and difficult step in our project is first to convert all the taxonomies into corresponding ontologies.

One of the advantages of using ontologies is that they naturally support large-scale distributed information. When information resources commit to the same ontology then the same meaning is anticipated for any term from that ontology. Even when information resources commit to different ontologies, there are still methods to integrate the information [8], as long as the ontologies have certain relationships, e.g. their concepts are defined in terms of a common ontology or an alignment is provided. Some other advantages of ontologies for the integration of heterogeneous and distributed classifications in e-commerce are discussed in [6].

The basis of the Semantic Web is that there are a number of ontologies, and different information resources commit to the definitions in these ontologies [1]. The problem considered in this paper can be solved by using the Semantic Web approaches. There are multiple semantic web languages with different features that have been intensively researched and designed. OWL is the W3C recommendation for a web ontology language and is an extension of the Resource Description Framework (RDF). The OWL class constructors (see Table 1) and axioms can be used to express rich semantics. In this paper, we focus on OWL DL, the sublanguage of OWL that most closely corresponds to description logics (DL). Most DLs are decidable and have sound and complete reasoning algorithms. The worst-case complexity of SHOIN(D), the DL corresponding to OWL DL, is NEXPTIME-complete [16], but modern tableaux-based reasoners typically have very good performance.

Creating domain ontologies from industrial products categorization standards is not straightforward as it appears. We have seen the RDF-S version [10] and the DAML+OIL version [12] of UNSPSC. There also exists a prototype of a RDF representation of eCl@ss 4.1 based on an OWL Full ontology [2]. These ontologies, however, were created automatically and have not been used for later integration. They do not reflect the semantics of the underlying standards precisely and specifically enough and are thus not suitable for the basis of inference. Hepp [9] points out that many classifications are created from a procurement viewpoint, and this results in hierarchies that do not have the strict "is-a" semantics of rdfs:subClassOf. For example, eCl@ss typically includes parts and accessories as subcategories of each product in its hierarchy. He proposed to represent a concept in the classification using two concepts in the ontology, one is a generic concept and the other is for the taxonomy category (representing the concept plus related goods). The taxonomy concepts are arranged in a strict rdfs:subClassOf hierarchy corresponding to the classification. A new "annotation class" is created for each concept, and is made as rdfs:subClassOf both the generic and the taxonomy class. Any specific products are asserted to be of this type.

Ontology matching is actively researched. There are mainly three categories [4]. The first is ontology mapping between an integrated global ontology and local ontologies, and typical tools include LSD and MOMIS. The second is mapping

Table 1. OWL class constructors

Constructor	DL Syntax	Example
intersectionOf	C1 ⊓ C2	GasTurbine ⊓ AircraftPart
unionOf	C1 ⊔ C2	Door ⊔ Airframe ⊔ TailSection
complementOf	¬C	¬Aircraft
oneOf	{x1,...,x2}	{F15, F16}
allValuesFrom	∀P.C	∀partOf.Airframe
someValuesFrom	∃P.C	∃hasPart.Door
maxCardinality	≤ nP	≤10hasPart
minCardinality	≥ nP	≥2hasPart

between local ontologies which is more suitable for cases with many ontologies and typical tools include GLUE, MAFRA and ONION. The third allows a single coherent merged ontology to be created when ontologies which have the same or overlapping domains; and tools include PROMPT [13] and Chimaera [11]. Automated and semi-automated methods typically use syntactic techniques, linguistic resources and/or ontology structure to identify matches. To the best of our knowledge, no automated technique can discover alignments more complicated than subsumption between two named classes.

Omelayenko [14] surveys various integration approaches for ecommerce. Most of these techniques focus on syntactic matching [3]. The approach proposed by Corcho et al. [5] is most similar to ours. In their approach they separate e-commerce standards into generic ontologies, which provide coarse-grained classifications of products, and regional ontologies whose concepts share a common root by all mapping to the concepts in the generic ontologies. Below those two levels, there are catalogs and optional local ontologies. This mechanism makes classifications into a multi-layered structure. However, since Corcho et al. do not map generic ontologies to other generic ontologies, each regional ontology must map to each concept in any overlapping portions of these ontologies. Also, since they do not map regional ontologies to each other, it is not possible to align the most specific concept in these ontologies. Finally, we note that they only use equivalent, subclass and union-of axioms in their mappings, limiting the types of mappings that they can express.

Compared to that proposal, the agreed common ontology in our approach is created based on common vocabularies from all incorporated e-commerce standards. Additionally all standards are on the same layer so that we do not need preliminary work to determine their roles. The mapping based on this structure is clear and easily understood. This simple structure has significant flexibility for changes of ontologies since each ontology only needs to map to a single interchange ontology. The additional advantage is that the user of a standard does not need to understand all partners' standards in order to integrate with them, because that standard only needs to be integrated with the common ontology. Obviously this costs less both when initially building mappings among them and when new ontologies are incorporated.

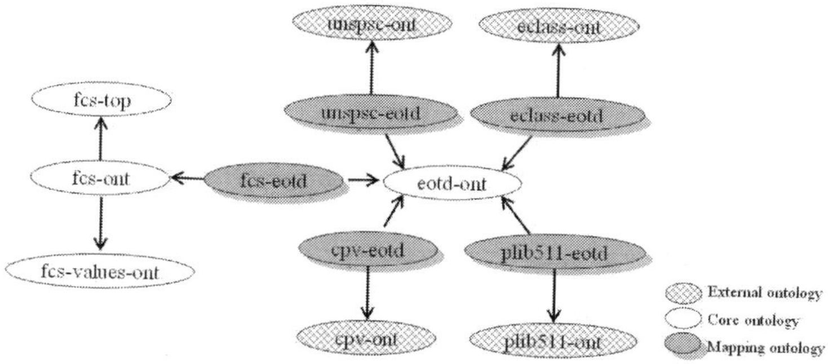

Fig. 1. The ontology network for the project. Each node is an ontology and the arrow points to another ontology which it extends from.

3 Approach

As introduced in the previous section, DL can be used to integrate taxonomies. To accomplish this, we must transform all taxonomies into OWL DL in order to describe concepts, terms and all expressive relationships in order to further embody higher level semantics. This step semi-automates the process of knowledge acquisition from the sources of information selected and adapts them to the ontological knowledge model. For all product ontologies, we assume that instances are "item of production," i.e. each instance is a specific part from a specific manufacturer. All created ontologies form an inter-related ontology network. Besides the FCS, the core of the network includes the eOTD which provides the means for comparing an external classification with the core FCS classification. It acts as a *lingua franca* whereby one classification can be mapped to another, which means classifications are not mapped directly to each other but by the eOTD intermediary. At the perimeter of the network (see Fig. 1) are all external ontologies derived from commercial taxonomies.

The next and critical step in our system is to formally define the terms in the FCS and external ontologies. We do this by writing axioms that ground the concepts defined in them with eOTD concepts. Therefore, the eOTD functions as the "standard" reference taxonomy against which other taxonomies are compared. Although these axioms could be included in the FCS and external ontologies themselves, for modularity reasons we place them in mapping ontologies. These mappings constitute "mediator" ontologies or navigational paths among different ontologies.

Assuming a sufficient intermediary ontology and axiomatization of the other ontologies, any concept can be translated to its subsuming concept in another ontology simply via logical entailment. We built a product classification translator that integrates all the taxonomies and analyzes potential reasonable relationships to search for "best available match" through multiple edges.

Using the translator's outcome we can solve a more practical requirement from the suppliers. The wrapper and compiler tools together fulfill this requirement to translate product descriptions to the FCS classification scheme and can be applied to heterogeneous formats, like XML, spreadsheets and other semi-structured or structured documents.

4 Ontology Construction

The core ontologies include four ontology files for the FCS taxonomy and one for the eOTD. In addition, there are four other ontologies for external taxonomies: eCl@ss, CPV, UNSPSC and PLIB-511. When we created these four external ontologies, we tried to automate the processes and keep their original hierarchical structures, because this can prove that our method does not need significant work to change or adapt other classifications for our approach. The semantic gap between those classifications and the FCS is shortened by ontology mapping which will be described in section 5. Therefore in this section we mainly discuss the construction of core ontologies and the characteristics and difficulties of external taxonomies we found during ontology construction.

4.1 Naming Convention

In order to build mappings systematically in later steps, it is necessary to establish certain naming conventions. In generating ontology identifiers we took the original name from the taxonomy and removed all spaces and punctuations converting it to Camelback notation. In accord with RDF convention, we start classes with an upper case letter and properties with a lower case letter (e.g. "taperIncludedAngle"). Properties that have the same name but apply to different classes are distinguished by numbers at the end of the class name. If a child uses the same name to refer to a more specific concept, we append "All" in the name of its super class. If a class actually includes things that are not is-a children of the named concept, we append "Etc" because this indicates that the more general concept is improperly named. We use the underscore where a class name begins with a number, such as "_12".

Sometimes, this naming convention fails to adequately denote the concept, because punctuation may aid in interpreting a name. We use a rdfs:label to preserve the original name. Thus for example, the class name: StudAssembly-TurnlockFastener could be supplemented with the label: "STUD ASSEMBLY, TURNLOCK FASTENER" for a more natural English interpretation.

4.2 FCS Ontology

The FCS is an extensive taxonomy for naming, classifying, and describing items of supply under inventory control by the Department of Defense (DoD). The Defense Logistics Information Service (DLIS) is responsible for the FCS and uses the taxonomy to catalog and differentiate items of supply. Each item in the FCS is assigned a four-digit code by the government to designate various groups of common use. The first two digits is the Federal Supply Groups (FSG) code identifying the group and the last two digits is the Federal Supply Class (FSC) code identifying the classes within each group. Each FSC has many Item Name Codes (INCs) which are five-digit numbers assigned by DLIS to each Approved Item Name (AIN). Although we don't discuss it in this paper, each INC also has many National Stock Numbers (NSNs) where each represents a different item of supply. The Master Requirement Code (MRC) is a four-character code assigned to a Federal Item Identification Guide (FIIG) requirement, and a set of MRCs can be used to indicate whether a requirement in a FIIG needs to be described for the item being identified.

Since the purpose of this project is to integrate external taxonomies with the FCS, our first step was to automatically create the FCS ontology from FCS database. The first ontology (fcs-top) for the FCS defines higher level terms in the FCS, such as the classes for FSGs and FSCs. Each FSC class is an rdfs:subClassOf its parent FSG class. This ontology contains all FSGs and FSCs. The second ontology (fcs-ont) for the FCS extends from fcs-top and defines classes for all the AINs in our scope. It includes properties for all mandatory MRCs of each AIN. Every AIN is an rdf:subClassOf its appropriate FSC class. The third ontology (fcs-values-ont) for FCS extends from fcs-ont and defines the values in the FIIGs of the FCS as instances in the ontology. To support the ontologies above, we created a meta ontology (fcs-meta) that defines annotation properties for recording the FSG, FSC, INC and MRC codes. This allows the ontology to be searched by DLIS personnel using their terminology.

Since this project was only a prototype, we focused on 128 AINs describing batteries, bearings, bushings, fasteners, gaskets and electronic circuits. In most cases, each AIN is represented in the ontology as a class and is identified by its INC in the fcs-ont. We use a minimum cardinality restriction to indicate properties defined as mandatory by the FCS. In principle, each item of production should have exactly one AIN, therefore we made all of the AINs under an FSC disjoint, and all FSCs under an FSG disjoint. In general disjointness axioms like these help in detecting errors in alignment axioms and can even be used to infer most specific relationships between classes in different ontologies. So we tried to make every disjoint relationship explicitly declared in every ontology we created. For instance, "fsc:BatteryNonrechargeable $\sqcap$ fsc:BatteryRechargeable $\equiv \bot$". Although we generally found the FCS to be a well-designed classification system, there were some defficiencies. For example, some different MRCs have the same syntactic words as name, e.g. MATT and MATL both have names "Material".

4.3 eOTD Ontology

The eOTD is an international standard for e-catalogs via the ISO 22745 designation used to create unambiguous language independent encoded descriptions of master data. It is designed to support industry classification by providing a classification neutral dictionary of names and attributes (or characteristics, properties) of items.

Since the eOTD is designed to be classification neutral, it does not contain a taxonomy. It only includes the most specific terms from each incorporated classification scheme. One of these schemes is the FCS. The eOTD includes a class for each AIN and a property for each MRC. In creating our OWL ontology, we only included those classes and properties that are in the scope of the project. In section 5.1, we discuss how we enrich the eOTD with a class hierarchy to better support our mappings.

4.4 External Ontologies

In this sub-section, we briefly introduce the four representative external ontologies we selected in this project.

Ecl@ss is a German initiative to create a standard classification of material and services for information exchange between suppliers and their customers. We use the 6.0 version of eCl@ss which has over 25,000 nodes arranged in a four-level hierarchy. We only created classes for those concepts that were in our project scope. We did not have access to a machine-readable copy, the only source we accessed is the website (http://www.eclass-online.com/). The eCl@ss ontology has a limited set of properties.

UNSPSC is a coding system to classify both products and services for use throughout the global eCommerce marketplace and its partners include 3M, AOL, Arthur Andersen, BT, Castrol and others. There are over 20,000 nodes in the hierarchy, which like eCl@ss is four levels deep. A limitation of the UNSPSC is that it does not define any properties, and no definitions are given for classes. We wrote a simple program to translate an Excel spreadsheet describing the catalogue into RDF.

CPV is the European-wide classification system for public procurement contracts. It has over 8,000 nodes arranged in a hierarchy with 7 levels. It employs two different methods for building the taxonomy. The first is a straightforward hierarchy proceeding from more general concepts to specific instances. The second method is based on a type relationship model. Like UNSPSC, CPV also does not have any properties nor are concept definitions provided. As above, we automated data extraction from Excel spreadsheets.

PLIB [15] is the Parts Library series of international standards, and is defined by ISO 13584. It is developed and maintained by the ISO technical committee. PLIB is a data model. In this project we use PLIB-511, the PLIB fasteners dictionary as an external ontology. Our ontology includes all of the classes and properties from this dictionary. PLIB-511 has a finer granularity of classes, but the FCS has a superior breadth of coverage. It is the only external ontology considered that has a relatively large number of properties. PLIB specifies an electronic interchange format which facilitates automatic conversion to OWL.

Table 2 summarizes all the ontologies we created so far. Most of our classifications have limited or no properties, for instance eCl@ss, UNSPSC and CPV described above. However properties are important components for a full ontology. Some properties do exist in taxonomies though they are at a fairly high level and lack detail. Furthermore, there are no definitions, at least in UNSPSC, CPV and the online version of eCl@ss that we were able to access over the Web. At best, there are lists of alternatives for specified terms. Also sometimes the modeling is inconsistent and fails to always observe implicit modeling conventions. These conditions introduce difficulties when mapping ontologies.

Table 2. Statistics of ontologies created

Ontology	Classes	Properties	Depth	Mean Siblings
fcs-top	735	365	2	14
fcs-ont	128	2	3	13
eOTD	194	180	5	10
eCl@ss	313	18	4	8
UNSPSC	228	0	4	13
CPV	208	0	7	5
PLIB-511	186	204	6	12

5 Ontology Mapping

This section discusses how to build mappings between ontologies for each corresponding taxonomy. All the original ontologies and mapping ontologies form an inter-connected ontology network (recall Fig. 1).

Our mapping construction is done iteratively as described in Fig. 2. As a preliminary step in the mapping process, we enriched the eOTD vocabulary to provide a solid basis for later steps. In the mapping process, we found and built a number of mappings manually, then used a reasoner to evaluate the validity of these constructs. If there were some errors in our previous manual construction, additional editing or correction of these mappings were needed. Otherwise, we repeated this process iteratively based on all of the preceding knowledge.

5.1 Enriching the eOTD

Given the wide variety of terms, attributes and properties across the taxonomies, we need to find a way to accommodate all of the differences. As we mentioned in section 4.3, one of the purposes of the eOTD is to serve as the common vocabulary that can be used to define concepts in all of other ontologies. However, this mapping is complicated by the fact that the eOTD does not have a hierarchy. Therefore we found it helpful to add more general concepts to the ontology. Not only are those concepts convenient when we are defining terms from more abstract external ontologies, but they enable definitions even when the eOTD terms provide incomplete coverage of the varieties of a product. We refer to these concepts as "abstract classes," and populate them by analyzing the FCS and, to a lesser extent, the external ontologies. Here we use two guidelines. The first is to remove one or more modifiers from class names. For instance, we created two abstract classes "Bearing" and "BearingRoller" for the AIN "Bearing, Roller, Self-Aligning." Another guideline is to identify "foundational" classes from FSGs and FSCs. For instance, we create four abstract classes "BearingsAntifriction," "BearingsPlain," "BearingsMounted" and "BearingsUnmounted" since we have

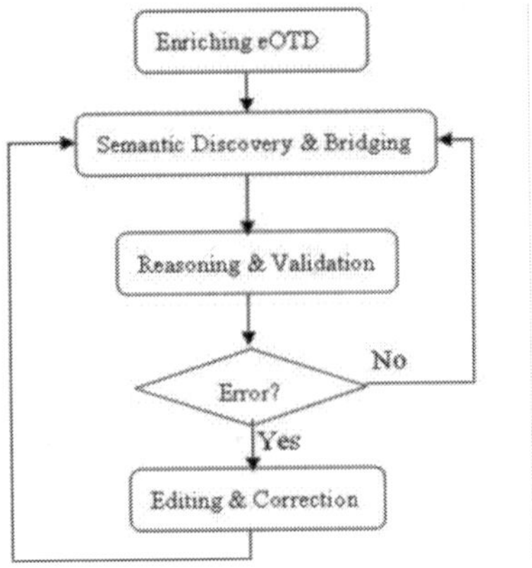

Fig. 2. The process of building mappings

FSCs "Bearings, Antifriction, Unmounted," "Bearings, Plain, Unmounted" and "Bearings, Mounted." The classes in the eOTD serve a similar role to Hepp's generic classes [9]. However, as we discuss in the next section, we provide explicit formal mappings to the original classes from commercial classification schemes.

5.2 Semantic Discovery and Bridging

Ontology mapping is the process of aligning the classes and properties of two ontologies. We do this by creating axioms in a mapping ontology that state equivalence, subclass, superclass relationships, exception, restriction, etc. In order to establish mappings between two ontologies, it is necessary to discover their logical and semantic relationships. Discovery of these relationships is the precondition to building the mappings.

We note that automated ontology alignment approaches [4] are not very helpful here. The differences between classification schemes are not as simple as the equivalence or subclass relations between named classes typically found by such systems. Even terms with identical names are often semantically different. For this reason, our alignment was a manual process that involved looking at names and ontology structures, and when possible we used definitions and actual product descriptions.

Our discovery focuses on two kinds of mapping axioms: owl:equivalentClass and rdf:subClassOf. In all mapping axioms, the left-hand side is a description composed of classes from one ontology, and the right-hand side is a description composed of classes from another ontology. Although equivalence is the most desirable matching, it is difficult to find it. Concepts that appear at one level in one classification frequently appear at different levels in the other taxonomies. When equivalence matches cannot be found, we tried to specify a most specific subsumer and most general subsumee. For example,

$$\text{cpv:PrimaryBatteries} \sqsubseteq \text{eOTD:BatteryAssemblyAll}$$
$$\text{eOTD:BatteryThermal} \sqsubseteq \text{cpv:PrimaryBatteries.}$$

To build these mappings, we used the most appropriate OWL constructs (see Table 1 for details) to express their relationship, such as those listed below.

1. Union ($A \equiv B \sqcup C$)

Sometimes a term name is the literal aggregation of two or more other terms from different classification. From the conventional English point of view, the phrase "A and B" usually does not mean the logical conjunction of A with B, but instead the disjunction of the concepts, e.g. "fsc:KnobsAndPointers $\equiv$ eOTD:Knob $\sqcup$ eOTD:Pointer."

2. Intersection ($A \equiv B \sqcap C$)

When a concept's name describes multiple independent characteristics, it is best defined as an intersection of classes representing these characteristics. For instance, "fsc:BearingAntifrictionUnmounted $\equiv$ eOTD:BearingAntifriction $\sqcap$ eOTD:BearingUnmounted."

3. Exclusion ($A \equiv B \sqcap \neg C$)

We often found that the different classifications have similar concepts with significant overlap, but there are often exceptions. In this case, we can correlate them by pointing out the difference between them. Exclusion can be used to specify that A and B are the same except for some C. This idiom is especially useful when trying to align a procurement-based taxonomy with formal "is-a" taxonomies. For instance, "eOTD:BearingPlain $\equiv$ eCl@ss:PlainBearing $\sqcap$ $\neg$ eCl@ss:PlainBearingParts."

4. Class vs. property distinction ($A \sqsubseteq \exists\, P.\{a, b, c\}$)

In some cases, the distinction made by a class in one ontology is made by a property in another. We found this to be true when mapping PLIB-511 to the FCS/eOTD, because it is a much more specialized taxonomy. We can use someValuesFrom or hasValue axioms to account for this. For example, "PLIB:HexagonHeadTappingScrewWithAFlatEnd" is classified partly according to two properties head style and point style. Meanwhile we can enumerate the values of both properties. So this item should have one of these values on the property. Thus "PLIB:HexagonHeadTappingScrewWithAFlatEnd $\sqsubseteq$ $\exists$eOTD:headStyle.{eOTD:Hexagon}" and "PLIB:HexagonHeadTappingScrew-WithAFlatEnd $\sqsubseteq$ $\exists$eOTD:pointStyle.{eOTD:Flat, eOTD:Flat2, eOTD:Flat3, eOTD:Flat4}." This restriction demonstrates that this concept in PLIB-511 has a determined value for the property head style, and simultaneously has a limited set of values on another property, point style.

5.3 Reasoning and Validation

Although domain experts and ontology developers can try to build all knowledge into an ontology, it is possible that they did not declare all assumptions or supplied incorrect axioms. Thus after each iteration of semantic discovery and bridging, a consistency check for the mapping of the ontology is necessary. A typical example of an inconsistency is when a class is a subclass of two disjoint classes. The most likely cause of this is an incorrect mapping axiom, but it is also possible that the disjointness condition is invalid.

Besides checking consistency, the second purpose of reasoning is to compute subsumption for the ontology. Subsumption determines which classes are necessarily subclasses of other classes based on the classes' descriptions and our bridging axioms. If insufficient subsumptions are discovered, then we need to add new mapping axioms or make the existing ones more specific.

6 Implementation

In previous sections, we have introduced the one-to-one ontology mappings. Now we will illustrate how they can be expanded to the mappings across the whole ontology network. The knowledge of domain experts is often limited to a few familiar classification schemes, and perhaps even only specific classes of product within these schemes. It is unrealistic to expect one person to know how to map to all other classification schemes. But if two classifications of different domains both have a correspondence (mapping) to a common third domain or more generally there is a correspondence path of multiple bridges between them, we can incorporate a DL reasoner to automatically infer a virtual correspondence between these two classifications. In other words, this method can implicitly "merge" two classifications via other ontologies.

Based on this idea we implemented a translator tool that intended to selects the most specific FCS concepts that include the terms appearing in the four external classification systems. It takes in a selected external ontology, core ontologies and the bridging ontologies between them together as input, and queries the DL reasoner about every kind of possible logical relationship we are interested in. Finally we visualize the quality of these mappings on the interface. We use HAWK[1], a homegrown Java Semantic Web API, to parse the ontologies and use a DIG interface to load these ontologies into FaCT++, although any DIG-compliant reasoner could be used.

We query the reasoner about equivalent and direct super class of each item in a selected external ontology. In the result we filter out those classes which are not in FCS ontologies, because we are only interested in those classes. Obviously, equivalent classes are preferred, because they imply an exact translation. Failing this, the most specific subsuming class is sufficient because this is an accurate (although more general) description of the original class. If the reasoner does not return any direct superclasses that are from the FCS, we then recursively query the classes that are returned until we find an FCS class. We call these results "indirect" subsumption. So the translator finally output three kinds of relations: equivalence, direct and indirect subsumption.

Table 3 gives a brief summary of translation results. We note that generic concepts rarely have matches due to classifications grouping things from different perspectives. As a result, very specialized schemes like PLIB-511 have a high degree of matches with FCS, while generic schemes like CPV have fewer matches.

[1] http://swat.cse.lehigh.edu/downloads/index.html#hawk

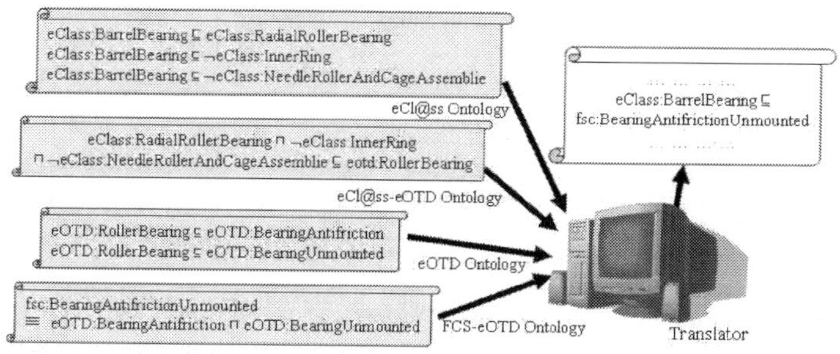

Fig. 3. An example of translation

For this reason, we recommend that suppliers use the most specific classification schemes available.

Table 3. Summary of computed matches. The scope means the number of concepts where we build mappings. The general scope includes all concepts in the scope and all concepts descended from them. The matching percentage is the sum of all three kinds of matches divided by the general scope.

Ontology	Scope	General Scope	Equivalence	Direct Subsumption	Indirect Subsumption	Matching Percentage
eCl@ss	86	191	13	21	78	58.64%
UNSPSC	79	103	7	55	18	77.67%
CPV	43	117	1	8	23	27.35%
PLIB-511	86	86	0	13	72	98.83%

Fig. 3 is an example of translation. A cataloger needs to register a new supplier who has used eCl@ss to describe their products, including a Barrel Bearing (code: 23.05.09.12). In order to determine how these items can be classified in the FCS, the cataloger can use our product classifciation translator. Given the eCl@ss ontology, its mapping to the eOTD ontology, and the mapping of the eOTD to the FCS ontology, the translator can determine that "eCl@ss:BarrelBearing $\sqsubseteq$ fsc:BearingAntifrictionUnmounted" is entailed. Thus the Barrel Bearing product should be clasified under this FSC.

To help both suppliers and customers to easily control and understand these translations, we developed a graphical user interface (see Fig. 4). Through it, users can select external classifications being mapped into the FCS classification. The results include the concept names and codes of both sides and the relationship mentioned above. The results of class and properties are separated in two pages. They are all listed in a tabular form which can be sorted by any column. Thus users can clearly read whether these mappings are satisfied.

In our prototype, the ontologies are of moderate size (typically hundreds of classes and a few thousand axioms) and as a result the translation is completed in tens of seconds including the reasoner's computation. However, in production

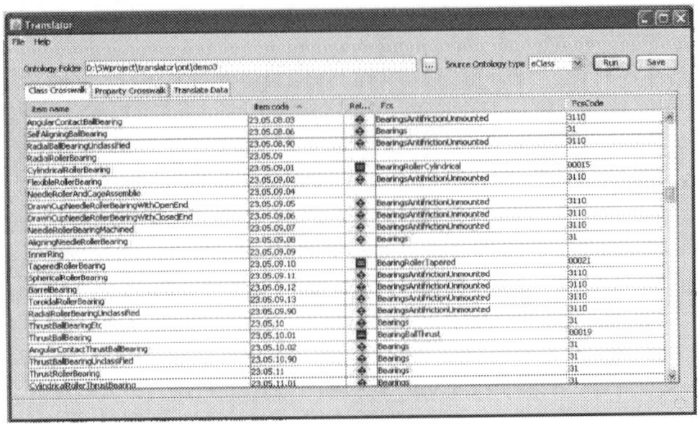

Fig. 4. The graphical user interface of the translator

the ontologies will be much larger. In order to save time, we provide a way to compile these mappings into persistent storage form, since these ontologies do not change very frequently (usually at the scale of months). This can make applications that use the translations much more efficient. The compiler can be rerun any time when source or target ontologies change. Therefore we can compare the mappings of different versions and more importantly subsequent steps can directly utilize these existing mappings rather than translating the ontologies "on the fly."

In order to translate data items from a target product description into FCS terms, we need to make the technical data items queryable by the OWL ontology language. So we provide a wrapper, i.e. a structure, for these data items to be translated into RDF Format. Via this process, the source data instances are transformed into standard form which uses terminology from the ontology version of the target classification.

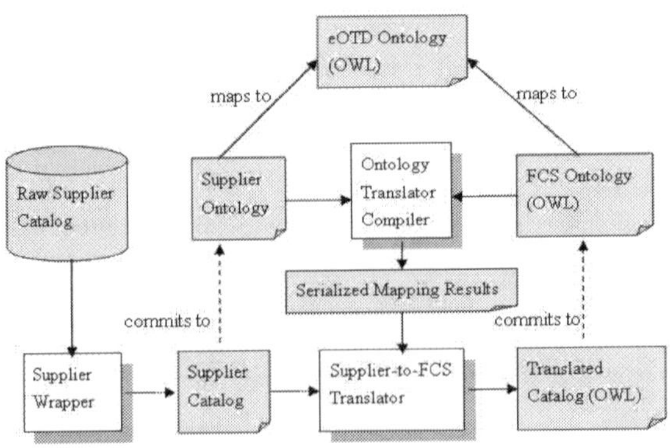

Fig. 5. The complete process of compilation. The white boxes are our prototype tools

As Fig. 5 shows, the wrapper takes in the supplier's raw catalog data which could be in the format of plain text, webpages or XML documents, and rewrites it into an RDF form that commits to the appropriate external ontology. Then the previously compiled translation results are incorporated to translate the wrapped catalog into a new catalog which commits to the FCS ontology. Specifically, the system first determines the category and type information of a catalog product description. After that it uses the previously compiled mapping relation (mentioned above) between this concept and the target concept in the FCS to rewrite the description of this product. With this tool, suppliers can easily find each product's corresponding descriptions in the terms of the FCS classification system.

7 Conclusion and Future Work

In this paper, we have demonstrated how Semantic Web technology can be used to ease integration of various e-commerce classification schemes. With some extensions, the eOTD provide an excellent common vocabulary and OWL DL is expressive enough to relate classification schemes that have different levels of specificity and different assumptions about what a parent-child relationship means. Ideally, we would like to see suppliers and maintainers of standard classification schemes create their own mappings to the eOTD, or have them use a standard product ontology that is so mapped, but in the meantime it is possible for these mappings to be developed by a third party. This would reduce the cost of integrating such ontology with the FCS or any other terminology. The clear advantage is less ambiguous classifications and the ability to automatically interchange with other schemes.

Future work includes extending the scope to include other kinds of products and improving our tools. We would like to refine the translator so that translating a source class to a property value works correctly and properties are translated in a more comprehensive way. We also believe that there are only a finite number of ways of modeling any given product, and would like to investigate the possibility of using automated methods to align new ontologies with the most similar ontology in our ontology network, achieving an economy of scale as the ontology network grows.

Acknowledgments

This work is supported by the Defense Supply Center Philadelphia, Philadelphia PA, and the Defense Logistics Agency, Ft. Belvoir, VA under contract SP4701-07-C-0006. We thank Peter Benson of the Electronic Commerce Code Management Association (ECCMA), Gerald Radack of Concurrent Technologies Corporation, Xinlei Wu and Ameet Chitnis for their help with the project.

References

1. Berners-Lee, T., Hendler, J., Lassila, O.: The Semantic Web. Scientific American 284, 34–43 (2001)
2. Bizer, C., Wolk, J.: RDF Version of the eClass 4.1 Product Classification Schema (retrieved June 2009), http://www.wiwiss.fu-berlin.de/suhl/bizer/ecommerce/eClass-4.1.rdf
3. Bowers, S., Delcambre, L.: Representing and Transforming Model-Based Information. In: Proceedings of the Workshop on the Semantic Web at ECDL 2000, Lisbon, Portugal, September 21 (2000)
4. Choi, N., Song, I.-Y., Han, H.: A Survey on Ontology Mapping Sigmod Record. ACM SIGMOD Record archive 35(3) (2006)
5. Corcho, O., Gomez-Perez, A.: Solving Integration Problems of Ecommerce Standards and Initiatives through Ontological Mappings. In: IJCAI 2001 Workshop on Ontologies and Information Sharing, pp. 131–140 (2001)
6. Fensel, D., McGuinness, D.L., Schulten, E., Ng, W.K., Lim, E.-P., Yan, G.: Ontologies and Electronic Commerce. IEEE Intelligent Systems 16(1), 8–14 (2001)
7. Guarino, N.: Formal Ontology in Information Systems. In: Proceedings of FOIS 1998, Trento, Italy, pp. 3–15. IOS Press, Amsterdam (1998)
8. Heflin, J., Hendler, J.: Dynamic Ontologies on the Web. In: Proceedings of the Seventeenth National Conference on Artificial Intelligence (AAAI 2000), pp. 443–449. AAAI/MIT Press (2000)
9. Hepp, M.: Products and Services Ontologies: A Methodology for Deriving OWL Ontologies from Industrial Categorization Standards. Int'l Journal on Semantic Web & Information Systems (IJSWIS) 2(1), 72–99 (2006)
10. Klein, M.: DAML+OIL and RDF Schema representation of UNSPSC (retrieved June 4, 2009), http://www.cs.vu.nl/~mcaklein/unspsc/
11. McGuinness, D., Fikes, R., Rice, J., Wilder, S.: An Environment for Merging and Testing Large Ontologies. In: Proceedings of the Seventh International Conference on Principles of Knowledge Representation and Reasoning (KR 2000), Breckenridge, Colorado, pp. 12–15 (April 2000)
12. McGuinness, D.L.: UNSPSC Ontology in DAML+OIL (retrieved from June 4, 2009) http://www.ksl.stanford.edu/projects/DAML/UNSPSC.daml
13. Noy, N., Musen, M.: PROMPT: Algorithm and Tool for Automated Ontology Merging and Alignment. In: Proceedings of the AAAI 2000 Conference, Austin, TX (2000)
14. Omelayenko, B.: Syntactic-Level Ontology Integration Rules for E-Commerce. In: FLAIRS Conference 2001, pp. 324–328 (2001)
15. Pierra, G.: Context-explication in conceptual ontologies: The PLIB approach. In: Proceedings of 10th ISPE International Conference on Concurrent Engineering: Research and Applications (ce 2003): Special Track on Data Integration in Engineering, pp. 243–254 (2003)
16. Tobies, S.: The complexity of reasoning with cardinality restrictions and nominals in expressive Description Logics. J. of Artificial Intelligence Research (JAIR) 12, 199–217 (2000)

Supporting Multi-view User Ontology to Understand Company Value Chains

Landong Zuo[1], Manuel Salvadores[1], SM Hazzaz Imtiaz[1], John Darlington[1], Nicholas Gibbins[1], Nigel R Shadbolt[1], and James Dobree[2]

[1] Intelligence, Agents, Multimedia (IAM) Group
School of Electronics and Computer Science
University of Southampton, UK
{lz,ms8,hsmi,jd,mng,nrs}@ecs.soton.ac.uk
[2] Semantric Ltd.
25 Landsdowne Gardens London, SW8 2EQ, UK
james@semantric.com

Abstract. The objective of the Market Blended Insight (MBI) project is to develop web based techniques to improve the performance of UK Business to Business (B2B) marketing activities. The analysis of company value chains is a fundamental task within MBI because it is an important model for understanding the market place and the company interactions within it. The project has aggregated rich data profiles of 3.7 million companies that form the active UK business community. The profiles are augmented by Web extractions from heterogeneous sources to provide unparalleled business insight. Advances by the Semantic Web in knowledge representation and logic reasoning allow flexible integration of data from heterogeneous sources, transformation between different representations and reasoning about their meaning. The MBI project has identified that the market insight and analysis interests of different types of users are difficult to maintain using a single domain ontology. Therefore, the project has developed a technique to undertake a plurality of analyses of value chains by deploying a distributed multi-view ontology to capture different user views over the classification of companies and their various relationships.

Keywords: Ontology Engineering, Data Management, Market Modeling, Semantic Web, Value Chain, Network Analysis.

1 Introduction

The MBI project [14] has the clear objective to make a significant performance improvement in UK business to business (B2B) marketing activities by providing unparalleled insight into UK business activity and support for intelligent decision-making processes. The project consortium includes the marketing departments of 5 influential companies each representing a different set of needs within the UK B2B market, 3M, AXA, British Gas Business, National Australia Group Europe (Clydesdale and York shire Bank) and Parcel Force Worldwide.

A. Bernstein et al. (Eds.): ISWC 2009, LNCS 5823, pp. 925–940, 2009.

A major issue for the project, and the market analysis techniques that have been researched, is how domain knowledge can be utilized to support the diverse analysis interests from all project partners. The issue is made complex by the need for rich semantics of the specific business scenarios that need to be modeled while at the same time supporting common forms of market analysis such as a value chain analysis. This paper argues that a single domain ontology is too rigid to resolve the problem and therefore, proposes a layered user viewpoint model which gives extra flexibility in terms of unique user views of information classification and relationship definition. The principle of multiple user view ontologies is generic but is illustrated in this paper with the latest achievement of value-chain analysis application.

The remainder of paper is structured as follows: section 2 reviews the related work, section 3 gives the background knowledge of value-chain analysis, section 4 describes the application architecture and processing work flow, section 5 includes the underlying data with extraction and integration processes, section 6 and section 7 explain the approach of multi-view ontology engineering and semantic annotation in detail, section 8 presents the execution result of value chain analysis, which is followed by the summary and future work in section 9.

2 Related Work

The information analysis over semantic integration of public information has attracted significant interest from different Semantic Web communities. The AKT PSI project [4] demonstrated the added-value of richer information insight by using Semantic Web technologies as an important means to integrate a large collection of public information from distributed legacy data sources. The underlying data, in less-standardized formats, was converted into RDF following local data-centric ontology. The different data sources were integrated using an ad-hoc ontology alignment approach.

The EDEN-IW project [12] integrated the water environmental information from heterogeneous databases to support cross-country decision-making processes at the European level. The semantic heterogeneity in terms of different data structures and multi-lingual representations was harmonized at the conceptual level by deploying a multi-layered ontology model. The conceptual viewpoints of local database ontology and user ontology were connected via rule-based mapping to the global ontology. The high-level user queries were translated into global terms and corresponding SQL syntax was constructed and executed over the legacy databases.

The first MBI prototype [14] extracted unstructured and semi-structured text from different web sources, transformed that into RDF and integrated the data with core structured and semantically annotated information. A global ontology was constructed from domain knowledge of the B2B market and was designed to be extensible in order to reuse additional conceptual schema extracted from external sources. The information analysis was performed using specific scenarios such as micro-segmentation and value-chain modeling. These require the application logic to follow either standard business classification schemes or ad-hoc operational logic of individual use cases.

An important goal for the project is to achieve a rich analysis insight over a wide range of data sources. The design of above applications was normally driven by specific key use cases in the closed domain in order to satisfy the mutual interests of particular user group. This could have led to a data-centric approach to save the cost of knowledge engineering. However, the potential diversity of operational logic in real business scenarios cannot be fully addressed in such an approach. The project identified a number of issues:

- The global specification of domain knowledge is too coarse and needs refinement to reflect the requirement of the individual analytical case.
- It can be too expensive to develop a complete global domain ontology that meets all requirements. This requires close collaboration between project stake-holders including domain experts, knowledge engineers and software developers.
- The representation of domain knowledge amongst different user group needs is rarely agreed as being common. In the extreme circumstance, the application development may have to repeat the knowledge learning process for every case resulting in separate solutions for each partner. The knowledge learning can become a very time-consuming process.
- There may be security concerns regarding sensitive information that cannot be shared.

The development process can become extended as new business scenarios emerge and changes to the analysis logic result in significant rework of the software implementation.

3 Background

A Value-Chain is defined by Porter [11] as a series of value-generating activities. Products pass through all activities of the chain in order, and at each activity the product gains some value. In MBI, the value chain is addressed as a network path consisting of relationship fragments between companies. The value-chain application is a semantic search and navigation tool featured with special interest in supporting the modeling of company relationships and the capability to identify key trading behavior and entities that have significant influence on a company's propensity to buy. The relationship is defined at a conceptual level as typed connections between two sub-types of company following specified patterns of trading behavior. In practice, the conceptual model regarding company classification scheme and trading behavior can vary with each individual case and can change frequently. The value-chain model needs to be analyzed in the context of a specific business scenario and user understanding of the industry domain.

4 Application Architecture

To achieve improved use of information the MBI project proposes a multi-view ontology approach to resolve the problem of varying user scenarios. It enables individual user conceptual views over a value-chain network as well as cross-domain knowledge manipulation by domain expertise. The approach is prototyped within a generic framework architecture, as shown in Fig. 1, which separates the definition or configuration of a form of marketing analysis from the execution or visualization of a specific analysis of the market.

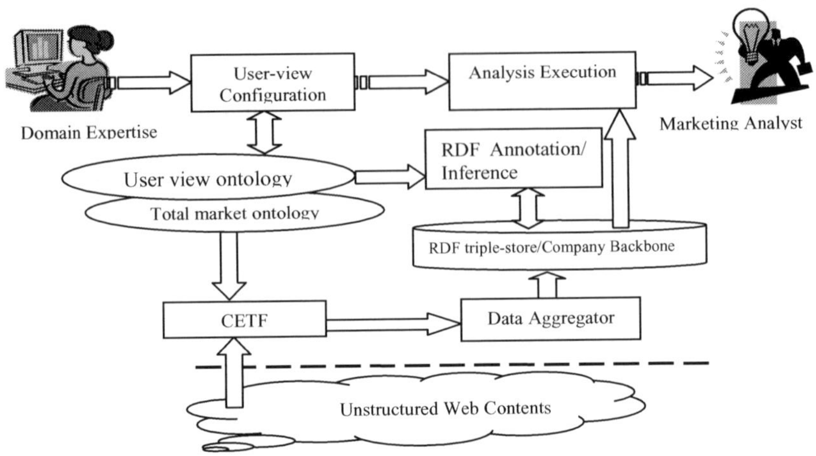

Fig. 1. Generic architecture of value-chain application

The Total Market ontology is a central knowledge base capturing the common conceptualization of a company profile. The common conceptualization includes the general definition of upper concepts and relationships common to all potential analysis projects.

Besides the common concepts and relationships, the refined viewpoint of an individual user scenario is addressed in the User View ontology that extends the Total Market ontology. The information access is enhanced by the augmentation insights of market including both common view and special view.

The user-view configuration is a web-based ontology editor that allows experts of industry domain to define their own knowledge view over the total market. The generated output is a user-view ontology that extends the general class and relationship definition from the Total Market ontology with additional finer definitions.

The Common Extraction and Translation Framework (CETF) is an ontology-driven process that extracts unstructured information from selected external web site. The web sites are of particular interests about trading relationships between companies. The output of web extractions is translated into RDF and follows the semantics of the Total Market ontology or its extensions.

The extracted RDF needs to be integrated with company profile backbone to establish a solid data foundation to support high level queries. The individual view of user conceptualization is not included in the extracted RDF because the extraction process is designed to be generic. The extraction process is completely independent of the analysis process. The application has an additional RDF annotation process that tags the extracted RDF triples into the specific user view. The details of multi-view ontology engineering and data annotation approach are further explained in Section 6 and Section 7. The marketing analyst is operating over the analysis execution interface where value-chain visualization and navigation is supported over high-level semantic query according to user-view conceptualization.

In this architecture, the knowledge representation is separated from operational process to achieve more flexible information access. The ontology is used here as an extensive knowledge vehicle that allows richer semantic definitions to represent complex analysis scenarios whilst diversity of user-views conceptualization is effectively maintained by introduction of user-defined concept and relationship over Total Market ontology. The concepts and relationships are restricted by semantic rules and mapped to super-classes in the Total Market ontology. The RDF annotation process loads the mapping rules, performs a forward-chained reasoning and generates SPARQL syntax to query the underlying integrated data. The logic consistency and answering correctness between domain expertises and marketing analyst is guaranteed.

Both types of users, the Marketing Analyst and the Domain Expert, interact with the system through a web portal. This portal is developed using the Django framework[1] and acts as a bridge to a collection of backend services. The collection of backend services work tightly with a set of JavaScript widgets in order to provide easy interaction and visualization functions over the semantic data. The list of functionalities is as follows:

1. Extract Graph Based Information: this data is extracted by a JavaScript library developed ad-hoc for this project[2]. The first prototype used Prefuse[3] but it did not meet the performance. Our implementation for visualizing graphs is Java Applet free and, it uses the HTML Canvas component to render the objects in the browser.
2. Represent Tree Based Structures: this service offers to the ontology editor the possibility of querying any graph structure and retrieve it in the shape of a tree. This functionality is flexible enough to allow users the creation of trees from RDF graph.
3. Extract Table and List Based Structures: the prototype uses extensively list and tables data structures. This service follows the schema definition for the data sets in order to retrieve the information in such a way as to be properly visualized in the front-end.
4. Schema Browsing Service: this service provides a unique point to query the ontology schema using T-Box reasoning. The front-end invokes this service to display a comprehensive view of the structure of RDF data (see Fig. 2).

[1] http://www.djangoproject.com/ (last accessed in April 2009).

[2] All the graphs-alike screenshots on this paper are examples of the mentioned JavaScript library.

[3] http://prefuse.org/ (last accessed in April 2009).

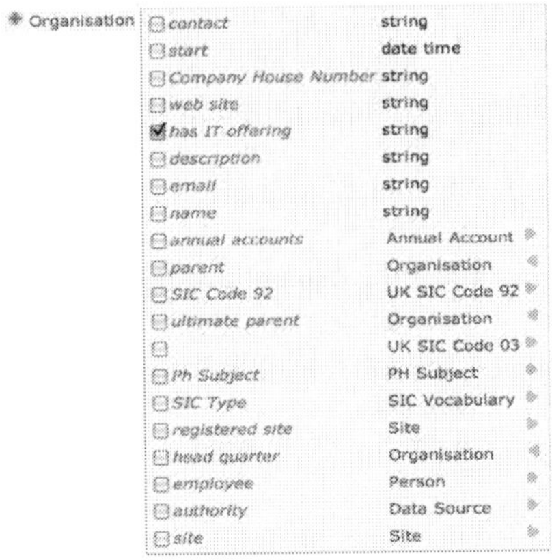

Fig. 2. Schema Browsing Service

5 Underlying Data

The MBI project has aggregated information from heterogeneous sources including structured data, semi-structured data and unstructured data. The company backbone is converted from an industry provided structured data source containing 3.7 million company profiles covering the total UK market. This is augmented with web-based content extracted from selective Web sites.

5.1 Company Backbone

The company backbone is created from an industry provided relational database dump. The company profile data is converted into RDF following concept and relationship definitions in the Total Market ontology. The company backbone is updated on a monthly basis providing an ongoing view of company profiles which can be analyzed over a historical period. In MBI, company backbone data is saved into persistent semantic storage of 4Store[4] RDF database and each update has more than 160 million triples.

5.2 Ontology-Driven Web Extraction

To exercise the value-chain model, the application prototype is used to create projects that focused on the domain of the Building and Construction (B&C) Industry using data that is regularly extracted from the trade web sites with rich contents about B&C projects and contracts in order to build a network of data depicting the historical trading relationship in the UK B&C industry. Two web sites are selected to support this prototype.

[4] The home page is at http://4store.org (last access in August 2009).

- Architect Journal Specification is a monthly UK-based magazine, which covers the issues of specifying products, and there use in building projects. The MBI project currently has extracted contract, supply and product information details including 632 building projects, 392 architects and 4318 suppliers.
- Barbour ABI is the leading provider of sales leads and construction contract data for the UK market. Their UK-based research team track 100% of all planning applications on every planned building project valued over £100,000 and publish contract update on the daily basis.

The CETF supports the generic ontology-driven solution dealing with unstructured or semi-structured web extractions. The CETF keeps a set of extraction patterns that is matched with the semantic definition of class, instance and properties in the ontology. This may include regular expressions, Gate jape grammar[5], natural languages expressions, concept instances, parsers etc. needed to build a gate pipeline and carry out any data extraction. The framework supports extraction from HTML pages with complex structures, such as nested tables or lists. The string similarity algorithm, for example Levenshtein Distance algorithm [8], is used to compute the similarity of table headers against its matching concepts in the ontology. The WordNet based semantic similarity measure [9] is also deployed. Some generic extraction logic is applied to translate semi-structured tabular data into RDF. For example, if at least one column maps to a concept attribute, instances are generated for each row with corresponding attributes. In the case of nested tables, a concept may be linked to other concepts through object properties. Then a column itself may contain concepts rather than just attributes of the main concept. The extraction task is performed on a scheduled or event-driven basis following the user's instruction. For example, the AJSPEC extraction is performed on the monthly basis and ABI extraction is on the daily basis and driven by any data update.

5.3 Semantic Integration for Companies

The output of a web extraction is identified in a different namespace connecting to the URL of extraction source in order to retain their provenance. The extraction output contains richer semantic information regarding company trading relations that need to be integrated with the company backbone to form the proper value chain definition. The comprehensive description of the approach to semantic integration is not part of this paper. The value-chain application has developed and deployed an ad-hoc integration model with fast mapping functions to match companies identified in web extractions to companies in the backbone data. The company matching can be declared as equivalent if the postcode of registered address of two companies is the same and their company names are closely matched. The Levenshtein Distance algorithm[8] is used here to measure string similarity between two company names. The algorithm output is normalized to the length of string and produces a real number between 0 and 1. The similarity is null if two strings are completely different. The similarity is 1 if two strings are exactly the same. If the similarity exceeds the acceptable threshold, the matching is successful and two companies are tagged with OWL:sameAs relation.

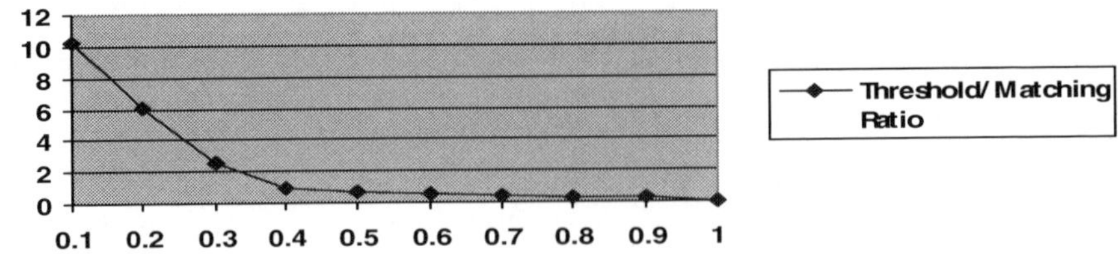

Fig. 3. Company matching using Normalized Levenshtein Distance algorithm

The instance matching result using AJSPEC data based on 4318 suppliers is shown in Fig. 3. Using the matching logic described above, the diagram shows how the matching ratio varies against the value change of similarity threshold. The matching ratio was calculated as the number of equivalent matching divided by total number of companies. The result shows the algorithm ends up with reasonable output if the threshold is between 0.4 and 0.6.

5.4 Data Storage and Query

The persistent storage of the company backbone is supported by 4Store, a large-scale RDF triple-store developed from previous work on the semantic database and query engine, 3Store [6]. The 4Store is capable of scaling up to 60 billion triples[5] with efficient supports of semantic query via SPARQL [3].

6 Ontology Engineering of Layered Model

The layered model is proposed and deployed in MBI value-chain application to support individual user-view, as shown in Fig. 4. The user-defined concepts and relationship presents the restricted view over the Total Market ontology. The user-defined concept is specified at schema level as classes that are mapped to the Total Market ontology using rdfs:subClassOf relationship plus filtering rules. Semantically, the user-defined company type is a special subset of total market *Organization* featured with unique attribute characteristics. The restriction of such subset are explicitly expressed in filtering rules that lead to a filtering process against *Organization* over underlying data using pre-defined filtering algebra. The user-defined company relationship is also explicitly specified in filtering rules to connect two company sub-types following the specific semantic patterns of trading behaviors in the Total Market ontology.

The data annotation process works as an off-line service to reason the semantic specification of user-defined concept and relationship with filtering rules and find satisfactory evidence from underlying data. The satisfying company individuals are annotated with additional triples in terms of user-defined class and relationship. The relating company profiles, the additional triples generated from annotation process are merged and saved into 4Store as user scenario data. Meta-data about this merging

[5] The number was cited from the page, http://www.garlik.com/press.php?id=136-GRLK_PRD

data set is created and linked to the individual user view. Thereafter, the knowledge representation of user conceptualization is transformed into underlying triples and the queries in user-defined terms are supported. The users can easily create new user-views, change and share the contents and make it quickly effective in analysis execution.

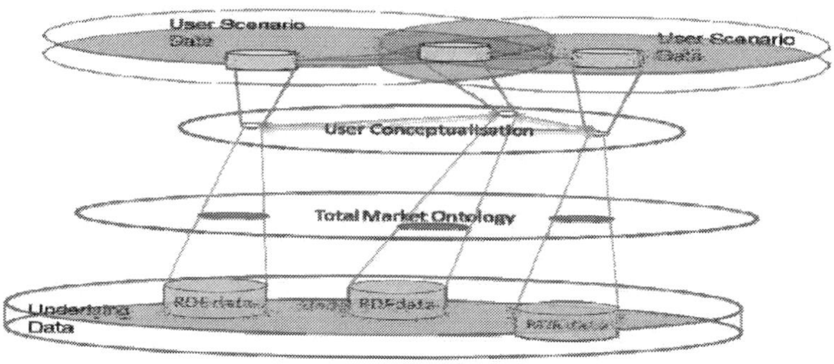

Fig. 4. Semantic adaptation crossing layered Ontology models

6.1 Total Market Ontology

The Total Market ontology was extended from the underlying schema of company backbone following a bottom-up approach, i.e. the ontology schema was derived from underlying database dump with neutral mappings to the conceptual structure of database tables and attributes. The ontology was developed in a generic purpose combining conceptual knowledge from relating domains including people, product, location and transactional information. The ontology has also adopted semantics from other established namespaces including Dublin core[1], OWL-time[7] and SKOS[2]. The ontology model benefits from this method in two extents: Firstly it supports query and reasoning request following the standard classes and attributes definition that would make the application extensible to connect to other applications via an open interface; Secondly, the company profile data is the major part of data aggregation. The straight mapping would save much of transformation efforts to convert relational database dumps into semantic forms.

The conceptualization of total market ontology is shown in Fig. 5, which provides an overall picture over the B2B trading domain with specific focus of *Organization*. The ontology contains many extending branches where further extension is conducted and created in separate ontologies, for examples the offering ontology extends the concept of *Product*, specifying standard taxonomic and part-of relationship in product domain and synonym representations, the business activity ontology describes the company classification in UKSIC Code 2003[10] standard and extends *UK SIC Code*. The extended semantics would allow richer data aggregation of public information and standards from different domains and thus supports finer company classification using comprehensive semantic restrictions.

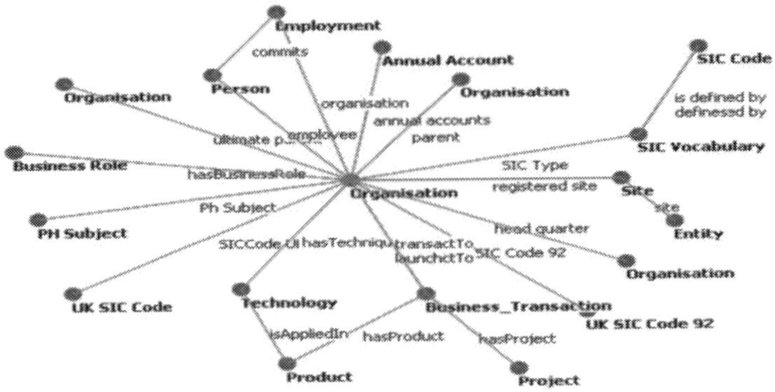

Fig. 5. Domain conceptualization of total market ontology

6.2 User-View Ontology

Information viewpoints model a given representation for some specific point of interest, among the set of possible representations [15]. The user-view ontology offers a particular perspective on the Total Market ontology, capturing explicit specification of additional concepts and relationships required for individual analysis scenarios. In value-chain analysis, the view ontology is developed in a top-down approach, i.e. specifying the user-defined company classification and relationships regardless of the existence of corresponding individuals in the underlying data. The user selects the Total Market ontology or additional conceptual knowledge derived from external Web Extraction to declare the restricted meaning of user-defined concept and relationship. The user-view specification is passed into reasoning service and the underlying RDF is annotated according to user views. The output of annotation service representing the value-chain relationship from different provenances can be further merged or compared to support higher order decision-making processes by the user.

The generic web-based ontology editing interface, user-view configuration, is developed to support user manipulation of view ontology. The underlying syntax of ontology language of OWL and filtering rules is hidden from the user with the ontology presented as a network graph following property domain and range relationships.

6.3 Common Ontology and Project Ontology

To improve the usability of user-view, the ontology representation is sub-typed into hierarchies regarding the common use and knowledge sharing of value-chain analysis. Two types of user-view are created in the prototype application: common view and project view. The common view contains the sharing knowledge, any concept agreed as being common amongst user groups is presented in the common view and to be shared amongst authorized users. The project view includes specific knowledge referring to the individual analysis scenario. It extends the common view ontology and inherits all common definition of company classification and relationship. If one concept or relationship in project view becomes popular and is recognized by most users, it can be inserted in common view making it visible to other project users. The user can easily create new individual project view from a sharing view of other users. The developing time of user-view conceptualization is saved. The view hierarchies can be further extended to include more layers according to the actual management hierarchy inside the organization, for example sales department and marketing department may share the common set of use scenario whilst the diverse view is still captured in the individual project view. In the value-chain model, the basic common view mainly defines general company classification scheme following UK SIC Code standards, for example, manufacturer, retailer and whole seller. The project includes the additional classification scheme based on the common definition plus semantic restrictions.

6.4 Filtering Rules to Map User-View and Total Market Ontology

The semantic meaning of user-defined concepts is specified as sub-class of total market concept and restricted with filtering rules, which is further interpreted as: the general class in total market ontology is filtered by attribute values of one or multiple connecting concepts following selective semantic path crossing the schema network of total market ontology. For example, a new concept "Manufacturer of Glazing" in individual user-view can be interpreted as: the existence of an *Organization* that has more than 2 historical transactions producing product "Glazing" between "01/01/2009" and "01/05/2009". In such statement, the general concept *Organization* is restricted by the semantic connection relating to concept *Product* and traverse concept *Business_Transaction* filtered by number of transaction and occurrence time period.

The MBI project has developed an ontology-driven, rich semantic interface to guide the user through the specification of such restriction logic and creation of a proper filtering template. The operation of new concept creation is conducted following the steps below,

1. The user needs to select the base concept to extend, in the case of value-chain it can be *Organization* or any user-defined sub-class of *Organization*.
2. Next, the user is supposed to select a restrictive concept in the Total Market ontology to apply the restriction over base concept, for example *Product*. The selection implies the semantics of new concept is limited by existing relations connecting base concept and *Product*.
3. The backend ontology service crosses the Total Market ontology graph to find any possible connecting paths between base concept and restrictive concept. The calculation of semantic path considers any class and relationship in the path taking into account of all domain/range properties being explicitly declared with the class or implicitly inherited from its super-classes.
4. The user selects one or multiple paths from the preliminary calculation result.
5. For each selected path, the user can select to restrict over any class or relation in the path. The setting-up of restriction is guided by pre-defined filtering algebra that works with ontological definitions including data-type property, class definition and object-property.

The preliminary result of ontology computing paths consists of any possible templates regarding the semantic restriction between the base concept and the restrictive concept. The path addresses the unique interpretation of semantic restriction over the base concept following the conceptual specification of the Total Market ontology. Although some paths may not make proper sense to the domain experts, it does comply with the logic structure defined in the underlying ontology. Therefore, the human knowledge of domain expertise is required to validate the preliminary result and make correct choice according to the analysis scenario. The significant flexibility and good data quality is made via combination of formal representation of domain knowledge with diverse understanding of individual scenario from domain expertise. Any multiple selections from the computing paths are regarded as logic AND in the annotation process.

The restriction over user-defined relationship has to follow the same procedure except that relationship is specified as connections between two company sub-types, for example, the user-defined relationship *Supply-To* may link *Manufacturer* and *Whole Seller* following a specific historical transaction pattern. Similarly, the user needs to select correct pattern from computing paths to declare the restriction. The user has options to select from two type of logic pattern, either explicit relationship or implicit relationship. The explicit relationship follows the single path to cross the semantic network and connect two companies. For example, the historical transaction pattern implies that two companies are connected in the relationship if there is at least one trading transaction between them. The implicit pattern defines a flexible model to handle implicit transaction information where interaction between buyer and seller is not explicitly specified. This requires user selection of the third concept to link two company types. For example, user specifies that a *Contractor* has *contactWith* connection to an *Architect*, if two companies have participated into the same building project. The user can also define other types of relationship in a similar pattern, for example the relation *competeTo* can be defined semantically to connect two companies producing the same type of *Product*. The implicit model can be further extended to cover more complex use cases.

The filtering rules are written in Jena-like[13] rule syntax plus additional extension supporting filtering algebra. Some functions are already supported by the Jena inference engine, the reason of re-definition in filtering algebra is explained in Section 7. The pre-defined filtering algebra includes following functions:

1. Data-type property functions:
   - String restriction functions: equals, starts with, contains and ends with;
   - Integer or Float type functions: less than, equals and greater than;
   - Date-time type functions: equals, starts from, ends by;
2. Class type restrictions: equals, all value from sub-type, union of;
3. Object Property restriction: occurrence (more than), occurrence(less than), occurrence(equals)

The selective combination of the filtering functions and computing paths gives the explicit meaning of user-defined concept and relationship in terms of how satisfied companies and relationships can be derived from underlying data.

7 Data Annotation via Rule-Based Reasoning

The value-chain application needs to query and reason over 200 million triples held in the 4Store. 4Store supports the SPARQL standard without any inference capability. Most inference engines, such as Pellet and Jena, are supposed to work over their native interface to access the underlying triple-stores, i.e. the reasoning engine is bounded with the query engine. To our knowledge, such generic inference engines do not support reasoning tasks by accessing an additional persistent triple-store via SPARQL interface. The MBI project has developed a computational model working as replacement for a generic logic inference engine to reason about filtering rules. The model takes into account of user-defined concept and relationship and filtering rules, to decompose the task, generate SPARQL syntax dynamically, and execute the queries over the underlying data. The underlying data of individual companies are annotated with additional triples generated from reasoning process. The model supports reasoning functions in terms of RDF forward-chained rule syntax to prove the supporting evidence that a company is an instance of classification type or the relationship exists between two companies. These are the primary pieces of information required for a value chain analysis.

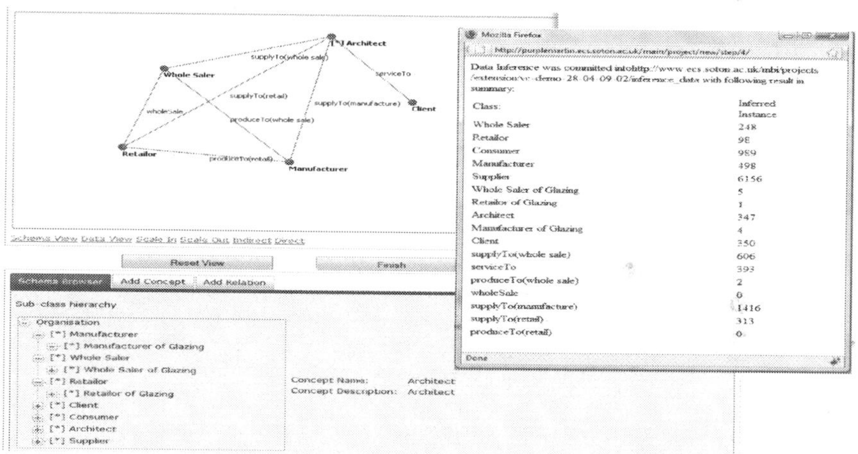

Fig. 6. Semantic reasoning over user-view conceptualization

The example of reasoning outcomes from a user conceptualization is shown in Fig. 6. The result was created from the AJSPEC data in the Building and Construction domain. The connecting network represents the individual user view over the industry domain. The user-defined company types are structured into a type hierarchy tree and the relationship types are named respectively following the historical transactions between two companies. The reasoning is conducted over iterations. The reasoning process tries to fire every satisfying rule in the iteration. The SPARQL syntax is constructed from filtering rules and executed against the underlying data to generate new triples for user-defined concept and relationship. The process continues until no more new triple is generated. The output in terms of generated triples are saved with

Web Extraction data and company profiles creating a new RDF data set, to support queries following the user-view conceptualization.

8 Value-Chain Execution over Project View

The value-chain execution is designed as a generic navigation service to visualize the value-chain relationships between companies according a specific user-view conceptualization. The value-chain model is constructed in a series of steps, each finding the potential connecting companies for a specific company following selected relationships or company types as defined in the user view. The analysis execution does not need to prove the underlying semantic meaning of each concept and relationship. It loads the user-view ontology and composes queries in the high-level user's terms. It executes the query against the user scenario data and visualizes the result in a network graph, see Fig. 7 .

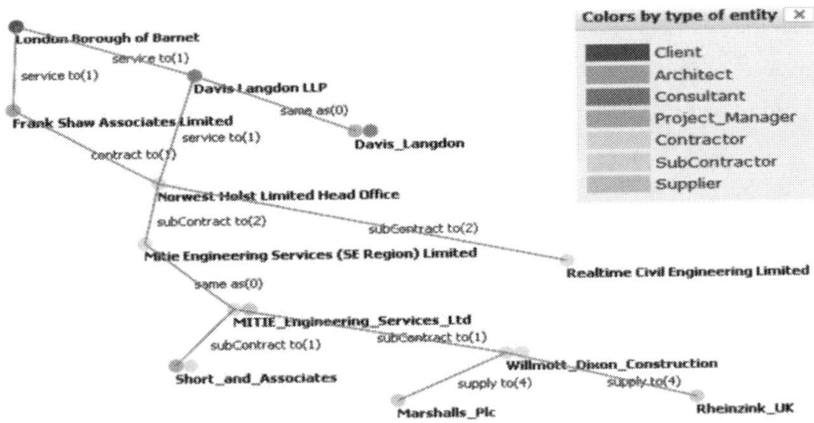

Fig. 7. Value-chain execution and visualization

The example shows the output of a value-chain execution. The construction starts from a client and follows the user-defined relationships to cross the trading network and reach other companies of user-defined types. The relationship closeness is defined by the weight of influential factors such as the number of transactions in the historical transactional data sets. The analysis execution enables quick construction of value-chain models following any user-defined logic view over different industry domains thus providing a generic service required by all project partners.

9 Conclusion and Future Work

The first MBI prototype showed the value of semantic data augmentation from heterogeneous sources. In this paper, we have considered an extension of the approach that covers the more important modeling and sharing of processing logic needed to analyze the underlying data. The extension creates flexibility in support of individual analytical user views to model the reality of specific business scenarios. The multi-view ontology approach was developed and deployed specifically for, but not limited to, value-chain analysis. It extends a single domain ontology to support modeling of varying analysis scenarios and fulfills the user requirements regarding different industry domains. The network analysis is separated at an abstract level into analysis configuration and analysis execution respectively supporting knowledge manipulation of user conceptualization and specific value-chain explorations. The processing logic of the data analysis is persisted in an ontology and aligned with underlying data via an annotation service following generic inference rules. The user-view ontology and corresponding annotated data are passed through two sub-phases in parallel, to separate high-level value-chain visualization and navigation functions from the scenario variation of analysis context. The ontology here is used as a super vehicle to convey rich semantic contents of analysis case from domain expertise to marketing analyst, and maintain the semantic consistency throughout annotation process. The information analysis benefits from explicit and extensible knowledge representation using Semantic Web technology. The application logic of analysis scenario is updated in an evolutionary rather revolutionary way by extending and reuse shared ontology views amongst user groups. Much time and effort is saved because user-view conceptualization is simply managed and validated by domain expert via the web-based ontological editing interface. The knowledge transferring is transparent to marketing analyst with less intervention required from software developer, thus performance of analysis project is improved in a long-term scale by using semantic web technology.

The future work is to continue to progress the inference service to provide a comprehensive analysis with more complex semantics, for example the value-chain analysis over an Offering ontology, to allow value chains analysis based on "potential product offering chains" as well as "historical transaction chains".

References

1. Dublin Core Metadata Element Set, Version 1.1, `http://dublincore.org/documents/2008/01/14/dces/`
2. SKOS simple Knowledge Organization System - Home Page, `http://www.w3.org/2004/02/skos/`
3. SPARQL Query Language for RDF, http://www.w3.org/TR/rdf-sparql-query/
4. Alani, H., Dupplaw, D., Sheridan, J., O'Hara, K., Darlington, J., Shadbolt, N.R., Tullo, C.: Unlocking the potential of public sector information with semantic web technology. In: Aberer, K., Choi, K.-S., Noy, N., Allemang, D., Lee, K.-I., Nixon, L.J.B., Golbeck, J., Mika, P., Maynard, D., Mizoguchi, R., Schreiber, G., Cudré-Mauroux, P. (eds.) ASWC 2007 and ISWC 2007. LNCS, vol. 4825, pp. 708–721. Springer, Heidelberg (2007)
5. Cunningham, H., et al.: JAPE: Regular Expressions Over Annotations, `http://gate.ac.uk/sale/tao/index.html#x1-1790007`
6. Harris, S., Gibbins, N.: 3store: Efficient Bulk RDF Storage. In: 1st International Workshop on Practical and Scalable Semantic Systems (PSSS 2003), Sanibel Island, Florida, USA (2003)
7. Hobbs, J.R., Pan, F.: Time Ontology in OWL, `http://www.w3.org/TR/owl-time/`

8. Levenshtein, V.: Binary Codes Capable of Correcting Deletions, Insertions, and Reversals, vol. 163, pp. 845–848 (1965)
9. Lin, D.: An Information-Theoretic Definition of Similarity. In: Proceedings of the 15th International Conference on Machine Learning, pp. 296–304 (1998)
10. National-Statistics, U.K.: Standard Industry Classification of Economic Activities pdf (2003), `http://www.statistics.gov.uk/methods_quality/sic/downloads/UK_SIC_Vol1.2003.pdf`
11. Porter, M.E.: Competitive Strategy: Techniques for Analyzing Industries and Competitors. Simon and Schuster (1980)
12. Poslad, S., Zuo, L.: An Adaptive Semantic Framework to Support Multiple User Viewpoints over Multiple Databases. In: Advances in Semantic Media Adaptation and Personalization, pp. 261–284 (2008)
13. Reynolds, D.: Jena 2 Inference Support, http://jena.sourceforge.net/inference/
14. Salvadores, M., Zuo, L., Imtiaz, S.H., Darlington, J., Gibbins, N., Shadbolt, N.R., Dobree, J.: Market blended insight: Modeling propensity to buy with the semantic web. In: Sheth, A.P., Staab, S., Dean, M., Paolucci, M., Maynard, D., Finin, T., Thirunarayan, K. (eds.) ISWC 2008. LNCS, vol. 5318, pp. 777–789. Springer, Heidelberg (2008)
15. Spaccapietra, S., Parent, C., Vangenot, C.: GIS databases: From multiscale to multiRepresentation. In: Choueiry, B.Y., Walsh, T. (eds.) SARA 2000. LNCS (LNAI), vol. 1864, pp. 57–70. Springer, Heidelberg (2000)

EXPRESS: EXPressing REstful Semantic Services Using Domain Ontologies

Areeb Alowisheq, David E. Millard, and Thanassis Tiropanis

Learning Societies Lab, School of Electronics and Computer Science
University of Southampton, Southampton, SO17 1BJ, UK
{aaa08r,dem,tt2}@ecs.soton.ac.uk

Abstract. Existing approaches to Semantic Web Services (SWS) require a domain ontology and a semantic description of the service. In the case of lightweight SWS approaches, such as SAWSDL, service description is achieved by semantically annotating existing web service interfaces. Other approaches such as OWL-S and WSMO describe services in a separate ontology. So, existing approaches separate service description from domain description, therefore increasing design efforts. We propose EXPRESS a lightweight approach to SWS that requires the domain ontology definition only. Its simplicity stems from the similarities between REST and the Semantic Web such as resource realization, self describing representations, and uniform interfaces. The semantics of a service is elicited from a resource's semantic description in the domain ontology and the semantics of the uniform interface, hence eliminating the need for ontologically describing services. We provide an example that illustrates EXPRESS and then discuss how it compares to SA-REST and WSMO.

Keywords: Semantic Web, Semantic Web Services, Ontologies, REST, SA-REST, WSMO.

1 Introduction

The emergence of Web Service technologies offers great business opportunities. Traditional Web Services, based on the SOAP/WSDL standards provide syntactic descriptions of services. Offering syntactic descriptions however, is insufficient for the automation or semi-automation of service discovery and composition, stating that a service accepts an integer and returns a string will not offer information on what the service does. In order to solve this problem research has been done to semantically, rather than syntactically, describe Web Services. The Semantic Web is a set of technologies enabling the semantic description of resources using standards such as RDF and OWL. Therefore it offers a solution to the lack of semantics in the Web Services world. The research community has introduced several approaches for Web Service semantic descriptions. These range from lightweight solutions like SAWSDL [1] to complex ones like OWL-S [2] and WSMO [3]. The complexity of these approaches stems from their heavy reliance on logical reasoning for the automation of discovery, matchmaking and composition. This complexity also means it will be very

A. Bernstein et al. (Eds.): ISWC 2009, LNCS 5823, pp. 941–948, 2009.

challenging for these features to be available at Web scale [4] [5] [6]. There is a trade-off between automation and scalability, and existing SWS approaches tend to focus on automation. Recently, there has been a rising interest in lightweight SWS for reasons of scalability and minimising complexity and design overhead [7] [8].

Another issue with these approaches, whether heavy or lightweight, is that in addition to semantically describing services they require a semantic description of the domain. This separation of domains and services descriptions stems from the SOA and RPC mindset these approaches are based on. This was the prevalent mindset in traditional Web Services when SWS research began. However another approach to Web Services came forward, known as RESTful Web Services. This approach is based on REST [9] where resources are key actors just as services are in SOA.

REST is an architectural style for network-based systems. It provides a set of constraints learnt from the Web's HTTP development and when applied can make systems scalable, reliable, reusable, resilient and other desired features of the Web as a network-based system. Constraints of REST are: identification of resources, manipulation of resources through representations, self descriptive messages, and hypermedia as the engine of application state. REST was not introduced as an approach to designing web services, yet it has been adopted by the non-corporate Web Service community as alternative to SOAP/WSDL. Although not always adhering to the all of REST's constraints [10] [11] [12], RESTful Web Services are gaining popularity and are adopted by major service providers like Google, Amazon and Yahoo.

The RESTful approach is a natural fit to the Semantic Web since the Semantic Web is based on resources and REST provides a uniform way to provide Web Services. In this paper we explain an approach we called EXPRESS [13] that offers Semantic RESTful Web Services by exploiting Semantic Web resources through a RESTful interface with the minimum of design and development overhead. EXPRESS uses the ontologies that describe classes, instances and relationships among them to create resources accessible via RESTful interfaces. Because the mapping between entities in an ontology and resources is direct, we created a tool that automatically creates a RESTful interface for the semantic resources, therefore simplifying the deployment process. The next section provides a brief overview of existing SWS approaches. In section 3 we discuss EXPRESS with an example. In section 4 we briefly compare EXPRESS with SA-REST and WSMO, then we conclude by highlighting the research questions and future directions.

2 Approaches to Semantic Web Services

We can classify SWS approaches into two main categories: in the first category are approaches that semantically enhance Web Services. The second are approaches that are based on manipulating semantic resources.

2.1 Semantically Enhancing Web Services Approaches

These approaches can be either based on SOA or based on REST.

2.1.1 SOA Based SWS

SAWSDL [1] is a lightweight solution and the only W3C SWS recommendation. It annotates WSDL components such as inputs and outputs with references to ontologies. More ambitious W3C submissions for SWS, such as OWL-S and WSMO, are more complex. OWL-S [2] based on OWL, defines an ontology describing 3 aspects of the service: profile, process and grounding. The profile is for advertising and discovery and contains non-functional and functional properties (inputs, outputs, preconditions and effects.) The service process describes how inputs relate to outputs and preconditions to effects. The grounding maps to a concrete service specification. The limitation of OWL-S is in using OWL as a language based on description logics. OWL-S is overcoming this by incorporating SWRL [14] for defining rules. WSMO [3], another approach, is based on 4 major elements for modelling: ontologies, web services, goals and mediators. Ontologies provide the terminology to describe the domain and services. Web services describe service capabilities (preconditions, assumptions, postconditions and effects) and interfaces (choreography and orchestration.) Goals model service requester's requirements which are used for matchmaking with service capabilities. Mediators handle heterogeneity. WSMO uses WSML[1] as the language for modelling ontologies and rules. It is more expressive and complex than OWL. A criticism of WSMO is its drifting from W3C standards. Efforts have been made to bridge between them.

2.1.2 REST Based SWS

RESTful WS are gaining more popularity, and interests in RESTful SWS are rising. SA-REST [8] is similar to SAWSDL, as it semantically annotates RESTful WS, but because there are no WSDL files for RESTful WS, it adds the annotations to web pages that describe the services. It uses GRDDL[2] or RDFa[3] to embed the annotations in HTML files. By adding semantics SA-REST aims to provide an easier way to create and coordinate mashups. hRESTs and microWSMO [7] are similar approaches to SA-REST. Another approach was introduced in [15] in their approach Semantic Bridge for Web Services (SBWS), they annotated WADL[4] documents linking them to ontologies.

2.2 Semantic Resources Based Approaches

Another part of the work in [15] involved providing a RESTful interface for Semantic data called Semantic REST. They mapped the HTTP methods into SPARQL commands that included proposed extensions for insertion, deletion and updating. In this way RDF datasets offering SPARQL endpoints can offer RESTful functionality integrating them with Web 2.0 clients.

Another approach that is based on semantic constructs is Triple Space Computing (TSC) [16]. It is based on Tuple Space Computing. The communication is shifted to reading and writing RDF triples in a shared triple space.

---

1 Web Service Modeling Language, http://www.w3.org/Submission/WSML/

2 Gleaning Resource Descriptions from Dialects of Languages, http://www.w3.org/TR/grddl/

3 RDFa in XHTML, http://www.w3.org/TR/rdfa-syntax/

4 Web Application Description Language, describes interfaces for RESTful WS, https://wadl.dev.java.net/

3 EXPRESS

EXPRESS eliminates the need for describing services separately because it provides resources with a uniform interface. The uniform interface is the HTTP methods GET, PUT, DELETE, POST and OPTIONS which define consistent operational semantics on all resources. The resources that EXPRESS exploits are entities described semantically in an OWL ontology. So by combining the expressivity and semantics in ontologies and providing a uniform interface to them, RESTful SWS can be created.

A service provider using EXPRESS provides an OWL file describing the resources in a Web Service. This is run through an EXPRESS deployment engine to generate URIs for classes, instances and properties. The service provider then specifies which of the HTTP methods can be applied to these resources and this can differ for different kinds of users, providing a role based access control (RBAC) at the resource methods level. The method is simple and generic and can be applied to any ontology. It builds upon existing standards and does not introduce additional complexity.

In this section we will describe how the method is applied in a simple example. We chose Amazon's Simple Storage Service S3[5], because it is a real service, it is simple so we describe how EXPRESS works in a limited space, and it is familiar to readers interested in REST[6]. S3 enables storing and managing data programmatically on Amazon's servers. It also provides the owner of the data with the ability to charge for downloads. There are two main concepts to manage users' data, Objects (data files) and Buckets (containers of these data files). S3 provides URIs for these objects. For example a file with a name -or key as S3 calls it- *doc* in a bucket *b1* would have the following URI *http://s3.amazon.com/b1/doc*. S3 also enables owners to control access to their data. S3 provides both REST and SOAP API.

3.1 A RESTful Semantic S3 Service

If Amazon wanted to provide a RESTful Semantic Web Service for S3, it should provide an ontology describing resources in S3 and relationships between them. We assume that this OWL file is provided. The next listing describes the relevant parts:

```
:User        a    owl:Class.
:Name        a    owl:DatatypeProperty;
                  rdfs:domain :User;       rdfs:range xsd:string.

:Bucket      a    owl:Class.
:Key         a    owl:DatatypeProperty;
                  rdfs:domain :Object;     rdfs:range xsd:string.
:Owner       a    owl:ObjectProperty;
                  rdfs:domain :Bucket;     rdfs:range :User.
:RequestPay  a    owl:DatatypeProperty;
                  rdfs:domain :Bucket;     rdfs:range xsd:boolean.
:CreationDate a   owl:DatatypeProperty;
                  rdfs:domain :Bucket;     rdfs:range xsd:string.
:Objects     a    owl:ObjectProperty;
                  rdfs:domain :Bucket;     rdfs:range :Object.
```

[5] Amazon Simple Storage Service (Amazon S3), http://aws.amazon.com/s3/

[6] This example, as a non-semantic RESTful Web Service is explained in [11].

```
:Object              a owl:Class.
:ContainingBucket    a owl:ObjectProperty;
              rdfs:domain :Object    rdfs:range :Bucket.
```

The OWL file is parsed; classes, properties and individuals are given URIs based on their names in the file. The following are examples of generated URIs.

http://s3.amazon.com/User (a class URI)
http://s3.amazon.com/Bucket/MyBucket (a bucket instance URI)

Properties also have URIs, for example the bucket's creation date has this URI

http://s3.amazon.com/Bucket/MyBucket/CreationDate.

An Amazon developer specifies which methods (GET, PUT, POST and DELETE) can be applied to each URI. The stubs are generated then the Amazon developer maps these stubs to existing services. Before providing an example, we will explain the differences between the URI structure in the existing S3 and our proposed S3. In the existing S3 the URIs of buckets and objects have the following forms respectively

http://s3.amazon.com/{bucket name}
http://s3.amazon.com/{bucket name}/{object name}

In our proposed S3 service the forms of the URIs are

http://s3.amazon.com/Bucket/{bucket name}
http://s3.amazon.com/Object/{object name}

The difference in the URI forms stems from design decisions. In the existing S3 there are only two types of resources: buckets and objects. The routing of requests to the processes dealing with each type is based on the structure of the URI. If a request was to *http://s3.amazon.com/**myspace*** then this will be considered a bucket and will be routed to the function that processes buckets. However if the request was to *http://s3.amazon.com/myspace/**m1*** it will be considered an object and routed to the processing function. EXPRESS however, is designed for a general purpose and in most cases there will be more than two resources in the system and therefore the routing decisions could not be made based on URI structure only. The URIs are designed to include the type of the requested resources as shown above and this also acts in accordance with the W3C note on cool URIs[7].

Now we will provide a simple scenario of how the service works by showing how a user can create a bucket, add objects to it and delete it. The interaction starts by the client accessing the OWL file. It can access it in the same way it GETs any other resource. The purpose of the OWL file is to show the resource representation -and thus the exchanged messages format-, relationships, and special instances. If the client wants to use the existing S3 services it will have to sign up with Amazon. The semantics of this action is creating a user. The OWL file contains the URIs of resources the client can manipulate. Restrictions in the OWL file such as $Bucket \sqsubseteq \exists Owner.User$ indicate that a user must be created before creating a bucket. As resources have a uniform interface - HTTP methods- the client knows how to create any resource, so the client will send a POST request to *http://s3.amazon.com/User*

[7] Cool URIs for the Semantic Web http://www.w3.org/TR/cooluris/

the message will contain required user information, specified by the OWL file as all the properties where user is the subject. In the excerpt of the S3 OWL file above the only property is required is the name, although other properties are required such as authentication information we did not discuss them due to space limitations. The service response will be creating a new user resource and returning its URI for example *http://s3.amazon.com/User/user1234*. The client can create a bucket in a similar approach. It sends a PUT request to *http://s3.amazon.com/Bucket/* and the message will be:

```
:MyBucket      a        :Bucket;
       :Key    "MyBucket"^^xsd:string;
       :Owner :user1234;
       :RequestPay   "false"^^xsd:boolean.
       :AccessControlPolicy "public-read"^^xsd:string.
```

Because the name of the bucket is specified by the client the PUT method is used instead of POST to create it. This complies with the HTTP standard. The server responds by creating the bucket at the requested URI which is *http://s3.amazon.com/Bucket/MyBucket*. In order to add an object to this bucket the client sends a similar PUT request to *http://s3.amazon.com/Object/* but with required Object properties and the file as a payload. This will make the files available at the S3 storage space and the client can then provide the URIs to its clients to download. Deleting a bucket is straightforward as a DELETE request is sent to a bucket's URI.

We can realize patterns in the way clients manipulate resources in EXPRESS, these patterns are further explained in [13]. To summarise, the Amazon developer needed to provide a domain ontology, specify control access on resources and then map the generated stubs to existing services. The resulting service will be semantic because: relationships between resources are described semantically, resources can be semantically associated to widely agreed on ontologies for example User could be defined as a subclass of foaf:person, and actions come down to adding, deleting and modifying assertions. The resulting service will also be RESTful, resources have URIs, resources have uniform interfaces, the exchanged messages are in OWL which is self-described, and the server guides the client by responding with URIs in which the client can follow where there are next states to go through.

4 Comparison to SA-REST and WSMO

In this section we will highlight the efforts in describing SWS in both SA-REST and WSMO. Creating the domain ontology is an effort that exists in all SWS approaches. We chose to compare to SA-REST, because like EXPRESS it recognises the increasing popularity of RESTful WS. SA-REST as explained in section 2 aims to integrate existing RESTful WS into the Semantic Web by semantically annotating their HTML documentations. Whereas EXPRESS is an approach of using OWL files and REST principles to describe and create RESTful SWS.

The efforts in SA-REST are creating the domain ontology then annotating the HTML documentation. A developer must annotate documentation pages with descriptions such as sarest:input, sarest:output, sarest:operation, sarest:lifting or sarest:lowering linking them to the domain ontology. For example in the case of S3 there are approximately 30 pages, the developer must decide which ones to annotate, and then annotate them with inputs, outputs, and actions. This can increase maintenance costs especially if documentation pages are scattered. In terms of RESTfulness SA-REST is not concerned if the services it describes are actually RESTful. As mentioned in the introduction not all RESTful Web Services adhere to REST's constraints. On the other hand services designed in EXPRESS are RESTful.

In the case of WSMO much more effort is needed. The developer needs to understand WSML and its variants. And for each web service a capability and interface have to be defined. The capability consists of axioms describing: preconditions, assumptions, postconditions, and effects. Preconditions and postconditions describe the Web Service information space state before and after the Web Service execution, whereas assumptions and effects describe the world's state. Furthermore the developer needs to describe the interface consisting of the choreography and orchestration of the service. In the choreography transitions rules that guide the interaction with this service must be specified. In the orchestration the rules guiding how this service uses other services to achieve its overall functionality is stated. For a simple service like S3 at least 15 service descriptions need to be created. It must be noted however, that the efforts included to describe a Web Service in WSMO are in the aim to automate or semi-automate the discovery, composition and invocation of Web Services. Criticisms to this approach and similar ones question whether the overhead is practical, and whether the automation's limited scalability justifies such efforts [4].

5 Conclusions and Future Work

In this paper we have explained EXPRESS, provided an example of how it works and briefly compared it to SA-REST and WSMO. The work done on EXPRESS is in its early stages. EXPRESS uses the OWL file as a description of a RESTful Semantic Service. The implemented deployment system parses OWL files and generates stubs to access the resources. It offers fine grained role based access control, controlling what can be accessed, how and who can access it. In order to understand EXPRESS better we would like to fully implement an existing system, develop the approach more, and analyse its applicability and constraints. EXPRESS's simplicity and nativity to both the Web and the Semantic Web, harvesting the strengths of both, and introducing the minimum level of complexity are features that motivate us to investigate it further. The research questions we would like to answer are:

1. How to perform automatic discovery and composition in EXPRESS?
2. How to facilitate transforming legacy systems to be Semantic and RESTful? Initial ideas are to start from ontologies derived from legacy DB schemas.
3. How to utilise the mapping between PI calculus and ROA Resource-Oriented Architecture [11] (an architecture influenced by REST) as described in [17] to answer the previous questions?

Our goal is provide pragmatic solutions that can contribute towards building infrastructure of the Semantic Web.

References

[1] Farrell, J., Lausen, H.: Semantic Annotations for WSDL and XML Schema. In: W3C Recommendation, World Wide Web Consortium, W3C (2007)
[2] Martin, D., Burstein, M., Hobbs, J., Lassila, O., McDermott, D., Mcllraith, S., Narayanan, S., Paulocci, M., Parsia, B., Payne, T.R., Sirin, E., Srinivasan, N., Sycara, K.: OWL-S: Semantic Markup for Web Services, W3C Member Submission, World Wide Web Consortium, W3C (2004)
[3] Bruijn, J.D., Bussler, C., Domingue, J., Fensel, D., Hepp, M., Keller, U., Kifer, M., König-Ries, B., Kopecky, J., Lara, R., Lausen, H., Oren, E., Polleres, A., Roman, D., Scicluna, J., Stollberg, M.: Web Service Modeling Ontology (WSMO), W3C Member Submission, W3C (2005)
[4] Klusch, M.: Semantic Web Service Description. In: CASCOM: Intelligent Service Coordination in the Semantic Web, Birkhäuser Verlag, Basel (2008)
[5] Fensel, D., Harmelen, F.V.: Unifying Reasoning and Search to Web Scale. IEEE Internet Computing 11(2), 96, 94–95 (2007)
[6] Hench, G., Simperl, E., Wahler, A., Fensel, D.: A Conceptual Roadmap for Scalable Semantic Computing. In: IEEE International Conference on Semantic Computing, pp. 562–568 (2008)
[7] Kopecky, J., Gomadam, K., Vitvar, T.: hRESTS: An HTML Microformat for Describing RESTful Web Services. In: IEEE/WIC/ACM International Conference on Web Intelligence and Intelligent Agent Technology, pp. 619–625 (2008)
[8] Sheth, A.P., Gomadam, K., Lathem, J.: SA-REST: Semantically Interoperable and Easier-to-Use Services and Mashups. IEEE Internet Computing 11, 91–94 (2007)
[9] Fielding, R.T.: Architectural Styles and the Design of Network-based Software Architectures. Doctoral dissertation, University of California (2000)
[10] Fielding, R.T.: A Little REST and Relaxation. In: Jazoon 2007 The International Conference for Java Technology (2007)
[11] Richardson, L., Ruby, S.: RESTful Web Services. O'Reilly Media, Sebastopol (2007)
[12] Vinoski, S.: RESTful Web Services Development Checklist. Internet Computing, IEEE 12, 96–95 (2008)
[13] Alowisheq, A., Millard, D.E.: EXPRESS: EXPressing REstful Semantic Services. To appear in: The 2nd Doctoral Workshop for 2009 IEEE/WIC/ACM International Joint Conference onWeb Intelligence and Intelligent Agent Technology, Milan, Italy (2009)
[14] Harrocks, I., Patel-Schneider, P.F., Boley, H., Tabet, S., Grosof, B., Dean, M.: SWRL: A Semantic Web Rule Language Combining OWL and RuleML, W3C Member Submission, World Wide Web Consortium, W3C (2004)
[15] Battle, R., Benson, E.: Bridging the semantic Web and Web 2.0 with Representational State Transfer (REST). Web Semantics 6, 61–69 (2008)
[16] Riemer, J., Martin-Recuerda, F., Ding, Y., Murth, M., Sapkota, B., Krummenacher, R., Shafiq, M.O., Fensel, D., Kühn, E.: Triple space computing: Adding semantics to space-based computing. In: Mizoguchi, R., Shi, Z.-Z., Giunchiglia, F. (eds.) ASWC 2006. LNCS, vol. 4185, pp. 300–306. Springer, Heidelberg (2006)
[17] Overdick, H.: The Resource-Oriented Architecture. In: IEEE Congress on Services, pp. 340–347 (2007)

A Lexical-Ontological Resource for Consumer Heathcare

Elena Cardillo

FBK-IRST, Via Sommarive 18, 38123 Trento, Italy
cardillo@fbk.eu

Abstract. In Consumer Healthcare Informatics it is still difficult for laypersons to understand and act on health information, due to the persistent communication gap between specialized medical terminology and that used by healthcare consumers. Furthermore, existing clinically-oriented terminologies cannot provide sufficient support when integrated into consumer-oriented applications, so there is a need to create consumer-friendly terminologies reflecting the different ways healthcare consumers express and think about health topics. Following this direction, this work suggests a way to support the design of an ontology-based system that mitigates this gap, using knowledge engineering and Semantic Web technologies. The system is based on the development of a consumer-oriented medical terminology which will be integrated with other existing domain ontologies/terminologies into a medical ontology repository. This will support consumer-oriented healthcare systems by providing many knowledge services to help users in accessing and managing their healthcare data.

Keywords: Medical Knowledge Acquisition, Knowledge Integration, Medical Ontologies, Consumer Medical Terminologies.

1 Introduction

With the advent of the Social Web and Healthcare Informatics technologies, we can recognize that a linguistic and semantic discrepancy still exists between specialized medical terminology used by healthcare providers or professionals, and the so called "lay" medical terminology used by healthcare consumers. The medical communication gap became more evident when consumers started to play an active role in healthcare information access, becoming more responsible for their personal healthcare, exploring health-related information sources on their own, consulting decision-support healthcare sites on the web, and using patient-oriented healthcare systems which allow them to directly read and interpret clinical notes or test results and to fill in their Personal Health Record (PHR). To help consumers fill this gap, the challenge is to sort out the different ways they communicate within distinct discourse groups and map the common, shared expressions and contexts to the more constrained, specialized language of healthcare professionals. In particular, medical Knowledge Integration in healthcare systems is facilitated by the use of Semantic Web technologies,

A. Bernstein et al. (Eds.): ISWC 2009, LNCS 5823, pp. 949–956, 2009.

helping consumers during their access to healthcare information and improving the exchange of their personal clinical data. Though much effort has been spent on the creation of these medical resources, used above all to help physicians in filling in Electronic Health Records (EHR), there is little work based on the use of consumer-oriented medical terminology, and in addition most existing studies have been done only for English.

Given this scenario, this work want to propose a methodology for the creation of a consumer-oriented lexical-ontological resource for Italian, and its integration with a coherent semantic medical resource, which could be used in healthcare systems, like Personal Health Records, to help consumers during the process of querying and accessing healthcare information. The present work will be structured as follows: In Section 2 is described the State of the Art in the field of medical terminologies/ontologies, both in consumer-oriented and clinically-oriented healthcare; in Section 3 are exposed the problem statement, our objectives and approach to reach them; in Section 4 are presented preliminary results; and finally in Section 5 are proposed concluding remarks and some future works.

2 State of the Art

2.1 Medical Terminologies and Ontologies

Over the last two decades the standardization efforts have established a number of medical terminologies and classification systems as well as conversion mappings between them to help medical professionals in managing and codifying their patients health care data, such as UMLS Metathesaurus, SNOMED, ICD-10 (International Classification of Diseases) and the ICPC-2 (International Classification of Primary Care). They concern *"the meaning, expression, and use of concepts in statements in the medical records or other clinical information systems"* [11]. In the presence of all these medical terminologies interoperability has become a significant problem. Content, structure, completeness, detail, cross-mapping, taxonomy, and definitions vary between existing vocabularies. So during the last few years, thanks to the Semantic Web perspective, the collaboration between the areas of Healthcare Informatics and Knowledge Representation generated a set of new methodologies and tools for improving healthcare systems, and in particular medical terminologies, which were translated into more formal representations using ontology languages (e.g. the logical formalization of SNOMED CT).

During the last few years much effort has also been spent on the creation of new Biomedical Ontologies (e.g. the Foundational Model Anatomy -FMA- [9]). Ontologies become relevant in healthcare if integrated into an EHRs, which manage an increasing volume of narrative data, to allow: structuring and semantics of the recorded information; and references to concepts from ontologies of the first kind, e.g. ICD 10/9 or SNOMED terms [5]. Two other important issues to take into account, given the presence of all these medical ontologies are: Ontology Mapping, to show how concepts of one ontology are semantically related to concepts of another ontology [6]; and Ontology Integration, which allows access to multiple heterogeneous ontologies. Much work has been done in this direction, for the alignment of different Biomedical Ontologies with concept overlap (here can be mentioned the works of Mork and Bernstein [8], and Zhang and Bodenreider [13]), and for their integration by means of medical ontology repositories, such as the creation of Bio-Portal, a Web-based system that serves as a repository for biomedical ontologies [10].

2.2 Consumer-Oriented Medical Vocabularies

In spite of these advantages reached by the integration of Healthcare Informatics and Semantic Web technologies, the vocabulary problem continues to plague health professionals and their information systems, and in particular laypersons who are the most damaged by the increased medical linguistic gap. To respond this healthcare consumers' needs, during the last few years, many researchers have labored over the creation of lexical resources that reflect the way consumers/patients express and think about health topics. One of the largest initiatives in this direction is the Consumer Health Vocabulary Initiative[1], resulted in the creation of the Open Access Collaborative Consumer Health Vocabulary (OAC CHV) for English. It includes lay medical terms and synonyms connected to their corresponding technical concepts in the UMLS Metathesaurus. They combined corpus-based text analysis with a human review approach, including the identification of consumer forms for "standard" health-related concepts. An overview of all these studies can be found elsewhere [7].

It is important to stress that there are only few examples of the real application of the most of initiatives. For example, in Zeng *et al.* [14] there is an attempt to face syntactic and semantic issues in the effort to improve PHRs readability, using the CHV to map content in EHRs and PHRs. On the other hand, Rosembloom *et al.* [12] developed a clinical interface terminology, a systematic collection of healthcare-related phrases (terms) to support clinicians' entries of patient-related information into computer programs such as clinical "note capture" and decision support tools, facilitating display of computer-stored patient information to clinician-users as simple human-readable texts.

3 Research Objectives and Directions

3.1 The Problem Statement

As mentioned in Section 1, healthcare consumers actually play an active role in accessing and managing their personal health care data. For this reason they need an easy and understandable access to medical information when using such healthcare systems, and at the same time physicians, from their side, need to understand patients reports on their conditions (severity pain, degree of discomfort). So communication terminology and understanding of medical concepts are

[1] `http://www.consumerhealthvocab.org`

serious barriers. This linguistic gap, in addition, also prevent full participation of consumers for example in shared health records, and often interferes in communication between patients and their health care providers. Furthermore, most of the existing "standards" medical terminologies and ontologies have been developed from the point of view of physicians, so they don't provide a sufficient support for their integration in all that applications designed for laypersons. This highlights the necessity of an intermediate consumer understandable terminology to be integrated with standard specific terminologies/ontologies in order to support the integration of consumer-oriented applications with that designed for experts. This thesis work will also focus on the nature of this medical communication gap in the Italian context, where there is a lack of consumer-oriented medical terminologies (as seen in Section 2, all previous works have been done for English), and where illiteracy, regional diversity, and the high presence of non-native speakers further intensify the problem.

3.2 The Objectives

The purpose of this work is to support the design of an ontology-based system that mitigates the language barrier between the healthcare consumer and professional medical domains. Knowing the forms used by laypersons and how such forms map to medical concepts is useful in assisting healthcare consumers to formulate queries and to understand retrieved medical documents, and also helps professionals and information systems to deal with patient inputs. The general aim can be divided into the following sub-objectives:

1. Development of a Consumer Medical Vocabulary for Italian, able to reflect the different ways consumers and patients express and think about health topics.
2. Integration of this "lay" terminology with other existing terminologies, in particular with the clinical ones relevant to reconstruct the process of care in General Practice.
3. Formal Representation in OWL language of these terminologies, and integration of them into a unique Medical Ontology Repository.
4. Implementation of Reasoning and Search services to support the development of semantic-based healthcare systems which need interchanges with patients and consumers.

3.3 Approach

The global approach followed for this research activity is divided in two macro phases. The first one includes the creation of a Consumer Health Vocabulary for Italian, for collecting common medical expressions and terms used by Italian speaking people. The second one focuses on the formal representation of medical terminologies which will be integrated with the developed consumers vocabulary, and the development of a Medical Ontology Repository in which all these ontologies and terminologies will be integrated. The activity will be characterized by the following tasks:

– Knowledge Acquisition/Terminology Extraction. Use of elicitation techniques to acquire all the lay terms, words, and expressions, used by laypeople to indicate specific medical concepts;
– Generation of the Italian Consumer Health Terminology. Selection of all the lay terms extracted that have been identified as good representatives of technical medical concepts, and consequent mapping analysis to a standard medical terminology.
– Formalization in terms of OWL. Medical terminologies such as ICD10 and ICPC2 will be formalized into OWL ontologies, and then will be integrated with the consumer-oriented medical vocabulary and other existing medical ontologies to guarantee semantic interoperability.
– Creation of a Medical Ontology Repository (MORe) and implementation of Knowledge Services. Some relevant resources will be integrated into MORe, an ontology collection that will be extended with a set of basic reasoning services to support the implementation of semantic based patient healthcare applications.

4 What Has Been Done So Far

4.1 Knowledge Acquisition Task

This first task aims at the acquisition of consumer-oriented terminology and knowledge about a specific subset of healthcare domain, and at the creation of the consumer-oriented medical vocabulary for Italian. A hybrid methodology was used for the identification of "lay" terms and expressions used by Italian speaking people to indicate "symptoms", "diseases", and "anatomical concepts". Three different target groups were considered: First Aid patients subjected to the Triage process; a community of high educated and middle age people; and finally a group of elderly people. In this methodology three different Elicitation Techniques were applied to the mentioned groups of people: 1) Collaborative wiki-based medical knowledge acquisition; 2) Nurse-assisted medical knowledge acquisition; and 3) Interactive medical knowledge acquisition combining traditional elicitation techniques (Focus Groups, Concepts Sorting and Games).

All the aquired knowledge was analysed by means of a term extraction tool (Text2Knowledge - T2K), which allowed to automatically extract terminology and to perform typical text processing techniques and statistical analyses (more details about the tool can be found in [1]). Term extracted were reviewed by two physicians to find incongruities done by laypeople in categorization of medical terms and in synonymy relations. Physicians have been also asked to map a term/medical concept pair by using a professional health classification system, the above mentioned International Classification for Primary Care 2nd Edition (ICPC2-E), which is used in particular by general practitioners for encoding symptoms, medical procedures and diagnosis. A more detailed description of the methodology for knowledge acquisition and of the term extraction process and mapping analysis can be found in Cardillo *et al.* [4].

First Results Evaluation. A variegated consumer-oriented terminology was acquired. From 225 Wiki pages 962 medical terms were extracted, and in particular were found 173 *Exact Mappings*, 80 *Related Mappings*, 94 *Hyperonyms*, 51 *Hypomyms* and, finally, 186 *Not Mapped* ICPC2 concepts. Most of the exact mappings to ICPC2 are related to anatomical concepts, and many synonyms were found for symptoms. Concerning the Nurse-assisted data set, from 2.000 Triage records 1108 relevant terms were extracted, providing mapping only for 726 terms. Here can be highlighted the high presence of lay terms used for expressing symptoms with exact mappings to ICPC2 (134 on a total of 240 exact mappings), but also many synonyms in lay terminology for ICPC2 concepts (386 *Related Mappings*). Finally, 321 medical terms were extracted by the Focus Group data set. Here all the symptoms extracted (79 terms) had corresponding medical concept in ICPC2 terminology (35 *Exact Mappings* and 44 *Related Mappings*).

The most profitable methodology for acquiring consumer-oriented medical terminology resulted the one assisted by Nurses. While Wiki-based method, even if not exploited for the collaborative characteristic, has demonstrated good qualitative and quantitative results. Comparing the three sets, the overlap is only of 60 relevant consumer medical terms. The overlap with ICPC2 is about 508 medical concepts on a total of 706 ICPC2 concepts. This means that all the other mapped terms can be considered synonyms or quasi synonyms of the ICPC2 concepts. The large number of not mapped terms and the low overlap between the three sets of extracted terms demonstrate that it was possible to extract a very variegated range of medical terms, many compound terms and expressions, which can be representative of the corresponding technical terms present in standard terminology, and which can be used as candidate for the construction of our consumer-oriented medical terminology for Italian.

4.2 OWL Encoding of Medical Classification Systems

A parallel activity to that of consumer-oriented terminology acquisition was performed to formalize two Medical Classification Systems into OWL ontologies [2]: the previously mentioned ICPC2 and ICD10, expressing the two ontologies according to the sublanguage DL (Description Logic). In the process of conversion of ICPC2-ICD10 to OWL formalism two important principles of classification have been preserved: the disjointness of terms (nodes) and the exhaustiveness of classification, by introducing the use of special groups of terms such as "other", "unspecified" and "not elsewhere classified", reflecting this property in OWL by the explicit definition of sibling classes as disjoint and by the closure definition of any subdivision class as to be equivalent to a disjunction of all its child classes (including other, unspecified and so on). In encoding ICD10 to OWL, every ICD10 chapter is a class and each section is a subclass, which contain in turn each three or four digit ICD items. So only the subsumption and the disjunction relations are defined, the concepts representing each ICD category are labelled by the ICD codes. In encoding ICPC2 we preserved its biaxal structure creating a class for each chapter (body system or problem area) and a class for each component (*Symptom and Complaint*; *Procedure*, and *Diagnosis and Disease*). We added disjoint statements between siblings and some objects and datatype properties (description, terms of inclusion, terms of exclusion, ICD10 corrispondence)[2].

A well-founded and medically sound mapping model between the two ontologies was constructed as well, by means of its formalization in terms of OWL axioms (686 in total) and the validation of its coherence using Semantic Web techniques. Standardly, given two heterogeneous representations, a mapping can be viewed as a triple $\langle e, e', r \rangle$, where e, e' are the entities (e.g., formula, terms, classes, etc.) belonging two the different representations, and r is the relation asserted by the mapping. Due to the idea of encoding ICPC-ICD mappings as OWL axioms, the entities in the mapping correspond to ICPC-2 and ICD-10 classes and expressions, while the relation r is given a set-theoretic meaning by using subsumption and equivalence. Details about methodology for formalization and results can be found in [3]. After the task of mapping analysis and the evaluation of the first results, the extracted "lay" terms considered as good synonyms for the ICPC2 symptoms and diseases have been added to the ICPC2 ontology to integrate it with the consumer-oriented terminology.

5 Concluding Remarks and Future Works

This paper proposed a thesis work aiming at the creation of a consumer-oriented lexical-ontological resource that would help fill in the medical linguistic gap between specialized and "lay" terminology, and which could be used in consumer-oriented halthcare systems to help consumers in accessing to and managing of their healthcare data. In particular, preliminary results have been presented for the task of consumer-oriented terminology acquisition, on the basis of statistical and mapping analyses, which helped to find overlaps between extracted "lay" terms and specialized medical concepts in the ICPC2 medical terminology. First results are encouraging because many consumer-oriented terms were acquired, and a low overlap with ICPC2 medical concepts and a high number of synonyms were found. The formalization in terms of OWL axioms of the ICP2 and ICD10 coding systems, and the existing clinical mappig between them, were provided, and results were very positive allowing the reduction of the efforts for upgrading mappings in view of the next publication of the two encoded systems, ICD11 and ICPC3; and to reuse Mapping Consistency, Debugging, and Entailment.

This thesis work will potentially contribute to the state of the art in several research areas, including Medical Terminology, Healthcare Informatics, Knowledge Acquisition and Representation, and it can have the following potential aspects: 1) the Cross-Domain Interdisciplinarity; 2) the Integration of specialized and consumer-oriented medical knowledge, which helps to fill the medical communication gap; and 3) new methodologies for integration tasks and for knowledge services, which this framework will offer in the application for example to a PHR, to improve its management and accessibility.

[2] These ontologies can be consulted at: `https://dkm.fbk.eu/index.php/Resources`

To improve the results of the knowledge acquisition process and to extract more variegated consumer-oriented terminology, a written corpus, which include forum postings of an Italian medical website for asking questions to on-line doctors[3] has been analyzing. This will allow extending our sample and cover a wider range of ages, people with different background and consequently different levels of health literacy.

References

1. Bartolini, R., Lenci, A., Marchi, S., Montemagni, S., Pirrelli, V.: Text-2-knowledge: Acquisizione semi-automatica di ontologie per l'indicizzazione semantica di documenti. Technical Report for the PEKITA Project, ILC. Pisa p.23 (2005)
2. Bechhofer, S., Van Harmelen, F., Hendler, J., Horrocks, I., McGuinness, D.L., Patel-Schneider, P.F., Stein, A.L.: OWL Web Ontology Language Reference, W3C Recommendation (2004)
3. Cardillo, E., Eccher, C., Tamilin, A., Serafini, L.: Logical Analysis of Mappings between Medical Classification Systems. In: Dochev, D., Pistore, M., Traverso, P. (eds.) AIMSA 2008. LNCS (LNAI), vol. 5253, pp. 311–321. Springer, Heidelberg (2008)
4. Cardillo, E., Serafini, L., Tamilin, A.: A Hybrid Methodology for Consumer-oriented Healthcare Knowledge Acquisition. In: proceedings of the KR4HC 2009 Workshop, Verona, July 19 (2009)
5. Ceusters, W., Smith, B., De Moor, G.: Ontology-Based Integration of Medical Coding Systems and Electronic Patient Records. In: MIE 2005 (2005)
6. Euzenat, J., Shvaiko, P.: Ontology Matching. Springer, Heidelberg (2007)
7. Keselman, A., Logan, R., Smith, C.A., Leroy, G., Zeng, Q.: Developing Informatics Tools and Strategies for Consumer-centered Health Communication. Journal of Am. Med. Inf. Assoc. 14(4), 473–483 (2008)
8. Mork, P., Bernstein, P.: Adapting a generic Match Algorithm to Align Ontologies of Human Anatomy. In: Proceedings of ICDE (2004)
9. Noy, N.F., Rubin, D.L.: Translating the Foundational Model of Anatomy into OWL, in Web Semantics: Science, Services and Agents on the World Wide Web. Elsevier Science 6(2), 133–136 (2008)
10. Noy, N.F., Musen, N., Shah, N., Dai, B., Dorf, M., Griffith, N., Jonquet, C., Montegut, M., Rubin, D., Youn, C.: BioPortal: A Web Repository for Biomedical Ontologies and Data Resources. In: The International Semantic Web Conference, ISWC 2008, Karlsruhe, Germany (2008)
11. Rector, A.: Clinical Terminology: Why is it so hard? Methods of Information in Medicine 38(4), 239–252 (1999)
12. Rosembloom, T.S., Miller, R.A., Johnson, K.B., Elkin, P.L., Brown, H.S.: Interface Terminologies: Facilitating Direct Entry of Clinical Data into Electronic Health Record Systems. Journal of Am. Med. Inf. Assoc. 13(3), 277–287 (2006)
13. Zhang, S., Bodenreiden, O.: Experience in aligning anatomical ontologies. International Journal on Semantic Web and Information Systems 3(2), 1–26 (2007)
14. Zeng, Q., Goryachev, S., Keselman, A., Rosendale, D.: Making Text in Electronic Health Records Comprehensible to Consumers: A Prototype Translator. In: The 31st American Medical Informatics Association's Annual Symposium, AMIA 2007, pp. 846–850 (2007)

[3] `http://medicitalia.it`

Semantic Web for Search*

Jessica Gronski**

UC Santa Cruz
jgronski@soe.ucsc.edu

Abstract. Semantic Web data seems like a promising source of information for improving search. While there is some literature about how semantic data should be used to enhance search, there are no positive conclusions about the best approach. This paper surveys existing approaches to semantic web search, describes adapting a TREC benchmark for evaluation, and proposes a learned representation algorithm for using semantic web data in search.

1 Introduction

The Semantic Web(SW)[2] aims to provide a common framework for publishing linked data on the web. While the SW has yet to become a fully adopted technology, the data that is published about web pages, can and should be exploited for improving web search. Improving web search with SW data is interesting not only because search is an important industry but because it provides another compelling reason for wide adoption of the SW.

This paper describes the beginning of my doctoral research on how to use SW data to improve traditional keyword search over documents. The next section describes the existing approaches to search using SW data and concludes with a summary of experimental results from the literature. Section 3 describes the work done to create a benchmark. Section 4 follows with the description of a proposed learned-representation search algorithm. Section 5 concludes with future work for the project.

2 Search Using Semantic Web Data

Regardless of whether the search algorithm retrieves web pages, RDF documents, RDF triples, or linked-data paths, SW search algorithms follow three information retrieval models: Boolean, Vector Space, or Link-based models.

2.1 Boolean Models

In boolean retrieval, documents are modeled as a set of boolean variables indicating whether the document contains a word, a query is modeled as a boolean

* All research is funded by Los Alamos National Labs.
** Thanks to my advisors Yi Zhang and Herbert Van de Sompel.

A. Bernstein et al. (Eds.): ISWC 2009, LNCS 5823, pp. 957–964, 2009.

proposition, and the retrieval algorithm returns all documents satisfying the boolean proposition. Given the query "cats AND dogs", a boolean model returns all documents containing both the terms "cats" and "dogs".

Boolean algorithms will miss documents that use words synonymous to the query terms and also fails to specify the order of documents returned. To address the latter problem, algorithms have ordered documents returned by the number of relevant terms contained [22] or used cover density, the distance between query terms within the document [5].

Many SW search algorithms employ boolean models [17,12,24,21,11] where documents are modeled as sets of variables which correspond to the `WHERE` clauses of a SPARQL query, and the boolean algorithm returns all query-satisfying items. Most do not address the ordering problem, an exception is K-search [3] which orders the boolean retrieved documents using the query's term frequency in the document text (ignoring the SW data). Others order the boolean retrieved documents using an adaptation of vector-space and link-based models to SW data and are grouped accordingly in subsequent sections.

2.2 Vector Space Models

The vector space model projects document and queries into a vector space with each dimension representing a different term. The ranking function returns documents ordered by measuring their distance to the query in vector space.

Using Euclidean distance for matching the query and document vectors discriminates against long documents so instead normalized cosine distance is used:

$$normcos(\vec{q}, \vec{d}) = \frac{\vec{q} \times \vec{d}}{||\vec{q}|| \cdot ||\vec{d}||}$$

where $\vec{q}$ and $\vec{d}$ are the vector representations of the query and document respectively. Alternatives include vector similarity measures which bias for length as it was discovered that in TREC, an established information retrieval benchmark, longer documents were more likely to be relevant[4,25].

The vector space models differ not only in how they compare vectors, but also in their approaches to mapping documents and queries into the term vector space. One approach, TFIDF[4,23], uses term frequency and inverse document frequency, calculates the document's weight in each of the vector space dimension (recall that each dimension represents a term) as the frequency of that term(t) in the document(d) divided by the frequency of documents in the corpus(Δ) containing the term.

$$TFIDF(d,t) = \frac{|t \in d|}{\sum_{\tau} |\tau \in d|} IDF(t)^{-1} \quad \text{where } IDF(t) = \frac{|\{\delta | t \in \delta \wedge \delta \in \Delta\}|}{|\Delta|}$$

The SW search papers which use vector-space models generally incorporate a variation of TFIDF [8] or use frequency directly [1,26] when ranking items. The approach taken by Vallet et. al[8] is to perform a boolean retrieval on the documents by translating the keyword query into a structured SW query. They then order the returned documents using an adaptation of TFIDF where instead of looking at the frequency of terms when calculating TFIDF they look at the frequency of the queried SW class instances. Two systems [1,26] that rank compound direct and indirect paths between SW instances, use the frequency of the constituent relationships to order the paths.

2.3 Link-Based Models

Link-based retrieval models rely on documents having links between them such as web pages or scholarly documents.

The Pagerank algorithm[19] models documents as nodes and HTML links as directed edges between nodes. Pagerank then models the user as a random walker that generally walks along edges and is equally likely to follow each outgoing link. With small probability, the walker will not follow a link but jump to any node with equal probability. Pagerank ranks the documents using the stationary distributions over documents which the random walker creates. Pagerank and its many variants[14,15,10] can be unified as part of the same framework and the variants have been shown to produce scores that are highly correlated[6].

Link-based SW algorithms adapt Pagerank to the SW setting by using the links between SW data [7,9,27,20]. The Swoogle search engine weighs different types of links more than others [7,9]. The SWRank algorithm[27] reverses the SW links when computing Pagerank. Most link-based algorithms use boolean retrieval to narrow the search to query relevant documents [7,9,27], and use their adapted Pagerank scores to order the boolean results. The exception is SWRank which incorporates a term-based TFIDF component to make the scoring query sensitive.

2.4 Experimental Evaluation in Semantic Web Search

Despite the large number of papers on using SW data in search, it is difficult to compare the retrieval performance of each algorithm as all but one of the papers report qualitative, positive results on different datasets and eschew established retrieval benchmarks. Vallet et. al is the notable exception as they used a TREC information retrieval benchmark for evaluation, however they reported negative results for their TFIDF algorithm. This lack of consensus on how to incorporate SW data in search is what motivates this work.

3 Experiment Setup

For evaluating any algorithm, we need an established benchmark ideally containing a set of documents with supporting SW data, a suite of queries, judgments for each document-query pair, and a baseline algorithm capturing the current best approach to the benchmark.

3.1 TREC Blog Track: Documents, Queries and Judgments

As we know of no ideal benchmark we choose to use the BLOG06 dataset created for the TREC benchmark conference's blog retrieval track because though it lacks explicit SW data, the results can be directly compared with competitive information retrieval algorithms.

The BLOG06 dataset was collected during late 2005 and early 2006 and consists of 100,649 feeds, 3,215,171 permalink documents (blog posts), and 324,880 homepages from both top-quality blogs, spam blogs, and manually picked blogs of unknown quality. The TREC blog track has two relevant tasks, the blog and post retrieval tasks (called the distillation and adhoc tasks in the literature). For each task TREC provides a set of questions, a dataset of documents and binary judgments of relevance for a subset of the blog-query or post-query pairs.

We shall evaluate an algorithm's performance using standard TREC metrics mean average precision (MAP), precision at ten documents (P@10), and R-Precision.

3.2 Semantic Web Data Creation

We extract SW data from BLOG06 documents and create SIOC[18] blog ontology classes and links between them which can be summarized on the right side of Figure 1. The two containment relationships `sioc:container_of` and `reply_of` shows that a post is contained in a blog and a comment is a reply of a post respectively. The `sioc:links_to` citation relationship indicates that the origin entity has an HTML link pointing to destination entity (all links, not in a comment or post are assumed in the blog). This data is distinct from ordinary HTML links as they originate from entities rather than web documents and thus the distinction between a informative link in a post and a spam link contained in a comment can be made. As blog-internal `sioc:links_to` data does not convey the same semantic information as links to external sites (often they are to the prior/next entry), we exclude this data.

For the learned graph approach we shall describe this SW data as a multi-relational graph $G = (V, \mathbb{E} = \{E \in V \times V\})$ where the nodes V are the instances of SIOC classes `weblog`, `post`, and `comment` and $\mathbb{E}$ is the set of directed, typed edge matrices E. Each E is a binary adjacency matrix which represents a connection that exists between different entities (either Blogs, posts, or comments) in the dataset. These edge sets are summarized on the right side of Figure 1.

3.3 TREC Baseline

Though the ideal baseline would be the best results reported in the TREC conference, we were unable to parse all blogs and thus cannot use these results. Instead we used the paper of the winners of the TREC 2007 blog distillation task (which used the same BLOG06 dataset) as a guideline to develop an algorithm which approximates their performance on the original data and apply it to our restricted dataset. The resulting baseline performance on both the adhoc

$E \in \mathbb{E}$	Description
$E_{\mathtt{w2w}}$	weblog `sioc:links_to` weblogs
$E_{\mathtt{p2w}}$	post `sioc:links_to` weblogs
$E_{\mathtt{p2p}}$	post `sioc:links_to` posts
$E_{\mathtt{c2w}}$	comment `sioc:links_to` weblogs
$E_{\mathtt{c2p}}$	comment `sioc:links_to` posts
$E_{\mathtt{p} \in \mathtt{w}}$	weblog `sioc:container_of` post
$E_{\mathtt{c} \in \mathtt{p}}$	post `sioc:reply_to` comment

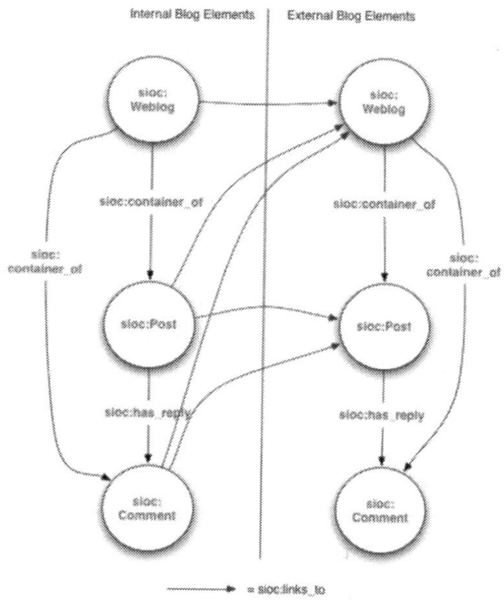

Fig. 1. Links between `(w)eblogs`, `(p)osts`, and `(c)omments` in the TREC dataset

	Data					
	BLOG06 (baseline/reported)			BLOG06 subset (baseline)		
Task	MAP	R-Prec.	P@10	MAP	R-Prec.	P@10
Blog Distillation	0.35/0.34	0.36/0.41	0.42/0.47	0.21	0.28	0.36
Adhoc Task	0.40/0.35	0.41/0.39	0.66/0.56	0.42	0.42	0.62

Fig. 2. On the left the results verify that the recreated TREC 2007 algorithm is close to the conference's reported values. The right column describes the performance of the recreated TREC algorithm on the parsed subset.

and distillation tasks is described in Figure 2 with the left side showing that it gives comparable performance on the entire BLOG06 dataset and the right side showing the baseline's results on the parsed subset.

4 Proposed Learned Representation Approach

We plan on using a learned representation approach for ranking documents with SW data. This link-based approach is based on recent work in the domain of document recommendations by Zhou et. al[28]. Their approach uses a multi-relational graph describing different kinds of links between entities (in their case: scholarly documents, authors and venues) to find latent representations of these entities and then finally produce document recommendations. In our dataset we have a multi-relational graph describing blog entities (blogs, posts, and comments) with HTML link relationships and structural relationships (a post is contained in a blog). Using Zhou's technique and the multi-relational graph, we shall find hidden representations of the blog entities and use the representations as features in a learning to rank algorithm to order the entities.

The approach used to discover the latent representations encodes graph edges as adjacency matrices, with each matrix describing the graph defined by one edge type. The latent representation of the nodes is found by defining a loss function between the adjacency matrices and the hidden representations of the nodes, and choosing the representations which minimize the loss function.

The loss function defined for the citation relationships, such as the post containing an HTML link to another post relationship $E_{\mathtt{p2p}}$, is a laplacian loss function:

$$Loss(X_P) = Tr(X_P^T L(\overline{E_{\mathtt{p2p}}}) X_P)$$

where X_P is the hidden representation of the post nodes and $L(E_{\mathtt{p2p}})$ is the Laplacian of the graph $E_{\mathtt{p2p}}$. The laplacian loss function smooths the difference between adjacent node values and captures the intuition that similar entities will cite one another so the values of their latent representations should be similar.

An undesirable way but effective way to minimize the laplacian loss function above is to use a representation where the values of every node is the same. This makes the function uninteresting and thus the loss function is regularized with $-log|X_P^T X_P|$ to prevent these kinds of simplistic representations.

Besides citation relationships between blogs there also exist containment relationships. For example, the posts in blogs containment graph $E_{\mathtt{p}\in\mathtt{w}}$ is a kind of containment relationship. We define a loss function for containment relationships which captures the intuition that post related by blogs will be close in the latent space:

$$Loss(X_W, X_P) = ||E_{\mathtt{p}\in\mathtt{w}} - X_P X_W^T||_F^2$$

where X_W and X_P are hidden representations of the blog and post entities respectively.

Using these two types of loss functions and taking a linear combination of all loss functions for each edge type we can construct a global loss function to minimize[1]:

$$\begin{aligned} Loss(X_W, X_P, X_C) = {} & k_1||E_{\mathtt{p}\in\mathtt{w}} - X_P X_W^T||_F^2 \\ & + k_2 \mathrm{Tr}(X_W^T L(E_{\mathtt{w2w}}) X_W) + k_3 \mathrm{Tr}(X_P^T L(E_{\mathtt{p2p}}) X_P) \\ & + k_4 \mathrm{Tr}(X_W^T L(E_{\mathtt{p}\in\mathtt{w}}^T E_{\mathtt{p2w}}) X_W) - log(|X_W^T X_W| * |X_P^T X_P|) \end{aligned}$$

where the k_i are tunable constants. As the equation is convex the entity representations minimizing the loss function can be found using a nonlinear conjugate gradient(CG) method given the derivatives (see appendix A). The minimal latent representations of the blog entities found will be used as features in a learning to rank algorithm such as Ranking SVM.

[1] For brevity, this loss functions omits relationships with the comment entity. The omitted terms mimic those contained in this representative loss function.

5 Future Work

In order to complete the evaluation of the learned representation approach we need to implement the loss function and apply the macopt[16] convex optimization package to optimize the function. Once our latent representation is found we plan on using Ranking SVM package provided by SVM^{light}[13], to incorporate the latent representation with the baseline algorithm. The algorithm will be considered effective only if the scores produced by ranking SVM will improve on the TREC-based baseline algorithm.

While blog search problem is not as general as web search, the features used by the proposed algorithm (containment and citation links) are not, and we expect will apply to other vertical search over SW data.

Another challenge we hope to address in the future is that the learned representation algorithm in its current form is ontology-sensitive and needs to know what kind of relationship a link-type is (containment or link relationship) to define the loss function. We hope to later develop an ontology-independent loss function or algorithm to deal with a more general environment where the meaning of the relationship is unknown.

References

1. Anyanwu, K., Maduko, A., Sheth, A.: Semrank: ranking complex relationship search results on the semantic web. In: WWW 2005, pp. 117–127. ACM Press, New York (2005)
2. Berners-Lee, T., Hendler, J., Lassila, O.: The semantic web: Scientific american. Scientific american (2001)
3. Bhagdev, R., Chapman, S., Ciravegna, F., Lanfranchi, V., Petrelli, D.: Hybrid search: Effectively combining keywords and semantic searches, pp. 554–568 (2008)
4. Buckley, C., Singhal, A., Mitra, M., Salton, G.: New retrieval approaches using smart: Trec
5. Clarke, C.L.A., Cormack, G.V., Tudhope, E.A.: Relevance ranking for one to three term queries. Inf. Process. Manage. 36(2), 291–311 (2000)
6. Ding, C., He, X., Husbands, P., Zha, H., Simon, H.: Pagerank, HITS and a unified framework for link analysis. Technical Report 49372, LBNL (2002)
7. Ding, L., Finin, T., Joshi, A., Peng, Y., Pan, R., Reddivari, P.: Search on the semantic web. Computer 38(10), 62–69 (2005)
8. Fernandez, M., Lopez, V., Sabou, M., Uren, V., Vallet, D., Motta, E., Castells, P.: Semantic search meets the web. In: IEEE Semantic Computing, pp. 253–260 (2008)
9. Finin, T., Mayfield, J., Joshi, A., Cost, R.S., Fink, C.: Information retrieval and the semantic web, p. 113a (2005)
10. Gevrey, J., Ruger, S.M.: Link-based approaches for text retrieval. In: Text REtrieval Conference (2001)
11. Guha, R., Mccool, R., Miller, E.: Semantic search. In: WWW 2003: Proceedings of the 12th international conference on World Wide Web, pp. 700–709. ACM Press, New York (2003)
12. Heflin, J., Hendler, J.: Searching the web with shoe. In: AAAI Workshop 2000, pp. 35–40 (2000)

13. Joachims, T.: Learning to Classify Text Using Support Vector Machines: Methods, Theory and Algorithms. Kluwer Academic Publishers, Norwell (2002)
14. Kleinberg, J.M.: Authoritative sources in a hyperlinked environment. Journal of the ACM 46(5), 604–632 (1999)
15. Lempel, R., Moran, S.: Salsa: the stochastic approach for link-structure analysis. ACM Trans. Inf. Syst. 19(2), 131–160 (2001)
16. Mackay, D.: Macopt, `http://www.inference.phy.cam.ac.uk/mackay/c/macopt.html`
17. Michalowski, M., Ambite, J.L., Thakkar, S., Tuchinda, R., Knoblock, C.A., Minton, S.: Retrieving and semantically integrating heterogeneous data from the web. Intelligent Systems, IEEE 19(3), 72–79 (2004)
18. Möller, K., Bojrs, U., Breslin, J.: Using semantics to enhance the blogging experience, pp. 679–696 (2006)
19. Page, L., Brin, S., Motwani, R., Winograd, T.: The pagerank citation ranking: Bringing order to the web. Technical report, Stanford Digital Library Technologies Project (1998)
20. Patel, C., Supekar, K., Lee, Y., Park, E.K.: Ontokhoj: a semantic web portal for ontology searching, ranking and classification. In: WIDM 2003, pp. 58–61. ACM Press, New York (2003)
21. Popov, B., Kiryakov, A., Ognyanoff, D., Manov, D., Kirilov, A.: Kim & ndash; a semantic platform for information extraction and retrieval. Nat. Lang. Eng. 10(3-4), 375–392 (2004)
22. Salton, G., Buckley, C.: Term-weighting approaches in automatic text retrieval. Inf. Process. Manage. 24(5), 513–523 (1988)
23. Salton, G., Fox, E.A., Wu, H.: Extended boolean information retrieval. Commun. ACM 26(11), 1022–1036 (1983)
24. Sheth, A., Bertram, C., Avant, D., Hammond, B., Kochut, K., Warke, Y.: Managing semantic content for the web. IEEE Internet Computing 6(4), 80–87 (2002)
25. Singhal, A., Buckley, C., Mitra, M.: Pivoted document length normalization. In: SIGIR 1996, pp. 21–29. ACM, New York (1996)
26. Stojanovic, N., Studer, R., Stojanovic, L.: An approach for the ranking of query results in the semantic web, pp. 500–516 (2003)
27. Wu, G., Li, J.: Swrank: An approach for ranking semantic web reversely and consistently. In: SKG 2007, pp. 116–121. IEEE Computer Society Press, Los Alamitos (2007)
28. Zhou, D., Zhu, S., Yu, K., Song, X., Tseng, B.L., Zha, H., Giles, L.C.: Learning multiple graphs for document recommendations. In: WWW 2008, pp. 141–150. ACM, New York (2008)

A Derivative of Loss Function

$$\frac{\partial Loss}{\partial X_W} = 2(X_P X_W^T - E_{\mathtt{p}\in\mathtt{W}})X_W + 2L(E_{\mathtt{W2W}})X_W + 2L(E_{\mathtt{p}\in\mathtt{W}}^T E_{\mathtt{p2W}})X_W + 2X_W(X_W^T X_W)^{-1}$$
$$\frac{\partial Loss}{\partial X_P} = 2(X_W X_P^T - E_{\mathtt{p}\in\mathtt{W}}^T)X_P + 2L(E_{\mathtt{p2p}})X_P + 2X_P(X_P^T X_P)^{-1}$$

Towards Agile Ontology Maintenance

Markus Luczak-Rösch

Freie Universität Berlin, Institute of Computer Science,
Corporate Semantic Web Workgroup, Berlin D-14195, Germany
markus.luczak-roesch@fu-berlin.de

Abstract. Ontologies are an appropriate means to represent knowledge on the Web. Research on ontology engineering reached practices for an integrative lifecycle support. However, a broader success of ontologies in Web-based information systems remains unreached while the more lightweight semantic approaches are rather successful. We assume, paired with the emerging trend of services and microservices on the Web, new dynamic scenarios gain momentum in which a shared knowledge base is made available to several dynamically changing services with disparate requirements. Our work envisions a step towards such a dynamic scenario in which an ontology adapts to the requirements of the accessing services and applications as well as the user's needs in an agile way and reduces the experts' involvement in ontology maintenance processes.

1 Introduction

Ontologies are an appropriate means to represent knowledge on the Web. Research on ontology engineering methodologies has come from describing the scratch development of ontologies and reached practices for an integrative lifecycle support. The ontology engineering discipline has changed from an individual art towards a collaborative and distributed process with disparate skilled users develop consensual models and distributed networks of ontologies. However, a broader success of ontologies in Web-based information systems remains unreached. They gained momentum in some characteristic and closed domains, such as health care and life sciences. On the every-day Web the more lightweight semantic approaches are rather successful which are based upon small vocabularies, e.g. the emerging linked data initiative. But also this lightweight semantic cannot deploy its full potential. The Web 2.0 resulted huge so called data silos. By use of wrappers or crawlers huge RDF datasets are derived from the relational databases of such silos. Consolidating and integrating the whole data of a specific application-dependent purpose or a specific individual remains a cumbersome task. Not to mention the control of the unintegratedly evolving knowledge in the silos.

As a next logical step one should await that, against the trend of the data silos, the user holds and controls her data on her own. Paired with the emerging trend of services and microservices on the Web [3] this results in a dynamic scenario in which a shared knowledge base is made available to several dynamically

A. Bernstein et al. (Eds.): ISWC 2009, LNCS 5823, pp. 965–972, 2009.

changing services with disparate requirements. This work envisions a step towards such a dynamic scenario in which an ontology adapts to the requirements of the accessing services and applications in an agile way. Therefore I design an innovative approach for agile ontology maintenance.

This paper starts with a presentation of the motivation for and the concrete problem statement of our work in Section 2. From our best knowledge we derived related approaches in the field of the research topic which we introduce briefly in the same section. Section 3 outlines the artifacts which we will contribute as the results of this work. The followed research methodology as well as the aimed evaluation are part of Section 4 before this paper ends with a summary of the goals, initial results, and the work in the nearest future in Section 5.

2 Motivation, Problem Statement and Related Work

The general and personal motivation for this work consists of three core parts. The first part is based upon our studies of the existing ontology engineering methodologies. It represents the fundamental direction of this work. As a second part, we derive from personal interviews with small and mid-sized enterprise (SME) partners of the project Corporate Semantic Web, that they look for a lightweight and dynamic process for ontology maintenance which minimizes the need for ontology experts to be present. We explicitly focus this scenario, however, we respect that there are enterprise settings as well which need and deal with heavyweight ontology engineering processes. On the whole, our idea meets the gap, which we identified as the result of the study of ontology engineering approaches and which has been also identified by others, such as [11]. That means concretely that research regards human-centered feedback as elementary part of the ontology lifecycle and ontology maintenance is more or less treated as the loop back to the beginning of the development process. Thus, ontology maintenance results as the specific direction of this work. The third part of the motivation is our personal vision of the next logical step of the Web from a social Web 2.0/3.0 towards a Web of services and alternative access devices. That means, that the next generation of the Web will be less driven by direct human access to contents and services by use of conventional client tools (e.g. Web browsers) but more by mobile devices and services. As a result of that the concepts of human-centered ontology engineering, such as argumentation to concepts and relations to reach the ontology consensus will loose impact.

2.1 A Running Example

To clarify this motivation we will briefly come up with a simple running example for the problem which we want to solve. Consider a company's knowledge base which includes information about the employees. In the beginning only personal information have been collected conforming the friend of a friend (FOAF) vocabulary. One service uses the knowledge base which generates lists of employees with certain interests. Each time a new service is bound to the knowledge base, such as a service for displaying absent employees or information about the income (e.g. for the accounting), the maintainer of the knowledge base has to find out which facts, in the sense of the T-box of the ontology, are missing and how she can easily adopt the current T-box and possibly the A-box as well to these new application requirements. It is also possible that separate services require the same information represented in different vocabularies (e.g. foaf:name vs. myvocabulary:name) which yields the conflict for the maintainer whether to replace the present representation or to model a mapping between both. The decision for either the first or the latter depends on several criteria, such as the computability of the ontology for complex reasoning or obsolete and unused information.

This yields the central research questions of our work:

1. *How does a methodology for ontology maintenance in an agile environment look like?* We search for a process which puts less emphasize on the initial development of an ontology but more on the ontology usage and evolution. That covers technical aspects of the modeling as well as aspects of the release management.
2. *Can we reduce the necessary influence of human experts in the ontology maintenance process by tracking feedback about ontology usage?* In this case our work searches for a formal model that allows the analysis of ontology usage for ontology evolution purposes.

2.2 Related Work

The methodologies and architectures for ontology engineering purposes mostly differ in details regarding to the composition of ontology engineering and application development, the range of users interacting in ontology engineering tasks, and the degree of lifecycle support. We studied the broad range of engineering methodologies, such as [6,15,7,9] beside others. While theses methodologies are from the perspective of the ontology engineers our approach changes this to the perspective of the ontology adopter which is a person with much less expertise in knowledge modeling and ontologies. We also border our approach from the recent one called *RapidOWL* [1], since RapidOWL introduces an idea of agile knowledge engineering. In contrast to us, Auer proposes a paradigm-based approach without any phase model and which is application-independent. The actually running NeOn project contributes the NeOn methodology for ontology engineering and the NeOn architecture for lifecycle support[16]. The NeOn methodology is well researched and adequate for projects with long iterations which are less dynamic than those which we address. The NeOn architecture proposes a feedback loop as part of the ontology lifecycle and adopts principles from service oriented architectures, thus it is an interesting base for our maintenance management framework and possibly reusable. However, it differs in detail since the feedback which it tracks is intended as explicit human feedback and an observation of informal background domain knowledge [18] while we want to adopt to the requirements of dynamically changing applications which use the knowledge base.

Ontology evolution has been researched under two scopes, that are (1) ontology versioning and change management as in [8,17] and (2) identification of informational extension, reduction, or reorganization as in [4,14,18]. Mostly, the representatives of both scopes end up in an expert-oriented perspective with focus to the initial ontology as the input and an evolved ontology as the output of the evolution process. The deployment of the new ontology including side effects to systems which apply it, such as updates of distributed stored instances because of schema changes, is not explicitly treated and the communication of the impact of ontology change from the ontology engineer to the ontology user are out of scope as well. We are also aware of the usage-oriented approaches presented in [12] and [13]. Compared to these we go one step further, since we focus both, user-oriented as well as application-oriented needs to an ontology.

As it is more or less usual for the ontology engineering discipline the whole field of software engineering methodologies is relevant work from which well-researched principles may be adopted. In this special case this is focused to maintenance management including bug and feature tracking [5] for software engineering processes. For sure, the field of database maintenance is related work, too. Here we focus on research on the incorporation of feedback into schema evolution and multi-tenant databases [2].

3 Contributions

We plan a multi layered result of this research activity. Altogether, from the bottom of theories to the top of method and tool support, we are working on (1) an ontology maintenance management methodology and (2) an ontology feedback tracking mechanism which both will be part of (3) an integrative ontology maintenance management framework.

COLM – An Agile Ontology Maintenance Methodology. The Corporate Ontology Lifecycle Methodology (COLM) reflects the agility of knowledge engineering processes and brings application dependency through the concrete definition of the application environment. We define it as an agile ontology maintenance methodology since it is focused on continuously evolving ontologies in an application-dependent context. To clarify which process steps are more expert-oriented and thus need higher human involvement and those which need less, COLM consists of two different cycles, namely the engineering cycle (high involvement) and the usage cycle (less involvement). The overall goal is to use an intuitive reporting of tracked usage information which indicates the necessity of change.

As depicted in Figure 1, the process starts at the *selection / development / integration* phase. The result of this phase is an ontology, which is *validated* within an application-dependent context. If it is approved that the ontology suites the requirements it is *deployed* to be in use and it is *populated.* Throughout the

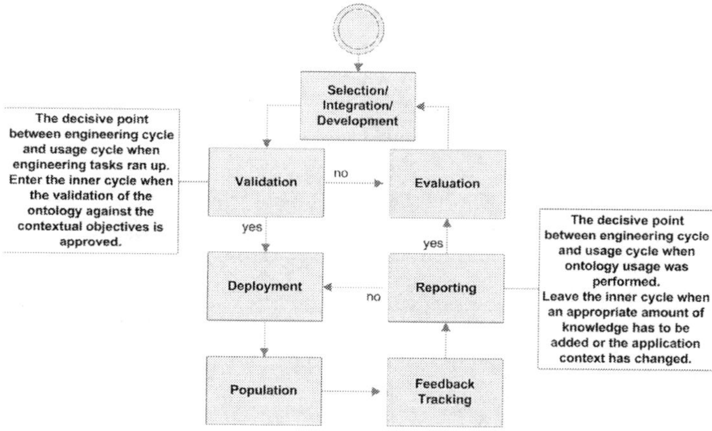

Fig. 1. The Corporate Ontology Lifecycle Methodology COLM

whole *feedback tracking* phase, formal statements about users' feedback and behavior are recorded and finally a *reporting* of this information is performed. The usage cycle is left if any necessary change has been detected and the knowledge engineers *evaluate* the weaknesses of the current ontology.

Ontology Change Recommendation by Feedback Tracking. Inspired by the approaches of argumentation-based ontology engineering and motivated by our viewpoint, that the human feedback within ontology engineering processes has to be reduced our integrative feedback tracking respects implicit feedback hidden in behaviors of users and applications. We winnow two categories of feedback. First, we regard the application-oriented feedback as the feedback which helps to fulfill the dynamically changing application requirements. Second, we regard the user-oriented feedback which helps to follow the evolution of dynamically changing and informal needs of the ontology user. In detail the feedback mechanisms which we design are *query observation* for the application-oriented feedback and *annotation behavior observation* for the user-oriented feedback.

The overall goal is to keep track of relevant information which help to create an ontology which properly describes a domain of interest that easy and rather small for better computation, so that the requirements of all accessing applications and the user's information needs are fulfilled.

Putting the Things Together. As we described it before we propose an evolution of ontologies by respecting their application-dependent context. Integrating new knowledge and evolving the existing by tracking the usage and distributing coexisting ontology versions to applications by intelligent version control are its central concepts. Our first draft proposal of this architecture is shown in Figure 2.

This depiction describes the various systems which access an ontology, respectively an ontology repository, for certain tasks, e.g. the ontology editors for

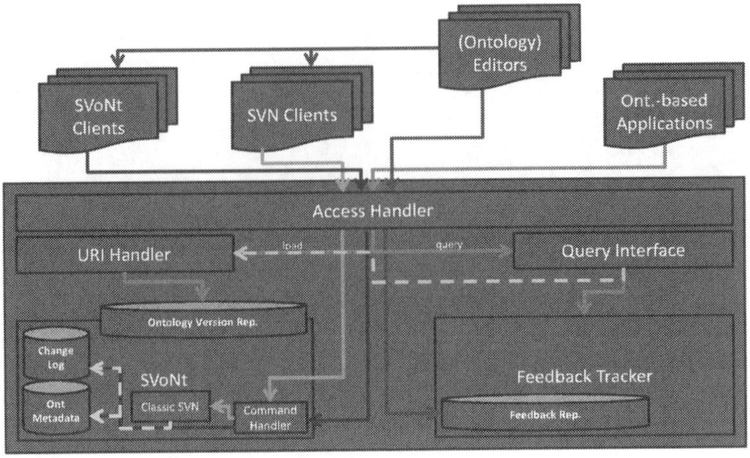

Fig. 2. High-level architecture of a proposed ontology maintenance management framework

manipulating an ontology or ontology-based applications for querying them. A central component is the feedback tracker which observes actions performed on the ontologies and supports the detection of new knowledge or potentially weak parts of the model.

The underlying technical essential of evolution in an agile environment is a flexible version management. At the moment we plan to integrate SVN for ontologies (SVoNt) into our maintenance management framework, which sets up on top of the well-known version control system SVN. To facilitate ontology versioning the text-based approach of SVN has to be extended to act on semantic structures of OWL ontologies. Internally, we add two major components which facilitate this additional functionality, namely consistency checks and generation of the ontology diff. From the set of differences we calculate a set of atomic change operations and store them into a log. In extend of SVoNt we also work on a process for an ontology release management which allows push and pull scenarios for the deployment of coexisting ontology versions. The model will provide a mechanism for the asynchronous evolution of a T-box and various central as well as distributed A-boxes which conform this schema.

4 Research Methodology and Aimed Evaluation

As we mentioned it in Section 2 the initial point for our work is based upon personal studies of the foundations of ontology engineering processes as well as a set of face to face interviews with representatives of small and mid-sized enterprises in Germany about the applicability of semantic technologies and ontologies in their corporate contexts.

Coming from theses requirements we aim at a multi layered contribution of this work, which touches fundamental research, instrumentalist research, and applied research. Thus it is necessary that we evaluate it multi-perspectively. We will choose methodologies which reflect the domain and the amount of the approach, range the results with reference to related work, and at least proof the applicability of the concepts in practice.

By today we see the chance to run and test COLM within two use cases. First, at the Ontonym GmbH which is a company that provides semantic search services supported by self-constructed and self-maintained ontologies in the background. Second, in cooperation with the DBpedia project, which supports a giant linked dataset based on s self-constructed and self-maintained ontology and crawled information from Wikipedia. Especially the latter use case seems to be valuable since our study will run in parallel to the development of a community-driven approach for an evolution of the DBpedia ontology. To set the approach in relation to other approaches and to measure its quality we will apply the ONTOCOM cost-estimation model for ontology engineering on it as it has been done for other methodologies, e.g. DILIGENT [10], as well.

5 Initial Results, Outlook and Conclusions

In this paper we presented our work towards an integrative ontology maintenance management for agile application-dependent scenarios. Based on the COLM methodology the SVoNt system for SVN-based version control of OWL ontologies and a feedback tracking mechanism will be combined into a framework application that supports the whole ontology lifecycle from the viewpoint of the ontology user.

The COLM methodology has been developed, published, and iteratively refined based on several valuable comments by experts. It is in a mature state right now. Based on the theories of COLM we proposed a high level architecture of an ontology maintenance management framework which will integrate the SVoNt server and the feedback tracking mechanism. The latter two components are in a preliminary design state right now. We envision to finish the fundamental work on these partial components until the end of 2009. The evaluation will be performed after the prototype implementation of our proposed framework is finished. We plan to finish this work until the beginning of 2011.

Altogether, our approach towards agile ontology maintenance should promote ontology engineering from the user's and the usage's perspective. In combination with several research trends, such as end-user generated microservices and the service Web 3.0, this is a promising step ahead to bring ontologies to broader success.

References

1. Auer, S., Herre, H.: RapidOWL — An Agile Knowledge Engineering Methodology. In: Virbitskaite, I., Voronkov, A. (eds.) PSI 2006. LNCS, vol. 4378, pp. 424–430. Springer, Heidelberg (2007)

2. Aulbach, S., Jacobs, D., Kemper, A., Seibold, M.: A comparison of flexible schemas for software as a service. In: SIGMOD 2009: Proceedings of the 35th SIGMOD international conference on Management of data, pp. 881–888. ACM, New York (2009)
3. Davies, M.: Towards a Semantic Infrastructure for User Generated Mobile Services. In: ASWC 2007 and ISWC 2007. LNCS, vol. 4825, pp. 924–928. Springer, Heidelberg (2009)
4. Djedidi, R., Aufaure, M.-A.: Ontological knowledge maintenance methodology. In: Lovrek, I., Howlett, R.J., Jain, L.C. (eds.) KES 2008, Part I. LNCS (LNAI), vol. 5177, pp. 557–564. Springer, Heidelberg (2008)
5. Erdil, K., Finn, E., Keating, K., Meattle, J., Park, S., Yoon, D.: Software Maintenance As Part of the Software Life Cycle. Comp180: Software Engineering Project, December 16 (2003)
6. Fernndez-Lpez, M., Gmez-Prez, A., Juristo, N.: METHONTOLOGY: From Ontological Art Towards Ontological Engineering. In: AAAI 1997 Spring Symposium on Ontological Engineering. AAAI Press, Menlo Park (1997)
7. Kotis, K., Vouros, A.: Human-centered ontology engineering: The HCOME methodology. Knowl. Inf. Syst. 10(1), 109–131 (2006)
8. Noy, N.F., Kunnatur, S., Klein, M., Musen, M.: Tracking Changes During Ontology Evolution. In: Proceedings of the Third International Semantic Web Conference, pp. 259–273. Springer, Berlin (2004)
9. Pinto, H.S., Tempich, C., Staab, S., Sure, Y.: Distributed Engineering of Ontologies (DILIGENT). Semantic Web and Peer-to-Peer. Springer, Heidelberg (2005)
10. Simperl, E., Tempich, C.: How Much Does It Cost? Applying ONTOCOM to DILIGENT. Technical Report, FU Berlin (2005)
11. Simperl, E., Tempich, C.: Ontology Engineering: A Reality Check. In: ASWC 2007 and ISWC 2007. LNCS, vol. 4825, pp. 836–854. Springer, Heidelberg (2006)
12. Stojanovic, N., Stojanovic, L.: Usage-Oriented Evolution of Ontology-Based Knowledge Management Systems. In: Meersman, R., Tari, Z., et al. (eds.) CoopIS 2002, DOA 2002, and ODBASE 2002. LNCS, vol. 2519. Springer, Heidelberg (2002)
13. Stojanovic, L., Stojanovic, N., Gonzalez, J., Studer, R.: OntoManager - A System for the Usage-Based Ontology Management. In: CoopIS/DOA/ODBASE, vol. 2888. Springer, Heidelberg (2003)
14. Stojanovic, L.: Methods and Tools for Ontology Evolution. PhD Thesis, University of Karlsruhe, Germany (2004)
15. Sure, Y., Studer, R.: On-To-Knowledge Methodology — Expanded Version. In: On-To-Knowledge deliverable, vol. 17, Institute AIFB, University of Karlsruhe (2002)
16. Tran, T., Haase, P., Lewen, H., Muñoz-García, Ó., Gómez-Pérez, A., Studer, R.: Lifecycle-Support in Architectures for Ontology-Based Information Systems. In: Aberer, K., Choi, K.-S., Noy, N., Allemang, D., Lee, K.-I., Nixon, L.J.B., Golbeck, J., Mika, P., Maynard, D., Mizoguchi, R., Schreiber, G., Cudré-Mauroux, P. (eds.) ASWC 2007 and ISWC 2007. LNCS, vol. 4825, pp. 508–522. Springer, Heidelberg (2007)
17. Völkel, M., Groza, T.: SemVersion: RDF-based Ontology Versioning System. In: Proceedings of the IADIS International Conference WWW / Internet 2006 (ICWI 2006), Murcia, Spain (2006)
18. Zablith, F.: Dynamic Ontology Evolution. In: International Semantic Web Conference (ISWC) Doctoral Consortium, Karlsruhe, Germany

Ontologies for User Interface Integration

Heiko Paulheim

SAP Research
heiko.paulheim@sap.com

Abstract. Application integration can be carried out on three different levels: the data source level, the business logic level, and the user interface level. With ontologies-based integration on the data source level dating back to the 1990s and semantic web services for integrating on the business logic level coming of age, it is time for the next logical step: employing ontologies for integration on the user interface level. Such an approach supports both the developer (in terms of reduced development times) and the user (in terms of better usability) of integrated applications. In this paper, we introduce a framework employing ontologies for integrating applications on the user interface level.

1 Introduction

Applications are often described in three layers: data, business logic, and user interface. Consequently, application integration can be performed on each of those three levels [1], as depicted in Fig. 1:

- Integrating the data sources, and developing common business logic and user interface layers above the integration layer,
- integrating the business logic, and developing a common user interface above the integration layer, and
- integrating the user interfaces.

Integrating applications on the user interface level means reusing existing user interfaces or parts thereof and coupling them in a way that *a user can interact with those interfaces as if they were a single application.* Such an integration may include that the applications share a common toolbar or menu, that one application reacts to user actions performed with another one, e.g. related objects are highlighted in other applications when selected in one application, objects can be dragged and dropped from on application to the other, etc. There are some approaches such as plugin-based systems [2], portals [3], and mashups [4] that propose integration on the user interface level. However, all of those approaches are either very limited concerning cross-application interaction or require deep changes of the applications in order to facilitate such interactions [1].

With semantic database integration [5] as well as ontology-based agents [6] and semantic web services [7,8], there have been considerable efforts to using ontologies in the integration on the database and business logic layer. However, little research has been conducted on ontology-based integration on the user

A. Bernstein et al. (Eds.): ISWC 2009, LNCS 5823, pp. 973–981, 2009.

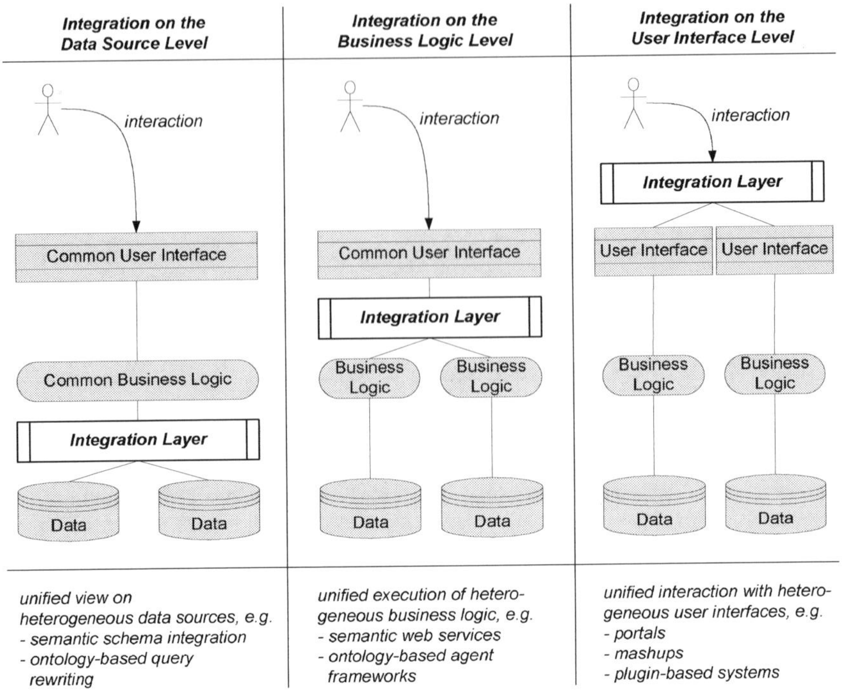

Fig. 1. Three layers of integration; based on [1]

interface level so far. In contrast to integration on the lower layers, user interface level integration has two significant advantages:

- The development of the user interface consumes up to 50% of the total efforts in developing an application [9]. Therefore, the benefit from reusing existing user interface components is significant.
- Users interacting with applications integrated on the user interface level will experience a decreased learning effort if already familiar with the applications' original interfaces, compared to interacting with a newly developed common user interface.

Yu et al. argue that user interface integration requires a description of the interfaces to be integrated that is *formal*, *human readable*, *modular*, and *simple* [10]. Ontologies perfectly meet the first two criteria and also provide the possibility for modularization [11]. As simplicity is a rather subjective criterion, and the description language must be flexible enough to cover all possible cases of integration, hence must not be too simple, we claim that ontologies are a suitable approach.

Since, as discussed above, *interaction* plays a crucial role in integrated user interfaces as well as it imposes problems in current approaches to user interface integration, the approach discussed in this paper aims at describing such interactions with ontologies. We introduce a prototype framework that allows run-time integration of user interfaces based on ontological descriptions of the interactions they support.

2 State of the Art

There are some approaches that employ ontologies for user interface integration in portals and mashups. The approach described in [12] uses semantic web services, i.e., web services described by means of ontologies. It rather focuses on communication between a portlet and its backend system than on inter-portlet communication and on user interaction. The work described in [13] uses ontologies to annotate the contents delivered by portlets. That approach is rather data-centric and has little focus on interaction. The work described in [14] shows how ontologies can help building mashup applications to integrate contents from diverse data sources in one mashup.

The work described in [15] and [16] shows how ontologies can be used to formalize user interfaces and to generate user interfaces with a model driven approach. The approach suggested in [17] also formalizes user interfaces with the help of ontologies, but with the aim of making user interfaces accessible to people with disabilities. Other approaches, such as [18] and [19], annotate software components in general (not necessarily user interface components) with ontologies to support the developer in searching and choosing appropriate components. The work described in [20] and [21] propose ontologies for describing different types of user interfaces on a rather general level, such as characterizing different input and output devices. Such formalizations are helpful, but so far, they have not been applied to *integrating* different user interfaces.

A research direction which comes close to ontologies-based user interface integration is the Semantic Desktop [22]. Here, data encapsulated in different applications is made accessible via a central query interface. In some semantic desktop systems, existing applications may be integrated as plugins [23]. The main focus of this direction, however, is to provide an integrated access to data in different applications rather than on cross-application interaction; it can therefore be regarded as an approach to data integration rather than to user interface integration.

3 Roadmap

3.1 Prototype

So far, a first prototype has been developed that shows how user interfaces can be integrated by using ontologies [24]. Three types of ontologies are used (see Fig. 2):

- An *ontology of the user interfaces and interactions domain*, which defines basic categories for describing applications. This ontology is part of the framework discussed in this paper.

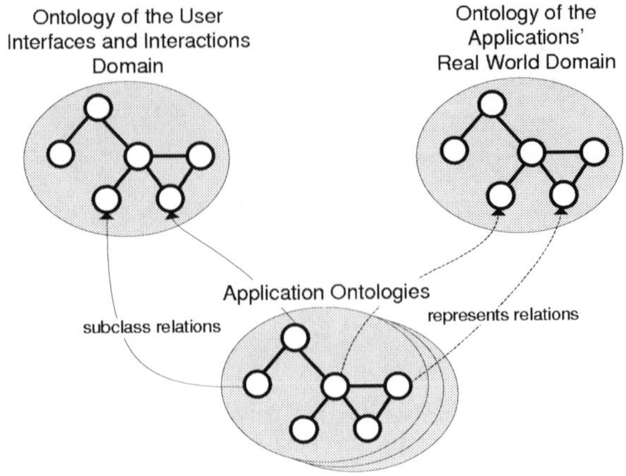

Fig. 2. Using two domain ontologies and several application ontologies for integration on the user interface level

- An *ontology of the application's real world domain*, which defines the categories of real world objects of the domain that the integrated application is built for (such as banking, travel, etc.). The information objects processed by the application *represent* those real worlds objects. When integrating applications from a given domain, an appropriate domain ontology has to be chosen or developed. A set of different real world domain ontologies may be used in case of modular domain ontologies or when developing cross-domain applications.
- One or more *application ontologies*, which use the user interfaces and interactions ontology's basic concepts to describe the applications to be integrated, and the interactions that are possible with them. The application ontologies may refer to the real world domain ontology for describing the types of objects that may be processed by the integrated applications. During the integration process, one application ontology per integrated application has to be developed.

This categorization resembles Guarino's classification [25] (without the top level layer, which may also be present in our framework, but its presence is not essential) – here, two kinds of domain ontologies are used. While the ontology of the user interfaces and interactions domain is a part of the integration framework, the real world domain ontology and application ontologies are dynamically added for each integrated application. It is particularly noteworthy that there is no direct connection from the user interfaces and interactions domain ontology to the real world domain ontology. Thus, the framework is domain independent.

Fig. 3 shows an overview of the framework prototype, which is based on Java and the OntoBroker reasoner [26]. Integrated applications consist of a class model for representing data, a user interface, and a business logic (the classical model view controller [27] components), and are described by an application ontology.

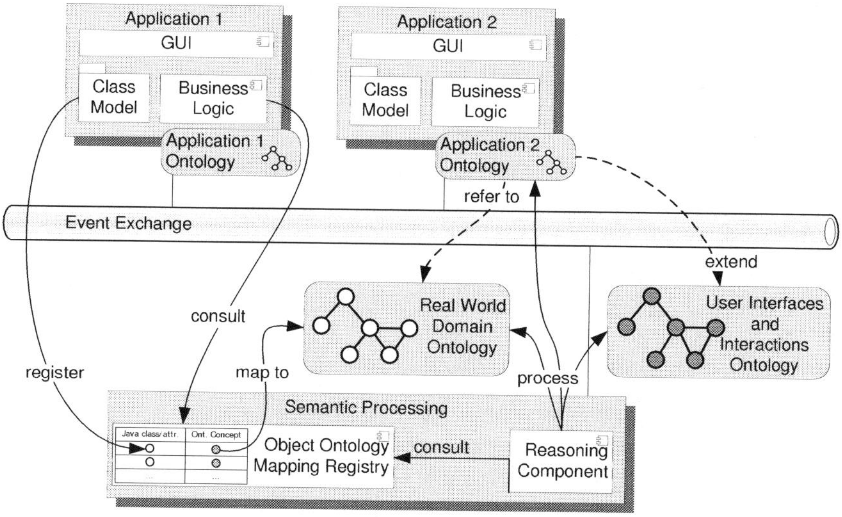

Fig. 3. Overview on the prototype's framework architecture. Applications are described by ontologies. A reasoning component evaluates those ontologies to facilitate integration at run time.

Applications communicate via events which are annotated using the respective application ontology. One key design decision is that no application sends events directly to any other application. Instead, events are processed by a reasoning component. For example, one application sends an event that a certain object is selected. The reasoner reads the event, queries the application ontologies to determine those applications which declare to react to that sort of event, and notifies the respective applications. Thus, no application has to directly react to other applications' events and only needs to process the event types defined in its own application ontology.

To allow mediation between different data models, an *object ontology mapping registry* is introduced. In this registry, each application stores annotations of its data model. Classes as well as properties may be annotated with concepts from the domain ontology. When the reasoner receives an object from an application or vice versa, the receiver consults the registry to analyse that object and convert it into a representation which can be processed by the reasoner.

3.2 Further Research Plan

While the prototype shows that the approach is valid and feasible, there are quite a few open research questions. We have shown that simple interactions (such as highlighting objects selected in different applications) are possible with our framework. More complex interactions will require extensions and refactoring of both the application ontology and the prototype implementation.

So far, we have only considered typical single-user WIMP (Windows, Icons, Mouse, Pointing) interfaces when modelling the application ontology. A more sophisticated ontology describing interactions would be flexible enough to allow different input and output devices (such as speech interfaces, gestures, and so on) [28], as well as multi-user interaction. Coupling the application ontology with a device ontology, such as the FIPA device ontology [29], could lead to a more universal framework.

Complex interactions may have conditions under which they may be performed, e.g. the visibility of a component or the presence of an application which is able to perform a certain task. To evaluate those conditions, an application's internal state has to be exposed to a certain extent. Thus, the relevant state information has to be identified and modelled in the user interfaces ontology. Furthermore, the state information has to be made known to the reasoner, either by the reasoner dynamically querying the applications or by the applications sending updates to the reasoner. A special case of preconditions are the users' rights to perform a certain interaction. Thus, adding an ontology of users' rights, such as the one proposed in [30], would be a feasible approach to account for that requirement.

The design decision that each application can use their own data model eases the reuse of existing components, but it comes with certain challenges in data integration. As there is a large variety of heterogeneities that may occur here [31], some methods to cope with those heterogeneities is needed. Such methods will probably impose certain restrictions on the data model used (such as a 1:1 mapping between elementary data types in the data model and data attributes in the domain ontology). We aim at finding a minimal set of such restrictions in order to allow a maximum degree of freedom in the applications' data models.

The application ontologies describing the interactions possible with applications may not only be used for integration. Another way of utilizing those ontologies is the provision of user assistance, such as automatic generation of help texts, or highlighting possible drop locations in different applications when dragging an object. The latter has already been successfully demonstrated in our prototype.

Finally, the approach requires validation beyond having a running prototype. We plan to conduct case studies where different example applications are to be integrated in a way providing a given set of interactions. Here, development efforts can be measured, e.g. in lines of code, and expert interviews with developers can reveal additional insights into the feasibility of the approach. In addition, studies with end users may be conducted to demonstrate the advantage of integrated user interfaces over non-integrated side-by-side use of applications.

4 Conclusion

In this paper, we have presented the idea of using ontologies for integrating applications on the user interface level. In our framework, applications are described by application ontologies, making use of two or more shared domain ontologies.

A first prototype shows that the approach is feasible. It has been successfully used in the SoKNOS project [32], where an integrated emergency management software has been built, consisting of twelve integrated single applications, serving purposes such as planning measures with resources, handling messages, or displaying relevant mission data on charts and geographic maps.

This prototype as well as the underlying ontology and algorithms are going to be subsequently extended. The aim is to enhance the framework in a way that it covers the most common interaction patterns between integrated applications, and that it can be enhanced in cases where more unusual interactions are to be implemented.

In summary, we believe that such a framework for integrating applications on the user interface level is a useful complement to existing integration efforts on the data and business logic level.

Acknowledgements

The work presented in this paper has been partly funded by the German Federal Ministry of Education and Research under grant no. 01ISO7009.

References

1. Daniel, F., Yu, J., Benatallah, B., Casati, F., Matera, M., Saint-Paul, R.: Understanding UI Integration: A Survey of Problems, Technologies, and Opportunities. IEEE Internet Computing 11(3), 59–66 (2007)
2. Birsan, D.: On plug-ins and extensible architectures. ACM Queue 3(2), 40–46 (2005)
3. Wege, C.: Portal Server Technology. IEEE Internet Computing 6(3), 73–77 (2002)
4. Abiteboul, S., Greenshpan, O., Milo, T.: Modeling the Mashup Space. In: WIDM 2008: Proceeding of the 10th ACM workshop on Web information and data management, pp. 87–94. ACM, New York (2008)
5. Doan, A., Halevy, A.Y.: Semantic Integration Research in the Database Community: A Brief Survey. AI Magazine 26(1), 83–94 (2005)
6. Sycara, K.P., Paolucci, M.: 17. International Handbooks on Information Systems. In: Ontologies in Agent Architectures, pp. 343–364. Springer, Heidelberg (2004)
7. Martin, D., Burstein, M., Hobbs, J., Lassila, O., McDermott, D., McIlraith, S., Narayanan, S., Paolucci, M., Parsia, B., Payne, T., Sirin, E., Srinivasan, N., Sycara, K.: OWL-S: Semantic Markup for Web Services (November 2004), `http://www.w3.org/Submission/OWL-S/`
8. Lausen, H., Polleres, A., Roman, D., de Bruijn, J., Bussler, C., Domingue, J., Fensel, D., Hepp, M., Keller, U., Kifer, M., König-Ries, B., Kopecky, J., Lara, R., Lausen, H., Oren, E., Polleres, A., Roman, D., Scicluna, J., Stollberg, M.: Web Service Modeling Ontology, WSMO (2005), `http://www.w3.org/Submission/WSMO/`
9. Myers, B.A., Rosson, M.B.: Survey on user interface programming. In: CHI 1992: Proceedings of the SIGCHI conference on Human factors in computing systems, pp. 195–202. ACM, New York (1992)
10. Yu, J., Benatallah, B., Saint-Paul, R., Casati, F., Daniel, F., Matera, M.: A framework for rapid integration of presentation components. In: WWW 2007: Proceedings of the 16th international conference on World Wide Web, pp. 923–932. ACM, New York (2007)

11. Gruber, T.R.: Toward Principles for the Design of Ontologies Used for Knowledge Sharing, Duluth, MN, USA, vol. 43, pp. 907–928. Academic Press, Inc., New York (1995)
12. Dettborn, T., König-Ries, B., Welsch, M.: Using Semantics in Portal Development. In: Proceedings of the 4th International Workshop on Semantic Web Enabled Software Engineering (2008)
13. Díaz, O., Iturrioz, J., Irastorza, A.: Improving portlet interoperability through deep annotation. In: WWW 2005: Proceedings of the 14th international conference on World Wide Web, pp. 372–381. ACM, New York (2005)
14. Ankolekar, A., Krötzsch, M., Tran, T., Vrandecic, D.: The Two Cultures: Mashing Up Web 2.0 and the Semantic Web. In: WWW 2007: Proceedings of the 16th International Conference on World Wide Web, pp. 825–834. ACM, New York (2007)
15. Sergevich, K.A., Viktorovna, G.V.: From an Ontology-Oriented Approach Conception to User Interface Development. International Journal Information Theories and Applications 10(1), 89–98 (2003)
16. Liu, B., Chen, H., He, W.: Deriving User Interface from Ontologies: A Model-Based Approach. In: ICTAI 2005: Proceedings of the 17th IEEE International Conference on Tools with Artificial Intelligence, pp. 254–259. IEEE Computer Society, Los Alamitos (2005)
17. W3C: WAI-ARIA Overview (2009), `http://www.w3.org/WAI/intro/aria`
18. Graubmann, P., Roshchin, M.: Semantic Annotation of Software Components. In: EUROMICRO 2006: Proceedings of the 32nd EUROMICRO Conference on Software Engineering and Advanced Applications, Washington, DC, USA, pp. 46–53. IEEE Computer Society, Los Alamitos (2006)
19. Happel, H.J., Korthaus, A., Seedorf, S., Tomczyk, P.: KOntoR: An Ontology-enabled Approach to Software Reuse. In: Zhang, K., Spanoudakis, G., Visaggio, G. (eds.) Proceedings of the Eighteenth International Conference on Software Engineering & Knowledge Engineering (SEKE), pp. 349–354 (2006)
20. Coutaz, J., Lachenal, C., Dupuy-Chessa, S.: Ontology for Multi-surface Interaction. In: Proceedings of IFIP INTERACT03: Human-Computer Interaction, IFIP Technical Committee No 13 on Human-Computer Interaction, pp. 447–454 (2003)
21. Eick, S.G., Wills, G.J.: High Interaction Graphics. European Journal of Operational Research 84, 445–459 (1995)
22. Sauermann, L., Bernardi, A., Dengel, A.: Overview and Outlook on the Semantic Desktop. In: Gil, Y., Motta, E., Benjamins, V.R., Musen, M.A. (eds.) ISWC 2005. LNCS, vol. 3729. Springer, Heidelberg (2005)
23. Cheyer, A., Park, J., Giuli, R.: Iris: Integrate. relate. infer. share. In: Workshop on the Semantic Desktop: Next Generation Personal Information Management and Collaboration Infrastructure (2005)
24. Paulheim, H.: Ontology-based Modularization of User Interfaces. In: Calvary, G., Graham, T.C.N., Gray, P. (eds.) Proceedings of The ACM SIGCHI Symposium on Engineering Interactive Computing Systems, pp. 23–28. ACM, New York (2009)
25. Guarino, N. (ed.): Formal Ontology and Information Systems. In: Guarino, N. (ed.) Proc. of the 1st Int'l Conf. on Formal Ontologies in Information Systems (FOIS 1998), Trento, Italy. IOS Press, Amsterdam (1998)
26. ontoPrise: OntoBroker Website (2009), `http://www.ontoprise.de/de/en/home/products/ontobroker.html`
27. Krasner, G.E., Pope, S.T.: A cookbook for using the model-view controller user interface paradigm in small-talk-80. Journal of Object Oriented Programming 1, 26–49 (1988)

28. van Dam, A.: Post-wimp user interfaces. Commun. ACM 40(2), 63–67 (1997)
29. Foundation for Intelligent Phyiscal Agents: FIPA Device Ontology Specification (December 2002), http://www.fipa.org/specs/fipa00091/index.html (accessed, 2008-01-22)
30. Kagal, L., Finin, T., Joshi, A.: A policy language for a pervasive computing environment. In: POLICY 2003: Proceedings of the IEEE 4th International Workshop on Policies for Distributed Systems and Networks (2003)
31. Klein, M.: Combining and relating ontologies: an analysis of problems and solutions. In: Gomez-Perez, A., Gruninger, M., Stuckenschmidt, H., Uschold, M. (eds.) Workshop on Ontologies and Information Sharing, IJCAI 2001, Seattle, USA (2001)
32. Doeweling, S., Probst, F., Ziegert, T., Manske, K.: SoKNOS - An Interactive Visual Emergency Management Framework. In: Amicis, R.D., Stojanovic, R., Conti, G. (eds.) GeoSpatial Visual Analytics. NATO Science for Peace and Security Series C: Environmental Security, pp. 251–262. Springer, Heidelberg (2009)

Semantic Usage Policies for Web Services*

Sebastian Speiser

Karlsruhe Institute of Technology (KIT), Karlsruhe, Germany
speiser@kit.edu

Abstract. Web Services provide standardized interfaces for accessing software systems and data sources over the Internet. Semantic descriptions of Web Services help to automate the discovery and invocation of new services and their integration into existing applications. However not all services are freely available for every purpose and not all data is in the public domain. Usage policies describe the terms and conditions under which services and data can be used. Current approaches to semantic Web Service description are mostly focused on functional properties and quality attributes, and do not cover usage policies. We plan to develop a formal language for usage policies with clearly defined semantics that relies on ontologies for representing domain specific terms. We will also extend service discovery and ranking algorithms to incorporate usage policies.

1 Introduction

The classical World Wide Web (WWW) makes services and data available through Web sites targeted at human users. The Web sites are interlinked but the integration of information and services from different sources is a manual effort. The Internet of Services as well as the Semantic Web both go one step further and aim at interoperable and machine-understandable descriptions and interfaces for services and data. It will therefore be easier to combine both types of resources to a network that fulfills a user need which is not satisfiable by any single component. Such networks can be created in various ways, e.g.

- Keyword searches are automatically translated into formal queries over multiple Web resources [1].
- Mashup editors which allow even unskilled users to combine Web services and data sources with minimal time effort [2].
- Creating a business process and orchestrate it with BPEL.

All approaches have in common that suitable services have to be selected that can be part of a network fulfilling the user need.

Users select services based on the net value they expect from the service consumption [3](p. 23ff). The net value is given by subtracting the costs from the

* This work was funded by the European project SOA4All and the Karlsruhe Service Research Institute (KSRI).

A. Bernstein et al. (Eds.): ISWC 2009, LNCS 5823, pp. 982–989, 2009.

experienced gross value. Both can only be determined after service consumption and therefore the selection is based on expected value and costs in which users incorporate their uncertainty about the result. In order to minimize the uncertainties and therefore increase the expected net value for users, two things have to be established: (i) services have to be described in such a way that users know which value and costs to expect, and (ii) these descriptions have to be trustworthy. This thesis deals with the first aspect.

The two most obvious factors influencing the value and cost of a service are its functionality that has to match the user's need and its price. Costs of a service are not restricted to money but also include other obligations and conditions the user has to fulfill, e.g. ensure that the output data is only stored on encrypted hard disks. Expected value is affected by uncertainties about the service experience which can be reduced by statements of the provider about the non-functional properties (NFPs) of the service which comprise the following areas:

- Quality of service (QoS) and data, e.g. availability of the service, are delivered stock prices real-time or delayed?
- Usage policy for input data. Data that is given by the user to the provider in order to consume a service can have restrictions on their usage, e.g. the user may specify that his e-mail address or credit card number are not allowed to be forwarded to other parties. This is also called privacy statement.
- Usage policy for output data, e.g. user is not allowed to display stock prices on public homepage, user has to attribute provider of weather data. We call this also data license.
- Usage policy of service, e.g. mobile data service must not be combined with VoIP service. We denote such policies also as service licenses. They also include customer obligations, e.g. pricing (every call costs 1 Euro, flat-fee for one month of 20 $).

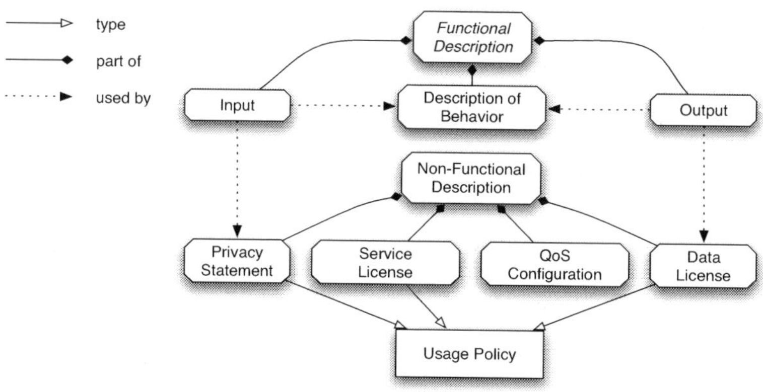

Fig. 1. Functional and Non-Functional Service Descriptions

In Figure 1 we show an overview of the service description, where we count input, output and behavioral description to the functional description and the above-mentioned aspects to the non-functional description.

As we consider a Web scenario with a large number of heterogeneous providers and an even larger number of services, we want to support the user with tools for service selection. Therefore we need formal descriptions of services. This is tackled by various semantic Web service approaches (e.g. WSMO[1], SAWSDL[2], OWL-S[3]). All approaches deal with functionality but non-functional properties are mostly restricted to quality of service and price.

This means that most usage policies are in natural language which requires a significant human effort to evaluate in terms of compatibility with user preferences. Especially in the mashup and keyword search scenario with their high user convenience this effort seems inappropriate. This makes it likely that natural language licenses are just ignored and possibly violated. Formal usage policies representing the licenses and user requirements would allow them to be integrated into service selection tools. In this way adhering to a license can be made easier for the user. As we mentioned above we consider a Web scenario with numerous, heterogeneous providers and therefore we need a license language with clearly defined formal semantics. Our goal is to create such a language and extend service selection and ranking algorithms to include usage policies.

In order to make the language applicable to a wide range of scenarios and different domains of Web resources we want to restrict it to a small core that has a well-defined meaning for permissions, prohibitions, obligations and constraints. The core elements can then be linked to domain-specific ontologies, e.g. for payments or subject and resource identification.

The rest of this paper is structured as follows. Section 2 discusses related work. In Section 3 we outline our planned solution for semantic usage policies and we describe in Section 4 how we want to achieve it. Finally we summarize in Section 5.

2 Related Work

O'Sullivan presents in his thesis a taxonomy of non-functional properties (NFPs) of services [4]. His results are based on the analysis of classic services and Web services from numerous domains. The taxonomy is very comprehensive and deals with various categories of NFPs, such as availability, price, discounts and quality. Usage policy related concepts that allow providers e.g. to forbid commercial usage of their service are not considered. The taxonomy itself is not concerned with the specification of service policies but delivers a vocabulary for common aspects of services.

The taxonomy of O'Sullivan was modeled as WSML ontologies and used to represent NFPs in WSMO [5]. Originally WSMO supported annotations like

[1] http://www.wsmo.org
[2] http://www.w3.org/TR/sawsdl
[3] http://www.w3.org/Submission/OWL-S

creator or title adapted from Dublin Core to describe non-functional properties of ontology elements. With the integration of the O'Sullivan NFP concepts and the extension of WSMO models to include rule-based definitions of NFPs for services, the approach is now able to describe expressive NFP offers in terms of the defined ontologies. Toma et al. present an algorithm that ranks WSMO services according to user specified preferences on NFPs [6]. The ranking is currently limited to NFPs that result in numerical values given the service specification and user provided context and input information.

Web service policy languages can be used to define quality guarantees and also user requirements. For example WS-Policy can specify with assertions which technologies users have to support, e.g. encryption[7]. Assertions are defined in different WS-* standards which cover technical domains but no usage rights and restrictions.

Formalized usage policies exist for digital objects such as e-books and music files. The RDF serialization of the Creative Commons licenses can be used for automatic determination of associated rights [8]. This is used by Google's advanced search[4], where users can specify their desired Creative Commons license and only documents that were annotated correspondingly by their owner are returned. However Creative Commons is mainly suitable for open contents, as the licenses are generally addressed to the public and do not specify rights for single users. Another point is that commercial usage can be allowed but no specific terms (e.g. price) can be expressed.

In the area of digital rights management (DRM), there exist several rights expression languages (REL), e.g. ODRL [9] and XrML [10]. These are XML based languages that allow more complex right specifications than Creative Commons. In theory existing RELs would be a good basis for service usage policies, however the mentioned RELs include language elements that go far beyond rights expression. They handle encryption, media encodings, user authentication and other domain-specific aspects. These are represented as syntactically defined XML elements and a meaning has to be given in the corresponding standard specifications (e.g. ODRL). Also they are lacking formal semantics which leads to ambiguities in scenarios with heterogeneous providers.

This problem is also recognized by Jamkhedkar et al. who present a DRM architecture that separates rights expression from other aspects [11]. They propose that there is a need for a core REL which is based on a mathematical foundation [12]. We share this thought and will further investigate if such a core REL could be reused for service usage policies.

Arnab and Hutchison present LiREL which is a formal REL [13]. They show what the difference to classical access control is and what the resulting requirements for an REL are. LiREL focuses on the expression of rights and requires an external standardized vocabulary for the definition of actions, constraints, etc. This is in contrast to our planned solution which relies on ontologies and ontology mappings rather than trying to build a standard vocabulary that covers all domains and has to be used by all providers and users.

---

[4] http://www.google.com/advanced_search

Gangadharan et al. developed an ODRL profile for service licenses (ODRL-S) [14]. As a service license they define regulations concerning the use of a service. This is a similar concept to our proposed usage policies, however their approach defines a static vocabulary of actions, including composition or attribution. This regulates general usage of the service but does not incorporate context information, e.g. this means that generally composition can be forbidden but there is no way to specify that composition is allowed with specific other services or providers. Also the policy does not refer to the output data of services. The lack of clear semantics is inherited from ODRL. The interpretation of licenses relies on information such as: composition is a special case of derivation. This hierarchy is however not modeled formally, so that it has to be implemented manually in every program that reasons about service licenses.

In general it can be said that existing rights expression approaches mostly rely on XML for syntax and specify the meaning of elements in human-readable documents. This makes it difficult to automatically match different terminologies that are likely to be used in the Web with its heterogeneous provider structure.

The thesis of Lamparter presents a policy-based approach for service offers, requests and contracts for Web service markets using ontologies to match different provider vocabularies [15]. The approach models functions ranging over non-functional properties of Web services. The functions are used to compactly represent policies that map configurations of NFPs to numerical values. Users specify their valuation of configurations and providers their prices. The policies are then used to find the optimal configuration with respect to the difference of user valuation and price. Lamparter identifies the allowed use of information returned by services as a relevant NFP and exemplifies it by showing how to restrict the disclosure of data. A general model for usage policies is however not given.

Kagal et al. developed the semantic policy language Rei[16]. It models the basic deontic concepts of permissions, prohibitions, obligations and dispensations. Concrete concepts have to be modeled in external domain-specific ontologies. Rei uses OWL as a serialization format but represents rules that have to be interpreted by an external rule engine based on logic programming. The usage of Rei was mainly for access control and we will investigate if it can also applied to usage policies.

The traditional access control model is the access control matrix, which specifies whether a specific action on a specific object by a specific subject is allowed or not [17]. Most access control approaches like role based access control do not explicitly state the matrix but can be reduced to it. Parker et al. break with this model and present the UCON_{ABC} approach for usage control [18] which not only regulates the access to resources but also their further usage [19]. Their model includes obligations on the user side and mutable subject and object attributes that have influence on user rights after the initial access to a resource was granted, e.g. an object can only be printed twice. Parker et al. observe that DRM, privacy policies and access control are developed independently despite their similarity and propose the UCON_{ABC} model as a theoretical foundation for the mentioned wide range of application areas [18]. The model has received broad attention and will be regarded in our formalization of usage policies.

3 Planned Contribution

In this section we list the components that are necessary for a solution that integrates formal usage policies in (semi-)automatic service selection.

The first required component is a language that represents the core notions of usage policies with clearly defined semantics. These notions include the ABC defined by $UCON_{ABC}$ as Authorizations (A), oBligations (B), Conditions (C). The core REL by Jamkhedkar et al., LiREL and ODRL have all slightly different terms such as for example constraints, permissions or prohibitions. We will investigate if these notions are substitutable and which are suitable for our purposes. We envision a solution that is in a similar relationship to ODRL as SAWSDL is to WSDL for Web service description. That means that the structure of the language itself remains the same, but individual elements of a document can reference concepts of an external domain ontology.

These domain ontologies form the second required component of our solution. We plan to reuse results from O'Sullivan [4] and their formalization as ontologies by Toma et al. [5]. For the domain of pricing we will consider the approach by Lamparter [15] which provides an extensive formalization for service pricing models depending on other non-functional attributes.

Being able to describe services is only one part of tool-supported service selection. The other one are algorithms which can be classified in two categories: discovery and ranking. Discovery means that the user formulates his functional and non-functional requirements for a service and the result is a list of matching services. We plan to first extend existing service discovery algorithms to include our proposed model for usage policies. However especially for non-functional properties users often have not only hard requirements but preferences (e.g. a user prefers generally cheaper services but also considers response time). Service ranking denotes the sorting of discovered services according to a preference structure. This structure can either be global for all users and built into the ranking algorithm or specified by each user in a preference formalism. Our goal is to extend existing service ranking algorithms such as the works by Toma et al. [6] and Lamparter [15] to be able to express preferences over usage policies.

In summary our planned contribution is a formal language for usage policies and extensions to service discovery and ranking algorithms that considers usage policy requirements and preferences of users.

4 Work Plan

Until now we have completed the review of the state of the art. We have analyzed different approaches to usage policies and related problems such as access control and DRM. We identified the requirement that the core language should be separated from domain-specific descriptions. We also decided on using ontologies for these descriptions as they support interoperability between different vocabularies likely to be used in a scenario with heterogeneous providers.

As a next step we plan to develop or adapt the core language for usage policies and preferences. As described above we will investigate the possibility to align the non-domain specific parts of ODRL with one of the formal models for usage control languages. The resulting language will integrate references to external ontologies. Based on a selection of Web service and information source licenses we will identify a set of required domain ontologies.

Afterwards we will develop a possibility for users to express their policy needs and preferences. These will serve as inputs to the discovery and ranking algorithms that will be the result of the next stage.

Finally the whole system is evaluated for three aspects: (i) expressivity of the language, (ii) efficiency, and (iii) added value for users. This will be done with the following experiments:

1. Take a list of popular services e.g. from Seekda (http://seekda.com) and distribute them to different users who have to create usage policies using our formalism. By assigning the same license to different users we can afterwards check the equivalence of their results and in this way verify their work. Users have to track which licenses or terms were not expressible in our framework. This experiment evaluates expressivity of the formalism. Additionally the users have to fill out questionnaires about the policy creation process. In this way we want to find out if the approach is practically usable or too complicated.
2. Based on the annotated services we will build a list of user requests and measure the time to complete discovery and ranking with and without usage policies. This will show us if the computational overhead from introducing more complex descriptions is reasonable. The added expressivity of user requirements can lead to a smaller number of services that have to be ranked, which can have a positive effect on the performance.
3. In order to test the added value for users, we will integrate our descriptions and algorithms in a tool, such as for example the keyword-driven search by Tran et al. [1]. With such an implementation we will be able to perform user studies where we can observe if users see a benefit in having tools that are aware of usage policies.

5 Conclusions

The Internet grows more and more into a system that provides us data and services that can easily be accessed and integrated, thanks to Semantic Web and Web Service technology. In such an environment it is natural for common users to build networks of services directly or indirectly and provide them to other users. This large number of providers and services needs formal and interoperable descriptions of services in order to let service location scale. Current approaches to this problem treat functionality and quality aspects. However not all services are freely usable and not all data is in the public domain. The policies of such services also include privacy statements and are currently mostly available in natural language. In this proposal we presented our planned research that includes a formalization of usage policies and their integration into service discovery and ranking algorithms. We believe that the possibility to express such policies will be a motivation for providers to open up their services and data to the Web as it reduces their risk of policy violation.

References

1. Tran, D.T., Wang, H., Rudolph, S., Cimiano, P.: Top-k Exploration of Query Candidates for Efficient Keyword Search on Graph-Shaped (RDF) Data. In: Proceedings of the 25th International Conference on Data Engineering, ICDE (2009)
2. Pautasso, C., Frisoni, M.: The Mashup Atelier. In: Second International Workshop on Web APIs and Services Mashups (Mashups 2008) at ICSOC (2008)
3. Ng, I.C.: The Pricing and Revenue Management of Services: a strategic approach. Routledge Advances in Management and Business Studies (2007)
4. O'Sullivan, J.: Towards a precise understanding of service properties. PhD thesis, Queensland University of Technology (2006)
5. Toma, I., Foxvog, D.: Non-Functional Properties in Web Services. Technical Report D28.4 v0.1, WSMO Working Draft (October 2006)
6. Toma, I., Roman, D., Fensel, D., Sapkota, B., Gómez, J.M.: A Multi-criteria Service Ranking Approach Based on Non-Functional Properties Rules Evaluation. In: ICSOC, pp. 435–441 (2007)
7. Vedamuthu, A., Orchard, D., Hirsch, F., Hondo, M., Yendluri, P., Bubez, T., Yacinalp, U.: Web Services Policy 1.5 - Framework. Technical report
8. Abelson, H., Adida, B., Linksvayer, M., Yergler, N.: ccREL: The Creative Commons Rights Expression Language. Technical report, Creative Commons (2008)
9. Iannella, R.: Open Digital Rights Language (ODRL) Version 1.1. W3C Note (2002)
10. Wang, X., Lao, G., Demartini, T., Reddy, H., Nguyen, M., Valenzuela, E.: XrML - eXtensible rights Markup Language. In: ACM workshop on XML security (2002)
11. Jamkhedkar, P.A., Heileman, G.L.: Digital rights management architectures. Comput. Electr. Eng. 35(2), 376–394 (2009)
12. Jamkhedkar, P.A., Heileman, G.L.: A formal conceptual model for rights. In: Digital Rights Management Workshop, pp. 29–38 (2008)
13. Arnab, A., Hutchison, A.: Persistent Access Control: A Formal Model for DRM. In: ACM workshop on Digital Rights Management, pp. 41–53 (2007)
14. Gangadharan, G.R., D'Andrea, V., Weiss, M.: Service Licensing Composition and Compatibility Analysis. Int. J. Cooperative Inf. Syst. 17(3), 301–317 (2008)
15. Lamparter, S.: Policy-based Contracting in Semantic Web Service Markets. PhD thesis, Universität Karlsruhe (TH), Institut AIFB (2007)
16. Kagal, L., Finin, T., Joshi, A.: A policy language for a pervasive computing environment. In: Policies for Distributed Systems and Networks, June 2003, pp. 63–74 (2003)
17. Lampson, B.W.: Protection. In: Proc. Firth Princeton Symposium on Information Sciences and Systems, pp. 437–443. Princeton University, Princeton (1971); reprinted in Operating Systems Review 8(1), 18–24 (1974)
18. Park, J., Sandhu, R.S.: The $UCON_{ABC}$ usage control model. ACM Trans. Inf. Syst. Secur. 7(1), 128–174 (2004)
19. Park, J., Sandhu, R.: Towards usage control models: beyond traditional access control. In: ACM Symp. on Access Control Models and Technologies SACMAT 2002 (2002)

Ontology-Driven Generalization of Cartographic Representations by Aggregation and Dimensional Collapse

Eric B. Wolf

Center of Excellence in GIScience
US Geological Survey
ebwolf@usgs.gov

Abstract. Automatic generalization of cartographic features has been recognized as a goal of Geographic Information Science (GIScience). Many successful algorithms have been introduced for generalization tasks such as point reduction and smoothing of linear features. Such algorithms operate well as a function of change in map scale or resolution. Other generalization tasks have proved considerably more difficult. Two of these operations, aggregation and dimensional collapse, are trivial to implement – replacing a set of points with an area feature or replacing an area feature with a single point – but have proven challenging to make operational. The decision to aggregate or collapse features is as much dependent on the context of the features as they are change in map scale. This dissertation proposes to show how ontologies can be used to inform automated generalization in these operations.

Keywords: ontologies, cartographic generalization, automated generalization.

1 Problem Statement

This dissertation proposes to link two significant areas of research: automated generalization and knowledge engineering with ontologies. Cartographic generalization is often split into two categories: feature modification based on geometry (e.g., sinuosity or area-perimeter ratios) and feature modification based on semantics. This study focuses on two generalization operations: aggregation and dimensional collapse. These two operations occur in response to constraints which can be measured empirically. But as Bertin [2] noted, altering the dimensionality of representation – what he refers to as implantation – is more of a conceptual change. It is the goal of this dissertation to demonstrate how ontologies can be used to constrain or enhance the conceptual changes during aggregation and dimensional collapse.

The intellectual merit of this effort lies in the application of current knowledge engineering methods to automated generalization. Generalization is well-documented as a process that is strongly dependent on the semantics of the information being modified. Ontologies have advanced knowledge in creating a means to encode such semantics in a form that is machine readable and that can guide automated processes. Efforts have been made to encode cartographic rules in ontologies [14] [7]. Other

A. Bernstein et al. (Eds.): ISWC 2009, LNCS 5823, pp. 990–997, 2009.

efforts have involved applying semantic relationships as an input into generalization operations [8]. No known effort has been made to extend existing cartographic task ontologies specifically toward generalization.

The broader impacts of this study lie in connecting the semantics of geographic information with the generalization process in a manner that can be automated. National Mapping Agencies (NMA), such as the United States Geological Survey (USGS), Ordnance Survey of the United Kingdom or Institut Geographique National of France, currently maintain multiple databases representing the same geographic extents at different levels of detail. The ability to derive all levels of detail from a single database would result in significant cost savings and reduction in update errors. The effort will directly involve the research goals of the USGS towards future versions of *The National Map*. Specifically, the effort could bridge existing research projects in the USGS in generalization and ontology development.

1.1 Generalization of Cartographic Representations

Data collected for a database with a scale of 1:5,000 would be too detailed for a map at a scale of 1:100,000. A representation appropriate for 1:5,000 scale would need to be generalized to a representation appropriate at 1:100,000 scale. Cartographic generalization is the process in which data is transformed to better represent geographic phenomena that are most significant for the map purpose or scale. Dimensional collapse and aggregation are related in that they alter the geometric dimensionality of the representations. Dimensional collapse involves reducing the geometric dimensionality of a representation: a two-dimensional polygon representing the borders of a city at fine scales may be replaced with a single zero-dimensional point symbol at coarser scales. Aggregation involves replacing collections of similar point symbols with a polygonal representation denoting a field of similar objects (Figure 1) [10]. These operations may be invoked due to constraints or for abstraction. Constraints are the limitations imposed by the communication medium (paper, screen, color palette, etc.) and human perception (semiotics) on how the abstracted data is communicated. Abstraction refers to the process in which the surveyor or data collector gather information and structure it according to demands of the desired end product. For instance, the location of wells, irrigation ditches and land parcels might be collected as locations and areas stored abstractly as points and polygons for the purpose of managing water rights [3].

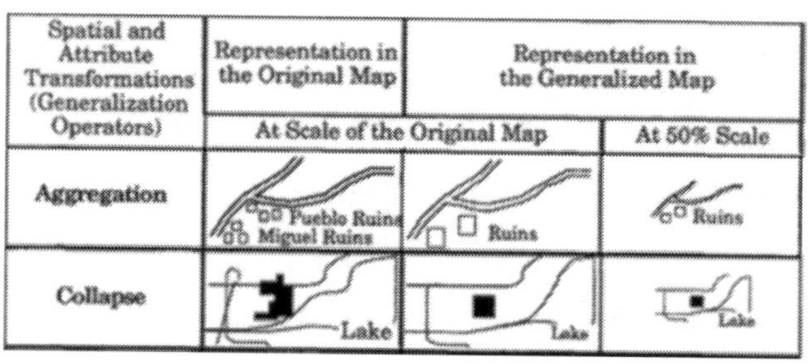

Fig. 1. Aggregation and Collapse generalization operations [10]

Bertin [2] divides cartographic generalization into two categories: structural generalization and conceptual generalization. Structural generalization involves geometric or symbolic changes to feature representation without impacting the semantic relationship with other features. Conceptual generalization requires not just symbolic or geometric changes, but a change in "implantation" [2]. A change in implantation means a change in representation from one primitive type, point, line or area, to another. Collapsing an area to a point or aggregating a collection of points into an area are examples of change in implantation [12]. Bertin provides the example of 'mines' symbolized by points and then aggregated into an area symbol for a 'coal field'. Structural generalization may alter the number of point symbols for mines but a conceptual generalization replaces the points entirely with an area symbol representing the coal field.

Similarly, Ratajski [11] divides generalization into the quantitative and the qualitative. Quantitative generalization is defined as "a gradual reduction of map contents depending on the reduction of the map scale" whereas qualitative generalization is given as "turning from the more elementary ideas to the more general ones". Continuing with Bertin's example, the application of quantitative generalization would result in a reduction of the number of mine symbols whereas the application of qualitative generalization would replace the concept of individual mines with a new concept of a 'coal field'.

Weibel [16] also creates a binary division describing generalization operators as either *context-independent* or *context-dependent*. Specifically, the generalization operators selection/elimination, simplification and smoothing are termed context-independent because they can be applied to individual feature representations without respect to surrounding features. Aggregation and displacement are examples of context-dependent operators that are only used when the surrounding spatial context necessitates it.

Mackaness [9] states "ultimately the results of generalisation do manifest themselves through the manipulation of geometric primitives. But current thinking argues that the reasoning behind those manipulations needs to be based on the analysis of the context." Thus, structural or quantitative generalization can be automated through the application of computational geometry but conceptual or qualitative generalization requires a context-oriented, semantic approach.

Dutton and Edwardes [4] declare "semantic properties and relations needed for generalization that cannot be inferred from geometry must be coded explicitly." In other words, the semantic relationships that govern generalization cannot always be determined through mathematical manipulation of the representations. These semantic relationships must be explicitly established. Further, in order to automate the generalization process the relationships must be encoded in a machine-readable format.

1.2 Ontologies for Cartographic Generalization

According to Smith [13], in the philosophical sense, ontology is the branch of metaphysics that "seeks to provide a definitive and exhaustive classification of entities in all spheres of being." In contrast to Smith, Gruber [5] defines an ontology as a formal, explicit specification of a conceptualization. In this definition, the term *conceptualization* refers to a model of some phenomenon constructed through the identification and enumeration of relevant concepts. The ontology should be *explicit* in that each concept and constraints on its use should be clearly delineated. This explicit conceptualization should be encoded *formally* – that is, the ontology should be machine-readable. The term *conceptualization* appears similar to both a map and a spatial database. Both provide an "abstract model of some phenomenon in the world" and both, through cartographic or model generalization identify "the relevant concepts of that phenomenon". In the case of a map, the conceptualization, or abstract model of some phenomenon in the world, is encoded in geometric shapes, patterns of color and texture. The kinds and categories in this abstract model are explicitly defined in the map legend but the constraints on their use are only loosely defined as visual variables and cartographic guidelines. But the conceptualization encoded as a map is not formal or machine-readable In the case of the spatial database, the data model nicely equates to the abstract model of some phenomenon in the world. The data model of a spatial database is also explicit and formal. The data model explicitly defines the concepts used and the constraints on their use. These data models are also machine readable but generally machine readable by specific database systems.

Gruber's definition of ontology has been made operational through formal languages like the Web Ontology Language OWL. By encoding conceptual relationships for a domain ontology in OWL, semantic relationships can be evaluated during automated processing. Automated generalization methods can use these linkages to maintain the semantic character of representations while reducing detail [15].

The application of ontologies to generalization and even the use of ontologies is not a new concept. Andrienko and Andrienko [1] develop what they term "the first system on automated mapping considering semantic relationships among data components." Their Descartes system automatically creates thematic maps based on semantic relationships of demographic statistics. Kulik et al. [8] present a novel algorithm for simplification of linear features that takes into account map purpose as encoded in the semantic weights. Other efforts focus on developing domain ontologies to facilitate cartographic generalization and representation in spatial databases. Dutton and Edwardes [4] state that "...without accounting for the *roles* of features in a landscape and in a map, it is difficult to select, simplify, displace and re-symbolize features appropriately." They develop an ontology based on the place names of features common between USGS topographic maps and NOAA navigational charts.

Torres et al. [14] describe domain ontologies for hydrological and topographic maps. The goal of their exercise is to explore representations of these maps as ontologies. Iosifesci-Enescu and Hurni [7] introduce ontologies for the formalization of cartographic knowledge and rules. They demonstrate both a cartographic domain ontology "centered on concepts of map, graphic element, visual variable and symbol" and point towards a task ontology that formalizes "the storage, access and exchange cartographic rules."

This dissertation will attempt tasks to help inform how semantic relationships drive the generalization operations of aggregation and dimensional collapse. While these operations are typically employed to maintain the presence of a feature representations in response to reduction in map scale, they are sometimes employed in response to changes in context according to the purpose of the map [4].

2 Research Questions

The questions this dissertation addresses are:

1. Can the semantic relationships encoded as ontologies be used to automate dimensional collapse and aggregation?
2. Separate domain ontologies must be created for each separate map series. Can a single task ontology for generalization be created to manage the semantics for all map scales? How is generalization modeled inside this ontology?
3. If multiple ontologies must be created, how can the ontologies be aligned to represent continuity across levels of detail?

3 Workplan

As a demonstration of the concepts in this dissertation, application ontologies will be developed for air navigation charts similar to the domain ontologies created by Torres et al. [14]. Air navigation charts are ideal for this case study because the database and scale remain fixed while representation is varied. The charts represent essentially the same information but with two unique perspectives. Visual Flight Rules (VFR) charts are designed for conditions in which pilots can navigate by visually surveying the landscape. Instrument Flight Rules (IFR) charts are designed for conditions in which pilots must rely solely on instrument readouts (Figure 2).

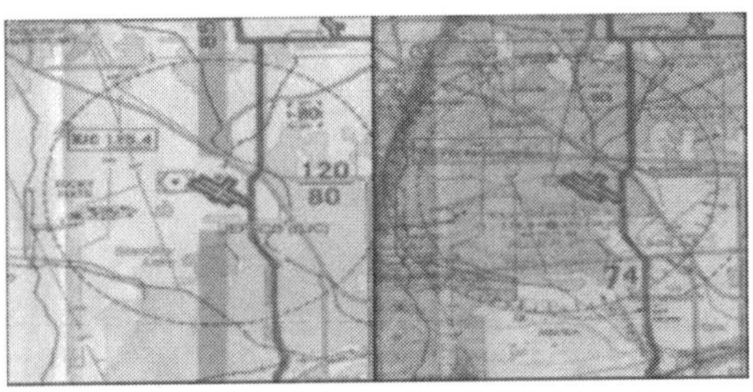

Fig. 2. Section of IFR and VFR Charts over Jefferson County Airport. Two cartographic ontologies based on one spatial database model.

Once the demonstration ontologies are completed, cartographic application ontologies will be created for multi-scale map series produced by the USGS (Figure 3). To limit the scope of the study, focus will be on feature complexes in and around selected international airports. A single ontology will be created for each multi-scale map series by examining the official data dictionaries. The availability of the official data dictionaries may further limit the scope of this task. A significant question arises from this process: Is it better to have one ontology that manages generalization and change in scale internally? Or is it better to have multiple ontologies with the generalization process becoming a process of creating semantic agreements between the ontologies?

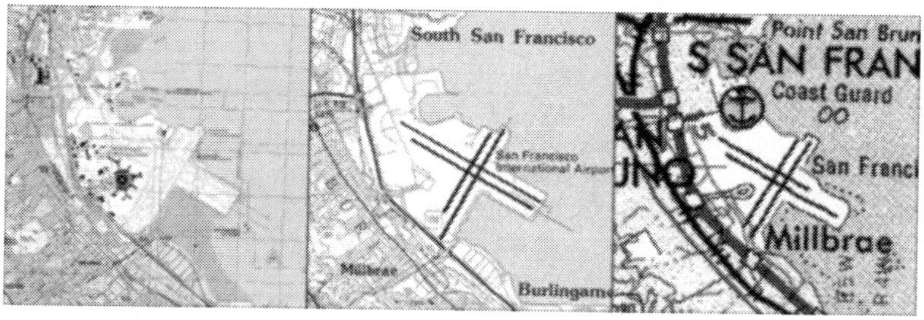

Fig. 3. San Francisco International Airport in USGS Topographic multi-scale map series. Left to right 1:24,000 7.5" Quandrangle, 1:100,000 30' x 60' Quadrangle, and 1:250,000 1 degree by 2 degree Quadrangle, presented as uniform-scale graphics.

Finally, task ontologies will be constructed for cartographic generalization operations used in each application ontology. Relationships among the domain ontologies developed in this study will be established. For instance, a domain ontology for the spatial databases underlying the USGS *National Map* could be aligned with different cartographic domain ontologies. The same will be done for the generalization task ontologies. Theoretically, such a system will allow for the generation of ontologically-dissimilar maps like the IFR and VFR charts as well as guiding the generalization of data for multi-scale map series.

In the first half of this research project, a series of domain and task ontologies are to be developed for creating similar maps. Several iterations of each ontology may be necessary to balance the requirements of each criterion. Gruber [6] gives objective criteria for evaluating the design of formal ontologies. The ontologies developed in this dissertation will be evaluated against these criteria

The demonstration task involves the flight navigation charts controls for scale and underlying database while varying map purpose. It is anticipated that a single domain ontology may be constructed for both VFR and IFR perspectives. Also, a task ontology for cartographic representation would be created based on a subset of the cartographic rules used on the flight navigation charts similar to Iosifescu-Enescu and Hurni's ontology [7]. The differences in the perspectives would be encoded in two application ontologies based on these domain and task ontologies.

The main research effort involves creating ontologies for multi-scale map series. This effort will be considerably more challenging than the demonstration, not just because of the number of maps involved, but because at least two different models are possible. In both models, it is anticipated that a single domain ontology will be needed for each map series. The cartographic task ontology from the demonstration would be expanded to include generalization operators. Task ontologies encode relationships among tasks but not specific sequences. Application ontologies rely on these relationships by the tasks to concepts in the domain ontology. Thus, application ontologies would be created for either each map or each map series depending on whether or not a single application ontology can be created for an entire map series.

Once completed, the final effort would be to utilize the hierarchy of domain and task ontologies to explore relationships between scale and ontology in order to answer the research questions. One method of exploring these relationships would be to map the successful ontologies onto heterogeneous database ontologies for the purpose of guiding the generalization process. The overall goal of the project is to guide the conceptual generalization of geographic representations through the semantic relationships encoded in the ontologies. In order to evaluate this goal, it will be necessary to develop programs that utilize ontologies in automating the generalization process.

4 Summary

This dissertation is designed to demonstrate how ontologies can be used to constrain or enhance the conceptual changes during aggregation and dimensional collapse. These two generalization operations modify the conceptual nature of representations and must be guided by semantics, not geometry [11] [2] [16] [9].

The expected outcomes of this research are the creation of a series of domain ontologies from data dictionaries published by NMAs as well as one or more task ontologies which model generalization within the context of cartography. These ontologies will be aligned as application ontologies that will be coupled with scripts to drive automated generalization for dimensional collapse and aggregation.

Acknowledgments. This research was supported by the United States Geological Survey as part of the project titled "Building an Ontology for *The National Map*".

References

1. Andrienko, G., Andrienko, N.: Knowledge Engineering for Automated Map Design in Descarates. In: ACM GIS 1999, Kansas City, MO USA (1999)
2. Bertin, J.: Semiology of Graphics, p. 415. The University of Wisconsin Press, Madison (1983)
3. Buttenfield, B.P., Mark, D.M.: Expert Systems in Cartographic Design. In: Geographic Information Systems: The Microcomputer and Modern Cartography, pp. 129–150. Pergamon Press, Oxford (1990)
4. Dutton, G., Edwardes, A.: Ontological Modeling of Geographical Relationships for Map Generalization. In: The 9th ICA Workshop on Generalization and Multiple Representations, Vancouver, Washington (2006)
5. Gruber, T.R.: A Translation Approach to Portable Ontology Specification. Knowledge Acquisition 5, 199–220 (1993)
6. Gruber, T.R.: Toward Principles for the Design of Ontologies Used for Knowledge Sharing. International Journal of Human-Computer Studies 43(5/6), 22 (1995)
7. Iosifescu-Enescu, I., Hurni, L.: Towards Cartographic Ontologies Or How Computers Learn Cartography. In: 23rd International Cartographic Conference, Moscow, Russia (2007)
8. Kulik, L., Duckham, M., et al.: Ontology-Driven Map Generalization. Journal of Visual Languages and Computing 16(2), 245–267 (2005)
9. Mackaness, W.A., Ruas, A., et al. (eds.): Generalisation of Geographic Information: Cartographic Modelling and Applications, p. 370. Elsevier, Oxford (2007)
10. McMaster, R.B., Shea, K.S.: Cartographic Generalization in a Digital Environment: When and How to Generalize. AutoCarto 9, Baltimore, MD (1989)

11. Ratajski, L.: Phenomenes Des Points De Generalization. International Yearbook of Cartography 7, 143–151 (1967)
12. Slocum, T.A., McMaster, R.B., et al.: Scale and Generalization. In: Thematic Cartography and Geographic Visualization, pp. 103–120. Prentice Hall, Upper Saddle River (2005)
13. Smith, B.: Ontology. In: Blackwell Guide to the Philosophy of Computing and Information, pp. 155–166. Blackwell, Malden (2003)
14. Torres, M., Quintero, R., et al.: Ontology-Driven Description of Spatial Data for Their Semantic Processing. In: GeoSpatial Semantics 2005. Springer, Heidelberg (2005)
15. Tran, V.X., Tsuji, H.: Owl-T: A Task Ontology Language for Automatic Service Composition. In: IEEE International Conference on Web Services. IEEE, Los Alamitos (2007)
16. Weibel, R.: Generalization of Spatial Data: Principles and Selected Algorithms. In: van Kreveld, M., Roos, T., Nievergelt, J., Widmayer, P. (eds.) CISM School 1996. LNCS, vol. 1340, pp. 99–152. Springer, Heidelberg (1997)

Populating the Semantic Web by Macro-reading Internet Text

Tom M. Mitchell[1], Justin Betteridge[1], Andrew Carlson[1], Estevam Hruschka[1,2], and Richard Wang[1]

[1] Carnegie Mellon University, Pittsburgh PA 15213, USA
Tom.Mitchell@cs.cmu.edu
http://www.cs.cmu.edu/~tom
[2] Federal University of Sao Carlos, Brazil

Abstract. A key question regarding the future of the semantic web is "how will we acquire structured information to populate the semantic web on a vast scale?" One approach is to enter this information manually. A second approach is to take advantage of pre-existing databases, and to develop common ontologies, publishing standards, and reward systems to make this data widely accessible. We consider here a third approach: developing software that automatically extracts structured information from unstructured text present on the web. We also describe preliminary results demonstrating that machine learning algorithms can learn to extract tens of thousands of facts to populate a diverse ontology, with imperfect but reasonably good accuracy.

1 The Problem

The future impact of the semantic web will depend critically on the breadth and depth of its content. One can imagine several approaches to constructing this content, including manual content entry by motivated teams of people, convincing owners of existing databases to publish them on the semantic web, and automatically extracting structured information from the vast quantity of unstructured online text. We consider here the third of these approaches, and argue both that it is feasible and that this kind of approach will be able to collect knowledge that is unlikely to be captured as easily by other approaches.

The feasibility of extracting structured information automatically from text will itself depend on the technical state-of-the-art of natural language processing (NLP) methods. We have witnessed significant progress in NLP over the past decade, on problems from sentence parsing [1] to named entity extraction [2], to question answering [3], to document classification [4]. Nevertheless, computer algorithms remain very far from being able to truly "understand" natural language text (e.g., to read and extract the full content of the paper you are currently reading). Given this shortcoming, why might we take the position that NLP algorithms offer a promising near-term approach to populating the semantic web?

We believe automatic methods offer a feasible near-term approach because the problem of automatically populating large databases from the internet can

A. Bernstein et al. (Eds.): ISWC 2009, LNCS 5823, pp. 998–1002, 2009.

be formulated so that it is much easier to solve than the problem of full natural language understanding. Our own formulation involves three key design choices:

1. Macro-reading instead of micro-reading. We use the term "micro-reading" to refer to the traditional NLP task where a single text document is input, and the desired output is the full information content of that document. In contrast, we define "macro-reading" as a task where the input is a large text collection (e.g., the web), and the desired output is a large collection of facts expressed by the text collection, without requiring that every fact be extracted. We argue that macro-reading is much easier than micro-reading, for two reasons. First, macro-reading does not require extracting every bit of information contained in the text collection. Second, in text corpora as large as the web, many important facts will be stated redundantly, thousands of times, using different wordings. A macro-reader can benefit from this redundancy by focusing on analyzing only the simple wordings of the fact, ignoring hopelessly complex sentences, and by statistically combining evidence from many text fragments in order to determine how strongly to believe a particular candidate hypothesis.

2. Ontology-driven reading. Much of the difficulty in truly understanding free-form text follows from the fact that it can say anything. In contrast, we formulate our macro-reading problem as a task of populating an ontology that is given as input, and that defines the categories (e.g., sport, person, team) and relations (e.g., plays-sport, plays-on-team) of interest. This is a natural way to frame a problem of populating some portion of the semantic web for which an ontology is available. It also makes our macro-reading problem easier in two ways. First, the system can focus only on a subset of text that is on-topic relative to the ontology. Second, the ontology itself can define meta-properties of its categories and relations that make extraction easier and more accurate (e.g., it can state that the relation 'plays-on-team' relates arguments of type 'person' and 'team').

3. Machine learning methods whose accuracy improves with ontology complexity. A third design choice is to use semi-supervised machine learning methods that automatically discover patterns of text and hypertext that support reliable fact extraction. Our machine learning approach acquires extraction patterns (e.g., "mayor of X" often implies X is a city) for each predicate (category or relation) in the input ontology. We build on earlier semi-supervised bootstrap learning methods [5,6,7,8] that learn from just a handful of labeled training examples, plus a large corpus of unlabeled text. While these earlier methods showed the feasibility of semi-supervised learning of extraction patterns, they were limited because accurate learning requires more constraints than are provided by a few dozen labeled training examples. Our algorithm achieves significantly higher accuracy by using the input ontology itself to provide additional constraints that guide the learner[9]. For example, when our algorithm learns extraction patterns for the predicates 'person', 'team' and 'plays-on-team', prior knowledge from the ontology requires that for any unlabeled sentence containing noun phrases A and B, the extractor for 'plays-on-team' can label $<A, B>$ a positive example of the relation only if the 'person' classifier labels A positive, and the 'team' classifier

Skype
isA: company
company_economic_sector: VoIP
competes_with: AOL, MSN, Yahoo, Google
acquired_by: Ebay

EBay
isA: company
company_CEO: Pierre Omidyar
competes_with: Dell, Google, Yahoo, Amazon, Amazon.com, Microsoft, AOL
acquired: PayPal, Skype

Fig. 1. Extracted facts for two companies discovered by our system. These two companies were extracted by the learned 'company' extractor, and the relations shown were extracted by learned relation extractors.

Table 1. Horn clause rules learned from extracted instances. Numbers indicate the conditional probability that the literal to the left of the ":-" will be satisfied if the literals to its right are satisfied.

```
0.84 playsSport(?x,?y) :- playsFor(?x,?z), teamPlaysSport(?z,?y)
0.70 playsSport(?x,baseball) :- playsFor(?x,yankees)
0.82 teamPlaysSport(?x,?y) :- playsFor(?x,?z), playsSport(?z,?y)
0.73 teamPlaysSport(?x,baseball) :- playsAgainst(?x,yankees)
```

labels B positive. As the ontology grows in complexity, the set of constraints on the learner also grows, resulting in even higher accuracy.

In summary, our approach uses a coupled semi-supervised learning algorithm to acquire extraction strategies for each predicate in the input ontology, and applies these to macro-read millions of web pages to populate that ontology.

2 The ReadTheWeb System

Our current system learns extraction patterns defined over free text and over HTML structure, starting from an initial ontology containing dozens of categories and relations, and 10-15 seed examples of each. The textual pattern learner, CBL [9], iteratively grows a set of extraction patterns while obeying mutual exclusion, subset, and type checking constraints given by the ontology. The HTML pattern learner, SEAL [10], learns patterns of HTML and text tokens that capture regularities such as HTML lists of predicate instances. Based on the belief that these techniques should make independent errors, our system only trusts instances that are extracted by both techniques. Such instances are added to the current beliefs at the end of each iteration, and the process repeats by invoking the subordinate techniques with the newly promoted instances. Figure 1 shows some facts extracted by a recent run of the system (see the complete results at http://rtw.ml.cmu.edu/readtheweb.html).

In a recent experiment involving 16 categories, our current system achieved an average precision of 97% while promoting 4224 category instances. In experiments with CBL which involved an additional 14 relations, it achieved an average precision of 83% for the categories and 84% for the relations while promoting 15520 category instances and 2674 relation instances.

Another component of our system mines the thousands of extracted beliefs, to learn probabilistic Horn clause rules that capture empirical regularities in this data. The resulting rules (Table 1) can then be used to infer additional beliefs to further populate the ontology. In this case the new beliefs are not extracted from text, but are instead inferred from the learned rules and other previously extracted beliefs. Note each learned rule contributes yet another constraint to couple the subsequent training of extractors for the predicates it mentions.

3 Conclusions

We argue that macro-reading the web to populate target ontologies is feasible in the near term, especially as progress continues on coupled semi-supervised learning methods, and intelligent approaches to lightly supervising them. Our preliminary results demonstrate that such an approach can successfully extract tens of thousands of beliefs to populate an input ontology, at imperfect but reasonable accuracy.

One key to our argument is that macro-reading is much easier than solving the full NLP problem. We note that micro-reading will also be important, especially for anotating individual web pages, and for extracting information that is stated only infrequently on the web (e.g., personal information that appears only on a person's home page). One direction for future work is to explore whether and how a macro-reader like ours can help train a micro-reader.

While we believe machine reading will play an important role in populating the semantic web, other approaches will be valuable too, and it is useful to understand different roles these different approaches can play. For example, publishing pre-existing databases may be especially useful for providing deep coverage over fairly narrow domains (e.g., the census data). In contrast, our approach of macro-reading the web may be better suited to populating more broad ontologies, especially with items that are mentioned frequently on the web (and hence easiest to macro-read). Because it can be retrained to fairly arbitrary ontologies, our approach might also be useful for more specialized applications where manual methods are prohibitively expensive.

Acknowledgements. This research has been supported by DARPA, Google, and the Brazilian Research Agency CAPES. Yahoo! Inc. has provided graduate fellowship support as well as access to their M45 computing cluster.

References

1. Nivre, J.: Incremental non-projective dependency parsing. HLT-NAACL, 396–403 (2007)
2. Tjong Kim Sang, E.F., De Meulder, F.: Introduction to the CoNLL-2003 shared task: Language-independent named entity recognition. In: Proceedings of CoNLL 2003, pp. 142–147 (2003)
3. Vorhees, E.: Overview of TREC 2007. In: TREC (2007)

4. Nigam, K., Andrew McCallum, S.T., Mitchell, T.: Text classification from labeled and unlabeled documents using em. Machine Learning 39, 103–134 (2000)
5. Yarowsky, D.: Unsupervised word sense disambiguation rivaling supervised methods. In: ACL, pp. 189–196 (1995)
6. Blum, A., Mitchell, T.: Combining labeled and unlabeled data with co-training. In: COLT (1998)
7. Riloff, E., Jones, R.: Learning dictionaries for information extraction by multi-level bootstrapping. In: AAAI (1999)
8. Brin, S.: Extracting patterns and relations from the world wide web. In: WebDB (1998)
9. Carlson, A., Betteridge, J., Hruschka Jr, E.R., Mitchell, T.M.: Coupling semi-supervised learning of categories and relations. In: Proceedings of the NAACL HLT 2009 Workshop on Semi-supervised Learning for NLP (2009)
10. Wang, R.C., Cohen, W.W.: Language-independent set expansion of named entities using the web. In: ICDM (2007)

Search 3.0: Present, Personal, Precise

Nova Spivack

Radar Networks, USA

The next generation of Web search is already beginning to emerge. With it we will see several shifts in the way people search, and the way major search engines provide search functionality to consumers.

To understand movement towards the next generation of Web search, it is helpful to first look at how it has developed to its current state. Beginning with Tim Berners-Lee's foundational proposal in 1989, the Web's evolution may be divided into three distinct phases.

Web 1.0, the first decade of the Web (1989 - 1999), was characterized by a distinctly desktop-like search paradigm. The overriding idea was that the Web is a collection of documents, not unlike the folder tree on the desktop, that must be searched and ranked hierarchically. Relevancy was considered to be how closely a document matched a given query string.

Web 2.0, the second decade of the Web (1999 - 2009), ushered in the beginnings of a shift towards social search. In particular, blogging tools, social bookmarking tools, social networks, social media sites, and microblogging services began to organize the Web around people and their relationships. This added the beginnings of a primitive "web of trust" to the search repertoire, enabling search engines to begin to take the social value of content (as evidenced by discussions, ratings, sharing, linking, referrals, etc.) as an additional measurement in the relevancy equation. Those items that were both most relevant on a keyword level, and most relevant in the social graph (closer and/or more popular in the graph), were considered to be more relevant. Thus results could be ranked according to their social value - how many people in the community liked them and current activity level - as well as by semantic relevancy measures.

In the coming third decade of the Web, Web 3.0 (2009 - 2019), there will be another shift in the search paradigm. This is a shift to from the past to the present, and from the social to the personal, and from the generic to the precise.

Established search engines like Google rank results primarily by keyword (semantic) relevancy. Social search engines rank results primarily by activity and social value (Digg, Twine 1.0, etc.). But the new search engines of the Web 3.0 era will also take into account two additional factors when determining relevancy: timeliness and personalization. The result will be a more precise and customized search capability.

Google returns the same results for everyone, but why should that be the case? In fact, when two different people search for the same information, they may want to elicit very different kinds of results. Someone who is a novice in

A. Bernstein et al. (Eds.): ISWC 2009, LNCS 5823, pp. 1003–1004, 2009.

a field may want beginner-level information to rank higher in the results than someone who is an expert. There may be a desire to emphasize things that are novel over things that have been seen before, or that have happened in the past; the more timely something is the more relevant it may be as well. These three themes - present, personal, and precise - will define the next great search experience.

To accomplish this, we need to make progress on a number of fronts. First of all, search engines need better ways to understand what content is, without having to do extensive computation. The best solution for this is to utilize metadata and the methods of the emerging semantic web.

Metadata reduces the need for computation in order to determine what content is about - it makes that information explicit and machine-understandable. To the extent that machine-understandable metadata is added or generated for the Web, it will become more precisely searchable and productive for searchers.

This applies especially to the area of the real-time Web, where, for example, short "tweets" of content contain very little context to support good natural-language processing. In such cases, a little metadata can go a long way. In addition, of course, metadata makes a dramatic difference in search of the larger non-real-time Web as well.

Beyond metadata, search engines need to modify their algorithms to be more personalized. Instead of a "one-size fits all" ranking for each query, the ranking may differ for different people depending on their varying interests and search histories.

Finally, to provide better search of the present, search has to become more realtime. To this end, rankings need to be developed that surface not only what just happened now, but what happened recently and is also trending upwards and/or of note. Realtime search has to be more than merely listing search results chronologically. There must be effective ways to filter the noise and surface what's most important effectively. Social graph analysis is a key tool for doing this, but in addition, powerful statistical analysis and new visualizations may also be required to make a compelling experience.

The pace at which Semantic technology is finally developing makes the shift into the third generation of Web search now realizable. Tools operating within the new paradigm of present (Oneriot, Topsy, Twitter), personalized (My6Sense, Siri, Twine), and precise (Bing/Powerset, Hakia, WolframAlpha) are already leading the way in expanding conceptions of how search can function. Search 3.0 will bring a breakthrough in the Semantic indexing of the entire Web, allowing vertical search to be taken to its ultimate conclusion.

Author Index

FSC
www.fsc.org
MIX
Papier | Fördert
gute Waldnutzung
FSC® C083411

Kolibri 360 GmbH
Ferdinand-Jühlke-Straße 7
99095 Erfurt, Deutschland
produktsicherheit@kolibri360.de